Properties for Design of Composite Structures

Properties for Design of Composite Structures

Theory and Implementation Using Software

Neil McCartney

National Physical Laboratory, Teddington, Middlesex, UK

WILEY

This edition first published 2022

Registered Office(s)
John Wiley & Sons, Inc., 111 River Street, Hoboken, NJ 07030, USA
John Wiley & Sons Ltd, The Atrium, Southern Gate, Chichester, West Sussex, PO19 8SQ, UK

Editorial Office
The Atrium, Southern Gate, Chichester, West Sussex, PO19 8SQ, UK

For details of our global editorial offices, customer services, and more information about Wiley products visit us at www.wiley.com.

Wiley also publishes its books in a variety of electronic formats and by print-on-demand. Some content that appears in standard print versions of this book may not be available in other formats.

Library of Congress Cataloging-in-Publication Data
Names: McCartney, Neil, author.
Title: Properties for design of composite structures : theory and implementation using software / Neil McCartney.
Description: Hoboken, NJ : John Wiley & Sons, 2022. | Includes bibliographical references and index.
Identifiers: LCCN 2021034688 (print) | LCCN 2021034689 (ebook) | ISBN 9781118485286 (hardback) | ISBN 9781118789681 (pdf) | ISBN 9781118789780 (epub) | ISBN 9781118789797 (ebook)
Subjects: LCSH: Composite materials--Design and construction.
Classification: LCC TA418.9.C6 M3495 2022 (print) | LCC TA418.9.C6 (ebook) | DDC 620.1/18--dc23
LC record available at https://lccn.loc.gov/2021034688
LC ebook record available at https://lccn.loc.gov/2021034689

Cover image: © NPL Management Ltd
Cover design by Wiley

Set in 9.5/12.5pt STIXTwoText by Integra Software Services Pvt. Ltd, Pondicherry, India

SKY6FC3FE0B-7599-4BFD-901F-6BCF0170F99E_061322

Contents

Preface

The effective properties of composite materials, such as thermoelastic and conductive, depend upon the nature of many materials' characteristics, which are needed to account for the composite structure, e.g. particulate or fibre reinforcement, unidirectional plies or multi-layered laminates. Account also must be taken of the resultant anisotropies that are often required when designing three-dimensional engineering components. Indeed, the properties of some constituents can be anisotropic themselves, such as carbon fibres and the individual plies of laminates. The accurate experimental determination of these properties is rarely, if ever, achieved in practice, and this leads on to the need for reliable estimation methods using analytical and/or numerical methods.

It is emphasised here that this book is not intended to be a reference text. The principal objective is to provide readers with a single source for a wealth of analytical methods and explicit formulae that can be used to estimate the effective properties of a wide range of composite types for both undamaged and damaged states. Readers wishing to track in more detail the historical development of composite theory will need to consult the literature, possibly starting with the limited number of references included in this book.

This book has been written over a number of years following the encouragement of the late Professor Anthony Kelly, FRS, CBE, to whom the author owes an enormous debt of gratitude. Professor Kelly was the Chairman of the Panel at the National Physical Laboratory (NPL) that decided, in 1969, that the author, following six months' probation, could be offered a permanent post. The author now has an Emeritus position at NPL (Senior NPL Fellow) that has enabled continued research on various aspects of material science, and the writing of this book. A much closer relationship developed when, in the early 1980s, we both independently solved a problem considered in Chapter 14. At that point and onwards Professor Kelly, who was then Vice-Chancellor of Surrey University, introduced the author to many theoretical aspects of composite technology, which are included in this book. Of particular note is the development of methods of analysing the effects of fibre fractures and matrix cracks in unidirectionally reinforced composites, extending the pioneering work of colleagues at NPL, which led to the very well-known ACK (Aveston, Cooper, Kelly) theory, developed in 1971, which continues to be widely cited in the literature.

Following the Introduction, Chapter 2 of this book introduces and describes in detail many continuum analyses that will be required in later chapters. Chapters 3–7 are concerned with a description of methods of estimating the effective properties of undamaged composite materials, including those reinforced with aligned spheroidal particles. Extensive use is made of the methodology developed by James Clerk Maxwell in his book on electromagnetism, where an ingenious analytical method is used to estimate the effective electrical conductivity of a uniformly distributed assembly of isotropic spherical particles. His result is thought to be the first expression ever developed for an effective property of a composite. His methodology is also extremely powerful and worthy of extensive use in this book.

During service, composite components are often exposed to loadings that are sufficient to cause the formation of microstructural damage such as fibre failures or matrix cracking, which are usually associated with the occurrence of fibre/matrix interface debonding. For laminates, delamination at ply interfaces is an additional form of load-induced damage. The initiation and growth of microstructural damage affects the effective properties of composite materials, which degrade. It is remarkable that the degradation of very many thermoelastic constants is characterised by a single damage dependent function, as will become clear in the book. Sometimes damage mechanisms lead to local stress-transfer within complex structures, which can delay the onset of structural failure during progressive loading. Chapters 8–14 of this book will focus on analytical and semi-analytical methods of predicting both the effect of damage on effective properties and, more importantly, the dependence of effective properties on local loading conditions when using an energy balance approach. Chapters 13 and 14 concern, respectively, a model of damage development during the fatigue loading of laminates and a model of composite degradation due to environmental damage (in fibres).

Chapters 15–19 comprise more advanced analyses, providing mathematical details of many complex derivations that arise when considering: i) the effective properties of aligned systems of spheroids, ii) interface debonding associated with a fibre fracture or matrix crack in a unidirectional composite, iii) the behaviour of bridged cracks of finite length in a unidirectional composite, iv) the behaviour of ply cracks of finite length in a cross-ply laminate, v) stress-transfer mechanics for general symmetric laminates subject to general in-plane loading and vi) stress-transfer mechanics associated with out-of-plane orthogonal bending of cross-ply laminates that do not need to be symmetric. The models considered can lead to analytical formulations that require the use of numerical methods to deal with the complexity arising from the use of analytical techniques. Software implementing the analysis described in this book is provided at the Wiley website link (www.wiley.com/go/mccartney/properties).

The author would like to take this opportunity to thank all colleagues at NPL, and in other institutions around the world, for all the technical interactions that have helped greatly to formulate and investigate the various aspects of composite materials that are described in this book. My thanks also extend to the staff at John Wiley, who have been very patient while the book has been completed, and for their considerable help in the final stages of book preparation.

This book is dedicated to my late wife, Irene, and all our family, who have encouraged the completion of this book despite the many years in preparation. The author cannot thank them enough for allowing him the time in their lives, first to undertake the research that is described, and then the writing of the book itself.

L N McCartney
April 2022

About the Companion Website

Properties for Design of Composite Structures: Theory and Implementation Using Software is accompanied by a companion website:

www.wiley.com/go/mccartney/properties

The website includes:

- Software Notes, where relevant by chapter

1

Introduction

Improving the properties and performance of materials by reinforcing them with different types of stiffer and/or stronger phases, such as particles or fibres, leads to the class of material known as composites. This approach was first exploited during natural developments (both living plants and organisms) and later by mankind, e.g. the use by Egyptians when making clay building bricks reinforced with straw to improve handling and performance, and when using gravel to reinforce cement forming a much stronger concrete material. Over the centuries, increasingly sophisticated composites have been developed, responding especially to the advent of higher-performance fibres (for high stiffness, strength and/or high temperature resistance). Even greater benefits can arise by the manufacture of composites reinforced with hollow and/or multicoated inclusions, and nanotubes.

The simplest inclusion geometry for matrix reinforcement is a set of spherical particles which might exhibit a range of radii. When the matrix and reinforcement is homogeneously mixed, the resulting material has improved properties which are usually considered to be isotropic. Another simple geometry uses aligned continuous fibres to reinforce an isotropic matrix forming a material that is anisotropic such that the properties in the fibre (or axial) direction differ from transverse properties in the plane normal to the fibre direction. This type of material is known as unidirectionally fibre-reinforced composite where the transverse stiffnesses and strengths are usually much lower than the stiffness and strength in the fibre direction. To overcome this significant practical problem, composite laminates are considered where a stack of unidirectional composites known as plies are bonded together where the fibres in each ply in the laminate are aligned in a direction that varies from ply to ply. Laminates are often weak under compression because of a damage mode known as delamination where debonding occurs at or near the interfaces between the various plies. To overcome this practical problem, woven or stitched fibre architectures are used. Such composites can be analysed effectively only if numerical methods are used. This topic will not, therefore, be considered in this book, as the intention is to focus here on the development and use of analytical methods. Composites, often made where the reinforcement is a set of short fibres which are either aligned in a given direction or are randomly oriented, are also not considered in this book.

The engineering application of composite materials for equilibrium calculations requires a full understanding the key materials properties which can be thought

Properties for Design of Composite Structures: Theory and Implementation Using Software, First Edition. Neil McCartney.

Companion Website: www.wiley.com/go/mccartney/properties

of as being classified as elastic, thermal, electrical and magnetic. In addition, when nonequilibrium transport phenomena occur, such as heat flow and electrical conduction, an additional type of property known as conductivities must also be considered. All these properties are well known, and for isotropic materials often encountered in engineering, a limited number of material properties are sufficient to characterise, fully, the physical behaviour of such materials. However, when composite materials are considered, the properties are usually no longer isotropic, and many more properties describing the directional behaviour of the material, and even of the fibres themselves, need to be considered in the engineering design process. This complexity also leads to needs for suitable measurement methods capable of determining the values of the multitude of properties required to fully characterise an anisotropic composite material, an important topic that will not be considered in this book. It is noted that the numerical analysis of three-dimensional composite components requires a very large number of materials properties, associated with the anisotropic nature of composite materials, some of which are extremely difficult to measure (e.g. through-thickness shear moduli). Analytical methods provide a pragmatic way of providing such data, but this leads on to needs to have access to reliable data for fibre properties, which are frequently anisotropic, and some properties are seldom known reliably (e.g. Poisson's ratios, shear moduli and transverse thermal expansion).

The effective properties of composite materials, which arise when samples are considered as homogeneous materials having anisotropic properties, depend in a complex way on the properties of the materials used in their manufacture (e.g. reinforcements and matrix) and on the geometrical arrangement of these materials. It is plainly not feasible to undertake an experimental programme designed to use measurement methods to determine the relationships between effective properties of composite materials and the constituent properties and structure. Instead, theoretical methods are used based on the well-established principles of continuum thermodynamics defined in its most general form so that both continuum mechanics and electrodynamics are considered in a thermodynamic context. It is indeed of interest to know that James Clerk Maxwell, developer of the famous Maxwell equations of electrodynamics, is believed to be the first scientist to develop a formula for an effective property of a composite material. He considered a cluster of spherical particles, all having the same isotropic permittivity value, embedded in an infinite matrix, having a different value for isotropic permittivity, and developed an elegant method of estimating the effective properties of the particle cluster. Although Maxwell argued that his neglect of particle interactions would limit the validity of his effective property to low volume fractions, it is known that results obtained using his methodology are in fact valid for much larger volume fractions. This important scientific contribution appeared in 1873 as part of Chapter 9 in his book entitled *A Treatise on Electricity and Magnetism*, published by Clarendon Press, Oxford. Maxwell's methodology will be used in this book to help understand the relationship of many effective composite properties to the properties of the reinforcements and their geometrical arrangements within a matrix.

In recent years, although many developments of materials and structures in the composites field have used a make-and-test philosophy, the scientific understanding that has now developed means that predictive methods of assessing composite performance are being used more widely. There is a wide spectrum of predictive techniques that can be used ranging from analytical models, which is the theme of this book, through to numerical simulations of engineering components, having complex geometries and loadings, which are based on numerical techniques such as the finite element and boundary element methods. There is a need to employ both analytical and numerical techniques. The former are models where predictions are possible through use of mathematical formulae that relate the important parameters that might be varied when designing a new material, whether a unidirectional ply or a complex laminate. The parameters normally encountered are the fibre volume fractions, the thermoelastic constants of both fibre and matrix, or of a ply or a laminate. When assessing damage resistance, other types of property are encountered such as fracture energies. Analytical methods provide a clear understanding of the key physical processes that are involved, and they provide methods of assessing whether, or not, candidate materials have good prospects of being used as improved engineering materials. Analytical models can also be used to develop exact solutions to relatively simple and amenable practical situations. These solutions can be used as special cases to validate the numerical methods which have much wider applicability.

The principal objectives of this book are to present, in a single publication, a description of the derivations of selected theoretical methods of predicting the effective properties of composite materials for situations where they are either undamaged or are subject to damage in the form of matrix cracking, in fibre-reinforced unidirectional composites, or in the plies of laminates, or to a lesser extent on the interfaces between neighbouring plies. The major focus of the book is on derivations of analytical formulae which can be the basis of software that is designed to predict composite behaviour, e.g. prediction of properties and growth of damage and its effect on composite properties. Software will be available from the John Wiley & Sons, Inc. website [1] including examples of software predictions associated with relevant chapters of this book.

The chapters of this book are grouped into three parts. The first group comprises Chapters 1 and 2, which provide the introduction and the fundamental relations for continuum models, and Chapters 3–7 that focus on preferred methods of estimating the thermoelastic properties of undamaged composites: particulate reinforced (both spheres and spheroids), fibre reinforced and laminates. The second group comprises Chapters 8–14 considering the fundamentals of ply cracking, and the predictions of ply crack formation in damaged composites, which are categorised into symmetric cross-plies and general symmetric laminates subject to general in-plane loading, and also nonsymmetric cross-ply laminates subject to combined biaxial bending and in-plane loading. A rigorous approach is developed that allows much theoretical development without having to know the detailed distributions of stress and strain within the laminates. Much effort is devoted to the development of very useful interrelationships between the effective properties of damaged laminates, and their use when using an energy balance approach to predict ply crack formation. Chapter 12 is concerned with an approach to the prediction of delaminations from preexisting ply cracks, whereas Chapters 13 and 14 consider ply crack formation under conditions of fatigue loading and under aggressive environmental conditions. The third and final group of chapters are more advanced texts where the mathematical details underpinning some of the earlier chapters are described in more detail. Spheroidal particle reinforcement for undamaged composites is considered in Chapter 15, and debonded fibre/matrix interfaces and crack bridging are described in Chapter 16, whereas crack bridging of ply cracks in laminates is described in Chapter 17. Stress transfer mechanics for ply cracks in general symmetric laminates is considered in Chapter 18, and Chapter 19 describes stress transfer mechanics for ply cracks in nonsymmetric cross-ply laminates subject to biaxial bending.

There is no attempt in this book to provide comprehensive accounts of relevant parts of the literature, although reference will be made to source publications related to the analytical methods described in the book. Some topics considered in this book, e.g. the chapters on particulate composites, delamination, fatigue damage and environmental damage, have been included to extend the range of applicability of the analytical methods described in the book. The content of these chapters is based essentially on specific publications by the author that are available in the literature.

Reference

1. John Wiley & Sons, Inc. website (www.wiley.com/go/mccartney/properties).

2

Fundamental Relations for Continuum Models

Overview: This chapter introduces the basic principles on which the mechanics of continua are based. Having defined the concepts of vectors and tensors, the physical quantities displacement and velocity are defined for continuous systems and then applied to the fundamental balance laws for mass, momentum (linear and angular) and energy. The principles of the thermodynamics of multicomponent fluid systems are first introduced. The strain tensor is then introduced so that the thermodynamic approach can be extended to solid systems for the single-component solids that will be considered in this book. The fundamental equations are then described for linear thermoelastic solids subject to infinitesimal deformations. The chapter then specifies the constitutive equations required for the analysis of anisotropic solids that will be encountered throughout the book, including the transformation of anisotropic properties following rotation about a given coordinate axis. The chapter concludes by considering bend formation applied to a homogeneous orthotropic plate.

2.1 Introduction

The principal objective of this book is to develop theoretical models and associated software that can predict the deformation behaviour of composite materials for situations where the system is first undamaged, but then develops progressively growing damage as the applied loading is gradually increased. Even though the composite systems are heterogeneous, continuum methods can be applied to each constituent of the composite, and to assist in the development of models for homogenised effectively continuous systems where the details of the reinforcement and damage distribution have been smoothed, and where effective properties may be defined.

The objective of this chapter is to describe the fundamental principles on which theoretical developments will be based. The topics to be covered are the nature of vectors and tensors, the definitions of displacement and velocity vectors, the balance laws for mass, momentum and energy, thermodynamics involving stress and strain state variables, and a thorough treatment of the linear elastic behaviour of anisotropic materials, including a contracted notation that is used widely in the composites field.

Properties for Design of Composite Structures: Theory and Implementation Using Software, First Edition. Neil McCartney.

Companion Website: www.wiley.com/go/mccartney/properties

2.2 Vectors

A vector $\boldsymbol{v}$ is a mathematical entity that possesses both a magnitude and a direction. The vector is usually a physical quantity that is independent of the coordinate system that will be used to describe its properties. A very convenient approach is first to define an orthonormal set of coordinates (x_1,x_2,x_3). Three axes are drawn in the x_1, x_2 and x_3 directions, which are all at right angles to one another. Such a system is often described as a Cartesian set of coordinates. The positive directions of the x_1, x_1 and x_3 axes are described by unit vectors $\boldsymbol{i}_1$ $\boldsymbol{i}_2$ and $\boldsymbol{i}_3$, respectively, where $\boldsymbol{i}_1=(1,0,0)$, $\boldsymbol{i}_2=(0,1,0)$, $\boldsymbol{i}_3=(0,0,1)$. The unit vectors are such that

$$\boldsymbol{i}_1.\boldsymbol{i}_1=1,\quad \boldsymbol{i}_2.\boldsymbol{i}_2=1,\quad \boldsymbol{i}_3.\boldsymbol{i}_3=1,\quad \boldsymbol{i}_1.\boldsymbol{i}_2=0,\quad \boldsymbol{i}_2.\boldsymbol{i}_3=0,\quad \boldsymbol{i}_1.\boldsymbol{i}_3=0. \tag{2.1}$$

The three coordinates (x_1,x_2,x_3) describe the location of a point $\boldsymbol{x}$ that is known as the position vector, which may be written as $\boldsymbol{x}=x_1\boldsymbol{i}_1+x_2\boldsymbol{i}_2+x_3\boldsymbol{i}_3$. In tensor theory based on Cartesian coordinates, this is written in the shorter form $\boldsymbol{x}=x_k\boldsymbol{i}_k$ where a summation over values k = 1, 2, 3 is implied when a suffix is repeated (k in this example). Any vector $\boldsymbol{v}$ may be written as $\boldsymbol{v}=v_1\boldsymbol{i}_1+v_2\boldsymbol{i}_2+v_3\boldsymbol{i}_3$, or as $\boldsymbol{v}=v_k\boldsymbol{i}_k$ when using tensor notation. The scalar quantities v_k, k = 1, 2, 3, are the components of the vector $\boldsymbol{v}$ with values depending on the choice of coordinates. The magnitude of the vector $\boldsymbol{v}$ is specified by

$$|\boldsymbol{v}|=\sqrt{v_1^2+v_2^2+v_3^2}, \tag{2.2}$$

and its value is independent of the system of coordinates that is selected. The magnitude is, thus, an invariant of the vector.

The unit vector in the direction of the vector $\boldsymbol{v}$ is specified by $\boldsymbol{v}/|\boldsymbol{v}|$. Examples of vectors that occur in the physical world are forces, displacements, velocities and tractions.

2.3 Tensors

The tensors to be used in the book are either second order or fourth order. Tensors are usually physical quantities that are independent of the coordinate system that is used to describe their properties. For the given coordinate system having unit vectors $\boldsymbol{i}_1$, $\boldsymbol{i}_2$ and $\boldsymbol{i}_3$, a second-order tensor $\boldsymbol{t}$ is expressed in terms of the unit vectors as follows:

$$\boldsymbol{t}=t_{11}\boldsymbol{i}_1\boldsymbol{i}_1+t_{12}\boldsymbol{i}_1\boldsymbol{i}_2+t_{13}\boldsymbol{i}_1\boldsymbol{i}_3+t_{21}\boldsymbol{i}_2\boldsymbol{i}_1+t_{22}\boldsymbol{i}_2\boldsymbol{i}_2+t_{23}\boldsymbol{i}_2\boldsymbol{i}_3+t_{31}\boldsymbol{i}_3\boldsymbol{i}_1+t_{32}\boldsymbol{i}_3\boldsymbol{i}_2+t_{33}\boldsymbol{i}_3\boldsymbol{i}_3, \tag{2.3}$$

or, more compactly, using tensor notation in the form

$$\boldsymbol{t}=t_{jk}\boldsymbol{i}_j\boldsymbol{i}_k, \tag{2.4}$$

where summation over values 1, 2, 3 is implied by the repeated suffices j and k. The quantities t_{jk} are known as the components of a second-order tensor with values depending on the choice of coordinates. There are three independent invariants of second-order tensors which can be expressed in a variety of forms, the simplest being

$$I_1=t_{kk},\qquad I_2=t_{jk}t_{kj},\qquad I_3=t_{ij}t_{jk}t_{ki}. \tag{2.5}$$

A fourth-order tensor $\boldsymbol{T}$ is expressed in terms of the unit vectors of the coordinate system as follows

$$\boldsymbol{T} = T_{ijkl}\,\boldsymbol{i}_i\,\boldsymbol{i}_j\,\boldsymbol{i}_k\,\boldsymbol{i}_l, \tag{2.6}$$

where summation over values 1, 2, 3, is implied by the repeated suffices i, j, k and l. The quantities T_{ijkl} are known as the components of a fourth-order tensor with values depending on the choice of coordinates.

2.3.1 Fourth-order Elasticity Tensors

Elastic stress-strain equations are often written in the following form (see, for example, (2.153) and (2.154) given later in the chapter which includes thermal terms).

$$\sigma_{ij} = C_{ijkl}\,\varepsilon_{kl}, \qquad \varepsilon_{ij} = S_{ijkl}\,\sigma_{kl}. \tag{2.7}$$

It is clear that

$$\sigma_{ij} = C_{ijkl}\,S_{klmn}\,\sigma_{mn}, \tag{2.8}$$

which may be written as

$$\sigma_{ij} = \mathrm{I}_{ijmn}\,\sigma_{mn}, \tag{2.9}$$

where

$$\mathrm{I}_{ijmn} = C_{ijkl}\,S_{klmn}. \tag{2.10}$$

The fourth-order tensor I_{ijmn} can be defined by

$$\mathrm{I}_{ijmn} = \delta_{im}\delta_{jn}, \tag{2.11}$$

where δ_{ij} denotes the Kronecker delta symbol which has the value unity when $i = j$ and the value zero otherwise. Clearly

$$\begin{aligned} C_{ijkl}\,\mathrm{I}_{klmn} &= C_{ijkl}\delta_{km}\delta_{ln} = C_{ijmn}, \\ \mathrm{I}_{klmn}C_{mnrs} &= \delta_{km}\delta_{ln}C_{mnrs} = C_{klrs}. \end{aligned} \tag{2.12}$$

The identity tensor defined by (2.11) does not exhibit the same symmetry as the stiffness and compliance tensors, which are such that

$$\begin{aligned} C_{ijkl} &= C_{ijlk} = C_{klij} = C_{jilk}, \\ S_{ijkl} &= S_{ijlk} = S_{klij} = S_{jilk}. \end{aligned} \tag{2.13}$$

It is noted that

$$\begin{aligned} \mathrm{I}_{1212} &= \delta_{11}\delta_{22} = 1, \\ \mathrm{I}_{2112} &= \delta_{21}\delta_{12} = 0, \\ \mathrm{I}_{1221} &= \delta_{12}\delta_{21} = 0, \end{aligned} \tag{2.14}$$

indicating that $\mathrm{I}_{ijmn} \neq \mathrm{I}_{jimn}$ and $\mathrm{I}_{ijmn} \neq \mathrm{I}_{jinm}$.

A symmetric fourth-order identity tensor may be defined by

$$\mathrm{I}^{\mathrm{sym}}_{ijmn} = \tfrac{1}{2}\left(\delta_{im}\delta_{jn} + \delta_{in}\delta_{jm}\right), \tag{2.15}$$

so that

$$
\begin{aligned}
\mathrm{I}^{\mathrm{sym}}_{1212} &= \tfrac{1}{2}\left(\delta_{11}\delta_{22} + \delta_{12}\delta_{21}\right) = \tfrac{1}{2},\\
\mathrm{I}^{\mathrm{sym}}_{2112} &= \tfrac{1}{2}\left(\delta_{21}\delta_{12} + \delta_{22}\delta_{11}\right) = \tfrac{1}{2},\\
\mathrm{I}^{\mathrm{sym}}_{1221} &= \tfrac{1}{2}\left(\delta_{12}\delta_{21} + \delta_{11}\delta_{22}\right) = \tfrac{1}{2}.
\end{aligned}
\tag{2.16}
$$

The definition (2.15) is used to define the fourth-order identity tensors used in this book which are denoted by I or **I**.

2.4 Displacement and Velocity Vectors

Consider a continuous elastic medium that is being deformed from some homogeneous initial state as a result of loading. At some time t, a material point at $\boldsymbol{x}$ will have moved from its initial location $\bar{\boldsymbol{x}}$ in the material. The motion of the medium can be described by the following transformation $\mathbf{g}$ and its inverse $\boldsymbol{G}$

$$\boldsymbol{x} = \mathbf{g}(\bar{\boldsymbol{x}}, t), \qquad \bar{\boldsymbol{x}} = \boldsymbol{G}(\boldsymbol{x}, t). \tag{2.17}$$

The vector $\bar{\boldsymbol{x}}$ defines 'material coordinates', associated with the motion of the medium that, together with the function $\mathbf{g}$, can be used to describe the spatial variation during deformation of any physical quantity with respect to its original configuration. The transformation (2.17) is assumed to be single-valued and possess continuous partial derivatives with respect to their arguments. It is also assumed that the inverse function $\boldsymbol{G}$ exists locally, and this is always the case when the Jacobian J is such that

$$J \equiv \det\left|\frac{\partial \mathbf{g}}{\partial \bar{\boldsymbol{x}}}\right| \neq 0. \tag{2.18}$$

The displacement of a material point $\bar{\boldsymbol{x}}$ is denoted by $\boldsymbol{u}(\bar{\boldsymbol{x}}, t)$ when using material coordinates, and is defined by

$$\boldsymbol{u} = \boldsymbol{x} - \bar{\boldsymbol{x}} = \mathbf{g}(\bar{\boldsymbol{x}}, t) - \bar{\boldsymbol{x}}. \tag{2.19}$$

The velocity $\boldsymbol{v}$ of a material point $\bar{\boldsymbol{x}}$ may be calculated using the relation

$$\boldsymbol{v} = \left.\frac{\partial \mathbf{g}(\bar{\boldsymbol{x}}, t)}{\partial t}\right|_{\bar{\boldsymbol{x}}} \equiv \bar{\boldsymbol{v}}(\bar{\boldsymbol{x}}, t) \equiv \boldsymbol{v}(\boldsymbol{x}, t). \tag{2.20}$$

2.5 Material Time Derivative

The concept of a material time derivative (often named the substantive derivative) associated with the motion of the material is fundamental to the mechanics of continuous media. If ϕ is any scalar, vector or tensor quantity such that $\phi(\boldsymbol{x},t) \equiv \bar{\phi}(\bar{\boldsymbol{x}},t)$, then its material time derivative is defined by

$$\frac{\mathrm{D}\phi}{\mathrm{D}t} \equiv \left.\frac{\partial \bar{\phi}(\bar{\boldsymbol{x}}, t)}{\partial t}\right|_{\bar{\boldsymbol{x}}} = \left.\frac{\partial \phi(\boldsymbol{x}, t)}{\partial t}\right|_{\boldsymbol{x}} + \left.\frac{\partial \phi(\boldsymbol{x}, t)}{\partial \boldsymbol{x}} \cdot \frac{\partial \boldsymbol{x}(\bar{\boldsymbol{x}}, t)}{\partial t}\right|_{\bar{\boldsymbol{x}}}. \tag{2.21}$$

On using (2.20) the material time derivative may then be written as

$$\frac{\mathrm{D}\phi}{\mathrm{D}t} \equiv \frac{\partial \phi}{\partial t} + \boldsymbol{v}\,.\,\nabla\phi, \text{ and is such that } \boldsymbol{v} \equiv \frac{\mathrm{D}\boldsymbol{x}}{\mathrm{D}t}, \tag{2.22}$$

where ∇ denotes the gradient with respect to the coordinates $\boldsymbol{x}$.

Consider now a region V of the system bounded by the closed surface S enclosing a sample of the medium which is moving such that the velocity distribution at time t is denoted by $\boldsymbol{v}(\boldsymbol{x},t)$. It follows that, for any extensive property $\phi(\boldsymbol{x},t)$ of the system, the material time derivative (associated with the mean motion) of the integral of $\phi(\boldsymbol{x},t)$ over the region fixed region V may be written as

$$\frac{\mathrm{d}}{\mathrm{d}t}\int_V \phi(\mathbf{x},t)\,\mathrm{d}V \equiv \int_V \left[\frac{\mathrm{D}\phi}{\mathrm{D}t}\,\mathrm{d}V + \phi\,\frac{\mathrm{D}(\mathrm{d}V)}{\mathrm{D}t}\right]. \tag{2.23}$$

From reference [1, Equation (2.4.9)], for example,

$$\frac{\mathrm{D}(\mathrm{d}V)}{\mathrm{D}t} = \nabla.\boldsymbol{v}\,\mathrm{d}V. \tag{2.24}$$

On substituting (2.24) into (2.23) it follows on using (2.22) that

$$\frac{\mathrm{d}}{\mathrm{d}t}\int_V \phi(\boldsymbol{x},t)\,\mathrm{d}V \equiv \int_V \left[\frac{\partial\phi}{\partial t} + \nabla.(\phi\boldsymbol{v})\right]\mathrm{d}V, \tag{2.25}$$

where use has been made of the following identity

$$\nabla.(\phi\boldsymbol{v}) \equiv \phi\nabla.\boldsymbol{v} + \boldsymbol{v}\,.\nabla\phi. \tag{2.26}$$

On using the divergence theorem, the identity (2.25) may then be written as

$$\frac{\mathrm{d}}{\mathrm{d}t}\int_V \phi(\boldsymbol{x},t)\,\mathrm{d}V \equiv \int_V \frac{\partial\phi}{\partial t}\,\mathrm{d}V + \int_S \boldsymbol{n}.\boldsymbol{v}\,\phi\,\mathrm{d}S. \tag{2.27}$$

The first term on the right-hand side accounts for any changes of the property ϕ locally at points within the region V, whereas the second term accounts for the mean advection of the property across the bounding surface S where $\boldsymbol{n}$ is the outward unit normal to the surface S bounding the region V. The important identities (2.22) and (2.27) are used repeatedly in the following analysis.

2.6 Continuity Equation

Consider at time t a moving sample of material occupying a fixed volume V bounded by a closed surface S. In the absence of mass source and sink terms, the total mass of material within the region V is fixed so that on setting $\phi = \rho$ in (2.27) the global form of the mass balance equation for the medium may be written as

$$\int_V \frac{\partial\rho}{\partial t}\,\mathrm{d}V = -\int_S \rho\boldsymbol{v}.\boldsymbol{n}\,\mathrm{d}S, \tag{2.28}$$

where ρ is the mass density. The term on the left-hand side of (2.28) is the rate of change of the total mass in the fixed region V. The term on the right-hand side is the rate at which the mass of the medium is transported across the surface S into the region V. On using the divergence theorem, the mass balance equation may be written as

$$\int_V \frac{\partial \rho}{\partial t} \mathrm{d}V = -\int_V \nabla.\left(\rho \boldsymbol{v}\right) \mathrm{d}V. \tag{2.29}$$

As (2.29) must be valid for any region V of the system, the following local form of the mass balance equation for the medium must be satisfied at all points in the system for all times t

$$\frac{\partial \rho}{\partial t} = -\nabla.(\rho \boldsymbol{v}). \tag{2.30}$$

On using (2.22) and the identity (2.26), the continuity equation (2.30) may be written in the equivalent form

$$\frac{\mathrm{D}\rho}{\mathrm{D}t} = -\rho \nabla.\boldsymbol{v}. \tag{2.31}$$

2.7 Equations of Motion and Equilibrium

The global form of the linear momentum balance equation for a fixed region V bounded by the closed surface S having outward unit normal $\boldsymbol{n}$ is written as

$$\frac{\mathrm{d}}{\mathrm{d}t}\int_V \rho \boldsymbol{v}\, \mathrm{d}V = \int_S \boldsymbol{n}.\boldsymbol{\sigma}\, dS + \int_V \rho \boldsymbol{b}\, \mathrm{d}V, \tag{2.32}$$

where $\boldsymbol{\sigma}$ is the stress tensor and where $\boldsymbol{b}$ is the body force per unit mass acting on the medium. Such body forces usually arise from the effects of gravity. On using (2.27) and the divergence theorem, the linear momentum balance equation (2.32) may be written as

$$\int_V \left[\frac{\partial\left(\rho \boldsymbol{v}\right)}{\partial t} + \nabla.\left(\rho \boldsymbol{v}\boldsymbol{v}\right)\right] \mathrm{d}V = \int_V \nabla.\boldsymbol{\sigma} \mathrm{d}V + \int_V \rho \boldsymbol{b} \mathrm{d}V. \tag{2.33}$$

As relation (2.33) must be satisfied for any region V of the system, it follows that the local form of the linear momentum balance equation has the form

$$\frac{\partial\left(\rho \boldsymbol{v}\right)}{\partial t} + \nabla.\left(\rho \boldsymbol{v}\boldsymbol{v}\right) = \nabla.\boldsymbol{\sigma} + \rho \boldsymbol{b}. \tag{2.34}$$

As

$$\nabla.\left(\rho \boldsymbol{v}\boldsymbol{v}\right) \equiv \rho \boldsymbol{v}\, \nabla.\boldsymbol{v} + \boldsymbol{v}.\nabla\left(\rho \boldsymbol{v}\right), \tag{2.35}$$

it follows from (2.34) on using (2.22) that

$$\rho \frac{\mathrm{D}\boldsymbol{v}}{\mathrm{D}t} + \boldsymbol{v}\frac{\mathrm{D}\rho}{\mathrm{D}t} + \rho \boldsymbol{v} \nabla.\boldsymbol{v} = \nabla.\boldsymbol{\sigma} + \rho \boldsymbol{b}. \tag{2.36}$$

On using (2.31), relation (2.36) reduces to the well-known equation of motion

$$\rho \frac{\mathrm{D}\boldsymbol{v}}{\mathrm{D}t} = \nabla.\boldsymbol{\sigma} + \rho \boldsymbol{b}, \tag{2.37}$$

which must be satisfied at every point in the medium for all times $t > 0$.

It is worth noting that by considering the balance of angular momentum it can be shown that in the absence of couple stresses, the stress tensor $\boldsymbol{\sigma}$ must be symmetric.

2.8 Energy Balance Equation

The energy balance equation is derived from the first law of thermodynamics which requires that, for a fixed region V bounded by a closed surface S, the rate of change of the sum of the internal energy and kinetic energy is balanced by the sum of the heat flowing across the external surface S, and the rate of working of the external tractions acting on S and of the body force acting in V. There are two types of stored energy that must be considered. The first is the internal energy that accounts for the strain energy stored owing to elastic deformation and the energy of the thermal agitations of atoms in the solid. The second is the kinetic energy arising from the local average motion of the medium. In addition, the heat flow across the external boundary and the mechanical work done by the applied tractions must be considered. Body forces such as that caused by gravity also need to be taken into account. A final type of energy is the local heating that can arise, for example, from the flow of electric currents when electrodynamic effects are otherwise neglected. However, this local heating term is a very useful theoretical device for imposing precise isothermal conditions in a simulation. Owing to various energy dissipation processes, heat will be generated locally in the medium, leading to temperature variations and heat flow. Imagine that it is possible to add or remove heat at every point of the medium, and that this may be controlled to maintain a constant uniform temperature at all points. This approach to achieving isothermal conditions requires the use of a distribution of local heat sources or sinks. An alternative method is to impose isothermal conditions only on the external boundary, and to assume that the thermal conductivity is effectively infinite so that the heat generated within the medium by dissipation processes can flow immediately out of the system, thus maintaining a uniform temperature.

In the absence of electrodynamic effects, the global form of the energy balance equation is written as

$$\begin{aligned}\frac{\mathrm{d}}{\mathrm{d}t}\int_V \rho\upsilon\,\mathrm{d}V + \frac{\mathrm{d}}{\mathrm{d}t}\int_V \tfrac{1}{2}\rho v^2\,\mathrm{d}V &= -\int_S \boldsymbol{n}.\boldsymbol{h}\,\mathrm{d}S + \int_S \boldsymbol{n}.(\boldsymbol{\sigma}.\boldsymbol{v})\,\mathrm{d}S \\ &\quad + \int_V \rho\boldsymbol{b}.\boldsymbol{v}\,\mathrm{d}V + \int_V \rho r\,\mathrm{d}V,\end{aligned} \tag{2.38}$$

where υ is the specific internal energy, $\boldsymbol{h}$ is the heat flux vector and r is a local rate of heat supply per unit mass. In (2.38), the symbol v^2 is used to denote the value of the scalar product $\boldsymbol{v}\,.\,\boldsymbol{v}$. On using (2.27) the energy balance equation (2.38) may be written as

$$\begin{aligned}\int_V \frac{\partial}{\partial t}\left(\rho\upsilon + \tfrac{1}{2}\rho v^2\right)\mathrm{d}V + \int_S \boldsymbol{n}.\rho\boldsymbol{v}\left(\upsilon + \tfrac{1}{2}v^2\right)\mathrm{d}S &= -\int_S \boldsymbol{n}.\boldsymbol{h}\,\mathrm{d}S \\ + \int_S \boldsymbol{n}.(\boldsymbol{\sigma}.\boldsymbol{v})\mathrm{d}S + \int_V \rho\boldsymbol{b}.\boldsymbol{v}\mathrm{d}V + \int_V \rho r\mathrm{d}V.\end{aligned} \tag{2.39}$$

On using the divergence theorem, it then follows that

$$\begin{aligned}\int_V \frac{\partial}{\partial t}\left(\rho\upsilon + \tfrac{1}{2}\rho v^2\right)\mathrm{d}V &= -\int_V \nabla.\left[\boldsymbol{h} - \boldsymbol{\sigma}.\boldsymbol{v} + \rho\boldsymbol{v}\left(\upsilon + \tfrac{1}{2}v^2\right)\right]\mathrm{d}S \\ &\quad + \int_V \rho\boldsymbol{b}.\boldsymbol{v}\mathrm{d}V + \int_V \rho r\mathrm{d}V.\end{aligned} \tag{2.40}$$

As relation (2.40) must be satisfied for all regions V, this leads to the following local form for the energy balance equation

$$\frac{\partial}{\partial t}\left(\rho \upsilon+\tfrac{1}{2}\rho \boldsymbol{v}^2\right)=-\nabla.\left[\boldsymbol{h}-\boldsymbol{\sigma}.\boldsymbol{v}+\rho\boldsymbol{v}\left(\upsilon+\tfrac{1}{2}\boldsymbol{v}^2\right)\right]+\rho\boldsymbol{b}.\boldsymbol{v}+\rho r, \tag{2.41}$$

which must be satisfied at every point in the medium for all times $t > 0$. On using the continuity equation (2.30), the local energy balance equation reduces to the form

$$\rho\frac{\mathrm{D}}{\mathrm{D}t}\left(\upsilon+\tfrac{1}{2}\boldsymbol{v}^2\right)=-\nabla.\left(\boldsymbol{h}-\boldsymbol{\sigma}.\boldsymbol{v}\right)+\rho\boldsymbol{b}.\boldsymbol{v}+\rho r. \tag{2.42}$$

From the equation of motion (2.37) it follows, on taking the scalar product with the velocity vector $\boldsymbol{v}$, that

$$\rho\frac{\mathrm{D}}{\mathrm{D}t}\left(\tfrac{1}{2}\boldsymbol{v}^2\right)=\nabla.\left(\boldsymbol{\sigma}.\boldsymbol{v}\right)-\boldsymbol{\sigma}:(\nabla\boldsymbol{v})^{\mathrm{T}}+\rho\boldsymbol{b}.\boldsymbol{v}, \tag{2.43}$$

where a superscript T denotes that the transpose of the tensor must be used. The symbol: is defined here such that for any second-order tensors $\boldsymbol{a}$ and $\boldsymbol{b}$ the double scalar product $\boldsymbol{a}:\boldsymbol{b}\equiv a_{ij}b_{ji}$. An alternative definition $\boldsymbol{a}:\boldsymbol{b}\equiv a_{ij}b_{ij}$ is sometimes used in the literature. On using (2.43), together with the symmetry of the stress tensor, (2.42) reduces to the form

$$\rho\frac{\mathrm{D}\upsilon}{\mathrm{D}t}=-\nabla.\boldsymbol{h}+\boldsymbol{\sigma}:\boldsymbol{d}+\rho r, \tag{2.44}$$

where $\boldsymbol{d}$ is the symmetric rate of deformation tensor defined by

$$\boldsymbol{d}=\frac{1}{2}\left[\nabla\boldsymbol{v}+\left(\nabla\boldsymbol{v}\right)^{\mathrm{T}}\right]=\boldsymbol{d}^{\mathrm{T}}. \tag{2.45}$$

Relation (2.44) is the well-known local form of the internal energy balance equation for a continuous medium.

2.8.1 Conservative Body Forces

Consider now the special case when the heat source per unit mass $r = 0$ and the body force is derivable from a scalar potential function ζ as follows

$$\boldsymbol{b}=-\nabla\zeta,\qquad \frac{\partial\zeta}{\partial t}\equiv 0. \tag{2.46}$$

The effects of the Earth's gravitational field can then be taken into account. The substitution of (2.46) into (2.41) leads to the relation

$$\frac{\partial}{\partial t}\left(\rho \upsilon+\tfrac{1}{2}\rho \boldsymbol{v}^2\right)=-\nabla.\left[\boldsymbol{h}-\boldsymbol{\sigma}.\boldsymbol{v}+\rho\boldsymbol{v}\left(\upsilon+\tfrac{1}{2}\boldsymbol{v}^2\right)\right]-\rho\boldsymbol{v}.\nabla\zeta. \tag{2.47}$$

As

$$\rho\boldsymbol{v}.\nabla\zeta=\nabla.\left(\rho\boldsymbol{v}\zeta\right)-\zeta\nabla.\left(\rho\boldsymbol{v}\right)=\nabla.\left(\rho\boldsymbol{v}\zeta\right)+\frac{\partial(\rho\zeta)}{\partial t}, \tag{2.48}$$

it then follows from (2.47) that the local form (2.44) for the total energy balance equation may also be expressed

$$\frac{\partial}{\partial t}\left(\rho\upsilon+\rho\zeta+\frac{1}{2}\rho\boldsymbol{v}^2\right)=-\nabla.\left[\boldsymbol{h}-\boldsymbol{\sigma}.\boldsymbol{v}+\rho\boldsymbol{v}\left(\upsilon+\zeta+\frac{1}{2}\boldsymbol{v}^2\right)\right]. \tag{2.49}$$

The energy balance equation may, therefore, be written in the more compact form

$$\frac{\partial \chi}{\partial t} = -\nabla . \boldsymbol{J}_{\chi}, \tag{2.50}$$

where the total energy per unit volume χ is given by

$$\chi = \rho \upsilon + \rho \zeta + \frac{1}{2} \rho \boldsymbol{v}^2, \tag{2.51}$$

which is the sum of the internal energy, potential energy and kinetic energy per unit volume, and where the total energy flux $\boldsymbol{J}_{\chi}$ is given by

$$\boldsymbol{J}_{\chi} = \boldsymbol{h} - \boldsymbol{\sigma} . \boldsymbol{v} + \rho \boldsymbol{v} \left(\upsilon + \zeta + \tfrac{1}{2} \boldsymbol{v}^2 \right). \tag{2.52}$$

On using (2.51), relation (2.52) may also be written in the form

$$\boldsymbol{J}_{\chi} = \chi \boldsymbol{v} + \boldsymbol{h} - \boldsymbol{\sigma} . \boldsymbol{v}, \tag{2.53}$$

which identifies the mechanisms whereby energy can enter or leave the system, namely advection (the term $\chi \boldsymbol{v}$), as heat (the term $\boldsymbol{h}$) and as external work (the term $-\boldsymbol{\sigma} . \boldsymbol{v}$).

The energy balance equation (2.50) implies energy conservation as the energy within any region V always remains fixed during any nonequilibrium process provided that there is no energy flow across the surface S bounding the region V. This result is easily established by integrating (2.50) over the region V, and then making use of the divergence theorem.

2.9 Equations of State for Hydrostatic Stress States

The balance equations of mass, momentum and energy introduce a variety of physical variables, but they do not involve in any way the properties of materials. An equation of state must now be introduced, implying the existence of a variety of material properties, depending upon the complexity of the material. If elastic effects are to be included in models, then it is through an equation of state that they must be introduced. Before using strain as a thermodynamic state variable, it is useful first to introduce some key thermodynamic relationships that can be applied when the stress field is hydrostatic, i.e. shear stresses are absent, characterised by a pressure p, such that the stress tensor has the form $\boldsymbol{\sigma} = -p\mathbf{I}$ where $\mathbf{I}$ is the second-order unit tensor.

2.9.1 Global Thermodynamic Relations

It is essential to understand fully the nature of thermodynamic principles that underpin modelling procedures. It is assumed now that the system studied is multi-component and uniform, and that the temperature and pressure are uniform as would be the case for many equilibrium states of the system. At equilibrium, the classical thermodynamic relations for a uniform multi-component mixture, where M_k, $k = 1, \ldots, n$ denote the total masses of the various species, are given by

$$\begin{aligned}
U &= \widehat{U}(S, V, \{M_k\}), \qquad & dU &= TdS - pdV + \sum_{k=1}^{n} \mu_k dM_k, \\
F &= \widehat{F}(T, V, \{M_k\}), \qquad & dF &= -SdT - pdV + \sum_{k=1}^{n} \mu_k dM_k, \\
G &= \widehat{G}(T, p, \{M_k\}), \qquad & dG &= -SdT + Vdp + \sum_{k=1}^{n} \mu_k dM_k,
\end{aligned} \tag{2.54}$$

where U is the internal energy of the system and F and G are the corresponding Helmholtz and Gibbs energies defined by

$$F = U - TS, \quad G = F + pV = U - TS + pV. \tag{2.55}$$

The brackets {} in (2.54) denote a set of state variables associated with the various species defined for $k = 1, \ldots, n$. The state variables S, T, V and p are the entropy, thermodynamic (i.e. absolute) temperature, volume and pressure, respectively, and the parameters μ_k, $k = 1, \ldots, n$, are the mass-based chemical potentials. Each of the relations (2.54) defines, for media having uniform state variables and hydrostatic stress states (i.e. in the absence of shear stresses), an equation of state for three different but equivalent selections of the independent state variables. It follows from the three differential relations in (2.54) that

$$\begin{gathered} T = \frac{\partial \hat{U}}{\partial S}, \quad p = -\frac{\partial \hat{U}}{\partial V} = -\frac{\partial \hat{F}}{\partial V}, \quad V = \frac{\partial \hat{G}}{\partial p}, \quad S = -\frac{\partial \hat{F}}{\partial T} = -\frac{\partial \hat{G}}{\partial T}, \\ \mu_k = \frac{\partial \hat{U}}{\partial M_k} = \frac{\partial \hat{F}}{\partial M_k} = \frac{\partial \hat{G}}{\partial M_k}, \ k = 1, \ldots, n. \end{gathered} \tag{2.56}$$

The total mass of the system denoted by M is given by

$$M = \sum_{k=1}^{n} M_k. \tag{2.57}$$

Any extensive state variable P depends directly on the total M of the system and is such that for *any* value of the dimensionless parameter λ

$$P(\lambda M) = \lambda P(M). \tag{2.58}$$

On applying this principle to the functions introduced in (2.54)

$$\begin{aligned} &\hat{U}(\lambda S, \lambda V, \{\lambda M_k\}) = \lambda \hat{U}(S, V, \{M_k\}), \\ &\hat{F}(T, \lambda V, \{\lambda M_k\}) = \lambda \hat{F}(T, V, \{M_k\}), \\ &\hat{G}(T, p, \{\lambda M_k\}) = \lambda \hat{G}(T, p, \{M_k\}). \end{aligned} \tag{2.59}$$

It follows that the functions $\hat{U}$, $\hat{F}$ or $\hat{G}$ are homogeneous of order unity in the extensive state variables. Provided that λ is independent of the state variables, on differentiating relations (2.59) with respect to λ, and then setting $\lambda = 1$, it follows on using (2.56) that

$$\begin{aligned} U &= \hat{U}(S, V, \{M_k\}) = S\frac{\partial \hat{U}}{\partial S} + V\frac{\partial \hat{U}}{\partial V} + \sum_{k=1}^{n} M_k \frac{\partial \hat{U}}{\partial M_k} \\ &= ST - pV + \sum_{k=1}^{n} M_k \mu_k, \\ F &= \hat{F}(T, V, \{M_k\}) = V\frac{\partial \hat{F}}{\partial V} + \sum_{k=1}^{n} M_k \frac{\partial \hat{F}}{\partial M_k} = -pV + \sum_{k=1}^{n} M_k \mu_k, \\ G &= \hat{G}(T, p, \{M_k\}) = \sum_{k=1}^{n} M_k \frac{\partial \hat{G}}{\partial M_k} = \sum_{k=1}^{n} M_k \mu_k. \end{aligned} \tag{2.60}$$

The quantities U, F, G, S, V and $\{M_k\}$ are the extensive variables of uniform multicomponent systems used in thermodynamics. The intensive variables, which do not depend on the total mass M of the system, are the temperature T, pressure p and the chemical potentials μ_k, $k = 1, \ldots, n$. The thermodynamic relations may only be used for uniform systems which are in a state of mechanical and thermal equilibrium. Their use needs to be extended to non-uniform states where gradients of state variables are encountered, and to dynamic states where the various species are subject to motion. In this book, diffusion mechanisms will be neglected, which means that all species at any given location in the material will move with the same velocity $\boldsymbol{v}$.

2.9.2 Local Thermodynamic Relations

The extensive quantities are not used directly in simulations, as local expressions of these thermodynamic quantities defined per mole, per unit mass or per unit volume need to be defined. The approach to be taken here is to define thermodynamic variables per unit mass by introducing the state variables υ, ψ, φ, Ω and η using the relations

$$\begin{gathered} U = M\upsilon, \qquad F = M\psi, \qquad G = M\varphi, \qquad V = M\Omega, \qquad S = M\eta, \\ M_k = M\omega_k,\ k = 1, \ldots, n, \text{ where } M = \sum_{k=1}^{n} M_k, \end{gathered} \tag{2.61}$$

implying that

$$\sum_{k=1}^{n} \omega_k = 1, \tag{2.62}$$

where υ,ψ and φ are the specific internal, specific Helmholtz and specific Gibbs energies, respectively, Ω is the volume per unit mass (i.e. $1/\rho$ where ρ is the mass density), whereas η and ω_k, $k = 1, \ldots, n$ are, respectively, the specific entropy and mass fractions for the n species. Substitution of (2.61) into (2.54) then leads to

$$\begin{aligned} \mathrm{d}\upsilon &= T\mathrm{d}\eta - p\mathrm{d}\Omega + \sum_{k=1}^{n} \mu_k \mathrm{d}\omega_k, \\ \mathrm{d}\psi &= -\eta \mathrm{d}T - p\mathrm{d}\Omega + \sum_{k=1}^{n} \mu_k \mathrm{d}\omega_k, \\ \mathrm{d}\varphi &= -\eta \mathrm{d}T + \Omega \mathrm{d}p + \sum_{k=1}^{n} \mu_k \mathrm{d}\omega_k, \end{aligned} \tag{2.63}$$

where use has been made of the following relation derived using (2.60) and (2.61)

$$\upsilon - T\eta + p\Omega = \psi + p\Omega = \varphi = \sum_{k=1}^{n} \omega_k \mu_k. \tag{2.64}$$

On setting $\lambda = 1/M$ in (2.59) and on using (2.61) the local equations of state are given by

$$\upsilon = \widehat{U}(\eta, \Omega, \{\omega_k\}), \qquad \psi = \widehat{F}(T, \Omega, \{\omega_k\}), \qquad \varphi = \widehat{G}(T, p, \{\omega_k\}). \tag{2.65}$$

The differential relations (2.63) imply that

$$\begin{gathered} T = \frac{\partial \widehat{U}}{\partial \eta}, \quad p = -\frac{\partial \widehat{U}}{\partial \Omega} = -\frac{\partial \widehat{F}}{\partial \Omega}, \quad \Omega = \frac{\partial \widehat{G}}{\partial p}, \quad \eta = -\frac{\partial \widehat{F}}{\partial T} = -\frac{\partial \widehat{G}}{\partial T}, \\ \mu_k = \frac{\partial \widehat{U}}{\partial \omega_k} = \frac{\partial \widehat{F}}{\partial \omega_k} = \frac{\partial \widehat{G}}{\partial \omega_k},\ k = 1, \ldots, n. \end{gathered} \tag{2.66}$$

It should be noted that the set of mass fractions $\{\omega_k\}$ appearing as state variables in (2.65) are not independent because their sum is unity as shown in (2.61). The partial derivatives with respect to mass fractions given in (2.66) must, however, be calculated under the assumption that the mass fractions are independent variables but that the relationship (2.62) is not yet assumed to be satisfied. This means that relation (2.62) cannot yet be used to simplify or modify the functions $\widehat{U}$, $\widehat{F}$ or $\widehat{G}$. It should also be noted that the functions $\widehat{U}$, $\widehat{F}$ and $\widehat{G}$ appearing in the local relations (2.66) are, in fact, defined for uniform systems at a global level, as shown in (2.54) where the set of state variables $\{M_k\}$, corresponding to $\{\omega_k\}$, are independent. As soon as all partial derivatives with respect to mass fractions have been calculated, it is then possible to simplify or modify results by making use the mass fraction relation (2.62).

The description of thermodynamic relations given here applies to multi-component fluids in equilibrium, and to solids provided the stress state is hydrostatic so that shear stresses are absent. In particular, the relations apply to the uniform initial reference state of elasticity theory where the background pressure and temperature are both uniform. Although the thermodynamic relations have been given for multi-component systems, it is emphasised that this is referring to mixtures of atoms and/or molecules of the various n species, which is an example of an 'atomistic' composite. Throughout this book there is no need to consider materials and associated models at the atomistic length scale, as all materials are considered homogenised into macroscopic continua. The composites to be considered are formed from a continuum representing the matrix and one or more types of reinforcement which are also modelled as continua. When applying thermodynamic principles to continua, it is sufficient to regard each material present as a thermodynamic system having just a single component so that $n = 1$. On using (2.62) it is then clear that $\omega_1 = 1$. The equations of state (2.65) may then be written in the form

$$v = \hat{v}(\eta, \Omega), \qquad \psi = \hat{\psi}(T, \Omega), \qquad \varphi = \hat{\varphi}(T, p), \tag{2.67}$$

and the differential relations (2.63) reduce to the simpler forms

$$\mathrm{d}v = T\mathrm{d}\eta - p\mathrm{d}\Omega, \quad \mathrm{d}\psi = -\eta\mathrm{d}T - p\mathrm{d}\Omega, \quad \mathrm{d}\varphi = -\eta\mathrm{d}T + \Omega\mathrm{d}p, \tag{2.68}$$

whereas relations (2.66) reduce to

$$T = \frac{\partial \hat{v}}{\partial \eta}, \quad p = -\frac{\partial \hat{v}}{\partial \Omega} = -\frac{\partial \hat{\psi}}{\partial \Omega}, \quad \Omega = \frac{\partial \hat{\varphi}}{\partial p}, \quad \eta = -\frac{\partial \hat{\psi}}{\partial T} = -\frac{\partial \hat{\varphi}}{\partial T}. \tag{2.69}$$

The analysis of this section has been presented as it establishes a basis for connecting composite modelling to the well-known principles of thermodynamics. The relationships (2.67)–(2.69) apply to single-component solids subject only to hydrostatic stress states leading to the possibility of specifying constitutive equations that depend on a limited set of material properties, e.g. bulk modulus and bulk thermal expansion coefficient. To extend the analysis to all types of thermoelastic material properties needed for the applications to be considered in this book, it is required to introduce in the next section strain tensors as state variables, rather than the specific volume Ω introduced in (2.65) and (2.67).

2.10 Strain Tensor

Unlike the situation for fluids, the deformation (i.e. motion) of solid media is defined relative to an initial homogeneous reference state, thus enabling the introduction of the concept of strain, which takes account of both hydrostatic deformation considered in Section 2.9 and shear deformation. Following the approach in [1], consider now a homogeneous solid body with an initial configuration that occupies a region $\overline{B}$, and which deforms into a different region B after the application of applied stresses and temperature changes. Let $\overline{\mathbf{p}}$ denote the position vector of some point in the undeformed body $\overline{B}$ referred to Cartesian coordinates $\overline{\boldsymbol{x}}$, that moves to the point $\mathbf{p}$ in the deformed body B referred to Cartesian coordinates $\boldsymbol{x}$. Let $\overline{\boldsymbol{i}}_K$ denote the orthogonal unit base vectors for the coordinate system $\overline{\boldsymbol{x}}$, and $\boldsymbol{i}_k$ denote the orthogonal unit base vectors for the coordinate system $\boldsymbol{x}$. The position vectors $\overline{\mathbf{p}}$ and $\mathbf{p}$ may then be expressed in the following form

$$\overline{\mathbf{p}} = \overline{x}_K \overline{\boldsymbol{i}}_K, \qquad \mathbf{p} = x_k \boldsymbol{i}_k, \tag{2.70}$$

where summations over values K, $k = 1, 2, 3$, are implied for repeated suffices. The corresponding infinitesimal vectors are written as

$$\mathrm{d}\overline{\mathbf{p}} = \mathrm{d}\overline{x}_K \overline{\boldsymbol{i}}_K, \qquad \mathrm{d}\mathbf{p} = \mathrm{d}x_k \boldsymbol{i}_k. \tag{2.71}$$

As

$$\overline{\boldsymbol{i}}_K . \overline{\boldsymbol{i}}_L = \delta_{KL} \qquad \text{and} \qquad \boldsymbol{i}_k . \boldsymbol{i}_l = \delta_{kl}, \tag{2.72}$$

it follows that

$$\begin{aligned} \mathrm{d}\overline{s}^2 &= \mathrm{d}\overline{\mathbf{p}}.\mathrm{d}\overline{\mathbf{p}} = \delta_{KL}\mathrm{d}\overline{x}_K \mathrm{d}\overline{x}_L = \mathrm{d}\overline{x}_K \mathrm{d}\overline{x}_K, \\ \mathrm{d}s^2 &= \mathrm{d}\mathbf{p}.\mathrm{d}\mathbf{p} = \delta_{kl}\mathrm{d}x_k \mathrm{d}x_l = \mathrm{d}x_k \mathrm{d}x_k. \end{aligned} \tag{2.73}$$

Any vector $\boldsymbol{v}$ may be written as

$$\boldsymbol{v} = \overline{v}_K \overline{\boldsymbol{i}}_K = v_k \boldsymbol{i}_k. \tag{2.74}$$

Define δ_{Kl} and δ_{kL} by the relation

$$\delta_{Kk} = \delta_{kK} = \overline{\boldsymbol{i}}_K . \boldsymbol{i}_k. \tag{2.75}$$

It then follows that

$$v_k = \delta_{kK} \overline{v}_K \qquad \text{and} \qquad \overline{v}_K = \delta_{Kk} v_k. \tag{2.76}$$

It is clear that

$$\delta_{Kk}\delta_{kL} = \delta_{KL} \qquad \text{and} \qquad \delta_{kK}\delta_{Kl} = \delta_{kl}. \tag{2.77}$$

The time-dependent deformation that transforms the undeformed region $\overline{B}$ into the region B(t) may be expressed as

$$\boldsymbol{x} = \boldsymbol{x}(\overline{\boldsymbol{x}}, t), \qquad \overline{\boldsymbol{x}} = \overline{\boldsymbol{x}}(\boldsymbol{x}, t). \tag{2.78}$$

It then follows that

$$\mathrm{d}\boldsymbol{x} = \frac{\partial \boldsymbol{x}}{\partial \overline{\boldsymbol{x}}} \mathrm{d}\overline{\boldsymbol{x}}, \quad \mathrm{d}\overline{\boldsymbol{x}} = \frac{\partial \overline{\boldsymbol{x}}}{\partial \boldsymbol{x}} \mathrm{d}\boldsymbol{x}, \tag{2.79}$$

where the increments are taken at some time t such that $\mathrm{d}t = 0$. In component form,

$$\mathrm{d}x_k = \frac{\partial x_k}{\partial \overline{x}_K} \mathrm{d}\overline{x}_K, \quad \mathrm{d}\overline{x}_K = \frac{\partial \overline{x}_K}{\partial x_k} \mathrm{d}x_k. \tag{2.80}$$

In the undeformed body, on using (2.71) and (2.73), the increment of arc length $d\bar{s}$ is such that

$$d\bar{s}^2 = d\bar{\boldsymbol{x}}.d\bar{\boldsymbol{x}} = \frac{\partial \bar{x}_K}{\partial x_k}\frac{\partial \bar{x}_K}{\partial x_l} dx_k dx_l = c_{kl} dx_k dx_l, \tag{2.81}$$

where c_{kl} is Cauchy's symmetric deformation tensor. Similarly, for the deformed body, the line increment $d\bar{s}$ deforms to an increment ds such that

$$ds^2 = d\boldsymbol{x}.d\boldsymbol{x} = \frac{\partial x_k}{\partial \bar{x}_K}\frac{\partial x_k}{\partial \bar{x}_L} d\bar{x}_K d\bar{x}_L = C_{KL} d\bar{x}_K d\bar{x}_L, \tag{2.82}$$

where C_{KL} is Green's symmetric deformation tensor. In dyadic form

$$ds^2 = d\boldsymbol{x}.d\boldsymbol{x} = \left(d\bar{\boldsymbol{x}}.\bar{\nabla}\boldsymbol{x}\right).\left(d\bar{\boldsymbol{x}}.\bar{\nabla}\boldsymbol{x}\right) = d\bar{\boldsymbol{x}}.\boldsymbol{C}.d\bar{\boldsymbol{x}}, \tag{2.83}$$

where $\bar{\nabla}$ denotes the gradient with respect to the material coordinates $\bar{\boldsymbol{x}}$, and where the symmetric Green deformation tensor $\boldsymbol{C}$ may be written as

$$C_{KL} = \frac{\partial x_i}{\partial \bar{x}_K}\frac{\partial x_i}{\partial \bar{x}_L}, \text{ i.e. } \boldsymbol{C} = \bar{\nabla}\boldsymbol{x}.\left(\bar{\nabla}\boldsymbol{x}\right)^{\mathrm{T}}. \tag{2.84}$$

From (2.81) and (2.82)

$$ds^2 - d\bar{s}^2 = \left(\delta_{kl} - c_{kl}\right) dx_k dx_l = \left(C_{KL} - \delta_{KL}\right) d\bar{x}_K d\bar{x}_L. \tag{2.85}$$

The Eulerian and Lagrange strain tensors are defined by

$$2e_{kl} = \delta_{kl} - c_{kl}, \qquad 2E_{KL} = C_{KL} - \delta_{KL}, \tag{2.86}$$

so that (2.85) may also be written as

$$ds^2 - d\bar{s}^2 = 2e_{kl} dx_k dx_l = 2E_{KL} d\bar{x}_K d\bar{x}_L. \tag{2.87}$$

On using the relation $\boldsymbol{u} = \boldsymbol{x} - \bar{\boldsymbol{x}}$ for the displacement vector and (2.84), the Lagrangian strain tensor may be written in terms of the displacement vector as follows

$$2\boldsymbol{E} \equiv \boldsymbol{C} - \mathbf{I} = \bar{\nabla}\boldsymbol{u} + \left(\bar{\nabla}\boldsymbol{u}\right)^{\mathrm{T}} + \bar{\nabla}\boldsymbol{u}.\left(\bar{\nabla}\boldsymbol{u}\right)^{\mathrm{T}}, \tag{2.88}$$

where $\mathbf{I}$ is the symmetric fourth-order identity tensor (see (2.15)). The quantity $\boldsymbol{E}$ is the strain tensor that is used in finite deformation theory where there is no restriction on the degree of deformation provided that the deformation is continuous and the condition (2.18) is satisfied at all points in the system.

The invariants of the strain tensors in terms of principal stretches are given by the relations

$$I_C = \lambda_1^2 + \lambda_2^2 + \lambda_3^2, \; I_c = \frac{1}{\lambda_1^2} + \frac{1}{\lambda_2^2} + \frac{1}{\lambda_3^2}, \tag{2.89}$$

$$II_C = \lambda_1^2\lambda_2^2 + \lambda_2^2\lambda_3^2 + \lambda_1^2\lambda_3^2, \qquad II_c = \frac{1}{\lambda_1^2\lambda_2^2} + \frac{1}{\lambda_2^2\lambda_3^2} + \frac{1}{\lambda_1^2\lambda_3^2}, \tag{2.90}$$

$$III_C = \lambda_1^2\lambda_2^2\lambda_3^2, \; III_c = \frac{1}{\lambda_1^2\lambda_2^2\lambda_3^2}. \tag{2.91}$$

It is clear that

$$I_c = \frac{II_C}{III_C}, \quad II_c = \frac{I_C}{III_C}, \quad III_c = \frac{1}{III_C}. \tag{2.92}$$

It will be very useful to introduce here the principal values C_J, $J = 1, 2, 3$, of Green's deformation tensor defined using the following relations

$$\left.\begin{aligned} \det\left(\boldsymbol{C} - C_J\mathbf{I}\right) = 0\,, \\ \left(\boldsymbol{C} - C_J\mathbf{I}\right).\boldsymbol{\nu}_J = 0\,, \end{aligned}\right\} \quad J = 1, 2, 3, \tag{2.93}$$

such that the symmetric tensor $\boldsymbol{C}$ may be written in the form

$$\boldsymbol{C} \equiv \sum_{J=1}^{3} C_J \boldsymbol{\nu}_J \boldsymbol{\nu}_J. \tag{2.94}$$

The quantities $\boldsymbol{\nu}_J$, $J = 1, 2, 3$, are orthogonal unit vectors defining the directions of the principal values. They have the following properties

$$\begin{gathered} \boldsymbol{\nu}_J.\boldsymbol{\nu}_J = 1, \;\; J = 1, 2, 3, \qquad \boldsymbol{\nu}_J.\boldsymbol{\nu}_K = 0 \;\; \text{if } J \neq K, \;\; \text{for } J, K = 1, 2, 3, \\ \boldsymbol{\nu}_J.\,\mathrm{d}\boldsymbol{\nu}_K = \mathrm{d}\boldsymbol{\nu}_J.\boldsymbol{\nu}_K = 0 \;\; \text{for } J, K = 1, 2, 3. \end{gathered} \tag{2.95}$$

The polar decomposition principle (see, for example, [2, Section 1.5]) states that the deformation gradient may be expressed in the following forms (dyadic and tensor)

$$\frac{\partial \boldsymbol{x}}{\partial \bar{\boldsymbol{x}}} \equiv \bar{\nabla}\boldsymbol{x} = \boldsymbol{R}.\boldsymbol{U} = \boldsymbol{V}.\boldsymbol{R}, \quad \frac{\partial x_i}{\partial \bar{x}_K} = R_{iL} U_{LK} = V_{ij} R_{jK}, \tag{2.96}$$

where $\boldsymbol{R}$ is the orthogonal rigid rotation tensor having the properties $\boldsymbol{R}.\boldsymbol{R}^{\mathrm{T}} = \boldsymbol{R}^{\mathrm{T}}.\boldsymbol{R} = \mathbf{I}$ with $\det(\boldsymbol{R}) = \pm 1$, and where $\boldsymbol{U}$ and $\boldsymbol{V}$ are positive-definite symmetric right and left stretch tensors.

It follows from (2.84) and (2.96) that in tensor form

$$\begin{aligned} C_{KL} &= \frac{\partial x_i}{\partial \bar{x}_K}\frac{\partial x_i}{\partial \bar{x}_L} = R_{iM} U_{MK} R_{iN} U_{NL} = R^{\mathrm{T}}_{Ni} R_{iM} U_{MK} U_{NL} \\ &= \delta_{MN} U_{MK} U_{NL} = U_{MK} U_{ML}, \end{aligned} \tag{2.97}$$

so that in dyadic form

$$\boldsymbol{C} = \boldsymbol{U}.\boldsymbol{U}. \tag{2.98}$$

The symmetric tensors $\boldsymbol{U}$ and $\boldsymbol{V}$ have common eigenvalues λ_J but different mutually orthogonal eigenvectors $\boldsymbol{\nu}_J$ such that

$$\boldsymbol{U}.\boldsymbol{\nu}_J = \lambda_J \boldsymbol{\nu}_J, \quad \boldsymbol{V}.\left(\boldsymbol{R}.\boldsymbol{\nu}_J\right) = \lambda_J\left(\boldsymbol{R}.\boldsymbol{\nu}_J\right), \;\; J = 1, 2, 3. \tag{2.99}$$

The eigenvalues λ_J, $J = 1, 2, 3$, are the principal stretches and

$$\boldsymbol{U} = \sum_{J=1}^{3} \lambda_J \boldsymbol{\nu}_J \boldsymbol{\nu}_J. \tag{2.100}$$

It then follows that the principal values C_J of the $\boldsymbol{C}$ may be written in terms of the principal stretches λ_J as follows

$$C_J = \lambda_J^2, \quad J = 1, 2, 3. \tag{2.101}$$

It follows from (2.94) that

$$\mathrm{d}\boldsymbol{C} = \sum_{J=1}^{3} \mathrm{d}C_J \boldsymbol{\nu}_J \boldsymbol{\nu}_J + \sum_{J=1}^{3} C_J \boldsymbol{\nu}_J \mathrm{d}\boldsymbol{\nu}_J + \sum_{J=1}^{3} C_J \mathrm{d}\boldsymbol{\nu}_J \boldsymbol{\nu}_J, \tag{2.102}$$

and on multiplying by $\boldsymbol{\nu}_K$ using the properties (2.95) it can be shown that

$$\boldsymbol{\nu}_K.\mathrm{d}\boldsymbol{C}.\boldsymbol{\nu}_K = \mathrm{d}C_K, \quad K = 1, 2, 3. \tag{2.103}$$

In terms of the principal values C_1, C_2 and C_3 of the tensor $\boldsymbol{C}$ and E_1, E_2 and E_3 of the tensor $\boldsymbol{E}$, the corresponding invariants may be written as (see [1, Section 1.10])

$$\begin{aligned} I_C &= C_1 + C_2 + C_3, \quad II_C = C_1C_2 + C_2C_3 + C_3C_1, \quad III_C = C_1C_2C_3, \\ I_E &= E_1 + E_2 + E_3, \quad II_E = E_1E_2 + E_2E_3 + E_3E_1, \quad III_E = E_1E_2E_3. \end{aligned} \tag{2.104}$$

It can be shown that

$$\frac{\rho}{\rho_0} = \frac{1}{\lambda_1\lambda_2\lambda_3} = \frac{1}{\sqrt{\det(\boldsymbol{C})}} = \frac{1}{\sqrt{III_C}}, \tag{2.105}$$

where ρ_0 is the uniform mass density before deformation has occurred at some reference temperature T_0 and reference pressure p_0.

2.11 Field Equations for Infinitesimal Deformations

Continuum mechanics is based upon conservation laws that lead to the basic field equations that are independent of the properties of material to which the laws are applied. The mathematical statement of these laws is now given for the case of infinitesimal deformations where there is no practical distinction between the use of so-called material coordinates and spatial coordinates.

For many practical applications the deformation gradients are sufficiently small for quadratic terms to be neglected when compared with linear terms leading to an infinitesimal deformation theory. The expression (2.88) for the Lagrangian strain tensor may then be approximated by $\boldsymbol{E} \cong \boldsymbol{\varepsilon}$ where

$$\boldsymbol{\varepsilon} \equiv \frac{1}{2}\left(\nabla\boldsymbol{u} + (\nabla\boldsymbol{u})^{\mathrm{T}}\right), \tag{2.106}$$

is known as the infinitesimal strain tensor and where ∇ denotes the gradient with respect to the coordinates $\boldsymbol{x}$. In addition, there is no need to distinguish between the initial and deformed states of the medium so that $\boldsymbol{x} \cong \bar{\boldsymbol{x}}$. Relation (2.106) may be written in component form so that

$$\varepsilon_{ij} \equiv \frac{1}{2}\left(\frac{\partial u_i}{\partial x_j} + \frac{\partial u_j}{\partial x_i}\right) = \varepsilon_{ji}, \tag{2.107}$$

which shows that the infinitesimal strain tensor $\boldsymbol{\varepsilon}$ is symmetric.

For a continuous medium having a uniform density distribution ρ_0 in its undeformed state, the principle of conservation of mass for infinitesimal deformations is expressed as

$$\rho = \rho_0, \quad \text{a constant}, \tag{2.108}$$

where ρ is the density of the medium during deformation. This equation simply states that the mass of a given set of material points remains constant during any deformation. The local density ρ of the medium measured relative to spatial coordinates will in fact vary because of non-uniform displacement gradients, but this change is negligible for infinitesimal deformation theory where the value of ρ corresponds to the initial density ρ_0 prior to deformation, as asserted by (2.108).

For the equilibrium situations considered in this book, body forces are neglected so that $\boldsymbol{b} = 0$ and the equation of motion (2.37) reduces in component form to

$$\rho_0 \frac{\partial v_j}{\partial t} = \frac{\partial \sigma_{ij}}{\partial x_i}, \tag{2.109}$$

where $v_j = \partial u_j / \partial t$ is the velocity vector, u_j is the displacement vector and σ_{ij} is the stress tensor. In the absence of body couples, the principle of the conservation of angular momentum leads to the symmetry of the stress tensor so that $\sigma_{ij} = \sigma_{ji}$.

The principle of the conservation of energy for infinitesimal deformations leads to the following local form of the internal energy balance equation that results from (2.44)

$$\frac{\partial(\rho_0 \upsilon)}{\partial t} = -\frac{\partial h_i}{\partial x_i} + \sigma_{ij}\frac{\partial \varepsilon_{ij}}{\partial t} + \rho_0 r, \tag{2.110}$$

where υ is the specific internal energy, h_i is the heat flux vector, r is the rate (per unit mass) at which heat from non-mechanical sources may be locally generated or lost that could arise, for example, from electrical heating and ε_{ij} is the infinitesimal strain tensor defined in terms of the displacement vector u_i by relation (2.107). Relations (2.109) and (2.110) are field equations that must always be satisfied for any type of homogeneous solid that deforms as a continuous medium and is subject to infinitesimal deformations.

When developing a theory of thermoelastic materials behaviour, it is useful to introduce an equation of state of the following form (a generalisation of $(2.67)_1$ to elastic materials subject to shear deformation)

$$\upsilon = \hat{\upsilon}(\eta, \varepsilon_{ij}), \tag{2.111}$$

where $\hat{\upsilon}$ is a prescribed function of the specific entropy η and the strain tensor ε_{ij}. The corresponding differential form is written as

$$\mathrm{d}\upsilon = T\mathrm{d}\eta + \frac{1}{\rho_0}\sigma_{ij}\,\mathrm{d}\varepsilon_{ij}, \tag{2.112}$$

and the thermodynamic temperature T (i.e. absolute temperature, which is always positive) is then defined by the relation

$$T = \frac{\partial \hat{\upsilon}}{\partial \eta}, \tag{2.113}$$

whereas the stress tensor is defined by

$$\sigma_{ij} = \rho_0 \frac{\partial \hat{\upsilon}}{\partial \varepsilon_{ij}}. \tag{2.114}$$

It follows from (2.110)–(2.114) that the local energy balance equation may be expressed in the form

$$T\frac{\partial(\rho_0 \eta)}{\partial t} = -\frac{\partial h_i}{\partial x_i} + \rho_0 r. \tag{2.115}$$

This relation is now rearranged in the form of an entropy balance equation

$$\frac{\partial(\rho_0 \eta)}{\partial t} = -\frac{\partial}{\partial x_i}\left(\frac{h_i}{T}\right) + \Delta + \frac{\rho_0 r}{T}, \tag{2.116}$$

where Δ is the rate of internal entropy production per unit volume given by

$$\Delta = h_i \frac{\partial}{\partial x_i}\left(\frac{1}{T}\right), \tag{2.117}$$

resulting from the thermodynamically irreversible process of heat conduction, and where in (2.116) the quantity h_i / T is the entropy flux vector. In words, (2.116) states that the rate of change of entropy in unit volume of material is balanced by the flow of entropy through its boundary, the rate of entropy production within the material resulting from heat flow and the rate of change of entropy resulting from localised non-mechanical heating effects. The second law of thermodynamics asserts that Δ must be non-negative for all possible local states of the system. This is assured if the following heat conduction law is assumed

$$h_i = -\kappa \frac{\partial T}{\partial x_i}, \tag{2.118}$$

where the thermal conductivity $\kappa \geq 0$ can be temperature dependent. The relation (2.118) linearly relating the heat flux to the temperature gradient is the well-known Fourier heat conduction law that demands that heat will always flow from regions of high temperature to regions of lower temperature.

2.12 Equilibrium Equations

When a continuous medium is in mechanical equilibrium, the equations of motion (2.109) reduce to the form

$$\frac{\partial \sigma_{ij}}{\partial x_i} = 0. \tag{2.119}$$

It is useful express the equilibrium equations (2.119) in terms of three coordinate systems that will be used later in this book. In terms of a set of Cartesian coordinates (x_1, x_2, x_3) the equilibrium equations (2.119) may be expressed as the following three independent equilibrium equations (see, for example, [3])

$$\frac{\partial \sigma_{11}}{\partial x_1} + \frac{\partial \sigma_{12}}{\partial x_2} + \frac{\partial \sigma_{13}}{\partial x_3} = 0, \tag{2.120}$$

$$\frac{\partial \sigma_{12}}{\partial x_1} + \frac{\partial \sigma_{22}}{\partial x_2} + \frac{\partial \sigma_{23}}{\partial x_3} = 0, \tag{2.121}$$

$$\frac{\partial \sigma_{13}}{\partial x_1} + \frac{\partial \sigma_{23}}{\partial x_2} + \frac{\partial \sigma_{33}}{\partial x_3} = 0, \tag{2.122}$$

where the stress tensor is symmetric such that

$$\sigma_{12} = \sigma_{21}, \sigma_{13} = \sigma_{31}, \sigma_{23} = \sigma_{32}. \tag{2.123}$$

When using cylindrical polar coordinates (r, θ, z) such that

$$x_1 = r\cos\theta, \quad x_2 = r\sin\theta, \quad x_3 = z, \tag{2.124}$$

the equilibrium equations (2.120)–(2.122) are written as

$$\frac{\partial \sigma_{rr}}{\partial r} + \frac{1}{r}\frac{\partial \sigma_{r\theta}}{\partial \theta} + \frac{\partial \sigma_{rz}}{\partial z} + \frac{\sigma_{rr} - \sigma_{\theta\theta}}{r} = 0, \tag{2.125}$$

$$\frac{\partial \sigma_{r\theta}}{\partial r} + \frac{1}{r}\frac{\partial \sigma_{\theta\theta}}{\partial \theta} + \frac{\partial \sigma_{\theta z}}{\partial z} + \frac{2\sigma_{r\theta}}{r} = 0, \tag{2.126}$$

$$\frac{\partial \sigma_{rz}}{\partial r}+\frac{1}{r}\frac{\partial \sigma_{\theta z}}{\partial \theta}+\frac{\partial \sigma_{zz}}{\partial z}+\frac{\sigma_{rz}}{r}=0, \tag{2.127}$$

where the stress tensor is symmetric such that

$$\sigma_{r\theta}=\sigma_{\theta r}, \sigma_{rz}=\sigma_{zr}, \sigma_{\theta z}=\sigma_{z\theta}. \tag{2.128}$$

When using spherical polar coordinates (r,θ,ϕ) such that

$$x_1=r\sin\theta\cos\phi,\quad x_2=r\sin\theta\sin\phi,\quad x_3=r\cos\theta, \tag{2.129}$$

the equilibrium equations (2.120)–(2.122) are written as

$$\frac{\partial \sigma_{rr}}{\partial r}+\frac{1}{r}\frac{\partial \sigma_{r\theta}}{\partial \theta}+\frac{1}{r\sin\theta}\frac{\partial \sigma_{r\phi}}{\partial \phi}+\frac{1}{r}\left(2\sigma_{rr}-\sigma_{\theta\theta}-\sigma_{\phi\phi}+\sigma_{r\theta}\cot\theta\right)=0, \tag{2.130}$$

$$\frac{\partial \sigma_{r\theta}}{\partial r}+\frac{1}{r}\frac{\partial \sigma_{\theta\theta}}{\partial \theta}+\frac{1}{r\sin\theta}\frac{\partial \sigma_{\theta\phi}}{\partial \phi}+\frac{1}{r}\left[3\sigma_{r\theta}+\left(\sigma_{\theta\theta}-\sigma_{\phi\phi}\right)\cot\theta\right]=0, \tag{2.131}$$

$$\frac{\partial \sigma_{r\phi}}{\partial r}+\frac{1}{r}\frac{\partial \sigma_{\theta\phi}}{\partial \theta}+\frac{1}{r\sin\theta}\frac{\partial \sigma_{\phi\phi}}{\partial \phi}+\frac{1}{r}\left(3\sigma_{r\phi}+2\sigma_{\theta\phi}\cot\theta\right)=0, \tag{2.132}$$

where the stress tensor is symmetric such that

$$\sigma_{r\theta}=\sigma_{\theta r}, \sigma_{r\phi}=\sigma_{\phi r}, \sigma_{\theta\phi}=\sigma_{\phi\theta}. \tag{2.133}$$

2.13 Strain–Displacement Relations

For the special case when the strain tensor ε_{ij} is uniform, and on using (2.107), the displacement fields

$$\begin{aligned} u_1(\boldsymbol{x}) &= \varepsilon_{11}x_1+\varepsilon_{12}x_2+\varepsilon_{13}x_3, \\ u_2(\boldsymbol{x}) &= \varepsilon_{12}x_1+\varepsilon_{22}x_2+\varepsilon_{23}x_3, \\ u_3(\boldsymbol{x}) &= \varepsilon_{13}x_1+\varepsilon_{23}x_2+\varepsilon_{33}x_3, \end{aligned} \tag{2.134}$$

and

$$\begin{aligned} u_1(\boldsymbol{x}) &= \varepsilon_{11}x_1+2\varepsilon_{12}x_2+2\varepsilon_{13}x_3, \\ u_2(\boldsymbol{x}) &= \varepsilon_{22}x_2+2\varepsilon_{23}x_3, \\ u_3(\boldsymbol{x}) &= \varepsilon_{33}x_3, \end{aligned} \tag{2.135}$$

both lead to the same strain field given by

$$\begin{aligned} &\varepsilon_{11}(\boldsymbol{x})\equiv\frac{\partial u_1}{\partial x_1}=\varepsilon_{11}, \quad \varepsilon_{22}(\boldsymbol{x})\equiv\frac{\partial u_2}{\partial x_2}=\varepsilon_{22}, \quad \varepsilon_{33}(\boldsymbol{x})\equiv\frac{\partial u_3}{\partial x_3}=\varepsilon_{33}, \\ &\varepsilon_{12}(\boldsymbol{x})\equiv\frac{1}{2}\left(\frac{\partial u_1}{\partial x_2}+\frac{\partial u_2}{\partial x_1}\right)=\varepsilon_{12}, \quad \varepsilon_{13}(\boldsymbol{x})\equiv\frac{1}{2}\left(\frac{\partial u_1}{\partial x_3}+\frac{\partial u_3}{\partial x_1}\right)=\varepsilon_{13}, \\ &\varepsilon_{23}(\boldsymbol{x})\equiv\frac{1}{2}\left(\frac{\partial u_2}{\partial x_3}+\frac{\partial u_3}{\partial x_2}\right)=\varepsilon_{23}. \end{aligned} \tag{2.136}$$

This implies that the displacement fields differ by a rigid rotation about some axis. Such a rigid rotation will not affect the values of the strain and stress fields. The form

(2.135) for the displacement field is preferred as it is then much easier to describe the meaning of the shear strains ε_{12}, ε_{13} and ε_{23}. When ε_{12} is the only non-zero strain component, it follows that u_1 is the only non-zero displacement component and that its value is proportional to x_2 so that the shear plane is normal to the x_2-axis. Similarly, when ε_{13} is the only non-zero strain component, it again follows that u_1 is the only non-zero displacement component and that its value is proportional to x_3 so that the shear plane is normal to the x_3-axis. When ε_{23} is the only non-zero strain component, it follows that u_2 is the only non-zero displacement component and that its value is proportional to x_3 so that the shear plane is again normal to the x_3-axis.

When using cylindrical polar coordinates (r,θ,z), the displacement components u_r, u_θ, u_z are related to the Cartesian components u_1, u_2, u_3 as follows

$$\begin{aligned} u_r &= u_1\cos\theta + u_2\sin\theta, \\ u_\theta &= u_2\cos\theta - u_1\sin\theta, \\ u_z &= u_3, \end{aligned} \tag{2.137}$$

having inverse

$$\begin{aligned} u_1 &= u_r\cos\theta - u_\theta\sin\theta, \\ u_2 &= u_\theta\cos\theta + u_r\sin\theta, \\ u_3 &= u_z, \end{aligned} \tag{2.138}$$

and the strain–displacement relations are given by

$$\begin{aligned} &\varepsilon_{rr} = \frac{\partial u_r}{\partial r}, \qquad \varepsilon_{\theta\theta} = \frac{1}{r}\left(u_r + \frac{\partial u_\theta}{\partial \theta}\right), \qquad \varepsilon_{zz} = \frac{\partial u_z}{\partial z}, \\ &\varepsilon_{r\theta} = \frac{1}{2}\left(\frac{1}{r}\frac{\partial u_r}{\partial \theta} + \frac{\partial u_\theta}{\partial r} - \frac{u_\theta}{r}\right), \ \varepsilon_{\theta z} = \frac{1}{2}\left(\frac{\partial u_\theta}{\partial z} + \frac{1}{r}\frac{\partial u_z}{\partial \theta}\right), \ \varepsilon_{rz} = \frac{1}{2}\left(\frac{\partial u_r}{\partial z} + \frac{\partial u_z}{\partial r}\right). \end{aligned} \tag{2.139}$$

When using spherical polar coordinates (r,θ,ϕ), the displacement components u_r, u_θ, u_ϕ are related to the Cartesian components u_1, u_2, u_3 as follows:

$$\begin{aligned} u_r &= \sin\theta\cos\phi\, u_1 + \sin\theta\sin\phi\, u_2 + \cos\theta\, u_3, \\ u_\theta &= \cos\theta\cos\phi\, u_1 + \cos\theta\sin\phi\, u_2 - \sin\theta\, u_3, \\ u_\phi &= \cos\phi\, u_2 - \sin\phi\, u_1, \end{aligned} \tag{2.140}$$

having inverse

$$\begin{aligned} u_1 &= \sin\theta\cos\phi\, u_r + \cos\theta\cos\phi\, u_\theta - \sin\phi\, u_\phi, \\ u_2 &= \sin\theta\sin\phi\, u_r + \cos\theta\sin\phi\, u_\theta + \cos\phi\, u_\phi, \\ u_3 &= \cos\theta\, u_r - \sin\theta\, u_\theta, \end{aligned} \tag{2.141}$$

and the strain–displacement relations are given by

$$\begin{aligned} &\varepsilon_{rr} = \frac{\partial u_r}{\partial r}, \qquad \varepsilon_{\theta\theta} = \frac{1}{r}\left(u_r + \frac{\partial u_\theta}{\partial \theta}\right), \\ &\varepsilon_{\phi\phi} = \frac{1}{r\sin\theta}\frac{\partial u_\phi}{\partial \phi} + \frac{u_r}{r} + \frac{u_\theta\cot\theta}{r}, \qquad \varepsilon_{r\theta} = \frac{1}{2}\left(\frac{1}{r}\frac{\partial u_r}{\partial \theta} + \frac{\partial u_\theta}{\partial r} - \frac{u_\theta}{r}\right), \\ &\varepsilon_{\theta\phi} = \frac{1}{2r}\left(\frac{1}{\sin\theta}\frac{\partial u_\theta}{\partial \phi} + \frac{\partial u_\phi}{\partial \theta} - u_\phi\cot\theta\right), \ \varepsilon_{r\phi} = \frac{1}{2}\left(\frac{1}{r\sin\theta}\frac{\partial u_r}{\partial \phi} + \frac{\partial u_\phi}{\partial r} - \frac{u_\phi}{r}\right). \end{aligned} \tag{2.142}$$

2.14 Constitutive Equations for Anisotropic Linear Thermoelastic Solids

As we are concerned in this book with various types of composite material, it is necessary to define a set of constitutive relations that will form the basis for the development of theoretical methods for predicting the behaviour of anisotropic materials. Consider a general homogeneous infinitesimal strain ε_{kl} (applied to a unit cube of the composite material) defined in terms of the displacement vector u_k and the position vector x_k by (2.107), namely,

$$\varepsilon_{kl} = \frac{1}{2}\left(\frac{\partial u_k}{\partial x_l} + \frac{\partial u_l}{\partial x_k}\right) = \varepsilon_{lk}. \tag{2.143}$$

With the assumption that the strain field is uniform in the composite, the displacement field is linear and may be written in the form

$$u_k = \varepsilon_{kl} x_l, \tag{2.144}$$

where it is assumed that the displacement vector is zero when $x_1 = x_2 = x_3 = 0$.

The local equation of state (2.111) is not of a form that can easily be related to experimental measurements as one of the state variables is assumed to be the specific entropy η. It is much more convenient, and much more practically useful, if the state variable η is replaced by the absolute temperature T. A local equation of state for the specific Helmoltz energy is assumed to have the following form (equivalent to (2.111) as implied by (2.67)–(2.69))

$$\psi = \hat{\psi}(T, \varepsilon_{ij}), \tag{2.145}$$

where ε_{ij} is the infinitesimal strain tensor introduced in Section 2.12 (see (2.107)). For infinitesimal deformations, and because $\psi \equiv e - T\eta$ from (2.64), the following differential form of (2.145) may be derived from (2.112)

$$\mathrm{d}\psi = -\,\eta\,\mathrm{d}T + \frac{1}{\rho_0}\sigma_{ij}\,\mathrm{d}\varepsilon_{ij}, \tag{2.146}$$

where the specific entropy η and the stress tensor σ_{ij} are now defined by

$$\eta = -\frac{\partial \hat{\psi}}{\partial T}, \qquad \sigma_{ij} = \rho_0 \frac{\partial \hat{\psi}}{\partial \varepsilon_{ij}}. \tag{2.147}$$

For infinitesimal deformations, a linear thermoelastic response can be assumed so that the Helmholtz energy per unit volume $\rho_0 \hat{\psi}$ has the form

$$\rho_0 \hat{\psi}(T, \varepsilon_{ij}) = \frac{1}{2} C_{ijkl}\,\varepsilon_{ij}\,\varepsilon_{kl} + \beta_{ij}\varepsilon_{ij}(T - T_0) + f(T), \tag{2.148}$$

where C_{ijkl} are the elastic constants having the dimensions of stress or modulus, β_{ij} are thermoelastic coefficients and where T_0 is a reference temperature. As the strain tensor is symmetric, it follows that $\beta_{ij} = \beta_{ji}$ and that

$$C_{ijkl}\,\varepsilon_{ij}\,\varepsilon_{kl} \equiv C_{ijkl}\,\varepsilon_{ji}\,\varepsilon_{kl} \equiv C_{jikl}\,\varepsilon_{ji}\,\varepsilon_{kl}, \text{ implying } C_{ijkl} = C_{jikl}, \tag{2.149}$$

$$C_{ijkl}\,\varepsilon_{ij}\,\varepsilon_{kl} \equiv C_{ijkl}\,\varepsilon_{ij}\,\varepsilon_{lk} \equiv C_{ijlk}\,\varepsilon_{ij}\,\varepsilon_{kl}, \text{ implying } C_{ijkl} = C_{ijlk}, \tag{2.150}$$

$$C_{ijkl}\,\varepsilon_{ij}\,\varepsilon_{kl} \equiv C_{klij}\,\varepsilon_{kl}\,\varepsilon_{ij} \equiv C_{klij}\,\varepsilon_{ij}\,\varepsilon_{kl}, \text{ implying } C_{ijkl} = C_{klij}. \tag{2.151}$$

On substituting (2.148) into (2.147), it follows that

$$\eta = -\beta_{ij}\varepsilon_{ij} + f'(T), \tag{2.152}$$

$$\sigma_{ij} = C_{ijkl}\varepsilon_{kl} + \beta_{ij}(T - T_0). \tag{2.153}$$

It should be noted that the stress components are zero everywhere when the strain is defined to be zero everywhere at the reference temperature T_0.

The inverse form of the linear stress-strain relations (2.153) is written as

$$\varepsilon_{ij} = S_{ijkl}\sigma_{kl} + \alpha_{ij}(T - T_0), \tag{2.154}$$

where the compliance tensor S_{ijkl} is such that

$$C_{ijkl}S_{klmn} = \tfrac{1}{2}\left(\delta_{im}\delta_{jn} + \delta_{in}\delta_{jm}\right), \tag{2.155}$$

where use has been made of (2.15), and where

$$\alpha_{ij} = S_{ijkl}\beta_{kl}, \tag{2.156}$$

are anisotropic thermal expansion coefficients.

2.14.1 Isotropic Materials

The situation simplifies when the material is linear thermoelastic and isotropic so that the stress-strain relations are

$$\varepsilon_{ij} = \frac{1}{E}\left[(1+\nu)\sigma_{ij} - \nu\delta_{ij}\sigma_{kk}\right] + \alpha\Delta T\delta_{ij}, \tag{2.157}$$

where E is Young's modulus, ν is Poisson's ratio, α is the thermal expansion coefficient and $\Delta T = T - T_0$. The inverse form is

$$\sigma_{ij} = \frac{E}{1+\nu}\left(\varepsilon_{ij} + \frac{\nu}{1-2\nu}\varepsilon_{kk}\delta_{ij}\right) - \frac{E\alpha\Delta T}{1-2\nu}\delta_{ij}. \tag{2.158}$$

When the stress field is hydrostatic so that $\sigma_{ij} = -p\delta_{ij}$, and because $\sigma_{kk} \equiv \sigma_{11} + \sigma_{22} + \sigma_{33}$, it follows from (2.157) that

$$\varepsilon_{ij} = \left(-\frac{1-2\nu}{E}p + \alpha\Delta T\right)\delta_{ij}. \tag{2.159}$$

The inverse form (2.158) may be written as

$$\sigma_{ij} = 2\mu\varepsilon_{ij} + \lambda\varepsilon_{kk}\delta_{ij} - (3\lambda + 2\mu)\alpha\Delta T\delta_{ij}, \tag{2.160}$$

where λ and μ are Lamé's constants, μ being the shear modulus, which can be calculated from Young's modulus and Poisson's ratio as follows:

$$\lambda = \frac{E\nu}{(1+\nu)(1-2\nu)}, \quad \mu = \frac{E}{2(1+\nu)}. \tag{2.161}$$

2.15 Introducing Contracted Notation

The general formulation for describing the elastic constants of anisotropic materials involves fourth-order tensors that are difficult to apply in many practical situations where analytical methods can be used. A simplified contracted notation is usually used for such analyses where the fourth-order tensors of elastic constants are replaced by a second-order matrix formulation that is now described. The matrix formulation makes use of the fact that the stress and strain tensors are symmetric. These symmetry properties enabled the derivation of the relationships (2.149)–(2.151).

The components of the stress and strain components are now assembled in column vectors of length six so that

$$\begin{bmatrix}\sigma_{11}\\ \sigma_{22}\\ \sigma_{33}\\ \sigma_{23}\\ \sigma_{13}\\ \sigma_{12}\end{bmatrix} \equiv \begin{bmatrix}\sigma_{1}\\ \sigma_{2}\\ \sigma_{3}\\ \sigma_{4}\\ \sigma_{5}\\ \sigma_{6}\end{bmatrix}, \quad \begin{bmatrix}\varepsilon_{11}\\ \varepsilon_{22}\\ \varepsilon_{33}\\ 2\varepsilon_{23}\\ 2\varepsilon_{13}\\ 2\varepsilon_{12}\end{bmatrix} \equiv \begin{bmatrix}\varepsilon_{1}\\ \varepsilon_{2}\\ \varepsilon_{3}\\ \varepsilon_{4}\\ \varepsilon_{5}\\ \varepsilon_{6}\end{bmatrix}. \tag{2.162}$$

It should be noted that a factor of two has been applied only to the shear terms of the relation involving the strains so that the quantities $2\varepsilon_{ij}$ for $i \neq j$ correspond to the widely used engineering shear strain values. General linear elastic stress-strain relations, including thermal expansion terms, have the contracted matrix form

$$\begin{bmatrix}\sigma_{1}\\ \sigma_{2}\\ \sigma_{3}\\ \sigma_{4}\\ \sigma_{5}\\ \sigma_{6}\end{bmatrix} = \begin{bmatrix} C_{11} & C_{12} & C_{13} & C_{14} & C_{15} & C_{16}\\ C_{21} & C_{22} & C_{23} & C_{24} & C_{25} & C_{26}\\ C_{31} & C_{32} & C_{33} & C_{34} & C_{35} & C_{36}\\ C_{41} & C_{42} & C_{43} & C_{44} & C_{45} & C_{46}\\ C_{51} & C_{52} & C_{53} & C_{54} & C_{55} & C_{56}\\ C_{61} & C_{62} & C_{63} & C_{64} & C_{65} & C_{66}\end{bmatrix} \begin{bmatrix}\varepsilon_{1}\\ \varepsilon_{2}\\ \varepsilon_{3}\\ \varepsilon_{4}\\ \varepsilon_{5}\\ \varepsilon_{6}\end{bmatrix} - \begin{bmatrix}U_{1}\\ U_{2}\\ U_{3}\\ U_{4}\\ U_{5}\\ U_{6}\end{bmatrix} \Delta T, \tag{2.163}$$

where C_{IJ} are symmetric elastic constants, which are components of the second-order matrix $\boldsymbol{C}$, and where U_I are thermoelastic constants associated with the tensor β_{ij}, which are components of the vector $\boldsymbol{U}$, the uppercase indices I and J ranging from 1 to 6. For orthotropic materials the stress-strain relations have the simpler matrix form

$$\begin{bmatrix}\sigma_{1}\\ \sigma_{2}\\ \sigma_{3}\\ \sigma_{4}\\ \sigma_{5}\\ \sigma_{6}\end{bmatrix} = \begin{bmatrix} C_{11} & C_{12} & C_{13} & 0 & 0 & 0\\ C_{21} & C_{22} & C_{23} & 0 & 0 & 0\\ C_{31} & C_{32} & C_{33} & 0 & 0 & 0\\ 0 & 0 & 0 & C_{44} & 0 & 0\\ 0 & 0 & 0 & 0 & C_{55} & 0\\ 0 & 0 & 0 & 0 & 0 & C_{66}\end{bmatrix} \begin{bmatrix}\varepsilon_{1}\\ \varepsilon_{2}\\ \varepsilon_{3}\\ \varepsilon_{4}\\ \varepsilon_{5}\\ \varepsilon_{6}\end{bmatrix} - \begin{bmatrix}U_{1}\\ U_{2}\\ U_{3}\\ 0\\ 0\\ 0\end{bmatrix} \Delta T. \tag{2.164}$$

The stress-strain relations (2.163) may be written, using a repeated summation convention for uppercase indices over the range 1, 2, ..., 6, as

$$\sigma_I = C_{IJ}\,\varepsilon_J - U_I \Delta T, \; I = 1, 2, \ldots, 6. \tag{2.165}$$

The inverse of the matrix C_{IJ} is denoted by the symmetric matrix S_{IJ} such that

$$C_{IJ}\,S_{JK} = S_{IJ} C_{JK} = \delta_{IK}, \tag{2.166}$$

where δ_{IK} is the Kronecker delta symbol having the value 1 when $I = J$ and the value 0 otherwise. On multiplying (2.165) on the left by S_{LI} and on using (2.166), it can be shown that

$$\varepsilon_I = S_{IJ}\sigma_J + V_I\Delta T,\ V_I = S_{IJ}U_J,\ I = 1, 2, \ldots, 6. \tag{2.167}$$

The quantities V_I are the components of the vector $\boldsymbol{V}$ which is associated with the thermal expansion tensor α_{ij}. The matrix form of (2.167) is given by

$$\begin{bmatrix} \varepsilon_1 \\ \varepsilon_2 \\ \varepsilon_3 \\ \varepsilon_4 \\ \varepsilon_5 \\ \varepsilon_6 \end{bmatrix} = \begin{bmatrix} S_{11} & S_{12} & S_{13} & S_{14} & S_{15} & S_{16} \\ S_{21} & S_{22} & S_{23} & S_{24} & S_{25} & S_{26} \\ S_{31} & S_{32} & S_{33} & S_{34} & S_{35} & S_{36} \\ S_{41} & S_{42} & S_{43} & S_{44} & S_{45} & S_{46} \\ S_{51} & S_{52} & S_{53} & S_{54} & S_{55} & S_{56} \\ S_{61} & S_{62} & S_{63} & S_{64} & S_{65} & S_{66} \end{bmatrix} \begin{bmatrix} \sigma_1 \\ \sigma_2 \\ \sigma_3 \\ \sigma_4 \\ \sigma_5 \\ \sigma_6 \end{bmatrix} + \begin{bmatrix} V_1 \\ V_2 \\ V_3 \\ V_4 \\ V_5 \\ V_6 \end{bmatrix} \Delta T, \tag{2.168}$$

and the corresponding orthotropic form is

$$\begin{bmatrix} \varepsilon_1 \\ \varepsilon_2 \\ \varepsilon_3 \\ \varepsilon_4 \\ \varepsilon_5 \\ \varepsilon_6 \end{bmatrix} = \begin{bmatrix} S_{11} & S_{12} & S_{13} & 0 & 0 & 0 \\ S_{21} & S_{22} & S_{23} & 0 & 0 & 0 \\ S_{31} & S_{32} & S_{33} & 0 & 0 & \\ 0 & 0 & 0 & S_{44} & 0 & 0 \\ 0 & 0 & 0 & 0 & S_{55} & 0 \\ 0 & 0 & 0 & 0 & 0 & S_{66} \end{bmatrix} \begin{bmatrix} \sigma_1 \\ \sigma_2 \\ \sigma_3 \\ \sigma_4 \\ \sigma_5 \\ \sigma_6 \end{bmatrix} + \begin{bmatrix} V_1 \\ V_2 \\ V_3 \\ 0 \\ 0 \\ 0 \end{bmatrix} \Delta T. \tag{2.169}$$

When expanded using the stress and strain tensor components and the symmetry of S_{IJ}, the stress-strain relations may be written as

$$\begin{aligned} \varepsilon_{11} &= S_{11}\sigma_{11} + S_{12}\sigma_{22} + S_{13}\sigma_{33} + V_1\Delta T, & 2\varepsilon_{23} &= S_{44}\sigma_{23}, \\ \varepsilon_{22} &= S_{12}\sigma_{11} + S_{22}\sigma_{22} + S_{23}\sigma_{33} + V_2\Delta T, & 2\varepsilon_{13} &= S_{55}\sigma_{13}, \\ \varepsilon_{33} &= S_{13}\sigma_{11} + S_{23}\sigma_{22} + S_{33}\sigma_{33} + V_3\Delta T, & 2\varepsilon_{12} &= S_{66}\sigma_{12}. \end{aligned} \tag{2.170}$$

2.16 Tensor Transformations

When considering laminated composite materials, where each ply is reinforced with aligned straight fibres that are inclined at various angles to a global set of coordinates, there is a need to define a set of local coordinates aligned with the fibres in each ply. There is also a need to determine the properties of each ply referred to the global coordinates. For a right-handed set of global coordinates x_1, x_2 and x_3, $\boldsymbol{i}_1$, $\boldsymbol{i}_2$ and $\boldsymbol{i}_3$ are unit vectors for the directions of the x_1-, x_2- and x_3-axes, respectively. For laminate models, the fibres are usually assumed to be in the x_1-direction and coordinate transformations involve rotations about the x_3-axis. When modelling unidirectional plies as transverse isotropic materials the rotations would need to be taken about the x_1-axis if the fibres are in the x_1-direction. Coordinate transformations involving rotations about the x_3-axis are now considered.

A right-handed second set of local coordinates x_1', x_2' and x_3' is obtained by rotating the reference set of coordinates about the x_3-axis by an angle ϕ as shown in Figure 2.1.

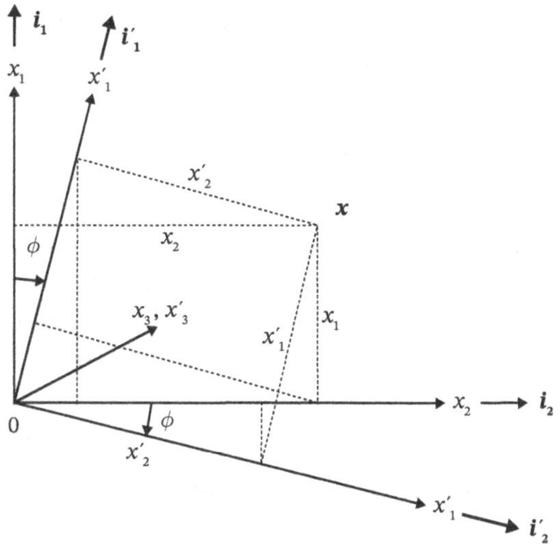

Figure 2.1 Transformation of right-handed Cartesian coordinates.

The rotation is clockwise when viewing along the positive direction of the x_3-axis. The unit vectors in the directions of the x_1', x_2' and x_3' axes are denoted by $\boldsymbol{i}_1'$, $\boldsymbol{i}_2'$ and $\boldsymbol{i}_3'$, respectively. Rotating about the x_3-axis enables account to be taken of the effects of off-axis plies in laminates, as considered in Chapters 6 and 7.

Any point in space can be represented by the vector $\boldsymbol{x}$ (a first-order tensor) the value of which is wholly independent of the coordinate system that is used to describe its components so that

$$\boldsymbol{x} = x_1\,\boldsymbol{i}_1 + x_2\,\boldsymbol{i}_2 + x_3\,\boldsymbol{i}_3 = \boldsymbol{x}_1'\,\boldsymbol{i}_1' + x_2'\,\boldsymbol{i}_2' + x_3'\,\boldsymbol{i}_3'. \tag{2.171}$$

It then follows on resolving vectors that

$$\begin{aligned} \boldsymbol{i}_1 &= \cos\phi\,\boldsymbol{i}_1' - \sin\phi\,\boldsymbol{i}_2', & \boldsymbol{i}_1' &= \cos\phi\,\boldsymbol{i}_1 + \sin\phi\,\boldsymbol{i}_2,\\ \boldsymbol{i}_2 &= \sin\phi\,\boldsymbol{i}_1' + \cos\phi\,\boldsymbol{i}_2', & \boldsymbol{i}_2' &= -\sin\phi\,\boldsymbol{i}_1 + \cos\phi\,\boldsymbol{i}_2,\\ \boldsymbol{i}_3 &= \boldsymbol{i}_3', & \boldsymbol{i}_3' &= \boldsymbol{i}_3. \end{aligned} \tag{2.172}$$

Transformation of a set of Cartesian coordinates (x_1,x_2,x_3) to (x_1',x_2',x_3') by a rotation of the x_1- and x_2-axes about the x_3-axis through an angle ϕ (as shown in Figure 2.1) leads to

$$\begin{aligned} x_1 &= x_1'\cos\phi - x_2'\sin\phi, & x_1' &= x_1\cos\phi + x_2\sin\phi,\\ x_2 &= x_1'\sin\phi + x_2'\cos\phi, & x_2' &= -x_1\sin\phi + x_2\cos\phi,\\ x_3 &= x_3', & x_3' &= x_3. \end{aligned} \tag{2.173}$$

These relationships can be established from the geometry shown in Figure 2.1 on making use of the various constructions shown as dotted lines.

The displacement vector $\boldsymbol{u}$ is a physical quantity that is wholly independent of the coordinate system that is used to describe its components. This vector may be written as (where summation over values 1, 2 and 3 is implied by repeated lower-case suffices)

$$\boldsymbol{u} = u_k\,\boldsymbol{i}_k = \boldsymbol{u}_k'\,\boldsymbol{i}_k', \tag{2.174}$$

where u_k and u'_k are the displacement components referred to the two coordinate systems being considered. It follows on using (2.172) that

$$\begin{aligned} u'_1 &= u_1\cos\phi + u_2\sin\phi, & u_1 &= u'_1\cos\phi - u'_2\sin\phi,\\ u'_2 &= u_2\cos\phi - u_1\sin\phi, & u_2 &= u'_1\sin\phi + u'_2\cos\phi,\\ u'_3 &= u_3, & u_3 &= u'_3. \end{aligned} \tag{2.175}$$

The stress and strain at any point in a material is a dyadic (an array of ordered vector pairs) or second-order tensor whose value is wholly independent of the coordinate system that is used to describe its components. The second-order stress tensor σ may, therefore, be written as (where summation over values 1, 2 and 3 is implied by repeated lower case suffices)

$$\boldsymbol{\sigma} = \sigma_{kl}\,\mathbf{i}_k\mathbf{i}_l = \sigma'_{kl}\,\mathbf{i}'_k\mathbf{i}'_l, \tag{2.176}$$

where σ_{kl} and σ'_{kl} are the stress components referred to the two coordinate systems being considered. On defining $m = \cos\phi$ and $n = \sin\phi$, it follows from (2.172) that

$$\begin{aligned}
\mathbf{i}'_1\mathbf{i}'_1 &= (m\mathbf{i}_1 + n\mathbf{i}_2)(m\mathbf{i}_1 + n\mathbf{i}_2)\\
&= m^2\mathbf{i}_1\mathbf{i}_1 + mn\mathbf{i}_1\mathbf{i}_2 + mn\mathbf{i}_2\mathbf{i}_1 + n^2\mathbf{i}_2\mathbf{i}_2,\\
\mathbf{i}'_2\mathbf{i}'_2 &= (-n\mathbf{i}_1 + m\mathbf{i}_2)(-n\mathbf{i}_1 + m\mathbf{i}_2)\\
&= n^2\mathbf{i}_1\mathbf{i}_1 - mn\mathbf{i}_1\mathbf{i}_2 - mn\mathbf{i}_2\mathbf{i}_1 + m^2\mathbf{i}_2\mathbf{i}_2,\\
\mathbf{i}'_3\mathbf{i}'_3 &= \mathbf{i}_3\mathbf{i}_3,\\
\mathbf{i}'_2\mathbf{i}'_3 &= -n\mathbf{i}_1\mathbf{i}_3 + m\mathbf{i}_2\mathbf{i}_3,\\
\mathbf{i}'_3\mathbf{i}'_2 &= -n\mathbf{i}_3\mathbf{i}_1 + m\mathbf{i}_3\mathbf{i}_2,\\
\mathbf{i}'_1\mathbf{i}'_3 &= m\mathbf{i}_1\mathbf{i}_3 + n\mathbf{i}_2\mathbf{i}_3,\\
\mathbf{i}'_3\mathbf{i}'_1 &= m\mathbf{i}_3\mathbf{i}_1 + n\mathbf{i}_3\mathbf{i}_2,\\
\mathbf{i}'_1\mathbf{i}'_2 &= (m\mathbf{i}_1 + n\mathbf{i}_2)(-n\mathbf{i}_1 + m\mathbf{i}_2)\\
&= -mn\mathbf{i}_1\mathbf{i}_1 + m^2\mathbf{i}_1\mathbf{i}_2 - n^2\mathbf{i}_2\mathbf{i}_1 + mn\mathbf{i}_2\mathbf{i}_2,\\
\mathbf{i}'_2\mathbf{i}'_1 &= (-n\mathbf{i}_1 + m\mathbf{i}_2)(m\mathbf{i}_1 + n\mathbf{i}_2)\\
&= -mn\mathbf{i}_1\mathbf{i}_1 - n^2\mathbf{i}_1\mathbf{i}_2 + m^2\mathbf{i}_2\mathbf{i}_1 + mn\mathbf{i}_2\mathbf{i}_2.
\end{aligned} \tag{2.177}$$

Thus, from (2.176) and (2.177), because $\sigma'_{12} = \sigma'_{21}$, $\sigma'_{13} = \sigma'_{31}$ and $\sigma'_{23} = \sigma'_{32}$

$$\begin{aligned}
\sigma_{11} &= m^2\sigma'_{11} + n^2\sigma'_{22} - 2mn\sigma'_{12},\\
\sigma_{22} &= n^2\sigma'_{11} + m^2\sigma'_{22} + 2mn\sigma'_{12},\\
\sigma_{33} &= \sigma'_{33},\\
\sigma_{23} &= m\sigma'_{23} + n\sigma'_{13} = \sigma_{32},\\
\sigma_{13} &= -n\sigma'_{23} + m\sigma'_{13} = \sigma_{31},\\
\sigma_{12} &= mn\sigma'_{11} - mn\sigma'_{22} + (m^2 - n^2)\sigma'_{12} = \sigma_{21}.
\end{aligned} \tag{2.178}$$

The inverse relationships are obtained by replacing ϕ by $-\phi$ (i.e. n is replaced by $-n$) so that

$$\begin{aligned}
\sigma'_{11} &= m^2\sigma_{11} + n^2\sigma_{22} + 2mn\sigma_{12},\\
\sigma'_{22} &= n^2\sigma_{11} + m^2\sigma_{22} - 2mn\sigma_{12},\\
\sigma'_{33} &= \sigma_{33},\\
\sigma'_{23} &= m\sigma_{23} - n\sigma_{13} = \sigma'_{32},\\
\sigma'_{13} &= n\sigma_{23} + m\sigma_{13} = \sigma'_{31},\\
\sigma'_{12} &= -mn\sigma_{11} + mn\sigma_{22} + \left(m^2 - n^2\right)\sigma_{12} = \sigma'_{21}.
\end{aligned} \tag{2.179}$$

The relationships (2.178) and (2.179) are the standard transformations, arising from tensor theory, for the rotation of stress components about one axis of a right-handed rectangular set of Cartesian coordinates. Identical transformations apply when considering the strain tensor so that

$$\begin{aligned}
\varepsilon_{11} &= m^2\varepsilon'_{11} + n^2\varepsilon'_{22} - 2mn\varepsilon'_{12},\\
\varepsilon_{22} &= n^2\varepsilon'_{11} + m^2\varepsilon'_{22} + 2mn\varepsilon'_{12},\\
\varepsilon_{33} &= \varepsilon'_{33},\\
\varepsilon_{23} &= m\varepsilon'_{23} + n\varepsilon'_{13} = \varepsilon_{32},\\
\varepsilon_{13} &= -n\varepsilon'_{23} + m\varepsilon'_{13} = \varepsilon_{31},\\
\varepsilon_{12} &= mn\varepsilon'_{11} - mn\varepsilon'_{22} + \left(m^2 - n^2\right)\varepsilon'_{12} = \varepsilon_{21},
\end{aligned} \tag{2.180}$$

with inverse relations

$$\begin{aligned}
\varepsilon'_{11} &= m^2\varepsilon_{11} + n^2\varepsilon_{22} + 2mn\varepsilon_{12},\\
\varepsilon'_{22} &= n^2\varepsilon_{11} + m^2\varepsilon_{22} - 2mn\varepsilon_{12},\\
\varepsilon'_{33} &= \varepsilon_{33},\\
\varepsilon'_{23} &= m\varepsilon_{23} - n\varepsilon_{13} = \varepsilon'_{32},\\
\varepsilon'_{13} &= n\varepsilon_{23} + m\varepsilon_{13} = \varepsilon'_{31},\\
\varepsilon'_{12} &= -mn\varepsilon_{11} + mn\varepsilon_{22} + \left(m^2 - n^2\right)\varepsilon_{12} = \varepsilon'_{21}.
\end{aligned} \tag{2.181}$$

2.17 Transformations of Elastic Constants

On substituting the stress-strain relations (2.170) into (2.181)

$$\begin{aligned}
\varepsilon'_{11} &= \left(m^2S_{11} + n^2S_{12}\right)\sigma_{11} + \left(m^2S_{12} + n^2S_{22}\right)\sigma_{22} + \left(m^2S_{13} + n^2S_{23}\right)\sigma_{33}\\
&\qquad + mnS_{66}\sigma_{12} + \left(m^2V_1 + n^2V_2\right)\Delta T,\\
\varepsilon'_{22} &= \left(n^2S_{11} + m^2S_{12}\right)\sigma_{11} + \left(n^2S_{12} + m^2S_{22}\right)\sigma_{22} + \left(n^2S_{13} + m^2S_{23}\right)\sigma_{33}\\
&\qquad - mnS_{66}\sigma_{12} + \left(n^2V_1 + m^2V_2\right)\Delta T,\\
\varepsilon'_{33} &= S_{13}\sigma_{11} + S_{23}\sigma_{22} + S_{33}\sigma_{33} + V_3\Delta T,\\
\varepsilon'_{23} &= \tfrac{1}{2}mS_{44}\sigma_{23} - \tfrac{1}{2}nS_{55}\sigma_{13} = \varepsilon'_{32},\\
\varepsilon'_{13} &= \tfrac{1}{2}nS_{44}\sigma_{23} + \tfrac{1}{2}mS_{55}\sigma_{13} = \varepsilon'_{31},\\
\varepsilon'_{12} &= mn\left[\left(S_{12} - S_{11}\right)\sigma_{11} + \left(S_{22} - S_{12}\right)\sigma_{22} + \left(S_{23} - S_{13}\right)\sigma_{33}\right]\\
&\qquad + \tfrac{1}{2}\left(m^2 - n^2\right)S_{66}\sigma_{12} - mn\left(V_1 - V_2\right)\Delta T = \varepsilon'_{21}.
\end{aligned} \tag{2.182}$$

Substitution of (2.178) into (2.182) leads to the relations

$$\begin{bmatrix} \varepsilon'_{11} \\ \varepsilon'_{22} \\ \varepsilon'_{33} \\ 2\varepsilon'_{23} \\ 2\varepsilon'_{13} \\ 2\varepsilon'_{12} \end{bmatrix} = \begin{bmatrix} S'_{11} & S'_{12} & S'_{13} & 0 & 0 & S'_{16} \\ S'_{12} & S'_{22} & S'_{23} & 0 & 0 & S'_{26} \\ S'_{13} & S'_{23} & S'_{33} & 0 & 0 & S'_{36} \\ 0 & 0 & 0 & S'_{44} & S'_{45} & 0 \\ 0 & 0 & 0 & S'_{45} & S'_{55} & 0 \\ S'_{16} & S'_{26} & S'_{36} & 0 & 0 & S'_{66} \end{bmatrix} \begin{bmatrix} \sigma'_{11} \\ \sigma'_{22} \\ \sigma'_{33} \\ \sigma'_{23} \\ \sigma'_{13} \\ \sigma'_{12} \end{bmatrix} + \begin{bmatrix} V'_1 \\ V'_2 \\ V'_3 \\ 0 \\ 0 \\ V'_6 \end{bmatrix} \Delta T, \tag{2.183}$$

where

$$\begin{aligned} S'_{11} &= m^4 S_{11} + n^4 S_{22} + m^2 n^2 \left(2S_{12} + S_{66}\right), \\ S'_{12} &= \left(m^4 + n^4\right) S_{12} + m^2 n^2 \left(S_{11} + S_{22} - S_{66}\right), \\ S'_{13} &= m^2 S_{13} + n^2 S_{23}, \\ S'_{22} &= n^4 S_{11} + m^4 S_{22} + m^2 n^2 \left(2S_{12} + S_{66}\right), \\ S'_{23} &= n^2 S_{13} + m^2 S_{23}, \\ S'_{33} &= S_{33}, \end{aligned} \tag{2.184}$$

$$\begin{aligned} S'_{44} &= m^2 S_{44} + n^2 S_{55}, \\ S'_{45} &= mn\left(S_{44} - S_{55}\right), \\ S'_{55} &= n^2 S_{44} + m^2 S_{55}, \end{aligned} \tag{2.185}$$

$$\begin{aligned} S'_{16} &= mn\left[\left(m^2 - n^2\right)\left(2S_{12} + S_{66}\right) - 2m^2 S_{11} + 2n^2 S_{22}\right], \\ S'_{26} &= mn\left[-\left(m^2 - n^2\right)\left(2S_{12} + S_{66}\right) - 2n^2 S_{11} + 2m^2 S_{22}\right], \\ S'_{36} &= 2mn\left(S_{23} - S_{13}\right), \\ S'_{66} &= \left(m^2 - n^2\right)^2 S_{66} + 4m^2 n^2 \left(S_{11} + S_{22} - 2S_{12}\right), \end{aligned} \tag{2.186}$$

and where

$$\begin{aligned} V'_1 &= m^2 V_1 + n^2 V_2, & V'_2 &= n^2 V_1 + m^2 V_2, \\ V'_3 &= V_3, & V'_6 &= -2mn\left(V_1 - V_2\right). \end{aligned} \tag{2.187}$$

2.17.1 Transverse Isotropic and Isotropic Solids

When considering unidirectionally reinforced fibre composites, as will be the case in Chapter 4, the effective composite properties are often assumed to be isotropic in the plane that is normal to the fibre direction taken here to be the x_3-direction as coordinate rotations considered previously have been about the x_3-axis. It is now assumed that $S_{11} = S_{22}$, $S_{44} = S_{55}$ and $S_{13} = S_{23}$. As $m^2 + n^2 = 1$ and

$$m^4 + n^4 = 1 - 2m^2 n^2, \ \left(m^2 - n^2\right)^2 = m^4 + n^4 - 2m^2 n^2 = 1 - 4m^2 n^2,$$

it then follows from (2.184)–(2.186) that

$$\begin{aligned}
&S'_{11} = S'_{22} = S_{11} + \left(S_{66} - 2S_{11} + 2S_{12}\right)m^2n^2,\\
&S'_{12} = S_{12} - \left(S_{66} - 2S_{11} + 2S_{12}\right)m^2n^2,\\
&S'_{13} = S'_{23} = S_{13}, \quad S'_{33} = S_{33},\\
&S'_{44} = S'_{55} = S_{44}, \quad S'_{45} = 0,\\
&S'_{16} = -S'_{26} = mn\left(m^2 - n^2\right)\left(S_{66} - 2S_{11} + 2S_{12}\right),\\
&S'_{36} = 0,\ S'_{66} = S_{66} - 4m^2n^2\left(S_{66} - 2S_{11} + 2S_{12}\right).
\end{aligned} \tag{2.188}$$

It should be noted that the factor $S_{66} - 2S_{11} + 2S_{12}$ appears repeatedly in these relations. When this factor is zero so that

$$S_{66} = 2\left(S_{11} - S_{12}\right), \tag{2.189}$$

it follows that

$$\begin{aligned}
&S'_{11} = S_{11}, \quad S'_{12} = S_{12}, \quad S'_{22} = S_{11},\\
&S'_{13} = S_{13}, \quad S'_{23} = S_{13},\\
&S'_{33} = S_{33}, \quad S'_{44} = S_{44}, \quad S'_{55} = S_{44}, \quad S'_{66} = S_{66},\\
&S'_{45} = 0, \quad S'_{16} = 0, \quad S'_{26} = 0, \quad S'_{36} = 0.
\end{aligned} \tag{2.190}$$

As it was assumed that $S_{11} = S_{22}$, $S_{44} = S_{55}$, $S_{13} = S_{23}$, it is clear that any rotation about the x_3-axis does not alter the value of the elastic constants on transformation. Thus, the material having the stress-strain relations (2.170) are transverse isotropic relative to the x_3-axis if the elastic constants are such that

$$S_{11} = S_{22},\ S_{44} = S_{55},\ S_{13} = S_{23},\ S_{66} = 2\left(S_{11} - S_{12}\right). \tag{2.191}$$

For *isotropic* materials, the elastic constants must satisfy the relations

$$S_{11} = S_{22} = S_{33},\ S_{44} = S_{55} = S_{66},\ S_{13} = S_{23} = S_{12},\ S_{66} = 2\left(S_{11} - S_{12}\right). \tag{2.192}$$

For a transverse isotropic solid the thermal expansion coefficients are such that $V_1 = V_2 = V^*$ and $V_3 = V$. It then follows from (2.187) that

$$V'_1 = V'_2 = V^*, \quad V'_3 = V, \quad V'_6 = 0. \tag{2.193}$$

For isotropic materials

$$V'_1 = V'_2 = V'_3 = V, \quad V'_6 = 0. \tag{2.194}$$

2.17.2 Introducing Familiar Thermoelastic Constants

It is useful to express the elastic constants S_{IJ} in terms of more familiar physical quantities such as the elastic constants, for linear elastic media, known as Young's moduli, shear moduli and Poisson's ratios. Consider a thin rectangular plate made of an orthotropic fibre reinforced material where the in-plane directions x_1 and x_2 are parallel to the edges of the plate and where the through-thickness direction is parallel to the

x_3-axis. The straight fibres in the plate are all parallel to the x_1-axis. For this situation, the elastic constants S_{IJ} in (2.170) are written in the form

$$\begin{aligned}
&S_{11} = \frac{1}{E_A}, \quad && S_{12} = -\frac{\nu_A}{E_A}, \quad && S_{13} = -\frac{\nu_a}{E_A},\\
&S_{21} = -\frac{\nu_A}{E_A}, && S_{22} = \frac{1}{E_T}, && S_{23} = -\frac{\nu_t}{E_T},\\
&S_{31} = -\frac{\nu_a}{E_A}, && S_{32} = -\frac{\nu_t}{E_T}, && S_{33} = \frac{1}{E_t},\\
&S_{44} = \frac{1}{\mu_t}, && S_{55} = \frac{1}{\mu_a}, && S_{66} = \frac{1}{\mu_A},\\
&V_1 = \alpha_A, && V_2 = \alpha_T, && V_3 = \alpha_t,
\end{aligned} \tag{2.195}$$

where Young's moduli are denoted by E, shear moduli by μ, Poisson's ratios by ν and thermal expansion coefficients by α. The stress-strain relations (2.170) may then be written as

$$\begin{aligned}
\varepsilon_{11} &= \frac{1}{E_A}\sigma_{11} - \frac{\nu_A}{E_A}\sigma_{22} - \frac{\nu_a}{E_A}\sigma_{33} + \alpha_A \Delta T, \quad & 2\varepsilon_{12} &= \frac{1}{\mu_A}\sigma_{12},\\
\varepsilon_{22} &= -\frac{\nu_A}{E_A}\sigma_{11} + \frac{1}{E_T}\sigma_{22} - \frac{\nu_t}{E_T}\sigma_{33} + \alpha_T \Delta T, & 2\varepsilon_{13} &= \frac{1}{\mu_a}\sigma_{13},\\
\varepsilon_{33} &= -\frac{\nu_a}{E_A}\sigma_{11} - \frac{\nu_t}{E_T}\sigma_{22} + \frac{1}{E_t}\sigma_{33} + \alpha_t \Delta T, & 2\varepsilon_{23} &= \frac{1}{\mu_t}\sigma_{23}.
\end{aligned} \tag{2.196}$$

The subscripts 'A' and 'T' refer to axial and transverse thermoelastic constants, respectively, involving in-plane stresses and deformations. The subscripts 'a' and 't' refer to axial and transverse constants, respectively, associated with out-of-plane stresses and deformations. The parameter ΔT is the difference between the current temperature of the material and the reference temperature for which all strains are zero when the sample is unloaded.

It is clear that when the plate is uniaxially loaded in the x_1-direction, the parameter ν_A is the Poisson's ratio determining the in-plane transverse deformation in the x_2-direction whereas ν_a is Poisson's ratio determining the transverse through-thickness deformation in the x_3-direction. When the plate is uniaxially loaded in the x_2-direction, the parameter ν_t is the Poisson's ratio determining the transverse through-thickness deformation in the x_3-direction.

It is useful, first, to show the form of the stress-strain equations (2.196) when the material is *transverse isotropic* about the x_3-axis, so that they may be used when considering the properties of unidirectional plies in a laminate where the fibres are aligned in the x_3-direction of the ply, and so that use can be made of analysis given in the previous section. It follows from (2.196) that when the material is transverse isotropic about the x_3-axis, the stress-strain relations are of the form

$$\begin{aligned}
\varepsilon_{11} &= \frac{1}{E_T}\sigma_{11} - \frac{\nu_t}{E_T}\sigma_{22} - \frac{\nu_A}{E_A}\sigma_{33} + \alpha_T \Delta T, \quad & 2\varepsilon_{12} &= \frac{1}{\mu_t}\sigma_{12},\\
\varepsilon_{22} &= -\frac{\nu_t}{E_T}\sigma_{11} + \frac{1}{E_T}\sigma_{22} - \frac{\nu_A}{E_A}\sigma_{33} + \alpha_T \Delta T, & 2\varepsilon_{13} &= \frac{1}{\mu_A}\sigma_{13},\\
\varepsilon_{33} &= -\frac{\nu_A}{E_A}\sigma_{11} - \frac{\nu_A}{E_A}\sigma_{22} + \frac{1}{E_A}\sigma_{33} + \alpha_A \Delta T, & 2\varepsilon_{23} &= \frac{1}{\mu_A}\sigma_{23}.
\end{aligned} \tag{2.197}$$

As $S_{11} = 1/E_T$, $S_{12} = -\nu_t/E_T$ and $S_{66} = 1/\mu_t$ it follows from (2.189) that for a transverse isotropic solid the following condition must be satisfied:

$$E_T = 2\mu_t\left(1+\nu_t\right). \tag{2.198}$$

In Chapter 4 considering fibre-reinforced materials, stress-strain relations are required for the cylindrical polar coordinates (r, θ, z) corresponding to the relations (2.197), which are given by

$$\begin{aligned}
\varepsilon_{rr} &= \frac{1}{E_T}\sigma_{rr} - \frac{\nu_t}{E_T}\sigma_{\theta\theta} - \frac{\nu_A}{E_A}\sigma_{zz} + \alpha_T\Delta T, & \varepsilon_{rz} &= \frac{\sigma_{rz}}{2\mu_A},\\
\varepsilon_{\theta\theta} &= -\frac{\nu_t}{E_T}\sigma_{rr} + \frac{1}{E_T}\sigma_{\theta\theta} - \frac{\nu_A}{E_A}\sigma_{zz} + \alpha_T\Delta T, & \varepsilon_{\theta z} &= \frac{\sigma_{\theta z}}{2\mu_A},\\
\varepsilon_{zz} &= -\frac{\nu_A}{E_A}\sigma_{rr} - \frac{\nu_A}{E_A}\sigma_{\theta\theta} + \frac{1}{E_A}\sigma_{zz} + \alpha_A\Delta T, & \varepsilon_{r\theta} &= \frac{\sigma_{r\theta}}{2\mu_t}.
\end{aligned} \tag{2.199}$$

When the fibres are aligned in a direction parallel to the x_1-axis, as required in Chapters 6 and 7 concerning laminates and their plies, the transverse isotropic stress-strain relations, resulting from the orthotropic form (2.196), are given by

$$\begin{aligned}
\varepsilon_{11} &= \frac{1}{E_A}\sigma_{11} - \frac{\nu_A}{E_A}\sigma_{22} - \frac{\nu_A}{E_A}\sigma_{33} + \alpha_A\Delta T, & 2\varepsilon_{12} &= \frac{1}{\mu_A}\sigma_{12},\\
\varepsilon_{22} &= -\frac{\nu_A}{E_A}\sigma_{11} + \frac{1}{E_T}\sigma_{22} - \frac{\nu_t}{E_T}\sigma_{33} + \alpha_T\Delta T, & 2\varepsilon_{13} &= \frac{1}{\mu_A}\sigma_{13},\\
\varepsilon_{33} &= -\frac{\nu_A}{E_A}\sigma_{11} - \frac{\nu_t}{E_T}\sigma_{22} + \frac{1}{E_T}\sigma_{33} + \alpha_T\Delta T, & 2\varepsilon_{23} &= \frac{1}{\mu_t}\sigma_{23}.
\end{aligned} \tag{2.200}$$

where, again, the relation (2.198) must be satisfied.

For plane strain conditions such that $\varepsilon_{11} \equiv 0$, it follows from (2.200) that

$$\begin{aligned}
\sigma_{11} &= \nu_A\left(\sigma_{22}+\sigma_{33}\right) - E_A\alpha_A\Delta T,\\
\varepsilon_{22}+\varepsilon_{33} &= \left(\frac{1-\nu_t}{E_T} - \frac{2\nu_A^2}{E_A}\right)\left(\sigma_{22}+\sigma_{33}\right) + 2\left(\alpha_T+\nu_A\alpha_A\right)\Delta T.
\end{aligned} \tag{2.201}$$

When $\Delta T = 0$, the term $\varepsilon_{22}+\varepsilon_{33}$ is the change in volume per unit volume $\Delta V/V$ for the plane strain conditions under discussion when an equiaxial transverse stress σ is applied such that $\sigma_2 = \sigma_3 = \sigma$. It then follows that a plane strain bulk modulus k_T can be defined by

$$\frac{1}{k_T} = \frac{2(1-\nu_t)}{E_T} - \frac{4\nu_A^2}{E_A}, \tag{2.202}$$

such that $\sigma = k_T\,\Delta V/V$ when $\Delta T = 0$.

For *isotropic* materials, $E_A = E_T = E$, $\nu_A = \nu_t = \nu$ and $\mu_A = \mu_t = \mu$ so that

$$\begin{aligned}
S_{11} &= \frac{1}{E}, & S_{12} &= -\frac{\nu}{E}, & S_{13} &= -\frac{\nu}{E},\\
S_{21} &= -\frac{\nu}{E}, & S_{22} &= \frac{1}{E}, & S_{23} &= -\frac{\nu}{E},\\
S_{31} &= -\frac{\nu}{E}, & S_{32} &= -\frac{\nu}{E}, & S_{33} &= \frac{1}{E},\\
S_{44} &= \frac{1}{\mu}, & S_{55} &= \frac{1}{\mu}, & S_{66} &= \frac{1}{\mu},\\
V_1 &= \alpha, & V_2 &= \alpha, & V_3 &= \alpha,
\end{aligned} \tag{2.203}$$

and so that (2.198) has the following form

$$E = 2\mu\left(1+\nu\right). \tag{2.204}$$

It is clear that the elastic constants of an isotropic material are fully characterised by just two independent elastic constants, such as one of the following combinations: (E, ν), (μ, ν) and (E, μ). One of Lamé's constants λ (the other is the shear modulus μ) and the bulk modulus k are often used as elastic constants for isotropic materials. These are related to Young's modulus E, the shear modulus μ and Poisson's ratio ν as follows (see (2.161)):

$$\lambda = \frac{E\nu}{\left(1+\nu\right)\left(1-2\nu\right)}, \quad k = \lambda + \tfrac{2}{3}\mu = \frac{E}{3(1-2\nu)}. \tag{2.205}$$

The inverse form is

$$E = \frac{\mu\left(3\lambda+2\mu\right)}{\lambda+\mu}, \qquad \nu = \frac{\lambda}{2\left(\lambda+\mu\right)}. \tag{2.206}$$

It is sometimes convenient to characterise an isotropic material using the two elastic constants μ and ν in which case, in addition to the relation (2.204),

$$\lambda = \frac{2\mu\nu}{1-2\nu}. \tag{2.207}$$

More frequently, and as required for Chapter 3, it is useful to express Young's modulus E and Poisson's ratio ν in terms of the bulk modulus k and the shear modulus μ. On using (2.204) and (2.205)

$$E = \frac{9k\mu}{3k+\mu}, \qquad \nu = \frac{3k-2\mu}{2(3k+\mu)}. \tag{2.208}$$

2.18 Analysis of Bend Deformation

For most engineering applications of composite components, the deformation experienced in service conditions will involve some degree of bending. As the effect of bending on ply crack formation in composite laminates is considered in Chapters 11 and 19, it is useful to describe here the essential fundamental aspects of an analysis of bend deformation for a uniform orthotropic plate.

2.18.1 Geometry and Basic Equations

A beam of rectangular cross section of length $2L$, width $2W$ and depth h is considered within a Cartesian coordinate system such that the x_1-axis is in the axial direction and the x_2-axis is in the in-plane transverse direction whereas the x_3-axis is in the through-thickness direction, as shown in Figure 2.2. The origin is selected to lie at the mid-point of the upper surface of the beam.

The modelling assumes that the beam is in equilibrium such that the equilibrium equations (2.120)–(2.122) are satisfied where the stress tensor is symmetric as in (2.123). The displacement components are denoted by u_i. The infinitesimal strains are then given by the relations (2.143). The stress-strain relations are assumed to be of the orthotropic form (2.196), namely,

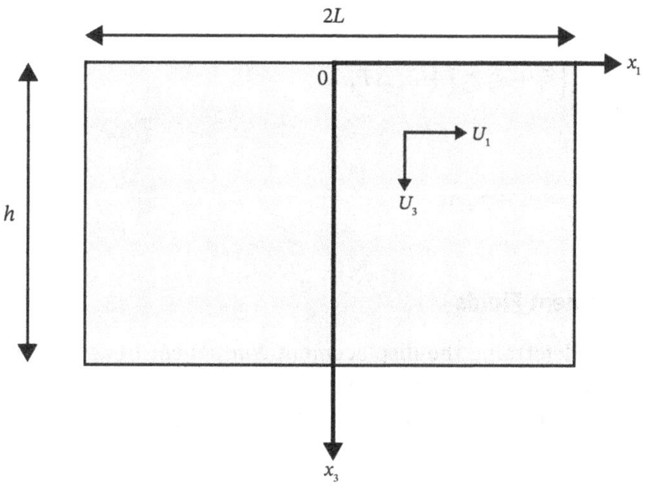

Figure 2.2 Schematic diagram of part of a rectangular orthotropic plate of length $2L$ and depth h, and coordinate system. The x_2-axis and u_2 displacement are directed out of the plane of the page, and the width is denoted by $2W$.

$$\begin{aligned}
\varepsilon_{11} &= \frac{1}{E_{\mathrm{A}}}\sigma_{11} - \frac{\nu_{\mathrm{A}}}{E_{\mathrm{A}}}\sigma_{22} - \frac{\nu_{\mathrm{a}}}{E_{\mathrm{A}}}\sigma_{33} + \alpha_{\mathrm{A}}\Delta T, & 2\varepsilon_{12} &= \frac{1}{\mu_{\mathrm{A}}}\sigma_{12}, \\
\varepsilon_{22} &= -\frac{\nu_{\mathrm{A}}}{E_{\mathrm{A}}}\sigma_{11} + \frac{1}{E_{\mathrm{T}}}\sigma_{22} - \frac{\nu_{\mathrm{t}}}{E_{\mathrm{T}}}\sigma_{33} + \alpha_{\mathrm{T}}\Delta T, & 2\varepsilon_{13} &= \frac{1}{\mu_{\mathrm{a}}}\sigma_{13}, \\
\varepsilon_{33} &= -\frac{\nu_{\mathrm{a}}}{E_{\mathrm{A}}}\sigma_{11} - \frac{\nu_{\mathrm{t}}}{E_{\mathrm{T}}}\sigma_{22} + \frac{1}{E_{\mathrm{t}}}\sigma_{33} + \alpha_{\mathrm{t}}\Delta T, & 2\varepsilon_{23} &= \frac{1}{\mu_{\mathrm{t}}}\sigma_{23}.
\end{aligned} \tag{2.209}$$

The beam is assumed to be in a state of orthogonal bending combined with uniform through-thickness loading such that

$$\sigma_{13} \equiv 0, \quad \sigma_{23} \equiv 0, \quad \sigma_{33} \equiv \sigma_{\mathrm{t}}, \tag{2.210}$$

and

$$u_1 = \left(\bar{\varepsilon}_{\mathrm{A}} + \hat{\varepsilon}_{\mathrm{A}} x_3\right) x_1, \quad u_2 = \left(\bar{\varepsilon}_{\mathrm{T}} + \hat{\varepsilon}_{\mathrm{T}} x_3\right) x_2, \tag{2.211}$$

where the strain parameters $\bar{\varepsilon}_{\mathrm{A}}$, $\bar{\varepsilon}_{\mathrm{T}}$, $\hat{\varepsilon}_{\mathrm{A}}$ and $\hat{\varepsilon}_{\mathrm{T}}$, the through-thickness stress σ_{t} and the temperature difference ΔT are assumed for the moment to be known. From (2.143), (2.209) and (2.211), the in-plane strains and shear stress are given by

$$\varepsilon_{11} = \bar{\varepsilon}_{\mathrm{A}} + \hat{\varepsilon}_{\mathrm{A}} x_3, \quad \varepsilon_{22} = \bar{\varepsilon}_{\mathrm{T}} + \hat{\varepsilon}_{\mathrm{T}} x_3, \quad \varepsilon_{12} \equiv 0, \quad \sigma_{12} \equiv 0. \tag{2.212}$$

The expression for ε_{22} in (2.209) is first solved for the stress component σ_{22} so that

$$\sigma_{22} = \nu_{\mathrm{A}} \frac{E_{\mathrm{T}}}{E_{\mathrm{A}}}\sigma_{11} + \nu_{\mathrm{t}}\sigma_{\mathrm{t}} + E_{\mathrm{T}}\left(\bar{\varepsilon}_{\mathrm{T}} + \hat{\varepsilon}_{\mathrm{T}} x_3\right) - E_{\mathrm{T}}\alpha_{\mathrm{T}}\Delta T. \tag{2.213}$$

It then follows from (2.209) that

$$\varepsilon_{11} = \frac{\sigma_{11}}{\tilde{E}_{\mathrm{A}}} - \frac{\tilde{\nu}_{\mathrm{a}}}{\tilde{E}_{\mathrm{A}}}\sigma_{\mathrm{t}} + \tilde{\alpha}_{\mathrm{A}}\Delta T - \nu_{\mathrm{A}}\frac{E_{\mathrm{T}}}{E_{\mathrm{A}}}\left(\bar{\varepsilon}_{\mathrm{T}} + \hat{\varepsilon}_{\mathrm{T}} x_3\right), \tag{2.214}$$

where

$$\frac{1}{\tilde{E}_{\mathrm{A}}} = \frac{1}{E_{\mathrm{A}}}\left(1 - (\nu_{\mathrm{A}})^2 \frac{E_{\mathrm{T}}}{E_{\mathrm{A}}}\right), \quad \frac{\tilde{\nu}_{\mathrm{a}}}{\tilde{E}_{\mathrm{A}}} = \frac{\nu_{\mathrm{a}} + \nu_{\mathrm{t}}\nu_{\mathrm{A}}}{E_{\mathrm{A}}}, \quad \tilde{\alpha}_{\mathrm{A}} = \alpha_{\mathrm{A}} + \nu_{\mathrm{A}}\frac{E_{\mathrm{T}}}{E_{\mathrm{A}}}\alpha_{\mathrm{T}}. \tag{2.215}$$

In addition, from (2.209)

$$\varepsilon_{33} = -\frac{\tilde{\nu}_a}{\tilde{E}_A}\sigma_{11} + \frac{\sigma_t}{\tilde{E}_t} - \nu_t\left(\bar{\varepsilon}_T + \hat{\varepsilon}_T x_3\right) + \tilde{\alpha}_t \Delta T, \tag{2.216}$$

where

$$\frac{1}{\tilde{E}_t} = \frac{1}{E_t} - \frac{(\nu_t)^2}{E_T}, \qquad \tilde{\alpha}_t = \alpha_t + \nu_t \alpha_T. \tag{2.217}$$

2.18.2 Stress and Displacement Fields

The immediate objective is to determine the displacement component u_3 and corresponding stresses σ_{11}, σ_{12} and σ_{12} in terms of the mechanical loading parameters $\bar{\varepsilon}_A, \hat{\varepsilon}_A, \bar{\varepsilon}_T, \hat{\varepsilon}_T, \sigma_t$ and the temperature difference ΔT. From (2.209), (2.210) and (2.212), the in-plane stresses must satisfy the relations

$$\sigma_{11} - \nu_A \sigma_{22} = E_A\left(\bar{\varepsilon}_A + \hat{\varepsilon}_A x_3 - \alpha_A \Delta T\right) + \nu_a \sigma_t, \tag{2.218}$$

$$-\frac{\nu_A E_T}{E_A}\sigma_{11} + \sigma_{22} = E_T\left(\bar{\varepsilon}_T + \hat{\varepsilon}_T x_3 - \alpha_T \Delta T\right) + \nu_t \sigma_t. \tag{2.219}$$

On solving (2.218) and (2.219) for the stresses, it follows that

$$\sigma_{11}(x_3) = \tilde{E}_A\left(\bar{\varepsilon}_A + \hat{\varepsilon}_A x_3\right) + \nu_A \tilde{E}_T\left(\bar{\varepsilon}_T + \hat{\varepsilon}_T x_3\right) + \tilde{\nu}_a \sigma_t - \tilde{E}_A \tilde{\alpha}_A \Delta T, \tag{2.220}$$

$$\sigma_{22}(x_3) = \nu_A \tilde{E}_T\left(\bar{\varepsilon}_A + \hat{\varepsilon}_A x_3\right) + \tilde{E}_T\left(\bar{\varepsilon}_T + \hat{\varepsilon}_T x_3\right) + \tilde{\nu}_t \sigma_t - \tilde{E}_T \tilde{\alpha}_T \Delta T, \tag{2.221}$$

where

$$\frac{1}{\tilde{E}_T} = \frac{1}{E_T}\left(1 - \nu_A^2 \frac{E_T}{E_A}\right), \qquad \frac{\tilde{\nu}_t}{\tilde{E}_T} = \frac{\nu_t}{E_T} + \frac{\nu_a \nu_A}{E_A}, \qquad \tilde{\alpha}_T = \alpha_T + \nu_A \alpha_A. \tag{2.222}$$

It is noted that the in-plane stresses are only linear functions of x_3. It then follows, on using (2.210), that all the equilibrium equations (2.120)–(2.122) are satisfied.

It now only remains to determine the value of the displacement component u_3 describing the deflection of the beam. From (2.216) and (2.220)

$$\varepsilon_{33} = -\tilde{\nu}_a\left(\bar{\varepsilon}_A + \hat{\varepsilon}_A x_3\right) - \left(\nu_t + \nu_A \frac{E_T}{E_A}\tilde{\nu}_a\right)\left(\bar{\varepsilon}_T + \hat{\varepsilon}_T x_3\right) + \left(\frac{1}{\tilde{E}_t} - \frac{(\tilde{\nu}_a)^2}{\tilde{E}_A}\right)\sigma_t + \left(\tilde{\alpha}_t + \tilde{\nu}_a \tilde{\alpha}_A\right)\Delta T. \tag{2.223}$$

On using (2.215) and (2.222) it can be shown that

$$\nu_t + \nu_A \frac{E_T}{E_A}\tilde{\nu}_a = \tilde{E}_T\left(\frac{\nu_t}{E_T} + \frac{\nu_a \nu_A}{E_A}\right) = \tilde{\nu}_t, \tag{2.224}$$

so that (2.223) may now be written as

$$\varepsilon_{33} \equiv \frac{\partial u_3}{\partial x_3} = -\tilde{\nu}_a\left(\bar{\varepsilon}_A + \hat{\varepsilon}_A x_3\right) - \tilde{\nu}_t\left(\bar{\varepsilon}_T + \hat{\varepsilon}_T x_3\right) + \left(\frac{1}{\tilde{E}_t} - \frac{(\tilde{\nu}_a)^2}{\tilde{E}_A}\right)\sigma_t + \left(\tilde{\alpha}_t + \tilde{\nu}_a \tilde{\alpha}_A\right)\Delta T. \tag{2.225}$$

Integration with respect to x_3 then leads to

$$u_3 = -\tilde{\nu}_a\left(\bar{\varepsilon}_A + \tfrac{1}{2}\hat{\varepsilon}_A x_3\right)x_3 - \tilde{\nu}_t\left(\bar{\varepsilon}_T + \tfrac{1}{2}\hat{\varepsilon}_T x_3\right)x_3 + \left(\frac{1}{\tilde{E}_t} - \frac{(\tilde{\nu}_a)^2}{\tilde{E}_A}\right)\sigma_t x_3 \\ + \left(\tilde{\alpha}_t + \tilde{\nu}_a\tilde{\alpha}_A\right)\Delta T x_3 + f(x_1, x_2). \tag{2.226}$$

As the shear stresses σ_{13} and σ_{23} are everywhere zero, it follows from (2.143), (2.209) and (2.211) that

$$\frac{\partial u_3}{\partial x_1} = -\frac{\partial u_1}{\partial x_3} = -\hat{\varepsilon}_A x_1, \quad \frac{\partial u_3}{\partial x_2} = -\frac{\partial u_2}{\partial x_3} = -\hat{\varepsilon}_T x_2. \tag{2.227}$$

Integration then yields

$$u_3 = -\tfrac{1}{2}\hat{\varepsilon}_A x_1^2 + g_1(x_2, x_3), \quad u_3 = -\tfrac{1}{2}\hat{\varepsilon}_T x_2^2 + g_2(x_1, x_3). \tag{2.228}$$

These relations must be consistent with (2.226) so that

$$u_3 = -\tilde{\nu}_a\left(\bar{\varepsilon}_A + \tfrac{1}{2}\hat{\varepsilon}_A x_3\right)x_3 - \tilde{\nu}_t\left(\bar{\varepsilon}_T + \tfrac{1}{2}\hat{\varepsilon}_T x_3\right)x_3 + \left(\frac{1}{\tilde{E}_t} - \frac{(\tilde{\nu}_a)^2}{\tilde{E}_A}\right)\sigma_t x_3 \\ + \left(\tilde{\alpha}_t + \tilde{\nu}_a\tilde{\alpha}_A\right)\Delta T x_3 - \tfrac{1}{2}\hat{\varepsilon}_A x_1^2 - \tfrac{1}{2}\hat{\varepsilon}_T x_2^2, \tag{2.229}$$

where the displacement component has been selected to be zero at the origin.

The through-thickness displacement of the top surface of the beam, at $x_3 = 0$, can be defined in terms of two lengths, R_1 and R_2, which are the radii of curvature of this surface in the x_1–x_3 plane and the x_2–x_3 plane, respectively. The exact relationships are given by the well-known formulae

$$\frac{1}{R_1} = -\frac{\dfrac{\partial^2 u_3}{\partial x_1^2}}{\left[1 + \left(\dfrac{\partial u_3}{\partial x_1}\right)^2\right]^{3/2}}, \quad \frac{1}{R_2} = -\frac{\dfrac{\partial^2 u_3}{\partial x_2^2}}{\left[1 + \left(\dfrac{\partial u_3}{\partial x_2}\right)^2\right]^{3/2}}. \tag{2.230}$$

For small deflections

$$\frac{1}{R_1} = -\frac{\partial^2 u_3}{\partial x_1^2}, \quad \frac{1}{R_2} = -\frac{\partial^2 u_3}{\partial x_2^2}. \tag{2.231}$$

Thus, it follows from (2.229) that

$$R_1 = \frac{1}{\hat{\varepsilon}_A}, \quad R_2 = \frac{1}{\hat{\varepsilon}_T}, \tag{2.232}$$

providing a useful physical interpretation of the strain parameters $\hat{\varepsilon}_A$ and $\hat{\varepsilon}_T$.

The final requirement is to determine the loading state that is consistent with the various strain parameter values. It is assumed that stresses within the beam can arise from an applied in-plane loading that is equivalent to an applied axial force F_A and a transverse force F_T acting in the mid-plane between the upper and lower surfaces of the beam, and an axial applied bending moment per unit area of cross section M_A and a transverse applied bending moment per unit area of cross section M_T. From mechanical equilibrium

$$F_A = 2hW\sigma_A = 2W\int_0^h \sigma_{11}(x_3)\mathrm{d}x_3, \qquad F_T = 2hL\sigma_T = 2L\int_0^h \sigma_{22}(x_3)\mathrm{d}x_3, \tag{2.233}$$

$$M_{\mathrm{A}}=\frac{1}{h}\int_0^h\left(x_3-\tfrac{1}{2}h\right)\sigma_{11}(x_3)\mathrm{d}x_3\,,\qquad M_{\mathrm{T}}=\frac{1}{h}\int_0^h\left(x_3-\tfrac{1}{2}h\right)\sigma_{22}(x_3)\mathrm{d}x_3,\tag{2.234}$$

where σ_{A} and σ_{T} are the effective axial and transverse applied stresses. On substituting (2.220) and (2.221) into (2.233), the following effective axial and transverse stresses are obtained

$$\sigma_{\mathrm{A}}=\tilde{E}_{\mathrm{A}}\left(\bar{\varepsilon}_{\mathrm{A}}+\tfrac{1}{2}h\hat{\varepsilon}_{\mathrm{A}}\right)+\nu_{\mathrm{A}}\tilde{E}_{\mathrm{T}}\left(\bar{\varepsilon}_{\mathrm{T}}+\tfrac{1}{2}h\hat{\varepsilon}_{\mathrm{T}}\right)+\tilde{\nu}_{\mathrm{a}}\sigma_{\mathrm{t}}-\tilde{E}_{\mathrm{A}}\tilde{\alpha}_{\mathrm{A}}\Delta T,\tag{2.235}$$

$$\sigma_{\mathrm{T}}=\nu_{\mathrm{A}}\tilde{E}_{\mathrm{T}}\left(\bar{\varepsilon}_{\mathrm{A}}+\tfrac{1}{2}h\hat{\varepsilon}_{\mathrm{A}}\right)+\tilde{E}_{\mathrm{T}}\left(\bar{\varepsilon}_{\mathrm{T}}+\tfrac{1}{2}h\hat{\varepsilon}_{\mathrm{T}}\right)+\tilde{\nu}_{\mathrm{t}}\sigma_{\mathrm{t}}-\tilde{E}_{\mathrm{T}}\tilde{\alpha}_{\mathrm{T}}\Delta T.\tag{2.236}$$

The relations (2.234) are now expressed in the form

$$hM_{\mathrm{A}}+\tfrac{1}{2}h^2\sigma_{\mathrm{A}}=\int_0^h x_3\sigma_{11}(x_3)\mathrm{d}x_3,\quad hM_{\mathrm{T}}+\tfrac{1}{2}h^2\sigma_{\mathrm{T}}=\int_0^h x_3\sigma_{22}(x_3)\mathrm{d}x_3.\tag{2.237}$$

On substituting (2.220) and (2.221) into (2.237) the following relations, enabling the determination of the effective axial and transverse bending moments per unit area of cross section, are obtained

$$M_{\mathrm{A}}+\tfrac{1}{2}h\sigma_{\mathrm{A}}=h\left[\tilde{E}_{\mathrm{A}}\left(\tfrac{1}{2}\bar{\varepsilon}_{\mathrm{A}}+\tfrac{1}{3}h\hat{\varepsilon}_{\mathrm{A}}\right)+\nu_{\mathrm{A}}\tilde{E}_{\mathrm{T}}\left(\tfrac{1}{2}\bar{\varepsilon}_{\mathrm{T}}+\tfrac{1}{3}h\hat{\varepsilon}_{\mathrm{T}}\right)+\tfrac{1}{2}\tilde{\nu}_{\mathrm{a}}\sigma_{\mathrm{t}}-\tfrac{1}{2}\tilde{E}_{\mathrm{A}}\tilde{\alpha}_{\mathrm{A}}\Delta T\right],\tag{2.238}$$

$$M_{\mathrm{T}}+\tfrac{1}{2}h\sigma_{\mathrm{T}}=h\left[\nu_{\mathrm{A}}\tilde{E}_{\mathrm{T}}\left(\tfrac{1}{2}\bar{\varepsilon}_{\mathrm{A}}+\tfrac{1}{3}h\hat{\varepsilon}_{\mathrm{A}}\right)+\tilde{E}_{\mathrm{T}}\left(\tfrac{1}{2}\bar{\varepsilon}_{\mathrm{T}}+\tfrac{1}{3}h\hat{\varepsilon}_{\mathrm{T}}\right)+\tfrac{1}{2}\tilde{\nu}_{\mathrm{t}}\sigma_{\mathrm{t}}-\tfrac{1}{2}\tilde{E}_{\mathrm{T}}\tilde{\alpha}_{\mathrm{T}}\Delta T\right].\tag{2.239}$$

On using (2.235) and (2.236) it follows that

$$M_{\mathrm{A}}=\tfrac{1}{12}h^2\tilde{E}_{\mathrm{A}}\hat{\varepsilon}_{\mathrm{A}}+\tfrac{1}{12}h^2\nu_{\mathrm{A}}\tilde{E}_{\mathrm{T}}\hat{\varepsilon}_{\mathrm{T}},\tag{2.240}$$

$$M_{\mathrm{T}}=\tfrac{1}{12}h^2\nu_{\mathrm{A}}\tilde{E}_{\mathrm{T}}\hat{\varepsilon}_{\mathrm{A}}+\tfrac{1}{12}h^2\tilde{E}_{\mathrm{T}}\hat{\varepsilon}_{\mathrm{T}}.\tag{2.241}$$

2.18.3 Some Special Cases

It is useful now to consider some important special cases that arise very often when considering the bending deformation of materials.

2.18.3.1 Four-point Bending Tests

The previous analysis can be used to determine the stress and strain state in beams subject to four-point bending. The analysis will apply near the mid-plane between the planes normal to the beam axis that contain the contact points of the inner rollers used in the experiments. For this case $\sigma_{\mathrm{A}}=\sigma_{\mathrm{T}}=\sigma_{\mathrm{t}}=M_{\mathrm{T}}=\Delta T=0$ and the relations (2.235), (2.236), (2.240) and (2.241) reduce to the form

$$\left(\bar{\varepsilon}_{\mathrm{A}}+\tfrac{1}{2}h\hat{\varepsilon}_{\mathrm{A}}\right)+\nu_{\mathrm{A}}\frac{E_{\mathrm{T}}}{E_{\mathrm{A}}}\left(\bar{\varepsilon}_{\mathrm{T}}+\tfrac{1}{2}h\hat{\varepsilon}_{\mathrm{T}}\right)=0,\tag{2.242}$$

$$\nu_{\mathrm{A}}\left(\bar{\varepsilon}_{\mathrm{A}}+\tfrac{1}{2}h\hat{\varepsilon}_{\mathrm{A}}\right)+\left(\bar{\varepsilon}_{\mathrm{T}}+\tfrac{1}{2}h\hat{\varepsilon}_{\mathrm{T}}\right)=0,\tag{2.243}$$

$$M_{\mathrm{A}}=\tfrac{1}{12}h^2\left(\tilde{E}_{\mathrm{A}}\hat{\varepsilon}_{\mathrm{A}}+\nu_{\mathrm{A}}\tilde{E}_{\mathrm{T}}\hat{\varepsilon}_{\mathrm{T}}\right),\tag{2.244}$$

$$\nu_{\mathrm{A}}\hat{\varepsilon}_{\mathrm{A}}+\hat{\varepsilon}_{\mathrm{T}}=0.\tag{2.245}$$

It is clear from (2.242), (2.243) and (2.245) that

$$\bar{\varepsilon}_{\mathrm{A}} + \tfrac{1}{2}h\hat{\varepsilon}_{\mathrm{A}} = \bar{\varepsilon}_{\mathrm{T}} + \tfrac{1}{2}h\hat{\varepsilon}_{\mathrm{T}} = 0, \qquad \hat{\varepsilon}_{\mathrm{T}} = -\nu_{\mathrm{A}}\hat{\varepsilon}_{\mathrm{A}}. \tag{2.246}$$

From $(2.215)_1$, (2.244) and (2.246)

$$\hat{\varepsilon}_{\mathrm{A}} = \frac{1}{R_1} = \frac{12M_{\mathrm{A}}}{h^2E_{\mathrm{A}}},\ \hat{\varepsilon}_{\mathrm{T}} = \frac{1}{R_2} = -\nu_{\mathrm{A}}\hat{\varepsilon}_{\mathrm{A}}, \tag{2.247}$$

$$\bar{\varepsilon}_{\mathrm{A}} = -\tfrac{1}{2}h\hat{\varepsilon}_{\mathrm{A}} = -\frac{6M_{\mathrm{A}}}{hE_{\mathrm{A}}},\ \bar{\varepsilon}_{\mathrm{T}} = -\tfrac{1}{2}h\hat{\varepsilon}_{\mathrm{T}} = \tfrac{1}{2}h\nu_{\mathrm{A}}\hat{\varepsilon}_{\mathrm{A}} = \frac{6\nu_{\mathrm{A}}M_{\mathrm{A}}}{hE_{\mathrm{A}}}. \tag{2.248}$$

It is noted from (2.211) that the quantities $\bar{\varepsilon}_{\mathrm{A}} + \frac{1}{2}\hat{\varepsilon}_{\mathrm{A}}h$ and $\bar{\varepsilon}_{\mathrm{T}} + \frac{1}{2}\hat{\varepsilon}_{\mathrm{T}}h$ appearing in (2.242), (2.243) and (2.246) are the in-plane strains on the mid-plane $x_3 = \frac{1}{2}h$ (i.e. the neutral plane) which are zero for the loading case under consideration.

2.18.3.2 Plane Strain Bending

Plane strain bending conditions are characterised by a zero transverse strain everywhere in the beam so that $\bar{\varepsilon}_{\mathrm{T}} = \hat{\varepsilon}_{\mathrm{T}} = 0$. It is also assumed that $\sigma_{\mathrm{A}} = \sigma_{\mathrm{t}} = \Delta T = 0$. It then follows that the relations (2.235), (2.236), (2.240) and (2.241) reduce to

$$\bar{\varepsilon}_{\mathrm{A}} = -\tfrac{1}{2}\hat{\varepsilon}_{\mathrm{A}}h,\ \sigma_{\mathrm{T}} = 0, \tag{2.249}$$

$$M_{\mathrm{A}} = \tfrac{1}{12}h^2\tilde{E}_{\mathrm{A}}\hat{\varepsilon}_{\mathrm{A}}, \tag{2.250}$$

$$M_{\mathrm{T}} = \tfrac{1}{12}h^2\nu_{\mathrm{A}}\tilde{E}_{\mathrm{T}}\hat{\varepsilon}_{\mathrm{A}} = \nu_{\mathrm{A}}\frac{E_{\mathrm{T}}}{E_{\mathrm{A}}}M_{\mathrm{A}}, \tag{2.251}$$

indicating the value of the transverse bending moment that must be applied to the beam to ensure plane strain conditions. It is noted from (2.250) that $\tilde{E}_{\mathrm{A}}$ may be interpreted as a plane strain bending modulus.

This completes the description of the fundamental theoretical concepts that will be required by subsequent chapters of this book.

References

1. Eringen, A.C. (1967). *Mechanics of Continua*. New York - London - Sydney: John Wiley & Sons, Inc.
2. Dunne, F.P.E. and Petrinic, N. (2005). *Introduction to Computational Plasticity*. Oxford University Press.
3. Love, A.E.H. (1944). *A Treatise on the Mathematical Theory of Elasticity*. Chapter XI, 4th ed.

3

Maxwell's Far-field Methodology Applied to the Prediction of Effective Properties of Multiphase Isotropic Particulate Composites

Overview: The far-field methodology developed by Maxwell, when estimating the effective electrical conductivity of isotropic particulate composites, is used to estimate effective thermoelastic properties of multiphase isotropic composites. In particular, Maxwell's methodology applied to the analogous thermal conduction problem is described, extending the approach to multiphase spherical particles having different sizes and properties. The methodology is also used to estimate the effective bulk and shear moduli, and the thermal expansion coefficients, of multiphase isotropic particulate composites. Results correspond with expressions derived in the literature, and coincide with, or lie between, variational bounds for all volume fractions. These characteristics, relating to isotropic effective properties, indicate that results obtained using the methodology, are not necessarily restricted to low volume fractions, as originally suggested by Maxwell. It is concluded that Maxwell's methodology is a unifying optimum technique to estimate the properties of multiphase isotropic particulate composites, because it provides closed form estimates that are fully consistent with other methods, without imposing restrictions, except that the particles must be spherical (but can have a range of size and properties) and the resultant effective properties must be isotropic.

3.1 Introduction

Estimating the effective thermoelastic and conduction properties of particulate composites is a very important engineering requirement that has been studied for many years. Examples of particulate-reinforced materials are concrete and the more recent developments in the metal matrix composites sector. A characteristic of both these materials is the irregular nature of particle geometry and their size. When modelling such materials, especially for analytical approaches, the particle geometries are usually idealised to spheres, spheroids (a good approximation to whiskers) or ellipsoids (useful when particles are in a platelet form). The immediate technical objective is to develop methods of estimating the effective thermoelastic properties of such materials. There are various methods of estimating the effective properties of particulate composites using analytical methods (see, for example, [1, 2]), but the approach that is considered here is that due to Maxwell [3], who is believed to have derived the first

Properties for Design of Composite Structures: Theory and Implementation Using Software, First Edition. Neil McCartney.

Companion Website: www.wiley.com/go/mccartney/properties

ever formula for an isotropic effective property (electrical conductivity) of a composite material (an isotropic medium reinforced with isotropic spherical particles).

Maxwell [3] provided an ingenious method of estimating the effective electrical conductivity of a cluster of spherical particles having spherical shape embedded in an infinite medium by considering the effect of the cluster on the far-field when the system is subject to a uniform electrical field. Maxwell asserted that the sizes and distribution of the particles must be such that particle interaction effects may be neglected, and he infers that his result will be valid only for small volume fractions of reinforcing particles. A detailed study [4] of the method used by Maxwell demonstrates that his methodology, focusing only on the far-field, can also be applied to the estimation of other properties of composite materials. As will become clear in this book (see Chapters 4 and 15), his methodology can easily be applied to composite materials reinforced with parallel fibres or aligned spheroids, the latter being used when considering short fibre or whisker reinforcements.

The principal objective of this chapter is to show, based on the approach described in [3], how Maxwell's methodology, first developed to estimate electrical conductivity, can also estimate the effective bulk modulus, shear modulus and thermal expansion coefficient of multiphase isotropic composites reinforced with homogeneous spherical particles. The methodology of Maxwell is naturally extended so that assemblies of multiphase spherical particles having a range of radii and/or properties may be considered. A second objective is to show that Maxwell's methodology is one reliable technique that provides closed-form estimates of effective properties and is not necessarily restricted to low volume fractions of particulate reinforcement. The first step is to describe the method developed by Maxwell [3] to estimate the effective isotropic conductivity of a composite reinforced with a uniform distribution of spherical particles. The analysis is presented in a thermal rather than electrical context, as it is a scenario more relevant to the other chapters in this book, and the thermal approach is, in fact, an exact analogue of the electrical counterpart.

3.2 General Description of Maxwell's Methodology Applied to Thermal Conductivity

To describe Maxwell's method [3], a more general situation is considered where a uniform cluster of isotropic spherical particles of various types (rather than a single type) are embedded in an infinite isotropic matrix. The multiphase approach to be adopted enables various particle materials and/or various sizes to be modelled. This is a key characteristic of Maxwell's methodology.

3.2.1 Description of Geometry

In a well-mixed cluster of N types of isotropic reinforcement embedded in and perfectly bonded to an infinite isotropic matrix (see Figure 3.1(a)), there are n_i spherical particles of radius a_i, $i = 1, \ldots N$. Particle properties of type i, which may differ from those of other types, are denoted by a superscript i. The cluster of all particle types may be *just* enclosed by a sphere of radius b and the particle distribution is homogeneous leading to isotropic effective properties of the composite formed by the cluster and

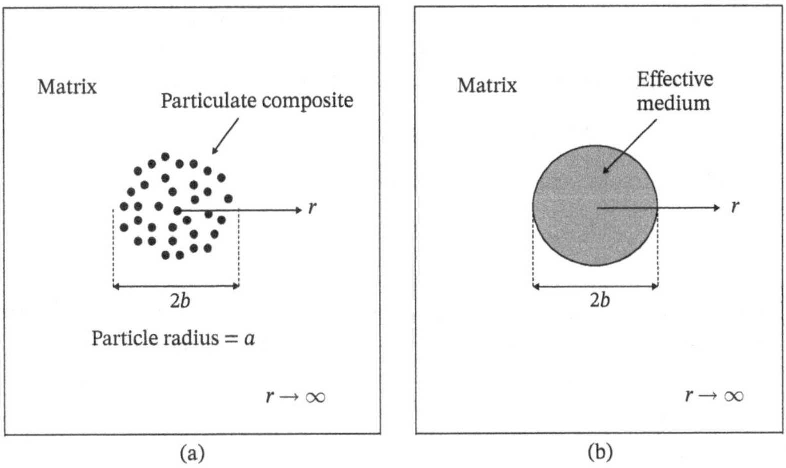

Figure 3.1 (a) Discrete particle model and (b) smoothed effective medium model of a particulate composite having spherical reinforcements embedded in infinite matrix material.

matrix lying within this sphere. The particles must not be distributed regularly or packed so densely that the effective properties are anisotropic. The volume fractions of particles of type i within the enclosing sphere of radius b are given by

$$V_{\mathrm{p}}^{i} = n_i a_i^3 / b^3, \quad i = 1, \ldots N, \quad \text{such that} \quad V_{\mathrm{m}} + \sum_{i=1}^{N} V_{\mathrm{p}}^{i} = 1, \tag{3.1}$$

where V_{m} is the volume fraction of matrix. For just one type of particle, with n particles of radius a within the enclosing sphere of radius b, the particulate volume fraction Vp is such that

$$V_{\mathrm{p}} = na^3 / b^3 = 1 - V_{\mathrm{m}}. \tag{3.2}$$

Whatever the nature and arrangement of spherical particles in the cluster, Maxwell's methodology considers the far-field when replacing the isotropic discrete particulate composite shown in Figure 3.1(a), that can just be enclosed by a sphere of radius b, by a homogeneous effective composite sphere of the same radius b embedded in the matrix as shown in Figure 3.1(b). There is no restriction on sizes, properties and locations of particles except that the equivalent effective medium is homogeneous and isotropic. Composites having statistical distributions of both size and properties can clearly be analysed.

3.2.2 Temperature Distribution for an Isolated Sphere Embedded in an Infinite Matrix

A set of spherical polar coordinates (r,θ,ϕ) is introduced having origin at the centre of the sphere of radius a. For steady-state conditions, the temperature distribution $T(r,\theta,\phi)$ in the particle and surrounding matrix must satisfy Laplace's equation, which is expressed in terms of spherical polar coordinates for the case where the temperature is independent of ϕ, namely,

$$\frac{\partial}{\partial r}\left(r^2 \frac{\partial T}{\partial r}\right) + \frac{1}{\sin\theta}\frac{\partial}{\partial \theta}\left(\sin\theta \frac{\partial T}{\partial \theta}\right) = 0. \tag{3.3}$$

On the external boundary $r \to \infty$ a temperature distribution is imposed that would lead, in a homogeneous matrix material under steady-state conditions, to the following linear temperature distribution having a uniform gradient α

$$T(r,\theta) = \alpha x_3, \quad \text{where} \quad x_3 = r\cos\theta. \tag{3.4}$$

At the particle/matrix interface $r = a$, the following conditions are imposed

$$T_\text{p} = T_\text{m}, \quad \kappa_\text{p}\frac{\partial T_\text{p}}{\partial r} = \kappa_\text{m}\frac{\partial T_\text{m}}{\partial r}, \tag{3.5}$$

where both the temperature and the normal heat flux are assumed to be continuous across the interface, and where the isotropic thermal conductivities of particles and matrix are denoted by κ_p and κ_m, respectively. The temperature distribution $T(r,\theta)$ in the particle and matrix, satisfying (3.3), the condition at infinity and the interface conditions (3.5), is given by

$$\left.\begin{aligned} T_\text{p}(r,\theta) &= \frac{3\kappa_\text{m}}{\kappa_\text{p}+2\kappa_\text{m}}\alpha r\cos\theta\,, \quad 0<r<a\,, \\ T_\text{m}(r,\theta) &= \alpha r\cos\theta + \left(\frac{3\kappa_\text{m}}{\kappa_\text{p}+2\kappa_\text{m}} - 1\right)\frac{\alpha a^3\cos\theta}{r^2}\,, \quad a<r<\infty\,. \end{aligned}\right\} \tag{3.6}$$

As $x_3 = r\cos\theta$, the temperature gradient is uniform within the particle. Any temperature T_0 can be added to (3.6) without affecting the satisfaction of the interface conditions (3.5). As an infinite medium having a uniform temperature gradient is considered, it is inevitable that temperatures lower than absolute zero will be encountered. This physical absurdity can be avoided by assuming that the system is taken as a spherical region having a very large but finite radius, in which case T_0 would be chosen large enough to ensure that the absolute temperature distribution is everywhere positive.

3.2.3 Maxwell's Methodology for Estimating Conductivity

The first step of Maxwell's approach considers the effect of embedding in the infinite isotropic matrix, an isolated cluster of isotropic spherical particles of different types that can be just contained within a sphere of radius b, as illustrated in Figure 3.1(a). The isotropic thermal conductivity of particles of type i is denoted by $\kappa_\text{p}^{(i)}$. The cluster is assumed to be effectively homogeneous regarding the distribution of particles, leading to an isotropic effective thermal conductivity κ_eff for the composite lying within the sphere of radius b. For a single particle, the matrix temperature distribution is perturbed from the distribution (3.4) to the distribution (3.6) that depends on particle geometry and properties. The perturbing effect in the matrix at large distances from the cluster of particles is estimated by superimposing the perturbations caused by each spherical particle, regarded as being isolated. The second step recognises that, at very large distances from the cluster, all the particles can be considered located at the origin that is chosen to be situated at the centre of one of the particles in the cluster. Thus, for the case of multiple phases, the approximate temperature distribution in the matrix at large distances from the cluster is given by the following generalisation of the second of relations (3.6):

$$T_\text{m} = \alpha r\cos\theta + \left[\sum_{i=1}^{N} n_i a_i^3\left(\frac{3\kappa_\text{m}}{\kappa_\text{p}^{(i)}+2\kappa_\text{m}} - 1\right)\right]\frac{\alpha\cos\theta}{r^2}, \quad r\to\infty. \tag{3.7}$$

On using (3.1), this relation may be expressed in terms of the volume fractions so that

$$T_{\mathrm{m}} = \alpha r \cos\theta + \left[\sum_{i=1}^{N} V_{\mathrm{p}}^{i} \frac{3\kappa_{\mathrm{m}}}{\kappa_{\mathrm{p}}^{(i)} + 2\kappa_{\mathrm{m}}} + V_{\mathrm{m}} - 1 \right] \frac{\alpha b^3 \cos\theta}{r^2}, \quad r \to \infty. \tag{3.8}$$

The third step involves replacing the composite having discrete particles lying within the sphere of radius b by a homogeneous spherical isotropic effective medium (see Figure 3.1(b)) having radius b and having the isotropic effective thermal conductivity κ_{eff} of the composite. On using (3.6), the temperature distribution in the matrix outside the sphere of effective medium having radius b is then given exactly, for a given value of κ_{eff}, by

$$T_{\mathrm{m}} = \alpha r \cos\theta + \left(\frac{3\kappa_{\mathrm{m}}}{\kappa_{\mathrm{eff}} + 2\kappa_{\mathrm{m}}} - 1 \right) \frac{\alpha b^3 \cos\theta}{r^2}, \quad b < r < \infty. \tag{3.9}$$

If the cluster in Figure 3.1(a) is represented accurately by the effective medium shown in Figure 3.1(b), then, at large distances from the cluster, the temperature distributions (3.8) and (3.9) should be identical, leading to the fourth step, where the perturbation terms in relations (3.8) and (3.9) are equated, so that

$$\frac{1}{\kappa_{\mathrm{eff}} + 2\kappa_{\mathrm{m}}} = \sum_{i=1}^{N} \frac{V_{\mathrm{p}}^{i}}{\kappa_{\mathrm{p}}^{(i)} + 2\kappa_{\mathrm{m}}} + \frac{V_{\mathrm{m}}}{\kappa_{\mathrm{m}} + 2\kappa_{\mathrm{m}}}, \tag{3.10}$$

which is a 'mixtures' relation for the quantity $1/(\kappa + 2\kappa_{\mathrm{m}})$. On using (3.1), the effective thermal conductivity may be estimated using

$$\kappa_{\mathrm{eff}} = \left(\sum_{i=1}^{N} \frac{V_{\mathrm{p}}^{i} \kappa_{\mathrm{p}}^{(i)}}{\kappa_{\mathrm{p}}^{(i)} + 2\kappa_{\mathrm{m}}} + \frac{V_{\mathrm{m}}}{3} \right) \Bigg/ \left(\sum_{i=1}^{N} \frac{V_{\mathrm{p}}^{i}}{\kappa_{\mathrm{p}}^{(i)} + 2\kappa_{\mathrm{m}}} + \frac{V_{\mathrm{m}}}{3\kappa_{\mathrm{m}}} \right). \tag{3.11}$$

For multiphase composites, Hashin and Shtrikman [5, Equations (3.21)–(3.23)] derived bounds for magnetic permeability, pointing out that they are analogous to bounds for effective thermal conductivity. Their conductivity bounds may be expressed in the following simpler form, having the same structure as the result (3.10) derived using Maxwell's methodology

$$\left.\begin{aligned} \frac{1}{\kappa_{\mathrm{eff}} + 2\kappa_{\min}} &\le \sum_{i=1}^{N} \frac{V_{\mathrm{p}}^{i}}{\kappa_{\mathrm{p}}^{(i)} + 2\kappa_{\min}} + \frac{V_{\mathrm{m}}}{\kappa_{\mathrm{m}} + 2\kappa_{\min}}, \\ \sum_{i=1}^{N} \frac{V_{\mathrm{p}}^{i}}{\kappa_{\mathrm{p}}^{(i)} + 2\kappa_{\max}} + \frac{V_{\mathrm{m}}}{\kappa_{\mathrm{m}} + 2\kappa_{\max}} &\le \frac{1}{\kappa_{\mathrm{eff}} + 2\kappa_{\max}}, \end{aligned}\right\} \tag{3.12}$$

where $\kappa_{\min}$ is the lowest value of conductivities for all phases, whereas $\kappa_{\max}$ is the highest value.

3.3 Bulk Modulus and Thermal Expansion Coefficient

3.3.1 Spherical Particle Embedded in Infinite Matrix Subject to Pressure and Thermal Loading

Consider an isolated particle of radius a perfectly bonded to an infinite surrounding matrix, subject to a pressure p applied at infinity and a uniform temperature change

ΔT from the stress-free temperature at which the stresses and strains in particle and matrix are zero. The displacement field in the particle and surrounding matrix is purely radial so that displacement components (u_r, u_θ, u_ϕ), referred to a set of spherical polar coordinates (r, θ, ϕ) with origin at the particle centre, are of the form

$$u_r = f(r), \quad u_\theta = 0, \quad u_\phi = 0. \tag{3.13}$$

The corresponding strain field obtained from (2.142) is then given by

$$\varepsilon_{rr} = \frac{\partial u_r}{\partial r}, \qquad \varepsilon_{\theta\theta} = \varepsilon_{\phi\phi} = \frac{u_r}{r}, \qquad \varepsilon_{r\theta} = \varepsilon_{r\phi} = \varepsilon_{\theta\phi} = 0. \tag{3.14}$$

The stress field follows from stress-strain relations expressed in the form (see (2.160) for the Cartesian equivalent)

$$\left.\begin{aligned}
\sigma_{rr} &= \lambda\left(\varepsilon_{rr} + \varepsilon_{\theta\theta} + \varepsilon_{\phi\phi}\right) + 2\mu\varepsilon_{rr} - (3\lambda + 2\mu)\alpha\Delta T, & \sigma_{r\theta} &= 2\mu\varepsilon_{r\theta},\\
\sigma_{\theta\theta} &= \lambda\left(\varepsilon_{rr} + \varepsilon_{\theta\theta} + \varepsilon_{\phi\phi}\right) + 2\mu\varepsilon_{\theta\theta} - (3\lambda + 2\mu)\alpha\Delta T, & \sigma_{r\phi} &= 2\mu\varepsilon_{r\phi},\\
\sigma_{\phi\phi} &= \lambda\left(\varepsilon_{rr} + \varepsilon_{\theta\theta} + \varepsilon_{\phi\phi}\right) + 2\mu\varepsilon_{\phi\phi} - (3\lambda + 2\mu)\alpha\Delta T, & \sigma_{\theta\phi} &= 2\mu\varepsilon_{\theta\phi},
\end{aligned}\right\} \tag{3.15}$$

where λ and μ are Lamé's constants and where α is now the linear coefficient of thermal expansion. On using the equilibrium equations (2.130)–(2.133), it can be shown that, within the spherical particle of radius a, the resulting bounded displacement and stress fields are given by

$$u_r^{\mathrm{p}} = \alpha_{\mathrm{p}}\Delta T r - \frac{p_0}{3k_{\mathrm{p}}} r, \qquad u_\theta^{\mathrm{p}} = u_\phi^{\mathrm{p}} = 0, \tag{3.16}$$

$$\sigma_{rr}^{\mathrm{p}} = \sigma_{\theta\theta}^{\mathrm{p}} = \sigma_{\phi\phi}^{\mathrm{p}} + \tfrac{2}{3}\mu_{\mathrm{p}} = -p_0, \quad \sigma_{r\theta}^{\mathrm{p}} = \sigma_{r\phi}^{\mathrm{p}} = \sigma_{\theta\phi}^{\mathrm{p}} = 0, \tag{3.17}$$

where $k_{\mathrm{p}} = \lambda_{\mathrm{p}} + \frac{2}{3}\mu_{\mathrm{p}}$ and μ_{p} are the bulk and shear moduli, respectively, for the particulate reinforcement, and where α_{p} is the corresponding thermal expansion coefficient. Clearly the strain and stress distributions within the particle are both uniform. For the matrix region it can be shown that

$$u_r^{\mathrm{m}} = \alpha_{\mathrm{m}}\Delta T r - \frac{p}{3k_{\mathrm{m}}} r + \frac{p_0 - p}{4\mu_{\mathrm{m}}}\frac{a^3}{r^2}, \quad u_\theta^{\mathrm{m}} = 0, \quad u_\phi^{\mathrm{m}} = 0, \tag{3.18}$$

$$\begin{gathered}
\sigma_{rr}^{\mathrm{m}} = -p - (p_0 - p)\frac{a^3}{r^3}, \quad \sigma_{\theta\theta}^{\mathrm{m}} = \sigma_{\phi\phi}^{\mathrm{m}} = -p + (p_0 - p)\frac{a^3}{2r^3},\\
\sigma_{r\theta}^{\mathrm{m}} = \sigma_{\theta\phi}^{\mathrm{m}} = \sigma_{r\phi}^{\mathrm{m}} = 0,
\end{gathered} \tag{3.19}$$

where $k_{\mathrm{m}} = \lambda_{\mathrm{m}} + \frac{4}{3}\mu_{\mathrm{m}}$ and μ_{m} are the bulk and shear moduli, respectively, for the matrix, and where α_{m} is the corresponding thermal expansion coefficient. The stress component σ_{rr} is automatically continuous across $r = a$ having the value $-p_0$. As the displacement component u_r must also be continuous across this interface, the value of p_0 must satisfy the relation

$$\left(\frac{1}{k_{\mathrm{p}}} + \frac{1}{\frac{4}{3}\mu_{\mathrm{m}}}\right) p_0 = 3\left(\alpha_{\mathrm{p}} - \alpha_{\mathrm{m}}\right)\Delta T + \left(\frac{1}{k_{\mathrm{m}}} + \frac{1}{\frac{4}{3}\mu_{\mathrm{m}}}\right) p. \tag{3.20}$$

3.3.2 Applying Maxwell's Methodology to Isotropic Multiphase Particulate Composites

Owing to the use of the far-field in Maxwell's methodology, it is again possible to consider multiple types of spherical reinforcement. The perturbing effect in the matrix at large distances from the cluster of particles is estimated by superimposing the perturbations caused by each particle, regarded as being isolated, and regarding all particles to be located at the origin. The properties of particles of type i are denoted by a superscript (i).

Relations (3.19) for nonzero stresses in the matrix are generalised to

$$\left.\begin{aligned} \sigma_{rr}^{\mathrm{m}} &= -p - \frac{1}{r^3}\sum_{i=1}^{N}\left(p_0^i - p\right) n_i\, a_i^3, \\ \sigma_{\theta\theta}^{\mathrm{m}} = \sigma_{\phi\phi}^{\mathrm{m}} &= -p + \frac{1}{2r^3}\sum_{i=1}^{N}\left(p_0^i - p\right) n_i\, a_i^3, \end{aligned}\right\} \tag{3.21}$$

where p_0^i is the pressure at the particle/matrix interface when an isolated particle of species i is placed in infinite matrix material. From (3.20) the following value of $p_0^i - p$ is obtained

$$\left(\frac{1}{k_{\mathrm{p}}^{(i)}} + \frac{1}{\frac{4}{3}\mu_{\mathrm{m}}}\right)\left(p_0^i - p\right) = 3\left(\alpha_{\mathrm{p}}^{(i)} - \alpha_{\mathrm{m}}\right)\Delta T + \left(\frac{1}{k_{\mathrm{m}}} - \frac{1}{k_{\mathrm{p}}^{(i)}}\right) p. \tag{3.22}$$

It then follows that the stress distribution in the matrix at large distances from the discrete cluster of particles shown in Figure 3.1(a) is approximately given by

$$\left.\begin{aligned} \sigma_{rr}^{\mathrm{m}} = &-p - \frac{pb^3}{r^3}\sum_{i=1}^{N}\left(\frac{1/k_{\mathrm{m}} + 3/(4\mu_{\mathrm{m}})}{1/k_{\mathrm{p}}^{(i)} + 3/(4\mu_{\mathrm{m}})} - 1\right) V_{\mathrm{p}}^i \\ &+ \frac{3\Delta T\, b^3}{r^3}\sum_{i=1}^{N}\frac{\alpha_{\mathrm{p}}^{(i)} - \alpha_{\mathrm{m}}}{1/k_{\mathrm{p}}^{(i)} + 3/(4\mu_{\mathrm{m}})} V_{\mathrm{p}}^i, \\ \sigma_{\theta\theta}^{\mathrm{m}} = \sigma_{\phi\phi}^{\mathrm{m}} = &-p + \frac{pb^3}{2r^3}\sum_{i=1}^{N}\left(\frac{1/k_{\mathrm{m}} + 3/(4\mu_{\mathrm{m}})}{1/k_{\mathrm{p}}^{(i)} + 3/(4\mu_{\mathrm{m}})} - 1\right) V_{\mathrm{p}}^i \\ &- \frac{3\Delta T\, b^3}{2r^3}\sum_{i=1}^{N}\frac{\alpha_{\mathrm{p}}^{(i)} - \alpha_{\mathrm{m}}}{1/k_{\mathrm{p}}^{(i)} + 3/(4\mu_{\mathrm{m}})} V_{\mathrm{p}}^i, \end{aligned}\right. \tag{3.23}$$

where the volume fractions V_{p}^i of particles of type i defined by (3.1) have been introduced.

When (3.23) is applied to a single sphere of radius b, having the effective properties of a composite representing the multiphase cluster of particles embedded in matrix material (see Figure 3.1(b)), the exact matrix stress distribution, for given values of k_{eff} and α_{eff}, is

$$\left.\begin{aligned} \sigma_{rr}^{\mathrm{m}} = &-p - \frac{pb^3}{r^3}\left(\frac{1/k_{\mathrm{m}} + 3/(4\mu_{\mathrm{m}})}{1/k_{\mathrm{eff}} + 3/(4\mu_{\mathrm{m}})} - 1\right) \\ &+ \frac{3\Delta T\, b^3}{r^3}\frac{\alpha_{\mathrm{eff}} - \alpha_{\mathrm{m}}}{1/k_{\mathrm{eff}} + 3/(4\mu_{\mathrm{m}})}, \\ \sigma_{\theta\theta}^{\mathrm{m}} = \sigma_{\phi\phi}^{\mathrm{m}} = &-p + \frac{pb^3}{2r^3}\left(\frac{1/k_{\mathrm{m}} + 3/(4\mu_{\mathrm{m}})}{1/k_{\mathrm{eff}} + 3/(4\mu_{\mathrm{m}})} - 1\right) \\ &- \frac{3\Delta T\, b^3}{2r^3}\frac{\alpha_{\mathrm{eff}} - \alpha_{\mathrm{m}}}{1/k_{\mathrm{eff}} + 3/(4\mu_{\mathrm{m}})}, \end{aligned}\right\} b < r < \infty. \tag{3.24}$$

Maxwell's methodology asserts that, at large distances from the cluster, the stress distributions (3.23) and (3.24), and hence the coefficients of p and ΔT, are identical leading to the following 'mixtures' rules for the functions $1/[1/k+3/(4\mu_{\mathrm{m}})]$ and $\alpha/[1/k+3/(4\mu_{\mathrm{m}})]$, respectively,

$$\frac{1}{1/k_{\mathrm{eff}}+3/(4\mu_{\mathrm{m}})}=\sum_{i=1}^{N}\frac{V_{\mathrm{p}}^{i}}{1/k_{\mathrm{p}}^{(i)}+3/(4\mu_{\mathrm{m}})}+\frac{V_{\mathrm{m}}}{1/k_{\mathrm{m}}+3/(4\mu_{\mathrm{m}})},\tag{3.25}$$

$$\frac{\alpha_{\mathrm{eff}}}{1/k_{\mathrm{eff}}+3/(4\mu_{\mathrm{m}})}=\sum_{i=1}^{N}\frac{V_{\mathrm{p}}^{i}\alpha_{\mathrm{p}}^{(i)}}{1/k_{\mathrm{p}}^{(i)}+3/(4\mu_{\mathrm{m}})}+\frac{V_{\mathrm{m}}\alpha_{\mathrm{m}}}{1/k_{\mathrm{m}}+3/(4\mu_{\mathrm{m}})}.\tag{3.26}$$

On using (3.1), the result (3.25) may be written as

$$\frac{1}{k_{\mathrm{eff}}+\frac{4}{3}\mu_{\mathrm{m}}}=\sum_{i=1}^{N}\frac{V_{\mathrm{p}}^{i}}{k_{\mathrm{p}}^{(i)}+\frac{4}{3}\mu_{\mathrm{m}}}+\frac{V_{\mathrm{m}}}{k_{\mathrm{m}}+\frac{4}{3}\mu_{\mathrm{m}}},\tag{3.27}$$

so that the effective bulk modulus of the multiphase particulate composite may instead be obtained from a 'mixtures' relation for the quantity $1/(k+\frac{4}{3}\kappa_{\mathrm{m}})$. On using (3.1) and (3.27), the effective bulk modulus may be estimated using

$$k_{\mathrm{eff}}=\left(\sum_{i=1}^{N}\frac{V_{\mathrm{p}}^{i}k_{\mathrm{p}}^{(i)}}{k_{\mathrm{p}}^{(i)}+\frac{4}{3}\mu_{\mathrm{m}}}+\frac{V_{\mathrm{m}}k_{\mathrm{m}}}{k_{\mathrm{m}}+\frac{4}{3}\mu_{\mathrm{m}}}\right)\Bigg/\left(\sum_{i=1}^{N}\frac{V_{\mathrm{p}}^{i}}{k_{\mathrm{p}}^{(i)}+\frac{4}{3}\mu_{\mathrm{m}}}+\frac{V_{\mathrm{m}}}{k_{\mathrm{m}}+\frac{4}{3}\mu_{\mathrm{m}}}\right).\tag{3.28}$$

It follows from (3.26) and (3.27) that the corresponding relation for effective thermal expansion is

$$\alpha_{\mathrm{eff}}=\left(\sum_{i=1}^{N}\frac{V_{\mathrm{p}}^{i}k_{\mathrm{p}}^{(i)}\alpha_{\mathrm{p}}^{(i)}}{k_{\mathrm{p}}^{(i)}+\frac{4}{3}\mu_{\mathrm{m}}}+\frac{V_{\mathrm{m}}k_{\mathrm{m}}\alpha_{\mathrm{m}}}{k_{\mathrm{m}}+\frac{4}{3}\mu_{\mathrm{m}}}\right)\Bigg/\left(\sum_{i=1}^{N}\frac{V_{\mathrm{p}}^{i}k_{\mathrm{p}}^{(i)}}{k_{\mathrm{p}}^{(i)}+\frac{4}{3}\mu_{\mathrm{m}}}+\frac{V_{\mathrm{m}}k_{\mathrm{m}}}{k_{\mathrm{m}}+\frac{4}{3}\mu_{\mathrm{m}}}\right).\tag{3.29}$$

The bounds for effective bulk modulus of multiphase isotropic composites derived by Hashin and Shtrikman [6, Equations (3.37)–(3.43)] and the bounds derived by Walpole [7, Equation (26)] are identical and may be expressed in the following simpler form having the same structure as the result (3.27) derived using Maxwell's methodology

$$\left.\begin{aligned}&\frac{1}{k_{\mathrm{eff}}+\frac{4}{3}\mu_{\mathrm{min}}}\le\sum_{i=1}^{N}\frac{V_{\mathrm{p}}^{i}}{k_{\mathrm{p}}^{(i)}+\frac{4}{3}\mu_{\mathrm{min}}}+\frac{V_{\mathrm{m}}}{k_{\mathrm{m}}+\frac{4}{3}\mu_{\mathrm{min}}},\\&\sum_{i=1}^{N}\frac{V_{\mathrm{p}}^{i}}{k_{\mathrm{p}}^{(i)}+\frac{4}{3}\mu_{\mathrm{max}}}+\frac{V_{\mathrm{m}}}{k_{\mathrm{m}}+\frac{4}{3}\mu_{\mathrm{max}}}\le\frac{1}{k_{\mathrm{eff}}+\frac{4}{3}\mu_{\mathrm{max}}},\end{aligned}\right\}\tag{3.30}$$

where the parameters k_{min} and μ_{min} are the lowest values of bulk and shear moduli of all phases in the composite, respectively, whereas k_{max} and μ_{max} are the highest values.

The bounds for effective thermal expansion involve the effective bulk modulus, and the specification of bounds is complex and beyond the scope of this chapter. An analysis has been undertaken showing numerically, for a very wide range of parameter values, that the effective thermal expansion obtained using Maxwell's methodology lies between the absolute bounds for all volume fractions of spherical particle distributions that are consistent with isotropic properties.

3.4 Shear Modulus

3.4.1 Spherical Particle Embedded in Infinite Matrix Material Subject to Pure Shear Loading

For a state of pure shear, and in the absence of thermal effects, the displacement field of a homogeneous sample of material referred to a set of Cartesian coordinates (x_1, x_2, x_3) has the form

$$u_1 = \gamma x_2, \quad u_2 = \gamma x_1, \quad u_3 = 0, \tag{3.31}$$

and the corresponding strain and stress components are given by

$$\varepsilon_{11} = 0, \ \varepsilon_{22} = 0, \ \varepsilon_{33} = 0, \qquad \varepsilon_{12} = \gamma, \ \varepsilon_{23} = 0, \ \varepsilon_{13} = 0, \tag{3.32}$$

$$\sigma_{11} = 0, \ \sigma_{22} = 0, \ \sigma_{33} = 0, \qquad \sigma_{12} = \tau, \ \sigma_{23} = 0, \ \sigma_{13} = 0. \tag{3.33}$$

The parameters γ and τ are the shear strain (half the engineering shear strain) and shear stress, respectively, such that $\tau = 2\mu\gamma$ where μ is the shear modulus of an isotropic material. The principal values of the stress field are along (tension) and perpendicular to (compression) the line $x_2 = x_1$.

A single spherical particle of radius a is now placed in, and perfectly bonded to, an infinite matrix, where the origin of spherical polar coordinates (r, θ, ϕ) is taken at the centre of the particle. The system is then subject only to a shear stress applied at infinity. At the particle/matrix interface the following perfect bonding boundary conditions must be satisfied:

$$\left.\begin{aligned} u_r^{\mathrm{p}}(a,\theta,\phi) &= u_r^{\mathrm{m}}(a,\theta,\phi), \quad \sigma_{rr}^{\mathrm{p}}(a,\theta,\phi) = \sigma_{rr}^{\mathrm{m}}(a,\theta,\phi), \\ u_\theta^{\mathrm{p}}(a,\theta,\phi) &= u_\theta^{\mathrm{m}}(a,\theta,\phi), \quad \sigma_{\theta\theta}^{\mathrm{p}}(a,\theta,\phi) = \sigma_{r\theta}^{\mathrm{m}}(a,\theta,\phi), \\ u_\phi^{\mathrm{p}}(a,\theta,\phi) &= u_\phi^{\mathrm{m}}(a,\theta,\phi), \quad \sigma_{\phi\phi}^{\mathrm{p}}(a,\theta,\phi) = \sigma_{r\phi}^{\mathrm{m}}(a,\theta,\phi). \end{aligned}\right\} \tag{3.34}$$

A displacement field equivalent to that used by Hashin [5], based on the analysis of Love [8, Equations (5)–(7)] that leads to a stress field satisfying the equilibrium equations and the stress-strain relations (3.15) with $\Delta T = 0$, can be used to solve the embedded isolated sphere problem (see Appendix A). The displacement and stress fields in the particle are bounded at $r = 0$ so that

$$\left.\begin{aligned} u_r^{\mathrm{p}} &= a\left[2A_{\mathrm{p}} r/a + 12\nu_{\mathrm{p}} C_{\mathrm{p}} (r/a)^3\right] \tfrac{1}{2}\sin^2\theta \sin 2\phi, \\ u_\theta^{\mathrm{p}} &= a\left[A_{\mathrm{p}} r/a + (7-4\nu_{\mathrm{p}}) C_{\mathrm{p}} (r/a)^3\right] \tfrac{1}{2}\sin 2\theta \sin 2\phi, \\ u_\phi^{\mathrm{p}} &= a\left[A_{\mathrm{p}} r/a + (7-4\nu_{\mathrm{p}}) C_{\mathrm{p}} (r/a)^3\right] \sin\theta \cos 2\phi, \end{aligned}\right\} 0 < r < a, \tag{3.35}$$

$$\left.\begin{aligned} \sigma_{rr}^{\mathrm{p}} &= 2\mu_{\mathrm{p}}\left[A_{\mathrm{p}} - 3\nu_{\mathrm{p}} C_{\mathrm{p}} (r/a)^2\right] \sin^2\theta \sin 2\phi, \\ \sigma_{r\theta}^{\mathrm{p}} &= \mu_{\mathrm{p}}\left[A_{\mathrm{p}} + (7+2\nu_{\mathrm{p}}) C_{\mathrm{p}} (r/a)^2\right] \sin 2\theta \sin 2\phi, \\ \sigma_{r\phi}^{\mathrm{p}} &= 2\mu_{\mathrm{p}}\left[A_{\mathrm{p}} + (7+2\nu_{\mathrm{p}}) C_{\mathrm{p}} (r/a)^2\right] \sin\theta \cos 2\phi, \end{aligned}\right\} 0 < r < a. \tag{3.36}$$

In the matrix the displacement field and stress field (stresses bounded as $r \to \infty$) have the form

$$\left.\begin{aligned}
u_r^{\mathrm{m}} &= a\left[2\gamma r/a - 3B_{\mathrm{m}}(a/r)^4 + (10-8\nu_{\mathrm{m}})D_{\mathrm{m}}(a/r)^2\right]\tfrac{1}{2}\sin^2\theta\,\sin 2\phi,\\
u_\theta^{\mathrm{m}} &= a\left[\gamma r/a + B_{\mathrm{m}}(a/r)^4 + (2-4\nu_{\mathrm{m}})D_{\mathrm{m}}(a/r)^2\right]\tfrac{1}{2}\sin 2\theta\,\sin 2\phi,\\
u_\phi^{\mathrm{m}} &= a\left[\gamma r/a + B_{\mathrm{m}}(a/r)^4 + (2-4\nu_{\mathrm{m}})D_{\mathrm{m}}(a/r)^2\right]\sin\theta\,\cos 2\phi,
\end{aligned}\right\} a<r<\infty, \tag{3.37}$$

$$\left.\begin{aligned}
\sigma_{rr}^{\mathrm{m}} &= 2\mu_{\mathrm{m}}\left[\gamma + 6B_{\mathrm{m}}(a/r)^5 - 2(5-\nu_{\mathrm{m}})D_{\mathrm{m}}(a/r)^3\right]\sin^2\theta\,\sin 2\phi,\\
\sigma_{r\theta}^{\mathrm{m}} &= \mu_{\mathrm{m}}\left[\gamma - 4B_{\mathrm{m}}(a/r)^5 + (2+2\nu_{\mathrm{m}})D_{\mathrm{m}}(a/r)^3\right]\sin 2\theta\,\sin 2\phi,\\
\sigma_{r\phi}^{\mathrm{m}} &= 2\mu_{\mathrm{m}}\left[\gamma - 4B_{\mathrm{m}}(a/r)^5 + (2+2\nu_{\mathrm{m}})D_{\mathrm{m}}(a/r)^3\right]\sin\theta\,\cos 2\phi,
\end{aligned}\right\} a<r<\infty. \tag{3.38}$$

The representation is identical in form to that used by Christensen and Lo [9] although they used a definition of ϕ that differs from that used here by an angle of $\pi/4$. This difference has no effect on the approach to be followed. It follows from (3.35)–(3.38) that the continuity conditions (3.34) are satisfied if the following four independent relations are satisfied

$$\left.\begin{aligned}
&2A_{\mathrm{p}} + 12\nu_{\mathrm{p}}C_{\mathrm{p}} = 2\gamma - 3B_{\mathrm{m}} + (10-8\nu_{\mathrm{m}})D_{\mathrm{m}},\\
&A_{\mathrm{p}} + (7-4\nu_{\mathrm{p}})C_{\mathrm{p}} = \gamma + B_{\mathrm{m}} + (2-4\nu_{\mathrm{m}})D_{\mathrm{m}},\\
&\mu_{\mathrm{p}}\left[A_{\mathrm{p}} - 3\nu_{\mathrm{p}}C_{\mathrm{p}}\right] = \mu_{\mathrm{m}}\left[\gamma + 6B_{\mathrm{m}} - 2(5-\nu_{\mathrm{m}})D_{\mathrm{m}}\right],\\
&\mu_{\mathrm{p}}\left[A_{\mathrm{p}} + (7+2\nu_{\mathrm{p}})C_{\mathrm{p}}\right] = \mu_{\mathrm{m}}\left[\gamma - 4B_{\mathrm{m}} + (2+2\nu_{\mathrm{m}})D_{\mathrm{m}}\right],
\end{aligned}\right\} \tag{3.39}$$

and it can then be shown that

$$\left.\begin{aligned}
&A_{\mathrm{p}} = \frac{15(1-\nu_{\mathrm{m}})\mu_{\mathrm{m}}}{(8-10\nu_{\mathrm{m}})\mu_{\mathrm{p}} + (7-5\nu_{\mathrm{m}})\mu_{\mathrm{m}}}\gamma, \quad B_{\mathrm{m}} = \frac{3(\mu_{\mathrm{m}}-\mu_{\mathrm{p}})}{(8-10\nu_{\mathrm{m}})\mu_{\mathrm{p}} + (7-5\nu_{\mathrm{m}})\mu_{\mathrm{m}}}\gamma,\\
&C_p = 0, \qquad D_{\mathrm{m}} = \frac{5}{2}\frac{(\mu_{\mathrm{m}}-\mu_{\mathrm{p}})}{(8-10\nu_{\mathrm{m}})\mu_{\mathrm{p}} + (7-5\nu_{\mathrm{m}})\mu_{\mathrm{m}}}\gamma = \frac{5}{6}B_{\mathrm{m}}.
\end{aligned}\right\} \tag{3.40}$$

As $C_{\mathrm{p}} = 0$, it follows from (3.35) and (3.36) that both the strain and stress distributions in the particle are uniform.

3.4.2 Application of Maxwell's Methodology

To apply Maxwell's methodology to a cluster of N particles embedded in an infinite matrix, the stress distribution in the matrix at large distances from the cluster is considered. The perturbing effect in the matrix at large distances from the cluster of particles is estimated by superimposing the perturbations caused by each particle, regarded as being isolated, and regarding all particles to be located at the origin. The properties of particles of type i will again be denoted by a superscript (i).

The stress distribution at very large distances from the cluster is then given by the following generalisation of relations (3.38)

$$\left.\begin{aligned}\sigma_{rr}^{\mathrm{m}} &\cong 2\mu_{\mathrm{m}}\left[\gamma-2(5-\nu_{\mathrm{m}})\sum_{i=1}^{N} D_{\mathrm{m}}^{i}\, n_i (a_i/r)^3\right]\sin^2\theta\,\sin 2\phi\,,\\ \sigma_{r\theta}^{\mathrm{m}} &\cong \mu_{\mathrm{m}}\left[\gamma+(2+2\nu_{\mathrm{m}})\sum_{i=1}^{N} D_{\mathrm{m}}^{i}\, n_i (a_i/r)^3\right]\sin 2\theta\,\sin 2\phi\,,\\ \sigma_{r\phi}^{\mathrm{m}} &\cong 2\mu_{\mathrm{m}}\left[\gamma+(2+2\nu_{\mathrm{m}})\sum_{i=1}^{N} D_{\mathrm{m}}^{i}\, n_i (a_i/r)^3\right]\sin\theta\,\cos 2\phi\,,\end{aligned}\right\} r\to\infty, \tag{3.41}$$

where from (3.40), for $i = 1, \ldots, N$,

$$\begin{aligned}D_{\mathrm{m}}^{i} &= \frac{5\gamma}{2}\,\frac{\mu_{\mathrm{m}}-\mu_{\mathrm{p}}^{(i)}}{(8-10\nu_{\mathrm{m}})\mu_{\mathrm{p}}^{(i)}+(7-5\nu_{\mathrm{m}})\mu_{\mathrm{m}}}\\ &= \frac{5\gamma}{2}\left[1-\frac{(8-10\nu_{\mathrm{m}})\mu_{\mathrm{m}}+(7-5\nu_{\mathrm{m}})\mu_{\mathrm{m}}}{(8-10\nu_{\mathrm{m}})\mu_{\mathrm{p}}^{(i)}+(7-5\nu_{\mathrm{m}})\mu_{\mathrm{m}}}\right].\end{aligned} \tag{3.42}$$

For the isolated sphere of radius b having the effective properties of the particulate composite cluster as illustrated in Figure 3.1(b), it follows that the stress field in the matrix at large distances is described exactly by relations of the type (3.38) where the coefficient D_{m} is replaced by $\bar{D}_{\mathrm{m}}$ having the value determined by the relation

$$\begin{aligned}\bar{D}_{\mathrm{m}} &= \frac{5\gamma}{2}\,\frac{\mu_{\mathrm{m}}-\mu_{\mathrm{eff}}}{(8-10\nu_{\mathrm{m}})\mu_{\mathrm{eff}}+(7-5\nu_{\mathrm{m}})\mu_{\mathrm{m}}}\\ &= \frac{5\gamma}{2}\left[1-\frac{(8-10\nu_{\mathrm{m}})\mu_{\mathrm{m}}+(7-5\nu_{\mathrm{m}})\mu_{\mathrm{m}}}{(8-10\nu_{\mathrm{m}})\mu_{\mathrm{eff}}+(7-5\nu_{\mathrm{m}})\mu_{\mathrm{m}}}\right],\end{aligned} \tag{3.43}$$

where μ_{eff} is the effective shear modulus of the isotropic particulate composite. It then follows from (3.38) and (3.40) that the exact matrix stress distribution, for a given value of μ_{eff}, is

$$\left.\begin{aligned}\sigma_{rr}^{\mathrm{m}} &= 2\mu_{\mathrm{m}}\left[\gamma+(36/5)\bar{D}_{\mathrm{m}}(b/r)^5-2(5-\nu_{\mathrm{m}})\,\bar{D}_{\mathrm{m}}(b/r)^3\right]\sin^2\theta\,\sin 2\phi,\\ \sigma_{r\theta}^{\mathrm{m}} &= \mu_{\mathrm{m}}\left[\gamma-(24/5)\bar{D}_{\mathrm{m}}(b/r)^5+2(1+\nu_{\mathrm{m}})\,\bar{D}_{\mathrm{m}}(b/r)^3\right]\sin 2\theta\,\sin 2\phi,\\ \sigma_{r\phi}^{\mathrm{m}} &= 2\mu_{\mathrm{m}}\left[\gamma-(24/5)\,\bar{D}_{\mathrm{m}}(b/r)^5+2(1+\nu_{\mathrm{m}})\,\bar{D}_{\mathrm{m}}(b/r)^3\right]\sin\theta\,\cos 2\phi,\end{aligned}\right\} b<r<\infty. \tag{3.44}$$

As the stress distribution given by (3.41) must be identical at large distances from the cluster with that specified by (3.44) it follows, from a consideration of terms proportional to r^{-3}, that

$$\bar{D}_{\mathrm{m}} = \sum_{i=1}^{N} V_{\mathrm{p}}^{i} D_{\mathrm{m}}^{i}, \tag{3.45}$$

where use has been made of (3.1). On substituting (3.42) and (3.43) into (3.45), it can be shown using (3.1) that the following 'mixtures' result is obtained for the function $1/(\mu+\mu_{\mathrm{m}}^{*})$

$$\left.\begin{aligned}\frac{1}{\mu_{\mathrm{eff}}+\mu_{\mathrm{m}}^{*}} &= \sum_{i=1}^{N}\frac{V_{\mathrm{p}}^{i}}{\mu_{\mathrm{p}}^{(i)}+\mu_{\mathrm{m}}^{*}}+\frac{V_{\mathrm{m}}}{\mu_{\mathrm{m}}+\mu_{\mathrm{m}}^{*}}\,,\\ \text{where}\quad \mu_{\mathrm{m}}^{*} &= \frac{7-5\nu_{\mathrm{m}}}{8-10\nu_{\mathrm{m}}}\mu_{\mathrm{m}} = \frac{9k_{\mathrm{m}}+8\mu_{\mathrm{m}}}{6(k_{\mathrm{m}}+2\mu_{\mathrm{m}})}\mu_{\mathrm{m}}\,.\end{aligned}\right\} \tag{3.46}$$

On using (3.1), the effective shear modulus may be estimated using the following relation

$$\mu_{\rm eff} = \left(\sum_{i=1}^{N} \frac{V_{\rm p}^i \mu_{\rm p}^{(i)}}{\mu_{\rm p}^{(i)} + \mu_{\rm m}^*} + \frac{V_{\rm m}\mu_{\rm m}}{\mu_{\rm m} + \mu_{\rm m}^*}\right) \Bigg/ \left(\sum_{i=1}^{N} \frac{V_{\rm p}^i}{\mu_{\rm p}^{(i)} + \mu_{\rm m}^*} + \frac{V_{\rm m}}{\mu_{\rm m} + \mu_{\rm m}^*}\right). \tag{3.47}$$

It can be shown that the bounds for the effective shear modulus derived by Hashin and Shtrikman [6, Equations (3.44)–(3.50)] and the bounds derived by Walpole [7, Equation (26)] are identical and may be expressed in the following form that has the same structure as the result (3.46) derived using Maxwell's methodology

$$\left.\begin{aligned} \frac{1}{\mu_{\rm eff} + \mu_{\rm min}^*} &\le \sum_{i=1}^{N} \frac{V_{\rm p}^i}{\mu_{\rm p}^{(i)} + \mu_{\rm min}^*} + \frac{V_{\rm m}}{\mu_{\rm m} + \mu_{\rm min}^*}, \\ \sum_{i=1}^{N} \frac{V_{\rm p}^i}{\mu_{\rm p}^{(i)} + \mu_{\rm max}^*} + \frac{V_{\rm m}}{\mu_{\rm m} + \mu_{\rm max}^*} &\le \frac{1}{\mu_{\rm eff} + \mu_{\rm max}^*}, \end{aligned}\right\} \tag{3.48}$$

$$\text{where} \quad \mu_{\rm min}^* = \mu_{\rm min} \frac{9k_{\rm min} + 8\mu_{\rm min}}{6\left(k_{\rm min} + 2\mu_{\rm min}\right)}, \quad \mu_{\rm max}^* = \mu_{\rm max} \frac{9k_{\rm max} + 8\mu_{\rm max}}{6\left(k_{\rm max} + 2\mu_{\rm max}\right)}. \tag{3.49}$$

The parameters $k_{\rm min}$ and $\mu_{\rm min}$ are the lowest values of the bulk and shear moduli of all phases in the composite, respectively, whereas $k_{\rm max}$ and $\mu_{\rm max}$ are the highest values. On writing

$$\begin{aligned} 6\mu_{\rm min}^* &= 4\mu_{\rm min} + 5/\left(2/k_{\rm min} + 1/\mu_{\rm min}\right), \\ 6\mu_{\rm max}^* &= 4\mu_{\rm max} + 5/\left(2/k_{\rm max} + 1/\mu_{\rm max}\right), \end{aligned} \tag{3.50}$$

it follows that $\mu_{\rm max}^* \ge \mu_{\rm min}^*$ for all values of the bulk and shear moduli, indicating that the 'max' and 'min' subscripts are used in an appropriate sense. It should be noted that $k_{\rm min}$ and $\mu_{\rm min}$ may be associated with different phases, and similarly for $k_{\rm max}$ and $\mu_{\rm max}$.

3.5 Summary of Results

3.5.1 Multiphase Composites

Key results, (3.10) for thermal conductivity, (3.27) for bulk modulus and (3.46) for shear modulus, derived using Maxwell's methodology have the following simple common structure, involving 'mixtures' formulae for the effective isotropic properties ϕ of the composite

$$\frac{1}{\phi_{\rm eff}^{(J)} + x_J} = \sum_{i=1}^{N} \frac{V_{\rm p}^i}{\phi_i^{(J)} + x_J} + \frac{V_{\rm m}}{\phi_{\rm m}^{(J)} + x_J}, \quad J = 1,2,3, \tag{3.51}$$

$$\text{where} \quad \left\{\begin{array}{lll} \phi^{(1)} \equiv \kappa, & \phi^{(2)} \equiv k, & \phi^{(3)} \equiv \mu, \\ x_1 = 2\kappa_m, & x_2 = \frac{4}{3}\mu_m, & x_3 = \mu_m^* = \dfrac{9k_{\rm m} + 8\mu_{\rm m}}{6\left(k_{\rm m} + 2\mu_{\rm m}\right)}\mu_{\rm m}. \end{array}\right. \tag{3.52}$$

The inequalities (3.12), (3.30) and (3.48), valid for all volume fractions, lead to rigorous bounds valid for any phase geometries that are statistically isotropic. They have the following common structure that is strongly related to the structure defined by (3.51) and (3.52) for effective properties determined using Maxwell's methodology

$$\left.\begin{aligned}&\frac{1}{\phi_{\text{eff}}^{(J)}+x_J^{\min}} \le \sum_{i=1}^{N}\frac{V_{\text{p}}^i}{\phi_i^{(J)}+x_J^{\min}}+\frac{V_{\text{m}}}{\phi_{\text{m}}^{(J)}+x_J^{\min}},\\ &\sum_{i=1}^{N}\frac{V_{\text{p}}^i}{\phi_i^{(J)}+x_J^{\max}}+\frac{V_{\text{m}}}{\phi_{\text{m}}^{(J)}+x_J^{\max}} \le \frac{1}{\phi_{\text{eff}}^{(J)}+x_J^{\max}},\end{aligned}\right\} \quad J=1,2,3, \tag{3.53}$$

$$\text{where}\quad \left.\begin{aligned}&x_1^{\min}=2\kappa_{\min}, \quad x_2^{\min}=\tfrac{4}{3}\mu_{\min}, \quad x_3^{\min}=\mu_{\min}^*,\\ &x_1^{\max}=2\kappa_{\max}, \quad x_2^{\max}=\tfrac{4}{3}\mu_{\max}, \quad x_3^{\max}=\mu_{\max}^*.\end{aligned}\right\} \tag{3.54}$$

By comparing the bounds (3.53) with (3.51) when $J = 1$, the result for thermal conductivity obtained using Maxwell's methodology is exactly the lower bound for κ_{eff} when $\kappa_{\min}=\kappa_{\text{m}}$, and the upper bound when $\kappa_{\max}=\kappa_{\text{m}}$. When $\kappa_{\min}<\kappa_{\text{m}}<\kappa_{\max}$, the result for effective thermal conductivity obtained using Maxwell's methodology lies between the bounds for all volume fractions. A comparison of (3.53) with (3.51) when $J = 2$ shows that the result obtained for bulk modulus using Maxwell's methodology leads exactly to the lower bound for k_{eff} when $\mu_{\min}=\mu_{\text{m}}$, and the upper bound when $\mu_{\max}=\mu_{\text{m}}$. When $\mu_{\min}<\mu_{\text{m}}<\mu_{\max}$, the result for effective bulk modulus obtained using Maxwell's methodology lies between the bounds for all volume fractions. A comparison of (3.53) with (3.51) when $J = 3$ shows that the result (3.51) obtained for shear modulus using Maxwell's methodology leads exactly to the lower bound for μ_{eff} when $\mu_{\min}^*=\mu_{\text{m}}^*$, and the upper bound when $\mu_{\max}^*=\mu_{\text{m}}^*$. When $\mu_{\min}^*<\mu_{\text{m}}^*<\mu_{\max}^*$, the result for effective shear modulus obtained using Maxwell's methodology lies between the bounds for all volume fractions. Thus, it has been shown that effective thermoelastic properties, obtained above using Maxwell's methodology, do not lie beyond rigorous bounds for properties for all volume fractions consistent with isotropic effective properties. This characteristic of Maxwell's methodology provides significant evidence that its validity is not confined to small volume fractions.

3.5.2 Two-phase Composites

When $N = 1$, it follows from (3.2) and (3.11) that the result first derived by Maxwell [3] for the analogous case of electrical conductivity is obtained, which may be expressed in the form of a mixtures estimate plus a correction term so that

$$\kappa_{\text{eff}} = V_{\text{p}}\kappa_{\text{p}}+V_{\text{m}}\kappa_{\text{m}}-\frac{\left(\kappa_{\text{p}}-\kappa_{\text{m}}\right)^2 V_{\text{p}}V_{\text{m}}}{V_{\text{p}}\kappa_{\text{m}}+V_{\text{m}}\kappa_{\text{p}}+2\kappa_{\text{m}}}. \tag{3.55}$$

It follows from (3.28), (3.29) and (3.47) derived using Maxwell's methodology, that the effective bulk modulus, thermal expansion coefficient and shear modulus may be expressed as a mixtures estimate plus a correction term, so that

$$\left.\begin{aligned}&k_{\text{eff}} = V_{\text{p}}k_{\text{p}}+V_{\text{m}}k_{\text{m}}-\frac{\left(k_{\text{p}}-k_{\text{m}}\right)^2 V_{\text{p}}V_{\text{m}}}{V_{\text{p}}k_{\text{m}}+V_{\text{m}}k_{\text{p}}+\frac{4}{3}\mu_{\text{m}}},\\ \text{or}\quad &\frac{1}{k_{\text{eff}}}=\frac{V_{\text{p}}}{k_{\text{p}}}+\frac{V_{\text{m}}}{k_{\text{m}}}-\frac{\left(1/k_{\text{p}}-1/k_{\text{m}}\right)^2 V_{\text{p}}V_{\text{m}}}{V_{\text{p}}/k_{\text{m}}+V_{\text{m}}/k_{\text{p}}+1/(\frac{4}{3}\mu_{\text{m}})},\end{aligned}\right\} \tag{3.56}$$

$$\alpha_{\text{eff}} = V_p\alpha_p + V_m\alpha_m + \frac{(k_p - k_m)(\alpha_p - \alpha_m)V_pV_m}{V_pk_p + V_mk_m + k_pk_m/(\frac{4}{3}\mu_m)}, \tag{3.57}$$

$$\mu_{\text{eff}} = V_p\mu_p + V_m\mu_m - \frac{(\mu_p - \mu_m)^2 V_pV_m}{V_p\mu_m + V_m\mu_p + \mu_m^*}, \tag{3.58}$$

where

$$\mu_m^* = \frac{7 - 5\nu_m}{8 - 10\nu_m}\mu_m = \frac{9k_m + 8\mu_m}{6(k_m + 2\mu_m)}\mu_m. \tag{3.59}$$

3.6 Bounds for Two-phase Isotropic Composites

It follows from Hashin and Shtrikman [5], and the review by Hashin [1], that bounds for the effective thermal conductivity of a two-phase composite, valid for arbitrary reinforcement geometries leading to statistically isotropic effective properties, may be expressed in the form

$$V_p\kappa_p + V_m\kappa_m - \frac{(\kappa_p - \kappa_m)^2 V_pV_m}{V_p\kappa_m + V_m\kappa_p + 2\kappa_{\min}} \le \kappa_{\text{eff}} \le V_p\kappa_p + V_m\kappa_m - \frac{(\kappa_p - \kappa_m)^2 V_pV_m}{V_p\kappa_m + V_m\kappa_p + 2\kappa_{\max}}, \tag{3.60}$$

where now $\kappa_{\max} = \max(\kappa_p, \kappa_m), \quad \kappa_{\min} = \min(\kappa_p, \kappa_m).$ (3.61)

Walpole [7, Equation (26)] has derived rigorous bounds for the effective bulk modulus, which can for a two-phase composite be expressed in the following two equivalent forms

$$V_pk_p + V_mk_m - \frac{(k_p - k_m)^2 V_pV_m}{V_pk_m + V_mk_p + \frac{4}{3}\mu_{\min}} \le k_{\text{eff}} \le V_pk_p + V_mk_m - \frac{(k_p - k_m)^2 V_pV_m}{V_pk_m + V_mk_p + \frac{4}{3}\mu_{\max}}, \tag{3.62}$$

$$\frac{V_p}{k_p} + \frac{V_m}{k_m} - \frac{(1/k_p - 1/k_m)^2 V_pV_m}{V_p/k_m + V_m/k_p + 3/(4\mu_{\max})} \le \frac{1}{k_{\text{eff}}} \le \frac{V_p}{k_p} + \frac{V_m}{k_m} - \frac{(1/k_p - 1/k_m)^2 V_pV_m}{V_p/k_m + V_m/k_p + 3/(4\mu_{\min})}, \tag{3.63}$$

where $\mu_{\max} = \max(\mu_p, \mu_m), \quad \mu_{\min} = \min(\mu_p, \mu_m).$ (3.64)

On using the bounds (3.62) for the bulk modulus, it can be shown that the bounds for the effective thermal expansion are such that if $\left(k_{\mathrm{p}}-k_{\mathrm{m}}\right)\left(\mu_{\mathrm{p}}-\mu_{\mathrm{m}}\right)\left(\alpha_{\mathrm{p}}-\alpha_{\mathrm{m}}\right) \geq 0$:

$$\bar{\alpha}+\frac{\left(k_{\mathrm{p}}-k_{\mathrm{m}}\right)\left(\alpha_{\mathrm{p}}-\alpha_{\mathrm{m}}\right) V_{\mathrm{p}} V_{\mathrm{m}}}{V_{\mathrm{p}} k_{\mathrm{p}}+V_{\mathrm{m}} k_{\mathrm{m}}+k_{\mathrm{m}} k_{\mathrm{p}} /\left(\frac{4}{3} \mu_{\mathrm{m}}\right)} \leq \alpha_{\mathrm{eff}} \leq \bar{\alpha}+\frac{\left(k_{\mathrm{p}}-k_{\mathrm{m}}\right)\left(\alpha_{\mathrm{p}}-\alpha_{\mathrm{m}}\right) V_{\mathrm{p}} V_{\mathrm{m}}}{V_{\mathrm{p}} k_{\mathrm{p}}+V_{\mathrm{m}} k_{\mathrm{m}}+k_{\mathrm{m}} k_{\mathrm{p}} /\left(\frac{4}{3} \mu_{\mathrm{p}}\right)}, \tag{3.65}$$

if $\left(k_{\mathrm{p}}-k_{\mathrm{m}}\right)\left(\mu_{\mathrm{p}}-\mu_{\mathrm{m}}\right)\left(\alpha_{\mathrm{p}}-\alpha_{\mathrm{m}}\right) \leq 0$:

$$\bar{\alpha}+\frac{\left(k_{\mathrm{p}}-k_{\mathrm{m}}\right)\left(\alpha_{\mathrm{p}}-\alpha_{\mathrm{m}}\right) V_{\mathrm{p}} V_{\mathrm{m}}}{V_{\mathrm{p}} k_{\mathrm{p}}+V_{\mathrm{m}} k_{\mathrm{m}}+k_{\mathrm{m}} k_{\mathrm{p}} /\left(\frac{4}{3} \mu_{\mathrm{p}}\right)} \leq \alpha_{\mathrm{eff}} \leq \bar{\alpha}+\frac{\left(k_{\mathrm{p}}-k_{\mathrm{m}}\right)\left(\alpha_{\mathrm{p}}-\alpha_{\mathrm{m}}\right) V_{\mathrm{p}} V_{\mathrm{m}}}{V_{\mathrm{p}} k_{\mathrm{p}}+V_{\mathrm{m}} k_{\mathrm{m}}+k_{\mathrm{m}} k_{\mathrm{p}} /\left(\frac{4}{3} \mu_{\mathrm{m}}\right)}. \tag{3.66}$$

Walpole [7, Equation (26)] has derived rigorous bounds for the effective shear modulus, which can, for an isotropic two-phase composite, be expressed in the following form having the same structure as the result (3.59)

$$V_{\mathrm{p}} \mu_{\mathrm{p}}+V_{\mathrm{m}} \mu_{\mathrm{m}}-\frac{\left(\mu_{\mathrm{p}}-\mu_{\mathrm{m}}\right)^{2} V_{\mathrm{p}} V_{\mathrm{m}}}{V_{\mathrm{p}} \mu_{\mathrm{m}}+V_{\mathrm{m}} \mu_{\mathrm{p}}+\mu_{\mathrm{min}}^{*}} \leq \mu_{\mathrm{eff}} \leq V_{\mathrm{p}} \mu_{\mathrm{p}}+V_{\mathrm{m}} \mu_{\mathrm{m}}-\frac{\left(\mu_{\mathrm{p}}-\mu_{\mathrm{m}}\right)^{2} V_{\mathrm{p}} V_{\mathrm{m}}}{V_{\mathrm{p}} \mu_{\mathrm{m}}+V_{\mathrm{m}} \mu_{\mathrm{p}}+\mu_{\mathrm{max}}^{*}}, \tag{3.67}$$

where μ_{min}^{*} and μ_{max}^{*} are defined by (3.49). The structure of (3.67) is identical to that given by Torquato [2, Equations (21.73)–(21.75)].

To conclude this section summarising results, it is useful to provide the relationships between the bulk and shear moduli and the elastic constants which are more frequently encountered in applications. It follows from (2.208) that the effective Young's modulus E_{eff} and effective Poisson's ratio ν_{eff} for an isotropic composite are given by

$$E_{\mathrm{eff}}=\frac{9 k_{\mathrm{eff}} \mu_{\mathrm{eff}}}{3 k_{\mathrm{eff}}+\mu_{\mathrm{eff}}}, \quad \nu_{\mathrm{eff}}=\frac{3 k_{\mathrm{eff}}-2 \mu_{\mathrm{eff}}}{2\left(3 k_{\mathrm{eff}}+\mu_{\mathrm{eff}}\right)}. \tag{3.68}$$

3.7 Comparison of Predictions with Known Results

When assessing the validity of undamaged particulate composites, it is particularly valuable to compare predictions using formulae for the relevant effective properties with those obtained from the use of alternative methods. This procedure can provide the confidence for use of the formulae in practical situations, and this approach to validation is now followed. When considering the effective bulk modulus, thermal expansion coefficient and thermal conductivity for two-phase composites having spherical particles of the same size, the results obtained using Maxwell's methodology are identical to the realistic bounds. They are also identical to estimates for effective properties obtained by applying the composite spheres assemblage model (see the review by Hashin [1]) for a particulate composite to a representative volume element comprising just one particle and a matrix region that is consistent with the volume fractions of the composite. In the case of shear modulus, the result (3.58) obtained using Maxwell's methodology, corresponds exactly to one of the variational bounds whenever $(k_p - k_m)(\mu_p - \mu_m) \geq 0$. For the case $(k_p - k_m)(\mu_p - \mu_m) \leq 0$, it can be shown that $\mu^*_{min} \leq \mu^*_m \leq \mu^*_{max}$ and a comparison of (3.59) and (3.67) then indicates that the result (3.58) for the effective shear modulus derived using Maxwell's methodology must lie between the bounds (3.67). In addition, the results (3.55–3.58), for two-phase composites arising from the use of Maxwell's methodology, are such that $\kappa_{eff} \rightarrow \kappa_p$, $k_{eff} \rightarrow k_p$, $\mu_{eff} \rightarrow \mu_p$ and $\alpha_{eff} \rightarrow \alpha_p$, respectively, when $V_p \rightarrow 1$, limits requiring V_p values attained only for a range of particle sizes, as for the composite spheres assembly model.

Effective properties of two-phase composites, derived using Maxwell's methodology, may be expressed as a mixtures estimate plus a correction term, as seen from (3.55)–(3.58). The correction is always proportional to the product V_pV_m, and it involves the square of property differences for the case of conductivity, bulk and shear moduli, and the product of differences of the bulk compressibility and expansion coefficient for the case of thermal expansion. These results are the preferred common form for effective properties, having the advantage that conditions governing whether an extreme value is an upper or lower bound are then easily determined. In addition, such conditions determine when both upper and lower bounds coincide with each other, and with predictions based on Maxwell's methodology, leading to exact nontrivial predictions for all volume fractions. For example, when $\mu_p = \mu_m$ the bounds for bulk modulus given by (3.62) are equal to the exact solution for any values of k_p, k_m and the volume fractions, and they are equal to the result (3.56) indicating that Maxwell's methodology leads, in this special nontrivial case, to an exact result for all volume fractions for which the composite is isotropic. For the case of thermal expansion, it follows from (3.65) and (3.66) that exact results are also obtained for any values of k_p, k_m, α_p, α_m and V_p, and they are equal to (3.57) indicating that Maxwell's methodology again leads, in a special nontrivial case, to an exact result for all volume fractions.

Results for effective properties of two-phase composites, are such that Maxwell's methodology, the composite spheres assemblage model when it can generate exact results, and the realistic variational bound, all lead to the same result. This suggests very strongly that the realistic bound is a much better estimate of effective properties for spherical particles than the other bound, which can be obtained simply by interchanging particle and matrix properties and the volume fractions. Further evidence of this phenomenon is provided in Figures 3.2–3.4 and Tables 3.1 and 3.2, where use has been made of the conductivity results of Sangani and Acrivos [10], based on the use of spherical harmonic expansions, and the results of Arridge [11] who used harmonics up to 11th order to estimate accurate values of bulk modulus and thermal expansion for body-centred cubic (b.c.c.) and face-centred cubic (f.c.c.) arrays of spherical particles having the same size. Whereas Sangani and Acrivos considered simple cubic, b.c.c. and f.c.c. arrays of spheres, only the f.c.c. results are shown in Figure 3.2, for the

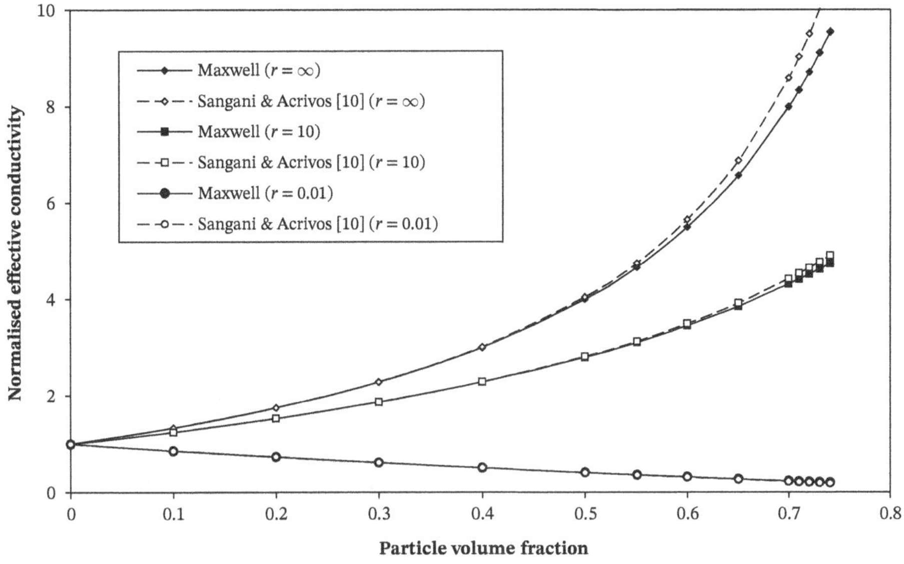

Figure 3.2 Dependence of ratio of effective and matrix thermal conductivities for a two-phase composite on particulate volume fraction for a face-centred cubic array of spherical particles, at various phase contrasts.

values $\kappa_p / \kappa_m = 0.01, 10, \infty$ of the phase contrast, as a larger range of volume fractions can be considered. It is seen that there is excellent agreement between predictions based on Maxwell's result (3.55) and the results of Sangani and Acrivos for a wide range of particulate volume fractions. Results (not shown) indicate that the agreement is less good at large volume fractions of particulate, when comparing Maxwell's result with the simple-cubic and b.c.c. results of Sangani and Acrivos.

It is worth noting from Bonnecaze and Brady [12, Tables 2–4], who used a multipole method to estimate the conductivity of cubic arrays of spherical particles, that their results for the case that retains only dipole–dipole interactions correspond almost exactly (to three significant figures) with the results shown in Figure 3.2 obtained using Maxwell's result (3.55). They did not make this comparison, considering only the results of Sangani and Acrivos [10]. This result suggests that Maxwell's result, which was derived assuming particles do not interact, is in fact valid also for the case when particle interactions are represented by dipole–dipole interactions, and this might explain why Maxwell's result is found to be a good approximation for a wide range of volume fractions. Further discussion of this issue is beyond the scope of this chapter. We note that for composites used in practice, the difference in the values of the thermomechanical properties (e.g. bulk modulus, thermal expansion coefficient) of the reinforcement and matrix, seldom lead to values of phase contrast that are greater than 10 or so. The phase contrast of the transport properties (such as electrical or thermal conductivity) can be very much greater. It follows that in practical situations, greater confidence may be placed in the Maxwell formulation being accurate at relatively large volume fractions for the thermomechanical properties when compared with the case of transport properties.

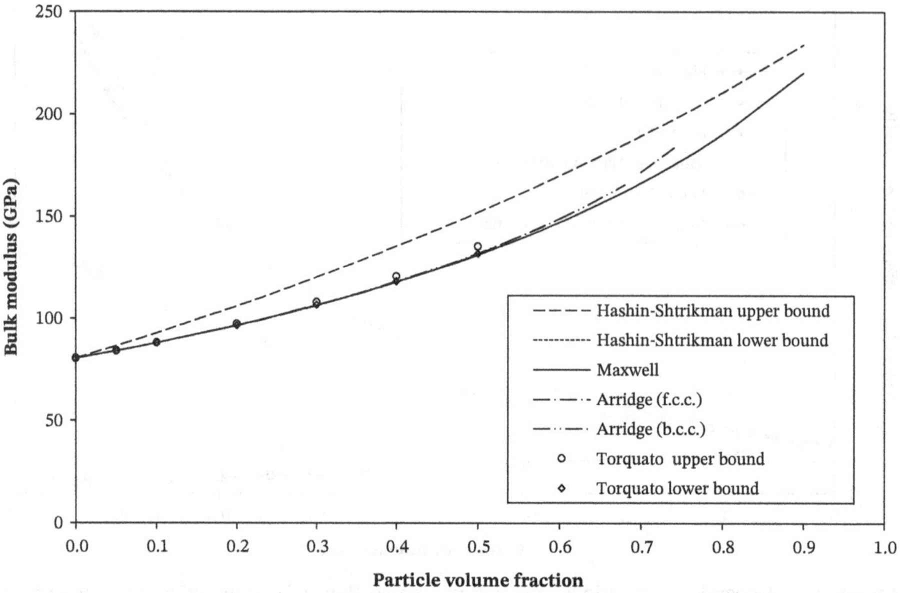

Figure 3.3 Dependence of effective bulk modulus for a two-phase composite on particulate volume fraction (see Table 3.1 for numerical values).

The results of Arridge [11] are based on the following properties for silicon carbide spheres in an aluminium matrix:

$$E_p = 483\text{GPa}, \quad E_m = 72.5\text{GPa}, \quad \nu_p = 0.19, \quad \nu_m = 0.35,$$

$$\alpha_p = 3.3\times10^{-6}\text{K}^{-1}, \quad \alpha_m = 22.5\times10^{-6}\text{K}^{-1}.$$

Table 3.1 Estimates of effective bulk modulus (GPa) for a two-phase particulate composite.

V_p	Maxwell's Methodology	Arridge [11] (f.c.c.)	Arridge [11] (b.c.c.)	Torquato [13] Lower bound
0	80.56	80.56	80.56	80.56
0.1	88.06	88.01	88.09	88.09
0.2	96.61	96.47	96.67	96.71
0.3	106.42	106.28	106.56	106.66
0.4	117.80	117.76	118.12	118.23
0.5	131.16	131.56	131.90	131.83
0.6	147.07	148.79	148.76	—
0.6802 (max.)	162.20	—	165.39	—
0.7	166.33	171.67	—	—
0.7405 (max.)	175.33	183.54	—	—
0.8	190.12	—	—	—
0.9	220.26	—	—	—

Table 3.2 Estimates of thermal expansion coefficient ($\times 10^6$ K $^{-1}$) of a two-phase particulate composite.

V_p	Maxwell's Methodology	Arridge [11] (f.c.c.)	Arridge [11] (b.c.c.)	Torquato [13] Upper bound
0	22.5	22.5	22.5	22.5
0.1	20.13	20.1	20.1	20.12
0.2	17.87	17.9	17.9	17.85
0.3	15.73	15.8	15.7	15.69
0.4	13.70	13.7	13.7	13.63
0.5	11.76	11.7	11.7	11.67
0.6	9.91	9.7	9.7	—
0.6802 (max.)	8.49	—	8.1	—
0.7	8.15	7.7	—	—
0.7405 (max.)	7.45	6.9	—	—
0.8	6.46	—	—	—
0.9	4.85	—	—	—

The corresponding values of bulk and shear moduli are $k_p = 259.68\text{GPa}$, $k_m = 80.56\text{GPa}$, $\mu_p = 202.94\text{GPa}$, $\mu_m = 26.85\text{GPa}$, and are such that $k_p > k_m$ and $\mu_p > \mu_m$. It should be noted that an array of spheres in a b.c.c. or in an f.c.c. arrangement possesses cubic symmetry. The thermal expansion coefficient of such an array is, therefore, isotropic. Arridge's results are given as mean values implying that the

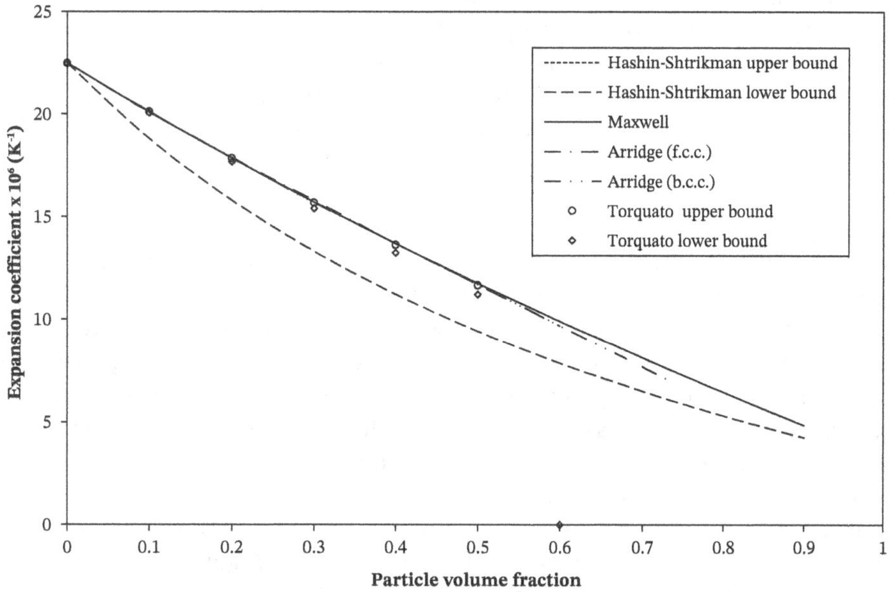

Figure 3.4 Dependence of the effective thermal expansion coefficient for a two-phase composite on particulate volume fraction (see Table 3.2 for numerical values).

expansion coefficients differ slightly in various directions, a situation that could arise because an insufficient number of harmonics has been included in the representation.

Additional evidence concerning the accuracy of realistic bounds is given by Torquato [13] who considered the effect on bounds of geometrical factors relating to the reinforcement, and developed three-point bounds that are more restrictive than the conventional two-point bounds (equivalent to the Hashin–Shtrikman [6] bounds) for the case of bulk modulus and thermal expansion of suspensions of spheres. The definition, (Torquato [13], Equation (25)), has in fact been replaced by $\zeta_2 = (\frac{5}{12} - \frac{3}{16}\ln 3)V_p$. Torquato's [13] three-point bounds are compared in Figures 3.3 and 3.4 with the almost exact results of Arridge [11], and results obtained from Maxwell's methodology and the two-point variational bounds. From (3.56) and (3.62), bulk modulus results using the Hashin–Shtrikman [6] lower bound and Maxwell's methodology are identical as $\mu_p > \mu_m$ for Arridge's properties. These results are very close to those obtained using the Arridge model for both f.c.c. and b.c.c. particle arrangements, and to the three-point lower bound estimate of Torquato [13]. The results of Arridge are shown for all volume fractions up to the closest packing value for f.c.c. and b.c.c. configurations of spherical particles. The f.c.c. and b.c.c. packing configurations lead to bulk moduli that are very close together for particulate volume fractions in the range $0 < V_p < 0.6$. Furthermore, the results obtained using Maxwell's methodology lie between the f.c.c. and b.c.c. estimates for volume fractions in the range $0 < V_p < 0.4$. For a significant range of volume fractions, the Hashin–Shtrikman upper bound is seen in Figure 3.3 to be significantly different to the corresponding lower bound, and to the three-point upper bound of Torquato.

For the case of thermal expansion, the Hashin–Shtrikman [6] upper bound and Maxwell's methodology result are identical as seen from (3.57) and (3.66), because for Arridge's properties $(k_p - k_m)(\mu_p - \mu_m)(\alpha_p - \alpha_m) \leq 0$. These results are seen in Figure 3.4 to be very close to those obtained using the Arridge model for both f.c.c. and b.c.c. particle arrangements, and to the three-point upper bound estimate of Torquato. The results of Arridge are again shown for all volume fractions up to the closest packing value for f.c.c. and b.c.c. configurations of spherical particles. The f.c.c. and b.c.c. packing configurations lead to expansion coefficients that are very close together, and very close to results obtained using Maxwell's methodology, for particulate volume fractions in the range $0 < V_p < 0.5$. For a significant range of volume fractions, the Hashin–Shtrikman lower bound is seen in Figure 3.4 to be significantly different to the corresponding upper bound, and to the three-point lower bound of Torquato. In view of the almost exact results of Arridge, and the observation that the three-point bounds for bulk modulus and thermal expansion derived by Torquato are reasonably close, it is deduced that Maxwell's methodology provides accurate estimates of bulk modulus and thermal expansion coefficient for a wide range of volume fractions.

For the case of a simple cubic array of spherical particles with volume fractions in the range $0 < V_p < 0.4$, Cohen and Bergman [14] (see Figure 3.4) have shown that bounds for shear modulus, obtained using a Fourier representation of an integrodifferential equation for the displacement field, are very close to the Hashin–Shtrikman lower bound when using properties for a glass–epoxy composite. The results of this chapter indicate that their bounds will also be very close to the result obtained using Maxwell's methodology, showing again that its validity is not restricted to low volume fractions, as might be expected from the approximations made.

For the cases of bulk modulus and thermal expansion, Maxwell's methodology is based on a stress distribution (3.24) in the matrix outside the sphere having radius b of effective medium, which is exact everywhere in the matrix (i.e. $b < r < \infty$) and involves an r-dependence only through terms proportional to r^{-3}. It follows from (3.23) that, for the discrete particle model (see Figure 3.1(a)), the asymptotic form for the stress field in the matrix as $r \to \infty$ has the same form as the exact solution for the equivalent effective medium model (see Figure 3.1(b)). The matching of the discrete and effective medium models at large distances, leading to an exact solution in the matrix ($b < r < \infty$) of the effective medium model, is thought to be one reason why estimates for bulk modulus and thermal expansion coefficient of two-phase composites are accurate for a wide range of volume fractions. When estimating thermal conductivity using Maxwell's methodology, relations (3.8) and (3.9) show that a similar situation arises. The r-dependence of the temperature gradient in the r-direction is through a term again proportional to r^{-3}, and as discussed previously, estimates of thermal conductivity based on Maxwell's methodology are again accurate for a wide range of volume fractions.

For the case of shear modulus, the exact solution for the stress field (3.44) in the matrix lying outside the sphere having radius b of effective medium (see Figure 3.1(b)) involves terms proportional to r^{-3} and r^{-5}, but only terms involving r^{-3} are used when applying Maxwell's methodology, as seen from (3.41). This means that, in contrast to the cases for the effective bulk modulus, thermal expansion coefficient and thermal conductivity, the resulting estimate for the effective shear modulus does not lead to an exact matrix stress distribution in the region $b < r < \infty$ outside the sphere of effective medium, and consequently estimates for effective shear modulus are likely to be less accurate than those for other effective properties.

The results, discussed previously for various effective properties, are remarkable as one might expect Maxwell's methodology to be accurate only for sufficiently low volume fractions of reinforcement. The reason is that the methodology involves the examination of the stress, displacement or temperature fields in the matrix at large distances from the cluster of particles, and assumes that the perturbing effect of each particle can be approximated by locating them at the same point. The nature of this approximation is such that interactions between particles are negligible, and it would be expected that resulting effective properties will be accurate only for low volume fractions, as originally suggested by Maxwell [3]. In view of compelling evidence presented in this chapter, based on a wide variety of considerations, a major conclusion is that results for two-phase composites derived using Maxwell's methodology are not limited to small particulate volume fractions and can be used with confidence using typical volume fractions often encountered in practice.

References

1. Hashin, Z. (1983). Analysis of composite materials - A survey. *Journal of Applied Mechanics* 50: 481–505.
2. Torquato, S. (2002). *Random Heterogeneous Materials.* New York: Springer-Verlag.
3. Maxwell, J.C. (1873). *A Treatise on Electricity and Magnetism*, 1st e (3rd e, 1892). Chapter 9, (Vol. 1, Art. 310-314, pp. 435–441). Oxford: Clarendon Press.
4. McCartney, L.N. and Kelly, A. (2008). Maxwell's far-field methodology applied to the prediction of properties of multi-phase isotropic particulate composites. *Proceedings of the Royal Society* A464: 423–446.
5. Hashin, Z. and Shtrikman, S. (1962). A variational approach to the theory of the effective magnetic permeability of multiphase materials. *Journal of Applied Physics* 33 (10): 3125–3131.
6. Hashin, Z. and Shtrikman, S. (1963). A variational approach to the theory of the elastic behaviour of multiphase composites. *Journal of the Mechanics and Physics of Solids* 11: 127–140.
7. Walpole, L.J. (1966). On bounds for the overall elastic moduli of inhomogeneous systems – I. *Journal of the Mechanics and Physics of Solids* 14: 151–162.
8. Love, A.E.H. (1944). *A Treatise on the Mathematical Theory of Elasticity*. Chapter XI, 4th ed. New York: Dover Publications.
9. Christensen, R.M. and Lo, K.H. (1979). Solutions for the effective shear properties in three phase sphere and cylinder models. *Journal of the Mechanics and Physics of Solids* 27: 315–330.
10. Sangani, A.S. and Acrivos, A. (1983). The effective conductivity of a periodic array of spheres. *Proceedings of the Royal Society of London* A386: 263–275.
11. Arridge, R.G.C. (1992). The thermal expansion and bulk modulus of composites consisting of arrays of spherical particles in a matrix, with body or face centred cubic symmetry. *Proceedings of the Royal Society of London* A438: 291–310.
12. Bonnecaze, R.T. and Brady, J.F. (1990). A method for determining the effective conductivity of dispersions of particles. *Proceedings of the Royal Society of London* A430: 285–313.
13. Torquato, S. (1990). Bounds on the thermoelastic properties of suspension of spheres. *Journal of Applied Physics* 67: 7223–7227.
14. Cohen, I. and Bergman, D.J. (2003). Effective elastic properties of periodic composite medium. *Journal of the Mechanics and Physics of Solids* 51: 1433–1457.

4

Maxwell's Methodology for the Prediction of Effective Properties of Unidirectional Multiphase Fibre-reinforced Composites

Overview: The methodology developed by Maxwell when estimating the effective electrical conductivity of an isotropic particulate composite is used to estimate many of the effective thermoelastic properties of a fibre-reinforced composite. A detailed description is first given of Maxwell's methodology applied to the thermal conduction problem, extending the approach to deal with multiphase fibres having different sizes and properties, and assuming perfect thermal contact at the fibre/matrix interfaces. It is noted that a published result, based on Maxwell's methodology, for two-phase systems having fibres of the same size, but including interfacial thermal resistance, is in error. A stress and displacement formulation that is radial in nature is used in conjunction with Maxwell's methodology to estimate values for many of the effective properties of a unidirectional fibre-reinforced composite. A method of applying Maxwell's methodology to the estimation of the effective shear modulus of a unidirectional fibre-reinforced composite is also given. A key aspect of the results obtained, when using Maxwell's methodology to estimate some of the thermoelastic properties of a unidirectional composite is that they correspond to those derived from the composite cylinders assemblage model, which, in turn, correspond to one of the bounds obtained when using variational methods. This correspondence indicates that the results obtained using Maxwell's methodology are not restricted to low fibre volume fractions. For each effective property, it is shown that the resulting formulae for the case of just two phases, having fibre reinforcements of the same size, may be expressed as a mixtures estimate plus a correction term that is used to derive the conditions that determine whether the extreme values of properties obtained by variational methods are upper or lower bounds. These conditions differ in some cases from those that have been given in the literature.

4.1 Introduction

Chapter 3 described how Maxwell's methodology [1, 2] can be applied to the estimation of the effective thermoelastic and conduction properties of isotropic spherical particulate composites. It was emphasised that Maxwell's methodology can provide good estimates of effective properties for a wide range of composites having practical interest. The objective of this chapter is to apply Maxwell's methodology to the prediction

Properties for Design of Composite Structures: Theory and Implementation Using Software, First Edition. Neil McCartney.

Companion Website: www.wiley.com/go/mccartney/properties

of good estimates of the effective properties of unidirectionally fibre-reinforced composites, where the fibres of circular cross section can be of different types and have different radii. Although some concepts regarding Maxwell's methodology are repeated, they are included here so that the chapter is more self-contained, and as the actual details of the method differ slightly. The analysis used differs significantly from that in Chapter 3.

The estimation of the thermoelastic and conduction properties of fibre-reinforced composites has been studied in detail in the literature over many years. Hashin [3] has given a very detailed review of many estimation methods, including the use of the well-known composite cylinders assemblage method, and methods based on variational techniques that lead to upper and lower bound estimates of properties. Key aspects are that the composite cylinders assemblage results correspond to one of the bounds obtained using variational methods, and that the other bound is obtained simply by interchanging the fibre and matrix properties and the volume fractions of fibre and matrix. This suggests that the bound that corresponds with the composite cylinders assemblage result is likely to be the nearest estimate to the actual value of the property that is being estimated.

A detailed study of the method used by Maxwell has revealed that his methodology can also be applied to the estimation of many other properties of fibre-reinforced composite materials that are reinforced with cylindrical fibres. The principal objective of this chapter is to show how Maxwell's methodology can be applied to the estimation of many of the effective properties of a composite reinforced with homogeneous cylindrical fibres. The methodology of Maxwell is easily extended so that assemblies of multiphase cylindrical fibres having a range of radii and properties may be considered.

Section 4.2 provides a detailed description of Maxwell's methodology applied to the thermal conduction problem, extending the approach to deal with multiphase cylindrical fibres having different sizes, and including the effects of thermal resistance at the fibre/matrix interfaces. It is noted that the results of Johnson and Hasselman [4] for two-phase systems having cylindrical fibres of the same size, but including interfacial thermal resistance, are in fact in error. Section 4.3 uses a radial stress and displacement formulation in conjunction with Maxwell's methodology to estimate values for the effective properties of a fibre-reinforced composite. Sections 4.4 and 4.5 consider the method of applying Maxwell's methodology to the estimation of the effective shear axial and transverse moduli of a fibre-reinforced composite. For each effective property, the resulting formulae for the case of just two phases, having cylindrical reinforcements of the same size, will be expressed as a mixtures estimate plus a correction term that is used to derive the conditions that determine whether the extreme values of properties are upper or lower bounds. These formulae differ in some cases from those that have been given in the literature.

4.2 General Description of Maxwell's Methodology Applied to Thermal Conductivity

Consider an isolated cluster of parallel fibre reinforcements, embedded in an isotropic infinite matrix, as shown in Figure 4.1(a). There are N different reinforcement types such that for $i = 1, \ldots, N$, there are n_i cylindrical fibres of radius a_i. The properties and geometry of the fibres of type i, that may differ from those of other types, are denoted

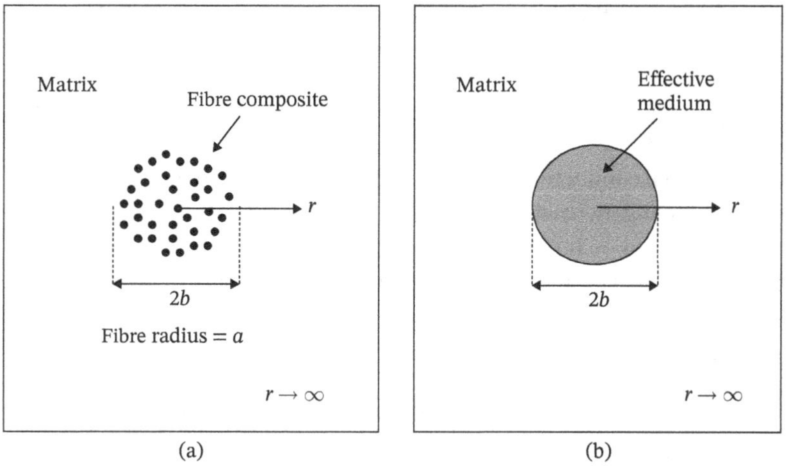

Figure 4.1 (a) Discrete fibre model and (b) smoothed effective medium model of a fibre-reinforced composite embedded in infinite matrix material.

by a superscript i. The cluster of all types of fibre is *just* enclosed by a cylinder of radius b and the parallel fibre distribution is sufficiently homogeneous for it to lead to transverse isotropic properties for the composite formed by the cluster of fibres and the matrix lying within this cylinder as shown in Figure 4.1(b). The volume fraction of fibres of type i within the cylinder of radius b is given by

$$V_{\rm f}^i = n_i a_i^2 \,/\, b^2, \quad \text{such that} \quad V_{\rm m} + \sum_{i=1}^{N} V_{\rm f}^i = 1, \tag{4.1}$$

where $V_{\rm m}$ is the corresponding volume fraction of matrix.

If there is only one type of fibre, and there are n fibres of radius a within the cylinder of radius b, then the fibre volume fraction of the composite is denoted by $V_{\rm f}$ such that

$$V_{\rm f} = \frac{na^2}{b^2} = 1 - V_{\rm m}. \tag{4.2}$$

It should be noted that by selecting all fibres to have the same properties but different sizes, composites having only a statistical distribution of fibre sizes can be considered. Similarly, by selecting all fibres to have the same size but different properties, composites having only a statistical distribution of properties can be considered. Clearly composites having statistical distributions of both size and properties can also be considered.

4.2.1 Temperature Distribution for an Isolated Fibre

It is useful to extend the methodology used by Maxwell [1] to estimate the electrical conductivity of a cluster of fibres embedded in an infinite matrix so that fibre-reinforced composites can be considered. The following description modifies the approach of Maxwell so that it applies to the analogous problem of thermal conductivity, and to a multiphase system of fibres. Hasselman and Johnson [4] considered the same problem where the fibres were all the same type size and where there is an interfacial thermal resistance between the fibers and the matrix.

Consider a single fibre of radius a embedded in an infinite matrix. A set of cylindrical polar coordinates (r, θ, z) is introduced having origin on the axis of the fibre. The temperature distribution in the fibre and surrounding matrix must satisfy Laplace's equation, namely

$$\frac{1}{r}\frac{\partial}{\partial r}\left(r\frac{\partial T}{\partial r}\right)+\frac{1}{r^2}\frac{\partial^2 T}{\partial \theta^2}=0. \tag{4.3}$$

The transverse thermal conductivity of the fibre is denoted by $\kappa_{\rm T}^{\rm f}$ whereas the matrix is assumed isotropic having thermal conductivity $\kappa_{\rm m}$. On the external boundary $r\to\infty$ a temperature distribution is imposed that would lead, in a homogeneous matrix material under steady-state conditions, to the following temperature distribution having a uniform gradient α

$$T(r,\theta)=\alpha z,\quad \text{where}\quad z=r\cos\theta. \tag{4.4}$$

At the fibre/matrix interface, the following boundary conditions are imposed

$$T_{\rm f}=T_{\rm m},\qquad \kappa_{\rm T}^{\rm f}\frac{\partial T_{\rm f}}{\partial r}=\kappa_{\rm m}\frac{\partial T_{\rm m}}{\partial r}\quad \text{on}\quad r=a. \tag{4.5}$$

The temperature distributions in the fibre and matrix, denoted by $T_{\rm f}$ and $T_{\rm m}$, respectively, must satisfy (4.3) and the boundary conditions (4.5) so that

$$T_{\rm f}\left(r,\theta\right)=\frac{2\kappa_{\rm m}}{\kappa_{\rm m}+\kappa_{\rm T}^{\rm f}}\alpha r\cos\theta,\quad 0<r<a, \tag{4.6}$$

$$T_{\rm m}(r,\theta)=\alpha r\cos\theta+\left(\frac{2\kappa_{\rm m}}{\kappa_{\rm m}+\kappa_{\rm T}^{\rm f}}-1\right)\frac{\alpha a^2\cos\theta}{r},\quad a<r<\infty. \tag{4.7}$$

It should be noted that any temperature T_0 can be added to the solution defined by (4.6) and (4.7) without affecting the satisfaction of the interface conditions (4.5). It should also be noted that when considering an infinite medium having a uniform temperature gradient, it is inevitable that temperatures lower than absolute zero will be encountered. This is not a matter for concern as the use of an infinite medium is simply a mathematical construct, designed to enable a specific method of estimating effective properties of the composite enclosed by the cylinder of radius b.

4.2.2 Maxwell's Methodology for Estimating Transverse Conductivity

The first stage is to consider the perturbing effect of an isolated cluster of parallel fibres, having different sizes and properties as described in Section 4.1, in the infinite matrix at very large distances from the cluster. The transverse thermal conductivity of the fibres of type i is denoted by $\kappa_{\rm T}^{{\rm f}(i)}$. The cluster is assumed to be homogeneous regarding the distribution of fibres, and leads to an effective transverse thermal conductivity $\kappa_{\rm T}^{\rm eff}$ for the composite. The perturbing effect of the fibre cluster is estimated by superimposing the perturbations caused by each fibre and recognising that, at large enough distances, all the fibres in the cluster can considered as located at the origin that is situated at the centre of one of the fibres in the cluster. Thus, for the case of multiple phases, the temperature distribution (4.7) for the matrix is generalised to the form

$$T_{\rm m}=\alpha r\cos\theta+\left[\sum_{i=1}^{N}n_i a_i^2\left(\frac{2\kappa_{\rm m}}{\kappa_{\rm m}+\kappa_{\rm T}^{{\rm f}(i)}}-1\right)\right]\frac{\alpha\cos\theta}{r}. \tag{4.8}$$

The cluster of *all* types of fibre is now considered to be enclosed in a cylinder of radius b such that the volume fraction $V_{\rm f}^i$ of fibres of type i within the cylinder of radius b is given by (4.1). On using (4.7) the temperature distribution in the matrix outside the cylinder of radius b is then given by

$$T_{\rm m} = \alpha r \cos\theta + \left[\frac{2\kappa_{\rm m}}{\kappa_{\rm m} + \kappa_{\rm T}^{\rm eff}} - 1\right]\frac{\alpha b^2 \cos\theta}{r}, \quad b < r < \infty. \tag{4.9}$$

The second stage is to equate the perturbation terms in (4.8) and (4.9) (i.e. the second terms on the right-hand side) so that, on using (4.1), the effective thermal conductivity of the multiphase cluster of fibres may be obtained from the relation

$$\frac{1}{\kappa_{\rm T}^{\rm eff} + \kappa_{\rm m}} = \sum_{i=1}^{N} \frac{V_{\rm f}^i}{\kappa_{\rm T}^{{\rm f}(i)} + \kappa_{\rm m}} + \frac{V_{\rm m}}{\kappa_{\rm m} + \kappa_{\rm m}}. \tag{4.10}$$

The effective thermal conductivity of the multiphase fibre composite is then given by

$$\kappa_{\rm T}^{\rm eff} = \left(\sum_{i=1}^{N} \frac{\kappa_{\rm T}^{{\rm f}(i)} V_{\rm f}^i}{\kappa_{\rm T}^{{\rm f}(i)} + \kappa_{\rm m}} + \frac{V_{\rm m}}{2} \right) \Bigg/ \left(\sum_{i=1}^{N} \frac{V_{\rm f}^i}{\kappa_{\rm T}^{{\rm f}(i)} + \kappa_{\rm m}} + \frac{V_{\rm m}}{2\kappa_{\rm m}} \right). \tag{4.11}$$

The bounds for the conductivity of a multiphase composite are given by Torquato [5] which may be written in the following simpler form, having the same structure as the result (4.10) derived using Maxwell's methodology

$$\begin{aligned} &\frac{1}{\kappa_{\rm T}^{\rm eff} + \kappa_{\rm T}^{\min}} \le \sum_{i=1}^{N} \frac{V_f^i}{\kappa_{\rm T}^{{\rm f}(i)} + \kappa_{\rm T}^{\min}} + \frac{V_m}{\kappa_{\rm m} + \kappa_{\rm T}^{\min}}, \\ &\sum_{i=1}^{N} \frac{V_f^i}{\kappa_{\rm T}^{{\rm f}(i)} + \kappa_{\rm T}^{\max}} + \frac{V_m}{\kappa_{\rm m} + \kappa_{\rm T}^{\max}} \le \frac{1}{\kappa_{\rm T}^{\rm eff} + \kappa_{\rm T}^{\max}}, \end{aligned} \tag{4.12}$$

where $\kappa_{\rm min}$ is the lowest value of conductivities for all phases, whereas $\kappa_{\rm max}$ is the highest value.

The effective axial thermal conductivity of the unidirectional composite is given by the following mixtures rule

$$\kappa_{\rm A}^{\rm eff} = \sum_{i=1}^{N} V_{\rm f}^i \kappa_{\rm A}^{f(i)} + V_{\rm m}\kappa_{\rm m}. \tag{4.13}$$

The validity of this relationship arises from the fact that the temperature distribution is such that there is no heat flow across the fibre/matrix interfaces. The temperature is linear in the axial direction having the same gradient in both fibre and matrix.

4.3 The Basic Equations for Thermoelastic Analysis

When considering an isolated cylindrical fibre embedded in matrix material, it is convenient to introduce a set of cylindrical polar coordinates (r, θ, z) where the origin lies on the axis of the fibre. All fibres are assumed to be made of transverse isotropic solids where properties are isotropic in the plane normal to the fibre axes. When using cylindrical polar coordinates, transverse isotropic solids (see (2.199)) are characterised by stress-strain relations of the form

$$\varepsilon_{rr} = \frac{1}{E_T}\sigma_{rr} - \frac{\nu_t}{E_T}\sigma_{\theta\theta} - \frac{\nu_A}{E_A}\sigma_{zz} + \alpha_T \Delta T, \tag{4.14}$$

$$\varepsilon_{\theta\theta} = -\frac{\nu_t}{E_T}\sigma_{rr} + \frac{1}{E_T}\sigma_{\theta\theta} - \frac{\nu_A}{E_A}\sigma_{zz} + \alpha_T \Delta T, \tag{4.15}$$

$$\varepsilon_{zz} = -\frac{\nu_A}{E_A}\sigma_{rr} - \frac{\nu_A}{E_A}\sigma_{\theta\theta} + \frac{1}{E_A}\sigma_{zz} + \alpha_A \Delta T, \tag{4.16}$$

$$\varepsilon_{rz} = \frac{\sigma_{rz}}{2\mu_A}, \qquad \varepsilon_{\theta z} = \frac{\sigma_{\theta z}}{2\mu_A}, \qquad \varepsilon_{r\theta} = \frac{\sigma_{r\theta}}{2\mu_t}, \tag{4.17}$$

$$\text{where} \quad E_T = 2\mu_t\left(1+\nu_t\right) \quad \text{but} \quad E_A \neq 2\mu_A\left(1+\nu_A\right). \tag{4.18}$$

Superscript 'f' is used to denote anisotropic fibre properties and subscript 'm' is used to denote the properties E_m, ν_m, μ_m and α_m of an isotropic matrix such that $E_m = 2\mu_m(1+\nu_m)$. If the fibres are also isotropic then

$$E_A^f = E_T^f = E_f, \ \mu_A^f = \mu_t^f = \mu_f, \ \nu_A^f = \nu_t^f = \nu_f, \ \alpha_A^f = \alpha_T^f = \alpha_f. \tag{4.19}$$

The equilibrium equations for the fibres and matrix in the absence of body forces are (see (2.125)–(2.127))

$$\frac{\partial \sigma_{rr}}{\partial r} + \frac{1}{r}\frac{\partial \sigma_{r\theta}}{\partial \theta} + \frac{\partial \sigma_{rz}}{\partial z} + \frac{\sigma_{rr} - \sigma_{\theta\theta}}{r} = 0, \tag{4.20}$$

$$\frac{\partial \sigma_{r\theta}}{\partial r} + \frac{1}{r}\frac{\partial \sigma_{\theta\theta}}{\partial \theta} + \frac{\partial \sigma_{\theta z}}{\partial z} + \frac{2\sigma_{\theta r}}{r} = 0, \tag{4.21}$$

$$\frac{\partial \sigma_{rz}}{\partial r} + \frac{1}{r}\frac{\partial \sigma_{\theta z}}{\partial \theta} + \frac{\partial \sigma_{zz}}{\partial z} + \frac{\sigma_{rz}}{r} = 0. \tag{4.22}$$

4.3.1 Properties Defined from Axisymmetric Distributions

The following analysis applies to an isolated cylindrical fibre of radius a that is perfectly bonded to an infinite matrix, subject to a uniform temperature change ΔT, where the system is subject to a uniform axial strain ε and a uniform transverse stress σ_T. The equilibrium equations (4.20)–(4.22) for the fibre and matrix, assuming symmetry about the fibre axis so that stress components are independent of θ and the shear stresses $\sigma_{r\theta}$ and $\sigma_{\theta z}$ are zero, then reduce to the form

$$\frac{\partial \sigma_{rr}}{\partial r} + \frac{\partial \sigma_{rz}}{\partial z} + \frac{\sigma_{rr} - \sigma_{\theta\theta}}{r} = 0, \tag{4.23}$$

$$\frac{\partial \sigma_{zz}}{\partial z} + \frac{\partial \sigma_{rz}}{\partial r} + \frac{\sigma_{rz}}{r} = 0. \tag{4.24}$$

In regions away from the loading mechanism it is reasonable to assume that

$$u_z^f \equiv u_z^m \equiv \varepsilon z, \tag{4.25}$$

where ε is the axial strain applied to the composite. A solution is now sought of the following classical Lamé form

$$u_r^{\mathrm{f}} = A_{\mathrm{f}} r, \quad u_\theta^{\mathrm{f}} \equiv 0, \tag{4.26}$$

$$u_r^{\mathrm{m}} = A_{\mathrm{m}} r + \frac{a^2 \phi}{2\mu_{\mathrm{m}} r}, \quad u_\theta^{\mathrm{m}} \equiv 0, \tag{4.27}$$

where A_{f}, A_{m} and ϕ are constants to be determined.

4.3.2 Solution for an Isolated Fibre Perfectly Bonded to the Matrix

The boundary and interface conditions that must be satisfied are

$$\left.\begin{aligned}
\sigma_{rr}^{\mathrm{m}}(\infty, z) &= \sigma_{\mathrm{T}}, \\
\sigma_{rz}^{\mathrm{m}}(\infty, z) &= 0, \\
\sigma_{rr}^{\mathrm{m}}(a, z) &= \sigma_{rr}^{\mathrm{f}}(a, z), \\
\sigma_{rz}^{\mathrm{m}}(a, z) &= \sigma_{rz}^{\mathrm{f}}(a, z), \\
u_r^{\mathrm{m}}(a, z) &= u_r^{\mathrm{f}}(a, z), \\
u_z^{\mathrm{m}}(R, z) &= u_z^{\mathrm{f}}(R, z).
\end{aligned}\right\} \tag{4.28}$$

On differentiating the displacement field, it follows, on using (2.139) on setting $u_\theta \equiv 0$, that.

$$\varepsilon_{rr}^{\mathrm{f}} = \frac{\partial u_r^{\mathrm{f}}}{\partial r} = A_{\mathrm{f}}, \quad \varepsilon_{\theta\theta}^{\mathrm{f}} = \frac{u_r^{\mathrm{f}}}{r} = A_{\mathrm{f}}, \quad \varepsilon_{zz}^{\mathrm{f}} = \frac{\partial u_z^{\mathrm{f}}}{\partial z} = \varepsilon, \tag{4.29}$$

$$\varepsilon_{rz}^{\mathrm{f}} = \varepsilon_{\theta z}^{\mathrm{f}} = \varepsilon_{r\theta}^{\mathrm{f}} \equiv 0, \tag{4.30}$$

$$\varepsilon_{rr}^{\mathrm{m}} = \frac{\partial u_r^{\mathrm{m}}}{\partial r} = A_{\mathrm{m}} - \frac{a^2 \phi}{2\mu_{\mathrm{m}} r^2}, \quad \varepsilon_{\theta\theta}^{\mathrm{m}} = \frac{u_r^{\mathrm{m}}}{r} = A_{\mathrm{m}} + \frac{a^2 \phi}{2\mu_{\mathrm{m}} r^2}, \quad \varepsilon_{zz}^{\mathrm{m}} = \frac{\partial u_z^{\mathrm{m}}}{\partial z} = \varepsilon, \tag{4.31}$$

$$\varepsilon_{rz}^{\mathrm{m}} = \varepsilon_{\theta z}^{\mathrm{m}} = \varepsilon_{r\theta}^{\mathrm{m}} \equiv 0. \tag{4.32}$$

It follows directly from (4.17), (4.30) and (4.32) that for fibre and matrix

$$\left.\begin{aligned}
\sigma_{rz}^{\mathrm{f}} &= \sigma_{\theta z}^{\mathrm{f}} = \sigma_{r\theta}^{\mathrm{f}} \equiv 0, \\
\sigma_{rz}^{\mathrm{m}} &= \sigma_{\theta z}^{\mathrm{m}} = \sigma_{r\theta}^{\mathrm{m}} \equiv 0.
\end{aligned}\right\} \tag{4.33}$$

On subtracting (4.14) and (4.15), it follows that for fibre and matrix

$$\varepsilon_{\theta\theta}^{\mathrm{f}} - \varepsilon_{rr}^{\mathrm{f}} = \frac{1}{2\mu_t^{\mathrm{f}}}(\sigma_{\theta\theta}^{\mathrm{f}} - \sigma_{rr}^{\mathrm{f}}), \tag{4.34}$$

$$\varepsilon_{\theta\theta}^{\mathrm{m}} - \varepsilon_{rr}^{\mathrm{m}} = \frac{1}{2\mu_{\mathrm{m}}}(\sigma_{\theta\theta}^{\mathrm{m}} - \sigma_{rr}^{\mathrm{m}}). \tag{4.35}$$

Relations (4.29) and (4.31) then assert that

$$\left.\begin{aligned}
\sigma_{\theta\theta}^{\mathrm{f}} &= \sigma_{rr}^{\mathrm{f}}, \\
\sigma_{\theta\theta}^{\mathrm{f}} &= \sigma_{rr}^{\mathrm{f}} + \frac{2a^2}{r^2}\phi.
\end{aligned}\right\} \tag{4.36}$$

On substituting (4.33) and (4.34) into the equilibrium equations (4.20)–(4.22) it follows that

$$\frac{\partial \sigma_{rr}^{\mathrm{f}}}{\partial r} = 0, \qquad \frac{\partial \sigma_{zz}^{\mathrm{f}}}{\partial z} = 0, \tag{4.37}$$

$$\frac{\partial \sigma_{rr}^{\mathrm{m}}}{\partial r} = \frac{2a^2\phi}{r^3}, \qquad \frac{\partial \sigma_{zz}^{\mathrm{m}}}{\partial z} = 0. \tag{4.38}$$

On integrating $(4.38)_1$ subject to the condition $(4.28)_1$,

$$\sigma_{rr}^{\mathrm{m}} = \sigma_{\mathrm{T}} - \frac{a^2}{r^2}\phi, \tag{4.39}$$

and from (4.36) it follows that

$$\sigma_{\theta\theta}^{\mathrm{m}} = \sigma_{\mathrm{T}} + \frac{a^2}{r^2}\phi. \tag{4.40}$$

On integrating $(4.37)_1$ using (4.39) together with the continuity condition $(4.28)_3$

$$\sigma_{rr}^{\mathrm{f}} = \sigma_{\mathrm{T}} - \phi, \tag{4.41}$$

and from (4.36) it follows that

$$\sigma_{\theta\theta}^{\mathrm{f}} = \sigma_{\mathrm{T}} - \phi. \tag{4.42}$$

The substitution of $(4.29)_3$, (4.41) and (4.42) into (4.16) leads to

$$\sigma_{zz}^{\mathrm{f}} = E_{\mathrm{A}}^{\mathrm{f}}(\varepsilon - \alpha_{\mathrm{A}}^{\mathrm{f}}\Delta T) + 2\nu_{\mathrm{A}}^{\mathrm{f}}\left(\sigma_{\mathrm{T}} - \phi\right). \tag{4.43}$$

It is clear from (4.41) that σ_{zz}^{f} is a constant, automatically satisfying $(4.37)_2$. The substitution of $(4.31)_3$, (4.39) and (4.40) into (4.16), applied to the matrix, leads to

$$\sigma_{zz}^{\mathrm{m}} = E_{\mathrm{m}}\left(\varepsilon - \alpha_{\mathrm{m}}\Delta T\right) + 2\nu_{\mathrm{m}}\sigma_{\mathrm{T}}, \tag{4.44}$$

thus automatically satisfying $(4.38)_2$. The substitution of (4.29), (4.31), (4.39)–(4.44) into relation (4.14), applied to the fibre and matrix, leads to

$$A_{\mathrm{f}} = -\nu_{\mathrm{A}}^{\mathrm{f}}\varepsilon + \left(\alpha_{\mathrm{T}}^{\mathrm{f}} + \nu_{\mathrm{A}}^{\mathrm{f}}\alpha_{\mathrm{A}}^{\mathrm{f}}\right)\Delta T + \frac{1}{2k_{\mathrm{T}}^{\mathrm{f}}}\left(\sigma_{\mathrm{T}} - \phi\right), \tag{4.45}$$

$$A_{\mathrm{m}} = -\nu_{\mathrm{m}}\varepsilon + \alpha_{\mathrm{m}}\left(1 + \nu_{\mathrm{m}}\right)\Delta T + \frac{\sigma_{\mathrm{T}}}{2k_{\mathrm{T}}^{\mathrm{m}}}, \tag{4.46}$$

where $k_{\mathrm{T}}^{\mathrm{f}}$ is the plane strain bulk modulus for the transverse isotropic fibre and $k_{\mathrm{T}}^{\mathrm{m}}$ is the plane strain bulk modulus of the isotropic matrix, defined by (see (2.202))

$$\frac{1}{k_{\mathrm{T}}^{\mathrm{f}}} = \frac{2(1-\nu_{\mathrm{t}}^{\mathrm{f}})}{E_{\mathrm{T}}^{\mathrm{f}}} - \frac{4(\nu_{\mathrm{A}}^{\mathrm{f}})^2}{E_{\mathrm{A}}^{\mathrm{f}}}, \tag{4.47}$$

$$\frac{1}{k_{\mathrm{T}}^{\mathrm{m}}} = \frac{2(1-\nu_{\mathrm{m}})}{E_{\mathrm{m}}} - \frac{4\nu_{\mathrm{m}}^2}{E_{\mathrm{m}}}. \tag{4.48}$$

It should be noted that

$$k_{\mathrm{T}}^{\mathrm{m}} = \frac{\mu_{\mathrm{m}}}{1-2\nu_{\mathrm{m}}} = k_{\mathrm{m}} + \frac{\mu_{\mathrm{m}}}{3}, \tag{4.49}$$

where $k_{\rm m}$ is the bulk modulus of the matrix defined by (see (2.205))

$$k_{\rm m} = \frac{E_{\rm m}}{3(1-2\nu_{\rm m})} = \frac{2\mu_{\rm m}(1+\nu_{\rm m})}{3(1-2\nu_{\rm m})}. \tag{4.50}$$

It now only remains to determine the constant ϕ, which can be specified on applying the remaining condition $(4.28)_5$, because the conditions $(4.28)_2$, $(4.28)_4$ and $(4.28)_6$ are automatically satisfied by (4.25) and (4.33). It follows from (4.26), (4.27), $(4.28)_5$, (4.45) and (4.46) that

$$\phi = \Lambda\left[(\nu_{\rm m} - \nu_{\rm A}^{\rm f})\varepsilon + \left(\alpha_{\rm T}^{\rm f} + \nu_{\rm A}^{\rm f}\alpha_{\rm A}^{\rm f}\right)\Delta T - \left(1+\nu_{\rm m}\right)\alpha_{\rm m}\Delta T + \frac{1}{2}\left(\frac{1}{k_{\rm T}^{\rm f}} - \frac{1}{k_{\rm T}^{\rm m}}\right)\sigma_{\rm T}\right], \tag{4.51}$$

where

$$\frac{1}{\Lambda} = \frac{1}{2}\left(\frac{1}{k_{\rm T}^{\rm f}} + \frac{1}{\mu_{\rm m}}\right). \tag{4.52}$$

The displacement distribution is specified by (4.25)–(4.27), and the corresponding stress distribution is specified by (4.33), (4.39)–(4.44). The stress-strain relations (4.14)–(4.17) and the equilibrium equations (4.20)–(4.22) are satisfied exactly. The boundary and interface conditions (4.28) are also satisfied exactly.

4.3.3 Solution in the Absence of Fibre

For the general loading conditions characterised by the parameters ε, σ_T and ΔT and applied to an infinite sample of matrix in the absence of fibre (filling the entire region of space) the solution is given by

$$u_r^{\rm m} = A_{\rm m} r, \quad u_\theta^{\rm m} = 0, \quad u_z^{\rm m} = \varepsilon, \tag{4.53}$$

$$\sigma_{rr}^{\rm m} = \sigma_{\theta\theta}^{\rm m} = \sigma_{\rm T}, \quad \sigma_{zz}^{\rm m} = E_{\rm m}\left(\varepsilon - \alpha_{\rm m}\Delta T\right) + 2\nu_{\rm m}\sigma_{\rm T}, \tag{4.54}$$

where the parameter A_m is given by (4.46). It should be noted that the axial stress in the matrix given by $(4.54)_2$ is identical to that in the matrix when the fibre is present (see (4.44)). It should also be noted that when the fibre is introduced the radial displacement function is perturbed as a second term appears in $(4.27)_1$ that is inversely proportional to the radial distance r. This additional term will now be considered when applying Maxwell's method of estimating the properties of a fibre-reinforced composite.

4.3.4 Applying Maxwell's Approach to Multiphase Fibre Composites

Owing to the use of the far-field in Maxwell's method for estimating the properties of fibre composites, it is possible to consider multiple fibre reinforcements. Suppose in a cluster of fibres that there are N different types such that for $i = 1, ..., N$ there are n_i fibres of radius a_i. The properties of the fibres of type i are denoted by a superscript i. The cluster is assumed to be homogeneous regarding the distribution of fibres, and leads to transverse isotropic effective properties.

For the case of multiple phases, relation $(4.27)_1$ is generalised to the following form

$$u_r^{\rm m} = A_{\rm m} r + \frac{1}{2\mu_{\rm m} r}\sum_{i=1}^{N} n_i a_i^2 \phi_i, \tag{4.55}$$

where

$$\phi_i = \frac{(\nu_{\rm m} - \nu_{\rm A}^{{\rm f}(i)})\varepsilon + \left(\alpha_{\rm T}^{{\rm f}(i)} + \nu_{\rm A}^{{\rm f}(i)}\alpha_{\rm A}^{{\rm f}(i)}\right)\Delta T - \alpha_{\rm m}\left(1+\nu_{\rm m}\right)\Delta T + \frac{1}{2}\left(\frac{1}{k_{\rm T}^{{\rm f}(i)}} - \frac{1}{k_{\rm T}^{\rm m}}\right)\sigma_{\rm T}}{\frac{1}{2}\left(\frac{1}{k_{\rm T}^{{\rm f}(i)}} + \frac{1}{\mu_{\rm m}}\right)}. \tag{4.56}$$

The cluster of all types of fibre is now considered to be enclosed in a cylinder of radius b such that the volume fraction of fibres of type i within the cylinder of radius b is given by $V_{\rm f}^i = n_i a_i^2 / b^2$. The volume fractions must satisfy relation $(4.1)_2$ namely

$$V_{\rm m} + \sum_{i=1}^{N} V_{\rm f}^i = 1. \tag{4.57}$$

It then follows that (4.27) may be written in the form

$$u_r^{\rm m} = A_{\rm m} r + \frac{b^2}{2\mu_{\rm m} r}\sum_{i=1}^{N} V_{\rm f}^i \phi_i. \tag{4.58}$$

When the result (4.55) is applied to a single fibre of radius b having effective properties corresponding to the multiphase cluster of fibres it follows that

$$u_r^{\rm m} = A_{\rm m} r + \frac{b^2\overline{\phi}}{2\mu_{\rm m} r}, \tag{4.59}$$

where the parameter $\overline{\phi}$ is defined by

$$\overline{\phi} = \frac{(\nu_{\rm m} - \nu_{\rm A}^{\rm eff})\varepsilon + \left(\alpha_{\rm T}^{\rm eff} + \nu_{\rm A}^{\rm eff}\alpha_{\rm A}^{\rm eff}\right)\Delta T - \alpha_{\rm m}\left(1+\nu_{\rm m}\right)\Delta T + \frac{1}{2}\left(\frac{1}{k_{\rm T}^{\rm eff}} - \frac{1}{k_{\rm T}^{\rm m}}\right)\sigma_{\rm T}}{\frac{1}{2}\left(\frac{1}{k_{\rm T}^{\rm eff}} + \frac{1}{\mu_{\rm m}}\right)}. \tag{4.60}$$

The coefficients of the $1/r$ terms in relations (4.59) and (4.58) must be identical so that

$$\overline{\phi} \equiv \sum_{i=1}^{N} V_{\rm f}^i \phi_i. \tag{4.61}$$

It then follows from (4.56) and (4.60) that for any values of the parameters ε, σ_T and ΔT

$$\begin{aligned} &\frac{(\nu_{\rm m} - \nu_{\rm A}^{\rm eff})\varepsilon + \left(\alpha_{\rm T}^{\rm eff} + \nu_{\rm A}^{\rm eff}\alpha_{\rm A}^{\rm eff}\right)\Delta T - \alpha_{\rm m}\left(1+\nu_{\rm m}\right)\Delta T + \frac{1}{2}\left(\frac{1}{k_{\rm T}^{\rm eff}} - \frac{1}{k_{\rm T}^{\rm m}}\right)\sigma_{\rm T}}{\frac{1}{k_{\rm T}^{eff}} + \frac{1}{\mu_m}} = \\ &\sum_{i=1}^{N} V_{\rm f}^i \frac{(\nu_{\rm m} - \nu_{\rm A}^{{\rm f}(i)})\varepsilon + \left(\alpha_{\rm T}^{{\rm f}(i)} + \nu_{\rm A}^{{\rm f}(i)}\alpha_{\rm A}^{{\rm f}(i)}\right)\Delta T - \alpha_{\rm m}\left(1+\nu_{\rm m}\right)\Delta T + \frac{1}{2}\left(\frac{1}{k_{\rm T}^{{\rm f}(i)}} - \frac{1}{k_{\rm T}^{\rm m}}\right)\sigma_{\rm T}}{\frac{1}{k_{\rm T}^{{\rm f}(i)}} + \frac{1}{\mu_{\rm m}}}. \end{aligned} \tag{4.62}$$

Clearly the following relations must be satisfied

$$\sum_{i=1}^{N} V_{\rm f}^i \frac{\frac{1}{k_{\rm T}^{{\rm f}(i)}} - \frac{1}{k_{\rm T}^{\rm m}}}{\frac{1}{k_{\rm T}^{{\rm f}(i)}} + \frac{1}{\mu_{\rm m}}} = \frac{\frac{1}{k_{\rm T}^{\rm eff}} - \frac{1}{k_{\rm T}^{\rm m}}}{\frac{1}{k_{\rm T}^{\rm eff}} + \frac{1}{\mu_{\rm m}}}, \tag{4.63}$$

$$\sum_{i=1}^{N} V_{\mathrm{f}}^{i} \frac{\nu_{\mathrm{A}}^{\mathrm{f}(i)} - \nu_{\mathrm{m}}}{\dfrac{1}{k_{\mathrm{T}}^{\mathrm{f}(i)}} + \dfrac{1}{\mu_{\mathrm{m}}}} = \frac{\nu_{\mathrm{A}}^{\mathrm{eff}} - \nu_{\mathrm{m}}}{\dfrac{1}{k_{\mathrm{T}}^{\mathrm{eff}}} + \dfrac{1}{\mu_{\mathrm{m}}}}, \tag{4.64}$$

$$\sum_{i=1}^{N} V_{\mathrm{f}}^{i} \frac{\left(\alpha_{\mathrm{T}}^{\mathrm{f}(i)} + \nu_{\mathrm{A}}^{\mathrm{f}(i)} \alpha_{\mathrm{A}}^{\mathrm{f}(i)}\right) - \alpha_{\mathrm{m}}\left(1+\nu_{\mathrm{m}}\right)}{\dfrac{1}{k_{\mathrm{T}}^{\mathrm{f}(i)}} + \dfrac{1}{\mu_{\mathrm{m}}}} = \frac{\left(\alpha_{\mathrm{T}}^{\mathrm{eff}} + \nu_{\mathrm{A}}^{\mathrm{eff}} \alpha_{\mathrm{A}}^{\mathrm{eff}}\right) - \alpha_{\mathrm{m}}\left(1+\nu_{\mathrm{m}}\right)}{\dfrac{1}{k_{\mathrm{T}}^{\mathrm{eff}}} + \dfrac{1}{\mu_{\mathrm{m}}}}. \tag{4.65}$$

On using (4.1), these relations are now expressed as

$$\frac{1}{\dfrac{1}{k_{\mathrm{T}}^{\mathrm{eff}}} + \dfrac{1}{\mu_{\mathrm{m}}}} = \sum_{i=1}^{N} \frac{V_{\mathrm{f}}^{i}}{\dfrac{1}{k_{\mathrm{T}}^{\mathrm{f}(i)}} + \dfrac{1}{\mu_{\mathrm{m}}}} + \frac{V_{\mathrm{m}}}{\dfrac{1}{k_{\mathrm{T}}^{\mathrm{m}}} + \dfrac{1}{\mu_{\mathrm{m}}}}, \tag{4.66}$$

$$\frac{\nu_{\mathrm{A}}^{\mathrm{eff}}}{\dfrac{1}{k_{\mathrm{T}}^{\mathrm{eff}}} + \dfrac{1}{\mu_{\mathrm{m}}}} = \sum_{i=1}^{N} \frac{V_{\mathrm{f}}^{i} \nu_{\mathrm{A}}^{\mathrm{f}(i)}}{\dfrac{1}{k_{\mathrm{T}}^{\mathrm{f}(i)}} + \dfrac{1}{\mu_{\mathrm{m}}}} + \frac{V_{\mathrm{m}} \nu_{\mathrm{m}}}{\dfrac{1}{k_{\mathrm{T}}^{\mathrm{m}}} + \dfrac{1}{\mu_{\mathrm{m}}}}, \tag{4.67}$$

$$\frac{\alpha_{\mathrm{T}}^{\mathrm{eff}} + \nu_{\mathrm{A}}^{\mathrm{eff}} \alpha_{\mathrm{A}}^{\mathrm{eff}}}{\dfrac{1}{k_{\mathrm{T}}^{\mathrm{eff}}} + \dfrac{1}{\mu_{\mathrm{m}}}} = \sum_{i=1}^{N} \frac{V_{\mathrm{f}}^{i}\left(\alpha_{\mathrm{T}}^{\mathrm{f}(i)} + \nu_{\mathrm{A}}^{\mathrm{f}(i)} \alpha_{\mathrm{A}}^{\mathrm{f}(i)}\right)}{\dfrac{1}{k_{\mathrm{T}}^{\mathrm{f}(i)}} + \dfrac{1}{\mu_{\mathrm{m}}}} + \frac{V_{\mathrm{m}} \alpha_{\mathrm{m}}\left(1+\nu_{\mathrm{m}}\right)}{\dfrac{1}{k_{\mathrm{T}}^{\mathrm{m}}} + \dfrac{1}{\mu_{\mathrm{m}}}}. \tag{4.68}$$

It should be noted that the quantity $\alpha_T + \nu_A \alpha_A$ is the transverse thermal expansion coefficient for plane strain conditions such that $u_z = 0$. On using (4.1), relation (4.66) may also be expressed as

$$\frac{1}{k_{\mathrm{T}}^{\mathrm{eff}} + \mu_{\mathrm{m}}} = \sum_{i=1}^{N} \frac{V_{\mathrm{f}}^{i}}{k_{\mathrm{T}}^{\mathrm{f}(i)} + \mu_{\mathrm{m}}} + \frac{V_{\mathrm{m}}}{k_{\mathrm{T}}^{\mathrm{m}} + \mu_{\mathrm{m}}}. \tag{4.69}$$

On using (4.1) and (4.66), relations (4.67) and (4.68) may be written as

$$\nu_{\mathrm{A}}^{\mathrm{eff}} = \left(\sum_{i=1}^{N} \frac{V_{\mathrm{f}}^{i} k_{\mathrm{T}}^{\mathrm{f}(i)} \nu_{\mathrm{A}}^{\mathrm{f}(i)}}{k_{\mathrm{T}}^{\mathrm{f}(i)} + \mu_{\mathrm{m}}} + \frac{V_{\mathrm{m}} k_{\mathrm{T}}^{\mathrm{m}} \nu_{\mathrm{m}}}{k_{\mathrm{T}}^{\mathrm{m}} + \mu_{\mathrm{m}}}\right) \bigg/ \left(\sum_{i=1}^{N} \frac{V_{\mathrm{f}}^{i} k_{\mathrm{T}}^{\mathrm{f}(i)}}{k_{\mathrm{T}}^{\mathrm{f}(i)} + \mu_{\mathrm{m}}} + \frac{V_{\mathrm{m}} k_{\mathrm{T}}^{\mathrm{m}}}{k_{\mathrm{T}}^{\mathrm{m}} + \mu_{\mathrm{m}}}\right), \tag{4.70}$$

$$\alpha_{\mathrm{T}}^{\mathrm{eff}} + \nu_{\mathrm{A}}^{\mathrm{eff}} \alpha_{\mathrm{A}}^{\mathrm{eff}} = \frac{\displaystyle\sum_{i=1}^{N} \frac{V_{\mathrm{f}}^{i} k_{\mathrm{T}}^{\mathrm{f}(i)}\left(\alpha_{\mathrm{T}}^{\mathrm{f}(i)} + \nu_{\mathrm{A}}^{\mathrm{f}(i)} \alpha_{\mathrm{A}}^{\mathrm{f}(i)}\right)}{k_{\mathrm{T}}^{\mathrm{f}(i)} + \mu_{\mathrm{m}}} + \frac{V_{\mathrm{m}} k_{\mathrm{T}}^{\mathrm{m}} \alpha_{\mathrm{m}}\left(1+\nu_{\mathrm{m}}\right)}{k_{\mathrm{T}}^{\mathrm{m}} + \mu_{\mathrm{m}}}}{\displaystyle\sum_{i=1}^{N} \frac{V_{\mathrm{f}}^{\mathrm{f}(i)} k_{\mathrm{T}}^{\mathrm{f}(i)}}{k_{\mathrm{T}}^{\mathrm{f}(i)} + \mu_{\mathrm{m}}} + \frac{V_{\mathrm{m}} k_{\mathrm{T}}^{\mathrm{m}}}{k_{\mathrm{T}}^{\mathrm{m}} + \mu_{\mathrm{m}}}}. \tag{4.71}$$

4.4 Axial Shear of Anisotropic Fibres

The following analysis applies to an isolated cylindrical fibre perfectly bonded to an infinite matrix, subject to a uniform temperature change ΔT and to a uniform axial shear stress. The fibre and matrix are regarded as transverse isotropic solids so that the stress-strain relations are given by (4.14)–(4.17).

4.4.1 Solution for an Isolated Fibre Perfectly Bonded to the Matrix

A solution is now sought of the following form

$$\begin{aligned} u_r^{\rm f} &\equiv 0, \quad u_\theta^{\rm f} \equiv 0, \quad u_z^{\rm f} \equiv Ar\cos\theta\,, \\ u_r^{\rm m} &\equiv 0, \quad u_\theta^{\rm m} \equiv 0, \quad u_z^{\rm m} \equiv \left(\alpha r + \frac{\beta}{r}\right)\cos\theta\,, \end{aligned} \tag{4.72}$$

where A, α and β are constants to be determined. On differentiating the displacement field using the strain-displacement relations (2.139), it follows that

$$\begin{aligned} \varepsilon_{rr}^{\rm f} &= \frac{\partial u_r^{\rm f}}{\partial r} = 0\,, \; \varepsilon_{\theta\theta}^{\rm f} = \frac{1}{r}\frac{\partial u_\theta^{\rm f}}{\partial \theta} + \frac{u_r^{\rm f}}{r} = 0\,, \; \varepsilon_{zz}^{\rm f} = \frac{\partial u_z^{\rm f}}{\partial z} = 0\,, \\ \varepsilon_{rz}^{\rm f} &\equiv \frac{1}{2}\left(\frac{\partial u_r^{\rm f}}{\partial z} + \frac{\partial u_z^{\rm f}}{\partial r}\right) = \frac{1}{2}A\cos\theta\,, \\ \varepsilon_{\theta z}^{\rm f} &\equiv \frac{1}{2}\left(\frac{\partial u_\theta^{\rm f}}{\partial z} + \frac{1}{r}\frac{\partial u_z^{\rm f}}{\partial \theta}\right) = -\frac{1}{2}A\sin\theta\,, \\ \varepsilon_{r\theta}^{f} &\equiv \frac{1}{2}\left(\frac{1}{r}\frac{\partial u_r^{\rm f}}{\partial \theta} + \frac{\partial u_\theta^{\rm f}}{\partial r} - \frac{u_\theta^{\rm f}}{r}\right) = 0\,, \end{aligned} \tag{4.73}$$

$$\begin{aligned} \varepsilon_{rr}^{\rm m} &= \frac{\partial u_r^{\rm m}}{\partial r} = 0\,, \; \varepsilon_{\theta\theta}^{\rm m} = \frac{1}{r}\frac{\partial u_\theta^{\rm m}}{\partial \theta} + \frac{u_r^{\rm m}}{r} = 0\,, \; \varepsilon_{zz}^{\rm m} = \frac{\partial u_z^{\rm m}}{\partial z} = 0\,, \\ \varepsilon_{rz}^{\rm m} &\equiv \frac{1}{2}\left(\frac{\partial u_r^{\rm m}}{\partial z} + \frac{\partial u_z^{\rm m}}{\partial r}\right) = \frac{1}{2}\left(\alpha - \frac{\beta}{r^2}\right)\cos\theta\,, \\ \varepsilon_{\theta z}^{\rm m} &\equiv \frac{1}{2}\left(\frac{\partial u_\theta^{\rm m}}{\partial z} + \frac{1}{r}\frac{\partial u_z^{\rm m}}{\partial \theta}\right) = -\frac{1}{2}\left(\alpha + \frac{\beta}{r^2}\right)\sin\theta\,, \\ \varepsilon_{r\theta}^{\rm m} &\equiv \frac{1}{2}\left(\frac{1}{r}\frac{\partial u_r^{\rm m}}{\partial \theta} + \frac{\partial u_\theta^{\rm m}}{\partial r} - \frac{u_\theta^{\rm m}}{r}\right) = 0\,. \end{aligned} \tag{4.74}$$

It follows directly from (4.14)–(4.17), (4.73) and (4.74) that

$$\begin{aligned} \sigma_{rr}^{\rm f} &= \sigma_{\theta\theta}^{\rm f} = \sigma_{zz}^{\rm f} = \sigma_{r\theta}^{\rm f} \equiv 0\,, \\ \sigma_{rr}^{\rm m} &= \sigma_{\theta\theta}^{\rm m} = \sigma_{zz}^{\rm m} = \sigma_{r\theta}^{\rm m} \equiv 0\,. \end{aligned} \tag{4.75}$$

In addition, it follows from (4.17), (4.73) and (4.74) that

$$\begin{aligned} \sigma_{rz}^{\rm f} &= 2\mu_{\rm A}^{\rm f}\varepsilon_{rz}^{\rm f} = \mu_{\rm A}^{\rm f}A\cos\theta\,, & \sigma_{\theta z}^{\rm f} &= 2\mu_{\rm A}^{\rm f}\varepsilon_{\theta z}^{\rm f} = -\,\mu_{\rm A}^{\rm f}A\sin\theta\,, \\ \sigma_{rz}^{\rm m} &= 2\mu_{\rm m}\varepsilon_{rz}^{\rm m} = \mu_{\rm m}\left(\alpha - \frac{\beta}{r^2}\right)\cos\theta\,, & \sigma_{\theta z}^{\rm m} &= 2\mu_{\rm m}\varepsilon_{\theta z}^{\rm m} = -\,\mu_{\rm m}\left(\alpha + \frac{\beta}{r^2}\right)\sin\theta\,. \end{aligned} \tag{4.76}$$

It is easily shown that the stress field automatically satisfies the equilibrium equations (4.20)–(4.22) for any values of the parameters A, α and β.

The following continuity conditions must be satisfied at the interface $r = a$ between fibre and matrix

$$\sigma_{rz}^{\rm m}(a,z) = \sigma_{rz}^{\rm f}(a,z), \tag{4.77}$$

$$u_z^{\rm m}(a,z) = u_z^{\rm f}(a,z). \tag{4.78}$$

It then follows from (4.76) that

$$\mu_{\rm A}^{\rm f}A = \mu_{\rm m}\left(\alpha - \frac{\beta}{a^2}\right), \tag{4.79}$$

and from (4.72) that

$$A = \alpha + \frac{\beta}{a^2}. \tag{4.80}$$

The substitution of (4.80) into (4.79) then leads to

$$\left.\begin{aligned} \left(\mu_A^f + \mu_m\right)\frac{\beta}{a^2} &= \left(\mu_m - \mu_A^f\right)\alpha\,, \\ A &= \frac{2\mu_m}{\mu_A^f + \mu_m}\,\alpha\,. \end{aligned}\right\} \tag{4.81}$$

4.4.2 Solution in the Absence of Fibre

For loading conditions characterised by the shear stress τ applied to an infinite sample of matrix in the absence of fibre (filling the entire region of space), the solution is given by

$$u_r^m \equiv 0, \quad u_\theta^m \equiv 0, \quad u_z^m \equiv \frac{1}{2}\tau\, r\cos\theta, \tag{4.82}$$

$$\sigma_{rz}^m = 2\mu_m \varepsilon_{rz}^m = \tau\cos\theta\,, \qquad \sigma_{\theta z}^m = 2\mu_m\varepsilon_{\theta z}^m = -\tau\sin\theta. \tag{4.83}$$

A comparison of (4.72) and (4.82) with (4.77) and (4.83) indicates that the identification $\alpha = \tau / 2$ can be made. It then follows from (4.81) that

$$A = \frac{\mu_m}{\mu_A^f + \mu_m}\,\tau\,, \qquad \frac{\beta}{a^2} = \frac{1}{2}\frac{\mu_m - \mu_A^f}{\mu_A^f + \mu_m}\,\tau. \tag{4.84}$$

Substitution into (4.72) leads to the following expression for the displacement component u_z:

$$u_z^m \equiv \frac{1}{2}\left(r + \frac{\mu_m - \mu_A^f}{\mu_A^f + \mu_m}\frac{a^2}{r}\right)\tau\cos\theta. \tag{4.85}$$

It should be noted that when fibre is introduced, the displacement components are perturbed as a second term appears in expression (4.85) for the matrix displacement that is inversely proportional to the radial distance r. This additional term will now be considered when applying Maxwell's method of estimating the effective properties of a fibre-reinforced composite.

4.4.3 Applying Maxwell's Approach to Multiphase Fibre Composites

Owing to the use of the far-field in Maxwell's method for estimating the properties of fibre composites, it is possible to consider multiple fibre reinforcements. Suppose in a cluster of fibres that there are N different types such that for $i = 1, \ldots, N$, there are n_i fibres of radius a_i. The properties of the fibres of type i are denoted by a superscript i. The cluster is assumed to be homogeneous regarding the distribution of fibres, and this leads to transverse isotropic effective properties.

For the case of multiple phases, relation (4.85) is generalised to the following form:

$$u_z^m = \frac{1}{2}\tau\cos\theta\, r + \frac{1}{2r}\tau\cos\theta\sum_{i=1}^{N} n_i a_i^2\,\frac{\mu_m - \mu_A^{f(i)}}{\mu_A^{f(i)} + \mu_m}. \tag{4.86}$$

When this result is applied to a single fibre of radius b having effective properties corresponding to the multiphase cluster of fibres, it follows that

$$u_z^{\mathrm{m}} = \frac{1}{2}\tau\cos\theta\, r + \frac{1}{2r}\tau\cos\theta\, b^2 \frac{\mu_{\mathrm{m}} - \mu_{\mathrm{A}}^{\mathrm{eff}}}{\mu_{\mathrm{A}}^{\mathrm{eff}} + \mu_{\mathrm{m}}}. \tag{4.87}$$

The cluster of all types of fibre is now considered to be enclosed in a cylinder of radius b such that the volume fraction of fibres of type i within the cylinder of radius b is given by $V_f^i = n_i a_i^2 / b^2$. The volume fractions must satisfy the relation

$$V_{\mathrm{m}} + \sum_{i=1}^{N} V_f^i = 1. \tag{4.88}$$

It then follows that (4.86) may be written in the form

$$u_z^{\mathrm{m}} = \frac{1}{2}\tau\cos\theta\, r + \frac{1}{2r}\tau\cos\theta\, b^2 \sum_{i=1}^{N} V_{\mathrm{f}}^i \frac{\mu_{\mathrm{m}} - \mu_{\mathrm{A}}^{\mathrm{f}(i)}}{\mu_{\mathrm{A}}^{\mathrm{f}(i)} + \mu_{\mathrm{m}}}. \tag{4.89}$$

The coefficients of the $1/r$ terms in relations (4.87) and (4.89) must be identical so that

$$\sum_{i=1}^{N} V_{\mathrm{f}}^i \frac{\mu_{\mathrm{A}}^{\mathrm{f}(i)} - \mu_{\mathrm{m}}}{\mu_{\mathrm{A}}^{\mathrm{f}(i)} + \mu_{\mathrm{m}}} = \frac{\mu_{\mathrm{A}}^{\mathrm{eff}} - \mu_{\mathrm{m}}}{\mu_{\mathrm{A}}^{\mathrm{eff}} + \mu_{\mathrm{m}}}. \tag{4.90}$$

It then follows on using (4.1) that the effective axial shear modulus for the multiphase composite may be written as

$$\frac{1}{\mu_{\mathrm{A}}^{\mathrm{eff}} + \mu_{\mathrm{m}}} = \sum_{i=1}^{N} \frac{V_{\mathrm{f}}^i}{\mu_{\mathrm{A}}^{\mathrm{f}(i)} + \mu_{\mathrm{m}}} + \frac{V_{\mathrm{m}}}{\mu_{\mathrm{m}} + \mu_{\mathrm{m}}}. \tag{4.91}$$

4.5 Transverse Shear of Multiphase Fibre Composites

Consider a cluster of n cylindrical fibres of the same radius a embedded in an infinite matrix having different properties. The cluster is just enclosed by a cylinder of radius b and the fibre distribution is sufficiently homogeneous for it to lead to transverse isotropic properties for the composite formed by the cluster of fibres and the matrix lying within this cylinder. If the fibre volume fraction of the composite is denoted by V_{f} then

$$V_{\mathrm{f}} = \frac{na^2}{b^2} = 1 - V_{\mathrm{m}}, \tag{4.92}$$

where V_{m} is the corresponding volume fraction of the matrix.

First, a single cylindrical fibre of radius a is placed in an infinite matrix and the origin of cylindrical polar coordinates (r, θ, z) is taken on the axis of the fibre. The system is then subject only to a transverse shear stress applied at infinity. The temperature change ΔT from the stress-free temperature, where the stresses and strains in fibre and matrix are everywhere zero, is also assumed to be everywhere zero.

4.5.1 Representation for Displacement Strain and Stress Distributions

Consider the displacement field having the form

$$\begin{aligned} u_r &= -\left(Ar + \frac{B}{r} + Cr^3 + \frac{D}{r^3}\right)\cos 2\theta\,, \qquad u_z = 0\,, \\ u_\theta &= \left(Ar + \frac{\mu_t}{k_T+\mu_t}\frac{B}{r} + \frac{2k_T+\mu_t}{k_T-\mu_t}Cr^3 - \frac{D}{r^3}\right)\sin 2\theta\,, \end{aligned} \tag{4.93}$$

where k_T and μ_t are the transverse bulk and shear moduli, respectively.

On using the strain–displacement relations (2.142), the corresponding strain field is given by

$$\begin{aligned} \varepsilon_{rr} &\equiv \frac{\partial u_r}{\partial r} = -\left(A - \frac{B}{r^2} + 3Cr^2 - \frac{3D}{r^4}\right)\cos 2\theta\,, \\ \varepsilon_{\theta\theta} &\equiv \frac{1}{r}\frac{\partial u_\theta}{\partial \theta} + \frac{u_r}{r} = \left(A - \frac{k_T-\mu_t}{k_T+\mu_t}\frac{B}{r^2} + \frac{k_T+\mu_t}{k_T-\mu_t}3Cr^2 - \frac{3D}{r^4}\right)\cos 2\theta\,, \\ \varepsilon_{r\theta} &\equiv \frac{1}{2}\left(\frac{1}{r}\frac{\partial u_r}{\partial \theta} + \frac{\partial u_\theta}{\partial r} - \frac{u_\theta}{r}\right) = \left(A + \frac{k_T}{k_T+\mu_t}\frac{B}{r^2} + \frac{3k_T}{k_T-\mu_t}Cr^2 + \frac{3D}{r^4}\right)\sin 2\theta\,, \\ \varepsilon_{zz} &\equiv \frac{\partial u_z}{\partial z} = 0\,, \quad \varepsilon_{rz} \equiv \frac{1}{2}\left(\frac{\partial u_r}{\partial z} + \frac{\partial u_z}{\partial r}\right) = 0\,, \quad \varepsilon_{\theta z} \equiv \frac{1}{2}\left(\frac{\partial u_\theta}{\partial z} + \frac{1}{r}\frac{\partial u_z}{\partial \theta}\right) = 0\,. \end{aligned} \tag{4.94}$$

When $\Delta T = 0$, the stress-strain relations (4.14)–(4.17) are written as

$$\varepsilon_{rr} = \frac{\sigma_{rr}}{E_T} - \frac{\nu_t}{E_T}\sigma_{\theta\theta} - \frac{\nu_A}{E_A}\sigma_{zz}, \tag{4.95}$$

$$\varepsilon_{\theta\theta} = -\frac{\nu_t}{E_T}\sigma_{rr} + \frac{\sigma_{\theta\theta}}{E_T} - \frac{\nu_A}{E_A}\sigma_{zz}, \tag{4.96}$$

$$\varepsilon_{zz} = -\frac{\nu_A}{E_A}\sigma_{rr} - \frac{\nu_A}{E_A}\sigma_{\theta\theta} + \frac{\sigma_{zz}}{E_A}, \tag{4.97}$$

$$\varepsilon_{rz} = \frac{\sigma_{rz}}{2\mu_A}, \quad \varepsilon_{\theta z} = \frac{\sigma_{\theta z}}{2\mu_A}, \quad \varepsilon_{r\theta} = \frac{\sigma_{r\theta}}{2\mu_t}, \tag{4.98}$$

where $E_T = 2\mu_t\left(1+\nu_t\right)$. It follows directly from (4.94) and (4.98) that

$$\sigma_{rz} = 0, \qquad \sigma_{\theta z} = 0. \tag{4.99}$$

From (4.94) and (4.97) it is clear that

$$\sigma_{zz} = \nu_A\left(\sigma_{rr} + \sigma_{\theta\theta}\right), \tag{4.100}$$

and on summing (4.95) and (4.96)

$$\varepsilon_{rr} + \varepsilon_{\theta\theta} = \frac{1-\nu_t}{E_T}\left(\sigma_{rr} + \sigma_{\theta\theta}\right) - \frac{2\nu_A}{E_A}\sigma_{zz}. \tag{4.101}$$

On substituting for σ_{zz} using (4.100), it then follows that

$$\varepsilon_{rr} + \varepsilon_{\theta\theta} = \left(\frac{1-\nu_t}{E_T} - \frac{2\nu_A^2}{E_A}\right)\left(\sigma_{rr} + \sigma_{\theta\theta}\right) = \frac{\sigma_{rr} + \sigma_{\theta\theta}}{2k_T}. \tag{4.102}$$

On subtracting (4.95) and (4.96),

$$\varepsilon_{rr} - \varepsilon_{\theta\theta} = \frac{\sigma_{rr} - \sigma_{\theta\theta}}{2\mu_t}. \tag{4.103}$$

It follows from (4.94), on addition and subtraction, that

$$\varepsilon_{rr} + \varepsilon_{\theta\theta} = 2\mu_t \left(\frac{1}{k_T + \mu_t} \frac{B}{r^2} + \frac{1}{k_T - \mu_t} 3Cr^2 \right) \cos 2\theta, \tag{4.104}$$

$$\varepsilon_{rr} - \varepsilon_{\theta\theta} = \left(-2A + \frac{2k_T}{k_T + \mu_t} \frac{B}{r^2} - \frac{2k_T}{k_T - \mu_t} 3Cr^2 + \frac{6D}{r^4} \right) \cos 2\theta. \tag{4.105}$$

Relations (4.102) and (4.104) then assert that

$$\sigma_{rr} + \sigma_{\theta\theta} = 2k_T \left(\varepsilon_{rr} + \varepsilon_{\theta\theta} \right) = 4\mu_t \left(\frac{k_T}{k_T + \mu_t} \frac{B}{r^2} + \frac{k_T}{k_T - \mu_t} 3Cr^2 \right) \cos 2\theta. \tag{4.106}$$

Relations (4.103) and (4.105) assert that

$$\begin{aligned} \sigma_{rr} - \sigma_{\theta\theta} &= 2\mu_t \left(\varepsilon_{rr} - \varepsilon_{\theta\theta} \right) \\ &= 4\mu_t \left(-A + \frac{k_T}{k_T + \mu_t} \frac{B}{r^2} - \frac{k_T}{k_T - \mu_t} 3Cr^2 + \frac{3D}{r^4} \right) \cos 2\theta. \end{aligned} \tag{4.107}$$

The addition and subtraction of (4.106) and (4.107) leads to the results

$$\sigma_{rr} = 2\mu_t \left(-A + \frac{2k_T}{k_T + \mu_t} \frac{B}{r^2} + \frac{3D}{r^4} \right) \cos 2\theta, \tag{4.108}$$

$$\sigma_{\theta\theta} = 2\mu_t \left(A + \frac{6k_T}{k_T - \mu_t} Cr^2 - \frac{3D}{r^4} \right) \cos 2\theta. \tag{4.109}$$

From (4.94) and (4.98), it follows that

$$\sigma_{r\theta} = 2\mu_t \left(A + \frac{k_T}{k_T + \mu_t} \frac{B}{r^2} + \frac{3k_T}{k_T - \mu_t} Cr^2 + \frac{3D}{r^4} \right) \sin 2\theta. \tag{4.110}$$

It is easily shown that the stress field given by relations (4.108)–(4.110) satisfies automatically the following equilibrium equations for any values of the parameters A, B, C and D (see (2.125)–(2.127))

$$\frac{\partial \sigma_{rr}}{\partial r} + \frac{1}{r} \frac{\partial \sigma_{r\theta}}{\partial \theta} + \frac{\sigma_{rr} - \sigma_{\theta\theta}}{r} = 0, \tag{4.111}$$

$$\frac{\partial \sigma_{r\theta}}{\partial r} + \frac{1}{r} \frac{\partial \sigma_{\theta\theta}}{\partial \theta} + \frac{2\sigma_{r\theta}}{r} = 0, \tag{4.112}$$

where use has been made of (4.99) and the fact that σ_{zz} is independent of z.

4.5.2 Stress Field in the Absence of Fibre

It follows from (4.93) and (4.108)–(4.110) that, when the matrix occupies the whole of space and is subject to a transverse shear stress τ at infinity, the resulting displacement and stress field is given by

$$\begin{aligned} u_r &= -\frac{\tau}{2\mu_m} r \cos 2\theta, \qquad u_\theta = \frac{\tau}{2\mu_m} r \sin 2\theta, \\ \sigma_{rr} &= -\tau \cos 2\theta, \quad \sigma_{\theta\theta} = \tau \cos 2\theta, \quad \sigma_{r\theta} = \tau \sin 2\theta. \end{aligned} \tag{4.113}$$

4.5.3 Displacement and Stress Fields in Fibre

The displacement and stress fields within the fibre must be bounded as $r \to 0$ and it follows from (4.93) and (4.108)–(4.110) that

$$u_r^{\rm f} = -\left(A_{\rm f} r + C_{\rm f} r^3\right)\cos 2\theta, \qquad u_z^{\rm f} = 0,$$
$$u_\theta^{\rm f} = \left(A_{\rm f} r + \frac{2k_{\rm T}^{\rm f} + \mu_{\rm t}^{\rm f}}{k_{\rm T}^{\rm f} - \mu_{\rm t}^{\rm f}} C_{\rm f} r^3\right)\sin 2\theta, \tag{4.114}$$

$$\sigma_{rr}^{\rm f} = -2\mu_{\rm t}^{\rm f} A_{\rm f}\cos 2\theta, \tag{4.115}$$

$$\sigma_{\theta\theta}^{\rm f} = 2\mu_{\rm t}^{\rm f}\left(A_{\rm f} + \frac{6k_{\rm T}^{\rm f}}{k_{\rm T}^{\rm f} - \mu_{\rm t}^{\rm f}} C_{\rm f} r^2\right)\cos 2\theta, \tag{4.116}$$

$$\sigma_{r\theta}^{\rm f} = 2\mu_{\rm t}^{\rm f}\left(A_{\rm f} + \frac{3k_{\rm T}^{\rm f}}{k_{\rm T}^{\rm f} - \mu_{\rm t}^{\rm f}} C_f r^2\right)\sin 2\theta. \tag{4.117}$$

4.5.4 Displacement and Stress Fields in Matrix

The displacement and stress fields within the matrix must be bounded as $r \to \infty$ and it follows from (4.93), (4.108)–(4.110) and (4.113) that

$$u_r^{\rm m} = -\left(\frac{\tau}{2\mu_{\rm m}} r + \frac{B_{\rm m}}{r} + \frac{D_{\rm m}}{r^3}\right)\cos 2\theta, \qquad u_z^{\rm m} = 0,$$
$$u_\theta^{\rm m} = \left(\frac{\tau}{2\mu_{\rm m}} r + \frac{\mu_{\rm m}}{k_{\rm T}^{\rm m} + \mu_{\rm m}} \frac{B_{\rm m}}{r} - \frac{D_{\rm m}}{r^3}\right)\sin 2\theta, \tag{4.118}$$

$$\sigma_{rr}^{\rm m} = 2\mu_{\rm m}\left(-\frac{\tau}{2\mu_{\rm m}} + \frac{2k_{\rm T}^{\rm m}}{k_{\rm T}^{\rm m} + \mu_{\rm m}} \frac{B_{\rm m}}{r^2} + \frac{3D_{\rm m}}{r^4}\right)\cos 2\theta, \tag{4.119}$$

$$\sigma_{\theta\theta}^{\rm m} = 2\mu_{\rm m}\left(\frac{\tau}{2\mu_{\rm m}} - \frac{3D_{\rm m}}{r^4}\right)\cos 2\theta, \tag{4.120}$$

$$\sigma_{r\theta}^{\rm m} = 2\mu_{\rm m}\left(\frac{\tau}{2\mu_{\rm m}} + \frac{k_{\rm T}^{\rm m}}{k_{\rm T}^{\rm m} + \mu_{\rm m}} \frac{B_{\rm m}}{r^2} + \frac{3D_{\rm m}}{r^4}\right)\sin 2\theta. \tag{4.121}$$

The unknown coefficients $A_{\rm f}, C_{\rm f}, B_{\rm m}, D_{\rm m}$ are found by imposing continuity conditions at the interface between the isolated fibre and the matrix. The continuity conditions are given by

$$u_r^{\rm f} = u_r^{\rm m} \text{ on } r = a, \tag{4.122}$$

$$u_\theta^{\rm f} = u_\theta^{\rm m} \text{ on } r = a, \tag{4.123}$$

$$\sigma_{rr}^{f} = \sigma_{rr}^{\rm m} \text{ on } r = a, \tag{4.124}$$

$$\sigma_{r\theta}^{\rm f} = \sigma_{r\theta}^{\rm m} \text{ on } r = a. \tag{4.125}$$

On imposing the continuity conditions (4.122)–(4.125), it follows that

$$A_f + C_f a^2 = \frac{\tau}{2\mu_m} + \frac{B_m}{a^2} + \frac{D_m}{a^4},$$
$$A_f + \frac{2k_T^f + \mu_t^f}{k_T^f - \mu_t^f} C_f a^2 = \frac{\tau}{2\mu_m} + \frac{\mu_m}{k_T^m + \mu_m}\frac{B_m}{a^2} - \frac{D_m}{a^4}, \tag{4.126}$$

and that

$$\mu_t^f A_f = \mu_m\left(\frac{\tau}{2\mu_t^m} - \frac{2k_T^m}{k_T^m + \mu_t^m}\frac{B_m}{a^2} - \frac{3D_m}{a^4}\right),$$
$$\mu_t^f\left(A_f + \frac{3k_T^f}{k_T^f - \mu_t^f} C_f a^2\right) = \mu_m\left(\frac{\tau}{2\mu_m} + \frac{k_T^m}{k_T^m + \mu_m}\frac{B_m}{a^2} + \frac{3D_m}{a^4}\right). \tag{4.127}$$

From (4.126), it follows on addition that

$$2A_f + \frac{3k_T^f}{k_T^f - \mu_t^f} C_f a^2 = \frac{\tau}{\mu_m} + \frac{k_T^m + 2\mu_m}{k_T^m + \mu_m}\frac{B_m}{a^2}, \tag{4.128}$$

and on subtraction that

$$C_f = -\frac{k_T^f - \mu_t^f}{k_T^f + 2\mu_t^f}\left(\frac{k_T^m}{k_T^m + \mu_m}\frac{B_m}{a^4} + \frac{2D_m}{a^6}\right). \tag{4.129}$$

The substitution of (4.129) into (4.128) then leads to

$$A_f = \frac{\tau}{2\mu_m} + \frac{k_T^m}{k_T^m + \mu_m}\left(1 + \frac{2\mu_m}{k_T^m} + \frac{3k_T^f}{k_T^f + 2\mu_t^f}\right)\frac{B_m}{2a^2} + \frac{3k_T^f}{k_T^f + 2\mu_t^f}\frac{D_m}{a^4}. \tag{4.130}$$

From (4.127) it follows on addition that

$$A_f = \frac{\mu_m}{2\mu_t^f}\left(\frac{\tau}{\mu_m} - \frac{k_T^m}{k_T^m + \mu_m}\frac{B_m}{a^2}\right) - \frac{3k_T^f}{2\left(k_T^f - \mu_t^f\right)} C_f a^2, \tag{4.131}$$

and on subtraction that

$$C_f = \frac{\mu_m}{\mu_t^f}\frac{k_T^f - \mu_t^f}{k_T^f}\left(\frac{k_T^m}{k_T^m + \mu_m}\frac{B_m}{a^4} + \frac{2D_m}{a^6}\right). \tag{4.132}$$

The substitution of (4.132) into (4.131) then leads to

$$A_f = \frac{\mu_m}{\mu_t^f}\left[\frac{\tau}{2\mu_m} - \frac{2k_T^m}{k_T^m + \mu_m}\frac{B_m}{a^2} - \frac{3D_m}{a^4}\right]. \tag{4.133}$$

It now follows from (4.129) and (4.132) that

$$B_m = -\frac{k_T^m + \mu_m}{k_T^m}\frac{2D_m}{a^2}, \quad C_f = 0, \tag{4.134}$$

and from (4.130), (4.133) and (4.134) that

$$\left(\frac{1}{\mu_m} + \frac{1}{\mu_t^f} + \frac{2}{k_T^m}\right)\frac{B_m}{a^2} = -\frac{k_T^m + \mu_m}{k_T^m}\left(\frac{1}{\mu_m} - \frac{1}{\mu_t^f}\right)\frac{\tau}{\mu_m}, \quad C_f = 0, \tag{4.135}$$

$$\left(\frac{1}{\mu_m} + \frac{1}{\mu_t^f} + \frac{2}{k_T^m}\right)\frac{D_m}{a^4} = \left(\frac{1}{\mu_m} - \frac{1}{\mu_t^f}\right)\frac{\tau}{2\mu_m}. \tag{4.136}$$

As $C_f = 0$, it follows from (4.115)–(4.117) that the stress and strain fields are uniform within the fibre. From (4.133) and (4.134)

$$A_f = \frac{\mu_m}{\mu_t^f}\left(\frac{\tau}{2\mu_m} + \frac{D_m}{a^4}\right). \tag{4.137}$$

On substituting (4.136) into (4.137) to eliminate D_m, it can be shown that

$$A_f = \frac{\mu_T^f k_T^m + \mu_t^f \mu_m}{\mu_t^f k_T^m + k_T^m \mu_m + 2\mu_t^f \mu_m}\frac{\tau}{\mu_t^f}. \tag{4.138}$$

It should be noted that the displacement and stress fields in the fibre and the matrix can now be calculated. It is clear from (4.118)–(4.121) that at large distances from the fibre the perturbations of the displacement and stress fields arising from the presence of the fibre are characterised by the values of the parameters B_m and D_m which are related according to relation (4.134) and depend on fibre properties. It is also clear that the far-field is insensitive to the actual location of the fibre. This means that a cluster of weakly interacting fibres can easily be considered, and this is the basis of Maxwell's method, which is now described.

4.5.5 Applying Maxwell's Approach to Multiphase Fibre Composites

Owing to the use of the far-field in Maxwell's method for estimating the properties of fibre composites, it is possible to consider multiple fibre reinforcements. Suppose in a cluster of fibres that there are N different types such that for $i = 1, \ldots, N$, there are n_i fibres of radius a_i. The properties of the fibres of type i are denoted by a superscript i. The cluster is assumed to be homogeneous regarding the distribution of fibres, and leads to transverse isotropic effective properties.

Consider the following asymptotic form of radial displacement field in the matrix as $r \to \infty$ that is derived from (4.118)

$$u_r^m = -\frac{\tau}{2\mu_m} r\cos 2\theta - \frac{k_T^m + \mu_m}{k_T^m \mu_m}\frac{a^2}{r}\frac{\dfrac{1}{\mu_t^f} - \dfrac{1}{\mu_m}}{\dfrac{1}{\mu_t^f} + \dfrac{1}{\mu_m} + \dfrac{2}{k_T^m}}\tau\cos 2\theta. \tag{4.139}$$

For the case of multiple phases, relation (4.139) is generalised to the following form

$$u_r^m = -\frac{\tau}{2\mu_m} r\cos 2\theta - \frac{k_T^m + \mu_m}{k_T^m \mu_m}\frac{1}{r}\tau\cos 2\theta \sum_{i=1}^{N} n_i a_i^2 \frac{\dfrac{1}{\mu_t^{f(i)}} - \dfrac{1}{\mu_m}}{\dfrac{1}{\mu_t^{f(i)}} + \dfrac{1}{\mu_m} + \dfrac{2}{k_T^m}}. \tag{4.140}$$

When the result (4.140) is applied to a single fibre of radius b having effective properties corresponding to the multiphase cluster of fibres, it follows that

$$u_r^m = -\frac{\tau}{2\mu_m} r\cos 2\theta - \frac{k_T^m + \mu_m}{k_T^m \mu_m}\frac{1}{r}\tau\cos 2\theta\, b^2 \frac{\dfrac{1}{\mu_t^{eff}} - \dfrac{1}{\mu_m}}{\dfrac{1}{\mu_t^{eff}} + \dfrac{1}{\mu_m} + \dfrac{2}{k_T^m}}. \tag{4.141}$$

The cluster of all types of fibre is now considered to be enclosed in a cylinder of radius b such that the volume fraction of fibres of type i within the cylinder of radius b is given by $V_f^i = n_i a_i^2 / b^2$. The volume fractions must satisfy the relation

$$V_m + \sum_{i=1}^{N} V_f^i = 1. \tag{4.142}$$

It then follows that (4.140) may be written in the form

$$u_r^m = -\frac{\tau}{2\mu_m} r\cos 2\theta - \frac{k_T^m + \mu_m}{k_T^m \mu_m} \frac{1}{r} \tau \cos 2\theta \, b^2 \sum_{i=1}^{N} V_f^i \frac{\frac{1}{\mu_t^{f(i)}} - \frac{1}{\mu_m}}{\frac{1}{\mu_t^{f(i)}} + \frac{1}{\mu_m} + \frac{2}{k_T^m}}. \tag{4.143}$$

The coefficients of the $1/r$ terms in relations (4.141) and (4.143) must be identical so that

$$\sum_{i=1}^{N} V_f^i \frac{\frac{1}{\mu_t^{f(i)}} - \frac{1}{\mu_m}}{\frac{1}{\mu_t^{f(i)}} + \frac{1}{\mu_m} + \frac{2}{k_T^m}} = \frac{\frac{1}{\mu_t^{eff}} - \frac{1}{\mu_m}}{\frac{1}{\mu_t^{eff}} + \frac{1}{\mu_m} + \frac{2}{k_T^m}}. \tag{4.144}$$

It then follows that the effective transverse shear modulus for the multiphase composite is given by

$$\frac{1}{\frac{1}{\mu_t^{eff}} + \frac{1}{\mu_m^*}} = \sum_{i=1}^{N} \frac{V_f^i}{\frac{1}{\mu_t^{f(i)}} + \frac{1}{\mu_m^*}} + \frac{V_m}{\frac{1}{\mu_m} + \frac{1}{\mu_m^*}}, \tag{4.145}$$

where

$$\mu_m^* = \frac{k_T^m \mu_m}{k_T^m + 2\mu_m}. \tag{4.146}$$

On using (4.1), the result (4.145) may also be written in the form

$$\frac{1}{\mu_t^{eff} + \mu_m^*} = \sum_{i=1}^{N} \frac{V_f^i}{\mu_t^{f(i)} + \mu_m^*} + \frac{V_m}{\mu_m + \mu_m^*}. \tag{4.147}$$

4.6 Other Effective Elastic Properties for Multiphase Fibre-reinforced Composites

Four independent effective elastic properties can now be estimated using relations (4.69), (4.67), (4.91) and (4.147), namely, ν_A^{eff}, k_T^{eff}, μ_A^{eff}, μ_t^{eff}. It is clear that Maxwell's methodology has not provided an expression for the axial modulus E_A^{eff} of a multiphase unidirectionally fibre-reinforced composite. This problem has, however, been overcome [6] by considering a special case of aligned spheroidal inclusions (see Chapter 15 for details and (15.100)) where it has been shown that the effective axial Young's modulus E_A^{eff} may be obtained from the following formula

$$E_A^{eff} + \frac{4k_T^{eff}\left(\nu_A^{eff}\right)^2 \mu_m}{k_T^{eff} + \mu_m} = \sum_{i=1}^{N} V_f^i \left(E_A^{f(i)} + \frac{4k_T^{f(i)}\left(\nu_A^{f(i)}\right)^2 \mu_m}{k_T^{f(i)} + \mu_m} \right) + V_m \left(E_m + \frac{4k_T^m \nu_m^2 \mu_m}{k_T^m + \mu_m} \right), \tag{4.148}$$

where

$$k_{\mathrm{T}}^{\mathrm{m}} = k_{\mathrm{m}} + \tfrac{1}{3}\mu_{\mathrm{m}}, \tag{4.149}$$

and where values of $\nu_{\mathrm{A}}^{\mathrm{eff}}$ and $k_{\mathrm{T}}^{\mathrm{eff}}$ have already been determined. The transverse Young's modulus $E_{\mathrm{T}}^{\mathrm{eff}}$ and transverse Poisson's ratio $\nu_{\mathrm{t}}^{\mathrm{eff}}$ can be estimated by making use of the following relations, corresponding to (4.18) and (4.47),

$$E_{\mathrm{T}}^{\mathrm{eff}} = 2\mu_{\mathrm{t}}^{\mathrm{eff}}\left(1+\nu_{\mathrm{t}}^{\mathrm{eff}}\right), \quad \frac{1}{k_{\mathrm{T}}^{\mathrm{eff}}} = \frac{2(1-\nu_{\mathrm{t}}^{\mathrm{eff}})}{E_{\mathrm{T}}^{\mathrm{eff}}} - \frac{4(\nu_{\mathrm{A}}^{\mathrm{eff}})^2}{E_{\mathrm{A}}^{\mathrm{eff}}}. \tag{4.150}$$

It follows that

$$\frac{1+\nu_{\mathrm{T}}^{\mathrm{eff}}}{E_{\mathrm{T}}^{\mathrm{eff}}} = \frac{1}{2\mu_{\mathrm{T}}^{\mathrm{eff}}}, \quad \frac{1-\nu_{\mathrm{T}}^{\mathrm{eff}}}{E_{\mathrm{T}}^{\mathrm{eff}}} = \frac{1}{2k_{\mathrm{T}}^{\mathrm{eff}}} + \frac{2(\nu_{\mathrm{A}}^{\mathrm{eff}})^2}{E_{\mathrm{A}}^{\mathrm{eff}}}, \tag{4.151}$$

so that

$$\frac{1}{E_{\mathrm{T}}^{\mathrm{eff}}} = \frac{1}{4\mu_{\mathrm{T}}^{\mathrm{eff}}} + \frac{1}{4k_{\mathrm{T}}^{\mathrm{eff}}} + \frac{(\nu_{\mathrm{A}}^{\mathrm{eff}})^2}{E_{\mathrm{A}}^{\mathrm{eff}}}, \quad \frac{\nu_{\mathrm{T}}^{\mathrm{eff}}}{E_{\mathrm{T}}^{\mathrm{eff}}} = \frac{1}{4\mu_{\mathrm{T}}^{\mathrm{eff}}} - \frac{1}{4k_{\mathrm{T}}^{\mathrm{eff}}} - \frac{(\nu_{\mathrm{A}}^{\mathrm{eff}})^2}{E_{\mathrm{A}}^{\mathrm{eff}}}, \tag{4.152}$$

enabling values of $E_{\mathrm{T}}^{\mathrm{eff}}$ and $\nu_{\mathrm{T}}^{\mathrm{eff}}$ to be determined.

4.7 Relationship between Two-phase and Multiphase Formulae

An interesting question is whether the formulae for the effective properties of multiphase composites can be derived from results that are valid only for two-phase composites. The results (4.10), (4.66)–(4.69), (4.91), (4.147) and (4.152) are mixtures relations of the type

$$P_J = V_{\mathrm{f}} P_J^{\mathrm{f}} + V_{\mathrm{m}} P_J^{\mathrm{m}}, \quad J=1,2,...,7, \tag{4.153}$$

for the following combinations of properties:

$$\begin{gathered} P_1 = \frac{1}{\kappa_{\mathrm{T}} + \kappa_{\mathrm{m}}},\ P_2 = \frac{1}{k_{\mathrm{T}} + \mu_{\mathrm{m}}},\ P_3 = \frac{\nu_{\mathrm{A}}}{\dfrac{1}{k_{\mathrm{T}}} + \dfrac{1}{\mu_{\mathrm{m}}}},\ P_4 = \frac{\alpha_{\mathrm{T}} + \nu_{\mathrm{A}}\alpha_{\mathrm{A}}}{\dfrac{1}{k_{\mathrm{T}}} + \dfrac{1}{\mu_{\mathrm{m}}}}, \\ P_5 = \frac{1}{\mu_{\mathrm{A}} + \mu_{\mathrm{m}}},\ P_6 = \frac{1}{\mu_{\mathrm{t}} + \mu_{\mathrm{m}}^*},\ P_7 = E_{\mathrm{A}} + \frac{4k_{\mathrm{T}}\left(\nu_{\mathrm{A}}\right)^2 \mu_{\mathrm{m}}}{k_{\mathrm{T}} + \mu_{\mathrm{m}}},\ \text{where } \mu_{\mathrm{m}}^* = \frac{k_{\mathrm{T}}^{\mathrm{m}}\mu_{\mathrm{m}}}{k_{\mathrm{T}}^{\mathrm{m}} + 2\mu_{\mathrm{m}}}. \end{gathered} \tag{4.154}$$

The properties P_J are effective properties of the unidirectional composite whereas P_J^{f} and P_J^{m} are the corresponding properties for the fibre and matrix, respectively. As $V_{\mathrm{f}} + V_{\mathrm{m}} = 1$, relation (4.153) may now be written in the form

$$P_J = P_J^{\mathrm{m}} + \left(P_J^{\mathrm{f}} - P_J^{\mathrm{m}}\right)V_f, \quad J=1,2,...,7, \tag{4.155}$$

implying that the effective property P_J of a two-phase composite is the sum of the matrix value P_J^{m} and the product of the property difference $P_J^{\mathrm{f}} - P_J^{\mathrm{m}}$ and the fibre volume fraction V_{f}. This approach can be extended to multiphase composites having N types of fibre reinforcement by generalising (4.155) to the following form:

$$P_J = P_J^{\mathrm{m}} + \sum_{i=1}^{N}\left(P_J^{\mathrm{f}(i)} - P_J^{\mathrm{m}}\right)V_{\mathrm{f}}^i, \quad J=1,2,...,7, \tag{4.156}$$

where $P_J^{f(i)}$ is the value of the property P_J for the fibre of type i and V_f^i is the corresponding fibre volume fraction. As, for the multiphase composite,

$$\sum_{i=1}^{N} V_f^i + V_m = 1, \tag{4.157}$$

relation (4.156) may be expressed in the mixtures form

$$P_J = \sum_{i=1}^{N} V_f^i P_J^{f(i)} + V_m P_J^m, \quad J=1,2,...,7. \tag{4.158}$$

When the result (4.158) is applied to the properties defined in (4.154), relations (4.10), (4.66)–(4.69), (4.91), (4.147) and (4.152) are generated. Thus, most multiphase properties can be derived from corresponding results for two-phase properties.

It should be noted that it has not been possible to derive the axial thermal expansion coefficient for a multiphase unidirectional composite, although the concentric cylinder model of a composite generates this property for a two-phase composite (see Appendix B). The challenge now is to determine the property combination P_8 ($\equiv \alpha_A^{eff}$) that enables the two-phase result (B.52) (see Appendix B) to be extended to multiphase composites using a relation of the form (4.158). Following inspection of the multiphase results so far obtained, one might expect that the required mixtures relation for a two-phase composite might have the following form:

$$\begin{aligned} E_A^{eff}\alpha_A^{eff} + \frac{4\nu_A^{eff}(\alpha_T^{eff}+\nu_A^{eff}\alpha_A^{eff})}{\dfrac{1}{k_T^{eff}}+\dfrac{1}{\mu_m}} &= V_f E_A^f \alpha_A^f + V_m E_m \alpha_m \\ &+ V_f \frac{4\nu_A^f(\alpha_T^f+\nu_A^f\alpha_A^f)}{\dfrac{1}{k_T^f}+\dfrac{1}{\mu_m}} + V_m \frac{4\nu_m(\alpha_m+\nu_m\alpha_m)}{\dfrac{1}{k_T^m}+\dfrac{1}{\mu_m}}. \end{aligned} \tag{4.159}$$

It can be shown that the conjectured relation (4.159) for just two phases is an equivalent form of the result (B.52) derived using the concentric cylinder model of a two-phase unidirectional composite. Relation (4.158) can then be used to generate the following corresponding expression which is conjectured to be valid for multiphase composites:

$$\begin{aligned} E_A^{eff}\alpha_A^{eff} + \frac{4\nu_A^{eff}(\alpha_T^{eff}+\nu_A^{eff}\alpha_A^{eff})}{\dfrac{1}{k_T^{eff}}+\dfrac{1}{\mu_m}} &= \sum_{i=1}^{N} V_f^i E_A^{f(i)} \alpha_A^{f(i)} + V_m E_m \alpha_m \\ &+ \sum_{i=1}^{N} V_f^i \frac{4\nu_A^{f(i)}(\alpha_T^{f(i)}+\nu_A^{f(i)}\alpha_A^{f(i)})}{\dfrac{1}{k_T^{f(i)}}+\dfrac{1}{\mu_m}} + V_m \frac{4\nu_m(\alpha_m+\nu_m\alpha_m)}{\dfrac{1}{k_T^m}+\dfrac{1}{\mu_m}}. \end{aligned} \tag{4.160}$$

4.8 Summary of Results for Multiphase Composites

It is useful to bring together the results that have been derived for multiphase fibre-reinforced composites. The values of the effective properties κ_A^{eff}, κ_T^{eff}, E_A^{eff}, E_T^{eff}, ν_A^{eff}, ν_t^{eff}, k_T^{eff}, μ_A^{eff}, μ_t^{eff}, α_A^{eff} and α_T^{eff} of the transverse isotropic unidirectionally fibre-reinforced composite may be obtained from the following set of formulae, obtained from relations (4.10), (4.13), (4.66)–(4.69), (4.91), (4.147), (4.152) and (4.160)

$$\kappa_{\mathrm{A}}^{\mathrm{eff}} = \sum_{i=1}^{N} V_{\mathrm{f}}^{i} \kappa_{\mathrm{A}}^{\mathrm{f}(i)} + V_{\mathrm{m}} \kappa_{\mathrm{m}}, \tag{4.161}$$

$$\frac{1}{\kappa_{\mathrm{T}}^{\mathrm{eff}} + \kappa_{\mathrm{m}}} = \sum_{i=1}^{N} \frac{V_{\mathrm{f}}^{i}}{\kappa_{\mathrm{T}}^{\mathrm{f}(i)} + \kappa_{\mathrm{m}}} + \frac{V_{\mathrm{m}}}{\kappa_{\mathrm{m}} + \kappa_{\mathrm{m}}}, \tag{4.162}$$

$$\frac{1}{\frac{1}{k_{\mathrm{T}}^{\mathrm{eff}}} + \frac{1}{\mu_{\mathrm{m}}}} = \sum_{i=1}^{N} \frac{V_{\mathrm{f}}^{i}}{\frac{1}{k_{\mathrm{T}}^{\mathrm{f}(i)}} + \frac{1}{\mu_{\mathrm{m}}}} + \frac{V_{\mathrm{m}}}{\frac{1}{k_{\mathrm{T}}^{\mathrm{m}}} + \frac{1}{\mu_{\mathrm{m}}}}, \tag{4.163}$$

$$\text{or} \quad \frac{1}{k_{\mathrm{T}}^{\mathrm{eff}} + \mu_{\mathrm{m}}} = \sum_{i=1}^{N} \frac{V_{\mathrm{f}}^{i}}{k_{\mathrm{T}}^{\mathrm{f}(i)} + \mu_{m}} + \frac{V_{\mathrm{m}}}{k_{\mathrm{T}}^{\mathrm{m}} + \mu_{\mathrm{m}}}, \tag{4.164}$$

$$\frac{\nu_{\mathrm{A}}^{\mathrm{eff}}}{\frac{1}{k_{\mathrm{T}}^{\mathrm{eff}}} + \frac{1}{\mu_{\mathrm{m}}}} = \sum_{i=1}^{N} \frac{V_{\mathrm{f}}^{i} \nu_{\mathrm{A}}^{\mathrm{f}(i)}}{\frac{1}{k_{\mathrm{T}}^{\mathrm{f}(i)}} + \frac{1}{\mu_{\mathrm{m}}}} + \frac{V_{\mathrm{m}} \nu_{\mathrm{m}}}{\frac{1}{k_{\mathrm{T}}^{\mathrm{m}}} + \frac{1}{\mu_{\mathrm{m}}}}, \tag{4.165}$$

$$\frac{1}{\mu_{\mathrm{A}}^{\mathrm{eff}} + \mu_{\mathrm{m}}} = \sum_{i=1}^{N} \frac{V_{\mathrm{f}}^{i}}{\mu_{\mathrm{A}}^{\mathrm{f}(i)} + \mu_{\mathrm{m}}} + \frac{V_{\mathrm{m}}}{\mu_{\mathrm{m}} + \mu_{\mathrm{m}}}, \tag{4.166}$$

$$\frac{1}{\mu_{\mathrm{t}}^{\mathrm{eff}} + \mu_{\mathrm{m}}^{*}} = \sum_{i=1}^{N} \frac{V_{\mathrm{f}}^{i}}{\mu_{\mathrm{t}}^{\mathrm{f}(i)} + \mu_{\mathrm{m}}^{*}} + \frac{V_{\mathrm{m}}}{\mu_{\mathrm{m}} + \mu_{\mathrm{m}}^{*}}, \quad \text{with} \quad \mu_{\mathrm{m}}^{*} = \frac{k_{\mathrm{T}}^{\mathrm{m}} \mu_{\mathrm{m}}}{k_{\mathrm{T}}^{\mathrm{m}} + 2\mu_{\mathrm{m}}}, \tag{4.167}$$

$$E_{\mathrm{A}}^{\mathrm{eff}} + \frac{4k_{\mathrm{T}}^{\mathrm{eff}} \left(\nu_{\mathrm{A}}^{\mathrm{eff}}\right)^{2} \mu_{\mathrm{m}}}{k_{\mathrm{T}}^{\mathrm{eff}} + \mu_{\mathrm{m}}} = \sum_{i=1}^{N} V_{\mathrm{f}}^{i} \left(E_{\mathrm{A}}^{\mathrm{f}(i)} + \frac{4k_{\mathrm{T}}^{\mathrm{f}(i)} \left(\nu_{\mathrm{A}}^{\mathrm{f}(i)}\right)^{2} \mu_{\mathrm{m}}}{k_{\mathrm{T}}^{\mathrm{f}(i)} + \mu_{\mathrm{m}}} \right) + V_{\mathrm{m}} \left(E_{\mathrm{m}} + \frac{4k_{\mathrm{T}}^{\mathrm{m}} \nu_{\mathrm{m}}^{2} \mu_{\mathrm{m}}}{k_{\mathrm{T}}^{\mathrm{m}} + \mu_{\mathrm{m}}} \right), \tag{4.168}$$

$$\frac{1}{E_{\mathrm{T}}^{\mathrm{eff}}} = \frac{1}{4\mu_{\mathrm{t}}^{\mathrm{eff}}} + \frac{1}{4k_{\mathrm{T}}^{\mathrm{eff}}} + \frac{(\nu_{\mathrm{A}}^{\mathrm{eff}})^{2}}{E_{\mathrm{A}}^{\mathrm{eff}}}, \quad \frac{\nu_{\mathrm{t}}^{\mathrm{eff}}}{E_{\mathrm{T}}^{\mathrm{eff}}} = \frac{1}{4\mu_{\mathrm{t}}^{\mathrm{eff}}} - \frac{1}{4k_{\mathrm{T}}^{\mathrm{eff}}} - \frac{4(\nu_{\mathrm{A}}^{\mathrm{eff}})^{2}}{E_{\mathrm{A}}^{\mathrm{eff}}}, \tag{4.169}$$

$$E_{\mathrm{A}}^{\mathrm{eff}} \alpha_{\mathrm{A}}^{\mathrm{eff}} + \frac{4\nu_{\mathrm{A}}^{\mathrm{eff}} (\alpha_{\mathrm{T}}^{\mathrm{eff}} + \nu_{\mathrm{A}}^{\mathrm{eff}} \alpha_{\mathrm{A}}^{\mathrm{eff}})}{\frac{1}{k_{\mathrm{T}}^{\mathrm{eff}}} + \frac{1}{\mu_{\mathrm{m}}}} = \sum_{i=1}^{N} V_{\mathrm{f}}^{i} E_{\mathrm{A}}^{\mathrm{f}(i)} \alpha_{\mathrm{A}}^{\mathrm{f}(i)} + V_{\mathrm{m}} E_{\mathrm{m}} \alpha_{\mathrm{m}} + \sum_{i=1}^{N} V_{f}^{i} \frac{4\nu_{\mathrm{A}}^{\mathrm{f}(i)} (\alpha_{\mathrm{T}}^{\mathrm{f}(i)} + \nu_{\mathrm{A}}^{\mathrm{f}(i)} \alpha_{\mathrm{A}}^{\mathrm{f}(i)})}{\frac{1}{k_{\mathrm{T}}^{\mathrm{f}(i)}} + \frac{1}{\mu_{\mathrm{m}}}} + V_{m} \frac{4\nu_{\mathrm{m}} (\alpha_{\mathrm{m}} + \nu_{\mathrm{m}} \alpha_{\mathrm{m}})}{\frac{1}{k_{\mathrm{T}}^{\mathrm{m}}} + \frac{1}{\mu_{\mathrm{m}}}}, \tag{4.170}$$

$$\frac{\alpha_{\mathrm{T}}^{\mathrm{eff}} + \nu_{\mathrm{A}}^{\mathrm{eff}} \alpha_{\mathrm{A}}^{\mathrm{eff}}}{\frac{1}{k_{\mathrm{T}}^{\mathrm{eff}}} + \frac{1}{\mu_{\mathrm{m}}}} = \sum_{i=1}^{N} \frac{V_{\mathrm{f}}^{i} \left(\alpha_{\mathrm{T}}^{\mathrm{f}(i)} + \nu_{\mathrm{A}}^{\mathrm{f}(i)} \alpha_{\mathrm{A}}^{\mathrm{f}(i)}\right)}{\frac{1}{k_{\mathrm{T}}^{\mathrm{f}(i)}} + \frac{1}{\mu_{\mathrm{m}}}} + \frac{V_{\mathrm{m}} \alpha_{\mathrm{m}} \left(1 + \nu_{\mathrm{m}}\right)}{\frac{1}{k_{\mathrm{T}}^{\mathrm{m}}} + \frac{1}{\mu_{\mathrm{m}}}}. \tag{4.171}$$

4.9 Results for Two-phase Fibre-reinforced Composites

For two-phase composites $N = 1$ so that, on writing $V_f^1 = V_f = 1 - V_m$, $\kappa_A^{f(1)} = \kappa_A^f$, $\kappa_T^{f(1)} = \kappa_T^f$, $k_T^{f(1)} = k_T^f$, $\mu_A^{f(1)} = \mu_A^f$, $\mu_t^{f(1)} = \mu_t^f$, $E_A^{f(1)} = E_A^f$, $E_T^{f(1)} = E_T^f$, $\nu_A^{f(1)} = \nu_A^f$, $\nu_t^{f(1)} = \nu_t^f$, $\alpha_A^{f(1)} = \alpha_A^f$ and $\alpha_T^{f(1)} = \alpha_T^f$, the results of Section 4.8 may be written as

$$\kappa_A^{eff} = V_f\,\kappa_A^f + V_m\kappa_m, \tag{4.172}$$

$$\frac{1}{\kappa_T^{eff} + \kappa_m} = \frac{V_f}{\kappa_T^f + \kappa_m} + \frac{V_m}{\kappa_m + \kappa_m}, \tag{4.173}$$

$$\frac{1}{\dfrac{1}{k_T^{eff}} + \dfrac{1}{\mu_m}} = \frac{V_f}{\dfrac{1}{k_T^f} + \dfrac{1}{\mu_m}} + \frac{V_m}{\dfrac{1}{k_T^m} + \dfrac{1}{\mu_m}}, \tag{4.174}$$

$$\text{or}\quad \frac{1}{k_T^{eff} + \mu_m} = \frac{V_f}{k_T^f + \mu_m} + \frac{V_m}{k_T^m + \mu_m}, \tag{4.175}$$

$$\frac{\nu_A^{eff}}{\dfrac{1}{k_T^{eff}} + \dfrac{1}{\mu_m}} = \frac{V_f\,\nu_A^f}{\dfrac{1}{k_T^f} + \dfrac{1}{\mu_m}} + \frac{V_m\nu_m}{\dfrac{1}{k_T^m} + \dfrac{1}{\mu_m}}, \tag{4.176}$$

$$\frac{1}{\mu_A^{eff} + \mu_m} = \frac{V_f}{\mu_A^f + \mu_m} + \frac{V_m}{\mu_m + \mu_m}, \tag{4.177}$$

$$\frac{1}{\mu_t^{eff} + \mu_m^*} = \frac{V_f}{\mu_t^f + \mu_m^*} + \frac{V_m}{\mu_m + \mu_m^*} \quad \text{with} \quad \mu_m^* = \frac{k_T^m \mu_m}{k_T^m + 2\mu_m}, \tag{4.178}$$

$$E_A^{eff} + \frac{4k_T^{eff}\left(\nu_A^{eff}\right)^2 \mu_m}{k_T^{eff} + \mu_m} = V_f\left(E_A^f + \frac{4k_T^f\left(\nu_A^f\right)^2 \mu_m}{k_T^f + \mu_m}\right) + V_m\left(E_m + \frac{4k_T^m \nu_m^2 \mu_m}{k_T^m + \mu_m}\right), \tag{4.179}$$

$$\frac{1}{E_T^{eff}} = \frac{1}{4\mu_t^{eff}} + \frac{1}{4k_T^{eff}} + \frac{(\nu_A^{eff})^2}{E_A^{eff}}, \quad \frac{\nu_t^{eff}}{E_T^{eff}} = \frac{1}{4\mu_t^{eff}} - \frac{1}{4k_T^{eff}} - \frac{4(\nu_A^{eff})^2}{E_A^{eff}}, \tag{4.180}$$

$$\begin{aligned} E_A^{eff}\alpha_A^{eff} + \frac{4\nu_A^{eff}(\alpha_T^{eff} + \nu_A^{eff}\alpha_A^{eff})}{\dfrac{1}{k_T^{eff}} + \dfrac{1}{\mu_m}} &= V_f\,E_A^f\alpha_A^f + V_m E_m\alpha_m \\ &\quad + V_f\,\frac{4\nu_A^f(\alpha_T^f + \nu_A^f\alpha_A^f)}{\dfrac{1}{k_T^f} + \dfrac{1}{\mu_m}} + V_m\frac{4\nu_m(\alpha_m + \nu_m\alpha_m)}{\dfrac{1}{k_T^m} + \dfrac{1}{\mu_m}}, \end{aligned} \tag{4.181}$$

$$\frac{\alpha_T^{eff} + \nu_A^{eff}\alpha_A^{eff}}{\dfrac{1}{k_T^{eff}} + \dfrac{1}{\mu_m}} = \frac{V_f\left(\alpha_T^f + \nu_A^f\alpha_A^f\right)}{\dfrac{1}{k_T^f} + \dfrac{1}{\mu_m}} + \frac{V_m(\alpha_m + \nu_m\alpha_m)}{\dfrac{1}{k_T^m} + \dfrac{1}{\mu_m}}. \tag{4.182}$$

These results (4.173)–(4.182) can also be written in the form of a mixtures estimate plus a correction term as follows

$$\kappa_T^{\text{eff}} = V_f \kappa_T^f + V_m \kappa_m - \frac{\left(\kappa_T^f - \kappa_m\right)^2 V_f V_m}{\left(1+V_f\right)\kappa_m + V_m \kappa_T^f}, \tag{4.183}$$

$$\frac{1}{k_T^{\text{eff}}} = \frac{V_f}{k_T^f} + \frac{V_m}{k_T^m} - \frac{\lambda}{2}\left(\frac{1}{k_T^f} - \frac{1}{k_T^m}\right)^2 V_f V_m, \tag{4.184}$$

$$\nu_A^{\text{eff}} = V_f \nu_A^f + V_m \nu_m - \frac{\lambda}{2}\left(\frac{1}{k_T^m} - \frac{1}{k_T^f}\right)\left(\nu_m - \nu_A^f\right) V_f V_m, \tag{4.185}$$

$$\mu_A^{\text{eff}} = V_f \mu_A^f + V_m \mu_m - \frac{(\mu_A^f - \mu_m)^2 V_f V_m}{\left(1+V_f\right)\mu_m + V_m \mu_A^f}, \tag{4.186}$$

$$\mu_t^{\text{eff}} = V_f \mu_t^f + V_m \mu_m - \frac{\left(\mu_t^f - \mu_m\right)^2 V_f V_m}{V_m \mu_t^f + V_f \mu_m + \dfrac{k_T^m \mu_m}{k_T^m + 2\mu_m}}, \tag{4.187}$$

$$E_A^{\text{eff}} = V_f E_A^f + V_m E_m + 2\lambda\left(\nu_m - \nu_A^f\right)^2 V_f V_m, \tag{4.188}$$

$$\frac{1}{E_T^{\text{eff}}} = \frac{1}{4\mu_t^{\text{eff}}} + \frac{1}{4k_T^{\text{eff}}} + \frac{(\nu_A^{\text{eff}})^2}{E_A^{\text{eff}}}, \quad \frac{\nu_t^{\text{eff}}}{E_T^{\text{eff}}} = \frac{1}{4\mu_t^{\text{eff}}} - \frac{1}{4k_T^{\text{eff}}} - \frac{4(\nu_A^{\text{eff}})^2}{E_A^{\text{eff}}}, \tag{4.189}$$

$$\begin{aligned} E_A^{\text{eff}} \alpha_A^{\text{eff}} &= V_f E_A^f \alpha_A^f + V_m E_m \alpha_m \\ &\quad + 2\lambda\left(\nu_m - \nu_A^f\right)\left(\alpha_m + \nu_m \alpha_m - \alpha_T^f - \nu_A^f \alpha_A^f\right) V_f V_m, \end{aligned} \tag{4.190}$$

$$\begin{aligned} \alpha_T^{\text{eff}} + \nu_A^{\text{eff}} \alpha_A^{\text{eff}} &= V_f(\alpha_T^f + \nu_A^f \alpha_A^f) + V_m(\alpha_m + \nu_m \alpha_m) \\ &\quad + \frac{\lambda}{2}\left(\frac{1}{k_T^f} - \frac{1}{k_T^m}\right)\left(\alpha_m + \nu_m \alpha_m - \alpha_T^f - \nu_A^f \alpha_A^f\right) V_f V_m, \end{aligned} \tag{4.191}$$

where the parameter λ is defined by

$$\frac{2}{\lambda} = \frac{1}{\mu_m} + \frac{V_f}{k_T^m} + \frac{V_m}{k_T^f}. \tag{4.192}$$

4.10 Bounds for Two-phase Fibre-reinforced Composites

Bounds for the effective properties of unidirectional fibre-reinforced composites have been derived using variational techniques by Hashin and Shtrikman [7], and in the review by Hashin [3]. In this section, these bounds are now associated with the results in this chapter derived using Maxwell's methodology. The results of Section 4.8, derived using Maxwell's methodology or using the concentric cylinder model of a composite, correspond exactly to one of the bounds. Furthermore, if one bound is known, then the other is obtained simply by interchanging the properties and volume fractions. It is of interest to determine the condition that specifies whether the extreme value is an upper or lower bound.

4.10.1 Thermal Conductivity

The thermal conductivity bounds are very easily determined from the result (4.183), which is such that the denominator of the third term on the right-hand side is the only term to change value when the properties and volume fractions are interchanged. It then follows that the bounds may be written as

$$V_f\kappa_T^f + V_m\kappa_m - \frac{\left(\kappa_T^f - \kappa_m\right)^2 V_f V_m}{\left(1+V_f\right)\kappa_m + V_m\kappa_T^f} \le \kappa_T^{eff} \le V_f\kappa_T^f + V_m\kappa_m - \frac{\left(\kappa_T^f - \kappa_m\right)^2 V_f V_m}{\left(1+V_m\right)\kappa_T^f + V_f\kappa_m}, \tag{4.193}$$

whenever $\kappa_m \le \kappa_T^f$. Whenever $\kappa_m \ge \kappa_T^f$, the bounds defined by (4.193) should be reversed.

4.10.2 Axial Young's Modulus

The bounds for the axial Young's modulus are given by

$$V_f E_A^f + V_m E_m + \frac{4\left(\nu_A^f - \nu_m\right)^2 V_f V_m}{\dfrac{V_m}{k_T^f} + \dfrac{V_f}{k_T^m} + \dfrac{1}{\mu_m}} \le E_A^{eff} \le V_f E_A^f + V_m E_m + \frac{4\left(\nu_A^f - \nu_m\right)^2 V_f V_m}{\dfrac{V_m}{k_T^f} + \dfrac{V_f}{k_T^m} + \dfrac{1}{\mu_T^f}}, \tag{4.194}$$

which are valid only if $\mu_T^f \ge \mu_m$, and the bounds are reversed if $\mu_T^f \le \mu_m$.

4.10.3 Axial Poisson's Ratio

The bounds for the axial Poisson's ratio are given by

$$V_f\nu_A^f + V_m\nu_m - \frac{\left(\nu_A^f - \nu_m\right)\left(\dfrac{1}{k_T^f} - \dfrac{1}{k_T^m}\right)V_f V_m}{\dfrac{V_m}{k_T^f} + \dfrac{V_f}{k_T^m} + \dfrac{1}{\mu_m}} \le \nu_A^{eff} \le V_f\nu_A^f + V_m\nu_m - \frac{\left(\nu_A^f - \nu_m\right)\left(\dfrac{1}{k_T^f} - \dfrac{1}{k_T^m}\right)V_f V_m}{\dfrac{V_m}{k_T^f} + \dfrac{V_f}{k_T^m} + \dfrac{1}{\mu_t^f}}, \tag{4.195}$$

which are valid only if $\left(\nu_A^f - \nu_m\right)\left(k_T^f - k_T^m\right)\left(\mu_t^f - \mu_m\right) \geq 0$, and the bounds are reversed if $\left(\nu_A^f - \nu_m\right)\left(k_T^f - k_T^m\right)\left(\mu_t^f - \mu_m\right) \leq 0$.

4.10.4 Transverse Bulk Modulus

The bounds for the transverse bulk modulus are given by

$$\frac{V_f}{k_T^f} + \frac{V_m}{k_T^m} - \frac{\left(\frac{1}{k_T^f} - \frac{1}{k_T^m}\right)^2 V_f V_m}{\frac{V_m}{k_T^f} + \frac{V_f}{k_T^m} + \frac{1}{\mu_m}} \leq \frac{1}{k_T^{eff}} \leq \frac{V_f}{k_T^f} + \frac{V_m}{k_T^m} - \frac{\left(\frac{1}{k_T^f} - \frac{1}{k_T^m}\right)^2 V_f V_m}{\frac{V_m}{k_T^f} + \frac{V_f}{k_T^m} + \frac{1}{\mu_t^f}}, \tag{4.196}$$

which are valid only if $\mu_t^f \leq \mu_m$, and the bounds are reversed if $\mu_t^f \geq \mu_m$.

4.10.5 Transverse Shear Modulus

The bounds for transverse shear modulus are given by

$$\begin{aligned} V_f\mu_t^f + V_m\mu_m - \frac{\left(\mu_t^f - \mu_m\right)^2 V_f V_m}{V_m\mu_t^f + V_f\mu_m + \frac{k_T^m\mu_m}{k_T^m + 2\mu_m}} &\leq \mu_t^{eff} \\ \leq V_f\mu_t^f + V_m\mu_m - \frac{\left(\mu_t^f - \mu_m\right)^2 V_f V_m}{V_m\mu_t^f + V_f\mu_m + \frac{k_T^f\mu_t^f}{k_T^f + 2\mu_t^f}}&, \end{aligned} \tag{4.197}$$

which are valid only if

$$\frac{1}{\mu_t^f} + \frac{2}{k_T^f} \leq \frac{1}{\mu_m} + \frac{2}{k_T^m}.$$

The bounds are reversed if

$$\frac{1}{\mu_t^f} + \frac{2}{k_T^f} \geq \frac{1}{\mu_m} + \frac{2}{k_T^m}.$$

4.10.6 Axial Shear Modulus

The bounds for the axial shear modulus are given by

$$\begin{aligned} V_f\mu_A^f + V_m\mu_m - \frac{\left(\mu_A^f - \mu_m\right)^2 V_f V_m}{V_m\mu_A^f + V_f\mu_m + \mu_m} &\leq \mu_A^{eff} \leq \\ V_f\mu_A^f + V_m\mu_m - \frac{\left(\mu_A^f - \mu_m\right)^2 V_f V_m}{V_m\mu_A^f + V_f\mu_m + \mu_A^f}&, \end{aligned} \tag{4.198}$$

which are valid only if $\mu_m \leq \mu_A^f$, and the bounds are reversed if $\mu_m \geq \mu_A^f$.

4.10.7 Axial Thermal Expansion

The bounds for the axial thermal expansion are given by

$$\begin{aligned} V_{\mathrm{f}}E_{\mathrm{A}}^{\mathrm{f}}\alpha_{\mathrm{A}}^{\mathrm{f}} + V_{\mathrm{m}}E_{\mathrm{m}}\alpha_{\mathrm{m}} + \frac{4\left(\nu_{\mathrm{A}}^{\mathrm{f}} - \nu_{\mathrm{m}}\right)\left(\hat{\alpha}_{\mathrm{T}}^{\mathrm{f}} - \hat{\alpha}_{\mathrm{T}}^{\mathrm{m}}\right)V_{\mathrm{f}}V_{\mathrm{m}}}{\dfrac{V_{\mathrm{m}}}{k_{\mathrm{T}}^{\mathrm{f}}} + \dfrac{V_{\mathrm{f}}}{k_{\mathrm{T}}^{\mathrm{m}}} + \dfrac{1}{\mu_{\mathrm{m}}}} &\le E_{\mathrm{A}}^{\mathrm{eff}}\alpha_{\mathrm{A}}^{\mathrm{eff}} \\ \le V_{\mathrm{f}}E_{\mathrm{A}}^{\mathrm{f}}\alpha_{\mathrm{A}}^{\mathrm{f}} + V_{\mathrm{m}}E_{\mathrm{m}}\alpha_{\mathrm{m}} + \frac{4\left(\nu_{\mathrm{A}}^{\mathrm{f}} - \nu_{\mathrm{m}}\right)\left(\hat{\alpha}_{\mathrm{T}}^{\mathrm{f}} - \hat{\alpha}_{\mathrm{T}}^{\mathrm{m}}\right)V_{\mathrm{f}}V_{\mathrm{m}}}{\dfrac{V_{\mathrm{m}}}{k_{\mathrm{T}}^{\mathrm{f}}} + \dfrac{V_{\mathrm{f}}}{k_{\mathrm{T}}^{\mathrm{m}}} + \dfrac{1}{\mu_{\mathrm{t}}^{\mathrm{f}}}}&, \end{aligned} \tag{4.199}$$

where

$$\hat{\alpha}_{\mathrm{T}} = \alpha_{\mathrm{T}} + \nu_{\mathrm{A}}\alpha_{\mathrm{A}}. \tag{4.200}$$

These bounds are valid only if $\left(\nu_{\mathrm{A}}^{\mathrm{f}} - \nu_{\mathrm{m}}\right)\left(\hat{\alpha}_{\mathrm{T}}^{\mathrm{f}} - \hat{\alpha}_{\mathrm{T}}^{\mathrm{m}}\right)\left(\mu_{\mathrm{t}}^{\mathrm{f}} - \mu_{\mathrm{m}}\right) \ge 0$, and the bounds are reversed if $\left(\nu_{\mathrm{A}}^{\mathrm{f}} - \nu_{\mathrm{m}}\right)\left(\hat{\alpha}_{\mathrm{T}}^{\mathrm{f}} - \hat{\alpha}_{\mathrm{T}}^{\mathrm{m}}\right)\left(\mu_{\mathrm{t}}^{\mathrm{f}} - \mu_{\mathrm{m}}\right) \le 0$.

4.10.8 Transverse Thermal Expansion

The bounds for the transverse thermal expansion are given by

$$\begin{aligned} V_{\mathrm{f}}\hat{\alpha}_{\mathrm{T}}^{\mathrm{f}} + V_{\mathrm{m}}\hat{\alpha}_{\mathrm{T}}^{\mathrm{m}} - \frac{\left(\dfrac{1}{k_{\mathrm{T}}^{\mathrm{f}}} - \dfrac{1}{k_{\mathrm{T}}^{\mathrm{m}}}\right)\left(\hat{\alpha}_{\mathrm{T}}^{\mathrm{f}} - \hat{\alpha}_{\mathrm{T}}^{\mathrm{m}}\right)V_{\mathrm{f}}V_{\mathrm{m}}}{\dfrac{V_{\mathrm{m}}}{k_{\mathrm{T}}^{\mathrm{f}}} + \dfrac{V_{\mathrm{f}}}{k_{\mathrm{T}}^{\mathrm{m}}} + \dfrac{1}{\mu_{\mathrm{m}}}} &\le \hat{\alpha}_{\mathrm{T}}^{\mathrm{eff}} \\ \le V_{\mathrm{f}}\hat{\alpha}_{\mathrm{T}}^{\mathrm{f}} + V_{\mathrm{m}}\hat{\alpha}_{\mathrm{T}}^{\mathrm{m}} - \frac{\left(\dfrac{1}{k_{\mathrm{T}}^{\mathrm{f}}} - \dfrac{1}{k_{\mathrm{T}}^{\mathrm{m}}}\right)\left(\hat{\alpha}_{\mathrm{T}}^{\mathrm{f}} - \hat{\alpha}_{\mathrm{T}}^{\mathrm{m}}\right)V_{\mathrm{f}}V_{\mathrm{m}}}{\dfrac{V_{\mathrm{m}}}{k_{\mathrm{T}}^{\mathrm{f}}} + \dfrac{V_{f}}{k_{\mathrm{T}}^{\mathrm{m}}} + \dfrac{1}{\mu_{\mathrm{t}}^{\mathrm{f}}}}&, \end{aligned} \tag{4.201}$$

which are valid only if $\left(k_{\mathrm{T}}^{\mathrm{f}} - k_{\mathrm{T}}^{\mathrm{m}}\right)\left(\hat{\alpha}_{\mathrm{T}}^{\mathrm{f}} - \hat{\alpha}_{\mathrm{T}}^{\mathrm{m}}\right)\left(\mu_{\mathrm{t}}^{\mathrm{f}} - \mu_{\mathrm{m}}\right) \le 0$, and the bounds are reversed if $\left(k_{\mathrm{T}}^{\mathrm{f}} - k_{\mathrm{T}}^{\mathrm{m}}\right)\left(\hat{\alpha}_{\mathrm{T}}^{\mathrm{f}} - \hat{\alpha}_{\mathrm{T}}^{\mathrm{m}}\right)\left(\mu_{\mathrm{t}}^{\mathrm{f}} - \mu_{\mathrm{m}}\right) \ge 0$.

4.11 Comparison of Predictions with Known Results

One of the more remarkable aspects of the results derived in this chapter is that Maxwell's methodology, when applied to two-phase fibre-reinforced composites where the reinforcing cylindrical fibres all have the same size, is that for the case of transverse thermal conductivity, bulk and shear moduli, transverse thermal expansion and transverse thermal conductivity, the effective properties correspond exactly to one of the bounds derived using variational techniques. This bound is also identical to estimates for the properties obtained using the composite cylinders assemblage method. These results are remarkable as one might expect Maxwell's methodology to be accurate only for sufficiently low volume fractions of reinforcement. The reason for this is that the methodology involves the examination of the stress and displacement fields in the matrix at large distances from the assembly of parallel fibres, and assuming that the perturbing effect of each fibre can be approximated by assuming all fibres are located at the same point. The nature of this approximation is such that one can expect that all interactions between fibres in the cluster are neglected, and that the resulting effective properties will be accurate only for low volume fractions. This appears to contradict the finding that the estimates correspond exactly with one of the variational bounds and with the results arising from the use of the composite cylinders assemblage model. The latter techniques are expected to lead to reasonable estimates of effective composite properties for all volume fractions. It should also be noted from (4.183) that $\kappa_{eff} \to \kappa_{\mathrm{f}}$ when $V_{\mathrm{f}} \to 1$. These facts lead to the conclusion that results derived using Maxwell's methodology are not restricted to low fibre volume fractions.

The effective properties of a two-phase composite, derived using Maxwell's methodology, may be expressed in the form of a mixtures estimate plus a correction term, as seen from the results in Section 4.8. It should be noted that the correction term is always proportional to the product $V_{\mathrm{f}}V_{\mathrm{m}}$ of volume fractions, and a term that involves the square of property differences for the case of conductivity, bulk and shear moduli, and the product of the bulk compressibility difference and expansion coefficient difference for the case of thermal expansion. The form of these results is the preferred common form for the effective properties, having the advantage that the conditions governing whether an extreme value is an upper or lower bound are easily determined.

It is useful now to compare predictions of effective composite properties with those in the literature obtained using numerical methods, and which are expected to be accurate. First, the transverse thermal conductivity is considered and the comparison of Maxwell's methodology with results of Perrins et al. [8] is shown in Figure 4.2. The normalised effective transverse conductivity is defined by $\kappa_{\mathrm{T}}^{\mathrm{eff}} / \kappa_{\mathrm{m}}$ and the three materials considered are for isotropic fibres and matrix such that $\kappa_{\mathrm{f}} / \kappa_{\mathrm{m}} = 2, 10, \infty$.

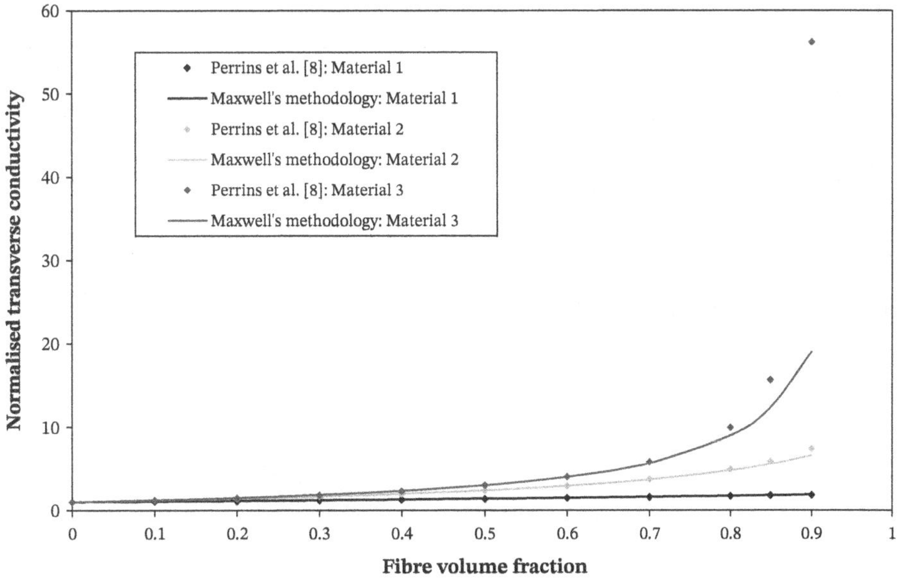

Figure 4.2 Comparison of results for normalised effective transverse thermal conductivity obtained using Maxwell's methodology with those of Modified from Perrins et al. [8] for three different materials.

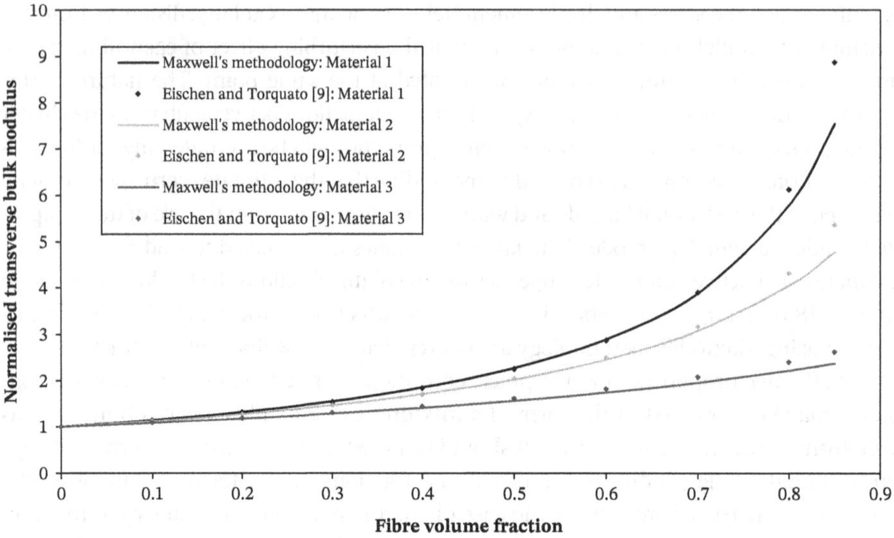

Figure 4.3 Comparison of results for normalised effective transverse bulk modulus obtained using Maxwell's methodology with those of Modified from Eischen and Torquato [9] for three different materials.

Figure 4.3 shows a comparison of transverse bulk modulus obtained using Maxwell's methodology with results of Eischen and Torquato [9]. The normalised effective transverse bulk modulus is defined by k_T^{eff} / k_m and the three materials considered are for isotropic fibres and matrix such that $\mu_f / \mu_m = 135, 22.5, 6.75$, $\mu_f / k_f = 0.75$ and $\mu_m / k_m = 0.33$.

Figure 4.4 shows a comparison of axial shear modulus obtained using Maxwell's methodology with results of Symm [10]. The normalised effective axial shear modulus

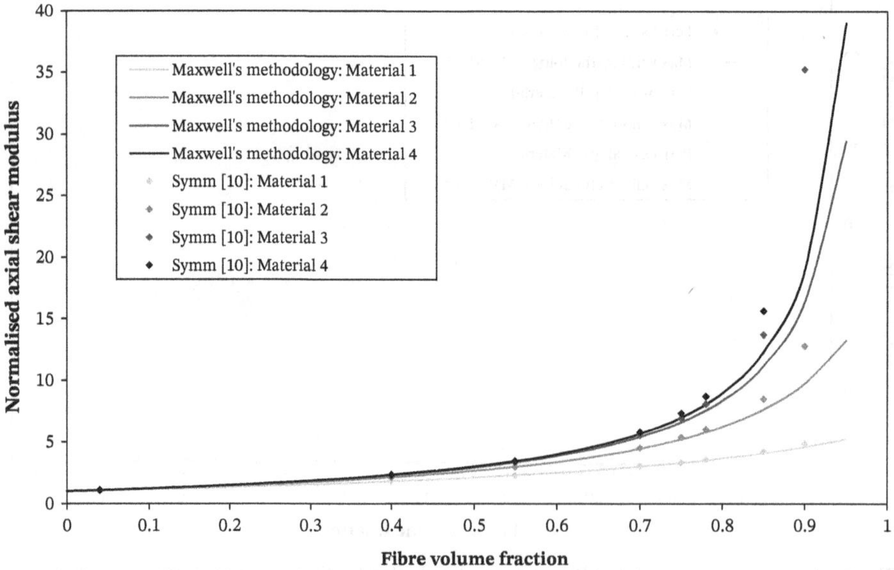

Figure 4.4 Comparison of results for normalised effective axial shear modulus obtained using Maxwell's methodology with those of Modified from Symm [10] for four different materials.

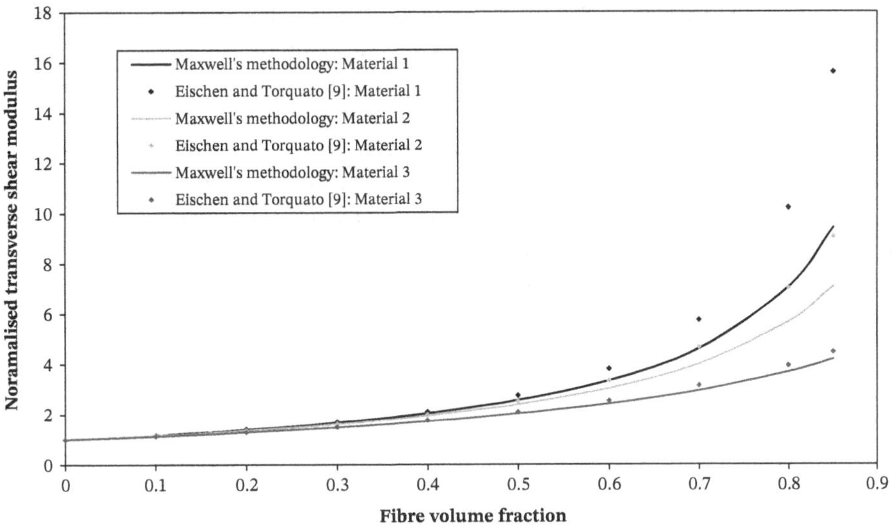

Figure 4.5 Comparison of results for normalised effective transverse shear modulus obtained using Maxwell's methodology with those of Modified from Eischen and Torquato [9] for three different materials.

is defined by k_A^{eff} / k_m and the four materials considered are for isotropic fibres and matrix such that $\mu_f / \mu_m = 6, 20, 120, \infty$.

Figure 4.5 shows a comparison of transverse shear modulus obtained using Maxwell's methodology with results of Eischen and Torquato [9]. The normalised effective transverse shear modulus is defined by $\mu_t^{\text{eff}} / \mu_m$ and the three materials considered are for isotropic fibres and matrix such that $\mu_f / \mu_m = 135, 22.5, 6.75$, $\mu_f / k_f = 0.75$ and $\mu_m / k_m = 0.33$.

References

1. Maxwell, J.C. (1873). *A Treatise on Electricity and Magnetism*, 1st ed. (3rd edition, 1892). Chapter 9, Vol. 1. Oxford: Clarendon Press.
2. McCartney, L.N. and Kelly, A. (2008). Maxwell's far-field methodology applied to the prediction of properties of multi-phase isotropic particulate composites. *Proceedings of the Royal Society* A464: 423–446.
3. Hashin, Z. (1983). Analysis of composite materials - A survey. *Journal of Applied Mechanics* 50: 481–505.
4. Hasselman, D.P.H. and Johnson, L.F. (1987). Effective thermal conductivity of composites with interfacial thermal barrier resistance. *Journal of Composite Materials* 21: 508–515.
5. Torquato, S. (2002). *Random Heterogeneous Materials*. New York: Springer-Verlag.
6. McCartney, L.N. (2010). Maxwell's far-field methodology predicting elastic properties of multiphase composites reinforced with aligned transversely isotropic spheroids. *Philosophical Magazine* 90: 4175–4207.

7. Hashin, Z. and Shtrikman, S. (1963). A variational approach to the theory of the elastic behaviour of multiphase composites. *Journal of the Mechanics and Physics of Solids* 11: 127–140.
8. Perrins, W.T., McKenzie, D.R., and McPhedran, R.C. (1979). Transport properties of regular arrays of cylinders. *Proceedings of the Royal Society of London* A369: 207–225.
9. Eischen, J.W. and Torquato, S. (1993). Determining elastic behaviour of composites by the boundary element method. *Journal of Applied Physics* 74 (1): 159–170.
10. Symm, G.T. (1970). The longitudinal shear modulus of a unidirectional fibrous composite. *Journal of Composite Materials* 4: 426–428.

5

Reinforcement with Ellipsoidal Inclusions

Overview: The objective of this chapter is to consider the effects on composite properties of reinforcing a matrix with ellipsoidal particles for cases where there is more than one species of reinforcement, which are assumed to be aligned and all species have the same shape. In the literature, most theoretical approaches have used a general approach involving fourth-order elastic constants and thermal stresses are not usually considered. In the following analysis of this chapter, bold characters denote second-order tensors such as stress $\boldsymbol{\sigma}$ and strain $\boldsymbol{\varepsilon}$ whereas uppercase symbols denote fourth-order tensors arising in elasticity theory. The elastic constants are denoted by the fourth-order tensor L.

5.1 Stress-Strain Relations

The stress-strain relations for any anisotropic material may be written in either of the following forms

$$\boldsymbol{\sigma} = L\boldsymbol{\varepsilon} = M^{-1}\boldsymbol{\varepsilon}, \qquad \boldsymbol{\varepsilon} = L^{-1}\boldsymbol{\sigma} = M\boldsymbol{\sigma}, \quad \text{so that} \quad M\,L \equiv \mathrm{I} \equiv L^{-1}M^{-1}, \tag{5.1}$$

where L denotes the symmetric fourth-order tensor of elastic moduli, M denotes the corresponding compliance tensor and where I denotes the symmetric fourth-order identity tensor defined by (2.15). In Cartesian component form, the stress-strain relations (5.1) may be written as

$$\sigma_{ij} = L_{ijkl}\,\varepsilon_{kl} = M_{ijkl}^{-1}\varepsilon_{kl}, \qquad \varepsilon_{ij} = L_{ijkl}^{-1}\sigma_{kl} = M_{ijkl}\sigma_{kl}. \tag{5.2}$$

5.2 Theorems for Mean Strain and Mean Stress

Consider a region R bounded by a closed surface Σ having unit outward normal $\boldsymbol{n}$ having Cartesian components defined by

$$\boldsymbol{n} \equiv n_1\boldsymbol{i}_1 + n_2\boldsymbol{i}_2 + n_3\boldsymbol{i}_3, \tag{5.3}$$

Properties for Design of Composite Structures: Theory and Implementation Using Software, First Edition. Neil McCartney.

Companion Website: www.wiley.com/go/mccartney/properties

where $\boldsymbol{i}_1, \boldsymbol{i}_2, \boldsymbol{i}_3$ are the orthogonal unit vectors in the coordinate directions x_1, x_2, x_3. For any scalar function ϕ

$$\int_R \nabla\phi \, \mathrm{d}V = \int_\Sigma \boldsymbol{n}\phi \mathrm{d}S. \tag{5.4}$$

This relation can be applied also to vector functions $\boldsymbol{v}$ so that

$$\int_R \nabla\boldsymbol{v} \mathrm{d}V = \int_\Sigma \boldsymbol{n}\boldsymbol{v} \mathrm{d}S. \tag{5.5}$$

The gradient of the vector $\boldsymbol{v}$ and the dyadic $\boldsymbol{n}\,\boldsymbol{v}$ are defined, respectively, by

$$\nabla\boldsymbol{v} \equiv \boldsymbol{i}_1\frac{\partial\boldsymbol{v}}{\partial x_1} + \boldsymbol{i}_2\frac{\partial\boldsymbol{v}}{\partial x_2} + \boldsymbol{i}_3\frac{\partial\boldsymbol{v}}{\partial x_3}, \tag{5.6}$$

$$\begin{aligned}\boldsymbol{n}\,\boldsymbol{v} &\equiv \left(n_1\boldsymbol{i}_1 + n_2\boldsymbol{i}_2 + n_3\boldsymbol{i}_3\right)\left(v_1\boldsymbol{i}_1 + v_2\boldsymbol{i}_2 + v_3\boldsymbol{i}_3\right), \\ &\equiv n_1v_1\boldsymbol{i}_1\boldsymbol{i}_1 + n_1v_2\boldsymbol{i}_1\boldsymbol{i}_2 + n_1v_3\boldsymbol{i}_1\boldsymbol{i}_3 + n_2v_1\boldsymbol{i}_2\boldsymbol{i}_1 + n_2v_2\boldsymbol{i}_2\boldsymbol{i}_2 + n_2v_3\boldsymbol{i}_2\boldsymbol{i}_3 \\ &\qquad + n_3v_1\boldsymbol{i}_3\boldsymbol{i}_1 + n_3v_2\boldsymbol{i}_3\boldsymbol{i}_2 + n_3v_3\boldsymbol{i}_3\boldsymbol{i}_3.\end{aligned} \tag{5.7}$$

Corresponding to (5.5) the following relation known as the divergence theorem can be applied to vector functions $\boldsymbol{v}$ so that

$$\int_R \nabla.\boldsymbol{v} \mathrm{d}V = \int_\Sigma \boldsymbol{n}.\boldsymbol{v} \mathrm{d}S. \tag{5.8}$$

5.2.1 Mean Strain

The closed surface Σ encloses a composite material containing a set of inclusions that can be of any shape and can have differing properties, which are embedded in a matrix material. It is assumed that the inclusions are perfectly bonded to the matrix so that the displacement vector $\boldsymbol{u}$ is continuous everywhere within Σ. The second-order infinitesimal strain tensor ε at any point within Σ is defined by

$$\varepsilon \equiv \frac{1}{2}\left[\nabla\boldsymbol{u} + \left(\nabla\boldsymbol{u}\right)^{\mathrm{T}}\right], \quad \text{where } \left(\nabla\boldsymbol{u}\right)^{\mathrm{T}} \equiv \frac{\partial\boldsymbol{u}}{\partial x_1}\boldsymbol{i}_1 + \frac{\partial\boldsymbol{u}}{\partial x_2}\boldsymbol{i}_2 + \frac{\partial\boldsymbol{u}}{\partial x_3}\boldsymbol{i}_3. \tag{5.9}$$

It follows that the strain component ε_{mn} is given by

$$\varepsilon_{mn} = \boldsymbol{i}_{\mathrm{m}} \cdot \frac{1}{2}\left[\nabla\boldsymbol{u} + \left(\nabla\boldsymbol{u}\right)^{\mathrm{T}}\right] . \boldsymbol{i}_n = \frac{1}{2}\left(\frac{\partial u_n}{\partial x_{\mathrm{m}}} + \frac{\partial u_{\mathrm{m}}}{\partial x_n}\right). \tag{5.10}$$

It follows from (5.4), (5.5) and (5.9) that

$$\int_R \left(\nabla\boldsymbol{v}\right)^{\mathrm{T}} \mathrm{d}V = \int_\Sigma \boldsymbol{v}\boldsymbol{n} \mathrm{d}S, \tag{5.11}$$

so that

$$\int_R \varepsilon \, \mathrm{d}V = \frac{1}{2}\int_\Sigma \left(\boldsymbol{u}\boldsymbol{n} + \boldsymbol{n}\boldsymbol{u}\right) \mathrm{d}S. \tag{5.12}$$

On denoting the volume of the region R enclosed by the surface Σ by Ω, the mean value of the strain tensor $\bar{\varepsilon}$ is given by

$$\bar{\varepsilon} \equiv \frac{1}{\Omega}\int_R \varepsilon \, \mathrm{d}V = \frac{1}{2\Omega}\int_\Sigma \left(\boldsymbol{u}\boldsymbol{n} + \boldsymbol{n}\boldsymbol{u}\right) \mathrm{d}S. \tag{5.13}$$

The mean value of the strain tensor, within an inhomogeneous region R where the displacement field is continuous, is determined by the values of the displacement vector on the bounding surface Σ. If a displacement field $\boldsymbol{u}_0$ applied on Σ to a medium that is homogeneous within R leads to a uniform strain field ε_0 within R, then it follows that, when the same displacement distribution is applied to a medium that is inhomogeneous within R, the mean strain $\bar{\varepsilon}$ in R has the value ε_0, i.e.

$$\bar{\varepsilon} \equiv \frac{1}{\Omega}\int_R \varepsilon \, \mathrm{d}V = \frac{1}{2\Omega}\int_\Sigma \left(\boldsymbol{u}_0\boldsymbol{n} + \boldsymbol{n}\boldsymbol{u}_0\right)\mathrm{d}S = \varepsilon_0. \tag{5.14}$$

This result applies to any inhomogeneous material and, in particular, it applies to any perfectly bonded composite, irrespective of the number, size, shape and properties of the reinforcing particles.

5.2.2 Mean Stress

The mean stress, in the possibly inhomogeneous region R having volume Ω, is defined by

$$\bar{\sigma} = \frac{1}{\Omega}\int_R \sigma \, \mathrm{d}V. \tag{5.15}$$

To estimate the mean value of the stress field in terms of boundary values for the stress field, use is made of the following identity, which is valid for equilibrium conditions in the absence of body forces,

$$\nabla.\left(\sigma \boldsymbol{r}\right) \equiv \sigma, \quad \text{where} \quad \boldsymbol{r} \equiv x_1\boldsymbol{i}_1 + x_2\boldsymbol{i}_2 + x_3\boldsymbol{i}_3. \tag{5.16}$$

It then follows, using (5.4), the tensor equivalent of (5.8), and (5.15), that the mean stress $\bar{\sigma}$, in region R and bounded by the surface Σ, is given by

$$\bar{\sigma} = \frac{1}{\Omega}\int_R \sigma \, \mathrm{d}V = \frac{1}{\Omega}\int_R \nabla.\left(\sigma \boldsymbol{r}\right)\mathrm{d}V = \frac{1}{\Omega}\int_\Sigma \boldsymbol{n}.\sigma \boldsymbol{r}\, \mathrm{d}S = \frac{1}{\Omega}\int_\Sigma \tau \boldsymbol{r}\, \mathrm{d}S, \tag{5.17}$$

where $\tau = \boldsymbol{n}.\sigma$ are the tractions applied to the external boundary Σ. The value of the mean stress is, thus, determined by values of the tractions τ on the external boundary Σ. For any homogeneous material, let σ_0 be a uniform stress field in the region R that arises from the application of tractions $\tau_0 = \boldsymbol{n}.\sigma_0$ on the external surface Σ. If the same tractions are applied to the composite, then provided the tractions are continuous across all interfaces, the mean value of the stress must have the value σ_0, i.e.

$$\bar{\sigma} = \frac{1}{\Omega}\int_R \sigma \, \mathrm{d}V = \frac{1}{\Omega}\int_\Sigma \boldsymbol{n}.\sigma_0 \boldsymbol{r}\, \mathrm{d}S = \frac{1}{\Omega}\int_\Sigma \tau_0 \boldsymbol{r}\, \mathrm{d}S = \sigma_0. \tag{5.18}$$

This result applies to any inhomogeneous material and, in particular, it applies to any perfectly bonded composite, irrespective of the number, size and shape of the reinforcing particles.

5.3 Eshelby Theory for an Isolated Particle

A summary is now given of the results for isolated inclusions given by Eshelby [1], which are recast in a form that is useful in this chapter. He considered an isolated inclusion of arbitrary shape bounded by a closed surface that is embedded in an infinite homogeneous isotropic linear elastic medium that is not loaded. The stress and strain are both zero everywhere in the system. The material in the inclusion is perfectly bonded to the material lying outside which is called the matrix. The inclusion then deforms by the addition a uniform strain ε_T, known as the transformation strain of the inclusion, whereas the material outside the inclusion remains bonded to the inclusion, and is not loaded at infinity. This process leads to nonuniform stresses and strains throughout the system. By using a particular cutting, transformation, rewelding and relaxation procedure, Eshelby derived an integral form for the displacement field in both the inclusion and matrix from which the strain and stress fields may in principle be calculated. He then considered the special case of ellipsoidal inclusions and proved that the resulting strain $\Delta\varepsilon_I$ arising from the strain transformation within the inclusion is uniform, being given by a relation of the form

$$\Delta\varepsilon_I = S\varepsilon_T, \tag{5.19}$$

where S is a fourth-order tensor that depends only on the properties of the matrix and the shape of the ellipsoidal inclusions. Values of the components of the Eshelby tensor S for various ellipsoidal geometries (including spheroids, cylinders, spheres) are given by Mura [2]. Eshelby then introduced the concept of an inhomogeneity, which is regarded as an inclusion having different isotropic linear elastic properties to those of the matrix, and considered the effects of applying at infinity a loading that would, in a homogeneous body, lead to a uniform strain field ε_A. He then considered how the inclusion (made of matrix material) in the presence of the applied strain field ε_A can be replaced by an inhomogeneity having different isotropic linear elastic properties without perturbing the stress and strain field in the matrix region. He showed that this procedure is possible only if the equivalent transformation strain ε_T is selected so that it satisfies the equation that is generalised here to the form

$$L_I\left(\varepsilon_A + \Delta\varepsilon_I\right) = L_m\left(\varepsilon_A + \Delta\varepsilon_I - \varepsilon_T\right), \tag{5.20}$$

where L_I and L_m are the fourth-order isotropic elastic constants for the inhomogeneity and matrix, respectively, and where $\Delta\varepsilon_I$ is now the additional uniform strain in the inhomogeneity arising from its presence when the matrix is loaded at infinity. The total strain in the inhomogeneity is clearly $\varepsilon_I = \varepsilon_A + \Delta\varepsilon_I$ and the expression on the left-hand side of (5.20) is clearly the stress in the inhomogeneity. The strain ε_T is now described as an equivalent transformation strain as it is a fictitious strain that must be applied to enable an inhomogeneity to be considered in place of the inclusion having matrix properties. It is emphasised that when considering inhomogeneities the stress and strain fields will be zero when the system is not loaded at infinity. When solving (5.20) for the equivalent transformation strain it must be remembered from (5.19) that the additional inclusion strain $\Delta\varepsilon_I$ is proportional to the equivalent transformation strain ε_T. The equation determining its value is, therefore,

$$L_I\left[\varepsilon_A + S\varepsilon_T\right] = L_m\left[\varepsilon_A + \left(S - I\right)\varepsilon_T\right], \tag{5.21}$$

where I is the symmetric fourth-order identity tensor (see (2.15)). It then follows that

$$\left[L_m - \left(L_m - L_I\right)S\right]\varepsilon_T = \left(L_m - L_I\right)\varepsilon_A, \tag{5.22}$$

indicating that $\varepsilon_T = 0$ when the inhomogeneity has the same properties as the matrix or the system is unloaded at infinity.

To summarise the results, derived by Eshelby [1], for an isolated ellipsoidal inhomogeneity embedded in an infinite isotropic matrix loaded at infinity by the uniform ε_{A}, the strain and stress in the inhomogeneity are uniform having the values

$$\varepsilon_{\mathrm{I}} = \varepsilon_{\mathrm{A}} + \Delta\varepsilon_{\mathrm{I}}, \qquad \sigma_{\mathrm{I}} = L_{\mathrm{I}}\left(\varepsilon_{\mathrm{A}} + \Delta\varepsilon_{\mathrm{I}}\right), \tag{5.23}$$

respectively, where the additional strain $\Delta\varepsilon_{\mathrm{I}}$ in the region of the inhomogeneity, arising from its presence, is given by (5.19) where the equivalent transformation strain ε_{T} required to operate Eshelby's analysis is obtained using (5.22). For the matrix, the strain and stress distributions are given by

$$\varepsilon_{\mathrm{m}} = \varepsilon_{\mathrm{A}} + \Delta\varepsilon_{\mathrm{m}}, \qquad \sigma_{\mathrm{m}} = L_{\mathrm{m}}\left(\varepsilon_{\mathrm{A}} + \Delta\varepsilon_{\mathrm{m}}\right), \tag{5.24}$$

where the additional strain $\Delta\varepsilon_{\mathrm{m}}$ in the matrix arising from the presence of the inhomogeneity will be nonuniform.

In a later paper, Eshelby [3] stated that the important relation (5.20), and hence (5.22), applies also to inhomogeneities that are anisotropic, but the matrix must always be isotropic. For the remainder of this book, the word inclusion is regarded as being an inhomogeneity having properties that are different to those of the embedding matrix.

5.4 Isolated Ellipsoidal Inclusion

An isolated ellipsoidal inclusion of species r is now considered that is embedded in an infinite homogeneous isotropic matrix such that in the absence of any applied loading the stress and strains within the matrix and an ellipsoidal inclusion are everywhere zero. When loading is applied to the matrix at infinity such that a uniform homogeneous strain field ε_{A} would result if the rth type of inclusion was made of matrix material, the total strain ε_r in the ellipsoidal inclusion is uniform and from (5.23) may be written as

$$\varepsilon_r = \varepsilon_{\mathrm{A}} + \Delta\varepsilon_r. \tag{5.25}$$

The uniform additional strain $\Delta\varepsilon_r$ resulting in the inclusion of species r, arising from the presence of the inclusion in the matrix, is given, on using (5.19), by

$$\Delta\varepsilon_r = S\varepsilon_T^{(r)}, \tag{5.26}$$

where $\varepsilon_T^{(r)}$ is the equivalent transformation strain associated with the inclusions of type r that is obtained using the following relation obtained from (5.22)

$$\left[L_{\mathrm{m}} - \left(L_{\mathrm{m}} - L_r\right)S\right]\varepsilon_T^{(r)} = \left(L_{\mathrm{m}} - L_r\right)\varepsilon_{\mathrm{A}}, \tag{5.27}$$

where L_r is the fourth-order tensor for the elastic constants (possible anisotropic) of the rth species, and where L_{m} is the fourth-order tensor of elastic constants for the matrix.

The corresponding stress distribution σ_r in the inclusion of species r is given by

$$\sigma_r = L_r\left(\varepsilon_{\mathrm{A}} + \Delta\varepsilon_r\right). \tag{5.28}$$

The total strain ε_{m} in the matrix may be written as

$$\varepsilon_{\mathrm{m}} = \varepsilon_{\mathrm{A}} + \Delta\varepsilon_{\mathrm{m}}, \tag{5.29}$$

so that $\Delta\varepsilon_{\mathrm{m}}$ is the additional strain field resulting in the matrix when the inclusion is present. The corresponding stress distribution σ_{m} in the matrix is simply given by

$$\sigma_{\mathrm{m}} = L_{\mathrm{m}}\left(\varepsilon_{\mathrm{A}} + \Delta\varepsilon_{\mathrm{m}}\right), \tag{5.30}$$

where the additional strain $\Delta\varepsilon_{\mathrm{m}}$ in the matrix, and hence the matrix stress, will be nonuniform. It follows from (5.20) that

$$L_r\left(\varepsilon_{\mathrm{A}} + \Delta\varepsilon_r\right) = L_{\mathrm{m}}\left(\varepsilon_{\mathrm{A}} + \Delta\varepsilon_r - \varepsilon_{\mathrm{T}}^{(r)}\right), \tag{5.31}$$

where $\varepsilon_{\mathrm{T}}^{(r)}$ is given by (5.27). On using (5.25), (5.31) may be written as

$$L_r\varepsilon_r = L_{\mathrm{m}}\left(\varepsilon_r - \varepsilon_{T}^{(r)}\right), \tag{5.32}$$

so that

$$\left(L_r - L_{\mathrm{m}}\right)\varepsilon_r = -L_{\mathrm{m}}\,\varepsilon_{\mathrm{T}}^{(r)}. \tag{5.33}$$

It then follows that

$$\varepsilon_{\mathrm{T}}^{(r)} = -L_{\mathrm{m}}^{-1}\left(L_r - L_{\mathrm{m}}\right)\varepsilon_r, \tag{5.34}$$

and substitution into (5.26) then leads to

$$\Delta\varepsilon_r = -S L_{\mathrm{m}}^{-1}\left(L_r - L_{\mathrm{m}}\right)\varepsilon_r. \tag{5.35}$$

On using (5.25), it follows that

$$\left[\mathrm{I} + S L_{\mathrm{m}}^{-1}\left(L_r - L_{\mathrm{m}}\right)\right]\varepsilon_r = \varepsilon_{\mathrm{A}}. \tag{5.36}$$

The uniform strain in an isolated inclusion of species r may then be written as

$$\varepsilon_r = \left[\mathrm{I} + S L_{\mathrm{m}}^{-1}\left(L_r - L_{\mathrm{m}}\right)\right]^{-1}\varepsilon_{\mathrm{A}} \equiv A_r\varepsilon_{\mathrm{A}}, \tag{5.37}$$

where A_r is the fourth-order strain concentration tensor for inclusions of species r. It is worth noting that $A_r = \mathrm{I}$ when $L_r = L_{\mathrm{m}}$ so that the properties of the inclusion are the same as those of the matrix.

5.5 Multiple Ellipsoidal Inclusions

A cluster of N species of aligned ellipsoidal inclusions is embedded in an infinite homogeneous isotropic linear elastic matrix that is bounded by the surface S_∞. The cluster itself is bounded by a closed surface Σ that lies in matrix material without touching any of the inclusions. The region enclosed by Σ is denoted by R and has volume Ω. The cluster of inclusions and the matrix material lying within Σ forms a composite having an effectively homogeneous distribution of n species of aligned ellipsoidal inclusions all having the same shape. The region R_r occupied by the inclusions of species r has volume Ω_r whereas the region R_{m} occupied by the matrix lying within the surface Σ has volume Ω_{m}. Clearly the region R is the union of the regions R_r, $r = 1, \ldots, N$, and R_{m}, and

$$\sum_{r=1}^{N}\Omega_r + \Omega_{\mathrm{m}} = \Omega. \tag{5.38}$$

It then follows that

$$\sum_{r=1}^{N} V_{\mathrm{p}}^{(r)} + V_{\mathrm{m}} = 1, \tag{5.39}$$

where volume fractions of species and of the matrix are defined respectively by

$$V_{\mathrm{p}}^{(r)} = \frac{\Omega_r}{\Omega}, \quad r = 1,\ldots,N, \quad V_{\mathrm{m}} = \frac{\Omega_{\mathrm{m}}}{\Omega}. \tag{5.40}$$

It is assumed that the inclusions are perfectly bonded to the matrix so that the displacement field is everywhere continuous. The composite is regarded as being subject on the external surface S_∞ to an applied displacement distribution that would generate a uniform strain field ε_{A} in a homogeneous material occupying the whole of the region S_∞. When the volume fractions of the inclusions are large enough there will be interaction effects between inclusions that are expected to lead to nonuniform stress and strain distributions within both the inclusions and the matrix.

Provided that the applied strain field ε_{A} will generate a uniform strain field within a homogeneous region R having volume Ω, the mean value $\bar{\varepsilon}$ of the strain tensor ε within R is simply ε_{A}. It then follows that

$$\bar{\varepsilon} \equiv \frac{1}{\Omega}\int_{R} \varepsilon \, \mathrm{d}V = \frac{1}{\Omega}\int_{R_{\mathrm{m}}} \varepsilon \, \mathrm{d}V + \frac{1}{\Omega}\sum_{r=1}^{N}\int_{R_r} \varepsilon \, \mathrm{d}V = \varepsilon_{\mathrm{A}}, \tag{5.41}$$

which may be written, on using (5.40), as

$$\varepsilon \equiv \frac{1}{\Omega}\int_{R} \varepsilon \, \mathrm{d}V = V_{\mathrm{m}}\,\bar{\varepsilon}_{\mathrm{m}} + \sum_{r=1}^{N} V_{\mathrm{p}}^{(r)}\,\bar{\varepsilon}_r = \varepsilon_{\mathrm{A}}, \tag{5.42}$$

where $\bar{\varepsilon}_{\mathrm{m}}$ and $\bar{\varepsilon}_r$, $r = 1, \ldots, N$, are mean strains for the matrix and inclusions, respectively, defined by

$$\bar{\varepsilon}_{\mathrm{m}} = \frac{1}{\Omega_{\mathrm{m}}}\int_{R_{\mathrm{m}}} \varepsilon \, \mathrm{d}V, \qquad \bar{\varepsilon}_r = \frac{1}{\Omega_r}\int_{R_r} \varepsilon \, \mathrm{d}V, \quad r = 1, \ldots, N. \tag{5.43}$$

An applied stress field σ_{A} that will generate a uniform stress field σ_{A} within a homogeneous region Ω, has the mean value $\bar{\sigma} = \sigma_{\mathrm{A}}$ of the stress tensor σ within Ω. It then follows that

$$\bar{\sigma} \equiv \frac{1}{\Omega}\int_{R} \sigma \, \mathrm{d}V = \frac{1}{\Omega}\int_{R_{\mathrm{m}}} \sigma \, \mathrm{d}V + \frac{1}{\Omega}\sum_{r=1}^{N}\int_{R_r} \sigma \, \mathrm{d}V = \sigma_{\mathrm{A}}, \tag{5.44}$$

which may be written as

$$\bar{\sigma} \equiv \frac{1}{\Omega}\int_{R} \sigma \, \mathrm{d}V = V_{\mathrm{m}}\,\bar{\sigma}_{\mathrm{m}} + \sum_{r=1}^{N} V_{\mathrm{p}}^{(r)}\,\bar{\sigma}_r = \sigma_{\mathrm{A}}, \tag{5.45}$$

where $\bar{\sigma}_{\mathrm{m}}$ and $\bar{\sigma}_r$, $r = 1, \ldots, N$, are mean stresses for the matrix and inclusions, respectively, defined by

$$\bar{\sigma}_{\mathrm{m}} = \frac{1}{\Omega_{\mathrm{m}}}\int_{R_{\mathrm{m}}} \sigma \, \mathrm{d}V, \qquad \bar{\sigma}_r = \frac{1}{\Omega_r}\int_{R_r} \sigma \, \mathrm{d}V, \quad r = 1, \ldots, N. \tag{5.46}$$

It is assumed that the matrix and inclusions have uniform properties so that

$$\bar{\sigma}_{\mathrm{m}} = L_{\mathrm{m}}\bar{\varepsilon}_{\mathrm{m}}, \quad \bar{\sigma}_r = L_r\bar{\varepsilon}_r, \quad r = 1, \ldots, N, \tag{5.47}$$

where L_m and L_r, $r = 1, \dots, N$ are the fourth-order elastic constant tensors for the matrix and inclusions, respectively.

5.6 Dilute Approximation

Displacements corresponding to the uniform strain field ε_A are now applied to the external surfaces of both the composite in region R and to the equivalent homogeneous effective composite material having a fourth-order elastic constant tensor that is denoted by L^*. The corresponding applied tractions are obtained from the uniform stress σ_A such that

$$\sigma_A = L^* \varepsilon_A. \tag{5.48}$$

On using the result (5.45), the mean stress in the inhomogeneous composite must also have the value σ_A so that

$$\sigma_A = V_m L_m \bar{\varepsilon}_m + \sum_{r=1}^{N} V_p^{(r)} L_r \bar{\varepsilon}_r, \tag{5.49}$$

where use has been made of (5.47). The mean strain in the matrix is obtained from relation (5.42) so that

$$V_m \bar{\varepsilon}_m = \varepsilon_A - \sum_{r=1}^{N} V_p^{(r)} \bar{\varepsilon}_r. \tag{5.50}$$

On substituting (5.48) and (5.50) into (5.49) it follows that

$$L^* \varepsilon_A = L_m \varepsilon_A + \sum_{r=1}^{N} V_p^{(r)} \left(L_r - L_m\right) \bar{\varepsilon}_r. \tag{5.51}$$

It is now assumed that the strain fields ε_r within the inclusions are all uniform, and that their values are given by relation (5.37). This means that subsequent results may be valid only for relatively low volume fractions of inclusions. It then follows from (5.43) that $\bar{\varepsilon}_r = \varepsilon_r$ and on substituting (5.37) into (5.51) it is found that the effective elastic moduli are given by

$$L^* = L_m + \sum_{r=1}^{N} V_p^{(r)} \left(L_r - L_m\right) A_r. \tag{5.52}$$

From (5.37) and (5.42) the average strain in the matrix $\bar{\varepsilon}_m$ may be found using the relation

$$V_m \bar{\varepsilon}_m = \left(\mathrm{I} - \sum_{r=1}^{N} V_p^{(r)} A_r \right) \varepsilon_A, \tag{5.53}$$

where I is the symmetric identity tensor. The mean strain in the matrix is now written as

$$\bar{\varepsilon}_m = A_m \varepsilon_A, \tag{5.54}$$

so that from (5.53) the relation defining the value of the strain concentration tensor for the matrix A_m is

$$V_m A_m + \sum_{r=1}^{N} V_p^{(r)} A_r = \mathrm{I}. \tag{5.55}$$

The principal result (5.52) may then be written on using (5.55) in the form of a mixtures relationship as follows

$$L^* = V_{\rm m} L_{\rm m} A_{\rm m} + \sum_{r=1}^{N} V_{\rm p}^{(r)} L_r A_r. \tag{5.56}$$

It is emphasised that the value of the fourth-order matrix strain concentration factor $A_{\rm m}$ is not necessarily equal to the symmetric identity tensor I.

5.7 General Case

Consider now the problem of attempting to take into account approximately the effects of particle interactions. Mori and Tanaka [4] originally calculated the average internal stress in the matrix of a material containing precipitates having eigenstrains. Their theory has been the starting point for a series of papers in the literature addressing the problem of estimating the effective properties of composite reinforced with ellipsoidal inclusions. The concepts of internal stress and eigenstrains need not concern us here as Benveniste [5] developed a new approach to Mori–Tanaka theory for the case of two-phase composites. Benveniste, Dvorak and Chen [6] extended the approach to multiple phases, included thermal expansion effects and considered both the case of dilute distributions of inclusions and the case when particle interactions are important. The Mori–Tanaka method, as described in [5, 6], replaces (5.37), i.e. $\varepsilon_r = A_r \varepsilon_{\rm A}$, for an isolated inclusion of species r, by the relation

$$\varepsilon_r = A_r \bar{\varepsilon}_{\rm m}, \tag{5.57}$$

where the strain concentration factor for the rth species is given, as before in (5.37), by

$$A_r = \left[{\rm I} + S L_{\rm m}^{-1} \left(L_r - L_{\rm m} \right) \right]^{-1}. \tag{5.58}$$

Their approach for interacting inclusions assumes that $\bar{\varepsilon}_r = A_r \bar{\varepsilon}_{\rm m}$, $r = 1, ..., N$. Substitution into (5.50) and (5.51) then leads to the relations

$$\bar{\varepsilon}_{\rm m} = \left(V_{\rm m} {\rm I} + \sum_{r=1}^{N} V_{\rm p}^{(r)} A_r \right)^{-1} \varepsilon_{\rm A}, \tag{5.59}$$

$$L^* \varepsilon_{\rm A} = L_{\rm m} \varepsilon_{\rm A} + \sum_{r=1}^{N} V_{\rm p}^{(r)} \left(L_r - L_{\rm m} \right) A_r \bar{\varepsilon}_{\rm m}. \tag{5.60}$$

It then follows directly that

$$L^* = L_{\rm m} + \sum_{r=1}^{N} V_{\rm p}^{(r)} \left(L_r - L_{\rm m} \right) A_r \left(V_{\rm m} {\rm I} + \sum_{r=1}^{N} V_{\rm p}^{(r)} A_r \right)^{-1}. \tag{5.61}$$

On using (5.39) this result is now written in the form

$$L^* = \left[L_{\rm m} V_{\rm m} + \sum_{r=1}^{N} V_{\rm p}^{(r)} L_r A_r \right] \left(V_{\rm m} {\rm I} + \sum_{r=1}^{N} V_{\rm p}^{(r)} A_r \right)^{-1}. \tag{5.62}$$

Clearly,

$$L^* V_{\rm m} + L^* \sum_{r=1}^{N} V_{\rm p}^{(r)} A_r = L_{\rm m} V_{\rm m} + \sum_{r=1}^{N} V_{\rm p}^{(r)} L_r A_r, \tag{5.63}$$

so that

$$\left(L^* - L_{\rm m}\right)V_{\rm m} + \sum_{r=1}^{N} V_{\rm p}^{(r)}\left(L^* - L_r\right)A_r = 0. \tag{5.64}$$

A fourth-order tensor L_m^* is now introduced so that

$$S L_{\rm m}^{-1} = \left(L_{\rm m} + L_{\rm m}^*\right)^{-1}, \quad \text{i.e.} \quad L_{\rm m}^* = L_{\rm m}\left(S^{-1} - {\rm I}\right). \tag{5.65}$$

It then follows from (5.58) that

$$\left[{\rm I} + \left(L_{\rm m} + L_{\rm m}^*\right)^{-1}\left(L_r - L_{\rm m}\right)\right] A_r = {\rm I}. \tag{5.66}$$

On multiplying (5.66) by $\left(L_{\rm m} + L_{\rm m}^*\right)$ it is easily shown that

$$A_r = \left(L_r + L_{\rm m}^*\right)^{-1}\left(L_{\rm m} + L_{\rm m}^*\right), \tag{5.67}$$

which is consistent with the identification $A_{\rm m} = {\rm I}$. It then follows that

$$\sum_{r=1}^{N} V_{\rm p}^{(r)} A_r = \left[\sum_{r=1}^{N} V_{\rm p}^{(r)}\left(L_r + L_{\rm m}^*\right)^{-1}\right]\left(L_{\rm m} + L_{\rm m}^*\right). \tag{5.68}$$

From (5.61)

$$\begin{aligned}\left(L^* + L_{\rm m}^*\right)\left(V_{\rm m}{\rm I} + \sum_{r=1}^{N} V_{\rm p}^{(r)} A_r\right) &= \left(L_{\rm m} + L_{\rm m}^*\right)\left(V_{\rm m}{\rm I} + \sum_{r=1}^{N} V_{\rm p}^{(r)} A_r\right) \\ &\quad + \sum_{r=1}^{N} V_{\rm p}^{(r)}\left(L_r - L_{\rm m}\right)A_r,\end{aligned} \tag{5.69}$$

which is first written as

$$\begin{aligned}\left(L^* + L_{\rm m}^*\right)\left(V_{\rm m}{\rm I} + \sum_{r=1}^{N} V_{p}^{(r)} A_r\right) &= \left(L_{\rm m} + L_{\rm m}^*\right)\left(V_{\rm m}{\rm I} + \sum_{r=1}^{N} V_{p}^{(r)} A_r\right) \\ &\quad + \sum_{r=1}^{N} V_{p}^{(r)}\left(L_r + L_{\rm m}^*\right)A_r - \left(L_{\rm m} + L_{\rm m}^*\right)\sum_{r=1}^{N} V_{p}^{(r)} A_r.\end{aligned} \tag{5.70}$$

On cancellation of terms

$$\left(L^* + L_{\rm m}^*\right)\left(V_{\rm m}{\rm I} + \sum_{r=1}^{N} V_{\rm p}^{(r)} A_r\right) = \left(L_{\rm m} + L_{\rm m}^*\right)V_{\rm m} + \sum_{r=1}^{N} V_{\rm p}^{(r)}\left(L_r + L_{\rm m}^*\right)A_r. \tag{5.71}$$

Now, on using (5.68),

$$\begin{aligned}V_{\rm m}{\rm I} + \sum_{r=1}^{N} V_{p}^{(r)} A_r &= V_{\rm m}{\rm I} + \left[\sum_{r=1}^{N} V_{p}^{(r)}\left(L_r + L_{\rm m}^*\right)^{-1}\right]\left(L_{\rm m} + L_{\rm m}^*\right) \\ &= \left[V_{\rm m}\left(L_{\rm m} + L_{\rm m}^*\right)^{-1} + \sum_{r=1}^{N} V_{p}^{(r)}\left(L_r + L_{\rm m}^*\right)^{-1}\right]\left(L_{\rm m} + L_{\rm m}^*\right),\end{aligned} \tag{5.72}$$

and, on using (5.67),

$$\sum_{r=1}^{N} V_{\rm p}^{(r)}\left(L_r + L_{\rm m}^*\right)A_r = \sum_{r=1}^{N} V_{\rm p}^{(r)}\left(L_{\rm m} + L_{\rm m}^*\right), \tag{5.73}$$

it follows from (5.71) and (5.39) that

$$\left(L^* + L^*_{\mathrm{m}}\right)\left[V_{\mathrm{m}}\left(L_{\mathrm{m}} + L^*_{\mathrm{m}}\right)^{-1} + \sum_{r=1}^{N} V_{\mathrm{p}}^{(r)}\left(L_r + L^*_{\mathrm{m}}\right)^{-1}\right] = I, \tag{5.74}$$

leading to the important result

$$\left(L^* + L^*_{\mathrm{m}}\right)^{-1} = V_{\mathrm{m}}\left(L_{\mathrm{m}} + L^*_{\mathrm{m}}\right)^{-1} + \sum_{r=1}^{N} V_{\mathrm{p}}^{(r)}\left(L_r + L^*_{\mathrm{m}}\right)^{-1}. \tag{5.75}$$

5.8 Walpole's Notation

Walpole's notation, as described in [7, 8], applies to any fourth-order tensor Q_{ijkl} having the following symmetry properties

$$Q_{ijkl} = Q_{jikl} = Q_{ijlk}. \tag{5.76}$$

In this case, the components of the fourth-order tensor Q_{ijkl} may be expressed in terms of six independent scalar quantities, denoted here by a, b, c, d, e, f so that the fourth-order tensor may be represented symbolically here as follows:

$$Q \equiv (a, b, c, d, e, f). \tag{5.77}$$

If $Q' \equiv (a', b', c', d', e', f')$, then the representation for the inner product AA' of two fourth-order tensors is written symbolically as follows (see [8], although notation has been changed here):

$$QQ' = \left(aa' + 2fe',\ bb' + 2ef',\ cc',\ dd',\ ea' + be',\ fb' + af'\right) \neq Q'Q. \tag{5.78}$$

The inverse Q^{-1} of the tensor Q is

$$Q^{-1} = \left(\frac{b}{\Delta}, \frac{a}{\Delta}, \frac{1}{c}, \frac{1}{d}, -\frac{e}{\Delta}, -\frac{f}{\Delta}\right), \quad \text{where} \quad \Delta = ab - 2ef. \tag{5.79}$$

It follows, on using (5.78) and (5.79), that

$$QQ^{-1} = Q^{-1}Q = \mathrm{I} = \left(1, 1, 1, 1, 0, 0\right), \tag{5.80}$$

defining symbolically the fourth-order unit tensor I. This approach represents the most recent formulation proposed by Walpole. It should be noted that it differs from that used by Qiu and Weng [9] because of the ordering of the terms when using the symbolic notation.

For a transverse isotropic solid, where the axial direction corresponds to the direction of the x_1-axis, the stress-strain relations (in the absence of thermal stresses) defining the elastic coefficients L_{ijkl} for the inclusion have the explicit form

$$\begin{aligned}
\sigma_{11} &= \left(E_{\mathrm{A}} + 4k_{\mathrm{T}}\nu_A^2\right)\varepsilon_{11} + 2\nu_A k_{\mathrm{T}}\,\varepsilon_{22} + 2\nu_A k_{\mathrm{T}}\,\varepsilon_{33},\\
\sigma_{22} &= 2\nu_{\mathrm{A}} k_{\mathrm{T}}\,\varepsilon_{11} + \left(k_{\mathrm{T}} + \mu_{\mathrm{t}}\right)\varepsilon_{22} + \left(k_{\mathrm{T}} - \mu_{\mathrm{t}}\right)\varepsilon_{33},\\
\sigma_{33} &= 2\nu_{\mathrm{A}} k_{\mathrm{T}}\,\varepsilon_{11} + \left(k_{\mathrm{T}} - \mu_{\mathrm{t}}\right)\varepsilon_{22} + \left(k_{\mathrm{T}} + \mu_{\mathrm{t}}\right)\varepsilon_{33},\\
\sigma_{12} &= 2\mu_{\mathrm{A}}\varepsilon_{12}, \quad \sigma_{13} = 2\mu_{\mathrm{A}}\varepsilon_{13}, \quad \sigma_{23} = 2\mu_{\mathrm{t}}\varepsilon_{23},
\end{aligned} \tag{5.81}$$

where there are five independent elastic constants given by E_{A} the axial Young's modulus, ν_{A} the axial Poisson's ratio, k_{T} the plane strain bulk modulus, μ_{A} and μ_{t}, which are the axial and transverse shear moduli, respectively. The corresponding transverse Young's modulus E_T and transverse Poisson's ratio ν_t are obtained from the following relations (obtained from (2.198) and (2.202)),

$$\frac{4}{E_T} = \frac{1}{k_T} + \frac{1}{\mu_t} + \frac{4\nu_A^2}{E_A}, \qquad \nu_t = \frac{E_T}{2\mu_t} - 1. \tag{5.82}$$

Qiu and Weng [9] use the following form of the stress-strain relations (5.81)

$$\begin{aligned} \sigma_{11} &= n\varepsilon_{11} + l(\varepsilon_{22} + \varepsilon_{33}), \\ \sigma_{22} + \sigma_{33} &= 2l'\varepsilon_{11} + 2k(\varepsilon_{22} + \varepsilon_{33}), \\ \sigma_{22} - \sigma_{33} &= 2m(\varepsilon_{22} - \varepsilon_{33}), \\ \sigma_{12} &= 2p\varepsilon_{12}, \ \sigma_{13} = 2p\varepsilon_{13}, \ \sigma_{23} = 2m\varepsilon_{23}, \end{aligned} \tag{5.83}$$

where there are six independent elastic constants denoted by k, l, l', m, n, p. If $l = l'$, it is stated in [8] that the matrix L_{ijkl} defined by $(5.2)_1$ is diagonally symmetric so that $L_{ijkl} = L_{klij}$, and the following identifications can be made

$$p \equiv \mu_A, \ \ m \equiv \mu_t, \ \ k \equiv k_T, \ \ l \equiv 2\nu_A k_T, \ \ n \equiv E_A + 4k_T\nu_A^2, \tag{5.84}$$

ensuring that the stress-strain relations (5.81) are equivalent to (5.83). On using (5.76), the full set of symmetry conditions are

$$L_{ijkl} = L_{jikl} = L_{ijlk} = L_{klij}. \tag{5.85}$$

The symmetric fourth-order tensor of elastic moduli L for the matrix may be expressed in the following symbolic form involving the well-known elastic constants

$$L \equiv \left(2k_T, E_A + 4k_T\nu_A^2, 2\mu_t, 2\mu_A, 2k_T\nu_A, 2k_T\nu_A\right), \tag{5.86}$$

having inverse

$$L^{-1} \equiv M \equiv \left(\frac{1}{2k_T} + \frac{2\nu_A^2}{E_A}, \frac{1}{E_A}, \frac{1}{2\mu_t}, \frac{1}{2\mu_A}, -\frac{\nu_A}{E_A}, -\frac{\nu_A}{E_A}\right), \tag{5.87}$$

which may also be written, on using (5.82), in the form

$$L^{-1} \equiv M \equiv \left(\frac{1-\nu_t}{E_T}, \frac{1}{E_A}, \frac{1}{2\mu_t}, \frac{1}{2\mu_A}, -\frac{\nu_A}{E_A}, -\frac{\nu_A}{E_A}\right). \tag{5.88}$$

Expressions (5.86) and (5.88), when used in conjunction with the symbolic form developed by Walpole [9], enable the fourth-order tensor L_{ijkl} of elastic constants to be related to the well-known elastic constants $E_A, E_T, \nu_A, \nu_T, \mu_A, \mu_t$ and k_T used when describing transverse isotropic materials. Walpole's approach is readily applied to the analysis presented in this chapter to determine the dependence of the effective properties of a transverse isotropic composite on the properties of the various reinforcements and matrix.

Spheroidal inclusions are considered further in Chapter 15 where it is reported that the analysis of Qiu and Weng [9] considered previously leads to predictions which appear to be numerically equivalent to results derived by applying Maxwell's methodology to aligned arrays of spheroidal particles embedded in an isotropic matrix.

References

1. Eshelby, J.D. (1957). The elastic field of an ellipsoidal inclusion. *Proceedings of the Royal Society of London* A241: 376–396.
2. Mura, T. (1982). *Micromechanics of Defects in Solids*. The Hague: Martinus Nijhoff.
3. Eshelby, J.D. (1959). The elastic field outside an ellipsoidal inclusion. *Proceedings of the Royal Society of London* A252: 561–569.
4. Mori, T. and Tanaka, K. (1973). Average stress in matrix and average elastic energy of materials with mis-fitting inclusions. *Acta Metallurgica* 21: 571–574.
5. Benveniste, Y. (1987). A new approach to the application of Mori-Tanaka theory in composite materials. *Mechanics of Materials* 6: 147–157.
6. Benveniste, Y., Dvorak, G.J., and Chen, T. (1991). On diagonal and elastic symmetry of the approximate effective stiffness tensor of heterogeneous media. *Journal of the Mechanics and Physics of Solids* 39: 927–946.
7. Walpole, L.J. (1969). On the overall elastic moduli of composite materials. *Journal of the Mechanics and Physics of Solids* 17: 235–251.
8. Walpole, L.J. (1981). Elastic behaviour of composite materials: Theoretical foundations. *Advances in Applied Mechanics* 21: 169–242.
9. Qiu, Y.P. and Weng, G.J. (1990). On the application of Mori-Tanaka's theory involving transversely isotropic spheroidal inclusions. *International Journal of Engineering Science* 28: 1121–1137.

6

Properties of an Undamaged Single Lamina

Overview: The building block for a laminated fibre-reinforced composite is an individual layer known as a lamina, or a ply, having a cuboid shape where one dimension (the thickness) is very much less than the other two, the length and width, which are the *in-plane* dimensions of the lamina. The length direction of the lamina defines the axial or longitudinal direction of the lamina while the width is measured in-plane at right angles to the axial direction, i.e. in the *in-plane transverse* direction. The lamina thickness is measured in the *through-thickness direction*, sometimes referred to as the out-of-plane transverse direction. The lamina is in its principal orientation when the parallel fibres are aligned in the axial direction. An undamaged laminate is formed when a set of laminae are perfectly bonded together having fibre orientations at various angles to the axial direction of the laminate. This chapter is concerned with defining and estimating the effective thermoelastic properties of an individual lamina, including for situations where the fibres in the lamina are at some nonzero angle to the axial direction.

6.1 Notation for the Properties of a Single Lamina

First, it is assumed that the fibres of the lamina are aligned exactly in the axial direction forming what is known as a unidirectional fibre-reinforced composite. Two different sets of notation are now introduced, which describe the properties of unidirectional fibre-reinforced composites. The first is a common notation based on right-handed Cartesian coordinates (x_1,x_2,x_3), whereas the second, used in this book, is a more compact notation that enables an immediate physical interpretation of each property. Consider a single lamina of a composite material, as shown in Figure 6.1, where the fibres are aligned in the x_1-direction. For this case, the fibre direction also corresponds with the axial direction of the lamina, whereas the x_2-direction corresponds with the in-plane transverse direction, and the x_3-direction corresponds with the through-thickness direction.

The thermoelastic constants are best defined with respect to stress-strain relations (see (6.1) and (6.2) below), and each material constant is described as follows.

Properties for Design of Composite Structures: Theory and Implementation Using Software, First Edition. Neil McCartney.

Companion Website: www.wiley.com/go/mccartney/properties

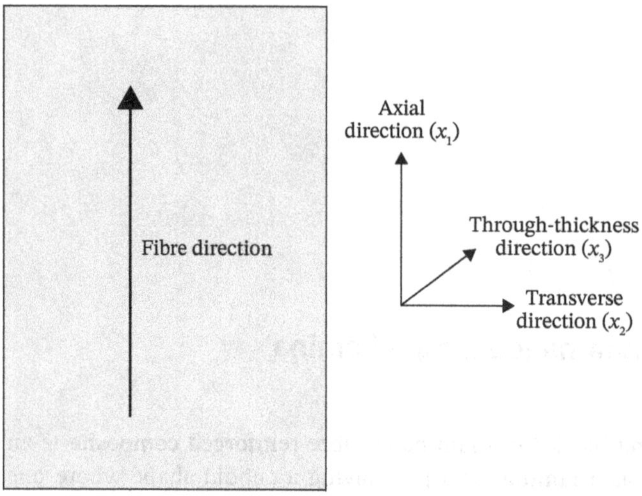

Figure 6.1 Method of defining principal directions and the coordinate system for a lamina where fibres are aligned in the axial direction.

	Common notation	Compact notation
Young's modulus in fibre direction (axial/longitudinal)	E_{11}	E_A
Young's modulus in in-plane transverse direction	E_{22}	E_T
Young's modulus in through-thickness direction	E_{33}	E_t
In-plane axial Poisson's ratio	ν_{12}	ν_A
Out-of-plane axial Poisson's ratio	ν_{13}	ν_a
Transverse Poisson's ratio	ν_{23}	ν_t
In-plane axial shear modulus	μ_{12}	μ_A
Out-of-plane axial shear modulus	μ_{13}	μ_a
Transverse shear modulus	μ_{23}	μ_t
Axial thermal expansion coefficient	α_{11}	α_A
In-plane transverse thermal expansion coefficient	α_{22}	α_T
Through-thickness thermal expansion coefficient	α_{33}	α_t

The thermal expansion coefficients are associated with a temperature difference $\Delta T = T - T_0$, where T is the uniform temperature of the lamina and T_0 is the uniform reference temperature of the lamina at which all stresses and strains are zero when the lamina is in an unloaded state.

6.2 Lamina Stress-Strain Relations

It is assumed that the loading of the lamina is such that the stress and strain distributions are uniform everywhere within the lamina. Such stress and deformation states occur when the external surfaces are subject to uniform applied tractions or linear displacements. The stress-strain relations referred to the coordinates (x_1,x_2,x_3) have the following orthotropic form:

$$\begin{aligned}
\varepsilon_{11} &= \frac{1}{E_{11}}\sigma_{11} - \frac{\nu_{21}}{E_{22}}\sigma_{22} - \frac{\nu_{31}}{E_{33}}\sigma_{33} + \alpha_{11}\Delta T, & \varepsilon_{12} &= \frac{\sigma_{12}}{2\mu_{12}}, \\
\varepsilon_{22} &= -\frac{\nu_{12}}{E_{11}}\sigma_{11} + \frac{1}{E_{22}}\sigma_{22} - \frac{\nu_{32}}{E_{33}}\sigma_{33} + \alpha_{22}\Delta T, & \varepsilon_{13} &= \frac{\sigma_{13}}{2\mu_{13}}, \\
\varepsilon_{33} &= -\frac{\nu_{13}}{E_{11}}\sigma_{11} - \frac{\nu_{23}}{E_{22}}\sigma_{22} + \frac{1}{E_{33}}\sigma_{33} + \alpha_{33}\Delta T, & \varepsilon_{23} &= \frac{\sigma_{23}}{2\mu_{23}}.
\end{aligned} \tag{6.1}$$

Using the compact notation to be used in this book, involving only independent thermoelastic constants, the stress-strain relations are written as

$$\begin{aligned}
\varepsilon_{\rm A} &= \frac{1}{E_{\rm A}}\sigma_{\rm A} - \frac{\nu_{\rm A}}{E_{\rm A}}\sigma_{\rm T} - \frac{\nu_{\rm a}}{E_{\rm A}}\sigma_{\rm t} + \alpha_{\rm A}\Delta T, & \gamma_{\rm A} &= \frac{\tau_{\rm A}}{\mu_{\rm A}}, \\
\varepsilon_{\rm T} &= -\frac{\nu_{\rm A}}{E_{\rm A}}\sigma_{\rm A} + \frac{1}{E_{\rm T}}\sigma_{\rm T} - \frac{\nu_{\rm t}}{E_{\rm T}}\sigma_{\rm t} + \alpha_{\rm T}\Delta T, & \gamma_{\rm a} &= \frac{\tau_{\rm a}}{\mu_{\rm a}}, \\
\varepsilon_{\rm t} &= -\frac{\nu_{\rm a}}{E_{\rm A}}\sigma_{\rm A} - \frac{\nu_{\rm t}}{E_{\rm T}}\sigma_{\rm T} + \frac{1}{E_{\rm t}}\sigma_{\rm t} + \alpha_{\rm t}\Delta T, & \gamma_{\rm t} &= \frac{\tau_{\rm t}}{\mu_{\rm t}}.
\end{aligned} \tag{6.2}$$

The subscripts 'A', 'T' and 't' attached to stresses, strains and properties indicate parameters associated with the axial, in-plane transverse and through-thickness directions of the lamina, respectively. It should be noted that the uppercase subscripts A and T are associated only with in-plane directions and parameters, whereas the lowercase subscripts are associated with the through-thickness direction and parameters. The above two sets of stress-strain relations are equivalent only if

$$\begin{aligned}
E_{\rm A} &= E_{11}, & E_{\rm T} &= E_{22}, & E_{\rm t} &= E_{33}, \\
\mu_{\rm A} &= \mu_{12}, & \mu_{\rm a} &= \mu_{13}, & \mu_{\rm t} &= \mu_{23}, \\
\alpha_{\rm A} &= \alpha_{11}, & \alpha_{\rm T} &= \alpha_{22}, & \alpha_{\rm t} &= \alpha_{33},
\end{aligned} \tag{6.3}$$

and

$$\begin{aligned}
&\frac{\nu_{\rm a}}{E_{\rm A}} = \frac{\nu_{13}}{E_{11}} = \frac{\nu_{31}}{E_{33}} \quad \text{so that } \nu_{13} = \nu_{\rm a}, \ \nu_{31} = \nu_{\rm a}\frac{E_{\rm t}}{E_{\rm A}}, \\
&\frac{\nu_{\rm t}}{E_{\rm T}} = \frac{\nu_{23}}{E_{22}} = \frac{\nu_{32}}{E_{33}} \quad \text{so that } \nu_{23} = \nu_{\rm t}, \ \nu_{32} = \nu_{\rm t}\frac{E_{\rm t}}{E_{\rm T}}, \\
&\frac{\nu_{\rm A}}{E_{\rm A}} = \frac{\nu_{21}}{E_{22}} = \frac{\nu_{12}}{E_{11}} \quad \text{so that } \nu_{12} = \nu_{\rm A}, \ \nu_{21} = \nu_{\rm A}\frac{E_{\rm T}}{E_{\rm A}}.
\end{aligned} \tag{6.4}$$

It should be noted that

$$\begin{aligned}
\gamma_{\rm A} &= 2\varepsilon_{12}, & \gamma_{\rm a} &= 2\varepsilon_{13}, & \gamma_{\rm t} &= 2\varepsilon_{23}, \\
\tau_{\rm A} &= \sigma_{12}, & \tau_{\rm a} &= \sigma_{13}, & \tau_{\rm t} &= \sigma_{23}.
\end{aligned} \tag{6.5}$$

The parameters $\gamma_{\rm A}$, $\gamma_{\rm a}$, $\gamma_{\rm t}$ are known as engineering shear strains which are twice the corresponding shear strains introduced when using tensor notation. It should be noted that just 9 lamina elastic properties appear in the compact version (6.2) of the stress-strain relations whereas 12 elastic properties appear when using the conventional approach shown in relations (6.1). Relations (6.4) indicate that the 9 lamina elastic properties of the compact notation are independent, adding a further reason for their use in this book. The thermoelastic constants of individual plies in a laminate are usually assumed to be transverse isotropic (see (2.197) and (2.198)) so that $E_t = E_T$, $\nu_a = \nu_A$, $\mu_a = \mu_A$, $\alpha_t = \alpha_T$ and $E_T = 2\mu_t(1+\nu_t)$. The number of independent thermoelastic constants then reduces from 12 to 7.

It is useful to give physical interpretations of the Young's moduli and Poisson's ratios by considering uniaxial stress states. First, for the case of uniaxial axial loading, where σ_A is the nonzero stress, it follows from (6.2) that the axial strain is $\varepsilon_A = \sigma_A / E_A$, the in-plane transverse strain is $\varepsilon_T = -(\nu_A / E_A)\sigma_A = -\nu_A\varepsilon_A$ and the through-thickness strain is $\varepsilon_t = -(\nu_a / E_A)\sigma_A = -\nu_a\varepsilon_A$. For the case of uniaxial in-plane transverse loading, where σ_T is the nonzero stress, it follows from (6.2) that the in-plane transverse strain $\varepsilon_T = \sigma_T / E_T$, the axial strain is $\varepsilon_A = -(\nu_A / E_A)\sigma_T = -(\nu_A E_T / E_A)\varepsilon_T$, and the through-thickness strain is $\varepsilon_t = -(\nu_t / E_T)\sigma_T = -\nu_t\varepsilon_T$. For the case of uniaxial through-thickness loading, where σ_t is the nonzero stress, it follows from (6.2) that the through-thickness strain is $\varepsilon_t = \sigma_t / E_t$, the axial strain is $\varepsilon_A = -(\nu_a / E_A)\sigma_t = -(\nu_a E_t / E_A)\varepsilon_t$ and the in-plane transverse strain is $\varepsilon_T = -(\nu_t / E_T)\sigma_t = -(\nu_t E_t / E_T)\varepsilon_t$.

6.3 Inverted Form of Lamina Stress-Strain Relations

It is useful to derive the inverted form of the stress-strain relations (6.2). From $(6.2)_3$

$$\sigma_t = E_t\,\varepsilon_t + \nu_a\frac{E_t}{E_A}\sigma_A + \nu_t\frac{E_t}{E_T}\sigma_T - E_t\,\alpha_t\,\Delta T. \tag{6.6}$$

Substitution into $(6.2)_1$ and $(6.2)_2$ then leads to

$$\left(1-\nu_a^2\frac{E_t}{E_A}\right)\frac{\sigma_A}{E_A} - \left(\nu_A+\nu_a\nu_t\frac{E_t}{E_T}\right)\frac{\sigma_T}{E_A} = \varepsilon_A + \nu_a\frac{E_t}{E_A}\varepsilon_t - \left(\alpha_A + \nu_a\frac{E_t}{E_A}\alpha_t\right)\Delta T, \tag{6.7}$$

$$-\left(\nu_A+\nu_a\nu_t\frac{E_t}{E_T}\right)\frac{\sigma_A}{E_A} + \left(1-\nu_t^2\frac{E_t}{E_T}\right)\frac{\sigma_T}{E_T} = \varepsilon_T + \nu_t\frac{E_t}{E_T}\varepsilon_t - \left(\alpha_T + \nu_t\frac{E_t}{E_T}\alpha_t\right)\Delta T. \tag{6.8}$$

On solving (6.7) and (6.8) for σ_A and σ_T, it can be shown that

$$\begin{aligned}\Lambda\frac{\sigma_A}{E_A} = {}& \left(1-\nu_t^2\frac{E_t}{E_T}\right)\left(\varepsilon_A - \alpha_A\Delta T\right) + \left(\nu_A+\nu_a\nu_t\frac{E_t}{E_T}\right)\frac{E_T}{E_A}\left(\varepsilon_T - \alpha_T\Delta T\right)\\ & + \left(\nu_a+\nu_t\nu_A\right)\frac{E_t}{E_A}\left(\varepsilon_t - \alpha_t\Delta T\right),\end{aligned} \tag{6.9}$$

$$\begin{aligned}\Lambda\frac{\sigma_T}{E_T} = {}& \left(\nu_A+\nu_a\nu_t\frac{E_t}{E_T}\right)\left(\varepsilon_A - \alpha_A\Delta T\right) + \left(1-\nu_a^2\frac{E_t}{E_A}\right)\left(\varepsilon_T - \alpha_T\Delta T\right)\\ & + \left(\nu_t+\nu_a\nu_A\frac{E_T}{E_A}\right)\frac{E_t}{E_T}\left(\varepsilon_t - \alpha_t\Delta T\right),\end{aligned} \tag{6.10}$$

where

$$\Lambda = \left(1-\nu_t^2\frac{E_t}{E_T}\right)\left(1-\nu_a^2\frac{E_t}{E_A}\right) - \frac{E_T}{E_A}\left(\nu_A+\nu_a\nu_t\frac{E_t}{E_T}\right)^2. \tag{6.11}$$

From (6.6), (6.9) and (6.10)

$$\Lambda \frac{\sigma_t}{E_t} = \left(\nu_a + \nu_t \nu_A\right)\left(\varepsilon_A - \alpha_A \Delta T\right) + \left(\nu_t + \nu_a \nu_A \frac{E_T}{E_A}\right)\left(\varepsilon_T - \alpha_T \Delta T\right) + \left(1 - \nu_A^2 \frac{E_T}{E_A}\right)\left(\varepsilon_t - \alpha_t \Delta T\right). \tag{6.12}$$

On assembling the required results, the inverted form of the stress-strain relations (6.2) may be written as

$$\begin{aligned}
\sigma_A &= \bar{E}_A \varepsilon_A + \bar{\nu}_A \bar{E}_T \varepsilon_T + \bar{\nu}_a \bar{E}_t \varepsilon_t - \bar{E}_A \bar{\alpha}_A \Delta T, \quad && \tau_A = \mu_A \gamma_A, \\
\sigma_T &= \bar{\nu}_A \bar{E}_T \varepsilon_A + \bar{E}_T \varepsilon_T + \bar{\nu}_t \bar{E}_t \varepsilon_t - \bar{E}_T \bar{\alpha}_T \Delta T, \quad && \tau_a = \mu_a \gamma_a, \\
\sigma_t &= \bar{\nu}_a \bar{E}_t \varepsilon_A + \bar{\nu}_t \bar{E}_t \varepsilon_T + \bar{E}_t \varepsilon_t - \bar{E}_t \bar{\alpha}_t \Delta T, \quad && \tau_t = \mu_t \gamma_t,
\end{aligned} \tag{6.13}$$

where

$$\begin{gathered}
\bar{E}_A = \frac{E_A}{\Lambda}\left(1 - \nu_t^2 \frac{E_t}{E_T}\right), \quad \bar{E}_T = \frac{E_T}{\Lambda}\left(1 - \nu_a^2 \frac{E_t}{E_A}\right), \quad \bar{E}_t = \frac{E_t}{\Lambda}\left(1 - \nu_A^2 \frac{E_T}{E_A}\right), \\
\bar{\nu}_A \bar{E}_T = \frac{E_T}{\Lambda}\left(\nu_A + \nu_a \nu_t \frac{E_t}{E_T}\right), \; \bar{\nu}_a \bar{E}_t = \frac{E_t}{\Lambda}\left(\nu_a + \nu_t \nu_A\right), \; \bar{\nu}_t \bar{E}_t = \frac{E_t}{\Lambda}\left(\nu_t + \nu_a \nu_A \frac{E_T}{E_A}\right),
\end{gathered} \tag{6.14}$$

$$\begin{aligned}
\bar{E}_A \bar{\alpha}_A &= \bar{E}_A \alpha_A + \bar{\nu}_A \bar{E}_T \alpha_T + \bar{\nu}_a \bar{E}_t \alpha_t, \\
\bar{E}_T \bar{\alpha}_T &= \bar{\nu}_A \bar{E}_T \alpha_A + \bar{E}_T \alpha_T + \bar{\nu}_t \bar{E}_t \alpha_t, \\
\bar{E}_t \bar{\alpha}_t &= \bar{\nu}_a \bar{E}_t \alpha_A + \bar{\nu}_t \bar{E}_t \alpha_T + \bar{E}_t \alpha_t,
\end{aligned} \tag{6.15}$$

$$\Lambda = 1 - \nu_a^2 \frac{E_t}{E_A} - \nu_t^2 \frac{E_t}{E_T} - \nu_A^2 \frac{E_T}{E_A} - 2\nu_a \nu_t \nu_A \frac{E_t}{E_A}. \tag{6.16}$$

It should be noted that the elastic coefficients of the three stress-strain relations (6.13) are symmetric as required (see (2.149)–(2.151)).

6.4 Generalised Plane Stress Conditions

In this book, *generalised plane stress conditions* are defined to be characterised by a through-thickness stress σ_t that has a uniform value everywhere in the system considered, whether a single ply or a laminate. In addition, the out-of-plane shear stresses τ_a and τ_t are both zero everywhere in the system. From $(6.13)_3$ it follows that

$$\varepsilon_t = \frac{\sigma_t}{\bar{E}_t} - \bar{\nu}_a \varepsilon_A - \bar{\nu}_t \varepsilon_T + \bar{\alpha}_t \Delta T, \tag{6.17}$$

and substitution into $(6.13)_{1,2}$ leads to

$$\begin{aligned}
\sigma_A &= \left(\bar{E}_A - \bar{\nu}_a^2 \bar{E}_t\right)\varepsilon_A + \left(\bar{\nu}_A \bar{E}_T - \bar{\nu}_a \bar{\nu}_t \bar{E}_t\right)\varepsilon_T + \bar{\nu}_a \sigma_t - \left(\bar{E}_A \bar{\alpha}_A - \bar{\nu}_a \bar{E}_t \bar{\alpha}_t\right)\Delta T, \\
\sigma_T &= \left(\bar{\nu}_A \bar{E}_T - \bar{\nu}_a \bar{\nu}_t \bar{E}_t\right)\varepsilon_A + \left(\bar{E}_T - \bar{\nu}_t^2 \bar{E}_t\right)\varepsilon_T + \bar{\nu}_t \sigma_t - \left(\bar{E}_T \bar{\alpha}_T - \bar{\nu}_t \bar{E}_t \bar{\alpha}_t\right)\Delta T.
\end{aligned} \tag{6.18}$$

It can be shown that

$$\begin{gathered}
\bar{E}_A - \bar{\nu}_a^2 \bar{E}_t = \tilde{E}_A = \frac{E_A}{1 - \nu_A^2 \dfrac{E_T}{E_A}}, \quad \bar{E}_T - \bar{\nu}_t^2 \bar{E}_t = \tilde{E}_T = \frac{E_T}{1 - \nu_A^2 \dfrac{E_T}{E_A}}, \\
\bar{\nu}_A \bar{E}_T - \bar{\nu}_a \bar{\nu}_t \bar{E}_t = \nu_A \tilde{E}_T = \frac{\nu_A E_T}{1 - \nu_A^2 \dfrac{E_T}{E_A}},
\end{gathered} \tag{6.19}$$

where use has been made of the identities

$$
\begin{aligned}
\Lambda &\equiv \left(1-\nu_t^2\frac{E_t}{E_T}\right)\left(1-\nu_a^2\frac{E_t}{E_A}\right)-\frac{E_T}{E_A}\left(\nu_A+\nu_a\nu_t\frac{E_t}{E_T}\right)^2,\\
&\equiv \left(1-\nu_t^2\frac{E_t}{E_T}\right)\left(1-\nu_A^2\frac{E_T}{E_A}\right)-\frac{E_t}{E_A}\left(\nu_a+\nu_t\nu_A\right)^2,\\
&\equiv \left(1-\nu_a^2\frac{E_t}{E_A}\right)\left(1-\nu_A^2\frac{E_T}{E_A}\right)-\frac{E_t}{E_T}\left(\nu_t+\nu_a\nu_A\frac{E_T}{E_A}\right)^2,\\
&\equiv \frac{1}{\nu_A}\left[\left(\nu_A+\nu_a\nu_t\frac{E_t}{E_T}\right)\left(1-\nu_A^2\frac{E_T}{E_A}\right)-\frac{E_t}{E_T}\left(\nu_a+\nu_t\nu_A\right)\left(\nu_t+\nu_a\nu_A\frac{E_T}{E_A}\right)\right],\\
&\equiv 1-\nu_a^2\frac{E_t}{E_A}-\nu_t^2\frac{E_t}{E_T}-\nu_A^2\frac{E_T}{E_A}-2\nu_a\nu_t\nu_A\frac{E_t}{E_A}.
\end{aligned}
\tag{6.20}
$$

From (6.14)

$$
\bar{\nu}_a=\frac{\nu_a+\nu_t\nu_A}{1-\nu_A^2\dfrac{E_T}{E_A}},\qquad \bar{\nu}_t=\frac{\nu_t+\nu_a\nu_A\dfrac{E_T}{E_A}}{1-\nu_A^2\dfrac{E_T}{E_A}}.
\tag{6.21}
$$

From (6.15)

$$
\begin{aligned}
\bar{E}_A\bar{\alpha}_A-\bar{\nu}_a\bar{E}_t\bar{\alpha}_t &=\left(\bar{E}_A-\bar{\nu}_a^2\bar{E}_t\right)\alpha_A+\left(\bar{\nu}_A\bar{E}_T-\bar{\nu}_a\bar{\nu}_t\bar{E}_t\right)\alpha_T=\frac{E_A\alpha_A+\nu_A E_T\alpha_T}{1-\nu_A^2\dfrac{E_T}{E_A}},\\
\bar{E}_T\bar{\alpha}_T-\bar{\nu}_t\bar{E}_t\bar{\alpha}_t &=\left(\bar{\nu}_A\bar{E}_T-\bar{\nu}_a\bar{\nu}_t\bar{E}_t\right)\alpha_A+\left(\bar{E}_T-\bar{\nu}_t^2\bar{E}_t\right)\alpha_T=\frac{\nu_A E_T\alpha_A+E_T\alpha_T}{1-\nu_A^2\dfrac{E_T}{E_A}}.
\end{aligned}
\tag{6.22}
$$

On using (6.19), (6.21) and (6.22) it then follows that for generalised plane stress conditions, the stress-strain relations (6.18) may be expressed in the form

$$
\begin{aligned}
\sigma_A &= \tilde{E}_A\varepsilon_A+\nu_A\tilde{E}_T\varepsilon_T+\tilde{\nu}_a\sigma_t-\tilde{E}_A\tilde{\alpha}_A\,\Delta T,\\
\sigma_T &= \nu_A\tilde{E}_T\varepsilon_A+\tilde{E}_T\varepsilon_T+\tilde{\nu}_t\sigma_t-\tilde{E}_T\tilde{\alpha}_T\Delta T,
\end{aligned}
\tag{6.23}
$$

where

$$
\tilde{E}_A=\frac{E_A}{1-\nu_A^2\dfrac{E_T}{E_A}},\quad \tilde{E}_T=\frac{E_T}{1-\nu_A^2\dfrac{E_T}{E_A}},\quad \tilde{\nu}_a=\frac{\nu_a+\nu_t\nu_A}{1-\nu_A^2\dfrac{E_T}{E_A}},\quad \tilde{\nu}_t=\frac{\nu_t+\nu_a\nu_A\dfrac{E_T}{E_A}}{1-\nu_A^2\dfrac{E_T}{E_A}},
$$
$$
\tilde{\alpha}_A=\alpha_A+\nu_A\frac{E_T}{E_A}\alpha_T,\qquad \tilde{\alpha}_T=\alpha_T+\nu_A\alpha_A.
\tag{6.24}
$$

The term *plane stress conditions* will refer to those generalised plane stress conditions for which $\sigma_t=\tau_a=\tau_t=0$. It is clear that when plane stress conditions prevail, so that $\sigma_t=0$, the stress-strain relations (6.23) reduce to the simpler form

$$
\begin{aligned}
\sigma_A &= \tilde{E}_A\varepsilon_A+\nu_A\tilde{E}_T\varepsilon_T-\tilde{E}_A\tilde{\alpha}_A\,\Delta T,\\
\sigma_T &= \nu_A\tilde{E}_T\varepsilon_A+\tilde{E}_T\varepsilon_T-\tilde{E}_T\tilde{\alpha}_T\Delta T.
\end{aligned}
\tag{6.25}
$$

On inverting these equations, the first two of the nonshear stress-strain relations in (6.2) are recovered for the case where $\sigma_t = 0$, as to be expected.

6.5 Generalised Plane Strain Conditions

In this book, *generalised plane strain conditions* are defined to be characterised by an in-plane transverse strain ε_T that has a uniform value everywhere in the system, whether a single ply or a laminate. In addition, the displacement components are independent of the coordinate x_2. From $(6.2)_2$

$$\sigma_T = \nu_A \frac{E_T}{E_A}\sigma_A + \nu_t\sigma_t + E_T\left(\varepsilon_T - \alpha_T\Delta T\right), \tag{6.26}$$

and on substituting into $(6.2)_1$ and $(6.2)_3$ using (6.24) leads to

$$\varepsilon_A - \alpha_A\Delta T = \frac{\sigma_A}{\tilde{E}_A} - \frac{\tilde{\nu}_a}{\tilde{E}_A}\sigma_t - \nu_A\frac{E_T}{E_A}\left(\varepsilon_T - \alpha_T\Delta T\right), \tag{6.27}$$

$$\varepsilon_t - \alpha_t\Delta T = -\frac{\tilde{\nu}_a}{\tilde{E}_A}\sigma_A + \frac{\sigma_t}{\tilde{E}_t} - \nu_t\left(\varepsilon_T - \alpha_T\Delta T\right), \tag{6.28}$$

where

$$\tilde{E}_t = \frac{E_t}{1 - \nu_t^2\dfrac{E_t}{E_T}}. \tag{6.29}$$

These relations are now written in the form

$$\varepsilon_A = \frac{\sigma_A}{\tilde{E}_A} - \frac{\tilde{\nu}_a}{\tilde{E}_A}\sigma_t - \nu_A\frac{E_T}{E_A}\varepsilon_T + \tilde{\alpha}_A\Delta T, \tag{6.30}$$

$$\varepsilon_t = -\frac{\tilde{\nu}_a}{\tilde{E}_A}\sigma_A + \frac{\tilde{\sigma}_t}{\tilde{E}_t} - \nu_t\varepsilon_T + \tilde{\alpha}_t\Delta T, \tag{6.31}$$

where

$$\tilde{\alpha}_t = \alpha_t + \nu_t\alpha_T. \tag{6.32}$$

The term *plane strain conditions* will refer to those generalised plane strain conditions for which $\varepsilon_T = \gamma_A = \gamma_t = 0$. This situation arises when the displacement in the x_2-direction is everywhere uniform and the remaining displacements depend only on x_1 and x_3. From (6.30) and (6.31) it is clear that, when plane strain conditions prevail, the stress-strain relations reduce to the form

$$\varepsilon_A = \frac{\sigma_A}{\tilde{E}_A} - \frac{\tilde{\nu}_a}{\tilde{E}_A}\sigma_t + \tilde{\alpha}_A\Delta T, \tag{6.33}$$

$$\varepsilon_t = -\frac{\tilde{\nu}_a}{\tilde{E}_A}\sigma_A + \frac{\sigma_t}{\tilde{E}_t} + \tilde{\alpha}_t\Delta T. \tag{6.34}$$

6.6 Extending the Contracted Notation for Tensors

The components of the stress and strain tensors defined by (2.162) are now extended so that

$$\begin{bmatrix}\sigma_A\\ \sigma_T\\ \sigma_t\\ \tau_t\\ \tau_a\\ \tau_A\end{bmatrix} \equiv \begin{bmatrix}\sigma_1\\ \sigma_2\\ \sigma_3\\ \sigma_4\\ \sigma_5\\ \sigma_6\end{bmatrix} \equiv \begin{bmatrix}\sigma_{11}\\ \sigma_{22}\\ \sigma_{33}\\ \sigma_{23}\\ \sigma_{13}\\ \sigma_{12}\end{bmatrix}, \quad \begin{bmatrix}\varepsilon_A\\ \varepsilon_T\\ \varepsilon_t\\ \gamma_t\\ \gamma_a\\ \gamma_A\end{bmatrix} \equiv \begin{bmatrix}\varepsilon_1\\ \varepsilon_2\\ \varepsilon_3\\ \varepsilon_4\\ \varepsilon_5\\ \varepsilon_6\end{bmatrix} \equiv \begin{bmatrix}\varepsilon_{11}\\ \varepsilon_{22}\\ \varepsilon_{33}\\ 2\varepsilon_{23}\\ 2\varepsilon_{13}\\ 2\varepsilon_{12}\end{bmatrix}. \tag{6.35}$$

General linear elastic stress-strain relations (2.163), including thermal expansion terms, are given by

$$\begin{bmatrix}\sigma_A\\ \sigma_T\\ \sigma_t\\ \tau_t\\ \tau_a\\ \tau_A\end{bmatrix} = \begin{bmatrix} C_{11} & C_{12} & C_{13} & C_{14} & C_{15} & C_{16}\\ C_{21} & C_{22} & C_{23} & C_{24} & C_{25} & C_{26}\\ C_{31} & C_{32} & C_{33} & C_{34} & C_{35} & C_{36}\\ C_{41} & C_{42} & C_{43} & C_{44} & C_{45} & C_{46}\\ C_{51} & C_{52} & C_{53} & C_{54} & C_{55} & C_{56}\\ C_{61} & C_{62} & C_{63} & C_{64} & C_{65} & C_{66}\end{bmatrix} \begin{bmatrix}\varepsilon_A\\ \varepsilon_T\\ \varepsilon_t\\ \gamma_t\\ \gamma_a\\ \gamma_A\end{bmatrix} - \begin{bmatrix}U_1\\ U_2\\ U_3\\ U_4\\ U_5\\ U_6\end{bmatrix} \Delta T. \tag{6.36}$$

For orthotropic materials these stress-strain relations have the simpler form (cf. (2.164))

$$\begin{bmatrix}\sigma_A\\ \sigma_T\\ \sigma_t\\ \tau_t\\ \tau_a\\ \tau_A\end{bmatrix} = \begin{bmatrix} C_{11} & C_{12} & C_{13} & 0 & 0 & 0\\ C_{21} & C_{22} & C_{23} & 0 & 0 & 0\\ C_{31} & C_{32} & C_{33} & 0 & 0 & 0\\ 0 & 0 & 0 & C_{44} & 0 & 0\\ 0 & 0 & 0 & 0 & C_{55} & 0\\ 0 & 0 & 0 & 0 & 0 & C_{66}\end{bmatrix} \begin{bmatrix}\varepsilon_A\\ \varepsilon_T\\ \varepsilon_t\\ \gamma_t\\ \gamma_a\\ \gamma_A\end{bmatrix} - \begin{bmatrix}U_1\\ U_2\\ U_3\\ 0\\ 0\\ 0\end{bmatrix} \Delta T. \tag{6.37}$$

By comparing (6.37) with (6.13) using (6.14)–(6.16), it follows that the nonzero components of the **C** matrix are related to the elastic constants defined by the stress-strain relations (6.2) as follows:

$$C_{11} = \frac{E_A}{\Lambda}\left(1 - \nu_t^2 \frac{E_t}{E_T}\right) = \bar{E}_A, \; C_{12} = \frac{E_T}{\Lambda}\left(\nu_A + \nu_a \nu_t \frac{E_t}{E_T}\right) = \bar{\nu}_A \bar{E}_T,$$
$$C_{13} = \frac{E_t}{\Lambda}(\nu_a + \nu_t \nu_A) = \bar{\nu}_a \bar{E}_t, \tag{6.38}$$

$$C_{21} = \frac{E_T}{\Lambda}\left(\nu_A + \nu_a \nu_t \frac{E_t}{E_T}\right) = \bar{\nu}_A \bar{E}_T, \; C_{22} = \frac{E_T}{\Lambda}\left(1 - \nu_a^2 \frac{E_t}{E_A}\right) = \bar{E}_T,$$
$$C_{23} = \frac{E_t}{\Lambda}\left(\nu_t + \nu_a \nu_A \frac{E_T}{E_A}\right) = \bar{\nu}_t \bar{E}_t, \tag{6.39}$$

$$C_{31} = \frac{E_t}{\Lambda}(\nu_a + \nu_t \nu_A) = \bar{\nu}_a \bar{E}_t, \; C_{32} = \frac{E_t}{\Lambda}\left(\nu_t + \nu_a \nu_A \frac{E_T}{E_A}\right) = \bar{\nu}_t \bar{E}_t,$$
$$C_{33} = \frac{E_t}{\Lambda}\left(1 - \nu_A^2 \frac{E_T}{E_A}\right) = \bar{E}_t, \tag{6.40}$$

$$C_{44} = \mu_t, \qquad C_{55} = \mu_a, \qquad C_{66} = \mu_A, \tag{6.41}$$

where

$$\Lambda = 1 - \nu_a^2 \frac{E_t}{E_A} - \nu_t^2 \frac{E_t}{E_T} - \nu_A^2 \frac{E_T}{E_A} - 2\nu_a \nu_t \nu_A \frac{E_t}{E_A}. \tag{6.42}$$

The nonzero components of the $\boldsymbol{U}$ vector are related to the thermoelastic constants defined by the stress-strain relations (6.2) as follows

$$U_1 = \bar{E}_A \bar{\alpha}_A, \quad U_2 = \bar{E}_T \bar{\alpha}_T, \quad U_3 = \bar{E}_t \bar{\alpha}_t. \tag{6.43}$$

The inverse matrix form of (6.37) is of the form

$$\begin{bmatrix} \varepsilon_A \\ \varepsilon_T \\ \varepsilon_t \\ \gamma_t \\ \gamma_a \\ \gamma_A \end{bmatrix} = \begin{bmatrix} S_{11} & S_{12} & S_{13} & 0 & 0 & 0 \\ S_{21} & S_{22} & S_{23} & 0 & 0 & 0 \\ S_{31} & S_{32} & S_{33} & 0 & 0 & \\ 0 & 0 & 0 & S_{44} & 0 & 0 \\ 0 & 0 & 0 & 0 & S_{55} & 0 \\ 0 & 0 & 0 & 0 & 0 & S_{66} \end{bmatrix} \begin{bmatrix} \sigma_A \\ \sigma_T \\ \sigma_t \\ \tau_t \\ \tau_a \\ \tau_A \end{bmatrix} + \begin{bmatrix} \alpha_A \\ \alpha_T \\ \alpha_t \\ 0 \\ 0 \\ 0 \end{bmatrix} \Delta T, \tag{6.44}$$

and a comparison with (6.2) shows that

$$S_{11} = \frac{1}{E_A}, \qquad S_{12} = -\frac{\nu_A}{E_A}, \quad S_{13} = -\frac{\nu_a}{E_A}, \tag{6.45}$$

$$S_{21} = -\frac{\nu_A}{E_A}, \quad S_{22} = \frac{1}{E_T}, \qquad S_{23} = -\frac{\nu_t}{E_T}, \tag{6.46}$$

$$S_{31} = -\frac{\nu_a}{E_A}, \quad S_{32} = -\frac{\nu_t}{E_T}, \quad S_{33} = \frac{1}{E_t}, \tag{6.47}$$

$$S_{44} = \frac{1}{\mu_t}, \qquad S_{55} = \frac{1}{\mu_a}, \qquad S_{66} = \frac{1}{\mu_A}. \tag{6.48}$$

The ordering of the components is selected to be consistent with that used in the next section when using the well-known contracted notation for anisotropic materials (see Section 6.5).

6.7 Thermoelastic Constants for Angled Laminae

Having considered in Chapter 4 the methods of estimating ply properties from those of the fibres and matrix it is required now to consider how these properties are used to derive those of angled plies. One very important parameter that must be clearly defined is the angle ϕ defining the fibre directions in an angled lamina. Figure 6.2 illustrates the definition that is used in this book, and it should be noted that this angle differs in sign from that used in some previous publications by the present author. The change of sign affects only the signs of shear coupling parameters (see Section 6.9).

Figure 6.2 also defines the three orthogonal principal directions of the laminate, and associates these directions with a right-handed system (x_1, x_2, x_3) of Cartesian coordinates defining a set of global axes. The angle ϕ defining the fibre direction of the lamina is measured from the global x_1-axis in a clockwise direction when viewing from a point situated on the negative part of the global x_3-axis (see Figure 6.2).

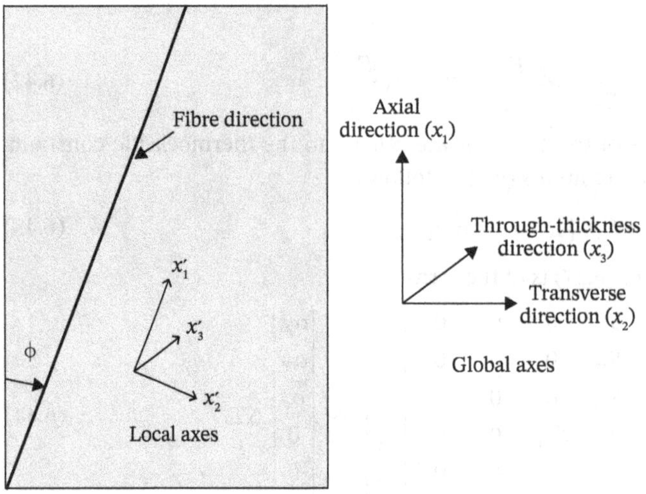

Figure 6.2 Method of defining principal directions and coordinate system for an angled lamina, and the angle ϕ specifying the fibre direction.

From (6.1) and (6.3), the strain components referred to local coordinates can be expressed in terms of local stress components using the following stress-strain relations for the lamina expressed in terms of stress and strain components referred to local axes (see Figure 6.2)

$$\varepsilon'_{11} = \frac{1}{E_A}\sigma'_{11} - \frac{\nu_A}{E_A}\sigma'_{22} - \frac{\nu_a}{E_A}\sigma'_{33} + \alpha_A \Delta T, \tag{6.49}$$

$$\varepsilon'_{22} = -\frac{\nu_A}{E_A}\sigma'_{11} + \frac{1}{E_T}\sigma'_{22} - \frac{\nu_t}{E_T}\sigma'_{33} + \alpha_T \Delta T, \tag{6.50}$$

$$\varepsilon'_{33} = -\frac{\nu_a}{E_A}\sigma'_{11} - \frac{\nu_t}{E_T}\sigma'_{22} + \frac{1}{E_t}\sigma'_{33} + \alpha_t \Delta T, \tag{6.51}$$

$$\varepsilon'_{23} = \frac{1}{2\mu_t}\sigma'_{23}, \quad \varepsilon'_{13} = \frac{1}{2\mu_a}\sigma'_{13}, \quad \varepsilon'_{12} = \frac{1}{2\mu_A}\sigma'_{12}. \tag{6.52}$$

The global strain components may be expressed in terms of global stress components by substituting (2.179) into (6.49)–(6.52) and then substituting the results into the transformation (2.180) that calculates the global strains in terms of the local values. The resulting stress-strain equations, including thermal expansion terms, relating global strain components to global stress components are given, using matrix notation, by

$$\begin{bmatrix} \varepsilon_A \\ \varepsilon_T \\ \varepsilon_t \\ \gamma_t \\ \gamma_a \\ \gamma_A \end{bmatrix} \equiv \begin{bmatrix} \varepsilon_{11} \\ \varepsilon_{22} \\ \varepsilon_{33} \\ 2\varepsilon_{23} \\ 2\varepsilon_{13} \\ 2\varepsilon_{12} \end{bmatrix} = \begin{bmatrix} S_{11} & S_{12} & S_{13} & 0 & 0 & S_{16} \\ S_{12} & S_{22} & S_{23} & 0 & 0 & S_{26} \\ S_{13} & S_{23} & S_{33} & 0 & 0 & S_{36} \\ 0 & 0 & 0 & S_{44} & S_{45} & 0 \\ 0 & 0 & 0 & S_{45} & S_{55} & 0 \\ S_{16} & S_{26} & S_{36} & 0 & 0 & S_{66} \end{bmatrix} \begin{bmatrix} \sigma_A \\ \sigma_T \\ \sigma_t \\ \tau_t \\ \tau_a \\ \tau_A \end{bmatrix} + \begin{bmatrix} V_1 \\ V_2 \\ V_3 \\ 0 \\ 0 \\ V_6 \end{bmatrix} \Delta T, \tag{6.53}$$

where it can be shown that the coefficients in these stress-strain relations are related to the thermoelastic constants of the lamina as follows:

$$S_{11} = m^2n^2\left(\frac{1}{\mu_A} - \frac{2\nu_A}{E_A}\right) + \frac{m^4}{E_A} + \frac{n^4}{E_T}, \tag{6.54}$$

$$S_{12} = m^2n^2\left(\frac{1}{E_A} + \frac{1}{E_T} - \frac{1}{\mu_A}\right) - \left(m^4 + n^4\right)\frac{\nu_A}{E_A}, \tag{6.55}$$

$$S_{13} = -m^2\frac{\nu_a}{E_A} - n^2\frac{\nu_t}{E_T}, \tag{6.56}$$

$$S_{16} = -mn\left[\left(m^2 - n^2\right)\left(\frac{1}{\mu_A} - \frac{2\nu_A}{E_A}\right) - \frac{2m^2}{E_A} + \frac{2n^2}{E_T}\right], \tag{6.57}$$

$$S_{22} = m^2n^2\left(\frac{1}{\mu_A} - \frac{2\nu_A}{E_A}\right) + \frac{m^4}{E_T} + \frac{n^4}{E_A}, \tag{6.58}$$

$$S_{23} = -m^2\frac{\nu_t}{E_T} - n^2\frac{\nu_a}{E_A}, \tag{6.59}$$

$$S_{26} = mn\left[\left(m^2 - n^2\right)\left(\frac{1}{\mu_A} - \frac{2\nu_A}{E_A}\right) - \frac{2m^2}{E_T} + \frac{2n^2}{E_A}\right], \tag{6.60}$$

$$S_{33} = \frac{1}{E_t}, \tag{6.61}$$

$$S_{36} = 2mn\left(\frac{\nu_t}{E_T} - \frac{\nu_a}{E_A}\right), \tag{6.62}$$

$$S_{44} = \frac{m^2}{\mu_t} + \frac{n^2}{\mu_a}, \tag{6.63}$$

$$S_{45} = mn\left(\frac{1}{\mu_a} - \frac{1}{\mu_t}\right), \tag{6.64}$$

$$S_{55} = \frac{m^2}{\mu_a} + \frac{n^2}{\mu_t}, \tag{6.65}$$

$$S_{66} = 4m^2n^2\left(\frac{1}{E_A} + \frac{1}{E_T} + \frac{2\nu_A}{E_A}\right) + \frac{\left(m^2 - n^2\right)^2}{\mu_A}, \tag{6.66}$$

$$\begin{aligned} V_1 &= m^2\alpha_A + n^2\alpha_T, \\ V_2 &= m^2\alpha_T + n^2\alpha_A, \\ V_3 &= \alpha_t, \ V_4 = 0, \ V_5 = 0, \\ V_6 &= 2mn\left(\alpha_A - \alpha_T\right), \end{aligned} \tag{6.67}$$

where $m = \cos\phi$ and $n = \sin\phi$. It should be noted that the expressions (6.53)–(6.67) reduce to (6.44)–(6.48) when $\phi = 0$ so that $m = 1$ and $n = 0$.

6.8 Inverse Approach

When considering laminated composites in Chapter 7, it is useful to have access to the inverted form of the stress-strain relations (6.53). On using (6.5), the inverted form (6.13) of the stress-strain relations for a lamina referred to local coordinates are first written in the form

$$\sigma'_{11} = \bar{E}_{\mathrm{A}}\varepsilon'_{11} + \bar{\nu}_{\mathrm{A}}\bar{E}_{\mathrm{T}}\varepsilon'_{22} + \bar{\nu}_{\mathrm{a}}\bar{E}_{\mathrm{t}}\varepsilon'_{33} - \bar{E}_{\mathrm{A}}\bar{\alpha}_{\mathrm{A}}\Delta T, \tag{6.68}$$

$$\sigma'_{22} = \bar{\nu}_{\mathrm{A}}\bar{E}_{\mathrm{T}}\varepsilon'_{11} + \bar{E}_{\mathrm{T}}\varepsilon'_{22} + \bar{\nu}_{\mathrm{t}}\bar{E}_{\mathrm{t}}\varepsilon'_{33} - \bar{E}_{\mathrm{T}}\bar{\alpha}_{\mathrm{T}}\Delta T, \tag{6.69}$$

$$\sigma'_{33} = \bar{\nu}_{\mathrm{a}}\bar{E}_{\mathrm{t}}\varepsilon'_{11} + \bar{\nu}_{\mathrm{t}}\bar{E}_{\mathrm{t}}\varepsilon'_{22} + \bar{E}_{\mathrm{t}}\varepsilon'_{33} - \bar{E}_{\mathrm{t}}\bar{\alpha}_{\mathrm{t}}\,\Delta T, \tag{6.70}$$

$$\sigma'_{23} = 2\mu_{\mathrm{t}}\,\varepsilon'_{23}, \quad \sigma'_{13} = 2\mu_{\mathrm{a}}\varepsilon'_{13}, \quad \sigma'_{12} = 2\mu_{\mathrm{A}}\varepsilon'_{12}. \tag{6.71}$$

The required inverted form for the stress-strain relations of an angled ply is obtained by substituting (2.181) into (6.68)–(6.71), and then substituting the results into (2.178) so that

$$\begin{bmatrix} \sigma_{\mathrm{A}} \\ \sigma_{\mathrm{T}} \\ \sigma_{\mathrm{t}} \\ \tau_{\mathrm{t}} \\ \tau_{\mathrm{a}} \\ \tau_{\mathrm{A}} \end{bmatrix} \equiv \begin{bmatrix} \sigma_{11} \\ \sigma_{22} \\ \sigma_{33} \\ \sigma_{23} \\ \sigma_{13} \\ \sigma_{12} \end{bmatrix} = \begin{bmatrix} C_{11} & C_{12} & C_{13} & 0 & 0 & C_{16} \\ C_{12} & C_{22} & C_{23} & 0 & 0 & C_{26} \\ C_{13} & C_{23} & C_{33} & 0 & 0 & C_{36} \\ 0 & 0 & 0 & C_{44} & C_{45} & 0 \\ 0 & 0 & 0 & C_{45} & C_{55} & 0 \\ C_{16} & C_{26} & C_{36} & 0 & 0 & C_{66} \end{bmatrix} \begin{bmatrix} \varepsilon_{\mathrm{A}} \\ \varepsilon_{\mathrm{T}} \\ \varepsilon_{\mathrm{t}} \\ \gamma_{\mathrm{t}} \\ \gamma_{\mathrm{a}} \\ \gamma_{\mathrm{A}} \end{bmatrix} - \begin{bmatrix} U_1 \\ U_2 \\ U_3 \\ 0 \\ 0 \\ U_6 \end{bmatrix} \Delta T, \tag{6.72}$$

where

$$\begin{aligned}
C_{11} &= m^4\bar{E}_{\mathrm{A}} + n^4\bar{E}_{\mathrm{T}} + 2m^2n^2\left(\bar{\nu}_{\mathrm{A}}\bar{E}_{\mathrm{T}} + 2\mu_{\mathrm{A}}\right), \\
C_{12} &= \left(m^4 + n^4\right)\bar{\nu}_{\mathrm{A}}\bar{E}_{\mathrm{T}} + m^2n^2\left(\bar{E}_{\mathrm{A}} + \bar{E}_{\mathrm{T}} - 4\mu_{\mathrm{A}}\right), \\
C_{13} &= \left(m^2\bar{\nu}_{\mathrm{a}} + n^2\bar{\nu}_{\mathrm{t}}\right)\bar{E}_{\mathrm{t}}, \quad C_{16} = mn\left[m^2\bar{E}_{\mathrm{A}} - n^2\bar{E}_{\mathrm{T}} - \left(m^2 - n^2\right)\left(\bar{\nu}_{\mathrm{A}}\bar{E}_{\mathrm{T}} + 2\mu_{\mathrm{A}}\right)\right], \\
C_{22} &= m^4\bar{E}_{\mathrm{T}} + n^4\bar{E}_{\mathrm{A}} + 2m^2n^2\left(\bar{\nu}_{\mathrm{A}}\bar{E}_{\mathrm{T}} + 2\mu_{\mathrm{A}}\right), \quad C_{23} = \left(m^2\bar{\nu}_{\mathrm{t}} + n^2\bar{\nu}_{\mathrm{a}}\right)\bar{E}_{\mathrm{t}}, \\
C_{26} &= mn\left[n^2\bar{E}_{\mathrm{A}} - m^2\bar{E}_{\mathrm{T}} + \left(m^2 - n^2\right)\left(\bar{\nu}_{\mathrm{A}}\bar{E}_{\mathrm{T}} + 2\mu_{\mathrm{A}}\right)\right], \quad C_{33} = \bar{E}_{\mathrm{t}}, \\
C_{36} &= mn\left(\bar{\nu}_{\mathrm{a}} - \bar{\nu}_{\mathrm{t}}\right)\bar{E}_{\mathrm{t}}, \quad C_{44} = m^2\mu_{\mathrm{t}} + n^2\mu_{\mathrm{a}}, \quad C_{45} = mn\left(\mu_{\mathrm{a}} - \mu_{\mathrm{t}}\right), \\
C_{55} &= m^2\mu_{\mathrm{a}} + n^2\mu_{\mathrm{t}}, \quad C_{66} = m^2n^2\left(\bar{E}_{\mathrm{A}} + \bar{E}_{\mathrm{T}} - 2\bar{\nu}_{\mathrm{A}}\bar{E}_{\mathrm{T}}\right) + \left(m^2 - n^2\right)^2\mu_{\mathrm{A}},
\end{aligned} \tag{6.73}$$

and

$$\begin{aligned}
U_1 &= m^2\bar{E}_{\mathrm{A}}\bar{\alpha}_{\mathrm{A}} + n^2\bar{E}_{\mathrm{T}}\bar{\alpha}_{\mathrm{T}}, \\
U_2 &= n^2\bar{E}_{\mathrm{A}}\bar{\alpha}_{\mathrm{A}} + m^2\bar{E}_{\mathrm{T}}\bar{\alpha}_{\mathrm{T}}, \\
U_3 &= \bar{E}_{\mathrm{t}}\bar{\alpha}_{\mathrm{t}}, \; U_6 = mn\left(\bar{E}_{\mathrm{A}}\bar{\alpha}_{\mathrm{A}} - \bar{E}_{\mathrm{T}}\bar{\alpha}_{\mathrm{T}}\right).
\end{aligned} \tag{6.74}$$

It should be noted that (6.73) and (6.74) reduce to (6.38)–(6.43) when $\phi = 0$ so that $m = 1$ and $n = 0$. In addition, it can be shown that

$$\boldsymbol{C} \cdot \boldsymbol{S} = \boldsymbol{S} \cdot \boldsymbol{C} = \mathbf{I}, \tag{6.75}$$

where $\boldsymbol{C}$ and $\boldsymbol{S}$ are the 6×6 matrices defined in (6.73) and (6.54)–(6.66), respectively, and where $\mathbf{I}$ is the second-order identity 6×6 matrix. In expanded form, relation (6.75) may be written as

$$\sum_{K=1}^{6} C_{IK} S_{KJ} = \sum_{K=1}^{6} S_{IK} C_{KJ} = \delta_{IJ}, \quad I,J=1,2,\ldots,6, \tag{6.76}$$

where $\delta_{IJ} = 1$ if $I = J$, $\delta_{IJ} = 0$ otherwise. The satisfaction of these relations by the elements of the C and S matrices has been checked numerically.

6.9 Shear Coupling Parameters and Reduced Stress-Strain Relations

The final task of this chapter is to introduce the concept of shear coupling and to define the thermoelastic constants that characterise the phenomenon. This concept is used extensively in subsequent chapters of this book. Readers should note that the definitions for the shear coupling parameters λ differ from those used in the literature in order that some results derived in this book are simplified.

Four of the stress-strain relations (6.53) are now expressed in the form

$$\begin{aligned}
\varepsilon_A &= \frac{1}{E_A^{(P)}}\sigma_A - \frac{\nu_A^{(P)}}{E_A^{(P)}}\sigma_T - \frac{\nu_a^{(P)}}{E_A^{(P)}}\sigma_t - \frac{\lambda_A^{(P)}}{\mu_A^{(P)}}\tau_A + \alpha_A^{(P)}\Delta T,\\
\varepsilon_T &= -\frac{\nu_A^{(P)}}{E_A^{(P)}}\sigma_A + \frac{1}{E_T^{(P)}}\sigma_T - \frac{\nu_t^{(P)}}{E_T^{(P)}}\sigma_t - \frac{\lambda_T^{(P)}}{\mu_A^{(P)}}\tau_A + \alpha_T^{(P)}\Delta T,\\
\varepsilon_t &= -\frac{\nu_a^{(P)}}{E_A^{(P)}}\sigma_A - \frac{\nu_t^{(P)}}{E_T^{(P)}}\sigma_T + \frac{1}{E_t^{(P)}}\sigma_t - \frac{\lambda_t^{(P)}}{\mu_A^{(P)}}\tau_A + \alpha_t^{(P)}\Delta T,\\
\gamma_A &= -\frac{\lambda_A^{(P)}}{\mu_A^{(P)}}\sigma_A - \frac{\lambda_T^{(P)}}{\mu_A^{(P)}}\sigma_T - \frac{\lambda_t^{(P)}}{\mu_A^{(P)}}\sigma_t + \frac{1}{\mu_A^{(P)}}\tau_A + \alpha_S^{(P)}\Delta T,
\end{aligned} \tag{6.77}$$

where the superscript '(P)' (P for ply) denotes that the thermoelastic constant refers to an angled ply. The remaining two stress-strain relations in (6.53) are concerned with out-of-plane shear loading, which is of very little practical relevance to single-ply behaviour. They are considered in Chapter 7 when considering undamaged laminates (see Section 7.6). By comparing (6.53) and (6.77) it is clear that

$$\begin{aligned}
&S_{11} = \frac{1}{E_A^{(P)}},\ S_{12} = -\frac{\nu_A^{(P)}}{E_A^{(P)}},\ S_{13} = -\frac{\nu_a^{(P)}}{E_A^{(P)}},\ S_{16} = -\frac{\lambda_A^{(P)}}{\mu_A^{(P)}},\ V_1 = \alpha_A^{(P)},\\
&S_{22} = \frac{1}{E_T^{(P)}},\ S_{23} = -\frac{\nu_t^{(P)}}{E_T^{(P)}},\ S_{26} = -\frac{\lambda_T^{(P)}}{\mu_A^{(P)}},\ V_2 = \alpha_T^{(P)},\\
&S_{33} = \frac{1}{E_t^{(P)}},\ S_{36} = -\frac{\lambda_t^{(P)}}{\mu_A^{(P)}},\ V_3 = \alpha_t^{(P)},\\
&S_{66} = \frac{1}{\mu_A^{(P)}},\ V_6 = \alpha_S^{(P)}.
\end{aligned} \tag{6.78}$$

The parameters λ_A, λ_T, λ_t are dimensionless shear coupling properties as they characterise the coupling of the shear stress τ_A to the nonshear strains ε_A, ε_T and ε_t. For the special case where the shear stress τ_A is the only nonzero stress and $\Delta T = 0$, it follows from (6.77) that $\varepsilon_A = -\lambda_A^{(P)}\gamma_A$, $\varepsilon_T = -\lambda_T^{(P)}\gamma_A$ and $\varepsilon_t = -\lambda_t^{(P)}\gamma_A$.

The parameter α_S characterises a shear deformation response to temperature changes. When the fibres are aligned in the axial and in-plane transverse directions, all four parameters have zero values. It should be noted that the sign of the shear coupling parameters $\lambda_A, \lambda_T, \lambda_t$ and the expansion coefficient α_S depend upon the sign of the orientation angle ϕ, indicating why it is essential to define exactly how this angle is defined (see Section 6.7).

From (6.77)$_4$

$$\tau_A = \lambda_A^{(P)}\sigma_A + \lambda_T^{(P)}\sigma_T + \lambda_t^{(P)}\sigma_t + \mu_A^{(P)}\gamma_A - \mu_A^{(P)}\alpha_S^{(P)}\Delta T, \tag{6.79}$$

and on substituting into the remaining relations of (6.77)

$$\begin{aligned}
\tilde{\varepsilon}_A &\equiv \varepsilon_A + \lambda_A^{(P)}\gamma_A = \frac{1}{\tilde{E}_A^{(P)}}\sigma_A - \frac{\tilde{\nu}_A^{(P)}}{\tilde{E}_A^{(P)}}\sigma_T - \frac{\tilde{\nu}_a^{(P)}}{\tilde{E}_A^{(P)}}\sigma_t + \tilde{\alpha}_A^{(P)}\Delta T,\\
\tilde{\varepsilon}_T &\equiv \varepsilon_T + \lambda_T^{(P)}\gamma_A = -\frac{\tilde{\nu}_A^{(P)}}{\tilde{E}_A^{(P)}}\sigma_A + \frac{1}{\tilde{E}_T^{(P)}}\sigma_T - \frac{\tilde{\nu}_t^{(P)}}{\tilde{E}_T^{(P)}}\sigma_t + \tilde{\alpha}_T^{(P)}\Delta T,\\
\tilde{\varepsilon}_t &\equiv \varepsilon_t + \lambda_t^{(P)}\gamma_A = -\frac{\tilde{\nu}_a^{(P)}}{\tilde{E}_A^{(P)}}\sigma_A - \frac{\tilde{\nu}_t^{(P)}}{\tilde{E}_T^{(P)}}\sigma_T + \frac{1}{\tilde{E}_t^{(P)}}\sigma_t + \tilde{\alpha}_t^{(P)}\Delta T,
\end{aligned} \tag{6.80}$$

where

$$\begin{aligned}
&\frac{1}{\tilde{E}_A^{(P)}} = \frac{1}{E_A^{(P)}} - \left(\lambda_A^{(P)}\right)^2 \frac{1}{\mu_A^{(P)}}, \qquad \frac{1}{\tilde{E}_T^{(P)}} = \frac{1}{E_T^{(P)}} - \left(\lambda_T^{(P)}\right)^2 \frac{1}{\mu_A^{(P)}},\\
&\frac{1}{\tilde{E}_t^{(P)}} = \frac{1}{E_t^{(P)}} - \left(\lambda_t^{(P)}\right)^2 \frac{1}{\mu_A^{(P)}}, \qquad \frac{\tilde{\nu}_t^{(P)}}{\tilde{E}_T^{(P)}} = \frac{\nu_t^{(P)}}{E_T^{(P)}} + \frac{\lambda_t^{(P)}\lambda_T^{(P)}}{\mu_A^{(P)}},\\
&\frac{\tilde{\nu}_A^{(P)}}{\tilde{E}_A^{(P)}} = \frac{\nu_A^{(P)}}{E_A^{(P)}} + \frac{\lambda_A^{(P)}\lambda_T^{(P)}}{\mu_A^{(P)}}, \qquad \frac{\tilde{\nu}_a^{(P)}}{\tilde{E}_A^{(P)}} = \frac{\nu_a^{(P)}}{E_A^{(P)}} + \frac{\lambda_t^{(P)}\lambda_A^{(P)}}{\mu_A^{(P)}},\\
&\tilde{\alpha}_A^{(P)} = \alpha_A^{(P)} + \lambda_A^{(P)}\alpha_S^{(P)}, \quad \tilde{\alpha}_T^{(P)} = \alpha_T^{(P)} + \lambda_T^{(P)}\alpha_S^{(P)}, \quad \tilde{\alpha}_t^{(P)} = \alpha_t^{(P)} + \lambda_t^{(P)}\alpha_S^{(P)}.
\end{aligned} \tag{6.81}$$

Relations (6.80) are known as the reduced stress-strain relations for the angled lamina as they have exactly the same form as three of the stress-strain relations (6.2) which apply when $\phi = 0$.

6.10 Mixed Form of Stress-Strain Relations

The final steps of this chapter are to manipulate the stress-strain equations (6.79) and (6.80) so that they are in a form that is useful when considering in Chapter 7 the effective properties of laminates. The objective is to express stresses and strains in terms of the parameters $\varepsilon_A, \varepsilon_T, \gamma_A, \sigma_t$ and ΔT which will have the same values in all plies of a laminate. The first two relations of (6.80) lead to

$$\begin{aligned}
\sigma_A = {} & \frac{\tilde{E}_A^{(P)}}{\Psi^{(P)}}\varepsilon_A + \frac{\tilde{\nu}_A^{(P)}\tilde{E}_T^{(P)}}{\Psi^{(P)}}\varepsilon_T + \frac{\hat{\nu}_a^{(P)}}{\Psi^{(P)}}\sigma_t\\
& + \left(\frac{\tilde{E}_A^{(P)}}{\Psi^{(P)}}\lambda_A^{(P)} + \frac{\tilde{\nu}_A^{(P)}\tilde{E}_T^{(P)}}{\Psi^{(P)}}\lambda_T^{(P)}\right)\gamma_A - \left(\frac{\tilde{E}_A^{(P)}}{\Psi^{(P)}}\tilde{\alpha}_A^{(P)} + \frac{\tilde{\nu}_A^{(P)}\tilde{E}_T^{(P)}}{\Psi^{(P)}}\tilde{\alpha}_T^{(P)}\right)\Delta T,\\
\sigma_T = {} & \frac{\tilde{\nu}_A^{(P)}\tilde{E}_T^{(P)}}{\Psi^{(P)}}\varepsilon_A + \frac{\tilde{E}_T^{(P)}}{\Psi^{(P)}}\varepsilon_T + \frac{\hat{\nu}_t^{(P)}}{\Psi^{(P)}}\sigma_t\\
& + \left(\frac{\tilde{\nu}_A^{(P)}\tilde{E}_T^{(P)}}{\Psi^{(P)}}\lambda_A^{(P)} + \frac{\tilde{E}_T^{(P)}}{\Psi^{(P)}}\lambda_T^{(P)}\right)\gamma_A - \left(\frac{\tilde{\nu}_A^{(P)}\tilde{E}_T^{(P)}}{\Psi^{(P)}}\tilde{\alpha}_A^{(P)} + \frac{\tilde{E}_T^{(P)}}{\Psi^{(P)}}\tilde{\alpha}_T^{(P)}\right)\Delta T,
\end{aligned} \tag{6.82}$$

where

$$\Psi^{(P)} = 1 - \left(\tilde{\nu}_A^{(P)}\right)^2 \frac{\tilde{E}_T^{(P)}}{\tilde{E}_A^{(P)}}, \; \hat{\nu}_a^{(P)} = \tilde{\nu}_a^{(P)} + \tilde{\nu}_t^{(P)}\tilde{\nu}_A^{(P)}, \; \hat{\nu}_t^{(P)} = \tilde{\nu}_t^{(P)} + \tilde{\nu}_a^{(P)}\tilde{\nu}_A^{(P)} \frac{\tilde{E}_T^{(P)}}{\tilde{E}_A^{(P)}}. \tag{6.83}$$

The third of relations (6.80) leads to the following expression for the through-thickness strain

$$\begin{aligned} \varepsilon_t = & -\frac{\hat{\nu}_a^{(P)}}{\Psi^{(P)}}\varepsilon_A - \frac{\hat{\nu}_t^{(P)}}{\Psi^{(P)}}\varepsilon_T + \frac{\Lambda^{(P)}}{\Psi^{(P)}\tilde{E}_t^{(P)}}\sigma_t \\ & - \left(\frac{\hat{\nu}_a^{(P)}}{\Psi^{(P)}}\lambda_A^{(P)} + \frac{\hat{\nu}_t^{(P)}}{\Psi^{(P)}}\lambda_T^{(P)} + \lambda_t^{(P)} \right)\gamma_A + \left(\frac{\hat{\nu}_a^{(P)}}{\Psi^{(P)}}\tilde{\alpha}_A^{(P)} + \frac{\hat{\nu}_t^{(P)}}{\Psi^{(P)}}\tilde{\alpha}_T^{(P)} + \tilde{\alpha}_t^{(P)} \right)\Delta T, \end{aligned} \tag{6.84}$$

where

$$\begin{aligned} \Lambda^{(P)} &= \Psi^{(P)} - \tilde{\nu}_a^{(P)} \frac{\tilde{E}_t^{(P)}}{\tilde{E}_A^{(P)}}\hat{\nu}_a^{(P)} - \tilde{\nu}_t^{(P)} \frac{\tilde{E}_t^{(P)}}{\tilde{E}_T^{(P)}}\hat{\nu}_t^{(P)}, \\ &= 1 - \left(\tilde{\nu}_a^{(P)}\right)^2 \frac{\tilde{E}_t^{(P)}}{\tilde{E}_A^{(P)}} - \left(\tilde{\nu}_t^{(P)}\right)^2 \frac{\tilde{E}_t^{(P)}}{\tilde{E}_T^{(P)}} - \left(\tilde{\nu}_A^{(P)}\right)^2 \frac{\tilde{E}_T^{(P)}}{\tilde{E}_A^{(P)}} - 2\tilde{\nu}_a^{(P)}\tilde{\nu}_t^{(P)}\tilde{\nu}_A^{(P)} \frac{\tilde{E}_t^{(P)}}{\tilde{E}_A^{(P)}}, \end{aligned} \tag{6.85}$$

consistent when $\phi = 0$ with relation (6.20) that identifies other forms for this expression. On using (6.82), the shear stress-strain relation (6.79) may be written as

$$\begin{aligned} \tau_A = & \frac{\tilde{E}_A^{(P)}}{\Psi^{(P)}}\hat{\lambda}_A^{(P)}\varepsilon_A + \frac{\tilde{E}_T^{(P)}}{\Psi^{(P)}}\hat{\lambda}_T^{(P)}\varepsilon_T + \left(\frac{\hat{\nu}_a^{(P)}}{\Psi^{(P)}}\lambda_A^{(P)} + \frac{\hat{\nu}_t^{(P)}}{\Psi^{(P)}}\lambda_T^{(P)} + \lambda_t^{(P)} \right)\sigma_t \\ & + \left(\frac{\tilde{E}_A^{(P)}}{\Psi^{(P)}}\lambda_A^{(P)}\hat{\lambda}_A^{(P)} + \frac{\tilde{E}_T^{(P)}}{\Psi^{(P)}}\lambda_T^{(P)}\hat{\lambda}_T^{(P)} + \mu_A^{(P)} \right)\gamma_A \\ & - \left(\frac{\tilde{E}_A^{(P)}}{\Psi^{(P)}}\hat{\lambda}_A^{(P)}\tilde{\alpha}_A^{(P)} + \frac{\tilde{E}_T^{(P)}}{\Psi^{(P)}}\hat{\lambda}_T^{(P)}\tilde{\alpha}_T^{(P)} + \mu_A^{(P)}\alpha_S^{(P)} \right)\Delta T, \end{aligned} \tag{6.86}$$

where

$$\hat{\lambda}_A^{(P)} = \lambda_A^{(P)} + \lambda_T^{(P)}\tilde{\nu}_A^{(P)} \frac{\tilde{E}_T^{(P)}}{\tilde{E}_A^{(P)}}, \qquad \hat{\lambda}_T^{(P)} = \lambda_T^{(P)} + \lambda_A^{(P)}\tilde{\nu}_A^{(P)}. \tag{6.87}$$

Relations (6.82), (6.84) and (6.86) are now collected together and written in the form

$$\begin{aligned} \sigma_A &= \Omega_{11}^{(P)}\varepsilon_A + \Omega_{12}^{(P)}\varepsilon_T + \Omega_{13}^{(P)}\sigma_t + \Omega_{16}^{(P)}\gamma_A - \omega_1^{(P)}\Delta T, \\ \sigma_T &= \Omega_{12}^{(P)}\varepsilon_A + \Omega_{22}^{(P)}\varepsilon_T + \Omega_{23}^{(P)}\sigma_t + \Omega_{26}^{(P)}\gamma_A - \omega_2^{(P)}\Delta T, \\ \varepsilon_t &= -\Omega_{13}^{(P)}\varepsilon_A - \Omega_{23}^{(P)}\varepsilon_T + \Omega_{33}^{(P)}\sigma_t - \Omega_{36}^{(P)}\gamma_A + \omega_3^{(P)}\Delta T, \\ \tau_A &= \Omega_{16}^{(P)}\varepsilon_A + \Omega_{26}^{(P)}\varepsilon_T + \Omega_{36}^{(P)}\sigma_t + \Omega_{66}^{(P)}\gamma_A - \omega_6^{(P)}\Delta T, \end{aligned} \tag{6.88}$$

where

$$\begin{aligned} &\Omega_{11}^{(P)} = \frac{\tilde{E}_A^{(P)}}{\Psi^{(P)}}, \; \Omega_{12}^{(P)} = \frac{\tilde{\nu}_A^{(P)}\tilde{E}_T^{(P)}}{\Psi^{(P)}}, \; \Omega_{13}^{(P)} = \frac{\hat{\nu}_a^{(P)}}{\Psi^{(P)}}, \quad \Omega_{16}^{(P)} = \frac{\tilde{E}_A^{(P)}}{\Psi^{(P)}}\lambda_A^{(P)} + \frac{\tilde{\nu}_A^{(P)}\tilde{E}_T^{(P)}}{\Psi^{(P)}}\lambda_T^{(P)}, \\ &\Omega_{22}^{(P)} = \frac{\tilde{E}_T^{(P)}}{\Psi^{(P)}}, \quad \Omega_{23}^{(P)} = \frac{\hat{\nu}_t^{(P)}}{\Psi^{(P)}}, \quad \Omega_{26}^{(P)} = \frac{\tilde{\nu}_A^{(P)}\tilde{E}_T^{(P)}}{\Psi^{(P)}}\lambda_A^{(P)} + \frac{\tilde{E}_T^{(P)}}{\Psi^{(P)}}\lambda_T^{(P)}, \\ &\Omega_{33}^{(P)} = \frac{\Lambda^{(P)}}{\Psi^{(P)}\tilde{E}_t^{(P)}}, \; \Omega_{36}^{(P)} = \frac{\hat{\nu}_a^{(P)}}{\Psi^{(P)}}\lambda_A^{(P)} + \frac{\hat{\nu}_t^{(P)}}{\Psi^{(P)}}\lambda_T^{(P)} + \lambda_t^{(P)}, \\ &\Omega_{66}^{(P)} = \frac{\tilde{E}_A^{(P)}}{\Psi^{(P)}}\lambda_A^{(P)}\hat{\lambda}_A^{(P)} + \frac{\tilde{E}_T^{(P)}}{\Psi^{(P)}}\lambda_T^{(P)}\hat{\lambda}_T^{(P)} + \mu_A^{(P)}, \end{aligned} \tag{6.89}$$

$$\omega_1^{(P)} = \frac{\tilde{E}_A^{(P)}}{\Psi^{(P)}}\tilde{\alpha}_A^{(P)} + \frac{\tilde{\nu}_A^{(P)}\tilde{E}_T^{(P)}}{\Psi^{(P)}}\tilde{\alpha}_T^{(P)}, \qquad \omega_2^{(P)} = \frac{\tilde{E}_T^{(P)}}{\Psi^{(P)}}\left(\tilde{\alpha}_T^{(P)} + \tilde{\nu}_A^{(P)}\tilde{\alpha}_A^{(P)}\right),$$
$$\omega_3^{(P)} = \frac{\hat{\nu}_a^{(P)}}{\Psi^{(P)}}\tilde{\alpha}_A^{(P)} + \frac{\hat{\nu}_t^{(P)}}{\Psi^{(P)}}\tilde{\alpha}_T^{(P)} + \tilde{\alpha}_t^{(P)}, \quad \omega_6^{(P)} = \frac{\tilde{E}_A^{(P)}}{\Psi^{(P)}}\hat{\lambda}_A^{(P)}\tilde{\alpha}_A^{(P)} + \frac{\tilde{E}_T^{(P)}}{\Psi^{(P)}}\hat{\lambda}_T^{(P)}\tilde{\alpha}_T^{(P)} + \mu_A^{(P)}\alpha_S^{(P)}, \tag{6.90}$$

and where $\Psi^{(P)}$ and $\Lambda^{(P)}$ are defined by (6.83) and (6.85), respectively.

It should be noted that the mixed form (6.88) of the stress-strain relations can be derived from (6.72) leading to the following alternative expressions for the coefficients:

$$\Omega_{11}^{(P)} = C_{11}^{(P)} - \frac{C_{13}^{(P)}C_{13}^{(P)}}{C_{33}^{(P)}}, \Omega_{12}^{(P)} = C_{12}^{(P)} - \frac{C_{13}^{(P)}C_{23}^{(P)}}{C_{33}^{(P)}}, \Omega_{13}^{(P)} = \frac{C_{13}^{(P)}}{C_{33}^{(P)}},$$
$$\Omega_{16}^{(P)} = C_{16}^{(P)} - \frac{C_{13}^{(P)}C_{36}^{(P)}}{C_{33}^{(P)}}, \tag{6.91}$$

$$\Omega_{22}^{(P)} = C_{22}^{(P)} - \frac{C_{23}^{(P)}C_{23}^{(P)}}{C_{33}^{(P)}}, \Omega_{23}^{(P)} = \frac{C_{23}^{(P)}}{C_{33}^{(P)}}, \Omega_{26}^{(P)} = C_{26}^{(P)} - \frac{C_{23}^{(P)}C_{36}^{(P)}}{C_{33}^{(P)}}, \tag{6.92}$$

$$\Omega_{33}^{(P)} = \frac{1}{C_{33}^{(P)}}, \Omega_{36}^{(P)} = \frac{C_{36}^{(P)}}{C_{33}^{(P)}}, \tag{6.93}$$

$$\Omega_{66}^{(P)} = C_{66}^{(P)} - \frac{C_{36}^{(P)}C_{36}^{(P)}}{C_{33}^{(P)}}, \tag{6.94}$$

$$\omega_1^{(P)} = U_1^{(P)} - \frac{C_{13}^{(P)}U_3^{(P)}}{C_{33}^{(P)}}, \omega_2^{(P)} = U_2^{(P)} - \frac{C_{23}^{(P)}U_3^{(P)}}{C_{33}^{(P)}}, \omega_3^{(P)} = \frac{U_3^{(P)}}{C_{33}^{(P)}},$$
$$\omega_6^{(P)} = U_6^{(P)} - \frac{C_{36}^{(P)}U_3^{(P)}}{C_{33}^{(P)}}. \tag{6.95}$$

6.11 Special Case of 0° and 90° Plies

To complete this chapter, it is useful to specify the stress-strain relations for the 0° and 90° plies in a laminate when referred to the global axes of a general symmetric laminate. When a 0° ply is considered so that $\phi = 0$ and $m = 1$ and $n = 0$, it follows from (6.54)–(6.67) that

$$\begin{aligned}
S_{11} &= \frac{1}{E_A}, & S_{12} &= -\frac{\nu_A}{E_A}, & S_{13} &= -\frac{\nu_a}{E_A},\\
S_{22} &= \frac{1}{E_T}, & S_{23} &= -\frac{\nu_t}{E_T}, & S_{33} &= \frac{1}{E_t},\\
S_{44} &= \frac{1}{\mu_t}, & S_{55} &= \frac{1}{\mu_a}, & S_{66} &= \frac{1}{\mu_A},\\
V_1 &= \alpha_A, & V_2 &= \alpha_T, & V_3 &= \alpha_t.
\end{aligned} \tag{6.96}$$

When a 90° ply is considered so that $\phi = \pi/2$ and $m = 0$ and $n = 1$, it follows from (6.54)–(6.67) that

$$
\begin{aligned}
S_{11} &= \frac{1}{E_T}, & S_{12} &= -\frac{\nu_A}{E_A}, & S_{13} &= -\frac{\nu_t}{E_T}, \\
S_{22} &= \frac{1}{E_A}, & S_{23} &= -\frac{\nu_a}{E_A}, & S_{33} &= \frac{1}{E_t}, \\
S_{44} &= \frac{1}{\mu_a}, & S_{55} &= \frac{1}{\mu_t}, & S_{66} &= \frac{1}{\mu_A}, \\
V_1 &= \alpha_T, & V_2 &= \alpha_A, & V_3 &= \alpha_t .
\end{aligned}
\tag{6.97}
$$

It then follows from (6.53) that the stress-strain relations, referred to global axes, for both a 0° ply and a 90° ply are of the orthotropic form

$$
\begin{bmatrix} \varepsilon_A \\ \varepsilon_T \\ \varepsilon_t \\ \gamma_t \\ \gamma_a \\ \gamma_A \end{bmatrix} \equiv \begin{bmatrix} \varepsilon_{11} \\ \varepsilon_{22} \\ \varepsilon_{33} \\ 2\varepsilon_{23} \\ 2\varepsilon_{13} \\ 2\varepsilon_{12} \end{bmatrix} = \begin{bmatrix} S_{11} & S_{12} & S_{13} & 0 & 0 & 0 \\ S_{12} & S_{22} & S_{23} & 0 & 0 & 0 \\ S_{13} & S_{23} & S_{33} & 0 & 0 & 0 \\ 0 & 0 & 0 & S_{44} & 0 & 0 \\ 0 & 0 & 0 & 0 & S_{55} & 0 \\ 0 & 0 & 0 & 0 & 0 & S_{66} \end{bmatrix} \begin{bmatrix} \sigma_A \\ \sigma_T \\ \sigma_t \\ \tau_t \\ \tau_a \\ \tau_A \end{bmatrix} + \begin{bmatrix} V_1 \\ V_2 \\ V_3 \\ 0 \\ 0 \\ 0 \end{bmatrix} \Delta T. \tag{6.98}
$$

On writing out the stress-strain relations explicitly, it follows that for a 0° ply

$$
\begin{aligned}
\varepsilon_A &= \frac{1}{E_A}\sigma_A - \frac{\nu_A}{E_A}\sigma_T - \frac{\nu_a}{E_A}\sigma_t + \alpha_A \Delta T, \\
\varepsilon_T &= -\frac{\nu_A}{E_A}\sigma_A + \frac{1}{E_T}\sigma_T - \frac{\nu_t}{E_T}\sigma_t + \alpha_T \Delta T, \\
\varepsilon_t &= -\frac{\nu_a}{E_A}\sigma_A - \frac{\nu_t}{E_T}\sigma_T + \frac{1}{E_t}\sigma_t + \alpha_t \Delta T, \\
\gamma_A &= \frac{1}{\mu_A}\tau_A, \quad \gamma_a = \frac{1}{\mu_a}\tau_a, \quad \gamma_t = \frac{1}{\mu_t}\tau_t,
\end{aligned}
\tag{6.99}
$$

and that for a 90° ply

$$
\begin{aligned}
\varepsilon_A &= \frac{1}{E_T}\sigma_A - \frac{\nu_A}{E_A}\sigma_T - \frac{\nu_t}{E_T}\sigma_t + \alpha_T \Delta T, \\
\varepsilon_T &= -\frac{\nu_A}{E_A}\sigma_A + \frac{1}{E_A}\sigma_T - \frac{\nu_a}{E_A}\sigma_t + \alpha_A \Delta T, \\
\varepsilon_t &= -\frac{\nu_t}{E_T}\sigma_A - \frac{\nu_a}{E_A}\sigma_T + \frac{1}{E_t}\sigma_t + \alpha_t \Delta T, \\
\gamma_A &= \frac{1}{\mu_A}\tau_A, \quad \gamma_a = \frac{1}{\mu_t}\tau_a, \quad \gamma_t = \frac{1}{\mu_a}\tau_t .
\end{aligned}
\tag{6.100}
$$

These special forms of stress-strain relations for the 0° and 90° plies are used throughout this book.

7

Effective Thermoelastic Properties of Undamaged Laminates

Overview: A key requirement when undertaking an analysis of engineering components made using composite laminates is knowledge of the values of the effective thermoelastic properties of a homogenised laminate that define the relationship between the effective stresses and strains of the laminate. Numerical simulation tools, such as finite element and boundary element methods, demand all the properties required to characterise three-dimensional stress states and the corresponding deformation. The objective of this chapter is to provide estimates of all the effective properties for *symmetric* undamaged laminates required by a three-dimensional component analysis. Although classical laminate theory addresses this issue (see, for example, [1]), the approach taken here focuses on the development of *explicit formulae* for effective thermoelastic properties, with the result that this chapter presents many results in forms that have not before been published in detail. The assumption is first made that the laminate is symmetric, which means that there is no need to consider the bending of the laminate. Later in the chapter, bend deformation is considered in detail in which case the laminates can be nonsymmetric. The analysis of this chapter defines the initial state of laminates before damage develops in the form of ply cracks: a topic that is considered in later chapters of this book.

7.1 Laminate Geometry (Symmetric Laminates)

Consider a laminate that is made by perfectly bonding together various laminae (i.e. plies) having different orientations so that there are no defects (i.e. the laminate is undamaged). The situation under consideration concerns first the deformation of a symmetric multilayered laminate of total thickness $2h$ constructed of $2n$ perfectly bonded plies that can have any combination of orientations, such that symmetry about the midplane of the laminate is preserved. The plies in each half of the laminate can be made of different materials and each can have a different thickness as illustrated in Figure 7.1.

As laminate symmetry is assumed, it is necessary to consider only the right-hand set of n layers (see Figure 7.1). A global right-handed set of Cartesian coordinates is chosen having the origin at the centre of the midplane of the laminate. The x_1-direction defines the longitudinal or axial direction, the x_2-direction defines the in-plane

Properties for Design of Composite Structures: Theory and Implementation Using Software, First Edition. Neil McCartney.

Companion Website: www.wiley.com/go/mccartney/properties

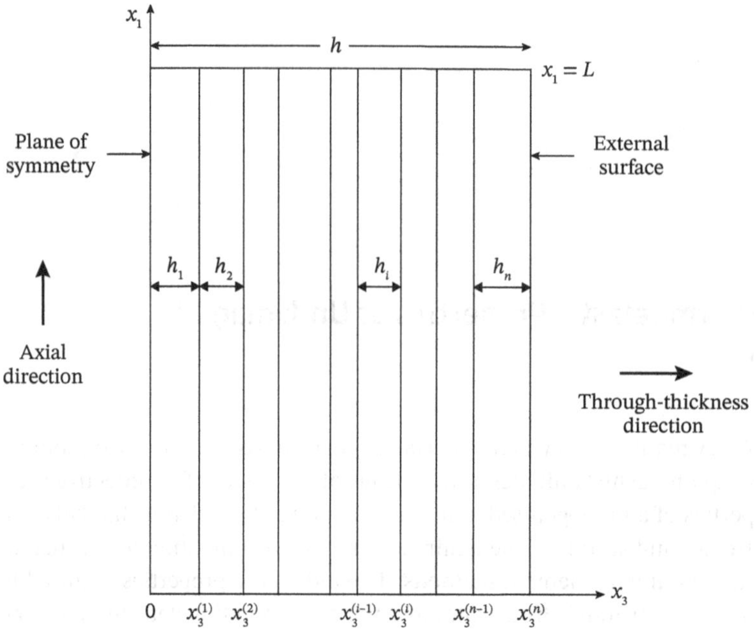

Figure 7.1 Schematic diagram of the geometry for one-half of a general symmetric laminate.

transverse direction and the x_3-direction defines the through-thickness direction. The locations of the n - 1 interfaces in one half of the laminate ($x_3 > 0$) are specified by $x_3 = x_3^{(i)}, i = 1, ..., n-1$. The midplane of the laminate is specified by $x_3 = x_3^{(0)} = 0$ and the external surface by $x_3 = x_3^{(n)} = h$ where h is the half-thickness of the laminate. The thickness of the ith layer is denoted by $h_i = x_3^{(i)} - x_3^{(i-1)}$ such that

$$h = \sum_{i=1}^{n} h_i. \tag{7.1}$$

Stress, strain and displacement components, and material properties associated with the ith layer are denoted by a superscript '(i)'. The orientation of the ith layer is specified by the angle ϕ_i (measured clockwise when looking in the direction of positive values of x_3) between the x_1-axis and the fibre direction of this layer (see Figure 6.2). The representative volume element of the laminate to be considered is specified by $0 \leq x_1 \leq L, 0 \leq x_2 \leq W, 0 \leq x_3 \leq h$.

7.2 Equilibrium Equations

The following equilibrium equations, for the stress components referred to global coordinates, must be satisfied for all values of x_1, x_2 in the region $x_3^{(i-1)} \leq x_3 \leq x_3^{(i)}$

$$\frac{\partial \sigma_{11}^{(i)}}{\partial x_1} + \frac{\partial \sigma_{12}^{(i)}}{\partial x_2} + \frac{\partial \sigma_{13}^{(i)}}{\partial x_3} = 0, \tag{7.2}$$

$$\frac{\partial \sigma_{12}^{(i)}}{\partial x_1} + \frac{\partial \sigma_{22}^{(i)}}{\partial x_2} + \frac{\partial \sigma_{23}^{(i)}}{\partial x_3} = 0, \tag{7.3}$$

$$\frac{\partial \sigma_{13}^{(i)}}{\partial x_1} + \frac{\partial \sigma_{23}^{(i)}}{\partial x_2} + \frac{\partial \sigma_{33}^{(i)}}{\partial x_3} = 0, \tag{7.4}$$

for values $i = 1, \ldots, n$, where it is understood that $x_3^{(0)} = 0$.

7.3 Interfacial and Boundary Conditions

As the plies of the laminate are assumed to be perfectly bonded, the interfacial tractions and displacements must be continuous throughout the layered system such that

$$\left.\begin{aligned} \sigma_{13}^{(i)}\left(x_1, x_2, x_3^{(i)}\right) &= \sigma_{13}^{(i+1)}\left(x_1, x_2, x_3^{(i)}\right) \\ \sigma_{23}^{(i)}\left(x_1, x_2, x_3^{(i)}\right) &= \sigma_{23}^{(i+1)}\left(x_1, x_2, x_3^{(i)}\right) \\ \sigma_{33}^{(i)}\left(x_1, x_2, x_3^{(i)}\right) &= \sigma_{33}^{(i+1)}\left(x_1, x_2, x_3^{(i)}\right) \end{aligned}\right\} \text{ for } i=1,\ldots,n-1, \left|x_1\right| \le L, \left|x_2\right| \le W, \tag{7.5}$$

and

$$\left.\begin{aligned} u_1^{(i)}\left(x_1, x_2, x_3^{(i)}\right) &= u_1^{(i+1)}\left(x_1, x_2, x_3^{(i)}\right) \\ u_2^{(i)}\left(x_1, x_2, x_3^{(i)}\right) &= u_2^{(i+1)}\left(x_1, x_2, x_3^{(i)}\right) \\ u_3^{(i)}\left(x_1, x_2, x_3^{(i)}\right) &= u_3^{(i+1)}\left(x_1, x_2, x_3^{(i)}\right) \end{aligned}\right\} \text{ for } i=1,\ldots,n-1, \left|x_1\right| \le L, \left|x_2\right| \le W. \tag{7.6}$$

The midplane $x_3 = 0$ and outer surface $x_3 = x_3^{(n)} = h$ of the system are assumed to be subject to uniform normal and shear tractions σ_t, τ_a and τ_t such that for *all* values $\left|x_1\right| \le L$, $\left|x_2\right| \le W$ and $\left|x_3\right| \le h$,

$$\sigma_{13}\left(x_1, x_2, x_3\right) = \tau_a, \quad \sigma_{23}\left(x_1, x_2, x_3\right) = \tau_t, \quad \sigma_{33}\left(x_1, x_2, x_3\right) = \sigma_t, \tag{7.7}$$

thus ensuring that the tractions continuity conditions (7.5) are satisfied. The displacement components are assumed to be zero at the origin of the coordinates.

7.4 Displacement and Strain Distributions

The parameters $u_k^{(i)}$, $k = 1, 2, 3$, denote the displacement components for the ith ply in the x_1, x_2 and x_3 directions, respectively, and $\varepsilon_{jk}^{(i)}$, $j, k = 1, 2, 3$, denote the strain components for infinitesimal deformation theory (see Chapter 2). For the ith ply in the region $x_3^{(i-1)} \le x_3 \le x_3^{(i)}$, assuming that the displacement components $u_j^{(1)}$, $j = 1, 2, 3$, are all zero at the origin of the Cartesian coordinates, the displacement field within this ply has the following form, based on (2.134), satisfying the displacement continuity conditions (7.6)

$$\left.\begin{aligned} u_1^{(i)}\left(x_1,x_2,x_3\right) &= \varepsilon_{11}x_1 + \varepsilon_{12}x_2 + 2\varepsilon_{13}^{(i)}\left(x_3 - x_3^{(i)}\right) + 2\sum_{j=1}^{i} h_j \varepsilon_{13}^{(j)}, \\ u_2^{(i)}\left(x_1,x_2,x_3\right) &= \varepsilon_{12}x_1 + \varepsilon_{22}x_2 + 2\varepsilon_{23}^{(i)}\left(x_3 - x_3^{(i)}\right) + 2\sum_{j=1}^{i} h_j \varepsilon_{23}^{(j)}, \\ u_3^{(i)}\left(x_1,x_2,x_3\right) &= \varepsilon_{33}^{(i)}\left(x_3 - x_3^{(i)}\right) + \sum_{j=1}^{i} h_j \varepsilon_{33}^{(j)}, \end{aligned}\right\} \quad i=1,\ldots,n, \tag{7.8}$$

where all the strain components are assumed to be uniform. Provided that the in-plane dimensions of the laminate are much greater than the laminate thickness, and provided that the laminate is undamaged, the in-plane strains of each ply in regions well away from the laminate edges will have the same values ε_{11}, ε_{22} and ε_{12} in each of the n plies. On using the compact notation of Chapter 4, it follows from (7.8) that for $x_3^{(i-1)} \le x_3 \le x_3^{(i)}$,

$$\left.\begin{aligned} u_1^{(i)}(x_1,x_2,x_3) &= \varepsilon_A x_1 + \tfrac{1}{2}\gamma_A x_2 + \gamma_a^{(i)}\left(x_3 - x_3^{(i)}\right) + \sum_{j=1}^{i} h_j \gamma_a^{(j)}, \\ u_2^{(i)}(x_1,x_2,x_3) &= \tfrac{1}{2}\gamma_A x_1 + \varepsilon_T x_2 + \gamma_t^{(i)}\left(x_3 - x_3^{(i)}\right) + \sum_{j=1}^{i} h_j \gamma_t^{(j)}, \\ u_3^{(i)}(x_1,x_2,x_3) &= \varepsilon_t^{(i)}\left(x_3 - x_3^{(i)}\right) + \sum_{j=1}^{i} h_j \varepsilon_t^{(j)}, \end{aligned}\right\} \quad i = 1,\ldots,n. \tag{7.9}$$

It follows immediately that the corresponding strains in the ith ply have the uniform values

$$\begin{aligned} &\varepsilon_{11}^{(i)} \equiv \frac{\partial u_1^{(i)}}{\partial x_1} = \varepsilon_A, \quad \varepsilon_{22}^{(i)} \equiv \frac{\partial u_2^{(i)}}{\partial x_2} = \varepsilon_T, \quad \varepsilon_{33}^{(i)} \equiv \frac{\partial u_3^{(i)}}{\partial x_3} = \varepsilon_t^{(i)}, \\ &2\varepsilon_{23}^{(i)} \equiv \left(\frac{\partial u_2^{(i)}}{\partial x_3} + \frac{\partial u_3^{(i)}}{\partial x_2}\right) = \gamma_t^{(i)}, \; 2\varepsilon_{13}^{(i)} \equiv \left(\frac{\partial u_1^{(i)}}{\partial x_3} + \frac{\partial u_3^{(i)}}{\partial x_1}\right) = \gamma_a^{(i)}, \\ &2\varepsilon_{12}^{(i)} \equiv \left(\frac{\partial u_1^{(i)}}{\partial x_2} + \frac{\partial u_2^{(i)}}{\partial x_1}\right) = \gamma_A. \end{aligned} \tag{7.10}$$

As $x_3^{(0)} = 0$ and $h_1 = x_3^{(1)}$ it follows from (7.9) that for the first ply adjacent to the midplane of the laminate

$$\begin{aligned} u_1^{(1)}(x_1,x_2,x_3) &= \varepsilon_A x_1 + \frac{1}{2}\gamma_A x_2 + \gamma_a^{(1)} x_3, \\ u_2^{(1)}(x_1,x_2,x_3) &= \frac{1}{2}\gamma_A x_1 + \varepsilon_T x_2 + \gamma_t^{(1)} x_3, \quad i = 1,\ldots,n, \\ u_3^{(1)}(x_1,x_2,x_3) &= \varepsilon_t^{(1)} x_3, \end{aligned} \tag{7.11}$$

satisfying the assumption that the displacement components are all zero at the origin of coordinates.

On the external surface $x_3 = x_3^{(n)} = h$ of the nth ply the displacement distribution is

$$\begin{aligned} u_1^{(n)}(x_1,x_2,h) &= \varepsilon_A x_1 + \frac{1}{2}\gamma_A x_2 + \sum_{i=1}^{n} h_i \gamma_a^{(i)}, \\ u_2^{(n)}(x_1,x_2,h) &= \frac{1}{2}\gamma_A x_1 + \varepsilon_T x_2 + \sum_{i=1}^{n} h_i \gamma_t^{(i)}, \\ u_3^{(n)}(x_1,x_2,h) &= \sum_{i=1}^{n} h_i \varepsilon_t^{(i)}. \end{aligned} \tag{7.12}$$

The first two of relations (7.12) are used to define the following effective out-of-plane shear strains for the laminate

$$\gamma_a = \frac{1}{h}\sum_{i=1}^{n} h_i \gamma_a^{(i)}, \quad \gamma_t = \frac{1}{h}\sum_{i=1}^{n} h_i \gamma_t^{(i)}. \tag{7.13}$$

On setting $n = 1$ and identifying γ_a, γ_t and h with $\gamma_a^{(1)}$, $\gamma_t^{(1)}$ and h_1, respectively, it follows that for a homogenised laminate, the displacement components may be written in the following form, derived from (7.11), valid for *all* values $|x_1| \le L$, $|x_2| \le W$ and $|x_3| \le h$:

$$\begin{aligned} u_1^{(n)}(x_1,x_2,x_3) &= \varepsilon_A x_1 + \frac{1}{2}\gamma_A x_2 + \gamma_a x_3, \\ u_2^{(n)}(x_1,x_2,x_3) &= \frac{1}{2}\gamma_A x_1 + \varepsilon_T x_2 + \gamma_t x_3, \\ u_3^{(n)}(x_1,x_2,x_3) &= \varepsilon_t x_3, \end{aligned} \tag{7.14}$$

where ε_t is the effective through-thickness strain defined by

$$\varepsilon_t = \frac{1}{h}\sum_{i=1}^{n} h_i \varepsilon_t^{(i)}. \tag{7.15}$$

7.5 Effective In-plane Properties for Laminate

The in-plane stresses in the ith ply and the through-thickness stress are obtained from four of the six stress-strain relations (6.72) so that on using $(7.7)_3$ and (7.10),

$$\begin{aligned} \sigma_A^{(i)} &= C_{11}^{(i)}\varepsilon_A + C_{12}^{(i)}\varepsilon_T + C_{13}^{(i)}\varepsilon_t^{(i)} + C_{16}^{(i)}\gamma_A - U_1^{(i)}\Delta T, \\ \sigma_T^{(i)} &= C_{12}^{(i)}\varepsilon_A + C_{22}^{(i)}\varepsilon_T + C_{23}^{(i)}\varepsilon_t^{(i)} + C_{26}^{(i)}\gamma_A - U_2^{(i)}\Delta T, \\ \sigma_t &= C_{13}^{(i)}\varepsilon_A + C_{23}^{(i)}\varepsilon_T + C_{33}^{(i)}\varepsilon_t^{(i)} + C_{36}^{(i)}\gamma_A - U_3^{(i)}\Delta T, \\ \tau_A^{(i)} &= C_{16}^{(i)}\varepsilon_A + C_{26}^{(i)}\varepsilon_T + C_{36}^{(i)}\varepsilon_t^{(i)} + C_{66}^{(i)}\gamma_A - U_6^{(i)}\Delta T, \end{aligned} \tag{7.16}$$

where the values of the components of the matrix C and vector U are known (see Chapter 6) in terms of ply properties and ply orientations. The components of U are associated with the temperature difference $\Delta T = T - T_0$, where T is the uniform temperature of the lamina and T_0 is the uniform reference temperature of the lamina at which all stresses and strains are zero when the lamina is in an unloaded state.

From $(7.16)_3$ it is clear that the through-thickness stress in each ply has the same uniform value σ_t thus ensuring that the traction continuity condition $(7.5)_3$ is satisfied. In addition, the through-thickness strain $\varepsilon_t^{(i)}$ for the ith ply can be found from $(7.16)_3$ and then substituted into the remaining equations of (7.16). This leads to ply stress-strain relations of the following mixed form, valid for values $i = 1, \ldots, n$, which are consistent with (6.88)

$$\begin{aligned} \sigma_A^{(i)} &= \Omega_{11}^{(i)}\varepsilon_A + \Omega_{12}^{(i)}\varepsilon_T + \Omega_{13}^{(i)}\sigma_t + \Omega_{16}^{(i)}\gamma_A - \omega_1^{(i)}\Delta T, \\ \sigma_T^{(i)} &= \Omega_{12}^{(i)}\varepsilon_A + \Omega_{22}^{(i)}\varepsilon_T + \Omega_{23}^{(i)}\sigma_t + \Omega_{26}^{(i)}\gamma_A - \omega_2^{(i)}\Delta T, \\ \varepsilon_t^{(i)} &= -\Omega_{13}^{(i)}\varepsilon_A - \Omega_{23}^{(i)}\varepsilon_T + \Omega_{33}^{(i)}\sigma_t - \Omega_{36}^{(i)}\gamma_A + \omega_3^{(i)}\Delta T, \\ \tau_A^{(i)} &= \Omega_{16}^{(i)}\varepsilon_A + \Omega_{26}^{(i)}\varepsilon_T + \Omega_{36}^{(i)}\sigma_t + \Omega_{66}^{(i)}\gamma_A - \omega_6^{(i)}\Delta T, \end{aligned} \tag{7.17}$$

where, consistent with (6.91)–(6.95),

$$\Omega_{11}^{(i)} = C_{11}^{(i)} - \frac{C_{13}^{(i)}C_{13}^{(i)}}{C_{33}^{(i)}},\ \Omega_{12}^{(i)} = C_{12}^{(i)} - \frac{C_{13}^{(i)}C_{23}^{(i)}}{C_{33}^{(i)}},\ \Omega_{13}^{(i)} = \frac{C_{13}^{(i)}}{C_{33}^{(i)}},\ \Omega_{16}^{(i)} = C_{16}^{(i)} - \frac{C_{13}^{(i)}C_{36}^{(i)}}{C_{33}^{(i)}}, \tag{7.18}$$

$$\Omega_{22}^{(i)} = C_{22}^{(i)} - \frac{C_{23}^{(i)} C_{23}^{(i)}}{C_{33}^{(i)}}, \quad \Omega_{23}^{(i)} = \frac{C_{23}^{(i)}}{C_{33}^{(i)}}, \quad \Omega_{26}^{(i)} = C_{26}^{(i)} - \frac{C_{23}^{(i)} C_{36}^{(i)}}{C_{33}^{(i)}}, \tag{7.19}$$

$$\Omega_{33}^{(i)} = \frac{1}{C_{33}^{(i)}}, \quad \Omega_{36}^{(i)} = \frac{C_{36}^{(i)}}{C_{33}^{(i)}}, \tag{7.20}$$

$$\Omega_{66}^{(i)} = C_{66}^{(i)} - \frac{C_{36}^{(i)} C_{36}^{(i)}}{C_{33}^{(i)}}, \tag{7.21}$$

$$\begin{aligned} \omega_1^{(i)} &= U_1^{(i)} - \frac{C_{13}^{(i)} U_3^{(i)}}{C_{33}^{(i)}}, \quad \omega_2^{(i)} = U_2^{(i)} - \frac{C_{23}^{(i)} U_3^{(i)}}{C_{33}^{(i)}}, \\ \omega_3^{(i)} &= \frac{U_3^{(i)}}{C_{33}^{(i)}}, \quad \omega_6^{(i)} = U_6^{(i)} - \frac{C_{36}^{(i)} U_3^{(i)}}{C_{33}^{(i)}}. \end{aligned} \tag{7.22}$$

The form (7.17) is such that the variables ε_A, ε_T, σ_t, γ_A and ΔT appear on the right-hand side for each ply of the laminate. The corresponding variables $\sigma_A^{(i)}$, $\sigma_T^{(i)}$, $\varepsilon_t^{(i)}$ and $\tau_A^{(i)}$ on the left-hand side of (7.17) vary from ply to ply but they are uniform within each ply.

When the laminate is considered as a homogenised plate, the effective stress-strain relations must be of the following form analogous to the corresponding ply relations (7.16)

$$\begin{aligned} \sigma_A &= C_{11}^{(L)} \varepsilon_A + C_{12}^{(L)} \varepsilon_T + C_{13}^{(L)} \varepsilon_t + C_{16}^{(L)} \gamma_A - U_1^{(L)} \Delta T, \\ \sigma_T &= C_{12}^{(L)} \varepsilon_A + C_{22}^{(L)} \varepsilon_T + C_{23}^{(L)} \varepsilon_t + C_{26}^{(L)} \gamma_A - U_2^{(L)} \Delta T, \\ \sigma_t &= C_{13}^{(L)} \varepsilon_A + C_{23}^{(L)} \varepsilon_T + C_{33}^{(L)} \varepsilon_t + C_{36}^{(L)} \gamma_A - U_3^{(L)} \Delta T, \\ \tau_A &= C_{16}^{(L)} \varepsilon_A + C_{26}^{(L)} \varepsilon_T + C_{36}^{(L)} \varepsilon_t + C_{66}^{(L)} \gamma_A - U_6^{(L)} \Delta T, \end{aligned} \tag{7.23}$$

where

$$\sigma_A = \frac{1}{h} \sum_{i=1}^{n} h_i \sigma_A^{(i)}, \quad \sigma_T = \frac{1}{h} \sum_{i=1}^{n} h_i \sigma_T^{(i)}, \quad \tau_A = \frac{1}{h} \sum_{i=1}^{n} h_i \tau_A^{(i)}, \quad \varepsilon_t = \frac{1}{h} \sum_{i=1}^{n} h_i \varepsilon_t^{(i)}. \tag{7.24}$$

The quantities σ_A, σ_T and τ_A are effective stresses applied to the laminate, whereas from (7.15) the quantity ε_t is the effective through-thickness strain. The superscript '(L)' is used to denote effective thermoelastic constants.

From $(7.23)_3$ it is clear that the effective through-thickness strain ε_t for the laminate can be found and then substituted into the remaining equations of (7.23) leading to laminate stress-strain relations of the following form analogous to the corresponding ply relations (7.17)

$$\begin{aligned} \sigma_A &= \Omega_{11}^{(L)} \varepsilon_A + \Omega_{12}^{(L)} \varepsilon_T + \Omega_{13}^{(L)} \sigma_t + \Omega_{16}^{(L)} \gamma_A - \omega_1^{(L)} \Delta T, \\ \sigma_T &= \Omega_{12}^{(L)} \varepsilon_A + \Omega_{22}^{(L)} \varepsilon_T + \Omega_{23}^{(L)} \sigma_t + \Omega_{26}^{(L)} \gamma_A - \omega_2^{(L)} \Delta T, \\ \varepsilon_t &= -\Omega_{13}^{(L)} \varepsilon_A - \Omega_{23}^{(L)} \varepsilon_T + \Omega_{33}^{(L)} \sigma_t - \Omega_{36}^{(L)} \gamma_A + \omega_3^{(L)} \Delta T, \\ \tau_A &= \Omega_{16}^{(L)} \varepsilon_A + \Omega_{26}^{(L)} \varepsilon_T + \Omega_{36}^{(L)} \sigma_t + \Omega_{66}^{(L)} \gamma_A - \omega_6^{(L)} \Delta T, \end{aligned} \tag{7.25}$$

where

$$\begin{aligned} \Omega_{11}^{(L)} &= C_{11}^{(L)} - \frac{C_{13}^{(L)} C_{13}^{(L)}}{C_{33}^{(L)}}, \quad \Omega_{12}^{(L)} = C_{12}^{(L)} - \frac{C_{13}^{(L)} C_{23}^{(L)}}{C_{33}^{(L)}}, \\ \Omega_{13}^{(L)} &= \frac{C_{13}^{(L)}}{C_{33}^{(L)}}, \quad \Omega_{16}^{(L)} = C_{16}^{(L)} - \frac{C_{13}^{(L)} C_{36}^{(L)}}{C_{33}^{(L)}}, \end{aligned} \tag{7.26}$$

$$\Omega_{22}^{(L)} = C_{22}^{(L)} - \frac{C_{23}^{(L)}C_{23}^{(L)}}{C_{33}^{(L)}}, \ \Omega_{23}^{(L)} = \frac{C_{23}^{(L)}}{C_{33}^{(L)}}, \ \Omega_{26}^{(L)} = C_{26}^{(L)} - \frac{C_{23}^{(L)}C_{36}^{(L)}}{C_{33}^{(L)}}, \tag{7.27}$$

$$\Omega_{33}^{(L)} = \frac{1}{C_{33}^{(L)}}, \ \Omega_{36}^{(L)} = \frac{C_{36}^{(L)}}{C_{33}^{(L)}}, \tag{7.28}$$

$$\Omega_{66}^{(L)} = C_{66}^{(L)} - \frac{C_{36}^{(L)}C_{36}^{(L)}}{C_{33}^{(L)}}, \tag{7.29}$$

$$\begin{aligned} \omega_1^{(L)} &= U_1^{(L)} - \frac{C_{13}^{(L)}U_3^{(L)}}{C_{33}^{(L)}}, \ \omega_2^{(L)} = U_2^{(L)} - \frac{C_{23}^{(L)}U_3^{(L)}}{C_{33}^{(L)}}, \\ \omega_3^{(L)} &= \frac{U_3^{(L)}}{C_{33}^{(L)}}, \ \omega_6^{(L)} = U_6^{(L)} - \frac{C_{36}^{(L)}U_3^{(L)}}{C_{33}^{(L)}}. \end{aligned} \tag{7.30}$$

On using (7.24) after multiplying (7.18)–(7.22) by the ply thickness h_i and summing over $i = 1, \ldots, n$, the effective stress-strain relations (7.25) are obtained where

$$\begin{aligned} &\Omega_{11}^{(L)} = \frac{1}{h}\sum_{i=1}^{n} h_i \Omega_{11}^{(i)}, \quad \Omega_{12}^{(L)} = \frac{1}{h}\sum_{i=1}^{n} h_i \Omega_{12}^{(i)}, \\ &\Omega_{22}^{(L)} = \frac{1}{h}\sum_{i=1}^{n} h_i \Omega_{22}^{(i)}, \quad \Omega_{33}^{(L)} = \frac{1}{h}\sum_{i=1}^{n} h_i \Omega_{33}^{(i)}, \\ &\Omega_{16}^{(L)} = \frac{1}{h}\sum_{i=1}^{n} h_i \Omega_{16}^{(i)}, \quad \Omega_{26}^{(L)} = \frac{1}{h}\sum_{i=1}^{n} h_i \Omega_{26}^{(i)}, \quad \Omega_{66}^{(L)} = \frac{1}{h}\sum_{i=1}^{n} h_i \Omega_{66}^{(i)}, \end{aligned} \tag{7.31}$$

$$\Omega_{13}^{(L)} = \frac{1}{h}\sum_{i=1}^{n} h_i \Omega_{13}^{(i)}, \quad \Omega_{23}^{(L)} = \frac{1}{h}\sum_{i=1}^{n} h_i \Omega_{23}^{(i)}, \quad \Omega_{36}^{(L)} = \frac{1}{h}\sum_{i=1}^{n} h_i \Omega_{36}^{(i)}, \tag{7.32}$$

$$\begin{aligned} &\omega_1^{(L)} = \frac{1}{h}\sum_{i=1}^{n} h_i \omega_1^{(i)}, \quad \omega_2^{(L)} = \frac{1}{h}\sum_{i=1}^{n} h_i \omega_2^{(i)}, \\ &\omega_3^{(L)} = \frac{1}{h}\sum_{i=1}^{n} h_i \omega_3^{(i)}, \quad \omega_6^{(L)} = \frac{1}{h}\sum_{i=1}^{n} h_i \omega_6^{(i)}. \end{aligned} \tag{7.33}$$

To calculate the effective laminate constants appearing in (7.23) the summations represented by the parameters $\Omega_{IJ}^{(L)}$ and $\omega_I^{(L)}$ are calculated using (7.31)–(7.33). The values of $C_{33}^{(L)}$ and $U_3^{(L)}$ are determined using the following relations derived from $(7.28)_1$ and $(7.30)_3$

$$C_{33}^{(L)} = \frac{1}{\Omega_{33}^{(L)}}, \qquad U_3^{(L)} = \frac{\omega_3^{(L)}}{\Omega_{33}^{(L)}}. \tag{7.34}$$

The effective elastic constants $C_{13}^{(L)}$, $C_{23}^{(L)}$ and $C_{36}^{(L)}$ are calculated next using the relations

$$C_{13}^{(L)} = \frac{\Omega_{13}^{(L)}}{\Omega_{33}^{(L)}}, \qquad C_{23}^{(L)} = \frac{\Omega_{23}^{(L)}}{\Omega_{33}^{(L)}}, \qquad C_{36}^{(L)} = \frac{\Omega_{36}^{(L)}}{\Omega_{33}^{(L)}}. \tag{7.35}$$

The remaining required components of the C matrix follow from (7.26), (7.27) and (7.29), and are given by

$$C_{11}^{(L)} = \Omega_{11}^{(L)} + \frac{(\Omega_{13}^{(L)})^2}{\Omega_{33}^{(L)}}, \quad C_{12}^{(L)} = \Omega_{12}^{(L)} + \frac{\Omega_{13}^{(L)}\Omega_{23}^{(L)}}{\Omega_{33}^{(L)}}, \quad C_{22}^{(L)} = \Omega_{22}^{(L)} + \frac{(\Omega_{23}^{(L)})^2}{\Omega_{33}^{(L)}}, \tag{7.36}$$

$$C_{16}^{(L)} = \Omega_{16}^{(L)} + \frac{\Omega_{13}^{(L)}\Omega_{36}^{(L)}}{\Omega_{33}^{(L)}}, \quad C_{26}^{(L)} = \Omega_{26}^{(L)} + \frac{\Omega_{23}^{(L)}\Omega_{36}^{(L)}}{\Omega_{33}^{(L)}}, \quad C_{66}^{(L)} = \Omega_{66}^{(L)} + \frac{(\Omega_{36}^{(L)})^2}{\Omega_{33}^{(L)}}. \tag{7.37}$$

The laminate constants $U_1^{(\mathrm{L})}$, $U_2^{(\mathrm{L})}$ and $U_6^{(\mathrm{L})}$ follow from (7.30), and are given by

$$U_1^{(\mathrm{L})} = \omega_1^{(\mathrm{L})} + \frac{\Omega_{13}^{(\mathrm{L})}}{\Omega_{33}^{(\mathrm{L})}}\omega_3^{(\mathrm{L})},\ U_2^{(\mathrm{L})} = \omega_2^{(\mathrm{L})} + \frac{\Omega_{23}^{(\mathrm{L})}}{\Omega_{33}^{(\mathrm{L})}}\omega_3^{(\mathrm{L})},\ U_6^{(\mathrm{L})} = \omega_6^{(\mathrm{L})} + \frac{\Omega_{36}^{(\mathrm{L})}}{\Omega_{33}^{(\mathrm{L})}}\omega_3^{(\mathrm{L})}. \quad (7.38)$$

It is now required to interpret the effective laminate properties $\boldsymbol{C}$ and $\boldsymbol{U}$ in terms of the Young's and in-plane shear moduli, Poisson's ratios and thermal expansion coefficients of the laminate. This is achieved by modifying relations (6.82), (6.84) and (6.86) so that they apply to a laminate rather than to an individual angled ply leading to the stress-strain relations

$$\begin{aligned}\sigma_{\mathrm{A}} = {} & \frac{\tilde{E}_{\mathrm{A}}^{(\mathrm{L})}}{\Psi^{(\mathrm{L})}}\varepsilon_{\mathrm{A}} + \frac{\tilde{\nu}_{\mathrm{A}}^{(\mathrm{L})}\tilde{E}_{\mathrm{T}}^{(\mathrm{L})}}{\Psi^{(\mathrm{L})}}\varepsilon_{\mathrm{T}} + \frac{\hat{\nu}_{\mathrm{a}}^{(\mathrm{L})}}{\Psi^{(\mathrm{L})}}\sigma_{\mathrm{t}} \\ & + \left(\frac{\tilde{E}_{\mathrm{A}}^{(\mathrm{L})}}{\Psi^{(\mathrm{L})}}\lambda_{\mathrm{A}}^{(\mathrm{L})} + \frac{\tilde{\nu}_{\mathrm{A}}^{(\mathrm{L})}\tilde{E}_{\mathrm{T}}^{(\mathrm{L})}}{\Psi^{(\mathrm{L})}}\lambda_{\mathrm{T}}^{(\mathrm{L})}\right)\gamma_{\mathrm{A}} - \left(\frac{\tilde{E}_{\mathrm{A}}^{(\mathrm{L})}}{\Psi^{(\mathrm{L})}}\tilde{\alpha}_{\mathrm{A}}^{(\mathrm{L})} + \frac{\tilde{\nu}_{\mathrm{A}}^{(\mathrm{L})}\tilde{E}_{\mathrm{T}}^{(\mathrm{L})}}{\Psi^{(\mathrm{L})}}\tilde{\alpha}_{\mathrm{T}}^{(\mathrm{L})}\right)\Delta T,\end{aligned} \quad (7.39)$$

$$\begin{aligned}\sigma_{\mathrm{T}} = {} & \frac{\tilde{\nu}_{\mathrm{A}}^{(\mathrm{L})}\tilde{E}_{\mathrm{T}}^{(\mathrm{L})}}{\Psi^{(\mathrm{L})}}\varepsilon_{\mathrm{A}} + \frac{\tilde{E}_{\mathrm{T}}^{(\mathrm{L})}}{\Psi^{(\mathrm{L})}}\varepsilon_{\mathrm{T}} + \frac{\hat{\nu}_{\mathrm{t}}^{(\mathrm{L})}}{\Psi^{(\mathrm{L})}}\sigma_{\mathrm{t}} \\ & + \left(\frac{\tilde{\nu}_{\mathrm{A}}^{(\mathrm{L})}\tilde{E}_{\mathrm{T}}^{(\mathrm{L})}}{\Psi^{(\mathrm{L})}}\lambda_{\mathrm{A}}^{(\mathrm{L})} + \frac{\tilde{E}_{\mathrm{T}}^{(\mathrm{L})}}{\Psi^{(\mathrm{L})}}\lambda_{\mathrm{T}}^{(\mathrm{L})}\right)\gamma_{\mathrm{A}} - \left(\frac{\tilde{\nu}_{\mathrm{A}}^{(\mathrm{L})}\tilde{E}_{\mathrm{T}}^{(\mathrm{L})}}{\Psi^{(\mathrm{L})}}\tilde{\alpha}_{\mathrm{A}}^{(\mathrm{L})} + \frac{\tilde{E}_{\mathrm{T}}^{(\mathrm{L})}}{\Psi^{(\mathrm{L})}}\tilde{\alpha}_{\mathrm{T}}^{(\mathrm{L})}\right)\Delta T,\end{aligned} \quad (7.40)$$

$$\begin{aligned}\varepsilon_{\mathrm{t}} = {} & -\frac{\hat{\nu}_{\mathrm{a}}^{(\mathrm{L})}}{\Psi^{(\mathrm{L})}}\varepsilon_{\mathrm{A}} - \frac{\hat{\nu}_{\mathrm{t}}^{(\mathrm{L})}}{\Psi^{(\mathrm{L})}}\varepsilon_{T} + \frac{\Lambda^{(\mathrm{L})}}{\Psi^{(\mathrm{L})}\tilde{E}_{\mathrm{t}}^{(\mathrm{L})}}\sigma_{\mathrm{t}} \\ & -\left(\frac{\hat{\nu}_{\mathrm{a}}^{(\mathrm{L})}}{\Psi^{(\mathrm{L})}}\lambda_{\mathrm{A}}^{(\mathrm{L})} + \frac{\hat{\nu}_{\mathrm{t}}^{(\mathrm{L})}}{\Psi^{(\mathrm{L})}}\lambda_{\mathrm{T}}^{(\mathrm{L})} + \lambda_{\mathrm{t}}^{(\mathrm{L})}\right)\gamma_{\mathrm{A}} + \left(\frac{\hat{\nu}_{\mathrm{a}}^{(\mathrm{L})}}{\Psi^{(\mathrm{L})}}\tilde{\alpha}_{\mathrm{A}}^{(\mathrm{L})} + \frac{\hat{\nu}_{\mathrm{t}}^{(\mathrm{L})}}{\Psi^{(\mathrm{L})}}\tilde{\alpha}_{\mathrm{T}}^{(\mathrm{L})} + \tilde{\alpha}_{\mathrm{t}}^{(\mathrm{L})}\right)\Delta T,\end{aligned} \quad (7.41)$$

$$\begin{aligned}\tau_{\mathrm{A}} = {} & \frac{\tilde{E}_{\mathrm{A}}^{(\mathrm{L})}}{\Psi^{(\mathrm{L})}}\hat{\lambda}_{\mathrm{A}}^{(\mathrm{L})}\varepsilon_{\mathrm{A}} + \frac{\tilde{E}_{\mathrm{T}}^{(\mathrm{L})}}{\Psi^{(\mathrm{L})}}\hat{\lambda}_{\mathrm{T}}^{(\mathrm{L})}\varepsilon_{\mathrm{T}} + \left(\frac{\hat{\nu}_{\mathrm{a}}^{(\mathrm{L})}}{\Psi^{(\mathrm{L})}}\lambda_{\mathrm{A}}^{(\mathrm{L})} + \frac{\hat{\nu}_{\mathrm{t}}^{(\mathrm{L})}}{\Psi^{(\mathrm{L})}}\lambda_{\mathrm{T}}^{(\mathrm{L})} + \lambda_{\mathrm{t}}^{(\mathrm{L})}\right)\sigma_{\mathrm{t}} \\ & + \left(\frac{\tilde{E}_{\mathrm{A}}^{(\mathrm{L})}}{\Psi^{(\mathrm{L})}}\lambda_{\mathrm{A}}^{(\mathrm{L})}\hat{\lambda}_{\mathrm{A}}^{(\mathrm{L})} + \frac{\tilde{E}_{\mathrm{T}}^{(\mathrm{L})}}{\Psi^{(\mathrm{L})}}\lambda_{\mathrm{T}}^{(\mathrm{L})}\hat{\lambda}_{\mathrm{T}}^{(\mathrm{L})} + \mu_{\mathrm{A}}^{(\mathrm{L})}\right)\gamma_{\mathrm{A}} \\ & - \left(\frac{\tilde{E}_{\mathrm{A}}^{(\mathrm{L})}}{\Psi^{(\mathrm{L})}}\hat{\lambda}_{\mathrm{A}}^{(\mathrm{L})}\tilde{\alpha}_{\mathrm{A}}^{(\mathrm{L})} + \frac{\tilde{E}_{\mathrm{T}}^{(\mathrm{L})}}{\Psi^{(\mathrm{L})}}\hat{\lambda}_{\mathrm{T}}^{(\mathrm{L})}\tilde{\alpha}_{\mathrm{T}}^{(\mathrm{L})} + \mu_{\mathrm{A}}^{(\mathrm{L})}\alpha_{\mathrm{S}}^{(\mathrm{L})}\right)\Delta T,\end{aligned} \quad (7.42)$$

where, on modifying the definitions (6.81), (6.83), (6.85) and (6.87), so that they apply to a laminate

$$\begin{aligned}&\frac{1}{\tilde{E}_{\mathrm{A}}^{(\mathrm{L})}} = \frac{1}{E_{\mathrm{A}}^{(\mathrm{L})}} - \left(\lambda_{\mathrm{A}}^{(\mathrm{L})}\right)^2\frac{1}{\mu_{\mathrm{A}}^{(\mathrm{L})}}, \qquad \frac{1}{\tilde{E}_{\mathrm{T}}^{(\mathrm{L})}} = \frac{1}{E_{\mathrm{T}}^{(\mathrm{L})}} - \left(\lambda_{\mathrm{T}}^{(\mathrm{L})}\right)^2\frac{1}{\mu_{\mathrm{A}}^{(\mathrm{L})}}, \\ &\frac{1}{\tilde{E}_{\mathrm{t}}^{(\mathrm{L})}} = \frac{1}{E_{\mathrm{t}}^{(\mathrm{L})}} - \left(\lambda_{\mathrm{t}}^{(\mathrm{L})}\right)^2\frac{1}{\mu_{\mathrm{A}}^{(\mathrm{L})}}, \qquad \frac{\tilde{\nu}_{\mathrm{t}}^{(\mathrm{L})}}{\tilde{E}_{\mathrm{T}}^{(\mathrm{L})}} = \frac{\nu_{\mathrm{t}}^{(\mathrm{L})}}{E_{\mathrm{T}}^{(\mathrm{L})}} + \frac{\lambda_{\mathrm{t}}^{(\mathrm{L})}\lambda_{\mathrm{T}}^{(\mathrm{L})}}{\mu_{\mathrm{A}}^{(\mathrm{L})}}, \\ &\frac{\tilde{\nu}_{\mathrm{A}}^{(\mathrm{L})}}{\tilde{E}_{\mathrm{A}}^{(\mathrm{L})}} = \frac{\nu_{\mathrm{A}}^{(\mathrm{L})}}{E_{\mathrm{A}}^{(\mathrm{L})}} + \frac{\lambda_{\mathrm{A}}^{(\mathrm{L})}\lambda_{\mathrm{T}}^{(\mathrm{L})}}{\mu_{\mathrm{A}}^{(\mathrm{L})}}, \qquad \frac{\tilde{\nu}_{\mathrm{a}}^{(\mathrm{L})}}{\tilde{E}_{\mathrm{A}}^{(\mathrm{L})}} = \frac{\nu_{\mathrm{a}}^{(\mathrm{L})}}{E_{\mathrm{A}}^{(\mathrm{L})}} + \frac{\lambda_{\mathrm{t}}^{(\mathrm{L})}\lambda_{\mathrm{A}}^{(\mathrm{L})}}{\mu_{\mathrm{A}}^{(\mathrm{L})}}, \\ &\tilde{\alpha}_{\mathrm{A}}^{(\mathrm{L})} = \alpha_{\mathrm{A}}^{(\mathrm{L})} + \lambda_{\mathrm{A}}^{(\mathrm{L})}\alpha_{\mathrm{S}}^{(\mathrm{L})},\ \tilde{\alpha}_{\mathrm{T}}^{(\mathrm{L})} = \alpha_{\mathrm{T}}^{(\mathrm{L})} + \lambda_{\mathrm{T}}^{(\mathrm{L})}\alpha_{\mathrm{S}}^{(\mathrm{L})},\ \tilde{\alpha}_{\mathrm{t}}^{(\mathrm{L})} = \alpha_{\mathrm{t}}^{(\mathrm{L})} + \lambda_{\mathrm{t}}^{(\mathrm{L})}\alpha_{\mathrm{S}}^{(\mathrm{L})},\end{aligned} \quad (7.43)$$

$$\Psi^{(L)} = 1 - \left(\tilde{\nu}_{\mathrm{A}}^{(\mathrm{L})}\right)^2\frac{\tilde{E}_{\mathrm{T}}^{(\mathrm{L})}}{\tilde{E}_{\mathrm{A}}^{(\mathrm{L})}},\ \hat{\nu}_{\mathrm{a}}^{(\mathrm{L})} = \tilde{\nu}_{\mathrm{a}}^{(\mathrm{L})} + \tilde{\nu}_{\mathrm{t}}^{(\mathrm{L})}\tilde{\nu}_{\mathrm{A}}^{(\mathrm{L})},\ \hat{\nu}_{\mathrm{t}}^{(\mathrm{L})} = \tilde{\nu}_{\mathrm{t}}^{(\mathrm{L})} + \tilde{\nu}_{\mathrm{a}}^{(\mathrm{L})}\tilde{\nu}_{\mathrm{A}}^{(\mathrm{L})}\frac{\tilde{E}_{\mathrm{T}}^{(\mathrm{L})}}{\tilde{E}_{\mathrm{A}}^{(\mathrm{L})}}, \quad (7.44)$$

$$\begin{aligned}\Lambda^{(\mathrm{L})} &= \Psi^{(\mathrm{L})} - \tilde{\nu}_{\mathrm{a}}^{(\mathrm{L})} \frac{\tilde{E}_{\mathrm{t}}^{(\mathrm{L})}}{\tilde{E}_{\mathrm{A}}^{(\mathrm{L})}} \hat{\nu}_{\mathrm{a}}^{(\mathrm{L})} - \tilde{\nu}_{\mathrm{t}}^{(\mathrm{L})} \frac{\tilde{E}_{\mathrm{t}}^{(\mathrm{L})}}{\tilde{E}_{\mathrm{T}}^{(\mathrm{L})}} \hat{\nu}_{\mathrm{t}}^{(\mathrm{L})}, \\ &= 1 - \left(\tilde{\nu}_{\mathrm{a}}^{(\mathrm{L})}\right)^2 \frac{\tilde{E}_{\mathrm{t}}^{(\mathrm{L})}}{\tilde{E}_{\mathrm{A}}^{(\mathrm{L})}} - \left(\tilde{\nu}_{\mathrm{t}}^{(\mathrm{L})}\right)^2 \frac{\tilde{E}_{\mathrm{t}}^{(\mathrm{L})}}{\tilde{E}_{\mathrm{T}}^{(\mathrm{L})}} - \left(\tilde{\nu}_{\mathrm{A}}^{(\mathrm{L})}\right)^2 \frac{\tilde{E}_{\mathrm{T}}^{(\mathrm{L})}}{\tilde{E}_{\mathrm{A}}^{(\mathrm{L})}} - 2\tilde{\nu}_{\mathrm{a}}^{(\mathrm{L})} \tilde{\nu}_{\mathrm{t}}^{(\mathrm{L})} \tilde{\nu}_{\mathrm{A}}^{(\mathrm{L})} \frac{\tilde{E}_{\mathrm{t}}^{(\mathrm{L})}}{\tilde{E}_{\mathrm{A}}^{(\mathrm{L})}},\end{aligned} \tag{7.45}$$

$$\hat{\lambda}_{\mathrm{A}}^{(\mathrm{L})} = \lambda_{\mathrm{A}}^{(\mathrm{L})} + \lambda_{\mathrm{T}}^{(\mathrm{L})} \tilde{\nu}_{\mathrm{A}}^{(\mathrm{L})} \frac{\tilde{E}_{\mathrm{T}}^{(\mathrm{L})}}{\tilde{E}_{\mathrm{A}}^{(\mathrm{L})}}, \qquad \hat{\lambda}_{\mathrm{A}}^{(\mathrm{L})} = \lambda_{\mathrm{T}}^{(\mathrm{L})} + \lambda_{\mathrm{A}}^{(\mathrm{L})} \tilde{\nu}_{\mathrm{A}}^{(\mathrm{L})}. \tag{7.46}$$

It then follows on comparing (7.25) with (7.39)–(7.42) that

$$\Omega_{11}^{(\mathrm{L})} = \frac{\tilde{E}_{\mathrm{A}}^{(\mathrm{L})}}{\Psi^{(\mathrm{L})}}, \qquad \Omega_{12}^{(\mathrm{L})} = \frac{\tilde{\nu}_{\mathrm{A}}^{(\mathrm{L})} \tilde{E}_{\mathrm{T}}^{(\mathrm{L})}}{\Psi^{(\mathrm{L})}}, \qquad \Omega_{13}^{(\mathrm{L})} = \frac{\tilde{\nu}_{\mathrm{a}}^{(\mathrm{L})} + \tilde{\nu}_{\mathrm{t}}^{(\mathrm{L})} \tilde{\nu}_{\mathrm{A}}^{(\mathrm{L})}}{\Psi^{(\mathrm{L})}}, \tag{7.47}$$

$$\Omega_{22}^{(\mathrm{L})} = \frac{\tilde{E}_{\mathrm{T}}^{(\mathrm{L})}}{\Psi^{(\mathrm{L})}}, \; \Omega_{23}^{(\mathrm{L})} = \frac{1}{\Psi^{(\mathrm{L})}} \left(\tilde{\nu}_{\mathrm{t}}^{(\mathrm{L})} + \tilde{\nu}_{\mathrm{a}}^{(\mathrm{L})} \tilde{\nu}_{\mathrm{A}}^{(\mathrm{L})} \frac{\tilde{E}_{\mathrm{T}}^{(\mathrm{L})}}{\tilde{E}_{\mathrm{A}}^{(\mathrm{L})}} \right), \; \Omega_{33}^{(\mathrm{L})} = \frac{\Lambda^{(\mathrm{L})}}{\Psi^{(\mathrm{L})} \tilde{E}_{\mathrm{t}}^{(\mathrm{L})}}, \tag{7.48}$$

$$\Omega_{16}^{(\mathrm{L})} = \frac{\tilde{E}_{\mathrm{A}}^{(\mathrm{L})}}{\Psi^{(\mathrm{L})}} \lambda_{\mathrm{A}}^{(\mathrm{L})} + \frac{\tilde{\nu}_{\mathrm{A}}^{(\mathrm{L})} \tilde{E}_{\mathrm{T}}^{(\mathrm{L})}}{\Psi^{(\mathrm{L})}} \lambda_{\mathrm{T}}^{(\mathrm{L})}, \quad \Omega_{26}^{(\mathrm{L})} = \frac{\tilde{\nu}_{\mathrm{A}}^{(\mathrm{L})} \tilde{E}_{\mathrm{T}}^{(\mathrm{L})}}{\Psi^{(\mathrm{L})}} \lambda_{\mathrm{A}}^{(\mathrm{L})} + \frac{\tilde{E}_{\mathrm{T}}^{(\mathrm{L})}}{\Psi^{(\mathrm{L})}} \lambda_{\mathrm{T}}^{(\mathrm{L})}, \tag{7.49}$$

$$\Omega_{36}^{(\mathrm{L})} = \frac{\lambda_{\mathrm{A}}^{(\mathrm{L})}}{\Psi^{(\mathrm{L})}} \left(\tilde{\nu}_{\mathrm{a}}^{(\mathrm{L})} + \tilde{\nu}_{\mathrm{t}}^{(\mathrm{L})} \tilde{\nu}_{\mathrm{A}}^{(\mathrm{L})} \right) + \frac{\lambda_{\mathrm{T}}^{(\mathrm{L})}}{\Psi^{(\mathrm{L})}} \left(\tilde{\nu}_{\mathrm{t}}^{(\mathrm{L})} + \tilde{\nu}_{\mathrm{a}}^{(\mathrm{L})} \tilde{\nu}_{\mathrm{A}}^{(\mathrm{L})} \frac{\tilde{E}_{\mathrm{T}}^{(\mathrm{L})}}{\tilde{E}_{\mathrm{A}}^{(\mathrm{L})}} \right) + \lambda_{\mathrm{t}}^{(\mathrm{L})}, \tag{7.50}$$

$$\Omega_{66}^{(\mathrm{L})} = \left(\lambda_{\mathrm{A}}^{(\mathrm{L})}\right)^2 \frac{\tilde{E}_{\mathrm{A}}^{(\mathrm{L})}}{\Psi^{(\mathrm{L})}} + 2\lambda_{\mathrm{A}}^{(\mathrm{L})} \lambda_{\mathrm{T}}^{(\mathrm{L})} \frac{\tilde{\nu}_{\mathrm{A}}^{(\mathrm{L})} \tilde{E}_{\mathrm{T}}^{(\mathrm{L})}}{\Psi^{(\mathrm{L})}} + \left(\lambda_{\mathrm{T}}^{(\mathrm{L})}\right)^2 \frac{\tilde{E}_{\mathrm{T}}^{(\mathrm{L})}}{\Psi^{(\mathrm{L})}} + \mu_{\mathrm{A}}^{(\mathrm{L})}, \tag{7.51}$$

$$\omega_1^{(\mathrm{L})} = \frac{\tilde{E}_{\mathrm{A}}^{(\mathrm{L})}}{\Psi^{(\mathrm{L})}} \tilde{\alpha}_{\mathrm{A}}^{(\mathrm{L})} + \frac{\tilde{\nu}_{\mathrm{A}}^{(\mathrm{L})} \tilde{E}_{\mathrm{T}}^{(\mathrm{L})}}{\Psi^{(\mathrm{L})}} \tilde{\alpha}_{\mathrm{T}}^{(\mathrm{L})}, \qquad \omega_2^{(L)} = \frac{\tilde{\nu}_{\mathrm{A}}^{(\mathrm{L})} \tilde{E}_{\mathrm{T}}^{(\mathrm{L})}}{\Psi^{(\mathrm{L})}} \tilde{\alpha}_{\mathrm{A}}^{(\mathrm{L})} + \frac{\tilde{E}_{\mathrm{T}}^{(\mathrm{L})}}{\Psi^{(\mathrm{L})}} \tilde{\alpha}_{\mathrm{T}}^{(\mathrm{L})}, \tag{7.52}$$

$$\omega_3^{(\mathrm{L})} = \frac{1}{\Psi^{(\mathrm{L})}} \left(\tilde{\nu}_{\mathrm{a}}^{(\mathrm{L})} + \tilde{\nu}_{\mathrm{t}}^{(\mathrm{L})} \tilde{\nu}_{\mathrm{A}}^{(\mathrm{L})} \right) \tilde{\alpha}_{\mathrm{A}}^{(\mathrm{L})} + \frac{1}{\Psi^{(\mathrm{L})}} \left(\tilde{\nu}_{\mathrm{t}}^{(\mathrm{L})} + \tilde{\nu}_{\mathrm{a}}^{(\mathrm{L})} \tilde{\nu}_{\mathrm{A}}^{(\mathrm{L})} \frac{\tilde{E}_{\mathrm{T}}^{(\mathrm{L})}}{\tilde{E}_{\mathrm{A}}^{(\mathrm{L})}} \right) \tilde{\alpha}_{\mathrm{T}}^{(\mathrm{L})} + \tilde{\alpha}_{\mathrm{t}}^{(\mathrm{L})}, \tag{7.53}$$

$$\omega_6^{(\mathrm{L})} = \left(\frac{\tilde{E}_{\mathrm{A}}^{(\mathrm{L})}}{\Psi^{(\mathrm{L})}} \lambda_{\mathrm{A}}^{(\mathrm{L})} + \frac{\tilde{\nu}_{\mathrm{A}}^{(\mathrm{L})} \tilde{E}_{\mathrm{T}}^{(\mathrm{L})}}{\Psi^{(\mathrm{L})}} \lambda_{\mathrm{T}}^{(\mathrm{L})} \right) \tilde{\alpha}_{\mathrm{A}}^{(\mathrm{L})} + \left(\frac{\tilde{E}_{\mathrm{T}}^{(\mathrm{L})}}{\Psi^{(\mathrm{L})}} \lambda_{\mathrm{T}}^{(\mathrm{L})} + \frac{\tilde{\nu}_{\mathrm{A}}^{(\mathrm{L})} \tilde{E}_{\mathrm{T}}^{(\mathrm{L})}}{\Psi^{(\mathrm{L})}} \lambda_{\mathrm{A}}^{(\mathrm{L})} \right) \tilde{\alpha}_{\mathrm{T}}^{(\mathrm{L})} + \mu_{\mathrm{A}}^{(\mathrm{L})} \alpha_{\mathrm{S}}^{(\mathrm{L})}. \tag{7.54}$$

It is emphasised that relations (7.31)–(7.33) will generate the same values for the quantities $\Omega_{ij}^{(\mathrm{L})}$ and $\omega_i^{(\mathrm{L})}$ as relations (7.47)–(7.54).

It can be shown using $(7.44)_1$, (7.47) and (7.48), that

$$\begin{aligned}&\tilde{E}_{\mathrm{A}}^{(\mathrm{L})} = \Omega_{11}^{(\mathrm{L})} - \frac{\left(\Omega_{12}^{(\mathrm{L})}\right)^2}{\Omega_{22}^{(\mathrm{L})}}, \quad \tilde{\nu}_{\mathrm{A}}^{(\mathrm{L})} = \frac{\Omega_{12}^{(\mathrm{L})}}{\Omega_{22}^{(\mathrm{L})}}, \quad \tilde{E}_{\mathrm{T}}^{(\mathrm{L})} = \Omega_{22}^{(\mathrm{L})} - \frac{\left(\Omega_{12}^{(\mathrm{L})}\right)^2}{\Omega_{11}^{(\mathrm{L})}}, \\ &\tilde{\nu}_{\mathrm{a}}^{(\mathrm{L})} = \Omega_{13}^{(\mathrm{L})} - \tilde{\nu}_{\mathrm{A}}^{(\mathrm{L})} \Omega_{23}^{(\mathrm{L})}, \qquad \tilde{\nu}_{\mathrm{t}}^{(\mathrm{L})} = \Omega_{23}^{(\mathrm{L})} - \tilde{\nu}_{\mathrm{A}}^{(\mathrm{L})} \frac{\tilde{E}_{\mathrm{T}}^{(\mathrm{L})}}{\tilde{E}_{\mathrm{A}}^{(\mathrm{L})}} \Omega_{13}^{(\mathrm{L})}, \\ &\frac{1}{\tilde{E}_{\mathrm{t}}^{(\mathrm{L})}} = \left(\frac{\tilde{\nu}_{\mathrm{a}}^{(\mathrm{L})}}{\tilde{E}_{\mathrm{A}}^{(\mathrm{L})}} \right)^2 \Omega_{11}^{(\mathrm{L})} + 2 \frac{\tilde{\nu}_{\mathrm{a}}^{(\mathrm{L})}}{\tilde{E}_{\mathrm{A}}^{(\mathrm{L})}} \frac{\tilde{\nu}_{\mathrm{t}}^{(\mathrm{L})}}{\tilde{E}_{\mathrm{T}}^{(\mathrm{L})}} \Omega_{12}^{(\mathrm{L})} + \left(\frac{\tilde{\nu}_{t}^{(\mathrm{L})}}{\tilde{E}_{\mathrm{T}}^{(\mathrm{L})}} \right)^2 \Omega_{22}^{(\mathrm{L})} + \Omega_{33}^{(\mathrm{L})}.\end{aligned} \tag{7.55}$$

On using (7.49)

$$\frac{\tilde{E}_{\mathrm{A}}^{(\mathrm{L})}}{\Psi^{(\mathrm{L})}} \lambda_{\mathrm{A}}^{(\mathrm{L})} + \frac{\tilde{\nu}_{\mathrm{A}}^{(\mathrm{L})} \tilde{E}_{\mathrm{T}}^{(\mathrm{L})}}{\Psi^{(\mathrm{L})}} \lambda_{\mathrm{T}}^{(\mathrm{L})} = \Omega_{16}^{(\mathrm{L})}, \qquad \frac{\tilde{\nu}_{\mathrm{A}}^{(\mathrm{L})} \tilde{E}_{\mathrm{T}}^{(\mathrm{L})}}{\Psi^{(\mathrm{L})}} \lambda_{\mathrm{A}}^{(\mathrm{L})} + \frac{\tilde{E}_{\mathrm{T}}^{(\mathrm{L})}}{\Psi^{(\mathrm{L})}} \lambda_{\mathrm{T}}^{(\mathrm{L})} = \Omega_{26}^{(\mathrm{L})}, \tag{7.56}$$

which may be written as the following simultaneous linear equations

$$\Omega_{11}^{(L)}\lambda_{A}^{(L)}+\Omega_{12}^{(L)}\lambda_{T}^{(L)}=\Omega_{16}^{(L)},\qquad \Omega_{12}^{(L)}\lambda_{A}^{(L)}+\Omega_{22}^{(L)}\lambda_{T}^{(L)}=\Omega_{26}^{(L)},\tag{7.57}$$

having the solution

$$\lambda_{A}^{(L)}=\frac{1}{\tilde{E}_{A}^{(L)}}\Omega_{16}^{(L)}-\frac{\tilde{\nu}_{A}^{(L)}}{\tilde{E}_{A}^{(L)}}\Omega_{26}^{(L)},\qquad \lambda_{T}^{(L)}=\frac{1}{\tilde{E}_{T}^{(L)}}\Omega_{26}^{(L)}-\frac{\tilde{\nu}_{A}^{(L)}}{\tilde{E}_{A}^{(L)}}\Omega_{16}^{(L)}.\tag{7.58}$$

It then follows from (7.47), (7.48), (7.50) and (7.51) that

$$\lambda_{t}^{(L)}=\Omega_{36}^{(L)}-\Omega_{13}^{(L)}\lambda_{A}^{(L)}-\Omega_{23}^{(L)}\lambda_{T}^{(L)},\tag{7.59}$$

$$\mu_{A}^{(L)}=\Omega_{66}^{(L)}-\Omega_{11}^{(L)}\left(\lambda_{A}^{(L)}\right)^{2}-2\Omega_{12}^{(L)}\lambda_{A}^{(L)}\lambda_{T}^{(L)}-\Omega_{22}^{(L)}\left(\lambda_{T}^{(L)}\right)^{2}.\tag{7.60}$$

Turning now to thermal expansion coefficients, it follows that (7.52) may be written as the following simultaneous linear equations

$$\Omega_{11}^{(L)}\tilde{\alpha}_{A}^{(L)}+\Omega_{12}^{(L)}\tilde{\alpha}_{T}^{(L)}=\omega_{1}^{(L)},\qquad \Omega_{12}^{(L)}\tilde{\alpha}_{A}^{(L)}+\Omega_{22}^{(L)}\tilde{\alpha}_{T}^{(L)}=\omega_{2}^{(L)},\tag{7.61}$$

having solution

$$\tilde{\alpha}_{A}^{(L)}=\frac{1}{\tilde{E}_{A}^{(L)}}\omega_{1}^{(L)}-\frac{\tilde{\nu}_{A}^{(L)}}{\tilde{E}_{A}^{(L)}}\omega_{2}^{(L)},\qquad \tilde{\alpha}_{T}^{(L)}=\frac{1}{\tilde{E}_{T}^{(L)}}\omega_{2}^{(L)}-\frac{\tilde{\nu}_{A}^{(L)}}{\tilde{E}_{A}^{(L)}}\omega_{1}^{(L)}.\tag{7.62}$$

On using (7.53)

$$\tilde{\alpha}_{t}^{(L)}=\omega_{3}^{(L)}-\Omega_{13}^{(L)}\tilde{\alpha}_{A}^{(L)}-\Omega_{23}^{(L)}\tilde{\alpha}_{T}^{(L)},\tag{7.63}$$

and from (7.54)

$$\alpha_{S}^{(L)}=\frac{1}{\mu_{A}^{(L)}}\left(\omega_{6}^{(L)}-\Omega_{16}^{(L)}\tilde{\alpha}_{A}^{(L)}-\Omega_{26}^{(L)}\tilde{\alpha}_{T}^{(L)}\right).\tag{7.64}$$

The final step in the determination of the remaining thermoelastic constants of the symmetric laminate involves rearranging the form of the laminate relations (7.43) so that

$$\begin{aligned}
&\frac{1}{E_{A}^{(L)}}=\frac{1}{\tilde{E}_{A}^{(L)}}+\frac{\left(\lambda_{A}^{(L)}\right)^{2}}{\mu_{A}^{(L)}},\qquad \frac{1}{E_{T}^{(L)}}=\frac{1}{\tilde{E}_{T}^{(L)}}+\frac{\left(\lambda_{T}^{(L)}\right)^{2}}{\mu_{A}^{(L)}},\\
&\frac{1}{E_{t}^{(L)}}=\frac{1}{\tilde{E}_{t}^{(L)}}+\frac{\left(\lambda_{t}^{(L)}\right)^{2}}{\mu_{A}^{(L)}},\qquad \nu_{t}^{(L)}=E_{T}^{(L)}\left(\frac{\tilde{\nu}_{t}^{(L)}}{\tilde{E}_{T}^{(L)}}-\frac{\lambda_{t}^{(L)}\lambda_{T}^{(L)}}{\mu_{A}^{(L)}}\right),\\
&\nu_{A}^{(L)}=E_{A}^{(L)}\left(\frac{\tilde{\nu}_{A}^{(L)}}{\tilde{E}_{A}^{(L)}}-\frac{\lambda_{A}^{(L)}\lambda_{T}^{(L)}}{\mu_{A}^{(L)}}\right),\qquad \nu_{a}^{(L)}=E_{A}^{(L)}\left(\frac{\tilde{\nu}_{a}^{(L)}}{\tilde{E}_{A}^{(L)}}-\frac{\lambda_{t}^{(L)}\lambda_{A}^{(L)}}{\mu_{A}^{(L)}}\right),\\
&\alpha_{A}^{(L)}=\tilde{\alpha}_{A}^{(L)}-\lambda_{A}^{(L)}\alpha_{S}^{(L)},\quad \alpha_{T}^{(L)}=\tilde{\alpha}_{T}^{(L)}-\lambda_{T}^{(L)}\alpha_{S}^{(L)},\quad \alpha_{t}^{(L)}=\tilde{\alpha}_{t}^{(L)}-\lambda_{t}^{(L)}\alpha_{S}^{(L)}.
\end{aligned}\tag{7.65}$$

By analogy with (6.77) and (6.78), the $\boldsymbol{S}$ matrix components for the laminate appearing in the stress-strain relations

$$\begin{aligned}
\varepsilon_{A}&=S_{11}^{(L)}\sigma_{A}+S_{12}^{(L)}\sigma_{T}+S_{13}^{(L)}\sigma_{t}+S_{16}^{(L)}\tau_{A}+V_{1}^{(L)}\Delta T,\\
\varepsilon_{T}&=S_{12}^{(L)}\sigma_{A}+S_{22}^{(L)}\sigma_{T}+S_{23}^{(L)}\sigma_{t}+S_{26}^{(L)}\tau_{A}+V_{2}^{(L)}\Delta T,\\
\varepsilon_{t}&=S_{13}^{(L)}\sigma_{A}+S_{23}^{(L)}\sigma_{T}+S_{33}^{(L)}\sigma_{t}+S_{36}^{(L)}\tau_{A}+V_{3}^{(L)}\Delta T,\\
\gamma_{A}&=S_{16}^{(L)}\sigma_{A}+S_{26}^{(L)}\sigma_{T}+S_{36}^{(L)}\sigma_{t}+S_{66}^{(L)}\tau_{A}+V_{6}^{(L)}\Delta T,
\end{aligned}\tag{7.66}$$

are given by

$$
\begin{aligned}
&S_{11}^{(\mathrm{L})}=\frac{1}{E_{\mathrm{A}}^{(\mathrm{L})}},\quad S_{12}^{(\mathrm{L})}=-\frac{\nu_{\mathrm{A}}^{(\mathrm{L})}}{E_{\mathrm{A}}^{(\mathrm{L})}},\quad S_{13}^{(\mathrm{L})}=-\frac{\nu_{\mathrm{a}}^{(\mathrm{L})}}{E_{\mathrm{A}}^{(\mathrm{L})}},\quad S_{16}^{(\mathrm{L})}=-\frac{\lambda_{\mathrm{A}}^{(\mathrm{L})}}{\mu_{\mathrm{A}}^{(\mathrm{L})}},\quad V_1^{(\mathrm{L})}=\alpha_{\mathrm{A}}^{(\mathrm{L})},\\
&\qquad S_{22}^{(\mathrm{L})}=\frac{1}{E_{\mathrm{T}}^{(\mathrm{L})}},\quad S_{23}^{(\mathrm{L})}=-\frac{\nu_{\mathrm{t}}^{(\mathrm{L})}}{E_{\mathrm{T}}^{(\mathrm{L})}},\quad S_{26}^{(\mathrm{L})}=-\frac{\lambda_{\mathrm{T}}^{(\mathrm{L})}}{\mu_{\mathrm{A}}^{(\mathrm{L})}},\quad V_2^{(\mathrm{L})}=\alpha_{\mathrm{T}}^{(\mathrm{L})},\\
&\qquad\qquad S_{33}^{(\mathrm{L})}=\frac{1}{E_{\mathrm{t}}^{(\mathrm{L})}},\quad S_{36}^{(\mathrm{L})}=-\frac{\lambda_{\mathrm{t}}^{(\mathrm{L})}}{\mu_{\mathrm{A}}^{(\mathrm{L})}},\quad V_3^{(\mathrm{L})}=\alpha_{\mathrm{t}}^{(\mathrm{L})},\\
&\qquad\qquad\qquad S_{66}^{(\mathrm{L})}=\frac{1}{\mu_{\mathrm{A}}^{(\mathrm{L})}},\quad V_6^{(\mathrm{L})}=\alpha_{\mathrm{S}}^{(\mathrm{L})}.
\end{aligned}
\tag{7.67}
$$

7.6 Out-of-Plane Shear Properties

From the stress-strain relations (6.53) it follows that for the ith ply of the laminate

$$
\begin{aligned}
\gamma_{\mathrm{t}}^{(i)}&=S_{44}^{(i)}\tau_{\mathrm{t}}+S_{45}^{(i)}\tau_{\mathrm{a}},\\
\gamma_{\mathrm{a}}^{(i)}&=S_{45}^{(i)}\tau_{\mathrm{t}}+S_{55}^{(i)}\tau_{\mathrm{a}},
\end{aligned}
\quad i=1,\ldots,n,
\tag{7.68}
$$

where the shear stresses τ_{a} and τ_{t} have the same uniform value in each of the n plies so that the traction continuity conditions $(7.5)_{1,2}$ are satisfied. The required elements of the $\boldsymbol{S}$ matrix are given by (6.63)–(6.65). On multiplying the shear stress-strain relations (7.68) by h_i, summing over $i = 1, \ldots, n$, and then dividing by h

$$
\begin{aligned}
\gamma_{\mathrm{t}}&=\frac{1}{h}\sum_{i=1}^{n}h_i\gamma_{\mathrm{t}}^{(i)}=\frac{1}{h}\left(\sum_{i=1}^{n}h_iS_{44}^{(i)}\right)\tau_{\mathrm{t}}+\frac{1}{h}\left(\sum_{i=1}^{n}h_iS_{45}^{(i)}\right)\tau_{\mathrm{a}},\\
\gamma_{\mathrm{a}}&=\frac{1}{h}\sum_{i=1}^{n}h_i\gamma_{\mathrm{a}}^{(i)}=\frac{1}{h}\left(\sum_{i=1}^{n}h_iS_{45}^{(i)}\right)\tau_{\mathrm{t}}+\frac{1}{h}\left(\sum_{i=1}^{n}h_iS_{55}^{(i)}\right)\tau_{\mathrm{a}},
\end{aligned}
\tag{7.69}
$$

where γ_{t} and γ_{a} are the effective out-of-plane shear strains applied to the laminate defined by (7.13). Relations (7.69) thus define the effective out-of-plane shear stress-strain relations for the laminate which are written in the form

$$
\begin{aligned}
\gamma_{\mathrm{t}}&=S_{44}^{(\mathrm{L})}\tau_{\mathrm{t}}+S_{45}^{(\mathrm{L})}\tau_{\mathrm{a}},\\
\gamma_{\mathrm{a}}&=S_{45}^{(\mathrm{L})}\tau_{\mathrm{t}}+S_{55}^{(\mathrm{L})}\tau_{\mathrm{a}}.
\end{aligned}
\tag{7.70}
$$

On comparing (7.69) and (7.70) it is clear that

$$
S_{44}^{(\mathrm{L})}=\frac{1}{h}\left(\sum_{i=1}^{n}h_iS_{44}^{(i)}\right),\quad S_{45}^{(\mathrm{L})}=\frac{1}{h}\left(\sum_{i=1}^{n}h_iS_{45}^{(i)}\right),\quad S_{55}^{(\mathrm{L})}=\frac{1}{h}\left(\sum_{i=1}^{n}h_iS_{55}^{(i)}\right).
\tag{7.71}
$$

On using the (6.63)–(6.65), relations (7.71) may also be written in the form

$$
\begin{aligned}
S_{44}^{(\mathrm{L})}&=\frac{1}{h}\sum_{i=1}^{n}h_i\left(\frac{m_i^2}{\mu_{\mathrm{t}}^{(i)}}+\frac{n_i^2}{\mu_{\mathrm{a}}^{(i)}}\right)=\frac{1}{\mu_{\mathrm{t}}^{(\mathrm{L})}},\\
S_{45}^{(\mathrm{L})}&=\frac{1}{h}\sum_{i=1}^{n}h_im_in_i\left(\frac{1}{\mu_{\mathrm{a}}^{(i)}}-\frac{1}{\mu_{\mathrm{t}}^{(i)}}\right)=\Phi^{(\mathrm{L})},\\
S_{55}^{(\mathrm{L})}&=\frac{1}{h}\sum_{i=1}^{n}h_i\left(\frac{m_i^2}{\mu_{\mathrm{a}}^{(i)}}+\frac{n_i^2}{\mu_{\mathrm{t}}^{(i)}}\right)=\frac{1}{\mu_{\mathrm{a}}^{(\mathrm{L})}},
\end{aligned}
\tag{7.72}
$$

defining three new effective elastic constants characterising the out-of-plane shear deformation of the laminate. These constants are denoted by $\mu_{\rm t}^{\rm (L)}$, $\Phi^{\rm (L)}$ and $\mu_{\rm a}^{\rm (L)}$, each having the dimension of stress (i.e. they are moduli). From (7.70) it is clear that the parameter $\Phi^{\rm (L)}$ is a measure of the interaction between the shear strain $\gamma_{\rm t}$ and the shear stress $\tau_{\rm a}$, and between the shear strain $\gamma_{\rm a}$ and the shear stress $\tau_{\rm t}$.

The inverse form of the stress-strain relations (7.68) for the ith ply is

$$\begin{aligned} \tau_{\rm t} &= C_{44}^{(i)}\gamma_{\rm t}^{(i)} + C_{45}^{(i)}\gamma_{\rm a}^{(i)} \\ \tau_{\rm a} &= C_{45}^{(i)}\gamma_{\rm t}^{(i)} + C_{55}^{(i)}\gamma_{\rm a}^{(i)} \end{aligned} \quad i=1,\ldots,n, \tag{7.73}$$

where, for $i = 1, \ldots, n$,

$$C_{44}^{(i)} = \frac{S_{55}^{(i)}}{S_{44}^{(i)}S_{55}^{(i)} - S_{45}^{(i)}S_{45}^{(i)}}, \quad C_{45}^{(i)} = -\frac{S_{45}^{(i)}}{S_{44}^{(i)}S_{55}^{(i)} - S_{45}^{(i)}S_{45}^{(i)}}, \quad C_{55}^{(i)} = \frac{S_{44}^{(i)}}{S_{44}^{(i)}S_{55}^{(i)} - S_{45}^{(i)}S_{45}^{(i)}}. \tag{7.74}$$

It can be shown that

$$C_{44}^{(i)}C_{55}^{(i)} - C_{45}^{(i)}C_{45}^{(i)} = \frac{1}{S_{44}^{(i)}S_{55}^{(i)} - S_{45}^{(i)}S_{45}^{(i)}}, \tag{7.75}$$

and

$$S_{44}^{(i)} = \frac{C_{55}^{(i)}}{C_{44}^{(i)}C_{55}^{(i)} - C_{45}^{(i)}C_{45}^{(i)}}, \quad S_{45}^{(i)} = -\frac{C_{45}^{(i)}}{C_{44}^{(i)}C_{55}^{(i)} - C_{45}^{(i)}C_{45}^{(i)}}, \quad S_{55}^{(i)} = \frac{C_{44}^{(i)}}{C_{44}^{(i)}C_{55}^{(i)} - C_{45}^{(i)}C_{45}^{(i)}}. \tag{7.76}$$

Similarly, the inverse form of the laminate stress-strain relations (7.70) is

$$\begin{aligned} \tau_{\rm t} &= C_{44}^{\rm (L)}\gamma_{\rm t} + C_{45}^{\rm (L)}\gamma_{\rm a}, \\ \tau_{\rm a} &= C_{45}^{\rm (L)}\gamma_{\rm t} + C_{55}^{\rm (L)}\gamma_{\rm a}, \end{aligned} \tag{7.77}$$

where

$$\begin{aligned} C_{44}^{\rm (L)} &= \frac{S_{55}^{\rm (L)}}{S_{44}^{\rm (L)}S_{55}^{\rm (L)} - S_{45}^{\rm (L)}S_{45}^{\rm (L)}}, \quad C_{45}^{\rm (L)} = -\frac{S_{45}^{\rm (L)}}{S_{44}^{\rm (L)}S_{55}^{\rm (L)} - S_{45}^{\rm (L)}S_{45}^{\rm (L)}}, \\ C_{55}^{\rm (L)} &= \frac{S_{44}^{\rm (L)}}{S_{44}^{\rm (L)}S_{55}^{\rm (L)} - S_{45}^{\rm (L)}S_{45}^{\rm (L)}}. \end{aligned} \tag{7.78}$$

On using the results (7.72),

$$\begin{aligned} C_{44}^{\rm (L)} &= \frac{\mu_{\rm t}^{\rm (L)}}{1-\mu_{\rm a}^{\rm (L)}\mu_{\rm t}^{\rm (L)}\left(\Phi^{\rm (L)}\right)^2}, \quad C_{45}^{\rm (L)} = -\frac{\mu_{\rm a}^{\rm (L)}\mu_{\rm t}^{\rm (L)}\Phi^{\rm (L)}}{1-\mu_{\rm a}^{\rm (L)}\mu_{\rm t}^{\rm (L)}\left(\Phi^{\rm (L)}\right)^2}, \\ C_{55}^{\rm (L)} &= \frac{\mu_{\rm a}^{\rm (L)}}{1-\mu_{\rm a}^{\rm (L)}\mu_{\rm t}^{\rm (L)}\left(\Phi^{\rm (L)}\right)^2}. \end{aligned} \tag{7.79}$$

7.7 Combined Stress-Strain Relations

On combining the stress-strain relations (7.23) and (7.77) it follows that the stress-strain relations for the laminate may be expressed in matrix form, analogous to (6.72) as follows

$$\begin{bmatrix}\sigma_{\mathrm{A}}\\ \sigma_{\mathrm{T}}\\ \sigma_{\mathrm{t}}\\ \tau_{\mathrm{t}}\\ \tau_{\mathrm{a}}\\ \tau_{\mathrm{A}}\end{bmatrix} \equiv \begin{bmatrix}\sigma_{11}\\ \sigma_{22}\\ \sigma_{33}\\ \sigma_{23}\\ \sigma_{13}\\ \sigma_{12}\end{bmatrix} = \begin{bmatrix} C_{11}^{(\mathrm{L})} & C_{12}^{(\mathrm{L})} & C_{13}^{(\mathrm{L})} & 0 & 0 & C_{16}^{(\mathrm{L})}\\ C_{12}^{(\mathrm{L})} & C_{22}^{(\mathrm{L})} & C_{23}^{(\mathrm{L})} & 0 & 0 & C_{26}^{(\mathrm{L})}\\ C_{13}^{(\mathrm{L})} & C_{23}^{(\mathrm{L})} & C_{33}^{(\mathrm{L})} & 0 & 0 & C_{36}^{(\mathrm{L})}\\ 0 & 0 & 0 & C_{44}^{(\mathrm{L})} & C_{45}^{(\mathrm{L})} & 0\\ 0 & 0 & 0 & C_{45}^{(\mathrm{L})} & C_{55}^{(\mathrm{L})} & 0\\ C_{16}^{(\mathrm{L})} & C_{26}^{(\mathrm{L})} & C_{36}^{(\mathrm{L})} & 0 & 0 & C_{66}^{(\mathrm{L})} \end{bmatrix} \begin{bmatrix}\varepsilon_{\mathrm{A}}\\ \varepsilon_{\mathrm{T}}\\ \varepsilon_{\mathrm{t}}\\ \gamma_{\mathrm{t}}\\ \gamma_{\mathrm{a}}\\ \gamma_{\mathrm{A}}\end{bmatrix} - \begin{bmatrix}U_1^{(\mathrm{L})}\\ U_2^{(\mathrm{L})}\\ U_3^{(\mathrm{L})}\\ 0\\ 0\\ U_6^{(\mathrm{L})}\end{bmatrix}\Delta T, \tag{7.80}$$

where the elastic constants $C_{IJ}^{(\mathrm{L})}$ are defined by (7.34)–(7.37) and (7.79), the thermal constants $U_I^{(\mathrm{L})}$ are defined by (7.38). Use needs to be made of relations (7.47)–(7.54) when calculating values of these elastic constants.

Similarly, on combining (7.66) and (7.70) the stress-strain relations for the laminate may be expressed in matrix form, analogous to (6.53), as follows

$$\begin{bmatrix}\varepsilon_{\mathrm{A}}\\ \varepsilon_{\mathrm{T}}\\ \varepsilon_{\mathrm{t}}\\ \gamma_{\mathrm{t}}\\ \gamma_{\mathrm{a}}\\ \gamma_{\mathrm{A}}\end{bmatrix} \equiv \begin{bmatrix}\varepsilon_{11}\\ \varepsilon_{22}\\ \varepsilon_{33}\\ 2\varepsilon_{23}\\ 2\varepsilon_{13}\\ 2\varepsilon_{12}\end{bmatrix} = \begin{bmatrix} S_{11}^{(\mathrm{L})} & S_{12}^{(\mathrm{L})} & S_{13}^{(\mathrm{L})} & 0 & 0 & S_{16}^{(\mathrm{L})}\\ S_{12}^{(\mathrm{L})} & S_{22}^{(\mathrm{L})} & S_{23}^{(\mathrm{L})} & 0 & 0 & S_{26}^{(\mathrm{L})}\\ S_{13}^{(\mathrm{L})} & S_{23}^{(\mathrm{L})} & S_{33}^{(\mathrm{L})} & 0 & 0 & S_{36}^{(\mathrm{L})}\\ 0 & 0 & 0 & S_{44}^{(\mathrm{L})} & S_{45}^{(\mathrm{L})} & 0\\ 0 & 0 & 0 & S_{45}^{(\mathrm{L})} & S_{55}^{(\mathrm{L})} & 0\\ S_{16}^{(\mathrm{L})} & S_{26}^{(\mathrm{L})} & S_{36}^{(\mathrm{L})} & 0 & 0 & S_{66}^{(\mathrm{L})} \end{bmatrix} \begin{bmatrix}\sigma_{\mathrm{A}}\\ \sigma_{\mathrm{T}}\\ \sigma_{\mathrm{t}}\\ \tau_{\mathrm{t}}\\ \tau_{\mathrm{a}}\\ \tau_{\mathrm{A}}\end{bmatrix} + \begin{bmatrix}V_1^{(\mathrm{L})}\\ V_2^{(\mathrm{L})}\\ V_3^{(\mathrm{L})}\\ 0\\ 0\\ V_6^{(\mathrm{L})}\end{bmatrix}\Delta T, \tag{7.81}$$

where

$$\begin{aligned}
&S_{11}^{(\mathrm{L})} = \frac{1}{E_{\mathrm{A}}^{(\mathrm{L})}},\quad S_{12}^{(\mathrm{L})} = -\frac{\nu_{\mathrm{A}}^{(\mathrm{L})}}{E_{\mathrm{A}}^{(\mathrm{L})}},\quad S_{13}^{(\mathrm{L})} = -\frac{\nu_{\mathrm{a}}^{(\mathrm{L})}}{E_{\mathrm{A}}^{(\mathrm{L})}},\quad S_{16}^{(\mathrm{L})} = -\frac{\lambda_{\mathrm{A}}^{(\mathrm{L})}}{\mu_{\mathrm{A}}^{(\mathrm{L})}},\quad V_1^{(\mathrm{L})} = \alpha_{\mathrm{A}}^{(\mathrm{L})},\\
&S_{22}^{(\mathrm{L})} = \frac{1}{E_{\mathrm{T}}^{(\mathrm{L})}},\quad S_{23}^{(\mathrm{L})} = -\frac{\nu_{\mathrm{t}}^{(\mathrm{L})}}{E_{\mathrm{T}}^{(\mathrm{L})}},\quad S_{26}^{(\mathrm{L})} = -\frac{\lambda_{\mathrm{T}}^{(\mathrm{L})}}{\mu_{\mathrm{A}}^{(\mathrm{L})}},\quad V_2^{(\mathrm{L})} = \alpha_{\mathrm{T}}^{(\mathrm{L})},\\
&S_{33}^{(\mathrm{L})} = \frac{1}{E_{\mathrm{t}}^{(\mathrm{L})}},\quad S_{36}^{(\mathrm{L})} = -\frac{\lambda_{\mathrm{t}}^{(\mathrm{L})}}{\mu_{\mathrm{A}}^{(\mathrm{L})}},\quad V_3^{(\mathrm{L})} = \alpha_{\mathrm{t}}^{(\mathrm{L})},\\
&S_{44}^{(\mathrm{L})} = \frac{1}{\mu_{\mathrm{t}}^{(\mathrm{L})}},\quad S_{45}^{(\mathrm{L})} = \Phi^{(\mathrm{L})},\quad S_{55}^{(\mathrm{L})} = \frac{1}{\mu_{\mathrm{a}}^{(\mathrm{L})}},\\
&S_{66}^{(\mathrm{L})} = \frac{1}{\mu_{\mathrm{A}}^{(\mathrm{L})}},\quad V_6^{(\mathrm{L})} = \alpha_{\mathrm{S}}^{(\mathrm{L})}.
\end{aligned} \tag{7.82}$$

It can be shown that

$$\boldsymbol{C}^{(\mathrm{L})}\cdot\boldsymbol{S}^{(\mathrm{L})} = \boldsymbol{S}^{(\mathrm{L})}\cdot\boldsymbol{C}^{(\mathrm{L})} = \mathbf{I}, \tag{7.83}$$

where $\boldsymbol{C}^{(\mathrm{L})}$ and $\boldsymbol{S}^{(\mathrm{L})}$ are the 6×6 matrices for the laminate appearing in (7.80) and (7.81), respectively, and where **I** is the second-order 6×6 identity matrix. In expanded form, relation (7.83) may be written as

$$\sum_{K=1}^{6} C_{IK}^{(\mathrm{L})} S_{KJ}^{(\mathrm{L})} = \sum_{K=1}^{6} S_{IK}^{(\mathrm{L})} C_{KJ}^{(\mathrm{L})} = \delta_{IJ},\quad I,J=1,\ldots,6, \tag{7.84}$$

where $\delta_{IJ}=1$ if $I=J$, $\delta_{IJ}=0$ otherwise. The satisfaction of these relations by the elements of the $\boldsymbol{C}^{(\mathrm{L})}$ and $\boldsymbol{S}^{(\mathrm{L})}$ matrices has been checked numerically.

7.8 Stress-Strain Equations for Transverse Isotropic Materials

When general symmetric laminates are used it is sometimes convenient to assume that the laminate is transverse isotropic about the axis describing the through-thickness direction. This special type of property is useful when solving laminate problems which are axisymmetric about this direction. Assuming transverse isotropy often enables two-dimensional solutions based on axisymmetry rather than three-dimensional solutions of a given problem. The purpose of this section is to derive conditions that must be satisfied by the effective properties of laminates in order that the laminate is transverse isotropic. Transverse isotropy was considered in Chapter 2 (see Section 2.17) when using stress-strain relations based on elastic compliances S_{ij}. It is useful to include here a discussion of transverse isotropy when using stress-strain relations based on elastic stiffnesses or moduli C_{ij}.

The transformation of a set of Cartesian coordinates (x_1,x_2,x_3) to (x_1',x_2',x_3') by a rotation of the x_1- and x_2-axes about the x_3-axis through an angle ϕ leads to

$$\begin{aligned} x_1 &= x_1'\cos\phi - x_2'\sin\phi, & x_1' &= \ \ x_1\cos\phi + x_2\sin\phi, \\ x_2 &= x_1'\sin\phi + x_2'\cos\phi, & x_2' &= -x_1\sin\phi + x_2\cos\phi, \\ x_3 &= x_3', & x_3' &= x_3. \end{aligned} \tag{7.85}$$

Stress-strain relations for an orthotropic material before transformation may be written (compare with (7.80)) as

$$\begin{bmatrix} \sigma_{11} \\ \sigma_{22} \\ \sigma_{33} \\ \sigma_{23} \\ \sigma_{13} \\ \sigma_{12} \end{bmatrix} = \begin{bmatrix} C_{11} & C_{12} & C_{13} & 0 & 0 & 0 \\ C_{12} & C_{22} & C_{23} & 0 & 0 & 0 \\ C_{13} & C_{23} & C_{33} & 0 & 0 & 0 \\ 0 & 0 & 0 & C_{44} & 0 & 0 \\ 0 & 0 & 0 & 0 & C_{55} & 0 \\ 0 & 0 & 0 & 0 & 0 & C_{66} \end{bmatrix} \begin{bmatrix} \varepsilon_{11} \\ \varepsilon_{22} \\ \varepsilon_{33} \\ 2\varepsilon_{23} \\ 2\varepsilon_{13} \\ 2\varepsilon_{12} \end{bmatrix}. \tag{7.86}$$

On setting $m = \cos\phi$ and $n = \sin\phi$, the transformation for stresses is given by

$$\begin{aligned} \sigma_{11}' &= m^2\sigma_{11} + n^2\sigma_{22} + 2mn\sigma_{12}, \\ \sigma_{22}' &= n^2\sigma_{11} + m^2\sigma_{22} - 2mn\sigma_{12}, \\ \sigma_{33}' &= \sigma_{33}, \\ \sigma_{23}' &= m\sigma_{23} - n\sigma_{13} = \sigma_{32}', \\ \sigma_{13}' &= n\sigma_{23} + m\sigma_{13} = \sigma_{31}', \\ \sigma_{12}' &= -mn\sigma_{11} + mn\sigma_{22} + \left(m^2 - n^2\right)\sigma_{12} = \sigma_{21}'. \end{aligned} \tag{7.87}$$

On substituting (7.86) into (7.87),

$$\begin{aligned} \sigma_{11}' &= \left(m^2C_{11} + n^2C_{12}\right)\varepsilon_{11} + \left(m^2C_{12} + n^2C_{22}\right)\varepsilon_{22} + \left(m^2C_{13} + n^2C_{23}\right)\varepsilon_{33} + 4mnC_{66}\varepsilon_{12}, \\ \sigma_{22}' &= \left(n^2C_{11} + m^2C_{12}\right)\varepsilon_{11} + \left(n^2C_{12} + m^2C_{22}\right)\varepsilon_{22} + \left(n^2C_{13} + m^2C_{23}\right)\varepsilon_{33} - 4mnC_{66}\varepsilon_{12}, \\ \sigma_{33}' &= C_{13}\varepsilon_{11} + C_{23}\varepsilon_{22} + C_{33}\varepsilon_{33}, \\ \sigma_{23}' &= 2mC_{44}\varepsilon_{23} - 2nC_{55}\varepsilon_{13} = \sigma_{32}', \\ \sigma_{13}' &= 2nC_{44}\varepsilon_{23} + 2mC_{55}\varepsilon_{13} = \sigma_{31}', \\ \sigma_{12}' &= mn\left(C_{12} - C_{11}\right)\varepsilon_{11} + mn\left(C_{22} - C_{12}\right)\varepsilon_{22} + mn\left(C_{23} - C_{13}\right)\varepsilon_{33} + 2\left(m^2 - n^2\right)C_{66}\varepsilon_{12} \\ &= \sigma_{21}'. \end{aligned} \tag{7.88}$$

The transformation for strains is given by

$$\begin{aligned}
\varepsilon_{11} &= m^2\varepsilon'_{11} + n^2\varepsilon'_{22} - 2mn\,\varepsilon'_{12}, \\
\varepsilon_{22} &= n^2\,\varepsilon'_{11} + m^2\varepsilon'_{22} + 2mn\,\varepsilon'_{12}, \\
\varepsilon_{33} &= \varepsilon'_{33}, \\
\varepsilon_{23} &= m\,\varepsilon'_{23} + n\,\varepsilon'_{13} = \varepsilon_{32}, \\
\varepsilon_{13} &= -\,n\,\varepsilon'_{23} + m\,\varepsilon'_{13} = \varepsilon_{31}, \\
\varepsilon_{12} &= mn\,\varepsilon'_{11} - mn\,\varepsilon'_{22} + \left(m^2 - n^2\right)\varepsilon'_{12} = \varepsilon_{21}.
\end{aligned} \tag{7.89}$$

On substituting (7.89) into (7.88),

$$\begin{bmatrix} \sigma'_{11} \\ \sigma'_{22} \\ \sigma'_{33} \\ \sigma'_{23} \\ \sigma'_{13} \\ \sigma'_{12} \end{bmatrix} = \begin{bmatrix} C'_{11} & C'_{12} & C'_{13} & 0 & 0 & C'_{16} \\ C'_{12} & C'_{22} & C'_{23} & 0 & 0 & C'_{26} \\ C'_{13} & C'_{23} & C'_{33} & 0 & 0 & C'_{36} \\ 0 & 0 & 0 & C'_{44} & C'_{45} & 0 \\ 0 & 0 & 0 & C'_{45} & C'_{55} & 0 \\ C'_{16} & C'_{26} & C'_{36} & 0 & 0 & C'_{66} \end{bmatrix} \begin{bmatrix} \varepsilon'_{11} \\ \varepsilon'_{22} \\ \varepsilon'_{33} \\ 2\varepsilon'_{23} \\ 2\varepsilon'_{13} \\ 2\varepsilon'_{12} \end{bmatrix}, \tag{7.90}$$

where

$$\begin{aligned}
C'_{11} &= m^4C_{11} + n^4C_{22} + 2m^2n^2\left(C_{12} + 2C_{66}\right), \\
C'_{22} &= n^4C_{11} + m^4C_{22} + 2m^2n^2\left(C_{12} + 2C_{66}\right), \\
C'_{12} &= \left(m^4 + n^4\right)C_{12} + m^2n^2\left(C_{11} + C_{22} - 4C_{66}\right), \\
C'_{13} &= m^2C_{13} + n^2C_{23},\ C'_{23} = n^2C_{13} + m^2C_{23},\ C'_{33} = C_{33}, \\
C'_{44} &= m^2C_{44} + n^2C_{55},\ C'_{45} = mn\left(C_{44} - C_{55}\right), C_{55} = m^2C_{55} + n^2\,C_{44}, \\
C'_{16} &= mn\left[\left(m^2 - n^2\right)\left(C_{12} + 2C_{66}\right) - m^2C_{11} + n^2C_{22}\right], \\
C'_{26} &= mn\left[\left(n^2 - m^2\right)\left(C_{12} + 2C_{66}\right) - n^2C_{11} + m^2C_{22}\right],\ C'_{36} = mn\left(C_{23} - C_{13}\right), \\
C'_{66} &= m^2n^2\left(C_{11} + C_{22} - 2C_{12}\right) + \left(m^2 - n^2\right)^2 C_{66}.
\end{aligned} \tag{7.91}$$

Assume now that $C_{11} = C_{22}$, $C_{44} = C_{55}$, $C_{13} = C_{23}$. As

$$m^4 + n^4 = 1 - 2m^2n^2, \qquad \left(m^2 - n^2\right)^2 = m^4 + n^4 - 2m^2n^2 = 1 - 4m^2n^2, \tag{7.92}$$

it then follows that

$$\begin{aligned}
C'_{11} &= C'_{22} = C_{11} - 2m^2n^2\left(C_{11} - C_{12} - 2C_{66}\right), \\
C'_{12} &= C_{12} + 2m^2n^2\left(C_{11} - C_{12} - 2C_{66}\right), \\
C'_{13} &= C'_{23} = C_{13},\ C'_{33} = C_{33},\ C'_{36} = 0, \\
C'_{44} &= C'_{55} = C_{44},\ C'_{45} = 0, \\
C'_{16} &= -C'_{26} = -mn\left(m^2 - n^2\right)\left(C_{11} - C_{12} - 2C_{66}\right), \\
C'_{66} &= C_{66} + 2m^2n^2\left(C_{11} - C_{12} - 2C_{66}\right).
\end{aligned} \tag{7.93}$$

It should be noted that the factor $C_{11} - C_{12} - 2C_{66}$ appears repeatedly in these relations. When this factor is zero so that

$$C_{66} = \frac{1}{2}\left(C_{11} - C_{12}\right), \tag{7.94}$$

it follows that

$$\begin{aligned} &C'_{11}=C'_{22}=C_{11},\quad C'_{12}=C_{12},\qquad C'_{13}=C'_{23}=C_{13},\\ &C'_{33}=C_{33},\qquad C'_{44}=C'_{55}=C_{44},\quad C'_{66}=C_{66},\\ &\qquad C'_{16}=C'_{26}=C'_{36}=C'_{45}=0. \end{aligned} \tag{7.95}$$

As it was assumed that $C_{11}=C_{22}$, $C_{44}=C_{55}$, $C_{13}=C_{23}$, it is clear that any rotation about the x_3-axis does not alter the value of the elastic constants on transformation. Thus, the material having the stress-strain relations (7.86) is transverse isotropic relative to the x_3-axis if the elastic constants are such that $C_{11}=C_{22}$, $C_{44}=C_{55}$, $C_{13}=C_{23}$, $C_{66}=\frac{1}{2}\left(C_{11}-C_{12}\right)$. It should be noted from (2.191) that the corresponding relation for elastic compliances was shown to be $S_{66}=2\left(S_{11}-S_{12}\right)$.

7.9 Accounting for Bend Deformation (Nonsymmetric Laminates)

An important deformation mode that has not yet been considered in this chapter concerns out-of-plane bending of laminates which can arise because of out-of-plane laminate loading, and because laminates are nonsymmetric. The approach to be taken here is to propose the form of a displacement field for the laminate and to show that it satisfies the required continuity equations, leading to a strain field that generates, from the required anisotropic stress-strain relations, a stress field that satisfies the equilibrium equations and continuity of tractions at ply interfaces. This avoids having to consider the more complex method of deriving the displacement field from strain–displacement relations, stress-strain relations, the equilibrium equations, the continuity conditions and associated boundary conditions.

Consider now a general laminate (possibly nonsymmetric), subject to general loading leading to in-plane and out-of-plane deformation, whose total thickness is now h rather than $2h$ as assumed in previous sections. It is useful to consider, for $i=1,\ldots,n$, the following generalisation of relations (7.9) that applied in the absence of out-of-plane bending

$$\begin{aligned} u_1^{(i)}(x_1,x_2,x_3)&=\left(\bar{\varepsilon}_A+\hat{\varepsilon}_A x_3\right)x_1+\tfrac{1}{2}\left(\bar{\gamma}_A+\hat{\gamma}_A x_3\right)x_2+\bar{\gamma}_a^{(i)}\left(x_3-x_3^{(i)}\right)+\sum_{j=1}^{i}h_j\bar{\gamma}_a^{(j)},\\ u_2^{(i)}(x_1,x_2,x_3)&=\left(\bar{\varepsilon}_T+\hat{\varepsilon}_T x_3\right)x_2+\tfrac{1}{2}\left(\bar{\gamma}_A+\hat{\gamma}_A x_3\right)x_1+\bar{\gamma}_t^{(i)}\left(x_3-x_3^{(i)}\right)+\sum_{j=1}^{i}h_j\bar{\gamma}_t^{(j)},\\ u_3^{(i)}(x_1,x_2,x_3)&=-\tfrac{1}{2}\hat{\varepsilon}_A x_1^2-\tfrac{1}{2}\hat{\varepsilon}_T x_2^2-\tfrac{1}{2}\hat{\gamma}_A x_1x_2+\bar{\varepsilon}_t^{(i)}\left(x_3-x_3^{(i)}\right)+\sum_{j=1}^{i}h_j\bar{\varepsilon}_t^{(j)}\\ &\quad+\tfrac{1}{2}\hat{\varepsilon}_t^{(i)}\left(x_3^2-(x_3^{(i)})^2\right)+\sum_{j=1}^{i}\hat{h}_j\hat{\varepsilon}_t^{(j)}, \end{aligned} \tag{7.96}$$

$$\text{with}\quad h_j=x_3^{(j)}-x_3^{(j-1)},\quad \hat{h}_j=\tfrac{1}{2}\left((x_3^{(j)})^2-(x_3^{(j-1)})^2\right),\quad \text{for } j=1,\ldots,n.$$

As $\hat{\varepsilon}_T$, $\hat{\varepsilon}_T$ and $\hat{\gamma}_A$ do not depend on the layer number i, this displacement distribution is continuous across all the layer interfaces. It follows that for all $i = 1, \ldots, n$,

$$\begin{aligned}
&\frac{\partial u_1^{(i)}}{\partial x_1}=\bar{\varepsilon}_{\mathrm{A}}+\hat{\varepsilon}_{\mathrm{A}}x_3, \quad \frac{\partial u_1^{(i)}}{\partial x_2}=\tfrac{1}{2}\left(\bar{\gamma}_{\mathrm{A}}+\hat{\gamma}_{\mathrm{A}}x_3\right), \quad \frac{\partial u_1^{(i)}}{\partial x_3}=\hat{\varepsilon}_{\mathrm{A}}x_1+\tfrac{1}{2}\hat{\gamma}_{\mathrm{A}}x_2+\bar{\gamma}_{\mathrm{a}}^{(i)},\\
&\frac{\partial u_2^{(i)}}{\partial x_1}=\tfrac{1}{2}\left(\bar{\gamma}_{\mathrm{A}}+\hat{\gamma}_{\mathrm{A}}x_3\right), \quad \frac{\partial u_2^{(i)}}{\partial x_2}=\bar{\varepsilon}_{\mathrm{T}}+\hat{\varepsilon}_{\mathrm{T}}x_3, \quad \frac{\partial u_2^{(i)}}{\partial x_3}=\hat{\varepsilon}_{\mathrm{T}}x_2+\tfrac{1}{2}\hat{\gamma}_{\mathrm{A}}x_1+\bar{\gamma}_{\mathrm{t}}^{(i)},\\
&\frac{\partial u_3^{(i)}}{\partial x_1}=-\hat{\varepsilon}_{\mathrm{A}}x_1-\tfrac{1}{2}\hat{\gamma}_{\mathrm{A}}x_2, \quad \frac{\partial u_3^{(i)}}{\partial x_2}=-\hat{\varepsilon}_{\mathrm{T}}x_2-\tfrac{1}{2}\hat{\gamma}_{\mathrm{A}}x_1, \quad \frac{\partial u_3^{(i)}}{\partial x_3}=\bar{\varepsilon}_{\mathrm{t}}^{(i)}+\hat{\varepsilon}_{\mathrm{t}}^{(i)}x_3.
\end{aligned} \tag{7.97}$$

The corresponding strain components are given by, for all $i = 1, \ldots, n$,

$$\begin{aligned}
&\varepsilon_{11}^{(i)}=\frac{\partial u_1^{(i)}}{\partial x_1}=\bar{\varepsilon}_{\mathrm{A}}+\hat{\varepsilon}_{\mathrm{A}}x_3, \quad \varepsilon_{22}^{(i)}=\frac{\partial u_2^{(i)}}{\partial x_2}=\bar{\varepsilon}_{\mathrm{T}}+\hat{\varepsilon}_{\mathrm{T}}x_3, \quad \varepsilon_{33}^{(i)}=\frac{\partial u_3^{(i)}}{\partial x_3}=\bar{\varepsilon}_{\mathrm{t}}^{(i)}+\hat{\varepsilon}_{\mathrm{t}}^{(i)}x_3,\\
&\varepsilon_{12}^{(i)}=\varepsilon_{21}^{(i)}=\frac{1}{2}\left(\frac{\partial u_1^{(i)}}{\partial x_2}+\frac{\partial u_2^{(i)}}{\partial x_1}\right)=\frac{1}{2}\left(\bar{\gamma}_{\mathrm{A}}+\hat{\gamma}_{\mathrm{A}}x_3\right),\\
&\varepsilon_{13}^{(i)}=\varepsilon_{31}^{(i)}=\frac{1}{2}\left(\frac{\partial u_1^{(i)}}{\partial x_3}+\frac{\partial u_3^{(i)}}{\partial x_1}\right)=\frac{1}{2}\bar{\gamma}_{\mathrm{a}}^{(i)}, \quad \varepsilon_{23}^{(i)}=\varepsilon_{32}^{(i)}=\frac{1}{2}\left(\frac{\partial u_2^{(i)}}{\partial x_3}+\frac{\partial u_3^{(i)}}{\partial x_2}\right)=\frac{1}{2}\bar{\gamma}_{\mathrm{t}}^{(i)}.
\end{aligned} \tag{7.98}$$

The equilibrium equations are given, for $i = 1, \ldots, n$, by

$$\frac{\partial \sigma_{11}^{(i)}}{\partial x_1}+\frac{\partial \sigma_{12}^{(i)}}{\partial x_2}+\frac{\partial \sigma_{13}^{(i)}}{\partial x_3}=0, \tag{7.99}$$

$$\frac{\partial \sigma_{12}^{(i)}}{\partial x_1}+\frac{\partial \sigma_{22}^{(i)}}{\partial x_2}+\frac{\partial \sigma_{23}^{(i)}}{\partial x_3}=0, \tag{7.100}$$

$$\frac{\partial \sigma_{13}^{(i)}}{\partial x_1}+\frac{\partial \sigma_{23}^{(i)}}{\partial x_2}+\frac{\partial \sigma_{33}^{(i)}}{\partial x_3}=0. \tag{7.101}$$

On using the contracted notation specified by

$$\begin{bmatrix}\sigma_{\mathrm{A}}\\ \sigma_{\mathrm{T}}\\ \sigma_{\mathrm{t}}\\ \tau_{\mathrm{t}}\\ \tau_{\mathrm{a}}\\ \tau_{\mathrm{A}}\end{bmatrix}\equiv\begin{bmatrix}\sigma_1\\ \sigma_2\\ \sigma_3\\ \sigma_4\\ \sigma_5\\ \sigma_6\end{bmatrix}\equiv\begin{bmatrix}\sigma_{11}\\ \sigma_{22}\\ \sigma_{33}\\ \sigma_{23}\\ \sigma_{13}\\ \sigma_{12}\end{bmatrix}, \qquad \begin{bmatrix}\varepsilon_{\mathrm{A}}\\ \varepsilon_{\mathrm{T}}\\ \varepsilon_{\mathrm{t}}\\ \gamma_{\mathrm{t}}\\ \gamma_{\mathrm{a}}\\ \gamma_{\mathrm{A}}\end{bmatrix}\equiv\begin{bmatrix}\varepsilon_1\\ \varepsilon_2\\ \varepsilon_3\\ \varepsilon_4\\ \varepsilon_5\\ \varepsilon_6\end{bmatrix}\equiv\begin{bmatrix}\varepsilon_{11}\\ \varepsilon_{22}\\ \varepsilon_{33}\\ 2\varepsilon_{23}\\ 2\varepsilon_{13}\\ 2\varepsilon_{12}\end{bmatrix}, \tag{7.102}$$

the equilibrium equations (7.99)–(7.101), may be written as

$$\frac{\partial \sigma_{\mathrm{A}}^{(i)}}{\partial x_1}+\frac{\partial \tau_{\mathrm{A}}^{(i)}}{\partial x_2}+\frac{\partial \tau_{\mathrm{a}}^{(i)}}{\partial x_3}=0, \tag{7.103}$$

$$\frac{\partial \tau_{\mathrm{A}}^{(i)}}{\partial x_1}+\frac{\partial \sigma_{\mathrm{T}}^{(i)}}{\partial x_2}+\frac{\partial \tau_{\mathrm{t}}^{(i)}}{\partial x_3}=0, \tag{7.104}$$

$$\frac{\partial \tau_{\mathrm{a}}^{(i)}}{\partial x_1}+\frac{\partial \tau_{\mathrm{t}}^{(i)}}{\partial x_2}+\frac{\partial \sigma_{\mathrm{T}}^{(i)}}{\partial x_3}=0. \tag{7.105}$$

The strains (7.98), when written using the contracted notation, are given, for all $i = 1, \ldots, n$, by

$$\begin{aligned} &\varepsilon_{\mathrm{A}}^{(i)} = \bar{\varepsilon}_{\mathrm{A}} + \hat{\varepsilon}_{\mathrm{A}} x_3, && \varepsilon_{\mathrm{T}}^{(i)} = \bar{\varepsilon}_{\mathrm{T}} + \hat{\varepsilon}_{\mathrm{T}} x_3, && \varepsilon_{\mathrm{t}}^{(i)} = \bar{\varepsilon}_{\mathrm{t}}^{(i)} + \hat{\varepsilon}_{\mathrm{t}}^{(i)} x_3, \\ &\gamma_{\mathrm{A}}^{(i)} = \bar{\gamma}_{\mathrm{A}} + \hat{\gamma}_{\mathrm{A}} x_3, && \gamma_{\mathrm{a}}^{(i)} = \bar{\gamma}_{\mathrm{a}}^{(i)}, && \gamma_{\mathrm{t}}^{(i)} = \bar{\gamma}_{\mathrm{t}}^{(i)}. \end{aligned} \tag{7.106}$$

The stress-strain relations, using contracted notation, are given, for all $i = 1, \ldots, n$, by

$$\begin{aligned} \sigma_{\mathrm{A}}^{(i)} &= C_{11}^{(i)} \varepsilon_{\mathrm{A}}^{(i)} + C_{12}^{(i)} \varepsilon_{\mathrm{T}}^{(i)} + C_{13}^{(i)} \varepsilon_{\mathrm{t}}^{(i)} + C_{16}^{(i)} \gamma_{\mathrm{A}}^{(i)} - U_1^{(i)} \Delta T, \\ \sigma_{\mathrm{T}}^{(i)} &= C_{12}^{(i)} \varepsilon_{\mathrm{A}}^{(i)} + C_{22}^{(i)} \varepsilon_{\mathrm{T}}^{(i)} + C_{23}^{(i)} \varepsilon_{\mathrm{t}}^{(i)} + C_{26}^{(i)} \gamma_{\mathrm{A}}^{(i)} - U_2^{(i)} \Delta T, \\ \sigma_{\mathrm{t}}^{(i)} &= C_{13}^{(i)} \varepsilon_{\mathrm{A}}^{(i)} + C_{23}^{(i)} \varepsilon_{\mathrm{T}}^{(i)} + C_{33}^{(i)} \varepsilon_{\mathrm{t}}^{(i)} + C_{36}^{(i)} \gamma_{\mathrm{A}}^{(i)} - U_3^{(i)} \Delta T, \\ \tau_{\mathrm{t}}^{(i)} &= C_{44}^{(i)} \gamma_{\mathrm{t}}^{(i)} + C_{45}^{(i)} \gamma_{\mathrm{a}}^{(i)}, \\ \tau_{\mathrm{a}}^{(i)} &= C_{45}^{(i)} \gamma_{\mathrm{t}}^{(i)} + C_{55}^{(i)} \gamma_{\mathrm{a}}^{(i)}, \\ \tau_{\mathrm{A}}^{(i)} &= C_{16}^{(i)} \varepsilon_{\mathrm{A}}^{(i)} + C_{26}^{(i)} \varepsilon_{\mathrm{T}}^{(i)} + C_{36}^{(i)} \varepsilon_{\mathrm{t}}^{(i)} + C_{66}^{(i)} \gamma_{\mathrm{A}}^{(i)} - U_6^{(i)} \Delta T. \end{aligned} \tag{7.107}$$

On substituting into (7.107) for the strains using (7.106), the stress field for the ith layer is given by

$$\begin{aligned} \sigma_{\mathrm{A}}^{(i)} = {} & C_{11}^{(i)} \bar{\varepsilon}_{\mathrm{A}} + C_{12}^{(i)} \bar{\varepsilon}_{\mathrm{T}} + C_{13}^{(i)} \bar{\varepsilon}_{\mathrm{t}}^{(i)} + C_{16}^{(i)} \bar{\gamma}_{\mathrm{A}} \\ & + \left(C_{11}^{(i)} \hat{\varepsilon}_{\mathrm{A}} + C_{12}^{(i)} \hat{\varepsilon}_{\mathrm{T}} + C_{13}^{(i)} \hat{\varepsilon}_{\mathrm{t}}^{(i)} + C_{16}^{(i)} \hat{\gamma}_{\mathrm{A}} \right) x_3 - U_1^{(i)} \Delta T, \end{aligned} \tag{7.108}$$

$$\begin{aligned} \sigma_{\mathrm{T}}^{(i)} = {} & C_{12}^{(i)} \bar{\varepsilon}_{\mathrm{A}} + C_{22}^{(i)} \bar{\varepsilon}_{\mathrm{T}} + C_{23}^{(i)} \bar{\varepsilon}_{\mathrm{t}}^{(i)} + C_{26}^{(i)} \bar{\gamma}_{\mathrm{A}} \\ & + \left(C_{12}^{(i)} \hat{\varepsilon}_{\mathrm{A}} + C_{22}^{(i)} \hat{\varepsilon}_{\mathrm{T}} + C_{23}^{(i)} \hat{\varepsilon}_{\mathrm{t}}^{(i)} + C_{26}^{(i)} \hat{\gamma}_{\mathrm{A}} \right) x_3 - U_2^{(i)} \Delta T, \end{aligned} \tag{7.109}$$

$$\begin{aligned} \sigma_{\mathrm{t}}^{(i)} = {} & C_{13}^{(i)} \bar{\varepsilon}_{\mathrm{A}} + C_{23}^{(i)} \bar{\varepsilon}_{\mathrm{T}} + C_{33}^{(i)} \bar{\varepsilon}_{\mathrm{t}}^{(i)} + C_{36}^{(i)} \bar{\gamma}_{\mathrm{A}} \\ & + \left(C_{13}^{(i)} \hat{\varepsilon}_{\mathrm{A}} + C_{23}^{(i)} \hat{\varepsilon}_{\mathrm{T}} + C_{33}^{(i)} \hat{\varepsilon}_{\mathrm{t}}^{(i)} + C_{36}^{(i)} \hat{\gamma}_{\mathrm{A}} \right) x_3 - U_3^{(i)} \Delta T, \end{aligned} \tag{7.110}$$

$$\tau_{\mathrm{t}}^{(i)} = C_{44}^{(i)} \bar{\gamma}_{\mathrm{t}}^{(i)} + C_{45}^{(i)} \bar{\gamma}_{\mathrm{a}}^{(i)}, \tag{7.111}$$

$$\tau_{\mathrm{a}}^{(i)} = C_{45}^{(i)} \bar{\gamma}_{\mathrm{t}}^{(i)} + C_{55}^{(i)} \bar{\gamma}_{\mathrm{a}}^{(i)}, \tag{7.112}$$

$$\begin{aligned} \tau_{\mathrm{A}}^{(i)} = {} & C_{16}^{(i)} \bar{\varepsilon}_{\mathrm{A}} + C_{26}^{(i)} \bar{\varepsilon}_{\mathrm{T}} + C_{36}^{(i)} \bar{\varepsilon}_{\mathrm{t}}^{(i)} + C_{66}^{(i)} \bar{\gamma}_{\mathrm{A}} \\ & + \left(C_{16}^{(i)} \hat{\varepsilon}_{\mathrm{A}} + C_{26}^{(i)} \hat{\varepsilon}_{\mathrm{T}} + C_{36}^{(i)} \hat{\varepsilon}_{\mathrm{t}}^{(i)} + C_{66}^{(i)} \hat{\gamma}_{\mathrm{A}} \right) x_3 - U_6^{(i)} \Delta T. \end{aligned} \tag{7.113}$$

The equilibrium equation (7.105) is satisfied by the stress field (7.108)–(7.113) derived from the displacement distribution (7.96) only if

$$C_{13}^{(i)} \hat{\varepsilon}_{\mathrm{A}} + C_{23}^{(i)} \hat{\varepsilon}_{\mathrm{T}} + C_{33}^{(i)} \hat{\varepsilon}_{\mathrm{t}}^{(i)} + C_{36}^{(i)} \hat{\gamma}_{\mathrm{A}} = 0, \quad \text{for all } i = 1, \ldots, n. \tag{7.114}$$

Relation (7.110) for the through-thickness stress may then be written as

$$\sigma_{\mathrm{t}}^{(i)} = C_{13}^{(i)} \bar{\varepsilon}_{\mathrm{A}} + C_{23}^{(i)} \bar{\varepsilon}_{\mathrm{T}} + C_{33}^{(i)} \bar{\varepsilon}_{\mathrm{t}}^{(i)} + C_{36}^{(i)} \bar{\gamma}_{\mathrm{A}} - U_3^{(i)} \Delta T, \quad \text{for all } i = 1, \ldots, n, \tag{7.115}$$

which are uniform in each ply. Relation (7.114), determining the value of the parameter $\hat{\varepsilon}_{\mathrm{t}}^{(i)}$ that ensures the equilibrium equations are satisfied, may be written as

$$\hat{\varepsilon}_{\mathrm{t}}^{(i)} = -\frac{C_{13}^{(i)}}{C_{33}^{(i)}} \hat{\varepsilon}_{\mathrm{A}} - \frac{C_{23}^{(i)}}{C_{33}^{(i)}} \hat{\varepsilon}_{\mathrm{T}} - \frac{C_{36}^{(i)}}{C_{33}^{(i)}} \hat{\gamma}_{\mathrm{A}}. \tag{7.116}$$

The stress-strain relations (7.107) are now manipulated so that they have the following equivalent form (an extended version of (7.17))

$$
\begin{aligned}
\sigma_{\mathrm{A}}^{(i)} &= \Omega_{11}^{(i)}\varepsilon_{\mathrm{A}}^{(i)} + \Omega_{12}^{(i)}\varepsilon_{\mathrm{T}}^{(i)} + \Omega_{13}^{(i)}\sigma_{\mathrm{t}}^{(i)} + \Omega_{16}^{(i)}\gamma_{\mathrm{A}}^{(i)} - \omega_1^{(i)}\Delta T,\\
\sigma_{\mathrm{T}}^{(i)} &= \Omega_{12}^{(i)}\varepsilon_{\mathrm{A}}^{(i)} + \Omega_{22}^{(i)}\varepsilon_{\mathrm{T}}^{(i)} + \Omega_{23}^{(i)}\sigma_{\mathrm{t}}^{(i)} + \Omega_{26}^{(i)}\gamma_{\mathrm{A}}^{(i)} - \omega_2^{(i)}\Delta T,\\
\varepsilon_{\mathrm{t}}^{(i)} &= -\Omega_{13}^{(i)}\varepsilon_{\mathrm{A}}^{(i)} - \Omega_{23}^{(i)}\varepsilon_{\mathrm{T}}^{(i)} + \Omega_{33}^{(i)}\sigma_{\mathrm{t}}^{(i)} - \Omega_{36}^{(i)}\gamma_{\mathrm{A}}^{(i)} + \omega_3^{(i)}\Delta T,\\
\gamma_{\mathrm{t}}^{(i)} &= S_{44}^{(i)}\tau_{\mathrm{t}}^{(i)} + S_{45}^{(i)}\tau_{\mathrm{a}}^{(i)},\\
\gamma_{\mathrm{a}}^{(i)} &= S_{45}^{(i)}\tau_{\mathrm{t}}^{(i)} + S_{55}^{(i)}\tau_{\mathrm{a}}^{(i)},\\
\tau_{\mathrm{A}}^{(i)} &= \Omega_{16}^{(i)}\varepsilon_{\mathrm{A}}^{(i)} + \Omega_{26}^{(i)}\varepsilon_{\mathrm{T}}^{(i)} + \Omega_{36}^{(i)}\sigma_{\mathrm{t}}^{(i)} + \Omega_{66}^{(i)}\gamma_{\mathrm{A}}^{(i)} - \omega_6^{(i)}\Delta T,
\end{aligned}
\tag{7.117}
$$

where from (7.18)–(7.22)

$$
\Omega_{11}^{(i)} = C_{11}^{(i)} - \frac{C_{13}^{(i)}C_{13}^{(i)}}{C_{33}^{(i)}},\ \Omega_{12}^{(i)} = C_{12}^{(i)} - \frac{C_{13}^{(i)}C_{23}^{(i)}}{C_{33}^{(i)}},\ \Omega_{13}^{(i)} = \frac{C_{13}^{(i)}}{C_{33}^{(i)}},\ \Omega_{16}^{(i)} = C_{16}^{(i)} - \frac{C_{13}^{(i)}C_{36}^{(i)}}{C_{33}^{(i)}},\tag{7.118}
$$

$$
\Omega_{22}^{(i)} = C_{22}^{(i)} - \frac{C_{23}^{(i)}C_{23}^{(i)}}{C_{33}^{(i)}},\quad \Omega_{23}^{(i)} = \frac{C_{23}^{(i)}}{C_{33}^{(i)}},\quad \Omega_{26}^{(i)} = C_{26}^{(i)} - \frac{C_{23}^{(i)}C_{36}^{(i)}}{C_{33}^{(i)}},\tag{7.119}
$$

$$
\Omega_{33}^{(i)} = \frac{1}{C_{33}^{(i)}},\quad \Omega_{36}^{(i)} = \frac{C_{36}^{(i)}}{C_{33}^{(i)}},\tag{7.120}
$$

$$
\Omega_{66}^{(i)} = C_{66}^{(i)} - \frac{C_{36}^{(i)}C_{36}^{(i)}}{C_{33}^{(i)}},\tag{7.121}
$$

$$
\omega_1^{(i)} = U_1^{(i)} - \frac{C_{13}^{(i)}U_3^{(i)}}{C_{33}^{(i)}},\ \omega_2^{(i)} = U_2^{(i)} - \frac{C_{23}^{(i)}U_3^{(i)}}{C_{33}^{(i)}},\ \omega_3^{(i)} = \frac{U_3^{(i)}}{C_{33}^{(i)}},\ \omega_6^{(i)} = U_6^{(i)} - \frac{C_{36}^{(i)}U_3^{(i)}}{C_{33}^{(i)}},\tag{7.122}
$$

and where

$$
S_{44}^{(i)} = \frac{C_{55}^{(i)}}{C_{44}^{(i)}C_{55}^{(i)} - C_{45}^{(i)}C_{45}^{(i)}},\ S_{45}^{(i)} = -\frac{C_{45}^{(i)}}{C_{44}^{(i)}C_{55}^{(i)} - C_{45}^{(i)}C_{45}^{(i)}},\ S_{55}^{(i)} = \frac{C_{44}^{(i)}}{C_{44}^{(i)}C_{55}^{(i)} - C_{45}^{(i)}C_{45}^{(i)}}.\tag{7.123}
$$

The through-thickness stress σ_{t} and the out-of-plane shear stresses τ_{a} and τ_{t} are assumed to have the same value in all plies ensuring that these stress components are continuous. The substitution of (7.106) into (7.117) using (7.116) then leads to the relations, for all $i = 1, \ldots, n$,

$$
\begin{aligned}
\sigma_{\mathrm{A}}^{(i)} &= \Omega_{11}^{(i)}\left(\bar{\varepsilon}_{\mathrm{A}} + \hat{\varepsilon}_{\mathrm{A}}x_3\right) + \Omega_{12}^{(i)}\left(\bar{\varepsilon}_{\mathrm{T}} + \hat{\varepsilon}_{\mathrm{T}}x_3\right) + \Omega_{13}^{(i)}\sigma_{\mathrm{t}} + \Omega_{16}^{(i)}\left(\bar{\gamma}_{\mathrm{A}} + \hat{\gamma}_{\mathrm{A}}x_3\right) - \omega_1^{(i)}\Delta T,\\
\sigma_{\mathrm{T}}^{(i)} &= \Omega_{12}^{(i)}\left(\bar{\varepsilon}_{\mathrm{A}} + \hat{\varepsilon}_{\mathrm{A}}x_3\right) + \Omega_{22}^{(i)}\left(\bar{\varepsilon}_{\mathrm{T}} + \hat{\varepsilon}_{\mathrm{T}}x_3\right) + \Omega_{23}^{(i)}\sigma_{\mathrm{t}} + \Omega_{26}^{(i)}\left(\bar{\gamma}_{\mathrm{A}} + \hat{\gamma}_{\mathrm{A}}x_3\right) - \omega_2^{(i)}\Delta T,\\
\bar{\varepsilon}_{\mathrm{t}}^{(i)} &= -\Omega_{13}^{(i)}\bar{\varepsilon}_{\mathrm{A}} - \Omega_{23}^{(i)}\bar{\varepsilon}_{\mathrm{T}} + \Omega_{33}^{(i)}\sigma_{\mathrm{t}} - \Omega_{36}^{(i)}\bar{\gamma}_{\mathrm{A}} + \omega_3^{(i)}\Delta T,\\
\gamma_{\mathrm{t}}^{(i)} &= S_{44}^{(i)}\tau_{\mathrm{t}} + S_{45}^{(i)}\tau_{\mathrm{a}},\\
\gamma_{\mathrm{a}}^{(i)} &= S_{45}^{(i)}\tau_{\mathrm{t}} + S_{55}^{(i)}\tau_{\mathrm{a}},\\
\tau_{\mathrm{A}}^{(i)} &= \Omega_{16}^{(i)}\left(\bar{\varepsilon}_{\mathrm{A}} + \hat{\varepsilon}_{\mathrm{A}}x_3\right) + \Omega_{26}^{(i)}\left(\bar{\varepsilon}_{\mathrm{T}} + \hat{\varepsilon}_{\mathrm{T}}x_3\right) + \Omega_{36}^{(i)}\sigma_{\mathrm{t}} + \Omega_{66}^{(i)}\left(\bar{\gamma}_{\mathrm{A}} + \hat{\gamma}_{\mathrm{A}}x_3\right) - \omega_6^{(i)}\Delta T.
\end{aligned}
\tag{7.124}
$$

The layer stresses may now be written, for all $i = 1, \ldots, n$, as

$$
\begin{aligned}
&\sigma_{\mathrm{A}}^{(i)} = \bar{\sigma}_{\mathrm{A}}^{(i)} + \hat{\sigma}_{\mathrm{A}}^{(i)}x_3, \quad && \sigma_{\mathrm{T}}^{(i)} = \bar{\sigma}_{\mathrm{T}}^{(i)} + \hat{\sigma}_{\mathrm{T}}^{(i)}x_3, \quad && \tau_{\mathrm{A}}^{(i)} = \bar{\tau}_{\mathrm{A}}^{(i)} + \hat{\tau}_{\mathrm{A}}^{(i)}x_3,\\
&\sigma_{\mathrm{t}}^{(i)} = \sigma_{\mathrm{t}}, && \tau_{\mathrm{a}}^{(i)} = \tau_{\mathrm{a}}, && \tau_{\mathrm{t}}^{(i)} = \tau_{\mathrm{t}},
\end{aligned}
\tag{7.125}
$$

where

$$\begin{aligned}
\bar{\sigma}_{\rm A}^{(i)} &= \Omega_{11}^{(i)}\bar{\varepsilon}_{\rm A} + \Omega_{12}^{(i)}\bar{\varepsilon}_{\rm T} + \Omega_{13}^{(i)}\sigma_{\rm t} + \Omega_{16}^{(i)}\bar{\gamma}_{\rm A} - \omega_1^{(i)}\Delta T,\\
\bar{\sigma}_{\rm T}^{(i)} &= \Omega_{12}^{(i)}\bar{\varepsilon}_{\rm A} + \Omega_{22}^{(i)}\bar{\varepsilon}_{\rm T} + \Omega_{23}^{(i)}\sigma_{\rm t} + \Omega_{26}^{(i)}\bar{\gamma}_{\rm A} - \omega_2^{(i)}\Delta T,\\
\bar{\varepsilon}_{\rm t}^{(i)} &= -\Omega_{13}^{(i)}\bar{\varepsilon}_{\rm A} - \Omega_{23}^{(i)}\bar{\varepsilon}_{\rm T} + \Omega_{33}^{(i)}\sigma_{\rm t} - \Omega_{36}^{(i)}\bar{\gamma}_{\rm A} + \omega_3^{(i)}\Delta T,\\
\bar{\tau}_{\rm A}^{(i)} &= \Omega_{16}^{(i)}\bar{\varepsilon}_{\rm A} + \Omega_{26}^{(i)}\bar{\varepsilon}_{\rm T} + \Omega_{36}^{(i)}\sigma_{\rm t} + \Omega_{66}^{(i)}\bar{\gamma}_{\rm A} - \omega_6^{(i)}\Delta T,
\end{aligned} \quad i=1,\ldots,n, \tag{7.126}$$

and, on including relation (7.116),

$$\begin{aligned}
\hat{\sigma}_{\rm A}^{(i)} &= \Omega_{11}^{(i)}\hat{\varepsilon}_{\rm A} + \Omega_{12}^{(i)}\hat{\varepsilon}_{\rm T} + \Omega_{16}^{(i)}\hat{\gamma}_{\rm A},\\
\hat{\sigma}_{\rm T}^{(i)} &= \Omega_{12}^{(i)}\hat{\varepsilon}_{\rm A} + \Omega_{22}^{(i)}\hat{\varepsilon}_{\rm T} + \Omega_{26}^{(i)}\hat{\gamma}_{\rm A},\\
\hat{\varepsilon}_{\rm t}^{(i)} &= -\Omega_{13}^{(i)}\hat{\varepsilon}_{\rm A} - \Omega_{23}^{(i)}\hat{\varepsilon}_{\rm T} - \Omega_{36}^{(i)}\hat{\gamma}_{\rm A},\\
\hat{\tau}_{\rm A}^{(i)} &= \Omega_{16}^{(i)}\hat{\varepsilon}_{\rm A} + \Omega_{26}^{(i)}\hat{\varepsilon}_{\rm T} + \Omega_{66}^{(i)}\hat{\gamma}_{\rm A}.
\end{aligned} \quad i=1,\ldots,n, \tag{7.127}$$

Effective applied stresses and applied bending moments per unit area are now defined by the following averages over the edges of the laminate having length $2L$, width $2W$ and total thickness h

$$\sigma_{\rm A} = \frac{1}{2hW}\int_{-W}^{W}\int_{0}^{h}\sigma_{11}(L,x_2,x_3)\,{\rm d}x_2{\rm d}x_3, \tag{7.128}$$

$$\sigma_{\rm T} = \frac{1}{2hL}\int_{-L}^{L}\int_{0}^{h}\sigma_{22}(x_1,W,x_3)\,{\rm d}x_1{\rm d}x_3, \tag{7.129}$$

$$\tau_{\rm A} = \frac{1}{2hW}\int_{-W}^{W}\int_{0}^{h}\sigma_{12}(L,x_2,x_3)\,{\rm d}x_2{\rm d}x_3 = \frac{1}{2hL}\int_{-L}^{L}\int_{0}^{h}\sigma_{21}(x_1,W,x_3)\,{\rm d}x_1{\rm d}x_3, \tag{7.130}$$

$$M_{\rm A} = \frac{1}{2hW}\int_{-W}^{W}\int_{0}^{h}\left(x_3 - \tfrac{1}{2}h\right)\sigma_{11}(L,x_2,x_3)\,{\rm d}x_2{\rm d}x_3, \tag{7.131}$$

$$M_{\rm T} = \frac{1}{2hL}\int_{-L}^{L}\int_{0}^{h}\left(x_3 - \tfrac{1}{2}h\right)\sigma_{22}(x_1,W,x_3)\,{\rm d}x_1{\rm d}x_3, \tag{7.132}$$

$$\begin{aligned}
M_{\rm S} &= \frac{1}{2hW}\int_{-W}^{W}\int_{0}^{h}\left(x_3 - \tfrac{1}{2}h\right)\sigma_{12}(L,x_2,x_3)\,{\rm d}x_2{\rm d}x_3\\
&= \frac{1}{2hL}\int_{-L}^{L}\int_{0}^{h}\left(x_3 - \tfrac{1}{2}h\right)\sigma_{21}(x_1,W,x_3)\,{\rm d}x_1{\rm d}x_3.
\end{aligned} \tag{7.133}$$

It is clear that

$$M_{\rm A} + \tfrac{1}{2}h\sigma_{\rm A} = \frac{1}{2hW}\int_{-W}^{W}\int_{0}^{h}x_3\sigma_{11}(L,x_2,x_3)\,{\rm d}x_2{\rm d}x_3, \tag{7.134}$$

$$M_{\rm T} + \tfrac{1}{2}h\sigma_{\rm T} = \frac{1}{2hL}\int_{-L}^{L}\int_{0}^{h}x_3\sigma_{22}(x_1,W,x_3)\,{\rm d}x_1{\rm d}x_3, \tag{7.135}$$

$$
\begin{aligned}
M_{\mathrm{S}}+\tfrac{1}{2}h\tau_{\mathrm{A}} &= \frac{1}{2hW}\int_{-W}^{W}\int_{0}^{h} x_3\sigma_{12}\left(L,x_2,x_3\right)\,\mathrm{d}x_2\mathrm{d}x_3 \\
&= \frac{1}{2h\mathrm{L}}\int_{-\mathrm{L}}^{\mathrm{L}}\int_{0}^{h} x_3\sigma_{21}\left(x_1,W,x_3\right)\,\mathrm{d}x_1\mathrm{d}x_3 .
\end{aligned}
\tag{7.136}
$$

On using (7.102) and (7.125)–(7.127), it follows from (7.128)–(7.136) that

$$
\begin{aligned}
\sigma_{\mathrm{A}} &= \frac{1}{h}\sum_{i=1}^{n}\int_{x_3^{(i-1)}}^{x_3^{(i)}}\left(\bar{\sigma}_{\mathrm{A}}^{(i)}+\hat{\sigma}_{\mathrm{A}}^{(i)}x_3\right)\mathrm{d}x_3 \\
&= \left(\frac{1}{h}\sum_{i=1}^{n}h_i\Omega_{11}^{(i)}\right)\bar{\varepsilon}_{\mathrm{A}}+\left(\frac{1}{h}\sum_{i=1}^{n}h_i\Omega_{12}^{(i)}\right)\bar{\varepsilon}_{\mathrm{T}}+\left(\frac{1}{h}\sum_{i=1}^{n}h_i\Omega_{13}^{(i)}\right)\sigma_t+\left(\frac{1}{h}\sum_{i=1}^{n}h_i\Omega_{16}^{(i)}\right)\bar{\gamma}_{\mathrm{A}} \\
&\quad+\left(\frac{1}{h}\sum_{i=1}^{n}\hat{h}_i\Omega_{11}^{(i)}\right)\hat{\varepsilon}_{\mathrm{A}}+\left(\frac{1}{h}\sum_{i=1}^{n}\hat{h}_i\Omega_{12}^{(i)}\right)\hat{\varepsilon}_{\mathrm{T}}+\left(\frac{1}{h}\sum_{i=1}^{n}\hat{h}_i\Omega_{16}^{(i)}\right)\hat{\gamma}_{\mathrm{A}}-\left(\frac{1}{h}\sum_{i=1}^{n}h_i\omega_1^{(i)}\right)\Delta T,
\end{aligned}
\tag{7.137}
$$

$$
\begin{aligned}
\sigma_{\mathrm{T}} &= \frac{1}{h}\sum_{i=1}^{n}\int_{x_3^{(i-1)}}^{x_3^{(i)}}\left(\bar{\sigma}_{\mathrm{T}}^{(i)}+\hat{\sigma}_{\mathrm{T}}^{(i)}x_3\right)\mathrm{d}x_3 \\
&= \left(\frac{1}{h}\sum_{i=1}^{n}h_i\Omega_{12}^{(i)}\right)\bar{\varepsilon}_{\mathrm{A}}+\left(\frac{1}{h}\sum_{i=1}^{n}h_i\Omega_{22}^{(i)}\right)\bar{\varepsilon}_{\mathrm{T}}+\left(\frac{1}{h}\sum_{i=1}^{n}h_i\Omega_{23}^{(i)}\right)\sigma_{\mathrm{t}}+\left(\frac{1}{h}\sum_{i=1}^{n}h_i\Omega_{26}^{(i)}\right)\bar{\gamma}_{\mathrm{A}} \\
&\quad+\left(\frac{1}{h}\sum_{i=1}^{n}\hat{h}_i\Omega_{12}^{(i)}\right)\hat{\varepsilon}_{\mathrm{A}}+\left(\frac{1}{h}\sum_{i=1}^{n}\hat{h}_i\Omega_{22}^{(i)}\right)\hat{\varepsilon}_{\mathrm{T}}+\left(\frac{1}{h}\sum_{i=1}^{n}\hat{h}_i\Omega_{26}^{(i)}\right)\hat{\gamma}_{\mathrm{A}}-\left(\frac{1}{h}\sum_{i=1}^{n}h_i\omega_2^{(i)}\right)\Delta T,
\end{aligned}
\tag{7.138}
$$

$$
\begin{aligned}
\tau_{\mathrm{A}} &= \frac{1}{h}\sum_{i=1}^{n}\int_{x_3^{(i-1)}}^{x_3^{(i)}}\left(\bar{\tau}_{\mathrm{A}}^{(i)}+\hat{\tau}_{\mathrm{A}}^{(i)}x_3\right)\mathrm{d}x_3 \\
&= \left(\frac{1}{h}\sum_{i=1}^{n}h_i\Omega_{16}^{(i)}\right)\bar{\varepsilon}_{\mathrm{A}}+\left(\frac{1}{h}\sum_{i=1}^{n}h_i\Omega_{26}^{(i)}\right)\bar{\varepsilon}_{\mathrm{T}}+\left(\frac{1}{h}\sum_{i=1}^{n}h_i\Omega_{36}^{(i)}\right)\sigma_{\mathrm{t}}+\left(\frac{1}{h}\sum_{i=1}^{n}h_i\Omega_{66}^{(i)}\right)\bar{\gamma}_{\mathrm{A}} \\
&\quad+\left(\frac{1}{h}\sum_{i=1}^{n}\hat{h}_i\Omega_{16}^{(i)}\right)\hat{\varepsilon}_{\mathrm{A}}+\left(\frac{1}{h}\sum_{i=1}^{n}\hat{h}_i\Omega_{26}^{(i)}\right)\hat{\varepsilon}_{\mathrm{T}}+\left(\frac{1}{h}\sum_{i=1}^{n}\hat{h}_i\Omega_{66}^{(i)}\right)\hat{\gamma}_{\mathrm{A}}-\left(\frac{1}{h}\sum_{i=1}^{n}h_i\omega_6^{(i)}\right)\Delta T,
\end{aligned}
\tag{7.139}
$$

$$
\begin{aligned}
M_{\mathrm{A}}+\tfrac{1}{2}h\sigma_{\mathrm{A}} &= \frac{1}{h}\sum_{i=1}^{n}\int_{x_3^{(i-1)}}^{x_3^{(i)}}\left(\bar{\sigma}_{\mathrm{A}}^{(i)}+\hat{\sigma}_{\mathrm{A}}^{(i)}x_3\right)x_3\mathrm{d}x_3 \\
&= \left(\frac{1}{h}\sum_{i=1}^{n}\hat{h}_i\Omega_{11}^{(i)}\right)\bar{\varepsilon}_{\mathrm{A}}+\left(\frac{1}{h}\sum_{i=1}^{n}\hat{h}_i\Omega_{12}^{(i)}\right)\bar{\varepsilon}_{\mathrm{T}}+\left(\frac{1}{h}\sum_{i=1}^{n}\hat{h}_i\Omega_{13}^{(i)}\right)\sigma_{\mathrm{t}}+\left(\frac{1}{h}\sum_{i=1}^{n}\hat{h}_i\Omega_{16}^{(i)}\right)\bar{\gamma}_{\mathrm{A}} \\
&\quad+\left(\frac{1}{h}\sum_{i=1}^{n}\tilde{h}_i\Omega_{11}^{(i)}\right)\hat{\varepsilon}_{\mathrm{A}}+\left(\frac{1}{h}\sum_{i=1}^{n}\tilde{h}_i\Omega_{12}^{(i)}\right)\hat{\varepsilon}_{\mathrm{T}}+\left(\frac{1}{h}\sum_{i=1}^{n}\tilde{h}_i\Omega_{16}^{(i)}\right)\hat{\gamma}_{\mathrm{A}}-\left(\frac{1}{h}\sum_{i=1}^{n}\hat{h}_i\omega_1^{(i)}\right)\Delta T,
\end{aligned}
\tag{7.140}
$$

$$
\begin{aligned}
M_{\mathrm{T}}+\tfrac{1}{2}h\sigma_{\mathrm{T}} &= \frac{1}{h}\sum_{i=1}^{n}\int_{x_3^{(i-1)}}^{x_3^{(i)}}\left(\bar{\sigma}_{\mathrm{T}}^{(i)}+\hat{\sigma}_{\mathrm{T}}^{(i)}x_3\right)x_3\mathrm{d}x_3 \\
&= \left(\frac{1}{h}\sum_{i=1}^{n}\hat{h}_i\Omega_{12}^{(i)}\right)\bar{\varepsilon}_{\mathrm{A}}+\left(\frac{1}{h}\sum_{i=1}^{n}\hat{h}_i\Omega_{22}^{(i)}\right)\bar{\varepsilon}_{\mathrm{T}}+\left(\frac{1}{h}\sum_{i=1}^{n}\hat{h}_i\Omega_{23}^{(i)}\right)\sigma_{\mathrm{t}}+\left(\frac{1}{h}\sum_{i=1}^{n}\hat{h}_i\Omega_{26}^{(i)}\right)\bar{\gamma}_{\mathrm{A}} \\
&\quad+\left(\frac{1}{h}\sum_{i=1}^{n}\tilde{h}_i\Omega_{12}^{(i)}\right)\hat{\varepsilon}_{\mathrm{A}}+\left(\frac{1}{h}\sum_{i=1}^{n}\tilde{h}_i\Omega_{22}^{(i)}\right)\hat{\varepsilon}_{\mathrm{T}}+\left(\frac{1}{h}\sum_{i=1}^{n}\tilde{h}_i\Omega_{26}^{(i)}\right)\hat{\gamma}_{\mathrm{A}}-\left(\frac{1}{h}\sum_{i=1}^{n}\hat{h}_i\omega_2^{(i)}\right)\Delta T,
\end{aligned}
\tag{7.141}
$$

$$
\begin{aligned}
M_{\mathrm{S}} + \tfrac{1}{2}h\tau_{\mathrm{A}} &= \frac{1}{h}\sum_{i=1}^{n}\int_{x_3^{(i-1)}}^{x_3^{(i)}}\left(\bar{\tau}_{\mathrm{A}}^{(i)} + \hat{\tau}_{\mathrm{A}}^{(i)}x_3\right)x_3\mathrm{d}x_3 \\
&= \left(\frac{1}{h}\sum_{i=1}^{n}\hat{h}_i\Omega_{16}^{(i)}\right)\bar{\varepsilon}_{\mathrm{A}} + \left(\frac{1}{h}\sum_{i=1}^{n}\hat{h}_i\Omega_{26}^{(i)}\right)\bar{\varepsilon}_{\mathrm{T}} + \left(\frac{1}{h}\sum_{i=1}^{n}\hat{h}_i\Omega_{36}^{(i)}\right)\sigma_{\mathrm{t}} + \left(\frac{1}{h}\sum_{i=1}^{n}\hat{h}_i\Omega_{66}^{(i)}\right)\bar{\gamma}_{\mathrm{A}} \\
&+ \left(\frac{1}{h}\sum_{i=1}^{n}\tilde{h}_i\Omega_{16}^{(i)}\right)\hat{\varepsilon}_{\mathrm{A}} + \left(\frac{1}{h}\sum_{i=1}^{n}\tilde{h}_i\Omega_{26}^{(i)}\right)\hat{\varepsilon}_{\mathrm{T}} + \left(\frac{1}{h}\sum_{i=1}^{n}\tilde{h}_i\Omega_{36}^{(i)}\right)\hat{\gamma}_{\mathrm{A}} - \left(\frac{1}{h}\sum_{i=1}^{n}\hat{h}_i\omega_6^{(i)}\right)\Delta T,
\end{aligned}
\tag{7.142}
$$

where, for $i = 1, \ldots, n$,

$$
h_i = x_3^{(i)} - x_3^{(i-1)},\ \hat{h}_i = \tfrac{1}{2}\left((x_3^{(i)})^2 - (x_3^{(i-1)})^2\right),\ \tilde{h}_i = \tfrac{1}{3}\left((x_3^{(i)})^3 - (x_3^{(i-1)})^3\right), \tag{7.143}
$$

and where $x_3^{(0)} = 0$ and $x_3^{(n)} = h$. Relations (7.137)–(7.142) are now written in the form

$$
\sigma_{\mathrm{A}} = \Omega_{11}^{(\mathrm{L})}\bar{\varepsilon}_{\mathrm{A}} + \Omega_{12}^{(\mathrm{L})}\bar{\varepsilon}_{\mathrm{T}} + \Omega_{13}^{(\mathrm{L})}\sigma_{\mathrm{t}} + \Omega_{16}^{(\mathrm{L})}\bar{\gamma}_{\mathrm{A}} + \hat{\Omega}_{11}^{(\mathrm{L})}\hat{\varepsilon}_{\mathrm{A}} + \hat{\Omega}_{12}^{(\mathrm{L})}\hat{\varepsilon}_{\mathrm{T}} + \hat{\Omega}_{16}^{(\mathrm{L})}\hat{\gamma}_{\mathrm{A}} - \omega_1^{(\mathrm{L})}\Delta T, \tag{7.144}
$$

$$
\sigma_{\mathrm{T}} = \Omega_{12}^{(\mathrm{L})}\bar{\varepsilon}_{\mathrm{A}} + \Omega_{22}^{(\mathrm{L})}\bar{\varepsilon}_{\mathrm{T}} + \Omega_{23}^{(\mathrm{L})}\sigma_{\mathrm{t}} + \Omega_{26}^{(\mathrm{L})}\bar{\gamma}_{\mathrm{A}} + \hat{\Omega}_{12}^{(\mathrm{L})}\hat{\varepsilon}_{\mathrm{A}} + \hat{\Omega}_{22}^{(\mathrm{L})}\hat{\varepsilon}_{\mathrm{T}} + \hat{\Omega}_{26}^{(\mathrm{L})}\hat{\gamma}_{\mathrm{A}} - \omega_2^{(\mathrm{L})}\Delta T, \tag{7.145}
$$

$$
\tau_{\mathrm{A}} = \Omega_{16}^{(\mathrm{L})}\bar{\varepsilon}_{\mathrm{A}} + \Omega_{26}^{(\mathrm{L})}\bar{\varepsilon}_{\mathrm{T}} + \Omega_{36}^{(\mathrm{L})}\sigma_{\mathrm{t}} + \Omega_{66}^{(\mathrm{L})}\bar{\gamma}_{\mathrm{A}} + \hat{\Omega}_{16}^{(\mathrm{L})}\hat{\varepsilon}_{\mathrm{A}} + \hat{\Omega}_{26}^{(\mathrm{L})}\hat{\varepsilon}_{\mathrm{T}} + \hat{\Omega}_{66}^{(\mathrm{L})}\hat{\gamma}_{\mathrm{A}} - \omega_6^{(\mathrm{L})}\Delta T, \tag{7.146}
$$

$$
\begin{aligned}
&M_{\mathrm{A}} + \tfrac{1}{2}h\sigma_{\mathrm{A}} = \\
&\quad \hat{\Omega}_{11}^{(\mathrm{L})}\bar{\varepsilon}_{\mathrm{A}} + \hat{\Omega}_{12}^{(\mathrm{L})}\bar{\varepsilon}_{\mathrm{T}} + \hat{\Omega}_{13}^{(\mathrm{L})}\sigma_{\mathrm{t}} + \hat{\Omega}_{16}^{(\mathrm{L})}\bar{\gamma}_{\mathrm{A}} + \tilde{\Omega}_{11}^{(\mathrm{L})}\hat{\varepsilon}_{\mathrm{A}} + \tilde{\Omega}_{12}^{(\mathrm{L})}\hat{\varepsilon}_{\mathrm{T}} + \tilde{\Omega}_{16}^{(\mathrm{L})}\hat{\gamma}_{\mathrm{A}} - \hat{\omega}_1^{(\mathrm{L})}\Delta T,
\end{aligned}
\tag{7.147}
$$

$$
\begin{aligned}
&M_{\mathrm{T}} + \tfrac{1}{2}h\sigma_{\mathrm{T}} = \\
&\quad \hat{\Omega}_{12}^{(\mathrm{L})}\bar{\varepsilon}_{\mathrm{A}} + \hat{\Omega}_{22}^{(\mathrm{L})}\bar{\varepsilon}_{\mathrm{T}} + \hat{\Omega}_{23}^{(\mathrm{L})}\sigma_{\mathrm{t}} + \hat{\Omega}_{26}^{(\mathrm{L})}\bar{\gamma}_{\mathrm{A}} + \tilde{\Omega}_{12}^{(\mathrm{L})}\hat{\varepsilon}_{\mathrm{A}} + \tilde{\Omega}_{22}^{(\mathrm{L})}\hat{\varepsilon}_{\mathrm{T}} + \tilde{\Omega}_{26}^{(\mathrm{L})}\hat{\gamma}_{\mathrm{A}} - \hat{\omega}_2^{(\mathrm{L})}\Delta T,
\end{aligned}
\tag{7.148}
$$

$$
\begin{aligned}
&M_{\mathrm{S}} + \tfrac{1}{2}h\tau_{\mathrm{A}} = \\
&\quad \hat{\Omega}_{16}^{(\mathrm{L})}\bar{\varepsilon}_{\mathrm{A}} + \hat{\Omega}_{26}^{(\mathrm{L})}\bar{\varepsilon}_{\mathrm{T}} + \hat{\Omega}_{36}^{(\mathrm{L})}\sigma_{\mathrm{t}} + \hat{\Omega}_{66}^{(\mathrm{L})}\bar{\gamma}_{\mathrm{A}} + \tilde{\Omega}_{16}^{(\mathrm{L})}\hat{\varepsilon}_{\mathrm{A}} + \tilde{\Omega}_{26}^{(\mathrm{L})}\hat{\varepsilon}_{\mathrm{T}} + \tilde{\Omega}_{66}^{(\mathrm{L})}\hat{\gamma}_{\mathrm{A}} - \hat{\omega}_6^{(\mathrm{L})}\Delta T,
\end{aligned}
\tag{7.149}
$$

where

$$
\begin{aligned}
&\Omega_{11}^{(\mathrm{L})} = \frac{1}{h}\sum_{i=1}^{n}h_i\Omega_{11}^{(i)},\ \Omega_{12}^{(\mathrm{L})} = \frac{1}{h}\sum_{i=1}^{n}h_i\Omega_{12}^{(i)},\ \Omega_{22}^{(\mathrm{L})} = \frac{1}{h}\sum_{i=1}^{n}h_i\Omega_{22}^{(i)},\ \Omega_{13}^{(\mathrm{L})} = \frac{1}{h}\sum_{i=1}^{n}h_i\Omega_{13}^{(i)}, \\
&\Omega_{16}^{(\mathrm{L})} = \frac{1}{h}\sum_{i=1}^{n}h_i\Omega_{16}^{(i)},\ \Omega_{26}^{(\mathrm{L})} = \frac{1}{h}\sum_{i=1}^{n}h_i\Omega_{26}^{(i)},\ \Omega_{66}^{(\mathrm{L})} = \frac{1}{h}\sum_{i=1}^{n}h_i\Omega_{66}^{(i)},\ \Omega_{23}^{(\mathrm{L})} = \frac{1}{h}\sum_{i=1}^{n}h_i\Omega_{23}^{(i)}, \\
&\Omega_{36}^{(\mathrm{L})} = \frac{1}{h}\sum_{i=1}^{n}h_i\Omega_{36}^{(i)},\ \omega_1^{(\mathrm{L})} = \frac{1}{h}\sum_{i=1}^{n}h_i\omega_1^{(i)},\ \omega_2^{(\mathrm{L})} = \frac{1}{h}\sum_{i=1}^{n}h_i\omega_2^{(i)},\ \omega_6^{(\mathrm{L})} = \frac{1}{h}\sum_{i=1}^{n}h_i\omega_6^{(i)},
\end{aligned}
\tag{7.150}
$$

$$
\begin{aligned}
&\hat{\Omega}_{11}^{(\mathrm{L})} = \frac{1}{h}\sum_{i=1}^{n}\hat{h}_i\Omega_{11}^{(i)},\ \hat{\Omega}_{12}^{(\mathrm{L})} = \frac{1}{h}\sum_{i=1}^{n}\hat{h}_i\Omega_{12}^{(i)},\ \hat{\Omega}_{22}^{(\mathrm{L})} = \frac{1}{h}\sum_{i=1}^{n}\hat{h}_{il}\Omega_{22}^{(i)},\ \hat{\Omega}_{13}^{(\mathrm{L})} = \frac{1}{h}\sum_{i=1}^{n}\hat{h}_i\Omega_{13}^{(i)}, \\
&\hat{\Omega}_{23}^{(\mathrm{L})} = \frac{1}{h}\sum_{i=1}^{n}\hat{h}_i\Omega_{23}^{(i)},\ \hat{\Omega}_{16}^{(\mathrm{L})} = \frac{1}{h}\sum_{i=1}^{n}\hat{h}_i\Omega_{16}^{(i)},\ \hat{\Omega}_{26}^{(\mathrm{L})} = \frac{1}{h}\sum_{i=1}^{n}\hat{h}_i\Omega_{26}^{(i)},\ \hat{\Omega}_{36}^{(\mathrm{L})} = \frac{1}{h}\sum_{i=1}^{n}\hat{h}_i\Omega_{36}^{(i)}, \\
&\hat{\Omega}_{66}^{(\mathrm{L})} = \frac{1}{h}\sum_{i=1}^{n}\hat{h}_i\Omega_{66}^{(i)},\ \hat{\omega}_1^{(\mathrm{L})} = \frac{1}{h}\sum_{i=1}^{n}\hat{h}_i\omega_1^{(i)},\ \hat{\omega}_2^{(\mathrm{L})} = \frac{1}{h}\sum_{i=1}^{n}\hat{h}_i\omega_2^{(i)},\ \hat{\omega}_6^{(\mathrm{L})} = \frac{1}{h}\sum_{i=1}^{n}\hat{h}_i\hat{\omega}_6^{(\mathrm{L})},
\end{aligned}
\tag{7.151}
$$

$$\tilde{\Omega}_{11}^{(L)} = \frac{1}{h}\sum_{i=1}^{n}\tilde{h}_i\,\Omega_{11}^{(i)}, \quad \tilde{\Omega}_{12}^{(L)} = \frac{1}{h}\sum_{i=1}^{n}\tilde{h}_i\,\Omega_{12}^{(i)}, \quad \tilde{\Omega}_{22}^{(L)} = \frac{1}{h}\sum_{i=1}^{n}\tilde{h}_i\,\Omega_{22}^{(i)},$$
$$\tilde{\Omega}_{16}^{(L)} = \frac{1}{h}\sum_{i=1}^{n}\tilde{h}_i\,\Omega_{16}^{(i)}, \quad \tilde{\Omega}_{26}^{(L)} = \frac{1}{h}\sum_{i=1}^{n}\tilde{h}_i\,\Omega_{26}^{(i)}, \quad \tilde{\Omega}_{36}^{(L)} = \frac{1}{h}\sum_{i=1}^{n}\tilde{h}_i\,\Omega_{36}^{(i)}. \tag{7.152}$$

The effective out-of-plane strain ε_{t} may be defined by the average

$$\begin{aligned}\varepsilon_t &= \left\langle \frac{\partial u_3}{\partial x_3} \right\rangle = \frac{1}{4hLW}\sum_{i=1}^{n}\int_{-L}^{L}\int_{-W}^{W}\int_{x_3^{(i-1)}}^{x_3^{(i)}} \frac{\partial u_3^{(i)}}{\partial x_3}\,\mathrm{d}x_1\mathrm{d}x_2\mathrm{d}x_3 \\ &= \frac{1}{4hLW}\sum_{i=1}^{n}\int_{-L}^{L}\int_{-W}^{W}\left[u_3^{(i)}\left(x_1,x_2,x_3^{(i)}\right) - u_3^{(i)}\left(x_1,x_2,x_3^{(i-1)}\right)\right]\mathrm{d}x_1\mathrm{d}x_2.\end{aligned} \tag{7.153}$$

It follows from (7.96) that

$$u_3^{(i)}\left(x_1,x_2,x_3^{(i)}\right) - u_3^{(i)}\left(x_1,x_2,x_3^{(i-1)}\right) = h_i\bar{\varepsilon}_{\mathrm{t}}^{(i)} + \frac{1}{2}\hat{h}_i\hat{\varepsilon}_{\mathrm{t}}^{(i)}, \tag{7.154}$$

and substitution into (7.153) then leads to

$$\varepsilon_{\mathrm{t}} = \frac{1}{h}\sum_{i=1}^{n}h_i\,\bar{\varepsilon}_{\mathrm{t}}^{(i)} + \frac{1}{h}\sum_{i=1}^{n}\hat{h}_i\,\hat{\varepsilon}_{\mathrm{t}}^{(i)}. \tag{7.155}$$

On using (7.116) and $(7.124)_3$

$$\begin{aligned}\varepsilon_{\mathrm{t}} =& -\Omega_{13}^{(\mathrm{L})}\bar{\varepsilon}_{\mathrm{A}} - \Omega_{23}^{(\mathrm{L})}\bar{\varepsilon}_{\mathrm{T}} + \Omega_{33}^{(\mathrm{L})}\sigma_{\mathrm{t}} - \Omega_{36}^{(\mathrm{L})}\bar{\gamma}_{\mathrm{A}} \\ &- \hat{\Omega}_{13}^{(\mathrm{L})}\hat{\varepsilon}_{\mathrm{A}} - \hat{\Omega}_{23}^{(\mathrm{L})}\hat{\varepsilon}_{\mathrm{T}} - \hat{\Omega}_{36}^{(\mathrm{L})}\hat{\gamma}_{\mathrm{A}} + \omega_3^{(\mathrm{L})}\Delta T,\end{aligned} \tag{7.156}$$

where the additional parameters introduced are defined by

$$\Omega_{33}^{(\mathrm{L})} = \frac{1}{h}\sum_{i=1}^{n}h_i\,\Omega_{33}^{(i)}, \quad \Omega_{36}^{(\mathrm{L})} = \frac{1}{h}\sum_{i=1}^{n}h_i\,\Omega_{36}^{(i)}, \quad \omega_3^{(\mathrm{L})} = \frac{1}{h}\sum_{i=1}^{n}h_i\,\omega_3^{(i)}. \tag{7.157}$$

When there is no bending deformation so that $\hat{\varepsilon}_{\mathrm{A}} = \hat{\varepsilon}_{\mathrm{T}} = \hat{\gamma}_{\mathrm{A}} = 0$, relations (7.144)–(7.146) and (7.156) reduce to a form identical with the results (7.25) for symmetric laminates.

The effective out-of-plane shear strains γ_{t} and γ_{a} are defined respectively by

$$\begin{aligned}\gamma_t =& \left\langle \frac{\partial u_2}{\partial x_3} + \frac{\partial u_3}{\partial x_2} \right\rangle = \frac{1}{4hLW}\sum_{i=1}^{n}\int_{-L}^{L}\int_{-W}^{W}\int_{x_3^{(i-1)}}^{x_3^{(i)}} \frac{\partial u_2}{\partial x_3}\mathrm{d}x_1\mathrm{d}x_2\mathrm{d}x_3 \\ &+ \frac{1}{4hLW}\sum_{i=1}^{n}\int_{-L}^{L}\int_{-W}^{W}\int_{x_3^{(i-1)}}^{x_3^{(i)}} \frac{\partial u_3}{\partial x_2}\mathrm{d}x_1\mathrm{d}x_2\mathrm{d}x_3 \\ =& \frac{1}{4hLW}\sum_{i=1}^{n}\int_{-L}^{L}\int_{-W}^{W}\left[u_2^{(i)}\left(x_1,x_2,x_3^{(i)}\right) - u_2^{(i)}\left(x_1,x_2,x_3^{(i-1)}\right)\right]\mathrm{d}x_1\mathrm{d}x_2 \\ &+ \frac{1}{4hLW}\sum_{i=1}^{n}\int_{-L}^{L}\int_{x_3^{(i-1)}}^{x_3^{(i)}}\left[u_3^{(i)}\left(x_1,W,x_3\right) - u_3^{(i)}\left(x_1,-W,x_3\right)\right]\mathrm{d}x_1\mathrm{d}x_3,\end{aligned} \tag{7.158}$$

and

$$\begin{aligned}\gamma_\mathrm{a} &= \left\langle \frac{\partial u_1}{\partial x_3} + \frac{\partial u_3}{\partial x_1} \right\rangle = \frac{1}{4hLW} \sum_{i=1}^{n} \int_{-L}^{L} \int_{-W}^{W} \int_{x_3^{(i-1)}}^{x_3^{(i)}} \frac{\partial u_1}{\partial x_3} \mathrm{d}x_1 \mathrm{d}x_2 \mathrm{d}x_3 \\ &+ \frac{1}{4hLW} \sum_{i=1}^{n} \int_{-L}^{L} \int_{-W}^{W} \int_{x_3^{(i-1)}}^{x_3^{(i)}} \frac{\partial u_3}{\partial x_1} \mathrm{d}x_1 \mathrm{d}x_2 \mathrm{d}x_3 \\ &= \frac{1}{4hLW} \sum_{i=1}^{n} \int_{-L}^{L} \int_{-W}^{W} \left[u_1^{(i)}\left(x_1, x_2, x_3^{(i)}\right) - u_1^{(i)}\left(x_1, x_2, x_3^{(i-1)}\right) \right] \mathrm{d}x_1 \mathrm{d}x_2 \\ &+ \frac{1}{4hLW} \sum_{i=1}^{n} \int_{-W}^{W} \int_{x_3^{(i-1)}}^{x_3^{(i)}} \left[u_3^{(i)}\left(L, x_2, x_3\right) - u_3^{(i)}\left(-L, x_2, x_3\right) \right] \mathrm{d}x_2 \mathrm{d}x_3 .\end{aligned} \tag{7.159}$$

From (7.96) it can be shown that, for $i = 1, \ldots, n$,

$$\begin{aligned} u_1^{(i)}\left(x_1, x_2, x_3^{(i)}\right) - u_1^{(i)}\left(x_1, x_2, x_3^{(i-1)}\right) &= h_i \hat{\varepsilon}_\mathrm{A} x_1 + \tfrac{1}{2} h_i \hat{\gamma}_\mathrm{A} x_2 + h_i \bar{\gamma}_a^{(i)}, \\ u_2^{(i)}\left(x_1, x_2, x_3^{(i)}\right) - u_2^{(i)}\left(x_1, x_2, x_3^{(i-1)}\right) &= h_i \hat{\varepsilon}_\mathrm{T} x_2 + \tfrac{1}{2} h_i \hat{\gamma}_\mathrm{A} x_1 + h_i \bar{\gamma}_\mathrm{t}^{(i)}, \\ u_3^{(i)}\left(x_1, W, x_3\right) - u_3^{(i)}\left(x_1, -W, x_3\right) &= -\hat{\gamma}_\mathrm{A} W x_1, \\ u_3^{(i)}\left(L, x_2, x_3\right) - u_3^{(i)}\left(-L, x_2, x_3\right) &= -\hat{\gamma}_\mathrm{A} L x_2 . \end{aligned} \tag{7.160}$$

Substitution into (7.158) and (7.159) then leads to the following definitions for the effective shear strains

$$\gamma_\mathrm{t} = \frac{1}{h} \sum_{i=1}^{n} h_i \bar{\gamma}_\mathrm{t}^{(i)}, \qquad \gamma_\mathrm{a} = \frac{1}{h} \sum_{i=1}^{n} h_i \bar{\gamma}_\mathrm{a}^{(i)}. \tag{7.161}$$

On using (7.106) and (7.124) it follows that

$$\begin{aligned} \gamma_\mathrm{t} &= \frac{1}{h} \sum_{i=1}^{n} h_i \gamma_\mathrm{t}^{(i)} = \frac{1}{h} \sum_{i=1}^{n} h_i S_{44}^{(i)} \tau_\mathrm{t} + \frac{1}{h} \sum_{i=1}^{n} h_i S_{45}^{(i)} \tau_\mathrm{a}, \\ \gamma_\mathrm{a} &= \frac{1}{h} \sum_{i=1}^{n} h_i \gamma_\mathrm{a}^{(i)} = \frac{1}{h} \sum_{i=1}^{n} h_i S_{45}^{(i)} \tau_\mathrm{t} + \frac{1}{h} \sum_{i=1}^{n} h_i S_{55}^{(i)} \tau_\mathrm{a}. \end{aligned} \tag{7.162}$$

These results are now written in the form

$$\begin{aligned} \gamma_\mathrm{t} &= S_{44}^{(\mathrm{L})} \tau_\mathrm{t} + S_{45}^{(\mathrm{L})} \tau_\mathrm{a}, \\ \gamma_\mathrm{a} &= S_{45}^{(\mathrm{L})} \tau_\mathrm{t} + S_{55}^{(\mathrm{L})} \tau_\mathrm{a}, \end{aligned} \tag{7.163}$$

where

$$S_{44}^{(\mathrm{L})} = \frac{1}{h} \sum_{i=1}^{n} h_i S_{44}^{(i)}, \qquad S_{45}^{(\mathrm{L})} = \frac{1}{h} \sum_{i=1}^{n} h_i S_{45}^{(i)}, \qquad S_{55}^{(\mathrm{L})} = \frac{1}{h} \sum_{i=1}^{n} h_i S_{55}^{(i)}. \tag{7.164}$$

On assembling all derived in-plane and through-thickness stress-strain relations for the laminate, it follows from (7.144)–(7.149), (7.156) and (7.163) that

$$\begin{aligned} \sigma_\mathrm{A} = {} & \Omega_{11}^{(\mathrm{L})} \bar{\varepsilon}_\mathrm{A} + \Omega_{12}^{(\mathrm{L})} \bar{\varepsilon}_\mathrm{T} + \Omega_{13}^{(\mathrm{L})} \sigma_\mathrm{t} + \Omega_{16}^{(\mathrm{L})} \bar{\gamma}_\mathrm{A} + \hat{\Omega}_{11}^{(\mathrm{L})} \hat{\varepsilon}_\mathrm{A} \\ & + \hat{\Omega}_{12}^{(\mathrm{L})} \hat{\varepsilon}_\mathrm{T} + \hat{\Omega}_{16}^{(\mathrm{L})} \hat{\gamma}_\mathrm{A} - \omega_1^{(\mathrm{L})} \Delta T, \end{aligned} \tag{7.165}$$

$$\begin{aligned} \sigma_\mathrm{T} = {} & \Omega_{12}^{(\mathrm{L})} \bar{\varepsilon}_\mathrm{A} + \Omega_{22}^{(\mathrm{L})} \bar{\varepsilon}_\mathrm{T} + \Omega_{23}^{(\mathrm{L})} \sigma_\mathrm{t} + \Omega_{26}^{(\mathrm{L})} \bar{\gamma}_\mathrm{A} + \hat{\Omega}_{12}^{(\mathrm{L})} \hat{\varepsilon}_\mathrm{A} \\ & + \hat{\Omega}_{22}^{(\mathrm{L})} \hat{\varepsilon}_\mathrm{T} + \hat{\Omega}_{26}^{(\mathrm{L})} \hat{\gamma}_\mathrm{A} - \omega_2^{(\mathrm{L})} \Delta T, \end{aligned} \tag{7.166}$$

$$\begin{aligned}\varepsilon_t = &-\Omega_{13}^{(L)}\bar{\varepsilon}_A - \Omega_{23}^{(L)}\bar{\varepsilon}_T + \Omega_{33}^{(L)}\sigma_t - \Omega_{36}^{(L)}\bar{\gamma}_A - \hat{\Omega}_{13}^{(L)}\hat{\varepsilon}_A \\ &- \hat{\Omega}_{23}^{(L)}\hat{\varepsilon}_T - \hat{\Omega}_{36}^{(L)}\hat{\gamma}_A + \omega_3^{(L)}\Delta T,\end{aligned} \quad (7.167)$$

$$\gamma_t = S_{44}^{(L)}\tau_t + S_{45}^{(L)}\tau_a, \quad (7.168)$$

$$\gamma_a = S_{45}^{(L)}\tau_t + S_{55}^{(L)}\tau_a, \quad (7.169)$$

$$\begin{aligned}\tau_A = &\ \Omega_{16}^{(L)}\bar{\varepsilon}_A + \Omega_{26}^{(L)}\bar{\varepsilon}_T + \Omega_{36}^{(L)}\sigma_t + \Omega_{66}^{(L)}\bar{\gamma}_A + \hat{\Omega}_{16}^{(L)}\hat{\varepsilon}_A \\ &+ \hat{\Omega}_{26}^{(L)}\hat{\varepsilon}_T + \hat{\Omega}_{66}^{(L)}\hat{\gamma}_A - \omega_6^{(L)}\Delta T,\end{aligned} \quad (7.170)$$

$$\begin{aligned}&M_A + \tfrac{1}{2}h\sigma_A = \\ &\hat{\Omega}_{11}^{(L)}\bar{\varepsilon}_A + \hat{\Omega}_{12}^{(L)}\bar{\varepsilon}_T + \hat{\Omega}_{13}^{(L)}\sigma_t + \hat{\Omega}_{16}^{(L)}\bar{\gamma}_A + \tilde{\Omega}_{11}^{(L)}\hat{\varepsilon}_A + \tilde{\Omega}_{12}^{(L)}\hat{\varepsilon}_T + \tilde{\Omega}_{16}^{(L)}\hat{\gamma}_A - \hat{\omega}_1^{(L)}\Delta T,\end{aligned} \quad (7.171)$$

$$\begin{aligned}&M_T + \tfrac{1}{2}h\sigma_T = \\ &\hat{\Omega}_{12}^{(L)}\bar{\varepsilon}_A + \hat{\Omega}_{22}^{(L)}\bar{\varepsilon}_T + \hat{\Omega}_{23}^{(L)}\sigma_t + \hat{\Omega}_{26}^{(L)}\bar{\gamma}_A + \tilde{\Omega}_{12}^{(L)}\hat{\varepsilon}_A + \tilde{\Omega}_{22}^{(L)}\hat{\varepsilon}_T + \tilde{\Omega}_{26}^{(L)}\hat{\gamma}_A - \hat{\omega}_2^{(L)}\Delta T,\end{aligned} \quad (7.172)$$

$$\begin{aligned}&M_S + \tfrac{1}{2}h\tau_A = \\ &\hat{\Omega}_{16}^{(L)}\bar{\varepsilon}_A + \hat{\Omega}_{26}^{(L)}\bar{\varepsilon}_T + \hat{\Omega}_{36}^{(L)}\sigma_t + \hat{\Omega}_{66}^{(L)}\bar{\gamma}_A + \tilde{\Omega}_{16}^{(L)}\hat{\varepsilon}_A + \tilde{\Omega}_{26}^{(L)}\hat{\varepsilon}_T + \tilde{\Omega}_{66}^{(L)}\hat{\gamma}_A - \hat{\omega}_6^{(L)}\Delta T.\end{aligned} \quad (7.173)$$

Relations (7.165)–(7.173) are not yet in the form of stress-strain relations owing to the presence of the through-thickness stress σ_t and the through-thickness shear stresses τ_t and τ_a on the right-hand sides. From (7.167) it can be shown that

$$\begin{aligned}\sigma_t = &\frac{\Omega_{13}^{(L)}}{\Omega_{33}^{(L)}}\bar{\varepsilon}_A + \frac{\Omega_{23}^{(L)}}{\Omega_{33}^{(L)}}\bar{\varepsilon}_T + \frac{1}{\Omega_{33}^{(L)}}\varepsilon_t + \frac{\hat{\Omega}_{36}^{(L)}}{\Omega_{33}^{(L)}}\bar{\gamma}_A + \frac{\hat{\Omega}_{13}^{(L)}}{\Omega_{33}^{(L)}}\hat{\varepsilon}_A \\ &+ \frac{\hat{\Omega}_{23}^{(L)}}{\Omega_{33}^{(L)}}\hat{\varepsilon}_T + \frac{\hat{\Omega}_{36}^{(L)}}{\Omega_{33}^{(L)}}\hat{\gamma}_A - \frac{\omega_3^{(L)}}{\Omega_{33}^{(L)}}\Delta T.\end{aligned} \quad (7.174)$$

On substitution into (7.165), (7.166) and (7.170)–(7.173), it follows that

$$\begin{aligned}\sigma_A = &\left(\Omega_{11}^{(L)} + \frac{\left(\Omega_{13}^{(L)}\right)^2}{\Omega_{33}^{(L)}}\right)\bar{\varepsilon}_A + \left(\Omega_{12}^{(L)} + \frac{\Omega_{13}^{(L)}\Omega_{23}^{(L)}}{\Omega_{33}^{(L)}}\right)\bar{\varepsilon}_T + \frac{\Omega_{13}^{(L)}}{\Omega_{33}^{(L)}}\varepsilon_t + \left(\Omega_{16}^{(L)} + \frac{\Omega_{13}^{(L)}\Omega_{36}^{(L)}}{\Omega_{33}^{(L)}}\right)\bar{\gamma}_A \\ &+ \left(\hat{\Omega}_{11}^{(L)} + \frac{\Omega_{13}^{(L)}\hat{\Omega}_{13}^{(L)}}{\Omega_{33}^{(L)}}\right)\hat{\varepsilon}_A + \left(\hat{\Omega}_{12}^{(L)} + \frac{\Omega_{13}^{(L)}\hat{\Omega}_{23}^{(L)}}{\Omega_{33}^{(L)}}\right)\hat{\varepsilon}_T + \left(\hat{\Omega}_{16}^{(L)} + \frac{\Omega_{13}^{(L)}\hat{\Omega}_{36}^{(L)}}{\Omega_{33}^{(L)}}\right)\hat{\gamma}_A \\ &- \left(\omega_1^{(L)} + \frac{\Omega_{13}^{(L)}\omega_3^{(L)}}{\Omega_{33}^{(L)}}\right)\Delta T,\end{aligned} \quad (7.175)$$

$$\begin{aligned}\sigma_T = &\left(\Omega_{12}^{(L)} + \frac{\Omega_{23}^{(L)}\Omega_{13}^{(L)}}{\Omega_{33}^{(L)}}\right)\bar{\varepsilon}_A + \left(\Omega_{22}^{(L)} + \frac{\left(\Omega_{23}^{(L)}\right)^2}{\Omega_{33}^{(L)}}\right)\bar{\varepsilon}_T + \frac{\Omega_{23}^{(L)}}{\Omega_{33}^{(L)}}\varepsilon_t + \left(\Omega_{26}^{(L)} + \frac{\Omega_{23}^{(L)}\Omega_{36}^{(L)}}{\Omega_{33}^{(L)}}\right)\bar{\gamma}_A \\ &+ \left(\hat{\Omega}_{12}^{(L)} + \frac{\Omega_{23}^{(L)}\hat{\Omega}_{13}^{(L)}}{\Omega_{33}^{(L)}}\right)\hat{\varepsilon}_A + \left(\hat{\Omega}_{22}^{(L)} + \frac{\Omega_{23}^{(L)}\hat{\Omega}_{23}^{(L)}}{\Omega_{33}^{(L)}}\right)\hat{\varepsilon}_T + \left(\hat{\Omega}_{26}^{(L)} + \frac{\Omega_{23}^{(L)}\hat{\Omega}_{36}^{(L)}}{\Omega_{33}^{(L)}}\right)\hat{\gamma}_A \\ &- \left(\omega_2^{(L)} + \frac{\Omega_{23}^{(L)}\omega_3^{(L)}}{\Omega_{33}^{(L)}}\right)\Delta T,\end{aligned} \quad (7.176)$$

$$\tau_{\rm A} = \left(\Omega_{16}^{(\rm L)} + \frac{\Omega_{36}^{(\rm L)}\Omega_{13}^{(\rm L)}}{\Omega_{33}^{(\rm L)}}\right)\bar{\varepsilon}_{\rm A} + \left(\Omega_{26}^{(\rm L)} + \frac{\Omega_{36}^{(\rm L)}\Omega_{23}^{(\rm L)}}{\Omega_{33}^{(\rm L)}}\right)\bar{\varepsilon}_{\rm T} + \frac{\Omega_{36}^{(\rm L)}}{\Omega_{33}^{(\rm L)}}\varepsilon_{\rm t} + \left(\Omega_{66}^{(\rm L)} + \frac{\left(\Omega_{36}^{(\rm L)}\right)^2}{\Omega_{33}^{(\rm L)}}\right)\bar{\gamma}_{\rm A}$$
$$+ \left(\hat{\Omega}_{16}^{(\rm L)} + \frac{\Omega_{36}^{(\rm L)}\hat{\Omega}_{13}^{(\rm L)}}{\Omega_{33}^{(\rm L)}}\right)\hat{\varepsilon}_{\rm A} + \left(\hat{\Omega}_{26}^{(\rm L)} + \frac{\Omega_{36}^{(\rm L)}\hat{\Omega}_{23}^{(\rm L)}}{\Omega_{33}^{(\rm L)}}\right)\hat{\varepsilon}_{\rm T} + \left(\hat{\Omega}_{66}^{(\rm L)} + \frac{\Omega_{36}^{(\rm L)}\hat{\Omega}_{36}^{(\rm L)}}{\Omega_{33}^{(\rm L)}}\right)\hat{\gamma}_{\rm A} \tag{7.177}$$
$$- \left(\omega_6^{(\rm L)} + \frac{\Omega_{36}^{(\rm L)}\omega_3^{(\rm L)}}{\Omega_{33}^{(\rm L)}}\right)\Delta T,$$

$$M_A + \tfrac{1}{2}h\sigma_A =$$
$$\left(\hat{\Omega}_{11}^{(\rm L)} + \frac{\hat{\Omega}_{13}^{(\rm L)}\Omega_{13}^{(\rm L)}}{\Omega_{33}^{(\rm L)}}\right)\bar{\varepsilon}_{\rm A} + \left(\hat{\Omega}_{12}^{(\rm L)} + \frac{\hat{\Omega}_{13}^{(\rm L)}\Omega_{23}^{(\rm L)}}{\Omega_{33}^{(\rm L)}}\right)\bar{\varepsilon}_{\rm T} + \frac{\hat{\Omega}_{13}^{(\rm L)}}{\Omega_{33}^{(\rm L)}}\varepsilon_{\rm t} + \left(\hat{\Omega}_{16}^{(\rm L)} + \frac{\hat{\Omega}_{13}^{(\rm L)}\Omega_{36}^{(\rm L)}}{\Omega_{33}^{(\rm L)}}\right)\bar{\gamma}_{\rm A}$$
$$+ \left(\tilde{\Omega}_{11}^{(\rm L)} + \frac{\left(\hat{\Omega}_{13}^{(\rm L)}\right)^2}{\Omega_{33}^{(\rm L)}}\right)\hat{\varepsilon}_{\rm A} + \left(\tilde{\Omega}_{12}^{(\rm L)} + \frac{\hat{\Omega}_{13}^{(\rm L)}\hat{\Omega}_{23}^{(\rm L)}}{\Omega_{33}^{(\rm L)}}\right)\hat{\varepsilon}_{\rm T} + \left(\tilde{\Omega}_{16}^{(\rm L)} + \frac{\hat{\Omega}_{13}^{(\rm L)}\hat{\Omega}_{36}^{(\rm L)}}{\Omega_{33}^{(\rm L)}}\right)\hat{\gamma}_{\rm A} \tag{7.178}$$
$$- \left(\hat{\omega}_1^{(\rm L)} + \frac{\hat{\Omega}_{13}^{(\rm L)}\omega_3^{(\rm L)}}{\Omega_{33}^{(\rm L)}}\right)\Delta T,$$

$$M_T + \tfrac{1}{2}h\sigma_T =$$
$$\left(\hat{\Omega}_{12}^{(\rm L)} + \frac{\hat{\Omega}_{23}^{(\rm L)}\Omega_{13}^{(\rm L)}}{\Omega_{33}^{(\rm L)}}\right)\bar{\varepsilon}_{\rm A} + \left(\hat{\Omega}_{22}^{(\rm L)} + \frac{\hat{\Omega}_{23}^{(\rm L)}\Omega_{23}^{(\rm L)}}{\Omega_{33}^{(\rm L)}}\right)\bar{\varepsilon}_{\rm T} + \frac{\hat{\Omega}_{23}^{(\rm L)}}{\Omega_{33}^{(\rm L)}}\varepsilon_{\rm t} + \left(\hat{\Omega}_{26}^{(\rm L)} + \frac{\hat{\Omega}_{23}^{(\rm L)}\Omega_{36}^{(\rm L)}}{\Omega_{33}^{(\rm L)}}\right)\bar{\gamma}_{\rm A}$$
$$+ \left(\tilde{\Omega}_{12}^{(\rm L)} + \frac{\hat{\Omega}_{23}^{(\rm L)}\hat{\Omega}_{13}^{(\rm L)}}{\Omega_{33}^{(\rm L)}}\right)\hat{\varepsilon}_{\rm A} + \left(\tilde{\Omega}_{22}^{(\rm L)} + \frac{\left(\hat{\Omega}_{23}^{(\rm L)}\right)^2}{\Omega_{33}^{(\rm L)}}\right)\hat{\varepsilon}_{\rm T} + \left(\tilde{\Omega}_{26}^{(\rm L)} + \frac{\hat{\Omega}_{23}^{(\rm L)}\hat{\Omega}_{36}^{(\rm L)}}{\Omega_{33}^{(\rm L)}}\right)\hat{\gamma}_{\rm A} \tag{7.179}$$
$$- \left(\hat{\omega}_2^{(\rm L)} + \frac{\hat{\Omega}_{23}^{(\rm L)}\omega_3^{(\rm L)}}{\Omega_{33}^{(\rm L)}}\right)\Delta T,$$

$$M_{\rm S} + \tfrac{1}{2}h\tau_{\rm A} =$$
$$\left(\hat{\Omega}_{16}^{(\rm L)} + \frac{\hat{\Omega}_{36}^{(\rm L)}\Omega_{13}^{(\rm L)}}{\Omega_{33}^{(\rm L)}}\right)\bar{\varepsilon}_{\rm A} + \left(\hat{\Omega}_{26}^{(\rm L)} + \frac{\hat{\Omega}_{36}^{(\rm L)}\Omega_{23}^{(\rm L)}}{\Omega_{33}^{(\rm L)}}\right)\bar{\varepsilon}_{\rm T} + \frac{\hat{\Omega}_{36}^{(\rm L)}}{\Omega_{33}^{(\rm L)}}\varepsilon_{\rm t} + \left(\hat{\Omega}_{66}^{(\rm L)} + \frac{\hat{\Omega}_{36}^{(\rm L)}\Omega_{36}^{(\rm L)}}{\Omega_{33}^{(\rm L)}}\right)\bar{\gamma}_{\rm A}$$
$$+ \left(\hat{\Omega}_{16}^{(\rm L)} + \frac{\hat{\Omega}_{36}^{(\rm L)}\hat{\Omega}_{13}^{(\rm L)}}{\Omega_{33}^{(\rm L)}}\right)\hat{\varepsilon}_{\rm A} + \left(\tilde{\Omega}_{26}^{(\rm L)} + \frac{\hat{\Omega}_{36}^{(\rm L)}\hat{\Omega}_{23}^{(\rm L)}}{\Omega_{33}^{(\rm L)}}\right)\hat{\varepsilon}_{\rm T} + \left(\tilde{\Omega}_{66}^{(\rm L)} + \frac{\left(\hat{\Omega}_{36}^{(\rm L)}\right)^2}{\Omega_{33}^{(\rm L)}}\right)\hat{\gamma}_{\rm A} \tag{7.180}$$
$$- \left(\hat{\omega}_6^{(\rm L)} + \frac{\hat{\Omega}_{36}^{(\rm L)}\omega_3^{(\rm L)}}{\Omega_{33}^{(\rm L)}}\right)\Delta T.$$

On using (7.70), (7.73) and (7.74), the inverse of relations (7.168) and (7.169) are given by (7.77) and (7.78), namely,

$$\begin{aligned} \tau_{\rm t} &= C_{44}^{(\rm L)}\gamma_{\rm t} + C_{45}^{(\rm L)}\gamma_{\rm a}, \\ \tau_{\rm a} &= C_{45}^{(\rm L)}\gamma_{\rm t} + C_{55}^{(\rm L)}\gamma_{\rm a}, \end{aligned} \tag{7.181}$$

where

$$C_{44}^{(\mathrm{L})}=\frac{S_{55}^{(\mathrm{L})}}{S_{44}^{(\mathrm{L})}S_{55}^{(\mathrm{L})}-S_{45}^{(\mathrm{L})}S_{45}^{(\mathrm{L})}},\quad C_{45}^{(\mathrm{L})}=-\frac{S_{45}^{(\mathrm{L})}}{S_{44}^{(\mathrm{L})}S_{55}^{(\mathrm{L})}-S_{45}^{(\mathrm{L})}S_{45}^{(\mathrm{L})}},$$
$$C_{55}^{(\mathrm{L})}=\frac{S_{44}^{(\mathrm{L})}}{S_{44}^{(\mathrm{L})}S_{55}^{(\mathrm{L})}-S_{45}^{(\mathrm{L})}S_{45}^{(\mathrm{L})}}. \tag{7.182}$$

On using relations (7.34)–(7.38), the results (7.174)–(7.181) may be written as

$$\sigma_{\mathrm{A}}=C_{11}^{(\mathrm{L})}\bar{\varepsilon}_{\mathrm{A}}+C_{12}^{(\mathrm{L})}\bar{\varepsilon}_{\mathrm{T}}+C_{16}^{(\mathrm{L})}\bar{\gamma}_{\mathrm{A}}+\hat{C}_{11}^{(\mathrm{L})}\hat{\varepsilon}_{\mathrm{A}}+\hat{C}_{12}^{(\mathrm{L})}\hat{\varepsilon}_{\mathrm{T}}+\hat{C}_{16}^{(\mathrm{L})}\hat{\gamma}_{\mathrm{A}}+C_{13}^{(\mathrm{L})}\varepsilon_{\mathrm{t}}-U_{1}^{(\mathrm{L})}\Delta T, \tag{7.183}$$

$$\sigma_{\mathrm{T}}=C_{12}^{(\mathrm{L})}\bar{\varepsilon}_{\mathrm{A}}+C_{22}^{(\mathrm{L})}\bar{\varepsilon}_{\mathrm{T}}+C_{26}^{(\mathrm{L})}\bar{\gamma}_{\mathrm{A}}+\hat{C}_{21}^{(\mathrm{L})}\hat{\varepsilon}_{\mathrm{A}}+\hat{C}_{22}^{(\mathrm{L})}\hat{\varepsilon}_{\mathrm{T}}+\hat{C}_{26}^{(\mathrm{L})}\hat{\gamma}_{\mathrm{A}}+C_{23}^{(\mathrm{L})}\varepsilon_{\mathrm{t}}-U_{2}^{(\mathrm{L})}\Delta T, \tag{7.184}$$

$$\sigma_{\mathrm{t}}=C_{13}^{(\mathrm{L})}\bar{\varepsilon}_{\mathrm{A}}+C_{23}^{(\mathrm{L})}\bar{\varepsilon}_{\mathrm{T}}+C_{36}^{(\mathrm{L})}\bar{\gamma}_{\mathrm{A}}+\hat{C}_{31}^{(\mathrm{L})}\hat{\varepsilon}_{\mathrm{A}}+\hat{C}_{32}^{(\mathrm{L})}\hat{\varepsilon}_{\mathrm{T}}+\hat{C}_{36}^{(\mathrm{L})}\hat{\gamma}_{\mathrm{A}}+C_{33}^{(\mathrm{L})}\varepsilon_{\mathrm{t}}-U_{3}^{(\mathrm{L})}\Delta T, \tag{7.185}$$

$$\tau_{\mathrm{A}}=C_{16}^{(\mathrm{L})}\bar{\varepsilon}_{\mathrm{A}}+C_{26}^{(\mathrm{L})}\bar{\varepsilon}_{\mathrm{T}}+C_{66}^{(\mathrm{L})}\bar{\gamma}_{\mathrm{A}}+\hat{C}_{61}^{(\mathrm{L})}\hat{\varepsilon}_{\mathrm{A}}+\hat{C}_{62}^{(\mathrm{L})}\hat{\varepsilon}_{\mathrm{T}}+\hat{C}_{66}^{(\mathrm{L})}\hat{\gamma}_{\mathrm{A}}+C_{36}^{(\mathrm{L})}\varepsilon_{\mathrm{t}}-U_{6}^{(\mathrm{L})}\Delta T, \tag{7.186}$$

$$M_{\mathrm{A}}+\tfrac{1}{2}h\sigma_{\mathrm{A}}=\hat{C}_{11}^{(\mathrm{L})}\bar{\varepsilon}_{\mathrm{A}}+\hat{C}_{21}^{(\mathrm{L})}\bar{\varepsilon}_{\mathrm{T}}+\hat{C}_{61}^{(\mathrm{L})}\bar{\gamma}_{\mathrm{A}}+\tilde{C}_{11}^{(\mathrm{L})}\hat{\varepsilon}_{\mathrm{A}}+\tilde{C}_{12}^{(\mathrm{L})}\hat{\varepsilon}_{\mathrm{T}}+\tilde{C}_{16}^{(\mathrm{L})}\hat{\gamma}_{\mathrm{A}}+\hat{C}_{31}^{(\mathrm{L})}\varepsilon_{\mathrm{t}}-\hat{U}_{1}^{(\mathrm{L})}\Delta T, \tag{7.187}$$

$$M_{\mathrm{T}}+\tfrac{1}{2}h\sigma_{\mathrm{T}}=\hat{C}_{12}^{(\mathrm{L})}\bar{\varepsilon}_{\mathrm{A}}+\hat{C}_{22}^{(\mathrm{L})}\bar{\varepsilon}_{\mathrm{T}}+\hat{C}_{62}^{(\mathrm{L})}\bar{\gamma}_{\mathrm{A}}+\tilde{C}_{12}^{(\mathrm{L})}\hat{\varepsilon}_{\mathrm{A}}+\tilde{C}_{22}^{(\mathrm{L})}\hat{\varepsilon}_{\mathrm{T}}+\tilde{C}_{26}^{(\mathrm{L})}\hat{\gamma}_{\mathrm{A}}+\hat{C}_{32}^{(\mathrm{L})}\varepsilon_{\mathrm{t}}-\hat{U}_{2}^{(\mathrm{L})}\Delta T, \tag{7.188}$$

$$M_{\mathrm{S}}+\tfrac{1}{2}h\tau_{\mathrm{A}}=\hat{C}_{16}^{(\mathrm{L})}\bar{\varepsilon}_{\mathrm{A}}+\hat{C}_{26}^{(\mathrm{L})}\bar{\varepsilon}_{\mathrm{T}}+\hat{C}_{66}^{(\mathrm{L})}\bar{\gamma}_{\mathrm{A}}+\tilde{C}_{16}^{(\mathrm{L})}\hat{\varepsilon}_{\mathrm{A}}+\tilde{C}_{26}^{(\mathrm{L})}\hat{\varepsilon}_{\mathrm{T}}+\tilde{C}_{66}^{(\mathrm{L})}\hat{\gamma}_{\mathrm{A}}+\hat{C}_{36}^{(\mathrm{L})}\varepsilon_{\mathrm{t}}-\hat{U}_{6}^{(\mathrm{L})}\Delta T, \tag{7.189}$$

$$\tau_{\mathrm{t}}=C_{44}^{(\mathrm{L})}\gamma_{\mathrm{t}}+C_{45}^{(\mathrm{L})}\gamma_{\mathrm{a}},\quad \tau_{\mathrm{a}}=C_{45}^{(\mathrm{L})}\gamma_{\mathrm{t}}+C_{55}^{(\mathrm{L})}\gamma_{\mathrm{a}}, \tag{7.190}$$

where the coefficients $C_{IJ}^{(\mathrm{L})}$ and $U_{I}^{(\mathrm{L})}$ are given by (7.34)–(7.38), and where

$$\left.\begin{aligned}
&\hat{C}_{11}^{(\mathrm{L})}=\hat{\Omega}_{11}^{(\mathrm{L})}+\frac{\Omega_{13}^{(\mathrm{L})}\hat{\Omega}_{13}^{(\mathrm{L})}}{\Omega_{33}^{(\mathrm{L})}},\quad \hat{C}_{12}^{(\mathrm{L})}=\hat{\Omega}_{12}^{(\mathrm{L})}+\frac{\Omega_{13}^{(\mathrm{L})}\hat{\Omega}_{23}^{(\mathrm{L})}}{\Omega_{33}^{(\mathrm{L})}},\quad \hat{C}_{16}^{(\mathrm{L})}=\hat{\Omega}_{16}^{(\mathrm{L})}+\frac{\Omega_{13}^{(\mathrm{L})}\hat{\Omega}_{36}^{(\mathrm{L})}}{\Omega_{33}^{(\mathrm{L})}},\\
&\hat{C}_{21}^{(\mathrm{L})}=\hat{\Omega}_{12}^{(\mathrm{L})}+\frac{\Omega_{23}^{(\mathrm{L})}\hat{\Omega}_{13}^{(\mathrm{L})}}{\Omega_{33}^{(\mathrm{L})}},\quad \hat{C}_{22}^{(\mathrm{L})}=\hat{\Omega}_{22}^{(\mathrm{L})}+\frac{\Omega_{23}^{(\mathrm{L})}\hat{\Omega}_{23}^{(\mathrm{L})}}{\Omega_{33}^{(\mathrm{L})}},\quad \hat{C}_{26}^{(\mathrm{L})}=\hat{\Omega}_{26}^{(\mathrm{L})}+\frac{\Omega_{23}^{(\mathrm{L})}\hat{\Omega}_{36}^{(\mathrm{L})}}{\Omega_{33}^{(\mathrm{L})}},\\
&\hat{C}_{61}^{(\mathrm{L})}=\hat{\Omega}_{16}^{(\mathrm{L})}+\frac{\Omega_{36}^{(\mathrm{L})}\hat{\Omega}_{13}^{(\mathrm{L})}}{\Omega_{33}^{(\mathrm{L})}},\quad \hat{C}_{62}^{(\mathrm{L})}=\hat{\Omega}_{26}^{(\mathrm{L})}+\frac{\Omega_{36}^{(\mathrm{L})}\hat{\Omega}_{23}^{(\mathrm{L})}}{\Omega_{33}^{(\mathrm{L})}},\quad \hat{C}_{66}^{(\mathrm{L})}=\hat{\Omega}_{66}^{(\mathrm{L})}+\frac{\Omega_{36}^{(\mathrm{L})}\hat{\Omega}_{36}^{(\mathrm{L})}}{\Omega_{33}^{(\mathrm{L})}},
\end{aligned}\right\} \tag{7.191}$$

$$\left.\begin{aligned}
&\tilde{C}_{11}^{(\mathrm{L})}=\tilde{\Omega}_{11}^{(\mathrm{L})}+\frac{(\hat{\Omega}_{13}^{(\mathrm{L})})^2}{\Omega_{33}^{(\mathrm{L})}},\quad \tilde{C}_{12}^{(\mathrm{L})}=\tilde{\Omega}_{12}^{(\mathrm{L})}+\frac{\hat{\Omega}_{13}^{(\mathrm{L})}\hat{\Omega}_{23}^{(\mathrm{L})}}{\Omega_{33}^{(\mathrm{L})}},\quad \tilde{C}_{16}^{(\mathrm{L})}=\tilde{\Omega}_{16}^{(\mathrm{L})}+\frac{\hat{\Omega}_{13}^{(\mathrm{L})}\hat{\Omega}_{36}^{(\mathrm{L})}}{\Omega_{33}^{(\mathrm{L})}},\\
&\tilde{C}_{22}^{(\mathrm{L})}=\tilde{\Omega}_{22}^{(\mathrm{L})}+\frac{(\hat{\Omega}_{23}^{(\mathrm{L})})^2}{\Omega_{33}^{(\mathrm{L})}},\quad \tilde{C}_{26}^{(\mathrm{L})}=\tilde{\Omega}_{26}^{(\mathrm{L})}+\frac{\hat{\Omega}_{23}^{(\mathrm{L})}\hat{\Omega}_{36}^{(\mathrm{L})}}{\Omega_{33}^{(\mathrm{L})}},\quad \tilde{C}_{66}^{(\mathrm{L})}=\tilde{\Omega}_{66}^{(\mathrm{L})}+\frac{(\hat{\Omega}_{36}^{(\mathrm{L})})^2}{\Omega_{33}^{(\mathrm{L})}},
\end{aligned}\right\} \tag{7.192}$$

$$\hat{C}_{31}^{(L)} = \frac{\hat{\Omega}_{13}^{(L)}}{\Omega_{33}^{(L)}}, \quad \hat{C}_{32}^{(L)} = \frac{\hat{\Omega}_{23}^{(L)}}{\Omega_{33}^{(L)}}, \quad \hat{C}_{36}^{(L)} = \frac{\hat{\Omega}_{36}^{(L)}}{\Omega_{33}^{(L)}}, \tag{7.193}$$

$$\hat{U}_1^{(L)} = \hat{\omega}_1^{(L)} + \frac{\hat{\Omega}_{13}^{(L)}\omega_3^{(L)}}{\Omega_{33}^{(L)}}, \quad \hat{U}_2^{(L)} = \hat{\omega}_2^{(L)} + \frac{\hat{\Omega}_{23}^{(L)}\omega_3^{(L)}}{\Omega_{33}^{(L)}}, \quad \hat{U}_6^{(L)} = \hat{\omega}_6^{(L)} + \frac{\hat{\Omega}_{36}^{(L)}\omega_3^{(L)}}{\Omega_{33}^{(L)}}. \tag{7.194}$$

It is noted that the elastic constants $\hat{C}_{12}^{(L)} \neq \hat{C}_{21}^{(L)}$, $\hat{C}_{16}^{(L)} \neq \hat{C}_{61}^{(L)}$, $\hat{C}_{26}^{(L)} \neq \hat{C}_{62}^{(L)}$ and that the components $\hat{C}_{13}^{(L)}$, $\hat{C}_{23}^{(L)}$, $\hat{C}_{63}^{(L)}$ and $\hat{U}_3^{(L)}$, $\hat{U}_4^{(L)}$, $\hat{U}_5^{(L)}$ have not been introduced.

The independent stresses considered in relations (7.183)–(7.190) are $\sigma_A, \sigma_T, \sigma_t, \tau_A, \tau_a, \tau_t$ and the independent bending moments per unit cross-sectional area are M_A, M_T, M_S, a total of nine loading parameters. The corresponding independent strains are $\bar{\varepsilon}_A, \bar{\varepsilon}_T, \varepsilon_t, \bar{\gamma}_A, \gamma_a, \gamma_t$ and the bending strains $\hat{\varepsilon}_A, \hat{\varepsilon}_T, \hat{\gamma}_A$, again nine in number. It should be noted from (7.98) that the quantities $\bar{\varepsilon}_A$, $\bar{\varepsilon}_T$ and $\bar{\gamma}_A$ denote the axial, transverse and in-plane shear strains, respectively, on the plane $x_3 = 0$ of the laminate (i.e. the lower free surface). The axial, transverse and in-plane shear strains on the midplane of the laminate are denoted by ε_A^*, ε_t^* and γ_A^*, respectively, where

$$\varepsilon_A^* = \bar{\varepsilon}_A + \tfrac{1}{2}h\hat{\varepsilon}_A, \qquad \varepsilon_T^* = \bar{\varepsilon}_T + \tfrac{1}{2}h\hat{\varepsilon}_T, \qquad \gamma_A^* = \bar{\gamma}_A + \tfrac{1}{2}h\hat{\gamma}_A.$$

It is usually preferred to replace the strain parameters $\bar{\varepsilon}_A$, $\bar{\varepsilon}_T$ and $\bar{\gamma}_A$ by ε_A^*, ε_T^* and γ_A^*, respectively. In principle, it is possible to manipulate the stress-strain relationships so that the nine strain parameters are expressed as a function of the nine loading parameters. Such an inversion, which is too involved to be performed analytically, needs to be undertaken numerically. The following relations, which form a subset of the nine inverse relations, are expressed using a notation that will be used elsewhere in this book (see Section 11.5 in Chapter 11)

$$\varepsilon_A^* = \frac{\sigma_A}{E_A} - \frac{\nu_A}{E_A}\sigma_T - \frac{\hat{\nu}_A}{\hat{E}_A}M_A - \frac{\hat{\eta}_A}{\hat{E}_T}M_T - \frac{\nu_a}{E_A}\sigma_t + \alpha_A\Delta T, \tag{7.195}$$

$$\varepsilon_t^* = -\frac{\nu_A}{E_A}\sigma_A + \frac{\sigma_T}{E_T} - \frac{\hat{\nu}_T}{\hat{E}_A}M_A - \frac{\hat{\eta}_T}{\hat{E}_T}M_T - \frac{\nu_t}{E_T}\sigma_t + \alpha_T\Delta T, \tag{7.196}$$

$$\hat{\varepsilon}_A = -\frac{\hat{\nu}_A}{\hat{E}_A}\sigma_A - \frac{\hat{\nu}_T}{\hat{E}_A}\sigma_T + \frac{M_A}{\hat{E}_A} - \frac{\hat{\delta}_A}{\hat{E}_A}M_T - \frac{\hat{\nu}_a}{\hat{E}_A}\sigma_t + \hat{\alpha}_A\Delta T, \tag{7.197}$$

$$\hat{\varepsilon}_T = -\frac{\hat{\eta}_A}{\hat{E}_T}\sigma_A - \frac{\hat{\eta}_T}{\hat{E}_T}\sigma_T - \frac{\hat{\delta}_A}{\hat{E}_A}M_A + \frac{M_T}{\hat{E}_T} - \frac{\hat{\eta}_a}{\hat{E}_T}\sigma_t + \hat{\alpha}_T\Delta T, \tag{7.198}$$

$$\varepsilon_t = -\frac{\nu_a}{E_A}\sigma_A - \frac{\nu_t}{E_T}\sigma_T - \frac{\hat{\nu}_a}{\hat{E}_A}M_A - \frac{\hat{\eta}_a}{\hat{E}_T}M_T + \frac{\sigma_t}{E_t} + \alpha_t\Delta T. \tag{7.199}$$

This subset of stress-strain relations arises when shear loading is absent so that $\tau_A = \tau_a = \tau_t = 0$ and $M_S = 0$.

To conclude this section concerning nonsymmetric laminates where bending modes of deformation arise, it is useful to consider the special case of *symmetric* laminates as bending is then absent and results derived in this section should reduce to those already given earlier in this chapter. For symmetric laminates, it is found, as expected, that $\hat{\varepsilon}_A = \hat{\varepsilon}_T = \hat{\gamma}_A = 0$. It then follows that relations (7.183)–(7.186) reduce to the simpler relations (7.23) provided use is made of relations (7.34)–(7.38).

7.10 A More Limited Explicit Formulation

In preparation for an analysis of damaged nonsymmetric cross-ply laminates in Chapters 11 and 19, it is now assumed that both in-plane and out-of-plane shear deformation are absent. It then follows that $\overline{\gamma}_A = \hat{\gamma}_A = 0$ and that, for values $i = 1, ..., n$, $\overline{\gamma}_a^{(i)} = \hat{\gamma}_a^{(i)} = 0$. The displacement distribution (7.96) then reduces to the simpler form for values $x_3^{(i-1)} \le x_3 < x_3^{(i)}, i = 1, \ldots, n,$

$$\begin{aligned} u_1^{(i)}(x_1,x_2,x_3) &= (\overline{\varepsilon}_A + \hat{\varepsilon}_A x_3)x_1, \\ u_2^{(i)}(x_1,x_2,x_3) &= (\overline{\varepsilon}_T + \hat{\varepsilon}_T x_3)x_2, \\ u_3^{(i)}(x_1,x_2,x_3) &= -\tfrac{1}{2}\hat{\varepsilon}_A x_1^2 - \tfrac{1}{2}\hat{\varepsilon}_T x_2^2 + \overline{\varepsilon}_t^{(i)}(x_3 - x_3^{(i)}) + \tfrac{1}{2}\hat{\varepsilon}_t^{(i)}\left(x_3^2 - (x_3^{(i)})^2\right) \\ &\quad + \sum_{j=1}^{i} h_j \overline{\varepsilon}_t^{(j)} + \sum_{j=1}^{i} \hat{h}_j \hat{\varepsilon}_t^{(j)}, \end{aligned} \tag{7.200}$$

where, from (7.143), $h_j = x_3^{(j)} - x_3^{(j-1)}$, $\hat{h}_j = \frac{1}{2}\left((x_3^{(j)})^2 - (x_3^{(j-1)})^2\right)$ for $j = 1, \ldots, N+1$.

From (7.106), the corresponding nonzero strains are

$$\varepsilon_A^{(i)} = \overline{\varepsilon}_A + \hat{\varepsilon}_A x_3, \quad \varepsilon_T^{(i)} = \overline{\varepsilon}_T + \hat{\varepsilon}_T x_3, \quad \varepsilon_t^{(i)} = \overline{\varepsilon}_t^{(i)} + \hat{\varepsilon}_t^{(i)} x_3. \tag{7.201}$$

7.10.1 The Stress Field

Owing to theses simplifications, rather than using the stress-strain relations (7.107) for each ply in the laminate, a more explicit approach is to use the following inverted form, where it is assumed that the through-thickness stress of each ply in the laminate has the same value so that $\sigma_{33}^{(i)} = \sigma_t$, $i = 1, \ldots, n$,

$$\varepsilon_A^{(i)} = \frac{\sigma_A^{(i)}}{E_A^{(i)}} - \frac{\nu_A^{(i)}}{E_A^{(i)}}\sigma_T^{(i)} - \frac{\nu_a^{(i)}}{E_A^{(i)}}\sigma_t + \alpha_A^{(i)}\Delta T, \tag{7.202}$$

$$\varepsilon_T^{(i)} = -\frac{\nu_A^{(i)}}{E_A^{(i)}}\sigma_A^{(i)} + \frac{\sigma_T^{(i)}}{E_T^{(i)}} - \frac{\nu_t^{(i)}}{E_T^{(i)}}\sigma_t + \alpha_T^{(i)}\Delta T, \tag{7.203}$$

$$\varepsilon_t^{(i)} = -\frac{\nu_a^{(i)}}{E_A^{(i)}}\sigma_A^{(i)} - \frac{\nu_t^{(i)}}{E_T^{(i)}}\sigma_T^{(i)} + \frac{\sigma_t}{E_t^{(i)}} + \alpha_t^{(i)}\Delta T. \tag{7.204}$$

On solving (7.203) for $\sigma_T^{(i)}$ and on using $(7.201)_2$, it can be shown that

$$\sigma_T^{(i)} = \nu_t^{(i)}\sigma_t + \nu_A^{(i)}\frac{E_T^{(i)}}{E_A^{(i)}}\sigma_A^{(i)} - E_T^{(i)}\alpha_T^{(i)}\Delta T + E_T^{(i)}(\overline{\varepsilon}_T + \hat{\varepsilon}_T x_3). \tag{7.205}$$

On substituting (7.205) into (7.202) using $(7.201)_1$, it can be shown that for $x_3^{(i-1)} \le x_3 \le x_3^{(i)}, i = 1, \ldots, n,$

$$\sigma_A^{(i)}(x_3) = \tilde{E}_A^{(i)}(\overline{\varepsilon}_A + \hat{\varepsilon}_A x_3) + \nu_A^{(i)}\tilde{E}_T^{(i)}(\overline{\varepsilon}_T + \hat{\varepsilon}_T x_3) + \tilde{\nu}_a^{(i)}\sigma_t - \tilde{E}_A^{(i)}\tilde{\alpha}_A^{(i)}\Delta T, \tag{7.206}$$

where

$$\begin{aligned} &\frac{1}{\tilde{E}_A^{(i)}} = \frac{1}{E_A^{(i)}}\left(1 - (\nu_A^{(i)})^2\frac{E_T^{(i)}}{E_A^{(i)}}\right), \qquad \frac{1}{\tilde{E}_T^{(i)}} = \frac{1}{E_T^{(i)}}\left(1 - (\nu_A^{(i)})^2\frac{E_T^{(i)}}{E_A^{(i)}}\right), \\ &\frac{\tilde{\nu}_a^{(i)}}{\tilde{E}_A^{(i)}} = \frac{\nu_a^{(i)} + \nu_t^{(i)}\nu_A^{(i)}}{E_A^{(i)}}, \qquad \tilde{\alpha}_A^{(i)} = \alpha_A^{(i)} + \nu_A^{(i)}\frac{E_T^{(i)}}{E_A^{(i)}}\alpha_T^{(i)}. \end{aligned} \tag{7.207}$$

On substituting (7.205) into (7.204),

$$\varepsilon_{\rm t}^{(i)} = -\frac{\tilde{\nu}_{\rm a}^{(i)}}{\tilde{E}_{\rm A}^{(i)}}\sigma_{\rm A}^{(i)} + \frac{1}{\tilde{E}_{\rm t}^{(i)}}\sigma_{\rm t} + \tilde{\alpha}_{\rm t}^{(i)}\Delta T - \nu_{\rm t}^{(i)}(\overline{\varepsilon}_{\rm T} + \hat{\varepsilon}_{\rm T}x_3), \tag{7.208}$$

where

$$\frac{1}{\tilde{E}_{\rm t}^{(i)}} = \frac{1}{E_{\rm t}^{(i)}} - \frac{(\nu_{\rm t}^{(i)})^2}{E_{\rm T}^{(i)}}, \qquad \tilde{\alpha}_{\rm t}^{(i)} = \alpha_{\rm t}^{(i)} + \nu_{\rm t}^{(i)}\alpha_{\rm T}^{(i)}. \tag{7.209}$$

On substituting (7.206) into (7.205), it can be shown that

$$\sigma_{\rm T}^{(i)}(x_3) = \tilde{\nu}_{\rm t}^{(i)}\sigma_{\rm t} + \nu_{\rm A}^{(i)}\tilde{E}_{\rm T}^{(i)}(\overline{\varepsilon}_{\rm A} + \hat{\varepsilon}_{\rm A}x_3) + \tilde{E}_{\rm T}^{(i)}(\overline{\varepsilon}_{\rm T} + \hat{\varepsilon}_{\rm T}x_3) - \tilde{E}_{\rm T}^{(i)}\tilde{\alpha}_{\rm T}^{(i)}\Delta T, \tag{7.210}$$

where

$$\frac{\tilde{\nu}_{\rm t}^{(i)}}{\tilde{E}_{\rm T}^{(i)}} = \frac{\nu_{\rm t}^{(i)}}{E_{\rm T}^{(i)}} + \frac{\nu_{\rm a}^{(i)}\nu_{\rm A}^{(i)}}{E_{\rm A}^{(i)}}, \qquad \tilde{\alpha}_{\rm T}^{(i)} = \alpha_{\rm T}^{(i)} + \nu_{\rm A}^{(i)}\alpha_{\rm A}^{(i)}. \tag{7.211}$$

It is assumed that stresses within the laminate can arise from any combination of mismatched thermal expansion coefficients, an applied in-plane loading that is equivalent to an applied axial force $F_{\rm A}$ and transverse force $F_{\rm T}$ acting at the mid-plane between the upper and lower surfaces of the laminate, and corresponding applied bending moments per unit area of laminate $M_{\rm A}$ and $M_{\rm T}$, respectively. In addition, the transverse strain parameters $\overline{\varepsilon}_{\rm T}$ and $\hat{\varepsilon}_{\rm T}$ are assumed for the moment to be given. From considerations of mechanical equilibrium,

$$F_{\rm A} = h\sigma_{\rm A} = \sum_{i=1}^{n} \int_{x_3^{(i-1)}}^{x_3^{(i)}} \sigma_{\rm A}^{(i)}(x_3)\,{\rm d}x_3, \tag{7.212}$$

$$F_{\rm T} = h\sigma_{\rm T} = \sum_{i=1}^{n} \int_{x_3^{(i-1)}}^{x_3^{(i)}} \sigma_{\rm T}^{(i)}(x_3)\,{\rm d}x_3, \tag{7.213}$$

$$M_{\rm A} = \frac{1}{h}\sum_{i=1}^{n} \int_{x_3^{(i-1)}}^{x_3^{(i)}} \left(x_3 - \frac{h}{2}\right)\sigma_{\rm A}^{(i)}(x_3)\,{\rm d}x_3, \tag{7.214}$$

$$M_{\rm T} = \frac{1}{h}\sum_{i=1}^{n} \int_{x_3^{(i-1)}}^{x_3^{(i)}} \left(x_3 - \frac{h}{2}\right)\sigma_{\rm T}^{(i)}(x_3)\,{\rm d}x_3. \tag{7.215}$$

On substituting (7.206) and (7.210) into (7.212)–(7.215), the following relations are obtained

$$\sigma_{\rm A} = \Omega_{11}^{(\rm L)}\overline{\varepsilon}_{\rm A} + \Omega_{12}^{(\rm L)}\overline{\varepsilon}_{\rm T} + \Omega_{13}^{(\rm L)}\sigma_{\rm t} + \hat{\Omega}_{11}^{(\rm L)}\hat{\varepsilon}_{\rm A} + \hat{\Omega}_{12}^{(\rm L)}\hat{\varepsilon}_{\rm T} - \omega_1^{(\rm L)}\Delta T, \tag{7.216}$$

$$\sigma_{\rm T} = \Omega_{12}^{(\rm L)}\overline{\varepsilon}_{\rm A} + \Omega_{22}^{(\rm L)}\overline{\varepsilon}_{\rm T} + \Omega_{23}^{(\rm L)}\sigma_{\rm t} + \hat{\Omega}_{12}^{(\rm L)}\hat{\varepsilon}_{\rm A} + \hat{\Omega}_{22}^{(\rm L)}\hat{\varepsilon}_{\rm T} - \omega_2^{(\rm L)}\Delta T, \tag{7.217}$$

$$M_{\rm A} + \tfrac{1}{2}h\sigma_{\rm A} = \hat{\Omega}_{11}^{(\rm L)}\overline{\varepsilon}_{\rm A} + \hat{\Omega}_{12}^{(\rm L)}\overline{\varepsilon}_{\rm T} + \hat{\Omega}_{13}^{(\rm L)}\sigma_{\rm t} + \tilde{\Omega}_{11}^{(\rm L)}\hat{\varepsilon}_{\rm A} + \tilde{\Omega}_{12}^{(\rm L)}\hat{\varepsilon}_{\rm T} - \hat{\omega}_1^{(\rm L)}\Delta T, \tag{7.218}$$

$$M_{\rm T} + \tfrac{1}{2}h\sigma_{\rm T} = \hat{\Omega}_{12}^{(\rm L)}\overline{\varepsilon}_{\rm A} + \hat{\Omega}_{22}^{(\rm L)}\overline{\varepsilon}_{\rm T} + \hat{\Omega}_{23}^{(\rm L)}\sigma_{\rm t} + \tilde{\Omega}_{12}^{(\rm L)}\hat{\varepsilon}_{\rm A} + \tilde{\Omega}_{22}^{(\rm L)}\hat{\varepsilon}_{\rm T} - \hat{\omega}_2^{(\rm L)}\Delta T, \tag{7.219}$$

where

$$\Omega_{11}^{(\mathrm{L})}=\frac{1}{h}\sum_{i=1}^{n}h_i\tilde{E}_{\mathrm{A}}^{(i)},\qquad \Omega_{22}^{(\mathrm{L})}=\frac{1}{h}\sum_{i=1}^{n}h_i\tilde{E}_{\mathrm{T}}^{(i)},\qquad \Omega_{12}^{(\mathrm{L})}=\frac{1}{h}\sum_{i=1}^{n}h_i\nu_{\mathrm{A}}^{(i)}\tilde{E}_{\mathrm{T}}^{(i)},$$
$$\hat{\Omega}_{11}^{(\mathrm{L})}=\frac{1}{h}\sum_{i=1}^{n}\hat{h}_i\tilde{E}_{\mathrm{A}}^{(i)},\qquad \hat{\Omega}_{22}^{(\mathrm{L})}=\frac{1}{h}\sum_{i=1}^{n}\hat{h}_i\tilde{E}_{\mathrm{T}}^{(i)},\qquad \hat{\Omega}_{12}^{(\mathrm{L})}=\frac{1}{h}\sum_{i=1}^{n}\hat{h}_i\nu_{\mathrm{A}}^{(i)}\tilde{E}_{\mathrm{T}}^{(i)},\tag{7.220}$$
$$\tilde{\Omega}_{11}^{(\mathrm{L})}=\frac{1}{h}\sum_{i=1}^{n}\tilde{h}_i\tilde{E}_{\mathrm{A}}^{(i)},\qquad \tilde{\Omega}_{22}^{(\mathrm{L})}=\frac{1}{h}\sum_{i=1}^{n}\tilde{h}_i\tilde{E}_{\mathrm{T}}^{(i)},\qquad \tilde{\Omega}_{12}^{(\mathrm{L})}=\frac{1}{h}\sum_{i=1}^{n}\tilde{h}_i\nu_{\mathrm{A}}^{(i)}\tilde{E}_{\mathrm{T}}^{(i)},$$

$$\Omega_{13}^{(\mathrm{L})}=\frac{1}{h}\sum_{i=1}^{n}h_i\,\tilde{\nu}_{\mathrm{a}}^{(i)},\qquad \Omega_{23}^{(\mathrm{L})}=\frac{1}{h}\sum_{i=1}^{n}h_i\,\tilde{\nu}_{\mathrm{t}}^{(i)},$$
$$\hat{\Omega}_{13}^{(\mathrm{L})}=\frac{1}{h}\sum_{i=1}^{n}\hat{h}_i\,\tilde{\nu}_{\mathrm{a}}^{(i)},\qquad \hat{\Omega}_{23}^{(\mathrm{L})}=\frac{1}{h}\sum_{i=1}^{n}\hat{h}_i\,\tilde{\nu}_{\mathrm{t}}^{(i)},\tag{7.221}$$

$$\omega_1^{(\mathrm{L})}=\frac{1}{h}\sum_{i=1}^{n}h_i\tilde{E}_{\mathrm{A}}^{(i)}\tilde{\alpha}_{\mathrm{A}}^{(i)},\qquad \omega_2^{(\mathrm{L})}=\frac{1}{h}\sum_{i=1}^{n}h_i\tilde{E}_{\mathrm{T}}^{(i)}\tilde{\alpha}_{\mathrm{T}}^{(i)},$$
$$\hat{\omega}_1^{(\mathrm{L})}=\frac{1}{h}\sum_{i=1}^{n}\hat{h}_i\tilde{E}_{\mathrm{A}}^{(i)}\tilde{\alpha}_{\mathrm{A}}^{(i)},\qquad \hat{\omega}_2^{(\mathrm{L})}=\frac{1}{h}\sum_{i=1}^{n}\hat{h}_i\tilde{E}_{\mathrm{T}}^{(i)}\tilde{\alpha}_{\mathrm{T}}^{(i)},\tag{7.222}$$

and where corresponding to (7.143)

$$h_i=x_3^{(i)}-x_3^{(i-1)},\quad \hat{h}_i=\tfrac{1}{2}\left((x_3^{(i)})^2-(x_3^{(i-1)})^2\right),\quad \tilde{h}_i=\tfrac{1}{3}\left((x_3^{(i)})^3-(x_3^{(i-1)})^3\right).\tag{7.223}$$

For an undamaged laminate with a specified uniform temperature difference ΔT, subject to a uniform through-thickness stress σ_{t} and applied strains $\bar{\varepsilon}_{\mathrm{A}}$, $\hat{\varepsilon}_{\mathrm{T}}$, $\bar{\varepsilon}_{\mathrm{T}}$ and $\hat{\varepsilon}_{\mathrm{T}}$, the resulting stresses σ_{A}, σ_{T} and bending moments per unit area M_{A}, M_{T} are given by relations (7.216)–(7.219). Substitution of the strains into (7.206) and (7.210) enables the longitudinal and transverse stress distributions in each ply to be found.

7.10.2 Calculation of Effective Through-thickness Strain

From (7.200) for values $x_3^{(i-1)}\le x_3<x_3^{(i)}$, $i=1,\ldots,n$, the through-thickness displacement is given by

$$\begin{aligned}u_3^{(i)}(x_1,x_2,x_3)=&-\tfrac{1}{2}\hat{\varepsilon}_{\mathrm{A}}x_1^2-\tfrac{1}{2}\hat{\varepsilon}_{\mathrm{T}}x_2^2+\bar{\varepsilon}_{\mathrm{t}}^{(i)}(x_3-x_3^{(i)})+\tfrac{1}{2}\hat{\varepsilon}_{\mathrm{t}}^{(i)}\left(x_3^2-(x_3^{(i)})^2\right)\\&+\sum_{j=1}^{i}h_j\bar{\varepsilon}_{\mathrm{t}}^{(j)}+\sum_{j=1}^{i}\hat{h}_j\hat{\varepsilon}_{\mathrm{t}}^{(j)},\end{aligned}\tag{7.224}$$

and, on differentiating with respect to x_3, it is clear the through-thickness strain in the ith ply is

$$\varepsilon_{\mathrm{t}}^{(i)}\equiv\frac{\partial u_3^{(i)}}{\partial x_3}=\bar{\varepsilon}_{\mathrm{t}}^{(i)}+\hat{\varepsilon}_{\mathrm{t}}^{(i)}x_3.\tag{7.225}$$

The values of the parameters $\bar{\varepsilon}_{\mathrm{t}}^{(i)}$, $\hat{\varepsilon}_{\mathrm{t}}^{(i)}$, $i=1,\ldots,n$, have yet to be determined. On using (7.204), (7.206) and (7.210), it can be shown that

$$\begin{aligned}\varepsilon_{\mathrm{t}}^{(i)}=&-\tilde{\nu}_{\mathrm{a}}^{(i)}(\bar{\varepsilon}_{\mathrm{A}}+\hat{\varepsilon}_{\mathrm{A}}x_3)-\tilde{\nu}_{\mathrm{t}}^{(i)}(\bar{\varepsilon}_{\mathrm{T}}+\hat{\varepsilon}_{\mathrm{T}}x_3)+\left(\frac{1}{\tilde{E}_{\mathrm{t}}^{(i)}}-\frac{(\tilde{\nu}_{\mathrm{a}}^{(i)})^2}{\tilde{E}_{\mathrm{A}}^{(i)}}\right)\sigma_{\mathrm{t}}\\&+\left(\tilde{\alpha}_{\mathrm{t}}^{(i)}+\tilde{\nu}_{\mathrm{a}}^{(i)}\tilde{\alpha}_{\mathrm{A}}^{(i)}\right)\Delta T,\end{aligned}\tag{7.226}$$

as from (7.207), (7.209) and (7.211)

$$\begin{aligned}
&\frac{\tilde{\nu}_{\rm a}^{(i)}}{\tilde{E}_{\rm A}^{(i)}}=\frac{\nu_{\rm a}^{(i)}+\nu_{\rm t}^{(i)}\nu_{\rm A}^{(i)}}{E_{\rm A}^{(i)}}, \quad \frac{\tilde{\nu}_{\rm t}^{(i)}}{\tilde{E}_{\rm T}^{(i)}}=\frac{\nu_{\rm t}^{(i)}}{E_{\rm T}^{(i)}}+\frac{\nu_{\rm a}^{(i)}\nu_{\rm A}^{(i)}}{E_{\rm A}^{(i)}},\\
&\frac{1}{E_{\rm t}^{(i)}}-\frac{\nu_{\rm a}^{(i)}}{E_{\rm A}^{(i)}}\tilde{\nu}_{\rm a}^{(i)}-\frac{\nu_{\rm t}^{(i)}}{E_{\rm T}^{(i)}}\tilde{\nu}_{\rm t}^{(i)}=\frac{1}{\tilde{E}_{\rm t}^{(i)}}-\frac{(\tilde{\nu}_{\rm a}^{(i)})^2}{\tilde{E}_{\rm A}^{(i)}},\\
&\alpha_{\rm t}^{(i)}+\frac{\nu_{\rm a}^{(i)}}{E_{\rm A}^{(i)}}\tilde{E}_{\rm A}^{(i)}\tilde{\alpha}_{\rm A}^{(i)}+\frac{\nu_{\rm t}^{(i)}}{E_{\rm T}^{(i)}}\tilde{E}_{\rm T}^{(i)}\tilde{\alpha}_{\rm T}^{(i)}=\tilde{\alpha}_{\rm t}^{(i)}+\tilde{\nu}_{\rm a}^{(i)}\tilde{\alpha}_{\rm A}^{(i)}.
\end{aligned} \tag{7.227}$$

By comparing relations (7.225) and (7.226), it is deduced that for values $i=1,\ldots,n$,

$$\bar{\varepsilon}_{\rm t}^{(i)}=-\tilde{\nu}_{\rm a}^{(i)}\bar{\varepsilon}_{\rm A}-\tilde{\nu}_{\rm t}^{(i)}\bar{\varepsilon}_{\rm T}+\left(\frac{1}{\tilde{E}_{\rm t}^{(i)}}-\frac{(\tilde{\nu}_{\rm a}^{(i)})^2}{\tilde{E}_{\rm A}^{(i)}}\right)\sigma_{\rm t}+\left(\tilde{\alpha}_{\rm t}^{(i)}+\tilde{\nu}_{\rm a}^{(i)}\tilde{\alpha}_{\rm A}^{(i)}\right)\Delta T, \tag{7.228}$$

$$\hat{\varepsilon}_{\rm t}^{(i)}=-\tilde{\nu}_{\rm a}^{(i)}\hat{\varepsilon}_{\rm A}-\tilde{\nu}_{\rm t}^{(i)}\hat{\varepsilon}_{\rm T}. \tag{7.229}$$

The top and bottom faces of the laminate are subject to a uniform applied stress $\sigma_{\rm t}$. The corresponding effective through-thickness strain $\varepsilon_{\rm t}$ for a laminate is defined by

$$\varepsilon_{\rm t}=\frac{1}{4hLW}\int_{-W}^{W}\int_{-L}^{L}\left[u_3(x_1,x_2,h)-u_3(x_1,x_2,0)\right]{\rm d}x_1\,{\rm d}x_2. \tag{7.230}$$

It follows from (7.224) that

$$\begin{aligned}
&u_3\left(x_1,x_2,0\right)=u_3^{(1)}\left(x_1,x_2,0\right)=-\frac{1}{2}\hat{\varepsilon}_{\rm A}x_1^2-\frac{1}{2}\hat{\varepsilon}_{\rm T}x_2^2,\\
&u_3\left(x_1,x_2,h\right)=u_3^{(n)}\left(x_1,x_2,h\right)=-\frac{1}{2}\hat{\varepsilon}_{\rm A}x_1^2-\frac{1}{2}\hat{\varepsilon}_{\rm T}x_2^2+\sum_{j=1}^{n}h_j\bar{\varepsilon}_{\rm t}^{(j)}+\sum_{j=1}^{n}\hat{h}_j\hat{\varepsilon}_{\rm t}^{(j)},
\end{aligned} \tag{7.231}$$

so that

$$\varepsilon_{\rm t}=\frac{1}{h}\sum_{j=1}^{n}h_j\bar{\varepsilon}_{\rm t}^{(j)}+\frac{1}{h}\sum_{j=1}^{n}\hat{h}_j\hat{\varepsilon}_{\rm t}^{(j)}. \tag{7.232}$$

On substituting (7.228) and (7.229) into (7.232), it can be shown that the effective through-thickness strain is given by

$$\varepsilon_{\rm t}=-\Omega_{13}^{({\rm L})}\bar{\varepsilon}_{\rm A}-\Omega_{23}^{({\rm L})}\bar{\varepsilon}_{\rm T}+\Omega_{33}^{({\rm L})}\sigma_{\rm t}-\hat{\Omega}_{13}^{({\rm L})}\hat{\varepsilon}_{\rm A}-\hat{\Omega}_{23}^{({\rm L})}\hat{\varepsilon}_{\rm T}+\omega_3^{({\rm L})}\Delta T, \tag{7.233}$$

where

$$\begin{aligned}
&\Omega_{13}^{({\rm L})}=\frac{1}{h}\sum_{i=1}^{n}h_i\tilde{\nu}_{\rm a}^{(i)}, \qquad \Omega_{23}^{({\rm L})}=\frac{1}{h}\sum_{i=1}^{n}h_i\tilde{\nu}_{\rm t}^{(i)},\\
&\hat{\Omega}_{13}^{({\rm L})}=\frac{1}{h}\sum_{i=1}^{n}\hat{h}_i\tilde{\nu}_{\rm a}^{(i)}, \qquad \hat{\Omega}_{23}^{({\rm L})}=\frac{1}{h}\sum_{i=1}^{n}\hat{h}_i\tilde{\nu}_{\rm t}^{(i)},
\end{aligned} \tag{7.234}$$

and

$$\begin{aligned}
&\Omega_{33}^{({\rm L})}=\frac{1}{h}\sum_{i=1}^{n}h_i\left(\frac{1}{E_{\rm t}^{(i)}}-\frac{\nu_{\rm a}^{(i)}}{E_{\rm A}^{(i)}}\tilde{\nu}_{\rm a}^{(i)}-\frac{\nu_{\rm t}^{(i)}}{E_{\rm T}^{(i)}}\tilde{\nu}_{\rm t}^{(i)}\right)=\frac{1}{h}\sum_{i=1}^{n}h_i\left(\frac{1}{\tilde{E}_{\rm t}^{(i)}}-\frac{(\tilde{\nu}_{\rm a}^{(i)})^2}{\tilde{E}_{\rm A}^{(i)}}\right),\\
&\omega_3^{({\rm L})}=\frac{1}{h}\sum_{i=1}^{n}h_i\left(\alpha_{\rm t}^{(i)}+\frac{\nu_{\rm a}^{(i)}}{E_{\rm A}^{(i)}}\tilde{E}_{\rm A}^{(i)}\tilde{\alpha}_{\rm A}^{(i)}+\frac{\nu_{\rm t}^{(i)}}{E_{\rm T}^{(i)}}\tilde{E}_{\rm T}^{(i)}\tilde{\alpha}_{\rm T}^{(i)}\right)=\frac{1}{h}\sum_{i=1}^{n}h_i\left(\tilde{\alpha}_{\rm t}^{(i)}+\tilde{\nu}_{\rm a}^{(i)}\tilde{\alpha}_{\rm A}^{(i)}\right).
\end{aligned} \tag{7.235}$$

It is noted that relations (7.221) are consistent with the results (7.234).

Explicit expressions for many of the matrices $C_{IJ}^{(L)}$, $\widehat{C}_{IJ}^{(L)}$ and $\tilde{C}_{IJ}^{(L)}$ and vectors $U_I^{(L)}$ and $\hat{U}_I^{(L)}$ appearing in relations (7.183)–(7.189) may then be obtained using some of relations (7.34)–(7.38) and (7.191)–(7.194).

References

1. Halpin, J.C. *Primer on Composite Analysis*, 2nd e. Lancaster, Penn., Basel: Technomic Publishing Co. Ltd., 1992.
2. Wiley web page (www.wiley.com/go/mccartney/properties).

8

Energy Balance Approach to Fracture in Anisotropic Elastic Material

Overview: This chapter describes the application of the well-known principles of nonequilibrium continuum thermodynamics to the prediction of fracture for isothermal conditions and infinitesimal deformations. By making use of the fact that kinetic energy is always positive, an inequality is derived for predicting the initiation of crack growth for isothermal conditions. This relationship, when used as an equality, is a necessary condition for quasistatic crack growth. A complex variable analysis is presented that enables elegant solutions, based on Chebyshev polynomials, to the special case of isolated cracks embedded in an infinite anisotropic medium.

8.1 Introduction

One objective of this chapter is to show how the classical energy method of Griffith [1], when modified to take account of temperature effects, should be modified to be consistent with the principles of continuum thermodynamics. This involves developing a theory of fracture in elastic monolithic materials for the case of quasistatic crack growth under isothermal conditions. It is achieved by making combined use of the well-established principles of nonequilibrium continuum mechanics *and* continuum thermodynamics [2–5], that will be shown to provide a rational framework for the prediction of fracture in continuous media. Nonequilibrium changes imply that state variables are time dependent contrasting with the classical approach where changes between equilibrium states are considered. The inclusion of time dependencies in the theory enables the rational consideration of fracture initiation and quasistatic crack growth. The approach taken inextricably links continuum mechanics to thermodynamics and leads to a treatment of fracture that provides a vehicle for wider application of the basic principles.

Properties for Design of Composite Structures: Theory and Implementation Using Software, First Edition. Neil McCartney.

Companion Website: www.wiley.com/go/mccartney/properties

8.2 Thermodynamics for Isothermal Deformations

The objective now is to apply some of the important principles of continuum thermodynamics to the fracture of homogeneous continuous media for the special case of isothermal conditions.

8.2.1 Local Energy Balance Equation Based on Helmholtz Energy

For many applications of continuum mechanics, isothermal conditions are assumed and it is useful to utilise an equation of state involving the temperature and strain as the state variables rather than specific entropy and strain. In addition, for infinitesimal deformations, it follows from (2.108) that the density ρ of the deforming medium may be approximated by the initial uniform density ρ_0. From (2.111)–(2.114) it follows that for infinitesimal deformations

$$\mathrm{d}(\rho_0 \upsilon) = T\mathrm{d}(\rho_0 \eta) + \sigma_{ij}\mathrm{d}\varepsilon_{ij}, \tag{8.1}$$

where the operator 'd' denotes the total derivative. The specific Helmholtz energy ψ is now introduced defined in terms of the specific internal energy υ by

$$\psi = \upsilon - \eta T. \tag{8.2}$$

On substituting for υ into (8.1) using (8.2), it follows that

$$\mathrm{d}(\rho_0 \psi) = -(\rho_0 \eta)\mathrm{d}T + \sigma_{ij}\,\mathrm{d}\varepsilon_{ij}, \tag{8.3}$$

which implies that the equation of state (2.111) can be replaced by one of the form

$$\psi = \hat{\psi}(T, \varepsilon_{ij}), \tag{8.4}$$

where the state variable η has been replaced by the temperature T. It follows from (8.3) and (8.4) that

$$\rho_0 \eta = -\frac{\partial(\rho_0 \hat{\psi})}{\partial T}, \tag{8.5}$$

and that

$$\sigma_{ij} = \frac{\partial(\rho_0 \hat{\psi})}{\partial \varepsilon_{ij}}. \tag{8.6}$$

To develop a model relevant for precise isothermal conditions, it is necessary to be able to control to temperature at every point in the system. This can be achieved by *imagining* that the rate of energy supply (or loss) per unit mass r appearing in the local energy balance equation (2.115) can be controlled at every point in the system in such a way that the temperature at every point in the system has a fixed uniform value T. On using (2.108), (8.4) and (8.5), the local energy balance equation (2.115) leads to

$$-T\frac{\partial^2(\rho_0 \hat{\psi})}{\partial T^2}\frac{\partial T}{\partial t} = -\frac{\partial h_i}{\partial x_i} + T\frac{\partial^2(\rho_0 \hat{\psi})}{\partial T \partial \varepsilon_{ij}}\frac{\partial \varepsilon_{ij}}{\partial t} + \rho_0 r. \tag{8.7}$$

As T = constant for isothermal conditions, it follows from (2.118) that $h_i = 0$, and from (8.7) that the rate of energy supply (or loss) would need to be controlled such that

$$\rho_0 r = -T\frac{\partial^2(\rho_0\hat{\psi})}{\partial T\,\partial \varepsilon_{ij}}\frac{\partial \varepsilon_{ij}}{\partial t} = \left.\frac{\partial(\rho_0\eta T)}{\partial t}\right|_T . \tag{8.8}$$

Relation (8.8) indicates how the rate of heat supply (or loss) must be regulated at every internal point of the system in order that isothermal conditions are achieved precisely at all points of the system. In practice, such control cannot be achieved, but if strain rates are small enough and thermal conductivities large enough, then isothermal conditions can be achieved approximately by heat (and entropy) transfer across the external boundary. For crack growth under these conditions, heat enters or leaves the system only through the external boundary and precise isothermal conditions cannot be achieved as nonuniform temperature distributions will occur leading to heat flow and to additional terms in the expression for the rate of entropy production that may be infinitesimally small. For this situation, the local rate of energy supply per unit volume $\rho_0 r = 0$ and relations (2.118) and (8.7) can be used to derive the heat transport equation. For subsequent theoretical developments it is assumed that isothermal conditions are achieved precisely by the imaginary control of the heat supply per unit mass (or loss) r at all internal points of the system.

On substituting (8.8) into (2.115), and on setting $h_i = 0$, it follows that, for isothermal conditions and infinitesimal deformations, the local energy balance equation may be written in the form

$$\frac{\partial(\rho_0\psi)}{\partial t} = \sigma_{ij}\frac{\partial \varepsilon_{ij}}{\partial t}, \tag{8.9}$$

where use has been made of relations (8.1) and (8.2).

8.2.2 Local Energy Balance Equation Based on Gibbs Energy

It is also useful to utilise an equation of state that involves temperature and stress as state variables. This is achieved by introducing the specific Gibbs energy g defined by

$$g = \psi - \frac{1}{\rho_0}\sigma_{ij}\varepsilon_{ij}. \tag{8.10}$$

It follows from (8.3) and (8.10) that

$$d(\rho_0 g) = -(\rho_0\eta)dT - \varepsilon_{ij}\,d\sigma_{ij}, \tag{8.11}$$

which implies that the equation of state (8.4) can be replaced by one of the form

$$g = \hat{g}(T, \sigma_{ij}), \tag{8.12}$$

where the state variable ε_{ij} has been replaced by the stress tensor σ_{ij}. It follows from (8.11) and (8.12) that

$$\rho_0\eta = -\frac{\partial(\rho_0\hat{g})}{\partial T}, \tag{8.13}$$

and that

$$\varepsilon_{ij} = -\frac{\partial(\rho_0 \hat{g})}{\partial \sigma_{ij}}. \tag{8.14}$$

From (2.108), (8.12) and (8.13) the local energy balance equation (2.115) leads to

$$-T\frac{\partial^2(\rho_0 \hat{g})}{\partial T^2}\frac{\partial T}{\partial t} = -\frac{\partial h_i}{\partial x_i} + T\frac{\partial^2(\rho_0 \hat{g})}{\partial T\,\partial \sigma_{ij}}\frac{\partial \sigma_{ij}}{\partial t} + \rho_0 r. \tag{8.15}$$

As $T =$ constant for isothermal conditions, it follows from (2.118) that $h_i = 0$, and from (8.15) that the rate of energy supply from nonmechanical sources may be written as

$$\rho_0 r = -T\frac{\partial^2(\rho_0 \hat{g})}{\partial T\,\partial \sigma_{ij}}\frac{\partial \sigma_{ij}}{\partial t} = \left.\frac{\partial(\rho_0 \eta T)}{\partial t}\right|_T. \tag{8.16}$$

Relation (8.16) indicates how the rate of heat supply (or loss) from nonmechanical sources must be regulated at every internal point of the system in order that isothermal conditions are achieved precisely at all points of the system. On substituting (8.16) into (2.115), it follows for isothermal conditions that the local energy balance equation has the form

$$\frac{\partial(\rho_0 g)}{\partial t} = -\varepsilon_{ij}\frac{\partial \sigma_{ij}}{\partial t}, \tag{8.17}$$

where use has been made of (8.2) and (8.10). The result (8.17) can also be derived directly from (8.9) using (8.10).

8.3 Linear Thermoelasticity

It is useful to introduce explicit expressions for the Helmholtz and Gibbs energy densities that correspond to continuous media that deform as linear anisotropic thermoelastic solids. The Helmholtz energy density function is written as

$$\rho_0 \hat{\psi} = \frac{1}{2} C_{ijkl}(\varepsilon_{ij} - \alpha_{ij}\Delta T)(\varepsilon_{kl} - \alpha_{kl}\Delta T) + f_0(\Delta T), \tag{8.18}$$

where $C_{ijkl}\,(= C_{klij})$ and $\alpha_{ij}\,(= \alpha_{ji})$ are the elastic constants and thermal expansion coefficients of an anisotropic solid, respectively, and where $\Delta T = T - T_0$ is the difference between the current thermodynamic temperature T and a reference temperature T_0 for which the stress and strain components are zero. The thermoelastic constants C_{ijkl} and α_{ij} are assumed here to be independent of the temperature. It follows from (8.6) and (8.18) that the corresponding linear stress-strain relations have the form

$$\sigma_{ij} = C_{ijkl}(\varepsilon_{kl} - \alpha_{kl}\Delta T), \tag{8.19}$$

which may be inverted into the form

$$\varepsilon_{ij} = S_{ijkl}\sigma_{kl} + \alpha_{ij}\Delta T, \tag{8.20}$$

where

$$S_{ijkl} = S_{klij}\,, \quad C_{ijkl}\,S_{klmn} = \delta_{im}\,\delta_{jn}, \tag{8.21}$$

and where δ_{ij} is the Kronecker delta symbol. It follows from (8.18) and (8.19) that $f_0(\Delta T)$ is the Helmholtz energy density at a temperature ΔT when the solid is stress-free. It follows from (8.10), (8.18), (8.20) and (8.21) that the corresponding expression for the Gibbs energy density is

$$\rho_0 g = -\frac{1}{2}\,S_{ijkl}\sigma_{ij}\sigma_{kl} - \sigma_{ij}\alpha_{ij}\,\Delta T + f_0(\Delta T), \tag{8.22}$$

which leads to the result (8.20) when using (8.14).

It is worth noting from (8.5)–(8.7) and (8.18) that the local rate of energy supply per unit volume from nonmechanical sources, needed to ensure precise isothermal conditions, may be expressed in the form

$$\rho_0 r = \rho_0 C_\varepsilon \frac{\partial T}{\partial t} + C_{ijkl}\alpha_{kl} T \frac{\partial \varepsilon_{ij}}{\partial t} + \frac{\partial h_i}{\partial x_i}, \quad C_\varepsilon = T \left.\frac{\partial \eta}{\partial T}\right|_\varepsilon, \tag{8.23}$$

where C_ε is the specific heat at constant strain (volume). From (8.13)–(8.15) and (8.22) it follows that the local rate of energy supply per unit volume, needed to ensure isothermal conditions, may also be expressed in the form

$$\rho_0 r = \rho_0\,C_\sigma \frac{\partial T}{\partial t} + \alpha_{ij}\,T \frac{\partial \sigma_{ij}}{\partial t} + \frac{\partial h_i}{\partial x_i}, \quad C_\sigma = T \left.\frac{\partial \eta}{\partial T}\right|_\sigma, \tag{8.24}$$

where C_σ is the specific heat at constant stress (pressure).

8.4 Global Energy Balance Equations

The global energy balance equation corresponding to the localised form (8.9) is obtained by integrating over the fixed region V bounded by the fixed closed surface S so that

$$\frac{\mathrm{d}}{\mathrm{d}t}\int_V \rho_0 \psi \,\mathrm{d}V = \int_V \sigma_{ij} \frac{\partial \varepsilon_{ij}}{\partial t}\,\mathrm{d}V. \tag{8.25}$$

On making use of (2.107), the symmetry of the stress tensor, the equations of motion (2.109) and the Gauss theorem, it can be shown that the global energy balance equation (8.25) for isothermal conditions may be written in the following form

$$\frac{\mathrm{d}}{\mathrm{d}t}\int_V \rho_0 \psi \,\mathrm{d}V + \frac{\mathrm{d}}{\mathrm{d}t}\int_V \frac{1}{2}\rho_0 v_j v_j \,\mathrm{d}V = \int_S \tau_j \frac{\partial u_j}{\partial t}\,\mathrm{d}S, \tag{8.26}$$

where $\tau_j = n_i\sigma_{ij}$ is the traction vector applied to the external surface S; n_i denoting the unit outward normal. The second term on the left-hand side of (8.26) is the rate of change of kinetic energy in the system. It is clear from (8.26) that if the external surface S of the system is subject to fixed displacements that may vary over the surface S, then the sum of the Helmholtz energy and kinetic energy for the system is always conserved (i.e. is always constant) for isothermal conditions. It follows that for quasistatic conditions, such that the acceleration vector and kinetic energy are negligible, the energy balance equation (8.26) reduces to the form

$$\frac{d}{dt}\int_V \rho_0\psi \, dV = \int_S \tau_j \frac{\partial u_j}{\partial t} dS, \tag{8.27}$$

whereas the equations of motion (2.109) are replaced by the following 'equilibrium equations'

$$\frac{\partial \sigma_{ij}}{\partial x_i} = 0. \tag{8.28}$$

The global energy balance equation corresponding to the localised form (8.17) is obtained by integrating over the region V bounded by the closed surface S so that

$$\frac{d}{dt}\int_V \rho_0 g \, dV = -\int_V \varepsilon_{ij} \frac{\partial \sigma_{ij}}{\partial t} dV. \tag{8.29}$$

On making use of (2.107), the symmetry of the stress tensor, the equations of motion (2.109) and the Gauss theorem it can be shown that the energy balance equation for isothermal conditions may be written in the following form

$$\frac{d}{dt}\int_V \rho_0 g \, dV = -\int_S u_j \frac{\partial \tau_j}{\partial t} dS + \int_V \rho_0 u_j \frac{\partial^2 v_j}{\partial t^2} dV,$$

which for quasistatic conditions reduces to the following more useful form

$$\frac{d}{dt}\int_V \rho_0 g \, dV = -\int_S u_j \frac{\partial \tau_j}{\partial t} dS. \tag{8.30}$$

It is clear from (8.30) that if the external surface S is subject to fixed applied tractions that may vary over the surface S, then the Gibbs energy for the system is conserved for quasistatic isothermal conditions.

8.5 Energy-based Global Fracture Criteria

The situation is now considered where the homogeneous linear thermoelastic medium that is subject to isothermal conditions for infinitesimal deformations contains a stress-free crack. Such problems lead to $r^{-1/2}$ singularities in both the stress and strain fields. The conditions for the growth of such a crack are now investigated to derive global fracture criteria. It is useful to consider a plate of unit thickness in the x_3-direction containing a semi-infinite crack occupying the region, $x_1 = 0$, $-\infty \leq x_2 \leq c(t)$, where the location of the crack tip $x_2 = c(t)$ is time dependent. Letting 2γ denote the fracture energy, i.e. the energy required to create unit area of stress-free surface from bulk material, the isothermal global energy balance equation (8.26) for a fixed region V and bounding surface S is modified, for the case of crack growth, to the form

$$2\gamma\dot{c} + \frac{d}{dt}\int_V \rho_0\psi \, dV + \frac{d}{dt}\int_V \frac{1}{2}\rho_0 v_j v_j \, dV = \int_S \tau_j \frac{\partial u_j}{\partial t} dS, \tag{8.31}$$

where the quantity 2γ appears in (8.31) rather than γ (the surface energy) because two surfaces are extending as the crack grows in length. When writing the energy balance equation in the form (8.31), it is implicitly assumed that the crack surfaces are everywhere stress-free, so that any cohesive force distributions are absent, where their effect on fracture is represented by the fracture energy term. Such an approach is reasonable only if the crack length is much longer than the length of the cohesive zone. Relationship of cohesive zones to the fracture energy term is considered later in Section 8.7. It should be noted that for a fixed region V that does not contain any crack tips the following relation is valid

$$\frac{\mathrm{d}}{\mathrm{d}t}\int_V \rho_0\psi \mathrm{d}V = \int_V \frac{\partial(\rho_0\psi)}{\partial t}\mathrm{d}V. \tag{8.32}$$

If, however, the region V contains at least one crack tip, then this relation is invalid for linear elastic materials as the integrand of the right-hand side has a $1/r^2$ singularity at the crack tip(s) leading to an unbounded integral.

Let the surface S_τ denote that portion of the external surface S that is subject to applied traction distributions that are held fixed in time. The remainder of the surface S_u is assumed to subjected to fixed displacement distributions. The energy balance equation (8.31) may be integrated over a finite time interval yielding the result

$$2\gamma\Delta c + \Delta F + \Delta K = \int_{S_\tau} \tau_j\,\Delta u_j\,\mathrm{d}S, \tag{8.33}$$

where $\Delta F = F - F_0$, $\Delta K = K - K_0$ and $\Delta u_j = u_j - u_j^0$ are respectively the changes in Helmholtz energy, kinetic energy and surface displacement that arise when the crack length changes by an amount $\Delta c = c - c_0$ under the action of fixed applied tractions and/or displacements, where

$$F = \int_V \rho_0\psi\mathrm{d}V, \quad K = \int_V \frac{1}{2}\rho_0 v_j v_j \mathrm{d}V, \tag{8.34}$$

and where the suffix '0' denotes the value of parameters in the initial equilibrium state.

When the initial state of a cracked solid is one of mechanical equilibrium, the velocity vector is zero everywhere in the system with the result that the kinetic energy for the system is also zero. A small departure from mechanical equilibrium, as would occur when crack growth is initiated, is characterised by an increase in the kinetic energy of the system. Consider now energy changes that must occur in a cracked body associated with the transition from a state of mechanical equilibrium to one where crack growth has been initiated.

As $K_0 = 0$ and $K > 0$, then $\Delta K > 0$ and it follows from (8.33) that

$$2\gamma\Delta c + \Delta F < \int_{S_\tau} \tau_j\Delta u_j\,\mathrm{d}S, \tag{8.35}$$

is a condition that must be satisfied for a transition to occur between an equilibrium state and one where crack length changes are occurring (either for values $\Delta c > 0$ or $\Delta c < 0$). The inequality (8.35) is a necessary condition for fracture or crack healing. For quasistatic changes in crack length, relation (8.33) reduces to one of equality, namely

$$2\gamma\Delta c + \Delta F = \int_{S_\tau} \tau_j \Delta u_j \, \mathrm{d}S. \tag{8.36}$$

For quasistatic conditions the Helmholtz energy for the region V depends upon crack length, which is time dependent, and on parameters defining the applied tractions and/or displacements which are assumed to be held fixed. It then follows from (8.35) that

$$\left(2\gamma + \frac{\partial F}{\partial c} - \int_{S_\tau} \tau_j \frac{\partial u_j}{\partial c} \mathrm{d}S\right)\delta c < 0, \tag{8.37}$$

for any infinitesimal change of crack length (positive or negative) subject to fixed applied tractions and/or displacements. Although negative changes of crack length (i.e. crack rewelding) are difficult to achieve in practice because of irreversible processes (including those that are chemical), such changes are allowed here as crack growth in elastic brittle materials can be thought of as occurring as a result of overcoming the interatomic cohesive forces that must be present near the tip of the crack. As such forces are reversible, it follows that crack rewelding is a possibility that should be taken into account. In practice, such rehealing is likely to be possible only when the solid that is cracking is placed in a vacuum. The work that must be done when overcoming these cohesive forces defines the surface energy γ appearing in the energy balance equation (8.31), which regards such cohesive forces as internal. If the cohesive forces were to be regarded as applied externally to the crack faces, then the first term on the left-hand side of (8.31) would be zero and an additional rate of working term would appear on the right-hand side involving the localised traction distributions, acting between the crack surfaces in the region of the crack tip, and the distribution of crack opening (see Section 8.7).

On defining G to be the Gibbs energy for a region V of a cracked body so that

$$G = \int_V \rho_0 \phi \, \mathrm{d}V, \tag{8.38}$$

it follows, from (2.107), (8.10) and (8.34), the 'equilibrium equations' (8.28), the symmetry of the stress tensor and the Gauss divergence theorem, that for the stress and displacement boundary conditions under discussion

$$\delta G = \delta F - \int_{S_\tau} \tau_j \delta u_j \, \mathrm{d}V - \int_{S_u} u_j \delta \tau_j \, \mathrm{d}V, \tag{8.39}$$

where δ denotes an infinitesimal change of state. Thus, relation (8.37) may also be written in the alternative form

$$\left(2\gamma + \frac{\partial G}{\partial c} + \int_{S_u} u_j \frac{\partial \tau_j}{\partial c} \mathrm{d}S\right)\delta c < 0, \tag{8.40}$$

which, if the tractions are held fixed over the entire surface S, reduces to the form

$$\left(2\gamma + \frac{\partial G}{\partial c}\right)\delta c < 0. \tag{8.41}$$

Thus, when tractions are held fixed on the entire external surface, crack growth $\delta c > 0$ can be initiated if and only if

$$2\gamma + \frac{\partial G}{\partial c} < 0, \tag{8.42}$$

and crack healing $\delta c < 0$ can be initiated if and only if

$$2\gamma + \frac{\partial G}{\partial c} > 0. \tag{8.43}$$

For the case

$$2\gamma + \frac{\partial G}{\partial c} = 0, \tag{8.44}$$

the crack is either in a state of mechanical equilibrium or the crack is propagating quasistatically. Relation (8.44) corresponds to the well-known Griffith fracture criterion [1].

8.6 Energy-based Local Fracture Criteria

For quasistatic crack growth, the energy balance equation (8.31) reduces to the form

$$2\gamma\dot{c} + \frac{\mathrm{d}}{\mathrm{d}t}\int_V \rho_0\psi \mathrm{d}V = \int_S \tau_j \frac{\partial u_j}{\partial t}\mathrm{d}S. \tag{8.45}$$

Relation (8.45) may also be applied locally at the crack tip. Select the region V to be a small region Ω containing the crack tip such that the crack tip remains within Ω during a short interval δt, and select the surface S to be the surface Σ enclosing Ω. If the region Ω and the interval δt are sufficiently small, it can be assumed that stress and displacement gradient fields within Ω are asymptotic to the singular parts of these distributions with the result that the stress and displacement fields can be regarded as moving 'rigidly' relative to the crack tip on which the singular stress field is centred. As in Ω

$$\psi(x_1,x_2,x_3,t) \approx \psi(x_1,x_2 - c(t),x_3),\ u_j(x_1,x_2,x_3,t) \approx u_j(x_1,x_2 - c(t),x_3), \tag{8.46}$$

it follows that as Ω shrinks to the crack tip

$$\frac{\partial u_j}{\partial t} = -\dot{c}\frac{\partial u_j}{\partial x_2}, \tag{8.47}$$

and that

$$\frac{\mathrm{d}}{\mathrm{d}t}\int_\Omega \rho_0\psi \mathrm{d}V = -\dot{c}\int_\Sigma n_2\rho_0\psi \mathrm{d}V. \tag{8.48}$$

It should be noted that (8.47) and (8.48) are exact results, rather than approximate, for the special case of self-similar crack growth for any region Ω.

On substituting (8.47) and (8.48) into (8.45) with $V = \Omega$ and $S = \Sigma$, and using $\tau_j = n_i\sigma_{ij}$, it follows that

$$\left[2\gamma - \int_\Sigma \left(n_2\rho_0\psi - n_i\sigma_{ij}\frac{\partial u_j}{\partial x_1}\right)\mathrm{d}S\right]\dot{c} = 0. \tag{8.49}$$

It then follows that nonzero crack velocities are possible if and only if

$$\int_{\Sigma} n_i \left(\rho_0 \psi \delta_{i2} - \sigma_{ij} \frac{\partial u_j}{\partial x_1} \right) \mathrm{d}S = 2\gamma. \tag{8.50}$$

The result (8.50) is clearly an energy-based fracture criterion for quasistatic isothermal conditions that is local to the crack tip region.

For quasistatic conditions of crack growth, the energy balance relation (8.29) is modified to the following form

$$2\gamma \dot{c} + \frac{\mathrm{d}}{\mathrm{d}t} \int_V \rho_0 \phi \, \mathrm{d}V + \int_S u_j \frac{\partial \tau_j}{\partial t} \mathrm{d}S = 0. \tag{8.51}$$

As in Ω

$$\phi(x_1,x_2,x_3,t) \approx \phi(x_1,x_2 - c(t),x_3),\ u_j(x_1,x_2,x_3,t) \approx u_j(x_1,x_2 - c(t),x_3), \tag{8.52}$$

it follows that as Ω shrinks to the crack tip

$$\frac{\mathrm{d}}{\mathrm{d}t} \int_{\Omega} \rho_0 \phi \mathrm{d}V = -\dot{c} \int_{\Sigma} n_2 \rho_0 \phi \mathrm{d}V. \tag{8.53}$$

and because $\tau_j = n_i \sigma_{ij}$, where n_i is time dependent, that

$$\frac{\partial \tau_j}{\partial t} = -\dot{c} n_i \frac{\partial \sigma_{ij}}{\partial x_2}. \tag{8.54}$$

Thus, on substituting (8.53) and (8.54) into (8.51) with $V = \Omega$ and $S = \Sigma$, it follows that

$$\left[2\gamma - \int_{\Sigma} \left(n_2 \rho_0 \phi + n_i u_j \frac{\partial \sigma_{ij}}{\partial x_2} \right) \mathrm{d}S \right] \dot{c} = 0. \tag{8.55}$$

It then follows that nonzero crack velocities are possible if and only if

$$\int_{\Sigma} n_i \left(\rho_0 g \, \delta_{i2} + u_j \frac{\partial \sigma_{ij}}{\partial x_2} \right) \mathrm{d}S = 2\gamma. \tag{8.56}$$

The result (8.56) is an alternative energy-based fracture criterion for quasistatic isothermal conditions that is local to the crack tip region.

8.7 Fracture Involving Cohesive Zones

When modelling cohesive forces explicitly, so that they are regarded as externally applied tractions, rather than introducing the concept of fracture energy, the energy balance equation (8.31) could be expressed in the form

$$\frac{\mathrm{d}}{\mathrm{d}t} \int_V \rho_0 \psi \mathrm{d}V + \frac{\mathrm{d}}{\mathrm{d}t} \int_V \frac{1}{2} \rho_0 v_j v_j \mathrm{d}V = \int_S \tau_j \frac{\partial u_j}{\partial t} \mathrm{d}S$$
$$- \int_{c(t)-r}^{c(t)} \sigma_{11}\big(c(t)-x_2\big) \frac{\partial \Delta u_1\big(c(t)-x_2\big)}{\partial t} \mathrm{d}x_2 - \int_{c(t)-r}^{c(t)} \sigma_{12}\big(c(t)-x_2\big) \frac{\partial \Delta u_2\big(c(t)-x_2\big)}{\partial t} \mathrm{d}x_2, \tag{8.57}$$

where Δu_α, $\alpha = 1, 2$, are the normal displacement discontinuities across the cohesive zone defined by

$$\Delta u_\alpha\left(c(t)-x_2\right) \equiv u_\alpha(x_1^+, x_2 - c(t), x_3) - u_\alpha(x_1^-, x_2 - c(t), x_3),\ \alpha = 1, 2, \\ \text{where}\quad x_1^\pm = \underset{\delta\to 0}{Lt}\left(x_1 \pm \delta\right), \tag{8.58}$$

and where the cohesive forces act over a region of length r at the crack tip and are assumed to be invariant in form as the crack propagates. This assumption is valid when the crack length is sufficiently long, but it will break down for very short crack lengths having the same order as the length r of the cohesive zone.

The second and third terms on the right-hand side of (8.57) may be written as

$$\int_{c(t)-r}^{c(t)} \sigma_{11}\left(c(t)-x_2\right)\frac{\partial \Delta u_1\left(c(t)-x_2\right)}{\partial t}\mathrm{d}x_2 + \int_{c(t)-r}^{c(t)} \sigma_{12}\left(c(t)-x_2\right)\frac{\partial \Delta u_2\left(c(t)-x_1\right)}{\partial t}\mathrm{d}x_2 \\ = \dot{c}\int_0^r \sigma_{11}\left(\xi\right)\frac{\partial \Delta u_1(\xi)}{\partial \xi}\,\mathrm{d}\xi + \dot{c}\int_0^r \sigma_{12}\left(\xi\right)\frac{\partial \Delta u_2(\xi)}{\partial \xi}\,\mathrm{d}\xi. \tag{8.59}$$

If it is assumed that the cohesive law is such that there exist energy functions A_1 and A_2 such that

$$\sigma_{11} = \frac{\mathrm{d}A_1\left(\Delta u_1\right)}{\mathrm{d}\Delta u_1},\ \sigma_{12} = \frac{\mathrm{d}A_2\left(\Delta u_2(u)\right)}{\mathrm{d}\Delta u_2},\ \text{where}\ A_1(0) = 0,\ A_2(0) = 0, \tag{8.60}$$

then (8.59) can also be written in the form

$$\int_{c(t)-r}^{c(t)} \sigma_{11}\left(c(t)-x_2\right)\frac{\partial \Delta u_1\left(c(t)-x_2\right)}{\partial t}\mathrm{d}x_2 + \int_{c(t)-r}^{c(t)} \sigma_{12}\left(c(t)-x_2\right)\frac{\partial \Delta u_2\left(c(t)-x_2\right)}{\partial t}\mathrm{d}x_2 \\ = \left[A_1\left(\Delta u_1(r)\right) + A_2\left(\Delta u_2(r)\right)\right]\dot{c} = 2\gamma\dot{c}. \tag{8.61}$$

Result (8.61) indicates the equivalence of the energy balance equations (8.31) and (8.57).

The quasistatic fracture criteria (8.50) and (8.56) are such that the left-hand sides are independent of the geometry selected for the surface Σ. Such 'path independence' is easily established by making use of the Gauss theorem, the symmetry of the stress tensor, the 'equilibrium equations' (8.28), together with relations (8.2) and (8.10), and the fact that the temperature distribution is uniform. Relations (8.50) and (8.56) are thus valid for *any* region Σ of the system containing the crack tip, even though their derivation assumed that Σ was small and centred on the crack tip. For linear elastic materials these relations demand that the stress and strain fields are $r^{-1/2}$ singular in the crack tip region.

In conclusion, an approach using both the principles of continuum mechanics and thermodynamics, has been described that enables the prediction of the initiation of crack growth of a single crack in a monolithic piece of material subject to isothermal conditions for thermoelastic materials, assuming that the material behaves as a continuum at *all* material points in the system.

8.8 Isolated Single Crack

An isolated crack of length $2c$ embedded in an infinite anisotropic linear elastic medium is now considered. The axial direction of the material is assumed to be normal to the crack plane. To be able to model bridged cracks, the crack faces are assumed to be subject to normal tractions that depend upon the opening of the crack faces. This section describes some useful methods of solving cracks problems of this type.

8.8.1 Anisotropic Stress-Strain Relations

As a very powerful complex variable method of dealing with isolated cracks in an anisotropic plate is to be used in this chapter, it is useful to introduce here an alternative rectangular set of Cartesian coordinate axes (x, y), such that $x \equiv x_2$ and $y \equiv x_1$ denote the in-plane transverse and axial directions, respectively, where a complex variable z may be written $z = x + iy$.

Consider a large rectangular homogeneous anisotropic plate that represents a homogenised cross-ply laminate material. The axes of material symmetry are parallel to the edges of the plate. It is assumed that there is no through-thickness loading ($\sigma_t = 0$) and no out-of-plane shear loading ($\tau_a = \tau_t = 0$), and that the axial loading direction is parallel to the y-axis. The plate is assumed to be orthotropic having the following in-plane stress-strain relations

$$\begin{aligned} \varepsilon_{xx} &= a_{11}\sigma_{xx} + a_{12}\sigma_{yy}, \\ \varepsilon_{yy} &= a_{12}\sigma_{xx} + a_{22}\sigma_{yy}, \\ 2\varepsilon_{xy} &= a_{66}\sigma_{xy}, \end{aligned} \tag{8.62}$$

where the compliances have the values (see (6.2) when $\sigma_t = 0$ and $\Delta T = 0$)

$$a_{11} = \frac{1}{E_T}, \quad a_{22} = \frac{1}{E_A}, \quad a_{12} = -\frac{\nu_A}{E_A}, \quad a_{66} = \frac{1}{\mu_A}. \tag{8.63}$$

For conditions of plane strain so that $\varepsilon_{zz} = 0$, the through-thickness stress is not zero, and on eliminating the through-thickness stress σ_{zz} it can easily be shown that (see (6.24), (6.29), (6.33) and (6.34) with $\Delta T = 0$)

$$a_{11} = \frac{1 - \nu_t^2 E_t / E_T}{E_T}, \quad a_{22} = \frac{1}{E_A}\left(1 - \nu_A^2 \frac{E_T}{E_A}\right), \quad a_{12} = -\frac{\nu_a + \nu_t \nu_A}{E_A}, \quad a_{66} = \frac{1}{\mu_A}. \tag{8.64}$$

For an homogenised 0°–90°–0° cross-ply laminate, the values of E_A, E_T and E_t (the axial, in-plane transverse and out-of-plane Young's moduli, respectively), ν_A (the principal axial Poisson's ratio), ν_a (the minor axial Poisson's ratio), ν_t (the transverse Poisson's ratio) and μ_A (the axial shear modulus) may be calculated from single-ply properties, as described in Chapter 7 using a general method.

8.8.2 A Representation for Stress and Displacement Fields

Consider a straight through-crack of length $2c$ embedded within an infinite transverse isotropic plate so that the crack plane is normal to one of the surfaces of the plate. It is assumed for the moment that the temperature difference $\Delta T = 0$. For conditions of generalised plane stress and plane strain, crack problems are two dimensional and it is possible to use complex variable techniques where the complex variable z is used, having real part x and imaginary part y, such that $x \equiv x_2$ and $y \equiv x_1$ enabling the introduction of a set of rectangular Cartesian coordinates (x, y) where the crack is parallel to the x-axis and occupies the region $|x - a| \leq c,\ y = b$ as shown in Figure 8.1. The complex variable z is defined by

$$z = x + i\,y. \tag{8.65}$$

The locations of the two crack tips are specified by

$$z = t_1 = a - c + ib, \tag{8.66}$$

$$z = t_2 = a + c + ib. \tag{8.67}$$

The stress and displacement representation developed by Sih and Liebowitz [6] is used as a basis for the subsequent analysis, which is valid for plane strain deformations in rectilinearly anisotropic bodies containing a crack. In the absence of body forces, as is assumed in the present model, this representation may also be applied to the conditions of generalised plane stress that are being considered.

The representation involves two analytic functions $\phi(z_1)$ and $\psi(z_2)$ of the complex variables

$$z_1 = x - a + s_1(y - b), \quad z_2 = x - a + s_2(y - b), \tag{8.68}$$

where s_1 and s_2 are the two roots having positive imaginary parts (described in the following) of the following quartic equation

$$a_{11}\,s^4 + (2a_{12} + a_{66})\,s^2 + a_{22} = 0. \tag{8.69}$$

The representation for the stress components is given by

$$\sigma_{xx} = \mathrm{Re}\left[s_1^2\phi'(z_1) + s_2^2\psi'(z_2)\right], \tag{8.70}$$

$$\sigma_{yy} = \mathrm{Re}\left[\phi'(z_1) + \psi'(z_2)\right], \tag{8.71}$$

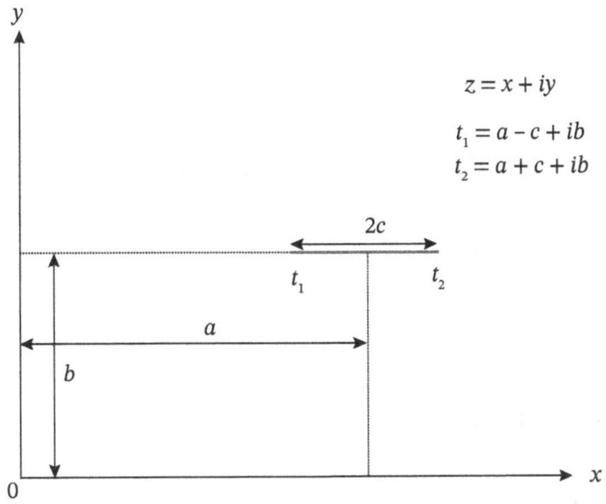

Figure 8.1 Geometry and loading of a transverse isotropic plate with a single crack.

$$\sigma_{xy} = \mathrm{Re}\left[s_1\phi'(z_1) + s_2\psi'(z_2)\right], \tag{8.72}$$

and that for the displacement components is (assuming $\Delta T = 0$)

$$u = \mathrm{Re}\left[p_1\phi(z_1) + p_2\psi(z_2)\right], \tag{8.73}$$

$$v = \mathrm{Re}\left[q_1\phi(z_1) + q_2\psi(z_2)\right], \tag{8.74}$$

where

$$\left.\begin{aligned} p_k &= a_{11}s_k^2 + a_{12} \\ q_k &= a_{12}s_k + a_{22}/s_k \end{aligned}\right\} \quad \text{for } k = 1, 2. \tag{8.75}$$

Putting

$$\gamma = \frac{a_{22}}{a_{11}}, \quad \delta = \frac{2a_{12} + a_{66}}{2a_{11}}, \tag{8.76}$$

it follows from (8.69) that

$$s^2 = -\delta \pm \sqrt{\delta^2 - \gamma}. \tag{8.77}$$

On substituting values from Table 3.1 into (8.63) or (8.64), it can be shown that for both glass-reinforced plastic (GRP) and carbon-fibre-reinforced polymer (CFRP), the values of s^2 are real and negative. The following distinct pure imaginary roots are, therefore, obtained

$$s_1 = i\sqrt{\delta - \sqrt{\delta^2 - \gamma}}, \quad s_2 = i\sqrt{\delta + \sqrt{\delta^2 - \gamma}}\,. \tag{8.78}$$

The other two roots of (8.69) are given by $\bar{s}_1 = -s_1$, $\bar{s}_2 = -s_2$. The stress and displacement representation automatically satisfies the equilibrium equations and the stress-strain relations (8.62) for any analytic functions $\phi(z)$ and $\psi(z)$ of the complex variable z. They are now assumed to take the following form:

$$\phi(z) \equiv \frac{1}{2\pi i}\int_{t_1}^{t_2} w(t)\,\hat{\rho}(t)\,\ln\frac{1}{z-t}\,dt, \tag{8.79}$$

$$\psi(z) \equiv \frac{1}{2\pi i}\int_{t_1}^{t_2} w(t)\,\hat{\sigma}(t)\,\ln\frac{1}{z-t}\,dt. \tag{8.80}$$

The density functions $\hat{\rho}(t)$ and $\hat{\sigma}(t)$ are assumed to be polynomials and

$$w(t) \equiv \frac{t_2 - t_1}{\left[(t-t_1)(t-t_2)\right]^{1/2}}. \tag{8.81}$$

In order that the displacement components u and v are single-valued functions at all points in the complex plane lying outside the crack, the following crack tip closure conditions must be satisfied:

$$\int_{t_1}^{t_2} w(t)\,\hat{\rho}(t)\,dt = 0, \quad \int_{t_1}^{t_2} w(t)\,\hat{\sigma}(t)\,dt = 0. \tag{8.82}$$

Differentiating (8.79) and (8.80) leads to

$$\phi'(z) \equiv \frac{1}{2\pi i}\int_{t_1}^{t_2} \frac{w(t)\,\hat{\rho}(t)}{t-z}\,dt, \tag{8.83}$$

$$\psi'(z) \equiv \frac{1}{2\pi i}\int_{t_1}^{t_2} \frac{w(t)\,\hat{\sigma}(t)}{t-z}\,dt. \tag{8.84}$$

Provided the conditions (8.82) are satisfied

$$\phi'(z) = O(z^{-2}), \quad \psi'(z) = O(z^{-2}) \;\text{ as } |z| \to \infty, \tag{8.85}$$

indicating that the stress field arising from the representation (8.70)–(8.72) has zero net force applied at infinity.

The algebra may be simplified by changing variables to

$$\zeta = \xi + i\eta, \tag{8.86}$$

where

$$\zeta = \zeta(z) = \frac{2z - (t_1 + t_2)}{t_2 - t_1}, \tag{8.87}$$

$$\xi = \frac{x-a}{c}, \quad \eta = \frac{y-b}{c}. \tag{8.88}$$

The crack is then described by $-1 < \xi < 1, \eta = 0$ and it follows from (8.83) and (8.84) that

$$\phi'(z) \equiv \frac{1}{\pi}\int_{-1}^{1} \frac{\rho(s)}{\sqrt{1-s^2}}\,\frac{ds}{\zeta(z)-s}, \tag{8.89}$$

$$\psi'(z) \equiv \frac{1}{\pi}\int_{-1}^{1} \frac{\sigma(s)}{\sqrt{1-s^2}}\,\frac{ds}{\zeta(z)-s}, \tag{8.90}$$

where $\rho(\xi) \equiv \hat{\rho}(t)$, $\sigma(\xi) \equiv \hat{\sigma}(t)$. From (8.82), the crack closure conditions are written as

$$\int_{-1}^{1} \frac{\rho(s)\,ds}{\sqrt{1-s^2}} = 0, \quad \int_{-1}^{1} \frac{\sigma(s)\,ds}{\sqrt{1-s^2}} = 0. \tag{8.91}$$

8.8.3 Chebyshev Polynomial Expansion

Chebyshev polynomials of the first kind are defined over the interval $[-1,1]$ by

$$T_n(\cos\alpha) \equiv \cos n\alpha, \; 0 \le \alpha \le \pi, \; n \ge 0. \tag{8.92}$$

Chebyshev polynomials of the second kind are defined by

$$U_n(\cos\alpha) = \frac{\sin(n+1)\alpha}{\sin\alpha}, \; 0 \le \alpha \le \pi, \; n \ge 0. \tag{8.93}$$

The functions $T_n(z)$ and $U_n(z)$ can both be analytically continued to the entire complex plane, simply by considering the usual analytic continuation of the functions sin(z) and cos(z). Use is made of the following two identities:

$$T_n\big((\lambda + 1/\lambda)/2\big) \equiv (\lambda^n + 1/\lambda^n)/2, \; n \ge 0, \tag{8.94}$$

$$U_{n-1}\left((\lambda+1/\lambda)/2\right) \equiv \left(\lambda^n - 1/\lambda^n\right)/\left(\lambda - 1/\lambda\right),\ n \geq 1. \tag{8.95}$$

The density functions $\rho(\xi)$ and $\sigma(\xi)$ are now assumed to be of the form

$$\rho(\xi) \equiv \sum_{n=1}^{N} A_n T_n(\xi), \quad \sigma(\xi) \equiv \sum_{n=1}^{N} B_n T_n(\xi), \tag{8.96}$$

where A_n and B_n are complex coefficients. The crack closure conditions (8.91) are automatically satisfied and substitution of (8.96) into (8.89) and (8.90) leads to

$$\phi'(z) \equiv \sum_{n=1}^{N} A_n H_n(\zeta), \quad \psi'(z) \equiv \sum_{n=1}^{N} B_n H_n(\zeta), \tag{8.97}$$

where, using a result established by Gladwell and England [7],

$$H_n(\zeta) \equiv \frac{1}{\pi}\int_{-1}^{1} \frac{T_n(s)}{\sqrt{1-s^2}}\frac{ds}{\zeta - s} \equiv \frac{\left[\zeta - (\zeta^2-1)^{1/2}\right]^n}{\left(\zeta^2-1\right)^{1/2}},\ n \geq 0. \tag{8.98}$$

By selecting the branch of $(\zeta^2-1)^{1/2}$ which is asymptotic to ζ as $|\zeta| \to \infty$, it can be shown that for $n \geq 1$,

$$H_n(\zeta) \equiv \frac{\left[\zeta - (\zeta^2-1)^{1/2}\right]^n}{\left(\zeta^2-1\right)^{1/2}} \equiv \frac{T_n(\zeta)}{\left(\zeta^2-1\right)^{1/2}} - U_{n-1}(\zeta). \tag{8.99}$$

8.8.4 Traction Distribution on the Crack

Let S_n^+, S_t^+, S_n^-, S_t^- be the normal and transverse tractions on the upper and lower surfaces of the crack. As

$$\left(\zeta^2-1\right)^{1/2} = \begin{cases} \sqrt{\xi^2-1}, & \xi>1,\ \eta=0, \\ i\sqrt{1-\xi^2}, & |\xi|<1,\ \eta=0+, \\ -i\sqrt{1-\xi^2}, & |\xi|<1,\ \eta=0-, \\ -\sqrt{\xi^2-1}, & \xi<-1,\ \eta=0, \end{cases} \tag{8.100}$$

it follows from (8.99) that

$$H_n^{\pm}(\xi) = \pm i\frac{T_n(\xi)}{\sqrt{1-\xi^2}} - U_{n-1}(\xi),\ |\xi|<1,\ n\geq 1, \tag{8.101}$$

where H_n^+ and H_n^- are the limiting values of $H_n(\zeta)$ on the positive and negative sides of the crack, respectively. Substituting (8.97) into (8.71) and (8.72) using (8.101) leads to the following relations valid only for $|\xi|<1$:

$$S_n^{\pm}(\xi) = \sum_{n=1}^{N}\left(\pm \mathrm{Im}\left[A_n + B_n\right]\frac{T_n(\xi)}{\sqrt{1-\xi^2}} - \mathrm{Re}\left[A_n + B_n\right]U_{n-1}(\xi)\right), \tag{8.102}$$

$$S_t^{\pm}(\xi) = \sum_{n=1}^{N}\left(\pm \mathrm{Im}\left[s_1 A_n + s_2 B_n\right]\frac{T_n(\xi)}{\sqrt{1-\xi^2}} + \mathrm{Re}\left[s_1 A_n + s_2 B_n\right]U_{n-1}(\xi)\right). \tag{8.103}$$

The tractions S_n^+, S_t^+, S_n^-, S_t^- must be bounded as $\xi \to \pm 1$, for values of ξ in the range $|\xi|<1$ leading to the following conditions for the complex coefficients A_n, B_n:

$$\mathrm{Im}\left[A_n + B_n\right] = \mathrm{Im}\left[s_1 A_n + s_2 B_n\right] = 0. \tag{8.104}$$

These conditions may easily be satisfied by choosing real α_n, β_n such that

$$\alpha_n = A_n + B_n, \quad \beta_n = -(s_1 A_n + s_2 B_n), \tag{8.105}$$

and

$$A_n = -\frac{\alpha_n s_2 + \beta_n}{s_1 - s_2}, \qquad B_n = \frac{\alpha_n s_1 + \beta_n}{s_1 - s_2}, \tag{8.106}$$

which, furthermore, yields the following simple expression for the tractions on the crack surfaces:

$$\left(S_n^{\pm} + i\, S_t^{\pm}\right)(\xi) = -\sum_{n=1}^{N} \left(\alpha_n + i\,\beta_n\right) U_{n-1}(\xi), \quad |\xi|<1. \tag{8.107}$$

8.8.5 Stress and Displacement Fields around the Crack

Define

$$G_n(\zeta) \equiv \left[\zeta - (\zeta^2 - 1)^{1/2}\right]^n, \quad n \geq 1, \tag{8.108}$$

noting that numerical calculation is convenient by making use of the fact that

$$G_n(\zeta) \equiv \exp\left[-n(\alpha + i\,\beta)\right], \quad \text{when } \zeta = \cosh(\alpha + i\beta). \tag{8.109}$$

It is easily shown that

$$\frac{\mathrm{d}}{\mathrm{d}\zeta} G_n(\zeta) = -n H_n(\zeta), \quad n \geq 1, \tag{8.110}$$

and it then follows from (8.97) that

$$\phi(z) = -c \sum_{n=1}^{N} \frac{A_n}{n} G_n(\zeta), \tag{8.111}$$

$$\psi(z) = -c \sum_{n=1}^{N} \frac{B_n}{n} G_n(\zeta). \tag{8.112}$$

Let

$$\zeta_1 = \xi + s_1 \eta \text{ and } \zeta_2 = \xi + s_2 \eta. \tag{8.113}$$

On substituting (8.97), (8.106), (8.111) and (8.112) into the representation (8.70)–(8.74), the stresses and displacement components may be expressed

$$\sigma_{xx} = -\sum_{n=1}^{N} \mathrm{Re}\left[\left(\alpha_n s_1 s_2 + \beta_n (s_1 + s_2)\right) H_n(\zeta_1) + s_2^2 (\alpha_n s_1 + \beta_n) \Delta H_n(\zeta_1, \zeta_2)\right], \tag{8.114}$$

$$\sigma_{yy} = \sum_{n=1}^{N} \mathrm{Re}\left[\alpha_n H_n(\zeta_1) - (\alpha_n s_1 + \beta_n) \Delta H_n(\zeta_1, \zeta_2)\right], \tag{8.115}$$

$$\sigma_{xy} = \sum_{n=1}^{N} \mathrm{Re}\left[\beta_n H_n(\zeta_1) + s_2(\alpha_n s_1 + \beta_n)\Delta H_n(\zeta_1, \zeta_2)\right], \tag{8.116}$$

$$\begin{aligned} u = c \sum_{n=1}^{N} \frac{1}{n} \mathrm{Re}\Big[\big(\alpha_n(a_{11}s_1s_2 - a_{12}) + \beta_n a_{11}(s_1+s_2)\big)G_n(\zeta_1) \\ + (a_{11}s_2^2 + a_{12})(\alpha_n s_1 + \beta_n)\Delta G_n(\zeta_1,\zeta_2)\Big], \end{aligned} \tag{8.117}$$

$$\begin{aligned} v = c \sum_{n=1}^{N} \frac{1}{n} \mathrm{Re}\left[\left(-\alpha_n a_{22}\left(\frac{s_1+s_2}{s_1 s_2}\right) + \beta_n\left(a_{12} - \frac{a_{22}}{s_1 s_2}\right)\right)G_n(\zeta_1)\right. \\ \left. + \left(a_{12}s_2 + \frac{a_{22}}{s_2}\right)(\alpha_n s_1 + \beta_n)\Delta G_n(\zeta_1,\zeta_2)\right], \end{aligned} \tag{8.118}$$

where

$$\Delta H_n(\zeta_1,\zeta_2) \equiv \begin{cases} \left[\dfrac{H_n(\zeta_1) - H_n(\zeta_2)}{s_1 - s_2}\right], & \text{for } s_1 \neq s_2, \\ \lim_{s_1 \to s_2}\left[\dfrac{H_n(\zeta_1) - H_n(\zeta_2)}{s_1 - s_2}\right] = -\eta\left(\dfrac{n}{(\zeta_1^2-1)^{1/2}} + \dfrac{\zeta_1}{\zeta_1^2 - 1}\right)H(\zeta_1), & \text{for } s_1 = s_2, \end{cases} \tag{8.119}$$

and

$$\Delta G_n(\zeta_1,\zeta_2) \equiv \begin{cases} \left[\dfrac{G_n(\zeta_1) - G_n(\zeta_2)}{s_1 - s_2}\right], & \text{for } s_1 \neq s_2, \\ \lim_{s_1 \to s_2}\left[\dfrac{G_n(\zeta_1) - G_n(\zeta_2)}{s_1 - s_2}\right] = -n\eta H(\zeta_1), & \text{for } s_1 = s_2. \end{cases} \tag{8.120}$$

One limiting situation, $s_1 = s_2 = i$, occurs when the material is isotropic. It should be noted that

$$\Delta G_n(\zeta_1,\zeta_2) \to 0, \quad \Delta H_n(\zeta_1,\zeta_2) \to 0 \text{ as } y \to 0.$$

8.8.6 Displacement Discontinuity across the Crack

From (8.99) and (8.108)

$$G_n(\zeta) \equiv H_n(\zeta)\left(\zeta^2 - 1\right)^{1/2} \equiv T_n(\xi) - \left(\zeta^2 - 1\right)^{1/2} U_{n-1}(\zeta). \tag{8.121}$$

It is deduced from (8.100) that the limiting values of $G_n(\zeta)$ on the crack faces are given by

$$G_n^{\pm}(\xi) = T_n(\xi) \pm i\sqrt{1 - \xi^2}\, U_{n-1}(\xi). \tag{8.122}$$

By considering the limiting distributions for the normal and tangential displacements along the upper and lower surfaces of the crack, which are denoted by V_{n}^{+}, V_{n}^{-}, U_{t}^{+}, U_{t}^{-}, use can be made of (8.117) and (8.118) together with (8.122) to obtain an expression for the displacement discontinuities across the crack:

$$\Delta v(\xi) \equiv (V_{\mathrm{n}}^{+} - V_{\mathrm{n}}^{-})(\xi) = 4c\sqrt{a_{22}g}\sqrt{1 - \xi^2} \sum_{n=1}^{N} \frac{\alpha_n}{n} U_{n-1}(\xi), \tag{8.123}$$

$$\Delta u(\xi) \equiv (U_t^+ - U_t^-)(\xi) = 4c\sqrt{a_{11}g}\sqrt{1-\xi^2}\sum_{n=1}^{N}\frac{\beta_n}{n}U_{n-1}(\xi), \tag{8.124}$$

where

$$g = \frac{1}{4}\left(2\sqrt{a_{11}a_{22}} + 2a_{12} + a_{66}\right). \tag{8.125}$$

It is useful to define, for an undamaged composite, an anisotropy factor Ω such that

$$\Omega^2 = \frac{E_A^2 a_{22}}{4}\left(2\sqrt{a_{11}a_{22}} + 2a_{12} + a_{66}\right) = E_A^2 a_{22} g. \tag{8.126}$$

It then follows that (8.123) and (8.124) may be written as

$$\Delta v(\xi) \equiv (V_n^+ - V_n^-)(\xi) = \frac{4c\Omega}{E_A}\sqrt{1-\xi^2}\sum_{n=1}^{N}\frac{\alpha_n}{n}U_{n-1}(\xi), \tag{8.127}$$

$$\Delta u(\xi) \equiv (U_t^+ - U_t^-)(\xi) = \frac{4c\Omega}{E_A}\sqrt{1-\xi^2}\sum_{n=1}^{N}\frac{\beta_n}{n}U_{n-1}(\xi). \tag{8.128}$$

On using (8.63) and (8.64), it follows that for conditions of generalised plane stress

$$\Omega^2 = \frac{1}{2}\left(\sqrt{\frac{E_A}{E_T}} - \nu_A\right) + \frac{E_A}{4\mu_A}, \tag{8.129}$$

and for plane strain conditions

$$\Omega^2 = \frac{E_A}{2}\left(1-\nu_A^2\frac{E_T}{E_A}\right)\left(\sqrt{\frac{1}{E_A E_T}\left(1-\nu_A^2\frac{E_T}{E_A}\right)\left(1-\nu_t^2\frac{E_t}{E_T}\right)} - \frac{\nu_a+\nu_t\nu_A}{E_A} + \frac{1}{2\mu_A}\right). \tag{8.130}$$

If the laminate is isotropic so that $E_A = E_T = E_t = E$, $\nu_A = \nu_a = \nu_t = \nu$ and $\mu_A = \mu$ where $E = 2\mu(1+\nu)$, the dimensionless parameter Ω has the value 1 for conditions of generalised plane stress and the value $1-\nu^2$ for conditions of plane strain.

8.8.7 Stress Intensity Factors

For the crack tip at t_1, the mode I and mode II stress intensity factors K_I^1, K_{II}^1 are defined by

$$K_I^1 + i K_{II}^1 = \lim_{\xi\to -1}\left[\sqrt{2\pi c(-\xi-1)}\left(S_n(\xi) + i S_t(\xi)\right)\right], \tag{8.131}$$

where $S_n(\xi)$ and $S_t(\xi)$ are the normal and tangential tractions acting on $\xi < -1, \eta = 0$. It follows from (8.71), (8.72), (8.97), (8.99) and (8.106) that on $\xi < -1, \eta = 0$ the tractions are given by

$$S_n(\xi) + i S_t(\xi) = -\sum_{n=1}^{N}(\alpha_n + i\beta_n)\left[\frac{T_n(\xi)}{\sqrt{\xi^2-1}} + U_{n-1}(\xi)\right]. \tag{8.132}$$

Substituting into (8.131) leads immediately to the simple result

$$K_I^1 + i K_{II}^1 = \sqrt{\pi c}\sum_{n=1}^{N}(-1)^{n+1}\left(\alpha_n + i\beta_n\right), \tag{8.133}$$

since $T_n(-1)=(-1)^n$. Similarly for the crack tip at t_2,

$$K_{\rm I}^2 + i K_{\rm II}^2 = \lim_{\xi \to 1} \left[\sqrt{2\pi c(\xi - 1)} \left(S_{\rm n}(\xi) + i S_{\rm t}(\xi) \right) \right], \tag{8.134}$$

where $S_{\rm n}(\xi)$ and $S_{\rm t}(\xi)$ are the normal and tangential tractions acting on $\xi > 1, \eta = 0$. For $\xi > 1$, $\eta = 0$,

$$S_{\rm n}(\xi) + i S_{\rm t}(\xi) = \sum_{n=1}^{N} (\alpha_n + i\beta_n) \left[\frac{T_n(\xi)}{\sqrt{\xi^2 - 1}} - U_{n-1}(\xi) \right]. \tag{8.135}$$

As $T_n(1)=1$, substitution into (8.134) yields the following corresponding simple result

$$K_{\rm I}^2 + i K_{\rm II}^2 = \sqrt{\pi c} \sum_{n=1}^{N} \left(\alpha_n + i\beta_n \right). \tag{8.136}$$

8.8.8 Integral Representations for the Solution of Matrix Crack Problems

It is now assumed that the temperature difference $\Delta T \neq 0$. The constitutive equations (8.62) describe the behaviour of a composite only when it is regarded as a homogeneous linear anisotropic material. These equations are describing the average stress-strain relations for the composite which in reality is inhomogeneous and has nonuniform stress and displacement distributions when viewed at a microscopic level. For nonzero values of ΔT, there exist thermal residual stresses at the microscopic level which average to zero when the composite is undamaged and is regarded as a homogeneous continuum. However, when considering the cracking of one phase in the matrix (fibres or matrix), the thermal residual stresses assume a greater importance even when cracking is viewed at a macroscopic level. To fully appreciate this point, it is useful to consider Figure 8.2.

In Figure 8.2(a) is shown the composite at some temperature such that $\Delta T \neq 0$ where all the applied tractions are zero. If the composite is unidirectional, then neither the fibres nor the matrix are stress-free. By applying a uniaxial stress $\sigma_{\rm A}$, as shown in Figure 8.2(b), it is possible to alter the values of the internal stresses so that the matrix stresses $\sigma_{yy}^{\rm m} (\equiv \sigma_{11}^{\rm m})$ and $\sigma_{xy}^{\rm m}$ $(\equiv \sigma_{12}^{\rm m})$ are everywhere zero in an undamaged composite. The value of the applied stress for this special case is denoted by $\sigma_{\rm th}$ which is regarded as a stress parameter characterising thermal residual stresses such that $\sigma_{\rm th} = 0$ when $\Delta T = 0$. For this special loading case, a matrix through-crack of length $2c$ can be introduced, as shown in Figure 8.2(b), without perturbing in any way the stress and displacement distributions within the composite. If the applied stress is then increased to the value $\sigma_{\rm A} > \sigma_{\rm th}$, then the matrix crack will open as shown in Figure 8.2(c). The matrix crack surfaces will be subject to a normal traction distribution denoted by $p(x)$, $|x| < c$, which restrains the opening of the crack as a result of the presence of intact fibres.

For the moment it is assumed that the thermal stress $\sigma_{\rm th}$ is known, and that the traction distribution $p(x)$ has been specified in advance. It then follows that the boundary conditions for the crack problem shown in Figure 8.2(c) can be imposed as follows:

$$\begin{aligned} \sigma_{yy} &= \sigma_{\rm A}, \quad \sigma_{xy} = 0, \quad \text{for } x^2 + y^2 \to \infty, \\ \sigma_{yy} &= p(x), \quad \sigma_{xy} = 0, \quad \text{on } y = 0 \text{ for } |x| < c. \end{aligned} \tag{8.137}$$

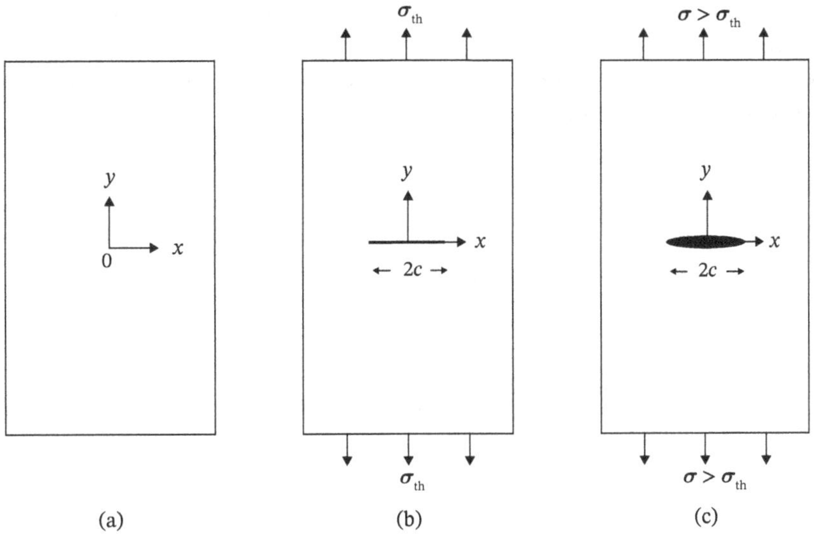

Figure 8.2 A composite plate (a) in a damage-free stress-free state, (b) subject to an applied stress σ_{th} such that the bridged crack is just closed and (c) subject to an applied stress $\sigma > \sigma_{th}$ leading to opening of the bridged crack.

This type of crack problem is easily solved by considering the auxiliary problem in which the external boundary is entirely stress-free and the crack is opened by the application of an internal pressure distribution. The boundary conditions for this auxiliary problem are

$$\begin{aligned} \sigma_{yy} &= 0, & \sigma_{xy} &= 0, \quad \text{for } x^2 + y^2 \to \infty, \\ \sigma_{yy} &= p(x) - \sigma_A, & \sigma_{xy} &= 0, \quad \text{on } y = 0 \text{ for } |x| < c. \end{aligned} \tag{8.138}$$

Clearly, the solution to the crack problem shown in Figure 8.2(c) is obtained by adding the stress field $\sigma_{yy} = \sigma_A, \sigma_{xy} = 0$, and the corresponding displacement field, to the solution of the auxiliary problem which can be obtained using the complex variable method described previously. The important results are that for a crack opened by a pressure distribution $\hat{p}(x)$ having the form (cf. (8.107))

$$\hat{p}(x) = \sum_{n=1}^{N} \alpha_n U_{n-1}(x/c), \quad |x| < c, \tag{8.139}$$

the stress intensity factors are given by (cf. (8.136))

$$K(x=-c) = \sqrt{\pi c} \sum_{n=1}^{N} (-1)^{n+1} \alpha_n, \quad K(x=c) = \sqrt{\pi c} \sum_{n=1}^{N} \alpha_n, \tag{8.140}$$

and the crack opening displacement (i.e. the discontinuity of the normal displacement across the crack) is given by (cf. (8.127))

$$\Delta v(x) = \frac{4\Omega}{E_A} \sqrt{c^2 - x^2} \sum_{n=1}^{N} \frac{\alpha_n}{n} U_{n-1}(x/c), \quad |x| < c, \tag{8.141}$$

where Ω is defined by (8.129) for generalised plane stress conditions and by (8.130) for plane strain conditions.

Using well-known properties of Chebyshev polynomials [8] the results (8.140) and (8.141) may be expressed in the forms

$$K(x=-c)=\sqrt{\frac{\pi}{c}}\int_{-c}^{c}\sqrt{\frac{c-x}{c+x}}\,\hat{p}(x)\mathrm{d}x,\quad K(x=c)=\sqrt{\frac{\pi}{c}}\int_{-c}^{c}\sqrt{\frac{c+x}{c-x}}\,\hat{p}(x)\mathrm{d}x,\tag{8.142}$$

$$\Delta v(x)=\frac{4\Omega}{E_{\mathrm{A}}}\sqrt{c^2-x^2}\int_{-c}^{c}\left(\int_{-c}^{t}\hat{p}(\xi)\mathrm{d}\xi\right)\frac{\mathrm{d}t}{(t-x)\sqrt{c^2-t^2}},\quad |x|<c,\tag{8.143}$$

where the principal value must be taken of the integral with respect to the variable t. For the special case of symmetric pressure distributions such that $\hat{p}(x)=\hat{p}(-x)$ for $|x|<c$, these relations may be written as

$$K(x=-c)=K(x=c)=K=2\sqrt{\frac{c}{\pi}}\int_{0}^{c}\frac{\hat{p}(x)}{\sqrt{c^2-x^2}}\,\mathrm{d}x\,,\tag{8.144}$$

$$\Delta v(x)=\frac{4\Omega}{\pi E_{\mathrm{A}}}\int_{0}^{c}\hat{p}(\xi)\ln\left|\frac{\sqrt{c^2-\xi^2}+\sqrt{c^2-x^2}}{\sqrt{c^2-\xi^2}-\sqrt{c^2-x^2}}\right|\mathrm{d}\xi,\quad |x|<c.\tag{8.145}$$

The result (8.145) is derived on reversing the orders of the integrations in (8.143), and then performing the integration with respect to the variable t.

8.8.9 Energy Balance Calculation

Following the approach described in [9], when the results (8.144) and (8.145) are applied to the auxiliary problem associated with the crack problem shown in Figure 8.2(c) it follows from (8.138) that the function $\hat{p}(x)$ must be defined by

$$\hat{p}(x)\equiv p(x)-\sigma_{\mathrm{A}},\quad |x|<c,\tag{8.146}$$

where it has been assumed that $p(x)=p(-x)$. Thus, for the crack problem shown in Figure 8.2(c)

$$K=2\sqrt{\frac{c}{\pi}}\int_{0}^{c}\frac{\sigma_{\mathrm{A}}-p(x)}{\sqrt{c^2-x^2}}\,\mathrm{d}x\,,\tag{8.147}$$

and the crack opening displacement may be written as

$$\Delta v(x)=\frac{4\Omega}{\pi E_{\mathrm{A}}}\int_{0}^{c}\left(\sigma_{\mathrm{A}}-p(\xi)\right)\ln\left|\frac{\sqrt{c^2-\xi^2}+\sqrt{c^2-x^2}}{\sqrt{c^2-\xi^2}-\sqrt{c^2-x^2}}\right|\mathrm{d}\xi,\quad |x|<c,\tag{8.148}$$

where use has been made of (8.125) and (8.126). It is worth noting that expression (8.147) for the stress intensity factor is not affected by anisotropy whereas expression (8.148) for the corresponding crack opening displacement involves the dimensionless anisotropy factor Ω defined in terms of elastic constants by (8.129) or (8.130).

The final step is to use these results to determine the value of the applied stress σ_A for which the embedded matrix crack of length $2c$ will grow unstably using the energy balance approach to fracture for bridged cracks, described in detail in Appendix J, which leads to the following fracture criterion (see (J.23) in Appendix J)

$$\int_0^c \left(\sigma_A - p(x,c)\right) \frac{\partial \Delta v(x,c)}{\partial c} \mathrm{d}x + \int_0^c \frac{\partial p(x,c)}{\partial c} \Delta v(x,c) \mathrm{d}x = 4\gamma, \tag{8.149}$$

where 2γ is the effective fracture energy, γ being the effective surface energy of the crack surfaces, and where the following identifications have been made

$$x_2 \equiv x, \quad \sigma_{22}^0(x_2) \equiv \sigma_A, \quad \sigma_{12}^0(x_2) \equiv 0, \quad \tau(x_2,c) \equiv 0,$$
$$\Delta u_1(x_2,c) \equiv \Delta v(x,c), \quad \Delta u_2(x_2,c) \equiv 0.$$

On differentiating (8.148) with respect to c, keeping σ_A fixed, it can be shown that

$$\begin{aligned} \frac{\pi E_A}{4\Omega} \frac{\partial \Delta v(x)}{\partial c} = {} & \frac{2c}{\sqrt{c^2 - x^2}} \int_0^c \frac{\sigma_A - p(\xi)}{\sqrt{c^2 - \xi^2}} \mathrm{d}\xi \\ & - \int_0^c \frac{\partial p(\xi,c)}{\partial c} \ln \left| \frac{\sqrt{c^2 - \xi^2} + \sqrt{c^2 - x^2}}{\sqrt{c^2 - \xi^2} - \sqrt{c^2 - x^2}} \right| \mathrm{d}\xi, \quad |x| < c. \end{aligned} \tag{8.150}$$

On multiplying (8.150) by the factor $\sigma_A - p(\xi)$ and then integrating, it follows from (8.147) and (8.148) that

$$\begin{aligned} & \int_0^c \left(\sigma_A - p(x,c)\right) \frac{\partial \Delta v(x,c)}{\partial c} \mathrm{d}x + \int_0^c \frac{\partial p(x,c)}{\partial c} \Delta v(x,c) \mathrm{d}x \\ & = \frac{8\Omega c}{\pi E_A} \left(\int_0^c \frac{\sigma_A - p(x,c)}{\sqrt{c^2 - x^2}} \mathrm{d}x \right)^2 = \frac{2\Omega}{E_A} K^2. \end{aligned} \tag{8.151}$$

Relations (8.149) and (8.151) then assert that the fracture criterion can be reduced to the simple form

$$K^2 = \frac{2E_A\gamma}{\Omega} = K_{Ic}^2, \tag{8.152}$$

which corresponds to the well-known Griffith fracture criterion for isotropic materials for which $\Omega = 1$ for generalised plane stress conditions or $\Omega = 1 - \nu^2$ for plane strain conditions, and $E_A = E$, where E is Young's modulus and ν is Poisson's ratio of an isotropic solid. The parameter K_{Ic} appearing in (8.152) is the mode I fracture toughness of an anisotropic solid. Relation (8.152) may also be expressed in the form

$$G_I = \frac{\Omega K^2}{E_A} = 2\gamma = G_{Ic}, \tag{8.153}$$

where G_{I} is the mode I energy release rate and G_{Ic} is its critical value at the point of fracture. It should be noted that mode II and mode III fracture refer to cracking under in-plane shear loading and out-of-plane shear loading, respectively.

8.8.10 Special Case of Long Ply Cracks

On applying the result (J.23) derived in Appendix J to the ply cracking problem under consideration, it follows from (8.151) and (8.152) that at the point of ply crack growth, when the axial applied stress has the value σ_{A},

$$\int_0^c \left(\sigma_{\mathrm{A}} - p(x,c)\right)\frac{\partial \Delta v(x,c)}{\partial c}\mathrm{d}x + \int_0^c \frac{\partial p(x,c)}{\partial c}\Delta v(x,c)\mathrm{d}x = 4\gamma. \tag{8.154}$$

For the case of very long cracks ($c \to \infty$), it is useful to introduce the parameter d, which is the length of the region $c - d < x < c$ adjacent to the crack tip at $x = c$ where the functions $p(x,c)$ and $\Delta v(x,c)$ are effectively dependent on the variables x and c. For the region $0 < x < c - d$ and for sufficiently long cracks, $p(x,c) = \sigma_\infty$ and $\Delta v(x,c) = \Delta v_\infty$ where σ_∞ and Δv_∞ are independent of x and c. For such long-crack conditions $\sigma_{\mathrm{A}} = \sigma_\infty$ so that the general expression (8.154) may then be replaced by

$$\int_{c-d}^c \left(\sigma_\infty - p(x-c)\right)\frac{\partial \Delta v(x-c)}{\partial c}\mathrm{d}x + \int_{c-d}^c \frac{\partial p(x-c)}{\partial c}\Delta v(x-c)\mathrm{d}x = 4\gamma. \tag{8.155}$$

The (x, c) dependence has been replaced by the $(x - c)$ dependence shown in (8.155) because the crack is very long such that the local deformation in the region of the crack tip at $x = c$ moves self-similarly as the crack propagates. It follows that (8.155) may now be written as

$$-\sigma_\infty \int_{c-d}^c \frac{\partial \Delta v(x-c)}{\partial x}\mathrm{d}x + \int_{c-d}^c p(x-c)\frac{\partial \Delta v(x-c)}{\partial x}\mathrm{d}x - \int_{c-d}^c \frac{\partial p(x-c)}{\partial x}\Delta v(x-c)\mathrm{d}x = 4\gamma. \tag{8.156}$$

On setting $x' = x - c$, it can then be shown that

$$\sigma_\infty \Delta v(-d) + \int_{-d}^0 p(x')\frac{\mathrm{d}\Delta v(x')}{\mathrm{d}x'}\mathrm{d}x' - \int_{-d}^0 \frac{\mathrm{d}p(x')}{\mathrm{d}x'}\Delta v(x')\mathrm{d}x' = 4\gamma. \tag{8.157}$$

On integrating by parts, it can be shown that

$$\int_{-d}^0 \frac{\mathrm{d}p(x')}{\mathrm{d}x'}\Delta v(x')\mathrm{d}x' = -p(-d)\Delta v(-d) - \int_{-d}^0 p(x')\frac{\mathrm{d}\Delta v(x')}{\mathrm{d}x'}\mathrm{d}x'. \tag{8.158}$$

It then follows that

$$\left[\sigma_\infty + p(-d)\right]\Delta v(-d) + 2\int_{-d}^0 p(x')\frac{\mathrm{d}\Delta v(x')}{\mathrm{d}x'}\mathrm{d}x' = 4\gamma. \tag{8.159}$$

The final step is to note that the definition of the parameter d asserts that $p(-d) = \sigma_\infty$ so that the condition for crack growth may be expressed

$$\sigma_\infty \Delta v(-d) + \int_{-d}^{0} p(x') \frac{\mathrm{d}\Delta v(x')}{\mathrm{d}x'} \mathrm{d}x' = 2\gamma. \tag{8.160}$$

This result may also be written as

$$\sigma_\infty \Delta v_\infty - \int_0^{\Delta v_\infty} p(\Delta v)\mathrm{d}(\Delta v) = \int_0^{\Delta v_\infty} \left[\sigma_\infty - p(\Delta v)\right] \mathrm{d}(\Delta v) = 2\gamma. \tag{8.161}$$

8.8.11 Concluding Remarks

The discussion of bridged cracks has used the matrix crack problem as the vehicle for the theoretical development of an energy-based fracture criterion. As described in [9], the approach developed for matrix cracks in unidirectionally fibre-reinforced composites is equally applicable to the study of ply cracking in cross-ply laminates. This topic is considered in Chapter 17 and the reader is also recommended to study [9].

References

1. Griffith, A.A. (1920). The phenomena of rupture and flow in solids. *Philosophical Transactions of the Royal Society of London*. A221: 163–198.
2. Eringen, A.C. (1967). Mechanics of Continua. New York-London-Sydney: John Wiley & Sons.
3. Truesdell, C. (1969). Rational Thermodynamics. New York-London-Sydney: McGraw-Hill Book Co.
4. Glansdorff, P. and Prigogine, I. (1971). Thermodynamic theory of structure, stability and fluctuations. London - New York – Sydney: Wiley – Interscience.
5. Eftis, J. and Liebowitz, H. (1976). On surface energy and the continuum thermodynamics of brittle fracture. *Engineering Fracture Mechanics* 8: 459–485.
6. Sih, G.C. and Liebowitz, H. (1968). Mathematical theories of brittle fracture in Fracture. Volume II, (ed. P. Liebowitz), 67. Academic Press.
7. Gladwell, G.M.L. and England, A.H. (1977). *Journal of Mechanics and Applied Mathematics* 30: 175.
8. Abramowitz, M. and Stegun, I.A. (1972). *Handbook of Mathematical Functions*. New York: Dover Publications. Chapter 22.
9. McCartney, L.N. (1992). Mechanics of growth of bridged cracks in composite materials: Part I. Basic principles. *Journal of Composites Technology and Research* 14: 133–146.

9

Ply Crack Formation in Symmetric Cross-ply Laminates

Overview: This chapter considers stress-transfer models that can be applied to simple $[0_m/90_n]_s$ cross-ply laminates in the presence of thermal residual stresses. A preferred shear-lag model is first developed to highlight various approximations that need to be made that can compromise the accuracy of predictions in some cases. To increase accuracy, a generalised plane strain model of stress transfer for uniformly distributed arrays of ply cracks is developed. Simple cross-ply laminates are considered to illustrate the methodology that is the basis of the generalised plane strain model. The predicted stress distributions for both shear-lag and generalised plane strain models are used to estimate the effective elastic constants of laminates having ply cracks in the 90° ply. The validity of very useful interrelationships between the thermoelastic constants is established analytically. The shear-lag and generalised plane stress predictions of the thermoelastic constants for damaged laminates are compared.

9.1 Fundamental Equations and Conditions

Consider the model of a simple $[0/90]_s$ cross-ply laminate illustrated in Figure 9.1 where two possible representative volume elements (RVEs) are shown. For the first, shown in Figure 9.1(a), one ply crack in the inner 90° ply is located on the plane $x_1 = 0$ and neighbouring ply cracks (not shown) are on the planes $x_1 = \pm 2L$. For the second RVE, shown in Figure 9.1(b), the plane $x_1 = 0$ is midway between two neighbouring ply cracks in the inner 90° ply on the planes $x_1 = \pm L$. The inner 90° ply has total thickness $2a$ and the two outer 0° plies each have thickness denoted by b so that the total thickness of the laminate is $2h$ where $h = a + b$. A set of Cartesian coordinates (x_1, x_2, x_3) is introduced such that the origin lies on the midplane of the laminate at the midpoint between two neighbouring cracks in the 90° ply. The x_1-axis is directed along the principal loading direction and the x_3-axis is directed in the through-thickness direction.

Properties for Design of Composite Structures: Theory and Implementation Using Software, First Edition. Neil McCartney.

Companion Website: www.wiley.com/go/mccartney/properties

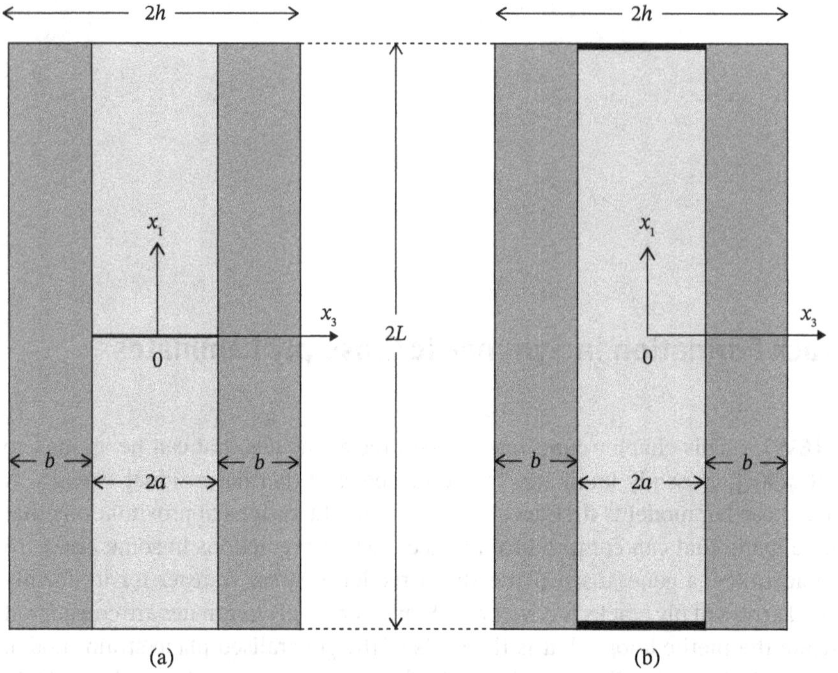

Figure 9.1 Representative volume elements for a cracked cross-ply laminate.

9.1.1 Basic Field Equations

The following equilibrium equations must be satisfied for both the 0° and 90° plies (see (2.120)–(2.122))

$$\frac{\partial \sigma_{11}}{\partial x_1} + \frac{\partial \sigma_{12}}{\partial x_2} + \frac{\partial \sigma_{13}}{\partial x_3} = 0, \tag{9.1}$$

$$\frac{\partial \sigma_{12}}{\partial x_1} + \frac{\partial \sigma_{22}}{\partial x_2} + \frac{\partial \sigma_{23}}{\partial x_3} = 0, \tag{9.2}$$

$$\frac{\partial \sigma_{13}}{\partial x_1} + \frac{\partial \sigma_{23}}{\partial x_2} + \frac{\partial \sigma_{33}}{\partial x_3} = 0, \tag{9.3}$$

where σ_{ij} are the stress components. The plies are regarded as transverse isotropic solids so that the stress-strain relations involve the axial and transverse values of Young's modulus E, Poisson's ratio ν, shear modulus μ and thermal expansion coefficient α. Superscripts '0' or '90' are used to denote the ply to which a stress, strain and displacement component refers. For the 0° plies (see (6.99))

$$\begin{aligned}
\varepsilon_{11}^0 &= \frac{\partial u_1^0}{\partial x_1} = \frac{1}{E_A^0}\sigma_{11}^0 - \frac{\nu_A^0}{E_A^0}\sigma_{22}^0 - \frac{\nu_a^0}{E_A^0}\sigma_{33}^0 + \alpha_A^0 \Delta T, & 2\varepsilon_{12}^0 &= \frac{\partial u_1^0}{\partial x_2} + \frac{\partial u_2^0}{\partial x_1} = \frac{\sigma_{12}^0}{\mu_A^0}, \\
\varepsilon_{22}^0 &= \frac{\partial u_2^0}{\partial x_2} = -\frac{\nu_A^0}{E_A^0}\sigma_{11}^0 + \frac{1}{E_T^0}\sigma_{22}^0 - \frac{\nu_t^0}{E_T^0}\sigma_{33}^0 + \alpha_T^0 \Delta T, & 2\varepsilon_{13}^0 &= \frac{\partial u_1^0}{\partial x_3} + \frac{\partial u_3^0}{\partial x_1} = \frac{\sigma_{13}^0}{\mu_a^0}, \\
\varepsilon_{33}^0 &= \frac{\partial u_3^0}{\partial x_3} = -\frac{\nu_a^0}{E_A^0}\sigma_{11}^0 - \frac{\nu_t^0}{E_T^0}\sigma_{22}^0 + \frac{1}{E_t^0}\sigma_{33}^0 + \alpha_t^0 \Delta T, & 2\varepsilon_{23}^0 &= \frac{\partial u_2^0}{\partial x_3} + \frac{\partial u_3^0}{\partial x_2} = \frac{\sigma_{23}^0}{\mu_t^0},
\end{aligned} \tag{9.4}$$

while for the 90° plies (see (6.100))

$$\begin{aligned}
\varepsilon_{11}^{90} &= \frac{\partial u_1^{90}}{\partial x_1} = \frac{1}{E_{\mathrm{T}}^{90}}\sigma_{11}^{90} - \frac{\nu_{\mathrm{A}}^{90}}{E_{\mathrm{A}}^{90}}\sigma_{22}^{90} - \frac{\nu_{\mathrm{t}}^{90}}{E_{\mathrm{T}}^{90}}\sigma_{33}^{90} + \alpha_{\mathrm{T}}^{90}\Delta T, \quad 2\varepsilon_{12}^{90} = \frac{\partial u_1^{90}}{\partial x_2} + \frac{\partial u_2^{90}}{\partial x_1} = \frac{\sigma_{12}^{90}}{\mu_{\mathrm{A}}^{90}}, \\
\varepsilon_{22}^{90} &= \frac{\partial u_2^{90}}{\partial x_2} = -\frac{\nu_{\mathrm{A}}^{90}}{E_{\mathrm{A}}^{90}}\sigma_{11}^{90} + \frac{1}{E_{\mathrm{A}}^{90}}\sigma_{22}^{90} - \frac{\nu_{\mathrm{a}}^{90}}{E_{\mathrm{A}}^{90}}\sigma_{33}^{90} + \alpha_{\mathrm{A}}^{90}\Delta T, \quad 2\varepsilon_{13}^{90} = \frac{\partial u_1^{90}}{\partial x_3} + \frac{\partial u_3^{90}}{\partial x_1} = \frac{\sigma_{13}^{90}}{\mu_t^{90}}, \\
\varepsilon_{33}^{90} &= \frac{\partial u_3^{90}}{\partial x_3} = -\frac{\nu_{\mathrm{t}}^{90}}{E_{\mathrm{T}}^{90}}\sigma_{11}^{90} - \frac{\nu_{\mathrm{a}}^{90}}{E_{\mathrm{A}}^{90}}\sigma_{22}^{90} + \frac{1}{E_{\mathrm{t}}^{90}}\sigma_{33}^{90} + \alpha_{\mathrm{t}}^{90}\Delta T, \quad 2\varepsilon_{23}^{90} = \frac{\partial u_2^{90}}{\partial x_3} + \frac{\partial u_3^{90}}{\partial x_2} = \frac{\sigma_{23}^{90}}{\mu_{\mathrm{a}}^{90}},
\end{aligned} \tag{9.5}$$

where the strain and displacement components are denoted by ε_{ij} and u_i, respectively. The subscripts A, T and t are attached to the properties to associate them with the axial, in-plane transverse and through-thickness directions of the lamina, respectively. It should be noted that the uppercase subscripts A and T are associated only with in-plane directions, whereas the lowercase subscripts are associated with the through-thickness direction. Relations (9.5) are either obtained by modifying directly relations (9.4) for the 0° plies, or by using the transformations considered in Chapter 2 to rotate the 0° ply by an angle $\pm 90^{\circ}$. The thermoelastic constants of individual plies in a laminate are usually assumed to be transverse isotropic (see Chapter 2) so that

$$E_{\mathrm{t}} = E_{\mathrm{T}}, \quad \nu_{\mathrm{a}} = \nu_{\mathrm{A}}, \quad \mu_{\mathrm{a}} = \mu_{\mathrm{A}}, \quad \alpha_{\mathrm{t}} = \alpha_{\mathrm{T}} \text{ and } E_{\mathrm{T}} = 2\mu_{\mathrm{t}}\left(1+\nu_{\mathrm{t}}\right).$$

9.1.2 Boundary and Interface Conditions

In order that the field equations can be solved uniquely, it is necessary to impose a sufficient number of boundary and interface conditions. The free surface conditions on $x_3 = \pm h$ and interface conditions on $x_3 = \pm a$ are first considered. On the free surfaces

$$\sigma_{33}^{0} = \sigma_{\mathrm{t}}, \quad \sigma_{13}^{0} = \sigma_{23}^{0} = 0, \quad \text{on} \quad x_3 = \pm h, \tag{9.6}$$

and on the interfaces

$$\begin{aligned}
\sigma_{33}^{0} &= \sigma_{33}^{90}, \quad \sigma_{13}^{0} = \sigma_{13}^{90}, \quad \sigma_{23}^{0} = \sigma_{23}^{90}, \\
u_1^{0} &= u_1^{90}, \quad u_2^{0} = u_2^{90}, \quad u_3^{0} = u_3^{90},
\end{aligned} \quad \text{on} \quad x_3 = \pm a. \tag{9.7}$$

The edges $x_2 = \pm W$ are such that in-plane transverse displacement is uniform having the following values

$$u_2^{0} = u_2^{90} = \pm W\varepsilon_{\mathrm{T}}, \qquad \text{on} \quad x_2 = \pm W, \tag{9.8}$$

where ε_{T} is the in-plane transverse strain that is uniform everywhere in the laminate when generalised plane strain conditions are imposed. The edges $x_2 = \pm W$ are assumed to have zero shear stresses so that

$$\sigma_{12}^{0} = \sigma_{12}^{90} = 0, \quad \sigma_{23}^{0} = \sigma_{23}^{90} = 0, \quad \text{on} \quad x_2 = \pm W. \tag{9.9}$$

For the above boundary conditions, and because of the symmetric nature of the laminate, there will be symmetry about $x_3 = 0$ of stress, strain and displacement components such that the following conditions are satisfied

$$\sigma_{13}^{90} = 0, \quad \sigma_{23}^{90} = 0, \quad u_3^{90} = 0, \quad \text{on} \quad x_3 = 0. \tag{9.10}$$

When applying laminate edge conditions applied on planes normal to the x_1-axis, two possible approaches can be made. Consider first the RVE shown in Figure 9.1(a) which can be used for undamaged laminates, and for damaged laminates where a ply crack in the 90° ply is located at $x_1 = 0$. The edges $x_1 = \pm L$ are such that in-plane axial displacement is uniform having the following values

$$u_1^0 = u_1^{90} = \pm L\varepsilon_{\rm A}, \quad \text{on} \quad x_1 = \pm L, \tag{9.11}$$

where $\varepsilon_{\rm A}$ is the effective axial applied strain. The edges $x_1 = \pm L$ are assumed to have zero shear stresses so that

$$\sigma_{12}^0 = \sigma_{12}^{90} = 0, \quad \sigma_{13}^0 = \sigma_{13}^{90} = 0, \quad \text{on} \quad x_1 = \pm L. \tag{9.12}$$

These conditions imply that there is symmetry about the plane $x_1 = 0$ so that

$$\sigma_{12}^0 = \sigma_{12}^{90} = 0, \quad \sigma_{13}^0 = \sigma_{13}^{90} = 0, \quad u_1^0 = u_1^{90} = 0, \quad \text{on} \quad x_1 = 0. \tag{9.13}$$

Consider now the RVE shown in Figure 9.1(b) where the 90° ply cracks are located on the planes $x_1 = \pm L$. The boundary conditions applied are given by

$$\sigma_{12}^0 = \sigma_{12}^{90} = 0, \quad \sigma_{13}^0 = \sigma_{13}^{90} = 0, \quad \sigma_{11}^{90} = 0, \quad u_1^0 = \pm L\varepsilon_{\rm A}, \quad \text{on} \quad x_1 = \pm L. \tag{9.14}$$

It is clear from both the RVEs shown in Figure 9.1 and the boundary conditions applied on planes normal to the x_1-axis that for infinite uniform of ply cracks there is symmetry about the planes $x_1 = 0$ and $x_1 = \pm L$.

9.1.3 Generalised Plane Strain Conditions

Generalised plane strain conditions are assumed so that the stress and strain distributions do not depend on the x_3-coordinate. This situation occurs when the displacement field is of the form

$$u_1 = u_1(x_1, x_3), \qquad u_2 = \varepsilon_{\rm T} x_2, \qquad u_3 = u_3(x_1, x_3), \tag{9.15}$$

where $\varepsilon_{22} = \varepsilon_{\rm T}$ is the uniform in-plane transverse strain in both the 0° and 90° plies. In addition, it is assumed that on the laminate faces $x_3 = \pm h$ that $\sigma_{33}^0 \equiv \sigma_{33}^{90} \equiv \sigma_{\rm t}$ where $\sigma_{\rm t}$ is a uniform applied through-thickness stress and $\sigma_{13} \equiv \sigma_{23} \equiv 0$. If $\varepsilon_{\rm T} = 0$, then the laminate is highly constrained in the transverse direction leading to well-known conditions of plane strain deformation. It should be noted that the solution for undamaged laminates derived in Section 9.2 is automatically one of generalised plane strain as the transverse strain $\varepsilon_{\rm T}$ is uniform everywhere in those regions of the laminate that are sufficiently well away from the edges.

From (9.4) it then follows that for the 0° ply

$$\sigma_{22}^0 = \frac{\nu_{\rm A}^0 E_{\rm T}^0}{E_{\rm A}^0}\sigma_{11}^0 + \nu_{\rm t}^0\sigma_{33}^0 - E_{\rm T}^0\alpha_{\rm T}^0\Delta T + E_{\rm T}^0\varepsilon_{\rm T}, \tag{9.16}$$

and

$$\varepsilon_{11}^0 = \frac{1}{\tilde{E}_{\rm A}^0}\sigma_{11}^0 - \frac{\tilde{\nu}_{\rm a}^0}{\tilde{E}_{\rm A}^0}\sigma_{33}^0 + \tilde{\alpha}_{\rm A}^0\Delta T - \nu_{\rm A}^0\frac{E_{\rm T}^0}{E_{\rm A}^0}\varepsilon_{\rm T}, \tag{9.17}$$

$$\varepsilon_{33}^{0} = -\frac{\tilde{\nu}_{a}^{0}}{\tilde{E}_{A}^{0}}\sigma_{11}^{0} + \frac{1}{\tilde{E}_{t}^{0}}\sigma_{33}^{0} + \tilde{\alpha}_{t}^{0}\Delta T - \nu_{t}^{0}\varepsilon_{T}, \tag{9.18}$$

where in addition to the relations (9.30) below

$$\frac{1}{\tilde{E}_{t}^{0}} = \frac{1}{E_{t}^{0}} - \frac{(\nu_{t}^{0})^{2}}{E_{T}^{0}}, \qquad \tilde{\alpha}_{t}^{0} = \alpha_{t}^{0} + \nu_{t}^{0}\alpha_{T}^{0}. \tag{9.19}$$

For the 90° ply it follows from (9.5) that

$$\sigma_{22}^{90} = \nu_{A}^{90}\sigma_{11}^{90} + \nu_{a}^{90}\sigma_{33}^{90} - E_{A}^{90}\alpha_{A}^{90}\Delta T + E_{A}^{90}\varepsilon_{T}, \tag{9.20}$$

and

$$\varepsilon_{11}^{90} = \frac{1}{\tilde{E}_{T}^{90}}\sigma_{11}^{90} - \frac{\tilde{\nu}_{t}^{90}}{\tilde{E}_{T}^{90}}\sigma_{33}^{90} + \tilde{\alpha}_{T}^{90}\Delta T - \nu_{A}^{90}\varepsilon_{T}, \tag{9.21}$$

$$\varepsilon_{33}^{90} = -\frac{\tilde{\nu}_{t}^{90}}{\tilde{E}_{T}^{90}}\sigma_{11}^{90} + \frac{1}{\tilde{E}_{t}^{90}}\sigma_{33}^{90} + \tilde{\alpha}_{t}^{90}\Delta T - \nu_{a}^{90}\varepsilon_{T}, \tag{9.22}$$

where in addition to the relations (9.34) below

$$\frac{1}{\tilde{E}_{t}^{90}} = \frac{1}{E_{t}^{90}} - \frac{(\nu_{a}^{90})^{2}}{E_{A}^{90}}, \qquad \tilde{\alpha}_{t}^{90} = \alpha_{t}^{90} + \nu_{a}^{90}\alpha_{A}^{90}. \tag{9.23}$$

It follows from (9.15) and the strain–displacement relations that $\varepsilon_{12} = \varepsilon_{23} = 0$ in both 0° and 90° plies and from the stress-strain relations (9.4) and (9.5) that $\sigma_{12} = \sigma_{23} = 0$. The equilibrium equation (9.2) is then automatically satisfied as, from (9.16), (9.20) and the stress-strain equations (9.4) and (9.5), σ_{22} is independent of x_2 as are both σ_{11} and σ_{13}. The remaining equilibrium equations (9.1) and (9.3) then reduce to the form

$$\frac{\partial \sigma_{11}}{\partial x_1} + \frac{\partial \sigma_{13}}{\partial x_3} = 0, \tag{9.24}$$

$$\frac{\partial \sigma_{13}}{\partial x_1} + \frac{\partial \sigma_{33}}{\partial x_3} = 0. \tag{9.25}$$

When developing stress-transfer models for application in conditions of generalised plane strain, it is very useful to consider averages through the thickness of the plies of various physical quantities. Assuming symmetry about the midplane $x_3 = 0$ of the laminate, the average of any quantities $f_0(x_1, x_3)$ and $f_{90}(x_1, x_3)$ associated with the 0° and 90° plies are defined, respectively, by

$$\bar{f}_0(x_1) = \frac{1}{b}\int_a^h f_0(x_1, x_3)\,\mathrm{d}x_3\,, \quad \bar{f}_{90}(x_1) = \frac{1}{a}\int_0^a f_{90}(x_1, x_3)\,\mathrm{d}x_3. \tag{9.26}$$

9.2 Solution for Undamaged Laminates

For undamaged laminates in regions away from laminate edges, the axial and transverse strains in both 0° and 90° plies have the uniform values $\hat{\varepsilon}_A$ and $\hat{\varepsilon}_T$, respectively. The 'hat' symbol is attached to the in-plane strains to distinguish them from the differing values ε_A and ε_T that will arise when the laminate is damaged. The through-thickness stress in both plies has the uniform value σ_t. For undamaged laminates the uniform stress and strain components in the plies will be denoted by $\hat\sigma_{ij}^0$, $\hat\sigma_{ij}^{90}$ and $\hat\varepsilon_{ij}^0$, $\hat\varepsilon_{ij}^{90}$, respectively. It follows from the stress-strain relations (9.4) that

$$\begin{aligned}
\hat\varepsilon_{\mathrm{A}} &= \frac{1}{E_{\mathrm{A}}^0}\hat\sigma_{11}^0 - \frac{\nu_{\mathrm{A}}^0}{E_{\mathrm{A}}^0}\hat\sigma_{22}^0 - \frac{\nu_{\mathrm{a}}^0}{E_{\mathrm{A}}^0}\sigma_{\mathrm{t}} + \alpha_{\mathrm{A}}^0\Delta T,\\
\hat\varepsilon_{\mathrm{T}} &= -\frac{\nu_{\mathrm{A}}^0}{E_{\mathrm{A}}^0}\hat\sigma_{11}^0 + \frac{1}{E_{\mathrm{T}}^0}\hat\sigma_{22}^0 - \frac{\nu_{\mathrm{t}}^0}{E_{\mathrm{T}}^0}\sigma_{\mathrm{t}} + \alpha_{\mathrm{T}}^0\Delta T,\\
\hat\varepsilon_{33} &= -\frac{\nu_{\mathrm{a}}^0}{E_{\mathrm{A}}^0}\hat\sigma_{11}^0 - \frac{\nu_{\mathrm{t}}^0}{E_{\mathrm{T}}^0}\hat\sigma_{22}^0 + \frac{1}{E_{\mathrm{t}}^0}\sigma_{\mathrm{t}} + \alpha_{\mathrm{t}}^0\Delta T,
\end{aligned}\tag{9.27}$$

so that on solving the first two equations for $\hat\sigma_{11}^0$ and $\hat\sigma_{22}^0$

$$\begin{aligned}
\frac{1}{\tilde E_{\mathrm{A}}^0}\hat\sigma_{11}^0 &= \hat\varepsilon_{\mathrm{A}} + \nu_{\mathrm{A}}^0\frac{E_{\mathrm{T}}^0}{E_{\mathrm{A}}^0}\hat\varepsilon_{\mathrm{T}} + \frac{\tilde\nu_{\mathrm{a}}^0}{\tilde E_{\mathrm{A}}^0}\sigma_{\mathrm{t}} - \tilde\alpha_{\mathrm{A}}^0\Delta T,\\
\frac{1}{\tilde E_{\mathrm{T}}^0}\hat\sigma_{22}^0 &= \nu_{\mathrm{A}}^0\hat\varepsilon_{\mathrm{A}} + \hat\varepsilon_{\mathrm{T}} + \frac{\tilde\nu_{\mathrm{t}}^0}{\tilde E_{\mathrm{T}}^0}\sigma_{\mathrm{t}} - \tilde\alpha_{\mathrm{T}}^0\Delta T,
\end{aligned}\tag{9.28}$$

and on substituting into the expression for $\hat\varepsilon_{33}^0$

$$\begin{aligned}
\hat\varepsilon_{33}^0 = &-\frac{\nu_{\mathrm{a}}^0+\nu_{\mathrm{t}}^0\nu_{\mathrm{A}}^0}{E_{\mathrm{A}}^0}\tilde E_{\mathrm{A}}^0\hat\varepsilon_{\mathrm{A}} - \left(\frac{\nu_{\mathrm{t}}^0}{E_{\mathrm{T}}^0} + \frac{\nu_{\mathrm{a}}^0\nu_{\mathrm{A}}^0}{E_{\mathrm{A}}^0}\right)\tilde E_{\mathrm{T}}^0\hat\varepsilon_{\mathrm{T}} + \left(\frac{1}{E_{\mathrm{t}}^0} - \frac{\nu_{\mathrm{a}}^0\tilde\nu_{\mathrm{a}}^0}{E_{\mathrm{A}}^0} - \frac{\nu_{\mathrm{t}}^0\tilde\nu_{\mathrm{t}}^0}{E_{\mathrm{T}}^0}\right)\sigma_{\mathrm{t}}\\
&+ \left(\alpha_{\mathrm{t}}^0 + \frac{\nu_{\mathrm{a}}^0}{E_{\mathrm{A}}^0}\tilde E_{\mathrm{A}}^0\tilde\alpha_{\mathrm{A}}^0 + \frac{\nu_{\mathrm{t}}^0}{E_{\mathrm{T}}^0}\tilde E_{\mathrm{T}}^0\tilde\alpha_{\mathrm{T}}^0\right)\Delta T,
\end{aligned}\tag{9.29}$$

where

$$\begin{aligned}
\frac{1}{\tilde E_{\mathrm{A}}^0} &= \frac{1}{E_{\mathrm{A}}^0}\left(1-(\nu_{\mathrm{A}}^0)^2\frac{E_{\mathrm{T}}^0}{E_{\mathrm{A}}^0}\right), &\quad \frac{1}{\tilde E_{\mathrm{T}}^0} &= \frac{1}{E_{\mathrm{T}}^0}\left(1-(\nu_{\mathrm{A}}^0)^2\frac{E_{\mathrm{T}}^0}{E_{\mathrm{A}}^0}\right),\\
\frac{\tilde\nu_{\mathrm{a}}^0}{\tilde E_{\mathrm{A}}^0} &= \frac{\nu_{\mathrm{a}}^0+\nu_{\mathrm{t}}^0\nu_{\mathrm{A}}^0}{E_{\mathrm{A}}^0}, &\quad \frac{\tilde\nu_{\mathrm{t}}^0}{\tilde E_{\mathrm{T}}^0} &= \frac{\nu_{\mathrm{t}}^0}{E_{\mathrm{T}}^0} + \frac{\nu_{\mathrm{a}}^0\nu_{\mathrm{A}}^0}{E_{\mathrm{A}}^0},\\
\tilde\alpha_{\mathrm{A}}^0 &= \alpha_{\mathrm{A}}^0 + \nu_{\mathrm{A}}^0\frac{E_{\mathrm{T}}^0}{E_{\mathrm{A}}^0}\alpha_{\mathrm{T}}^0, &\quad \tilde\alpha_{\mathrm{T}}^0 &= \alpha_{\mathrm{T}}^0 + \nu_{\mathrm{A}}^0\alpha_{\mathrm{A}}^0.
\end{aligned}\tag{9.30}$$

Similarly, from (9.5)

$$\begin{aligned}
\hat\varepsilon_{\mathrm{A}} &= \frac{1}{E_{\mathrm{T}}^{90}}\hat\sigma_{11}^{90} - \frac{\nu_{\mathrm{A}}^{90}}{E_{\mathrm{A}}^{90}}\hat\sigma_{22}^{90} - \frac{\nu_{\mathrm{t}}^{90}}{E_{\mathrm{T}}^{90}}\sigma_{\mathrm{t}} + \alpha_{\mathrm{T}}^{90}\Delta T,\\
\hat\varepsilon_{\mathrm{T}} &= -\frac{\nu_{\mathrm{A}}^{90}}{E_{\mathrm{A}}^{90}}\hat\sigma_{11}^{90} + \frac{1}{E_{\mathrm{A}}^{90}}\hat\sigma_{22}^{90} - \frac{\nu_{\mathrm{a}}^{90}}{E_{\mathrm{A}}^{90}}\sigma_{\mathrm{t}} + \alpha_{\mathrm{A}}^{90}\Delta T,\\
\hat\varepsilon_{33}^{90} &= -\frac{\nu_{\mathrm{t}}^{90}}{E_{\mathrm{T}}^{90}}\hat\sigma_{11}^{90} - \frac{\nu_{\mathrm{a}}^{90}}{E_{\mathrm{t}}^{90}}\hat\sigma_{22}^{90} + \frac{1}{E_{\mathrm{t}}^{90}}\sigma_{\mathrm{t}} + \alpha_{\mathrm{t}}^{90}\Delta T,
\end{aligned}\tag{9.31}$$

so that on solving the first two equations for $\hat\sigma_{11}^{90}$ and $\hat\sigma_{22}^{90}$

$$\begin{aligned}
\frac{1}{\tilde E_{\mathrm{T}}^{90}}\hat\sigma_{11}^{90} &= \hat\varepsilon_{\mathrm{A}} + \nu_{\mathrm{A}}^{90}\hat\varepsilon_{\mathrm{T}} + \frac{\tilde\nu_{\mathrm{t}}^{90}}{\tilde E_{\mathrm{T}}^{90}}\sigma_{\mathrm{t}} - \tilde\alpha_{\mathrm{T}}^{90}\Delta T,\\
\frac{1}{\tilde E_{\mathrm{A}}^{90}}\hat\sigma_{22}^{90} &= \nu_{\mathrm{A}}^{90}\frac{E_{\mathrm{T}}^{90}}{E_{\mathrm{A}}^{90}}\hat\varepsilon_{\mathrm{A}} + \hat\varepsilon_{\mathrm{T}} + \frac{\tilde\nu_{\mathrm{a}}^{90}}{\tilde E_{\mathrm{A}}^{90}}\sigma_{\mathrm{t}} - \tilde\alpha_{\mathrm{A}}^{90}\Delta T,
\end{aligned}\tag{9.32}$$

and on substituting into the expression for $\hat{\varepsilon}_{33}^{90}$

$$\hat{\varepsilon}_{33}^{90} = -\left(\frac{\nu_t^{90}}{E_T^{90}} + \frac{\nu_a^{90}\nu_A^{90}}{E_A^{90}}\right)\tilde{E}_T^{90}\hat{\varepsilon}_A - \frac{\nu_a^{90} + \nu_t^{90}\nu_A^{90}}{E_A^{90}}\tilde{E}_A^{90}\hat{\varepsilon}_T \\ + \left(\frac{1}{E_t^{90}} - \frac{\nu_t^{90}\tilde{\nu}_t^{90}}{E_T^{90}} - \frac{\nu_a^{90}\tilde{\nu}_a^{90}}{E_A^{90}}\right)\sigma_t + \left(\alpha_t^{90} + \frac{\nu_a^{90}}{E_A^{90}}\tilde{E}_A^{90}\tilde{\alpha}_A^{90} + \frac{\nu_t^{90}}{E_T^{90}}\tilde{E}_T^{90}\tilde{\alpha}_T^{90}\right)\Delta T, \tag{9.33}$$

where

$$\frac{1}{\tilde{E}_A^{90}} = \frac{1}{E_A^{90}}\left(1 - (\nu_A^{90})^2\frac{E_T^{90}}{E_A^{90}}\right), \qquad \frac{1}{\tilde{E}_T^{90}} = \frac{1}{E_T^{90}}\left(1 - (\nu_A^{90})^2\frac{E_T^{90}}{E_A^{90}}\right), \\ \frac{\tilde{\nu}_a^{90}}{\tilde{E}_A^{90}} = \frac{\nu_a^{90}}{E_A^{90}} + \frac{\nu_t^{90}\nu_A^{90}}{E_A^{90}}, \qquad \frac{\tilde{\nu}_t^{90}}{\tilde{E}_T^{90}} = \frac{\nu_t^{90}}{E_T^{90}} + \frac{\nu_a^{90}\nu_A^{90}}{E_A^{90}}, \\ \tilde{\alpha}_A^{90} = \alpha_A^{90} + \nu_A^{90}\frac{E_T^{90}}{E_A^{90}}\alpha_T^{90}, \qquad \tilde{\alpha}_T^{90} = \alpha_T^{90} + \nu_A^{90}\alpha_A^{90}. \tag{9.34}$$

From a consideration of mechanical equilibrium, the uniform ply stresses can be used to define, for an undamaged laminate, the effective axial stress $\hat{\sigma}_A$ and the effective in-plane transverse stress $\hat{\sigma}_T$ as follows

$$h\hat{\sigma}_A = b\hat{\sigma}_{11}^0 + a\hat{\sigma}_{11}^{90}, \tag{9.35}$$

$$h\hat{\sigma}_T = b\hat{\sigma}_{22}^0 + a\hat{\sigma}_{22}^{90}. \tag{9.36}$$

The 'hat' symbol is used to distinguish these effective stresses from those that will result when the laminate is damaged. Corresponding to the uniform through-thickness stress σ_t, an effective through-thickness strain $\hat{\varepsilon}_t$ can be defined by the relation

$$h\hat{\varepsilon}_t = b\hat{\varepsilon}_{33}^0 + a\hat{\varepsilon}_{33}^{90}. \tag{9.37}$$

It should be noted that the value σ_t for the through-thickness stress of an undamaged laminate corresponds to the effective value when the laminate is damaged. On using relations (9.35) and (9.36) together with the results (9.28) and (9.32) it can be shown that

$$\hat{\sigma}_A = A\hat{\varepsilon}_A + B\hat{\varepsilon}_T + C\sigma_t - P\Delta T, \\ \hat{\sigma}_T = B\hat{\varepsilon}_A + F\hat{\varepsilon}_T + G\sigma_t - Q\Delta T, \tag{9.38}$$

where

$$A = \frac{b}{h}\tilde{E}_A^0 + \frac{a}{h}\tilde{E}_T^{90}, \quad B = \frac{b}{h}\nu_A^0\tilde{E}_T^0 + \frac{a}{h}\nu_A^{90}\tilde{E}_T^{90}, \quad C = \frac{b}{h}\tilde{\nu}_a^0 + \frac{a}{h}\tilde{\nu}_t^{90}, \\ F = \frac{b}{h}\tilde{E}_T^0 + \frac{a}{h}\tilde{E}_A^{90}, \quad G = \frac{b}{h}\tilde{\nu}_t^0 + \frac{a}{h}\tilde{\nu}_a^{90}, \\ P = \frac{b}{h}\tilde{E}_A^0\tilde{\alpha}_A^0 + \frac{a}{h}\tilde{E}_T^{90}\tilde{\alpha}_T^{90}, \quad Q = \frac{b}{h}\tilde{E}_T^0\tilde{\alpha}_T^0 + \frac{a}{h}\tilde{E}_A^{90}\tilde{\alpha}_A^{90}. \tag{9.39}$$

On solving (9.38) for the in-plane strains $\hat{\varepsilon}_A$ and $\hat{\varepsilon}_T$

$$\hat{\varepsilon}_A = \frac{1}{E_A^{(L)}}\hat{\sigma}_A - \frac{\nu_A^{(L)}}{E_A^{(L)}}\hat{\sigma}_T - \frac{\nu_a^{(L)}}{E_A^{(L)}}\sigma_t + \alpha_A^{(L)}\Delta T, \\ \hat{\varepsilon}_T = -\frac{\nu_A^{(L)}}{E_T^{(L)}}\hat{\sigma}_T + \frac{1}{E_T^{(L)}}\hat{\sigma}_T - \frac{\nu_t^{(L)}}{E_T^{(L)}}\sigma_t + \alpha_T^{(L)}\Delta T, \tag{9.40}$$

where

$$E_{\mathrm{A}}^{(\mathrm{L})}=A-\frac{B^2}{F},\qquad E_{\mathrm{T}}^{(\mathrm{L})}=F-\frac{B^2}{A},\qquad \nu_{\mathrm{A}}^{(\mathrm{L})}=\frac{B}{F},$$
$$\nu_{\mathrm{a}}^{(\mathrm{L})}=C-\frac{BG}{F},\qquad \nu_{\mathrm{t}}^{(\mathrm{L})}=G-\frac{BC}{A},\tag{9.41}$$
$$\alpha_{\mathrm{A}}^{(\mathrm{L})}=\frac{1}{E_{\mathrm{A}}^{(\mathrm{L})}}\left(P-\nu_{\mathrm{A}}^{(\mathrm{L})}Q\right),\quad \alpha_{\mathrm{T}}^{(\mathrm{L})}=\frac{1}{E_{\mathrm{T}}^{(\mathrm{L})}}\left(Q-\nu_{\mathrm{A}}^{(\mathrm{L})}\frac{E_{\mathrm{T}}^{(\mathrm{L})}}{E_{\mathrm{A}}^{(\mathrm{L})}}P\right).$$

It should be noted that

$$\tilde{E}_{\mathrm{A}}^{(\mathrm{L})}=A=\frac{b}{h}\tilde{E}_{\mathrm{A}}^{0}+\frac{a}{h}\tilde{E}_{\mathrm{T}}^{90},\qquad \nu_{\mathrm{A}}^{(\mathrm{L})}\frac{E_{\mathrm{T}}^{(\mathrm{L})}}{E_{\mathrm{A}}^{(\mathrm{L})}}=\frac{B}{A}=\frac{b\nu_{\mathrm{A}}^{0}\tilde{E}_{\mathrm{T}}^{0}+a\nu_{\mathrm{A}}^{90}\tilde{E}_{\mathrm{T}}^{90}}{b\tilde{E}_{\mathrm{A}}^{0}+a\tilde{E}_{\mathrm{T}}^{90}},$$
$$\tilde{E}_{\mathrm{T}}^{(\mathrm{L})}=F=\frac{b}{h}\tilde{E}_{\mathrm{T}}^{0}+\frac{a}{h}\tilde{E}_{\mathrm{A}}^{90},\qquad \nu_{\mathrm{A}}^{(\mathrm{L})}=\frac{B}{F}=\frac{b\nu_{\mathrm{A}}^{0}\tilde{E}_{\mathrm{T}}^{0}+a\nu_{\mathrm{A}}^{90}\tilde{E}_{\mathrm{T}}^{90}}{b\tilde{E}_{\mathrm{T}}^{0}+a\tilde{E}_{\mathrm{A}}^{90}},\tag{9.42}$$

where

$$\tilde{E}_{\mathrm{A}}^{(\mathrm{L})}=\frac{E_{\mathrm{A}}^{(\mathrm{L})}}{1-(\nu_{\mathrm{A}}^{(\mathrm{L})})^2E_{\mathrm{T}}^{(\mathrm{L})}/E_{\mathrm{A}}^{(\mathrm{L})}},\qquad \tilde{E}_{\mathrm{T}}^{(\mathrm{L})}=\frac{E_{\mathrm{T}}^{(\mathrm{L})}}{1-(\nu_{\mathrm{A}}^{(\mathrm{L})})^2E_{\mathrm{T}}^{(\mathrm{L})}/E_{\mathrm{A}}^{(\mathrm{L})}}.\tag{9.43}$$

On using (9.37) together with the results (9.29) and (9.33) it can be shown that

$$\hat{\varepsilon}_{\mathrm{t}}=-A'\hat{\varepsilon}_{\mathrm{A}}-B'\hat{\varepsilon}_{\mathrm{T}}+C'\sigma_{\mathrm{t}}+P'\Delta T,\tag{9.44}$$

where

$$A'=\frac{b}{h}\frac{\nu_{\mathrm{a}}^{0}+\nu_{\mathrm{t}}^{0}\nu_{\mathrm{A}}^{0}}{E_{\mathrm{A}}^{0}}\tilde{E}_{\mathrm{A}}^{0}+\frac{a}{h}\left(\frac{\nu_{\mathrm{t}}^{90}}{E_{\mathrm{T}}^{90}}+\frac{\nu_{\mathrm{a}}^{90}\nu_{\mathrm{A}}^{90}}{E_{\mathrm{A}}^{90}}\right)\tilde{E}_{\mathrm{T}}^{90}=\frac{b}{h}\tilde{\nu}_{\mathrm{a}}^{0}+\frac{a}{h}\tilde{\nu}_{\mathrm{t}}^{90},$$
$$B'=\frac{b}{h}\left(\frac{\nu_{\mathrm{t}}^{0}}{E_{\mathrm{T}}^{0}}+\frac{\nu_{\mathrm{a}}^{0}\nu_{\mathrm{A}}^{0}}{E_{\mathrm{A}}^{0}}\right)\tilde{E}_{\mathrm{T}}^{0}+\frac{a}{h}\frac{\nu_{\mathrm{a}}^{90}+\nu_{\mathrm{t}}^{90}\nu_{\mathrm{A}}^{90}}{E_{\mathrm{A}}^{90}}\tilde{E}_{\mathrm{A}}^{90}=\frac{b}{h}\tilde{\nu}_{\mathrm{t}}^{0}+\frac{a}{h}\tilde{\nu}_{\mathrm{a}}^{90},$$
$$C'=\frac{b}{h}\left(\frac{1}{E_{\mathrm{t}}^{0}}-\frac{\nu_{\mathrm{a}}^{0}\tilde{\nu}_{\mathrm{a}}^{0}}{E_{\mathrm{A}}^{0}}-\frac{\nu_{\mathrm{t}}^{0}\tilde{\nu}_{\mathrm{t}}^{0}}{E_{\mathrm{T}}^{0}}\right)+\frac{a}{h}\left(\frac{1}{E_{\mathrm{t}}^{90}}-\frac{\nu_{\mathrm{t}}^{90}\tilde{\nu}_{\mathrm{t}}^{90}}{E_{\mathrm{T}}^{90}}-\frac{\nu_{\mathrm{a}}^{90}\tilde{\nu}_{\mathrm{a}}^{90}}{E_{\mathrm{A}}^{90}}\right),\tag{9.45}$$
$$P'=\frac{b}{h}\left(\alpha_{\mathrm{t}}^{0}+\frac{\nu_{\mathrm{a}}^{0}}{E_{\mathrm{A}}^{0}}\tilde{E}_{\mathrm{A}}^{0}\tilde{\alpha}_{\mathrm{A}}^{0}+\frac{\nu_{\mathrm{t}}^{0}}{E_{\mathrm{T}}^{0}}\tilde{E}_{\mathrm{T}}^{0}\tilde{\alpha}_{\mathrm{T}}^{0}\right)+\frac{a}{h}\left(\alpha_{\mathrm{t}}^{90}+\frac{\nu_{\mathrm{a}}^{90}}{E_{\mathrm{A}}^{90}}\tilde{E}_{\mathrm{A}}^{90}\tilde{\alpha}_{\mathrm{A}}^{90}+\frac{\nu_{\mathrm{t}}^{90}}{E_{\mathrm{T}}^{90}}\tilde{E}_{\mathrm{T}}^{90}\tilde{\alpha}_{\mathrm{T}}^{90}\right).$$

On substituting the expressions (9.40) into (9.44) it follows that

$$\hat{\varepsilon}_{\mathrm{t}}=-\frac{\nu_{\mathrm{a}}^{(\mathrm{L})}}{E_{\mathrm{A}}^{(\mathrm{L})}}\hat{\sigma}_{\mathrm{A}}-\frac{\nu_{\mathrm{t}}^{(\mathrm{L})}}{E_{\mathrm{T}}^{(\mathrm{L})}}\hat{\sigma}_{\mathrm{T}}+\frac{1}{E_{\mathrm{t}}^{(\mathrm{L})}}\sigma_{\mathrm{t}}+\alpha_{\mathrm{t}}^{(\mathrm{L})}\Delta T,\tag{9.46}$$

where

$$\nu_{\mathrm{a}}^{(\mathrm{L})}=A'-B'\nu_{\mathrm{A}}^{(\mathrm{L})},\qquad \nu_{\mathrm{t}}^{(\mathrm{L})}=B'-A'\nu_{\mathrm{A}}^{(\mathrm{L})}\frac{E_{\mathrm{T}}^{(\mathrm{L})}}{E_{\mathrm{A}}^{(\mathrm{L})}},$$
$$\frac{1}{E_{\mathrm{t}}^{(\mathrm{L})}}=C'+A'\frac{\nu_{\mathrm{a}}^{(\mathrm{L})}}{E_{\mathrm{A}}^{(\mathrm{L})}}+B'\frac{\nu_{\mathrm{t}}^{(\mathrm{L})}}{E_{\mathrm{T}}^{(\mathrm{L})}},\qquad \alpha_{\mathrm{t}}^{(\mathrm{L})}=P'-A'\alpha_{\mathrm{A}}^{(\mathrm{L})}-B'\alpha_{\mathrm{T}}^{(\mathrm{L})}.\tag{9.47}$$

It can be shown that the same expressions result for the minor Poisson's ratios $\nu_{\mathrm{a}}^{(\mathrm{L})}$ and $\nu_{\mathrm{t}}^{(\mathrm{L})}$ when using relations (9.41) or (9.47). It should be noted that the results (9.40) and (9.46) for the laminate have exactly the same form as the stress-strain relations (9.4) for the 0° ply.

9.3 Shear-lag Theory for Cross-ply Laminates

Stress-transfer phenomena in cross-ply laminates have been examined extensively using an approximate method of stress analysis, first developed for UD composites, that is known as 'shear-lag theory'. The objective now is to describe the nature of this methodology and to identify the various approximations that have to be made. It is assumed in this section that the through-thickness stress σ_{33} is everywhere zero. The approach to be adopted follows the work of Nuismer and Tan [1], as in previous work by the present author [2]. Nairn and Mendels [3] have critically examined approaches to shear-lag theory, including comparison of predictions with the results of finite element analysis, and concluded that Nuismer and Tan's approach, which is now to be followed, is the most reliable shear-lag methodology, although it is by no means a satisfactory solution to stress-transfer problems in cross-ply laminates.

The shear-lag approach is based on averages defined over the regions occupied by the 0° and 90° plies. Averaging (9.24) over the 0° ply using $(9.26)_1$ leads to the result

$$\frac{\mathrm{d}\bar{\sigma}_{11}^{0}}{\mathrm{d}x_1} = \frac{\tau(x_1)}{b}, \tag{9.48}$$

assuming that $\sigma_{13}(x_1, h) = 0$, where $\tau(x_1) = \sigma_{13}(x_1, a)$ is the value of the shear stress on the interface $x_2 = a$. On averaging (9.24) over the 90° ply using $(9.26)_2$ it can be shown that

$$\frac{\mathrm{d}\bar{\sigma}_{11}^{90}}{\mathrm{d}x_1} = -\frac{\tau(x_1)}{a}, \tag{9.49}$$

since $\sigma_{13}(x_1, 0) = 0$ because of symmetry about the plane $x_3 = 0$. The result (9.49) also assumes that σ_{13} is continuous across the interface $x_2 = a$. It follows from (9.48) and (9.49) that

$$\frac{\mathrm{d}}{\mathrm{d}x_1}\left(b\bar{\sigma}_{11}^{0} + a\bar{\sigma}_{11}^{90}\right) = 0. \tag{9.50}$$

The integrated form of (9.50) is

$$b\bar{\sigma}_{11}^{0}(x_1) + a\bar{\sigma}_{11}^{90}(x_1) = h\sigma_{\mathrm{A}}, \tag{9.51}$$

where σ_{A} is the effective stress applied in the axial direction. It is emphasised that relation (9.51) is valid for any value of x_1.

From (9.4) and (9.5) the stress-strain relations involving the shear stress σ_{13} may be written for the 0° and 90° plies as follows

$$\frac{\partial u_1^{0}}{\partial x_3} + \frac{\partial u_3^{0}}{\partial x_1} = \frac{\sigma_{13}^{0}}{\mu_{\mathrm{a}}^{0}}, \qquad \frac{\partial u_1^{90}}{\partial x_3} + \frac{\partial u_3^{90}}{\partial x_1} = \frac{\sigma_{13}^{90}}{\mu_{\mathrm{t}}^{90}}. \tag{9.52}$$

To develop a shear-lag model these relations are now approximated by

$$\frac{\partial u_1^0}{\partial x_3} = \frac{\sigma_{13}^0}{\mu_a^0}, \qquad \frac{\partial u_1^{90}}{\partial x_3} = \frac{\sigma_{13}^{90}}{\mu_t^{90}}. \tag{9.53}$$

The shear stress distribution is now assumed to have the following form

$$\sigma_{13}^0(x_1,x_3) = \tau(x_1)\frac{h-x_3}{b}, \qquad \sigma_{13}^{90}(x_1,x_3) = \tau(x_1)\frac{x_3}{a}. \tag{9.54}$$

It is clear that $\sigma_{13}^0 = \sigma_{13}^{90}$ on $x_3 = a$, and that $\sigma_{13}^0(x_1,h) = 0$. On multiplying $(9.53)_1$ by $(h - x_3)$, on substituting the expression $(9.54)_1$ for σ_{13}^0 and then integrating with respect to x_3 over the 0° ply, it can be shown that

$$\int_a^h (h-x_3)\frac{\partial u_1^0}{\partial x_3}\mathrm{d}x_3 = \frac{b^2\tau(x_1)}{3\mu_a^0}. \tag{9.55}$$

On integrating the left-hand side of (9.55) by parts it can be shown that

$$\bar{u}_1^0(x_1) - u_1^0(x_1,a) = \frac{b\tau(x_1)}{3\mu_a^0}. \tag{9.56}$$

On multiplying $(9.53)_2$ by x_3, on substituting expression $(9.54)_2$ for σ_{13} and then integrating with respect to x_3 over the 90° ply, it can be shown that

$$\int_0^a x_3\frac{\partial u_1^{90}}{\partial x_3}\mathrm{d}x_3 = \frac{a^2\tau(x_1)}{3\mu_t^{90}}. \tag{9.57}$$

On integrating the left-hand side of (9.57) by parts, it can be shown that

$$u_1^{90}(x_1,a) - \bar{u}_1^{90}(x_1) = \frac{a\tau(x_1)}{3\mu_t^{90}}. \tag{9.58}$$

As there is perfect bonding on the interface $x_3 = a$ so that $u_1^0(x_1,a) = u_1^{90}(x_1,a)$, the addition of (9.56) and (9.58) leads to the important result

$$\bar{u}_1^0(x_1) - \bar{u}_1^{90}(x_1) = \frac{1}{3}\left(\frac{b}{\mu_a^0} + \frac{a}{\mu_t^{90}}\right)\tau(x_1). \tag{9.59}$$

Clearly the interfacial shear stress is proportional to the difference between the average axial displacements in the 0° and 90° plies. It should be noted that a relationship of this form was derived by Nuismer and Tan [1].

As it has been assumed in this section that the through-thickness stress σ_{33} is zero, relations (9.17) and (9.21) are now averaged using (9.26) so that

$$\bar{\varepsilon}_{11}^0 = \frac{\mathrm{d}\bar{u}_1^0}{\mathrm{d}x_1} = \frac{1}{\tilde{E}_A^0}\bar{\sigma}_{11}^0 + \tilde{\alpha}_A^0\Delta T - \nu_A^0\frac{E_T^0}{E_A^0}\varepsilon_T, \tag{9.60}$$

$$\bar{\varepsilon}_{11}^{90} = \frac{\mathrm{d}\bar{u}_1^{90}}{\mathrm{d}x_1} = \frac{1}{\tilde{E}_T^{90}}\bar{\sigma}_{11}^{90} + \tilde{\alpha}_T^{90}\Delta T - \nu_A^{90}\varepsilon_T. \tag{9.61}$$

On differentiating these results with respect to x_1 and on making using of relations (9.48) and (9.49)

$$\frac{\mathrm{d}^2\bar{u}_1^0}{\mathrm{d}x_1^2} = \frac{1}{\tilde{E}_A^0}\frac{\mathrm{d}\bar{\sigma}_{11}^0}{\mathrm{d}x_1} = \frac{1}{\tilde{E}_A^0}\frac{\tau(x_1)}{b}, \tag{9.62}$$

$$\frac{d^2\bar{u}_1^{90}}{dx_1^2} = \frac{1}{\tilde{E}_T^{90}} \frac{d\bar{\sigma}_{11}^{90}}{dx_1} = -\frac{1}{\tilde{E}_T^{90}} \frac{\tau(x_1)}{a}. \tag{9.63}$$

On subtracting these relations making use the of the result (9.59), the following second-order ordinary differential equation for the interfacial shear stress distribution $\tau(x_1)$ may be derived

$$\frac{d^2\tau(x_1)}{dx_1^2} = k^2\tau(x_1), \tag{9.64}$$

where

$$k^2 = \frac{3\mu_a^0\mu_t^{90}\left(b\tilde{E}_A^0 + a\tilde{E}_T^{90}\right)}{ab\tilde{E}_A^0\tilde{E}_T^{90}\left(b\mu_t^{90} + a\mu_a^0\right)}. \tag{9.65}$$

As $\tau(0)=0$, the shear stress distribution satisfying (9.64) has the form

$$\tau(x_1) = \tilde{A}\sinh(kx_1), \tag{9.66}$$

where $\tilde{A}$ is a constant to be determined. It follows from (9.48), (9.49) and (9.66) that

$$\frac{d\bar{\sigma}_{11}^0}{dx_1} = \frac{\tilde{A}}{b}\sinh(kx_1), \qquad \frac{d\bar{\sigma}_{11}^{90}}{dx_1} = -\frac{\tilde{A}}{a}\sinh(kx_1). \tag{9.67}$$

On integration, and on making use of the requirements that $b\bar{\sigma}_{11}^0(L) = h\sigma_A, \bar{\sigma}_{11}^{90}(L) = 0$, it follows that

$$\bar{\sigma}_{11}^0(x_1) = \frac{h\sigma_A}{b} - \frac{\tilde{A}}{kb}\left[\cosh(kL) - \cosh(kx_1)\right], \; \bar{\sigma}_{11}^{90}(x_1) = \frac{\tilde{A}}{ka}\left[\cosh(kL) - \cosh(kx_1)\right], \tag{9.68}$$

satisfying relation (9.51) for any value of $\tilde{A}$. An estimate of the value of $\tilde{A}$ can be obtained by considering the situation where $L \to \infty$. As the plane $x_1 = 0$ is then remote from the ply cracks at $x_1 = \pm L$, the axial stresses σ_{11}^0 and σ_{11}^{90} on $x_1 = 0$ would have the uniform values $\hat{\sigma}_{11}^0$ and $\hat{\sigma}_{11}^{90}$ that occur in an undamaged laminate subject to an axial stress σ_A and transverse strain ε_T implying that $\bar{\sigma}_{11}^0(0) = \hat{\sigma}_{11}^0$ and $\bar{\sigma}_{11}^{90}(0) = \hat{\sigma}_{11}^{90}$. The stress values $\hat{\sigma}_{11}^0$ and $\hat{\sigma}_{11}^{90}$ must be such that

$$b\hat{\sigma}_{11}^0 + a\hat{\sigma}_{11}^{90} = h\sigma_A. \tag{9.69}$$

It then follows from (9.66) and $(9.68)_2$ that, because L is very large for this case, the condition $\bar{\sigma}_{11}^{90}(0) = \hat{\sigma}_{11}^{90}$ leads to

$$\tilde{A} = \frac{ka\hat{\sigma}_{11}^{90}}{\cosh(kL)}, \quad \text{and} \quad \tau(x_1) = ka\hat{\sigma}_{11}^{90}\frac{\sinh(kx_1)}{\cosh(kL)}, \tag{9.70}$$

so that on using (9.68) and (9.69)

$$\begin{aligned} \bar{\sigma}_{11}^0(x_1) &= \frac{h\sigma_A}{b} - \frac{a}{b}\hat{\sigma}_{11}^{90}\left(1 - \frac{\cosh(kx_1)}{\cosh(kL)}\right), \\ \bar{\sigma}_{11}^{90}(x_1) &= \hat{\sigma}_{11}^{90}\left(1 - \frac{\cosh(kx_1)}{\cosh(kL)}\right). \end{aligned} \tag{9.71}$$

The fact that the stress field (9.71) can be specified only by imposing a boundary condition that is valid as $L \to \infty$ is regarded as a significant disadvantage of shear-lag models. It is not clear to what extent ply interactions are being considered.

As σ_{33} has been assumed to be zero, on averaging (9.16) and (9.20) using (9.26), and on using (9.71), it can be shown that

$$\bar{\sigma}_{22}^{0}(x_1) = -\frac{\nu_A^0 E_T^0}{E_A^0}\frac{a}{b}\hat{\sigma}_{11}^{90}\left(1 - \frac{\cosh(kx_1)}{\cosh(kL)}\right) + \frac{\nu_A^0 E_T^0}{E_A^0}\frac{h\sigma_A}{b} - E_T^0\alpha_T^0\Delta T + E_T^0\varepsilon_T, \quad (9.72)$$

$$\bar{\sigma}_{22}^{90}(x_1) = \nu_A^{90}\hat{\sigma}_{11}^{90}\left(1 - \frac{\cosh(kx_1)}{\cosh(kL)}\right) - E_A^{90}\alpha_A^{90}\Delta T + E_A^{90}\varepsilon_T. \quad (9.73)$$

For an undamaged laminate, the axial strains in both the 0° and 90° plies have the same uniform value $\hat{\varepsilon}_A$. It then follows from (9.60) and (9.61) that

$$\hat{\varepsilon}_A = \frac{1}{\tilde{E}_A^0}\sigma_\infty^0 + \tilde{\alpha}_A^0\Delta T - \nu_A^0\frac{E_T^0}{E_A^0}\varepsilon_T = \frac{1}{\tilde{E}_T^{90}}\sigma_\infty^{90} + \tilde{\alpha}_T^{90}\Delta T - \nu_A^{90}\varepsilon_T. \quad (9.74)$$

On using (9.69) it then follows that

$$\left(b\tilde{E}_A^0 + a\tilde{E}_T^{90}\right)\frac{\hat{\sigma}_{11}^{0}}{\tilde{E}_A^0} = h\sigma_A - a\tilde{E}_T^{90}\left(\tilde{\alpha}_A^0 - \tilde{\alpha}_T^{90}\right)\Delta T - a\tilde{E}_T^{90}\left(\nu_A^{90} - \nu_A^0\frac{E_T^0}{E_A^0}\right)\varepsilon_T, \quad (9.75)$$

$$\left(b\tilde{E}_A^0 + a\tilde{E}_T^{90}\right)\frac{\hat{\sigma}_{11}^{90}}{\tilde{E}_T^{90}} = h\sigma_A + b\tilde{E}_A^0\left(\tilde{\alpha}_A^0 - \tilde{\alpha}_T^{90}\right)\Delta T + b\tilde{E}_A^0\left(\nu_A^{90} - \nu_A^0\frac{E_T^0}{E_A^0}\right)\varepsilon_T. \quad (9.76)$$

The axial average displacement field, which satisfies the conditions $\bar{u}_1^0(0) = \bar{u}_1^{90}(0) = 0$, is derived on integrating (9.60) and (9.61) using (9.69) and (9.71) so that

$$\bar{u}_1^0(x_1) = \frac{\hat{\sigma}_{11}^{90}}{k\tilde{E}_A^0}\frac{a}{b}\frac{\sinh(kx_1)}{\cosh(kL)} + \hat{\varepsilon}_A x_1, \quad \bar{u}_1^{90}(x_1) = -\frac{\hat{\sigma}_{11}^{90}}{k\tilde{E}_T^{90}}\frac{\sinh(kx_1)}{\cosh(kL)} + \hat{\varepsilon}_A x_1. \quad (9.77)$$

Given that the 90° ply is cracked, an effective axial applied strain for a cracked laminate ε_A is defined by

$$\varepsilon_A = \frac{\bar{u}_1^0(L)}{L} = \frac{a\Phi\hat{\sigma}_{11}^{90}}{b\tilde{E}_A^0} + \hat{\varepsilon}_A, \quad \text{where} \quad \Phi = \frac{\tanh(kL)}{kL}. \quad (9.78)$$

So far, the analysis has assumed generalised plane strain conditions where the transverse strain ε_T has the same prescribed uniform value for both 0° and 90° plies. In most applications it is more convenient to prescribe the effective transverse stress σ_T which is defined by

$$hL\sigma_T = \int_0^h\int_0^L \sigma_{22}(x_1,x_3)\,dx_1 dx_3 = \int_0^L \left(b\bar{\sigma}_{22}^0(x_1) + a\bar{\sigma}_{22}^{90}(x_1)\right)dx_1. \quad (9.79)$$

On using (9.72) and (9.73)

$$\begin{aligned} b\bar{\sigma}_{22}^0(x_1) + a\bar{\sigma}_{22}^{90}(x_1) = {} & a\hat{\sigma}_{11}^{90}\left(\nu_A^{90} - \frac{\nu_A^0 E_T^0}{E_A^0}\right)\left(1 - \frac{\cosh(kx_1)}{\cosh(kL)}\right) + \frac{\nu_A^0 E_T^0}{E_A^0}h\sigma_A \\ & - \left(bE_T^0\alpha_T^0 + aE_A^{90}\alpha_A^{90}\right)\Delta T + \left(bE_T^0 + aE_A^{90}\right)\varepsilon_T. \end{aligned} \quad (9.80)$$

The substitution of (9.80) into (9.79) then leads to

$$h\sigma_{\mathrm{T}} = \left(\nu_{\mathrm{A}}^{90} - \nu_{\mathrm{A}}^{0}\frac{E_{\mathrm{T}}^{0}}{E_{\mathrm{A}}^{0}}\right)\left(1 - \frac{\tanh(kL)}{kL}\right)a\hat{\sigma}_{11}^{90}$$
$$+ \frac{\nu_{\mathrm{A}}^{0}E_{\mathrm{T}}^{0}}{E_{\mathrm{A}}^{0}}h\sigma_{\mathrm{A}} - \left(bE_{\mathrm{T}}^{0}\alpha_{\mathrm{T}}^{0} + aE_{\mathrm{A}}^{90}\alpha_{\mathrm{A}}^{90}\right)\Delta T + \left(bE_{\mathrm{T}}^{0} + aE_{\mathrm{A}}^{90}\right)\varepsilon_{\mathrm{T}}. \tag{9.81}$$

From (9.16), (9.20), (9.36) and (9.69) it can be shown that for an undamaged laminate the effective transverse stress is given by

$$h\hat{\sigma}_{\mathrm{T}} = \left(\nu_{\mathrm{A}}^{90} - \frac{\nu_{\mathrm{A}}^{0}E_{\mathrm{T}}^{0}}{E_{\mathrm{A}}^{0}}\right)a\hat{\sigma}_{11}^{90} + \frac{\nu_{\mathrm{A}}^{0}E_{\mathrm{T}}^{0}}{E_{\mathrm{A}}^{0}}h\sigma_{\mathrm{A}} - \left(bE_{\mathrm{T}}^{0}\alpha_{\mathrm{T}}^{0} + aE_{\mathrm{A}}^{90}\alpha_{\mathrm{A}}^{90}\right)\Delta T + \left(bE_{\mathrm{T}}^{0} + aE_{\mathrm{A}}^{90}\right)\varepsilon_{\mathrm{T}}. \tag{9.82}$$

The result (9.81) may then be written in the more compact form

$$\sigma_{\mathrm{T}} = \left(\nu_{\mathrm{A}}^{0}\frac{E_{\mathrm{T}}^{0}}{E_{\mathrm{A}}^{0}} - \nu_{\mathrm{A}}^{90}\right)\frac{a}{h}\Phi\hat{\sigma}_{11}^{90} + \hat{\sigma}_{\mathrm{T}}, \quad \text{where} \quad \Phi = \frac{\tanh(kL)}{kL}. \tag{9.83}$$

It should be noted that $\sigma_{\mathrm{T}} \to \hat{\sigma}_{\mathrm{T}}$ as $L \to \infty$.

This completes the development of useful relations for a shear-lag model that can be applied to uniform distributions of ply cracks in a simple cross-ply laminate. In the next section, a more accurate stress-transfer model is developed and discussed in detail where an alternative more accurate definition of the damage dependent parameter Φ is derived and used to predict the dependence of the thermoelastic constants of a cracked laminate on ply crack density. Equivalent shear-lag predictions can be made simply by replacing the function Φ used in Section 9.4 by the simpler but less accurate shear-lag function defined by (9.83).

The preferred shear-lag approach described previously is based on various approximations that are cause for concern from a mechanics point of view. It is useful now to identify these approximations clearly. The first approximation that is made concerns the neglect of the terms $\partial u_3 / \partial x_1$ in the expression for the shear strain ε_{13} (see (9.52) and (9.53)). This assumption is expected to lead to significant errors. It is observed that the equilibrium equation (9.25) and the stress-strain relations $(9.4)_{2,3}$ and $(9.5)_{2,3}$ have not been considered, indicating further sources of significant error. It is concluded that although the preferred shear-lag approach is relatively easy to develop and understand, it does suffer from having to make a number of assumptions that are expected to lead to unacceptable errors.

9.4 Generalised Plane Strain Theory for Cross-ply Laminates

As mentioned in Section 9.3, stress-transfer phenomena in cross-ply laminates have been examined extensively using an approximate method of stress analysis, first developed for unidirectional composites, that is known as 'shear-lag theory'. The objective now is to improve the nature of this methodology by removing the various approximations that had to be made when developing shear-lag solutions. The approach to be adopted extends previous work by the present author [4].

Consider a symmetric damaged laminate of length $2L$, width $2W$ and total thickness $2h$, for which the plies are uniformly thick and perfectly bonded together. In-plane loading is applied by imposing uniform axial and transverse displacements, denoted by $\pm U_{\mathrm{A}}$ and $\pm U_{\mathrm{T}}$, on the edges of the laminate thus defining an effective axial strain $\varepsilon_{\mathrm{A}} = U_{\mathrm{A}} / L$ and an effective transverse strain $\varepsilon_{\mathrm{T}} = U_{\mathrm{T}} / W$. The faces of the laminate are subjected to a uniform applied stress denoted by σ_{t}. Corresponding to the effective in-plane strains ε_{A} and ε_{T}, axial and transverse effective applied stresses can be defined for damaged laminates by

$$\sigma_{\mathrm{A}} = \frac{1}{4Wh} \int_{-W}^{W} \int_{-h}^{h} \sigma_{11} \, \mathrm{d}x_2 \, \mathrm{d}x_3, \qquad \sigma_{\mathrm{T}} = \frac{1}{4Lh} \int_{-L}^{L} \int_{-h}^{h} \sigma_{22} \, \mathrm{d}x_1 \, \mathrm{d}x_3. \tag{9.84}$$

Corresponding to the applied through-thickness stress σ_{t} an effective through-thickness strain ε_{t} can be defined for damaged and undamaged laminates by

$$\varepsilon_{\mathrm{t}} = \frac{1}{8LWh} \int_{-W}^{W} \int_{-L}^{L} \left[u_3(x_1,x_2,h) - u_3(x_1,x_2,-h) \right] \mathrm{d}x_1 \, \mathrm{d}x_2. \tag{9.85}$$

For an undamaged laminate subject to these displacements, the axial and transverse strains will be uniform throughout all plies of the laminate having the values ε_{A} and ε_{T}, respectively, and the through-thickness stress will be uniform throughout the laminate having the value σ_{t} in each ply. It then follows from (9.84) and (9.85) that $\sigma_{\mathrm{A}} = \hat{\sigma}_{\mathrm{A}}$, $\sigma_{\mathrm{T}} = \hat{\sigma}_{\mathrm{T}}$ and $\varepsilon_{\mathrm{t}} = \hat{\varepsilon}_{\mathrm{t}}$ where $\hat{\sigma}_{\mathrm{A}}$, $\hat{\sigma}_{\mathrm{T}}$ and $\hat{\varepsilon}_{\mathrm{t}}$ are defined by relations (9.35)–(9.37).

For generalised plane strain conditions, the shear stress distribution is assumed to have the following form

$$\sigma_{13}^{0}(x_1,x_3) = C'(x_1)(h - x_3), \qquad \sigma_{13}^{90}(x_1,x_3) = \frac{b}{a} C'(x_1) x_3, \tag{9.86}$$

where $C(x_1)$is a stress-transfer function that is to be determined. Relations (9.86) ensure that the shear stress σ_{13} is continuous across the interface $x = a$. On substituting (9.86) into (9.24) and then integrating with respect to x_3 it can be shown that

$$\sigma_{11}^{0}(x_1,x_3) = C(x_1) + \hat{\sigma}_{11}^{0}, \qquad \sigma_{11}^{90}(x_1,x_3) = -\frac{b}{a} C(x_1) + \hat{\sigma}_{11}^{90}, \tag{9.87}$$

where, from (9.28) and (9.32), $\hat{\sigma}_{11}^{0}$ and $\hat{\sigma}_{11}^{90}$ are the values of the stress component σ_{11} in the 0° and 90° plies, respectively, when the laminate is undamaged so that $C(x_1) \equiv 0$. On using (9.35) it follows from (9.87) that the effective applied axial stress σ_{A} defined by $(9.84)_1$ is given by

$$h\sigma_{\mathrm{A}} = b\sigma_{11}^{0}(x_1,x_3) + a\sigma_{11}^{90}(x_1,x_3) = b\hat{\sigma}_{11}^{0} + a\hat{\sigma}_{11}^{90} = h\hat{\sigma}_{\mathrm{A}}. \tag{9.88}$$

It follows that the effective applied axial stress for damaged laminate σ_{A} is equal to the effective axial stress $\hat{\sigma}_{\mathrm{A}}$ for the corresponding undamaged laminate.

On substituting (9.86) into (9.25) and then integrating with respect to x_3 it can be shown that

$$\sigma_{33}^{0}(x_1,x_3) = \frac{1}{2} C''(x_1) \left(h - x_3 \right)^2 + \sigma_{\mathrm{t}}, \quad \sigma_{33}^{90}(x_1,x_3) = \frac{b}{2a} C''(x_1) \left(ah - x_3^2 \right) + \sigma_{\mathrm{t}}, \tag{9.89}$$

where the stress component σ_{33} is continuous across the interface $x = a$ and has zero value on the free surfaces $x_3 = \pm h$.

It follows from (9.16) and (9.20) that the stress σ_{22} is given by

$$\sigma_{22}^{0}(x_1,x_3)=\frac{\nu_{A}^{0}E_{T}^{0}}{E_{A}^{0}}C(x_1)+\frac{1}{2}\nu_{t}^{0}C''(x_1)\big(h-x_3\big)^2+\hat{\sigma}_{22}^{0},\tag{9.90}$$

$$\sigma_{22}^{90}(x_1,x_3)=-\frac{b}{a}\nu_{A}^{90}C(x_1)+\frac{b}{2a}\nu_{a}^{90}C''(x_1)\big(ah-x_3^2\big)+\hat{\sigma}_{22}^{90},\tag{9.91}$$

where the values of $\hat{\sigma}_{22}^{0}$ and $\hat{\sigma}_{22}^{90}$ are given by (9.28) and (9.32), respectively.

The stress representation (9.86)–(9.91) is now substituted into the stress-strain relation (9.5) for the strain component ε_{33}^{90} leading to the following result, on integration with respect to $x_3$and on using (9.23)

$$u_3^{90}(x_1,x_3)=\frac{b}{a}\frac{\tilde{\nu}_{t}^{90}}{\tilde{E}_{T}^{90}}C(x_1)x_3+\frac{b}{6a}\frac{1}{\tilde{E}_{t}^{90}}C''(x_1)\big(3ah-x_3^2\big)x_3+\hat{\varepsilon}_{33}^{90}x_3,\tag{9.92}$$

where the value of $\hat{\varepsilon}_{33}^{90}$ is given by (9.31) and where the symmetry condition $u_3^{90}(x_1,0)=0$ has been imposed. The stress representation (9.86)–(9.91) is then substituted into the stress-strain relation (9.4) for the strain component ε_{33}^{0} leading to, on integration with respect to x_3 and on using (9.19),

$$\begin{aligned}u_3^{0}(x_1,x_3)=\hat{\varepsilon}_{33}^{0}(x_3-a)+\hat{\varepsilon}_{33}^{90}a+\left(b\frac{\tilde{\nu}_{t}^{90}}{\tilde{E}_{T}^{90}}-\frac{\tilde{\nu}_{a}^{0}}{\tilde{E}_{A}^{0}}(x_3-a)\right)C(x_1)\\+\frac{1}{6}\left[\frac{1}{\tilde{E}_{t}^{0}}\Big(b^3-(h-x_3)^3\Big)+\frac{ab}{\tilde{E}_{t}^{90}}(2h+b)\right]C''(x_1),\end{aligned}\tag{9.93}$$

where the value of $\hat{\varepsilon}_{33}^{0}$ is given by (9.29) and where the continuity condition $u_3^{0}(x_1,a)=u_3^{90}(x_1,a)$ has been applied.

The next step is to use shear stress-strain relations to determine the form of the distributions of the displacement component u_1 in terms of the stress-transfer function $C(x_1)$. On substituting the stress representation (9.86)–(9.91) and (9.93) into the stress-strain relation (9.4) for the strain component ε_{13}^{0} leading to the following result, on integration with respect to x_3

$$\begin{aligned}u_1^{0}(x_1,x_3)=&\left(\frac{\tilde{\nu}_{a}^{0}}{2\tilde{E}_{A}^{0}}(x_3-a)^2-\frac{\tilde{\nu}_{t}^{90}}{\tilde{E}_{T}^{90}}b(x_3-a)+\frac{b^2-(h-x_3)^2}{2\mu_{a}^{0}}\right)C'(x_1)\\&-\left(\frac{(h-x_3)^4-b^4+4b^3(x_3-a)}{24\tilde{E}_{t}^{0}}+\frac{ab(2h+b)(x_3-a)}{6\tilde{E}_{t}^{90}}\right)C'''(x_1)+\bar{A}(x_1),\end{aligned}\tag{9.94}$$

where $\bar{A}(x_1)\equiv u_1^{0}(x_1,a)$ is for the moment an arbitrary function of x_1. Similarly, on substituting the stress representation (9.86)–(9.91) and (9.92) into the stress-strain relation (9.5) for the strain component ε_{13}^{90}, and on integration with respect to x_3, it can be shown that

$$\begin{aligned}u_1^{90}(x_1,x_3)=&\frac{b}{2a}\left(\frac{1}{\mu_{t}^{90}}-\frac{\tilde{\nu}_{t}^{90}}{\tilde{E}_{T}^{90}}\right)(x_3^2-a^2)C'(x_1)\\&+\frac{b}{24a}\frac{1}{\tilde{E}_{t}^{90}}\Big(x_3^4-a^4-6ah(x_3^2-a^2)\Big)C'''(x_1)+\bar{A}(x_1),\end{aligned}\tag{9.95}$$

where the continuity condition $u_1^0(x_1,a)=u_1^{90}(x_1,a)$ has been imposed.

The solution specified by relations (9.86)–(9.95) automatically satisfies the equilibrium equations (9.24) and (9.25), and all the stress-strain relations (9.4) and (9.5), except for the two relations $(9.4)_1$ and $(9.5)_1$ involving strains ε_{11}^0 and ε_{11}^{90}, for any functions $C(x_1)$ and $\bar{A}(x_1)$. It is possible, however, to satisfy these relations after they are averaged through the thickness of the 0° and 90° plies, respectively, as is now described.

On averaging (9.17) and (9.21) using (9.26)

$$\frac{\partial \bar{u}_1^0}{\partial x_1}=\frac{1}{\tilde{E}_{\mathrm{A}}^0}\bar{\sigma}_{11}^0-\frac{\tilde{\nu}_{\mathrm{a}}^0}{\tilde{E}_{\mathrm{A}}^0}\bar{\sigma}_{33}^0+\tilde{\alpha}_{\mathrm{A}}^0\Delta T-\nu_{\mathrm{A}}^0\frac{E_{\mathrm{T}}^0}{E_{\mathrm{A}}^0}\varepsilon_{\mathrm{T}}, \tag{9.96}$$

$$\frac{\partial \bar{u}_1^{90}}{\partial x_1}=\frac{1}{\tilde{E}_{\mathrm{T}}^{90}}\bar{\sigma}_{11}^{90}-\frac{\tilde{\nu}_{\mathrm{t}}^{90}}{\tilde{E}_{\mathrm{T}}^{90}}\bar{\sigma}_{33}^{90}+\tilde{\alpha}_{\mathrm{T}}^{90}\Delta T-\nu_{\mathrm{A}}^{90}\varepsilon_{\mathrm{T}}. \tag{9.97}$$

On averaging (9.87) and (9.89) over the 0° and 90° plies, respectively, it can be shown that

$$\bar{\sigma}_{11}^0(x_1)=C(x_1)+\hat{\sigma}_{11}^0,\qquad \bar{\sigma}_{11}^{90}(x_1)=-\frac{b}{a}C(x_1)+\hat{\sigma}_{11}^{90}, \tag{9.98}$$

$$\bar{\sigma}_{33}^0(x_1)=\frac{b^2}{6}C''(x_1)+\sigma_{\mathrm{t}},\quad \bar{\sigma}_{33}^{90}(x_1)=\frac{b(2a+3b)}{6}C''(x_1)+\sigma_{\mathrm{t}}. \tag{9.99}$$

On averaging (9.94) and (9.95) over the 0° and 90° plies, respectively, it can be shown that

$$\bar{u}_1^0(x_1)=\left(\frac{\tilde{\nu}_a^0}{6\tilde{E}_{\mathrm{A}}^0}-\frac{\tilde{\nu}_{\mathrm{t}}^{90}}{2\tilde{E}_{\mathrm{T}}^{90}}+\frac{1}{3\mu_{\mathrm{a}}^0}\right)b^2C'(x_1)-\left(\frac{1}{20\tilde{E}_{\mathrm{t}}^0}+\frac{a(2a+3b)}{12b^2\tilde{E}_{\mathrm{t}}^{90}}\right)b^4C'''(x_1)+\bar{A}(x_1), \tag{9.100}$$

$$\bar{u}_1^{90}(x_1)=\frac{a}{3b}\left(\frac{\tilde{\nu}_{\mathrm{t}}^{90}}{\tilde{E}_{\mathrm{T}}^{90}}-\frac{1}{\mu_{\mathrm{t}}^{90}}\right)b^2C'(x_1)+\frac{a^2(5b+4a)}{30b^3}\frac{1}{\tilde{E}_{\mathrm{t}}^{90}}b^4C'''(x_1)+\bar{A}(x_1). \tag{9.101}$$

On substituting (9.98)–(9.101) into relations (9.96) and (9.97), on eliminating the terms $\hat{\sigma}_{11}^0$ and $\hat{\sigma}_{11}^{90}$ using (9.28) and (9.32), it follows that

$$\begin{aligned}\left(\frac{1}{20\tilde{E}_{\mathrm{t}}^0}+\frac{a(2a+3b)}{12b^2\tilde{E}_{\mathrm{t}}^{90}}\right)b^4C''''(x_1)-\left(\frac{\tilde{\nu}_a^0}{3\tilde{E}_{\mathrm{A}}^0}-\frac{\tilde{\nu}_{\mathrm{t}}^{90}}{2\tilde{E}_{\mathrm{T}}^{90}}+\frac{1}{3\mu_{\mathrm{a}}^0}\right)b^2C''(x_1)\\+\frac{1}{\tilde{E}_{\mathrm{A}}^0}C(x_1)=-\hat{\varepsilon}_{\mathrm{A}}+\bar{A}'(x_1),\end{aligned} \tag{9.102}$$

$$\begin{aligned}\frac{a^2(5b+4a)}{30b^3\tilde{E}_{\mathrm{t}}^{90}}b^4C''''(x_1)+\frac{a}{3b}\left(\frac{4a+3b}{2a}\frac{\tilde{\nu}_{\mathrm{t}}^{90}}{\tilde{E}_{\mathrm{T}}^{90}}-\frac{1}{\mu_{\mathrm{t}}^{90}}\right)b^2C''(x_1)\\+\frac{b}{a}\frac{1}{\tilde{E}_{\mathrm{T}}^{90}}C(x_1)=\hat{\varepsilon}_{\mathrm{A}}-\bar{A}'(x_1).\end{aligned} \tag{9.103}$$

On adding (9.102) and (9.103) the term $\hat{\varepsilon}_{\mathrm{A}}-\bar{A}'(x_1)$ is eliminated and the following homogeneous fourth-order ordinary differential equation is obtained

$$\tilde{F} b^4 C''''(x_1) - \tilde{G} b^2 C''(x_1) + \tilde{H} C(x_1) = 0, \tag{9.104}$$

where

$$\begin{aligned}
\tilde{F} &= \frac{1}{20\tilde{E}_t^0} + \frac{2}{15\tilde{E}_t^{90}}\frac{a}{b}\left(\frac{a^2}{b^2} + \frac{5}{2}\frac{a}{b} + \frac{15}{8}\right) > 0, \\
\tilde{G} &= \frac{1}{3}\left(\frac{1}{\mu_a^0} + \frac{1}{\mu_t^{90}}\frac{a}{b} + \frac{\tilde{\nu}_a^0}{\tilde{E}_A^0} - \frac{\tilde{\nu}_t^{90}}{\tilde{E}_T^{90}}\frac{2a+3b}{b}\right), \\
\tilde{H} &= \frac{1}{\tilde{E}_A^0} + \frac{1}{\tilde{E}_T^{90}}\frac{b}{a} > 0.
\end{aligned} \tag{9.105}$$

From (9.102) and (9.103) it is clear that the function $\bar{A}(x_1)$ can be calculated using either of the following relations

$$\begin{aligned}
\bar{A}(x_1) = &\left(\frac{1}{20\tilde{E}_t^0} + \frac{a(2a+3b)}{12b^2\tilde{E}_t^{90}}\right) b^4 C'''(x_1) - \left(\frac{\tilde{\nu}_a^0}{3\tilde{E}_A^0} - \frac{\tilde{\nu}_t^{90}}{2\tilde{E}_T^{90}} + \frac{1}{3\mu_a^0}\right) b^2 C'(x_1) \\
&+ \frac{1}{\tilde{E}_A^0}\bar{C}(x_1) + \hat{\varepsilon}_A x_1,
\end{aligned} \tag{9.106}$$

$$\begin{aligned}
\bar{A}(x_1) = &-\frac{a^2(5b+4a)}{30b^3\tilde{E}_t^{90}} b^4 C'''(x_1) - \frac{a}{3b}\left(\frac{4a+3b}{2a}\frac{\tilde{\nu}_t^{90}}{\tilde{E}_T^{90}} - \frac{1}{\mu_t^{90}}\right) b^2 C'(x_1) \\
&- \frac{b}{a}\frac{1}{\tilde{E}_T^{90}}\bar{C}(x_1) + \hat{\varepsilon}_A x_1,
\end{aligned} \tag{9.107}$$

where

$$\bar{C}(x_1) \equiv \int_0^{x_1} C(x)\,\mathrm{d}x. \tag{9.108}$$

When there is symmetry about $x_1 = 0$, it follows that $C'(x_1) = 0$ and $C'''(x_1) = 0$ and the integration constant has been selected so that $A(x_1) = 0$. On substituting (9.106) into (9.100)

$$\bar{u}_1^0(x_1) = \frac{1}{\tilde{E}_A^0}\bar{C}(x_1) - \frac{\tilde{\nu}_a^0}{6\tilde{E}_A^0} b^2 C'(x_1) + \hat{\varepsilon}_A x_1, \tag{9.109}$$

and on substituting (9.107) into (9.101)

$$\bar{u}_1^{90}(x_1) = -\frac{b}{a}\frac{1}{\tilde{E}_T^{90}}\bar{C}(x_1) - \frac{2a+3b}{6b}\frac{\tilde{\nu}_t^{90}}{\tilde{E}_T^{90}} b^2 C'(x_1) + \hat{\varepsilon}_A x_1. \tag{9.110}$$

It should be noted that $\bar{u}_1^0(0) = \bar{u}_1^{90}(0) = 0$ consistent with the conditions $(9.13)_3$. Because the function $\bar{A}(x_1)$ is now known in terms of $C(x_1)$, the u_1 displacement distributions (9.94) and (9.95) are fully specified in terms of the stress-transfer function $C(x_1)$.

On defining dimensionless parameters r and s so that

$$r = \frac{\tilde{G}}{2\tilde{F}}, \qquad s = \sqrt{\frac{\tilde{H}}{\tilde{F}}}, \tag{9.111}$$

the most general solution of the differential equation (9.104) satisfying the symmetry condition $C(x_1) \equiv C(-x_1)$ is given by:
if $s > r$,

$$C(x_1) = \tilde{P}\cosh\frac{px_1}{b}\cos\frac{qx_1}{b} + \tilde{Q}\sinh\frac{px_1}{b}\sin\frac{qx_1}{b}, \tag{9.112}$$

if $s = r$,

$$C(x_1) = \tilde{P}\cosh\frac{px_1}{b} + \tilde{Q}\frac{x_1}{b}\sinh\frac{px_1}{b}, \tag{9.113}$$

if $s < r$,

$$C(x_1) = \tilde{P}\cosh\frac{px_1}{b}\cosh\frac{qx_1}{b} + \tilde{Q}\sinh\frac{px_1}{b}\sinh\frac{qx_1}{b}, \tag{9.114}$$

where p and q are dimensionless parameters defined by

$$p = \sqrt{\tfrac{1}{2}(r+s)}\,, \qquad q = \sqrt{\tfrac{1}{2}\left|r-s\right|}. \tag{9.115}$$

Rather than consider the three cases separately, it is more convenient to select the form (9.114) of the solution for the case $r > s$ and then allow the parameter q to be real, zero or pure imaginary. This situation can be dealt with using software that includes the use of complex arithmetic. The solution (9.114) is now expressed in the equivalent form

$$C(x_1) = \hat{A}\cosh\frac{(p+q)x_1}{b} + \hat{B}\cosh\frac{(p-q)x_1}{b}, \tag{9.116}$$

where

$$\begin{aligned} \hat{P} &= \hat{A} + \hat{B}, & \hat{Q} &= \hat{A} - \hat{B}, \\ p &= \sqrt{\tfrac{1}{2}(r+s)}\,, & q &= \sqrt{\tfrac{1}{2}(r-s)}\,. \end{aligned} \tag{9.117}$$

This completes the development of the general solution for symmetric situations. Further progress depends upon the selection of appropriate boundary conditions for specific types of problem.

9.4.1 Solution for Ply Cracks

Consider now a uniform array of ply cracks, having density $\rho = 1/(2L)$, in the 90° ply of the cross-ply laminate as shown in Figure 9.1(b). The tractions on the ply crack surfaces must be zero so that from (9.86) and (9.87)

$$\sigma_{13}^{90}(L,x_3) = \frac{b}{a}C'(L)x_3 = 0, \quad \text{implying} \quad C'(L) = 0, \tag{9.118}$$

$$\sigma_{11}^{90}(L,x_3) = -\frac{b}{a}C(L) + \hat{\sigma}_{11}^{90} = 0, \quad \text{implying} \quad C(L) = \frac{a}{b}\hat{\sigma}_{11}^{90}. \tag{9.119}$$

On applying these conditions to the solution (9.116) the parameters $\hat{A}$ and $\hat{B}$ must be selected so that

$$\hat{A}\cosh\frac{(p+q)L}{b} + \hat{B}\cosh\frac{(p-q)L}{b} = \frac{a}{b}\hat{\sigma}_{11}^{90}, \tag{9.120}$$

$$\hat{A}\,\frac{p+q}{b}\sinh\frac{(p+q)L}{b} + \hat{B}\,\frac{p-q}{b}\sinh\frac{(p-q)L}{b} = 0. \tag{9.121}$$

On solving (9.120) and (9.121)

$$\hat{A} = -\frac{a}{b}\,\frac{(p-q)\tanh\dfrac{(p-q)L}{b}}{\cosh\dfrac{(p+q)L}{b}}\,\Gamma\hat{\sigma}_{11}^{90}, \quad \hat{B} = \frac{a}{b}\,\frac{(p+q)\tanh\dfrac{(p+q)L}{b}}{\cosh\dfrac{(p-q)L}{b}}\,\Gamma\hat{\sigma}_{11}^{90}, \tag{9.122}$$

where

$$\frac{1}{\Gamma} = (q+p)\tanh\frac{(p+q)L}{b} + (q-p)\tanh\frac{(p-q)L}{b}. \tag{9.123}$$

The only boundary condition for a damaged laminate that has not been satisfied is given by $(9.14)_1$. It is clear from (9.94), (9.95) and (9.106) or (9.107) that it is not possible for this boundary condition to be satisfied by the approximate solution derived. The boundary condition (9.11) is now replaced by the following averaged condition

$$\bar{u}_1^0 = \pm L\varepsilon_A, \quad \text{on} \quad x_1 = \pm L. \tag{9.124}$$

It then follows from (9.109) and (9.118) that

$$\varepsilon_A = \frac{1}{\tilde{E}_A^0}\frac{\bar{C}(L)}{L} + \hat{\varepsilon}_A. \tag{9.125}$$

It follows from (9.108) and (9.116) that

$$\bar{C}(x_1) = \frac{b\hat{A}}{p+q}\sinh\frac{(p+q)x_1}{b} + \frac{b\hat{B}}{p-q}\sinh\frac{(p-q)x_1}{b}, \tag{9.126}$$

so that on using (9.122)

$$\bar{C}(L) = \frac{a}{b}\hat{\sigma}_{11}^{90}L\Phi, \tag{9.127}$$

where Φ is a dimensionless parameter defined by

$$\Phi = \frac{4\Gamma pq}{p^2-q^2}\,\frac{b}{L}\tanh\frac{(p+q)L}{b}\tanh\frac{(p-q)L}{b}. \tag{9.128}$$

From (9.125) it follows that

$$\varepsilon_A = \frac{a\Phi\hat{\sigma}_{11}^{90}}{b\tilde{E}_A^0} + \hat{\varepsilon}_A. \tag{9.129}$$

On using (9.36) it follows from (9.90) and (9.91) that the effective applied transverse stress σ_T defined by $(9.84)_2$ is given by

$$\begin{aligned}\sigma_T &= \frac{1}{4Lh}\int_{-L}^{L}\int_{-h}^{h}\sigma_{22}\,dx_1\,dx_3 = \frac{1}{Lh}\int_0^L\int_0^a\sigma_{22}^{90}\,dx_1\,dx_3 + \frac{1}{Lh}\int_0^L\int_a^h\sigma_{22}^{0}\,dx_1\,dx_3\\ &= \frac{b}{Lh}\left(\nu_A^0\frac{E_T^0}{E_A^0} - \nu_A^{90}\right)\bar{C}(L) + \hat{\sigma}_T,\end{aligned} \tag{9.130}$$

where use has been made of the relations $C'(0)=0$ and $C'(L)=0$. It follows that the effective applied transverse stress for damaged laminate $\sigma_{\rm T}$ is not equal to the effective axial stress $\hat{\sigma}_{\rm T}$ for the corresponding undamaged laminate. On using (9.127) it follows that

$$\sigma_{\rm T}=\frac{a}{h}\left(\nu_{\rm A}^{0}\frac{E_{\rm T}^{0}}{E_{\rm A}^{0}}-\nu_{\rm A}^{90}\right)\Phi\hat{\sigma}_{11}^{90}+\hat{\sigma}_{\rm T}. \tag{9.131}$$

9.5 Calculation of In-plane Thermoelastic Constants for Damaged Laminate

There are two approaches, involving complex algebra, which can be made to the estimation of the effective thermoelastic constants of a damaged laminate. In order that the reader can understand how the analysis is derived, the key steps of the complex algebra are shown for each approach.

9.5.1 Approach 1

It is assumed that the values of $\sigma_{\rm A}=\hat{\sigma}_{\rm A}$, $\varepsilon_{\rm T}=\hat{\varepsilon}_{\rm T}$, $\sigma_{\rm t}$ and ΔT are known so that the stress-strain relations (9.40) for an undamaged laminate may be written as

$$\begin{aligned}\hat{\varepsilon}_{\rm A}&=\frac{1}{E_{\rm A}^{(\rm L)}}\sigma_{\rm A}-\frac{\nu_{\rm A}^{(\rm L)}}{E_{\rm A}^{(\rm L)}}\hat{\sigma}_{\rm T}-\frac{\nu_{\rm a}^{(\rm L)}}{E_{\rm A}^{(\rm L)}}\sigma_{\rm t}+\alpha_{\rm A}^{(\rm L)}\Delta T,\\ \varepsilon_{\rm T}&=-\frac{\nu_{\rm A}^{(\rm L)}}{E_{\rm A}^{(\rm L)}}\sigma_{\rm A}+\frac{1}{E_{\rm T}^{(\rm L)}}\hat{\sigma}_{\rm T}-\frac{\nu_{\rm t}^{(\rm L)}}{E_{\rm T}^{(\rm L)}}\sigma_{\rm t}+\alpha_{\rm T}^{(\rm L)}\Delta T.\end{aligned} \tag{9.132}$$

On using (9.131) to eliminate $\hat{\sigma}_{\rm T}$ in (9.132), it follows that

$$\hat{\varepsilon}_{\rm A}=\frac{1}{E_{\rm A}^{(\rm L)}}\sigma_{\rm A}-\frac{\nu_{\rm A}^{(\rm L)}}{E_{\rm A}^{(\rm L)}}\sigma_{\rm T}-\frac{\nu_{\rm a}^{(\rm L)}}{E_{\rm A}^{(\rm L)}}\sigma_{\rm t}+\frac{\nu_{\rm A}^{(\rm L)}}{E_{\rm A}^{(\rm L)}}\left(\nu_{\rm A}^{0}\frac{E_{\rm T}^{0}}{E_{\rm A}^{0}}-\nu_{\rm A}^{90}\right)\frac{a}{h}\Phi\hat{\sigma}_{11}^{90}+\alpha_{\rm A}^{(\rm L)}\Delta T, \tag{9.133}$$

$$\varepsilon_{\rm T}=-\frac{\nu_{\rm A}^{(\rm L)}}{E_{\rm A}^{(\rm L)}}\sigma_{\rm A}+\frac{1}{E_{\rm T}^{(\rm L)}}\sigma_{\rm T}-\frac{\nu_{\rm t}^{(\rm L)}}{E_{\rm T}^{(\rm L)}}\sigma_{\rm t}-\frac{1}{E_{\rm T}^{(\rm L)}}\left(\nu_{\rm A}^{0}\frac{E_{\rm T}^{0}}{E_{\rm A}^{0}}-\nu_{\rm A}^{90}\right)\frac{a}{h}\Phi\hat{\sigma}_{11}^{90}+\alpha_{\rm T}^{(\rm L)}\Delta T. \tag{9.134}$$

From $(9.32)_1$, on eliminating $\varepsilon_{\rm T}$ using (9.134),

$$\begin{aligned}&\left[\frac{1}{\tilde{E}_{\rm T}^{90}}+\frac{\nu_{\rm A}^{90}}{E_{\rm T}^{(\rm L)}}\left(\nu_{\rm A}^{0}\frac{E_{\rm T}^{0}}{E_{\rm A}^{0}}-\nu_{\rm A}^{90}\right)\frac{a}{h}\Phi\right]\hat{\sigma}_{11}^{90}=\hat{\varepsilon}_{\rm A}-\nu_{\rm A}^{90}\frac{\nu_{\rm A}^{(\rm L)}}{E_{\rm A}^{(\rm L)}}\sigma_{\rm A}+\frac{\nu_{\rm A}^{90}}{E_{\rm T}^{(\rm L)}}\sigma_{\rm T}\\ &\qquad+\left(\frac{\tilde{\nu}_{\rm t}^{90}}{\tilde{E}_{\rm T}^{90}}-\nu_{\rm A}^{90}\frac{\nu_{\rm t}^{(\rm L)}}{E_{\rm T}^{(\rm L)}}\right)\sigma_{\rm t}+\left(\nu_{\rm A}^{90}\alpha_{\rm T}^{(\rm L)}-\tilde{\alpha}_{\rm T}^{90}\right)\Delta T.\end{aligned} \tag{9.135}$$

On using (9.133) to eliminate $\hat{\varepsilon}_{\rm A}$, and on using (9.34)

$$\begin{aligned}&\left[1-\frac{\tilde{E}_{\rm T}^{90}}{E_{\rm T}^{(\rm L)}}\left(\nu_{\rm A}^{(\rm L)}\frac{E_{\rm T}^{(\rm L)}}{E_{\rm A}^{(\rm L)}}-\nu_{\rm A}^{90}\right)\left(\nu_{\rm A}^{0}\frac{E_{\rm T}^{0}}{E_{\rm A}^{0}}-\nu_{\rm A}^{90}\right)\frac{a}{h}\Phi\right]\frac{\hat{\sigma}_{11}^{90}}{\tilde{E}_{\rm T}^{90}}=\left(1-\nu_{\rm A}^{90}\nu_{\rm A}^{(\rm L)}\right)\frac{\sigma_{\rm A}}{E_{\rm A}^{(\rm L)}}\\ &\quad-\left(\nu_{\rm A}^{(\rm L)}\frac{E_{\rm T}^{(\rm L)}}{E_{\rm A}^{(\rm L)}}-\nu_{\rm A}^{90}\right)\frac{\sigma_{\rm T}}{E_{\rm T}^{(\rm L)}}-\left(\frac{\nu_{\rm a}^{(\rm L)}}{E_{\rm A}^{(\rm L)}}-\frac{\nu_{\rm t}^{90}}{E_{\rm T}^{90}}-\frac{\nu_{\rm a}^{90}\nu_{\rm A}^{90}}{E_{\rm A}^{90}}+\nu_{\rm A}^{90}\frac{\nu_{\rm t}^{(\rm L)}}{E_{\rm T}^{(\rm L)}}\right)\sigma_{\rm t}\\ &\quad+\left(\alpha_{\rm A}^{(\rm L)}+\nu_{\rm A}^{90}\alpha_{\rm T}^{(\rm L)}-\tilde{\alpha}_{\rm T}^{90}\right)\Delta T.\end{aligned} \tag{9.136}$$

On using (9.42) it can be shown that

$$\nu_{A}^{(L)}\frac{E_{T}^{(L)}}{E_{A}^{(L)}}-\nu_{A}^{90}=\left(\nu_{A}^{0}\frac{E_{T}^{0}}{E_{A}^{0}}-\nu_{A}^{90}\right)\frac{b}{h}\frac{\tilde{E}_{A}^{0}}{\tilde{E}_{A}^{(L)}}, \tag{9.137}$$

so that relation (9.136) may now be written as

$$\begin{aligned}\xi\frac{\hat{\sigma}_{11}^{90}}{\tilde{E}_{T}^{90}}=&\left(1-\nu_{A}^{90}\nu_{A}^{(L)}\right)\frac{\sigma_{A}}{E_{A}^{(L)}}-\left(\nu_{A}^{0}\frac{E_{T}^{0}}{E_{A}^{0}}-\nu_{A}^{90}\right)\frac{\tilde{E}_{A}^{0}}{\tilde{E}_{A}^{(L)}}\frac{b}{h}\frac{\sigma_{T}}{E_{T}^{(L)}}\\&-\left\{\frac{\nu_{a}^{(L)}}{E_{A}^{(L)}}-\frac{\nu_{t}^{90}}{E_{T}^{90}}+\nu_{A}^{90}\left(\frac{\nu_{t}^{(L)}}{E_{T}^{(L)}}-\frac{\nu_{a}^{90}}{E_{A}^{90}}\right)\right\}\sigma_{t}+\left(\alpha_{A}^{(L)}+\nu_{A}^{90}\alpha_{T}^{(L)}-\tilde{\alpha}_{T}^{90}\right)\Delta T,\end{aligned} \tag{9.138}$$

where

$$\xi=1-\left(\nu_{A}^{0}\frac{E_{T}^{0}}{E_{A}^{0}}-\nu_{A}^{90}\right)^{2}\frac{ab}{h^{2}}\frac{\tilde{E}_{A}^{0}\tilde{E}_{T}^{90}}{\tilde{E}_{A}^{(L)}E_{T}^{(L)}}\Phi. \tag{9.139}$$

The next step of the first approach is to eliminate $\hat{\varepsilon}_{A}$ and $\hat{\sigma}_{11}^{90}$ in (9.133) and (9.134) using (9.129) and (9.138). On eliminating $\hat{\varepsilon}_{A}$ in (9.133)

$$\varepsilon_{A}=\frac{1}{E_{A}^{(L)}}\sigma_{A}-\frac{\nu_{A}^{(L)}}{E_{A}^{(L)}}\sigma_{T}-\frac{\nu_{a}^{(L)}}{E_{A}^{(L)}}\sigma_{t}+\alpha_{A}^{(L)}\Delta T+\left(1-\nu_{A}^{(L)}\nu_{A}^{90}\right)\frac{a}{b}\frac{\tilde{E}_{A}^{(L)}}{E_{A}^{(L)}\tilde{E}_{A}^{0}}\Phi\hat{\sigma}_{11}^{90}. \tag{9.140}$$

On eliminating $\hat{\sigma}_{11}^{90}$ in (9.134) and (9.140) using (9.138) the final form of the stress-strain relations for a damaged laminate are obtained, namely,

$$\varepsilon_{A}=\frac{1}{E_{A}}\sigma_{A}-\frac{\nu_{A}}{E_{A}}\sigma_{T}-\frac{\nu_{a}}{E_{A}}\sigma_{t}+\alpha_{A}\Delta T, \tag{9.141}$$

$$\varepsilon_{T}=-\frac{\nu_{A}}{E_{A}}\sigma_{A}+\frac{1}{E_{T}}\sigma_{T}-\frac{\nu_{t}}{E_{T}}\sigma_{t}+\alpha_{T}\Delta T, \tag{9.142}$$

where the thermoelastic constants of the damaged laminate are given by

$$\frac{1}{E_{A}}=\frac{1}{E_{A}^{(L)}}+\left(1-\nu_{A}^{(L)}\nu_{A}^{90}\right)^{2}\frac{a}{b}\frac{\tilde{E}_{A}^{(L)}\tilde{E}_{T}^{90}}{(E_{A}^{(L)})^{2}\tilde{E}_{A}^{0}}\frac{\Phi}{\xi}, \tag{9.143}$$

$$\frac{1}{E_{T}}=\frac{1}{E_{T}^{(L)}}\left[1+\left(\nu_{A}^{0}\frac{E_{T}^{0}}{E_{A}^{0}}-\nu_{A}^{90}\right)^{2}\frac{ab}{h^{2}}\frac{\tilde{E}_{A}^{0}\tilde{E}_{T}^{90}}{\tilde{E}_{A}^{(L)}E_{T}^{(L)}}\frac{\Phi}{\xi}\right],\ \text{implying } E_{T}=\xi E_{T}^{(L)}, \tag{9.144}$$

$$\frac{\nu_{A}}{E_{A}}=\frac{\nu_{A}^{(L)}}{E_{A}^{(L)}}+\left(1-\nu_{A}^{(L)}\nu_{A}^{90}\right)\left(\nu_{A}^{0}\frac{E_{T}^{0}}{E_{A}^{0}}-\nu_{A}^{90}\right)\frac{a}{h}\frac{\tilde{E}_{T}^{90}}{E_{A}^{(L)}E_{T}^{(L)}}\frac{\Phi}{\xi}, \tag{9.145}$$

$$\frac{\nu_{a}}{E_{A}}=\frac{\nu_{a}^{(L)}}{E_{A}^{(L)}}+\left\{\frac{\nu_{a}^{(L)}}{E_{A}^{(L)}}-\frac{\nu_{t}^{90}}{E_{T}^{90}}+\nu_{A}^{90}\left(\frac{\nu_{t}^{(L)}}{E_{T}^{(L)}}-\frac{\nu_{a}^{90}}{E_{A}^{90}}\right)\right\}\left(1-\nu_{A}^{(L)}\nu_{A}^{90}\right)\frac{a}{b}\frac{\tilde{E}_{A}^{(L)}}{E_{A}^{(L)}}\frac{\tilde{E}_{T}^{90}}{\tilde{E}_{A}^{0}}\frac{\Phi}{\xi}, \tag{9.146}$$

$$\frac{\nu_{t}}{E_{T}}=\frac{\nu_{t}^{(L)}}{E_{T}^{(L)}}-\left\{\frac{\nu_{a}^{(L)}}{E_{A}^{(L)}}-\frac{\nu_{t}^{90}}{E_{T}^{90}}+\nu_{A}^{90}\left(\frac{\nu_{t}^{(L)}}{E_{T}^{(L)}}-\frac{\nu_{a}^{90}}{E_{A}^{90}}\right)\right\}\left(\nu_{A}^{0}\frac{E_{T}^{0}}{E_{A}^{0}}-\nu_{A}^{90}\right)\frac{a}{h}\frac{\tilde{E}_{T}^{90}}{E_{T}^{(L)}}\frac{\Phi}{\xi}, \tag{9.147}$$

$$\alpha_{A}=\alpha_{A}^{(L)}+\left(1-\nu_{A}^{(L)}\nu_{A}^{90}\right)\left(\alpha_{A}^{(L)}+\nu_{A}^{90}\alpha_{T}^{(L)}-\tilde{\alpha}_{T}^{90}\right)\frac{a}{b}\frac{\tilde{E}_{A}^{(L)}}{E_{A}^{(L)}}\frac{\tilde{E}_{T}^{90}}{\tilde{E}_{A}^{0}}\frac{\Phi}{\xi}, \tag{9.148}$$

$$\alpha_{\mathrm{T}}=\alpha_{\mathrm{T}}^{(\mathrm{L})}-\left(\alpha_{\mathrm{A}}^{(\mathrm{L})}+\nu_{\mathrm{A}}^{90}\alpha_{\mathrm{T}}^{(\mathrm{L})}-\tilde{\alpha}_{\mathrm{T}}^{90}\right)\left(\nu_{\mathrm{A}}^{0}\frac{E_{\mathrm{T}}^{0}}{E_{\mathrm{A}}^{0}}-\nu_{\mathrm{A}}^{90}\right)\frac{a}{h}\frac{\tilde{E}_{\mathrm{T}}^{90}}{E_{\mathrm{T}}^{(\mathrm{L})}}\frac{\Phi}{\xi}. \tag{9.149}$$

The results (9.141)–(9.149) show that the stress-strain relations of a damaged laminate are exactly of the same form as those for an undamaged laminate. The formation of damage affects only the values of the thermoelastic constants, and not the form of the stress-strain relations.

9.5.2 Approach 2

On using $(9.32)_1$ to substitute for $\hat{\sigma}_{11}^{90}$ into both (9.129) and (9.131), assuming that the values of $\sigma_{\mathrm{A}}=\hat{\sigma}_{\mathrm{A}}$, $\varepsilon_{\mathrm{T}}=\hat{\varepsilon}_{\mathrm{T}}$, σ_{t} and ΔT are known,

$$\varepsilon_{\mathrm{A}}-\hat{\varepsilon}_{\mathrm{A}}=\left(\hat{\varepsilon}_{\mathrm{A}}+\nu_{\mathrm{A}}^{90}\varepsilon_{\mathrm{T}}+\frac{\tilde{\nu}_{\mathrm{t}}^{90}}{\tilde{E}_{\mathrm{T}}^{90}}\sigma_{\mathrm{t}}-\tilde{\alpha}_{\mathrm{T}}^{90}\Delta T\right)\frac{a}{b}\frac{\tilde{E}_{\mathrm{T}}^{90}}{\tilde{E}_{\mathrm{A}}^{0}}\Phi, \tag{9.150}$$

$$\sigma_{\mathrm{T}}-\hat{\sigma}_{\mathrm{T}}=\left(\hat{\varepsilon}_{\mathrm{A}}+\nu_{\mathrm{A}}^{90}\varepsilon_{\mathrm{T}}+\frac{\tilde{\nu}_{\mathrm{t}}^{90}}{\tilde{E}_{\mathrm{T}}^{90}}\sigma_{\mathrm{t}}-\tilde{\alpha}_{\mathrm{T}}^{90}\Delta T\right)\left(\nu_{\mathrm{A}}^{0}\frac{E_{\mathrm{T}}^{0}}{E_{\mathrm{A}}^{0}}-\nu_{\mathrm{A}}^{90}\right)\frac{a}{h}\tilde{E}_{\mathrm{T}}^{90}\Phi, \tag{9.151}$$

so that

$$\sigma_{\mathrm{T}}-\hat{\sigma}_{\mathrm{T}}=\lambda\tilde{E}_{\mathrm{A}}^{(\mathrm{L})}\left(\varepsilon_{\mathrm{A}}-\hat{\varepsilon}_{\mathrm{A}}\right), \tag{9.152}$$

where λ is a useful laminate constant defined by (see also (9.137))

$$\lambda=\nu_{\mathrm{A}}^{(\mathrm{L})}\frac{E_{\mathrm{T}}^{(\mathrm{L})}}{E_{\mathrm{A}}^{(\mathrm{L})}}-\nu_{\mathrm{A}}^{90}=\left(\nu_{\mathrm{A}}^{0}\frac{E_{\mathrm{T}}^{0}}{E_{\mathrm{A}}^{0}}-\nu_{\mathrm{A}}^{90}\right)\frac{b}{h}\frac{\tilde{E}_{\mathrm{A}}^{0}}{\tilde{E}_{\mathrm{A}}^{(\mathrm{L})}}. \tag{9.153}$$

From the stress-strain relations (9.40) for an undamaged laminate, because it is assumed that the values of $\sigma_{\mathrm{A}}=\hat{\sigma}_{\mathrm{A}}$, $\varepsilon_{\mathrm{T}}=\hat{\varepsilon}_{\mathrm{T}}$, σ_{t} and ΔT are known,

$$\hat{\sigma}_{\mathrm{T}}=\nu_{\mathrm{A}}^{(\mathrm{L})}\frac{E_{\mathrm{T}}^{(\mathrm{L})}}{E_{\mathrm{A}}^{(\mathrm{L})}}\sigma_{\mathrm{A}}+\nu_{\mathrm{t}}^{(\mathrm{L})}\sigma_{\mathrm{t}}+E_{\mathrm{T}}^{(\mathrm{L})}\varepsilon_{\mathrm{T}}-E_{\mathrm{T}}^{(\mathrm{L})}\alpha_{\mathrm{T}}^{(\mathrm{L})}\Delta T, \tag{9.154}$$

$$\hat{\varepsilon}_{\mathrm{A}}=\frac{1}{\tilde{E}_{\mathrm{A}}^{(\mathrm{L})}}\sigma_{\mathrm{A}}-\frac{\tilde{\nu}_{\mathrm{a}}^{(\mathrm{L})}}{\tilde{E}_{\mathrm{A}}^{(\mathrm{L})}}\sigma_{\mathrm{t}}-\nu_{\mathrm{A}}^{(\mathrm{L})}\frac{E_{\mathrm{T}}^{(\mathrm{L})}}{E_{\mathrm{A}}^{(\mathrm{L})}}\varepsilon_{\mathrm{T}}+\tilde{\alpha}_{\mathrm{A}}^{(\mathrm{L})}\Delta T, \tag{9.155}$$

where $\tilde{E}_{\mathrm{A}}^{(\mathrm{L})}$ is defined by (9.42) and where

$$\frac{\tilde{\nu}_{\mathrm{a}}^{(\mathrm{L})}}{\tilde{E}_{\mathrm{A}}^{(\mathrm{L})}}=\frac{\nu_{\mathrm{a}}^{(\mathrm{L})}+\nu_{\mathrm{t}}^{(\mathrm{L})}\nu_{\mathrm{A}}^{(\mathrm{L})}}{E_{\mathrm{A}}^{(\mathrm{L})}},\qquad\tilde{\alpha}_{\mathrm{A}}^{(\mathrm{L})}=\alpha_{\mathrm{A}}^{(\mathrm{L})}+\nu_{\mathrm{A}}^{(\mathrm{L})}\frac{E_{\mathrm{T}}^{(\mathrm{L})}}{E_{\mathrm{A}}^{(\mathrm{L})}}\alpha_{\mathrm{T}}^{(\mathrm{L})}. \tag{9.156}$$

Relations (9.150)–(9.155), on eliminating the terms $\hat{\varepsilon}_{\mathrm{A}}$ and $\hat{\sigma}_{\mathrm{T}}$, assert that for a damaged laminate

$$\varepsilon_{\mathrm{A}}=\frac{1}{\tilde{E}_{\mathrm{A}}}\sigma_{\mathrm{A}}-\frac{\tilde{\nu}_{\mathrm{a}}}{\tilde{E}_{\mathrm{A}}}\sigma_{\mathrm{t}}-\nu_{\mathrm{A}}\frac{E_{\mathrm{T}}}{E_{\mathrm{A}}}\varepsilon_{\mathrm{T}}+\tilde{\alpha}_{\mathrm{A}}\Delta T, \tag{9.157}$$

$$\sigma_{\mathrm{T}}=\nu_{\mathrm{A}}\frac{E_{\mathrm{T}}}{E_{\mathrm{A}}}\sigma_{\mathrm{A}}+E_{\mathrm{T}}\varepsilon_{\mathrm{T}}+\nu_{\mathrm{t}}\sigma_{\mathrm{t}}-E_{\mathrm{T}}\alpha_{\mathrm{T}}\Delta T, \tag{9.158}$$

where

$$\frac{1}{\tilde{E}_A} = \frac{1}{E_A}\left(1 - \nu_A^2 \frac{E_T}{E_A}\right), \qquad \frac{\tilde{\nu}_a}{\tilde{E}_A} = \frac{\nu_a + \nu_t \nu_A}{E_A}, \qquad \tilde{\alpha}_A = \alpha_A + \nu_A \frac{E_T}{E_A}\alpha_T, \tag{9.159}$$

and where

$$\frac{1}{\tilde{E}_A} = \frac{1}{\tilde{E}_A^{(L)}}\left(1 + \frac{a}{b}\frac{\tilde{E}_T^{90}}{\tilde{E}_A^0}\Phi\right), \tag{9.160}$$

$$\frac{\tilde{\nu}_a}{\tilde{E}_A} = \frac{\tilde{\nu}_a^{(L)}}{\tilde{E}_A^{(L)}} + \left(\frac{\tilde{\nu}_a^{(L)}}{\tilde{E}_A^{(L)}} - \frac{\tilde{\nu}_t^{90}}{\tilde{E}_T^{90}}\right)\frac{a}{b}\frac{\tilde{E}_T^{90}}{\tilde{E}_A^0}\Phi, \tag{9.161}$$

$$\nu_A \frac{E_T}{E_A} = \nu_A^{(L)}\frac{E_T^{(L)}}{E_A^{(L)}} + \nu_A^{(L)}\frac{E_T^{(L)}}{E_A^{(L)}} - \nu_A^{90} = \left(\nu_A^0 \frac{E_T^0}{E_A^0} - \nu_A^{90}\right)\frac{a}{h}\frac{\tilde{E}_T^{90}}{\tilde{E}_A^{(L)}}\Phi, \tag{9.162}$$

$$\tilde{\alpha}_A = \tilde{\alpha}_A^{(L)} + \left(\tilde{\alpha}_A^{(L)} - \tilde{\alpha}_T^{90}\right)\frac{a}{b}\frac{\tilde{E}_T^{90}}{\tilde{E}_A^0}\Phi, \tag{9.163}$$

$$\nu_A \frac{E_T}{E_A} = \nu_A^{(L)}\frac{E_T^{(L)}}{E_A^{(L)}} + \left(\nu_A^0 \frac{E_T^0}{E_A^0} - \nu_A^{90}\right)\frac{a}{h}\frac{\tilde{E}_T^{90}}{\tilde{E}_A^{(L)}}\Phi, \tag{9.164}$$

$$E_T = E_T^{(L)} - \left(\nu_A^0 \frac{E_T^0}{E_A^0} - \nu_A^{90}\right)^2 \frac{ab}{h^2}\frac{\tilde{E}_A^0 \tilde{E}_T^{90}}{\tilde{E}_A^{(L)}}\Phi, \tag{9.165}$$

$$\nu_t = \nu_t^{(L)} - \left(\frac{\tilde{\nu}_a^{(L)}}{\tilde{E}_A^{(L)}} - \frac{\tilde{\nu}_t^{90}}{\tilde{E}_T^{90}}\right)\left(\nu_A^0 \frac{E_T^0}{E_A^0} - \nu_A^{90}\right)\frac{a}{h}\tilde{E}_T^{90}\Phi, \tag{9.166}$$

$$E_T\alpha_T = E_T^{(L)}\alpha_T^{(L)} - \left(\tilde{\alpha}_A^{(L)} - \tilde{\alpha}_T^{90}\right)\left(\nu_A^0 \frac{E_T^0}{E_A^0} - \nu_A^{90}\right)\frac{a}{h}\tilde{E}_T^{90}\Phi. \tag{9.167}$$

On using (9.42) it can be shown that the expressions (9.162) and (9.164) are equivalent because of relation (9.137).

From (9.160)–(9.167) it can be shown on eliminating Φ that

$$\frac{\dfrac{\tilde{\nu}_a}{\tilde{E}_A} - \dfrac{\tilde{\nu}_a^{(L)}}{\tilde{E}_A^{(L)}}}{\dfrac{1}{\tilde{E}_A} - \dfrac{1}{\tilde{E}_A^{(L)}}} = \left(\frac{\tilde{\nu}_a^{(L)}}{\tilde{E}_A^{(L)}} - \frac{\tilde{\nu}_t^{90}}{\tilde{E}_T^{90}}\right)\tilde{E}_A^{(L)}, \tag{9.168}$$

$$\frac{\tilde{\alpha}_A - \tilde{\alpha}_A^{(L)}}{\dfrac{1}{\tilde{E}_A} - \dfrac{1}{\tilde{E}_A^{(L)}}} = \left(\tilde{\alpha}_A^{(L)} - \tilde{\alpha}_T^{90}\right)\tilde{E}_A^{(L)}, \tag{9.169}$$

$$\frac{E_T\alpha_T - E_T^{(L)}\alpha_T^{(L)}}{\nu_A \dfrac{E_T}{E_A} - \nu_A^{(L)}\dfrac{E_T^{(L)}}{E_A^{(L)}}} = -\left(\tilde{\alpha}_A^{(L)} - \tilde{\alpha}_T^{90}\right)\tilde{E}_A^{(L)}, \tag{9.170}$$

$$\frac{\nu_t - \nu_t^{(L)}}{\nu_A \dfrac{E_T}{E_A} - \nu_A^{(L)} \dfrac{E_T^{(L)}}{E_A^{(L)}}} = \left(\frac{\tilde{\nu}_a^{(L)}}{\tilde{E}_A^{(L)}} - \frac{\tilde{\nu}_t^{90}}{\tilde{E}_T^{90}} \right) \tilde{E}_A^{(L)}, \tag{9.171}$$

$$\frac{E_T \alpha_T - E_T^{(L)} \alpha_T^{(L)}}{\tilde{\alpha}_A - \tilde{\alpha}_A^{(L)}} = -\left(\nu_A^0 \frac{E_T^0}{E_A^0} - \nu_A^{90} \right) \frac{b}{h} \tilde{E}_A^0, \tag{9.172}$$

$$\frac{E_T - E_T^{(L)}}{\nu_A \dfrac{E_T}{E_A} - \nu_A^{(L)} \dfrac{E_T^{(L)}}{E_A^{(L)}}} = -\left(\nu_A^0 \frac{E_T^0}{E_A^0} - \nu_A^{90} \right) \frac{b}{h} \tilde{E}_A^0. \tag{9.173}$$

This completes the descriptions of two self-consistent approaches to the prediction of the effective in-plane thermoelastic constants for a simple cross-ply model having a uniform array of ply cracks in the 90° ply.

9.6 Through-thickness Properties of Damaged Laminates

On applying (9.85) to the ply crack problem being considered

$$\varepsilon_t = \frac{1}{Lh} \int_0^L u_3^0(x_1, h) \, dx_1. \tag{9.174}$$

On using (9.93) together with the conditions $C'(0) = 0$ and $C'(L) = 0$

$$\varepsilon_t = \hat{\varepsilon}_t + \left(\frac{\tilde{\nu}_t^{90}}{\tilde{E}_T^{90}} - \frac{\tilde{\nu}_a^0}{\tilde{E}_A^0} \right) \frac{b}{h} \frac{\bar{C}(L)}{L}, \tag{9.175}$$

where use has been made of relation (9.37). As it has been assumed that $\sigma_A = \hat{\sigma}_A$, it follows on using (9.46) and (9.127) that

$$\varepsilon_t = -\frac{\nu_a^{(L)}}{E_A^{(L)}} \sigma_A - \frac{\nu_t^{(L)}}{E_T^{(L)}} \hat{\sigma}_T + \frac{1}{E_t^{(L)}} \sigma_t + \alpha_t^{(L)} \Delta T + \left(\frac{\tilde{\nu}_t^{90}}{\tilde{E}_T^{90}} - \frac{\tilde{\nu}_a^0}{\tilde{E}_A^0} \right) \frac{a}{h} \Phi \, \hat{\sigma}_{11}^{90}. \tag{9.176}$$

On using (9.131)

$$\varepsilon_t = -\frac{\nu_a^{(L)}}{E_A^{(L)}} \sigma_A - \frac{\nu_t^{(L)}}{E_T^{(L)}} \sigma_T + \frac{1}{E_t^{(L)}} \sigma_t + \alpha_t^{(L)} \Delta T - \Omega \frac{a}{h} \Phi \hat{\sigma}_{11}^{90}, \tag{9.177}$$

where

$$\Omega = \frac{\tilde{\nu}_a^0}{\tilde{E}_A^0} - \frac{\tilde{\nu}_t^{90}}{\tilde{E}_T^{90}} - \frac{\nu_t^{(L)}}{E_T^{(L)}} \left(\nu_A^0 \frac{E_T^0}{E_A^0} - \nu_A^{90} \right). \tag{9.178}$$

On substituting for $\hat{\sigma}_{11}^{90}$ into (9.177) using (9.138) the following through-thickness stress-strain relation is obtained

$$\varepsilon_t = -\frac{\nu_a}{E_A} \sigma_A - \frac{\nu_t}{E_T} \sigma_T + \frac{1}{E_t} \sigma_t + \alpha_t \Delta T, \tag{9.179}$$

where

$$\frac{\nu_{\mathrm{a}}}{E_{\mathrm{A}}}=\frac{\nu_{\mathrm{a}}^{(\mathrm{L})}}{E_{\mathrm{A}}^{(\mathrm{L})}}+\Omega\left(1-\nu_{\mathrm{A}}^{90}\nu_{\mathrm{A}}^{(\mathrm{L})}\right)\frac{a}{h}\frac{\tilde{E}_{\mathrm{T}}^{90}}{E_{\mathrm{A}}^{(\mathrm{L})}}\frac{\Phi}{\xi}, \tag{9.180}$$

$$\frac{\nu_{\mathrm{t}}}{E_{\mathrm{T}}}=\frac{\nu_{\mathrm{t}}^{(L)}}{E_{\mathrm{T}}^{(\mathrm{L})}}-\Omega\left(\nu_{\mathrm{A}}^{0}\frac{E_{\mathrm{T}}^{0}}{E_{\mathrm{A}}^{0}}-\nu_{\mathrm{A}}^{90}\right)\frac{ab}{h^2}\frac{\tilde{E}_{\mathrm{A}}^{0}\tilde{E}_{\mathrm{T}}^{90}}{\tilde{E}_{\mathrm{A}}^{(\mathrm{L})}E_{\mathrm{T}}^{(\mathrm{L})}}\frac{\Phi}{\xi}, \tag{9.181}$$

$$\frac{1}{E_{\mathrm{t}}}=\frac{1}{E_{\mathrm{t}}^{(\mathrm{L})}}+\Omega\left\{\frac{\nu_{\mathrm{a}}^{(\mathrm{L})}}{E_{\mathrm{A}}^{(\mathrm{L})}}-\frac{\nu_{\mathrm{t}}^{90}}{E_{\mathrm{T}}^{90}}+\nu_{\mathrm{A}}^{90}\left(\frac{\nu_{\mathrm{t}}^{(\mathrm{L})}}{E_{\mathrm{T}}^{(\mathrm{L})}}-\frac{\nu_{\mathrm{a}}^{90}}{E_{\mathrm{A}}^{90}}\right)\right\}\frac{a}{h}\tilde{E}_{\mathrm{T}}^{90}\frac{\Phi}{\xi}, \tag{9.182}$$

$$\alpha_{\mathrm{t}}=\alpha_{\mathrm{t}}^{(\mathrm{L})}-\Omega\left(\alpha_{\mathrm{A}}^{(\mathrm{L})}+\nu_{\mathrm{A}}^{90}\alpha_{\mathrm{T}}^{(\mathrm{L})}-\tilde{\alpha}_{\mathrm{T}}^{90}\right)\frac{a}{h}\tilde{E}_{\mathrm{T}}^{90}\frac{\Phi}{\xi}. \tag{9.183}$$

Relations (9.146) and (9.147) are equivalent to the results (9.180) and (9.181) only if the following relation is satisfied

$$\frac{\nu_{\mathrm{a}}^{(\mathrm{L})}}{E_{\mathrm{A}}^{(\mathrm{L})}}-\frac{\nu_{\mathrm{t}}^{90}}{E_{\mathrm{T}}^{90}}+\nu_{\mathrm{A}}^{90}\left(\frac{\nu_{\mathrm{t}}^{(\mathrm{L})}}{E_{\mathrm{T}}^{(\mathrm{L})}}-\frac{\nu_{\mathrm{a}}^{90}}{E_{\mathrm{A}}^{90}}\right)=\Omega\frac{b}{h}\frac{\tilde{E}_{\mathrm{A}}^{0}}{\tilde{E}_{\mathrm{A}}^{(\mathrm{L})}}. \tag{9.184}$$

On using (9.178) this relation may be expressed

$$\left\{\frac{\nu_{\mathrm{a}}^{(\mathrm{L})}}{E_{\mathrm{A}}^{(\mathrm{L})}}-\frac{\nu_{\mathrm{t}}^{90}}{E_{\mathrm{T}}^{90}}+\nu_{\mathrm{A}}^{90}\left(\frac{\nu_{\mathrm{t}}^{(L)}}{E_{\mathrm{T}}^{(\mathrm{L})}}-\frac{\nu_{\mathrm{a}}^{90}}{E_{\mathrm{A}}^{90}}\right)\right\}\tilde{E}_{\mathrm{A}}^{(\mathrm{L})}=\frac{b}{h}\left\{\frac{\tilde{\nu}_{\mathrm{a}}^{0}}{\tilde{E}_{\mathrm{A}}^{0}}-\frac{\tilde{\nu}_{\mathrm{t}}^{90}}{\tilde{E}_{\mathrm{T}}^{90}}-\frac{\nu_{\mathrm{t}}^{(\mathrm{L})}}{E_{\mathrm{T}}^{(\mathrm{L})}}\left(\nu_{\mathrm{A}}^{0}\frac{E_{\mathrm{T}}^{0}}{E_{\mathrm{A}}^{0}}-\nu_{\mathrm{A}}^{90}\right)\right\}\tilde{E}_{\mathrm{A}}^{0}. \tag{9.185}$$

From (9.30) and (9.34)

$$\frac{\tilde{\nu}_{\mathrm{a}}^{0}}{\tilde{E}_{\mathrm{A}}^{0}}-\frac{\tilde{\nu}_{\mathrm{t}}^{90}}{\tilde{E}_{\mathrm{T}}^{90}}=\frac{\nu_{\mathrm{a}}^{0}}{E_{\mathrm{A}}^{0}}+\frac{\nu_{\mathrm{t}}^{0}\nu_{\mathrm{A}}^{0}}{E_{\mathrm{A}}^{0}}-\frac{\nu_{\mathrm{t}}^{90}}{E_{\mathrm{T}}^{90}}-\frac{\nu_{\mathrm{a}}^{90}\nu_{\mathrm{A}}^{90}}{E_{\mathrm{A}}^{90}}, \tag{9.186}$$

and from (9.42)

$$\tilde{E}_{\mathrm{A}}^{(\mathrm{L})}=\frac{b}{h}\tilde{E}_{\mathrm{A}}^{0}+\frac{a}{h}\tilde{E}_{\mathrm{T}}^{90}. \tag{9.187}$$

On substituting these values into (9.185)

$$\frac{\nu_{\mathrm{a}}^{(\mathrm{L})}\tilde{E}_{\mathrm{A}}^{(\mathrm{L})}}{E_{\mathrm{A}}^{(\mathrm{L})}}=\left\{\frac{\nu_{\mathrm{a}}^{0}}{E_{\mathrm{A}}^{0}}+\frac{\nu_{\mathrm{t}}^{0}\nu_{\mathrm{A}}^{0}}{E_{\mathrm{A}}^{0}}\right\}\frac{b}{h}\tilde{E}_{\mathrm{A}}^{0}+\left\{\frac{\nu_{\mathrm{t}}^{90}}{E_{\mathrm{T}}^{90}}+\frac{\nu_{\mathrm{a}}^{90}\nu_{\mathrm{A}}^{90}}{E_{\mathrm{A}}^{90}}\right\}\frac{a}{h}\tilde{E}_{\mathrm{T}}^{90}-\frac{\nu_{\mathrm{t}}^{(\mathrm{L})}}{E_{\mathrm{T}}^{(\mathrm{L})}}\frac{1}{h}\left(b\nu_{\mathrm{A}}^{0}\tilde{E}_{\mathrm{T}}^{0}+a\nu_{\mathrm{A}}^{90}\tilde{E}_{\mathrm{T}}^{90}\right). \tag{9.188}$$

From (9.42) it can be shown that $h\tilde{E}_{\mathrm{T}}^{(L)}\nu_{\mathrm{A}}^{(\mathrm{L})}=b\nu_{\mathrm{A}}^{0}\tilde{E}_{\mathrm{T}}^{0}+a\nu_{\mathrm{A}}^{90}\tilde{E}_{\mathrm{T}}^{90}$ so that (9.188) may be written as

$$\nu_{\mathrm{a}}^{(\mathrm{L})}\frac{\tilde{E}_{\mathrm{A}}^{(\mathrm{L})}}{E_{\mathrm{A}}^{(\mathrm{L})}}+\nu_{\mathrm{t}}^{(\mathrm{L})}\nu_{\mathrm{A}}^{(\mathrm{L})}\frac{\tilde{E}_{\mathrm{T}}^{(\mathrm{L})}}{E_{\mathrm{T}}^{(\mathrm{L})}}=\frac{b}{h}\tilde{\nu}_{\mathrm{a}}^{0}+\frac{a}{h}\tilde{\nu}_{\mathrm{t}}^{90}. \tag{9.189}$$

It then follows that

$$\frac{\nu_{\mathrm{a}}^{(\mathrm{L})}+\nu_{\mathrm{t}}^{(\mathrm{L})}\nu_{\mathrm{A}}^{(\mathrm{L})}}{1-(\nu_{\mathrm{A}}^{(\mathrm{L})})^2E_{\mathrm{T}}^{(\mathrm{L})}/E_{\mathrm{A}}^{(\mathrm{L})}}=\frac{b}{h}\tilde{\nu}_{\mathrm{a}}^{0}+\frac{a}{h}\tilde{\nu}_{\mathrm{t}}^{90}. \tag{9.190}$$

The validity of relation (9.190), and hence of relation (9.184), is easily established by making use of relations (9.45) and (9.47) to show that

$$\nu_{\mathrm{a}}^{(\mathrm{L})}+\nu_{\mathrm{t}}^{(\mathrm{L})}\nu_{\mathrm{A}}^{(\mathrm{L})}=\left(1-(\nu_{\mathrm{A}}^{(\mathrm{L})})^2\frac{E_{\mathrm{T}}^{(\mathrm{L})}}{E_{\mathrm{A}}^{(\mathrm{L})}}\right)A'=\left(1-(\nu_{\mathrm{A}}^{(\mathrm{L})})^2\frac{E_{\mathrm{T}}^{(\mathrm{L})}}{E_{\mathrm{A}}^{(\mathrm{L})}}\right)\left(\frac{b}{h}\tilde{\nu}_{\mathrm{a}}^0+\frac{a}{h}\tilde{\nu}_{\mathrm{t}}^{90}\right). \tag{9.191}$$

9.7 Consideration of Ply Crack Closure

A damaged laminate having cracks in the 90° ply can always be loaded so that the ply cracks just close such that the stress normal to the ply cracks is zero everywhere in the 90° ply. To understand this, it is necessary only to consider an undamaged laminate and to determine loading conditions for which the stress component $\hat{\sigma}_{11}^{90}=0$. When an undamaged laminate is in this state, any number of ply cracks can be formed without changing in any way the stress or displacement distributions in the laminate. Let $\varepsilon_{\mathrm{A}}^{\mathrm{c}}$, $\varepsilon_{\mathrm{T}}^{\mathrm{c}}$ and $\sigma_{\mathrm{t}}^{\mathrm{c}}$ be the in-plane laminate strains and through-thickness stress that generate a ply crack closure condition.

From (9.31) at the point of ply crack closure $\hat{\sigma}_{11}^{90}=0$

$$\begin{aligned}
\varepsilon_{\mathrm{A}}^{\mathrm{c}}&=-\frac{\nu_{\mathrm{A}}^{90}}{E_{\mathrm{A}}^{90}}\hat{\sigma}_{22}^{90}-\frac{\nu_{\mathrm{t}}^{90}}{E_{\mathrm{T}}^{90}}\sigma_{\mathrm{t}}^{\mathrm{c}}+\alpha_{\mathrm{T}}^{90}\Delta T,\\
\varepsilon_{\mathrm{T}}^{\mathrm{c}}&=\frac{1}{E_{E}^{90}}\hat{\sigma}_{22}^{90}-\frac{\nu_{\mathrm{a}}^{90}}{E_{\mathrm{A}}^{90}}\sigma_{\mathrm{t}}^{\mathrm{c}}+\alpha_{\mathrm{A}}^{90}\Delta T.
\end{aligned} \tag{9.192}$$

On eliminating $\hat{\sigma}_{22}^{90}$ in (9.192) it follows that

$$\varepsilon_{\mathrm{A}}^{\mathrm{c}}=-\nu_{\mathrm{A}}^{90}\varepsilon_{\mathrm{T}}^{\mathrm{c}}-\left(\frac{\nu_{\mathrm{t}}^{90}}{E_{\mathrm{T}}^{90}}+\frac{\nu_{\mathrm{a}}^{90}\nu_{\mathrm{A}}^{90}}{E_{\mathrm{A}}^{90}}\right)\sigma_{\mathrm{t}}^{\mathrm{c}}+\left(\alpha_{\mathrm{T}}^{90}+\nu_{\mathrm{A}}^{90}\alpha_{\mathrm{A}}^{90}\right)\Delta T. \tag{9.193}$$

9.7.1 Uniaxial Loading in Axial Direction

Consider the uniaxial loading of the laminate in the axial direction to the point where the ply cracks just close. Clearly $\hat{\sigma}_{\mathrm{T}}=\hat{\sigma}_{\mathrm{t}}=0$ so that from (9.40)

$$\begin{aligned}
\bar{\varepsilon}_{\mathrm{A}}^{\mathrm{c}}&=\frac{1}{E_{\mathrm{A}}^{(\mathrm{L})}}\bar{\sigma}_{\mathrm{A}}^{\mathrm{c}}+\alpha_{\mathrm{A}}^{(\mathrm{L})}\Delta T,\\
\bar{\varepsilon}_{\mathrm{T}}^{\mathrm{c}}&=-\frac{\nu_{\mathrm{A}}^{(\mathrm{L})}}{E_{\mathrm{A}}^{(\mathrm{L})}}\bar{\sigma}_{\mathrm{A}}^{\mathrm{c}}+\alpha_{\mathrm{T}}^{(\mathrm{L})}\Delta T,
\end{aligned} \tag{9.194}$$

where $\bar{\varepsilon}_{\mathrm{A}}^{\mathrm{c}}$ and $\bar{\varepsilon}_{\mathrm{T}}^{\mathrm{c}}$ are the in-plane laminate strains that arise when the uniaxially applied axial stress $\bar{\sigma}_{\mathrm{A}}^{\mathrm{c}}$ generates the ply crack closure condition. For this uniaxial loading case, the closure relation (9.193) reduces to

$$\bar{\varepsilon}_{\mathrm{A}}^{\mathrm{c}}=-\nu_{\mathrm{A}}^{90}\bar{\varepsilon}_{\mathrm{T}}^{\mathrm{c}}+\left(\alpha_{\mathrm{T}}^{90}+\nu_{\mathrm{A}}^{90}\alpha_{\mathrm{A}}^{90}\right)\Delta T, \tag{9.195}$$

and on using (9.194)

$$\frac{\bar{\sigma}_{\mathrm{A}}^{\mathrm{c}}}{\Delta T}=E_{\mathrm{A}}^{(\mathrm{L})}\frac{\alpha_{\mathrm{T}}^{90}-\alpha_{\mathrm{A}}^{(\mathrm{L})}+\nu_{\mathrm{A}}^{90}(\alpha_{\mathrm{A}}^{90}-\alpha_{\mathrm{T}}^{(\mathrm{L})})}{1-\nu_{\mathrm{A}}^{90}\nu_{\mathrm{A}}^{(\mathrm{L})}}=-k_1. \tag{9.196}$$

9.7.2 Uniaxial Loading in In-plane Transverse Direction

Consider now the uniaxial loading of the laminate in the in-plane transverse direction to the point where the ply cracks just close. Clearly $\hat{\sigma}_{\mathrm{A}} = \hat{\sigma}_{\mathrm{t}} = 0$ for this case so that from (9.40)

$$\begin{aligned} \tilde{\varepsilon}_{\mathrm{A}}^{\mathrm{c}} &= -\frac{\nu_{\mathrm{A}}^{(\mathrm{L})}}{E_{\mathrm{A}}^{(\mathrm{L})}} \tilde{\sigma}_{\mathrm{T}}^{\mathrm{c}} + \alpha_{\mathrm{A}}^{(\mathrm{L})} \Delta T, \\ \tilde{\varepsilon}_{\mathrm{T}}^{\mathrm{c}} &= \frac{1}{E_{\mathrm{T}}^{(\mathrm{L})}} \tilde{\sigma}_{\mathrm{T}}^{\mathrm{c}} + \alpha_{\mathrm{T}}^{(\mathrm{L})} \Delta T, \end{aligned} \tag{9.197}$$

where $\tilde{\varepsilon}_{\mathrm{A}}^{\mathrm{c}}$ and $\tilde{\varepsilon}_{\mathrm{T}}^{\mathrm{c}}$ are the in-plane laminate strains that arise when the uniaxially applied in-plane transverse stress $\tilde{\sigma}_{\mathrm{T}}^{\mathrm{c}}$ generates the ply crack closure condition. For this uniaxial loading case, the closure relation (9.193) reduces to

$$\tilde{\varepsilon}_{\mathrm{A}}^{\mathrm{c}} = -\nu_{\mathrm{A}}^{90} \tilde{\varepsilon}_{\mathrm{T}}^{\mathrm{c}} + \left(\alpha_{\mathrm{T}}^{90} + \nu_{\mathrm{A}}^{90} \alpha_{\mathrm{A}}^{90} \right) \Delta T, \tag{9.198}$$

and on using (9.197)

$$\frac{\tilde{\sigma}_{\mathrm{T}}^{\mathrm{c}}}{\Delta T} = \frac{\alpha_{\mathrm{T}}^{90} - \alpha_{\mathrm{A}}^{(\mathrm{L})} + \nu_{\mathrm{A}}^{90} (\alpha_{\mathrm{A}}^{90} - \alpha_{\mathrm{T}}^{(\mathrm{L})})}{\dfrac{\nu_{\mathrm{A}}^{90}}{E_{\mathrm{T}}^{(\mathrm{L})}} - \dfrac{\nu_{\mathrm{A}}^{(\mathrm{L})}}{E_{\mathrm{A}}^{(\mathrm{L})}}} = -k_2. \tag{9.199}$$

9.7.3 Uniaxial Loading in Through-thickness Direction

Consider now the uniaxial loading of the laminate in the through-thickness direction to the point where the ply cracks just close. Clearly $\hat{\sigma}_{\mathrm{A}} = \hat{\sigma}_{\mathrm{T}} = 0$ for this case so that from (9.40)

$$\begin{aligned} \hat{\varepsilon}_{\mathrm{A}}^{\mathrm{c}} &= -\frac{\nu_{\mathrm{a}}^{(\mathrm{L})}}{E_{\mathrm{A}}^{(\mathrm{L})}} \hat{\sigma}_{\mathrm{t}}^{\mathrm{c}} + \alpha_{\mathrm{A}}^{(\mathrm{L})} \Delta T, \\ \hat{\varepsilon}_{\mathrm{T}}^{\mathrm{c}} &= -\frac{\nu_{\mathrm{t}}^{(\mathrm{L})}}{E_{\mathrm{T}}^{(\mathrm{L})}} \hat{\sigma}_{\mathrm{t}}^{\mathrm{c}} + \alpha_{\mathrm{T}}^{(\mathrm{L})} \Delta T, \end{aligned} \tag{9.200}$$

where $\hat{\varepsilon}_{\mathrm{A}}^{\mathrm{c}}$ and $\hat{\varepsilon}_{\mathrm{T}}^{\mathrm{c}}$ are the in-plane laminate strains that arise when the uniaxially applied through-thickness stress $\hat{\sigma}_{\mathrm{t}}^{\mathrm{c}}$ generates the ply crack closure condition. For this uniaxial loading case, the closure relation (9.193) is written as

$$\hat{\varepsilon}_{\mathrm{A}}^{\mathrm{c}} = -\nu_{\mathrm{A}}^{90} \hat{\varepsilon}_{\mathrm{T}}^{\mathrm{c}} - \left(\frac{\nu_{\mathrm{t}}^{90}}{E_{\mathrm{T}}^{90}} + \frac{\nu_{\mathrm{a}}^{90} \nu_{\mathrm{A}}^{90}}{E_{\mathrm{A}}^{90}} \right) \hat{\sigma}_{\mathrm{t}}^{\mathrm{c}} + \left(\alpha_{\mathrm{T}}^{90} + \nu_{\mathrm{A}}^{90} \alpha_{\mathrm{A}}^{90} \right) \Delta T, \tag{9.201}$$

and on using (9.200)

$$\frac{\hat{\sigma}_{\mathrm{t}}^{\mathrm{c}}}{\Delta T} = \frac{\alpha_{\mathrm{T}}^{90} - \alpha_{\mathrm{A}}^{(\mathrm{L})} + \nu_{\mathrm{A}}^{90} (\alpha_{\mathrm{A}}^{90} - \alpha_{\mathrm{T}}^{(\mathrm{L})})}{\dfrac{\nu_{\mathrm{t}}^{90}}{E_{\mathrm{T}}^{90}} + \dfrac{\nu_{\mathrm{a}}^{90} \nu_{\mathrm{A}}^{90}}{E_{\mathrm{A}}^{90}} - \dfrac{\nu_{\mathrm{a}}^{(\mathrm{L})}}{E_{\mathrm{A}}^{(\mathrm{L})}} - \dfrac{\nu_{\mathrm{A}}^{90} \nu_{\mathrm{t}}^{(\mathrm{L})}}{E_{\mathrm{T}}^{(\mathrm{L})}}} = -k_3. \tag{9.202}$$

9.7.4 Derivation of Important Interrelationships

First, from (9.196) and (9.199) it follows that

$$\frac{\bar{\sigma}_{\mathrm{A}}^{\mathrm{c}}}{\tilde{\sigma}_{\mathrm{T}}^{\mathrm{c}}} = \frac{E_{\mathrm{A}}^{(\mathrm{L})}}{E_{\mathrm{T}}^{(\mathrm{L})}} \frac{\nu_{\mathrm{A}}^{90} - \nu_{\mathrm{A}}^{(\mathrm{L})} E_{\mathrm{T}}^{(\mathrm{L})} / E_{\mathrm{A}}^{(\mathrm{L})}}{1 - \nu_{\mathrm{A}}^{90} \nu_{\mathrm{A}}^{(\mathrm{L})}} = \frac{k_1}{k_2} = k, \tag{9.203}$$

and from (9.196) and (9.202)

$$\frac{\bar{\sigma}_{\rm A}^{\rm c}}{\bar{\sigma}_{\rm t}^{\rm c}} = \frac{E_{\rm A}^{\rm (L)}}{1-\nu_{\rm A}^{90}\nu_{\rm A}^{\rm (L)}}\left\{\frac{\nu_{\rm t}^{90}}{E_{\rm T}^{90}} - \frac{\nu_{\rm a}^{\rm (L)}}{E_{\rm A}^{\rm (L)}} + \nu_{\rm A}^{90}\left(\frac{\nu_{\rm a}^{90}}{E_{\rm A}^{90}} - \frac{\nu_{\rm t}^{\rm (L)}}{E_{\rm T}^{\rm (L)}}\right)\right\} = \frac{k_1}{k_3} = k'. \tag{9.204}$$

It should be noted that the values of the constants k_1, k_2, k_3, k and k' depend on the properties of the 90° ply, and of effective properties of the laminate when in an undamaged state. Clearly their values are independent of any damage such as ply cracking. It should also be noted that the constants k and k' are independent of thermal expansion coefficients.

Consider now relation (9.143) which is written as

$$\Psi = \frac{1}{E_{\rm A}} - \frac{1}{E_{\rm A}^{\rm (L)}} = \left(1-\nu_{\rm A}^{\rm (L)}\nu_{\rm A}^{90}\right)^2 \frac{a}{b}\frac{\tilde{E}_{\rm A}^{\rm (L)}\tilde{E}_{\rm T}^{90}}{(E_{\rm A}^{\rm (L)})^2\tilde{E}_{\rm A}^{0}}\frac{\Phi}{\xi}, \tag{9.205}$$

so that

$$\frac{\Phi}{\xi} = \frac{b(E_{\rm A}^{\rm (L)})^2\tilde{E}_{\rm A}^{0}\,\Psi}{a\tilde{E}_{\rm A}^{\rm (L)}\tilde{E}_{\rm T}^{90}\left(1-\nu_{\rm A}^{\rm (L)}\nu_{\rm A}^{90}\right)^2}. \tag{9.206}$$

The ratio Φ/ξ is now eliminated in relations (9.144)–(9.149), (9.182) and (9.183) so that on using (9.34), (9.153), (9.184), (9.196) and (9.203)–(9.206)

$$\begin{aligned}
&\frac{1}{E_{\rm A}} - \frac{1}{E_{\rm A}^{\rm (L)}} = \Psi, \quad && \frac{1}{E_{\rm T}} - \frac{1}{E_{\rm T}^{\rm (L)}} = k^2\Psi, \quad && \frac{1}{E_{\rm t}} - \frac{1}{E_{\rm t}^{\rm (L)}} = (k')^2\Psi,\\
&\frac{\nu_{\rm A}^{\rm (L)}}{E_{\rm A}^{\rm (L)}} - \frac{\nu_{\rm A}}{E_{\rm A}} = k\Psi, && \frac{\nu_{\rm a}^{\rm (L)}}{E_{\rm A}^{\rm (L)}} - \frac{\nu_{\rm a}}{E_{\rm A}} = k'\Psi, && \frac{\nu_{\rm t}^{\rm (L)}}{E_{\rm T}^{\rm (L)}} - \frac{\nu_{\rm t}}{E_{\rm T}} = kk'\Psi,\\
&\alpha_{\rm A} - \alpha_{\rm A}^{\rm (L)} = k_1\Psi, && \alpha_{\rm T} - \alpha_{\rm T}^{\rm (L)} = kk_1\Psi, && \alpha_{\rm t} - \alpha_{\rm t}^{\rm (L)} = k'k_1\Psi.
\end{aligned} \tag{9.207}$$

To summarise the results obtained so far, it is clear from (9.207) that the degradation of all the thermoelastic constants of a damaged laminate are governed by the three undamaged laminate constants k_1, k and k' defined by (9.196), (9.203) and (9.204), respectively, and a single function Ψ accounting for the damage state that is defined by (9.205).

On using (9.196), (9.199), (9.202) and (9.207) it can be shown that

$$\frac{\bar{\sigma}_{\rm A}^{\rm c}}{\Delta T} = -\frac{\alpha_{\rm A} - \alpha_{\rm A}^{\rm (L)}}{\dfrac{1}{E_{\rm A}} - \dfrac{1}{E_{\rm A}^{\rm (L)}}} = -\frac{\alpha_{\rm T} - \alpha_{\rm T}^{\rm (L)}}{\dfrac{\nu_{\rm A}^{\rm (L)}}{E_{\rm A}^{\rm (L)}} - \dfrac{\nu_{\rm A}}{E_{\rm A}}} = -\frac{\alpha_{\rm t} - \alpha_{\rm t}^{\rm (L)}}{\dfrac{\nu_{\rm a}^{\rm (L)}}{E_{\rm A}^{\rm (L)}} - \dfrac{\nu_{\rm a}}{E_{\rm A}}} = -k_1, \tag{9.208}$$

$$\frac{\tilde{\sigma}_{\rm T}^{\rm c}}{\Delta T} = -\frac{\alpha_{\rm A} - \alpha_{\rm A}^{\rm (L)}}{\dfrac{\nu_{\rm A}^{\rm (L)}}{E_{\rm A}^{\rm (L)}} - \dfrac{\nu_{\rm A}}{E_{\rm A}}} = -\frac{\alpha_{\rm T} - \alpha_{\rm T}^{\rm (L)}}{\dfrac{1}{E_{\rm T}} - \dfrac{1}{E_{\rm T}^{\rm (L)}}} = -\frac{\alpha_{\rm t} - \alpha_{\rm t}^{\rm (L)}}{\dfrac{\nu_{\rm t}^{\rm (L)}}{E_{\rm T}^{\rm (L)}} - \dfrac{\nu_{\rm t}}{E_{\rm T}}} = -k_2, \tag{9.209}$$

$$\frac{\bar{\sigma}_{\rm t}^{\rm c}}{\Delta T} = -\frac{\alpha_{\rm A} - \alpha_{\rm A}^{\rm (L)}}{\dfrac{\nu_{\rm a}^{\rm (L)}}{E_{\rm A}^{\rm (L)}} - \dfrac{\nu_{\rm a}}{E_{\rm A}}} = -\frac{\alpha_{\rm T} - \alpha_{\rm T}^{\rm (L)}}{\dfrac{\nu_{\rm t}^{\rm (L)}}{E_{\rm T}^{\rm (L)}} - \dfrac{\nu_{\rm t}}{E_{\rm T}}} = -\frac{\alpha_{\rm t} - \alpha_{\rm t}^{\rm (L)}}{\dfrac{1}{E_{\rm t}} - \dfrac{1}{E_{\rm t}^{\rm (L)}}} = -k_3. \tag{9.210}$$

On using (9.203), (9.204) and (9.207) it can also be shown that

$$\frac{\dfrac{\nu_{\rm t}^{(\rm L)}}{E_{\rm T}^{(\rm L)}}-\dfrac{\nu_{\rm t}}{E_{\rm T}}}{\dfrac{\nu_{\rm a}^{(\rm L)}}{E_{\rm A}^{(\rm L)}}-\dfrac{\nu_{\rm a}}{E_{\rm A}}}=\frac{\dfrac{\nu_{\rm A}^{(\rm L)}}{E_{\rm A}^{(\rm L)}}-\dfrac{\nu_{\rm A}}{E_{\rm A}}}{\dfrac{1}{E_{\rm A}}-\dfrac{1}{E_{\rm A}^{(\rm L)}}}=\frac{\dfrac{1}{E_{\rm T}}-\dfrac{1}{E_{\rm T}^{(\rm L)}}}{\dfrac{\nu_{\rm A}^{(\rm L)}}{E_{\rm A}^{(\rm L)}}-\dfrac{\nu_{\rm A}}{E_{\rm A}}}=\frac{\alpha_{\rm T}-\alpha_{\rm T}^{(\rm L)}}{\alpha_{\rm A}-\alpha_{\rm A}^{(\rm L)}}=k, \tag{9.211}$$

$$\frac{\dfrac{1}{E_{\rm t}}-\dfrac{1}{E_{\rm t}^{(\rm L)}}}{\dfrac{\nu_{\rm a}^{(\rm L)}}{E_{\rm A}^{(\rm L)}}-\dfrac{\nu_{\rm a}}{E_{\rm A}}}=\frac{\dfrac{\nu_{\rm a}^{(\rm L)}}{E_{\rm A}^{(\rm L)}}-\dfrac{\nu_{\rm a}}{E_{\rm A}}}{\dfrac{1}{E_{\rm A}}-\dfrac{1}{E_{\rm A}^{(\rm L)}}}=\frac{\dfrac{\nu_{\rm t}^{(\rm L)}}{E_{\rm T}^{(\rm L)}}-\dfrac{\nu_{\rm t}}{E_{\rm T}}}{\dfrac{\nu_{\rm A}^{(\rm L)}}{E_{\rm A}^{(\rm L)}}-\dfrac{\nu_{\rm A}}{E_{\rm A}}}=\frac{\alpha_{\rm t}-\alpha_{\rm t}^{(\rm L)}}{\alpha_{\rm A}-\alpha_{\rm A}^{(\rm L)}}=k'. \tag{9.212}$$

9.7.5 An Alternative Derivation of Interrelations

The results (9.207) were derived using the approximate stress-transfer model developed in Sections 9.4–9.6. They can also be derived without having to consider a stress-transfer model. The effective stress-strain relations for a damaged laminate are given by (9.141), (9.142) and (9.179), namely,

$$\begin{aligned}
\varepsilon_{\rm A}&=\frac{1}{E_{\rm A}}\,\sigma_{\rm A}-\frac{\nu_{\rm A}}{E_{\rm A}}\,\sigma_{\rm T}-\frac{\nu_{\rm a}}{E_{\rm A}}\,\sigma_{\rm t}+\alpha_{\rm A}\,\Delta T,\\
\varepsilon_{\rm T}&=-\frac{\nu_{\rm A}}{E_{\rm A}}\,\sigma_{\rm A}+\frac{1}{E_{\rm T}}\,\sigma_{\rm T}-\frac{\nu_{\rm t}}{E_{\rm T}}\,\sigma_{\rm t}+\alpha_{\rm T}\,\Delta T,\\
\varepsilon_{\rm t}&=-\frac{\nu_{\rm a}}{E_{\rm A}}\,\sigma_{\rm A}-\frac{\nu_{\rm t}}{E_{\rm T}}\,\sigma_{\rm T}+\frac{1}{E_{\rm t}}\,\sigma_{\rm t}+\alpha_{\rm t}\,\Delta T.
\end{aligned} \tag{9.213}$$

For an undamaged laminate the stress-strain relations are given by

$$\begin{aligned}
\varepsilon_{\rm A}&=\frac{1}{E_{\rm A}^{(\rm L)}}\,\sigma_{\rm A}-\frac{\nu_{\rm A}^{(\rm L)}}{E_{\rm A}^{(\rm L)}}\,\sigma_{\rm T}-\frac{\nu_{\rm a}^{(\rm L)}}{E_{\rm A}^{(\rm L)}}\,\sigma_{\rm t}+\alpha_{\rm A}^{(\rm L)}\Delta T,\\
\varepsilon_{\rm T}&=-\frac{\nu_{\rm A}^{(\rm L)}}{E_{\rm A}^{(\rm L)}}\,\sigma_{\rm A}+\frac{1}{E_{\rm T}^{(\rm L)}}\,\sigma_{\rm T}-\frac{\nu_{\rm t}^{(\rm L)}}{E_{\rm T}^{(\rm L)}}\,\sigma_{\rm t}+\alpha_{\rm T}^{(\rm L)}\Delta T,\\
\varepsilon_{\rm t}&=-\frac{\nu_{\rm a}^{(\rm L)}}{E_{\rm A}^{(\rm L)}}\,\sigma_{\rm A}-\frac{\nu_{\rm t}^{(\rm L)}}{E_{\rm T}^{(\rm L)}}\,\sigma_{\rm T}+\frac{1}{E_{\rm t}^{(\rm L)}}\,\sigma_{\rm t}+\alpha_{\rm t}^{(\rm L)}\Delta T.
\end{aligned} \tag{9.214}$$

9.7.5.1 Uniaxial Loading in Axial Direction

Just at the point of ply crack closure, when the loading is uniaxial in the axial direction so that $\sigma_{\rm A}=\bar{\sigma}_{\rm A}^{\rm c}$, $\sigma_{\rm T}=0$ and $\sigma_{\rm t}=0$, it follows that the stress and strain states in a damaged laminate are identical to those for an undamaged laminate subject to the same axial stress $\bar{\sigma}_{\rm A}^{\rm c}$. It then follows from (9.213) that for damaged laminates the closure strains are given by the relations

$$\begin{aligned}
\bar{\varepsilon}_{\rm A}^{\rm c}&=\frac{1}{E_{\rm A}}\,\bar{\sigma}_{\rm A}^{\rm c}+\alpha_{\rm A}\,\Delta T,\\
\bar{\varepsilon}_{\rm T}^{\rm c}&=-\frac{\nu_{\rm A}}{E_{\rm A}}\,\bar{\sigma}_{\rm A}^{\rm c}+\alpha_{\rm T}\,\Delta T,\\
\bar{\varepsilon}_{\rm t}^{\rm c}&=-\frac{\nu_{\rm a}}{E_{\rm A}}\,\bar{\sigma}_{\rm A}^{\rm c}+\alpha_{\rm t}\,\Delta T,
\end{aligned} \tag{9.215}$$

and that they are also given by the stress-strain relations of an undamaged laminate (9.214) so that

$$\begin{aligned}
\bar{\varepsilon}_{\mathrm{A}}^{\mathrm{c}} &= \frac{1}{E_{\mathrm{A}}^{(\mathrm{L})}}\,\bar{\sigma}_{\mathrm{A}}^{\mathrm{c}} + \alpha_{\mathrm{A}}^{(\mathrm{L})}\Delta T,\\
\bar{\varepsilon}_{\mathrm{T}}^{\mathrm{c}} &= -\frac{\nu_{\mathrm{A}}^{(\mathrm{L})}}{E_{\mathrm{A}}^{(\mathrm{L})}}\,\bar{\sigma}_{\mathrm{A}}^{\mathrm{c}} + \alpha_{\mathrm{T}}^{(\mathrm{L})}\Delta T,\\
\bar{\varepsilon}_{\mathrm{t}}^{\mathrm{c}} &= -\frac{\nu_{\mathrm{a}}^{(\mathrm{L})}}{E_{\mathrm{A}}^{(\mathrm{L})}}\,\bar{\sigma}_{\mathrm{A}}^{\mathrm{c}} + \alpha_{\mathrm{t}}^{(\mathrm{L})}\Delta T.
\end{aligned} \tag{9.216}$$

It then follows, on eliminating the strains $\bar{\varepsilon}_{\mathrm{A}}^{\mathrm{c}}$, $\bar{\varepsilon}_{\mathrm{T}}^{\mathrm{c}}$ and $\bar{\varepsilon}_{\mathrm{t}}^{\mathrm{c}}$ and on using (9.196), that

$$\frac{\bar{\sigma}_{\mathrm{A}}^{\mathrm{c}}}{\Delta T} = -\frac{\alpha_{\mathrm{A}} - \alpha_{\mathrm{A}}^{(\mathrm{L})}}{\dfrac{1}{E_{\mathrm{A}}} - \dfrac{1}{E_{\mathrm{A}}^{(\mathrm{L})}}} = -\frac{\alpha_{\mathrm{T}} - \alpha_{\mathrm{T}}^{(\mathrm{L})}}{\dfrac{\nu_{\mathrm{A}}^{(\mathrm{L})}}{E_{\mathrm{A}}^{(\mathrm{L})}} - \dfrac{\nu_{\mathrm{A}}}{E_{\mathrm{A}}}} = -\frac{\alpha_{\mathrm{t}} - \alpha_{\mathrm{t}}^{(\mathrm{L})}}{\dfrac{\nu_{\mathrm{a}}^{(\mathrm{L})}}{E_{\mathrm{A}}^{(\mathrm{L})}} - \dfrac{\nu_{\mathrm{a}}}{E_{\mathrm{A}}}} = -\,k_1, \tag{9.217}$$

which agrees with the earlier result (9.208).

9.7.5.2 Uniaxial Loading in In-plane Transverse Direction

Similarly, just at the point of ply crack closure, when the loading is uniaxial in the in-plane transverse direction so that $\sigma_{\mathrm{A}} = 0$, $\sigma_{\mathrm{T}} = \tilde{\sigma}_{\mathrm{T}}^{\mathrm{c}}$ and $\sigma_{\mathrm{t}} = 0$, it follows that the stress and strain states in a damaged laminate are again identical to those for an undamaged laminate subject to the same uniaxial stress $\tilde{\sigma}_{\mathrm{T}}^{\mathrm{c}}$. It then follows from (9.213) that the closure strains are

$$\begin{aligned}
\tilde{\varepsilon}_{\mathrm{A}}^{\mathrm{c}} &= -\frac{\nu_{\mathrm{A}}}{E_{\mathrm{A}}}\,\tilde{\sigma}_{\mathrm{T}}^{\mathrm{c}} + \alpha_{\mathrm{A}}\,\Delta T,\\
\tilde{\varepsilon}_{\mathrm{T}}^{\mathrm{c}} &= \frac{1}{E_{\mathrm{T}}}\,\tilde{\sigma}_{\mathrm{T}}^{\mathrm{c}} + \alpha_{\mathrm{T}}\,\Delta T,\\
\tilde{\varepsilon}_{\mathrm{t}}^{\mathrm{c}} &= -\frac{\nu_{\mathrm{t}}}{E_{\mathrm{T}}}\,\tilde{\sigma}_{\mathrm{T}}^{\mathrm{c}} + \alpha_{\mathrm{t}}\,\Delta T,
\end{aligned} \tag{9.218}$$

and that they are also given by the stress-strain relations of undamaged laminate (9.214) so that

$$\begin{aligned}
\tilde{\varepsilon}_{\mathrm{A}}^{\mathrm{c}} &= -\frac{\nu_{\mathrm{A}}^{(\mathrm{L})}}{E_{\mathrm{A}}^{(\mathrm{L})}}\,\tilde{\sigma}_{\mathrm{T}}^{\mathrm{c}} + \alpha_{\mathrm{A}}^{(\mathrm{L})}\Delta T,\\
\tilde{\varepsilon}_{\mathrm{T}}^{\mathrm{c}} &= \frac{1}{E_{\mathrm{T}}^{(\mathrm{L})}}\,\tilde{\sigma}_{\mathrm{T}}^{\mathrm{c}} + \alpha_{\mathrm{T}}^{(\mathrm{L})}\Delta T,\\
\tilde{\varepsilon}_{\mathrm{t}}^{\mathrm{c}} &= -\frac{\nu_{\mathrm{t}}^{(\mathrm{L})}}{E_{\mathrm{T}}^{(\mathrm{L})}}\,\tilde{\sigma}_{\mathrm{T}}^{\mathrm{c}} + \alpha_{\mathrm{t}}^{(\mathrm{L})}\Delta T.
\end{aligned} \tag{9.219}$$

It then follows, on eliminating the closure strains and on using (9.199), that

$$\frac{\tilde{\sigma}_{\mathrm{T}}^{\mathrm{c}}}{\Delta T} = -\frac{\alpha_{\mathrm{A}} - \alpha_{\mathrm{A}}^{(\mathrm{L})}}{\dfrac{\nu_{\mathrm{A}}^{(\mathrm{L})}}{E_{\mathrm{A}}^{(\mathrm{L})}} - \dfrac{\nu_{\mathrm{A}}}{E_{\mathrm{A}}}} = -\frac{\alpha_{\mathrm{T}} - \alpha_{\mathrm{T}}^{(\mathrm{L})}}{\dfrac{1}{E_{\mathrm{T}}} - \dfrac{1}{E_{\mathrm{T}}^{(\mathrm{L})}}} = -\frac{\alpha_{\mathrm{t}} - \alpha_{\mathrm{t}}^{(\mathrm{L})}}{\dfrac{\nu_{\mathrm{t}}^{(\mathrm{L})}}{E_{\mathrm{T}}^{(\mathrm{L})}} - \dfrac{\nu_{\mathrm{t}}}{E_{\mathrm{T}}}} = -\,k_2, \tag{9.220}$$

which agrees with the earlier result (9.209).

9.7.5.3 Uniaxial Loading in Through-thickness Direction

In addition, just at the point of ply crack closure, when the loading is uniaxial in the through-thickness direction so that $\sigma_A = 0$, $\sigma_T = 0$ and $\sigma_t = \hat{\sigma}_t^c$, it follows that the stress and strain states in a damaged laminate are identical to those for an undamaged laminate subject to the same uniaxial stress $\hat{\sigma}_t^c$. It then follows from (9.213) that the closure strains are

$$\begin{aligned} \hat{\varepsilon}_A^c &= -\frac{\nu_a}{E_A}\,\hat{\sigma}_t^c + \alpha_A\,\Delta T, \\ \hat{\varepsilon}_T^c &= -\frac{\nu_t}{E_T}\,\hat{\sigma}_t^c + \alpha_T\,\Delta T, \\ \hat{\varepsilon}_t^c &= \frac{1}{E_t}\,\hat{\sigma}_t^c + \alpha_t\,\Delta T, \end{aligned} \tag{9.221}$$

and that they are also given by the stress-strain relations of undamaged laminate (9.214) so that

$$\begin{aligned} \hat{\varepsilon}_A^c &= -\frac{\nu_a^{(L)}}{E_A^{(L)}}\,\hat{\sigma}_t^c + \alpha_A^{(L)}\Delta T, \\ \hat{\varepsilon}_T^c &= -\frac{\nu_t^{(L)}}{E_T^{(L)}}\,\hat{\sigma}_t^c + \alpha_T^{(L)}\Delta T, \\ \hat{\varepsilon}_t^c &= \frac{1}{E_t^{(L)}}\,\hat{\sigma}_t^c + \alpha_t^{(L)}\Delta T. \end{aligned} \tag{9.222}$$

It then follows, on eliminating the closure strains and on using (9.202), that

$$\frac{\hat{\sigma}_t^c}{\Delta T} = -\frac{\alpha_A - \alpha_A^{(L)}}{\dfrac{\nu_a^{(L)}}{E_A^{(L)}} - \dfrac{\nu_a}{E_A}} = -\frac{\alpha_T - \alpha_T^{(L)}}{\dfrac{\nu_t^{(L)}}{E_T^{(L)}} - \dfrac{\nu_t}{E_T}} = -\frac{\alpha_t - \alpha_t^{(L)}}{\dfrac{1}{E_t} - \dfrac{1}{E_t^{(L)}}} = -k_3, \tag{9.223}$$

which agrees with the earlier result (9.210).

9.7.6 An Observation

The various interrelationships described here have been derived using two distinct methods. The first is based on an approximate model of stress transfer whereas the second considers ply crack closure, which applies more generally and would include the exact solution to the stress-transfer problem if it could be found. It is clear from (9.207) that it is required only to calculate the value of the parameter Ψ from values of the axial Young's modulus for a damaged laminate predicted by a model or from experimental data, as the dependence of all other thermoelastic laminate constants on ply crack damage is given in terms of known properties of undamaged laminates and this value of Ψ.

9.8 Example Predictions

The key results of this chapter are expressions for the various effective thermoelastic properties of a simple cross-ply laminate having an array of uniformly spaced ply cracks. For the example to be considered the following ply properties, typical of a transverse isotropic carbon-fibre-reinforced composite, are used:

$$\begin{aligned} &E_A = 140.77\,\text{GPa}, && E_T = 8.85\,\text{GPa}, && E_t = 8.85\,\text{GPa}, \\ &\nu_A = 0.28, && \nu_a = 0.28, && \nu_t = 0.43, \\ &\mu_A = 4.59\,\text{GPa}, && \mu_a = 4.59\,\text{GPa}, && \mu_t = 3.09441\,\text{GPa}, \\ &\alpha_A = 0.245 \times 10^{-6}\,\text{K}^{-1}, && \alpha_T = 45.6 \times 10^{-6}\,\text{K}^{-1}, && \alpha_t = 45.6 \times 10^{-6}\,\text{K}^{-1}. \end{aligned}$$

When these ply properties are used in conjunction with the formulae (9.143)–(9.149) and (9.180)–(9.183), for a set of ply crack densities in the range 0–4 cracks mm^{-1}, the results shown in Figure 9.2 are obtained when $a = b = 0.5$ mm The results shown assume the following identifications:

$$\begin{aligned} &E_A \equiv \text{EA}, && E_T \equiv \text{ET}, && E_t \equiv \text{Est}, \\ &\nu_A \equiv \text{nuA}, && \nu_a \equiv \text{nusa}, && \nu_t \equiv \text{nust}, \\ &\mu_A \equiv \text{muA}, && \mu_a \equiv \text{musa}, && \mu_t \equiv \text{must}, \\ &\alpha_A \equiv \text{alA}, && \alpha_T \equiv \text{alT}, && \alpha_t \equiv \text{alst}. \end{aligned}$$

It is noted that for ply crack densities exceeding 2 mm^{-1}, the effective properties no longer depend on the crack density. This is because the ply discount limit has been reached where the axial Young's and shear moduli of the 90° ply are assumed to be effectively zero. In addition, it is seen that the effective in-plane transverse modulus

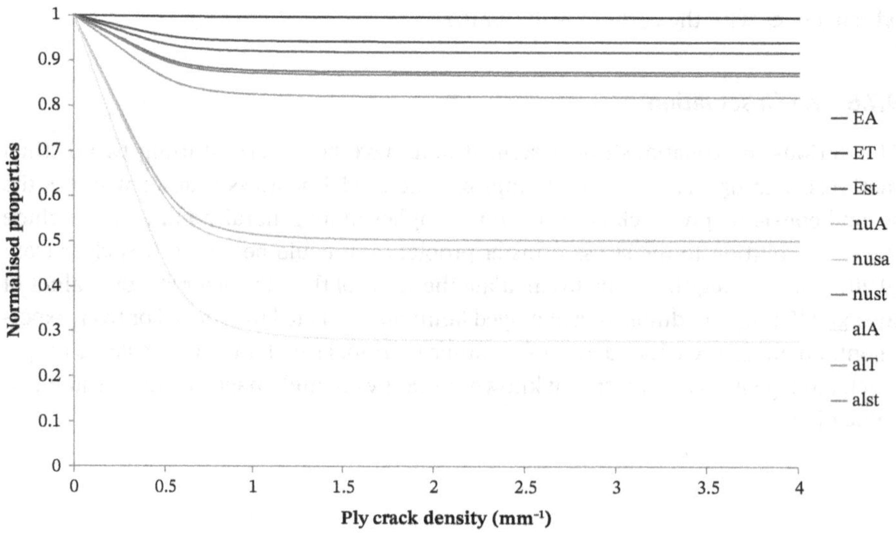

Figure 9.2 Predictions of the normalised effective properties of a simple cross-ply laminate as a function of the density of a uniform distribution of ply cracks in the 90° ply.

E_T is hardly affected by ply cracking, and that the effective axial thermal expansion coefficient is affected a great deal by ply cracking.

An important characteristic of the detailed analysis provided in this chapter is that results depend upon the uniformity of the through-thickness distributions of both stress and displacement distributions. One way of developing more accurate solutions is to extend the model so that it is applicable to any number of both 0° and 90° plies. Each ply considered in the model described in this chapter can then be replaced by a set of identical plies having a common orientation, either 0° or 90°. The thicknesses of the replacement plies for each orientation can vary but their total must correspond to the thicknesses of the original 0° and 90° plies. This procedure is known as ply refinement and has been used, for example, in the author contributions to the international Worldwide Failure Exercise WWFE III [5] concerned with the assessment of damage models for composite laminates. Ply refinement is considered in Chapter 10 where cross-ply laminates are regarded as a special case of the model that is designed for general symmetrical laminates.

Additional comment:
In this chapter it should be noted that, because there are only two plies of orientation 0° and 90° which could be made of different materials, the properties used for the 90° ply have taken account of the required 90° rotation. This means that, when both the 0° and 90° plies are made of the same material, there is no need to refer the ply property to a particular ply orientation. In other words, for this case the superscripts '0' and '90' can be removed, their presence being required only if the 0° and 90° plies are made of a different material.

References

1. Nuismer, R.J. and Tan, S.C. (1988). Constitutive relations of a cracked composite lamina. Journal of Composite Materials 22 (4): 306–321.
2. McCartney, L.N., 'Analytical models of stress transfer in unidirectional composites and cross-ply laminates, and their application to the prediction of matrix/transverse cracking', Proc. IUTAM Symposium on 'Local mechanics concepts for composite materials systems', Blacksburg, Va., (1991), 251–282.
3. Nairn, J.A. and Mendels, D.A. (2001). On the use of planar shear-lag methods for stress-transfer analysis of multi-layered composites. Mechanics of Materials 33: 335–362.
4. McCartney, L.N. (1992). Theory of stress transfer in a 0-90-0 cross-ply laminate containing a parallel array of transverse cracks. Journal of the Mechanics and Physics of Solids 40: 27–68.
5. Kaddour, A.S., Hinton, M.J., Smith, P.A., and Li, S. (2013). The background to the third world-wide failure exercise. Journal of Composite Materials 47 (20-21): 2417–2426.

Further Reading

Kaddour, A.S., Hinton, M.J., Smith, P.A., and Li, S. (2013). A comparison between the predictive capability of matrix cracking, damage and failure criteria for fibre reinforced

laminates: Part A of the third world-wide failure exercise. Journal of Composite Materials 47 (20–21): 2749–2779.
McCartney, L.N. (2013). Derivations of energy-based modelling for ply cracking in general symmetric laminates. Journal of Composite Materials 47 (20–21): 2641–2673.
McCartney, L.N. (2013). Energy methods for modelling damage in laminates. Journal of Composite Materials 47 (20–21): 2613–2640.

10

Theoretical Basis for a Model of Ply Cracking in General Symmetric Laminates

Overview: This chapter is concerned with a general theoretical framework that enables a thorough understanding of how energy-based methods can be used to predict first ply cracking and the progressive formation of uniform distributions of ply cracks when a general symmetric laminate, subject to general in-plane loading and through-thickness loading in the presence of thermal residual stresses. Numerous interrelationships between the thermoelastic constants of damage laminates are derived by considering the closure of ply cracks for special loading situations. These relationships enable rather complex expressions for the energy per unit volume of a damaged laminate to be simplified to a very useful form involving an effective applied stress that captures the effect on damage formation of three principal loading parameters, the effective axial, transverse and through-thickness stresses. An approximate, but accurate methodology is then described that enables the prediction of progressive ply crack formation leading to nonuniform ply crack distributions which are observed in practice.

10.1 Introduction

The cross-ply laminates considered in Chapter 9, having reinforcement in the 0° and 90° directions, are of limited practical utility as reinforcement is often required in additional directions. To meet this need, general symmetric laminates are now considered where reinforcement in any number of directions is possible provided that the laminate is symmetric so that unwanted bending modes of deformation can be avoided. This extends the modelling requirement to the consideration of general symmetric laminates having multiple plies, or to symmetric cross-ply laminates having multiple 0° and 90° plies. The objective now is to develop a methodology for understanding the effects of ply crack formation on the effective properties of a general symmetric laminate where through-thickness loading and in-plane general loading can be considered in the presence of thermal residual stresses.

When solving ply cracking problems for general symmetric laminates the first step is to rotate the laminate and loading so that the plies in which ply cracks will be allowed to form have a 90° orientation. The subsequent analysis assumes that such a transformation (if needed) has already been made. To achieve the challenging objectives while avoiding numerical methods such as finite element and boundary element

Properties for Design of Composite Structures: Theory and Implementation Using Software,
First Edition. Neil McCartney.

Companion Website: www.wiley.com/go/mccartney/properties

methods, some simplifications must be made to enable a tractable analytical development which is nevertheless complex. The first approximation is to assume that ply cracks form in only a single orientation that must correspond to that of the fibres in a 90° ply. If the laminate and loading have to be rotated so that the orientation of the ply in which ply cracks are required to form becomes a 90° ply following the rotation, general in-plane loading must be considered. The second approximation is to assume that ply cracks can form in a discrete number of planes parallel to the fibres in a 90° ply. The ply crack distribution in each cracked ply must be the same; however, not all 90° plies need to be cracked. These assumptions are made so that a relatively simple representative volume can be used. It should be noted that the approach to be developed cannot consider arrays of staggered cracks, as would arise in the two 90° plies of a $[90/0]_s$ laminate where localised bending deformation could arise: a case that is not treated by the model.

10.2 Geometry and Basic Field Equations

The problem under consideration concerns the triaxial deformation of a symmetric, possibly unbalanced, multilayered laminate having width $2W$ and total thickness $2h$ constructed of $2N + 2$ perfectly bonded layers which can have any combination of orientations provided that there is at least one 90° ply in the laminate and that symmetry about the midplane of the laminate is preserved. Some of the layers can have the same properties (as part of a ply refinement scheme) so that through-thickness variations in the stress and displacement fields can be modelled adequately. It is assumed that the ply crack distribution in damaged 90° plies is uniform having a separation $2L$, and that the cracks in each damaged 90° ply of the laminate lie in the same plane.

A set of rectangular Cartesian coordinates (x_1, x_2, x_3) is introduced having origin located at the intersection of the midplane of the laminate and the plane midway between the planes containing two neighbouring ply cracks. The RVE of the laminate to be considered is specified by $0 \le x_1 \le L$, $0 \le x_2 \le W$, $0 \le x_3 \le h$. Ply cracks in some or all of the 90° plies are assumed to form on the plane $x_1 = L$. The plane $x_1 = 0$ is a symmetry plane implying that neighbouring ply cracks would be on the plane $x_1 = -L$. The planes $x_2 = 0$ and $x_3 = 0$ are also planes of symmetry.

10.3 Boundary Conditions for Uniformly Cracked Laminates

Boundary conditions need to be imposed on the external edges of the uniformly cracked laminate. It is assumed that in-plane biaxial and shear loads are applied to the cracked laminate by imposing displacements on the edges of the laminate that would, in an undamaged laminate, lead to a state of combined uniform biaxial strain and uniform shear strain. As the laminate is assumed to have length $2L$ in the x_1-direction and width $2W$ in the x_2-direction, the edge conditions for the *uncracked* layers of a damaged laminate are given by

$$\begin{aligned} &\sigma_{13} = 0, \quad u_1 = \varepsilon_A L + \tfrac{1}{2}\gamma_A x_2, \quad u_3 = \tfrac{1}{2}\gamma_A L + \varepsilon_T x_2, \qquad \text{on } x_1 = L, \\ &\sigma_{23} = 0, \quad u_2 = \varepsilon_A x_1 + \tfrac{1}{2}\gamma_A W, \quad u_3 = \tfrac{1}{2}\gamma_A x_1 + \varepsilon_T W, \qquad \text{on } x_2 = W, \end{aligned} \tag{10.1}$$

where ε_A, ε_T and γ_A define the imposed in-plane axial, transverse and *engineering* shear strains for a uniformly *cracked* laminate. It should be noted that these edge conditions do not prevent sliding of the ply crack surfaces, but they do apply some constraint to the uncracked layers. Any cracked layer elements on $x_1 = L$ will be stress-free so that

$$\sigma_{11}^{(i)}(L,x_2,x_3)=0,\quad \sigma_{12}^{(i)}(L,x_2,x_3)=0,\quad \sigma_{13}^{(i)}(L,x_2,x_3)=0,\ \text{for all}\ x_2,x_3. \tag{10.2}$$

The external face on $x_3 = x_3^{(N+1)} = h$ of the damaged laminate is assumed to be subject to a uniform normal traction and zero shear tractions so that

$$\sigma_{13}^{(N+1)}(x_1,x_2,h)=0,\quad \sigma_{23}^{(N+1)}(x_1,x_2,h)=0,\quad \sigma_{33}^{(N+1)}(x_1,x_2,h)=\sigma_t,\ \text{for all}\ x_1,x_2. \tag{10.3}$$

By symmetry about the plane $x_3 = 0$, the same conditions apply on the external face $x_3 = -h$ of the laminate.

Because applied displacements of the form (10.1), involving applied strains ε_A, ε_T and γ_A, have been imposed on the edges of the cracked laminate, the corresponding stresses σ_{11}, σ_{22} and σ_{12}, resulting from an exact elasticity analysis, will be nonuniform. Corresponding effective in-plane axial, transverse and shear stresses are, therefore, defined by

$$\begin{aligned}
\sigma_A &= \frac{1}{4hW}\int_{-W}^{W}\int_{-h}^{h}\sigma_{11}(L,x_2,x_3)\,\mathrm{d}x_2\mathrm{d}x_3, \quad \sigma_T = \frac{1}{4hL}\int_{-L}^{L}\int_{-h}^{h}\sigma_{22}(x_1,W,x_3)\,\mathrm{d}x_1\mathrm{d}x_3,\\
\tau_A &= \frac{1}{4hW}\int_{-W}^{W}\int_{-h}^{h}\sigma_{12}(L,x_2,x_3)\,\mathrm{d}x_2\mathrm{d}x_3 = \frac{1}{4hL}\int_{-L}^{L}\int_{-h}^{h}\sigma_{12}(x_1,W,x_3)\,\mathrm{d}x_1\mathrm{d}x_3,
\end{aligned} \tag{10.4}$$

where $2h$ is the total thickness of the laminate. The effective out-of-plane strain ε_t is defined for a damaged laminate by

$$\varepsilon_t = \frac{1}{4hLW}\int_{-L}^{L}\int_{-W}^{W} u_3(x_1,x_2,h)\,\mathrm{d}x_1\mathrm{d}x_2, \tag{10.5}$$

where use has been made of symmetry about the plane $x_3 = 0$.

The loading parameters σ_t, σ_A, τ_A and ε_T are to be applied to both cracked and undamaged laminates, whereas the values of the corresponding parameters ε_t, ε_A, γ_A and σ_T, describe the response of the damaged laminate to loading and will have differing values when the laminate is undamaged.

10.4 Generalised Plane Strain Conditions

The fundamental assumption for generalised plane strain conditions is that the displacement field is of the following form

$$u_1 = f_1(x_1,x_3)+\varepsilon_A^0 x_1+\tfrac{1}{2}\gamma_A^0 x_2,\quad u_2 = f_2(x_1,x_3)+\tfrac{1}{2}\gamma_A^0 x_1+\varepsilon_T x_2,\quad u_3 = f_3(x_1,x_3), \tag{10.6}$$

where ε_A^0 and γ_A^0 are the effective axial and shear strains in the undamaged laminate when the stresses σ_t, σ_A and τ_A are applied, when the in-plane transverse strain has the same value ε_T in all layers of the laminate, and when the temperature difference is ΔT. The parameters σ_t and ε_T are regarded as input values for both cracked and undamaged laminates. For an undamaged laminate, the corresponding through-thickness strain will be denoted by ε_t^0 and the corresponding effective transverse stress will be denoted by σ_T^0. It should be noted that the functions f_k, $k = 1, 2, 3$, appearing in (10.6) are all independent of x_2.

The representation (10.6) is valid only if ply cracks form in planes normal to the x_1-direction, and only if well away from the edges of the laminate where the local stress and deformation distributions depend on how the loading is applied. The regions of laminate in which the edge effects are found depend on laminate geometry, properties and loading. This issue can be examined using models of the type to be described here, and experience suggests that these effects are found in narrow strips at the edges of the laminate having thicknesses that are about the half-thickness h of the laminate. The representation (10.6) for the displacement field differs from an earlier approach [1] by a rigid rotation that does not affect the stress or strain fields.

It follows from (7.66) and (7.67) that the stress-strain relations for a general symmetric laminate have the form (the superscript (L) denoting a laminate property is now suppressed)

$$\varepsilon_A = \frac{\sigma_A}{E_A} - \frac{\nu_A}{E_A}\sigma_T - \frac{\nu_a}{E_A}\sigma_t - \frac{\lambda_A}{\mu_A}\tau_A + \alpha_A \Delta T, \tag{10.7}$$

$$\varepsilon_T = -\frac{\nu_A}{E_A}\sigma_A + \frac{\sigma_T}{E_T} - \frac{\nu_t}{E_T}\sigma_t - \frac{\lambda_T}{\mu_A}\tau_A + \alpha_T \Delta T, \tag{10.8}$$

$$\varepsilon_t = -\frac{\nu_a}{E_A}\sigma_A - \frac{\nu_t}{E_T}\sigma_T + \frac{\sigma_t}{E_t} - \frac{\lambda_t}{\mu_A}\tau_A + \alpha_t \Delta T, \tag{10.9}$$

$$\gamma_A = -\frac{\lambda_A}{\mu_A}\sigma_A - \frac{\lambda_T}{\mu_A}\sigma_T - \frac{\lambda_t}{\mu_A}\sigma_t + \frac{\tau_A}{\mu_A} + \alpha_S \Delta T, \tag{10.10}$$

where, for specified values of $\sigma_A, \sigma_T, \sigma_t$ and τ_A and ΔT, the quantities $\varepsilon_A, \varepsilon_T, \varepsilon_t$ and γ_A are the effective through-thickness, axial, in-plane transverse and shear strains of the damaged laminate, respectively, and where $E_A, E_T, E_t, \nu_A, \nu_a, \nu_t, \lambda_A, \lambda_T, \lambda_t, \alpha_A, \alpha_T, \alpha_t$, and α_S denote the corresponding effective thermoelastic constants of the damaged laminate. The parameters λ_A, λ_T, and λ_t are ratios indicating the degree of shear coupling, and it should be noted that their definition differs from that used in the original publication by the present author [1]. The parameter α_S is an expansion coefficient governing the amount of shear strain that can arise when the temperature is changed.

10.5 Reduced Stress-Strain Relations for a Cracked Laminate

It follows from (10.10) that the in-plane shear stress is

$$\tau_A = \mu_A \gamma_A + \lambda_A \sigma_A + \lambda_T \sigma_T + \lambda_t \sigma_t - \mu_A \alpha_S \Delta T. \tag{10.11}$$

Substitution into (10.7)–(10.9) then leads to the following reduced stress-strain equations that are of the same form as those for cross-ply laminates subject to triaxial loading modes without shear

$$\hat{\varepsilon}_A \equiv \varepsilon_A + \lambda_A \gamma_A = \frac{\sigma_A}{\tilde{E}_A} - \frac{\tilde{\nu}_A}{\tilde{E}_A}\sigma_T - \frac{\tilde{\nu}_a}{\tilde{E}_A}\sigma_t + \tilde{\alpha}_A \Delta T, \tag{10.12}$$

$$\hat{\varepsilon}_T \equiv \varepsilon_T + \lambda_T\gamma_A = -\frac{\tilde{\nu}_A}{\tilde{E}_A}\sigma_A + \frac{\sigma_T}{\tilde{E}_T} - \frac{\tilde{\nu}_t}{\tilde{E}_T}\sigma_t + \tilde{\alpha}_T\Delta T, \tag{10.13}$$

$$\hat{\varepsilon}_t \equiv \varepsilon_t + \lambda_t\gamma_A = -\frac{\tilde{\nu}_a}{\tilde{E}_A}\sigma_A - \frac{\tilde{\nu}_t}{\tilde{E}_T}\sigma_T + \frac{\sigma_t}{\tilde{E}_t} + \tilde{\alpha}_t\Delta T, \tag{10.14}$$

where $\hat{\varepsilon}_A$, $\hat{\varepsilon}_T$ and $\hat{\varepsilon}_t$ can be interpreted as the axial strain, the in-plane transverse strain and the through-thickness strain for a damaged laminate, respectively, that is constrained so that the shear strain is zero, and where the reduced thermoelastic constants are defined by

$$\frac{1}{\tilde{E}_A} = \frac{1}{E_A} - \frac{\lambda_A^2}{\mu_A}, \tag{10.15}$$

$$\frac{1}{\tilde{E}_T} = \frac{1}{E_T} - \frac{\lambda_T^2}{\mu_A}, \tag{10.16}$$

$$\frac{\tilde{\nu}_t}{\tilde{E}_T} = \frac{\nu_t}{E_T} + \frac{\lambda_T\lambda_t}{\mu_A}, \tag{10.17}$$

$$\frac{\tilde{\nu}_a}{\tilde{E}_A} = \frac{\nu_a}{E_A} + \frac{\lambda_A\lambda_t}{\mu_A}, \tag{10.18}$$

$$\frac{\tilde{\nu}_A}{\tilde{E}_A} = \frac{\nu_A}{E_A} + \frac{\lambda_A\lambda_T}{\mu_A}, \tag{10.19}$$

$$\tilde{\alpha}_t = \alpha_t + \lambda_t\alpha_S, \tag{10.20}$$

$$\tilde{\alpha}_A = \alpha_A + \lambda_A\alpha_S, \tag{10.21}$$

$$\tilde{\alpha}_T = \alpha_T + \lambda_T\alpha_S. \tag{10.22}$$

10.6 Interrelationships for Thermoelastic Constants

Consider now the stress field in the 90° plies of an uncracked general symmetric laminate for which each ply has the same in-plane strains ε_{11}, ε_{22} and ε_{12}, but differing out-of-plane strains ε_{13}, ε_{23} and ε_{33}. The objective is to determine the values of the effective applied stresses for which $\sigma_{11} = \sigma_A^{(i)} = 0$ and $\sigma_{12} = \tau_A^{(i)} = 0$ where $\sigma_A^{(i)}$ and $\tau_A^{(i)}$ are the axial and shear stresses in any one of the 90° plies that is labelled the ith ply. This applied stress state corresponds also to the closure condition for cracks that are present in any of the 90° plies of a cracked laminate. At laminate level, the values for the through-thickness, axial, transverse and shear stresses at the point of closure are denoted by $\sigma_t^c, \sigma_A^c, \sigma_T^c$ and τ_A^c, respectively. For 90° plies only, relations (6.100) imply that $\gamma_A = \tau_A^{(i)} / \mu_A^{(i)}$. The shear stress $\tau_A^{(i)}$ in the 90° plies is clearly zero only if the in-plane shear strain $\gamma_A = 0$. Thus, the point of closure of any cracks in the 90° plies is determined by considering the values of the applied stresses such that the stress component $\sigma_{11} = \sigma_A^{(i)} = 0$ in each of the 90° plies of an uncracked laminate for which $\gamma_A = 0$. It follows from (6.100), on setting $\sigma_A^{(i)} = \tau_A^{(i)} = 0$, that the axial strain ε_A^c and the transverse strain ε_T^c at the point of crack closure must satisfy the relation

$$\varepsilon_{\mathrm{A}}^{\mathrm{c}} = -A_0\sigma_{\mathrm{t}}^{\mathrm{c}} - B_0\varepsilon_{\mathrm{T}}^{\mathrm{c}} + C_0\Delta T, \tag{10.23}$$

where (see (D.13) in Appendix D)

$$A_0 = \frac{\nu_{\mathrm{t}}^{(90)}}{E_{\mathrm{T}}^{(90)}} + \frac{\nu_{\mathrm{a}}^{(90)}\nu_{\mathrm{A}}^{(90)}}{E_{\mathrm{A}}^{(90)}}, \quad B_0 = \nu_{\mathrm{A}}^{(90)}, \quad C_0 = \alpha_{\mathrm{T}}^{(90)} + \nu_{\mathrm{A}}^{(90)}\alpha_{\mathrm{A}}^{(90)}, \tag{10.24}$$

where $E_{\mathrm{A}}^{(90)}, E_{\mathrm{T}}^{(90)}, \nu_{\mathrm{a}}^{(90)}, \nu_{\mathrm{t}}^{(90)}, \nu_{\mathrm{A}}^{(90)}, \alpha_{\mathrm{A}}^{(90)}, \alpha_{\mathrm{T}}^{(90)}$ are the moduli, Poisson's ratios and thermal expansion coefficients for the 90° plies, and where the axial direction corresponds to the fibre direction of the 0° plies. Thus, there is a multitude of stress states for which crack closure occurs. Subsequent theoretical developments in this chapter will allow any loading choice satisfying this crack closure relationship. The superscript '(90)' is attached to the properties used in (10.24) just in case the 90° ply is made of a material that differs from one or more of the other plies in the laminate.

10.6.1 Ply Crack Closure for Constrained Uniaxial Loading in Axial Direction

Consider ply crack closure for uniaxial loading in the axial direction such that $\sigma_{\mathrm{t}} = \sigma_{\mathrm{T}} = 0$ where the laminate is constrained so that γ_{A}. It follows from (10.12)–(10.14), on setting $\hat{\varepsilon}_{\mathrm{A}} = \bar{\varepsilon}_{\mathrm{A}}^{\mathrm{c}}$, $\hat{\varepsilon}_{\mathrm{T}} = \bar{\varepsilon}_{\mathrm{T}}^{\mathrm{c}}$, $\hat{\varepsilon}_{\mathrm{t}} = \bar{\varepsilon}_{\mathrm{t}}^{\mathrm{c}}$ and $\sigma_{\mathrm{A}} = \bar{\sigma}_{\mathrm{A}}^{\mathrm{c}}$, that

$$\bar{\varepsilon}_{\mathrm{A}}^{\mathrm{c}} = \frac{\bar{\sigma}_{\mathrm{A}}^{\mathrm{c}}}{\tilde{E}_{\mathrm{A}}} + \tilde{\alpha}_{\mathrm{A}}\Delta T, \tag{10.25}$$

$$\bar{\varepsilon}_{\mathrm{T}}^{\mathrm{c}} = -\frac{\tilde{\nu}_{\mathrm{A}}}{\tilde{E}_{\mathrm{A}}}\bar{\sigma}_{\mathrm{A}}^{\mathrm{c}} + \tilde{\alpha}_{\mathrm{T}}\Delta T, \tag{10.26}$$

$$\bar{\varepsilon}_{\mathrm{t}}^{\mathrm{c}} = -\frac{\tilde{\nu}_{\mathrm{a}}}{\tilde{E}_{\mathrm{A}}}\bar{\sigma}_{\mathrm{A}}^{\mathrm{c}} + \tilde{\alpha}_{\mathrm{t}}\Delta T. \tag{10.27}$$

Similarly, for undamaged laminates it follows that

$$\bar{\varepsilon}_{\mathrm{A}}^{\mathrm{c}} = \frac{\bar{\sigma}_{\mathrm{A}}^{\mathrm{c}}}{\tilde{E}_{\mathrm{A}}^{0}} + \tilde{\alpha}_{\mathrm{A}}^{0}\Delta T, \tag{10.28}$$

$$\bar{\varepsilon}_{\mathrm{T}}^{\mathrm{c}} = -\frac{\tilde{\nu}_{\mathrm{A}}^{0}}{\tilde{E}_{\mathrm{A}}^{0}}\bar{\sigma}_{\mathrm{A}}^{\mathrm{c}} + \tilde{\alpha}_{\mathrm{T}}^{0}\Delta T, \tag{10.29}$$

$$\bar{\varepsilon}_{\mathrm{t}}^{\mathrm{c}} = -\frac{\tilde{\nu}_{\mathrm{a}}^{0}}{\tilde{E}_{\mathrm{A}}^{0}}\bar{\sigma}_{\mathrm{A}}^{\mathrm{c}} + \tilde{\alpha}_{\mathrm{t}}^{0}\Delta T, \tag{10.30}$$

where the superscript '0' is used here to denote the properties of the laminate when it is undamaged. It should be noted that relations (10.25)–(10.27) and (10.28)–(10.30) have made use of the fact that at the point where ply cracks *just* close the strains in a cracked laminate must be the same as those in an undamaged laminate. This is the case because at the point of closure, the stress normal to the ply crack within the cracked ply is zero so that it does not matter whether this ply is cracked or not. It then follows from (10.25)–(10.30), on eliminating $\bar{\varepsilon}_{\mathrm{A}}^{\mathrm{c}}$, $\bar{\varepsilon}_{\mathrm{T}}^{\mathrm{c}}$ and $\bar{\varepsilon}_{\mathrm{t}}^{\mathrm{c}}$, that

$$\bar{\sigma}_{\mathrm{A}}^{\mathrm{c}} = -\frac{\tilde{\alpha}_{\mathrm{A}} - \tilde{\alpha}_{\mathrm{A}}^{0}}{\dfrac{1}{\tilde{E}_{\mathrm{A}}} - \dfrac{1}{\tilde{E}_{\mathrm{A}}^{0}}}\Delta T = -\frac{\tilde{\alpha}_{\mathrm{T}} - \tilde{\alpha}_{\mathrm{T}}^{0}}{\dfrac{\tilde{\nu}_{\mathrm{A}}^{0}}{\tilde{E}_{\mathrm{A}}^{0}} - \dfrac{\tilde{\nu}_{\mathrm{A}}}{\tilde{E}_{\mathrm{A}}}}\Delta T = -\frac{\tilde{\alpha}_{\mathrm{t}} - \tilde{\alpha}_{\mathrm{t}}^{0}}{\dfrac{\tilde{\nu}_{\mathrm{a}}^{0}}{\tilde{E}_{\mathrm{A}}^{0}} - \dfrac{\tilde{\nu}_{\mathrm{a}}}{\tilde{E}_{\mathrm{A}}}}\Delta T. \tag{10.31}$$

It follows fom (10.23) that for the case of constrained uniaxial loading of the laminate under discussion

$$\bar{\varepsilon}_{\rm A}^{\rm c} = -B_0\bar{\varepsilon}_{\rm T}^{\rm c} + C_0\,\Delta T. \tag{10.32}$$

Thus, on substituting (10.28) and (10.29) into (10.32) and solving for $\bar{\sigma}_{\rm A}^{\rm c}$ the following alternative expression for the crack closure stress is obtained

$$\bar{\sigma}_{\rm A}^{\rm c} = -\frac{\tilde{E}_{\rm A}^0\left[\tilde{\alpha}_{\rm A}^0 + \tilde{\alpha}_{\rm T}^0 B_0 - C_0\right]}{1-\tilde{\nu}_{\rm A}^0 B_0}\Delta T. \tag{10.33}$$

It follows from (10.31) and (10.33) that the following interrelationships between the effective thermoelastic constants of damaged laminates are valid

$$\frac{\tilde{\alpha}_{\rm A}-\tilde{\alpha}_{\rm A}^0}{\dfrac{1}{\tilde{E}_{\rm A}}-\dfrac{1}{\tilde{E}_{\rm A}^0}} = \frac{\tilde{\alpha}_{\rm T}-\tilde{\alpha}_{\rm T}^0}{\dfrac{\tilde{\nu}_{\rm A}^0}{\tilde{E}_{\rm A}^0}-\dfrac{\tilde{\nu}_{\rm A}}{\tilde{E}_{\rm A}}} = \frac{\tilde{\alpha}_{\rm t}-\tilde{\alpha}_{\rm t}^0}{\dfrac{\tilde{\nu}_{\rm a}^0}{\tilde{E}_{\rm A}^0}-\dfrac{\tilde{\nu}_{\rm a}}{\tilde{E}_{\rm A}}} = k_1, \tag{10.34}$$

where

$$k_1 = \frac{\tilde{E}_{\rm A}^0\left(\tilde{\alpha}_{\rm A}^0 + \tilde{\alpha}_{\rm T}^0 B_0 - C_0\right)}{1-\tilde{\nu}_{\rm A}^0 B_0}. \tag{10.35}$$

From (10.33) and (10.35), the parameter k_1 is a constant for an *undamaged* laminate such that

$$\bar{\sigma}_{\rm A}^{\rm c} = -k_1\Delta T. \tag{10.36}$$

10.6.2 Ply Crack Closure for Constrained Uniaxial Loading in In-plane Transverse Direction

Consider now ply crack closure for the case of constrained uniaxial transverse loading such that $\sigma_{\rm t} = \sigma_{\rm A} = 0$ and $\gamma_{\rm A} = 0$. By following the approach of the previous section, it can be shown using (10.12)–(10.14) that, on setting $\hat{\varepsilon}_{\rm A} = \hat{\varepsilon}_{\rm A}^{\rm c}$, $\hat{\varepsilon}_{\rm T} = \hat{\varepsilon}_{\rm T}^{\rm c}$, $\hat{\varepsilon}_{\rm t} = \hat{\varepsilon}_{\rm t}^{\rm c}$ and $\sigma_{\rm T} = \hat{\sigma}_{\rm T}^{\rm c}$, that

$$\hat{\varepsilon}_{\rm A}^{\rm c} = -\frac{\tilde{\nu}_{\rm A}}{\tilde{E}_{\rm A}}\hat{\sigma}_{\rm T}^{\rm c} + \tilde{\alpha}_{\rm A}\Delta T, \tag{10.37}$$

$$\hat{\varepsilon}_{\rm T}^{\rm c} = \frac{\hat{\sigma}_{\rm T}^{\rm c}}{\tilde{E}_{\rm T}} + \tilde{\alpha}_{\rm T}\Delta T, \tag{10.38}$$

$$\hat{\varepsilon}_{\rm t}^{\rm c} = -\frac{\tilde{\nu}_{\rm t}}{\tilde{E}_{\rm T}}\hat{\sigma}_{\rm T}^{\rm c} + \tilde{\alpha}_{\rm t}\Delta T. \tag{10.39}$$

Similarly, for undamaged laminates it follows that

$$\hat{\varepsilon}_{\rm A}^{\rm c} = -\frac{\tilde{\nu}_{\rm A}^0}{\tilde{E}_{\rm A}^0}\hat{\sigma}_{\rm T}^{\rm c} + \tilde{\alpha}_{\rm A}^0\Delta T, \tag{10.40}$$

$$\hat{\varepsilon}_{\rm T}^{\rm c} = \frac{\hat{\sigma}_{\rm T}^{\rm c}}{\tilde{E}_{\rm T}^0} + \tilde{\alpha}_{\rm T}^0\Delta T, \tag{10.41}$$

$$\hat{\varepsilon}_{\rm t}^{\rm c} = -\frac{\tilde{\nu}_{\rm t}^0}{\tilde{E}_{\rm T}^0}\hat{\sigma}_{\rm T}^{\rm c} + \tilde{\alpha}_{\rm t}^0\Delta T. \tag{10.42}$$

It should be noted that relations (10.37)–(10.39) and (10.40)–(10.42) have again made use of the fact that at the point where ply cracks *just* close the strains in a cracked laminate must be the same as those in an undamaged laminate. As stated previously, this is the case because at the point of closure, the stress normal to the ply crack within the cracked ply is zero so that it does not matter whether this ply is cracked or not. It follows from (10.39)–(10.42), on eliminating $\hat{\varepsilon}_{\mathrm{A}}^{\mathrm{c}}$, $\hat{\varepsilon}_{\mathrm{T}}^{\mathrm{c}}$ and $\hat{\varepsilon}_{\mathrm{t}}^{\mathrm{c}}$, that

$$\hat{\sigma}_{\mathrm{T}}^{\mathrm{c}} = -\frac{\tilde{\alpha}_{\mathrm{A}} - \tilde{\alpha}_{\mathrm{A}}^{0}}{\dfrac{\tilde{\nu}_{\mathrm{A}}^{0}}{\tilde{E}_{\mathrm{A}}^{0}} - \dfrac{\tilde{\nu}_{\mathrm{A}}}{\tilde{E}_{\mathrm{A}}}}\Delta T = -\frac{\tilde{\alpha}_{\mathrm{T}} - \tilde{\alpha}_{\mathrm{T}}^{0}}{\dfrac{1}{\tilde{E}_{\mathrm{T}}} - \dfrac{1}{\tilde{E}_{\mathrm{T}}^{0}}}\Delta T = -\frac{\tilde{\alpha}_{\mathrm{t}} - \tilde{\alpha}_{\mathrm{t}}^{0}}{\dfrac{\tilde{\nu}_{\mathrm{t}}^{0}}{\tilde{E}_{\mathrm{T}}^{0}} - \dfrac{\tilde{\nu}_{\mathrm{t}}}{\tilde{E}_{\mathrm{T}}}}\Delta T. \tag{10.43}$$

It follows from (10.23) that for the case of transverse uniaxial loading under discussion

$$\hat{\varepsilon}_{\mathrm{A}}^{\mathrm{c}} = -B\varepsilon_{\mathrm{T}}^{\mathrm{c}} + C_0\Delta T. \tag{10.44}$$

Thus, on using the stress-strain relations for an undamaged laminate the following alternative expression for the crack closure stress is obtained

$$\hat{\sigma}_{\mathrm{T}}^{\mathrm{c}} = -\frac{\tilde{E}_{\mathrm{T}}^{0}\left(\tilde{\alpha}_{\mathrm{A}}^{0} + \tilde{\alpha}_{\mathrm{T}}^{0}B_0 - C_0\right)}{B_0 - \tilde{\nu}_{\mathrm{A}}^{0}\tilde{E}_{\mathrm{T}}^{0}/\tilde{E}_{\mathrm{A}}^{0}}\Delta T. \tag{10.45}$$

It then follows from (10.43) and (10.45) that

$$\frac{\tilde{\alpha}_{\mathrm{A}} - \tilde{\alpha}_{\mathrm{A}}^{0}}{\dfrac{\tilde{\nu}_{\mathrm{A}}^{0}}{\tilde{E}_{\mathrm{A}}^{0}} - \dfrac{\tilde{\nu}_{\mathrm{A}}}{\tilde{E}_{\mathrm{A}}}} = \frac{\tilde{\alpha}_{\mathrm{T}} - \tilde{\alpha}_{\mathrm{T}}^{0}}{\dfrac{1}{\tilde{E}_{\mathrm{T}}} - \dfrac{1}{\tilde{E}_{\mathrm{T}}^{0}}} = \frac{\tilde{\alpha}_{\mathrm{t}} - \tilde{\alpha}_{\mathrm{t}}^{0}}{\dfrac{\tilde{\nu}_{\mathrm{t}}^{0}}{\tilde{E}_{\mathrm{T}}^{0}} - \dfrac{\tilde{\nu}_{\mathrm{t}}}{\tilde{E}_{\mathrm{T}}}} = k_2, \tag{10.46}$$

where

$$k_2 = \frac{\tilde{E}_{\mathrm{T}}^{0}\left(\tilde{\alpha}_{\mathrm{A}}^{0} + \tilde{\alpha}_{\mathrm{T}}^{0}B_0 - C_0\right)}{B_0 - \tilde{\nu}_{\mathrm{A}}^{0}\tilde{E}_{\mathrm{T}}^{0}/\tilde{E}_{\mathrm{A}}^{0}}. \tag{10.47}$$

The constant k_2 is another constant for an *undamaged* laminate and is such that $\hat{\sigma}_{\mathrm{T}}^{\mathrm{c}} = -k_2\Delta T$.

It follows from (10.35) and (10.47) that

$$k = \frac{k_1}{k_2} = \frac{\tilde{E}_{\mathrm{A}}^{0}}{\tilde{E}_{\mathrm{T}}^{0}}\frac{B_0 - \tilde{\nu}_{\mathrm{A}}^{0}\tilde{E}_{\mathrm{T}}^{0}/\tilde{E}_{\mathrm{A}}^{0}}{1 - \tilde{\nu}_{\mathrm{A}}^{0}B_0}, \tag{10.48}$$

where the constant k for an *undamaged* laminate is independent of the thermal expansion coefficients. It then follows from (10.34), (10.46) and (10.48) that the following interrelationships between the effective thermoelastic constants of damaged laminates must also be valid

$$\frac{\dfrac{\tilde{\nu}_{\mathrm{t}}^{0}}{\tilde{E}_{\mathrm{T}}^{0}} - \dfrac{\tilde{\nu}_{\mathrm{t}}}{\tilde{E}_{\mathrm{T}}}}{\dfrac{\tilde{\nu}_{\mathrm{a}}^{0}}{\tilde{E}_{\mathrm{A}}^{0}} - \dfrac{\tilde{\nu}_{\mathrm{a}}}{\tilde{E}_{\mathrm{A}}}} = \frac{\dfrac{\tilde{\nu}_{\mathrm{A}}^{0}}{\tilde{E}_{\mathrm{A}}^{0}} - \dfrac{\tilde{\nu}_{\mathrm{A}}}{\tilde{E}_{\mathrm{A}}}}{\dfrac{1}{\tilde{E}_{\mathrm{A}}} - \dfrac{1}{\tilde{E}_{\mathrm{A}}^{0}}} = \frac{\dfrac{1}{\tilde{E}_{\mathrm{T}}} - \dfrac{1}{\tilde{E}_{\mathrm{T}}^{0}}}{\dfrac{\tilde{\nu}_{\mathrm{A}}^{0}}{\tilde{E}_{\mathrm{A}}^{0}} - \dfrac{\tilde{\nu}_{\mathrm{A}}}{\tilde{E}_{\mathrm{A}}}} = \frac{\tilde{\alpha}_{\mathrm{T}} - \tilde{\alpha}_{\mathrm{T}}^{0}}{\tilde{\alpha}_{\mathrm{A}} - \tilde{\alpha}_{\mathrm{A}}^{0}} = k. \tag{10.49}$$

10.6.3 Ply Crack Closure for Constrained Uniaxial Loading in Through-thickness Direction

Consider now ply crack closure for the case of constrained uniaxial through-thickness loading such that $\sigma_A = \sigma_T = 0$ and $\gamma_A = 0$. By following the approach of the previous two sections, it can be shown using (10.12)–(10.14) that, on setting $\hat{\varepsilon}_A = \hat{\varepsilon}_A^c$, $\hat{\varepsilon}_T = \hat{\varepsilon}_T^c$, $\hat{\varepsilon}_t = \hat{\varepsilon}_t^c$ and $\sigma_T = \hat{\sigma}_T^c$, that

$$\hat{\varepsilon}_A^c = -\frac{\tilde{\nu}_a}{\tilde{E}_A}\hat{\sigma}_t^c + \tilde{\alpha}_A \Delta T, \tag{10.50}$$

$$\hat{\varepsilon}_T^c = -\frac{\tilde{\nu}_t}{\tilde{E}_T}\hat{\sigma}_t^c + \tilde{\alpha}_T \Delta T, \tag{10.51}$$

$$\hat{\varepsilon}_t^c = \frac{\hat{\sigma}_t^c}{\tilde{E}_t} + \tilde{\alpha}_t \Delta T. \tag{10.52}$$

Similarly, for undamaged laminates it follows that

$$\hat{\varepsilon}_A^c = -\frac{\tilde{\nu}_a^0}{\tilde{E}_A^0}\hat{\sigma}_t^c + \tilde{\alpha}_A^0 \Delta T, \tag{10.53}$$

$$\hat{\varepsilon}_T^c = -\frac{\tilde{\nu}_t^0}{\tilde{E}_T^0}\hat{\sigma}_t^c + \tilde{\alpha}_T^0 \Delta T, \tag{10.54}$$

$$\hat{\varepsilon}_t^c = \frac{\hat{\sigma}_t^c}{\tilde{E}_t^0} + \tilde{\alpha}_t^0 \Delta T. \tag{10.55}$$

It should be noted that relations (10.50)–(10.52) and (10.53)–(10.55) have again made use of the fact that at the point where ply cracks *just* close the strains in a cracked laminate must be the same as those in an undamaged laminate. As stated previously, this is the case because at the point of closure the stress normal to the ply crack within the cracked ply is zero so that it does not matter whether this ply is cracked. It then follows from (10.50)–(10.55), on eliminating $\hat{\varepsilon}_A^c$, $\hat{\varepsilon}_T^c$ and $\hat{\varepsilon}_t^c$, that

$$\hat{\sigma}_t^c = -\frac{\tilde{\alpha}_A - \tilde{\alpha}_A^0}{\dfrac{\tilde{\nu}_a^0}{\tilde{E}_A^0} - \dfrac{\tilde{\nu}_a}{\tilde{E}_A}}\Delta T = -\frac{\tilde{\alpha}_T - \tilde{\alpha}_T^0}{\dfrac{\tilde{\nu}_t^0}{\tilde{E}_T^0} - \dfrac{\tilde{\nu}_t}{\tilde{E}_T}}\Delta T = -\frac{\tilde{\alpha}_t - \tilde{\alpha}_t^0}{\dfrac{1}{\tilde{E}_t} - \dfrac{1}{\tilde{E}_t^0}}\Delta T. \tag{10.56}$$

It follows from (10.23) that for the case of constrained uniaxial loading under discussion

$$\hat{\varepsilon}_A^c = -A_0 \hat{\sigma}_t^c - B_0 \hat{\varepsilon}_T^c + C_0 \Delta T. \tag{10.57}$$

Thus, on using the stress-strain relations for an undamaged laminate, the following alternative expression for the crack closure stress is obtained

$$\hat{\sigma}_t^c = -\frac{\tilde{\alpha}_A^0 + \tilde{\alpha}_T^0 B_0 - C_0}{A_0 - \dfrac{\tilde{\nu}_a^0}{\tilde{E}_A^0} - \dfrac{\tilde{\nu}_t^0}{\tilde{E}_T^0}B_0}\Delta T. \tag{10.58}$$

It then follows from (10.56) and (10.58) that

$$\frac{\tilde{\alpha}_A - \tilde{\alpha}_A^0}{\dfrac{\tilde{\nu}_a^0}{\tilde{E}_A^0} - \dfrac{\tilde{\nu}_a}{\tilde{E}_A}} = \frac{\tilde{\alpha}_T - \tilde{\alpha}_T^0}{\dfrac{\tilde{\nu}_t^0}{\tilde{E}_T^0} - \dfrac{\tilde{\nu}_t}{\tilde{E}_T}} = \frac{\tilde{\alpha}_t - \tilde{\alpha}_t^0}{\dfrac{1}{\tilde{E}_t} - \dfrac{1}{\tilde{E}_t^0}} = k_3, \tag{10.59}$$

where

$$k_3 = \frac{\tilde{\alpha}_{\mathrm{A}}^0 + \tilde{\alpha}_{\mathrm{T}}^0 B_0 - C_0}{A_0 - \dfrac{\tilde{\nu}_{\mathrm{a}}^0}{\tilde{E}_{\mathrm{A}}^0} - \dfrac{\tilde{\nu}_{\mathrm{t}}^0}{\tilde{E}_{\mathrm{T}}^0} B_0}. \tag{10.60}$$

The constant k_3 is thus another undamaged laminate constant such that $\tilde{\sigma}_{\mathrm{t}}^{\mathrm{c}} = -k_3 \Delta T$.
It follows from (10.35) and (10.60) that

$$k' = \frac{k_1}{k_3} = \frac{\tilde{E}_{\mathrm{A}}^0 A_0 - \tilde{\nu}_{\mathrm{a}}^0 - \left(\tilde{\nu}_{\mathrm{t}}^0 \tilde{E}_{\mathrm{A}}^0 / \tilde{E}_{\mathrm{T}}^0\right) B_0}{1 - \tilde{\nu}_{\mathrm{A}}^0 B_0}, \tag{10.61}$$

where the constant k' is independent of the thermal expansion coefficients for undamaged laminates. It then follows from (10.34), (10.59) and (10.61) that the following interrelationships between effective thermoelastic constants of damaged laminates must also be valid:

$$\frac{\dfrac{1}{\tilde{E}_{\mathrm{t}}} - \dfrac{1}{\tilde{E}_{\mathrm{t}}^0}}{\dfrac{\tilde{\nu}_{\mathrm{a}}^0}{\tilde{E}_{\mathrm{A}}^0} - \dfrac{\tilde{\nu}_{\mathrm{a}}}{\tilde{E}_{\mathrm{A}}}} = \frac{\dfrac{\tilde{\nu}_{\mathrm{a}}^0}{\tilde{E}_{\mathrm{A}}^0} - \dfrac{\tilde{\nu}_{\mathrm{a}}}{\tilde{E}_{\mathrm{A}}}}{\dfrac{1}{\tilde{E}_{\mathrm{A}}} - \dfrac{1}{\tilde{E}_{\mathrm{A}}^0}} = \frac{\dfrac{\tilde{\nu}_{\mathrm{t}}^0}{\tilde{E}_{\mathrm{T}}^0} - \dfrac{\tilde{\nu}_{\mathrm{t}}}{\tilde{E}_{\mathrm{T}}}}{\dfrac{\tilde{\nu}_{\mathrm{A}}^0}{\tilde{E}_{\mathrm{A}}^0} - \dfrac{\tilde{\nu}_{\mathrm{A}}}{\tilde{E}_{\mathrm{A}}}} = \frac{\tilde{\alpha}_{\mathrm{t}} - \tilde{\alpha}_{\mathrm{t}}^0}{\tilde{\alpha}_{\mathrm{A}} - \tilde{\alpha}_{\mathrm{A}}^0} = k'. \tag{10.62}$$

On using (10.24) the expressions (10.48) and (10.61) for the laminate constants k and k' may be written in terms of undamaged laminate properties as follows:

$$k = \frac{\tilde{E}_{\mathrm{A}}^0}{\tilde{E}_{\mathrm{T}}^0} \frac{\nu_{\mathrm{A}}^{(90)} - \tilde{\nu}_{\mathrm{A}}^0 \tilde{E}_{\mathrm{T}}^0 / \tilde{E}_{\mathrm{A}}^0}{1 - \nu_{\mathrm{A}}^{(90)} \tilde{\nu}_{\mathrm{A}}^0}, \tag{10.63}$$

$$k' = \frac{\tilde{E}_{\mathrm{A}}^0 \left(\nu_{\mathrm{t}}^{(90)} / E_{\mathrm{T}}^{(90)} + \nu_{\mathrm{a}}^{(90)} \nu_{\mathrm{A}}^{(90)} / E_{\mathrm{A}}^{(90)}\right) - \tilde{\nu}_{\mathrm{a}}^0 - \nu_{\mathrm{A}}^{(90)} \tilde{\nu}_{\mathrm{t}}^0 \tilde{E}_{\mathrm{A}}^0 / \tilde{E}_{\mathrm{T}}^0}{1 - \nu_{\mathrm{A}}^{(90)} \tilde{\nu}_{\mathrm{A}}^0}. \tag{10.64}$$

10.6.4 Useful Independent Interrelationships

Relations (10.34), (10.46), (10.49), (10.59) and (10.62) are not of course all independent. It will, therefore, be very useful to identify a set of independent interrelationships. To achieve this objective, it is convenient to introduce the parameter D defined by

$$D = \frac{1}{\tilde{E}_{\mathrm{A}}} - \frac{1}{\tilde{E}_{\mathrm{A}}^0}. \tag{10.65}$$

The quantity D can be regarded as a definition of a macroscopic property degradation parameter arising in the field of continuum damage mechanics. It can be shown, from (10.34), (10.49) and (10.62), that the following set of relations form the required independent interrelationships

$$\frac{\tilde{\nu}_{\mathrm{A}}^0}{\tilde{E}_{\mathrm{A}}^0} - \frac{\tilde{\nu}_{\mathrm{A}}}{\tilde{E}_{\mathrm{A}}} = kD, \tag{10.66}$$

$$\frac{1}{\tilde{E}_{\mathrm{T}}} - \frac{1}{\tilde{E}_{\mathrm{T}}^0} = k^2 D, \tag{10.67}$$

$$\frac{\tilde{\nu}_{\mathrm{a}}^0}{\tilde{E}_{\mathrm{A}}^0} - \frac{\tilde{\nu}_{\mathrm{a}}}{\tilde{E}_{\mathrm{A}}} = k' D, \tag{10.68}$$

$$\frac{1}{\tilde{E}_{\rm t}} - \frac{1}{\tilde{E}_{\rm t}^0} = \left(k'\right)^2 D, \tag{10.69}$$

$$\frac{\tilde{\nu}_{\rm t}^0}{\tilde{E}_{\rm T}^0} - \frac{\tilde{\nu}_{\rm t}}{\tilde{E}_{\rm T}} = kk'\,D, \tag{10.70}$$

$$\tilde{\alpha}_{\rm A} - \tilde{\alpha}_{\rm A}^0 = k_1\,D, \tag{10.71}$$

$$\tilde{\alpha}_{\rm T} - \tilde{\alpha}_{\rm T}^0 = kk_1 D, \tag{10.72}$$

$$\tilde{\alpha}_{\rm t} - \tilde{\alpha}_{\rm t}^0 = k'k_1 D. \tag{10.73}$$

In these interrelationships, the parameters k, k' and k_1 are constants for undamaged laminates defined by relations (10.48), (10.61) and (10.35), respectively. Thus, the values of these constants are readily calculated using (10.15)–(10.22) and (10.24) applied to undamaged laminates, and the property definitions given in Section 10.5 and Appendix B. On using relations (10.66)–(10.73), the effective thermoelastic constants of a damaged laminate can be calculated in terms of the damage dependent parameter D defined by (10.65) and the corresponding values for undamaged laminates.

10.7 Predicting Crack Formation under Fixed Applied Stresses

Consider a general symmetric laminate that already contains m cracks in the 90° plies, and then the simultaneous formation of $n - m$ new cracks such that there are n cracks in total. The locations of the ply cracks are assumed to be such that the overall deformation of the laminate is governed by the effective stress-strain equations (10.7)–(10.10). This means that the crack distribution is regarded as being sufficiently uniformly distributed for both in-plane and out-of-plane bending deformations to be negligible. The new ply cracks are assumed to form quasistatically under conditions of fixed temperature. Consider the case of crack formation under fixed applied tractions. This means that the applied stresses $\sigma_{\rm A}$, $\sigma_{\rm T}$, $\sigma_{\rm t}$ and $\tau_{\rm A}$ are considered to be held fixed during the formation of each crack, but that progressively increasing stresses will be required to form new cracks. From energy balance considerations and the fact that kinetic energy is never negative, the criterion for crack formation under these conditions has the form (see Section 8.5)

$$\Delta\Gamma + \Delta G < 0, \tag{10.74}$$

where ΔG is the change in Gibbs energy, and where the energy absorbed in volume V of laminate by the formation of the new cracks is given by

$$\Delta\Gamma = V\left(\Gamma_n - \Gamma_m\right). \tag{10.75}$$

In (10.75), the parameter Γ_k denotes the energy absorbed in unit volume (averaged over a sufficiently large region V of laminate) during the formation of $k = m, n$ ply cracks in the 90° plies. The corresponding change of Gibbs energy in the region V of the laminate is

$$\Delta G = \int_V \left(g_n - g_m\right) \mathrm{d}V. \tag{10.76}$$

In (10.76), g_k denotes the Gibbs energy per unit volume when there are k cracks in the laminate. It follows from (10.74)–(10.76) that the formation of n ply cracks in a laminate already having m ply cracks is governed by the inequality

$$P_n < P_m, \tag{10.77}$$

where P_k is the total energy per unit volume for a laminate having k cracks defined by

$$P_k \equiv \Gamma_k + \bar{g}_k - \bar{g}_0, \quad \bar{g} = \frac{1}{V}\int_V g\,\mathrm{d}V, \quad k = 0, 1, \ldots, n, \tag{10.78}$$

where $\bar{g}_0$ is the value of $\bar{g}$ when the laminate is undamaged and where $\Gamma_0 = 0$. The energy P has thus been defined so that it is zero when the laminate is uncracked. Appendix C shows that the average Gibbs energy (equivalent to the complementary energy) for region V may be written in the form

$$\begin{aligned}\bar{g} = &-\frac{\sigma_A^2}{2E_A} - \frac{\sigma_T^2}{2E_T} - \frac{\sigma_t^2}{2E_t} - \frac{\tau_A^2}{2\mu_A} + \frac{\nu_A}{E_A}\sigma_A\sigma_T + \frac{\nu_a}{E_A}\sigma_A\sigma_t + \frac{\nu_t}{E_T}\sigma_T\sigma_t \\ &+ \frac{\lambda_A}{\mu_A}\sigma_A\tau_A + \frac{\lambda_T}{\mu_A}\sigma_T\tau_A + \frac{\lambda_t}{\mu_A}\sigma_t\tau_A - \left[\sigma_A\alpha_A + \sigma_T\alpha_T + \sigma_t\alpha_t + \tau_A\alpha_S\right]\Delta T \\ &+ \frac{1}{2}\left[\sigma_A^c\alpha_A + \sigma_T^c\alpha_T + \sigma_t^c\alpha_t + \tau_A^c\alpha_S\right]\Delta T - \frac{1}{2V}\int_V \sigma_{ij}^c\alpha_{ij}\Delta T\,\mathrm{d}V + g_0(\Delta T),\end{aligned} \tag{10.79}$$

where σ_A^c, σ_T^c, σ_t^c and τ_A^c denote any combination of the applied stresses for which the ply cracks are just closed. On using (10.11) and (10.15)–(10.22) to eliminate the shear stress τ_A in (10.79) it can be shown that

$$\begin{aligned}\bar{g} = &-\frac{\sigma_A^2}{2\tilde{E}_A} - \frac{\sigma_T^2}{2\tilde{E}_T} - \frac{\sigma_t^2}{2\tilde{E}_t} + \frac{\tilde{\nu}_A}{\tilde{E}_A}\sigma_A\sigma_T + \frac{\tilde{\nu}_a}{\tilde{E}_A}\sigma_A\sigma_t + \frac{\tilde{\nu}_t}{\tilde{E}_T}\sigma_T\sigma_t \\ &- \left[\sigma_A\tilde{\alpha}_A + \sigma_T\tilde{\alpha}_T + \sigma_t\tilde{\alpha}_t\right]\Delta T - \frac{1}{2}\mu_A\left(\gamma_A\right)^2 + \frac{1}{2}\mu_A\left(\alpha_S\Delta T\right)^2 \\ &+ \frac{1}{2}\left[\sigma_A^c\alpha_A + \sigma_T^c\alpha_T + \sigma_t^c\alpha_t + \tau_A^c\alpha_S\right]\Delta T - \frac{1}{2V}\int_V \sigma_{ij}^c\alpha_{ij}\Delta T\,\mathrm{d}V + g_0(\Delta T).\end{aligned} \tag{10.80}$$

At the point of closure, for a damaged laminate with $\gamma_A = 0$, it follows from (10.11) that

$$\tau_A^c = \lambda_A\sigma_A^c + \lambda_A\sigma_T^c + \lambda_t\sigma_t^c - \mu_A\alpha_S\Delta T. \tag{10.81}$$

Substitution into (10.80) then leads to the result

$$\begin{aligned}\bar{g} = &-\frac{\sigma_A^2}{2\tilde{E}_A} - \frac{\sigma_T^2}{2\tilde{E}_T} - \frac{\sigma_t^2}{2\tilde{E}_t} + \frac{\tilde{\nu}_A}{\tilde{E}_A}\sigma_A\sigma_T + \frac{\tilde{\nu}_a}{\tilde{E}_A}\sigma_A\sigma_t + \frac{\tilde{\nu}_t}{\tilde{E}_T}\sigma_T\sigma_t \\ &- \left[\sigma_A\tilde{\alpha}_A + \sigma_T\tilde{\alpha}_T + \sigma_t\tilde{\alpha}_t\right]\Delta T - \frac{1}{2}\mu_A\left(\gamma_A\right)^2 \\ &+ \frac{1}{2}\left[\sigma_A^c\tilde{\alpha}_A + \sigma_T^c\tilde{\alpha}_T + \sigma_t^c\tilde{\alpha}_t\right]\Delta T - \frac{1}{2V}\int_V \sigma_{ij}^c\alpha_{ij}\Delta T\,\mathrm{d}V + g_0(\Delta T).\end{aligned} \tag{10.82}$$

For the case of an undamaged laminate, (10.82) takes the form

$$\begin{aligned}\bar{g}_0 = &-\frac{\sigma_A^2}{2\tilde{E}_A^0}-\frac{\sigma_T^2}{2\tilde{E}_T^0}-\frac{\sigma_t^2}{2\tilde{E}_t^0}+\frac{\tilde{\nu}_A^0}{\tilde{E}_A^0}\sigma_A\sigma_T+\frac{\tilde{\nu}_a^0}{\tilde{E}_A^0}\sigma_A\sigma_t+\frac{\tilde{\nu}_t^0}{\tilde{E}_T^0}\sigma_T\sigma_t\\ &-\left[\sigma_A\tilde{\alpha}_A^0+\sigma_T\tilde{\alpha}_T^0+\sigma_t\tilde{\alpha}_t^0\right]\Delta T-\frac{1}{2}\mu_A^0\left(\gamma_A^0\right)^2 \\ &+\frac{1}{2}\left[\sigma_A^c\tilde{\alpha}_A^0+\tilde{\sigma}_T^c\alpha_T^0+\sigma_t^c\tilde{\alpha}_t^0\right]\Delta T-\frac{1}{2V}\int_V \sigma_{ij}^c\alpha_{ij}\Delta T\,dV+g_0(\Delta T),\end{aligned}\tag{10.83}$$

where γ_A^0 is the engineering in-plane shear strain of the laminate in an undamaged state subject to the same stresses and temperature difference. On subtracting (10.82) and (10.83), it then follows that

$$\begin{aligned}\bar{g}-\bar{g}_0 = &-\left(\frac{1}{\tilde{E}_A}-\frac{1}{\tilde{E}_A^0}\right)\frac{\sigma_A^2}{2}-\left(\frac{1}{\tilde{E}_T}-\frac{1}{\tilde{E}_T^0}\right)\frac{\sigma_T^2}{2}-\left(\frac{1}{\tilde{E}_t}-\frac{1}{\tilde{E}_t^0}\right)\frac{\sigma_t^2}{2}\\ &-\left(\frac{\tilde{\nu}_A^0}{\tilde{E}_A^0}-\frac{\tilde{\nu}_A}{\tilde{E}_A}\right)\sigma_A\sigma_T-\left(\frac{\tilde{\nu}_a^0}{\tilde{E}_A^0}-\frac{\tilde{\nu}_a}{\tilde{E}_A}\right)\sigma_A\sigma_t-\left(\frac{\tilde{\nu}_t^0}{\tilde{E}_T^0}-\frac{\tilde{\nu}_t}{\tilde{E}_T}\right)\sigma_T\sigma_t\\ &-\sigma_A\left(\tilde{\alpha}_A-\tilde{\alpha}_A^0\right)\Delta T-\sigma_T\left(\tilde{\alpha}_T-\tilde{\alpha}_T^0\right)\Delta T-\sigma_t\left(\tilde{\alpha}_t-\tilde{\alpha}_t^0\right)\Delta T\\ &+\frac{1}{2}\sigma_A^c\left(\tilde{\alpha}_A-\tilde{\alpha}_A^0\right)\Delta T+\frac{1}{2}\sigma_T^c\left(\tilde{\alpha}_T-\tilde{\alpha}_T^0\right)\Delta T+\frac{1}{2}\sigma_t^c\left(\tilde{\alpha}_t-\tilde{\alpha}_t^0\right)\Delta T\\ &-\frac{1}{2}\mu_A\left(\gamma_A\right)^2+\frac{1}{2}\mu_A^0\left(\gamma_A^0\right)^2.\end{aligned}\tag{10.84}$$

Result (10.84) is now expressed in terms of the dimensionless parameter D defined by (10.65) using relations (10.66)–(10.73)

$$\begin{aligned}\bar{g}-\bar{g}_0 = &-\frac{1}{2}D\Big[\sigma_A^2+k^2\sigma_T^2+k'^2\sigma_t^2+2k\sigma_A\sigma_T+2k'\sigma_A\sigma_t+2kk'\sigma_T\sigma_t\\ &+2k_1\sigma_A\Delta T+2kk_1\sigma_T\Delta T+2k'k_1\sigma_t\Delta T-k_1\sigma_A^c\Delta T-kk_1\sigma_T^c\Delta T-k'k_1\sigma_t^c\Delta T\Big]\\ &-\frac{1}{2}\mu_A\left(\gamma_A\right)^2+\frac{1}{2}\mu_A^0\left(\gamma_A^0\right)^2.\end{aligned}\tag{10.85}$$

On using (10.12)–(10.14) for the undamaged and damaged states and subtracting

$$\hat{\varepsilon}_A-\hat{\varepsilon}_A^0=\left(\frac{1}{\tilde{E}_A}-\frac{1}{\tilde{E}_A^0}\right)\sigma_A+\left(\frac{\tilde{\nu}_A^0}{\tilde{E}_A^0}-\frac{\tilde{\nu}_A}{\tilde{E}_A}\right)\sigma_T+\left(\frac{\tilde{\nu}_a^0}{\tilde{E}_A^0}-\frac{\tilde{\nu}_a}{\tilde{E}_A}\right)\sigma_t+\left(\tilde{\alpha}_A-\tilde{\alpha}_A^0\right)\Delta T,\tag{10.86}$$

$$\hat{\varepsilon}_T-\hat{\varepsilon}_T^0=\left(\frac{\tilde{\nu}_A^0}{\tilde{E}_A^0}-\frac{\tilde{\nu}_A}{\tilde{E}_A}\right)\sigma_A+\left(\frac{1}{\tilde{E}_T}-\frac{1}{\tilde{E}_T^0}\right)\sigma_T+\left(\frac{\tilde{\nu}_t^0}{\tilde{E}_T^0}-\frac{\tilde{\nu}_t}{\tilde{E}_T}\right)\sigma_t+\left(\tilde{\alpha}_T-\tilde{\alpha}_T^0\right)\Delta T,\tag{10.87}$$

$$\hat{\varepsilon}_t-\hat{\varepsilon}_t^0=\left(\frac{\tilde{\nu}_a^0}{\tilde{E}_A^0}-\frac{\tilde{\nu}_a}{\tilde{E}_A}\right)\sigma_A+\left(\frac{\tilde{\nu}_t^0}{\tilde{E}_T^0}-\frac{\tilde{\nu}_t}{\tilde{E}_T}\right)\sigma_T+\left(\frac{1}{\tilde{E}_t}-\frac{1}{\tilde{E}_t^0}\right)\sigma_t+\left(\tilde{\alpha}_t-\tilde{\alpha}_t^0\right)\Delta T,\tag{10.88}$$

where $\hat{\varepsilon}_A^0$, $\hat{\varepsilon}_T^0$ and $\hat{\varepsilon}_t^0$ are the strains for an undamaged laminate subject to the applied stresses σ_A, σ_T and σ_t and the temperature difference ΔT. On using (10.65)–(10.73), relations (10.86)–(10.88) may be written in the form

$$\hat{\varepsilon}_A-\hat{\varepsilon}_A^0 = D\left(\sigma_A+k\sigma_T+k'\sigma_t-\bar{\sigma}_A^c\right),\tag{10.89}$$

$$\hat{\varepsilon}_{\mathrm{T}} - \hat{\varepsilon}_{\mathrm{T}}^{0} = k\,D\left(\sigma_{\mathrm{A}} + k\sigma_{\mathrm{T}} + k'\sigma_{\mathrm{t}} - \bar{\sigma}_{\mathrm{A}}^{\mathrm{c}}\right), \tag{10.90}$$

$$\hat{\varepsilon}_{\mathrm{t}} - \hat{\varepsilon}_{\mathrm{t}}^{0} = k'D\left(\sigma_{\mathrm{A}} + k\sigma_{\mathrm{T}} + k'\sigma_{\mathrm{t}} - \bar{\sigma}_{\mathrm{A}}^{\mathrm{c}}\right). \tag{10.91}$$

At any critical point of crack closure, such that $\sigma_{\mathrm{A}} = \sigma_{\mathrm{A}}^{\mathrm{c}}$, $\sigma_{\mathrm{T}} = \sigma_{\mathrm{T}}^{\mathrm{c}}$ and $\sigma_{\mathrm{t}} = \sigma_{\mathrm{t}}^{\mathrm{c}}$, it is clear that the strains in a damaged laminate will be identical to those for an undamaged laminate subject to the same stresses, that is, $\hat{\varepsilon}_{\mathrm{A}} = \hat{\varepsilon}_{\mathrm{A}}^{0} = \hat{\varepsilon}_{\mathrm{A}}^{\mathrm{c}}$, $\hat{\varepsilon}_{\mathrm{T}} = \hat{\varepsilon}_{\mathrm{T}}^{0} = \hat{\varepsilon}_{\mathrm{T}}^{\mathrm{c}}$ and $\hat{\varepsilon}_{\mathrm{t}} = \hat{\varepsilon}_{\mathrm{t}}^{0} = \hat{\varepsilon}_{\mathrm{t}}^{\mathrm{c}}$. It then follows from (10.89)–(10.91) and (10.36) that

$$\sigma_{\mathrm{A}}^{\mathrm{c}} + k\sigma_{\mathrm{T}}^{\mathrm{c}} + k'\sigma_{\mathrm{t}}^{\mathrm{c}} = \bar{\sigma}_{\mathrm{A}}^{\mathrm{c}} = -k_1\Delta T, \tag{10.92}$$

for any critical state of crack closure (i.e. ply cracks are just closed having zero normal and shear tractions). Result (10.85) may then be written in the following form

$$\begin{aligned}\bar{g} - \bar{g}_0 = &-\frac{1}{2}D\Big[\sigma_{\mathrm{A}}^2 + k^2\sigma_{\mathrm{T}}^2 + k'^2\sigma_{\mathrm{t}}^2 + 2k\sigma_{\mathrm{A}}\sigma_{\mathrm{T}} + 2k'\sigma_{\mathrm{A}}\sigma_{\mathrm{t}} + 2k'k\sigma_{\mathrm{T}}\sigma_{\mathrm{t}} \\ &-2\sigma_{\mathrm{A}}\bar{\sigma}_{\mathrm{A}}^{\mathrm{c}} - 2k\sigma_{\mathrm{T}}\bar{\sigma}_{\mathrm{A}}^{\mathrm{c}} - 2k'\sigma_{\mathrm{t}}\bar{\sigma}_{\mathrm{A}}^{\mathrm{c}} + \left(\bar{\sigma}_{\mathrm{A}}^{\mathrm{c}}\right)^2\Big] - \frac{1}{2}\mu_{\mathrm{A}}\left(\gamma_{\mathrm{A}}\right)^2 + \frac{1}{2}\mu_{\mathrm{A}}^{0}\left(\gamma_{\mathrm{A}}^{0}\right)^2,\end{aligned} \tag{10.93}$$

which may be expressed in the following compact form where the applied stress dependence is a perfect square involving a linear combination of the applied stresses σ_{A}, σ_{T} and σ_{t}, and the uniaxial closure stress $\bar{\sigma}_{\mathrm{A}}^{\mathrm{c}}$

$$\bar{g} - \bar{g}_0 = -\frac{1}{2}D\left(\sigma_{\mathrm{A}} + k\sigma_{\mathrm{T}} + k'\sigma_{\mathrm{t}} - \bar{\sigma}_{\mathrm{A}}^{\mathrm{c}}\right)^2 - \frac{1}{2}\mu_{\mathrm{A}}\left(\gamma_{\mathrm{A}}\right)^2 + \frac{1}{2}\mu_{\mathrm{A}}^{0}\left(\gamma_{\mathrm{A}}^{0}\right)^2. \tag{10.94}$$

It is clear from (10.94) that a key loading parameter s for ply cracking, in the absence of shear deformation, may be defined by

$$s = \sigma_{\mathrm{A}} + k\sigma_{\mathrm{T}} + k'\sigma_{\mathrm{t}}, \tag{10.95}$$

so that

$$\bar{g} - \bar{g}_0 = -\frac{1}{2}D\left(s - \bar{\sigma}_{\mathrm{A}}^{\mathrm{c}}\right)^2 - \frac{1}{2}\mu_{\mathrm{A}}\left(\gamma_{\mathrm{A}}\right)^2 + \frac{1}{2}\mu_{\mathrm{A}}^{0}\left(\gamma_{\mathrm{A}}^{0}\right)^2. \tag{10.96}$$

On using (10.65), (10.67), (10.69) and (10.89)–(10.91), result (10.94) may also be written in one of the following three compact equivalent forms involving applied strains

$$\begin{aligned}&\bar{g} - \bar{g}_0 + \tfrac{1}{2}\mu_{\mathrm{A}}\left(\gamma_{\mathrm{A}}\right)^2 - \tfrac{1}{2}\mu_{\mathrm{A}}^{0}\left(\gamma_{\mathrm{A}}^{0}\right)^2 \\ &= -\frac{\tfrac{1}{2}\left(\hat{\varepsilon}_{\mathrm{A}} - \hat{\varepsilon}_{\mathrm{A}}^{0}\right)^2}{\dfrac{1}{\tilde{E}_{\mathrm{A}}} - \dfrac{1}{\tilde{E}_{\mathrm{A}}^{0}}} = -\frac{\tfrac{1}{2}\left(\hat{\varepsilon}_{\mathrm{T}} - \hat{\varepsilon}_{\mathrm{T}}^{0}\right)^2}{\dfrac{1}{\tilde{E}_{\mathrm{T}}} - \dfrac{1}{\tilde{E}_{\mathrm{T}}^{0}}} = -\frac{\tfrac{1}{2}\left(\hat{\varepsilon}_{\mathrm{t}} - \hat{\varepsilon}_{\mathrm{t}}^{0}\right)^2}{\dfrac{1}{\tilde{E}_{\mathrm{t}}} - \dfrac{1}{\tilde{E}_{\mathrm{t}}^{0}}}.\end{aligned} \tag{10.97}$$

It is emphasised that the exceedingly simple results (10.96) and (10.97) are exact, and that they take full account of the effect of multiaxial loading and of the effect of thermal residual stresses. The results involve only macroscopic properties defined at the laminate level even though they have been derived from a detailed consideration of the stress and deformation distributions at the ply level. The results are based on the single assumption of damage homogeneity where damage formation in the laminate is such that the stress/strain relationships (10.7) and (10.10) are satisfied for all states of damage, including for both embedded and fully developed ply cracks in the 90° plies.

For crack initiation, the use of (10.65), (10.77), (10.78), (10.96) and (10.97) leads to the following equivalent criteria ((10.99) is one of three possible equivalent forms)

$$\left(\frac{1}{\tilde{E}_{\mathrm{A}}} - \frac{1}{\tilde{E}_{\mathrm{A}}^{0}}\right)\left(s - \bar{\sigma}_{\mathrm{A}}^{\mathrm{c}}\right)^{2} + \mu_{\mathrm{A}}\left(\gamma_{\mathrm{A}}\right)^{2} - \mu_{\mathrm{A}}^{0}\left(\gamma_{\mathrm{A}}^{0}\right)^{2} - 2\Gamma > 0, \tag{10.98}$$

$$\frac{\left(\hat{\varepsilon}_{\mathrm{A}} - \hat{\varepsilon}_{\mathrm{A}}^{0}\right)^{2}}{\dfrac{1}{\tilde{E}_{\mathrm{A}}} - \dfrac{1}{\tilde{E}_{\mathrm{A}}^{0}}} + \mu_{\mathrm{A}}\left(\gamma_{\mathrm{A}}\right)^{2} - \mu_{\mathrm{A}}^{0}\left(\gamma_{\mathrm{A}}^{0}\right)^{2} - 2\Gamma > 0, \tag{10.99}$$

where Γ is energy absorption per unit volume specified later (see (10.188)).

It is emphasised that the approach described does not provide any information that indicates how the effective reduced axial Young's modulus $\tilde{E}_{\mathrm{A}}$ and axial shear modulus μ_{A} depend upon the damage in the laminate. The detailed stress analysis described in Chapter 18 provides this information for the case of uniformly spaced arrays of ply cracks.

10.8 Accounting for Nonuniform Cracking during Ply Crack Simulation

It is emphasised that the results, based on the consideration of crack closure, presented in Section 10.6 do, in fact, apply to laminates having both uniform and nonuniform distributions of ply cracks in the 90° plies. They are expected to apply also to other forms of damage provided that it is possible to apply a loading to the laminate that leads to the closure of all cracks present. This property is now exploited when considering ply crack distributions that are nonuniform.

When a laminate has a nonuniform distribution of n ply cracks, as shown in Figure 10.1, it is useful to develop an approximate method of estimating the effective properties of the whole laminate which is assumed to have length $2L$ The laminate is divided into a set of n regions $V_1, \ldots, V_n$ such that the undeformed length of region V_I is $2L_I$. Clearly

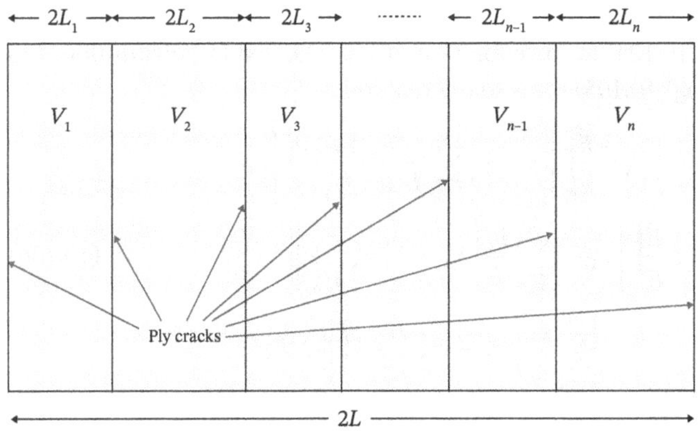

Figure 10.1 Schematic diagram illustrating the geometry of a nonuniformly cracked laminate.

$$L = \sum_{I=1}^{n} L_I, \tag{10.100}$$

and it follows that the effective or mean ply crack density has the value $\overline{\rho} = n/(2L)$. If the laminate had a uniform distribution of ply cracks having separation $2L_I$, then the thermoelastic constants for such a laminate can be estimated. The effective stress-strain relations for the region V_I referred to laminate axes are now written in the following form corresponding to the form (10.7) and (10.10) of the ply relations

$$\varepsilon_{\rm A}^I = \frac{\sigma_{\rm A}}{E_{\rm A}^I} - \frac{\nu_{\rm A}^I}{E_{\rm A}^I}\sigma_{\rm T}^I - \frac{\nu_{\rm a}^I}{E_{\rm A}^I}\sigma_{\rm t} - \frac{\lambda_{\rm A}^I}{\mu_{\rm A}^I}\tau_{\rm A}^I + \alpha_{\rm A}^I \Delta T, \tag{10.101}$$

$$\varepsilon_{\rm T} = -\frac{\nu_{\rm A}^I}{E_{\rm A}^I}\sigma_{\rm A} + \frac{\sigma_{\rm T}^I}{E_{\rm T}^I} - \frac{\nu_{\rm t}^I}{E_{\rm T}^I}\sigma_{\rm t} - \frac{\lambda_{\rm T}^I}{\mu_{\rm A}^I}\tau_{\rm A}^I + \alpha_{\rm T}^I \Delta T, \tag{10.102}$$

$$\varepsilon_{\rm t}^I = -\frac{\nu_{\rm a}^I}{E_{\rm A}^I}\sigma_{\rm A} - \frac{\nu_{\rm t}^I}{E_{\rm T}^I}\sigma_{\rm T}^I + \frac{\sigma_{\rm t}}{E_{\rm t}^I} - \frac{\lambda_{\rm t}^I}{\mu_{\rm A}^I}\tau_{\rm A}^I + \alpha_{\rm t}^I \Delta T, \tag{10.103}$$

$$\gamma_{\rm A} = -\frac{\lambda_{\rm A}^I}{\mu_{\rm A}^I}\sigma_{\rm A} - \frac{\lambda_{\rm T}^I}{\mu_{\rm A}^I}\sigma_{\rm T}^I - \frac{\lambda_{\rm t}^I}{\mu_{\rm A}^I}\sigma_{\rm t} + \frac{\tau_{\rm A}^I}{\mu_{\rm A}^I} + \alpha_{\rm S}^I \Delta T, \tag{10.104}$$

where the thermoelastic constants are labelled with a superscript I to denote that they refer to those values that would arise for a uniformly cracked laminate having a crack separation $2L_I$.

The effective stress-strain relations (10.101) and (10.104) are applied to the region V_I of the nonuniformly cracked laminate. The effective strains $\varepsilon_{\rm A}^I$ and ε_t^I and the effective transverse and shear stresses $\sigma_{\rm T}^I$ and $\tau_{\rm A}^I$, have a superscript I attached as their values will depend upon the region V_I being considered. The effective axial stress $\sigma_{\rm A}$, the through-thickness stress $\sigma_{\rm t}$ and the transverse and shear strains $\varepsilon_{\rm T}$ and $\gamma_{\rm A}$, all have the same value for each of the regions V_I , $I = 1, ..., n$. On considering the displacement field in the absence of bend deformation, the effective through-thickness strain, axial strain, transverse and shear stresses applied to the nonuniformly cracked laminate are given by

$$\varepsilon_{\rm A} = \sum_{I=1}^{n} \frac{L_I}{L}\varepsilon_{\rm A}^I, \quad \sigma_{\rm T} = \sum_{I=1}^{n} \frac{L_I}{L}\sigma_{\rm T}^I, \quad \varepsilon_{\rm t} = \sum_{I=1}^{n} \frac{L_I}{L}\varepsilon_{\rm t}^I, \quad \tau_{\rm A} = \sum_{I=1}^{n} \frac{L_I}{L}\tau_{\rm A}^I. \tag{10.105}$$

On using (10.102) and (10.104) to solve for $\sigma_{\rm T}^i$ and $\tau_{\rm A}^i$, followed by substitution into (10.103) and (10.104) using (10.105) leads to the relations

$$\begin{aligned}\varepsilon_{\rm A} = {} & \left[\sum_{I=1}^{n} \frac{L_I}{L}\frac{1}{\tilde{E}_{\rm A}^I}\left(1 - \left(\tilde{\nu}_{\rm A}^I\right)^2 \frac{\tilde{E}_{\rm T}^I}{\tilde{E}_{\rm A}^I}\right)\right]\sigma_{\rm A} - \left[\sum_{I=1}^{n} \frac{L_I}{L}\frac{\tilde{\nu}_{\rm a}^I + \tilde{\nu}_{\rm A}^I\tilde{\nu}_{\rm t}^I}{\tilde{E}_{\rm A}^I}\right]\sigma_{\rm t} \\ & - \left(\sum_{I=1}^{n} \frac{L_I}{L}\tilde{\nu}_{\rm A}^I \frac{\tilde{E}_{\rm T}^I}{\tilde{E}_{\rm A}^I}\right)\varepsilon_{\rm T} - \left[\sum_{I=1}^{n} \frac{L_I}{L}\left(\lambda_{\rm A}^I + \lambda_{\rm T}^I\tilde{\nu}_{\rm A}^I \frac{\tilde{E}_{\rm T}^I}{\tilde{E}_{\rm A}^I}\right)\right]\gamma_{\rm A} \\ & + \left[\sum_{I=1}^{n} \frac{L_I}{L}\left(\tilde{\alpha}_{\rm A}^I + \tilde{\nu}_{\rm A}^I \frac{\tilde{E}_{\rm T}^I}{\tilde{E}_{\rm A}^I}\tilde{\alpha}_{\rm T}^I\right)\right]\Delta T,\end{aligned} \tag{10.106}$$

$$\left(\sum_{I=1}^{n}\frac{L_I}{L}\tilde{E}_{\rm T}^I\right)\varepsilon_{\rm T}+\left(\sum_{I=1}^{n}\frac{L_I}{L}\lambda_{\rm T}^I\tilde{E}_{\rm T}^I\right)\gamma_{\rm A}=$$
$$-\left(\sum_{I=1}^{n}\frac{L_I}{L}\tilde{\nu}_{\rm A}^I\frac{\tilde{E}_{\rm T}^I}{\tilde{E}_{\rm A}^I}\right)\sigma_{\rm A}+\sigma_{\rm T}-\left(\sum_{I=1}^{n}\frac{L_I}{L}\tilde{\nu}_{\rm t}^I\right)\sigma_{\rm t}+\left(\sum_{I=1}^{n}\frac{L_I}{L}\tilde{E}_{\rm T}^I\tilde{\alpha}_{\rm T}^I\right)\Delta T, \tag{10.107}$$

$$\varepsilon_{\rm t}=-\left[\sum_{I=1}^{n}\frac{L_I}{L}\frac{\tilde{\nu}_{\rm a}^I+\tilde{\nu}_{\rm A}^I\tilde{\nu}_{\rm t}^I}{\tilde{E}_{\rm A}^I}\right]\sigma_{\rm A}+\left[\sum_{I=1}^{n}\frac{L_I}{L}\left(\frac{1}{\tilde{E}_{\rm t}^I}-\frac{(\tilde{\nu}_{\rm t}^I)^2}{\tilde{E}_{\rm T}^I}\right)\right]\sigma_{\rm t}-\left[\sum_{I=1}^{n}\frac{L_I}{L}\tilde{\nu}_{\rm t}^I\right]\varepsilon_{\rm T}$$
$$-\left[\sum_{I=1}^{n}\frac{L_I}{L}\left(\lambda_{\rm t}^I+\lambda_{\rm T}^I\tilde{\nu}_{\rm t}^I\right)\right]\gamma_{\rm A}+\left[\sum_{I=1}^{n}\frac{L_I}{L}\left(\tilde{\alpha}_{\rm t}^I+\tilde{\nu}_{\rm t}^I\tilde{\alpha}_{\rm T}^I\right)\right]\Delta T, \tag{10.108}$$

$$\tau_{\rm A}=\left[\sum_{I=1}^{n}\frac{L_I}{L}\left(\lambda_{\rm A}^I+\lambda_{\rm T}^I\tilde{\nu}_{\rm A}^I\frac{\tilde{E}_{\rm T}^I}{\tilde{E}_{\rm A}^I}\right)\right]\sigma_{\rm A}+\left[\sum_{I=1}^{n}\frac{L_I}{L}\left(\lambda_{\rm t}^I+\lambda_{\rm T}^I\tilde{\nu}_{\rm t}^I\right)\right]\sigma_{\rm t}$$
$$+\left(\sum_{I=1}^{n}\frac{L_I}{L}\lambda_{\rm T}^I\tilde{E}_{\rm T}^I\right)\varepsilon_{\rm T}+\left[\sum_{I=1}^{n}\frac{L_I}{L}\left(\mu_{\rm A}^I+\left(\lambda_{\rm T}^I\right)^2\tilde{E}_{\rm T}^I\right)\right]\gamma_{\rm A}$$
$$-\left[\sum_{I=1}^{n}\frac{L_I}{L}\left(\mu_{\rm A}^I\alpha_{\rm S}^I+\lambda_{\rm T}^I\tilde{E}_{\rm T}^I\tilde{\alpha}_{\rm T}^I\right)\right]\Delta T. \tag{10.109}$$

For the case of uniform ply crack distributions it follows directly from (10.106)–(10.109) that

$$\varepsilon_{\rm A}=\left(1-\tilde{\nu}_{\rm A}^2\frac{\tilde{E}_{\rm T}}{\tilde{E}_{\rm A}}\right)\frac{\sigma_{\rm A}}{\tilde{E}_{\rm A}}-\frac{\tilde{\nu}_{\rm a}+\tilde{\nu}_{\rm A}\tilde{\nu}_{\rm t}}{\tilde{E}_{\rm A}}\sigma_{\rm t}$$
$$-\tilde{\nu}_{\rm A}\frac{\tilde{E}_{\rm T}}{\tilde{E}_{\rm A}}\varepsilon_{\rm T}-\left(\lambda_{\rm A}+\lambda_{\rm T}\tilde{\nu}_{\rm A}\frac{\tilde{E}_{\rm T}}{\tilde{E}_{\rm A}}\right)\gamma_{\rm A}+\left(\tilde{\alpha}_{\rm A}+\tilde{\nu}_{\rm A}\frac{\tilde{E}_{\rm T}}{\tilde{E}_{\rm A}}\tilde{\alpha}_{\rm T}\right)\Delta T, \tag{10.110}$$

$$\tilde{E}_{\rm T}\varepsilon_{\rm T}+\lambda_{\rm T}\tilde{E}_{\rm T}\gamma_{\rm A}=-\tilde{\nu}_{\rm A}\frac{\tilde{E}_{\rm T}}{\tilde{E}_{\rm A}}\sigma_{\rm A}+\sigma_{\rm T}-\tilde{\nu}_{\rm t}\sigma_{\rm t}+\tilde{E}_{\rm T}\tilde{\alpha}_{\rm T}\Delta T, \tag{10.111}$$

$$\varepsilon_{\rm t}=-\frac{\tilde{\nu}_{\rm a}+\tilde{\nu}_{\rm A}\tilde{\nu}_{\rm t}}{\tilde{E}_{\rm A}}\sigma_{\rm A}+\left(\frac{1}{\tilde{E}_{\rm t}}-\frac{\tilde{\nu}_{\rm t}^2}{\tilde{E}_{\rm T}}\right)\sigma_{\rm t}-\tilde{\nu}_{\rm t}\varepsilon_{\rm T}$$
$$-\left(\lambda_{\rm t}+\lambda_{\rm T}\tilde{\nu}_{\rm t}\right)\gamma_{\rm A}+\left(\tilde{\alpha}_{\rm t}+\tilde{\nu}_{\rm t}\tilde{\alpha}_{\rm T}\right)\Delta T, \tag{10.112}$$

$$\tau_{\rm A}=\left(\lambda_{\rm A}+\lambda_{\rm T}\tilde{E}_{\rm T}\frac{\tilde{\nu}_{\rm A}}{\tilde{E}_{\rm A}}\right)\sigma_{\rm A}+\left(\lambda_{\rm t}+\lambda_{\rm T}\tilde{\nu}_{\rm t}\right)\sigma_{\rm t}+\lambda_{\rm T}\tilde{E}_{\rm T}\varepsilon_{\rm T}$$
$$+\left[\mu_{\rm A}+\left(\lambda_{\rm T}\right)^2\tilde{E}_{\rm T}\right]\gamma_{\rm A}-\left(\mu_{\rm A}\alpha_{\rm S}+\lambda_{\rm T}\tilde{E}_{\rm T}\tilde{\alpha}_{\rm T}\right)\Delta T. \tag{10.113}$$

The basis of the approximate method of dealing with nonuniform ply crack distributions is to assume that (10.110)–(10.113) derived for uniform distributions, are valid also for nonuniform distributions. A comparison of (10.106)–(10.109) and (10.110)–(10.113) then leads to the following relationships showing how the effective thermoelastic constants for the laminate as a whole having a nonuniform crack distribution may be calculated from those of the regions V_I, $I = 1,\ldots,n$ between neighbouring ply cracks whose values are estimated for the case of uniform crack distributions having crack separations $2L_I$, $I = 1,\ldots,n$,

$$\tilde{E}_{\mathrm{T}} = \sum_{I=1}^{n} \frac{L_I}{L} \tilde{E}_{\mathrm{T}}^I, \tag{10.114}$$

$$\tilde{\nu}_{\mathrm{t}} = \sum_{I=1}^{n} \frac{L_I}{L} \tilde{\nu}_{\mathrm{t}}^I, \tag{10.115}$$

$$\frac{\tilde{\nu}_{\mathrm{A}}}{\tilde{E}_{\mathrm{A}}} = \frac{1}{\tilde{E}_{\mathrm{T}}} \sum_{I=1}^{n} \frac{L_I}{L} \frac{\tilde{\nu}_{\mathrm{A}}^I}{\tilde{E}_{\mathrm{A}}^I} \tilde{E}_{\mathrm{T}}^I, \tag{10.116}$$

$$\lambda_{\mathrm{T}} \frac{\mu_{\mathrm{A}}}{E_{\mathrm{A}}} = \frac{1}{\tilde{E}_{\mathrm{T}}} \sum_{I=1}^{n} \frac{L_I}{L} \frac{\mu_{\mathrm{A}}^I}{E_{\mathrm{A}}^I} \tilde{E}_{\mathrm{T}}^I, \tag{10.117}$$

$$\tilde{\alpha}_{\mathrm{T}} = \frac{1}{\tilde{E}_{\mathrm{T}}} \sum_{I=1}^{n} \frac{L_I}{L} \tilde{E}_{\mathrm{T}}^I \tilde{\alpha}_{\mathrm{T}}^I, \tag{10.118}$$

$$\frac{1}{\tilde{E}_{\mathrm{A}}} = \left(\frac{\tilde{\nu}_{\mathrm{A}}}{\tilde{E}_{\mathrm{A}}}\right)^2 \tilde{E}_{\mathrm{T}} + \sum_{I=1}^{n} \frac{L_I}{L} \frac{1}{\tilde{E}_{\mathrm{A}}^I} \left(1 - (\tilde{\nu}_{\mathrm{A}}^I)^2 \frac{\tilde{E}_{\mathrm{T}}^I}{\tilde{E}_{\mathrm{A}}^I}\right), \tag{10.119}$$

$$\frac{1}{\tilde{E}_{\mathrm{t}}} = \frac{\tilde{\nu}_{\mathrm{t}}^2}{\tilde{E}_{\mathrm{T}}} + \sum_{I=1}^{n} \frac{L_I}{L} \left[\frac{1}{\tilde{E}_{\mathrm{t}}^I} - \frac{(\tilde{\nu}_{\mathrm{t}}^I)^2}{\tilde{E}_{\mathrm{T}}^I}\right], \tag{10.120}$$

$$\frac{\tilde{\nu}_{\mathrm{a}}}{\tilde{E}_{\mathrm{A}}} = -\frac{\tilde{\nu}_{\mathrm{A}}}{\tilde{E}_{\mathrm{A}}} \tilde{\nu}_{\mathrm{t}} + \sum_{I=1}^{n} \frac{L_I}{L} \frac{1}{\tilde{E}_{\mathrm{A}}^I} \left(\tilde{\nu}_{\mathrm{a}}^I + \tilde{\nu}_{\mathrm{A}}^I \tilde{\nu}_{\mathrm{t}}^I\right), \tag{10.121}$$

$$\tilde{\alpha}_{\mathrm{A}} = -\frac{\tilde{\nu}_{\mathrm{A}}}{\tilde{E}_{\mathrm{A}}} \tilde{E}_{\mathrm{T}} \tilde{\alpha}_{\mathrm{T}} + \sum_{I=1}^{n} \frac{L_I}{L} \left(\tilde{\alpha}_{\mathrm{A}}^I + \frac{\tilde{\nu}_{\mathrm{A}}^I}{\tilde{E}_{\mathrm{A}}^I} \tilde{E}_{\mathrm{T}}^I \tilde{\alpha}_{\mathrm{T}}^I\right), \tag{10.122}$$

$$\tilde{\alpha}_{\mathrm{t}} = -\tilde{\nu}_{\mathrm{t}} \tilde{\alpha}_{\mathrm{T}} + \sum_{I=1}^{n} \frac{L_I}{L} \left(\tilde{\alpha}_{\mathrm{t}}^I + \tilde{\nu}_{\mathrm{t}}^I \tilde{\alpha}_{\mathrm{T}}^I\right), \tag{10.123}$$

$$\lambda_{\mathrm{A}} + \lambda_{\mathrm{T}} \frac{\tilde{\nu}_{\mathrm{A}}}{\tilde{E}_{\mathrm{A}}} \tilde{E}_{\mathrm{T}} = \sum_{I=1}^{n} \frac{L_I}{L} \left(\lambda_{\mathrm{A}}^I + \lambda_{\mathrm{T}}^I \frac{\tilde{\nu}_{\mathrm{A}}^I}{\tilde{E}_{\mathrm{A}}^I} \tilde{E}_{\mathrm{T}}^I\right), \tag{10.124}$$

$$\lambda_{\mathrm{t}} + \lambda_{\mathrm{T}} \tilde{\nu}_{\mathrm{t}} = \sum_{I=1}^{n} \frac{L_I}{L} \left(\lambda_{\mathrm{t}}^I + \lambda_{\mathrm{T}}^I \tilde{\nu}_{\mathrm{t}}^I\right), \tag{10.125}$$

$$\mu_{\mathrm{A}} = -\left(\lambda_{\mathrm{T}}\right)^2 \tilde{E}_{\mathrm{T}} + \sum_{I=1}^{n} \frac{L_I}{L} \left(\mu_{\mathrm{A}}^I + \left(\lambda_{\mathrm{T}}^I\right)^2 \tilde{E}_{\mathrm{T}}^I\right), \tag{10.126}$$

$$\alpha_{\mathrm{S}} = -\frac{\lambda_{\mathrm{T}}}{\mu_{\mathrm{A}}} \tilde{E}_{\mathrm{T}} \tilde{\alpha}_{\mathrm{T}} + \frac{1}{\mu_{\mathrm{A}}} \sum_{I=1}^{n} \frac{L_I}{L} \left(\mu_{\mathrm{A}}^I \alpha_{\mathrm{S}}^I + \lambda_{\mathrm{T}}^I \tilde{E}_{\mathrm{T}}^I \tilde{\alpha}_{\mathrm{T}}^I\right). \tag{10.127}$$

Relations (10.114)–(10.127) are sufficient to determine values of the following parameters

$$\tilde{E}_{\mathrm{A}}, \tilde{E}_{\mathrm{T}}, \tilde{E}_{\mathrm{t}}, \tilde{\nu}_{\mathrm{A}}, \tilde{\nu}_{\mathrm{a}}, \tilde{\nu}_{\mathrm{t}}, \mu_{\mathrm{A}}, \tilde{\alpha}_{\mathrm{A}}, \tilde{\alpha}_{\mathrm{T}}, \tilde{\alpha}_{\mathrm{t}}, \alpha_{\mathrm{S}}, \frac{\lambda_{\mathrm{A}}}{\mu_{\mathrm{A}}}, \frac{\lambda_{\mathrm{T}}}{\mu_{\mathrm{A}}}, \frac{\lambda_{\mathrm{t}}}{\mu_{\mathrm{A}}}.$$

It is convenient to define the parameters k_I by

$$k_I = \frac{L_I}{L}\frac{\tilde{E}_T^I}{\tilde{E}_T}, \quad I = 1, \ldots, n. \tag{10.128}$$

It then follows from (10.114) that

$$\sum_{I=1}^{n} k_I = 1, \tag{10.129}$$

$$\frac{1}{\tilde{E}_T} = \sum_{I=1}^{n} \frac{k_I}{\tilde{E}_T^I}, \tag{10.130}$$

and from (10.115)–(10.118) that

$$\frac{\tilde{\nu}_t}{\tilde{E}_T} = \sum_{I=1}^{n} k_I \frac{\tilde{\nu}_t^I}{\tilde{E}_T^I}, \tag{10.131}$$

$$\frac{\tilde{\nu}_A}{\tilde{E}_A} = \sum_{I=1}^{n} k_I \frac{\tilde{\nu}_A^I}{\tilde{E}_A^I}, \tag{10.132}$$

$$\lambda_T = \sum_{I=1}^{n} k_I \lambda_T^I, \tag{10.133}$$

$$\tilde{\alpha}_T = \sum_{I=1}^{n} k_I \tilde{\alpha}_T^I. \tag{10.134}$$

Owing to (10.129), it follows from (10.130)–(10.132) and (10.134) that

$$\frac{1}{\tilde{E}_T} - \frac{1}{\tilde{E}_T^0} = \sum_{I=1}^{n} k_I \left(\frac{1}{\tilde{E}_T^I} - \frac{1}{\tilde{E}_T^0} \right), \tag{10.135}$$

$$\frac{\tilde{\nu}_t}{\tilde{E}_T} - \frac{\tilde{\nu}_t^0}{\tilde{E}_T^0} = \sum_{I=1}^{n} k_I \left(\frac{\tilde{\nu}_t^I}{\tilde{E}_T^I} - \frac{\tilde{\nu}_t^0}{\tilde{E}_T^0} \right), \tag{10.136}$$

$$\frac{\tilde{\nu}_A}{\tilde{E}_A} - \frac{\tilde{\nu}_A^0}{\tilde{E}_A^0} = \sum_{I=1}^{n} k_I \left(\frac{\tilde{\nu}_A^I}{\tilde{E}_A^I} - \frac{\tilde{\nu}_A^0}{\tilde{E}_A^0} \right), \tag{10.137}$$

$$\tilde{\alpha}_T - \tilde{\alpha}_T^0 = \sum_{I=1}^{n} k_I \left(\tilde{\alpha}_T^I - \tilde{\alpha}_T^0 \right). \tag{10.138}$$

Use can now be made of the interrelationships (10.49) and (10.62) applied to each region V_I, $I = 1, \ldots, n$, and to the nonuniformly cracked laminate as a whole, to derive the following expressions:

$$\frac{1}{\tilde{E}_A} - \frac{1}{\tilde{E}_A^0} = \sum_{I=1}^{n} k_I \left(\frac{1}{\tilde{E}_A^I} - \frac{1}{\tilde{E}_A^0} \right), \tag{10.139}$$

$$\frac{1}{\tilde{E}_t} - \frac{1}{\tilde{E}_t^0} = \sum_{I=1}^{n} k_I \left(\frac{1}{\tilde{E}_t^I} - \frac{1}{\tilde{E}_t^0} \right), \tag{10.140}$$

$$\frac{\tilde{\nu}_a}{\tilde{E}_A} - \frac{\tilde{\nu}_a^0}{\tilde{E}_A^0} = \sum_{I=1}^{n} k_I \left(\frac{\tilde{\nu}_a^I}{\tilde{E}_A^I} - \frac{\tilde{\nu}_a^0}{\tilde{E}_A^0} \right), \tag{10.141}$$

$$\tilde{\alpha}_{\rm A} - \tilde{\alpha}_{\rm A}^0 = \sum_{I=1}^{n} k_I \left(\tilde{\alpha}_{\rm A}^I - \tilde{\alpha}_{\rm A}^0 \right), \tag{10.142}$$

$$\tilde{\alpha}_{\rm t} - \tilde{\alpha}_{\rm t}^0 = \sum_{I=1}^{n} k_I \left(\tilde{\alpha}_{\rm t}^I - \tilde{\alpha}_{\rm t}^0 \right). \tag{10.143}$$

It is useful to show how the relations of the type (10.139)–(10.143) are derived. From (10.49),

$$\frac{\tilde{\nu}_{\rm t}}{\tilde{E}_{\rm T}} - \frac{\tilde{\nu}_{\rm t}^0}{\tilde{E}_{\rm T}^0} = k \left(\frac{\tilde{\nu}_{\rm a}}{\tilde{E}_{\rm A}} - \frac{\tilde{\nu}_{\rm a}^0}{\tilde{E}_{\rm A}^0} \right) \quad \text{and} \quad \frac{\tilde{\nu}_{\rm t}^I}{\tilde{E}_{\rm T}^I} - \frac{\tilde{\nu}_{\rm t}^0}{\tilde{E}_{\rm T}^0} = k \left(\frac{\tilde{\nu}_{\rm a}^I}{\tilde{E}_{\rm A}^I} - \frac{\tilde{\nu}_{\rm a}^0}{\tilde{E}_{\rm A}^0} \right). \tag{10.144}$$

Substitution into (10.136) then leads to relation (10.141). From (10.62),

$$\frac{1}{\tilde{E}_{\rm t}} - \frac{1}{\tilde{E}_{\rm t}^0} = k' \left(\frac{\tilde{\nu}_{\rm a}^0}{\tilde{E}_{\rm A}^0} - \frac{\tilde{\nu}_{\rm a}}{\tilde{E}_{\rm A}} \right) \quad \text{and} \quad \frac{1}{\tilde{E}_{\rm t}^I} - \frac{1}{\tilde{E}_{\rm t}^0} = k' \left(\frac{\tilde{\nu}_{\rm a}^0}{\tilde{E}_{\rm A}^0} - \frac{\tilde{\nu}_{\rm a}^I}{\tilde{E}_{\rm A}^I} \right). \tag{10.145}$$

It then follows from (10.141) that relation (10.139) is derived. The remaining relations are similarly derived. On using (10.129) the following results are obtained from (10.139)–(10.143) that are much simpler than those given by (10.119)–(10.123)

$$\frac{1}{\tilde{E}_{\rm A}} = \sum_{I=1}^{n} \frac{k_I}{\tilde{E}_{\rm A}^I}, \tag{10.146}$$

$$\frac{1}{\tilde{E}_{\rm t}} = \sum_{I=1}^{n} \frac{k_I}{\tilde{E}_{\rm t}^I}, \tag{10.147}$$

$$\frac{\tilde{\nu}_{\rm a}}{\tilde{E}_{\rm A}} = \sum_{I=1}^{n} k_I \frac{\tilde{\nu}_{\rm a}^I}{\tilde{E}_{\rm A}^I}, \tag{10.148}$$

$$\tilde{\alpha}_{\rm A} = \sum_{I=1}^{n} k_I \tilde{\alpha}_{\rm A}^I, \tag{10.149}$$

$$\tilde{\alpha}_{\rm t} = \sum_{I=1}^{n} k_I \tilde{\alpha}_{\rm t}^I. \tag{10.150}$$

The equivalence of relations (10.119)–(10.123) with relations (10.146)–(10.150) has been confirmed numerically (i.e. predictions agree to eight or more significant figures when using double length arithmetic).

It follows from (10.124), (10.125) and (10.128) that

$$\lambda_{\rm A} \frac{1}{\tilde{E}_{\rm T}} + \lambda_{\rm T} \frac{\tilde{\nu}_{\rm A}}{\tilde{E}_{\rm A}} = \sum_{I=1}^{n} k_I \left(\frac{\lambda_{\rm A}^I}{\tilde{E}_{\rm T}^I} + \lambda_{\rm T}^I \frac{\tilde{\nu}_{\rm A}^I}{\tilde{E}_{\rm A}^I} \right), \tag{10.151}$$

$$\lambda_t \frac{1}{\tilde{E}_{\rm T}} + \lambda_{\rm T} \frac{\tilde{\nu}_{\rm t}}{\tilde{E}_{\rm T}} = \sum_{I=1}^{n} k_I \left(\frac{\lambda_{\rm t}^I}{\tilde{E}_{\rm T}^I} + \lambda_{\rm T}^I \frac{\tilde{\nu}_{\rm t}^I}{\tilde{E}_{\rm T}^I} \right). \tag{10.152}$$

As the values of $\lambda_{\rm T}$, $\tilde{E}_{\rm T}$, $\tilde{\nu}_{\rm A} / \tilde{E}_{\rm A}$ and $\tilde{\nu}_{\rm t} / \tilde{E}_{\rm T}$ are known from relations (10.130)–(10.133), relations (10.151) and (10.152) can be used to determine values of $\lambda_{\rm A}$ and $\lambda_{\rm t}$, respectively. The value of $\mu_{\rm A}$ may be determined using relation (10.126), which may be written in the form

$$\frac{\mu_{\rm A}}{\tilde{E}_{\rm T}} + \left(\lambda_{\rm T}\right)^2 = \sum_{I=1}^{n} k_I \left[\frac{\mu_{\rm A}^I}{\tilde{E}_{\rm T}^I} + \left(\lambda_{\rm T}^I\right)^2\right]. \tag{10.153}$$

The value of $\alpha_{\rm S}$ may be obtained from result (10.127), which can be written as

$$\alpha_{\rm S} = \frac{\tilde{E}_{\rm T}}{\mu_{\rm A}}\left[-\lambda_{\rm T}\tilde{\alpha}_{\rm T} + \sum_{I=1}^{n} k_I \left(\frac{\mu_{\rm A}^I}{\tilde{E}_{\rm T}^I}\alpha_{\rm S}^I + \lambda_{\rm T}^I\tilde{\alpha}_{\rm T}^I\right)\right]. \tag{10.154}$$

It follows from (10.15)–(10.19) that

$$\frac{1}{E_{\rm A}} = \frac{1}{\tilde{E}_{\rm A}} + \frac{\lambda_{\rm A}^2}{\mu_{\rm A}}, \tag{10.155}$$

$$\frac{1}{E_{\rm T}} = \frac{1}{\tilde{E}_{\rm T}} + \frac{\lambda_{\rm T}^2}{\mu_{\rm A}}, \tag{10.156}$$

$$\frac{1}{E_{\rm t}} = \frac{1}{\tilde{E}_{\rm t}} + \frac{\lambda_{\rm t}^2}{\mu_{\rm A}}, \tag{10.157}$$

$$\frac{\nu_{\rm A}}{E_{\rm A}} = \frac{\tilde{\nu}_{\rm A}}{\tilde{E}_{\rm A}} - \frac{\lambda_{\rm A}\lambda_{\rm T}}{\mu_{\rm A}}, \tag{10.158}$$

$$\frac{\nu_{\rm a}}{E_{\rm A}} = \frac{\tilde{\nu}_{\rm a}}{\tilde{E}_{\rm A}} - \frac{\lambda_{\rm t}\lambda_{\rm A}}{\mu_{\rm A}}, \tag{10.159}$$

$$\frac{\nu_{\rm t}}{E_{\rm T}} = \frac{\tilde{\nu}_{\rm t}}{\tilde{E}_{\rm T}} - \frac{\lambda_{\rm t}\lambda_{\rm T}}{\mu_{\rm A}}. \tag{10.160}$$

Thus, it is now possible to determine the values of the thermoelastic constants

$$E_{\rm A}, E_{\rm T}, E_{\rm t}, \nu_{\rm A}, \nu_{\rm a}, \nu_{\rm t}, \lambda_{\rm A}, \lambda_{\rm T}, \lambda_{\rm t}.$$

The through-thickness, axial and transverse thermal expansion coefficients may then be obtained from relations (10.20)–(10.22) so that

$$\alpha_{\rm A} = \tilde{\alpha}_{\rm A} - \lambda_{\rm A}\alpha_{\rm S}, \tag{10.161}$$

$$\alpha_{\rm T} = \tilde{\alpha}_{\rm T} - \lambda_{\rm T}\alpha_{\rm S}, \tag{10.162}$$

$$\alpha_{\rm t} = \tilde{\alpha}_{\rm t} - \lambda_{\rm t}\alpha_{\rm S}. \tag{10.163}$$

with the result that all the thermoelastic constants for nonuniformly cracked laminates appearing in the stress-strain relations (10.7)–(10.10) have now been determined.

It is useful to assemble all the key relations that are required to estimate the effective thermoelastic constants of a nonuniformly cracked general symmetric laminate. First, the values of the following summations are required:

$$S_1 = \sum_{I=1}^{n} \frac{L_I}{L}\tilde{E}_{\rm T}^I, \quad \text{so that} \quad k_I = \frac{L_I}{L}\frac{\tilde{E}_{\rm T}^I}{S_1}, \quad I = 1,\ldots,n, \tag{10.164}$$

$$S_2 = \sum_{I=1}^{n} k_I \frac{\tilde{\nu}_{\rm t}^I}{\tilde{E}_{\rm T}^I}, \tag{10.165}$$

$$S_3 = \sum_{I=1}^{n} k_I \frac{\tilde{\nu}_A^I}{\tilde{E}_A^I}, \tag{10.166}$$

$$S_4 = \sum_{I=1}^{n} k_I \lambda_T^I, \tag{10.167}$$

$$S_5 = \sum_{I=1}^{n} k_I \tilde{\alpha}_T^I, \tag{10.168}$$

$$S_6 = \sum_{I=1}^{n} \frac{k_I}{\tilde{E}_t^I}, \tag{10.169}$$

$$S_7 = \sum_{I=1}^{n} \frac{k_I}{\tilde{E}_A^I}, \tag{10.170}$$

$$S_8 = \sum_{I=1}^{n} k_I \frac{\tilde{\nu}_a^I}{\tilde{E}_A^I}, \tag{10.171}$$

$$S_9 = \sum_{I=1}^{n} k_I \tilde{\alpha}_t^I, \tag{10.172}$$

$$S_{10} = \sum_{I=1}^{n} k_I \tilde{\alpha}_A^I, \tag{10.173}$$

$$S_{11} = \sum_{I=1}^{n} k_I \left[\frac{\mu_A^I}{\tilde{E}_T^I} + \left(\lambda_T^I\right)^2 \right], \tag{10.174}$$

$$S_{12} = \sum_{I=1}^{n} k_I \left(\frac{\lambda_A^I}{\tilde{E}_T^I} + \lambda_T^I \frac{\tilde{\nu}_A^I}{\tilde{E}_A^I} \right), \tag{10.175}$$

$$S_{13} = \sum_{I=1}^{n} k_I \left(\frac{\lambda_t^I}{\tilde{E}_T^I} + \lambda_T^I \frac{\tilde{\nu}_t^I}{\tilde{E}_T^I} \right), \tag{10.176}$$

$$S_{14} = \sum_{I=1}^{n} k_I \left(\frac{\mu_A^I}{\tilde{E}_T^I} \alpha_S^I + \lambda_T^I \tilde{\alpha}_T^I \right), \tag{10.177}$$

such that

$$\begin{gathered}
\tilde{E}_T = S_1, \quad \frac{\tilde{\nu}_t}{\tilde{E}_T} = S_2, \quad \frac{\tilde{\nu}_A}{\tilde{E}_A} = S_3, \quad \lambda_T = S_4, \quad \tilde{\alpha}_T = S_5, \\
\frac{1}{\tilde{E}_t} = S_6, \quad \frac{1}{\tilde{E}_A} = S_7, \quad \frac{\tilde{\nu}_a}{\tilde{E}_A} = S_8, \quad \tilde{\alpha}_t = S_9, \quad \tilde{\alpha}_A = S_{10}, \\
\mu_A = \left(S_{11} - S_4^2\right) S_1, \quad \lambda_A = \left(S_{12} - S_3 S_4\right) S_1, \quad \lambda_t = \left(S_{13} - S_2 S_4\right) S_1, \\
\alpha_S = \frac{S_1}{\mu_A} \left(S_{14} - S_4 S_5\right).
\end{gathered} \tag{10.178}$$

It then follows from (10.155)–(10.163) that

$$\frac{1}{E_A} = \frac{1}{\tilde{E}_A} + \left(S_{12} - S_3 S_4\right)^2 S_1^2 \frac{1}{\mu_A}, \tag{10.179}$$

$$\frac{1}{E_{\mathrm{T}}} = \frac{1}{\tilde{E}_{\mathrm{T}}} + S_4^2 \frac{1}{\mu_{\mathrm{A}}}, \tag{10.180}$$

$$\frac{1}{E_{\mathrm{t}}} = S_6 + \frac{\lambda_{\mathrm{t}}^2}{\mu_{\mathrm{A}}}, \tag{10.181}$$

$$\frac{\nu_{\mathrm{A}}}{E_{\mathrm{A}}} = \frac{\tilde{\nu}_{\mathrm{A}}}{\tilde{E}_{\mathrm{A}}} - \left(S_{12} - S_3 S_4\right) S_1 S_4 \frac{1}{\mu_{\mathrm{A}}}, \tag{10.182}$$

$$\frac{\nu_{\mathrm{a}}}{E_{\mathrm{A}}} = \frac{\tilde{\nu}_{\mathrm{a}}}{\tilde{E}_{\mathrm{A}}} - \frac{\lambda_{\mathrm{t}}\lambda_{\mathrm{A}}}{\mu_{\mathrm{A}}}, \tag{10.183}$$

$$\frac{\nu_{\mathrm{t}}}{E_{\mathrm{T}}} = \frac{\tilde{\nu}_{\mathrm{t}}}{\tilde{E}_{\mathrm{T}}} - \left(S_{13} - S_2 S_4\right) S_1 S_4 \frac{1}{\mu_{\mathrm{A}}}, \tag{10.184}$$

$$\alpha_{\mathrm{A}} = \tilde{\alpha}_{\mathrm{A}} - \left(S_{12} - S_3 S_4\right) S_1 \alpha_{\mathrm{S}}, \tag{10.185}$$

$$\alpha_{\mathrm{T}} = \tilde{\alpha}_{\mathrm{T}} - S_4 \alpha_{\mathrm{S}}, \tag{10.186}$$

$$\alpha_{\mathrm{t}} = \tilde{\alpha}_{\mathrm{t}} - \left(S_{13} - S_2 S_4\right) S_1 \alpha_{\mathrm{S}}. \tag{10.187}$$

Results (10.164)–(10.187) enable the effective thermoelastic constants of a nonuniformly cracked laminate to be estimated approximately from corresponding values that assume uniform distributions of ply cracks for various ply crack separations (i.e. densities).

10.9 Progressive Ply Cracking

It is useful to discuss progressive cracking in a length $2L$ of laminate where the damage is defined by $\{L_1, L_2, \ldots, L_n\}$ where n is the number of cracks in length $2L$ of the laminate and where $2L_I$, $I = 1, \ldots, n$, denote the distances between neighbouring cracked planes. For this case, the energy absorption per unit volume for length $2L$ of laminate is given by

$$\Gamma = \frac{h^{(90)}}{hL} \sum_{j=1}^{M} \delta_j^{(90)} \Gamma_j^{(90)}, \tag{10.188}$$

where $2h^{(90)}$ is the total thickness of all 90° plies in the laminate having total thickness $2h$, and where M is the number of potential cracking sites in 90° plies which are ordered in a regular way, e.g. from top to bottom in the plies which are taken in order from the centre of the laminate to the outside, symmetry about the midplane of the laminate being assumed. The quantity $2\Gamma_j^{(90)}$ is the fracture energy for the jth potential cracking site of the 90° plies. For conditions of general in-plane loading, as ply cracking will occur under conditions of mixed mode loading, the fracture energies $2\Gamma_j^{(90)}$ take account of such mixed mode effects, and will for many laminates differ from the mode I fracture energies normally measured. The parameters $\delta_j^{(90)}$ describe the crack pattern in the laminate such that

$$\delta_j^{(90)} = \begin{cases} 0 \text{ if } j\text{th site of the } 90^\circ \text{ ply is uncracked,} \\ 1 \text{ if } j\text{th site of the } 90^\circ \text{ ply is cracked.} \end{cases} \tag{10.189}$$

It follows from this definition that the number of ply cracks in length $2L$ of laminate is given by

$$n = \sum_{j=1}^{M} \delta_j^{(90)}. \tag{10.190}$$

In simulations where statistical scatter is to be included, the fracture energies $2\Gamma_j^{(90)}$ are selected at random from a normal distribution having mean value a and variance b. This can be achieved numerically (as when using PREDICT [2]), using the Box–Muller method [3] where two independent random numbers z_1 and z_2 lying in the range $0 \leq z_1, z_2 \leq 1$ are provided by a random number generating subroutine and then applied to the following formula that generates a random values v belonging to a normal distribution

$$v = a + b\sqrt{2\ln(1/z_1)}\cos(2\pi z_2).$$

When $b = 0$, all fracture energies selected have the same value. In simulations the value of M is recommended to lie in the range 128–1024 (i.e. 2^7–2^{10}) where the use of powers of 2 enable, in principle but not numerically, the precise progressive doubling of crack densities during loading for cases when the fracture energies of all potential fracture sites have the same value.

For progressive cracking consider the two damage states $\{L_i\}$ and $\{L_j\}$, where the superscript i and j attached to stresses, strains and thermoelastic constants will be used to denote the number of laminate segments defined by the ply crack planes, and thus which damage state is being considered. It is easily shown that the interrelationships (10.49) and (10.62), also valid for nonuniform crack spacings, may be generalised to the forms

$$\frac{\dfrac{\tilde{\nu}_t^j}{\tilde{E}_T^j} - \dfrac{\tilde{\nu}_t^i}{\tilde{E}_T^i}}{\dfrac{\tilde{\nu}_a^j}{\tilde{E}_A^j} - \dfrac{\tilde{\nu}_a^i}{\tilde{E}_A^i}} = \frac{\dfrac{\tilde{\nu}_A^j}{\tilde{E}_A^j} - \dfrac{\tilde{\nu}_A^i}{\tilde{E}_A^i}}{\dfrac{1}{\tilde{E}_A^i} - \dfrac{1}{\tilde{E}_A^j}} = \frac{\dfrac{1}{\tilde{E}_T^i} - \dfrac{1}{\tilde{E}_T^j}}{\dfrac{\tilde{\nu}_A^j}{\tilde{E}_A^j} - \dfrac{\tilde{\nu}_A^i}{\tilde{E}_A^i}} = \frac{\tilde{\alpha}_T^i - \tilde{\alpha}_T^j}{\tilde{\alpha}_A^i - \tilde{\alpha}_A^j} = k, \tag{10.191}$$

$$\frac{\dfrac{1}{\tilde{E}_t^i} - \dfrac{1}{\tilde{E}_t^j}}{\dfrac{\tilde{\nu}_a^j}{\tilde{E}_A^j} - \dfrac{\tilde{\nu}_a^i}{\tilde{E}_A^i}} = \frac{\dfrac{\tilde{\nu}_a^j}{\tilde{E}_A^j} - \dfrac{\tilde{\nu}_a^i}{\tilde{E}_A^i}}{\dfrac{1}{\tilde{E}_A^i} - \dfrac{1}{\tilde{E}_A^j}} = \frac{\dfrac{\tilde{\nu}_t^j}{\tilde{E}_T^j} - \dfrac{\tilde{\nu}_t^i}{\tilde{E}_T^i}}{\dfrac{\tilde{\nu}_A^j}{\tilde{E}_A^j} - \dfrac{\tilde{\nu}_A^i}{\tilde{E}_A^i}} = \frac{\tilde{\alpha}_t^i - \tilde{\alpha}_t^j}{\tilde{\alpha}_A^i - \tilde{\alpha}_A^j} = k'. \tag{10.192}$$

On using (10.65), (10.77), (10.78), (10.96), (10.191) and (10.192) it can be shown that the criterion for progressive cracking may be written in the compact form

$$\left(\frac{1}{\tilde{E}_A^j} - \frac{1}{\tilde{E}_A^i}\right)\left(s - \overline{\sigma}_A^c\right)^2 + \mu_A^j\left(\gamma_A^j\right)^2 - \mu_A^i\left(\gamma_A^i\right)^2 - 2\left(\Gamma_j - \Gamma_i\right) > 0, \tag{10.193}$$

where γ_A^i is the in-plane shear strain for the damage state i. This result is consistent with the initiation condition (10.98). It must be emphasised that the result (10.193) takes full account of anisotropy and thermal residual stresses, and is valid for combined biaxial and shear loading conditions. The result is exact and simple in form. It is concluded that stress-transfer models are needed only to estimate the effective reduced axial Young's modulus and axial shear modulus of a cracked laminate. All other aspects of crack formation can be dealt with using the general framework described here.

It is also emphasised that the criterion (10.193) is valid *only* if new cracks form for fixed values of $\sigma_A, \sigma_T, \sigma_t, \tau_A$ and ΔT. The criteria (10.98) and (10.193), derived for initiation and progressive ply cracking, respectively, would have rather complicated forms if expressed directly in terms of $\sigma_A, \sigma_T, \sigma_t, \tau_A$ and ΔT. The left-hand sides of (10.98) and (10.193) would lead to quadratic forms in the variables $\sigma_A, \sigma_T, \sigma_t, \tau_A$ and ΔT that are difficult to solve in general. To simplify the problem, the effective applied stresses are now assumed to be such that during loading

$$\sigma_T = \hat{A}\,\sigma_A, \quad \sigma_t = \hat{B}\,\sigma_A, \quad \tau_A = \hat{C}\,\sigma_A, \tag{10.194}$$

where $\hat{A}$, $\hat{B}$ and $\hat{C}$ are specified values.

Consider the progressive ply cracking criterion (10.193). On substituting in terms of $\sigma_A, \sigma_T, \sigma_t, \tau_A$ and ΔT for the shear strains and the effective stress s using (10.10) and (10.95), the criterion (10.193) may be expressed in the following quadratic form involving only the axial applied stress σ_A:

$$f(\sigma_A) \equiv p\sigma_A^2 + q\sigma_A + r > 0, \tag{10.195}$$

where the coefficients p, q and r (dependent on the loading parameters $\hat{A}, \hat{B}, \hat{C}$ and ΔT) can be defined from the left-hand side of (10.193), which is denoted by $f(\sigma_A)$, using the following relations

$$p = \frac{1}{2}\left[f(1) + f(-1)\right] - f(0), \tag{10.196}$$

$$q = \frac{1}{2}\left[f(1) - f(-1)\right], \tag{10.197}$$

$$r = f(0). \tag{10.198}$$

Let σ^+ and σ^- be the roots of the quadratic equation

$$p\sigma_A^2 + q\sigma_A + r = 0. \tag{10.199}$$

If $q^2 < 4pr$, then the values given by (10.199) are complex and clearly ply cracking cannot occur. If $q^2 \geq 4pr$, then ply cracking can be predicted and results to date indicate that $p > 0$. Just one positive root occurs whenever $r < 0$. Extensive predictions to date have been such that there is only one positive root σ^+ that is the required solution, and ply cracking occurs only if $\sigma_A > \sigma^+$. The results given by (10.195) can also be applied to the case of ply crack initiation governed by (10.98) instead of (10.193).

References

1. McCartney, L.N. (2000). Model to predict effects of triaxial loading on ply cracking in general symmetric laminates. *Composites Science and Technology* 60: 2255–2279. (See errata in (2002) *Composites Science and Technology* 62: 1273–1274).
2. Software system 'PREDICT' which is a specific module of CoDA (see http://www.npl.co.uk/npl/cmmt/cog/coda.html).
3. Box, G.E.P. and Muller, M.E. (1958). A note on the generation of random normal deviates. *Annals of Mathematical Statistics* 29: 610–611.

11

Ply Cracking in Cross-ply Laminates Subject to Biaxial Bending

Overview: This chapter is concerned with a general theoretical framework that enables a thorough understanding of how energy-based methods can be used to predict first ply cracking and the progressive formation of uniform distributions of ply cracks when a unsymmetric cross-ply laminate is subject to general in-plane loading, biaxial bending and through-thickness loading in the presence of thermal residual stresses. Numerous interrelationships between the thermoelastic constants of damaged laminates are derived by considering the closure of ply cracks for special loading situations. These relationships enable rather complex expressions for the energy per unit volume of a damaged laminate to be simplified to a very useful form involving an effective applied stress that captures the effect on damage formation of three principal loading parameters, the effective axial, transverse and through-thickness stresses. The general analysis is then applied to some simple loading cases.

11.1 Introduction

Predicting the initiation and progressive formation of ply cracks in laminated composites subject to the complex loading that is expected in composite components is an important technological objective. This goal has been achieved [1,2] for the case of general symmetric laminates that are subject to general in-plane loading. A practically important deformation mode, that has to date been rather neglected in the literature, concerns the prediction of ply crack formation in laminates that are subject to out-of-plane bending. A precursor to the development of such prediction methods is the need for a model to predict the effective thermoelastic constants for a cracked laminate that is subject to combined biaxial loading and out-of-plane bending. Such a model is developed in this chapter, although a brief description of the work (valid for plane strain conditions only) has appeared elsewhere [3].

Li et al. [4] have developed a finite strip analysis of cracked laminates that is capable of handling multiple-ply laminates and out-of-plane bending deformation modes. Their technique is based on a series representation for the displacement field combined with the use of minimisation principles applied to the strain energy. Their analysis does not take account of the effects of thermal residual stresses.

Properties for Design of Composite Structures: Theory and Implementation Using Software, First Edition. Neil McCartney.

Companion Website: www.wiley.com/go/mccartney/properties

In Chapter 10, it was shown how a great deal of insight and fundamental understanding could be obtained when considering the effects of ply crack formation in general symmetric laminates subject to general in-plane loading, through-thickness loading and thermal residual stresses. The objective here is to develop a similar approach when considering nonsymmetrical cross-ply laminates subject to biaxial in-plane loading, combined with biaxial out-of-plane bending loads, in the presence of thermal residual stresses.

11.2 Geometry and Basic Equations

While a composite laminate is normally considered to comprise a set of plies perfectly bonded together, it is useful to introduce the concept of perfectly bonded composite layers where there may be more than one layer allocated to each ply of the laminate. The refinement of plies into such layers will enable through-thickness stress, displacement and temperature variations in each ply to be investigated, leading to more accurate modelling of the stress and displacement fields. A laminate consisting of $n = N+1$ perfectly bonded layers of 0° and 90° plies, each of length $2L$, is considered within a Cartesian coordinate system such that the x_1-axis is in the axial direction whereas the x_2-axis is in the in-plane transverse direction, and the x_3-axis is in the through-thickness direction, as shown in Figure 11.1. In Chapters 18 and 19, the symbol N is used to denote the number of interfaces where unknown stress-transfer functions are defined and need to be determined. The origin is selected to lie at the midpoint of the upper surface of the laminate, i.e. it lies in the first of the layers of the laminate that are labelled $i=1,\ldots,n$ from top to bottom. The layer interfaces are located on the planes $x_3 = x_3^{(i)}, i=1,\ldots,N$ where $x_3^{(0)} = 0$ and $x_3^{(n)} = h$, the total thickness of the laminate. The thickness of the ith layer is denoted by $h_i = x_3^{(i)} - x_3^{(i-1)}$.

The modelling assumes that the laminate composite is in mechanical equilibrium such that the following equilibrium equations are satisfied for each layer in the laminate (see (2.120)–(2.122))

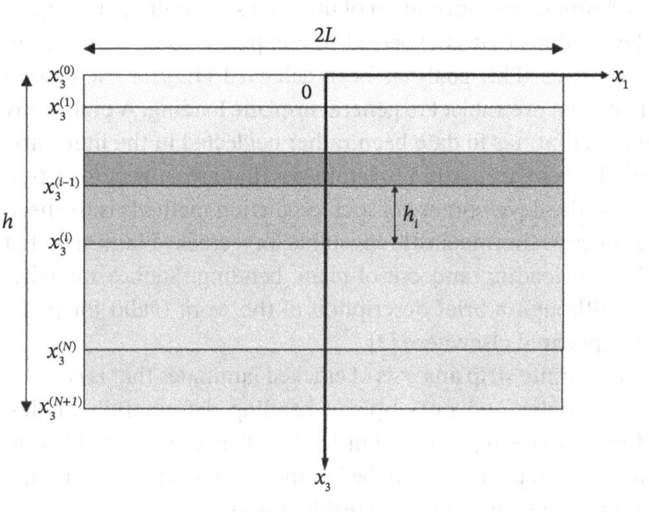

Figure 11.1 Geometry of a multilayered laminate showing a crack.

$$\frac{\partial \sigma_{11}}{\partial x_1}+\frac{\partial \sigma_{12}}{\partial x_2}+\frac{\partial \sigma_{13}}{\partial x_3}=0, \tag{11.1}$$

$$\frac{\partial \sigma_{12}}{\partial x_1}+\frac{\partial \sigma_{22}}{\partial x_2}+\frac{\partial \sigma_{23}}{\partial x_3}=0, \tag{11.2}$$

$$\frac{\partial \sigma_{13}}{\partial x_1}+\frac{\partial \sigma_{23}}{\partial x_2}+\frac{\partial \sigma_{33}}{\partial x_3}=0, \tag{11.3}$$

where the stress tensor is symmetric such that

$$\sigma_{12}=\sigma_{21}, \sigma_{13}=\sigma_{31}, \sigma_{23}=\sigma_{32}. \tag{11.4}$$

The displacements within the ith layer in the x_1, x_2 and x_3 directions are denoted by $u_1^{(i)}$, $u_2^{(i)}$ and $u_3^{(i)}$, respectively. The strains in the ith layer are then given by (see (2.136))

$$\varepsilon_{11}^{(i)}=\frac{\partial u_1^{(i)}}{\partial x_1}, \qquad \varepsilon_{22}^{(i)}=\frac{\partial u_2^{(i)}}{\partial x_2}, \qquad \varepsilon_{33}^{(i)}=\frac{\partial u_3^{(i)}}{\partial x_3}, \tag{11.5}$$

$$\begin{aligned}\varepsilon_{12}^{(i)}=\varepsilon_{21}^{(i)}=\frac{1}{2}\left(\frac{\partial u_1^{(i)}}{\partial x_2}+\frac{\partial u_2^{(i)}}{\partial x_1}\right), \varepsilon_{13}^{(i)}=\varepsilon_{31}^{(i)}=\frac{1}{2}\left(\frac{\partial u_1^{(i)}}{\partial x_3}+\frac{\partial u_3^{(i)}}{\partial x_1}\right),\\ \varepsilon_{23}^{(i)}=\varepsilon_{32}^{(i)}=\frac{1}{2}\left(\frac{\partial u_2^{(i)}}{\partial x_3}+\frac{\partial u_3^{(i)}}{\partial x_2}\right).\end{aligned} \tag{11.6}$$

The stress-strain relations for the ith layer are assumed to be of the form (see (2.196))

$$\varepsilon_{11}^{(i)}=\frac{\sigma_{11}^{(i)}}{E_{\mathrm{A}}^{(i)}}-\frac{\nu_{\mathrm{A}}^{(i)}}{E_{\mathrm{A}}^{(i)}}\sigma_{22}^{(i)}-\frac{\nu_{\mathrm{a}}^{(i)}}{E_{\mathrm{A}}^{(i)}}\sigma_{33}^{(i)}+\alpha_{\mathrm{A}}^{(i)}\Delta T, \tag{11.7}$$

$$\varepsilon_{22}^{(i)}=-\frac{\nu_{\mathrm{A}}^{(i)}}{E_{\mathrm{A}}^{(i)}}\sigma_{11}^{(i)}+\frac{\sigma_{22}^{(i)}}{E_{\mathrm{T}}^{(i)}}-\frac{\nu_{\mathrm{t}}^{(i)}}{E_{\mathrm{T}}^{(i)}}\sigma_{33}^{(i)}+\alpha_{\mathrm{T}}^{(i)}\Delta T, \tag{11.8}$$

$$\varepsilon_{33}^{(i)}=-\frac{\nu_{\mathrm{a}}^{(i)}}{E_{\mathrm{A}}^{(i)}}\sigma_{11}^{(i)}-\frac{\nu_{\mathrm{t}}^{(i)}}{E_{\mathrm{T}}^{(i)}}\sigma_{22}^{(i)}+\frac{\sigma_{33}^{(i)}}{E_{\mathrm{t}}^{(i)}}+\alpha_{\mathrm{t}}^{(i)}\Delta T, \tag{11.9}$$

$$2\varepsilon_{12}^{(i)}=\frac{\sigma_{12}^{(i)}}{\mu_{\mathrm{A}}^{(i)}}, \tag{11.10}$$

$$2\varepsilon_{13}^{(i)}=\frac{\sigma_{13}^{(i)}}{\mu_{\mathrm{a}}^{(i)}}, \tag{11.11}$$

$$2\varepsilon_{23}^{(i)}=\frac{\sigma_{23}^{(i)}}{\mu_{\mathrm{t}}^{(i)}}. \tag{11.12}$$

The parameters $E^{(i)}$, $\nu^{(i)}$, $\mu^{(i)}$ and $\alpha^{(i)}$ denote Young's modulus, Poisson's ratio, shear modulus and thermal expansion coefficient in the ith layer of the laminate. The subscripts 'A' and 'a' refer to axial thermoelastic constants involving in-plane and out-of-plane stresses and deformations, respectively. The subscripts 'T' and 't' refer to transverse constants for in-plane and out-of-plane stresses and deformations, respectively. The parameter ΔT is the difference between the current temperature (assumed uniform throughout the laminate) and the stress-free temperature of the laminate.

11.3 In-plane Transverse Loading and Bending

To allow for the anticlastic bending associated with thick laminates, bending is considered to occur in both the x_1–x_3 plane and in the x_2–x_3 plane. It is assumed that edge effects associated with the laminate external surfaces on $x_2 = \pm W$ are neglected, implying that the ensuing analysis will apply in all regions of the laminate except near these edges. The bending in the x_1–x_3 plane will be analysed in detail; bending in the x_2–x_3 plane is modelled by assuming that $u_1^{(i)}$ is independent of x_2, that $u_2^{(i)}$ is linear in x_3 across the whole laminate, and that there is no shear deformation in the x_2–x_3 plane throughout the laminate, i.e.

$$\left.\begin{aligned} &u_2^{(i)} \equiv (\bar{\varepsilon}_T + \hat{\varepsilon}_T x_3)x_2, \\ &\frac{\partial u_1^{(i)}}{\partial x_2} \equiv 0, \\ &\varepsilon_{23}^{(i)} \equiv 0, \end{aligned}\right\}, \quad i = 1,\ldots,N+1. \tag{11.13}$$

It is assumed for the moment that the transverse strain parameters $\bar{\varepsilon}_T$ and $\hat{\varepsilon}_T$ are given. They will, however, be regarded as in-plane transverse loading parameters for both undamaged and damaged laminates. From the assumptions (11.13), it follows using (11.5) and (11.6) that

$$\varepsilon_{22}^{(i)} \equiv \bar{\varepsilon}_T + \hat{\varepsilon}_T x_3, \ \varepsilon_{12}^{(i)} \equiv 0. \tag{11.14}$$

Relations (11.12), (11.10), $(11.13)_3$ and $(11.14)_2$ lead to

$$\sigma_{12}^{(i)} \equiv 0, \sigma_{23}^{(i)} \equiv 0, \tag{11.15}$$

so that $\sigma_{13}^{(i)}$ and $\varepsilon_{13}^{(i)}$ are the only nonzero shear stress and shear strain components. On using (11.8) and $(11.14)_1$

$$\sigma_{22}^{(i)} = \nu_t^{(i)}\sigma_{33}^{(i)} + \nu_A^{(i)}\frac{E_T^{(i)}}{E_A^{(i)}}\sigma_{11}^{(i)} - E_T^{(i)}\alpha_T^{(i)}\Delta T + E_T^{(i)}(\bar{\varepsilon}_T + \hat{\varepsilon}_T x_3), \tag{11.16}$$

thus satisfying (11.2) provided that σ_{11} and σ_{33} are independent of x_2 for all plies, implying that $\sigma_{11}^{(i)} = \sigma_{11}^{(i)}(x_1, x_3)$ and $\sigma_{33}^{(i)} = \sigma_{33}^{(i)}(x_1, x_3)$ for the ith ply. This situation is now assumed and is confirmed in Chapter 19 (see (19.7) and (19.8)).

On eliminating $\sigma_{22}^{(i)}$ in (11.7) and (11.9) using (11.16) it can be shown that

$$\varepsilon_{11}^{(i)} = \frac{\sigma_{11}^{(i)}}{\tilde{E}_A^{(i)}} - \frac{\tilde{\nu}_a^{(i)}}{\tilde{E}_A^{(i)}}\sigma_{33}^{(i)} + \tilde{\alpha}_A^{(i)}\Delta T - \nu_A^{(i)}\frac{E_T^{(i)}}{E_A^{(i)}}(\bar{\varepsilon}_T + \hat{\varepsilon}_T x_3), \tag{11.17}$$

$$\varepsilon_{33}^{(i)} = -\frac{\tilde{\nu}_a^{(i)}}{\tilde{E}_A^{(i)}}\sigma_{11}^{(i)} + \frac{1}{\tilde{E}_t^{(i)}}\sigma_{33}^{(i)} + \tilde{\alpha}_t^{(i)}\Delta T - \nu_t^{(i)}\left(\bar{\varepsilon}_T + \hat{\varepsilon}_T x_3\right), \tag{11.18}$$

where

$$\frac{1}{\tilde{E}_{\mathrm{t}}^{(i)}}=\frac{1}{E_{\mathrm{t}}^{(i)}}-\frac{\left(\nu_{\mathrm{t}}^{(i)}\right)^2}{E_{\mathrm{T}}^{(i)}},\quad \frac{\tilde{\nu}_a^{(i)}}{\tilde{E}_{\mathrm{A}}^{(i)}}=\frac{\nu_{\mathrm{a}}^{(i)}+\nu_{\mathrm{A}}^{(i)}\nu_{\mathrm{t}}^{(i)}}{E_{\mathrm{A}}^{(i)}}, \tag{11.19}$$

$$\frac{1}{\tilde{E}_{\mathrm{A}}^{(i)}}=\frac{1}{E_{\mathrm{A}}^{(i)}}\left[1-\left(\nu_{\mathrm{A}}^{(i)}\right)^2\frac{E_{\mathrm{T}}^{(i)}}{E_{\mathrm{A}}^{(i)}}\right], \tag{11.20}$$

$$\tilde{\alpha}_{\mathrm{t}}^{(i)}=\alpha_{\mathrm{t}}^{(i)}+\nu_{\mathrm{t}}^{(i)}\alpha_{\mathrm{T}}^{(i)},\quad \tilde{\alpha}_{\mathrm{A}}^{(i)}=\alpha_{\mathrm{A}}^{(i)}+\nu_{\mathrm{A}}^{(i)}\frac{E_{\mathrm{T}}^{(i)}}{E_{\mathrm{A}}^{(i)}}\alpha_{\mathrm{T}}^{(i)}. \tag{11.21}$$

Thus, when considering the anticlastic bending of a composite laminate, the following stress-strain relations and equilibrium equations should be satisfied as closely as possible:

$$\varepsilon_{11}^{(i)}\equiv\frac{\partial u_1^{(i)}}{\partial x_1}=\frac{\sigma_{11}^{(i)}}{\tilde{E}_{\mathrm{A}}^{(i)}}-\frac{\tilde{\nu}_a^{(i)}}{\tilde{E}_{\mathrm{A}}^{(i)}}\sigma_{33}^{(i)}+\tilde{\alpha}_{\mathrm{A}}^{(i)}\Delta T-\nu_{\mathrm{A}}^{(i)}\frac{E_{\mathrm{T}}^{(i)}}{E_{\mathrm{A}}^{(i)}}\left(\bar{\varepsilon}_{\mathrm{T}}+\hat{\varepsilon}_{\mathrm{T}}x_3\right), \tag{11.22}$$

$$\varepsilon_{22}^{(i)}\equiv\frac{\partial u_2^{(i)}}{\partial x_2}=\bar{\varepsilon}_{\mathrm{T}}+\hat{\varepsilon}_{\mathrm{T}}x_3, \tag{11.23}$$

$$\varepsilon_{33}^{(i)}\equiv\frac{\partial u_3^{(i)}}{\partial x_3}=-\frac{\tilde{\nu}_a^{(i)}}{\tilde{E}_{\mathrm{A}}^{(i)}}\sigma_{11}^{(i)}+\frac{1}{\tilde{E}_{\mathrm{t}}^{(i)}}\sigma_{33}^{(i)}+\tilde{\alpha}_{\mathrm{t}}^{(i)}\ \Delta T-\nu_{\mathrm{t}}^{(i)}\left(\bar{\varepsilon}_{\mathrm{T}}+\hat{\varepsilon}_{\mathrm{T}}x_3\right), \tag{11.24}$$

$$\varepsilon_{13}^{(i)}\equiv\frac{1}{2}\left(\frac{\partial u_1^{(i)}}{\partial x_3}+\frac{\partial u_3^{(i)}}{\partial x_1}\right)=\frac{\sigma_{13}^{(i)}}{2\mu_{\mathrm{a}}^{(i)}}, \tag{11.25}$$

$$\frac{\partial\sigma_{11}^{(i)}}{\partial x_1}+\frac{\partial\sigma_{13}^{(i)}}{\partial x_3}=0, \tag{11.26}$$

$$\frac{\partial\sigma_{13}^{(i)}}{\partial x_1}+\frac{\partial\sigma_{33}^{(i)}}{\partial x_3}=0. \tag{11.27}$$

Relations (11.26) and (11.27) result from the equilibrium equations (11.1) and (11.3) because from (11.15), the shear stresses $\sigma_{12}^{(i)}$ and $\sigma_{23}^{(i)}$ are zero.

11.4 Interfacial and Boundary Conditions

The origin for coordinates is selected to be midway between two transverse cracks. As the layers are assumed to be perfectly bonded the displacements and interfacial tractions must be continuous through the laminate such that

$$\left.\begin{aligned}\sigma_{13}^{(i)}(x_1,x_3^{(i)})&=\sigma_{13}^{(i+1)}(x_1,x_3^{(i)}),\\ \sigma_{33}^{(i)}(x_1,x_3^{(i)})&=\sigma_{33}^{(i+1)}(x_1,x_3^{(i)}),\\ u_1^{(i)}(x_1,x_3^{(i)})&=u_1^{(i+1)}(x_1,x_3^{(i)}),\\ u_3^{(i)}(x_1,x_2,x_3^{(i)})&=u_3^{(i+1)}(x_1,x_2,x_3^{(i)}),\end{aligned}\right\}\ \text{for } i=1,\ldots,N,\ |x_1|\le L. \tag{11.28}$$

Symmetry about the plane $x_1 = 0$ implies that

$$\sigma_{13}(0,x_3)=0,\ u_1(0,x_3)=0,\ \text{for } 0 \le x_3 \le h, \tag{11.29}$$

and the outer surfaces of the laminate are assumed to be subject only to a uniform normal stress σ_t so that for $|x_1| \le L$,

$$\sigma_{33}^{(1)}(x_1,0)=\sigma_{33}^{(N+1)}(x_1,x_3^{(N+1)})=\sigma_t,\ \sigma_{13}^{(1)}(x_1,0)=0,\ \sigma_{13}^{(N+1)}(x_1,x_3^{(N+1)})=0. \tag{11.30}$$

Because the surfaces of the transverse cracks are stress-free,

$$\sigma_{11}^{(i)}(\pm L, x_3)=0,\ \ \sigma_{13}^{(i)}(\pm L, x_3)=0,\ \ \text{for } x_3^{(i-1)} \le x_3 \le x_3^{(i)}, \tag{11.31}$$

for cracked layers only. For a uniform distribution of transverse cracks symmetry requires that for uncracked layers

$$\sigma_{13}^{(i)}(\pm L, x_3)=0, \text{for } x_3^{(i-1)} \le x_3 \le x_3^{(i)}. \tag{11.32}$$

The axial displacement for all the uncracked layers is assumed to have the form

$$u_1^{(i)}(\pm L, x_3)=\pm(\bar{\varepsilon}_A + \hat{\varepsilon}_A x_3)L, \tag{11.33}$$

where $\bar{\varepsilon}_A$ and $\hat{\varepsilon}_A$ are parameters whose values characterise the effective axial strain and bending of a cracked laminate. The corresponding values for an uncracked laminate are denoted by $\bar{\varepsilon}_A^0$ and $\hat{\varepsilon}_A^0$.

11.5 Effective Stress-Strain Relations

The bending moments per unit area of loading cross section for the axial and transverse directions are defined respectively by

$$\begin{aligned} M_A &= \frac{1}{2hW}\int_0^h\int_{-W}^W \left(x_3 - \frac{h}{2}\right)\sigma_{11}(L,x_2,x_3)\,dx_2 dx_3, \\ M_T &= \frac{1}{2hL}\int_0^h\int_{-L}^L \left(x_3 - \frac{h}{2}\right)\sigma_{22}(x_1,W,x_3)\,dx_1 dx_3. \end{aligned} \tag{11.34}$$

The moments are taken about the midplane of the laminate that might not correspond to the neutral axis if the laminate is geometrically nonsymmetrical and/or nonsymmetrically damaged. The corresponding effective applied axial and transverse stresses are defined respectively by

$$\sigma_A = \frac{1}{2hW}\int_0^h\int_{-W}^W \sigma_{11}(L,x_2,x_3)\,dx_2 dx_3,\ \ \sigma_T = \frac{1}{2hL}\int_0^h\int_{-L}^L \sigma_{22}(x_1,W,x_3)\,dx_1 dx_3. \tag{11.35}$$

The faces of the laminate are assumed to be subject to a uniform applied tensile traction σ_t. The corresponding effective through-thickness strain ε_t for a laminate is defined by

$$\varepsilon_{\rm t} = \frac{1}{4hLW}\int_{-W}^{W}\int_{-L}^{L}\left[u_3(x_1,x_2,h) - u_3(x_1,x_2,0)\right]\,{\rm d}x_1{\rm d}x_2, \tag{11.36}$$

and the effective applied axial and transverse strains $\varepsilon_{\rm A}^*$ and $\varepsilon_{\rm T}^*$ are defined by

$$\begin{aligned}\varepsilon_{\rm A}^* &= \frac{1}{4hLW}\int_{0}^{h}\int_{-W}^{W}\left[u_1(L,x_2,x_3) - u_1(-L,x_2,x_3)\right]\,{\rm d}x_2{\rm d}x_3,\\ \varepsilon_{\rm T}^* &= \frac{1}{4hLW}\int_{0}^{h}\int_{-L}^{L}\left[u_2(x_1,W,x_3) - u_2(x_1,-W,x_3)\right]\,{\rm d}x_1{\rm d}x_3,\end{aligned} \tag{11.37}$$

where u_k, $k = 1, 2, 3$, are the components of the displacement vector.

To apply bending of the laminate for the axial and transverse directions, the following edge boundary conditions are imposed for the displacement components u_1 and u_2

$$u_1 = \begin{cases} L(\overline{\varepsilon}_{\rm A} + \hat{\varepsilon}_{\rm A} x_3) \text{ on } x_1 = L,\\ -L(\overline{\varepsilon}_{\rm A} + \hat{\varepsilon}_{\rm A} x_3) \text{ on } x_1 = -L,\end{cases} \quad u_2 = \begin{cases} W(\overline{\varepsilon}_{\rm T} + \hat{\varepsilon}_{\rm T} x_3) \text{ on } x_2 = W,\\ -W(\overline{\varepsilon}_{\rm T} + \hat{\varepsilon}_{\rm T} x_3) \text{ on } x_2 = -W,\end{cases} \tag{11.38}$$

where $\overline{\varepsilon}_{\rm A}$ and $\overline{\varepsilon}_{\rm T}$ denote the effective in-plane axial and transverse strains on the free surface $x_3 = 0$. These effective strain parameters are defined by

$$\begin{aligned}\overline{\varepsilon}_{\rm A} &= \frac{1}{4LW}\int_{-W}^{W}\left[u_1(L,x_2,0) - u_1(-L,x_2,0)\right]\,{\rm d}x_2,\\ \overline{\varepsilon}_{\rm T} &= \frac{1}{4LW}\int_{-L}^{L}\left[u_2(x_1,W,0) - u_2(x_1,-W,0)\right]\,{\rm d}x_1.\end{aligned} \tag{11.39}$$

For infinitesimal deformations, the principal radii of curvature of the surface $x_3 = 0$ of the deformed laminate are given by (see (2.232))

$$R_1 = \frac{1}{|\hat{\varepsilon}_{\rm A}|}, \quad R_2 = \frac{1}{|\hat{\varepsilon}_{\rm T}|},$$

so that $\hat{\varepsilon}_{\rm A}$ and $\hat{\varepsilon}_{\rm T}$ are respectively the curvatures of the surface $x_3 = 0$ of the deformed laminate in the axial and transverse directions. Substituting the edge boundary conditions (11.38) into (11.37) and performing the integrations leads to

$$\varepsilon_{\rm A}^* = \overline{\varepsilon}_{\rm A} + \frac{1}{2}h\hat{\varepsilon}_{\rm A}, \quad \varepsilon_{\rm T}^* = \overline{\varepsilon}_{\rm T} + \frac{1}{2}h\hat{\varepsilon}_{\rm T}. \tag{11.40}$$

It is clear that $\varepsilon_{\rm A}^*$ and $\varepsilon_{\rm T}^*$ are the axial and transverse strains on the midplane of the laminate subject to the edge conditions (11.38).

For the moment it is assumed that the stress-strain relations for a damaged cross-ply laminate may be expressed in the following form, generalising results given in Chapter 7 for undamaged laminates,

$$\varepsilon_{\rm A}^* = \frac{\sigma_{\rm A}}{E_{\rm A}} - \frac{\nu_{\rm A}}{E_{\rm A}}\sigma_{\rm T} - \frac{\hat{\nu}_{\rm A}}{\hat{E}_{\rm A}}M_{\rm A} - \frac{\hat{\eta}_{\rm A}}{\hat{E}_{\rm T}}M_{\rm T} - \frac{\nu_{\rm a}}{E_{\rm A}}\sigma_{\rm t} + \alpha_{\rm A}\Delta T, \tag{11.41}$$

$$\varepsilon_{\mathrm{T}}^{*}=-\frac{\nu_{\mathrm{A}}}{E_{\mathrm{A}}}\sigma_{\mathrm{A}}+\frac{\sigma_{\mathrm{T}}}{E_{\mathrm{T}}}-\frac{\hat{\nu}_{\mathrm{T}}}{\hat{E}_{\mathrm{A}}}M_{\mathrm{A}}-\frac{\hat{\eta}_{\mathrm{T}}}{\hat{E}_{\mathrm{T}}}M_{\mathrm{T}}-\frac{\nu_{\mathrm{t}}}{E_{\mathrm{T}}}\sigma_{\mathrm{t}}+\alpha_{\mathrm{T}}\Delta T, \tag{11.42}$$

$$\hat{\varepsilon}_{\mathrm{A}}=-\frac{\hat{\nu}_{\mathrm{A}}}{\hat{E}_{\mathrm{T}}}\sigma_{\mathrm{A}}-\frac{\hat{\nu}_{\mathrm{T}}}{\hat{E}_{\mathrm{A}}}\sigma_{\mathrm{T}}+\frac{M_{\mathrm{A}}}{\hat{E}_{\mathrm{A}}}-\frac{\hat{\delta}_{\mathrm{A}}}{\hat{E}_{\mathrm{A}}}M_{\mathrm{T}}-\frac{\hat{\nu}_{\mathrm{a}}}{\hat{E}_{\mathrm{A}}}\sigma_{\mathrm{t}}+\hat{\alpha}_{\mathrm{A}}\Delta T, \tag{11.43}$$

$$\hat{\varepsilon}_{\mathrm{T}}=-\frac{\hat{\eta}_{\mathrm{A}}}{\hat{E}_{\mathrm{T}}}\sigma_{\mathrm{A}}-\frac{\hat{\eta}_{\mathrm{T}}}{\hat{E}_{\mathrm{T}}}\sigma_{\mathrm{T}}-\frac{\hat{\delta}_{\mathrm{A}}}{\hat{E}_{\mathrm{T}}}M_{\mathrm{A}}+\frac{M_{\mathrm{T}}}{\hat{E}_{\mathrm{T}}}-\frac{\hat{\eta}_{\mathrm{a}}}{\hat{E}_{\mathrm{T}}}\sigma_{\mathrm{t}}+\hat{\alpha}_{\mathrm{T}}\Delta T, \tag{11.44}$$

$$\varepsilon_{\mathrm{t}}=-\frac{\nu_{\mathrm{a}}}{E_{\mathrm{A}}}\sigma_{\mathrm{A}}-\frac{\nu_{\mathrm{t}}}{E_{\mathrm{T}}}\sigma_{\mathrm{T}}-\frac{\hat{\nu}_{\mathrm{a}}}{\hat{E}_{\mathrm{A}}}M_{\mathrm{A}}-\frac{\hat{\eta}_{\mathrm{a}}}{\hat{E}_{\mathrm{T}}}M_{\mathrm{T}}+\frac{\sigma_{\mathrm{t}}}{E_{\mathrm{t}}}+\alpha_{\mathrm{t}}\Delta T, \tag{11.45}$$

which defines the various thermoelastic constants that characterise the properties of a damaged laminate subject to combined in-plane biaxial loading, and out-of-plane through-thickness loading and bending. Clearly the stress-strain relations (11.41)–(11.45) are of the same form as those for undamaged laminates (see (7.195)–(7.199)) and a superscript '0' is used in the following analysis to denote that the value of a thermoelastic constant refers to its undamaged state. In Chapter 19, it will become clear that the assumed form of the laminate stress-strain relations is fully justified.

11.6 Reduced Stress-Strain Relations for Constrained Triaxial Loading

The damage-dependent stress-strain relations (11.43) and (11.44) are treated as a linear system of algebraic equations with unknowns M_{A} and M_{T}. For convenience, the following parameters are defined

$$\hat{\delta}_{\mathrm{T}}=\hat{\delta}_{\mathrm{A}}\frac{\hat{E}_{\mathrm{T}}}{\hat{E}_{\mathrm{A}}},\quad \Lambda=1-\hat{\delta}_{\mathrm{A}}\,\hat{\delta}_{\mathrm{T}}. \tag{11.46}$$

The solution of (11.43) and (11.44) can then be shown to be

$$M_{\mathrm{A}}=\frac{1}{\Lambda}\left[\begin{array}{l}\hat{E}_{\mathrm{A}}(\hat{\varepsilon}_{\mathrm{A}}+\hat{\delta}_{\mathrm{T}}\hat{\varepsilon}_{\mathrm{T}})-\hat{E}_{\mathrm{A}}(\hat{\alpha}_{\mathrm{A}}+\hat{\delta}_{\mathrm{T}}\hat{\alpha}_{\mathrm{T}})\Delta T\\ \quad+(\hat{\nu}_{\mathrm{a}}+\hat{\delta}_{\mathrm{A}}\hat{\eta}_{\mathrm{a}})\sigma_{\mathrm{t}}+(\hat{\nu}_{\mathrm{A}}+\hat{\delta}_{\mathrm{A}}\hat{\eta}_{\mathrm{A}})\sigma_{\mathrm{A}}+(\hat{\nu}_{\mathrm{T}}+\hat{\delta}_{\mathrm{A}}\hat{\eta}_{\mathrm{T}})\sigma_{\mathrm{T}}\end{array}\right], \tag{11.47}$$

$$M_{\mathrm{T}}=\frac{1}{\Lambda}\left[\begin{array}{l}\hat{E}_{\mathrm{T}}(\hat{\varepsilon}_{\mathrm{T}}+\hat{\delta}_{\mathrm{A}}\hat{\varepsilon}_{\mathrm{A}})-\hat{E}_{\mathrm{T}}(\hat{\alpha}_{\mathrm{T}}+\hat{\delta}_{\mathrm{A}}\hat{\alpha}_{\mathrm{A}})\Delta T\\ \quad+(\hat{\eta}_{\mathrm{a}}+\hat{\delta}_{\mathrm{T}}\hat{\nu}_{\mathrm{a}})\sigma_{\mathrm{t}}+(\hat{\eta}_{\mathrm{A}}+\hat{\delta}_{\mathrm{T}}\hat{\nu}_{\mathrm{A}})\sigma_{\mathrm{A}}+(\hat{\eta}_{\mathrm{T}}+\hat{\delta}_{\mathrm{T}}\hat{\nu}_{\mathrm{T}})\sigma_{\mathrm{T}}\end{array}\right]. \tag{11.48}$$

These results can be substituted into the remaining damage-dependent stress-strain relations (11.41)–(11.45) leading to reduced stress-strain relations of the same form as those for a cross-ply laminate subject only to triaxial loading, without shear or bend. The reduced equations are

$$\bar{\varepsilon}_{\mathrm{A}}\equiv\varepsilon_{\mathrm{A}}^{*}+\frac{\hat{\nu}_{\mathrm{A}}}{\Lambda}\left(\hat{\varepsilon}_{\mathrm{A}}+\hat{\delta}_{\mathrm{T}}\hat{\varepsilon}_{\mathrm{T}}\right)+\frac{\hat{\eta}_{\mathrm{A}}}{\Lambda}\left(\hat{\varepsilon}_{\mathrm{T}}+\hat{\delta}_{\mathrm{A}}\hat{\varepsilon}_{\mathrm{A}}\right)=\frac{\sigma_{\mathrm{A}}}{\bar{E}_{\mathrm{A}}}-\frac{\bar{\nu}_{\mathrm{A}}}{\bar{E}_{\mathrm{A}}}\sigma_{\mathrm{T}}-\frac{\bar{\nu}_{\mathrm{a}}}{\bar{E}_{\mathrm{A}}}\sigma_{\mathrm{t}}+\bar{\alpha}_{\mathrm{A}}\Delta T, \tag{11.49}$$

$$\hat{\varepsilon}_{\mathrm{T}} \equiv \varepsilon_{\mathrm{T}}^{*} + \frac{\hat{\nu}_{\mathrm{T}}}{\Lambda}\left(\hat{\varepsilon}_{\mathrm{A}} + \hat{\delta}_{\mathrm{T}}\hat{\varepsilon}_{\mathrm{T}}\right) + \frac{\hat{\eta}_{\mathrm{T}}}{\Lambda}\left(\hat{\varepsilon}_{\mathrm{T}} + \hat{\delta}_{\mathrm{A}}\hat{\varepsilon}_{\mathrm{A}}\right) = -\frac{\widehat{\nu}_{\mathrm{A}}}{\widehat{E}_{\mathrm{A}}}\sigma_{\mathrm{A}} + \frac{\sigma_{\mathrm{T}}}{\widehat{E}_{\mathrm{T}}} - \frac{\widehat{\nu}_{\mathrm{t}}}{\widehat{E}_{\mathrm{T}}}\sigma_{\mathrm{t}} + \widehat{\alpha}_{\mathrm{T}}\Delta T, \tag{11.50}$$

$$\hat{\varepsilon}_{\mathrm{t}} \equiv \varepsilon_{\mathrm{t}} + \frac{\hat{\nu}_{\mathrm{a}}}{\Lambda}\left(\hat{\varepsilon}_{\mathrm{A}} + \hat{\delta}_{\mathrm{T}}\hat{\varepsilon}_{\mathrm{T}}\right) + \frac{\hat{\eta}_{\mathrm{a}}}{\Lambda}\left(\hat{\varepsilon}_{\mathrm{T}} + \hat{\delta}_{\mathrm{A}}\hat{\varepsilon}_{\mathrm{A}}\right) = -\frac{\widehat{\nu}_{\mathrm{a}}}{\widehat{E}_{\mathrm{A}}}\sigma_{\mathrm{A}} - \frac{\widehat{\nu}_{\mathrm{t}}}{\widehat{E}_{\mathrm{T}}}\sigma_{\mathrm{T}} + \frac{\sigma_{\mathrm{t}}}{\widehat{E}_{\mathrm{t}}} + \widehat{\alpha}_{\mathrm{t}}\Delta T, \tag{11.51}$$

where the reduced strains $\widehat{\varepsilon}_{\mathrm{A}}$, $\widehat{\varepsilon}_{\mathrm{T}}$ and $\widehat{\varepsilon}_{\mathrm{t}}$ can be interpreted as strains for a damaged laminate, subject to triaxial loading and constrained so that shear and bending strains are zero, and where the reduced thermoelastic constants are defined by

$$\frac{1}{\widehat{E}_{\mathrm{A}}} = \frac{1}{E_{\mathrm{A}}} - \frac{\hat{\nu}_{\mathrm{A}}\overline{\nu}_{\mathrm{A}}}{\widehat{E}_{\mathrm{A}}} - \frac{\hat{\eta}_{\mathrm{A}}\overline{\eta}_{\mathrm{A}}}{\widehat{E}_{\mathrm{T}}}, \tag{11.52}$$

$$\frac{1}{\widehat{E}_{\mathrm{T}}} = \frac{1}{E_{\mathrm{T}}} - \frac{\hat{\nu}_{\mathrm{T}}\overline{\nu}_{\mathrm{T}}}{\widehat{E}_{\mathrm{A}}} - \frac{\hat{\eta}_{\mathrm{T}}\overline{\eta}_{\mathrm{T}}}{\widehat{E}_{\mathrm{T}}}, \tag{11.53}$$

$$\frac{1}{\widehat{E}_{\mathrm{t}}} = \frac{1}{E_{\mathrm{t}}} - \frac{\hat{\nu}_{\mathrm{a}}\overline{\nu}_{\mathrm{a}}}{\widehat{E}_{\mathrm{A}}} - \frac{\hat{\eta}_{\mathrm{a}}\overline{\eta}_{\mathrm{a}}}{\widehat{E}_{\mathrm{T}}}, \tag{11.54}$$

$$\frac{\widehat{\nu}_{\mathrm{A}}}{\widehat{E}_{\mathrm{A}}} = \frac{\nu_{\mathrm{A}}}{E_{\mathrm{A}}} + \frac{\hat{\nu}_{\mathrm{A}}\overline{\nu}_{\mathrm{T}}}{\widehat{E}_{\mathrm{A}}} + \frac{\hat{\eta}_{\mathrm{A}}\overline{\eta}_{\mathrm{T}}}{\widehat{E}_{\mathrm{T}}}, \tag{11.55}$$

$$\frac{\widehat{\nu}_{\mathrm{a}}}{\widehat{E}_{\mathrm{A}}} = \frac{\nu_{\mathrm{a}}}{E_{\mathrm{A}}} + \frac{\hat{\nu}_{\mathrm{a}}\overline{\nu}_{\mathrm{A}}}{\widehat{E}_{\mathrm{A}}} + \frac{\hat{\eta}_{\mathrm{a}}\overline{\eta}_{\mathrm{A}}}{\widehat{E}_{\mathrm{T}}}, \tag{11.56}$$

$$\frac{\widehat{\nu}_{\mathrm{t}}}{\widehat{E}_{\mathrm{T}}} = \frac{\nu_{\mathrm{t}}}{E_{\mathrm{T}}} + \frac{\hat{\nu}_{\mathrm{T}}\overline{\nu}_{\mathrm{a}}}{\widehat{E}_{\mathrm{A}}} + \frac{\hat{\eta}_{\mathrm{T}}\overline{\eta}_{\mathrm{a}}}{\widehat{E}_{\mathrm{T}}}, \tag{11.57}$$

$$\widehat{\alpha}_{\mathrm{A}} = \alpha_{\mathrm{A}} + \overline{\nu}_{\mathrm{A}}\widehat{\alpha}_{\mathrm{A}} + \overline{\eta}_{\mathrm{A}}\widehat{\alpha}_{\mathrm{T}}, \tag{11.58}$$

$$\widehat{\alpha}_{\mathrm{T}} = \alpha_{\mathrm{T}} + \overline{\nu}_{\mathrm{T}}\widehat{\alpha}_{\mathrm{A}} + \overline{\eta}_{\mathrm{T}}\widehat{\alpha}_{\mathrm{T}}, \tag{11.59}$$

$$\widehat{\alpha}_{\mathrm{t}} = \alpha_{\mathrm{t}} + \overline{\nu}_{\mathrm{a}}\widehat{\alpha}_{\mathrm{A}} + \overline{\eta}_{\mathrm{a}}\widehat{\alpha}_{\mathrm{T}}, \tag{11.60}$$

where

$$\begin{aligned} \overline{\nu}_{\mathrm{A}} &= \frac{\hat{\nu}_{\mathrm{A}} + \hat{\delta}_{\mathrm{A}}\hat{\eta}_{\mathrm{A}}}{\Lambda}, & \overline{\nu}_{T} &= \frac{\hat{\nu}_{\mathrm{T}} + \hat{\delta}_{\mathrm{A}}\hat{\eta}_{\mathrm{T}}}{\Lambda}, & \overline{\nu}_{\mathrm{a}} &= \frac{\hat{\nu}_{\mathrm{a}} + \hat{\delta}_{\mathrm{A}}\hat{\eta}_{\mathrm{a}}}{\Lambda}, \\ \overline{\eta}_{\mathrm{A}} &= \frac{\hat{\eta}_{\mathrm{A}} + \hat{\delta}_{\mathrm{T}}\hat{\nu}_{\mathrm{A}}}{\Lambda}, & \overline{\eta}_{\mathrm{T}} &= \frac{\hat{\eta}_{\mathrm{T}} + \hat{\delta}_{\mathrm{T}}\hat{\nu}_{\mathrm{T}}}{\Lambda}, & \overline{\eta}_{\mathrm{t}} &= \frac{\hat{\eta}_{\mathrm{a}} + \hat{\delta}_{\mathrm{T}}\hat{\nu}_{\mathrm{a}}}{\Lambda}. \end{aligned} \tag{11.61}$$

The corresponding reduced stress-strain relations for an undamaged symmetric laminate are as follows:

$$\widehat{\varepsilon}_{\mathrm{A}}^{0} = \frac{1}{\widehat{E}_{\mathrm{A}}^{0}}\sigma_{\mathrm{A}} - \frac{\widehat{\nu}_{\mathrm{A}}^{0}}{\widehat{E}_{\mathrm{A}}^{0}}\sigma_{\mathrm{T}} - \frac{\widehat{\nu}_{\mathrm{a}}^{0}}{\widehat{E}_{\mathrm{A}}^{0}}\sigma_{\mathrm{t}} + \widehat{\alpha}_{\mathrm{A}}^{0}\Delta T, \tag{11.62}$$

$$\widehat{\varepsilon}_{\mathrm{T}}^{0} = -\frac{\widehat{\nu}_{\mathrm{A}}^{0}}{\widehat{E}_{\mathrm{A}}^{0}}\sigma_{\mathrm{A}} + \frac{\sigma_{\mathrm{T}}}{\widehat{E}_{\mathrm{T}}^{0}} - \frac{\widehat{\nu}_{\mathrm{t}}^{0}}{\widehat{E}_{\mathrm{T}}^{0}}\sigma_{\mathrm{t}} + \widehat{\alpha}_{\mathrm{T}}^{0}\Delta T, \tag{11.63}$$

$$\widehat{\varepsilon}_{\mathrm{t}}^{0} = -\frac{\widehat{\nu}_{\mathrm{a}}^{0}}{\widehat{E}_{\mathrm{A}}^{0}}\sigma_{\mathrm{A}} - \frac{\widehat{\nu}_{\mathrm{t}}^{0}}{\widehat{E}_{\mathrm{T}}^{0}}\sigma_{\mathrm{T}} + \frac{\sigma_{\mathrm{t}}}{\widehat{E}_{\mathrm{t}}^{0}} + \widehat{\alpha}_{\mathrm{t}}^{0}\Delta T, \tag{11.64}$$

where a superscript '0' denotes that the laminate property refers to its value for the undamaged state of the laminate. It should be noted that, if the laminate is symmetric and undamaged, then the parameters $\widehat{\nu}_{\mathrm{A}}$, $\widehat{\nu}_{\mathrm{T}}$, $\widehat{\nu}_{\mathrm{a}}$, $\widehat{\eta}_{\mathrm{A}}$, $\widehat{\eta}_{\mathrm{T}}$, $\widehat{\eta}_{\mathrm{a}}$, $\widehat{\alpha}_{\mathrm{A}}$ and $\widehat{\alpha}_{\mathrm{T}}$ are all zero so that $\widehat{E}_{\mathrm{t}} = E_{\mathrm{t}}$, and similarly for the other thermoelastic constants defined by (11.52)–(11.60).

11.7 Ply Crack Closure for Uniaxial Loading

Having reduced the stress-strain relations to the form (11.49)–(11.51), various interrelationships between the coefficients may be derived. For a cross-ply laminate with ply cracks in the 90° plies only, it is shown in Appendix D (see (D.9)) that, at the point of crack closure the axial strain, through-thickness stress, in-plane transverse strain and the curvatures must satisfy

$$\bar{\varepsilon}_{\mathrm{A}}^{\mathrm{c}} = -A\sigma_{\mathrm{t}}^{\mathrm{c}} - B\bar{\varepsilon}_{\mathrm{T}}^{\mathrm{c}} + C\Delta T, \quad \hat{\varepsilon}_{\mathrm{A}}^{\mathrm{c}} = -B\hat{\varepsilon}_{\mathrm{T}}^{\mathrm{c}}, \tag{11.65}$$

where

$$A = \frac{\nu_{\mathrm{t}}^{(90)}}{E_{\mathrm{T}}^{(90)}} + \frac{\nu_{\mathrm{a}}^{(90)}\nu_{\mathrm{A}}^{(90)}}{E_{\mathrm{A}}^{(90)}}, \quad B = \nu_{\mathrm{A}}^{(90)}, \quad C = \alpha_{\mathrm{T}}^{(90)} + \nu_{\mathrm{A}}^{(90)}\alpha_{\mathrm{A}}^{(90)}, \tag{11.66}$$

where a superscript '(90)' indicates a stress-strain coefficient for a single 90° ply (hence, it is necessary in this case that all the 90° plies be made of the same material).
Consider now ply crack closure for the case of uniaxial loading such that $\sigma_{\mathrm{t}} = \sigma_{\mathrm{T}} = 0$ and $\hat{\varepsilon}_{\mathrm{A}} = \hat{\varepsilon}_{\mathrm{T}} = 0$. It follows from (11.49)–(11.51) on setting $\bar{\varepsilon}_{\mathrm{A}} = \bar{\varepsilon}_{\mathrm{A}}^{\mathrm{c}}, \bar{\varepsilon}_{\mathrm{T}} = \bar{\varepsilon}_{\mathrm{T}}^{\mathrm{c}}, \bar{\varepsilon}_{\mathrm{t}} = \bar{\varepsilon}_{\mathrm{t}}^{\mathrm{c}}$ and $\sigma_{\mathrm{A}} = \bar{\sigma}_{\mathrm{A}}^{\mathrm{c}}$, that

$$\bar{\varepsilon}_{\mathrm{A}}^{\mathrm{c}} = \frac{\bar{\sigma}_{\mathrm{A}}^{\mathrm{c}}}{\widehat{E}_{\mathrm{A}}} + \widehat{\alpha}_{\mathrm{A}}\Delta T, \tag{11.67}$$

$$\bar{\varepsilon}_{\mathrm{T}}^{\mathrm{c}} = -\frac{\widehat{\nu}_{\mathrm{A}}}{\widehat{E}_{\mathrm{A}}}\bar{\sigma}_{\mathrm{A}}^{\mathrm{c}} + \widehat{\alpha}_{\mathrm{T}}\Delta T, \tag{11.68}$$

$$\bar{\varepsilon}_{\mathrm{t}}^{\mathrm{c}} = -\frac{\widehat{\nu}_{\mathrm{a}}}{\widehat{E}_{\mathrm{A}}}\bar{\sigma}_{\mathrm{A}}^{\mathrm{c}} + \widehat{\alpha}_{\mathrm{t}}\Delta T. \tag{11.69}$$

Similarly, for undamaged laminates at the point of ply crack closure, the same uniaxial stress and axial strain must result, and it follows that

$$\bar{\varepsilon}_{\mathrm{A}}^{\mathrm{c}} = \frac{\bar{\sigma}_{\mathrm{A}}^{\mathrm{c}}}{\widehat{E}_{\mathrm{A}}^{0}} + \widehat{\alpha}_{\mathrm{A}}^{0}\Delta T, \tag{11.70}$$

$$\bar{\varepsilon}_{\mathrm{T}}^{\mathrm{c}} = -\frac{\hat{\nu}_{\mathrm{A}}^{0}}{\hat{E}_{\mathrm{A}}^{0}}\,\bar{\sigma}_{\mathrm{A}}^{\mathrm{c}} + \hat{\alpha}_{\mathrm{T}}^{0}\,\Delta T, \tag{11.71}$$

$$\bar{\varepsilon}_{\mathrm{t}}^{\mathrm{c}} = -\frac{\hat{\nu}_{\mathrm{a}}^{0}}{\hat{E}_{\mathrm{A}}^{0}}\,\bar{\sigma}_{\mathrm{A}}^{\mathrm{c}} + \hat{\alpha}_{\mathrm{t}}^{0}\,\Delta T. \tag{11.72}$$

Using (11.67) and (11.70) to eliminate $\bar{\varepsilon}_{\mathrm{A}}^{\mathrm{c}}$ leads to

$$\bar{\sigma}_{\mathrm{A}}^{\mathrm{c}} = -\frac{\hat{\alpha}_{\mathrm{A}} - \hat{\alpha}_{\mathrm{A}}^{0}}{\dfrac{1}{\hat{E}_{\mathrm{A}}} - \dfrac{1}{\hat{E}_{\mathrm{A}}^{0}}}\,\Delta T. \tag{11.73}$$

Similarly, the elimination of $\bar{\varepsilon}_{\mathrm{T}}^{\mathrm{c}}$ using relations (11.68) and (11.71) leads to

$$\bar{\sigma}_{\mathrm{A}}^{\mathrm{c}} = -\frac{\hat{\alpha}_{\mathrm{T}} - \hat{\alpha}_{\mathrm{T}}^{0}}{\dfrac{\hat{\nu}_{\mathrm{A}}^{0}}{\hat{E}_{\mathrm{A}}^{0}} - \dfrac{\hat{\nu}_{\mathrm{A}}}{\hat{E}_{\mathrm{A}}}}\,\Delta T. \tag{11.74}$$

It follows from (11.69) and (11.72), on eliminating $\bar{\varepsilon}_{\mathrm{t}}^{\mathrm{c}}$ that

$$\bar{\sigma}_{\mathrm{A}}^{\mathrm{c}} = -\frac{\hat{\alpha}_{\mathrm{t}} - \hat{\alpha}_{\mathrm{t}}^{0}}{\dfrac{\hat{\nu}_{\mathrm{a}}^{0}}{\hat{E}_{\mathrm{A}}^{0}} - \dfrac{\hat{\nu}_{\mathrm{a}}}{\hat{E}_{\mathrm{A}}}}\,\Delta T. \tag{11.75}$$

It follows from (11.65), for the case of uniaxial loading under discussion where $\hat{\varepsilon}_{\mathrm{A}} = \hat{\varepsilon}_{\mathrm{T}} = 0$, that

$$\bar{\varepsilon}_{\mathrm{A}}^{\mathrm{c}} = -B\bar{\varepsilon}_{\mathrm{T}}^{\mathrm{c}} + C\Delta T. \tag{11.76}$$

Thus, on substituting (11.70) and (11.71) into (11.76) and solving for $\bar{\sigma}_{\mathrm{A}}^{\mathrm{c}}$, the following alternative expression for the crack closure stress is obtained

$$\bar{\sigma}_{\mathrm{A}}^{\mathrm{c}} = -\frac{\hat{E}_{\mathrm{A}}^{0}(\hat{\alpha}_{\mathrm{A}}^{0} + \hat{\alpha}_{\mathrm{T}}^{0}B - C)}{1 - \hat{\nu}_{\mathrm{A}}^{0}B}\,\Delta T = -\hat{k}_1\,\Delta T, \tag{11.77}$$

where

$$\hat{k}_1 = \frac{\hat{E}_{\mathrm{A}}^{0}(\hat{\alpha}_{\mathrm{A}}^{0} + \hat{\alpha}_{\mathrm{T}}^{0}B - C)}{1 - \hat{\nu}_{\mathrm{A}}^{0}B}. \tag{11.78}$$

It follows from (11.73)–(11.75) that

$$\frac{\hat{\alpha}_{\mathrm{A}} - \hat{\alpha}_{\mathrm{A}}^{0}}{\dfrac{1}{\hat{E}_{\mathrm{A}}} - \dfrac{1}{\hat{E}_{\mathrm{A}}^{0}}} = \frac{\hat{\alpha}_{\mathrm{T}} - \hat{\alpha}_{\mathrm{T}}^{0}}{\dfrac{\hat{\nu}_{\mathrm{A}}^{0}}{\hat{E}_{\mathrm{A}}^{0}} - \dfrac{\hat{\nu}_{\mathrm{A}}}{\hat{E}_{\mathrm{A}}}} = \frac{\hat{\alpha}_{\mathrm{t}} - \hat{\alpha}_{\mathrm{t}}^{0}}{\dfrac{\hat{\nu}_{\mathrm{a}}^{0}}{\hat{E}_{\mathrm{A}}^{0}} - \dfrac{\hat{\nu}_{\mathrm{a}}}{\hat{E}_{\mathrm{A}}}} = \hat{k}_1. \tag{11.79}$$

Similarly, by considering ply crack closure during uniaxial loading in the in-plane transverse and through-thickness directions, it can be shown that

$$\frac{\widehat{\alpha}_{\mathrm{A}}-\widehat{\alpha}_{\mathrm{A}}^{0}}{\dfrac{\widehat{\nu}_{\mathrm{A}}^{0}}{\widehat{E}_{\mathrm{A}}^{0}}-\dfrac{\widehat{\nu}_{\mathrm{A}}}{\widehat{E}_{\mathrm{A}}}}=\frac{\widehat{\alpha}_{\mathrm{T}}-\widehat{\alpha}_{\mathrm{T}}^{0}}{\dfrac{1}{\widehat{E}_{\mathrm{T}}}-\dfrac{1}{\widehat{E}_{\mathrm{T}}^{0}}}=\frac{\widehat{\alpha}_{\mathrm{t}}-\widehat{\alpha}_{\mathrm{t}}^{0}}{\dfrac{\widehat{\nu}_{\mathrm{t}}^{0}}{\widehat{E}_{\mathrm{T}}^{0}}-\dfrac{\widehat{\nu}_{\mathrm{t}}}{\widehat{E}_{\mathrm{T}}}}=\widehat{k}_{2},\tag{11.80}$$

$$\frac{\widehat{\alpha}_{\mathrm{A}}-\widehat{\alpha}_{\mathrm{A}}^{0}}{\dfrac{\widehat{\nu}_{\mathrm{a}}^{0}}{\widehat{E}_{\mathrm{A}}^{0}}-\dfrac{\widehat{\nu}_{\mathrm{a}}}{\widehat{E}_{\mathrm{A}}}}=\frac{\widehat{\alpha}_{\mathrm{T}}-\widehat{\alpha}_{\mathrm{T}}^{0}}{\dfrac{\widehat{\nu}_{\mathrm{t}}^{0}}{\widehat{E}_{\mathrm{T}}^{0}}-\dfrac{\widehat{\nu}_{\mathrm{t}}}{\widehat{E}_{\mathrm{T}}}}=\frac{\widehat{\alpha}_{\mathrm{t}}-\widehat{\alpha}_{\mathrm{t}}^{0}}{\dfrac{1}{\widehat{E}_{\mathrm{t}}}-\dfrac{1}{\widehat{E}_{\mathrm{t}}^{0}}}=\widehat{k}_{3},\tag{11.81}$$

where

$$\widehat{k}_{2}=\frac{\widehat{E}_{\mathrm{T}}^{0}(\widehat{\alpha}_{\mathrm{A}}^{0}+\widehat{\alpha}_{\mathrm{T}}^{0}B-C)}{B-\widehat{\nu}_{\mathrm{A}}^{0}\dfrac{\widehat{E}_{\mathrm{T}}^{0}}{\widehat{E}_{\mathrm{A}}^{0}}},\tag{11.82}$$

$$\widehat{k}_{3}=\frac{\widehat{\alpha}_{\mathrm{A}}^{0}+\widehat{\alpha}_{\mathrm{T}}^{0}B-C}{A-\dfrac{\widehat{\nu}_{\mathrm{a}}^{0}}{\widehat{E}_{\mathrm{A}}^{0}}-\dfrac{\widehat{\nu}_{\mathrm{t}}^{0}}{\widehat{E}_{\mathrm{T}}^{0}}B}.\tag{11.83}$$

Results (11.79)–(11.81) are equivalent to the following two sets of independent interrelationships

$$\frac{\dfrac{\widehat{\nu}_{\mathrm{t}}^{0}}{\widehat{E}_{\mathrm{T}}^{0}}-\dfrac{\widehat{\nu}_{\mathrm{t}}}{\widehat{E}_{\mathrm{T}}}}{\dfrac{\widehat{\nu}_{\mathrm{a}}^{0}}{\widehat{E}_{\mathrm{A}}^{0}}-\dfrac{\widehat{\nu}_{\mathrm{a}}}{\widehat{E}_{\mathrm{A}}}}=\frac{\dfrac{\widehat{\nu}_{\mathrm{A}}^{0}}{\widehat{E}_{\mathrm{A}}^{0}}-\dfrac{\widehat{\nu}_{\mathrm{A}}}{\widehat{E}_{\mathrm{A}}}}{\dfrac{1}{\widehat{E}_{\mathrm{A}}}-\dfrac{1}{\widehat{E}_{\mathrm{A}}^{0}}}=\frac{\dfrac{1}{\widehat{E}_{\mathrm{T}}}-\dfrac{1}{\widehat{E}_{\mathrm{T}}^{0}}}{\dfrac{\widehat{\nu}_{\mathrm{A}}^{0}}{\widehat{E}_{\mathrm{A}}^{0}}-\dfrac{\widehat{\nu}_{\mathrm{A}}}{\widehat{E}_{\mathrm{A}}}}=\frac{\widehat{\alpha}_{\mathrm{T}}-\widehat{\alpha}_{\mathrm{T}}^{0}}{\widehat{\alpha}_{\mathrm{A}}-\widehat{\alpha}_{\mathrm{A}}^{0}}=\frac{\widehat{k}_{1}}{\widehat{k}_{2}}=\widehat{k},\tag{11.84}$$

$$\frac{\dfrac{1}{\widehat{E}_{\mathrm{t}}}-\dfrac{1}{\widehat{E}_{\mathrm{t}}^{0}}}{\dfrac{\widehat{\nu}_{\mathrm{a}}^{0}}{\widehat{E}_{\mathrm{A}}^{0}}-\dfrac{\widehat{\nu}_{\mathrm{a}}}{\widehat{E}_{\mathrm{A}}}}=\frac{\dfrac{\widehat{\nu}_{\mathrm{a}}^{0}}{\widehat{E}_{\mathrm{A}}^{0}}-\dfrac{\widehat{\nu}_{\mathrm{a}}}{\widehat{E}_{\mathrm{A}}}}{\dfrac{1}{\widehat{E}_{\mathrm{A}}}-\dfrac{1}{\widehat{E}_{\mathrm{A}}^{0}}}=\frac{\dfrac{\widehat{\nu}_{\mathrm{t}}^{0}}{\widehat{E}_{\mathrm{T}}^{0}}-\dfrac{\widehat{\nu}_{\mathrm{t}}}{\widehat{E}_{\mathrm{T}}}}{\dfrac{\widehat{\nu}_{\mathrm{A}}^{0}}{\widehat{E}_{\mathrm{A}}^{0}}-\dfrac{\widehat{\nu}_{\mathrm{A}}}{\widehat{E}_{\mathrm{A}}}}=\frac{\widehat{\alpha}_{\mathrm{t}}-\widehat{\alpha}_{\mathrm{t}}^{0}}{\widehat{\alpha}_{\mathrm{A}}-\widehat{\alpha}_{\mathrm{A}}^{0}}=\frac{\widehat{k}_{1}}{\widehat{k}_{3}}=\widehat{k}',\tag{11.85}$$

where

$$\widehat{k}=\frac{\widehat{k}_{1}}{\widehat{k}_{2}}=\frac{\widehat{E}_{\mathrm{A}}^{0}}{\widehat{E}_{\mathrm{T}}^{0}}\,\frac{B-\widehat{\nu}_{\mathrm{A}}^{0}\dfrac{\widehat{E}_{\mathrm{T}}^{0}}{\widehat{E}_{\mathrm{A}}^{0}}}{1-\widehat{\nu}_{\mathrm{A}}^{0}B},\tag{11.86}$$

$$\widehat{k}'=\frac{\widehat{k}_{1}}{\widehat{k}_{3}}=\frac{\widehat{E}_{\mathrm{A}}^{0}A-\widehat{\nu}_{\mathrm{a}}^{0}-\widehat{\nu}_{\mathrm{t}}^{0}\dfrac{\widehat{E}_{\mathrm{A}}^{0}}{\widehat{E}_{\mathrm{T}}^{0}}B}{1-\widehat{\nu}_{\mathrm{A}}^{0}B}.\tag{11.87}$$

Thus, $\widehat{k}_1$, $\widehat{k}_2$, $\widehat{k}_3$, $\widehat{k}$ and $\widehat{k}'$ are laminate constants which are independent of the state of damage, and $\widehat{k}$ and $\widehat{k}'$ are also independent of the thermal expansion coefficients. On using (11.66) the expressions (11.86) and (11.87) for the laminate constants $\widehat{k}$ and $\widehat{k}'$ may be written in terms of undamaged laminate properties as follows:

$$\widehat{k}=\frac{1}{\widehat{E}_{\mathrm{T}}^{0}}\frac{\widehat{E}_{\mathrm{A}}^{0}\nu_{\mathrm{A}}^{(90)}-\widehat{\nu}_{\mathrm{A}}^{0}\widehat{E}_{\mathrm{T}}^{0}}{1-\nu_{\mathrm{A}}^{(90)}\widehat{\nu}_{\mathrm{A}}^{0}}, \tag{11.88}$$

$$\widehat{k}'=\frac{\widehat{E}_{\mathrm{A}}^{0}(\nu_{\mathrm{t}}^{(90)}/E_{\mathrm{T}}^{(90)}+\nu_{\mathrm{a}}^{(90)}\nu_{\mathrm{A}}^{(90)}/E_{\mathrm{A}}^{(90)})-\widehat{\nu}_{\mathrm{a}}^{0}-\nu_{\mathrm{A}}^{(90)}\widehat{\nu}_{\mathrm{t}}^{0}\widehat{E}_{\mathrm{A}}^{0}/\widehat{E}_{\mathrm{T}}^{0}}{1-\nu_{\mathrm{A}}^{(90)}\widehat{\nu}_{\mathrm{A}}^{0}}. \tag{11.89}$$

11.7.1 Useful Independent Relationships

To form a useful set of independent interrelationships, define

$$\Phi=\frac{1}{\widehat{E}_{\mathrm{A}}}-\frac{1}{\widehat{E}_{\mathrm{A}}^{0}}. \tag{11.90}$$

It can then be shown that the required relationships are

$$\frac{1}{\widehat{E}_{\mathrm{T}}}-\frac{1}{\widehat{E}_{\mathrm{T}}^{0}}=\widehat{k}^{2}\Phi, \tag{11.91}$$

$$\frac{1}{\widehat{E}_{\mathrm{t}}}-\frac{1}{\widehat{E}_{\mathrm{t}}^{0}}=(\widehat{k}')^{2}\Phi, \tag{11.92}$$

$$\frac{\widehat{\nu}_{\mathrm{A}}^{0}}{\widehat{E}_{\mathrm{A}}^{0}}-\frac{\widehat{\nu}_{\mathrm{A}}}{\widehat{E}_{\mathrm{A}}}=\widehat{k}\,\Phi, \tag{11.93}$$

$$\frac{\widehat{\nu}_{\mathrm{a}}^{0}}{\widehat{E}_{\mathrm{A}}^{0}}-\frac{\widehat{\nu}_{\mathrm{a}}}{\widehat{E}_{\mathrm{A}}}=\widehat{k}'\Phi, \tag{11.94}$$

$$\frac{\widehat{\nu}_{\mathrm{t}}^{0}}{\widehat{E}_{\mathrm{T}}^{0}}-\frac{\widehat{\nu}_{\mathrm{t}}}{\widehat{E}_{\mathrm{T}}}=\widehat{k}\,\widehat{k}'\Phi, \tag{11.95}$$

$$\widehat{\alpha}_{\mathrm{A}}-\widehat{\alpha}_{\mathrm{A}}^{0}=\widehat{k}_{1}\Phi, \tag{11.96}$$

$$\widehat{\alpha}_{\mathrm{T}}-\widehat{\alpha}_{\mathrm{T}}^{0}=\widehat{k}\,\widehat{k}_{1}\Phi, \tag{11.97}$$

$$\widehat{\alpha}_{\mathrm{t}}-\widehat{\alpha}_{\mathrm{t}}^{0}=\widehat{k}'\,\widehat{k}_{1}\Phi. \tag{11.98}$$

The results (11.90)–(11.98) indicate that the degradation of all the thermoelastic constants of a damaged laminate arising from ply cracking in the 90° plies can be characterised by a single parameter Φ that is defined at the macroscopic ply level.

11.8 Energy for a Cracked Laminate Subject to Biaxial Bending

It is shown in Appendix D (see (D.9)) that, if there are cracks only in the 90° plies, then

$$\hat{\varepsilon}_{\mathrm{A}}+\nu_{\mathrm{A}}^{(90)}\hat{\varepsilon}_{\mathrm{T}}=0, \tag{11.99}$$

must be satisfied at the point of ply crack closure. Hence, any combination of the curvatures satisfying (11.99) may be used as the basis for a closure condition. For simplicity it is natural to choose $\hat{\varepsilon}_{\rm A} = \hat{\varepsilon}_{\rm T} = 0$, in which case (11.47) and (11.48) lead to the following expressions for the moments per unit area $M_{\rm A}^{\rm c}$ and $M_{\rm T}^{\rm c}$ at the point of closure

$$M_{\rm A}^{\rm c} = \frac{1}{\Lambda}\left[\begin{array}{l}(\hat{\nu}_{\rm A} + \hat{\delta}_{\rm A}\hat{\eta}_{\rm A})\sigma_{\rm A}^{\rm c} + (\hat{\nu}_{\rm T} + \hat{\delta}_{\rm T}\hat{\eta}_{\rm T})\sigma_{\rm T}^{\rm c} \\ + (\hat{\nu}_{\rm a} + \hat{\delta}_{\rm A}\hat{\eta}_{\rm a})\sigma_{\rm t}^{\rm c} - \hat{E}_{\rm A}(\hat{\alpha}_{\rm A} + \hat{\delta}_{\rm T}\hat{\alpha}_{\rm T})\Delta T\end{array}\right], \tag{11.100}$$

$$M_{\rm T}^{\rm c} = \frac{1}{\Lambda}\left[\begin{array}{l}(\hat{\eta}_{\rm A} + \hat{\delta}_{\rm T}\hat{\nu}_{\rm A})\sigma_{\rm A}^{\rm c} + (\hat{\eta}_{\rm T} + \hat{\delta}_{\rm T}\hat{\nu}_{\rm T})\sigma_{\rm T}^{\rm c} \\ + (\hat{\eta}_{\rm a} + \hat{\delta}_{\rm T}\hat{\nu}_{\rm a})\sigma_{\rm t}^{\rm c} - \hat{E}_{\rm T}(\hat{\alpha}_{\rm T} + \hat{\delta}_{\rm A}\hat{\alpha}_{\rm A})\Delta T\end{array}\right]. \tag{11.101}$$

Substituting (11.47), (11.48), (11.100) and (11.101) into (C.32) (see Appendix C), followed by an elimination of the strains using (11.49)–(11.51) leads to the compact result for the Gibbs energy (equivalent to the complementary energy) per unit volume of a damaged laminate

$$\begin{aligned}\hat{g} = &-\frac{\sigma_{\rm t}^2}{2\hat{E}_{\rm t}} - \frac{\sigma_{\rm A}^2}{2\hat{E}_{\rm A}} - \frac{\sigma_{\rm T}^2}{2\hat{E}_{\rm T}} + \frac{\hat{\nu}_{\rm a}}{\hat{E}_{\rm A}}\sigma_{\rm t}\sigma_{\rm A} + \frac{\hat{\nu}_{\rm t}}{\hat{E}_{\rm T}}\sigma_{\rm t}\sigma_{\rm T} + \frac{\hat{\nu}_{\rm A}}{\hat{E}_{\rm A}}\sigma_{\rm A}\sigma_{\rm T} \\ &-(\sigma_{\rm t}\hat{\alpha}_{\rm t} + \sigma_{\rm A}\hat{\alpha}_{\rm A} + \sigma_{\rm T}\hat{\alpha}_{\rm T})\Delta T + \frac{1}{2}(\sigma_{\rm t}^{\rm c}\hat{\alpha}_{\rm t} + \sigma_{\rm A}^{\rm c}\hat{\alpha}_{\rm A} + \sigma_{\rm T}^{\rm c}\hat{\alpha}_{\rm T})\Delta T \\ &- F(\hat{\varepsilon}_{\rm A}, \hat{\varepsilon}_{\rm T}) - \frac{1}{2V}\int_V \sigma_{ij}^{\rm c}\alpha_{ij}\Delta T\, {\rm d}V + g_0(\Delta T),\end{aligned} \tag{11.102}$$

where

$$F(\hat{\varepsilon}_{\rm A}, \hat{\varepsilon}_{\rm T}) = \frac{1}{2\Lambda}\left[\hat{E}_{\rm A}\hat{\varepsilon}_{\rm A}(\hat{\varepsilon}_{\rm A} + \hat{\delta}_{\rm T}\hat{\varepsilon}_{\rm T}) + \hat{E}_{\rm T}\hat{\varepsilon}_{\rm T}(\hat{\varepsilon}_{\rm T} + \hat{\delta}_{\rm A}\hat{\varepsilon}_{\rm A})\right], \quad \Lambda = 1 - \hat{\delta}_{\rm A}\hat{\delta}_{\rm T}. \tag{11.103}$$

The corresponding expression for an undamaged laminate is

$$\begin{aligned}\hat{g}_0 = &-\frac{\sigma_{\rm t}^2}{2\hat{E}_{\rm t}^0} - \frac{\sigma_{\rm A}^2}{2\hat{E}_{\rm A}^0} - \frac{\sigma_{\rm T}^2}{2\hat{E}_{\rm T}^0} + \frac{\hat{\nu}_{\rm a}^0}{\hat{E}_{\rm A}^0}\sigma_{\rm t}\sigma_{\rm A} + \frac{\hat{\nu}_{\rm t}^0}{\hat{E}_{\rm A}^0}\sigma_{\rm t}\sigma_{\rm T} + \frac{\hat{\nu}_{\rm A}^0}{\hat{E}_{\rm A}^0}\sigma_{\rm A}\sigma_{\rm T} \\ &-(\sigma_{\rm t}\hat{\alpha}_{\rm t}^0 + \sigma_{\rm A}\hat{\alpha}_{\rm A}^0 + \sigma_{\rm T}\hat{\alpha}_{\rm T}^0)\Delta T + \frac{1}{2}(\sigma_{\rm t}^{\rm c}\hat{\alpha}_{\rm t}^0 + \sigma_{\rm A}^{\rm c}\hat{\alpha}_{\rm A}^0 + \sigma_{\rm T}^{\rm c}\hat{\alpha}_{\rm T}^0)\Delta T \\ &- F_0(\hat{\varepsilon}_{\rm A}^0, \hat{\varepsilon}_{\rm T}^0) - \frac{1}{2V}\int_V \sigma_{ij}^{\rm c}\alpha_{ij}\Delta T\, {\rm d}V + g_0(\Delta T),\end{aligned} \tag{11.104}$$

where

$$F_0(\hat{\varepsilon}_{\rm A}, \hat{\varepsilon}_{\rm T}) = \frac{1}{2\Lambda_0}\left[\hat{E}_{\rm A}^0\hat{\varepsilon}_{\rm A}(\hat{\varepsilon}_{\rm A} + \hat{\delta}_{\rm T}^0\hat{\varepsilon}_{\rm T}) + \hat{E}_{\rm T}^0\hat{\varepsilon}_{\rm T}(\hat{\varepsilon}_{\rm T} + \hat{\delta}_{\rm A}^0\hat{\varepsilon}_{\rm A})\right], \quad \Lambda_0 = 1 - \hat{\delta}_{\rm A}^0\hat{\delta}_{\rm T}^0. \tag{11.105}$$

Subtracting (11.104) from (11.102) leads to the relation

$$
\begin{aligned}
\widehat{g}-\widehat{g}_0 = & -\frac{\sigma_t^2}{2}\left(\frac{1}{\widehat{E}_t}-\frac{1}{\widehat{E}_t^0}\right)-\frac{\sigma_A^2}{2}\left(\frac{1}{\widehat{E}_A}-\frac{1}{\widehat{E}_A^0}\right)-\frac{\sigma_T^2}{2}\left(\frac{1}{\widehat{E}_T}-\frac{1}{\widehat{E}_T^0}\right) \\
& -\left(\frac{\widehat{\nu}_a^0}{\widehat{E}_A^0}-\frac{\widehat{\nu}_a}{\widehat{E}_A}\right)\sigma_t\sigma_A+\left(\frac{\widehat{\nu}_t^0}{\widehat{E}_T^0}-\frac{\widehat{\nu}_t}{\widehat{E}_T}\right)\sigma_t\sigma_T+\left(\frac{\widehat{\nu}_A^0}{\widehat{E}_A^0}-\frac{\widehat{\nu}_A}{\widehat{E}_A}\right)\sigma_A\sigma_T \\
& -\sigma_t(\widehat{\alpha}_t-\widehat{\alpha}_t^0)\Delta T-\sigma_A(\widehat{\alpha}_A-\widehat{\alpha}_A^0)\Delta T-\sigma_T(\widehat{\alpha}_T-\widehat{\alpha}_T^0)\Delta T \\
& +\frac{1}{2}\sigma_t^0(\widehat{\alpha}_t-\widehat{\alpha}_t^0)\Delta T+\frac{1}{2}\sigma_A^0(\widehat{\alpha}_A-\widehat{\alpha}_A^0)\Delta T+\frac{1}{2}\sigma_T^0(\widehat{\alpha}_T-\widehat{\alpha}_T^0)\Delta T \\
& -F(\hat{\varepsilon}_A,\hat{\varepsilon}_T)+F_0(\hat{\varepsilon}_A^0,\hat{\varepsilon}_T^0).
\end{aligned}
\tag{11.106}
$$

Use can now be made of (11.90) and the independent interrelationships (11.91)–(11.98) to express result (11.106) in terms of the damage function Φ:

$$
\begin{aligned}
\widehat{g}-\widehat{g}_0 = & -\frac{\Phi}{2}\Big[(\widehat{k}')^2\sigma_t^2+\sigma_A^2+\widehat{k}^2\sigma_T^2+2\widehat{k}'\sigma_t\sigma_A+2\widehat{k}\widehat{k}'\sigma_t\sigma_T+2\widehat{k}\sigma_A\sigma_T \\
& +2\widehat{k}_1\widehat{k}'\sigma_t\Delta T+2\widehat{k}_1\sigma_A\Delta T+2\widehat{k}\widehat{k}_1\sigma_T\Delta T \\
& -\widehat{k}_1\widehat{k}'\sigma_t^c\Delta T-\widehat{k}_1\sigma_A^c\Delta T-\widehat{k}\widehat{k}_1\sigma_T^c\Delta T\Big] \\
& -F(\hat{\varepsilon}_A,\hat{\varepsilon}_T)+F_0(\hat{\varepsilon}_A^0,\hat{\varepsilon}_T^0).
\end{aligned}
\tag{11.107}
$$

It follows from the reduced stress-strain relations, by subtraction of the damage-independent forms (11.62)–(11.64) from the damage-dependent forms (11.49)–(11.51), that

$$
\widehat{\varepsilon}_t-\widehat{\varepsilon}_t^0=\left(\frac{1}{\widehat{E}_t}-\frac{1}{\widehat{E}_t^0}\right)\sigma_t+\left(\frac{\widehat{\nu}_a^0}{\widehat{E}_A^0}-\frac{\widehat{\nu}_a}{\widehat{E}_A}\right)\sigma_A+\left(\frac{\widehat{\nu}_t^0}{\widehat{E}_T^0}-\frac{\widehat{\nu}_t}{\widehat{E}_T}\right)\sigma_T+(\widehat{\alpha}_t-\widehat{\alpha}_t^0)\Delta T, \tag{11.108}
$$

$$
\widehat{\varepsilon}_A-\widehat{\varepsilon}_A^0=\left(\frac{\widehat{\nu}_a^0}{\widehat{E}_A^0}-\frac{\widehat{\nu}_a}{\widehat{E}_A}\right)\sigma_t+\left(\frac{1}{\widehat{E}_A}-\frac{1}{\widehat{E}_A^0}\right)\sigma_A+\left(\frac{\widehat{\nu}_A^0}{\widehat{E}_A^0}-\frac{\widehat{\nu}_A}{\widehat{E}_A}\right)\sigma_T+(\widehat{\alpha}_A-\widehat{\alpha}_A^0)\Delta T, \tag{11.109}
$$

$$
\widehat{\varepsilon}_T-\widehat{\varepsilon}_T^0=\left(\frac{\widehat{\nu}_t^0}{\widehat{E}_T^0}-\frac{\widehat{\nu}_t}{\widehat{E}_T}\right)\sigma_t+\left(\frac{\widehat{\nu}_A^0}{\widehat{E}_A^0}-\frac{\widehat{\nu}_A}{\widehat{E}_A}\right)\sigma_A+\left(\frac{1}{\widehat{E}_T}-\frac{1}{\widehat{E}_T^0}\right)\sigma_T+(\widehat{\alpha}_T-\widehat{\alpha}_T^0)\Delta T. \tag{11.110}
$$

Using (11.90)–(11.98), these equations reduce to

$$
\widehat{\varepsilon}_t-\widehat{\varepsilon}_t^0=\widehat{k}'\Phi(\widehat{k}'\sigma_t+\sigma_A+\widehat{k}\sigma_T-\overline{\sigma}_A^c), \tag{11.111}
$$

$$
\widehat{\varepsilon}_A-\widehat{\varepsilon}_A^0=\Phi(\widehat{k}'\sigma_t+\sigma_A+\widehat{k}\sigma_T-\overline{\sigma}_A^c), \tag{11.112}
$$

$$
\widehat{\varepsilon}_T-\widehat{\varepsilon}_T^0=\widehat{k}\Phi(\widehat{k}'\sigma_t+\sigma_A+\widehat{k}\sigma_T-\overline{\sigma}_A^c), \tag{11.113}
$$

where $\bar{\sigma}_A^c = -\hat{k}_1 \Delta T$ is clearly the closure stress for ply cracking during the uniaxial loading of the laminate in the axial direction. It is clear that at the point of crack closure, the strains in a damaged laminate will be identical to those in an undamaged laminate subject to the same imposed stresses. Thus, (11.111)–(11.113) all reduce to the single equation

$$\hat{k}'\sigma_t^c + \sigma_A^c + \hat{k}\sigma_T^c = \bar{\sigma}_A^c = -\hat{k}_1 \Delta T, \tag{11.114}$$

which must hold for any stress state that just closes the ply cracks. Substitution into (11.107) to eliminate ΔT leads to

$$\begin{aligned} \hat{g} - \hat{g}_0 = -\frac{\Phi}{2}\Big[&(\hat{k}')^2\sigma_t^2 + \sigma_A^2 + \hat{k}^2\sigma_T^2 + 2\hat{k}'\sigma_t\sigma_A + 2\hat{k}\hat{k}'\sigma_t\sigma_T + 2\hat{k}\sigma_A\sigma_T \\ &- 2\hat{k}'\sigma_t\bar{\sigma}_A^c - 2\sigma_A\bar{\sigma}_A^c - 2\hat{k}\sigma_T\bar{\sigma}_A^c + (\bar{\sigma}_A^c)^2\Big] \\ &- F(\hat{\varepsilon}_A, \hat{\varepsilon}_T) + F_0(\hat{\varepsilon}_A^0, \hat{\varepsilon}_T^0), \end{aligned} \tag{11.115}$$

which may be written in the compact form

$$\hat{g} - \hat{g}_0 = -\frac{\Phi}{2}\left[\hat{k}'\sigma_t + \sigma_A + \hat{k}\sigma_T - \bar{\sigma}_A^c\right]^2 - F(\hat{\varepsilon}_A, \hat{\varepsilon}_T) + F_0(\hat{\varepsilon}_A^0, \hat{\varepsilon}_T^0). \tag{11.116}$$

11.9 Predicting First Ply Cracking

Consider the special case where, during the formation of every ply crack, the fracture energy for ply crack formation has a unique value 2γ. The first objective is to determine the conditions for which it is energetically favourable for an array of equally spaced ply cracks having density ρ_0 to form quasistatically in an undamaged laminate subject to fixed applied loads and temperature. Consider a macroscopic region V of the laminate. From energy balance considerations and the fact that kinetic energy is never negative, the criterion for crack formation under these conditions has the form [2]

$$\Delta\Gamma + \Delta G < 0, \tag{11.117}$$

where the energy absorbed in a macroscopic volume V of laminate by the formation of the new cracks is given by

$$\Delta\Gamma = V\Gamma. \tag{11.118}$$

In (11.118) the parameter Γ denotes the energy absorbed per unit volume of laminate during the formation of new ply crack surfaces in the 90° plies that have led to the initial damage state denoted by the ply crack density ρ_0. It is easily shown that

$$\Gamma = \frac{2\gamma\rho_0 h^{(90)}}{h}, \tag{11.119}$$

where $h^{(90)}$ is the total thickness of the 90° plies in which the ply cracks have formed. The corresponding change of Gibbs energy in the region V of the laminate is

$$\Delta G = \int_V (g - g_0) \mathrm{d}V. \tag{11.120}$$

In (11.120), g denotes the Gibbs energy per unit volume when the damage in the laminate is characterised by the ply crack density ρ_0, and where g_0 denotes the corresponding value of g in the undamaged state. It follows from (11.117)–(11.120) that ply crack formation is governed by the inequality

$$P < 0, \tag{11.121}$$

where P is the total energy defined by

$$P \equiv \Gamma + \frac{1}{V}\int_V (g - g_0) \mathrm{d}V \equiv \Gamma + \hat{g} - \hat{g}_0, \tag{11.122}$$

where use has been made of the definition of the average Gibbs energy given by (C4) (see Appendix C) and denoted here by $\hat{g}$, and where from (11.116),

$$\hat{g} - \hat{g}_0 = -\frac{\Phi(\rho_0)}{2}(\hat{k}'\sigma_t + \sigma_A + \hat{k}\sigma_T - \overline{\sigma}_A^c)^2 - F(\hat{\varepsilon}_A, \hat{\varepsilon}_T, \rho_0) + F_0(\hat{\varepsilon}_A^0, \hat{\varepsilon}_T^0), \tag{11.123}$$

where from (11.90),

$$\Phi(\rho_0) = \frac{1}{\hat{E}_A(\rho_0)} - \frac{1}{\hat{E}_A^0}, \tag{11.124}$$

and from (11.103),

$$F(\hat{\varepsilon}_A, \hat{\varepsilon}_T, \rho) = \frac{1}{2\Lambda(\rho)}\left[\hat{E}_A(\rho)\hat{\varepsilon}_A(\hat{\varepsilon}_A + \hat{\delta}_T(\rho)\hat{\varepsilon}_T) + \hat{E}_T(\rho)\hat{\varepsilon}_T(\hat{\varepsilon}_T + \hat{\delta}_A(\rho)\hat{\varepsilon}_A)\right],$$
$$\Lambda(\rho) = 1 - \hat{\delta}_A(\rho)\hat{\delta}_T(\rho). \tag{11.125}$$

On using (11.119) and (11.121)–(11.123), it then follows that first ply crack formation is governed by the inequality

$$\frac{\Phi(\rho_0)}{2}(\hat{k}'\sigma_t + \sigma_A + \hat{k}\sigma_T - \overline{\sigma}_A^c)^2 + F(\hat{\varepsilon}_A, \hat{\varepsilon}_T, \rho_0) - F_0(\hat{\varepsilon}_A^0, \hat{\varepsilon}_T^0) > \frac{2\gamma\rho_0 h^{(90)}}{h}. \tag{11.126}$$

It is emphasised that the approach described here does not provide any information that indicates how the effective thermoelastic constants depend upon ply crack density. A detailed stress analysis is required to provide this information (see Section 11.7.1 and Chapter 19).

11.10 Progressive Cracking

Because the fracture energy for ply cracking has a precise value (i.e. it is not statistically distributed among the various potential ply crack formation sites in the laminate), and because the initial ply cracking is in the form of a uniformly spaced array of ply cracks, progressive ply crack formation will occur only by the formation of new ply cracks at the midpoints between the existing ply cracks. Thus, if ρ_0 is the ply crack density that forms initially in an undamaged laminate, then successive progressive ply crack formation will be such that uniform ply crack densities will be described by the sequence $\rho_i = 2^i \rho_0,\ i \geq 1$. Consider now the energy balance involved when the ply crack density changes from ρ_i to ρ_{i+1} for a laminate subject to fixed applied loads and temperature.

The energy absorbed in a macroscopic volume V of laminate by the formation of the new ply cracks is given, for $i = 1, 2, 3, \ldots,$ by

$$\Delta\Gamma = V(\Gamma_{i+1} - \Gamma_i), \tag{11.127}$$

where, consistent with (11.119),

$$\Gamma_i = \frac{2^{i+1}\gamma\rho_0 h^{(90)}}{h}, \tag{11.128}$$

denotes the energy absorbed per unit volume of laminate (averaged over the macroscopic region V) during the formation of new crack surfaces in the 90° plies that have led to a damage state denoted by the ply crack density ρ_i. The corresponding change of Gibbs energy in the region V of the laminate is

$$\Delta G = \int_V (g_{i+1} - g_i)\mathrm{d}V. \tag{11.129}$$

In (11.129), g_i denotes the Gibbs energy per unit volume when the damage in the laminate is characterised by the ply crack density ρ_i. It follows from (11.117), (11.127) and (11.129) that crack formation associated with the change in ply crack density from ρ_i to ρ_{i+1} is governed by the inequality

$$P_{i+1} < P_i, \tag{11.130}$$

where P_i is the total energy defined using (11.122) as follows

$$P_i \equiv \Gamma_i + \frac{1}{V}\int_V (g_i - g_0)\,\mathrm{d}V \equiv \Gamma_i + \hat{g}_i - \hat{g}_0. \tag{11.131}$$

It follows from (11.123) that when the ply crack density is ρ_i

$$\begin{aligned} \hat{g}_{i+1} - \hat{g}_i = -\frac{1}{2}\left[\Phi(\rho_{i+1}) - \Phi(\rho_i)\right](\hat{k}'\sigma_t + \sigma_A + \hat{k}\sigma_T - \bar{\sigma}_A^c)^2 \\ - F(\hat{\varepsilon}_A^{i+1}, \hat{\varepsilon}_T^{i+1}, \rho_{i+1}) + F(\hat{\varepsilon}_A^i, \hat{\varepsilon}_T^i, \rho_i). \end{aligned} \tag{11.132}$$

It then follows from (11.130)–(11.132) that the criterion for progressive ply crack formation may be written, for $i = 0, 1, 2, \ldots,$ as

$$\begin{aligned} &\frac{1}{2}\left[\Phi(\rho_{i+1}) - \Phi(\rho_i)\right](\hat{k}'\sigma_t + \sigma_A + \hat{k}\sigma_T - \bar{\sigma}_A^c)^2 \\ &\quad + F(\hat{\varepsilon}_A^{i+1}, \hat{\varepsilon}_T^{i+1}, \rho_{i+1}) - F(\hat{\varepsilon}_A^i, \hat{\varepsilon}_T^i, \rho_i) > \frac{2^{i+1}\gamma\rho_0 h^{(90)}}{h}. \end{aligned} \tag{11.133}$$

11.10.1 Ply Crack Formation during Simple Bending

To explore the characteristics of the energy balance model it is useful to consider ply cracking in the simplest possible situation involving bend deformation. It is now assumed that $\sigma_t = \sigma_A = \sigma_T = M_T = 0$ and that $\Delta T = 0$ so that the thermal residual stresses are zero. It follows from (11.43), (11.44) and (11.46) that

$$\hat{\varepsilon}_A = \frac{M_A}{\hat{E}_A}, \qquad \hat{\varepsilon}_T = -\frac{\hat{\delta}_A}{\hat{E}_A} M_A. \tag{11.134}$$

As, from (11.114), $\overline{\sigma}_A^c = 0$, it follows using (11.124) and (11.125) that

$$F(\hat{\varepsilon}_A, \hat{\varepsilon}_T, \rho_0) = \frac{M_A}{2\hat{E}_A(\rho_0)}. \tag{11.135}$$

The criterion (11.126) for ply crack formation may then be written in the following form where the right-hand side is independent of ρ_0

$$\frac{1}{\rho_0}\left(\frac{1}{\hat{E}_A(\rho_0)} - \frac{1}{\hat{E}_A^0}\right) M_A^2 > \frac{4\gamma h^{(90)}}{h}, \tag{11.136}$$

that is,

$$M_A > \sqrt{\frac{4\gamma h^{(90)}/h}{\frac{1}{\rho_0}\left(\frac{1}{\hat{E}_A(\rho_0)} - \frac{1}{\hat{E}_A^0}\right)}}. \tag{11.137}$$

For the other case of simple bending where $\sigma_t = \sigma_A = \sigma_T = M_A = 0$ and $\Delta T = 0$, it follows similarly that

$$\hat{\varepsilon}_A = -\frac{\hat{\delta}_A}{\hat{E}_A} M_T, \qquad \hat{\varepsilon}_T = \frac{M_T}{\hat{E}_T}, \tag{11.138}$$

$$F(\hat{\varepsilon}_A, \hat{\varepsilon}_T, \rho_0) = \frac{M_T}{2\hat{E}_T(\rho_0)}. \tag{11.139}$$

The criterion (11.126) for initial ply crack formation in the 90° plies may then be written as

$$\frac{1}{\rho_0}\left(\frac{1}{\hat{E}_T(\rho_0)} - \frac{1}{\hat{E}_T^0}\right) M_T^2 > \frac{4\gamma h^{(90)}}{h}, \tag{11.140}$$

that is,

$$M_T > \sqrt{\frac{4\gamma h^{(90)}/h}{\frac{1}{\rho_0}\left(\frac{1}{\hat{E}_T(\rho_0)} - \frac{1}{\hat{E}_T^0}\right)}}. \tag{11.141}$$

11.10.2 Simple Bending with Thermal Residual Stresses

It is now assumed that $\sigma_t = \sigma_A = \sigma_T = M_T = 0$, but that $\Delta T \neq 0$ so that the thermal residual stresses are nonzero. On substituting (11.43) and (11.44) into (11.126) it can be shown that the relation governing ply crack formation may be written in the form

$$aM_A^2 + bM_A\Delta T + c(\Delta T)^2 > \frac{2\gamma\rho_0 h^{(90)}}{h}, \tag{11.142}$$

where a, b and c are parameters that depend on the thermoelastic constants in both damaged and undamaged states whose values are most easily determined by numerical methods. It follows directly from (11.142) that

$$M_A > \frac{1}{2a}\left[-b\Delta T + \sqrt{b^2(\Delta T)^2 - 4a\left(c(\Delta T)^2 - 2\gamma\rho_0 h^{(90)}/h\right)}\right]. \tag{11.143}$$

11.11 An Alternative Approximate Approach to Ply Cracking

The approach to the prediction of progressive ply crack formation described in Section 11.11 leads to the situation that there is a characteristic initial ply crack density ρ_0 that successively doubles as the applied loading is increased. Crack density versus loading plots then have unrealistic step profiles as so few points defining the plot can be generated. This idealised situation will not occur in practice, as statistical materials property variability will lead to ply cracks forming one at a time during loading creating arrays of cracks that will not be distributed uniformly, as is usually observed. Such statistical variability was considered in Chapter 10 when considering ply crack formation in general symmetric laminates subject to general in-plane loading and through-thickness loading but in the absence of bend deformation. The ply cracks were assumed to form one at a time at locations which led to the need to consider the effective thermoelastic constants of laminates having nonuniform arrays of ply cracks. Dealing with such complexity for cross-ply laminates in the presence of bend deformation, whilst technically feasible, would lead to very complex analysis that has not been attempted. An approximate approach is, however, now developed where ply cracks are progressively formed one at a time enabling a mean ply crack density to be calculated. The effective properties of the cracked laminate are estimated *approximately* by assuming that at each stage the ply cracks present are uniformly distributed thus enabling the use of the results in Section 11.10.

Consider a length $2L$ of laminate in which there are m ply cracks that might not be uniformly spaced. The mean ply crack density can then be defined by $\rho_m = m/2L$. Assuming that $m-1$ ply cracks have already formed in a laminate it is now required to determine the conditions for which the mth ply crack can form. By applying the principles developed in Section 11.10 to this problem, it is clear that relation (11.133), developed for crack densities that progressively double during loading, must be modified to the form

$$\frac{1}{2}\left[\Phi(\rho_m)-\Phi(\rho_{m-1})\right](\widehat{k'}\sigma_t+\sigma_A+\widehat{k}\sigma_T+\widehat{k}_1\Delta T)^2 \\ +F(\hat{\varepsilon}_A^m,\hat{\varepsilon}_T^m,\rho_m)-F(\hat{\varepsilon}_A^{m-1},\hat{\varepsilon}_T^{m-1},\rho_{m-1})>\frac{\gamma h^{(90)}}{Lh}. \tag{11.144}$$

This relation can be used to determine whether the laminate loading is sufficient to form the mth ply crack, given that $m-1$ ply cracks have already formed. Although the ply crack densities are easily defined using the relation $\rho_m=m/2L$ and the undamaged laminate constants $\widehat{k'}$, $\widehat{k}$ and $\widehat{k}_1$ are known, it is necessary also to be able to estimate the effective thermoelastic constants of the damaged laminate appearing in relations (11.124) and (11.125). The required effective properties of the damaged laminate are estimated approximately by assuming that the ply crack densities ρ_i required by the criterion (11.144) are uniform and can, thus, be calculated using results derived earlier in this chapter.

The simplest way of predicting progressive ply crack formation is to assume that the loading is proportional so that at any stage of loading

$$\sigma_A=K_1 s,\quad M_A=K_2 s,\quad \sigma_T=K_3 s,\quad M_T=K_4 s,\quad \sigma_t=K_5 s, \tag{11.145}$$

where K_i, $i=1,\ldots,5$, are dimensionless load ratios and s is a stress with value characterising the state of loading. It follows from (11.124), (11.125) and (11.144) that, for a given state of ply cracking, the next ply crack will form when the characteristic stress s has the value satisfying an algebraically complicated quadratic equation which has the following simple form

$$P(s,\Delta T)\equiv As^2+Bs\Delta T+C(\Delta T)^2=\frac{\gamma h^{(90)}}{Lh}, \tag{11.146}$$

where the function P is determined using the left-hand side of (11.144), and where A, B and C can then be determined using

$$P(1,0)=A,\quad P(0,1)=C,\quad P(1,1)=A+B+C. \tag{11.147}$$

Given a value of the temperature difference ΔT and a value of the fracture energy for ply cracking 2γ, it is then possible to solve the quadratic equation (11.146) to determine a value of the applied stress parameter s. This calculation can be performed for every ply crack that forms enabling the construction of ply crack density versus stress or strain plots, and the stress-strain relation that results when ply cracks progressively form.

To conclude this chapter, it is noted that all relations derived are expected to be exact results. The analysis given does not lead to a method of estimating the effective thermoelastic constants of a damaged laminate, but it does generate sets of relationships (11.91)–(11.98) between constants, for damaged and undamaged states, that all depend on a single damage parameter Φ defined by (11.90). Values of Φ can be obtained using a stress-transfer analysis that is described in Chapter 19.

References

1. McCartney, L.N. and Pierse, C. (1997). Stress transfer mechanics for multiple-ply laminates subject to bending. NPL Report CMMT(A)55, February.
2. McCartney, L.N. and Pierse, C. (1997). Stress transfer mechanics for multiple ply laminates for axial loading and bending. *Proc. 11th Int. Conf. on Composite Materials, Gold Coast*, Australia. July 14–18, vol. V, 662–671.
3. Li, S., Reid, S.R., and Soden, P.D. (1994). A finite strip analysis of cracked laminates. *Mechanics of Materials* 18: 289–311.
4. Smith, P.A. and Ogin, S.L. (2000). Characterisation and modelling of matrix cracking in a $(0/90)_{2s}$ GFRP laminate loaded in flexure. *Proceedings of the Royal Society of London* A 456: 2755–2770.

Further Reading

Smith, P.A. and Ogin, S.L. (1999). On transverse matrix cracking in cross-ply laminates loaded in simple bending. *Composites Part A* 30: 1003–1008.

McCartney, L.N. Stress transfer mechanics for multiple ply cross-ply laminates subject to bending. (2001). *Summary in Proceedings of 6th International Conference on the Deformation and Fracture of Composites*, Manchester. April.

McCartney, L N. and Byrne, M.J.W. (2001). Energy balance method for predicting cracking in cross-ply laminates during bend deformation. *Proc. 10th Int. Conf. on Fracture (ICF-10).* Honolulu: Advances in Fracture Research. 2–6 Dec.

McCartney, L.N. (1998). Prediction of microcracking in composite materials. In: *FRACTURE: A Topical Encyclopaedia of Current Knowledge Dedicated to Alan Arnold Griffith*, (ed. G.P. Cherepanov), Melbourne, USA: Krieger Publishing Company. 905–916.

12

Energy-based Delamination Theory for Biaxial Loading in the Presence of Thermal Stresses

Overview: This chapter addresses the issue of using energy balance methods and crack closure concepts to predict the growth of delaminations associated with ply cracks during the progressive loading of cross-ply laminates subject to a combination of in-plane biaxial stresses and thermal residual stresses. When the effective applied stresses and the temperature are held fixed during delamination growth, and there is negligible interaction of the delamination tips with the ply cracks, very simple analytical formulae for the energy release rate can be derived for unconstrained and generalised plane strain conditions, which are exact when the ply crack separation tends to infinity. In some practical applications, such as for wide plates, delamination growth is constrained transversely by surrounding undamaged laminate and this has an effect on energy release rates. An analytical method is described that determines the energy release rate for delamination growth under such constrained conditions. This result is also exact in the limit of ply crack separations tending to infinity.

12.1 Introduction

Progressive ply crack formation in cross-ply laminates is often accompanied during the later stages of loading by the appearance of delaminations that are understood to have initiated from ply cracks. It is of interest, therefore, to be able to predict the degree of loading required to enable delaminations, which have already initiated at ply crack tips, to grow along the interfaces between the 0° and 90° plies and meet delaminations growing in the opposite direction.

O'Brien [1] developed, for cross-ply laminates, an analytical model based on energy balance principles that determines the value of the applied uniaxial stress for 'steady-state' delamination for the special case where there is no interaction between ply cracks that form in the 90° plies. In addition, thermal residual stresses were neglected but considered in O'Brien's later paper [2]. Nairn and Hu [3] used a variational technique to investigate the behaviour of delaminations that have initiated at ply cracks in $[(S)/90_n]_s$ laminates where S is any orthotropic sublaminate that is much stiffer than the 90° plies. The properties of the S plies might be homogenised thermoelastic constants that represent the effect of off-axis plies. The interactions between neighbouring ply cracks in the 90° plies were taken into account, as were the effects of thermal

Properties for Design of Composite Structures: Theory and Implementation Using Software, First Edition. Neil McCartney.

Companion Website: www.wiley.com/go/mccartney/properties

stresses. For the special case where thermal stresses are ignored and the ply crack density is very small, so that there is negligible interaction between the ply cracks, the energy release rate for the delamination was shown to be identical to that predicted by O'Brien [1]. Takeda and Ogihara [4] used a shear-lag method of accounting for stress transfer when investigating the effects, on ply cracking and delamination damage in CFRP laminates, of the inclusion in the model of interlaminar shear layers, and of thermal residual stresses. Their result for the energy release rate for delamination reduces to that of O'Brien [1] when thermal stresses are ignored, shear layers are absent, and the separation of ply cracks is large enough for ply crack interaction effects to be negligible.

The delamination models described in [1–4], all assume that the transverse stresses in both 0° and 90° plies are zero everywhere, even for an undamaged laminate. This situation can never occur in practice as bonded 0° and 90° plies always exhibit the same nonzero in-plane transverse strain that is shared by the 0° and 90° plies. The corresponding values for the transverse in-plane stresses in the 0° and 90° plies differ, and they can be combined to calculate the effective transverse stress applied to the laminate that might be zero. One objective of this chapter is to show how the O'Brien model can be extended by considering generalised plane strain deformations and showing how to take account of biaxial in-plane loading combined with thermal residual stresses characterised by the difference between the laminate temperature and its stress-free temperature.

A further objective of this chapter is to develop analytical methods of estimating the delamination energy release rates for three different types of laminate loading. To achieve this, a special case is assumed where the uniform ply crack density is so low that neighbouring ply cracks are widely separated. In addition, the delamination crack tips are well away from the ply cracks so that they do not interact with the ply cracks, nor with the other delaminations that are present in the same section of the 0°/90° interface between neighbouring ply cracks. For this case, the stress and deformation in regions well away from the delamination crack tips will not be affected by their growth because delamination growth is self-similar. When neighbouring ply cracks are wide apart and the delamination length is large enough, the self-similar region of the laminate moves with the delamination tip during growth. Self-similar growth conditions lead to limiting values of the energy release rate, which are of importance when assessing the threat of significant laminate damage in the form of interply delaminations.

When taking proper account of the transverse strains in the 0° and 90° plies of a laminate, it is necessary to consider very carefully the transverse boundary conditions that need to be applied when developing models of delamination growing from ply cracks in the 90° plies. For a uniaxially loaded laminate, the values of the transverse strains in the 0° and 90° plies that have delaminated will be different, and they will both differ from that in the undamaged part of the laminate. Such a situation will be encountered when testing coupons in the laboratory. However, when delaminations form at ply cracks in a wide laminated plate that is part of an engineering structure, much of the plate will be undamaged and this must constrain the transverse strains that occur in the regions of delamination. This chapter develops a delamination model that can be applied in this practically important situation, by assuming generalised plane strain conditions where the transverse in-plane strain in the damaged laminate is constrained everywhere to have the value that would result in an undamaged laminate subject to the same loading. The approach will account for the effects of thermal residual stresses but ignores interaction effects between neighbouring ply cracks in the 90° plies.

The analytical methodology for self-similar delamination growth to be described in this chapter is an exact approach that is ideal for assessing the quality of numerical methods of solution such as those based on the boundary element or the finite element method. The assumptions made, which enable effective stresses and strains to be defined, are readily applied to numerical methods if axial loading is applied by subjecting the damaged laminate to a prescribed axial strain. The results of this chapter will, therefore, provide exact values of energy release rates that can be used to assess the errors of numerical methods applied to relatively simple laminates and loading.

12.2 Geometry and Mode of Loading

Consider a $[0_m / 90_n]_s$ cross-ply laminate having an array of equally spaced ply cracks in the 90° ply having separation $2L$. A set of Cartesian coordinates (x_1, x_2, x_3) are introduced having origin at the intersection of the midplane between two neighbouring ply cracks, and the midplanes of the laminate in the longitudinal and transverse directions, such that x_1 is in the axial direction, x_2 is in the in-plane transverse direction and x_3 is in the through-thickness direction. The geometry of a representative volume element of the laminate is shown in Figure 12.1.

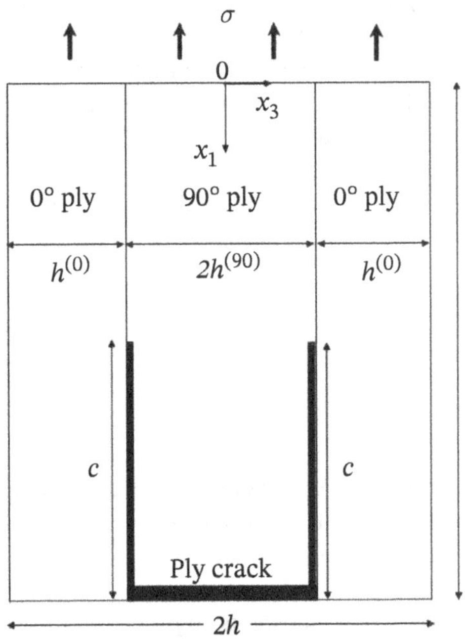

Figure 12.1 Schematic diagram of a damaged laminate illustrating the geometry and loading, where x_2 is normal to the plane of the page.

The ply thicknesses of the 0° and 90° plies are denoted by $h^{(0)}$ and $h^{(90)}$, respectively, so that the total laminate thickness $2h = 2(h^{(0)} + h^{(90)})$, and the laminate width is $2W$. Four delaminations of equal length c are growing in both the interfaces between the 90° ply and the 0° plies, having been initiated at the tips of the ply cracks. In each of the interfaces the total delaminated length in the region between two neighbouring ply cracks is thus $2c$. The external faces $x_3 = \pm h$ of the laminate are stress-free and the temperature T is everywhere uniform and fixed in time, having a value that differs from the uniform stress-free temperature T_0 of the laminate. The temperature difference is denoted by $\Delta T = T - T_0$. It is assumed that there are no applied shear tractions on the laminate edges. The laminate is now assumed to be subject to imposed uniform axial displacements $\pm\delta_A$ on $x_1 = \pm L$ leading to an effective applied axial strain $\varepsilon_A = \delta_A / L$. Similarly, the laminate is subject to imposed uniform transverse displacements $\pm\delta_T$ on $x_2 = \pm W$ leading to an effective applied transverse in-plane strain $\varepsilon_T = \delta_T / W$.

Owing to the presence of damage, the traction distributions on $x_1 = \pm L$ and on $x_2 = \pm W$ will be nonuniform and complex so that it is necessary to introduce an *effective* applied axial stress σ_A and an *effective* applied in-plane transverse stress σ_T defined respectively by

$$\sigma_A = \frac{1}{4Wh}\int_{-h}^{h}\int_{-W}^{W} \sigma_{22}(L, x_2, x_3)\, dx_2\, dx_3, \tag{12.1}$$

$$\sigma_T = \frac{1}{4Lh}\int_{-h}^{h}\int_{-L}^{L} \sigma_{33}(x_1, W, x_3)\, dx_1\, dx_3. \tag{12.2}$$

Because there are no normal or shear tractions acting on the laminate faces and edges, it follows that the through-thickness stress $\sigma_t = 0$ whereas the effective axial stress acting on the plane at any location x_1 is σ_A and the effective transverse stress acting on the plane at any location x_2 is σ_T. It is emphasised that the word 'effective' could be replaced by 'average', and that the imposition of an effective applied stress does not imply that the tractions applied are uniform.

As it has been assumed that $\sigma_t = 0$, the effective stress-strain relations for the damaged laminate, relating linearly the quantities ε_A, ε_T, σ_A, σ_T and ΔT, are expected to be of the form (see Chapter 7)

$$\varepsilon_A = \frac{\sigma_A}{E_A} - \frac{\nu_A}{E_A}\sigma_T + \alpha_A \Delta T, \tag{12.3}$$

$$\varepsilon_T = -\frac{\nu_A}{E_A}\sigma_A + \frac{\sigma_T}{E_T} + \alpha_T \Delta T, \tag{12.4}$$

where for a damaged laminate the axial and transverse Young's moduli are denoted by E_A and E_T, respectively, the major axial Poisson's ratio is denoted by ν_A, and the axial and transverse thermal expansion coefficients are denoted by α_A and α_T, respectively.

There is an assumed dependence of the thermoelastic constants on ply crack density $\rho = 1/(2L)$ and the length c of the delaminations, indicating that these thermoelastic constants are effective values for a damaged laminate. The form (12.3) and (12.4) is known to be valid when ply cracking is the only form of damage. It is expected that they will be valid also for laminates with both ply cracking and delaminations provided that any crack contacts are friction-free. To illustrate this point, relations (12.3) and (12.4) have been used to predict the applied strains for which the ply cracks and delaminations are expected to just close (see Section 12.4 and Appendix D). When these strains are applied, unpublished boundary element methods solutions (not presented here) have confirmed that both ply cracks and delaminations are indeed just closed (i.e. no gaps and zero tractions on fractured surfaces).

12.3 Undamaged Laminates

The effective in-plane stress-strain relations for an undamaged laminate are obtained from (12.3) and (12.4) so that

$$\varepsilon_{\mathrm{A}}^{0}=\frac{\sigma_{\mathrm{A}}}{E_{\mathrm{A}}^{0}}-\frac{\nu_{\mathrm{A}}^{0}}{E_{\mathrm{A}}^{0}}\sigma_{\mathrm{T}}+\alpha_{\mathrm{A}}^{0}\Delta T, \tag{12.5}$$

$$\varepsilon_{\mathrm{T}}^{0}=-\frac{\nu_{\mathrm{A}}^{0}}{E_{\mathrm{A}}^{0}}\sigma_{\mathrm{A}}+\frac{\sigma_{\mathrm{T}}}{E_{\mathrm{T}}^{0}}+\alpha_{\mathrm{T}}^{0}\Delta T, \tag{12.6}$$

where $\varepsilon_{\mathrm{A}}^{0}$ and $\varepsilon_{\mathrm{T}}^{0}$ are the in-plane axial and transverse strains for the 0° and 90° plies, respectively, and where σ_{A} and σ_{T} are the corresponding effective applied stresses in the axial and transverse directions, respectively. The superscript '0' is used to denote that the thermoelastic constant refers to an undamaged laminate. The values of the undamaged thermoelastic constants appearing in (12.5) and (12.6) can be calculated using analysis given, for example, in [5]. The following explicit formulae can be derived when the 0° and 90° plies are made of the same material

$$\left(1-\widehat{\nu}_{\mathrm{A}}^{2}\frac{\widehat{E}_{\mathrm{T}}}{\widehat{E}_{\mathrm{A}}}\right)E_{\mathrm{A}}^{0}=\frac{h^{(0)}\widehat{E}_{\mathrm{A}}+h^{(90)}\widehat{E}_{\mathrm{T}}}{h}-\frac{h\widehat{\nu}_{\mathrm{A}}^{2}\widehat{E}_{\mathrm{T}}^{2}}{h^{(0)}\widehat{E}_{\mathrm{T}}+h^{(90)}\widehat{E}_{\mathrm{A}}}, \tag{12.7}$$

$$\left(1-\widehat{\nu}_{\mathrm{A}}^{2}\frac{\widehat{E}_{\mathrm{T}}}{\widehat{E}_{\mathrm{A}}}\right)E_{\mathrm{T}}^{0}=\frac{h^{(0)}\widehat{E}_{\mathrm{T}}+h^{(90)}\widehat{E}_{\mathrm{A}}}{h}-\frac{h\widehat{\nu}_{\mathrm{A}}^{2}\widehat{E}_{\mathrm{T}}^{2}}{h^{(0)}\widehat{E}_{\mathrm{A}}+h^{(90)}\widehat{E}_{\mathrm{T}}}, \tag{12.8}$$

$$\nu_{\mathrm{A}}^{0}=\frac{h\widehat{\nu}_{\mathrm{A}}\widehat{E}_{\mathrm{T}}}{h^{(0)}\widehat{E}_{\mathrm{T}}+h^{(90)}\widehat{E}_{\mathrm{A}}},\quad \frac{E_{\mathrm{T}}^{0}}{E_{\mathrm{A}}^{0}}=\frac{h^{(0)}\widehat{E}_{\mathrm{T}}+h^{(90)}\widehat{E}_{\mathrm{A}}}{h^{(0)}\widehat{E}_{\mathrm{A}}+h^{(90)}\widehat{E}_{\mathrm{T}}},\quad \nu_{\mathrm{A}}^{0}\frac{E_{\mathrm{T}}^{0}}{E_{\mathrm{A}}^{0}}=\frac{h\widehat{\nu}_{\mathrm{A}}\widehat{E}_{\mathrm{T}}}{h^{(0)}\widehat{E}_{\mathrm{A}}+h^{(90)}\widehat{E}_{\mathrm{T}}}, \tag{12.9}$$

$$\begin{aligned}\left(1-\widehat{\nu}_{\mathrm{A}}^{2}\frac{\widehat{E}_{\mathrm{T}}}{\widehat{E}_{\mathrm{A}}}\right)E_{\mathrm{A}}^{0}\alpha_{\mathrm{A}}^{0}&=\left[\frac{h^{(0)}}{h}-\widehat{\nu}_{\mathrm{A}}\frac{h^{(90)}\widehat{E}_{\mathrm{T}}}{h^{(0)}\widehat{E}_{\mathrm{T}}+h^{(90)}\widehat{E}_{\mathrm{A}}}\right]\left(\widehat{E}_{\mathrm{A}}\widehat{\alpha}_{\mathrm{A}}+\widehat{\nu}_{\mathrm{A}}\widehat{E}_{\mathrm{T}}\widehat{\alpha}_{\mathrm{T}}\right)\\&+\left[\frac{h^{(90)}}{h}-\widehat{\nu}_{\mathrm{A}}\frac{h^{(0)}\widehat{E}_{\mathrm{T}}}{h^{(0)}\widehat{E}_{\mathrm{T}}+h^{(90)}\widehat{E}_{\mathrm{A}}}\right]\left(\widehat{E}_{\mathrm{T}}\widehat{\alpha}_{\mathrm{T}}+\widehat{\nu}_{\mathrm{A}}\widehat{E}_{\mathrm{T}}\widehat{\alpha}_{\mathrm{A}}\right),\end{aligned} \tag{12.10}$$

$$\begin{aligned}\left(1-\widehat{\nu}_{\mathrm{A}}^{2}\frac{\widehat{E}_{\mathrm{T}}}{\widehat{E}_{\mathrm{A}}}\right)E_{\mathrm{T}}^{0}\alpha_{\mathrm{T}}^{0}&=\left[\frac{h^{(0)}}{h}-\widehat{\nu}_{\mathrm{A}}\frac{h^{(90)}\widehat{E}_{\mathrm{T}}}{h^{(0)}\widehat{E}_{\mathrm{A}}+h^{(90)}\widehat{E}_{\mathrm{T}}}\right]\left(\widehat{E}_{\mathrm{T}}\widehat{\alpha}_{\mathrm{T}}+\widehat{\nu}_{\mathrm{A}}\widehat{E}_{\mathrm{T}}\widehat{\alpha}_{\mathrm{A}}\right)\\&+\left[\frac{h^{(90)}}{h}-\widehat{\nu}_{\mathrm{A}}\frac{h^{(0)}\widehat{E}_{\mathrm{T}}}{h^{(0)}\widehat{E}_{\mathrm{A}}+h^{(90)}\widehat{E}_{\mathrm{T}}}\right]\left(\widehat{E}_{\mathrm{A}}\widehat{\alpha}_{\mathrm{A}}+\widehat{\nu}_{\mathrm{A}}\widehat{E}_{\mathrm{T}}\widehat{\alpha}_{\mathrm{T}}\right),\end{aligned} \tag{12.11}$$

where, for any thermoelastic property p of a single ply, the symbol $\widehat{p}$ denotes that for the transverse isotropic material used to make the 0° plies. It is of interest to consider the special case when $\widehat{\nu}_A = 0$ so that the thermoelastic constants defined by (12.7)–(12.11) for an undamaged laminate reduce to the following 'mixtures' forms

$$\begin{aligned} hE_A^0 &= h^{(0)}\widehat{E}_A + h^{(90)}\widehat{E}_T, & hE_T^0 &= h^{(0)}\widehat{E}_T + h^{(90)}\widehat{E}_A, \quad \nu_A^0 = 0, \\ hE_A^0\alpha_A^0 &= h^{(0)}\widehat{E}_A\widehat{\alpha}_A + h^{(90)}\widehat{E}_T\widehat{\alpha}_T, & hE_T^0\alpha_T^0 &= h^{(0)}\widehat{E}_T\widehat{\alpha}_T + h^{(90)}\widehat{E}_A\widehat{\alpha}_A. \end{aligned} \tag{12.12}$$

An examination of the models described in [1–4] shows that they are based on relations of the form (12.12), indicating that there is an implied unphysical assumption that $\widehat{\nu}_A = 0$. The analysis of this chapter overcomes this deficiency.

12.4 Consideration of Crack Closure

The stress-free temperature T_0 of a laminate is a very important reference state for the approach to be used in this chapter. When the temperature T has this value and the laminate is unloaded, it follows from (12.3) and (12.4) that the stress field within the laminate is everywhere zero, and this reference state is used to define the zero for the corresponding strain field. In addition, when the laminate is in this state, ply cracks and delaminations of any type can be introduced in an undamaged laminate without perturbing in any way the stress and strain fields. This approach can also be taken for a damaged laminate having both ply cracks and delaminations if the laminate is suitably loaded so that all cracks are *just* closed, i.e. crack faces are touching and stress-free.

First, consider the special case when the loading is uniaxial in the *axial* direction (so that $\sigma_A = \sigma_A^c$, $\sigma_T = 0$) and the axial load is such that the ply cracks and delaminations are all *just* closed. For this state, the surfaces of the cracks are just touching and the tractions acting on the crack surfaces are zero. As already mentioned, for cases where any crack contacts are frictionless this situation has been confirmed to within numerical error using an accurate boundary element method, that the delaminations also will just close. Another way of understanding such closure is to consider the uniaxial axial loading of an undamaged laminate where the axial stress in the 90° ply is everywhere zero and where the tractions on the free surfaces $x_3 = \pm h$ and the ply interfaces are also zero. In this state, ply cracks in the 90° plies and delaminations can be introduced anywhere in the laminate without the stress and displacement fields being perturbed. Thus, at the point of closure, the axial stress and the axial and in-plane transverse strains have the same values in both damaged and undamaged laminates so that, from (12.3)–(12.6),

$$\varepsilon_A^c = \frac{\sigma_A^c}{E_A} + \alpha_A \Delta T, \qquad \varepsilon_T^c = -\frac{\nu_A}{E_A}\sigma_A^c + \alpha_T \Delta T, \tag{12.13}$$

$$\varepsilon_A^c = \frac{\sigma_A^c}{E_A^0} + \alpha_A^0 \Delta T, \qquad \varepsilon_T^c = -\frac{\nu_A^0}{E_A^0}\sigma_A^c + \alpha_T^0 \Delta T, \tag{12.14}$$

where σ_A^c is interpreted as the axial closure stress and where ε_A^c and ε_T^c are the axial and transverse closure strains, respectively, for the case of uniaxial loading in the axial direction. Elimination of the closure strains leads to the relations

$$\left(\frac{1}{E_A} - \frac{1}{E_A^0}\right)\sigma_A^c = -\left(\alpha_A - \alpha_A^0\right)\Delta T, \tag{12.15}$$

$$\left(\frac{\nu_A^0}{E_A^0} - \frac{\nu_A}{E_A}\right)\sigma_A^c = -\left(\alpha_T - \alpha_T^0\right)\Delta T. \tag{12.16}$$

Consider now another special case when the loading is uniaxial in the *in-plane transverse* direction and the transverse load is such that the ply cracks are *just* closed so that $\sigma_A = 0$ and $\sigma_T = \sigma_T^c$. For cases where crack contacts are frictionless, it is again expected that the delaminations also will just close. It then follows that at the point of closure the transverse stress and both the axial and in-plane transverse strains will have the same values in both the damaged and undamaged laminates so that, from (12.3)–(12.6),

$$\bar{\varepsilon}_A^c = -\frac{\nu_A}{E_A}\sigma_T^c + \alpha_A \Delta T, \qquad \bar{\varepsilon}_T^c = \frac{\sigma_T^c}{E_T} + \alpha_T \Delta T, \tag{12.17}$$

$$\bar{\varepsilon}_A^c = -\frac{\nu_A^0}{E_A^0}\sigma_T^c + \alpha_A^0 \Delta T, \qquad \bar{\varepsilon}_T^c = \frac{\sigma_T^c}{E_T^0} + \alpha_T^0 \Delta T, \tag{12.18}$$

where σ_T^c is interpreted as the transverse closure stress and where $\bar{\varepsilon}_A^c$ and $\bar{\varepsilon}_T^c$ are the axial and transverse closure strains, respectively, for the case of uniaxial loading in the transverse direction. Elimination of these closure strains leads to the relations

$$\left(\frac{\nu_A^0}{E_A^0} - \frac{\nu_A}{E_A}\right)\sigma_T^c = -\left(\alpha_A - \alpha_A^0\right)\Delta T, \tag{12.19}$$

$$\left(\frac{1}{E_T} - \frac{1}{E_T^0}\right)\sigma_T^c = -\left(\alpha_T - \alpha_T^0\right)\Delta T. \tag{12.20}$$

It follows from (12.15), (12.16), (12.19) and (12.20), and crack closure analysis described in Appendix D (Section 3) that the crack closure stresses σ_A^c and σ_T^c are given by

$$\sigma_A^c = -\frac{\alpha_A - \alpha_A^0}{\dfrac{1}{E_A} - \dfrac{1}{E_A^0}}\Delta T = -\frac{\alpha_T - \alpha_T^0}{\dfrac{\nu_A^0}{E_A^0} - \dfrac{\nu_A}{E_A}}\Delta T = -k_1\,\Delta T, \tag{12.21}$$

$$\sigma_T^c = -\frac{\alpha_A - \alpha_A^0}{\dfrac{\nu_A^0}{E_A^0} - \dfrac{\nu_A}{E_A}}\Delta T = -\frac{\alpha_T - \alpha_T^0}{\dfrac{1}{E_T} - \dfrac{1}{E_T^0}}\Delta T = -k_2\,\Delta T, \tag{12.22}$$

where k_1 and k_2 are *undamaged* laminate constants defined by

$$k_1 = \frac{h^{(0)}\widehat{E}_A}{h}\,\frac{h^{(0)}\widehat{E}_T + h^{(90)}\widehat{E}_A - h\widehat{\nu}_A\widehat{E}_T}{h^{(0)}\widehat{E}_T + h^{(90)}\widehat{E}_A - h\widehat{\nu}_A^2\widehat{E}_T}\left(\widehat{\alpha}_A - \widehat{\alpha}_T\right), \tag{12.23}$$

$$k_2 = \left[h^{(0)}\widehat{E}_T + h^{(90)}\widehat{E}_A - h\widehat{\nu}_A\widehat{E}_T\right]\frac{\widehat{E}_A\left(\widehat{\alpha}_A - \widehat{\alpha}_T\right)}{h\widehat{\nu}_A\left(\widehat{E}_A - \widehat{E}_T\right)}, \tag{12.24}$$

where any property p for a 0° ply is shown as a symbol $\hat{p}$. It follows from (12.21) and (12.22) that the following interrelationships must be satisfied for all delamination lengths and ply crack densities:

$$\frac{\dfrac{\nu_A^0}{E_A^0}-\dfrac{\nu_A}{E_A}}{\dfrac{1}{E_A}-\dfrac{1}{E_A^0}}=\frac{\dfrac{1}{E_T}-\dfrac{1}{E_T^0}}{\dfrac{\nu_A^0}{E_A^0}-\dfrac{\nu_A}{E_A}}=\frac{\alpha_T-\alpha_T^0}{\alpha_A-\alpha_A^0}=\frac{k_1}{k_2}=k,\qquad \frac{\dfrac{1}{E_T}-\dfrac{1}{E_T^0}}{\dfrac{1}{E_A}-\dfrac{1}{E_A^0}}=k^2, \tag{12.25}$$

where k is an *undamaged* laminate constant independent of thermal expansion coefficients, derived from (12.23)–(12.25), given by

$$k=\frac{\hat{\nu}_A h^{(0)}\left(\hat{E}_A-\hat{E}_T\right)}{h^{(0)}\hat{E}_T+h^{(90)}\hat{E}_A-h\hat{\nu}_A^2\hat{E}_T}=\frac{\sigma_A^c}{\sigma_T^c}. \tag{12.26}$$

12.5 Analysis for Unconstrained Conditions

Consider first the case where the damaged laminate is loaded such that σ_A and σ_T are the effective applied axial and transverse stresses applied to the laminate. When W has the same order as L, the debonded 90° ply can relax in the fibre direction leading to the situation where the transverse strain in this ply has a uniform value that will differ from the value of the transverse strains in the loaded 0° ply and in the fully bonded region. This would be the case when testing a delaminating cross-ply coupon by applying a uniaxial load. This loading case is described by the term *unconstrained* delamination growth. Three distinct regions are identified in Figure 12.2(a) (showing one half of a laminate) which are to be considered separately.

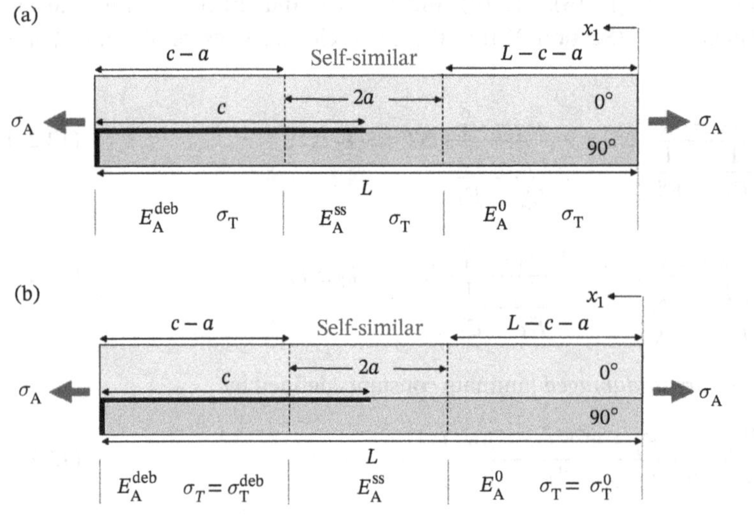

Figure 12.2 Schematic diagram showing geometry for self-similar growth under conditions of: (a) unconstrained delamination in the presence of a uniform transverse stress σ_T; (b) generalised plane strain.

12.5.1 Bonded Region of Laminate

The undamaged region of the laminate (see Figure 12.2(a)) occupies the region $0 < x_1 < L - c - a$, where $a > 0$. The stress and strain distributions are uniform in each of the plies. In this region generalised plane strain conditions will prevail in regions away from the laminate edges such that the effective axial and transverse stresses are σ_A and σ_T. The stress-strain relations (12.5) and (12.6) apply in this region where ε_A^0 and ε_T^0 are the axial and transverse strains, respectively, acting everywhere in the bonded region of the laminate. The required values of the thermoelastic constants for the fully bonded part of the laminate are given by relations (12.7)–(12.11).

12.5.2 Debonded Region of Laminate

Now consider the region $L - c + a < x_1 < L$ of length $c - a$ in which the interfaces are entirely debonded owing to the presence of the delamination crack. In this region the stress and strain distributions are again uniform in each of the plies. In this region the 0° and 90° plies are separated and each is subject to the in-plane transverse stress σ_T, whereas the 0° ply is subject to the axial stress σ_A and the 90° ply has a zero axial stress. The stress-strain relations for the 0° plies are given by

$$\varepsilon_A^{\rm deb} = \varepsilon_A^{(0)} = \frac{1}{\widehat{E}_A}\sigma_A^{(0)} - \frac{\widehat{\nu}_A}{\widehat{E}_A}\sigma_T + \widehat{\alpha}_A \Delta T, \tag{12.27}$$

$$\varepsilon_T^{(0)} = -\frac{\widehat{\nu}_A}{\widehat{E}_A}\sigma_A^{(0)} + \frac{1}{\widehat{E}_T}\sigma_T + \widehat{\alpha}_T \Delta T. \tag{12.28}$$

Only the 0° plies support the axial load in this case so that the uniform stress supported by each 0° ply is

$$\sigma_A^{(0)} = \frac{h\sigma_A}{h^{(0)}}. \tag{12.29}$$

On substituting (12.29) into (12.27), the uniform axial strain for the debonded region is

$$\varepsilon_A^{\rm deb} = \frac{\sigma_A}{E_A^{\rm deb}} - \frac{\nu_A^{\rm deb}}{E_A^{\rm deb}}\sigma_T + \alpha_A^{\rm deb}\Delta T, \tag{12.30}$$

where

$$\frac{1}{E_A^{\rm deb}} = \frac{h}{h^{(0)}}\frac{1}{\widehat{E}_A}, \qquad \frac{\nu_A^{\rm deb}}{E_A^{\rm deb}} = \frac{\widehat{\nu}_A}{\widehat{E}_A}, \qquad \alpha_A^{\rm deb} = \widehat{\alpha}_A. \tag{12.31}$$

12.5.3 Self-similar Region

In the remaining section of the region between two neighbouring ply cracks, occupying the region $L - c - a < x_1 < L - c + a$ having length $2a$, the stress and deformation distributions are nonuniform, but they do not vary as the delaminations increase their length provided this region moves 'rigidly' with the delamination tip, i.e. delamination growth is self-similar. This region is regarded as being subject to boundary conditions on $x_1 = L - c - a$ and $x_1 = L - c + a$ that match those of the neighbouring regions. The stress and deformation distributions are complex but are not required here. The effective axial Young's modulus of this region is required, denoted by $E_{\mathrm{A}}^{\mathrm{ss}}$ and independent of the delamination length c. Similarly, the effective axial Poisson's ratio and axial thermal expansion coefficient of this region, denoted by $\nu_{\mathrm{A}}^{\mathrm{ss}}$ and $\alpha_{\mathrm{A}}^{\mathrm{ss}}$ respectively, are also independent of the delamination length c. In this region of length $2a$ moving with the debond tip, where self-similar conditions prevail, the effective axial and transverse stresses are σ_{A} and σ_{T}, respectively.

12.5.4 Laminate Stress-Strain Relations

The three regions of the laminate are now considered in turn. With reference to Figure 12.2(a), in the bonded region of length $L-c-a$ where the local effective transverse stress has the value σ_{T}, the effective axial strain is given by (12.5) leading to a corresponding axial displacement

$$(L-c-a)\left(\frac{\sigma_{\mathrm{A}}}{E_{\mathrm{A}}^{0}}-\frac{\nu_{\mathrm{A}}^{0}}{E_{\mathrm{A}}^{0}}\sigma_{\mathrm{T}}+\alpha_{\mathrm{A}}^{0}\Delta T\right).$$

In the debonded region of length $c-a$, the effective axial strain is given by (12.30) leading to a corresponding axial displacement

$$(c-a)\left(\frac{\sigma_{\mathrm{A}}}{E_{\mathrm{A}}^{\mathrm{deb}}}-\frac{\nu_{\mathrm{A}}^{\mathrm{deb}}}{E_{\mathrm{A}}^{\mathrm{deb}}}\sigma_{\mathrm{T}}+\alpha_{\mathrm{A}}^{\mathrm{deb}}\Delta T\right).$$

In the region of length $2a$ moving with the debond tip where self-similar conditions prevail, the effective axial displacement is fixed as the debond grows in length having the value

$$2a\left(\frac{\sigma_{\mathrm{A}}}{E_{\mathrm{A}}^{\mathrm{ss}}}-\frac{\nu_{\mathrm{A}}^{\mathrm{ss}}}{E_{\mathrm{A}}^{\mathrm{ss}}}\sigma_{\mathrm{T}}+\alpha_{\mathrm{A}}^{\mathrm{ss}}\Delta T\right).$$

It then follows, by adding the above displacement contributions and then taking the limit $L\to\infty$ such that $c=\eta L$, that the overall effective axial strain ε_{A} may be obtained from the relation

$$\varepsilon_{\mathrm{A}}=(1-\eta)\left(\frac{\sigma_{\mathrm{A}}}{E_{\mathrm{A}}^{0}}-\frac{\nu_{\mathrm{A}}^{0}}{E_{\mathrm{A}}^{0}}\sigma_{\mathrm{T}}+\alpha_{\mathrm{A}}^{0}\Delta T\right)+\eta\left(\frac{\sigma_{\mathrm{A}}}{E_{\mathrm{A}}^{\mathrm{deb}}}-\frac{\nu_{\mathrm{A}}^{\mathrm{deb}}}{E_{\mathrm{A}}^{\mathrm{deb}}}\sigma_{\mathrm{T}}+\alpha_{\mathrm{A}}^{\mathrm{deb}}\Delta T\right). \tag{12.32}$$

This result has the form of (12.3) where on using (12.31)

$$\frac{1}{E_{\mathrm{A}}}=\frac{1-\eta}{E_{\mathrm{A}}^{0}}+\eta\frac{h}{h^{(0)}}\frac{1}{\widehat{E}_{\mathrm{A}}},\quad \frac{\nu_{\mathrm{A}}}{E_{\mathrm{A}}}=(1-\eta)\frac{\nu_{\mathrm{A}}^{0}}{E_{\mathrm{A}}^{0}}+\eta\frac{\widehat{\nu}_{\mathrm{A}}}{E_{\mathrm{A}}},\quad \alpha_{\mathrm{A}}=(1-\eta)\alpha_{\mathrm{A}}^{0}+\eta\widehat{\alpha}_{\mathrm{A}}. \tag{12.33}$$

It is clear that the terms involving the unknown quantities $E_{\mathrm{A}}^{\mathrm{ss}}$, $\nu_{\mathrm{A}}^{\mathrm{ss}}$ and $\alpha_{\mathrm{A}}^{\mathrm{ss}}$ have been eliminated.

12.6 Analysis for Generalised Plane Strain Conditions

The case is now considered where there is some lateral constraint in the laminate that leads to the situation where the in-plane transverse strain may be assumed to be uniform everywhere in the laminate having the value $\varepsilon_{\mathrm{T}}^{0}$. This lateral strain is unknown at the outset and its value is found from solutions by selecting its value so that the effective transverse stress has a prescribed value σ_{T}. This type of situation is known as *generalised plane strain* and will apply in those regions of a damaged laminate which are not influenced by stress-transfer effects near the laminate edges. For conditions of plane strain, the value of $\varepsilon_{\mathrm{T}}^{0}$ is zero.

Consider the case where the damaged laminate is loaded such that $\sigma_{\mathrm{A}} \neq 0$, $\varepsilon_{\mathrm{T}} = \varepsilon_{\mathrm{T}}^{0}$, as illustrated in Figure 12.2(b) (showing one half of a laminate), which shows three distinct regions of the laminate for which the effective axial applied stress and transverse strain have the same values σ_{A} and $\varepsilon_{\mathrm{T}}^{0}$, respectively.

12.6.1 Bonded Region of Laminate

The effective in-plane transverse stress in the undamaged region is $\sigma_{\mathrm{T}} = \sigma_{\mathrm{T}}^{0}$, so that from (12.5) and (12.6), it follows that

$$\sigma_{\mathrm{T}}^{0} = \nu_{\mathrm{A}}^{0} \frac{E_{\mathrm{T}}^{0}}{E_{\mathrm{A}}^{0}} \sigma_{\mathrm{A}} + E_{\mathrm{T}}^{0} \varepsilon_{\mathrm{T}}^{0} - E_{\mathrm{T}}^{0} \alpha_{\mathrm{T}}^{0} \Delta T, \tag{12.34}$$

$$\varepsilon_{\mathrm{A}}^{0} = \frac{\sigma_{\mathrm{A}}}{\tilde{E}_{\mathrm{A}}^{0}} - \nu_{\mathrm{A}}^{0} \frac{E_{\mathrm{T}}^{0}}{E_{\mathrm{A}}^{0}} \varepsilon_{\mathrm{T}}^{0} + \tilde{\alpha}_{\mathrm{A}}^{0} \Delta T, \tag{12.35}$$

where

$$\frac{1}{\tilde{E}_{\mathrm{A}}^{0}} = \frac{1}{E_{\mathrm{A}}^{0}} \left(1 - \left(\nu_{\mathrm{A}}^{0} \right)^{2} \frac{E_{\mathrm{T}}^{0}}{E_{\mathrm{A}}^{0}} \right), \qquad \tilde{\alpha}_{\mathrm{A}}^{0} = \alpha_{\mathrm{A}}^{0} + \nu_{\mathrm{A}}^{0} \frac{E_{\mathrm{T}}^{0}}{E_{\mathrm{A}}^{0}} \alpha_{\mathrm{T}}^{0}. \tag{12.36}$$

12.6.2 Debonded Region of Laminate

The axial stress in the 0° plies is again given by (12.29) so that from (D.23) (see Appendix D),

$$\sigma_{\mathrm{T}}^{(0)} = \frac{h}{h^{(0)}} \widehat{\nu}_{\mathrm{A}} \frac{\widehat{E}_{\mathrm{T}}}{\widehat{E}_{\mathrm{A}}} \sigma_{\mathrm{A}} + \widehat{E}_{\mathrm{T}} \varepsilon_{\mathrm{T}}^{0} - \widehat{E}_{\mathrm{T}} \widehat{\alpha}_{\mathrm{T}} \Delta T. \tag{12.37}$$

On substituting (12.29) and (12.37) into (D.22) (see Appendix D), the uniform axial strain in the debonded region is

$$\varepsilon_{\mathrm{A}}^{\mathrm{deb}} = \frac{1}{\widehat{E}_{\mathrm{A}}} \left(1 - \widehat{\nu}_{\mathrm{A}}^{2} \frac{\widehat{E}_{\mathrm{T}}}{\widehat{E}_{\mathrm{A}}} \right) \frac{h \sigma_{\mathrm{A}}}{h^{(0)}} - \widehat{\nu}_{\mathrm{A}} \frac{\widehat{E}_{\mathrm{T}}}{\widehat{E}_{\mathrm{A}}} \varepsilon_{\mathrm{T}}^{0} + \left(\widehat{\alpha}_{\mathrm{A}} + \widehat{\nu}_{\mathrm{A}} \frac{\widehat{E}_{\mathrm{T}}}{\widehat{E}_{\mathrm{A}}} \widehat{\alpha}_{\mathrm{T}} \right) \Delta T. \tag{12.38}$$

As the axial stress in the 90° ply is zero in the debonded region, i.e. $\sigma_{\mathrm{A}}^{(90)} = 0$, the transverse stress and axial strain in the 90° ply are then obtained from (D.24) and (D.25) (see Appendix D) so that

$$\sigma_{\mathrm{T}}^{(90)} = \widehat{E}_{\mathrm{A}} \left(\varepsilon_{\mathrm{T}}^{0} - \widehat{\alpha}_{\mathrm{A}} \Delta T \right), \tag{12.39}$$

$$\varepsilon_{\mathrm{A}}^{(90)} = -\widehat{\nu}_{\mathrm{A}} \varepsilon_{\mathrm{T}}^{0} + \left(\widehat{\alpha}_{\mathrm{T}} + \widehat{\nu}_{\mathrm{A}} \widehat{\alpha}_{\mathrm{A}} \right) \Delta T. \tag{12.40}$$

The effective transverse stress in the debonded region $\sigma_{\mathrm{T}}^{\mathrm{deb}}$ is obtained from (12.37), (12.39) and

$$h\sigma_{\rm T}^{\rm deb} = h^{(0)}\sigma_{\rm T}^{(0)} + h^{(90)}\sigma_{\rm T}^{(90)}, \tag{12.41}$$

leading to

$$\sigma_{\rm T}^{\rm deb} = \widehat{\nu}_{\rm A}\frac{\widehat{E}_{\rm T}}{\widehat{E}_{\rm A}}\sigma_{\rm A} + \left(h^{(0)}\widehat{E}_{\rm T} + h^{(90)}\widehat{E}_{\rm A}\right)\frac{\varepsilon_{\rm T}^0}{h} - \left(h^{(0)}\widehat{E}_{\rm T}\widehat{\alpha}_{\rm T} + h^{(90)}\widehat{E}_{\rm A}\widehat{\alpha}_{\rm A}\right)\frac{\Delta T}{h}. \tag{12.42}$$

On solving (12.42) for the transverse strain $\varepsilon_{\rm T}^0$,

$$\varepsilon_{\rm T}^0 = -\frac{\nu_{\rm A}^{\rm deb}}{E_{\rm A}^{\rm deb}}\sigma_{\rm A} + \frac{\sigma_{\rm T}^{\rm deb}}{E_{\rm T}^{\rm deb}} + \alpha_{\rm T}^{\rm deb}\Delta T, \tag{12.43}$$

where

$$E_{\rm T}^{\rm deb} = \frac{h^{(0)}}{h}\widehat{E}_{\rm T} + \frac{h^{(90)}}{h}\widehat{E}_{\rm A}, \tag{12.44}$$

$$\nu_{\rm A}^{\rm deb}\frac{E_{\rm T}^{\rm deb}}{E_{\rm A}^{\rm deb}} = \widehat{\nu}_{\rm A}\frac{\widehat{E}_{\rm T}}{\widehat{E}_{\rm A}}, \tag{12.45}$$

$$E_{\rm T}^{\rm deb}\alpha_{\rm T}^{\rm deb} = \frac{h^{(0)}}{h}\widehat{E}_{\rm T}\widehat{\alpha}_{\rm T} + \frac{h^{(90)}}{h}\widehat{E}_{\rm A}\widehat{\alpha}_{\rm A}. \tag{12.46}$$

From (12.38), on using (12.43) and (12.45), it can be shown that

$$\varepsilon_{\rm A}^{\rm deb} = \frac{\sigma_{\rm A}}{E_{\rm A}^{\rm deb}} - \frac{\nu_{\rm A}^{\rm deb}}{E_{\rm A}^{\rm deb}}\sigma_{\rm T}^{\rm deb} + \alpha_{\rm A}^{\rm deb}\Delta T, \tag{12.47}$$

where

$$\frac{1}{E_{\rm A}^{\rm deb}} = \frac{1}{\widehat{E}_{\rm A}}\left(1 - \widehat{\nu}_{\rm A}^2\frac{\widehat{E}_{\rm T}}{\widehat{E}_{\rm A}}\right)\frac{h}{h^{(0)}} + \left(\widehat{\nu}_{\rm A}\frac{\widehat{E}_{\rm T}}{\widehat{E}_{\rm A}}\right)^2\frac{1}{E_{\rm T}^{\rm deb}}, \tag{12.48}$$

$$\alpha_{\rm A}^{\rm deb} = \widehat{\alpha}_{\rm A} + \widehat{\nu}_{\rm A}\frac{\widehat{E}_{\rm T}}{\widehat{E}_{\rm A}}\left(\widehat{\alpha}_{\rm T} - \alpha_{\rm T}^{\rm deb}\right). \tag{12.49}$$

On substituting (12.42) into (12.47) it can be shown using (12.44)–(12.46) that

$$\varepsilon_{\rm A}^{\rm deb} = \frac{\sigma_{\rm A}}{\tilde{E}_{\rm A}^{\rm deb}} - \nu_{\rm A}^{\rm deb}\frac{E_{\rm T}^{\rm deb}}{E_{\rm A}^{\rm deb}}\varepsilon_{\rm T}^0 + \tilde{\alpha}_{\rm A}^{\rm deb}\Delta T, \tag{12.50}$$

where

$$\frac{1}{\tilde{E}_{\rm A}^{\rm deb}} = \frac{1}{E_{\rm A}^{\rm deb}} - \left(\nu_{\rm A}^{\rm deb}\frac{E_{\rm T}^{\rm deb}}{E_{\rm A}^{\rm deb}}\right)^2\frac{1}{E_{\rm T}^{\rm deb}} = \frac{1}{\widehat{E}_{\rm A}}\left(1 - \widehat{\nu}_{\rm A}^2\frac{\widehat{E}_{\rm T}}{\widehat{E}_{\rm A}}\right)\frac{h}{h^{(0)}}, \tag{12.51}$$

$$\tilde{\alpha}_{\rm A}^{\rm deb} = \alpha_{\rm A}^{\rm deb} + \nu_{\rm A}^{\rm deb}\frac{E_{\rm T}^{\rm deb}}{E_{\rm A}^{\rm deb}}\alpha_{\rm T}^{\rm deb} = \widehat{\alpha}_{\rm A} + \widehat{\nu}_{\rm A}\frac{\widehat{E}_{\rm T}}{\widehat{E}_{\rm A}}\widehat{\alpha}_{\rm T}. \tag{12.52}$$

12.6.3 Self-similar Region

As in Section 12.5.3 the effective thermoelastic constants for this self-similar region, which are again denoted using the superscript 'ss', are independent of the delamination length c. In the region of length $2a$ moving with the debond tip, where self-similar conditions prevail, the effective transverse stress is denoted by σ_T^{ss} corresponding to the imposed uniform value ε_T^0 assumed for the transverse strain.

12.6.4 Laminate Stress-Strain Relations

In the bonded region of length $L - c - a$ (see Figure 12.2(b)), the effective transverse stress has the value σ_T^0 given by (12.34). In the debonded region of length $c - a$, the effective transverse stress is σ_T^{deb} given by (12.42). It then follows that the overall effective transverse stress σ_T for the region of length L is obtained from the relation

$$hL\sigma_T = h(L-c-a)\sigma_T^0 + 2ha\sigma_T^{ss} + h(c-a)\sigma_T^{deb}. \tag{12.53}$$

Solving (12.4) for σ_T with $\varepsilon_T = \varepsilon_T^0$ and then substituting the result into (12.3) leads to

$$\sigma_T = \nu_A \frac{E_T}{E_A}\sigma_A + E_T\varepsilon_T^0 - E_T\alpha_T \Delta T, \tag{12.54}$$

$$\varepsilon_A = \frac{\sigma_A}{\tilde{E}_A} - \nu_A \frac{E_T}{E_A}\varepsilon_T^0 + \tilde{\alpha}_A \Delta T, \tag{12.55}$$

where

$$\frac{1}{\tilde{E}_A} = \frac{1}{E_A}\left(1 - \nu_A^2 \frac{E_T}{E_A}\right) = \frac{1}{E_A} - \left(\nu_A \frac{E_T}{E_A}\right)^2 \frac{1}{E_T}, \qquad \tilde{\alpha}_A = \alpha_A + \nu_A \frac{E_T}{E_A}\alpha_T. \tag{12.56}$$

In the limit $L \to \infty$, keeping a fixed and assuming $c = \eta L$ where $0 < \eta < 1$, it follows from (12.53) that

$$\sigma_T \to (1-\eta)\sigma_T^0 + \eta\sigma_T^{deb}. \tag{12.57}$$

On using (12.34)

$$\sigma_T^{deb} = \frac{\sigma_T}{\eta} - \frac{1-\eta}{\eta}\left(\nu_A^0 \frac{E_T^0}{E_A^0}\sigma_A + E_T^0\varepsilon_T^0 - E_T^0\alpha_T^0 \Delta T\right). \tag{12.58}$$

Substitution into (12.42) then leads to

$$\begin{aligned}\left[(1-\eta)E_T^0 + \eta\left(\frac{h^{(0)}}{h}\widehat{E}_T + \frac{h^{(90)}}{h}\widehat{E}_A\right)\right]\varepsilon_T^0 &= -\left((1-\eta)\nu_A^0 \frac{E_T^0}{E_A^0} + \eta\widehat{\nu}_A \frac{\widehat{E}_T}{\widehat{E}_A}\right)\sigma_A + \sigma_T \\ &+ \left[(1-\eta)E_T^0\alpha_T^0 + \eta\left(\frac{h^{(0)}}{h}\widehat{E}_T\widehat{\alpha}_T + \frac{h^{(90)}}{h}\widehat{E}_A\widehat{\alpha}_A\right)\right]\Delta T.\end{aligned} \tag{12.59}$$

By comparing (12.4) (with $\varepsilon_T = \varepsilon_T^0$) and (12.59), it is deduced that

$$E_T = (1-\eta)E_T^0 + \eta\left(\frac{h^{(0)}}{h}\widehat{E}_T + \frac{h^{(90)}}{h}\widehat{E}_A\right), \tag{12.60}$$

$$\nu_{\mathrm{A}}\frac{E_{\mathrm{T}}}{E_{\mathrm{A}}}=(1-\eta)\nu_{\mathrm{A}}^{0}\frac{E_{\mathrm{T}}^{0}}{E_{\mathrm{A}}^{0}}+\eta\widehat{\nu}_{\mathrm{A}}\frac{\widehat{E}_{\mathrm{T}}}{\widehat{E}_{\mathrm{A}}},\tag{12.61}$$

$$E_{\mathrm{T}}\alpha_{\mathrm{T}}=(1-\eta)E_{\mathrm{T}}^{0}\alpha_{\mathrm{T}}^{0}+\eta\left(\frac{h^{(0)}}{h}\widehat{E}_{\mathrm{T}}\widehat{\alpha}_{\mathrm{T}}+\frac{h^{(90)}}{h}\widehat{E}_{\mathrm{A}}\widehat{\alpha}_{\mathrm{A}}\right),\tag{12.62}$$

enabling the calculation of the values of E_{T}, $\nu_{\mathrm{A}}\,/\,E_{\mathrm{A}}$ and α_{T}.

Three regions of the laminate are now considered in turn for conditions of generalised plane strain deformation characterised by the transverse strain $\varepsilon_{\mathrm{T}}^{0}$. With reference to Figure 12.2(b), in the bonded region of length $L - c - a$ where the local effective transverse stress is σ_{T}^{0}, the effective axial strain is given by (12.35) leading to a corresponding axial displacement

$$(L-c-a)\left(\frac{\sigma_{\mathrm{A}}}{\tilde{E}_{\mathrm{A}}^{0}}-\nu_{\mathrm{A}}^{0}\frac{E_{\mathrm{T}}^{0}}{E_{\mathrm{A}}^{0}}\varepsilon_{\mathrm{T}}^{0}+\tilde{\alpha}_{\mathrm{A}}^{0}\Delta T\right),$$

where the effective thermoelastic constants of the undamaged laminate are given by (12.7)–(12.11) and (12.36). In the debonded region of length $c - a$, the effective axial strain is given by (12.50) leading to a corresponding axial displacement

$$(c-a)\left(\frac{\sigma_{\mathrm{A}}}{\tilde{E}_{\mathrm{A}}^{\mathrm{deb}}}-\nu_{\mathrm{A}}^{\mathrm{deb}}\frac{E_{\mathrm{T}}^{\mathrm{deb}}}{E_{\mathrm{A}}^{\mathrm{deb}}}\varepsilon_{\mathrm{T}}^{0}+\tilde{\alpha}_{\mathrm{A}}^{\mathrm{deb}}\Delta T\right),$$

where the effective laminate constants of the debonded part of the laminate are given by relations (12.45), (12.51) and (12.52). In the region of length $2a$ moving with the debond tip where self-similar conditions prevail, the effective axial displacement is fixed as the debond grows in length having the value

$$2a\left(\frac{\sigma_{\mathrm{A}}}{\tilde{E}_{\mathrm{A}}^{\mathrm{ss}}}-\nu_{\mathrm{A}}^{\mathrm{ss}}\frac{E_{\mathrm{T}}^{\mathrm{ss}}}{E_{\mathrm{A}}^{\mathrm{ss}}}\varepsilon_{\mathrm{T}}^{0}+\tilde{\alpha}_{\mathrm{A}}^{\mathrm{ss}}\Delta T\right),$$

involving the effective thermoelastic constants of the self-similar region. It then follows, on adding these displacement contributions and then taking the limit $L\to\infty$ such that $c=\eta L$, that the overall effective axial strain is given by

$$\varepsilon_{\mathrm{A}}=\left(\frac{1-\eta}{\tilde{E}_{\mathrm{A}}^{0}}+\frac{\eta}{\tilde{E}_{\mathrm{A}}^{\mathrm{deb}}}\right)\sigma_{\mathrm{A}}-\left((1-\eta)\frac{\nu_{\mathrm{A}}^{0}}{E_{\mathrm{A}}^{0}}+\eta\frac{\nu_{\mathrm{A}}^{\mathrm{deb}}}{E_{\mathrm{A}}^{\mathrm{deb}}}\right)\varepsilon_{\mathrm{T}}^{0}+\left[(1-\eta)\tilde{\alpha}_{\mathrm{A}}^{0}+\eta\tilde{\alpha}_{\mathrm{A}}^{\mathrm{deb}}\right]\Delta T.\tag{12.63}$$

The terms involving the unknown thermoelastic constants for the self-similar region have been eliminated. It then follows on comparing (12.63) with (12.55) that

$$\frac{1}{\tilde{E}_{\mathrm{A}}}=\frac{1-\eta}{\tilde{E}_{\mathrm{A}}^{0}}+\frac{\eta}{\tilde{E}_{\mathrm{A}}^{\mathrm{deb}}},\tag{12.64}$$

$$\nu_{\mathrm{A}}\frac{E_{\mathrm{T}}}{E_{\mathrm{A}}}=(1-\eta)\nu_{\mathrm{A}}^{0}\frac{E_{\mathrm{T}}^{0}}{E_{\mathrm{A}}^{0}}+\eta\nu_{\mathrm{A}}^{\mathrm{deb}}\frac{E_{\mathrm{T}}^{\mathrm{deb}}}{E_{\mathrm{A}}^{\mathrm{deb}}},\tag{12.65}$$

$$\tilde{\alpha}_{\mathrm{A}}=(1-\eta)\tilde{\alpha}_{\mathrm{A}}^{0}+\eta\tilde{\alpha}_{\mathrm{A}}^{\mathrm{deb}}.\tag{12.66}$$

On using (12.45), (12.51) and (12.52), these relations may be written as

$$\frac{1}{\tilde{E}_A}=\frac{1-\eta}{\tilde{E}_A^0}+\frac{\eta}{\widehat{E}_A}\left(1-\widehat{\nu}_A^2\frac{\widehat{E}_T}{\widehat{E}_A}\right)\frac{h}{h^{(0)}}, \tag{12.67}$$

$$\nu_A\frac{E_T}{E_A}=(1-\eta)\nu_A^0\frac{E_T^0}{E_A^0}+\eta\widehat{\nu}_A\frac{\widehat{E}_T}{\widehat{E}_A}, \tag{12.68}$$

$$\tilde{\alpha}_A=(1-\eta)\tilde{\alpha}_A^0+\eta\left(\widehat{\alpha}_A+\widehat{\nu}_A\frac{\widehat{E}_T}{\widehat{E}_A}\widehat{\alpha}_T\right). \tag{12.69}$$

Relation (12.68) is consistent with (12.61) as to be expected. The axial modulus E_A and thermal expansion coefficient α_A are obtained using the results (12.67)–(12.69) in the following relations derived from (12.56)

$$\frac{1}{E_A}=\frac{1}{\tilde{E}_A}+\left(\nu_A\frac{E_T}{E_A}\right)^2\frac{1}{E_T}, \qquad \alpha_A=\tilde{\alpha}_A-\left(\nu_A\frac{E_T}{E_A}\right)\alpha_T. \tag{12.70}$$

It should be noted that the results (12.60)–(12.62) and (12.70) are, in fact, consistent with the interrelationships (12.25) derived by considering the closure of the ply cracks and delaminations.

12.7 Calculation of the Gibbs Energy

Delamination growth is governed by energy balance principles which are described in Appendix I. On using (12.3), (12.4) and (I.13) (see Appendix I), it follows that the effective thermoelastic constants of the damaged laminate will depend on the ply crack density $\rho=1/(2L)$ and on the length c of the delamination cracks, and that

$$\frac{\partial\hat{G}}{\partial\sigma_A}=-\frac{\sigma_A}{E_A}+\frac{\nu_A}{E_A}\sigma_T-\alpha_A\Delta T, \tag{12.71}$$

$$\frac{\partial\hat{G}}{\partial\sigma_T}=\frac{\nu_A}{E_A}\sigma_A-\frac{\sigma_T}{E_T}-\alpha_T\Delta T. \tag{12.72}$$

It then follows on integration that the Gibbs energy averaged per unit volume $\hat{G}$ for a damaged cross-ply laminate must have the form

$$\hat{G}=-\frac{\sigma_A^2}{2E_A}-\frac{\sigma_T^2}{2E_T}+\frac{\nu_A}{E_A}\sigma_A\sigma_T-\sigma_A\alpha_A\Delta T-\sigma_T\alpha_T\Delta T+\overline{G}, \tag{12.73}$$

where the term $\overline{G}$ represents the Gibbs energy per unit volume when the damaged laminate is mechanically unloaded. The value of $\overline{G}$ is thus independent of the loading parameters σ_A, σ_T but it will depend on the damage if thermal residual stresses are present. In the absence of thermal stresses, $\overline{G}$ would be independent of the loading and damage parameters, and could then be treated as a material constant for a given fixed temperature difference ΔT.

To avoid the problem of estimating $\overline{G}$ when thermal stresses are present, it is useful to consider the crack closure state when it is expected that all ply cracks and delaminations will be just closed, assuming frictionless contacts of crack faces, as discussed in Section 12.4. For such closure conditions, the stress and deformation distributions in the laminate are identical to those in the corresponding undamaged laminate subject to the same applied stresses and temperature. Consider the crack closure situation that arises when the laminate is uniaxially loaded in the axial direction so that $\sigma_A = \sigma_A^c$, $\sigma_T = 0$. It follows from (12.73) that, on introducing the Gibbs energy per unit volume G_c for such closure conditions, the following expression for $\bar{G}$ is derived

$$\bar{G} = \sigma_A^c \left(\frac{\sigma_A^c}{2E_A} + \alpha_A \Delta T \right) + G_c. \tag{12.74}$$

The value of G_c could easily be calculated as its value does not depend on the damage present although its value will not in fact be required. It follows from (12.73) and (12.74) that for general loading conditions the Gibbs energy per unit volume is given by

$$\begin{aligned}\hat{G} = -\frac{\sigma_A^2}{2E_A} - \frac{\sigma_T^2}{2E_T} + \frac{\nu_A}{E_A}\sigma_A\sigma_T - \sigma_A\alpha_A\Delta T - \sigma_T\alpha_T\Delta T \\ + \sigma_A^c \left[\frac{\sigma_A^c}{2E_A} + \alpha_A \Delta T \right] + G_c.\end{aligned} \tag{12.75}$$

The corresponding value for the Gibbs energy per unit volume when the laminate is undamaged G_0 is given by

$$\begin{aligned}G_0 = -\frac{\sigma_A^2}{2E_A^0} - \frac{\sigma_T^2}{2E_T^0} + \frac{\nu_A^0}{E_A^0}\sigma_A\sigma_T - \sigma_A\alpha_A^0\Delta T - \sigma_T\alpha_T^0\Delta T \\ + \sigma_A^c \left[\frac{\sigma_A^c}{2E_A^0} + \alpha_A^0 \Delta T \right] + G_c.\end{aligned} \tag{12.76}$$

On subtracting, so that G_c is eliminated, it then follows that

$$\begin{aligned}\hat{G} - G_0 = -\frac{\sigma_A^2}{2}\left(\frac{1}{E_A} - \frac{1}{E_A^0} \right) - \frac{\sigma_T^2}{2}\left(\frac{1}{E_T} - \frac{1}{E_T^0} \right) + \left(\frac{\nu_A}{E_A} - \frac{\nu_A^0}{E_A^0} \right)\sigma_A\sigma_T \\ -\sigma_A\left(\alpha_A - \alpha_A^0\right)\Delta T - \sigma_T\left(\alpha_T - \alpha_T^0\right)\Delta T + \sigma_A^c\left[\frac{\sigma_A^c}{2}\left(\frac{1}{E_A} - \frac{1}{E_A^0} \right) + \left(\alpha_A - \alpha_A^0\right)\Delta T \right].\end{aligned} \tag{12.77}$$

On using the interrelations (12.21) and (12.25) it then follows that the Gibbs energy per unit volume of damaged laminate may be written in the compact form

$$\hat{G} = -\frac{1}{2}\left(\frac{1}{E_A} - \frac{1}{E_A^0} \right)\left[\sigma_A + k\sigma_T - \sigma_A^c\right]^2 + G_0, \quad \text{where} \quad \sigma_A^c = -k_1\Delta T. \tag{12.78}$$

It should be noted that when $\sigma_A = \sigma_A^c$, $\sigma_T = 0$, it follows that $\hat{G} = G_0$ the undamaged value of the Gibbs energy per unit volume, as at closure the stress and deformation distributions for both damaged and undamaged laminates are identical. Result (12.78) is valid for all delamination lengths and ply crack densities.

12.8 Calculation of Energy Release Rates for Delamination Growth

The energy release rate for delamination is considered in detail in Appendix I where it is shown that delamination growth (i.e. $\dot{c}(t) > 0$) is possible only if (see (I.15) in Appendix I)

$$\frac{16\gamma W}{V} + \frac{\partial \hat{G}}{\partial c} = 0, \quad (\Delta T, \rho \text{ held fixed}). \tag{12.79}$$

This relation will be the basis of a study of delamination governed by energy balance considerations for conditions of self-similar delamination growth. Three situations will now be considered: (i) unconstrained delamination, (ii) delamination for generalised plane strain conditions associated with a fixed value of the effective transverse stress and (iii) constrained delamination associated with a fixed uniform value of the transverse strain.

12.8.1 Unconstrained Delamination ($\sigma_{\mathrm{A}}, \sigma_{\mathrm{T}}$ Fixed)

For the case of unconstrained delamination, the laminate is assumed to be subject to fixed values of the applied stresses σ_{A} and σ_{T}, but the laminate strains are allowed to relax in the transverse direction as described in Section 12.5. Use is now made of the results (12.78) and (12.79) to calculate the total energy release rates associated with the growth of the delaminations in the interfaces between the 0° and 90° plies. The damage in the laminate is characterised by the uniform and fixed ply crack density $\rho = 1/(2L)$ and the variable length c of the four delaminations associated with each ply crack. The axial modulus E_{A} will depend on both ρ and c, but the quantities E_{A}^0, k, k_1 and G_0 will all be independent of these damage parameters. On applying the energy-based fracture criterion (12.79) to a representative region of the laminate having volume $V = 8LWh$, it follows that

$$2\gamma + Lh\frac{\partial \hat{G}}{\partial c} = 0, \quad (\Delta T, \rho \text{ held fixed}). \tag{12.80}$$

As the effective applied stresses $\sigma_{\mathrm{A}}, \sigma_{\mathrm{T}}$ are held fixed, and on using (12.78), the energy release rate for delamination is given by

$$ERR(\sigma_{\mathrm{A}}, \sigma_{\mathrm{T}}, \Delta T) \equiv -Lh\frac{\partial \hat{G}}{\partial c} = \frac{1}{2}Lh[\sigma_{\mathrm{A}} + k\sigma_{\mathrm{T}} + k_1\Delta T]^2 \frac{\partial}{\partial c}\left(\frac{1}{E_{\mathrm{A}}}\right), \tag{12.81}$$

which assumes that the temperature difference ΔT, and the ply crack density ρ both remain fixed during delamination growth. The critical value of the energy release rate, enabling delamination growth, is clearly 2γ.

Relation (12.81), which is valid for all delamination lengths and ply crack densities, provides a compliance method of estimating the energy release rate for debonding. For conditions of self-similar delamination growth, it follows from (12.33) and (12.81) that

$$ERR(\sigma_{\mathrm{A}}, \sigma_{\mathrm{T}}, \Delta T) = \frac{1}{2}h[\sigma_{\mathrm{A}} + k\sigma_{\mathrm{T}} + k_1\Delta T]^2 \left(\frac{h}{h^{(0)}}\frac{1}{\hat{E}_{\mathrm{A}}} - \frac{1}{E_{\mathrm{A}}^0}\right). \tag{12.82}$$

This result does not depend on the value of $\eta = c/L$ so that delamination under conditions of self-similar growth can be considered as steady-state cracking under fixed applied stresses σ_A and σ_T.

It is noted from the more general relation (12.81) that if the energy release rate is known for uniaxial stress states in the absence of thermal stresses, then its value for any biaxial loading in the presence of thermal residual stresses is given by

$$ERR(\sigma_A, \sigma_T, \Delta T) = \left[\frac{\sigma_A + k\sigma_T + k_1\Delta T}{\sigma_A}\right]^2 ERR(\sigma_A, 0, 0). \tag{12.83}$$

12.8.2 Generalised Plane Strain (σ_A, σ_T Fixed)

The damaged laminate is now assumed to be in a state of generalised plane strain so that the effective transverse strain ε_T is uniform for all states of axial stress and temperature, as described in Section 12.6. The effective transverse strain is chosen using (12.4) so that the corresponding effective transverse stress σ_T has a prescribed fixed value for a given value of the axial applied stress σ_A and the temperature difference ΔT.

For these generalised plane strain conditions, the expression (12.81) for the energy release rate for delamination growth remains valid and the effective transverse strain ε_T will depend on delamination length. On using (12.70),

$$\frac{\partial}{\partial c}\left(\frac{1}{E_A}\right) = \frac{\partial}{\partial c}\left(\frac{1}{\tilde{E}_A}\right) + \frac{2\nu_A}{E_A}\frac{\partial}{\partial c}\left(\nu_A \frac{E_T}{E_A}\right) - \left(\frac{\nu_A}{E_A}\right)^2 \frac{\partial E_T}{\partial c}. \tag{12.84}$$

From (12.60), (12.61) and (12.67),

$$L\frac{\partial E_T}{\partial c} = \frac{h^{(0)}}{h}\widehat{E}_T + \frac{h^{(90)}}{h}\widehat{E}_A - E_T^0 = P, \tag{12.85}$$

$$L\frac{\partial}{\partial c}\left(\nu_A \frac{E_T}{E_A}\right) = \widehat{\nu}_A \frac{\widehat{E}_T}{\widehat{E}_A} - \nu_A^0 \frac{E_T^0}{E_A^0} = Q, \tag{12.86}$$

$$L\frac{\partial}{\partial c}\left(\frac{1}{\tilde{E}_A}\right) = \frac{1}{\widehat{E}_A}\left(1 - \widehat{\nu}_A^2 \frac{\widehat{E}_T}{\widehat{E}_A}\right)\frac{h}{h^{(0)}} - \frac{1}{\overline{E}_A^0} = R. \tag{12.87}$$

Substitution into (12.84) then yields

$$L\frac{\partial}{\partial c}\left(\frac{1}{E_A}\right) = R + \frac{2\nu_A}{E_A}Q - \left(\frac{\nu_A}{E_A}\right)^2 P. \tag{12.88}$$

Substitution into (12.81) leads to the following expression for the energy release rate for delamination under self-similar growth and generalised plane strain conditions

$$ERR(\sigma_A, \sigma_T, \Delta T) = \frac{1}{2}h[\sigma_A + k\sigma_T + k_1\Delta T]^2 \left(R + \frac{2\nu_A}{E_A}Q - \left(\frac{\nu_A}{E_A}\right)^2 P\right), \tag{12.89}$$

where the critical value, enabling delamination growth, is 2γ.

The parameters P, Q and R are constants which are independent of the debond length c, and it follows from (12.60) and (12.61) that

$$\frac{\nu_{\mathrm{A}}}{E_{\mathrm{A}}}=\frac{(1-\eta)\nu_{\mathrm{A}}^{0}\dfrac{E_{\mathrm{T}}^{0}}{E_{\mathrm{A}}^{0}}+\eta\widehat{\nu}_{\mathrm{A}}\dfrac{\widehat{E}_{\mathrm{T}}}{\widehat{E}_{\mathrm{A}}}}{(1-\eta)E_{\mathrm{T}}^{0}+\eta\left(\dfrac{h^{(0)}}{h}\widehat{E}_{\mathrm{T}}+\dfrac{h^{(90)}}{h}\widehat{E}_{\mathrm{A}}\right)},\tag{12.90}$$

depending on the value of $\eta=c/L$ so that delamination under conditions of self-similar growth cannot be considered as steady-state cracking under fixed applied stresses.

12.8.3 Constrained Delamination (ε_{A}, ε_{T} Fixed)

For many practical applications of composites, where damage occurs in the form of ply cracks and associated delaminations, the engineering structure involves the use of large laminates (i.e. $W >> L$). In such laminates the occurrence of a ply crack and delamination is a very local event and the deformation resulting from the damage is constrained by other undamaged parts of the structure so that: (i) the effective axial strain is constrained to have a prescribed value $\varepsilon_{\mathrm{A}}^{0}$ that leads in an undamaged laminate to an effective axial stress σ_{A}^{0}, (ii) the transverse strain is constrained to be everywhere uniform having the value $\varepsilon_{\mathrm{T}}^{0}$ that leads in an undamaged laminate to a given value σ_{T}^{0} for the effective applied transverse stress. This special case of generalised plane strain deformation is described by the term *constrained delamination*. As shown in [6] for the case of boundary elements, the direct form of imposing generalised plane strain conditions is by means of establishing and imposing the value of ε_{T}. This situation also leads to the need to develop a method of estimating energy release rates for constrained delamination. For this case the effective transverse strain is uniform everywhere in the damaged laminate having the fixed value $\varepsilon_{\mathrm{T}}^{0}$ during delamination. For given prescribed values of the effective axial strain $\varepsilon_{\mathrm{A}}^{0}$ and the temperature difference ΔT, the value of $\varepsilon_{\mathrm{T}}^{0}$ is selected so that an *undamaged* laminate is subject to a prescribed value σ_{T}^{0} for the effective transverse stress. The axial stress $\sigma_{\mathrm{A}}^{\mathrm{app}}$ that must be applied when $\sigma_{\mathrm{T}}=\sigma_{\mathrm{T}}^{0}$ is obtained from the stress-strain relation (12.5) for an undamaged laminate so that

$$\sigma_{\mathrm{A}}^{\mathrm{app}}=E_{\mathrm{A}}^{0}\varepsilon_{\mathrm{A}}^{0}+\nu_{\mathrm{A}}^{0}\sigma_{\mathrm{T}}^{0}-E_{\mathrm{A}}^{0}\alpha_{\mathrm{A}}^{0}\Delta T.\tag{12.91}$$

The corresponding value of $\varepsilon_{\mathrm{T}}^{0}$ is then obtained from (12.6) so that

$$\varepsilon_{\mathrm{T}}^{0}=-\frac{\nu_{\mathrm{A}}^{0}}{E_{\mathrm{A}}^{0}}\sigma_{\mathrm{A}}^{\mathrm{app}}+\frac{\sigma_{\mathrm{T}}^{0}}{E_{\mathrm{T}}^{0}}+\alpha_{\mathrm{T}}^{0}\Delta T.\tag{12.92}$$

For a damaged laminate subject to the fixed effective transverse strain $\varepsilon_{\mathrm{T}}^{0}$ the value of the effective transverse applied stress $\sigma_{\mathrm{T}}^{\mathrm{app}}$ is obtained using (12.57) so that

$$\sigma_{\mathrm{T}}^{\mathrm{app}}=(1-\eta)\sigma_{\mathrm{T}}^{0}+\eta\sigma_{\mathrm{T}}^{\mathrm{deb}},\tag{12.93}$$

where the value of σ_{T}^{0} is given by (12.34) and the value of $\sigma_{\mathrm{T}}^{\mathrm{deb}}$ is obtained using (12.42) with the value $\sigma_{\mathrm{A}}=\sigma_{\mathrm{A}}^{\mathrm{app}}$.

As the state of deformation is one of generalised plane strain, the analysis of Section 12.6 is relevant and the result (12.89) can also be used for the constrained case so that the energy release rate is given by

$$ERR(\sigma_A, \sigma_T, \Delta T) = \frac{1}{2} h \left[\sigma_A^{app} + k \sigma_T^{app} + k_1 \Delta T \right]^2 \left(R + \frac{2\nu_A}{E_A} Q - \left(\frac{\nu_A}{E_A} \right)^2 P \right), \quad (12.94)$$

where the critical value, enabling delamination growth, is 2γ. Relation (12.90) is valid for this case so that the result (12.94) depends on the value of $\eta = c / L$ indicating that delamination under conditions of self-similar growth cannot be considered as steady-state cracking under fixed applied stresses.

It is emphasised that results in this chapter obtained entirely by analytical methods, are exact in the limit $L \to \infty$, and that they can be used to assess the accuracy of all types of numerical method based on the assumed boundary conditions.

12.9 Results

For the investigation a simple $[0/90]_s$ laminate, where the ply thickness is 0.25 mm, has been selected that has the following transverse isotropic properties typical of a carbon-fibre-reinforced epoxy material:

$E_A = 141.3$ GPa $\quad E_T = E_t = 9.58$ GPa $\quad \nu_A = \nu_a = 0.3 \quad \nu_T = 0.32$

$\mu_A = \mu_a = 5.0$ GPa $\quad \mu_t = 3.629$ GPa $\quad \alpha_A = -1 \times 10^{-6}$ /°C $\quad \alpha_T = 2.6 \times 10^{-5}$ /°C.

The shear moduli μ are required only for input into boundary element method analysis, as the analytical results presented in this chapter do not depend on these elastic constants. Both the analytical and boundary element models require a value for the temperature difference $\Delta T = T - T_0$. When $\Delta T \neq 0$, thermal residual stresses which are present will affect the energy release rates for delamination. The laminate is assumed either to be free of thermal stresses (i.e. $\Delta T = 0$) or simulated at a temperature T that is 100 °C lower than the stress-free temperature of the laminate so that $\Delta T = -100$ °C.

As an important objective of this chapter is to compare predictions based on analytical results with accurate results from numerical studies based on boundary element method analysis, it is useful to choose a set of boundary conditions that is convenient for the comparison. The comparison case chosen applies specified axial and transverse displacements to the damaged laminate. The axial displacement is chosen so that the effective axial applied strain for the damaged laminate has the value $\varepsilon_A = \varepsilon_A^0$ where $\varepsilon_A^0 = 0.01$. The transverse displacement is uniform having a value the leads to the effective transverse strain ε_T^0 that arises in an undamaged laminate subject to the prescribed effective axial strain ε_A and to a zero effective transverse stress.

12.10 Discussion

From Table 12.1, for constrained conditions with $\varepsilon_A^{app} = 0.01$ and a small nonzero transverse stress σ_T^{app} in the absence of thermal stresses (i.e. when $\Delta T = 0$°C), as the ply crack separation $2L$ increases the energy release rate estimated using the boundary

Table 12.1 Predictions of loading parameters and energy release rates for delamination in a typical damaged CFRP estimated using the boundary element method ($\varepsilon_A^{app} = 0.01$).

ΔT (°C)	L (mm)	ε_T^0	σ_A^{app} (GPa)	σ_T^{app} (GPa)	ERR[a] $\Delta c = 1.22 \times 10^{-6}$ mm (J m^{-2})
0	5	−0.000380	0.7314	−0.007961	115.22
0	10	−0.000380	0.7324	−0.007671	115.49
0	20	−0.000380	0.7328	−0.007526	115.65
0	50	−0.000380	0.7331	−0.007439	115.71
−100	5	−0.000500	0.7333	−0.010002	181.87
−100	10	−0.000500	0.7345	−0.009637	182.29
−100	20	−0.000500	0.7351	−0.009455	182.54
−100	50	−0.000500	0.7355	−0.009346	182.64

[a]Twelve elements exactly.

element method gradually increases such that ERR = 115.71 J m^{-2} when L = 50 mm The corresponding asymptotic limiting value predicted using the analytical result (12.94) is given in Table 12.2 as ERR = 117.62 J m^{-2} which is consistent with the boundary element method results of Table 12.1. When thermal stresses are present characterised by the value $\Delta T = -100^o$C for the temperature difference, the values of energy release rate predicted by the boundary element method again increase as $2l$ increases such that ERR = 182.64 J m^{-2} when L = 50 mm The corresponding asymptotic limiting value predicted by the analytical result (12.94) is given in Table 12.2 as ERR = 185.66 J m^{-2} which is again consistent with the trend of the boundary element method results given in Table 12.1. The differences between the boundary element method results when L = 50 mm and the limiting analytical result valid when $L \to \infty$ are indicating that the size of the self-similar zones (see Figure 12.2) might be significant. Despite this, these differences are only 1.6% when $\Delta T = 0^o$C and when $\Delta T = -100^o$C, thus indicating the high accuracy of the boundary element method for delamination problems. The trends of the boundary element method results in Table 12.1 indicate that smaller percentage differences would result if ply crack separations greater than L = 50 mm were to be investigated. It is observed from Tables 12.1 and 12.2 that the limiting values of the effective applied axial stress σ_A^{app} in Table 12.2 obtained using analytical methods are very close indeed to the values predicted by the boundary element method for the case L = 50 mm

In Table 12.2 it is of interest to compare results for the case of constrained delamination with those predicted by (12.82) for the case of unconstrained delamination where $\sigma_T = 0$. The energy release rate for unconstrained conditions is significantly greater than the values for constrained conditions, independent of whether thermal stresses are present. The energy release rate results predicted by (12.89) for generalised plane strain conditions are also given in Table 12.2. It is seen that when thermal stresses are absent so that $\Delta T = 0^o$C the generalised plane strain result is very close to that for constrained delamination. A significant difference is, however, observed when thermal stresses are present characterised by the temperature difference $\Delta T = -100^o$C. It is seen from all

Table 12.2 Limiting loading parameters and energy release rates for delamination in a typical damaged CFRP estimated using analytical methods ($\varepsilon_A^{app} = 0.01$, $L \to \infty$).

ΔT (°C)	σ_A^{app} (GPa)	σ_T^{app} (GPa)	ε_T^0	*ERR* Constrained ($\sigma_T \neq 0$) (J m^{-2})	*ERR* Unconstrained ($\sigma_T = 0$) (J m^{-2})	*ERR* GPS[a] ($\sigma_T = 0$) (J m^{-2})	*ERR* O'Brien ($\sigma_T = 0$) (J m^{-2})
0	0.7333	-0.00666	-0.000380	117.62	129.12	118.32	120.83
-100	0.7357	-0.00837	-0.000500	185.66	203.82	186.76	—

[a]Generalised plane strain.

the results given in both Tables 12.1 and 12.2 that thermal residual stresses lead to a very large increase in the energy release rates for delamination for all loading cases.

It is seen from (12.82), (12.89) and (12.94) that the energy release rate is influenced by the loading parameters σ_A, σ_T and the temperature difference ΔT through the square of an effective stress term $s = \sigma_A + k\sigma_T + k_1\Delta T$, where the parameters k_1 and k are defined by (12.23) and (12.26), respectively. For the ply properties (carbon-fibre-reinforced epoxy) used previously, the value of $k = 0.265$ and $k_1 = -0.00186$ GPa. This indicates that, for the material used in the investigation, the effect of the transverse stress σ_T and the temperature difference ΔT on the energy release rate for delamination is of practical significance. Clearly, the occurrence of tensile transverse stresses and negative temperature differences increases the energy release rate and, thus, increase the likelihood of delamination.

It is useful to compare result (12.82) with that predicted using O'Brien's model [1], which applies only for uniaxial stress states and neglects thermal residual stresses. Furthermore, the model assumes that the transverse stresses are everywhere zero in both the 0° and 90° plies, a situation that cannot in general arise in the fully bonded part of a damaged laminate. The energy release rate for steady state delamination in a $[0_n/90_m]_s$ cross-ply predicted by O'Brien's model [1] is given by the following simple formula

$$ERR = \frac{h}{2}\sigma_A^2\left(\frac{h}{h^{(0)}\hat{E}_A} - \frac{1}{E_c}\right), \tag{12.95}$$

where

$$hE_c = h^{(0)}\hat{E}_A + h^{(90)}\hat{E}_T. \tag{12.96}$$

Expression (12.95) can be derived from (12.7) and (12.82) on substituting $\sigma_T = 0$, $\Delta T = 0$°C and $\nu_A = 0$. Result (12.82) shows how O'Brien's result may be improved by allowing nonzero values of ν_A, and generalised so that biaxial loading and thermal residual stresses can be considered. Using the value of σ_A^{app} given in Table 12.2 for the temperature difference $\Delta T = 0$°C that leads to an effective axial applied strain $\varepsilon_A = 0.01$, the energy release rate when using O'Brien's formula (12.95) given in Table 12.2 is 120.83 J m^{-2}. When nonzero values of ν_A are taken into account, the energy release rate is given by (12.82) with $\sigma_T = 0$ leading to the result *ERR* = 129.12 J m^{-2}. These results show that, when thermal residual stresses are absent, O'Brien's result

(12.95) predicts a value for the energy release rate that is significantly less than the unconstrained limiting value predicted by (12.82), but that it is a much better estimate when constrained delamination takes place as the implied approximation $\nu_A = 0$ is almost equivalent to the condition $\varepsilon_T = 0$ applied to the undamaged part of the laminate.

Takeda and Ogihara [4, (26)] have given a result for the debonding energy release rate that includes the effects of thermal residual stresses and reduces, using the notation of this chapter, to the following form, when $L \to \infty$ and $\eta = 0.5$,

$$ERR = \frac{hh^{(90)}}{2h^{(0)}} \frac{\hat{E}_T}{\hat{E}_A E_c} \left(E_c \varepsilon_0 - \frac{h^{(0)} \hat{E}_A}{h} (\hat{\alpha}_T - \hat{\alpha}_A) \Delta T \right)^2, \tag{12.97}$$

although there is some ambiguity regarding the meaning of the strain ε_0. A corresponding result of Nairn and Hu [3, (37)] can be written in the form

$$ERR = \frac{hh^{(90)}}{2h^{(0)}} \frac{\hat{E}_T}{\hat{E}_A E_c} \left(\sigma_A - \frac{h^{(0)} \hat{E}_A}{h} (\hat{\alpha}_T - \hat{\alpha}_A) \Delta T \right)^2. \tag{12.98}$$

Relations (12.97) and (12.98) are identical only if $\sigma_A = E_c \varepsilon_0$, where E_c is defined by (12.96), suggesting that ε_0 should be interpreted as the mechanical axial strain rather than total axial strain. It is worth noting from (D.36) (see Appendix D) that when $\hat{\nu}_A = 0$

$$k_1 = \frac{h^{(0)} \hat{E}_A}{h} (\hat{\alpha}_A - \hat{\alpha}_T), \tag{12.99}$$

so that (12.98) may be written as

$$ERR = \frac{hh^{(90)}}{2h^{(0)}} \frac{\hat{E}_T}{\hat{E}_A E_c} (\sigma_A + k_1 \Delta T)^2. \tag{12.100}$$

It is easily shown that this result is identical to (12.82) only if $\hat{\nu}_A = 0$. It is clear that the result (12.82) should be used in preference to published results as it is valid for any value of $\hat{\nu}_A$.

To conclude the discussion it is worthwhile emphasising that the approach to predicting delamination growth based on the use of the Gibbs energy function $\hat{G}(\sigma_A, \sigma_T, \Delta T, \rho, c)$ could be replaced by an equivalent approach based on the Helmholtz energy function $\hat{F}(\varepsilon_A, \varepsilon_T, \Delta T, \rho, c)$. Using the stress-strain relations (12.3) and (12.4) for a damaged laminate and expression (12.73) for $\hat{G}$, it can be shown that

$$\begin{aligned} \hat{F} \equiv \hat{G} + \sigma_A \varepsilon_A + \sigma_T \varepsilon_T = {} & \frac{1}{2} \tilde{E}_A \varepsilon_A^2 + \frac{1}{2} \tilde{E}_T \varepsilon_T^2 + \nu_A \tilde{E}_T \varepsilon_A \varepsilon_T - \tilde{E}_A \tilde{\alpha}_A \varepsilon_A \Delta T \\ & - \tilde{E}_T \tilde{\alpha}_T \varepsilon_T \Delta T + \left(\frac{1}{2} \tilde{E}_A \alpha_A^2 + \nu_A \tilde{E}_T \alpha_A \alpha_T + \frac{1}{2} \tilde{E}_T \alpha_T^2 \right) (\Delta T)^2 + \bar{G}, \end{aligned} \tag{12.101}$$

where

$$\frac{E_A}{\tilde{E}_A} = \frac{E_T}{\tilde{E}_T} = 1 - \nu_A^2 \frac{E_T}{E_A}, \quad \tilde{\alpha}_A = \alpha_A + \nu_A \frac{E_T}{E_A} \alpha_T, \quad \tilde{\alpha}_T = \alpha_T + \nu_A \alpha_A. \tag{12.102}$$

The inverse form of the stress-strain relations for the damaged laminate are then given by

$$\sigma_{\mathrm{A}} \equiv \frac{\partial \widehat{F}}{\partial \varepsilon_{\mathrm{A}}} = \tilde{E}_{\mathrm{A}}\varepsilon_{\mathrm{A}} + \nu_{\mathrm{A}}\tilde{E}_{\mathrm{T}}\varepsilon_{\mathrm{T}} - \tilde{E}_{\mathrm{A}}\tilde{\alpha}_{\mathrm{A}}\Delta T, \tag{12.103}$$

$$\sigma_{\mathrm{T}} \equiv \frac{\partial \widehat{F}}{\partial \varepsilon_{\mathrm{T}}} = \nu_{\mathrm{A}}\tilde{E}_{\mathrm{T}}\varepsilon_{\mathrm{A}} + \tilde{E}_{\mathrm{T}}\varepsilon_{\mathrm{T}} - \tilde{E}_{\mathrm{T}}\tilde{\alpha}_{\mathrm{T}}\Delta T. \tag{12.104}$$

By considering crack closure when the laminate is uniaxially strained in the axial and transverse directions, the following additional interrelationships can be derived using similar methods to those described in Section 12.4:

$$\frac{\nu_{\mathrm{A}}\tilde{E}_{\mathrm{T}} - \nu_{\mathrm{A}}^0\tilde{E}_{\mathrm{T}}^0}{\tilde{E}_{\mathrm{A}} - \tilde{E}_{\mathrm{A}}^0} = \frac{\tilde{E}_{\mathrm{T}} - \tilde{E}_{\mathrm{T}}^0}{\nu_{\mathrm{A}}\tilde{E}_{\mathrm{T}} - \nu_{\mathrm{A}}^0\tilde{E}_{\mathrm{T}}^0} = \frac{\tilde{E}_{\mathrm{T}}\tilde{\alpha}_{\mathrm{T}} - \tilde{E}_{\mathrm{T}}^0\tilde{\alpha}_{\mathrm{T}}^0}{\tilde{E}_{\mathrm{A}}\tilde{\alpha}_{\mathrm{A}} - \tilde{E}_{\mathrm{A}}^0\tilde{\alpha}_{\mathrm{A}}^0} = \widehat{\nu}_{\mathrm{A}}, \quad \frac{\tilde{E}_{\mathrm{T}} - \tilde{E}_{\mathrm{T}}^0}{\tilde{E}_{\mathrm{A}} - \tilde{E}_{\mathrm{A}}^0} = \widehat{\nu}_{\mathrm{A}}^2. \tag{12.105}$$

Expression (12.101) for the Helmholtz energy per unit volume $\hat{F}$ may then be expressed in the following simple form

$$\widehat{F} = \frac{1}{2}\left(\tilde{E}_{\mathrm{A}} - \tilde{E}_{\mathrm{A}}^0\right)\left[\varepsilon_{\mathrm{A}} + \widehat{\nu}_{\mathrm{A}}\varepsilon_{\mathrm{T}} - \hat{\varepsilon}_{\mathrm{A}}^{\mathrm{c}}\right]^2 + F_0, \tag{12.106}$$

where

$$\hat{\varepsilon}_{\mathrm{A}}^{\mathrm{c}} = (\widehat{\alpha}_{\mathrm{T}} + \widehat{\nu}_{\mathrm{A}}\widehat{\alpha}_{\mathrm{A}})\Delta T, \tag{12.107}$$

is the axial crack closure strain for a laminate under uniaxial straining conditions such that $\varepsilon_{\mathrm{T}} = 0$, and where F_0 is the Helmholtz energy per unit volume for an undamaged laminate in a stress-free state.

On applying the energy balance fracture criterion (I.2) (see Appendix I) to the constrained delamination case of Section 12.8, the energy release rate is given by

$$ERR\left(\varepsilon_{\mathrm{A}}^{\mathrm{app}}, \varepsilon_{\mathrm{T}}^0, \Delta T\right) = -hL\frac{\partial \widehat{F}}{\partial c}. \qquad (\Delta T, \rho \text{ held fixed}) \tag{12.108}$$

When the applied effective strain $\varepsilon_{\mathrm{A}}^{\mathrm{app}}$ and temperature difference ΔT are given, the energy release rate for delamination, for conditions of transverse constraint $\varepsilon_{\mathrm{T}} = \varepsilon_{\mathrm{T}}^0$, is then given by

$$ERR\left(\varepsilon_{\mathrm{A}}^{\mathrm{app}}, \varepsilon_{\mathrm{T}}^0, \Delta T\right) \equiv -hL\frac{\partial \widehat{F}}{\partial c} = \frac{1}{2}h\tilde{E}_{\mathrm{A}}^2 R\left[\varepsilon_{\mathrm{A}}^{\mathrm{app}} + \widehat{\nu}_{\mathrm{A}}\varepsilon_{\mathrm{T}}^0 - \hat{\varepsilon}_{\mathrm{A}}^{\mathrm{c}}\right]^2, \tag{12.109}$$

requiring a critical value 2γ for delamination growth. It has been shown that the result (12.94) based on the Gibbs energy and the result (12.108) based on the Helmholtz energy predict identical values for the energy release rate (to 17 significant figures using high-precision arithmetic).

Although the main purpose of this chapter is to determine energy release rates for delamination under self-similar growth conditions, the following questions arise: are such conditions ever realised in practice and does ply crack saturation intervene? As the results listed in Table 12.1 indicate that very long delamination lengths are required for the *ERR* to approach the self-similar growth limit, it seems this limit is unlikely to be achieved in practical applications. However, and most importantly, it is noted that the self-similar limit is an upper bound for the ERR, so that its use in design situations will be conservative, i.e. it will lead to pessimistic (and, therefore, safe) predictions of delamination resistance.

As cross-ply laminates are of limited practical use, it is worthwhile pointing out that the methodology of this chapter could be applied to more general laminates provided they are balanced and symmetric. Off-axis plies and 0° plies would need first to be homogenised into an effective 0° ply having properties that will then differ from those of the 90° plies, as suggested by Nairn and Hu [3].

References

1. O'Brien, T.K. (1985). Analysis of local delaminations and their influence on composite laminate behaviour. *ASTM STP* 876: 282–297.
2. O'Brien, T.K. (1992). Residual thermal and moisture influences on the analysis of local delaminations. *Journal of Composites Technology and Research* 14: 86–94.
3. Nairn, J.A. and Hu, S. (1992). The initiation and growth of delaminations induced by matrix microcracks in laminated-composites. *International Journal of Fracture* 57: 1–24.
4. Takeda, N. and Ogihara, S. (1994). Initiation and growth of delamination from the tips of transverse cracks in CFRP cross-ply laminates. *Composites Science and Technology* 52: 309–318.
5. McCartney, L.N. (2003). Physically based damage models for laminated composites. *Proceedings of the Institution of Mechanical Engineers* 217 (Part L, *Journal of Materials: Design and Applications*): 163–199.
6. Blázquez, A., Mantič, V., Paris, F., and McCartney, L.N. (2008). Stress state characterization of delamination cracks in [0/90] symmetric laminates by BEM. *International Journal of Solids and Structures* 45: 1632–1662.

The analysis and results presented in this Chapter are based on the publication

McCartney, L.N., Blazquez, A., and Paris, F. (2012). Energy-based delamination theory for biaxial loading in presence of thermal stresses. *Composites Science and Technology* 72: 1753–1766.

13

Energy Methods for Fatigue Damage Modelling of Laminates

Overview: The initiation and growth of damage in composite materials are phenomena that precede the catastrophic failure event where a material sample or component fragments or separates into two pieces. During fatigue loading, the damage grows stably owing to cyclic stressing and leads to a gradual deterioration of mechanical properties and ultimately to failure. For cross-ply laminates, the estimation is necessary of effective stress intensity factors or energy release rates, for statically loaded ply cracks in 90° plies that are bridged by the uncracked 0° plies, particularly when considering the early stages of property degradation. Such relations are used in conjunction with a fatigue crack growth law to predict the progressive development of damage during cyclic loading. For such loading, this chapter justifies, on the basis of detailed physical modelling using energy methods rather than empiricism, the stress range intensity factor as the correlating parameter for fatigue crack growth data, rather than energy release rates or differences of energy release rate.

Use is made of an accurate stress-transfer model for multiple-ply cross-ply laminates to predict the dependence of energy release rates and stress intensity factors for long bridged ply cracks on the applied stress and ply crack separation. Two methods of analysis are considered. The first uses a method that can be extended to deal with small laminate defects where the energy release rate and stress intensity factor depend on the size of the defect, but the laminate is subject only to a uniaxial load. The second method applies to multiaxial loading but assumes that the stress intensity factor or energy release rate is independent of the defect size. Both methods are, however, shown to lead to the same energy release rate and stress intensity factor for long ply cracks subject to uniaxial loading. The two methods also take full account of the effects of thermal residual stresses through the use of a crack closure concept.

Simplifying assumptions are made when developing a model that predicts the degradation of most of the thermoelastic constants of a fatigue damaged cross-ply laminate as a function of the number of fatigue cycles. Such data are needed to predict the fatigue behaviour of structures having complex stress states using finite element analysis. Work carried out to validate the fatigue model, based on simplifying assumptions, has led to pessimistic predictions of performance that have the advantage that they can be exploited in the form of conservative design methods.

Properties for Design of Composite Structures: Theory and Implementation Using Software, First Edition. Neil McCartney.

Companion Website: www.wiley.com/go/mccartney/properties

13.1 Introduction

The estimation of the life of a composite structural element, when subjected to fatigue or creep deformations, is a notoriously difficult task to attempt. If material failure is defined to be the separation of a composite into two or more separate pieces, as in the case of metal components, then it must be recognised that composite failure is generally preceded by a multitude of localised damage effects, such as ply cracking, delamination, fibre/matrix debonding and fibre fracture, that are strongly interactive, and subject to statistical uncertainty. These damage mechanisms lead to a progressive deterioration in performance of the composite to the point of catastrophic failure.

The complexity is well described in [1], which describes contributions from recognised experts in the field, and is apparent from earlier work in the field (see, for example, Talreja [2,3], Ogin et al. [4] and Boniface et al. [5]). Harris [6] has given a very useful historical review of the fatigue behaviour of fibre-reinforced composites, whereas Beaumont [7] has developed physical models of fatigue-induced damage in the form of ply cracks, splits and delaminations. In [8–10] additional approaches to modelling of relevance to some aspects of this chapter are described.

To develop a practically useful life prediction design methodology, it is vital that the initiation and growth of damage preceding failure is capable of being predicted with confidence. As the prediction of the initiation of fatigue damage in composites is exceedingly difficult, even for metals, it is usual to consider fatigue damage growth by assuming that preexisting defects of known size are already present in the material or structure. The overall objective of this chapter is, therefore, to outline some useful methods that can be recommended for use when predicting the growth of preexisting ply defects, and the effects of fatigue induced damage on the thermoelastic performance of $[0_m/90_n]_s$ cross-ply laminates.

Carbon fibres are well known for not exhibiting significant degradation during fatigue loading. In fact, a good approximation is to assume that the fibres do not degrade at all. This means that, when modelling the fatigue degradation of unidirectional CFRP subject to uniaxial loading, it is highly unlikely that any model will be able to predict the failure of the composite. The only exception is for composites subject to sufficiently large loads, where load transfer from the degraded matrix to the fibres leads to progressive fibre failure. This type of behaviour is often observed in *S–N* curves for CFRP. The modelling of matrix degradation in the composite is exceedingly difficult and not amenable to meaningful analysis. This difficulty applies to both unidirectional and multidirectional CFRP laminates. The modelling discussed in this chapter does not consider the failure of CFRP, but concentrates on the development of a model to predict the degradation of laminate mechanical properties as a result of ply crack growth. In contrast to carbon fibres, glass fibres degrade with fatigue cycling, and the mechanism is thought to result from the stress-corrosion cracking at fibre defects, which also occurs when the fibres are subjected to a static load. This type of degradation has been considered in another publication [11] and in Chapter 14.

During fatigue loading (at modest stress levels) of cross-ply laminates, ply crack growth normal to the loading direction, is an important mode of fatigue damage that is encountered. This leads to the presence of cracks in the 90° plies of the composite, which are effectively bridged by 0° plies. The presence of these cracks leads to a progressive degradation of the properties of the composite, such as Young's moduli,

Poisson's ratios and thermal expansion coefficients. When attempting to develop a realistic fatigue model of laminate degradation it is necessary to distinguish between three different types of ply crack. The first type concerns fully developed ply cracks that traverse the entire cross section of the 90° ply of the laminate. This type of ply crack either forms instantaneously when the laminate is first loaded, or when progressive fatigue-induced growth of preexisting defects leads eventually to fully developed ply cracks. The key aspect concerning fully developed cracks is that their lengths cannot increase in the through-thickness directions during subsequent fatigue cycling although interply delaminations may initiate and grow (see, for example, [10]). The second type of ply crack starts near one of the laminate edges and has grown part of the way across the laminate in the width direction (i.e. in fibre direction of the 90° plies) rather than the through-thickness direction. During fatigue cycling ply crack growth can occur only at the interior crack tip. The third type of ply crack is fully embedded, as both ply crack tips are in the interior of a 90° ply in the laminate and both can grow in the fibre direction of the 90° ply during fatigue cycling. Realistic fatigue damage modelling needs to take account of the effects of fully developed cracks (whether formed during initial loading or progressively as a result of fatigue loading), and of the possibility that new defects may be continually initiated. During fatigue loading, delamination growth from the ply cracks that form is another important damage mode [10] that will affect the degradation of laminate properties especially during the later stages of damage development. A satisfactory treatment of this phenomenon will require the use of reliable stress-transfer models. This aspect of fatigue damage development is beyond the scope of this chapter, although a simple model of delamination has already been considered in Chapter 12.

Understanding the factors that determine the formation and growth of ply cracks during fatigue loading is a key objective. The stress intensity factors and energy release rates at the tips of bridged ply cracks are influenced strongly by the bridging 0° plies and it is vital to be able to predict the dependence of the local stress intensity factors and energy release rates on crack bridging stresses. The concept of crack bridging stress and crack opening displacement has been discussed in detail [12,13]. It has been shown [12] how crack bridging relations derived from sublaminate models of ply cracking can be used to develop an equivalent continuum model that can be treated using the methods of fracture mechanics suitably modified to account for bridging (see also Chapter 8). The methods developed will be extended here so that the growth of fatigue cracks can also be treated. The approach to be adopted involves relating the effective stress p carried by the bridging 0° plies to the effective crack opening displacement Δv for the equivalent continuum model (following notation and (x, y) coordinates adopted in Chapter 8, Section 8.8). One objective of this chapter is to use the p–Δv relation to predict the stress intensity factors and energy release rates of ply cracks for the case when 0° plies remain intact during ply crack growth. For cross-ply laminates, where ply cracking in the 90° plies is the only form of damage, and where the interfaces between the 0° and plies 90° remain perfectly bonded, the p–Δv relation is linear owing to the assumption that the stress analysis is based on linear elasticity theory. The application of nonlinear crack bridging relations to fatigue crack growth has been considered in detail for the case of fibre-reinforced titanium composites [14]. The approach of this chapter is to extend the same general method that was described in [14] by developing a simplified approach to mathematical modelling that is valid for the linear p–Δv relations that arise in most applications when considering ply crack growth in laminates.

An alternative approach for estimating the stress intensities and energy release rates for ply cracking is considered, based on rigorous ply cracking theory that includes the effect of multiaxial loading, that confirms the validity of the first approach for the special case of uniaxial loading of the laminate. The stress intensity factor is used to define the effective stress range intensity factor that can be used in conjunction with a fatigue crack growth law to predict the growth of ply cracks in laminates subject to cyclic loading.

13.2 Defining Preexisting Damage

The prediction of the number of cycles needed to initiate a ply crack, having a size that is measurable, is exceedingly difficult because of complex microscopic mechanisms that contribute to the formation of small ply cracks. Such mechanisms operate at the fibre/matrix interfaces, and lead to debonded areas on fibre/matrix interfaces that can grow in the through-thickness and width directions, and between plies leading to ply cracks and delaminations. The variability of fibre/matrix interface quality and inhomogeneous fibre spacings lead to ply crack initiation times that are subject to significant scatter, and to the need for a statistical approach. The pragmatic approach to this problem is to regard the ply crack initiation phase of fatigue damage as a known parameter so that predictive modelling can concentrate on the growth of damage and the onset of catastrophic failure. Figure 13.1 illustrates the type of ply crack that is considered in this chapter. The ply crack having length $2c$ is normal to the axial direction (in y-direction) and it traverses the entire thickness of the 90° ply but not necessarily the full width of the laminate in the x-direction.

Here, the approach taken is to consider small preexisting ply cracks within the model, where their length has grown during the initiation stage to a fixed critical size c_i that has the order of the ply thickness. The value of c_i can be determined using Figure 13.2 that is a master plot (see [13] for details) that applies to all cross-ply laminates.

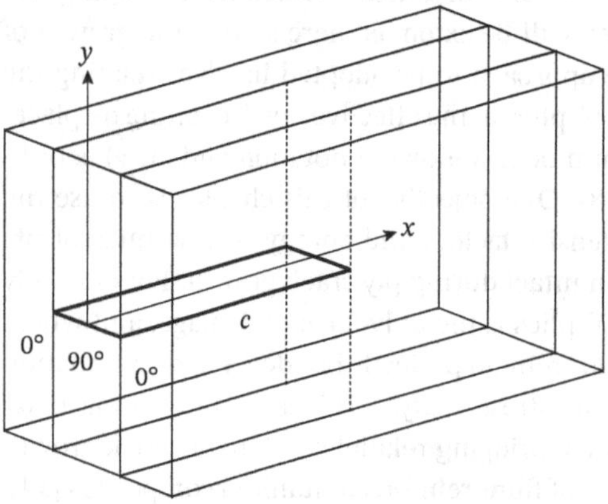

Figure 13.1 Schematic diagram of a ply crack in a cross-ply laminate representing one half of a laminate having an embedded crack or the whole of a laminate with an edge crack.

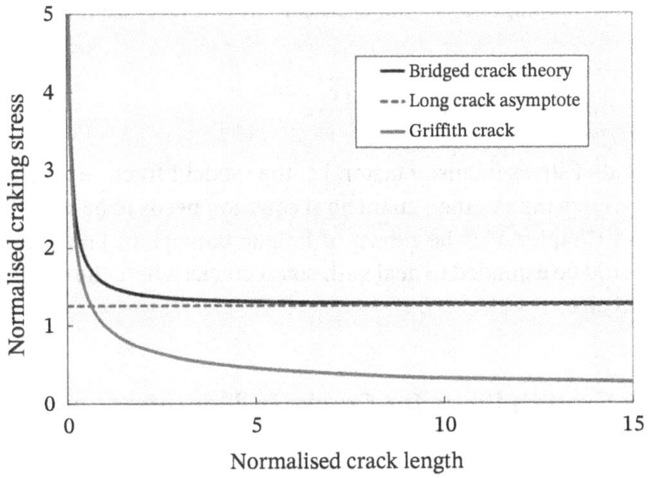

Figure 13.2 Master curve for cross-ply laminates showing how the normalised ply cracking stress is related to the normalised length of an embedded ply crack.

In Figure 13.2 the continuous black line denotes the curve relating the axial stress for ply cracking as a function of ply crack length. The cracking stress is normalised by dividing the difference $\sigma_A - \sigma_c$ between the uniaxial applied axial stress σ_A and the crack closure stress σ_c by the parameter σ_0 whereas the half-crack length c is normalised by dividing by c_0 where

$$\sigma_0^2 = \frac{8\Lambda\gamma}{\pi}, \quad c_0 = \frac{E_A}{4\Lambda\Omega}, \tag{13.1}$$

and where the significance of the parameters Λ and Ω are described in Sections 13.4 and 13.5, respectively. The quantity E_A is the effective axial Young's modulus of the laminate in an undamaged state and 2γ is the effective fracture energy (i.e. critical value of the energy release rate) for ply cracking in a homogenised laminate model. The crack closure stress σ_c is that value of the uniaxial applied stress where the ply crack is just closed. It depends upon the temperature difference $\Delta T = T - T_0$ where T is the laminate temperature and where T_0 is the stress-free temperature for the laminate. The crack closure stress σ_c is, in fact, proportional to the temperature difference ΔT.

For small enough crack lengths, the cracking stress depends on the ply crack length as shown in Figure 13.2. As the ply crack length increases, the curve approaches an asymptote that characterises steady-state ply cracking where the energy release rate is independent of crack length. The limiting value, given by the relation (see [13])

$$\frac{\sigma_A - \sigma_c}{\sigma_0} = \sqrt{\frac{\pi}{2}} \cong 1.253, \tag{13.2}$$

defines a simple lower bound for the ply cracking stress that can be used as a design parameter. It is seen that if the normalised crack length is greater than three, then the energy release rate can be regarded as being independent of ply crack length, thus defining the initial defect size c_i assumed for the fatigue model. The gray line of Figure 13.2 denotes the curve that would result if the bridging effect of the 0° plies were to be absent. This curve corresponds to the classical Griffith fracture criterion described by the relation

$$K \equiv \sigma_A \sqrt{\pi c} = K_{Ic} = \sigma_0 \sqrt{\pi c_0}, \text{ so that } \frac{\sigma_A}{\sigma_0} = \sqrt{\frac{c_0}{c}}, \tag{13.3}$$

where K_{Ic} is the critical mode I stress intensity factor, i.e. the model I fracture toughness. If defects are smaller in size than c_i, then an integral equation needs to be solved, as described in [12,13] and Chapter 17. The model of fatigue damage in laminates described in this chapter could be extended to deal with small cracks where the cracking stress depends on defect size.

13.3 Fatigue Crack Growth Laws for Cracks in Homogeneous Anisotropic Materials

A key issue regarding the prediction of fatigue damage in laminated composites, concerns the selection of the correlating parameter for uniform amplitude fatigue crack growth data, and the method for taking account of the load ratio $R = \sigma_{min}/\sigma_A$, where σ_{min} is the minimum stress of the fatigue cycle and σ_A is the maximum axial load. When carrying out fatigue lifing calculations for monolithic unreinforced materials three essential ingredients are needed. The first is a crack growth law that, for homogeneous materials, has often been assumed to have the following empirical form (the well-known Paris equation)

$$\frac{dc}{dN} = C(\Delta K)^n, \tag{13.4}$$

where C and n are material constants, and where ΔK is the stress range intensity factor for the crack having the form,

$$\Delta K \equiv Y \Delta\sigma \sqrt{\pi c}, \tag{13.5}$$

where $\Delta\sigma = \sigma_A - \sigma_{min}$ is the applied stress range of uniform amplitude fatigue loading, where $2c$ denotes the embedded crack length after N fatigue cycles, and where Y is a factor that takes geometrical effects into account. The parameter Y is often dependent on crack length, but for convenience it is treated as a constant in the following analysis.

The second ingredient, for loading cases where small-scale crack tip yielding conditions prevail, is knowledge of the fracture toughness of the material K_{Ic} governing the occurrence of unstable fracture. Following on from (13.3), the unstable fracture condition is often written in the form

$$K \equiv Y \sigma_A \sqrt{\pi c} = K_{Ic} \quad \text{or} \quad G \equiv Y^2 \sigma_A^2 \pi c = G_{Ic}, \tag{13.6}$$

where σ_A denotes the maximum stress that is applied during the fatigue cycle, where G is the energy release rate corresponding to the stress intensity factor K, and where G_{Ic} is the critical value of the energy release rate needed to initiate conditions of unstable crack growth. For anisotropic materials, the fracture criterion can be written in the form

$$G \equiv \frac{\Omega}{E_A} K^2 = G_{Ic} . \tag{13.7}$$

The parameter Ω characterises the anisotropy of the material, as discussed in detail in Section 13.4. The third ingredient is knowledge of the initial defect size c_i (as discussed in Section 13.2) that can be used, together with the crack growth law (13.4) and instability condition (13.6) or (13.7), to predict fatigue life and residual strength.

When attempting to take account of mean load effects, more complex forms for the crack growth law (13.4) are possible, for example,

$$\frac{dc}{dN} = \frac{C(\Delta K - \Delta K_0)^n}{K_{Ic}^2 - K^2}, \tag{13.8}$$

where C and n are material constants. The term ΔK_0 is a fatigue crack growth threshold that the stress range intensity factor must exceed if crack propagation is to occur. The denominator in (13.8) ensures that as $K \rightarrow K_{Ic}$ the crack growth rate tends to infinity; a phenomenon often observed from fatigue data relating to homogeneous materials. More general forms of (13.8) have been proposed, but the specific form given by (13.8) is preferred here as the crack growth relation can be written in the following form

$$\left(G_{Ic} - G\right)\frac{dc}{dN} = C'(\Delta K - \Delta K_0)^n, \tag{13.9}$$

where C' is another material constant. This type of relationship arises when carrying out an energy balance analysis for fatigue loading when treated quasistatically (see, for example, [15]). The left-hand side is associated with the strain energy release rate, whereas the right-hand side can be regarded as arising from localised energy dissipation at the crack tip contributing directly to stable crack growth through damage formation during fatigue loading, rather than to localised heating.

Empirical fatigue crack growth laws have been proposed of the form (see, for example, [8] when considering delamination growth)

$$\frac{dc}{dN} = CG^n \frac{1 - \dfrac{G_0^m}{G^m}}{1 - \dfrac{G^p}{G_{Ic}^p}}, \tag{13.10}$$

that is one possible empirical generalisation of (13.8) where the energy release rate has been used in place of the stress range intensity factor and the stress intensity factor. The form of (13.10) does not have a direct dependence on the R ratio of uniform amplitude loading (i.e. the ratio of the minimum and maximum stresses of the stress cycle). As written, assuming C is independent of the R ratio, it predicts that a *significant* growth rate will occur even if the applied stress range tends to zero, i.e. there is almost no stress cycling. One would expect that a fatigue crack growth law should have a direct dependence on the R ratio, and that the growth rate would tend to zero as $R \rightarrow 1$. It should be noted that the form (13.8) involves both the stress maximum and the stress range (and, hence, the R ratio) through the definitions of the stress intensity factors (13.5) and (13.6), and is such that the growth rate is zero when $\Delta\sigma = 0$. The form of (13.8) suggests that (13.10) might be generalised to the following form (just one of many empirical possibilities)

$$\frac{\mathrm{d}c}{\mathrm{d}N} = C\Delta G^n \frac{1 - \dfrac{G_0^{\mathrm{m}}}{G^{\mathrm{m}}}}{1 - \dfrac{G^{\mathrm{p}}}{G_{\mathrm{Ic}}^{\mathrm{p}}}}. \tag{13.11}$$

This leads on to the problem of uniquely defining ΔG. One approach is to write

$$\Delta G = G - G_{\min}, \tag{13.12}$$

where G is the strain energy release rate at the maximum load of the cycle and $G_{\min}$ is the energy release rate at the minimum load of the stress cycle. On using (13.6) and (13.7), it is clear that ΔG is then proportional to $\sigma_{\mathrm{A}}^2 - \sigma_{\min}^2$. No physical reasoning has ever been given to justify the assumption that fatigue crack growth data can be correlated using a ΔG parameter defined by (13.12). As $\Delta K = (1 - R)K$, where R is the stress ratio defined by $R = \sigma_{\min} / \sigma_{\mathrm{A}}$, relation (13.7) for the strain energy release rate can, however, be used to define an equally plausible alternative relation for ΔG as follows:

$$\Delta G \equiv \frac{\Omega}{E_{\mathrm{A}}} \Delta K^2 = (1 - R)^2 G = \left(\sqrt{G} - \sqrt{G_{\min}}\right)^2 = \frac{\Omega}{E_{\mathrm{A}}} \left(K - K_{\min}\right)^2. \tag{13.13}$$

The suggested possible definitions (13.12) and (13.13) for ΔG both have the desirable property that $\Delta G \to 0$ as $\sigma_{\mathrm{A}} \to \sigma_{\min}$. Another method of discriminating between these choices is, therefore, needed. This issue is discussed later (see Section 13.8).

13.4 Ply Crack Instability Criteria for Monotonic Loading

Consider a large rectangular homogeneous anisotropic plate that represents a homogenised cross-ply laminate material. The axes of material symmetry are parallel to the edges of the plate. Following the approach of Chapter 8, introduce a rectangular set of Cartesian coordinate axes (x, y) such that $x \equiv x_2$ and $y \equiv x_1$ denote the in-plane transverse and axial directions, respectively, so that the in-plane stress-strain relations for the laminate are of the form (8.62) (see also [12])

$$\begin{aligned} \varepsilon_{xx} &= a_{11}\sigma_{xx} + a_{12}\sigma_{yy}, \\ \varepsilon_{yy} &= a_{12}\sigma_{xx} + a_{22}\sigma_{yy}, \\ 2\varepsilon_{xy} &= a_{66}\sigma_{xy}, \end{aligned} \tag{13.14}$$

The parameters a_{11}, a_{12}, a_{22} and a_{66} are elastic constants for an orthotropic solid having the following values, for generalised plane stress conditions (thin sheets) (see (8.63)),

$$a_{11} = \frac{1}{E_{\mathrm{T}}}, \quad a_{22} = \frac{1}{E_{\mathrm{A}}}, \quad a_{12} = -\frac{\nu_{\mathrm{A}}}{E_{\mathrm{A}}}, \quad a_{66} = \frac{1}{\mu_{\mathrm{A}}}. \tag{13.15}$$

In (13.15), the quantities E, ν and μ denote Young's modulus, Poisson's ratio and shear modulus, respectively, whereas α denotes thermal expansion coefficients in (13.14). The subscripts A and T refer the thermoelastic constants to the axial and in-plane transverse directions, respectively. As already indicated, in this chapter the same coordinates (x, y) and corresponding notation are used as in the crack problems described in Chapter 8 (Section 8.8).

The first two diagrams of Figure 13.3 are edge views of a $[0_m/90_n]_s$ laminate that show the formation of a ply crack under fixed axial stress p, leading to an increase in length of the laminate by the amount Δv. The stress distribution for the laminate shown in the central diagram in Figure 13.3 is highly nonuniform but can be estimated accurately [16]. The axial stress distribution in the ply crack plane is also highly nonuniform being zero in the 90° ply and exhibiting stress singularities in the 0° plies at the two crack tips. A continuum model of the ply crack is constructed by smearing out the nonuniform axial stress distribution in the crack plane into a uniform effective stress distribution having value p that acts across the entire plane of the crack, as shown in the right-hand diagram of Figure 13.3, which is the edge view of the equivalent homogeneous continuum model of the laminate. The effective stress p acting across the crack is a function of the crack opening displacement, i.e. $p = p(\Delta v)$.

Based on analysis given in [12], it has been shown in Chapter 8 that the relationship, valid for nonlinear p–Δv relations, determining the stress for the instability of sufficiently long ply cracks in the 90° ply of a $[0_m/90_n]_s$ laminate subject to monotonic loading, is given by relation (8.149), namely

$$\int_0^c \left(\sigma_A - p(x,c)\right) \frac{\partial \Delta v(x,c)}{\partial c}\, dx + \int_0^c \frac{\partial p(x,c)}{\partial c} \Delta v(x,c)\, dx = 4\gamma. \tag{13.16}$$

It has also been shown in Chapter 8 that the fracture criterion for the unstable growth of ply cracks has the following simple form (consistent with relations (13.6) and (13.7))

$$K^2 = \frac{2E_A\gamma}{\Omega} \equiv K_{Ic}^2 \quad \text{or} \quad G = 2\gamma \equiv G_{Ic}, \tag{13.17}$$

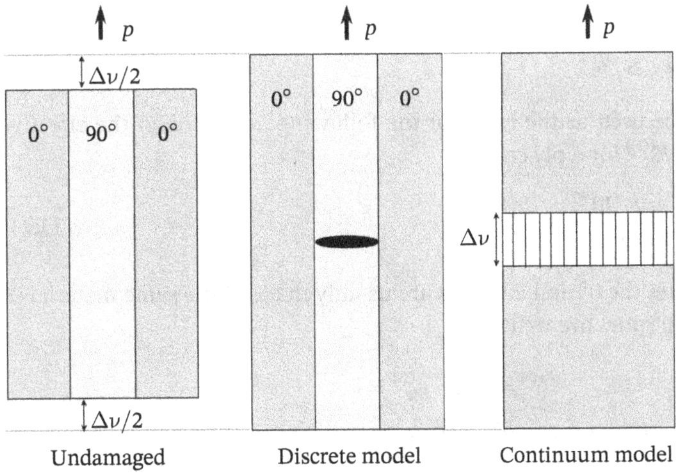

Figure 13.3 The continuum model of ply cracking in a cross-ply laminate.

where 2γ is the effective fracture energy for ply crack growth in a homogenised laminate model. For generalised plane stress conditions, the dimensionless parameter Ω is defined by (8.129), namely,

$$\Omega^2 = \frac{1}{2}\left(\sqrt{\frac{E_A}{E_T}} - \nu_A\right) + \frac{E_A}{4\mu_A}. \tag{13.18}$$

The methodology described in Chapter 8 enables the calculation of the stress intensity factor K for the continuum model of a ply crack in a laminate, and the master curve shown in Figure 13.2 indicates how large the initiation crack length $2c_i$ must be for the stress intensity factor to be independent of the crack length.

The parameter K_{Ic} given in (13.17) is the effective mode I fracture toughness of the composite. This parameter (together with K) is effective in the sense that its role is defined to be identical to that for unbridged cracks whereas its implementation takes full account of the fact that the ply crack is bridged by the 0° plies, such that ply cracking is allowed only in the 90° ply. Relation (13.17) corresponds to the well-known Griffith fracture criterion for unbridged cracks when the material is isotropic so that $\Omega = 1$ (generalised plane stress) or $\Omega = 1 - \nu^2$ (plane strain), and when $E_A = E$, where ν is Poisson's ratio and E is Young's modulus for an isotropic solid (see (8.129) and (8.130)). Relation (13.17) shows how K_{Ic} is related to the effective fracture energy 2γ. For ply cracks in a $[0_m/90_n]_s$ cross-ply laminate $2\gamma = 2\gamma^{(90)}h^{(90)}/h$ where $2h$ is the laminate thickness, $2h^{(90)}$ is the total thickness of the 90° ply, and where $2\gamma^{(90)}$ is the fracture energy for cracking parallel to the fibres in a monolithic sample of 90° ply material.

For monolithic 90° ply material the relation corresponding to (13.17) is written as

$$\left[K_{Ic}^{(90)}\right]^2 = \frac{2\gamma^{(90)}E_T^{(90)}}{\Omega^{(90)}}, \tag{13.19}$$

where the superscript $^{(90)}$ denotes that the quantity is associated with the property of the 90° ply. It follows from (13.17), (13.19) and the relationship between γ and $\gamma^{(90)}$ that the effective fracture toughness for the continuum model of a ply crack in a laminate may be calculated using the relation

$$K_{Ic}^2 = \frac{h^{(90)}}{h}\frac{\Omega^{(90)}}{\Omega}\frac{E_A}{E_T^{(90)}}\left[K_{Ic}^{(90)}\right]^2. \tag{13.20}$$

Relation (13.20) can be used as the basis for the following definition of the effective stress intensity factor $K^{(90)}$ for a ply crack

$$\left[K^{(90)}\right]^2 = \frac{h}{h^{(90)}}\frac{\Omega}{\Omega^{(90)}}\frac{E_T^{(90)}}{E_A}K^2. \tag{13.21}$$

In cross-ply laminates the 0° and 90° plies are usually made of the same material so that the properties of 0° plies are written as

$$E_A^{(0)} = \widehat{E}_A,\quad E_T^{(0)} = \widehat{E}_T,\quad \nu_A^{(0)} = \widehat{\nu}_A,\quad \nu_T^{(0)} = \widehat{\nu}_T,$$
$$\mu_A^{(0)} = \widehat{\mu}_A,\quad \alpha_A^{(0)} = \widehat{\alpha}_A,\quad \alpha_T^{(0)} = \widehat{\alpha}_T,$$

where $\widehat{E}_{\mathrm{A}}, \widehat{E}_{\mathrm{T}}, \widehat{\nu}_{\mathrm{A}}, \widehat{\nu}_{\mathrm{T}}, \widehat{\mu}_{\mathrm{A}}, \widehat{\alpha}_{\mathrm{A}}, \widehat{\alpha}_{\mathrm{T}}$ are the seven independent in-plane thermoelastic constants of the plies of the laminate. For the 90° plies the following identifications must be made

$$E_{\mathrm{A}}^{(90)} = \widehat{E}_{\mathrm{T}}, \quad E_{\mathrm{T}}^{(90)} = \widehat{E}_{\mathrm{A}}, \quad \nu_{\mathrm{A}}^{(90)} = \widehat{\nu}_{\mathrm{A}} \frac{\widehat{E}_{\mathrm{T}}}{\widehat{E}_{\mathrm{A}}}, \quad \nu_{\mathrm{T}}^{(90)} = \widehat{\nu}_{\mathrm{A}},$$
$$\mu_{\mathrm{A}}^{(90)} = \widehat{\mu}_{\mathrm{A}}, \quad \alpha_{\mathrm{A}}^{(90)} = \widehat{\alpha}_{\mathrm{T}}, \quad \alpha_{\mathrm{T}}^{(90)} = \widehat{\alpha}_{\mathrm{A}}.$$

These relations enable the parameter $\Omega^{(90)}$ to be calculated using (13.18). It now remains to show how the effective stress intensity factor K for the continuum model of bridged ply cracking may be estimated.

13.5 Stress Intensity Factors and Energy Release Rates for Long Ply Cracks

Consider a long ply crack in the 90° ply of a cross-ply laminate, as shown in Figure 13.1 now considered to show the entire laminate. The crack is of length c and the growth of the right-hand crack tip is in the direction of the x-axis. The left-hand crack tip is regarded as being at the edge of the laminate and cannot grow. Consider the case of linear p–Δv crack bridging relationships that can be expressed [12,13] in the form

$$p = \Lambda \Delta v + \sigma_{\mathrm{c}}, \tag{13.22}$$

where σ_{c} is the value of the applied stress for which the axial stress in the 90° plies of an undamaged composite would be zero so that the ply crack is just closed. The value of σ_{c} is zero when thermal expansion mismatches between the 0° and 90° plies are negligible, which is seldom the case. The crack bridging parameters Λ and σ_{c} are specified in Section 13.6. As Δv_0 is the crack opening displacement when the applied stress is σ_{A}, it follows from (13.22) that

$$\sigma_{\mathrm{A}} = \Lambda \Delta v_0 + \sigma_{\mathrm{c}}. \tag{13.23}$$

On substituting (13.22) into (13.16), making use of (13.23), it is easily shown that the instability condition, for sufficiently long ply cracks, is given by the simple relation

$$G = \frac{1}{2\Lambda} (\sigma_{\mathrm{A}} - \sigma_{\mathrm{c}})^2 = 2\gamma, \tag{13.24}$$

which is independent of crack length as to be expected for sufficiently long ply cracks. As (13.24) is equivalent to (13.17) it follows that the effective stress intensity factor (continuum model) for a sufficiently long ply crack is given by the simple relation

$$K = \sqrt{\frac{E_{\mathrm{A}}}{2\Omega\Lambda}} (\sigma_{\mathrm{A}} - \sigma_{\mathrm{c}}). \tag{13.25}$$

It is emphasised that the value of this effective stress intensity factor, valid for a sufficiently long ply crack, is independent of crack length. On substituting (13.25) into

(13.21), the corresponding stress intensity factor for a sufficiently long ply crack in a 90° ply is given by

$$K^{(90)} = \sqrt{\frac{hE_{\mathrm{T}}^{(90)}}{2\Lambda h^{(90)}\Omega^{(90)}}}(\sigma_{\mathrm{A}} - \sigma_{\mathrm{c}}). \tag{13.26}$$

It is worth noting that the energy release rate (13.24) and the stress intensity factors defined by (13.25) and (13.26) are all zero when the applied stress is equal to the crack closure stress σ_{c}.

13.6 Determination of Crack Bridging Parameters

Relation (13.22) or (13.23) defining the effective crack bridging law, for a single fully developed ply crack in a cross-ply laminate, can be derived from the axial stress-strain relation for a cracked laminate with an array of equally spaced fully developed ply cracks having density $\rho = 1/2L$ where $2L$ is the crack separation. The crack bridging law for an isolated crack will result when $L \to \infty$. The axial stress-strain relation for a uniaxially loaded undamaged cross-ply laminate is

$$\varepsilon_{\mathrm{A}} = \frac{\sigma_{\mathrm{A}}}{E_{\mathrm{A}}} + \alpha_{\mathrm{A}}\Delta T, \tag{13.27}$$

where ε_{A} is the axial strain for an undamaged laminate. The corresponding relation for a damaged laminate subject to uniaxial loading having an effective ply crack density ρ is

$$\varepsilon_{\mathrm{A}}(\rho) = \frac{\sigma_{\mathrm{A}}}{E_{\mathrm{A}}(\rho)} + \alpha_{\mathrm{A}}(\rho)\Delta T, \tag{13.28}$$

where now the effective axial modulus and effective thermal expansion coefficient depend on the crack density ρ. As the ply crack distribution in the laminate is regarded as effectively uniform such that the separation of the ply cracks is $2L$, the strains defined by (13.27) and (13.28) are directly related to the crack opening displacement Δv appearing in the crack-bridging relation (13.23) as follows:

$$\frac{\Delta v}{2L} = \varepsilon_{\mathrm{A}}(\rho) - \varepsilon_{\mathrm{A}}, \quad \text{i.e.} \quad \Delta v = \frac{1}{\rho}\left[\varepsilon_{\mathrm{A}}(\rho) - \varepsilon_{\mathrm{A}}\right], \quad \text{because} \quad \rho = \frac{1}{2L}. \tag{13.29}$$

On substituting (13.27) and (13.28) into (13.29) it follows that

$$\Delta v = \frac{1}{\rho}\left[\left(\frac{1}{E_{\mathrm{A}}(\rho)} - \frac{1}{E_{\mathrm{A}}}\right)\sigma_{\mathrm{A}} + \left(\alpha_{\mathrm{A}}(\rho) - \alpha_{\mathrm{A}}\right)\Delta T\right]. \tag{13.30}$$

As this relation must be identical in form to (13.23) it is deduced that the parameters Λ and σ_{c} are specified by

$$\frac{1}{\Lambda} = \frac{1}{\rho}\left(\frac{1}{E_{\mathrm{A}}(\rho)} - \frac{1}{E_{\mathrm{A}}}\right), \quad \sigma_{\mathrm{c}} = -\frac{\alpha_{\mathrm{A}}(\rho) - \alpha_{\mathrm{A}}}{\dfrac{1}{E_{\mathrm{A}}(\rho)} - \dfrac{1}{E_{\mathrm{A}}}}\Delta T. \tag{13.31}$$

The values of Λ and σ_c defined by (13.31) can be estimated from an accurate stress-transfer model [16,17]. It has been shown [16,17] that the ratio of the thermoelastic constants defining σ_c is independent of the ply crack density. The expression for σ_c given by (13.31) may be written as

$$\sigma_c = -k_1 \Delta T, \tag{13.32}$$

where k_1 is a laminate constant that is easily calculated from the undamaged properties and geometry of the plies in the cracked laminate, as shown in [16,17]. It should be noted that as $\rho \rightarrow 0$ the parameter Λ will tend to a finite value that is the result for an isolated crack. It should be noted also that the crack closure stress σ_c accounts for the effects on ply cracking of thermal residual stresses in the laminate.

13.7 Alternative Method of Calculating Energy Release Rates

Consider a length $2L$ of a cross-ply laminate containing a long single ply crack in the 90° ply, as shown in Figure 13.1 (three-dimensional view) and Figure 13.4 (side view). The laminate is divided into three regions defining a transition zone, between cracked and uncracked regions, in which the stress and displacement fields are complex and will move in a self-similar way with the crack tip as it grows. For sufficiently long ply cracks it is not necessary to consider this region when carrying out an energy balance calculation to estimate the energy release rate at the ply crack tip.

The stress and displacement distributions in the uncracked region, having volume V, ahead of the growing crack is easily found using laminate theory. For the cracked region on the other side of the transition region, having volume V as shown in Figure 13.4, the ply crack can be treated as a fully developed ply crack (i.e. traversing the entire cross section of the 90° ply in this region). Methods are available for accurately determining the stress and displacement distributions in this region (see, for example, [16,17]). The new ply crack surfaces are assumed to

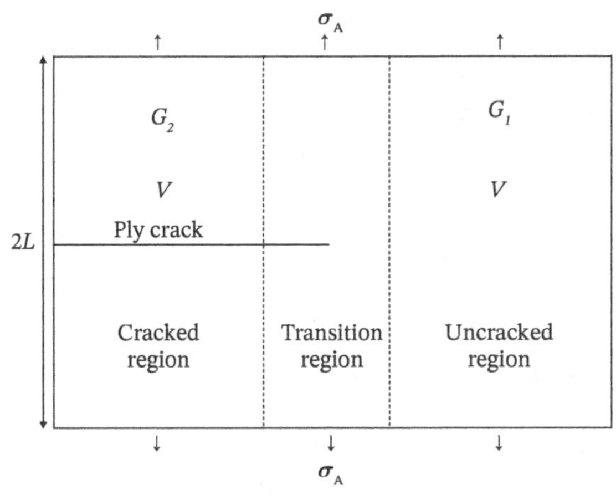

Figure 13.4 Side view of a long ply crack growing in the 90° ply of a cross-ply laminate.

form isothermally and quasistatically under conditions of fixed effective triaxial applied tractions, σ_A in the axial direction, σ_T in the in-plane transverse direction and σ_t in the through-thickness direction (although only the axial stress σ_A is shown in Figure 13.4).

From energy balance considerations and the fact that kinetic energy is never negative, the criterion for ply crack formation under these conditions has the form [16,17]

$$\Delta\Gamma + \Delta G < 0, \tag{13.33}$$

where the energy absorbed in a volume V of laminate by the formation of the new ply crack is given by

$$\Delta\Gamma = V\Gamma = V\frac{h^{(90)}\gamma^{(90)}}{hL}, \tag{13.34}$$

and where ΔG is the change in Gibbs energy resulting from the formation of the ply crack. In (13.34) the parameter Γ denotes the energy absorbed in unit volume of laminate during the formation of the ply crack surfaces in the 90° plies and it depends upon the effective fracture energy for ply cracking $2\gamma^{(90)}$, the half-thickness $h^{(90)}$ of the 90° ply and the half-thickness h of the laminate. The energy absorbed per unit volume is taken as the total area of crack surface formed in unit volume of laminate multiplied by the effective fracture energy for ply cracking. The corresponding change of Gibbs energy in the region V of the laminate is

$$\Delta G = G_2 - G_1 = \int_V \left[g - g_0\right] dV\,, \tag{13.35}$$

where G_1 and G_2 are the Gibbs energies in the uncracked and cracked regions, respectively, as shown in Figure 13.4, and where g and g_0 are the nonuniform distributions of Gibbs energy per unit volume for the uncracked and cracked regions, respectively. The Gibbs energy is used in (13.35) because the applied stresses and temperature are being regarded as the independent state variables. If external displacements were specified rather than applied stresses, then the Helmholtz energy would replace the Gibbs energy. The Gibbs energy does, in fact, correspond to the complementary energy that is often referred to when carrying out energy balance calculations in the absence of thermal residual stresses.

The cracked region of the laminate can be thought of as a RVE of a laminate where a series of ply cracks are growing which all have the same separation $2L$ The effective ply crack density in the cracked region is then $\rho = 1/(2L)$. It can be shown from the analysis given in [17], when applied to cross-ply laminates, that

$$\langle g\rangle \equiv \frac{1}{V}\int_V g\,dV = \langle g_0\rangle - \frac{1}{2}\left(\frac{1}{E_A(\rho)} - \frac{1}{E_A}\right)\left[k'\sigma_t + \sigma_A + k\sigma_T - \sigma_c\right]^2, \tag{13.36}$$

where $\langle g\rangle$ is the average Gibbs energy in region V for a damaged laminate, $\langle g_0\rangle$ is the corresponding value for an undamaged laminate and where σ_c (see (13.32)) is the ply crack closure stress when the laminate is subject to uniaxial loading in the axial direction normal to the planes of the ply cracks. The applied stress dependence appears as a perfect square involving a linear combination of the applied stresses.

It follows from (13.33)–(13.36) that the criterion for ply crack growth, i.e. ply crack formation in a previously undamaged laminate, is given by

$$\frac{hL}{h^{(90)}}\left(\frac{1}{E_A(\rho)}-\frac{1}{E_A}\right)\left[k'\sigma_t+\sigma_A+k\sigma_T-\sigma_c\right]^2 > 2\gamma^{(90)}. \tag{13.37}$$

This result implies that the energy release rate $G^{(90)}$ for the ply crack is

$$G^{(90)} = \frac{h}{2h^{(90)}}\frac{1}{\rho}\left(\frac{1}{E_A(\rho)}-\frac{1}{E_A}\right)\left[k'\sigma_t+\sigma_A+k\sigma_T-\sigma_c\right]^2, \tag{13.38}$$

where use has been made of the relation $\rho = 1/(2L)$, and that the critical value for crack growth $G_c^{(90)}$ is given by

$$G_c^{(90)} = 2\gamma^{(90)}. \tag{13.39}$$

The stress intensity factor may be obtained from the following relation valid for monolithic 90° ply material

$$G^{(90)} = \frac{\Omega^{(90)}\left[K^{(90)}\right]^2}{E_T^{(90)}}, \quad \text{so that } K^{(90)} = \sqrt{\frac{E_T^{(90)}G^{(90)}}{\Omega^{(90)}}}. \tag{13.40}$$

It then follows from (13.38) that

$$K^{(90)} = \sqrt{\frac{hE_T^{(90)}}{2h^{(90)}\Omega^{(90)}}\frac{1}{\rho}\left(\frac{1}{E_A(\rho)}-\frac{1}{E_A}\right)}\left(k'\sigma_t+\sigma_A+k\sigma_T-\sigma_c\right). \tag{13.41}$$

For the special case of uniaxial loading such that $\sigma_t = \sigma_T = 0$, the stress intensity factor reduces to the form

$$K^{(90)} = \sqrt{\frac{hE_T^{(90)}}{2h^{(90)}\Omega^{(90)}}\frac{1}{\rho}\left(\frac{1}{E_A(\rho)}-\frac{1}{E_A}\right)}\left(\sigma_A-\sigma_c\right). \tag{13.42}$$

This result is identical to relation (13.26) derived in Section 13.5 on substituting the expression for Λ given by (13.31).

Result (13.41), although applying only for sufficiently long ply cracks, is valid for conditions of triaxial loading. This contrasts with result (13.26) that, although derived only for uniaxial loading conditions, was derived using a method that can be extended to deal with embedded ply cracks of any size. As already mentioned, the extension to ply cracks of any size involves the need to solve an integral equation (see [12,13]) and discussion of this aspect is beyond the scope of this chapter.

13.8 Parameters Defining Cyclic Crack Tip Deformation

Although the theory for the prediction of the unstable growth of a ply crack is well understood and can be characterised either by using stress intensity factors or energy release rates, the corresponding situation for the prediction of fatigue crack growth remains to be resolved. The approach described in Sections 13.6 and 13.7, based on linear elasticity theory that leads to $r^{-1/2}$ stress and strain singularities at the ply crack tips, can be used for elastic–plastic crack tip deformations provided that small-scale crack tip yielding conditions prevail. Although such localised plastic yielding prevents the development of the singularities, the stress field in the neighbourhood of the ply crack tips continues to be characterised by the stress intensity factor to the extent that it continues to control the energy flow to the crack tip that enables crack growth to occur. A useful concept is the existence of a fracture process zone local to the ply crack tip in which micromechanisms of damage leading to crack growth are considered to occur. Continuum modelling does not apply in this region, but its size is considered small enough for it to be neglected when predicting stress and displacement distributions in cracked laminates. Outside this region there might be a plastic zone in which continuum modelling applies and in which energy will be dissipated as heat due to plastic deformation that is irreversible in a thermodynamic sense. Such energy dissipation will need to be considered when carrying out an energy balance calculation for a growing ply crack tip. For small-scale yielding conditions the plastic zone can be ignored when carrying out a stress analysis, but the effective fracture energy included in the energy balance calculation will include a contribution arising from localised plastic flow in the crack tip region. For conditions of small scale yielding the form of the criterion for the unstable growth of the ply crack is indistinguishable from that which applies when a perfectly elastic material is considered (including cohesive fracture models).

The concept of using an energy balance approach for the prediction of unstable crack growth can be extended to the case of fatigue crack growth. It is first necessary to consider the energy balance that must be applied when cyclic deformation is occurring in the crack tip region. Much of the cyclic deformation will be contributing to localised heating, but some will lead to the degradation of the material lying within the fracture process zone. Any energy balance approach to the prediction of fatigue crack growth must take account of this partition of energy. This approach has been made when developing a model for fatigue crack growth in homogeneous metals for conditions of small-scale crack tip yielding [15]. The analysis presented in [15] is based on the Dugdale [18] model of plasticity that assumes there is a yielding zone in strip form ahead of the crack tip, and that within the strip yield zone the axial stress is uniform, having a value related to the uniaxial yield stress for the material. This constraint in shape of the plastic zone enables analytical solutions to elastic–plastic crack problems. For metals such a model is relevant only for thin sheets where plastic flow is governed by the Tresca yield criterion. However, when considering any plastic flow that might occur in the matrix of any 90° ply containing a ply crack that is growing owing to fatigue loading, the Dugdale model (if suitably modified to take account of anisotropy) could be expected to apply more generally. This may be possible because for small-scale yielding conditions the presence of the fibres in the 90° is likely to restrict plastic flow to the region directly ahead of the crack. One of the beneficial aspects of the Dugdale model is that it enables a convenient linear elastic analysis of the stress distribution at any stage during the load cycle. The length of the yield zone is selected so that the $r^{-1/2}$ singularity normally arising from an elastic analysis is removed. During unloading a reversed plastic flow zone develops (smaller than the tensile yield zone) in which energy is dissipated as heat or contributes to the progressive failure of the material in the fracture process zone where localised damage formation leads to stable incremental crack growth. The stress field at any stage of unloading is predicted by subtracting two different solutions to the Dugdale model. The first is the solution that is predicted at the maximum applied stress σ_A of the fatigue cycle using a yield stress σ_Y. The second is the solution to the Dugdale model where the yield stress has the value $2\sigma_Y$ and the applied stress has the value $\sigma_A - \sigma'$ where σ' is the stress applied at any other stage of unloading during the fatigue cycle. The reversed plastic zone always lies within the plastic zone whenever $|\sigma'| < \sigma_A$. Because the axial stresses in the plastic and reversed plastic zones are uniform having the values σ_Y and $-\sigma_Y$, respectively, it is an easy matter to calculate the local energy dissipated either as heat or within the fracture process zone. It is worth noting that for fully reversed fatigue cycling ($R = -1$) the reversed plastic zone has the same length as the plastic zone. The details of this analysis for isotropic metals are given in [14,15]. The important conclusion of relevance here is that the cyclic stress field in the crack tip region and the rate of fatigue crack growth are characterised by the stress range intensity factor, even when using an energy balance approach. The results are, thus, fully consistent with definition (13.13) for ΔG and inconsistent with definition (13.12).

The Dugdale model can be criticised on the grounds that it is based on the assumption that the axial stress in the plastic and reversed plastic zones is uniform. A similar type of assumption has also been made when considering craze zones at the tips of cracks in homogeneous samples of viscoelastic materials [19]. For small-scale crack tip crazing, crack growth can again be characterised by the stress intensity factor. Uniform stresses in these zones enable a relatively simple approach to the analysis of the stress problems that arise, but in practice such uniform stress distributions may not occur. Nonuniform craze stress distributions in viscoelastic materials have been considered [20] and again lead to formulations involving the stress intensity factor.

As fatigue crack growth has been argued as resulting from energy dissipation (or energy absorption) in the crack tip failure process zones, energy concepts are used to define the effective stress range intensity factor ΔK. Relationships (13.25) and (13.26), which were derived for statically loaded brittle matrix composites by energy balance arguments, are the basis for the following definitions of the effective stress range intensity factor ΔK for the continuum model of a bridged ply crack and the corresponding stress range intensity factor $\Delta K^{(90)}$ for a ply crack in the 90° ply

$$\Delta K = \sqrt{\frac{E_A}{2\Omega\Lambda}}\,\Delta\sigma, \tag{13.43}$$

$$\Delta K^{(90)} = \sqrt{\frac{hE_T^{(90)}}{2\Lambda h^{(90)}\Omega^{(90)}}}\,\Delta\sigma, \tag{13.44}$$

where $\Delta\sigma$ is the stress range for uniform amplitude stress cycling. These definitions are consistent with the ΔG concept provided it is defined by (13.13).

It should be noted that definitions (13.43) and (13.44) do not involve thermal stresses. This is to be expected as thermal stresses are fixed during stress cycling, assuming isothermal conditions. Definitions (13.43) and (13.44) are consistent with the following relationship whose derivation, on the basis of linear elastic stress analysis applied to the Dugdale model of plasticity, is given in the note at the end of this chapter (see also [14])

$$\Delta K(\Delta\sigma, \Delta T) = K(\Delta\sigma, 0). \tag{13.45}$$

The final issue that needs to be discussed is whether a fatigue crack growth threshold exists for ply cracks and whether it should be modelled. The threshold for fatigue crack growth in homogeneous metals has been associated with the occurrence of crack closure during load cycling. Owing to plastic flow in the near-crack-tip region, a crack tip that is growing owing to fatigue loading leaves behind a wake of plastically deformed (stretched) material on the surfaces of the growing crack. Such stretching leads to the likelihood of the upper and lower crack surfaces coming into contact during load cycling, leading to compressive surface tractions for those parts of the stress cycle where contact occurs. This phenomenon reduces the magnitude of the effective stress intensity factor and in consequence the fatigue crack growth rate. The Dugdale model has been used [15] to analyse this situation, and an approximate method has been developed for taking such crack closure effects into account. A similar situation will arise for a ply crack growing owing to fatigue loading in the 90° ply of a laminate. This aspect will introduce significant R-ratio effects on fatigue crack growth, but this issue will not be considered here. The major objective is to concentrate on modelling the essential (first order) rather than secondary features governing the growth of fatigue cracks in cross-ply laminates.

13.9 Fatigue Crack Growth Law for Ply Cracks in Cross-ply Laminates

For ply cracks in the 90° plies of cross-ply laminates, a fatigue crack growth law is assumed of the form

$$\frac{\mathrm{d}c}{\mathrm{d}N} = C\left[\Delta K^{(90)}\right]^n, \tag{13.46}$$

where $\Delta K^{(90)}$ is the energy-based effective stress range intensity factor defined by (13.44) for the fatigue crack growth of a ply crack in the 90° ply of the laminate. The value of this parameter will be influenced by the 0° plies that bridge the ply crack and by the fact that only the 90° plies are cracking during fatigue loading. The corresponding instability condition for a ply crack in the 90° ply is assumed to be of the form

$$K^{(90)} = K_{\mathrm{Ic}}^{(90)}, \tag{13.47}$$

where $K^{(90)}$ and $K_{\mathrm{Ic}}^{(90)}$ are defined by (13.26) and (13.19), respectively.

For ply cracking under fixed values of the maximum axial stress σ_{A}, the stress intensity factors K and $K^{(90)}$ are independent of crack length. Thus, unstable ply crack growth during fatigue loading will not occur in the laminate until the crack tip region interacts with the edge affected regions of the laminate. Relation (13.47) is relevant only if the maximum axial stress σ_{A} were to be increasing with the number of fatigue cycles so that instability occurs before the crack has grown across the laminate to the far edge.

13.10 Predicting Fatigue Damage and Property Degradation (First-order Model)

The principal difficulty when predicting the occurrence of fatigue damage in composites is dealing with the initiation stage of damage development where micromechanisms are occurring that are difficult to observe and whose effects on properties are negligible. The only pragmatic way of overcoming this difficulty is to assume that a laminate has a preexisting distribution of defects, and that the nucleation of new defects during the fatigue process is a material characteristic that is known. In this chapter, the continual nucleation of defects is, however, neglected. To develop the model, the laminate is assumed to be such that ply cracks can form only in a set of parallel equally spaced sites (traversing the full width of the laminate) of which there are a large number denoted by m. These sites are to be referred to as 'potential ply crack formation sites' (see the dotted lines in Figure 13.5) that will usually be much larger than the number of ply cracks that are actually present.

Following manufacture of the laminate it is assumed that there are $s < m$ preexisting randomly distributed defects of the same size $2c_i$, and that the defects are located away from the edges, each occupying a different potential ply crack formation site so that each ply crack formation site has at most just one small ply defect. The size of the defect is just large enough for the stress intensity factor to be independent of crack length (see Figures 13.1 and 13.2).

The ply defects are assumed to be distributed uniformly in the laminate (when viewed at the macroscopic level) and the total initial length of defects in the laminate is thus $2sc_i$. The effective initial ply crack density ρ_0 arising from the many preexisting defects is defined by

$$\rho_0 = \frac{sc_i}{aW}, \tag{13.48}$$

where W is the total width of the laminate and $2a$ is the total length of laminate considered. As a result of fatigue loading, it is assumed that each ply crack will experience the same stress field such that the stress intensity factor and energy release rates for the crack have the same value and are independent of their length. Ply crack

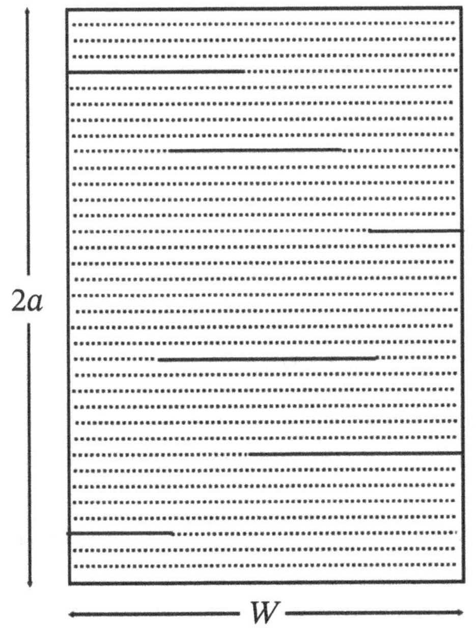

Figure 13.5 Schematic diagram of the uniformly spaced potential crack formation sites (dotted lines) in a cross-ply laminate where some fatigue ply cracks are present.

interaction effects are neglected, although it is possible to take these into account (see [21]). As a ply crack tip reaches the edge of the laminate during fatigue crack growth it is regarded as jumping to the other edge of the laminate, i.e. cyclic edge conditions are applied and edge stress effects are ignored. It is as though the laminate behaves as a tube without edges and curvature but having a finite circumference.

After N fatigue cycles having a fixed uniform amplitude $\Delta\sigma$ and peak stress σ_A, the initial defects of length $2c_i$ will have grown to a length $2c$ so that the effective ply crack density is then given by

$$\rho = \frac{sc}{aW}. \tag{13.49}$$

Relation (13.49) implies that the gradual growth of the large number of small ply defects is equivalent (when viewed at the macroscopic level) to a uniform density ρ of fully developed ply cracks. There can be a continuous variation of the ply crack density by altering the separation of the representative fully developed cracks. This approach, the validity of which is very difficult to test, is a practical solution to the very complex problem (see [21]) of being able to analyse the interaction of randomly distributed cracks that will have significant interaction as the effective crack density increases (see Chapter 17 for preliminary results concerning ply crack interaction effects). It follows from (13.49) and these continuity arguments that

$$\frac{\mathrm{d}c}{\mathrm{d}N} = \frac{aW}{s}\frac{\mathrm{d}\rho}{\mathrm{d}N}. \tag{13.50}$$

The maximum crack density that can arise is given by $\rho_{\max} = m/2a$ occurring when each potential cracking site has in fact cracked. Relations (13.31) and (13.44) lead to

$$\Delta K^{(90)} = \beta\Delta\sigma\sqrt{\frac{1}{\rho}\left(\frac{E_A}{E_A(\rho)} - 1\right)}, \tag{13.51}$$

where β is a dimensionless parameter defined by

$$\beta = \sqrt{\frac{hE_T^{(90)}}{2h^{(90)}E_A\Omega^{(90)}}}. \tag{13.52}$$

On substituting (13.50) and (13.51) into the defect growth law (13.46) it can be shown on integration that

$$\frac{N}{N_0} = \int_{\rho_0}^{\rho}\left(\frac{E_A}{E_A(\rho)} - 1\right)^{-\frac{n}{2}}\frac{\rho^{n/2}\,\mathrm{d}\rho}{\rho_{\max}^{1+n/2}}, \qquad N_0 = \frac{aW\rho_{\max}^{1+n/2}}{sC(\beta\Delta\sigma)^n}, \tag{13.53}$$

where N_0 is a dimensionless normalising constant that depends upon the stress range $\Delta\sigma$ of the uniform amplitude fatigue cycling. Relation (13.53) implies that for $\rho_0 \le \rho \le \rho_{\max}$

$$\rho = \rho(N/N_0), \qquad E_A = E_A(\rho(N/N_0)) \equiv E_A(N). \tag{13.54}$$

Now define a fatigue damage parameter by

$$D(N) = \frac{E_A}{E_A(N)} - 1 > 0. \tag{13.55}$$

This damage parameter can be accurately calculated in terms of the representative ply crack density using the stress-transfer models described in [16,17,22] and the PREDICT module of the software system known as CoDA [23] (which can deal with general symmetric laminates for mode I loading conditions). It follows from [16,17,22] that the thermoelastic constants for a fatigue damaged cross-ply laminate must be of the following form:

$$\frac{1}{E_A(N)} = \frac{1}{E_A} + \frac{D(N)}{E_A}, \tag{13.56}$$

$$\frac{1}{E_T(N)} = \frac{1}{E_T} + \frac{k^2 D(N)}{E_A}, \tag{13.57}$$

$$\frac{1}{E_t(N)} = \frac{1}{E_t} + \frac{(k')^2 D(N)}{E_A}, \tag{13.58}$$

$$\frac{\nu_A(N)}{E_A(N)} = \frac{\nu_A}{E_A} - \frac{k D(N)}{E_A}, \tag{13.59}$$

$$\frac{\nu_a(N)}{E_A(N)} = \frac{\nu_a}{E_A} - \frac{k' D(N)}{E_A}, \tag{13.60}$$

$$\frac{\nu_t(N)}{E_T(N)} = \frac{\nu_t}{E_T} - \frac{k k' D(N)}{E_A}, \tag{13.61}$$

$$\alpha_A(N) = \alpha_A + \frac{k_1 D(N)}{E_A}, \tag{13.62}$$

$$\alpha_T(N) = \alpha_T + \frac{k k_1 D(N)}{E_A}, \tag{13.63}$$

$$\alpha_t(N) = \alpha_t + \frac{k' k_1 D(N)}{E_A}. \tag{13.64}$$

The parameters k, k' and k_1 are laminate constants specified, in terms of ply properties and laminate geometry, in [16,17]. The thermoelastic constants given by (13.56)–(13.64) appear in the stress-strain relations for a damaged cross-ply laminate, namely

$$\varepsilon_A = \frac{\sigma_A}{E_A(N)} - \frac{\nu_A(N)}{E_A(N)}\sigma_T - \frac{\nu_a(N)}{E_A(N)}\sigma_t + \alpha_A(N)\Delta T, \tag{13.65}$$

$$\varepsilon_T = -\frac{\nu_A(N)}{E_A(N)}\sigma_A + \frac{\sigma_T}{E_T(N)} - \frac{\nu_t(N)}{E_T(N)}\sigma_t + \alpha_T(N)\Delta T, \tag{13.66}$$

$$\varepsilon_t = -\frac{\nu_a(N)}{E_A(N)}\sigma_A - \frac{\nu_t(N)}{E_T(N)}\sigma_T + \frac{\sigma_t}{E_t(N)} + \alpha_t(N)\Delta T, \tag{13.67}$$

where σ_A, σ_T, σ_t are the effective through-thickness, axial and transverse stresses applied to the laminate, and where ε_A, ε_T, ε_t are the corresponding effective strains. Thus, all the thermoelastic constants of the fatigue damaged laminate can be determined as a function of the number of fatigue cycles executed, except for the shear moduli. Such information is vital for input into structural calculations of fatigue damage accumulation where load transfer in a structure can gradually occur owing to the localised evolution of fatigue damage.

13.11 Material Properties for Example Simulations

The dependence of axial Young's modulus on ply crack density during static uniaxial loading has been measured [24], for a range of both CFRP and GRP cross-ply laminates having various thicknesses for the 90° plies. For the model predictions [17], undamaged ply properties were estimated (see [23]) from fibre and matrix properties using a concentric cylinder model assuming that the fibre/matrix interfaces were perfectly bonded. The properties assumed for the carbon fibre (HTA) reinforced epoxy (922) are as follows.

	Fibre	*Matrix*
E_A (GPa)	208.0	3.75
E_T (GPa)	16.7	3.75
μ_A (GPa)	18.0	1.32042
ν_A	0.25	0.42
ν_T	0.35	0.42
α_A (/°C × 10^6)	−1.1	50.0
α_T (/°C × 10^6)	22.1	50.0

For the CFRP the fibre (HTA) and matrix (922) properties used assume that the fibre volume fraction is 0.575 leading to the following ply properties estimated by PREDICT [23]:

$E_A = 121.23$ GPa $\quad E_T = 9.11$ GPa $\quad \nu_A = 0.320 \quad \nu_T = 0.552$

$\mu_A = 3.924$ GPa $\quad \alpha_A = -0.349 \times 10^{-6}$/°C $\quad \alpha_T = 42.21 \times 10^{-6}$/°C.

The properties assumed for the E-glass fibre-reinforced epoxy (922) are as follows.

	Fibre	*Matrix*
E_A (GPa)	72.0	3.75
E_T (GPa)	72.0	3.75
μ_A (GPa)	29.5082	1.32042
ν_A	0.22	0.42
ν_T	0.22	0.42
α_A (/°C × 10^6)	5.1	50.0
α_T (/°C × 10^6)	5.1	50.0

Assuming a fibre volume fraction of 0.6, the GRP ply properties have the following values estimated by PREDICT [23]:

$\hat{E}_A = 121.23$ GPa $\quad \hat{E}_T = 9.11$ GPa $\quad \hat{\nu}_A = 0.320 \quad \hat{\nu}_T = 0.552$

$\hat{\mu}_A = 3.924$ GPa $\quad \hat{\alpha}_A = -0.349 \times 10^{-6/°C} \quad \hat{\alpha}_T = 42.21 \times 10^{-6}$ /°C.

It is worth noting that a comparison between model predictions and the experimental data for static loading [24] is given in [17], where very good agreement has been achieved for both CFRP and GRP materials and all 90° ply thicknesses. Corresponding Poisson's ratio data is subject to much more scatter, but a comparison with the model is not necessary when a good correlation with the axial modulus has been achieved, as theoretical work [16,17,22] has shown that the ply crack density dependence of many other laminate properties is easily estimated from the axial modulus data (see, for example, (13.56)–(13.64)). The properties given previously are now used for preliminary simulations of fatigue damage in a cross-ply laminate.

13.12 Relationship of the Model to Experimental Data

One of the key parameters controlling the growth of fatigue damage in the form of ply cracking is the exponent n appearing in the ply crack growth law (13.46). It is useful first to investigate the effect of its value on fatigue damage growth. When estimating the properties of the fatigue damaged laminate, use is made of the stress-transfer model [16,17,22] that allows each ply to be refined into elemental layers so that through-thickness variations in the stress and displacement fields can be modelled accurately. Each ply was subdivided into five elements of equal thickness, and the elements next to the interface between the plies were successively subdivided three times thus ensuring that the stress singularity at the ply crack tips can be resolved. The analysis and software neglect the effects of any ply cracks that might form during the initial loading of the laminate.

For the case of a CFRP (HTA fibres in 922 epoxy matrix having a volume fraction of 0.575 and a ply thickness of 0.25 mm) whose properties are given in Section 13.11, Figure 13.6 shows, for values of n in the range 2–8, the result of carrying out the integration specified in (13.53) in order to relate the loss in modulus of a $[0/90]_s$ laminate to the number of fatigue cycles that have been carried out. The predictions assumed that the number of potential ply cracking sites $m = 1000$, the maximum ply crack density $\rho_{\max} = 3$ cracks/mm, and that $\rho_0 = \rho_{\max} / m$.

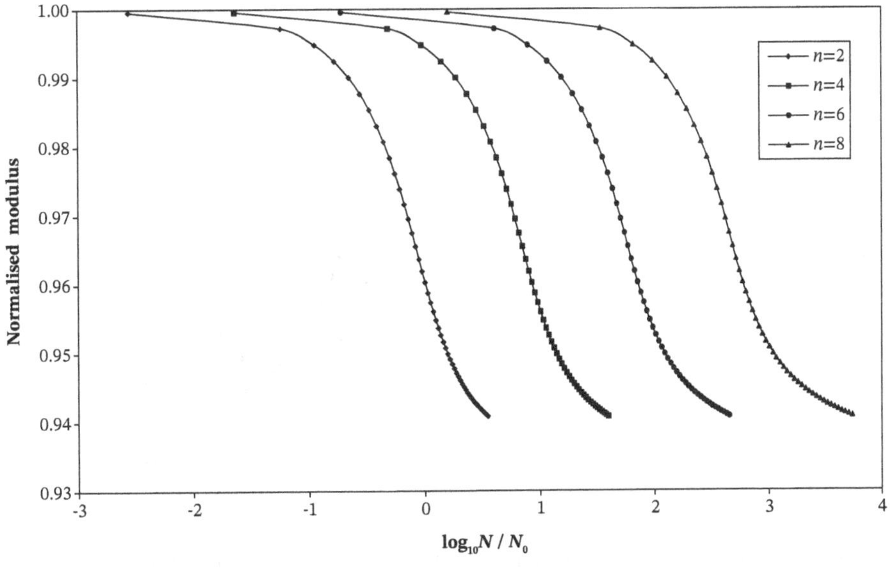

Figure 13.6 Dependence of axial modulus of CFRP $[0/90]_s$ laminate (HTA fibre in 922 matrix) for various values of the fatigue exponent n.

It is seen that the exponent n has a significant effect on the number of cycles to achieve a specified loss in axial modulus, but hardly any effect on the actual shape of the curves: an unexpected result. At first sight the curves appear identical in shape and their location is merely shifted. There are, in fact, minor differences, as seen at the right-hand side ends of the curves. The curves are seen to be sigmoidal in shape and would lead to a maximum modulus reduction of at most 93% of the undamaged modulus. It is emphasised that the curves plotted are independent of the stress range and maximum stress used to generate the data. It should be noted that the normalising parameter N_0 defined by (13.53) is inversely proportional to $(\Delta\sigma)^n$. Changes in the value of $\Delta\sigma$ merely leads to a horizontal shift of the curves, each translating by an amount that depends on the value of the exponent n. Corresponding results for the case of a GRP $[0/90]_s$ laminate (plies have thickness 0.25 mm and are made of E-glass in 913 epoxy having a fibre volume fraction of 0.6) are given in Figure 13.7. It is seen that the same sigmoidal shape of curve results having similar characteristics to those for the CFRP. For GRP, it is seen that the maximum modulus reduction is at most 75% of the undamaged value. Despite the similarities in the behaviour of CFRP and GRP, there is clearly one important difference. The spread of the curves for CFRP shown in Figure 13.6, as n varies from 2 to 8, ranges over three decades, whereas the spread in the corresponding curves for GRP shown in Figure 13.7 ranges over less than one decade.

It should be noted that

$$\log_{10}(N/N_0) = \log_{10} N - \log_{10} N_0,$$

so that the data in Figure 13.7 for GRP can be compared directly with experimental data [24] shown in Figure 13.8 apart from a rigid shift along the abscissa. Thus, the shape of the curves in Figures 13.7 and 13.8 would be identical if the experimental data

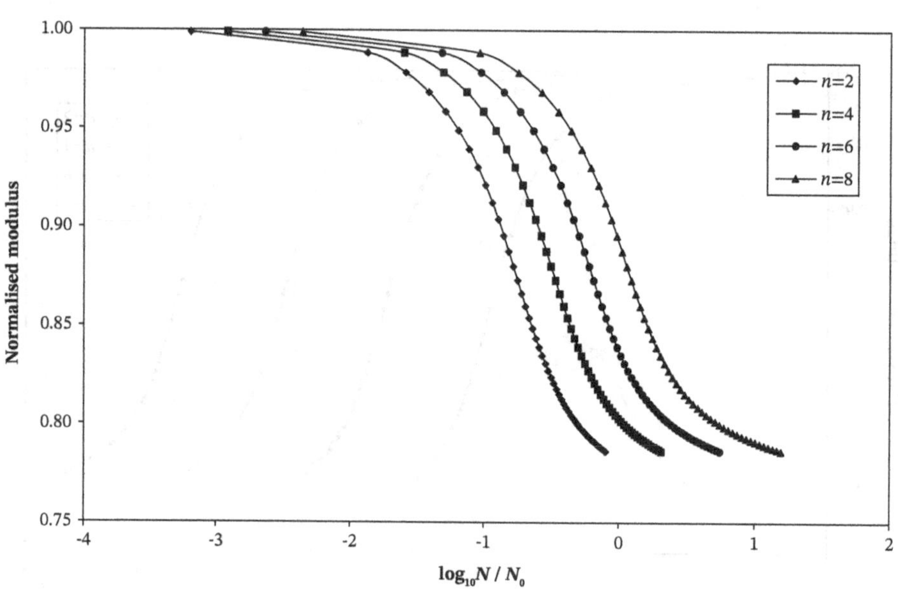

Figure 13.7 Dependence of axial modulus of GRP $[0/90]_s$ laminate (E-glass fibre in 913 matrix) for various values of the fatigue exponent n.

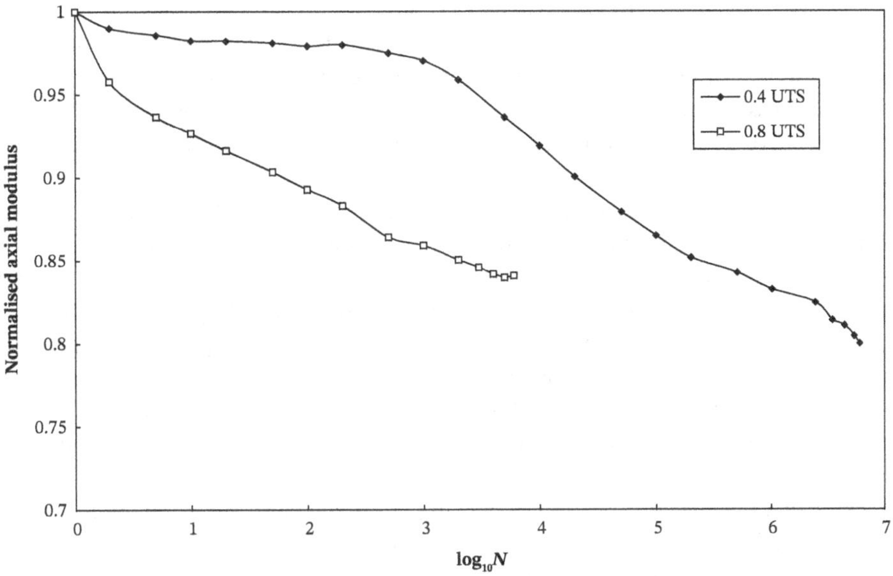

Figure 13.8 Experimental fatigue [24] data for $[0_2/90_2]_s$ GRP (E-glass in 913) at two different maximum stresses during cycling.

were equivalent to that predicted by the model. In practice, the value of n for a GRP composite has been given as 5.6 (see [4]).

When comparing Figure 13.7 with Figure 13.8, it is clear that the modulus degrades much less rapidly than model predictions, although it is seen that the actual total loss in modulus is very similar, as is the cycle range over which the degradation occurs. There are three possible explanations for the difference in behaviour. First, the form (13.46) assumed for the fatigue crack growth law may not be correct. The inclusion of threshold and instability effects in the defect growth law, as in (13.9), could have a significant effect on the shape of the curves shown in Figure 13.7. Second, the assumption that all defects contributing to the degradation are preexisting may not be correct. It is indeed likely that ply defects are continually nucleating throughout the fatigue process, but observing the behaviour would be very difficult. Such observations would be essential for the development of a reliable model of such nucleation phenomena. An important conclusion of this investigation is that the basis of the fatigue growth law (13.46) needs to be investigated by observations of fatigue crack rates in translucent GRP and their correlation to the stress intensity factor. The third reason for differences between the model and experimental data concerns the assumption, which is difficult to test, that the effect of the total area of many partially developed ply cracks may be modelled by a uniform density of fully developed ply cracks having the same area (see (13.49)).

The experimental data shown in Figure 13.8 exhibit an initial loss in modulus that is not predicted by the model. This is indicating that the ply cracking during initial loading may be having an observable effect on the axial modulus. The model, which has ignored this aspect that appears to be important (especially at large values of the stress range), needs to be modified to take account of the initial ply cracking. It is clear from the data shown in Figures 13.7 and 13.8 that the first-order model is predicting a pessimistic loss of properties resulting from fatigue loading. From a design point of view, this pessimism can be converted into the positive advantage of having achieved a conservative approach to the design of composite laminates that are resistant to fatigue damage.

To conclude this investigation into the modelling of fatigue damage in composite laminates, it is useful to show model predictions for degradation of all the thermoelastic constants. Figure 13.9 shows the cycle dependence of the in-plane thermoelastic constants of a GRP laminate that have been normalised by dividing by their values for the uncracked state of the laminate. The data have been plotted up to a ply crack density of 3 cracks mm^{-1} and an exponent of $n = 6$ was used. It is seen that as the number of cycles tends to zero all the thermoelastic constants will tend to unity as to be expected. As the number of cycles increases the thermoelastic constants are seen to be tending to asymptotes that are defined by the ply discount limit where the 90° plies have so much cracking that the mechanical effect of the 90° plies in the laminate is negligible. Figure 13.10 shows the corresponding out-of-plane thermoelastic constants that exhibit a similar behaviour to the results for in-plane properties shown in Figure 13.9.

As ply cracking owing to fatigue loading is the only damage mode that is being modelled, the failure of the laminate owing to fatigue loading cannot be modelled using the methods of this chapter. Provided the applied maximum load in the fatigue cycle is large enough, the progressive growth of ply cracking in the 90° ply will lead to fibre fracture in the 0° plies although this is not modelled here, nor are the initial fully developed ply cracks that will be inevitable in such cases. Fibre fracture occurs because the axial stress in the 0° plies increases sufficiently (owing to ply cracking) causing progressive fibre damage in the 0° plies and the eventual failure of the laminate. For

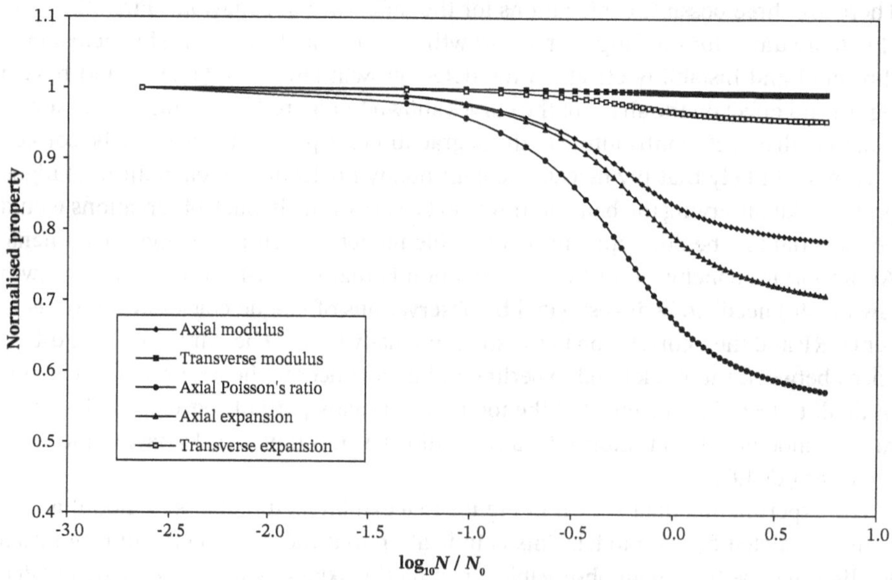

Figure 13.9 Dependence of in-plane properties of a GRP $[0/90]_S$ laminate on the number of cycles.

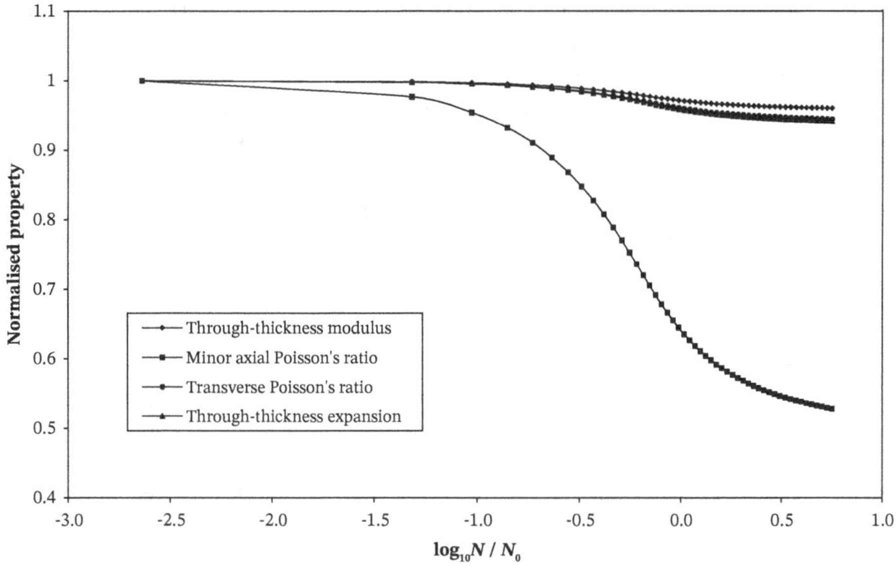

Figure 13.10 Dependence of through-thickness properties of a GRP $[0/90]_S$ laminate on the number of cycles.

these cases it should be possible to predict the lifetime of the laminate assuming some given initial damage state. For lower levels of the applied maximum stress in the fatigue cycle, the laminate will never fail as load transfer from the 90° ply to the 0° plies is insufficient to initiate fibre fracture. The laminate will have an effectively infinite lifetime for such cases. There is an intermediate regime where load transfer owing to ply cracking in the 90° ply leads to a few fibre fractures in the 0° plies that act as initiation sites for subsequent damage development. The modelling of this case is very complex and beyond the scope of this chapter.

13.13 Discussion

This chapter has described the features of a preliminary model for ply crack formation in a cross-ply laminate during fatigue loading. The model can be applied to multiple-ply cross-ply laminates that are cracked in the transverse plies and which are subject to thermal residual stresses. By adopting a simplifying approach, it has been possible to develop a convenient method of assessing the degradation of the properties of cross-ply laminates that will result from ply crack formation during fatigue loading. An attempt to validate the model has shown that the first-order approach leads to a conservative estimate of fatigue performance that can be exploited in the design of fatigue resistant composite laminates. By making use of established results from ply crack formation theory for monotonic loading of cross-ply laminates it has been possible to predict the degradation of all nonshear properties of the laminate as a function of the number of fatigue cycles executed. This data is required by structural analyses that attempt to predict the growth of damage in laminated structures arising from fatigue loading. Shear properties can in fact be considered using the methods [22] developed for general symmetric laminates.

The model described is a first-order approach where there is focus only on the ply cracking mechanism for fatigue damage formation in laminates. There are many other aspects for which predictive schemes will need to be developed if the reliable prediction of fatigue lifetimes is to become a reality. It is useful to indicate the nature of these aspects so that they can be considered in any future developments of the approach. These aspects are as follows.

- The nature of the initiation of microcracks from ply defects needs thorough study, and the development of predictive models is needed for this contribution (often the longest) to fatigue life.
- The progressive formation of new microcracks during stress cycling needs to be modelled. The current model merely assumes that there are preexisting defects that all have the same size. The model needs to be developed so that a statistical distribution of defects can be considered. There is also a need to include the effect on properties of the formation of fully developed ply cracks during the initial loading of the laminate.
- The form of fatigue crack growth law to be used for ply crack formation needs to be critically examined by carrying out careful experimental studies and by analysing the measured data using the methods described here. The work should concentrate on characterising the nature of fatigue threshold (if it exists), and on the study of the crack growth acceleration that occurs if fatigue crack instability approaches.
- The effects of R ratio on fatigue crack growth rate needs careful study, as do the effects of crack closure (if it occurs) on R-ratio effects.
- Cross-ply laminates are hardly ever used in practice, but they are very useful for developing an understanding of the important physical processes that determine the fatigue performance of laminates. In order that lifetime prediction procedures can be applied to laminated structures, a methodology needs to be developed that will enable fatigue damage growth to be modelled in general symmetric laminates. The methods of this chapter are easily extended to deal with more general laminates provided that ply crack growth is mode I (see [17,22]).
- If fatigue crack growth laws are to be applied in structures, then it is vital that multiaxial loading effects are modelled. This includes the study of mixed mode fatigue crack growth and the accompanying instability preceding catastrophic failure. The effects of out-of-plane bending on fatigue damage growth also need to be understood when considering the fatigue performance of structures. The methods of this chapter can be extended to deal with the bending of cross-ply laminates using analyses described in [17]. One way of dealing with the fatigue of balanced general symmetric laminates subject to bending is to homogenise the 0° plies and off-axis plies and then use the effective homogenised thermoelastic properties as modified properties for use in a cross-ply model.
- Other damage modes need to be considered such as interply delamination initiating at edges and ply cracks (as in [10,25]). Crack growth here occurs between layers of different materials, and will be mixed mode in nature. In addition, fibre fracture needs to be considered leading on to the occurrence of laminate failure. All damage

phenomena will be highly interactive and their interrelationship needs to be fully understood.

- Cyclic loads applied to structures in practice are very seldom uniform amplitude in nature. Random loading is often encountered leading to the need to predict property degradation and lifetime by integrating over each cycle in the random sequence. Some progress can be made by assuming periodic–random loading and making use of 'rain-flow' techniques [26] to decompose the random sequences into equivalent complete cycles of various amplitudes. Overloads can have an important effect of fatigue damage development and their effect will need to be analysed. In metals, overloads can lead to the deceleration of fatigue cracks because of crack tip blunting and associated closure effects [27].
- The viscoelastic nature of the matrix in polymer composites leads to energy dissipation in bulk regions where deformation is occurring (whether reversible or irreversible). The energy dissipation during fatigue loading leads to heating (autogenous) which can be large when the frequency of stress cycling is high. The increases in temperature that arise during fatigue loading will alter the thermal residual stresses, the physical properties of the matrix and the properties of the plies and their degradation. Local heating effects can also occur locally in the crack tip region with the result that fracture energies can also be expected to be affected by temperature and frequency. Thus, cycling frequency rates can be an important factor that needs to be considered, especially when validating models by comparison with experimental data obtained under rapid test conditions in the laboratory. Practical applications will in most cases be at relatively low cycling rates so that autogenous heating may not then be a problem.
- Models predicting the degradation of laminate properties owing to fatigue loading need further careful validation by comparison of model predictions with experimental data.

This list of important issues that affect the prediction of the fatigue performance of laminates structures, including lifetime prediction and residual strength prediction, leads to the following general conclusions.

1. The *reliable* prediction of the fatigue performance of laminated structures is very much in its infancy.
2. The complex nature of fatigue damage mechanisms in composite laminates, and their relation to the laminate structure and type of loading, means that engineers can have confidence in predictive schemes only if a proper understanding of influencing factors is available. Developing this understanding in detail, and increasing confidence, will take a significant time and resource.
3. Short-term solutions based on simplifications will have very limited applicability, but they do have an important role to play as dealing with fatigue damage development is becoming an increasingly important requirement for both users and designers of composite laminate structures.
4. Putting fatigue performance on one side, there remain significant problems with the prediction of the static strength of laminates structures, as shown by the International Exercise on Failure Criteria [28]. We must be able to predict static

failure reliably if we are ever to achieve reliable life prediction and residual strength in fatigue.

References

1. Harris, B. (ed.) (2003). Fatigue in composite materials. Cambridge: Woodhead Publishing Ltd.
2. Talreja, R. (1981). Fatigue of composite materials: Damage mechanisms and fatigue life diagrams. *Proceedings of the Royal Society of London* A378: 461–475.
3. Talreja, R. (1987). Fatigue of composite materials. Lancaster, PA, USA.: Technomic Publishing Inc.
4. Ogin, S.L., Smith, P.A., and Beaumont, P.W.R. (1985). A stress intensity factor approach fatigue growth of transverse ply cracks. *Composites Science and Technology* 24: 47–59.
5. Boniface, L., Ogin, S.L., and Smith, P.A. (1991). Strain Energy Release Rates and the Fatigue Growth of Matrix Cracks in Model Arrays in Composite Laminates. *Proceedings of the Royal Society of London* A432: 427–444.
6. Harris, B. (2003). 'A historical review of the fatigue behaviour of fibre-reinforced plastics', Chapter 1 in 'Fatigue in composite materials', Cambridge: Woodhead Publishing Ltd.
7. Beaumont, P.W.R. (2003). Physical modelling of damage development in structural composite materials under stress. Chapter 13 in 'Fatigue in composite materials', Cambridge: Woodhead Publishing Ltd.
8. Martin, R.. (2003). Delamination fatigue. Chapter 6 in 'Fatigue in composite materials', Cambridge: Woodhead Publishing Ltd.
9. Ladeveze, P. and Lubineau, G. (2003). A computational mesodamage model for life prediction for laminates. Chapter 15 in 'Fatigue in composite materials', Cambridge: Woodhead Publishing Ltd.
10. Kashtalyan, M. and Soutis, C. (2003). Analysis of matrix crack induced delamination in composite laminates under static and fatigue loading. Chapter 17 in 'Fatigue in composite materials', Cambridge: Woodhead Publishing Ltd.
11. McCartney, L.N. (2001). 'Model of composite degradation due to environmentally assisted fatigue damage', NPL Report MATC(A) 26, May.
12. McCartney, L.N. (1992). Mechanics for the growth of bridged cracks in composite materials: I - Basic principles. *Journal Of Composites Technology And Research* 14: 133–146.
13. McCartney, L.N. (1992). Mechanics for the growth of bridged cracks in composite materials: II - Applications. *Journal Of Composites Technology And Research* 14: 147–154.
14. McCartney, L.N. (1996). Stress transfer mechanics: Models that should be the basis of life prediction methodology. *ASTM STP 1253* on 'Life Prediction Methodology for Titanium Matrix Composites', pp.85–113.
15. Lindley, T.C. and McCartney, L.N. (1981). Mechanics and mechanisms of fatigue crack growth. Chapter 5 in In: *Developments in Fracture Mechanics - 2*, (ed., G.G. Chell), London - New Jersey: Applied Science Publishers.

16. McCartney, L.N. (1998). Predicting transverse crack formation in cross-ply laminates. *Composites Science and Technology* 58: 1069–1081.
17. McCartney, L.N. (2003). Physically based damage models for laminated composites. *Proceedings of the Institution of Mechanical Engineers* 217(Part L: J. Materials: Design & Applications): 163–199.
18. Dugdale, D.S. (1960). *Journal of the Mechanics and Physics of Solids.* 8: 100.
19. McCartney, L.N. (1980). Derivation of crack growth laws for linear viscoelastic solids based upon the concept of a fracture process zone. *International Journal of Fracture* 16: 375–382.
20. McCartney, L.N. (1988). Crack growth predictions for viscoelastic materials exhibiting non-uniform craze deformation. *International Journal of Fracture* 37: 279–301.
21. McCartney, L.N. (1995). 'Interaction of transverse cracks of finite size in a cross-ply laminate', Proc. 10th Int. Conf. on Composite Materials, Whistler, Canada, August Vol.1, pp. 399–406.
22. McCartney, L.N. (2000). Model to predict effects of triaxial loading on ply cracking in general symmetric laminates. *Composites Science and Technology* 60: 2255–2279.
23. Software system 'PREDICT' which is a specific module of CoDA. (see http://www.npl.co.uk/npl/cmmt/cog/coda.html)
24. Broughton, W.R. and Lodeiro, M.J. (2000).Fatigue testing of composite laminates. NPL Report CMMT(A) 252, November.
25. Nairn, J.A. and Hu, S. (1992). The initiation and growth of delaminations induced by matrix microcracks in laminated composites. *International Journal of Fracture* 57(1): 1–24.
26. McCartney, L.N. (1976). The effect of periodic-random loading on fatigue crack growth. *International Journal of Fracture* 12(2): 273.
27. McCartney, L.N. (1978). A theoretical explanation of the delaying effects of overloads on fatigue crack propagation. *International Journal of Fracture* 14(2): 213.
28. Soden, P.D., Hinton, M.J., and Kaddour, A.S. (1998). Lamina properties, lay-up configurations and loading conditions for a range of fibre reinforced composite laminates. *Composites Science and Technology* 58(7): 1011–1022. and other papers in the same special issue. Part B of the exercise is published in *Comp. Sci. Tech.* 2002, **62**(12 and 13) and Part C is published in Comp. Sci. Tech. 2004, **64** (3 and 4).

Note: Modelling reversed plasticity using the Dugdale model

Consider the stress field, predicted by the Dugdale model [18] described in Section 13.8, that arises when the uniaxial axial stress applied to a bridged ply crack in a laminate is reduced from the value σ_A to the value $\sigma_A - \Delta\sigma$, as would occur during fatigue cycling. The yield and fracture process zones are assumed to be so small that their presence has negligible effect on the stress field. Suppose that the stress field at the maximum stress σ_A of the fatigue cycle is given by the function $\sigma_{ij}(y, z, \sigma_A, \Delta T)$ where y is the coordinate for the axial direction of the laminate and z denotes the through-thickness coordinate. The parameter ΔT denotes the difference between the current temperature and the stress-free temperature of the laminate. The corresponding stress intensity factor for the bridged ply crack is denoted by $K(\sigma_A, \Delta T)$. It is convenient to decompose the stress field σ_{ij} as follows:

$$\sigma_{ij} = \hat{\sigma}_{ij}(y, z, \sigma_A) + f_{ij}(z)\Delta T,$$

where the first term is the stress field that results when $\Delta T = 0$, and where the second term represents the thermal stress distribution characterised by the function f_{ij} that is a function only of the through-thickness coordinate z. The thermal part of the stress field will remain unchanged during any stress change provided the temperature is fixed. Consider now a stress reduction from σ_A to the value $\sigma_A - \Delta\sigma$. The stress field following any reduction in the applied stress may be determined by using the superposition principle. The stress field following the stress reduction $\sigma_A \rightarrow \sigma_A - \Delta\sigma$ is given by

$$\sigma_{ij} = \hat{\sigma}_{ij}(y, z, \sigma_A) + f_{ij}(z)\Delta T - \hat{\sigma}_{ij}(y, z, \Delta\sigma),$$

where the term subtracted corresponds to the linear elastic solution for a bridged crack. The stress intensity factor following the stress reduction $\sigma_A \rightarrow \sigma_A - \Delta\sigma$ has the value $K(\sigma_A, \Delta T) - K(\Delta\sigma, 0)$ so that the stress range intensity factor is given by $\Delta K = K(\Delta\sigma, 0)$. It is noted that, because the stress field for a ply crack in a uniaxially loaded cross-ply laminate is determined using the methods of linear elasticity theory, the following relationship must be satisfied by the function $\hat{\sigma}_{ij}$ characterising the stress field when $\Delta T = 0$:

$$\hat{\sigma}_{ij}(y, z, \sigma_A) \equiv \sigma_A \hat{\sigma}_{ij}(y, z, 1).$$

Note: This chapter is based on the publication:
L. N. McCartney, 'Energy methods for fatigue damage modelling of laminates', Composites Science and Technology, **68**, (2008), 2601–2615.

14

Model of Composite Degradation Due to Environmental Damage

Overview: This chapter describes a model that can predict the time-dependent axial strength of a unidirectional fibre-reinforced composite arising from environmental exposure. The model assumes that the environmental exposure leads to interfaces that are very weak so that a parallel two bar model can be developed where one bar represents the behaviour of the fibres, which are regarded as a loose bundle, and where the second bar represents the matrix. The principal effect of the environment is assumed to be the time-dependent strength of individual fibres arising from stress corrosion cracking of small defects within the fibres. The initial strength of the fibres is assumed to be governed by the Weibull distribution which, when used in conjunction with fracture mechanics, defines an initial statistical size distribution of fibre defects. Defect growth arising from environmental exposure is assumed to be governed by a fracture-mechanics-based growth law where the defect growth rate is controlled by the stress intensity factor. The model described extends previous work in the 1980s for loose fibre bundles to the case where the fibres are embedded in a composite so that the matrix has an effect. Preliminary predictions are given which show the time dependence of the fibre stress as a function of exposure time under applied load. The time dependence of the residual bundle strength is also given showing a significant effect of the matrix. By normalising the residual bundle strength with respect to the corresponding static strength, it is shown that the normalised residual strength of the fibre bundle is virtually independent of matrix behaviour. The residual strength of the composite is obtained by multiplying the bundle strength by the fibre volume fraction.

14.1 Introduction

The axial strength of unidirectional fibre-reinforced composites is controlled by the strength of the fibres. In cross-ply laminates, the axial strength of the laminate is controlled to a large degree by the strength of the fibres in the 0° plies. Fibre strength is statistical in nature owing to the presence of defects both on fibre surfaces and in their interior. This aspect has been described in the review [1]. The effect of interface properties on axial strength are of secondary importance, and modelling their effect on

Properties for Design of Composite Structures: Theory and Implementation Using Software,
First Edition. Neil McCartney.

Companion Website: www.wiley.com/go/mccartney/properties

axial strength requires the use of sophisticated stress-transfer models and Monte Carlo simulation techniques. For unprotected glass fibres, it is well known (see, for example, [2]) that the environmental exposure of the composite leads to time-dependent reductions in fibre strength. The strength reduction of the fibres results because of the progressive growth of fibre defects caused by stress corrosion cracking at a microscopic level.

Environmental exposure, provided that it is saturated, can lead to a deterioration in interface properties. Given that the axial strength of a unidirectional composite is not affected to a great extent by interface properties, it is reasonable to assume, when modelling the axial behaviour of a unidirectional composite, that the interfaces in the composite following prolonged exposure have no 'strength'. This enables a relatively simple approach to be taken that will provide good insight into the axial behaviour of a composite when exposed to an aggressive environment.

Owing to the dominance of fibre behaviour, earlier modelling work applied to glass-fibre composites [3, 4] regarded the unidirectional composite as a loose bundle of parallel fibres having equal length, so that the relatively low load-carrying capacity of the matrix was ignored. The glass fibres in the bundle were assumed to be attached to rigid supports which were able to share the applied load equally between all surviving fibres. The objective of this chapter is to describe the extension of the existing loose bundle model so that the load-carrying capacity of the matrix is taken into account when considering the axial behaviour of glass-fibre-reinforced composites subject to environmental exposure.

14.2 Model Geometry

Consider a unidirectional fibre-reinforced composite having a fibre volume fraction V_f and matrix volume fraction V_m such that $V_f + V_m = 1$. The composite has been wholly immersed in an aggressive environment for a sufficient time for the composite to be fully saturated. The application of axial load to the composite leads to the environmental growth of defects in the fibres: a phenomenon well known to afflict glass fibres [2]. The interfaces between the fibres and matrix are regarded as being significantly weakened by the environment to the extent that it can be assumed that the fibres and matrix behave independently in regions of the composite that are well away from the uniaxial loading mechanism where clamping effects become important. This assumption means that the composite can be modelled as a parallel bar model, as shown in Figure 14.1. The fibres in the composite are regarded as acting as a loose bundle forming one bar of the model. The matrix material in the composite is considered as being gathered together to form the other bar of the model, which is regarded as homogeneous, i.e. the bar is solid. When a fibre fails, the load it carried is shared between the surviving fibres in the bundle and the matrix in such a way that all surviving fibres and matrix experience the same axial strain increment. The fibres are assumed to have the same length so that each surviving fibre has the same stress throughout the progressive failure process. The composite is subject to a fixed applied load F for all times $t > 0$, where $t = 0$ corresponds to the time when the fixed load F is first applied. Environmental defect growth in the fibres leads to progressive fibre failure until the bundle collapses. It is assumed that bundle collapse corresponds to the catastrophic failure of the composite,

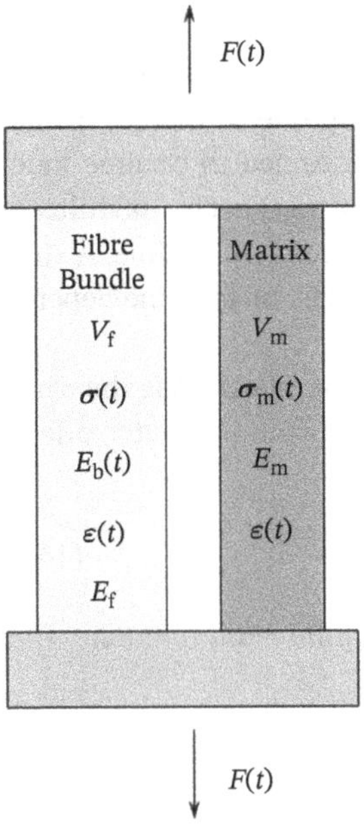

Figure 14.1 Schematic diagram of the parallel bar model of a unidirectional composite that is used to predict the effects of environmental exposure on axial composite properties.

i.e. the matrix strength is insufficient to maintain the load when all the fibres have failed. The objective is to develop the parallel bar model of a composite so that it can predict the dependence of composite life t_f on the fixed applied load F, and the dependence of the residual strength of the composite on elapsed time t from the instant of first loading.

The behaviour of bundles of loose fibres subject to environmental degradation has been modelled by Kelly and McCartney [3] and McCartney [4] for the case when the load applied to the bundle is fixed in time. For the parallel bar model of the composite that is subject to a fixed load F, the progressive failure of fibres in the bundle leads to a time dependence of the effective bundle stiffness, and consequently to a time dependence of the load applied to the fibre bundle. Thus, the earlier modelling [3, 4] requires modification if it is to be applied to the prediction of the behaviour of a uniaxially loaded unidirectional composite material having weak interfaces.

14.3 Basic Mechanics for the Parallel Bar Model of a Composite

The analysis of the parallel bar model shown in Figure 14.1 will neglect any axial thermal stresses arising from a mismatch of the thermal expansion coefficients of the fibres and the matrix. The area fraction of all fibres in the bundle is denoted by A_b and that of the matrix is denoted by A_m. It follows that

$$V_f = \frac{A_b}{A_b + A_m}, \qquad V_m = \frac{A_m}{A_b + A_m} = 1 - V_f. \tag{14.1}$$

The load applied to the fibre bundle at time t is denoted by $F_b(t)$, the stress in each surviving fibre being denoted by $\sigma(t)$. The cross-sectional area of each of the fibres in the bundle is denoted by A, and there are N_0 fibres in the composite so that $A_b = N_0A$. The axial modulus of each fibre is denoted by E_f, which is assumed to be time independent. The axial stress at time t in the matrix is denoted by $\sigma_m(t)$. The modulus of the matrix is denoted by E_m, which is also assumed to be independent of time. A time dependence could be included to account for viscoelastic effects, or for time-dependence arising from matrix ageing.

The axial strain in all surviving fibres of the bundle and the matrix has the same time-dependent value that is denoted by $\varepsilon(t)$. As thermal expansion mismatch effects are neglected it follows that

$$\varepsilon(t) = \frac{\sigma(t)}{E_f} = \frac{\sigma_m(t)}{E_m}. \tag{14.2}$$

The balance of forces in the parallel bar model leads to the equilibrium relation

$$F_b(t) + A_m\sigma_m(t) = F. \tag{14.3}$$

The number of surviving fibres in the bundle at time t is denoted by $N(t)$ so that the load applied to the bundle at time t may be written as

$$F_b(t) = N(t)A\sigma(t). \tag{14.4}$$

On substituting (14.4) into (14.3), followed by the elimination of $\sigma_m(t)$ using (14.2) it is easily shown that the number of fibres surviving at time t is related to the fibre stress $\sigma(t)$ through the following relation that quantitatively characterises the load sharing that occurs when fibres in the composite fail

$$\left(\frac{N(t)}{N_0} + \alpha\right)\sigma(t) = \frac{F}{N_0A}, \quad \text{where} \quad \alpha = \frac{V_mE_m}{V_fE_f}, \tag{14.5}$$

and where $N(0) = N_0$. This is the generalisation to a composite material of the relation used in the modelling a loose bundle of fibres subject to environmental degradation [3, 4], which is recovered from (14.5) on letting $\alpha \to 0$.

It is useful to relate the number of fibres surviving in the bundle at time t to the effective axial modulus of the bundle $E_b(t)$. The effective stress applied to the bundle is defined by

$$\sigma_b(t) = \frac{F_b(t)}{N_0A}, \tag{14.6}$$

and the axial strain of the bundle and the individual fibres has the value,

$$\varepsilon(t) = \frac{\sigma(t)}{E_f} = \frac{\sigma_b(t)}{E_b(t)} = \frac{N(t)}{N_0}\frac{\sigma(t)}{E_b(t)}, \tag{14.7}$$

where use has been made of (14.4) and (14.6). Clearly the effective axial modulus of the fibre bundle is given by

$$E_b(t) = \frac{N(t)}{N_0}E_f. \tag{14.8}$$

The effective axial stress σ_A applied to the composite is defined by

$$\sigma_A = \frac{F}{A_b + A_m}, \tag{14.9}$$

and it can be shown from (14.1) and (14.5), together with the fact that $A_b = N_0 A$, that

$$\sigma_A(t) = E_c(t)\varepsilon(t), \quad \text{where} \quad E_c(t) = V_f E_b(t) + V_m E_m, \tag{14.10}$$

where $E_c(t)$ is the effective axial modulus of the composite defined by the rule of mixtures, as to be expected.

14.4 Accounting for Defect Growth

The objective here is to show how the analysis of Kelly and McCartney [3], developed for loose bundles exposed to an aggressive environment, must be modified for application to a unidirectional composite having weak interfaces. The analysis is based on the assumption that the strength of individual fibres is determined by surface defects whose effective size and distribution along the fibre surface is statistically distributed.

Fibre failure is assumed to be governed by a Griffith type of failure criterion having the form

$$K^2 = y^2\sigma^2 a = K_{Ic}^2, \tag{14.11}$$

where K is an effective stress intensity factor for a fibre defect of effective size a subject to a fibre stress σ, K_{Ic} is the effective fracture toughness of the fibre material and y is a dimensionless parameter designed to account for defect geometry. The aggressive environment leads to defect growth when the fibre is under load. Such defect growth is assumed to be governed by a growth law of the form

$$\frac{da}{dt} = CK^n, \tag{14.12}$$

where C and n are material constants.

When a constant load is applied to a unidirectional composite, exposed to an aggressive environment to the point of saturation, the fibre defects grow in size according to the growth law (14.12) eventually leading to fibre failure when the failure criterion (14.11) is satisfied. Thus, fibres progressively fail and the load carried by failed fibres is, for the parallel bar model under discussion, transferred to the surviving fibres and matrix using the load-sharing rule (14.5). The stress in each fibre of the system is, thus, time dependent. It is useful to present here the relationship that determines the initial defect size $X_0(t)$ that requires a time t to grow to the critical size $a_c(t)$, at which the fibre stress is $\sigma(t)$, under the influence of a time-dependent fibre stress history $\sigma(\tau)$; $0 < \tau < t$. The critical defect size at time t is predicted by (14.11) to be

$$a_c(t) = \left[\frac{K_{Ic}}{y\sigma(t)}\right]^2, \tag{14.13}$$

and it can be shown on integrating (14.12) between the limits $X_0(t)$ and $a_c(t)$ that

$$X_0(t) = \frac{K_{Ic}^2}{y^2}\left[\sigma^{n-2}(t) + (n-2)\lambda\int_0^t \sigma^n(\tau)\mathrm{d}\tau\right]^{\frac{2}{2-n}}, \tag{14.14}$$

where

$$\lambda = \frac{1}{2} C K_{\mathrm{Ic}}^{n-2} y^2. \tag{14.15}$$

On using (14.11), it follows from (14.14) that the initial strength $\sigma_{\mathrm{i}}(t)$ of the fibres, that fail at time t when their stress is $\sigma(t)$, is given by

$$\sigma_{\mathrm{i}}(t) = \left[\sigma^{n-2}(t) + (n-2)\,\lambda \int_0^t \sigma^n(\tau)\mathrm{d}\tau \right]^{\frac{1}{n-2}}. \tag{14.16}$$

The cross-sectional area of the sample of unidirectional composite is assumed to be large enough for there to be a very large number of fibres. It can then be assumed that the bundle of fibres used in the parallel bar model contains, in effect, every possible fibre strength that can arise in the statistical distribution. It is assumed that the strength distribution of the fibres is given by the two-parameter Weibull distribution [5] so that, for a large bundle of N_0 fibres, the expected number of fibres N surviving when the stress in each fibre is σ is given by

$$N = N_0 \exp\left[-\left(\frac{\sigma}{\sigma_0} \right)^m \right], \tag{14.17}$$

where σ_0 is a scaling parameter that will depend on the length of the composite.

14.5 Prediction of Maximum load

It is useful to investigate the prediction of the maximum load that can be carried by a unidirectional composite assuming that the parallel bar model is valid. When using (14.17) in conjunction with (14.5) it is easily shown that the total load carried by the composite, when the stress in surviving fibres has the value σ, is obtained from

$$\hat{F} = \left(\alpha + \mathrm{e}^{-\hat{\sigma}^m} \right) \hat{\sigma}, \tag{14.18}$$

where $\hat{F}$ and $\hat{\sigma}$ are a dimensionless normalised load and normalised stress, respectively, defined by

$$\hat{F} = \frac{F}{N_0 \sigma_0 A}, \qquad \hat{\sigma} = \frac{\sigma}{\sigma_0}. \tag{14.19}$$

The maximum value of the load that can be carried by the composite occurs when $\hat{F}$ has a local maximum when plotted as a function of $\hat{\sigma}$. The maximum fibre stress $\hat{\sigma}_{\max}$ satisfies the transcendental equation

$$m \hat{\sigma}_{\max}^m = 1 + \alpha \mathrm{e}^{\hat{\sigma}_{\max}^m}. \tag{14.20}$$

The corresponding maximum load $F_{\max}$ for the composite is then obtained using

$$\frac{F_{\max}}{N_0 A \sigma_0} = \hat{F}_{\max} = \left(\alpha + \mathrm{e}^{-\hat{\sigma}_{\max}^m} \right) \hat{\sigma}_{\max} = \alpha \frac{m \hat{\sigma}_{\max}^{m+1}}{m \hat{\sigma}_{\max}^m - 1}, \tag{14.21}$$

which is consistent with the known result for a loose bundle [3] when the limit $\alpha \to 0$ is taken.

Equation (14.20) governing the maximum fibre stress does not always have a solution as is easily seen by examining the form of the left- and right-hand side of (14.20). On letting $x = \hat{\sigma}_{\max}^m$ the critical conditions defining the limit of solutions to (14.20) may be written as

$$mx = 1 + \alpha e^x, \qquad m = \alpha e^x. \tag{14.22}$$

These conditions correspond to the touching of the curves $y = mx - 1$ and $y = \alpha e^x$. It is easily seen that the critical condition occurs when

$$x = \ln \frac{m}{\alpha} = \frac{1+m}{m}. \tag{14.23}$$

It is concluded that (14.20) has a solution only if

$$\alpha < m\, e^{-\left(1+\frac{1}{m}\right)}. \tag{14.24}$$

If this condition is not satisfied, then it is deduced that the fibres progressively fail until there is just one surviving fibre which will then fail, i.e. the bundle does not suddenly collapse. The value of the Weibull modulus m for fibres of interest is usually such that condition (14.24) is satisfied so that bundle collapse is always expected in practice.

14.6 Prediction of Progressive Damage

At time t, the fibres that survive in the composite are those whose initial strengths were greater than $\sigma_i(t)$ defined by (14.16). It then follows from (14.17) that the expected number of surviving fibres $N(t)$ at time t is given by

$$\frac{N(t)}{N_0} = \hat{N}(t) = \exp\left[-\left(\frac{\sigma_i(t)}{\sigma_0}\right)^m\right]. \tag{14.25}$$

On substituting (14.16) into (14.25),

$$\left(\ln \frac{1}{\hat{N}(t)}\right)^{\frac{n-2}{m}} = \hat{\sigma}^{n-2}(t) + (n-2)\eta \int_0^t \hat{\sigma}^n(\tau)\,d\tau, \tag{14.26}$$

where

$$\eta = \lambda \sigma_0^2 = \tfrac{1}{2} C K_{Ic}^{n-2} y^2 \sigma_0^2, \tag{14.27}$$

and where use has been made of definitions (14.19), which when applied to the load sharing rule (14.5) lead to

$$\hat{N}(t) = \frac{\hat{F}}{\hat{\sigma}(t)} - \alpha. \tag{14.28}$$

On substituting (14.28) into (14.26),

$$\left(\ln \frac{\hat{\sigma}(t)}{\hat{F} - \alpha\hat{\sigma}(t)}\right)^{\frac{n-2}{m}} = \hat{\sigma}^{n-2}(t) + (n-2)\eta \int_0^t \hat{\sigma}^n(\tau)\,\mathrm{d}\tau. \tag{14.29}$$

On differentiating (14.29) with respect to t, the following differential equation governing the time dependence of the normalised fibre stress $\hat{\sigma}(t)$ is obtained

$$\left[\frac{1}{m}\frac{\hat{F}}{\hat{F} - \alpha\hat{\sigma}(t)}\left(\ln \frac{\hat{\sigma}(t)}{\hat{F} - \alpha\hat{\sigma}(t)}\right)^{\frac{n-m-2}{m}} - \hat{\sigma}^{n-2}(t)\right]\frac{\mathrm{d}\hat{\sigma}(t)}{\mathrm{d}(\eta t)} = \hat{\sigma}^{n+1}(t). \tag{14.30}$$

This differential equation is solved by standard numerical techniques subject to the initial condition

$$\hat{\sigma}(0) = s_0, \tag{14.31}$$

where s_0 is the solution of the transcendental equation

$$\hat{F} = \left(\alpha + \mathrm{e}^{-s_0^m}\right) s_0, \tag{14.32}$$

corresponding to (14.18), that must be solved numerically.

14.7 Predicting the Failure Stress and Time to Failure

The structure of the differential equation is such that $\mathrm{d}\hat{\sigma}/\mathrm{d}t \to \infty$ when $\hat{\sigma}(t) \to \hat{\sigma}_\mathrm{f}$ where

$$\frac{1}{m}\frac{\hat{F}}{\hat{F} - \alpha\hat{\sigma}_\mathrm{f}}\left(\ln \frac{\hat{\sigma}_\mathrm{f}}{\hat{F} - \alpha\hat{\sigma}_\mathrm{f}}\right)^{\frac{n-m-2}{m}} = \hat{\sigma}_\mathrm{f}^{n-2}. \tag{14.33}$$

The stress $\hat{\sigma}_\mathrm{f}$ in the surviving fibres when the composite fails can, thus, be determined using numerical methods without having to solve the differential equation (14.30). It should be noted that when $\hat{F} = \hat{F}_{\max}$, the solution of (14.33) is given by $\hat{\sigma}_\mathrm{f} = \hat{\sigma}_{\max}$ where $\hat{\sigma}_{\max}$ and $\hat{F}_{\max}$ are given by (14.20) and (14.21), respectively. The time to failure for the composite is denoted by t_f.

The number of surviving fibres just before composite failure is obtained using (14.25) and is given by

$$\frac{N(t_\mathrm{f})}{N_0} = \hat{N}_\mathrm{f} = \mathrm{e}^{-\hat{\sigma}_\mathrm{i}^m}, \qquad \hat{\sigma}_\mathrm{i} = \frac{\sigma_\mathrm{i}(t_\mathrm{f})}{\sigma_0}. \tag{14.34}$$

The transcendental equation (14.33), that usually must be solved numerically, involves the dimensionless loading parameter $\hat{F}$ in a complicated way. It is useful to unravel the dependence on this parameter by using the load sharing rule (14.28) to express (14.33) in terms of N_f as follows

$$\hat{F}^{n-2} = \frac{1}{m}\frac{\left[\hat{N}_\mathrm{f} + \alpha\right]^{n-1}}{\hat{N}_\mathrm{f}}\left(\ln \frac{1}{\hat{N}_\mathrm{f}}\right)^{\frac{n-m-2}{m}}. \tag{14.35}$$

Having solved (14.35) to find $\hat{N}_f$ using numerical methods, the normalised failure stress is obtained, on making use of (14.5), from the relation

$$\hat{\sigma}_f = \frac{\hat{F}}{\hat{N}_f + \alpha}. \tag{14.36}$$

The time to failure t_f can be predicted only by solving the differential equation (14.30) in the normalised stress range $s_0 \leq \hat{\sigma}(t) \leq \hat{\sigma}_f$.

14.8 Predicting Residual Strength

A key requirement concerning the effects of environment on composite degradation is the prediction of the time dependence of the residual strength of a composite. This has already been considered for the case of a loose bundle of fibres [4]. The objective now is to extend the analysis, and simplify it as far as possible, so that the residual strength of a unidirectional composite with weak interfaces can be predicted. After an elapsed time t from the application of a fixed load F, the load is instantaneously increased until the composite fails catastrophically. Just before the load is suddenly increased, the stress in the surviving fibres has the value $\sigma(t)$ and at any stage during the subsequent instantaneous load increase the value of the stress in the fibres is denoted by s. When the fibre stress has the value s, the critical defect size has the following value specified by (14.11)

$$a_c^* = \left(\frac{K_{Ic}}{ys}\right)^2. \tag{14.37}$$

It is necessary to calculate the original size X^* of the critical defect using relations (14.11) and (14.12). It is easily shown that

$$\left(X^*\right)^{\frac{2-n}{2}} = \left(a_c^*\right)^{\frac{2-n}{2}} + \frac{1}{2}C(n-2)y^n \int_0^t \sigma^n(\tau)\,d\tau. \tag{14.38}$$

On using (14.11) the initial strength of the fibres that are critical at time t when the fibre stress has the value s is denoted by s_i and is given, on using (14.38), by the relation

$$\hat{s}_i^{n-2} = \hat{s}^{n-2} + (n-2)\eta \int_0^t \hat{\sigma}^n(\tau)\,d\tau, \tag{14.39}$$

where use has been made of (14.27) and where

$$\hat{s}_i = \frac{s_i}{\sigma_0}, \qquad \hat{s} = \frac{s}{\sigma_0}. \tag{14.40}$$

On using (14.29), relation (14.39) may be written in the form

$$\hat{s}_i^{n-2} = \hat{s}^{n-2} - k(t), \tag{14.41}$$

where

$$k(t) = \hat{\sigma}^{n-2}(t) - \left[\ln \frac{\hat{\sigma}(t)}{\hat{F} - \alpha\hat{\sigma}(t)}\right]^{\frac{n-2}{m}}. \tag{14.42}$$

The load applied to the composite F_s, when the fibre stress has the value s, is obtained from (14.19) and (14.28) so that

$$\frac{F_s}{N_0\sigma_0 A} = \hat{F}_s = \left(\alpha + \hat{N}_s\right)\hat{s}, \tag{14.43}$$

where $\hat{F}_s$ is the normalised applied load and where $\hat{N}_s$ is the normalised number of surviving fibres when the load on the composite is such that the fibre stress has the value s. It follows from (14.17) that

$$\hat{N}_s = \mathrm{e}^{-\hat{s}_i^m}. \tag{14.44}$$

On substituting into (14.43), the following expression is derived for the normalised load applied to the composite during a residual strength test

$$\hat{F}_s = \left(\alpha + \mathrm{e}^{-\hat{s}_i^m}\right)\hat{s}. \tag{14.45}$$

A normalised residual strength $S(t)$ of the composite at time t is defined to be the maximum value of F_s when s is varied, or alternatively the maximum value of $\hat{F}_s$ when $\hat{s}$ is varied. Noting that $k(t)$ is independent of $\hat{s}$, the maximum value of $\hat{F}_s$ occurs when $\hat{s}_i = x(t)$, where $x(t)$ satisfies the transcendental equation

$$1 + \alpha\mathrm{e}^{x^m(t)} = mx^m(t)\left(1 + \frac{k(t)}{x^{n-2}(t)}\right). \tag{14.46}$$

On using (14.41), the stress $\sigma_{\max}(t)$ in the surviving fibres just before the composite fails during a residual strength test is obtained from

$$\frac{\sigma_{\max}(t)}{\sigma_0} = \left[x^{n-2}(t) + k(t)\right]^{\frac{1}{n-2}} = \hat{\sigma}_{\max}(t). \tag{14.47}$$

It then follows from (14.41) and (14.45) that the normalised residual strength of the composite $S(t)$ is obtained using

$$\frac{S(t)}{\sigma_0} = \hat{\sigma}_{\max}(t)\left[\alpha + \mathrm{e}^{-x^m(t)}\right] = \hat{S}(t). \tag{14.48}$$

When $t = 0$, it can be shown using (14.29) that $k(0) = 0$ in which case the transcendental equation (14.46) reduces to the form (14.20), which needs to be solved when calculating the maximum load carried by the composite. The actual residual strength of the composite, defined as maximum load divided by total cross-sectional area, is in fact given by $V_f S(t)$.

14.9 Example Results

To assess the properties of the model, some predictions have been made to illustrate the principal characteristics of the model. There are four parameters that need to be specified in order that example predictions are obtained. The first is the Weibull exponent m characterising the strength distribution of the fibres before environmental exposure. This parameter, which often has values in the range 4–8, appears in relation (14.17) defining the expected number of fibre failures for a given fibre stress. Most predictions will assume that $m = 8$, although lifetime predictions will consider the case when $m = 4$. The value of m is usually obtained from single-fibre strength tests. The second parameter is the exponent n appearing in the defect growth law (14.12). This parameter, which usually has values in the range 3–30, is obtained from stress corrosion cracking tests carried out using monolithic glass test pieces. Predictions of behaviour will be made for the three values $n = 3$, 15 and 30. The third parameter is the ratio α defined by (14.5), which takes account of the properties of the fibre and matrix, and also of the volume fractions. For many glass-fibre composites of interest, the value of α lies in the range 0–0.1. Predictions will be made for the three values $\alpha =$ 0, 0.05 and 0.1. When $\alpha = 0$ the composite is being modelled in exactly the same way as a loose bundle of fibres. The fourth and final parameter that needs to be specified is the level of loading applied to the composite. The model assumes that the ratio $F/F_{\max}$ is given where F is the axial load applied to the composite and $F_{\max}$ is the maximum load carried by the composite, i.e. the failure load before environmental exposure. The value of the ratio $F/F_{\max}$ always lies in the range 0–1.

The Euler–Richardson solution technique [6] is used to solve the differential equation (14.30) where the dimensionless time ηt is regarded as the unknown function of $\hat{\sigma}$. In other words, the differential equation can be used directly to determine an increment in the value of ηt for any given increment in $\hat{\sigma}$. The initial condition is specified by (14.31) and (14.32) and the range $s_0 \leq \hat{\sigma} \leq \hat{\sigma}_{\mathrm{f}}$ is subdivided into 1000 equal intervals when solving the differential equation. The upper bound $\hat{\sigma}_{\mathrm{f}}$ is determined by relations (14.35) and (14.36). Figure 14.2 shows the result of solving the differential equation (14.30) to find the dimensionless lifetime ηt_{f} for the case when $\alpha = 0.1$ and $n = 15$ for various values of loading parameter $F/F_{\max}$, and for values $m = 4$, 8 of the Weibull exponent. The parameter η is defined by (14.27). It is seen that as $F/F_{\max}$ tends to unity the lifetime tends to zero (the logarithm tending to $-\infty$). In addition, it is clear that the effect of varying the Weibull exponent in the range $4 \leq m \leq 8$ leads to only slight differences in predictions. For this reason, all further predictions assume the single value $m = 8$.

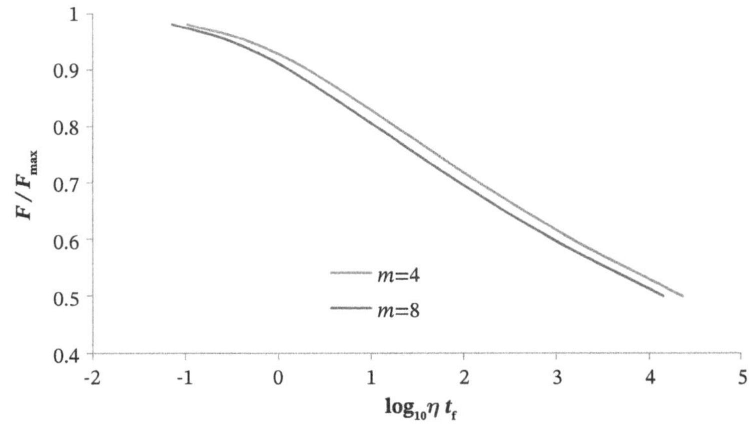

Figure 14.2 The dependence of composite lifetime (ηt_{f}) on the axial loading parameter ($F/F_{\max}$) for the case when $\alpha = 0.1$ and $n = 15$.

For the cases $m = 8$, $F/F_{max} = 0.6$ and $n = 3, 15, 30$, Figures 14.3–14.5 show the predictions for the normalised fibre stress $\hat{\sigma}$, defined by (14.19), as a function of elapsed time ratio t/t_f. Each figure shows predictions for the three values $\alpha = 0, 0.05$ and 0.1. On comparing Figures 14.3–14.5, it is seen that increasing the value of the parameter n leads to larger values for the normalised fibre stress as composite failure is approached. In addition, larger values of n lead to a more sudden failure. The effect of varying the parameter $\hat{\sigma}$ is not very significant, i.e. the matrix does not have much of an effect on the stress history of the fibres in the composite.

For the cases $m = 8$, $F/F_{max} = 0.6$ and $n = 3, 15, 30$, Figures 14.6–14.8 show the predictions for the normalised residual strength α, defined by (14.48), as a function of elapsed time ratio t/t_f. Each figure shows predictions for the three values $\alpha = 0, 0.05$ and 0.1. On comparing Figures 14.6–14.8, it is seen that increasing the value of the parameter n leads to curves (Figure 14.6) which vary from almost linear having slightly

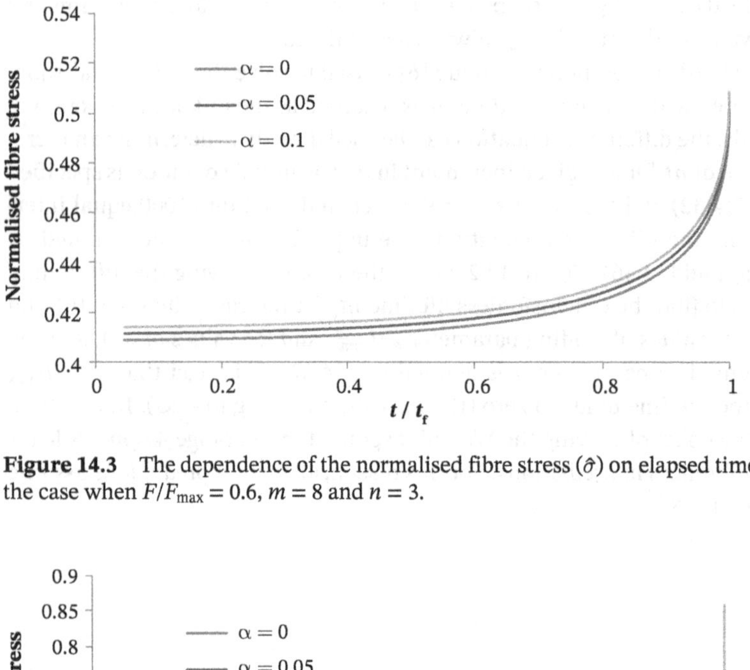

Figure 14.3 The dependence of the normalised fibre stress ($\hat{\sigma}$) on elapsed time ratio (t/t_f) for the case when $F/F_{max} = 0.6$, $m = 8$ and $n = 3$.

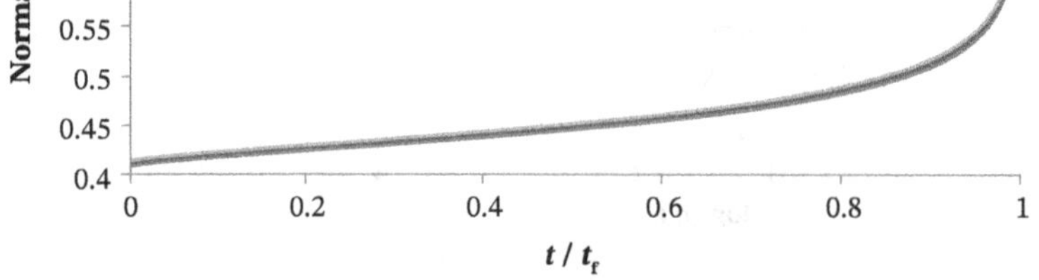

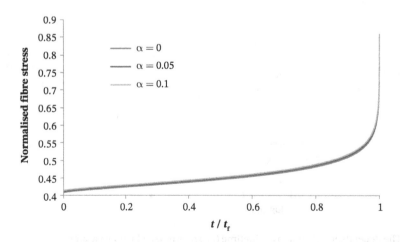

Figure 14.4 The dependence of the normalised fibre stress ($\hat{\sigma}$) on elapsed time ratio (t/t_f) for the case when $F/F_{max} = 0.6$, $m = 8$ and $n = 15$.

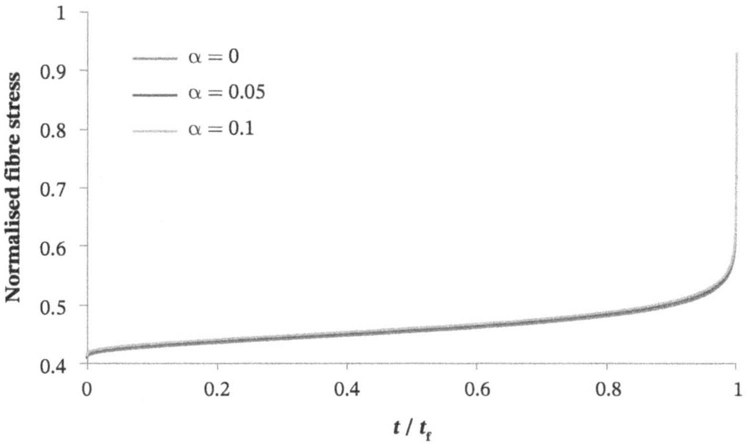

Figure 14.5 The dependence of the normalised fibre stress ($\hat{\sigma}$) on elapsed time ratio (t/t_f) for the case when $F/F_{max} = 0.6$, $m = 8$ and $n = 30$.

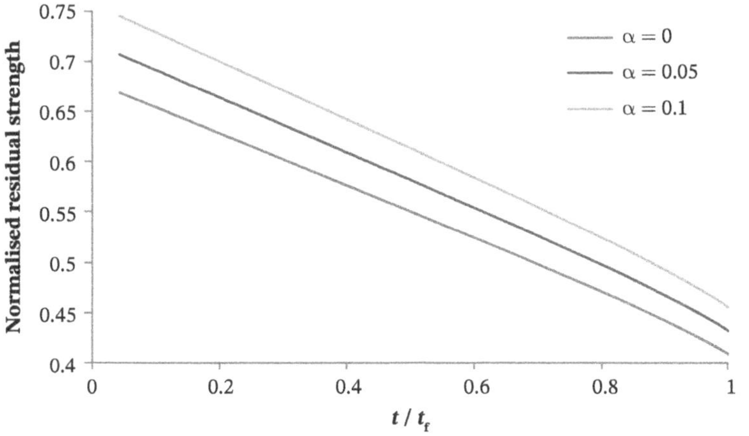

Figure 14.6 The dependence of the normalised residual strength ($\hat{s}$) on elapsed time ratio (t/t_f) for the case when $F/F_{max} = 0.6$, m = 8 and $n = 3$.

different gradients for the three values of α, to curves (Figure 14.8) which are almost parallel for each value of α, but which exhibit a sudden loss of strength as $t/t_f \rightarrow 1$. The effect of varying the parameter α is significant, i.e. the matrix is having an effect on the time dependence of the residual strength of the composite.

On replotting the results shown in Figures 14.6–14.8, so that in effect the residual strength at time t is divided by the static strength of the composite, it is seen from Figures 14.9–14.11 that there is hardly any dependence of the residual strength ratio $S/F_{max} = \hat{S}/\hat{F}_{max}$ on the parameter α. This result is found for all the values of $n = 3$, 15 and 30.

The results shown in Figure 14.12 are an attempt to illustrate many of the most important features of the model, where the values $\alpha = 0.025$, $m = 8$ and $n = 12$, have been assumed. The black line shows the relationship between composite lifetime and the loading parameter F/F_{max} where F_{max} is the initial strength of the composite, i.e. before exposure to the environment, corresponding to a zero exposure time. The three

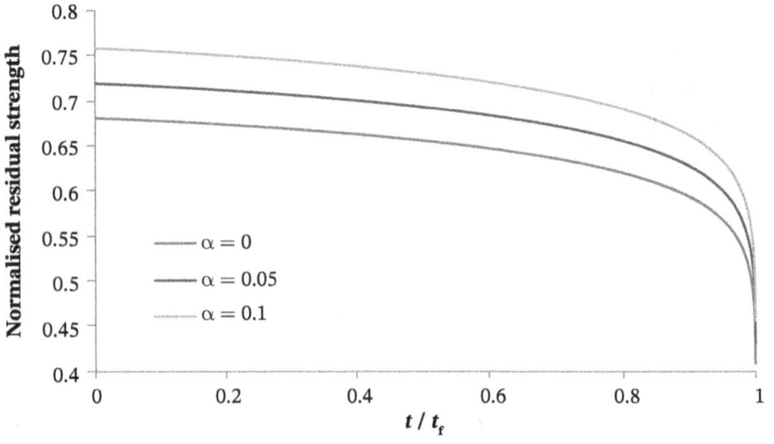

Figure 14.7 The dependence of the normalised residual strength ($\hat{s}$) on elapsed time ratio (t/t_f) for the case when $F/F_{max} = 0.6$, $m = 8$ and $n = 15$.

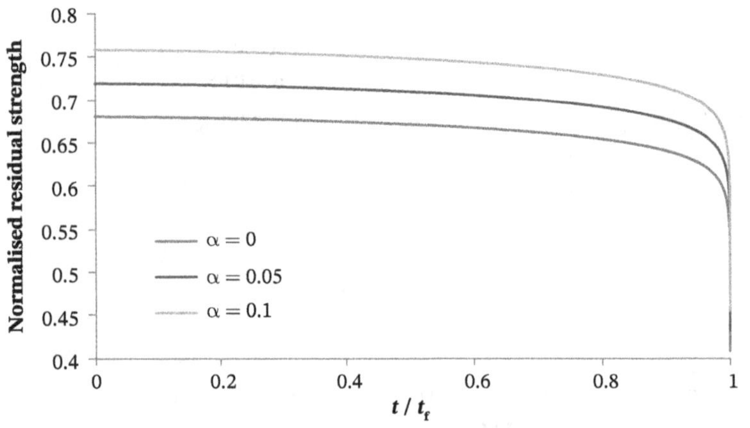

Figure 14.8 The dependence of the normalised residual strength ($\hat{s}$) on elapsed time ratio (t/t_f) for the case when $F/F_{max} = 0.6$, $m = 8$ and $n = 30$.

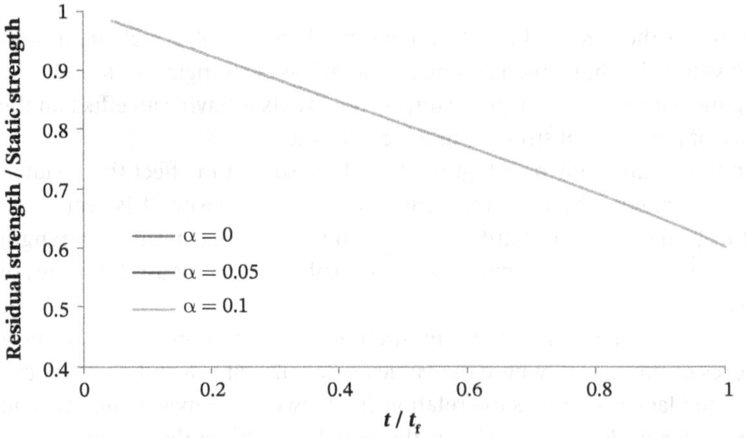

Figure 14.9 The dependence of the residual strength ratio on elapsed time ratio (t/t_f) for the case when $F/F_{max} = 0.6$, $m = 8$ and $n = 3$.

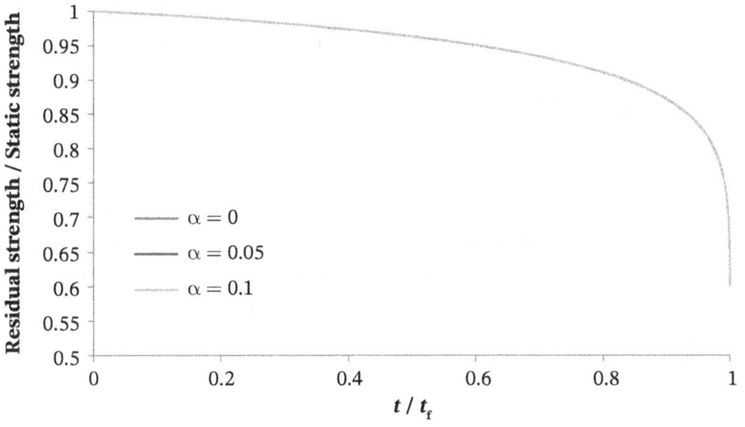

Figure 14.10 The dependence of the residual strength ratio on elapsed time ratio (t/t_f) for the case when $F/F_{max} = 0.6$, $m = 8$ and $n = 15$.

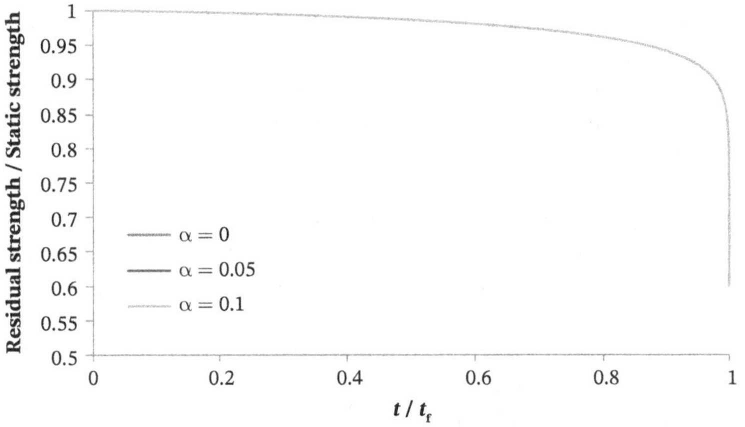

Figure 14.11 The dependence of the residual strength ratio on elapsed time ratio (t/t_f) for the case when $F/F_{max} = 0.6$, $m = 8$ and $n = 30$.

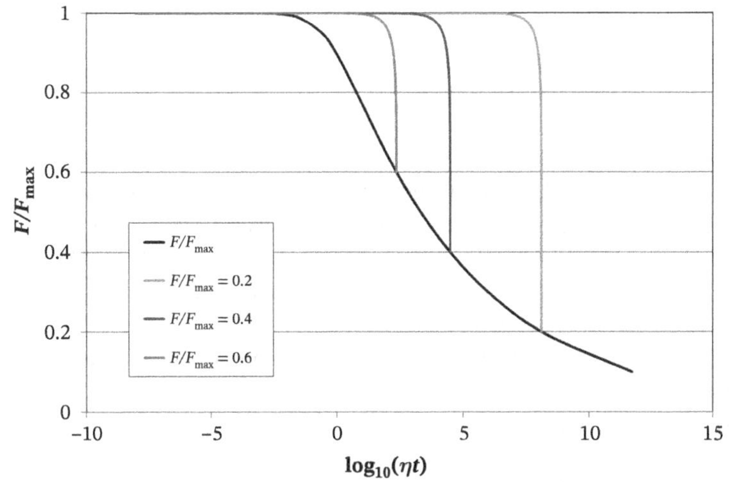

Figure 14.12 The dependence of composite lifetime (ηt_f) on the axial loading parameter (F/F_{max}) for the case when $\alpha = 0.025$, $m = 8$ and $n = 12$, and corresponding residual strength curves.

gray lines in the figure are the corresponding residual strength plots, where the composite is assumed to have been exposed to the environment for a specified time after which the load is increased until the composite fails. The exposure has been undertaken at the three different fixed loads specified by $F / F_{max} = 0.2, 0.4, 0.6$. All three plots start from the same point where, without any exposure, the residual strength must be equal to the static strength of the composite. When the composite is exposed to the environment at the three fixed load levels, it is seen that the residual strength is hardly affected by the environment for a good proportion of the lifetimes. As exposure at the specified loading progresses further, the residual strength starts to reduce until a very fast decrease occurs leading quickly to the final failure of the composite. The final failure point does, in fact, lie on the black line representing the lifetimes following exposure to a fixed loads F.

14.10 Conclusion

The principal conclusion to be drawn from the preliminary results presented is that the time dependence of the axial properties of a unidirectional fibre-reinforced glass composite subject to environmental exposure under fixed load can be predicted using a parallel bar model of the composite where interface bonding is neglected. The model enables the prediction of the stress history of the fibre stress in surviving fibres from the point of first loading to the occurrence of catastrophic failure. The fibre stress is found to be almost independent of the matrix properties.

The model can also be used to predict the time dependence of the residual strength of the fibre bundle and of the composite, a property which does show some dependence on matrix properties. However, it has been shown that when the residual strength is divided by the static strength, the resulting residual strength ratio is virtually independent of the matrix properties. It is concluded that the residual strength ratio for a unidirectional composite is predictable (and, therefore, measurable) from the static strength of the composite, and the time dependence of the residual strength of a loose bundle of fibres. This major result of the modelling identifies an efficient method for predicting and measuring the time dependence of the residual axial strength of unidirectional composites.

References

1. Broughton, W.R. and McCartney, L.N. (April 1998), Predictive models for assessing long-term performance of polymer matrix composites. *NPL Report CMMT(A)95.*
2. Metcalfe, A.G., Gulden, M.E., and Schmitz, G.K. (1971). Glass technology 12: 15–23.
3. Kelly, A. and McCartney, L.N. (1981). Failure by stress corrosion of bundles of fibres. *Proceedings of the Royal Society of London* A374: 475–489.
4. McCartney, L.N. (1982). Time dependent strength of large bundles of fibres loaded in corrosive environments. *Fibre Science & Technology* 16: 95–109.
5. Weibull, W. (1951). A statistical distribution function of wide applicability. *Journal of Applied Mechanics* 19: 293–297.
6. Churchhouse, R.F. (1981). Handbook of Applicable Mathematics. Vol. 3. In: *Numerical Methods* (ed. W. Ledermann), 319–321. Chichester - New York -Brisbane - Toronto: John Wiley and Sons.

15

Maxwell's Far-field Methodology Predicting Elastic Properties of Multiphase Composites Reinforced with Aligned Transversely Isotropic Spheroids

Overview: Maxwell's methodology, developed to estimate the effective electrical conductivity of isotropic particulate composites, is used with a far-field elasticity result of Eshelby to derive closed-form formulae for effective transversely isotropic elastic properties of *multiphase* composites comprising aligned *transversely isotropic* spheroidal inclusions embedded in an *isotropic* matrix. Very simple expressions are derived for the effective shear moduli. Closed-form analytical results for all elastic constants are shown, using exact numerical methods, to be identical to more complex results derived by Qiu and Weng on applying Mori–Tanaka theory to spheroidal reinforcements. This is a contradictory result as Maxwell's approach neglects inclusion interactions whereas Mori–Tanaka theory is designed, to some extent, to take such interactions into account. The rational conclusion is that inclusion interaction effects for volume fractions of practical relevance do not affect the far-field to any significant degree so that Maxwell's methodology, when combined with Eshelby's analysis, has much wider applicability than expected. Results for isotropic composites having distributions of spherical particles, and transversely isotropic composites having distributions of aligned fibres, correspond with known expressions, and can coincide with, or lie between, variational bounds for all volume fractions. A new simple expression having a 'mixtures' structure is obtained for the axial modulus of multiphase fibre-reinforced composites that reduces to concentric cylinders estimates when there are just two phases. To demonstrate accuracy, property results for a variety of composites are compared with accurate numerical results in the literature for two-phase composites having reinforcement volume fractions in the range 0–0.7.

15.1 Introduction

The development of methods to estimate the effective properties of multiphase composites reinforced by spheroidal reinforcements has a long history. For example, Mori–Tanaka theory [1], estimating the average stress and elastic energy for composites reinforced with misfitting inclusions, has been applied by Weng [2] to develop an approximate method of estimating effective elastic properties of general multiphase anisotropic composites reinforced with arbitrarily oriented anisotropic inclusions. Explicit formulae for effective elastic properties were derived for suspensions of

Properties for Design of Composite Structures: Theory and Implementation Using Software, First Edition. Neil McCartney.

Companion Website: www.wiley.com/go/mccartney/properties

uniformly distributed, multiphase isotropic spherical particles in an isotropic matrix. Norris [3] emphasised the relationship of effective properties, estimated using Mori–Tanaka theory, to general bounds, showing that predicted properties for two-phase composites always satisfy the Hashin–Shtrikman bounds [4], a result that does not generalise to multiphase composites so that caution should be used in this case. Weng [2] reformulated Mori–Tanaka theory so that it is recast into a form that has an identical structure to that used when deriving the Hashin–Shtrikman bounds, and he used a notation based on the treatment of fourth-order tensors for elastic constants developed and used by Walpole [6–9]. Qiu and Weng [10] applied the modified Mori–Tanaka theory to composites having transversely isotropic spheroidal inclusions and derived explicit but complex formulae that can be used to estimate all elastic properties. Benveniste et al. [11] investigated the diagonal and elastic symmetry of the fourth-order effective elastic property tensor for heterogeneous media and showed that the Mori–Tanaka and self-consistent methods lead to diagonal and symmetric property tensors for all two-phase composites. If, however, all the inclusions have a similar shape and are aligned, then they showed that the symmetry properties apply also to multiphase composites. Chen and Dvorak [12] have applied the Mori–Tanaka method to estimate explicit formulae for the effective elastic properties of composites reinforced with aligned or randomly oriented, transversely isotropic fibres or platelets, and for fibrous composites reinforced with cylindrical orthotropic fibres. For the general case of transversely isotropic spheroidal inclusions embedded in an isotropic matrix, the effective elastic properties presented as formulae for fourth-order tensors are very difficult to interpret in terms of the five independent elastic constants, as shown by Qiu and Weng [10]. This has meant that the practical application in an engineering context of the valuable results has been difficult.

As a result of collaborating with the late Professor Anthony Kelly, regarding methods of estimating the effective thermoelastic constants and conductivities of composite materials, the author was introduced to the pioneering work of James Clerk Maxwell [13] who provided an ingenious method of estimating the effective electrical conductivity of a cluster of spherical particles, having the same size and embedded in an infinite medium, by considering the effect of the cluster on the far-field, when the system is subject to a uniform electrical field. Maxwell modestly asserted that the sizes and distribution of the particles must be such that particle interaction effects may be neglected, and he infers that his result will be valid only for small volume fractions of reinforcing particles. A result for effective permittivity that is analogous to Maxwell's result for electrical conductivity is known as the 'Maxwell–Garnett mixing formula', and it has a microscopic analogue that is known as the 'Clausius–Mossotti' (or 'Lorentz–Lorenz') formula, which has been related to effective elastic property estimation by Felderhof and Iske [14] and Cohen and Bergman [15, 16].

More recently, McCartney and Kelly [17] studied the method used by Maxwell and demonstrated that his methodology, focusing only on the far-field, can also be applied to the estimation of other properties of composite materials. The principal objective was to show how Maxwell's methodology could be used to estimate explicit formulae for the effective bulk modulus, shear modulus and thermal expansion coefficient of multiphase isotropic composites reinforced with homogeneous spherical particles (see Chapter 3). The methodology of Maxwell was naturally extended so that assemblies of multiphase spherical particles having a range of radii and/or properties could be considered. A second objective was to show that Maxwell's methodology is one reliable technique that provides closed-form estimates of effective properties and is not necessarily restricted to low volume fractions of particulate reinforcement as has often been claimed in the literature.

More recent unpublished work (now included here as Chapter 4) has shown that, when Maxwell's methodology is applied to clusters of aligned transversely isotropic cylindrical fibres of different types, embedded in an infinite isotropic matrix material, a similar situation arises to that described in [17]. For the fibre case, the effective thermoelastic properties of the composite are transversely isotropic, but it is not known how to estimate the axial Young's modulus and axial thermal expansion coefficient for multiphase fibre-reinforced composites. This leads on to the idea of considering clusters of aligned spheroidal inclusions having various sizes and properties that can represent aligned short fibres and particulate composites of various types, and of making use of a classical analysis due to Eshelby [18, 19]. He considered the elastic field for isolated ellipsoidal inclusions (both isotropic and anisotropic) that are embedded in an infinite isotropic matrix and subjected to loading that would in a homogenous material lead to uniform stress and strain fields. Of particular relevance is an expression for the far-field displacement field that provides a very convenient method of extending Maxwell's methodology for spheres to the case of aligned transversely isotropic spheroids and fibres embedded in an isotropic matrix, the investigation of which is the principal objective here. It is noted that Torquato [20] has observed: (i) this correspondence for the case of isotropic ellipsoidal inclusions when using a formulation based on the fourth-order elastic constants described previously, and (ii) that the Mori and Tanaka results, and hence those in this chapter based on Maxwell's methodology, coincide with one of the bounds due to Willis [21] depending on whether the matrix is stiffer or more compliant than all of the inclusions, as shown by Weng [22].

This chapter first steps back in time combining the pioneering work of Maxwell [13] and Eshelby [18] to develop a new method of estimating explicit closed-form formulae for the effective elastic properties of composites reinforced by aligned transversely isotropic spheroidal inclusions embedded in an isotropic matrix. The new approach is shown to generate very simple expressions for the shear moduli, and lead to a new result for the effective axial Young's modulus for a multiphase composite reinforced with aligned fibres. The approach will be shown capable of generating many of the well-known results for effective properties that have been derived in the literature using a variety of other methods. It is thought that the new method of estimating effective properties will enable engineers to understand more readily, and calculate more efficiently, the transversely isotropic effective properties of composites reinforced with aligned spheroidal inclusions.

15.2 General Description of Maxwell's Methodology Applied to Spheroidal Inclusions

The following description is based on Maxwell's [13] far-field approach, when estimating the electrical conductivity of a cluster of isotropic spherical particles embedded in an infinite isotropic matrix. The methodology is generalised here so that multiphase composites having aligned transversely isotropic spheroidal inclusions of various types can be considered.

15.2.1 Description of Geometry

In a well-mixed cluster of N types of aligned spheroidal reinforcement embedded in and perfectly bonded to an infinite isotropic matrix, there are n_i spheroidal inclusions having major axes a_i and minor axes b_i, $i = 1, \ldots, N$. The centres of the spheroids representing the inclusions are assumed to be homogeneously and isotropically distributed within the cluster. Inclusion properties of type i, which may differ from those of other types, are denoted by a subscript or superscript i, and they are assumed to be transversely isotropic with the principal direction aligned with the major axes of the spheroids. The cluster of all inclusion types may be just enclosed by a spheroid of major axis $\bar{a}$ and minor axis $\bar{b}$ having the same alignment as the inclusions. The homogeneous inclusion distribution leads to transversely isotropic effective properties of the composite formed by the cluster of aligned transversely isotropic spheroidal inclusions and isotropic matrix lying within this enclosing spheroid. The volume fractions of inclusions of type i within the enclosing spheroid of radii $\bar{a}$ and $\bar{b}$ are given by

$$V_{\mathrm{p}}^{i} = \frac{n_i a_i b_i^2}{\bar{a}\bar{b}^2}, \quad i = 1, \ldots, N, \quad \text{such that} \quad V_{\mathrm{m}} + \sum_{i=1}^{N} V_{\mathrm{p}}^{i} = 1, \tag{15.1}$$

where V_{m} is the volume fraction of matrix. For just one type of inclusion, as shown in Figure 15.1, with n inclusions having major axis a and minor axis b, the particulate volume fraction V_{p} is such that

$$V_{\mathrm{p}} = \frac{nab^2}{\bar{a}\bar{b}^2} = 1 - V_{\mathrm{m}}. \tag{15.2}$$

Whatever the nature and arrangement of the aligned spheroidal transversely isotropic inclusions in the cluster, Maxwell's methodology considers the far-field when replacing the discrete particulate composite, that can be enclosed by a spheroid having axes $\bar{a}$ and $\bar{b}$, by a homogeneous effective composite spheroid having the same axes $\bar{a}$

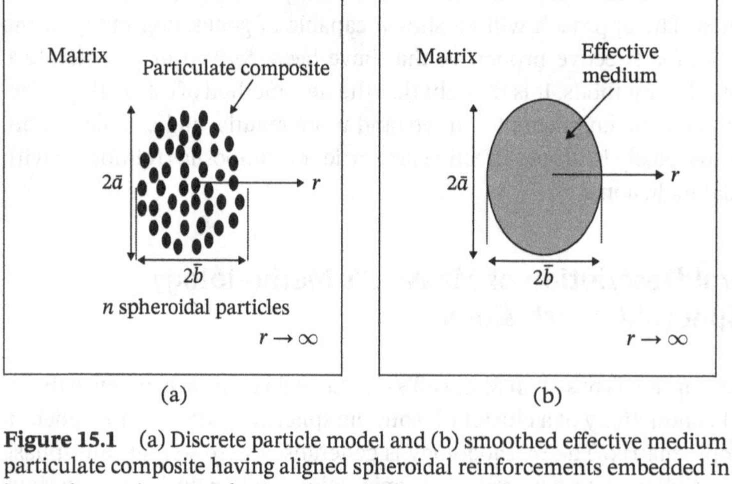

Figure 15.1 (a) Discrete particle model and (b) smoothed effective medium model of a particulate composite having aligned spheroidal reinforcements embedded in an infinite isotropic matrix material.

and $\bar{b}$ embedded in the matrix. There is no restriction on sizes, properties and locations of inclusions provided that the equivalent effective medium is homogeneous and transversely isotropic. Composites having statistical distributions of both inclusion size and properties can clearly be analysed, but they must all be aligned in the same direction.

15.2.2 Maxwell's Methodology for Estimating Elastic Constants

The first step considers the effect of embedding in the infinite matrix, an isolated cluster of aligned spheroidal transversely isotropic inclusions of different types that can be just contained within the spheroid having the same alignment. At infinity a stress is applied that would lead in matrix material alone to uniform stress and strain fields. For a single isolated inclusion embedded in an infinite matrix, the matrix displacement distribution is perturbed by the presence of the inclusion, and the perturbation from the uniform strain field depends on inclusion geometry and properties. The analysis of Eshelby [18, 19] enables the perturbations of the matrix displacement field to be determined at large distances from the inclusion. According to Maxwell's methodology (see [17]), the perturbing effect in the matrix at large distances from all the inclusions in the cluster is estimated by superimposing the perturbations caused by each inclusion, regarded as being isolated. The second step recognises that, at very large distances from the cluster, all the inclusions can be considered to be located at the origin that is chosen to be situated at the centre of one of the inclusions in the cluster. The third step replaces the composite having discrete inclusions lying within the bounding spheroid by the homogeneous spheroidal effective medium having axes $\bar{a}$ and $\bar{b}$, and having the transversely isotropic effective elastic properties of the composite.

Maxwell's methodology [13] is now combined with that of Eshelby [18, 19] so that a method can be developed that enables the estimation of the transversely isotropic effective elastic properties of a multiphase particulate composite reinforced with aligned spheroidal transversely isotropic inclusions having different sizes. An additional assumption is now made where it is assumed that the aspect ratios a_i / b_i, $i = 1, \ldots, N$, of the spheroidal inclusions, and that of the enclosing spheroid $\bar{a} / \bar{b}$, all have the same value. The principal reasons for this assumption are that it is useful to simplify the analysis that will be rather complex, and as it is one objective of this chapter to apply the methodology to aligned cylindrical fibres, these can be enclosed only by a surface having a similar geometry.

15.3 Isolated Spheroidal Inclusion

Eshelby [18] considered the elastic field outside an isolated ellipsoidal inclusion embedded in and perfectly bonded to an infinite matrix for the case where the inclusion and the matrix are isotropic materials. Eshelby [18] stated how the case of an anisotropic inclusion in an isotropic matrix could be solved. The application of Maxwell's methodology to a system of aligned spheroidal inclusions embedded in an infinite isotropic matrix is considered in this chapter, which requires knowledge of the stress or displacement distribution at large distances from a transversely isotropic spheroidal inclusion.

From Eshelby [18] the relation determining the transformation strain tensor $\varepsilon_{ij}^{\mathrm{T}}$ within the inclusion in terms of the strain tensor $\varepsilon_{ij}^{\mathrm{A}}$ applied to the matrix at infinity is

$$C^{ijkl}\left(\varepsilon_{kl}^{\mathrm{C}}+\varepsilon_{kl}^{\mathrm{A}}\right)=\lambda\left(\varepsilon_{kk}^{\mathrm{C}}+\varepsilon_{kk}^{\mathrm{A}}-\varepsilon_{kk}^{\mathrm{T}}\right)\delta_{ij}+2\mu\left(\varepsilon_{ij}^{\mathrm{C}}+\varepsilon_{ij}^{\mathrm{A}}-\varepsilon_{ij}^{\mathrm{T}}\right), \tag{15.3}$$

where C_{ijkl} are the anisotropic elastic constants for the inclusion, λ and μ (the shear modulus) are Lamé's constants for the isotropic matrix, and where $\varepsilon_{ij}^{\mathrm{C}}$ is the 'constrained strain' within the inclusion when it transforms while embedded in the matrix. From Eshelby [18],

$$\varepsilon_{ij}^{\mathrm{C}}=S_{ijkl}\varepsilon_{kl}^{\mathrm{T}}, \tag{15.4}$$

where the Eshelby tensor S_{ijkl} has dimensionless components depending only on Poisson's ratio of the isotropic matrix and the aspect ratio of the ellipsoid. The elastic constants C_{ijkl} are such that $C_{ijkl}=C_{jikl}=C_{ijlk}=C_{jilk}$. The substitution of (15.4) into (15.3) leads to

$$C_{ijkl}\left(S_{klmn}\varepsilon_{mn}^{\mathrm{T}}+\varepsilon_{kl}^{\mathrm{A}}\right)=\lambda\left(S_{kkmn}\varepsilon_{mn}^{\mathrm{T}}-\varepsilon_{kk}^{\mathrm{T}}+\varepsilon_{kk}^{\mathrm{A}}\right)\delta_{ij}+2\mu\left(S_{ijkl}\varepsilon_{kl}^{\mathrm{T}}-\varepsilon_{ij}^{\mathrm{T}}+\varepsilon_{ij}^{\mathrm{A}}\right). \tag{15.5}$$

For a transversely isotropic solid, where the axial direction corresponds to the direction of the x_1-axis, the stress-strain relations defining the elastic coefficients C_{ijkl} for the inclusion have the explicit form

$$\sigma_{11}=\left(E_{\mathrm{A}}+4k_{\mathrm{T}}\nu_{\mathrm{A}}^{2}\right)\varepsilon_{11}+2\nu_{\mathrm{A}}k_{\mathrm{T}}\varepsilon_{22}+2\nu_{\mathrm{A}}k_{\mathrm{T}}\varepsilon_{33}, \tag{15.6}$$

$$\sigma_{22}=2\nu_{\mathrm{A}}k_{\mathrm{T}}\varepsilon_{11}+\left(k_{\mathrm{T}}+\mu_{\mathrm{t}}\right)\varepsilon_{22}+\left(k_{\mathrm{T}}-\mu_{\mathrm{t}}\right)\varepsilon_{33}, \tag{15.7}$$

$$\sigma_{33}=2\nu_{\mathrm{A}}k_{\mathrm{T}}\varepsilon_{11}+\left(k_{\mathrm{T}}-\mu_{\mathrm{t}}\right)\varepsilon_{22}+\left(k_{\mathrm{T}}+\mu_{\mathrm{t}}\right)\varepsilon_{33}, \tag{15.8}$$

$$\sigma_{12}=2\mu_{\mathrm{A}}\varepsilon_{12},\qquad \sigma_{13}=2\mu_{\mathrm{A}}\varepsilon_{13},\qquad \sigma_{23}=2\mu_{\mathrm{t}}\varepsilon_{23}, \tag{15.9}$$

where E_{A} is the axial Young's modulus, ν_{A} is the axial Poisson's ratio, k_{T} is the plane strain bulk modulus and where μ_{A} and μ_{t} are the axial and transverse shear moduli, respectively. The corresponding transverse Young's modulus E_{T} and transverse Poisson's ratio ν_{t} are obtained using the following relations

$$\frac{4}{E_{\mathrm{T}}}=\frac{1}{k_{\mathrm{T}}}+\frac{1}{\mu_{\mathrm{t}}}+\frac{4\nu_{\mathrm{A}}^{2}}{E_{\mathrm{A}}},\qquad \nu_{\mathrm{t}}=\frac{E_{\mathrm{T}}}{2\mu_{\mathrm{t}}}-1. \tag{15.10}$$

The stress-strain relations (15.6)–(15.9) are used to characterise both the reinforcing inclusions and the effective medium representing the composite. For the isotropic matrix the elastic constants are the Young's modulus E, Poisson's ratio ν, the shear modulus μ and the bulk modulus k, which satisfy the relations

$$E=2\mu(1+\nu),\qquad k=\lambda+\frac{2}{3}\mu=\frac{E}{3(1-2\nu)}. \tag{15.11}$$

It is assumed that the major axes of the various spheroids in the composite are aligned with the x_1-axis. It should be noted that for spheroidal inclusions the tensor components S_{ijkl} are such that

$$S_{2222} = S_{3333},\quad S_{2211} = S_{3311},\quad S_{1212} = S_{2121} = S_{1313} = S_{3131},\quad S_{2323} = S_{3232},$$
$$S_{2222} - S_{2233} = 2S_{2323},\quad S_{2211} - S_{1122} = \nu\left(S_{2222} + S_{2233} - S_{1111} - S_{1122}\right). \tag{15.12}$$

Expanding (15.5) using (15.6)–(15.9) leads to the following six linear equations that determine the components of the strain tensor $\varepsilon_{ij}^{\mathrm{T}}$

$$\begin{aligned}
&\left[\left(E_{\mathrm{A}} + 4k_{\mathrm{T}}\nu_{\mathrm{A}}^{2} - \lambda - 2\mu\right)S_{1111} + \left(2\nu_{\mathrm{A}}k_{\mathrm{T}} - \lambda\right)S_{2211} + \left(2\nu_{\mathrm{A}}k_{\mathrm{T}} - \lambda\right)S_{3311}\right]\varepsilon_{11}^{\mathrm{T}}\\
&+\left[\left(E_{\mathrm{A}} + 4k_{\mathrm{T}}\nu_{\mathrm{A}}^{2} - \lambda - 2\mu\right)S_{1122} + \left(2\nu_{\mathrm{A}}k_{\mathrm{T}} - \lambda\right)S_{2222} + \left(2\nu_{\mathrm{A}}k_{\mathrm{T}} - \lambda\right)S_{3322}\right]\varepsilon_{22}^{\mathrm{T}}\\
&+\left[\left(E_{\mathrm{A}} + 4k_{\mathrm{T}}\nu_{\mathrm{A}}^{2} - \lambda - 2\mu\right)S_{1133} + \left(2\nu_{\mathrm{A}}k_{\mathrm{T}} - \lambda\right)S_{2233} + \left(2\nu_{\mathrm{A}}k_{\mathrm{T}} - \lambda\right)S_{3333}\right]\varepsilon_{33}^{\mathrm{T}}\\
&+\lambda\left(\varepsilon_{11}^{\mathrm{T}} + \varepsilon_{22}^{\mathrm{T}} + \varepsilon_{33}^{\mathrm{T}}\right) + 2\mu\varepsilon_{11}^{\mathrm{T}}\\
&\quad = -\left(E_{\mathrm{A}} + 4k_{\mathrm{T}}\nu_{\mathrm{A}}^{2} - \lambda - 2\mu\right)\varepsilon_{11}^{\mathrm{A}} - \left(2\nu_{\mathrm{A}}k_{\mathrm{T}} - \lambda\right)\varepsilon_{22}^{\mathrm{A}} - \left(2\nu_{\mathrm{A}}k_{\mathrm{T}} - \lambda\right)\varepsilon_{33}^{\mathrm{A}},
\end{aligned} \tag{15.13}$$

$$\begin{aligned}
&\left[\left(2\nu_{\mathrm{A}}k_{\mathrm{T}} - \lambda\right)S_{1111} + \left(k_{\mathrm{T}} + \mu_{\mathrm{T}} - \lambda - 2\mu\right)S_{2211} + \left(k_{\mathrm{T}} - \mu_{\mathrm{T}} - \lambda\right)S_{3311}\right]\varepsilon_{11}^{\mathrm{T}}\\
&+\left[\left(2\nu_{\mathrm{A}}k_{\mathrm{T}} - \lambda\right)S_{1122} + \left(k_{\mathrm{T}} + \mu_{\mathrm{T}} - \lambda - 2\mu\right)S_{2222} + \left(k_{\mathrm{T}} - \mu_{\mathrm{T}} - \lambda\right)S_{3322}\right]\varepsilon_{22}^{\mathrm{T}}\\
&+\left[\left(2\nu_{\mathrm{A}}k_{\mathrm{T}} - \lambda\right)S_{1133} + \left(k_{\mathrm{T}} + \mu_{\mathrm{T}} - \lambda - 2\mu\right)S_{2233} + \left(k_{\mathrm{T}} - \mu_{\mathrm{T}} - \lambda\right)S_{3333}\right]\varepsilon_{33}^{\mathrm{T}}\\
&+\lambda\left(\varepsilon_{11}^{\mathrm{T}} + \varepsilon_{22}^{\mathrm{T}} + \varepsilon_{33}^{\mathrm{T}}\right) + 2\mu\varepsilon_{22}^{\mathrm{T}}\\
&\quad = -\left(2\nu_{\mathrm{A}}k_{\mathrm{T}} - \lambda\right)\varepsilon_{11}^{\mathrm{A}} - \left(k_{\mathrm{T}} + \mu_{\mathrm{T}} - \lambda - 2\mu\right)\varepsilon_{22}^{\mathrm{A}} - \left(k_{\mathrm{T}} - \mu_{\mathrm{T}} - \lambda\right)\varepsilon_{33}^{\mathrm{A}},
\end{aligned} \tag{15.14}$$

$$\begin{aligned}
&\left[\left(2\nu_{\mathrm{A}}k_{\mathrm{T}} - \lambda\right)S_{1111} + \left(k_{\mathrm{T}} - \mu_{\mathrm{T}} - \lambda\right)S_{2211} + \left(k_{\mathrm{T}} + \mu_{\mathrm{T}} - \lambda - 2\mu\right)S_{3311}\right]\varepsilon_{11}^{\mathrm{T}}\\
&+\left[\left(2\nu_{\mathrm{A}}k_{\mathrm{T}} - \lambda\right)S_{1122} + \left(k_{\mathrm{T}} - \mu_{\mathrm{T}} - \lambda\right)S_{2222} + \left(k_{\mathrm{T}} + \mu_{\mathrm{T}} - \lambda - 2\mu\right)S_{3322}\right]\varepsilon_{22}^{\mathrm{T}}\\
&+\left[\left(2\nu_{\mathrm{A}}k_{\mathrm{T}} - \lambda\right)S_{1133} + \left(k_{\mathrm{T}} - \mu_{\mathrm{T}} - \lambda\right)S_{2233} + \left(k_{\mathrm{T}} + \mu_{\mathrm{T}} - \lambda - 2\mu\right)S_{3333}\right]\varepsilon_{33}^{\mathrm{T}}\\
&+\lambda\left(\varepsilon_{11}^{\mathrm{T}} + \varepsilon_{22}^{\mathrm{T}} + \varepsilon_{33}^{\mathrm{T}}\right) + 2\mu\varepsilon_{33}^{\mathrm{T}}\\
&\quad = -\left(2\nu_{A}k_{T} - \lambda\right)\varepsilon_{11}^{A} - \left(k_{T} - \mu_{T} - \lambda\right)\varepsilon_{22}^{A} - \left(k_{T} + \mu_{T} - \lambda - 2\mu\right)\varepsilon_{33}^{A},
\end{aligned} \tag{15.15}$$

$$\varepsilon_{12}^{\mathrm{T}} = \frac{\varepsilon_{12}^{\mathrm{A}}}{\dfrac{\mu}{\mu - \mu_{\mathrm{A}}} - 2S_{1212}},\quad \varepsilon_{13}^{\mathrm{T}} = \frac{\varepsilon_{13}^{\mathrm{A}}}{\dfrac{\mu}{\mu - \mu_{\mathrm{A}}} - 2S_{1313}},\quad \varepsilon_{23}^{\mathrm{T}} = \frac{\varepsilon_{23}^{\mathrm{A}}}{\dfrac{\mu}{\mu - \mu_{\mathrm{T}}} - 2S_{2323}}. \tag{15.16}$$

15.4 Far-field Displacement Distribution

Consider an isolated ellipsoidal inclusion having axes a, b, c perfectly bonded to an infinite isotropic matrix having elastic properties λ, μ. Eshelby [18] derived the following expression for the perturbation displacement field at large distances from the ellipsoidal inclusion:

$$u_{i}^{\mathrm{C}} = \frac{abc}{6(1-\nu)}\left[(1-2\nu)\left(\varepsilon_{ik}^{\mathrm{T}}l_{k} + \varepsilon_{ki}^{\mathrm{T}}l_{k} - \varepsilon_{kk}^{\mathrm{T}}l_{i}\right) + 3\varepsilon_{jk}^{\mathrm{T}}l_{i}l_{j}l_{k}\right]\frac{1}{r^{2}}, \tag{15.17}$$

where $\nu = \frac{1}{2}\lambda/(\lambda+\mu)$ is Poisson's ratio for the matrix and where l_i are direction cosines of the point $r = (x_1, x_2, x_3)$ relative to the origin of spherical coordinates (r, θ, ϕ) defined by

$$l_1 = \sin\theta\cos\phi\,, \quad l_2 = \sin\theta\sin\phi\,, \quad l_3 = \cos\theta\,. \tag{15.18}$$

Eshelby [18] justified the use of (15.17) for the case of anisotropic inclusions embedded in an isotropic matrix. For the case where $\varepsilon_{11}^{\mathrm{T}}$, $\varepsilon_{22}^{\mathrm{T}}$, $\varepsilon_{33}^{\mathrm{T}}$ are the only nonzero strains, (15.17) may be written, for $k = 1, 2, 3$ (no summation over repeated suffices), as

$$u_k^{\mathrm{C}} = \frac{abc}{6(1-\nu)}\left[(1-2\nu)\left(2\varepsilon_{kk}^{\mathrm{T}} - \varepsilon_{11}^{\mathrm{T}} - \varepsilon_{22}^{\mathrm{T}} - \varepsilon_{33}^{\mathrm{T}}\right) + 3\varepsilon_{11}^{\mathrm{T}} l_1^2 + 3\varepsilon_{22}^{\mathrm{T}} l_2^2 + 3\varepsilon_{33}^{\mathrm{T}} l_3^2\right]\frac{l_k}{r^2}. \tag{15.19}$$

For the case when $\varepsilon_{12}^{\mathrm{T}}$ is the only nonzero strain, (15.17) may be written as

$$u_1^{\mathrm{C}} = \frac{abc}{3(1-\nu)}\left[1-2\nu+3l_1^2\right]\varepsilon_{12}^{\mathrm{T}}\frac{l_2}{r^2}, \quad u_2^{\mathrm{C}} = \frac{abc}{3(1-\nu)}\left[1-2\nu+3l_2^2\right]\varepsilon_{12}^{\mathrm{T}}\frac{l_1}{r^2},$$
$$u_3^{\mathrm{C}} = \frac{abc}{1-\nu}\varepsilon_{12}^{\mathrm{T}}\frac{l_1 l_2 l_3}{r^2}. \tag{15.20}$$

For the case when $\varepsilon_{13}^{\mathrm{T}}$ is the only nonzero strain

$$u_1^{\mathrm{C}} = \frac{abc}{3(1-\nu)}\left[1-2\nu+3l_1^2\right]\varepsilon_{13}^{\mathrm{T}}\frac{l_3}{r^2}, \quad u_2^{\mathrm{C}} = \frac{abc}{1-\nu}\varepsilon_{13}^{\mathrm{T}}\frac{l_1 l_2 l_3}{r^2},$$
$$u_3^{\mathrm{C}} = \frac{abc}{3(1-\nu)}\left[1-2\nu+3l_3^2\right]\varepsilon_{13}^{\mathrm{T}}\frac{l_1}{r^2}. \tag{15.21}$$

and for the case when $\varepsilon_{23}^{\mathrm{T}}$ is the only nonzero strain

$$u_1^{\mathrm{C}} = \frac{abc}{1-\nu}\varepsilon_{23}^{\mathrm{T}}\frac{l_1 l_2 l_3}{r^2}, \quad u_2^{\mathrm{C}} = \frac{abc}{3(1-\nu)}\left[1-2\nu+3l_2^2\right]\varepsilon_{23}^{\mathrm{T}}\frac{l_3}{r^2},$$
$$u_3^{\mathrm{C}} = \frac{abc}{3(1-\nu)}\left[1-2\nu+3l_3^2\right]\varepsilon_{23}^{\mathrm{T}}\frac{l_2}{r^2}. \tag{15.22}$$

The subsequent analysis in this chapter assumes that $b = c$ so that the inclusions are spheroidal. In addition, matrix properties are denoted by a suffix 'm' as there will not now be confusion with the tensor notation.

15.5 Estimating Shear Properties

On combining $(15.16)_1$ with (15.20), it follows that, for a strain field $\varepsilon_{12}^{\mathrm{A}}$ applied to a single transversely isotropic spheroid of type i embedded in an infinite isotropic matrix, the far-field displacement distribution has the form

$$u_1^{\mathrm{C}} = \frac{a_i b_i^2}{3(1-\nu_{\mathrm{m}})}\left[1-2\nu_{\mathrm{m}}+3l_1^2\right]\frac{\varepsilon_{12}^{\mathrm{A}}}{\dfrac{\mu_{\mathrm{m}}}{\mu_{\mathrm{m}}-\mu_{\mathrm{A}}^i} - 2S_{1212}}\frac{l_2}{r^2}, \tag{15.23}$$

$$u_2^{\mathrm{C}} = \frac{a_i b_i^2}{3(1-\nu_{\mathrm{m}})}\left[1-2\nu_{\mathrm{m}}+3l_2^2\right]\frac{\varepsilon_{12}^{\mathrm{A}}}{\dfrac{\mu_{\mathrm{m}}}{\mu_{\mathrm{m}}-\mu_A^i}-2S_{1212}}\frac{l_1}{r^2}, \tag{15.24}$$

$$u_3^{\mathrm{C}} = \frac{a_i b_i^2}{1-\nu_{\mathrm{m}}}\frac{\varepsilon_{12}^{\mathrm{A}}}{\dfrac{\mu_{\mathrm{m}}}{\mu_{\mathrm{m}}-\mu_{\mathrm{A}}^i}-2S_{1212}}\frac{l_1 l_2 l_3}{r^2}. \tag{15.25}$$

As it is assumed that the aspect ratios of all types of spheroid in the cluster are identical and equal to that for the enclosing spheroid, Maxwell's methodology asserts that the far-field displacement distribution for a cluster of N spheroidal inclusions is given by

$$u_1^{\mathrm{C}} = \frac{1-2\nu_{\mathrm{m}}+3l_1^2}{3(1-\nu_{\mathrm{m}})}\frac{l_2}{r^2}\varepsilon_{12}^{\mathrm{A}}\sum_{i=1}^{N}\frac{n_i a_i b_i^2}{\dfrac{\mu_{\mathrm{m}}}{\mu_{\mathrm{m}}-\mu_{\mathrm{A}}^i}-2S_{1212}}, \tag{15.26}$$

$$u_2^{\mathrm{C}} = \frac{1-2\nu_{\mathrm{m}}+3l_2^2}{3(1-\nu_{\mathrm{m}})}\frac{l_1}{r^2}\varepsilon_{12}^{\mathrm{A}}\sum_{i=1}^{N}\frac{n_i a_i b_i^2}{\dfrac{\mu_{\mathrm{m}}}{\mu_{\mathrm{m}}-\mu_{\mathrm{A}}^i}-2S_{1212}}, \tag{15.27}$$

$$u_3^{\mathrm{C}} = \frac{\varepsilon_{12}^{\mathrm{A}}}{1-\nu_{\mathrm{m}}}\frac{l_1 l_2 l_3}{r^2}\sum_{i=1}^{N}\frac{n_i a_i b_i^2}{\dfrac{\mu_{\mathrm{m}}}{\mu_{\mathrm{m}}-\mu_{\mathrm{A}}^i}-2S_{1212}}. \tag{15.28}$$

When the equivalent single spheroidal transversely isotropic inclusion representing the cluster of aligned spheroids in the matrix is subject to the same applied strain field, the far-field displacement distribution will have the form

$$u_1^{\mathrm{C}} = \frac{1-2\nu_{\mathrm{m}}+3l_1^2}{3(1-\nu_{\mathrm{m}})}\frac{l_2}{r^2}\varepsilon_{12}^{\mathrm{A}}\frac{\bar{a}\bar{b}^2}{\dfrac{\mu_{\mathrm{m}}}{\mu_{\mathrm{m}}-\mu_{\mathrm{A}}^{\mathrm{eff}}}-2S_{1212}}, \tag{15.29}$$

$$u_2^{\mathrm{C}} = \frac{1-2\nu_{\mathrm{m}}+3l_2^2}{3(1-\nu_{\mathrm{m}})}\frac{l_1}{r^2}\varepsilon_{12}^{\mathrm{A}}\frac{\bar{a}\bar{b}^2}{\dfrac{\mu_{\mathrm{m}}}{\mu_{\mathrm{m}}-\mu_{\mathrm{A}}^{\mathrm{eff}}}-2S_{1212}}, \tag{15.30}$$

$$u_3^{\mathrm{C}} = \frac{\varepsilon_{12}^{\mathrm{A}}}{1-\nu_{\mathrm{m}}}\frac{l_1 l_2 l_3}{r^2}\frac{\bar{a}\bar{b}^2}{\dfrac{\mu_{\mathrm{m}}}{\mu_{\mathrm{m}}-\mu_{\mathrm{A}}^{\mathrm{eff}}}-2S_{1212}}. \tag{15.31}$$

The application of Maxwell's methodology demands that the far-field displacements defined by (15.26)–(15.28) and (15.29)–(15.31) are identical, and this is leads to the following simple relationship that can be used to estimate the effective *axial* modulus μ_A^{eff} of the composite

$$\frac{1}{\dfrac{\mu_{\mathrm{m}}}{\mu_{\mathrm{A}}^{\mathrm{eff}}-\mu_{\mathrm{m}}}+2S_{1212}} = \sum_{i=1}^{N}\frac{V_{\mathrm{p}}^i}{\dfrac{\mu_{\mathrm{m}}}{\mu_{\mathrm{A}}^i-\mu_{\mathrm{m}}}+2S_{1212}}, \tag{15.32}$$

where the inclusion volume fractions $V_{\rm p}^i$ have been introduced using (15.1). As $S_{1212} = S_{1313}$, it follows that (15.32) can be obtained also from relations $(15.16)_2$ and (15.21), which are relevant when the applied strain field is given by $\varepsilon_{13}^{\rm A}$. However, on using $(15.16)_3$ in conjunction with (15.22), it can be shown that, when the applied strain field is given by $\varepsilon_{23}^{\rm A}$, the effective *transverse* shear modulus $\mu_{\rm t}^{\rm eff}$ may be found from the simple relation

$$\frac{1}{\dfrac{\mu_{\rm m}}{\mu_{\rm t}^{\rm eff} - \mu_{\rm m}} + 2S_{2323}} = \sum_{i=1}^{N} \frac{V_{\rm p}^i}{\dfrac{\mu_{\rm m}}{\mu_{\rm t}^i - \mu_{\rm m}} + 2S_{2323}}. \tag{15.33}$$

It can be shown, on using (15.1), that the results (15.32) and (15.33) may be also be expressed as the following simple 'mixtures' relationships:

$$\frac{1}{\mu_{\rm A}^{\rm eff} + \overset{*}{\mu}_{\rm A}} = \sum_{i=1}^{N} \frac{V_{\rm p}^i}{\mu_{\rm A}^i + \overset{*}{\mu}_{\rm A}} + \frac{V_{\rm m}}{\mu_{\rm m} + \overset{*}{\mu}_{\rm A}}, \quad \text{where} \quad \overset{*}{\mu}_{\rm A} = \left(\frac{1}{2S_{1212}} - 1\right)\mu_{\rm m}, \tag{15.34}$$

$$\frac{1}{\mu_{\rm T}^{\rm eff} + \overset{*}{\mu}_{\rm T}} = \sum_{i=1}^{N} \frac{V_{\rm p}^i}{\mu_{\rm T}^i + \overset{*}{\mu}_{\rm T}} + \frac{V_{\rm m}}{\mu_{\rm m} + \overset{*}{\mu}_{\rm T}}, \quad \text{where} \quad \overset{*}{\mu}_{\rm T} = \left(\frac{1}{2S_{2323}} - 1\right)\mu_{\rm m}. \tag{15.35}$$

15.6 Far-field Solution for Nonshear Case

For a strain field $\varepsilon_{11}^{\rm A}$, $\varepsilon_{22}^{\rm A}$ and $\varepsilon_{33}^{\rm A}$ applied to a single transversely isotropic spheroidal inclusion of type i embedded in an infinite matrix, it follows from (15.17) that the far-field displacement distribution has the form, for $k = 1, 2, 3$, (no summation over repeated suffices)

$$u_k^{\rm C} = \frac{a_i b_i^2}{6(1-\nu_{\rm m})} \left[\begin{array}{l} (1-2\nu_{\rm m})\left(2\varepsilon_{kk}^{{\rm T}(i)} - \varepsilon_{11}^{{\rm T}(i)} - \varepsilon_{22}^{{\rm T}(i)} - \varepsilon_{33}^{{\rm T}(i)}\right) \\ \qquad + 3\varepsilon_{11}^{{\rm T}(i)} l_1^2 + 3\varepsilon_{22}^{{\rm T}(i)} l_2^2 + 3\varepsilon_{33}^{{\rm T}(i)} l_3^2 \end{array} \right] \frac{l_k}{r^2}. \tag{15.36}$$

Maxwell's methodology asserts that the far-field displacement field for the various sets of spheroidal inclusions in the cluster are then given by, for $k = 1, 2, 3$,

$$u_k^{\rm C} = \frac{1}{6(1-\nu_{\rm m})} \sum_{i=1}^{N} a_i b_i^2 \left[\begin{array}{l} (1-2\nu_{\rm m})\left(2\varepsilon_{kk}^{{\rm T}(i)} - \varepsilon_{11}^{{\rm T}(i)} - \varepsilon_{22}^{{\rm T}(i)} - \varepsilon_{33}^{{\rm T}(i)}\right) \\ \qquad + 3\varepsilon_{11}^{{\rm T}(i)} l_1^2 + 3\varepsilon_{22}^{{\rm T}(i)} l_2^2 + 3\varepsilon_{33}^{{\rm T}(i)} l_3^2 \end{array} \right] \frac{l_k}{r^2}. \tag{15.37}$$

When the equivalent single spheroidal transversely isotropic inclusion representing the cluster of aligned spheroids is subject to the same applied strain field, the far-field displacement distribution will have the form, for $k = 1, 2, 3$,

$$u_k^{\rm C} = \frac{l_k \bar{a} \bar{b}^2}{6(1-\nu_{\rm m}) r^2} \left[\begin{array}{l} (1-2\nu_{\rm m})\left(2\bar{\varepsilon}_{kk}^{\rm T} - \bar{\varepsilon}_{11}^{\rm T} - \bar{\varepsilon}_{22}^{\rm T} - \bar{\varepsilon}_{33}^{\rm T}\right) \\ \qquad + 3\bar{\varepsilon}_{11}^{\rm T} l_1^2 + 3\bar{\varepsilon}_{22}^{\rm T} l_2^2 + 3\bar{\varepsilon}_{33}^{\rm T} l_3^2 \end{array} \right]. \tag{15.38}$$

The application of Maxwell's methodology demands that the far-field displacements defined by (15.37) and (15.38) are identical, and this leads to the following relationships that will be used to estimate the effective nonshear effective elastic constants of the composite

$$\bar{a}\bar{b}^2\bar{\varepsilon}_{11}^{\mathrm{T}}=\sum_{i=1}^{N}a_ib_i^2\,\varepsilon_{11}^{\mathrm{T}(i)},\quad \bar{a}\bar{b}^2\bar{\varepsilon}_{22}^{\mathrm{T}}=\sum_{i=1}^{N}a_ib_i^2\,\varepsilon_{22}^{\mathrm{T}(i)},\quad \bar{a}\bar{b}^2\bar{\varepsilon}_{33}^{\mathrm{T}}=\sum_{i=1}^{N}a_ib_i^2\,\varepsilon_{33}^{\mathrm{T}(i)}. \tag{15.39}$$

On using (15.1), these relationships may be written as

$$\bar{\varepsilon}_{11}^{\mathrm{T}}=\sum_{i=1}^{N}V_{\mathrm{p}}^{i}\,\varepsilon_{11}^{\mathrm{T}(i)},\quad \bar{\varepsilon}_{22}^{\mathrm{T}}=\sum_{i=1}^{N}V_{\mathrm{p}}^{i}\,\varepsilon_{22}^{\mathrm{T}(i)},\quad \bar{\varepsilon}_{33}^{\mathrm{T}}=\sum_{i=1}^{N}V_{\mathrm{p}}^{i}\,\varepsilon_{33}^{\mathrm{T}(i)}. \tag{15.40}$$

It is, thus, first necessary to determine the transformation strains $\varepsilon_{11}^{\mathrm{T}(i)}$, $\varepsilon_{22}^{\mathrm{T}(i)}$ and $\varepsilon_{33}^{\mathrm{T}(i)}$ associated with the applied strain field $\varepsilon_{11}^{\mathrm{A}}$, $\varepsilon_{22}^{\mathrm{A}}$ and $\varepsilon_{33}^{\mathrm{A}}$ for each type of inclusion in the cluster, and the strains $\bar{\varepsilon}_{11}^{\mathrm{T}}$, $\bar{\varepsilon}_{22}^{\mathrm{T}}$ and $\bar{\varepsilon}_{33}^{\mathrm{T}}$ associated with the equivalent single inclusion.

Consider the linear equations (15.13)–(15.15) for a single inclusion that must be solved for the transformation strains $\varepsilon_{11}^{\mathrm{T}}$, $\varepsilon_{22}^{\mathrm{T}}$ and $\varepsilon_{33}^{\mathrm{T}}$ in each inclusion of the cluster and in the effective medium representing the cluster. It can be shown that when $\varepsilon_{33}^{\mathrm{A}}=\varepsilon_{22}^{\mathrm{A}}$,

$$\begin{aligned}
\varepsilon_{11}^{\mathrm{T}} &= \left[(QR-PS)(S_{2222}+S_{2233})-(3k_{\mathrm{m}}+\mu_{\mathrm{m}})P-\mu_{\mathrm{m}}Q\right]\frac{\varepsilon_{11}^{\mathrm{A}}}{\Delta}\\
&\quad +2\left[(PS-QR)S_{1122}-(3k_{\mathrm{m}}+\mu_{\mathrm{m}})R-\mu_{\mathrm{m}}S\right]\frac{\varepsilon_{22}^{\mathrm{A}}}{\Delta},\\
\varepsilon_{22}^{\mathrm{T}} &= \left[(PS-QR)S_{2211}+\left(\tfrac{3}{2}k_{\mathrm{m}}-\mu_{\mathrm{m}}\right)P-\mu_{\mathrm{m}}Q\right]\frac{\varepsilon_{11}^{\mathrm{A}}}{\Delta}\\
&\quad +\left[(QR-PS)S_{1111}+(3k_{\mathrm{m}}-2\mu_{\mathrm{m}})R-2\mu_{\mathrm{m}}S\right]\frac{\varepsilon_{22}^{\mathrm{A}}}{\Delta}=\varepsilon_{33}^{\mathrm{T}},
\end{aligned} \tag{15.41}$$

and when $\varepsilon_{33}^{\mathrm{A}}=-\varepsilon_{22}^{\mathrm{A}}$,

$$\begin{aligned}
\varepsilon_{11}^{\mathrm{T}} &= \left[(QR-PS)(S_{2222}+S_{2233})-(3k_{\mathrm{m}}+\mu_{\mathrm{m}})P-\mu_{\mathrm{m}}Q\right]\frac{\varepsilon_{11}^{\mathrm{A}}}{\Delta},\\
\varepsilon_{22}^{\mathrm{T}} &= \left[(PS-QR)S_{2211}+\tfrac{1}{2}(3k_{\mathrm{m}}-2\mu_{\mathrm{m}})P-\mu_{\mathrm{m}}Q\right]\frac{\varepsilon_{11}^{\mathrm{A}}}{\Delta}-\frac{\varepsilon_{22}^{\mathrm{A}}}{\Lambda},\\
\varepsilon_{33}^{\mathrm{T}} &= \left[(PS-QR)S_{2211}+\tfrac{1}{2}(3k_{\mathrm{m}}-2\mu_{\mathrm{m}})P-\mu_{\mathrm{m}}Q\right]\frac{\varepsilon_{11}^{\mathrm{A}}}{\Delta}+\frac{\varepsilon_{22}^{\mathrm{A}}}{\Lambda},
\end{aligned} \tag{15.42}$$

where

$$\begin{aligned}
P &= E_{\mathrm{A}}+2\nu_{\mathrm{A}}(2\nu_{\mathrm{A}}-1)k_{\mathrm{T}}-2\mu_{\mathrm{m}},\\
Q &= 6\nu_{\mathrm{A}}k_{\mathrm{T}}-3k_{\mathrm{m}}+2\mu_{\mathrm{m}},\\
R &= (2\nu_{\mathrm{A}}-1)k_{\mathrm{T}}+\mu_{\mathrm{m}},\\
S &= 3k_{\mathrm{T}}-3k_{\mathrm{m}}-\mu_{\mathrm{m}}=Q-3R,
\end{aligned} \tag{15.43}$$

$$\Lambda = S_{2222}-S_{2233}+\frac{\mu_{\mathrm{m}}}{\mu_{\mathrm{T}}-\mu_{\mathrm{m}}}, \tag{15.44}$$

$$\Delta = L_1P+L_2Q+L_3R+\Psi(QR-PS)+9k_{\mathrm{m}}\mu_{\mathrm{m}}, \tag{15.45}$$

and where

$$\begin{aligned}
L_1 &= (3k_m + \mu_m)S_{1111} - (3k_m - 2\mu_m)S_{1122}\,,\\
L_2 &= \left[S_{1111} + 2(S_{1122} + S_{2211} + S_{2222} + S_{2233})\right]\mu_m\,,\\
L_3 &= 2(3k_m - 2\mu_m)S_{2211} - (3k_m + 4\mu_m)(S_{2222} + S_{2233})\,,\\
\Psi &= 2S_{1122}S_{2211} - S_{1111}(S_{2222} + S_{2233})\,.
\end{aligned} \tag{15.46}$$

15.7 Solving for Parameters Defining Properties of the Effective Medium

It is useful to consider two special loading cases, which reduce the complexity of the approach. The first case is for uniaxial axial applied strains whereas the second considers plane strain equibiaxial transverse loading.

15.7.1 Uniaxial Axial Loading

For the special case when ε_{11}^{A} is the only nonvanishing applied strain it follows from (15.41) or (15.42) that the strains $\varepsilon_{11}^{T(i)}$, $\varepsilon_{22}^{T(i)}$ and $\varepsilon_{33}^{T(i)}$ appearing in (15.40) are given by

$$\begin{aligned}
\varepsilon_{11}^{T(i)} &= \left[(Q_iR_i - P_iS_i)(S_{2222} + S_{2233}) - (3k_m + \mu_m)P_i - \mu_m Q_i\right]\frac{\varepsilon_{11}^{A}}{\Delta_i}\,,\\
\varepsilon_{22}^{T(i)} &= \left[(P_iS_i - Q_iR_i)S_{2211} + \frac{1}{2}(3k_m - 2\mu_m)P_i - \mu_m Q_i\right]\frac{\varepsilon_{11}^{A}}{\Delta_i} = \varepsilon_{33}^{T}\,,
\end{aligned} \tag{15.47}$$

where

$$\begin{aligned}
P_i &= E_A^i + 2\nu_A^i(2\nu_A^i - 1)k_T^i - 2\mu_m\,,\\
Q_i &= 6\nu_A^i k_T^i - 3k_m + 2\mu_m\,,\\
R_i &= (2\nu_A^i - 1)k_T^i + \mu_m\,,\\
S_i &= 3k_T^i - 3k_m - \mu_m = Q_i - 3R_i\,,
\end{aligned} \tag{15.48}$$

$$\Delta_i = L_1P_i + L_2Q_i + L_3R_i + \Psi(Q_iR_i - P_iS_i) + 9k_m\mu_m. \tag{15.49}$$

Similarly, for the effective composite

$$\begin{aligned}
\bar{\varepsilon}_{11}^{T} &= \left[(\bar{Q}\bar{R} - \bar{P}\bar{S})(S_{2222} + S_{2233}) - (3k_m + \mu_m)\bar{P} - \mu_m\bar{Q}\right]\frac{\varepsilon_{11}^{A}}{\bar{\Delta}}\,,\\
\bar{\varepsilon}_{22}^{T} &= \left[(\bar{P}\bar{S} - \bar{Q}\bar{R})S_{2211} + \frac{1}{2}(3k_m - 2\mu_m)\bar{P} - \mu_m\bar{Q}\right]\frac{\varepsilon_{11}^{A}}{\bar{\Delta}} = \bar{\varepsilon}_{33}^{T}\,,
\end{aligned} \tag{15.50}$$

where

$$\begin{aligned}
\bar{P} &= E_A^{eff} + 2\nu_A^{eff}(2\nu_A^{eff} - 1)k_T^{eff} - 2\mu_m\,,\\
\bar{Q} &= 6\nu_A^{eff}k_T^{eff} - 3k_m + 2\mu_m\,,\\
\bar{R} &= (2\nu_A^{eff} - 1)k_T^{eff} + \mu_m\,,\\
\bar{S} &= 3k_T^{eff} - 3k_m - \mu_m = \bar{Q} - 3\bar{R}\,,
\end{aligned} \tag{15.51}$$

$$\bar{\Delta} = L_1\bar{P} + L_2\bar{Q} + L_3\bar{R} + \Psi\left(\bar{Q}\bar{R} - \bar{P}\bar{S}\right) + 9k_{\mathrm{m}}\mu_{\mathrm{m}}. \tag{15.52}$$

It then follows from (15.40) that

$$\begin{aligned}
&\left[\left(\bar{Q}\bar{R} - \bar{P}\bar{S}\right)\left(S_{2222} + S_{2233}\right) - \left(3k_{\mathrm{m}} + \mu_{\mathrm{m}}\right)\bar{P} - \mu_{\mathrm{m}}\bar{Q}\right]\frac{1}{\bar{\Delta}} = A \\
&\qquad = \sum_{i=1}^{N}\left[\left(Q_iR_i - P_iS_i\right)\left(S_{2222} + S_{2233}\right) - \left(3k_{\mathrm{m}} + \mu_{\mathrm{m}}\right)P_i - \mu_{\mathrm{m}}Q_i\right]\frac{V_p^i}{\Delta_i}, \\
&\left[\left(\bar{Q}\bar{R} - \bar{P}\bar{S}\right)S_{2211} - \tfrac{1}{2}\left(3k_{\mathrm{m}} - 2\mu_{\mathrm{m}}\right)\bar{P} + \mu_{\mathrm{m}}\bar{Q}\right]\frac{1}{\bar{\Delta}} = B \\
&\qquad = \sum_{i=1}^{N}\left[\left(Q_iR_i - P_iS_i\right)S_{2211} - \tfrac{1}{2}\left(3k_{\mathrm{m}} - 2\mu_{\mathrm{m}}\right)P_i + \mu_{\mathrm{m}}Q_i\right]\frac{V_p^i}{\Delta_i}.
\end{aligned} \tag{15.53}$$

The values of the dimensionless parameters A and B are known as they can be calculated from inclusion and matrix parameters, and the volume fractions.

15.7.2 Plane-strain Equibiaxial Transverse Loading

For the special case when $\varepsilon_{11}^{\mathrm{A}} = 0$ and $\varepsilon_{22}^{\mathrm{A}} = \varepsilon_{33}^{\mathrm{A}}$, it follows from (15.41) that the strains $\varepsilon_{11}^{\mathrm{T}(i)}$, $\varepsilon_{22}^{\mathrm{T}(i)}$ and $\varepsilon_{33}^{\mathrm{T}(i)}$ appearing in (15.40) are given by

$$\begin{aligned}
\varepsilon_{11}^{\mathrm{T}(i)} &= 2\left[\left(P_iS_i - Q_iR_i\right)S_{1122} - \left(3k_{\mathrm{m}} + \mu_{\mathrm{m}}\right)R_i - \mu_{\mathrm{m}}S_i\right]\frac{\varepsilon_{22}^{\mathrm{A}}}{\Delta_i}, \\
\varepsilon_{22}^{\mathrm{T}(i)} &= \left[\left(Q_iR_i - P_iS_i\right)S_{1111} + \left(3k_{\mathrm{m}} - 2\mu_{\mathrm{m}}\right)R_i - 2\mu_{\mathrm{m}}S_i\right]\frac{\varepsilon_{22}^{\mathrm{A}}}{\Delta_i} = \varepsilon_{33}^{\mathrm{T}}.
\end{aligned} \tag{15.54}$$

Similarly, for the effective composite

$$\begin{aligned}
\bar{\varepsilon}_{11}^{\mathrm{T}} &= 2\left[\left(\bar{P}\bar{S} - \bar{Q}\bar{R}\right)S_{1122} - \left(3k_{\mathrm{m}} + \mu_{\mathrm{m}}\right)\bar{R} - \mu_{\mathrm{m}}\bar{S}\right]\frac{\varepsilon_{22}^{\mathrm{A}}}{\bar{\Delta}}, \\
\bar{\varepsilon}_{22}^{\mathrm{T}} &= \left[\left(\bar{Q}\bar{R} - \bar{P}\bar{S}\right)S_{1111} + \left(3k_{\mathrm{m}} - 2\mu_{\mathrm{m}}\right)\bar{R} - 2\mu_{\mathrm{m}}\bar{S}\right]\frac{\varepsilon_{22}^{\mathrm{A}}}{\bar{\Delta}} = \bar{\varepsilon}_{33}^{\mathrm{T}}.
\end{aligned} \tag{15.55}$$

As $S_i = Q_i - 3R_i$ and $\bar{S} = \bar{Q} - 3\bar{R}$, it then follows on using (15.40) that

$$\begin{aligned}
&\left[\left(\bar{Q}\bar{R} - \bar{P}\bar{S}\right)S_{1122} + \mu_{\mathrm{m}}\bar{Q} + \left(3k_{\mathrm{m}} - 2\mu_{\mathrm{m}}\right)\bar{R}\right]\frac{1}{\bar{\Delta}} = C \\
&\qquad = \sum_{i=1}^{N}\left[\left(Q_iR_i - P_iS_i\right)S_{1122} + \mu_{\mathrm{m}}Q_i + \left(3k_{\mathrm{m}} - 2\mu_{\mathrm{m}}\right)R_i\right]\frac{V_p^i}{\Delta_i}, \\
&\left[\left(\bar{Q}\bar{R} - \bar{P}\bar{S}\right)S_{1111} - 2\mu_{\mathrm{m}}\bar{Q} + \left(3k_{\mathrm{m}} + 4\mu_{\mathrm{m}}\right)\bar{R}\right]\frac{1}{\bar{\Delta}} = D \\
&\qquad = \sum_{i=1}^{N}\left[\left(Q_iR_i - P_iS_i\right)S_{1111} - 2\mu_{\mathrm{m}}Q_i + \left(3k_{\mathrm{m}} + 4\mu_{\mathrm{m}}\right)R_i\right]\frac{V_p^i}{\Delta_i}.
\end{aligned} \tag{15.56}$$

The values of the dimensionless parameters C and D are again known as they can be calculated from inclusion and matrix parameters, and the volume fractions. It is worth noting that if relation (15.42) is used for shear states, then no new information is provided as results obtained correspond exactly to results already given.

15.7.3 Defining a Soluble Set of Nonlinear Algebraic Equations

The problem that now remains is to find values of $\bar{P}$, $\bar{Q}$, $\bar{R}$ and $\bar{S} = \bar{Q} - 3\bar{R}$ satisfying the nonlinear relations (15.53) and (15.56), which are written as

$$(S_{2222} + S_{2233})\Omega - (3k_{\rm m} + \mu_{\rm m})\bar{P} - \mu_{\rm m}\bar{Q} = A\bar{\Delta}, \tag{15.57}$$

$$S_{2211}\Omega - \frac{1}{2}(3k_{\rm m} - 2\mu_{\rm m})\bar{P} + \mu_{\rm m}\bar{Q} = B\bar{\Delta}, \tag{15.58}$$

$$S_{1122}\Omega + \mu_{\rm m}\bar{Q} + (3k_{\rm m} - 2\mu_{\rm m})\bar{R} = C\bar{\Delta}, \tag{15.59}$$

$$S_{1111}\Omega - 2\mu_{\rm m}\bar{Q} + (3k_{\rm m} + 4\mu_{\rm m})\bar{R} = D\bar{\Delta}, \tag{15.60}$$

where

$$\Omega = \bar{Q}\bar{R} - \bar{P}\bar{S} = \bar{Q}\bar{R} - \bar{P}\bar{Q} + 3\bar{P}\bar{R}, \tag{15.61}$$

and where, from (15.52),

$$\bar{\Delta} = L_1\bar{P} + L_2\bar{Q} + L_3\bar{R} + \Psi\Omega + 9k_{\rm m}\mu_{\rm m}. \tag{15.62}$$

It can be shown that

$$(3k_{\rm m} - 2\mu_{\rm m})A - 2(3k_{\rm m} + \mu_{\rm m})B + (3k_{\rm m} + 4\mu_{\rm m})C - (3k_{\rm m} - 2\mu_{\rm m})D \equiv 0, \tag{15.63}$$

$$\begin{aligned}(3k_{\rm m} - 2\mu_{\rm m})(S_{2222} + S_{2233}) - 2(3k_{\rm m} + \mu_{\rm m})S_{2211}\\ + (3k_{\rm m} + 4\mu_{\rm m})S_{1122} - (3k_{\rm m} - 2\mu_{\rm m})S_{1111} \equiv 0,\end{aligned} \tag{15.64}$$

$$\begin{aligned}&(3k_{\rm m} - 2\mu_{\rm m})\left[-(3k_{\rm m} + \mu_{\rm m})\bar{P} - \mu_{\rm m}\bar{Q}\right]\\ &-2(3k_{\rm m} + \mu_{\rm m})\left[-\tfrac{1}{2}(3k_{\rm m} - 2\mu_{\rm m})\bar{P} + \mu_{\rm m}\bar{Q}\right]\\ &+(3k_{\rm m} + 4\mu_{\rm m})\left[\mu_{\rm m}\bar{Q} + (3k_{\rm m} - 2\mu_{\rm m})\bar{R}\right]\\ &-(3k_{\rm m} - 2\mu_{\rm m})\left[-2\mu_{\rm m}\bar{Q} + (3k_{\rm m} + 4\mu_{\rm m})\bar{R}\right] \equiv 0.\end{aligned} \tag{15.65}$$

Relations (15.63)–(15.65) show that the four equations (15.57)–(15.60) are linearly dependent. It should be noted that (15.64) is an alternative to the last of relations (15.12). The following three independent equations are derived from (15.58)–(15.60) and (15.62), which are then to be solved for the unknowns $\bar{P}$, $\bar{Q}$ and $\bar{R}$ in terms of Ω defined by (15.61), which is a nonlinear function of $\bar{P}$, $\bar{Q}$ and $\bar{R}$,

$$\left(L_1 + \frac{3k_{\rm m} - 2\mu_{\rm m}}{2B}\right)\bar{P} + \left(L_2 - \frac{\mu_{\rm m}}{B}\right)\bar{Q} + L_3\bar{R} = -\left(\Psi - \frac{S_{2211}}{B}\right)\Omega - 9k_{\rm m}\mu_{\rm m}, \tag{15.66}$$

$$L_1\bar{P} + \left(L_2 - \frac{\mu_{\rm m}}{C}\right)\bar{Q} + \left(L_3 - \frac{3k_{\rm m} - 2\mu_{\rm m}}{C}\right)\bar{R} = -\left(\Psi - \frac{S_{1122}}{C}\right)\Omega - 9k_{\rm m}\mu_{\rm m}, \tag{15.67}$$

$$L_1\overline{P}+\left(L_2+\frac{2\mu_m}{D}\right)\overline{Q}+\left(L_3-\frac{3k_m+4\mu_m}{D}\right)\overline{R}=-\left(\Psi-\frac{S_{1111}}{D}\right)\Omega-9k_m\mu_m. \quad (15.68)$$

It follows from (15.53) and (15.56) that

$$\begin{aligned}
A&=(S_{2222}+S_{2233})W-(3k_m+\mu_m)X-\mu_m Y,\\
B&=S_{2211}W-\frac{1}{2}(3k_m-2\mu_m)X+\mu_m Y,\\
C&=S_{1122}W+\mu_m Y+(3k_m-2\mu_m)Z,\\
D&=S_{1111}W-2\mu_m Y+(3k_m+4\mu_m)Z,
\end{aligned} \quad (15.69)$$

where W, X, Y and Z are known constants defined by

$$W=\sum_{i=1}^{N}V_p^i\frac{Q_iR_i-P_iS_i}{\Delta_i},\quad X=\sum_{i=1}^{N}V_p^i\frac{P_i}{\Delta_i},\quad Y=\sum_{i=1}^{N}V_p^i\frac{Q_i}{\Delta_i},\quad Z=\sum_{i=1}^{N}V_p^i\frac{R_i}{\Delta_i}. \quad (15.70)$$

Following a great deal of complex and laborious algebra (see Appendix E for details), (15.66)–(15.68) may be expressed as

$$\Phi\overline{P}=(\beta_1\Psi+\alpha_1)\hat{\Omega}-\beta_1, \quad (15.71)$$

$$\Phi\overline{Q}=(\beta_2\Psi+\alpha_2)\hat{\Omega}-\beta_2, \quad (15.72)$$

$$\Phi\overline{R}=(\beta_3\Psi+\alpha_3)\hat{\Omega}-\beta_3, \quad (15.73)$$

where $\Phi=L_1X+L_2Y+L_3Z+2\Psi W-1,$ (15.74)

and where the dimensionless constant Ψ is given by (15.46). Using (15.61), $\hat{\Omega}$ is defined by

$$\hat{\Omega}=\frac{\Omega}{9k_m\mu_m}=\frac{\overline{Q}\overline{R}-\overline{P}\overline{Q}+3\overline{P}\overline{R}}{9k_m\mu_m}, \quad (15.75)$$

and the remaining coefficients in (15.71)–(15.73) are defined by

$$\begin{aligned}
\alpha_1&=2\mu_m(S_{2211}+S_{2222}+S_{2233})\Phi,\\
\beta_1&=9k_m\mu_m X-2\mu_m(S_{2211}+S_{2222}+S_{2233})W,\\
\alpha_2&=-\lambda\Phi,\\
\beta_2&=9k_m\mu_m Y+\lambda W,\\
\alpha_3&=-\mu_m(S_{1111}+2S_{1122})\Phi,\\
\beta_3&=9k_m\mu_m Z+\mu_m(S_{1111}+2S_{1122})W,
\end{aligned} \quad (15.76)$$

where

$$\begin{aligned}
\lambda&=(3k_m+4\mu_m)S_{1122}-(3k_m-2\mu_m)S_{1111}\\
&=2(3k_m+\mu_m)S_{2211}-(3k_m-2\mu_m)(S_{2222}+S_{2233}).
\end{aligned} \quad (15.77)$$

15.8 Determination of Effective Composite Properties

The elimination of the parameters $\overline{P}$, $\overline{Q}$ and $\overline{R}$ using (15.71)–(15.73) and (15.75) leads to the following quadratic equation for $\hat{\Omega}$

$$\alpha\hat{\Omega}^2 - \beta\hat{\Omega} - \gamma = 0, \tag{15.78}$$

$$\text{where} \quad \begin{cases} \alpha = \Psi\beta + \Psi^2\gamma, \\ \beta = \{-(\alpha_1\beta_2 + \alpha_2\beta_1) + (\alpha_2\beta_3 + \alpha_3\beta_2) + 3(\alpha_1\beta_3 + \alpha_3\beta_1)\} \\ \qquad + (-\alpha_1\alpha_2 + \alpha_2\alpha_3 + 3\alpha_1\alpha_3)/\Psi - 2\Psi\gamma, \\ \gamma = \beta_1\beta_2 - \beta_2\beta_3 - 3\beta_1\beta_3. \end{cases} \tag{15.79}$$

The solutions of (15.78) are given by

$$\hat{\Omega} = \frac{\Omega}{9k_m\mu_m} = \frac{1}{2\alpha}\left[\beta \pm \sqrt{\beta^2 + 4\alpha\gamma}\,\right]. \tag{15.80}$$

It has been noted from numerical calculations that the solutions for $\hat{\Omega}$ are either equal or they differ in value such that one of the roots leads to a zero value for $\overline{\Delta}$, indicating that the solutions (15.50) and (15.55) for special cases break down when applied to the single effective spheroid that represents the composite. When one of the roots leads to the result $\overline{\Delta} = 0$, the required value of $\hat{\Omega}$ is that value given by the other root of (15.80). It has also been observed numerically that the root leading to the result $\overline{\Delta} = 0$ does not depend on any of the elastic properties of the inclusions. For the case of cylinders, it has been observed that the roots of (15.80) are always equal. These characteristics of the solution strongly suggest that simpler analytical solutions should exist.

It can be shown that

$$\Phi\overline{\Delta} = \left[(L_1\beta_1 + L_2\beta_2 + L_3\beta_3) - 9k_m\mu_m\Phi\right]\left(\Psi\hat{\Omega} - 1\right), \tag{15.81}$$

indicating that the solution $\hat{\Omega}$ that leads to the situation $\overline{\Delta} = 0$ is simply

$$\hat{\Omega} = \frac{\Omega}{9k_m\mu_m} = \frac{1}{\Psi}. \tag{15.82}$$

From (15.46), it is clear that Ψ is independent of inclusion elastic properties, as observed numerically. It can then be shown that the required solution of (15.78) such that $\overline{\Delta} \neq 0$ is given by

$$\hat{\Omega} = \frac{\Omega}{9k_m\mu_m} = -\frac{\Psi\gamma}{\alpha} = \frac{\beta}{\alpha} - \frac{1}{\Psi}. \tag{15.83}$$

It should be noted that the quantity Φ defined by (15.74) may also be expressed in the form

$$\Phi = \sqrt{\frac{3\alpha_1\alpha_3 - \alpha_1\alpha_2 + \alpha_2\alpha_3}{9k_m\mu_m\Psi}}. \tag{15.84}$$

It follows from (15.72) and (15.73) and the relation $\overline{S} = \overline{Q} - 3\overline{R}$ that

$$\Phi\overline{S} = \Phi\left(\overline{Q} - 3\overline{R}\right) = \left[(\beta_2 - 3\beta_3)\Psi + \alpha_2 - 3\alpha_3\right]\hat{\Omega} - (\beta_2 - 3\beta_3). \tag{15.85}$$

It is now possible to calculate the required solution $\overline{P}$, $\overline{Q}$, $\overline{R}$ and $\overline{S}$ of the nonlinear algebraic equations (15.66)–(15.68) and (15.61) using (15.71)–(15.73), (15.83) and (15.85). Relations (15.51) are then used to derive the following relationships for the effective nonshear properties of the composite

$$\begin{aligned} k_{\mathrm{T}}^{\mathrm{eff}} &= \tfrac{1}{3}\left(\overline{S}+3k_{\mathrm{m}}+\mu_{\mathrm{m}}\right), \\ \nu_{\mathrm{A}}^{\mathrm{eff}} &= \frac{1}{2}\left(\frac{\overline{R}-\mu_{\mathrm{m}}}{k_{\mathrm{T}}^{\mathrm{eff}}}+1\right) = \frac{\overline{Q}+3k_{\mathrm{m}}-2\mu_{\mathrm{m}}}{6k_{\mathrm{T}}^{\mathrm{eff}}}, \\ E_{\mathrm{A}}^{\mathrm{eff}} &= \overline{P}+2\nu_{\mathrm{A}}^{\mathrm{eff}}\left(1-2\nu_{\mathrm{A}}^{\mathrm{eff}}\right)k_{\mathrm{T}}^{\mathrm{eff}}+2\mu_{\mathrm{m}}. \end{aligned} \tag{15.86}$$

The corresponding shear properties $\mu_{\mathrm{A}}^{\mathrm{eff}}$ and $\mu_{\mathrm{T}}^{\mathrm{eff}}$ have already been obtained and are given by (15.34) and (15.35). The corresponding values of the transverse Young's modulus $E_{\mathrm{T}}^{\mathrm{eff}}$ and transverse Poisson's ratio $\nu_{\mathrm{T}}^{\mathrm{eff}}$ are obtained using (15.10).

It has, thus, been shown that it is possible to apply the analysis of Eshelby [18, 19] together with Maxwell's methodology [13] to develop a method of estimating the effective elastic properties of a multiphase distribution of aligned spheroidal transversely isotropic inclusions having the same aspect ratio that reinforce an isotropic matrix. The effective elastic properties may be estimated using the results (15.34), (15.35) and (15.86), and the various associated relationships that have been derived.

15.9 Composites Reinforced with Isotropic Spherical Inclusions

It is useful to consider the special case when the reinforcing inclusions are spherical and isotropic, and distributed so that the composite is also isotropic. Rather than developing the required solution using (15.71)–(15.73), it is simpler to derive the results using the following method. When the inclusions are isotropic

$$\begin{aligned} &E_{\mathrm{A}}^{i}=E_{\mathrm{T}}^{i}=E_{\mathrm{p}}^{i}=2\mu_{\mathrm{p}}^{i}\left(1+\nu_{\mathrm{p}}^{i}\right), && \mu_{\mathrm{A}}^{i}=\mu_{\mathrm{T}}^{i}=\mu_{\mathrm{p}}^{i}, && \nu_{\mathrm{A}}^{i}=\nu_{\mathrm{T}}^{i}=\nu_{\mathrm{p}}^{i}, \\ &E_{\mathrm{A}}^{\mathrm{eff}}=E_{\mathrm{T}}^{\mathrm{eff}}=E_{\mathrm{p}}^{\mathrm{eff}}=2\mu_{\mathrm{p}}^{\mathrm{eff}}\left(1+\nu_{\mathrm{p}}^{\mathrm{eff}}\right), && \mu_{\mathrm{A}}^{\mathrm{eff}}=\mu_{\mathrm{T}}^{\mathrm{eff}}=\mu_{\mathrm{p}}^{\mathrm{eff}}, && \nu_{\mathrm{A}}^{\mathrm{eff}}=\nu_{\mathrm{T}}^{\mathrm{eff}}=\nu_{\mathrm{p}}^{\mathrm{eff}}. \end{aligned} \tag{15.87}$$

The bulk modulus k_{p} and plane strain bulk modulus k_{T} are now introduced, defined by

$$k_{\mathrm{p}}=\frac{E_{\mathrm{p}}}{3\left(1-2\nu_{\mathrm{p}}\right)}=\frac{2\mu_{\mathrm{p}}\left(1+\nu_{\mathrm{p}}\right)}{3\left(1-2\nu_{\mathrm{p}}\right)}, \qquad k_{\mathrm{T}}=\frac{\mu_{\mathrm{p}}}{1-2\nu_{\mathrm{p}}}=\frac{3k_{\mathrm{p}}}{2\left(1+\nu_{\mathrm{p}}\right)}=k_{\mathrm{p}}+\frac{1}{3}\mu_{\mathrm{p}}. \tag{15.88}$$

It then follows from (15.48) and (15.51) that

$$\begin{aligned} P_i &= 2\left(\mu_{\mathrm{p}}^{i}-\mu_{\mathrm{m}}\right), & \overline{P} &= 2\left(\mu_{\mathrm{p}}^{\mathrm{eff}}-\mu_{\mathrm{m}}\right), \\ Q_i &= 3\left(k_{\mathrm{p}}^{i}-k_{\mathrm{m}}\right)-2\left(\mu_{\mathrm{p}}^{i}-\mu_{\mathrm{m}}\right), & \overline{Q} &= 3\left(k_{\mathrm{p}}^{\mathrm{eff}}-k_{\mathrm{m}}\right)-2\left(\mu_{\mathrm{p}}^{\mathrm{eff}}-\mu_{\mathrm{m}}\right), \\ R_i &= -\left(\mu_{\mathrm{p}}^{i}-\mu_{\mathrm{m}}\right), & \overline{R} &= -\left(\mu_{\mathrm{p}}^{\mathrm{eff}}-\mu_{\mathrm{m}}\right), \\ S_i &= 3\left(k_{\mathrm{p}}^{i}-k_{\mathrm{m}}\right)+\left(\mu_{\mathrm{p}}^{i}-\mu_{\mathrm{m}}\right) & \overline{S} &= 3\left(k_{\mathrm{p}}^{\mathrm{eff}}-k_{\mathrm{m}}\right)+\left(\mu_{\mathrm{p}}^{\mathrm{eff}}-\mu_{\mathrm{m}}\right) \\ &= Q_i-3R_i, & &= \overline{Q}-3\overline{R}. \end{aligned} \tag{15.89}$$

It can be shown from (15.46) that

$$\Psi = 2S_{1122}\,S_{2211} - S_{1111}\left(S_{2222} + S_{2233}\right) = -\frac{8-10\nu_{\mathrm{m}}}{15(1-\nu_{\mathrm{m}})}\frac{1+\nu_{\mathrm{m}}}{3(1-\nu_{\mathrm{m}})}, \tag{15.90}$$

and it follows from (15.49) and (15.52) that

$$\begin{aligned}
\Delta_i &= \frac{1}{5\left(1-\nu_{\mathrm{m}}\right)^2}\left[\left(1+\nu_{\mathrm{m}}\right)k_{\mathrm{p}}^i + 2\left(1-2\nu_{\mathrm{m}}\right)k_{\mathrm{m}}\right]\left[\left(8-10\nu_{\mathrm{m}}\right)\mu_{\mathrm{p}}^i + \left(7-5\nu_{\mathrm{m}}\right)\mu_{\mathrm{m}}\right],\\
\bar{\Delta} &= \frac{1}{5\left(1-\nu_{\mathrm{m}}\right)^2}\left[\left(1+\nu_{\mathrm{m}}\right)k_{\mathrm{p}}^{\mathrm{eff}} + 2\left(1-2\nu_{\mathrm{m}}\right)k_{\mathrm{m}}\right]\left[\left(8-10\nu_{\mathrm{m}}\right)\mu_{\mathrm{p}}^{\mathrm{eff}} + \left(7-5\nu_{\mathrm{m}}\right)\mu_{\mathrm{m}}\right],
\end{aligned} \tag{15.91}$$

and from (15.41) that

$$\begin{aligned}
\varepsilon_{11}^{\mathrm{T}(i)} = \varepsilon_{22}^{\mathrm{T}(i)} = \varepsilon_{33}^{\mathrm{T}(i)} &= -\frac{3\left(k_{\mathrm{p}}^i - k_{\mathrm{m}}\right)}{5\left(1-\nu_{\mathrm{m}}\right)}\left[\left(8-10\nu_{\mathrm{m}}\right)\mu_{\mathrm{p}}^i + \left(7-5\nu_{\mathrm{m}}\right)\mu_{\mathrm{m}}\right]\frac{\varepsilon}{\Delta_i},\\
\bar{\varepsilon}_{11}^{\mathrm{T}} = \bar{\varepsilon}_{22}^{\mathrm{T}} = \bar{\varepsilon}_{33}^{\mathrm{T}} &= -\frac{3\left(k_{\mathrm{p}}^{\mathrm{eff}} - k_{\mathrm{m}}\right)}{5\left(1-\nu_{\mathrm{m}}\right)}\left[\left(8-10\nu_{\mathrm{m}}\right)\mu_{\mathrm{p}}^{\mathrm{eff}} + \left(7-5\nu_{\mathrm{m}}\right)\mu_{\mathrm{m}}\right]\frac{\varepsilon}{\bar{\Delta}},
\end{aligned} \tag{15.92}$$

where ε is the equiaxial applied strain. The substitution of (15.91) into (15.92), using the relation $\left(1-2\nu_{\mathrm{m}}\right)k_{\mathrm{m}} = \frac{2}{3}\left(1+\nu_{\mathrm{m}}\right)\mu_{\mathrm{m}}$, leads to

$$\begin{aligned}
\varepsilon_{11}^{\mathrm{T}(i)} = \varepsilon_{22}^{\mathrm{T}(i)} = \varepsilon_{33}^{\mathrm{T}(i)} &= -3\,\frac{1-\nu_{\mathrm{m}}}{1+\nu_{\mathrm{m}}}\,\frac{k_{\mathrm{p}}^i - k_{\mathrm{m}}}{k_{\mathrm{p}}^i + \frac{4}{3}\mu_{\mathrm{m}}}\,\varepsilon,\\
\bar{\varepsilon}_{11}^{\mathrm{T}} = \bar{\varepsilon}_{22}^{\mathrm{T}} = \bar{\varepsilon}_{33}^{\mathrm{T}} &= -3\,\frac{1-\nu_{\mathrm{m}}}{1+\nu_{\mathrm{m}}}\,\frac{k_{\mathrm{p}}^{\mathrm{eff}} - k_{\mathrm{m}}}{k_{\mathrm{p}}^{\mathrm{eff}} + \frac{4}{3}\mu_{\mathrm{m}}}\,\varepsilon.
\end{aligned} \tag{15.93}$$

On substituting (15.93) into (15.40), the bulk modulus of an isotropic particulate composite is obtained (see, for example, [17]), namely,

$$\frac{1}{k_{\mathrm{p}}^{\mathrm{eff}} + \frac{4}{3}\mu_{\mathrm{m}}} = \sum_{i=1}^{N}\frac{V_{\mathrm{p}}^i}{k_{\mathrm{p}}^i + \frac{4}{3}\mu_{\mathrm{m}}} + \frac{V_{\mathrm{m}}}{k_{\mathrm{m}} + \frac{4}{3}\mu_{\mathrm{m}}}. \tag{15.94}$$

The corresponding expression for the shear modulus is obtained from (15.34) or (15.35), leading to the following result

$$\begin{aligned}
&\frac{1}{\mu_{\mathrm{p}}^{\mathrm{eff}} + \mu_{\mathrm{m}}^*} = \sum_{i=1}^{N}\frac{V_{\mathrm{p}}^i}{\mu_{\mathrm{p}}^i + \mu_{\mathrm{m}}^*} + \frac{V_{\mathrm{m}}}{\mu_{\mathrm{m}} + \mu_{\mathrm{m}}^*},\\
&\text{where } \mu_{\mathrm{m}}^* = \frac{7-5\nu_{\mathrm{m}}}{8-10\nu_{\mathrm{m}}}\mu_{\mathrm{m}} = \frac{9k_{\mathrm{m}} + 8\mu_{\mathrm{m}}}{6\left(k_{\mathrm{m}} + 2\mu_{\mathrm{m}}\right)}\mu_{\mathrm{m}}.
\end{aligned} \tag{15.95}$$

The results (15.94) and (15.95) have been shown [17] to predict effective properties that are in close agreement with accurate results that have been given in the literature for volume fractions of practical interest. It was shown that the f.c.c. and b.c.c. packing configurations considered by Arridge [23] lead to bulk moduli that are very close together for particulate volume fractions in the range $0 < V_p < 0.6$ (see [17]).

Furthermore, the results for spherical isotropic particles obtained using Maxwell's methodology lie between the f.c.c. and b.c.c. estimates for volume fractions in the range $0 < V_p < 0.4$. For the case of a simple cubic array of spherical particles with volume fractions in the range $0 < V_p < 0.4$, it was shown that bounds for shear modulus, obtained by Cohen and Bergman [15] (see also Figure 15.4) using a Fourier representation of an integrodifferential equation for the displacement field, are very close to results obtained using Maxwell's methodology.

Further evidence that Maxwell's methodology can provide useful predictions for the bulk and shear moduli of isotropic composites reinforced with isotropic spherical particles is provided by the results of Sangani and Mo [24] for the extreme cases where the particles are either rigid for behave as cavities. Tables 15.1 and 15.2 compare the results of Sangani and Mo (labelled Accurate) with those based on the formulae (15.94) and (15.95) (labelled Maxwell). It can be seen from Table 15.1 that predictions for the effective bulk modulus based on Maxwell's methodology applied to the extreme case of rigid particles become more accurate as the matrix Poisson's ratio increases and becomes exact when Poisson's ratio is 0.5 (i.e. an incompressible matrix) for all particle volume fractions considered, whereas predictions for the effective shear modulus become much less accurate, especially as Poisson's ratio for the matrix tends to the incompressible value 0.5. Significant differences in shear modulus are seen for this extreme case when the volume fraction has the value 0.6. As to be expected at low volume fractions there is good agreement for all values of the matrix Poisson's ratio ≤ 0.4.

From Table 15.2 it can be seen that predictions for the effective bulk modulus based on Maxwell's methodology applied to the other extreme case of cavities become more accurate as the matrix Poisson's ratio increases and becomes the exact value 0 when Poisson's ratio is 0.5 for all particle volume fractions considered, whereas predictions for the effective shear modulus become much less accurate. At low volume fractions

Table 15.1 Effective properties for random distributions of rigid isotropic spherical particles.

V_p	ν_m	k_{eff}/k_m Accurate	k_{eff}/k_m Maxwell	k_{eff}/k_m Dipole	μ_{eff}/μ_m Accurate	μ_{eff}/μ_m Maxwell	μ_{eff}/μ_m Dipole
0.1	0.2	1.23	1.222	1.226	1.236	1.222	1.228
0.1	0.3	1.183	1.179	1.18	1.242	1.233	1.234
0.1	0.4	1.145	1.143	1.143	1.257	1.25	1.249
0.1	0.5	1.111	1.111	1.111	1.311	1.278	1.291
0.45	0.2	2.93	2.636	—	3.21	2.636	—
0.45	0.3	2.55	2.322	—	3.43	2.718	—
0.45	0.4	2.18	2.052	—	3.9	2.841	—
0.45	0.5	1.818	1.818	—	5.7	3.045	—
0.6	0.2	5.28	4	—	6	4	—
0.6	0.3	4.4	3.423	—	6.7	4.15	—
0.6	0.4	3.59	2.929	—	8.2	4.375	—
0.6	0.5	2.502	2.5	—	17.6	4.75	—

Table 15.2 Effective properties for random distributions of spherical cavities.

V_p	ν_m	k_{eff}/k_m Accurate	k_{eff}/k_m Maxwell	k_{eff}/k_m Dipole	μ_{eff}/μ_m Accurate	μ_{eff}/μ_m Maxwell	μ_{eff}/μ_m Dipole
0.1	0.2	0.8156	0.818	0.8169	0.813	0.818	0.816
0.1	0.3	0.771	0.774	0.772	0.82	0.825	0.823
0.1	0.4	0.662	0.667	0.664	0.829	0.833	0.832
0.1	0.5	0	0	0	0.856	0.844	0.859
0.45	0.2	0.362	0.379	0.377	0.346	0.379	0.379
0.45	0.3	0.299	0.3177	0.3143	0.355	0.39	0.391
0.45	0.4	0.1971	0.2136	0.2106	0.367	0.4044	0.405
0.45	0.5	0	0	0	0.399	0.423	0.427
0.6	0.2	0.223	0.25	—	0.198	0.25	—
0.6	0.3	0.177	0.2025	—	0.205	0.259	—
0.6	0.4	0.11	0.129	—	0.212	0.27	—
0.6	0.5	0	0	—	0.337	0.286	—

there is again good agreement for all values of the matrix Poisson's ratio ≤ 0.4. From Tables 15.1 and 15.2 it is clear that estimates of properties based on Maxwell's methodology are more accurate for cavities than they are for the case of rigid particles. Also listed in Tables 15.1 and 15.2 are predictions based on a dipole approximation that was considered by Sangani and Mo [24]. It is observed that corresponding predictions of effective elastic properties agree very closely with those obtained using Maxwell's methodology. This agreement suggests that, although particle interactions are neglected when using Maxwell's methodology, the approach appears to take account of dipole interactions at least.

Kushch [25] has developed a numerical method of estimating accurately the effective elastic properties of regular arrays of aligned spheroidal particles having the same geometry. Simple cubic arrays have been considered for isotropic particles and matrix for the case when the particle volume fraction is 0.1 such that the centroids of the particles are distributed on a cubic lattice. Extreme values of the particle shear modulus are considered and Poisson's ratio for both the particles and matrix has the value 0.3. It is useful to compare in Tables 15.3 and 15.4 the predictions of the elastic constants C_{1111} and C_{3333} given by Kushch [25] with those obtained using the values calculated using the methodology described in this chapter. It follows from (15.6) and (15.8) that these elastic constants are defined by

$$C_{1111} = E_A + 4k_T\nu_A^2 , \qquad C_{3333} = k_T + \mu_t = C_{2222}.$$

When the particle shear modulus ratio $\mu_p / \mu_m = 0$ (a cavity-like spheroid), and for all values $0.25 \leq b / a \leq 2.25$, the results for C_{1111} based on Maxwell's methodology differ from those of Kushch by less than 9%. Differences less than 1.7% are found when $\mu_p / \mu_m = 10$ and differences less than 2.4% when $\mu_p / \mu_m = 1000$.

For the effective property C_{3333} differences are less than 42% when $\mu_p / \mu_m = 0$, less than 14% when $\mu_p / \mu_m = 10$ and less than 35% when $\mu_p / \mu_m = 1000$. However, if

Table 15.3 Predictions for C_{1111} for distributions of spheroidal particles.

	Random (Maxwell)	Simple cubic	Random (Maxwell)	Simple cubic	Random (Maxwell)	Simple cubic
μ_p / μ_m	0	0	10	10	1000	1000
a / b						
0.25	2.963	2.856	4.691	4.645	5.590	5.482
0.5	2.913	2.877	4.297	4.267	4.589	4.536
0.75	2.849	2.839	4.148	4.155	4.320	4.336
1.0	2.777	2.799	4.071	4.102	4.199	4.248
1.25	2.703	2.762	4.025	4.070	4.130	4.199
1.5	2.629	2.732	3.995	4.049	4.086	4.167
1.75	2.556	2.705	3.973	4.035	4.056	4.145
2.0	2.485	2.681	3.957	4.023	4.033	4.128
2.25	2.417	2.653	3.944	4.012	4.016	4.113

Table 15.4 Predictions for C_{3333} for distributions of spheroidal particles.

	Random (Maxwell)	Simple cubic	Random (Maxwell)	Simple cubic	Random (Maxwell)	Simple cubic
μ_p / μ_m	0	0	10	10	1000	1000
a / b						
0.25	2.611	1.843	3.991	3.889	4.081	3.944
0.5	2.678	2.474	4.015	3.956	4.116	4.032
0.75	2.735	2.693	4.043	4.026	4.156	4.132
1.0	2.777	2.799	4.071	4.102	4.199	4.248
1.25	2.809	2.859	4.100	4.185	4.243	4.388
1.5	2.8315	2.897	4.129	4.281	4.289	4.565
1.75	2.848	2.923	4.157	4.397	4.335	4.810
2.0	2.859	2.941	4.184	4.555	4.3814	5.225
2.25	2.867	2.954	4.211	4.868	4.428	6.765

spheroids are considered such that $0.5 \leq b / a \leq 2$ the differences for C_{1111} are less than 7.4% and those for C_{3333} are less than 16.2%. A contribution to the largest differences at the extreme values of the aspect ratio *b/a* will certainly be due to differences in the properties of composites reinforced with random and simple cubic arrays of particles. It is concluded from the results given in Tables 15.3 and 15.4 that there is reasonable agreement between the results of Kushch [25] for simple cubic arrays of spheroidal particles and those based on Maxwell's methodology for random distributions of aligned particles for cases of practical relevance (e.g. $\mu_p / \mu_m \approx 10$ including cavities having low volume fractions).

15.10 Composites Reinforced with Aligned Transversely Isotropic Cylindrical Fibres

For the case of transversely isotropic cylindrical fibres, rather than developing the required solution using (15.71)–(15.73), it is again simpler to derive the results using another method that involves less laborious algebra (see Appendix F for details). The principal results are given by

$$\Delta_i = 9k_{\mathrm{m}}\mu_{\mathrm{m}}\frac{k_{\mathrm{T}}^i + \mu_{\mathrm{m}}}{k_{\mathrm{T}}^{\mathrm{m}} + \mu_{\mathrm{m}}}, \tag{15.96}$$

$$\overline{\Delta} = 9k_{\mathrm{m}}\mu_{\mathrm{m}}\frac{k_{\mathrm{T}}^{\mathrm{eff}} + \mu_{\mathrm{m}}}{k_{\mathrm{T}}^{\mathrm{m}} + \mu_{\mathrm{m}}}, \tag{15.97}$$

$$\frac{1}{k_{\mathrm{T}}^{\mathrm{eff}} + \mu_{\mathrm{m}}} = \sum_{i=1}^{N}\frac{V_{\mathrm{p}}^i}{k_{\mathrm{T}}^i + \mu_{\mathrm{m}}} + \frac{V_{\mathrm{m}}}{k_{\mathrm{T}}^{\mathrm{m}} + \mu_{\mathrm{m}}}, \tag{15.98}$$

$$\frac{\nu_{\mathrm{A}}^{\mathrm{eff}}k_{\mathrm{T}}^{\mathrm{eff}}}{k_{\mathrm{T}}^{\mathrm{eff}} + \mu_{\mathrm{m}}} = \sum_{i=1}^{N}V_{\mathrm{p}}^i\frac{\nu_{\mathrm{A}}^i k_{\mathrm{T}}^i}{k_{\mathrm{T}}^i + \mu_{\mathrm{m}}} + V_{\mathrm{m}}\frac{\nu_{\mathrm{m}}k_{\mathrm{T}}^{\mathrm{m}}}{k_{\mathrm{T}}^{\mathrm{m}} + \mu_{\mathrm{m}}}, \tag{15.99}$$

$$\begin{aligned} E_{\mathrm{A}}^{\mathrm{eff}} + \frac{4k_{\mathrm{T}}^{\mathrm{eff}}\left(\nu_{\mathrm{A}}^{\mathrm{eff}}\right)^2\mu_{\mathrm{m}}}{k_{\mathrm{T}}^{\mathrm{eff}} + \mu_{\mathrm{m}}} &= \sum_{i=1}^{N}V_{\mathrm{p}}^i\left(E_{\mathrm{A}}^i + \frac{4k_{\mathrm{T}}^i\left(\nu_{\mathrm{A}}^i\right)^2\mu_{\mathrm{m}}}{k_{\mathrm{T}}^i + \mu_{\mathrm{m}}}\right) \\ &\quad + V_{\mathrm{m}}\left(E_{\mathrm{m}} + \frac{4k_{\mathrm{T}}^{\mathrm{m}}\nu_{\mathrm{m}}^2\mu_{\mathrm{m}}}{k_{\mathrm{T}}^{\mathrm{m}} + \mu_{\mathrm{m}}}\right), \end{aligned} \tag{15.100}$$

$$\text{where } k_{\mathrm{T}}^{\mathrm{m}} = k_{\mathrm{m}} + \frac{1}{3}\mu_{\mathrm{m}}. \tag{15.101}$$

The results (15.98) and (15.99) can also be derived using similar methods to those described for spherical inclusions in [17]. The result (15.100) for the axial Young's modulus for multiphase unidirectional composite is thought to be a new formula. It follows from (15.34) and (15.35) that the corresponding effective shear moduli are given by

$$\frac{1}{\mu_{\mathrm{A}}^{\mathrm{eff}} + \mu_{\mathrm{m}}} = \sum_{i=1}^{N}\frac{V_{\mathrm{p}}^i}{\mu_{\mathrm{A}}^i + \mu_{\mathrm{m}}} + \frac{V_{\mathrm{m}}}{\mu_{\mathrm{m}} + \mu_{\mathrm{m}}}, \tag{15.102}$$

$$\frac{1}{\mu_{\mathrm{t}}^{\mathrm{eff}} + \mu_{\mathrm{t}}^*} = \sum_{i=1}^{N}\frac{V_{\mathrm{p}}^i}{\mu_{\mathrm{t}}^i + \mu_{\mathrm{t}}^*} + \frac{V_{\mathrm{m}}}{\mu_{\mathrm{m}} + \mu_{\mathrm{t}}^*}, \quad \text{where } \mu_{\mathrm{t}}^* = \frac{k_{\mathrm{T}}^{\mathrm{m}}\mu_{\mathrm{m}}}{k_{\mathrm{T}}^{\mathrm{m}} + 2\mu_{\mathrm{m}}}. \tag{15.103}$$

Five independent effective elastic properties can thus be estimated, namely, $k_{\mathrm{T}}^{\mathrm{eff}}$, $\mu_{\mathrm{t}}^{\mathrm{eff}}$, $\mu_{\mathrm{A}}^{\mathrm{eff}}$, $\nu_{\mathrm{A}}^{\mathrm{eff}}$ and $E_{\mathrm{A}}^{\mathrm{eff}}$. The transverse Young's modulus $E_{\mathrm{T}}^{\mathrm{eff}}$ and Poisson's ratio $\nu_{\mathrm{t}}^{\mathrm{eff}}$ can be estimated using relations (15.10). For fibre-reinforced composites subject to plane strain deformations so that the axial displacement and strain are everywhere zero, two other elastic constants E_{T}^* and ν_{t}^* are relevant defined by the relations

$$\frac{4}{E_{\mathrm{T}}^{*}}=\frac{1}{\mu_{\mathrm{t}}^{\mathrm{eff}}}+\frac{1}{k_{\mathrm{T}}^{\mathrm{eff}}},\quad \nu_{\mathrm{t}}^{*}=\frac{k_{\mathrm{T}}^{\mathrm{eff}}-\mu_{\mathrm{T}}^{\mathrm{eff}}}{k_{\mathrm{T}}^{\mathrm{eff}}+\mu_{\mathrm{T}}^{\mathrm{eff}}}. \tag{15.104}$$

The constant E_{T}^{*} is the transverse Young's modulus for plane strain conditions in the axial direction, whereas ν_{T}^{*} is the corresponding transverse Poisson's ratio. It is useful to note that

$$E_{\mathrm{T}}^{*}=2\mu_{\mathrm{t}}^{\mathrm{eff}}\left(1+\nu_{\mathrm{t}}^{*}\right). \tag{15.105}$$

Predictions of effective properties for fibre-reinforced composites are now compared with those in the literature for two phases obtained using numerical methods, and which are expected to be accurate. Eischen and Torquato [26] have considered, using a boundary element method, the estimation of elastic constants for hexagonal arrays of aligned fibres subject to plane strain conditions. Three different materials systems were included having isotropic fibres and matrix properties given by $\mu_{\mathrm{f}}\,/\,\mu_{\mathrm{m}}=135, 22.5, 6.75$, $\mu_{\mathrm{f}}\,/\,k_{\mathrm{f}}=0.75$, $\mu_{\mathrm{m}}\,/\,k_{\mathrm{m}}=0.33$, $\nu_{\mathrm{m}}=0.35$, $\nu_{\mathrm{f}}=0.2$ and the value $\mu_{\mathrm{m}}\,/\,k_{\mathrm{T}}^{\mathrm{m}}=0.3$ is also given. The matrix values $\mu_{\mathrm{m}}\,/\,k_{\mathrm{m}}=0.33$, $\nu_{\mathrm{m}}=0.35$ and $\mu_{\mathrm{m}}\,/\,k_{\mathrm{T}}^{\mathrm{m}}=0.3$ are not consistent with relation (15.101) due to rounding errors. To avoid this problem the matrix values assumed here are $\mu_{\mathrm{m}}\,/\,k_{\mathrm{m}}=1/3$, $\nu_{\mathrm{m}}=0.35$ and $\mu_{\mathrm{m}}\,/\,k_{\mathrm{T}}^{\mathrm{m}}=0.3$ which are consistent with relation (15.101).

Figure 15.2 shows a comparison of the effective plane strain bulk modulus $k_{\mathrm{T}}^{\mathrm{eff}}$, obtained using relation (15.98) resulting from Maxwell's methodology, with the accurate boundary element results [26]. The normalised plane strain bulk modulus is defined by $k_{\mathrm{T}}^{\mathrm{eff}}\,/\,k_{\mathrm{T}}^{\mathrm{m}}$ where $k_{\mathrm{T}}^{\mathrm{m}}$ is defined by (15.101). For fibre volume fractions in the range $0\leq V_{\mathrm{f}}\leq 0.7$, the results predicted using Maxwell's methodology agree

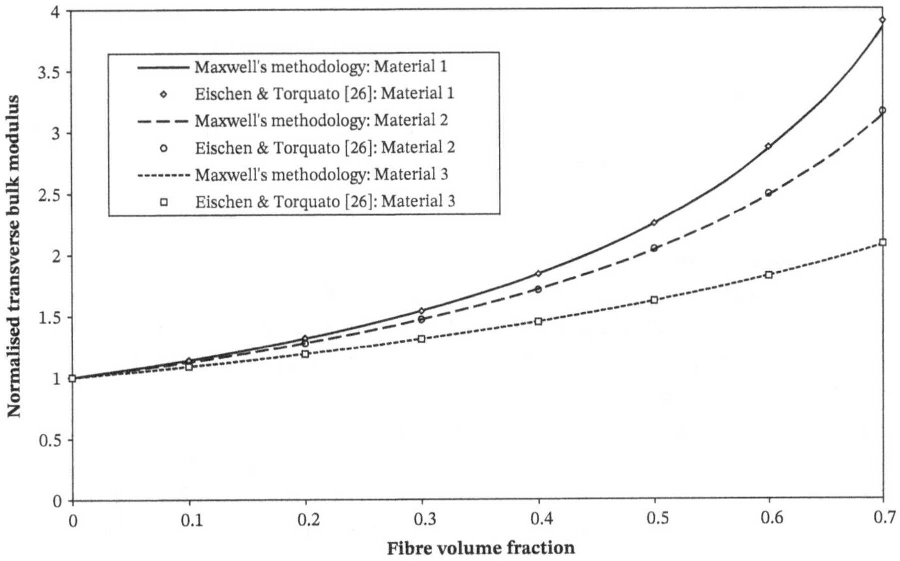

Figure 15.2 Comparison of results for normalised effective plane strain bulk $k_{\mathrm{T}}^{\mathrm{eff}}$ modulus obtained using Maxwell's methodology with those of Eischen and Torquato [26] for three different materials.

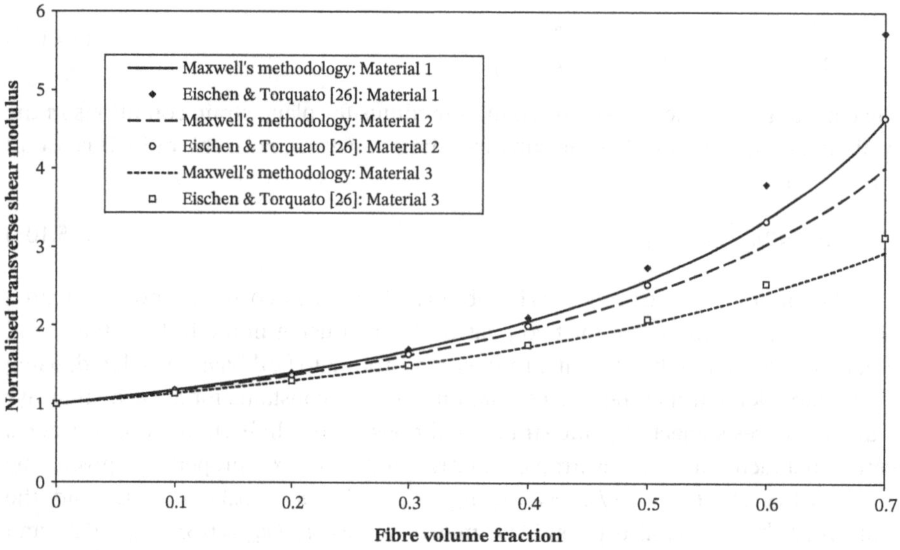

Figure 15.3 Comparison of results for normalised effective transverse shear modulus μ_t^{eff} obtained using Maxwell's methodology with those of Eischen and Torquato [26] for three different materials.

exceedingly well (errors less than 1.4%) with the results of Eischen and Torquato. For larger volume fractions significant differences arise especially when the fibres have a much larger shear modulus than the matrix.

Figure 15.3 shows a comparison of transverse shear modulus μ_t^{eff}, obtained using relation (15.103) resulting from Maxwell's methodology, with results of Eischen and Torquato [26]. The normalised effective transverse shear modulus is defined by μ_t^{eff} / μ_m. For fibre volume fractions in the range $0 \leq V_f \leq 0.4$, the results predicted using Maxwell's methodology agree well (errors less than 2.3%) with the results of Eischen and Torquato. For larger volume fractions, significant differences arise especially when the fibres have a much larger shear modulus than the matrix.

Figure 15.4 shows a comparison of the plane strain transverse Poisson's ratio ν_t^*, defined by (15.104) and obtained from relations (15.98) and (15.103) resulting from Maxwell's methodology, with the results of Eischen and Torquato [26]. For fibre volume fractions in the range $0 \leq V_f \leq 0.4$, the results predicted using Maxwell's methodology agree well (errors less than 1.8%) with the results of Eischen and Torquato. For larger volume fractions, significant differences arise especially when the fibres have a much larger shear modulus than the matrix.

Figure 15.5 shows a comparison of the plane strain transverse Young's modulus E_T^*, defined by (15.104) and obtained from relations (15.98) and (15.103) resulting from Maxwell's methodology, with the results of Eischen and Torquato [26]. For fibre volume fractions in the range $0 \leq V_f \leq 0.4$, the results predicted using Maxwell's methodology agree well (errors less than 2.1%) with the results of Eischen and Torquato. For larger volume fractions, significant differences arise especially when the fibres have a much larger shear modulus than the matrix.

Figure 15.6 shows a comparison of axial shear modulus μ_A^{eff}, obtained using relation (15.102) resulting from Maxwell's methodology, with the results of Symm [27]. The

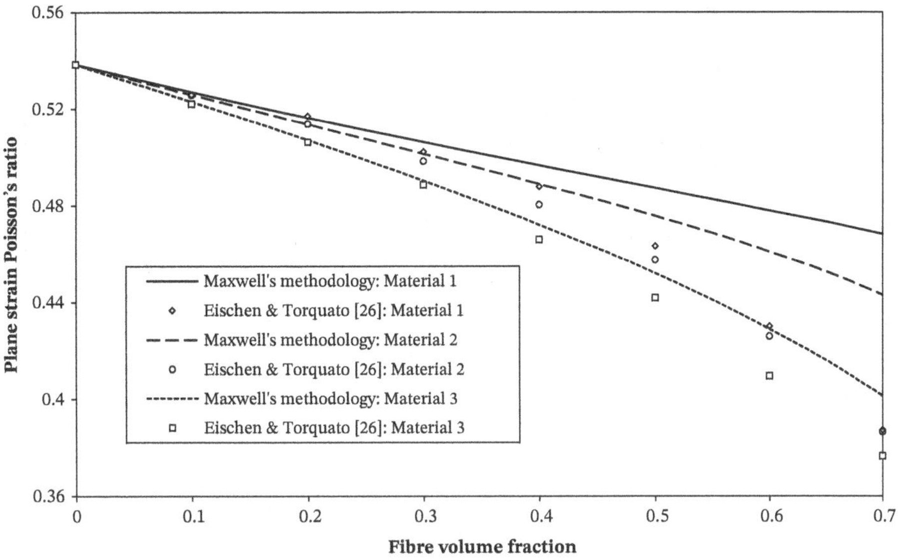

Figure 15.4 Comparison of results for the effective plane strain transverse Poisson's ratio ν_t^* obtained using Maxwell's methodology with those of Eischen and Torquato [26] for three different materials.

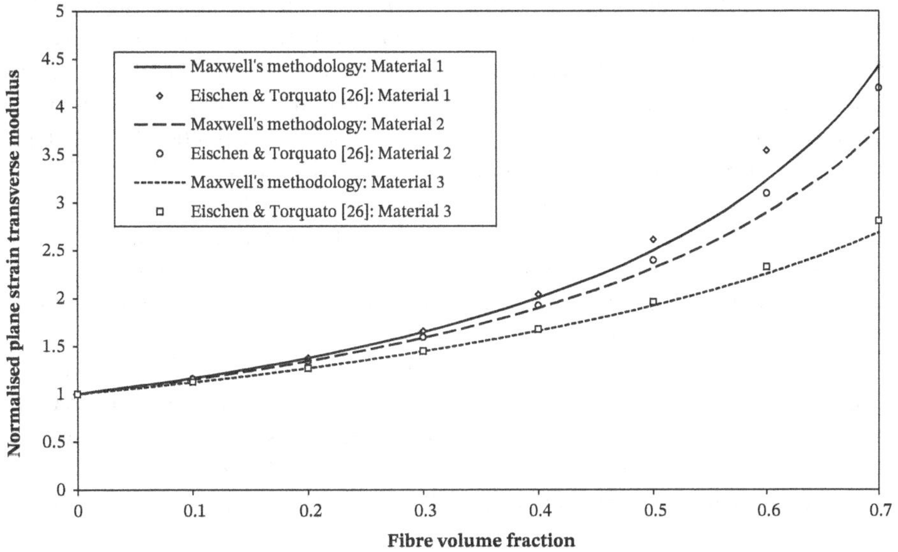

Figure 15.5 Comparison of results for the normalised effective plane strain transverse Young's modulus E_T^* obtained using Maxwell's methodology with those of Eischen and Torquato [26] for three different materials.

normalised effective axial shear modulus is defined by $\mu_A^{\text{eff}} / \mu_m$, and the four materials considered are for isotropic fibres and matrix such that $\mu_f / \mu_m = 6, 20, 120, \infty$. For fibre volume fractions in the range $0 \leq V_f \leq 0.7$, the results predicted using Maxwell's methodology agree very well (errors less than 2.6%) with the results of Symm [27]. For larger volume fractions significant differences arise especially when the fibres have a much larger shear modulus than the matrix.

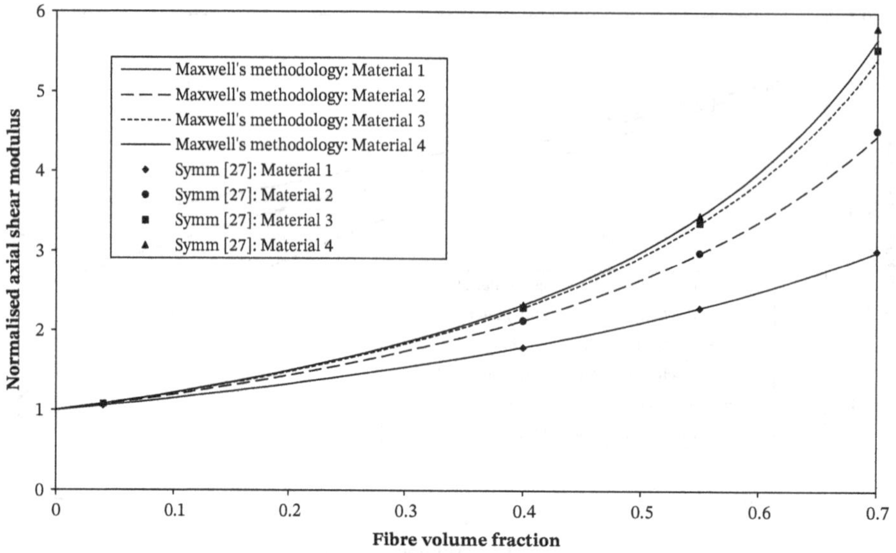

Figure 15.6 Comparison of results for normalised effective axial shear modulus μ_A^{eff} obtained using Maxwell's methodology with those of Symm [27] for four different materials.

These results have not tested the validity of the values of the elastic constants E_A^{eff} and ν_A^{eff} that can be estimated using (15.99) and (15.100). In [28], values were given for all the elastic constants calculated using a finite element analysis (Li and Zou, Private communication, 2001) for the two cases of hexagonal arrays of aligned carbon and glass fibres embedded in an isotropic epoxy matrix. The volume fraction for both types of fibre was taken to be 0.6. The fibre and matrix properties are given in [27, Tables 15.1 and 15.3]. The carbon fibres were transversely isotropic while the glass fibres were assumed to be isotropic. Table 15.5 compares the finite element results to the various properties of two-phase composites that can be estimated using the results given in Section 15.10. It is seen that the effective properties E_A^{eff}, ν_A^{eff}, μ_A^{eff} and k_T^{eff} are estimated very accurately (errors less 0.5%) using Maxwell's methodology for both carbon- and glass-fibre composites. For the effective properties E_T^{eff}, ν_t^{eff} and μ_t^{eff}, the percentage errors for the carbon-fibre composite are less than 3.7% whereas those for the glass-fibre composite are less than 9.6%. These trends are consistent with those observed when comparing in Figures 15.2–15.6 predictions of results based on Maxwell's methodology with the boundary element results given by Eischen and Torquato [26] and by Symm [27].

15.11 Discussion of Results

It is first noted that the Mori–Tanaka mean field result for the fourth-order effective elastic property tensor L may be obtained (see, for example, [2]), in terms of those for the matrix L_m and the N phases of inclusion having properties $L_r, r=1,\ldots,N$, using a relation of the form

Table 15.5 Comparison of effective properties estimated using Maxwell's methodology with finite element results for carbon- and glass-fibre-reinforced composites having a volume fraction of 0.6.

Property	CFRP (FEA)	CFRP (Maxwell)	% Error	GRP (FEA)	GRP (Maxwell)	% Error
E_A^{eff}	136.70	136.7032	0.0023	45.76	45.7625	0.0055
E_T^{eff}	8.901	8.7564	1.62	11.80	11.0193	6.62
ν_A^{eff}	0.2526	0.2526	0	0.2515	0.2517	0.08
ν_t^{eff}	0.3082	0.3194	3.63	0.4020	0.4405	9.58
μ_A^{eff}	4.550	4.5365	0.30	4.339	4.3179	0.49
μ_t^{eff}	3.4020	3.3184	2.46	4.2083	3.8249	9.11
k_T^{eff}	6.5114	6.5110	0.0066	10.4355	10.4146	0.20

$$L^* = \left[L_m V_m + \sum_{r=1}^{N} V_p^r L_r \left[I + S L_m^{-1} \left(L_r - L_m \right) \right]^{-1} \right] \cdot \left(V_m I + \sum_{r=1}^{N} V_p^r \left[I + S L_m^{-1} \left(L_r - L_m \right) \right]^{-1} \right)^{-1}, \tag{15.106}$$

where S is the fourth-order tensor derived by Eshelby [18] and where I is the symmetric fourth-order unit tensor. When a fourth-order tensor L_m^* is defined by the relation

$$L_m^* = L_m \left(S^{-1} - I \right), \tag{15.107}$$

it can be shown that relation (15.106) may be expressed in the following form that exhibits a mixtures structure for a fourth-order tensor of the form $\left(L^* + L_m^* \right)^{-1}$

$$\left(L^* + L_m^* \right)^{-1} = V_m \left(L_m + L_m^* \right)^{-1} + \sum_{r=1}^{N} V_p^r \left(L_r + L_m^* \right)^{-1}. \tag{15.108}$$

Relation (15.108) follows directly from relations given in the literature (e.g. Norris [3], Weng [5], Walpole [6, 7], Qiu and Weng [10], and Benveniste et al. [11]). Its form is indicative of many of the results that have been derived in this chapter (see (15.34), (15.35), (15.94), (15.95), (15.102) and (15.103)).

When using Maxwell's methodology combined with Eshelby's far-field solution for the displacement in the matrix, the results (15.34) and (15.35) for the effective axial and shear moduli are easily derived for a composite having uniform distributions of aligned transversely isotropic spheroidal inclusions embedded in an isotropic matrix. These results are very simple in form showing that the effective shear moduli can be estimated using a mixtures relationship. It is noted that their form is a scalar equivalent of the result (15.108) defining the fourth-order tensor L^* that is based on Mori–Tanaka theory. The situation regarding the nonshear elastic constants is far more complex. The analysis in Sections 15.7 and 15.8 leads to the results (15.86) for the effective elastic constants of the composite that depend on the parameters $\overline{P}, \overline{Q}, \overline{R}$ and $\overline{S} = \overline{Q} - 3\overline{R}$, which are complicated functions of the properties and volume fractions of the inclusions, and of matrix properties.

For the case when the aspect ratios of all types of spheroidal inclusion are the same, Qiu and Weng [10] derived explicit and very complex expressions for the independent effective elastic constants p, m, k, l and n describing the properties of a transversely isotropic composite, such that

$$p \equiv \mu_A, \quad \mathrm{m} \equiv \mu_t, \quad k \equiv k_T, \quad l \equiv 2\nu_A k_T, \quad n \equiv E_A + 4k_T\nu_A^2. \tag{15.109}$$

The complexity of the expressions for the effective elastic constants arises because 12 different summations over the phases have to be performed. The approach derived in this chapter involves only single summations when estimating the elastic constants μ_A and μ_t as seen from very simple results (15.34) and (15.35), and the four summations given by (15.70) when estimating the remaining elastic constants E_A, ν_A and k_T using (15.86), i.e. six summations in total. In spite of their complexity, the results (15.86) and associated relations are, however, simpler than those given by Qiu and Weng [10].

As comparing algebraically the possible equivalence of the results of Qiu and Weng [10] for the elastic constants (15.109) with the corresponding results of this chapter would be very laborious, values have been compared using *exact* numerical methods provided by the open-source algebraic programming system REDUCE [29]. Exact agreement has been obtained for a wide range of parameter values. The numerical methods are based on the representation of the values of physical quantities and values of associated expressions by rational numbers having integer denominators and numerators. When performing the comparison, rational values are assumed for the tensor components S_{1111}, S_{2222}, S_{1122}, S_{2233} and S_{1212}. The remaining components of the S_{ijkl} tensor are then calculated using (15.12). It has been shown using these methods that the results of this chapter are exactly equivalent to the more complex results arising from the analysis of Qiu and Weng [10] based on Mori and Tanaka theory [1]. The principal result is thus obtained, namely, that the approximate Mori–Tanaka model, as implemented by Qiu and Weng [10] when estimating effective elastic properties, leads to results that are identical to the approximate relations obtained using Maxwell's methodology combined with Eshelby's far-field result for the displacement field. It is now useful to discuss other relevant results from the literature.

For the case of distributions of various types of isotropic spherical particles uniformly distributed in a matrix, having properties that are less than those of all reinforcements, it was shown [17] that the predictions of effective properties based on Maxwell's methodology were identical to the lower Hashin–Shtrikman bound [4], and that this bound was very close to accurate effective properties estimated by other methods. For the case of multiphase fibre-reinforced composites there is a similar situation. The lower bounds for the plane strain transverse bulk modulus and both the transverse and axial shear moduli of multiphase composites derived by Hashin [30] can be manipulated so that they correspond exactly to relations (15.98), (15.102) and (15.103) derived here.

For two-phase composites, predictions of many of the effective properties based on Maxwell's methodology are identical to those generated by the composite sphere assemblage and composite cylinder models, and they also correspond exactly to one of the variational bounds. These results strongly suggest that Maxwell's methodology is not restricted to dilute distributions of reinforcing inclusions. It can be shown for two-phase fibre-reinforced composites that relation (15.100) for the axial Young's modulus is identical to that which is obtained when using the concentric cylinders model of a unidirectional composite (Hashin and Rosen [31] for isotropic constituents, Hashin [32] for anisotropic constituents). Although the comparisons of results shown in Figures 15.2–15.6 indicate that estimates based on Maxwell's methodology become less accurate as the volume fraction of reinforcement increases, it is worth noting that as the volume fraction tends to unity, so that the system becomes almost homogenous without any matrix, the general relations (15.34), (15.35) for shear, and the specific relations (15.94), (15.95), (15.98)–(15.100), (15.102) and (15.103) all predict the expected result that the effective properties tend to the properties of the reinforcement. Distributions of very small inclusions would of course be needed to realise this limit in practice.

For the special case where the matrix and spherical inclusions are isotropic and have the same shear modulus, the effective bulk modulus of an isotropic composite based on Mori–Tanaka theory [1] was shown by Weng [2] to correspond to the exact solution of Hill [33]. As Maxwell's methodology has been shown to predict elastic moduli that correspond exactly with Mori and Tanaka's results, it follows that for this special case the results given in this Chapter for the effective bulk modulus must be exact for all volume fractions. For the special case where transversely isotropic spheroids have the form of aligned very thin circular discs, Weng [5] has shown that the Mori–Tanaka moduli are exact, implying that corresponding results derived for this limit using Maxwell's methodology will also be exact for all volume fractions.

It is indeed remarkable that the pioneering methodology developed by Maxwell [13], for predicting the effective properties of a composite, when combined with Eshelby's method [18] for predicting the far-field in the matrix for an isolated ellipsoid, are capable of predicting formulae for effective elastic properties that have been shown to correspond with many of those derived subsequently in the literature using alternative methods. A key characteristic of the approach is that a single method can be used to generate estimates for *all* the elastic constants of multiphase composites having an isotropic matrix and reinforced with aligned transversely isotropic inclusions of the same aspect ratio, or with aligned fibres, or with spherical particles. The nature of the methodology is such that it has good potential for application to other situations of practical interest. For example, composites reinforced with inclusions having imperfect interfaces or one or more coatings of uniform thickness could easily be treated provided that the solution for an isolated inclusion is available. In the field of nanocomposites, nanoinclusions (e.g. nanoparticles and carbon nanotubes) are associated with interphase matrix regions adjacent to the inclusions having different properties to those of the bulk matrix. The interphase regions of these systems, which form owing to nanoscale interactions between the embedded nanoinclusions and adjacent polymer chains, can occupy a substantial portion of the volume fraction of the composite due to the very large surface area of nanoinclusion per unit volume available for interaction with local polymer chains. This effect results in significant changes to the effective properties of the polymer composite when compared with those of the bulk polymer (see, for example, Fisher et al. [34] who considered nanotubes with adjacent interphase layers using the Mori–Tanaka method). Surface tension effects (see, for example, Duan et al. [35] and Mogilevskaya et al. [36, 37]), which lead to discontinuities in the normal traction distributions at the nanoinclusion boundaries, can also modify the effective properties of the nanocomposite.

A common, and in fact incorrect, view of Maxwell's methodology is that it is expected to yield results valid only for very dilute concentrations of inclusions. However, it has been shown in this chapter that, for spherical, fibrous and spheroidal inclusions, results based on Maxwell's methodology are, in fact, valid for a much wider range of volume fractions. An apparent contradiction has, thus, been identified. Maxwell's methodology implicitly neglects interactions between inclusions and yet it predicts accurate values (sometimes exact) for effective elastic constants for a wide range of volume fractions, up to 0.7 for some properties of fibre composites. In addition, for the case of spheroidal inclusions having the same aspect ratio, Maxwell's methodology has been shown to lead to expressions for all elastic constants that are apparently identical to the more complex results derived by Qiu and Weng [10], which are based on a mean field method of taking inclusion interactions into account. The only rational conclusion is that inclusion interaction effects for volume fractions of practical interest may not affect the far-field to any significant degree with the result that Maxwell's methodology, when combined with Eshelby's analysis, has much wider applicability than expected.

It is noted from the discussion in [17] that Bonnecaze and Brady [38] used a method of estimating the conductivity of a composite reinforced with cubic arrays of spherical particles that captures both far-field and near-field particle interactions. They compared their results with those of Sangani and Acrivos [39] providing numerical values of results in various tables. One type of estimate takes account only of dipole interactions, ignoring higher-order terms, and it leads to results that appear coincident with results obtained using Maxwell's formula (agreement to three significant figures in most cases) for all volume fractions up to closest packing. This agreement was not noticed in their paper. They may have discovered a method that is showing why Maxwell's methodology works so well as the effect of a distribution of interacting dipoles at large distances may be identical to that for the case when they are all located at the same point, as assumed by Maxwell. Further support of this idea is given in this chapter for the case of rigid spherical particles and cavities (Tables 15.1 and 15.2). It is noted that for the case of aligned cylindrical fibres, Mogilevskaya and Crouch [40–42] have developed a complex variable technique that can be used to investigate numerically the effect of fibre interactions at large distances from a cluster of fibres in an infinite matrix, including cases for homogeneously imperfect interfaces and uniform interphases.

References

1. Mori, T. and Tanaka, K. (1973). Average stress in matrix and average elastic energy of materials with mis-fitting inclusions. *Acta Metallurgica* 21: 571–574.
2. Weng, G.J. (1984). Some elastic properties of reinforced solids with special reference to isotropic ones containing spherical inclusions. *International Journal* of *Engineering Science* 22: 845–856.
3. Norris, A.N. (1989). An examination of the Mori–Tanaka effective medium approximation for multi-phase composites. *Journal of Applied Mechanics* 56: 83–88.
4. Hashin, Z. and Shtrikman, S. (1963). A variational approach to the theory of the elastic behaviour of multiphase composites. *Journal of the Mechanics and Physics of Solids* 11: 127–140.
5. Weng, G.J. (1990). The theoretical connection between Mori–Tanaka's theory and the Hashin–Shtrikman-Walpole bounds. *International Journal of Engineering Science* 28 (11): 1111–1120.
6. Walpole, L.J. (1966). On bounds for the overall elastic moduli of inhomogeneous systems -I. *Journal of the Mechanics and Physics of Solids* 14: 151–162.
7. Walpole, L.J. (1966). On bounds for the overall elastic moduli of inhomogeneous systems -II. *Journal of the Mechanics and Physics of Solids* 14: 289–301.
8. Walpole, L.J. (1969). On the overall elastic moduli of composite materials. *Journal of the Mechanics and Physics of Solids* 17: 235–251.
9. Walpole, L.J. (1981). Elastic behaviour of composite materials: Theoretical foundations. *Advances in Applied Mechanics* 21: 169–242.
10. Qiu, Y.P. and Weng, G.J. (1990). On the application of Mori–Tanaka's theory involving transversely isotropic spheroidal inclusions. *International Journal of Engineering Science* 28: 1121–1137.
11. Benveniste, Y., Dvorak, G.J., and Chen, T. (1991). On diagonal and elastic symmetry of the approximate effective stiffness tensor of heterogeneous media. *Journal of the Mechanics and Physics of Solids* 39: 927–946.
12. Chen, T. and Dvorak, G.J. (1992). Mori–Tanaka estimates of the overall elastic moduli of certain composite materials. *Journal of Applied Mechanics* 59: 539–546.
13. Maxwell, J.C. (1873). *A Treatise on Electricity and Magnetism.* Chapter 9, (Vol. 1, Art. 310–314, pp. 435–441), 1st e (3rd e, 1892). Oxford: Clarendon Press.
14. Felderhof, B.U. and Iske, P.L. (1992). Mean-field approximation to the effective elastic moduli of a solid suspension of spheres. *Physical Review* A45 (2): 611–617.
15. Cohen, I. and Bergman, D.J. (2003). Effective elastic properties of periodic composite medium. *Journal of the Mechanics and Physics of Solids* 51: 1433–1457.
16. Cohen, I. and Bergman, D.J. (2003). Clausius-Mossotti-type approximation for elastic moduli of a cubic array of spheres. *Physical Review B* 68: 024104.
17. McCartney, L.N. and Kelly, A. (2008). Maxwell's far-field methodology applied to the prediction of the properties of multi-phase isotropic particulate composites. *Proceedings of the Royal Society of London* A464: 423–446.
18. Eshelby, J.D. (1957). The elastic field of an ellipsoidal inclusion. *Proceedings of the Royal Society of London* A241: 376–396.
19. Eshelby, J.D. (1959). The elastic field outside an ellipsoidal inclusion. *Proceedings of the Royal Society of London* A252: 561–569.

20. Torquato, S. (2002). *Random Heterogeneous Materials*. New York: Springer.
21. Willis, J.R. (1977). Bounds and self-consistent estimates for the overall properties of anisotropic composites. *Journal of the Mechanics and Physics of Solids* 25: 185–202.
22. Weng, G.J. (1992). Explicit evaluation of Willis' bounds with ellipsoidal inclusions. *International Journal of Engineering Science* 30: 83–92.
23. Arridge, R.G.C. (1992). The thermal expansion and bulk modulus of composites consisting of arrays of spherical particles in a matrix, with body or face centred cubic symmetry. *Proceedings of the Royal Society of London* A438: 291–310.
24. Sangani, A.S. and Mo, G. (1997). Elastic interactions in particulate composites with perfect as well as imperfect interfaces. *Journal of the Mechanics and Physics of Solids* 45: 2001–2031.
25. Kushch, V.I. (1997). Microstresses and effective elastic moduli of a solid reinforced with periodically distributed spheroidal particles. *International Journal of Solids and Structures* 34: 1353–1366.
26. Eischen, J.W. and Torquato, S. (1993). Determining elastic behaviour of composites by the boundary element method. *Journal of Applied Physics* 74 (1): 159–170.
27. Symm, G.T. (1970). The longitudinal shear modulus of a unidirectional fibrous composite. *Journal of Composite Materials* 4: 426–428.
28. McCartney, L.N. (2003). Physically based damage models for laminated composites. *Proceedings of the Institution of Mechanical Engineers* 217 (Part L, J. Materials: Design and Applications): 163–199.
29. Algebraic programming system REDUCE version 3.8 (February 2004). http://www.uni-koeln.de/REDUCE.
30. Hashin, Z. (1965). On elastic behaviour of fibre reinforced materials of arbitrary transverse phase geometry. *Journal of the Mechanics and Physics of Solids* 13: 119–134.
31. Hashin, Z. and Rosen, B.W. (1964). The elastic moduli of fiber-reinforced materials. *Journal of Applied Mechanics* 31: 223–232.
32. Hashin, Z. (1979). Analysis of properties of fiber composites with anisotropic constituents. *Journal of Applied Mechanics* 46: 543–550.
33 Hill, R. (1963). Elastic properties of reinforced solids: Some theoretical principles. *Journal of the Mechanics and Physics of Solids* 11: 357–372.
34. Fisher, F.T., Lee, K.-C., and Brinson, L.C. (2005). Elastic and viscoelastic properties of non-bulk polymer interphases in nanotube reinforced polymers. *SEM 22005 Annual Conference on Experimental and Applied Mechanics, June 7–9*, Portland, OR. http://personal.stevens.edu/~ffisher/pubs/Fisher_SEMconf_2005.pdf
35. Duan, H.L., Wang, J., Huang, Z.P., and Karihaloo, B.L. (2005). Eshelby formalism for nano-homogeneities. *Proceedings of the Royal Society* A461: 3335–3353.
36. Mogilevskaya, S.G., Crouch, S.L., Stolarski, H.K., and Benusiglio, A. (2010). Equivalent homogeneity method for evaluating the effective elastic properties of unidirectional multi-phase composites with surface/interface effects. *International Journal of Solids and Structures* 47: 407–418.
37. Mogilevskaya, S.G., Crouch, S.L., La Grotta, A., and Stolarski, H.K. (2009). The effects of surface elasticity and surface tension on the transverse overall elastic behaviour of unidirectional nano-composites. *Composites Science and Technology* doi:10.1016/j.compscitech.2009.11.012.
38. Bonnecaze, R.T. and Brady, J.F. (1990). A method for determining the effective conductivity of dispersions of particles. *Proceedings of the Royal Society of London* A430: 285–313.

39. Sangani, A.S. and Acrivos, A. (1983). The effective conductivity of a periodic array of spheres. *Proceedings of the Royal Society of London* A386: 263–275.
40. Mogilevskaya, S.G. and Crouch, S.L. (2001). A Galerkin boundary integral equation method for multiple circular elastic inclusions. *International Journal for Numerical Methods in Engineering* 52: 1069–1106.
41. Mogilevskaya, S.G. and Crouch, S.L. (2002). A Galerkin boundary integral equation method for multiple circular elastic inclusions with homogeneously imperfect interfaces. *International Journal of Solids and Structures* 39: 4723–4746.
42. Mogilevskaya, S.G. and Crouch, S.L. (2004). A Galerkin boundary integral equation method for multiple circular elastic inclusions with uniform interphase layers. *International Journal of Solids and Structures* 41: 1285–1311.

NOTE: This chapter is based on the publication

McCartney, L.N. (2010). Maxwell's far-field methodology predicting elastic properties of multiphase composites reinforced with aligned transversely isotropic spheroids. *Philosophical Magazine* 90: 4175–4207.

16

Debonding Models and Application to Fibre Fractures and Matrix Cracks

Overview: Analytical stress-transfer models are described that enable estimates to be made of the stress and displacement fields associated with fibre fractures or matrix cracks in unidirectional fibre-reinforced composites. The models represent a clear improvement on popular shear-lag-based methodologies, which are described for completeness. The model takes account of thermal residual stresses, and is based on simplifying assumptions that the axial stress in the fibre is independent of the radial coordinate, and similarly for the matrix. A representation for both the stress and displacement fields is derived that satisfies exactly the equilibrium equations, the required interface continuity equations for displacement and tractions, and all stress-strain equations except for that which relates to axial deformation. In addition, the representation is such that the Reissner energy functional has a stationary value provided that averaged axial stress-strain relations for the fibre and matrix are satisfied. The improved representation is fully consistent with variational mechanics and provides both the stress and displacement distributions in the fibre and the matrix.

For isolated or interacting fibre fractures or matrix cracks, interface debonding is considered where two types of condition are investigated. First, it is assumed that the shear stress is uniform within the debonded region, and a small transition zone is included in the model in order that essential zero traction conditions can be satisfied on the crack surfaces. Second, it is assumed that stress transfer in the debonded region is controlled by Coulomb friction. Predictions are made for carbon-fibre-reinforced epoxy composites, and the model is then used to consider fibre bridging of isolated matrix cracks.

16.1 Introduction

For unidirectional polymer composites, the fibre fracture is perhaps the most important damage mode, as it can be the precursor to progressive damage growth that leads on to the catastrophic failure of the composite. Fibre fracture is often accompanied by the formation of fibre/matrix debonds that affect the stress transfer between fibre and matrix, and the performance of the composite in directions normal to the fibre direction. For high-temperature brittle-matrix composites, the most important damage mode is matrix cracking in a direction normal to the fibre direction. The matrix cracks

Properties for Design of Composite Structures: Theory and Implementation Using Software, First Edition. Neil McCartney.

Companion Website: www.wiley.com/go/mccartney/properties

can be bridged by a number of intact fibres and the opening of the cracks is affected by the degree of fibre/matrix debonding.

To understand these damage modes, it is essential that a reliable analysis is undertaken so that stress transfer between fibres and matrix is modelled adequately. A concentric cylinder model is often used to analyse fibre fractures and matrix cracking, where stress transfer is estimated between two concentric cylinders and where the inner cylinder represents the fibre and the outer cylinder represents the matrix. Historically, shear-lag theories were first used for this purpose, as discussed in detail by Nairn [1] for the special case where the matrix remains perfectly bonded to the fibres. He critically assessed shear-lag methods which have frequently been used in the literature to analyse fibre/matrix stress transfer. He concluded that shear-lag methods provide poor estimates of shear stresses and energy release rates, and cannot be used for low fibre volume fractions. For brittle-matrix composites, a shear-lag model was the basis of Aveston–Cooper–Kelly (ACK) theory [2], when using an energy balance method to predict the conditions for matrix cracking when it is accompanied by fibre/matrix debonding characterised by a uniform interfacial shear stress (see also [3–5]). Hutchinson and Jensen [6] avoided the use of a shear-lag model by using the well-known Lamé solution, together with mode II fracture mechanics principles, to consider matrix cracking for cases where frictional slip occurs at the fibre/matrix interfaces characterised by either a uniform interfacial shear stress or by the Coulomb friction law. Their approach estimates a 'steady-state' energy release rate using a method that does not consider the variation of the stress field in the direction of the fibres, but only the stress states far 'upstream' and far 'downstream' relative to the debond location. To obtain more accurate solutions than shear-lag methods, Nairn [7] used a variational calculation based on the principle of minimising the complementary energy to develop a solution for the stress field only for the special case where the fibre/matrix interface remains perfectly bonded. He applied the model to the analysis of single-fibre fibre-pull-out tests and microdrop debond tests.

In this chapter, an axisymmetric model is considered subject to applied axial load distributions that gives rise to axial stress transfer between the concentric cylinders through the action of distributed shear stresses on the interface between the two cylinders. The model is constructed so that the cylinders can be made of different transversely isotropic materials. The model takes full account of the effects of the thermal residual stresses induced in the system during manufacture as a result of the difference in thermal expansion behaviour of the fibre and matrix. The solution technique, briefly described in [8], is an improvement of an earlier stress-transfer model [9]. The solution of the stress-transfer problem involves the development of an ordinary differential equation that can be solved by analytical methods, and enables subsequent calculation of the stress and displacement distributions throughout the system. The axisymmetric model of stress transfer leads to stationary values of the Reissner energy functional [10] so that the stress and displacement distribution derived from the model would also result from carrying out a corresponding variational calculation. Thus, the stress-transfer model is the best that can be developed based upon the single fundamental assumption that axial stresses in each cylinder are independent of the radial coordinate. This assumption is not expected to lead to reasonable predictions when the fibre volume fraction is small, as in a single-fibre pull-out or fragmentation test, but it is expected to lead to very useful and reasonably accurate predictions for use in fibre failure or matrix cracking models applied to unidirectional composites of practical interest.

Stress transfer between fibre and matrix will be considered for the idealised case when either a fibre of a composite is uniformly fragmented, or when the matrix has cracked uniformly so that the matrix cracks are normal to the fibre axis. The analysis to be presented here focuses on a simplifying situation where the fibre/matrix interface is assumed to be frictionally bonded. This means that there is no physical or chemical bonding between fibre and matrix. It is assumed that thermal residual stresses are such that the matrix clamps the fibre so that the composite will behave as a perfectly bonded composite provided that there is no slippage between the fibre and matrix. This would certainly be the situation for an undamaged composite. If a fibre fracture or matrix crack is present, then some fibre/matrix slippage is expected near the fracture plane. An indication will be given of how physically or chemically bonded interfaces may be analysed.

16.2 Field Equations

Consider the concentric cylinder model of a unidirectionally fibre-reinforced composite illustrated in Figure 16.1. The fibre, of radius R, is perfectly bonded to a surrounding annulus of matrix having inner radius R and outer radius a. The model is representative of a unidirectional composite having a fibre volume fraction V_{f}, which means that the outer radius of the matrix must have the value

$$a = \frac{R}{\sqrt{V_{\text{f}}}}, \tag{16.1}$$

ensuring that the fibre volume fraction for the concentric cylinder model is the same as that of the composite being modelled. The volume fraction of the matrix is given by $V_{\text{m}} = 1 - V_{\text{f}}$.

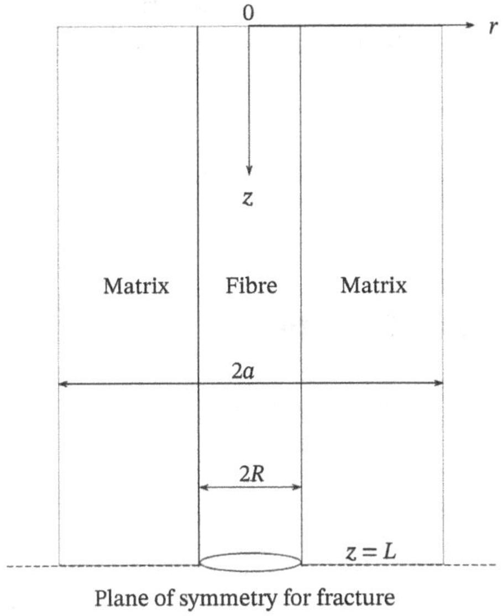

Figure 16.1 Schematic diagram of the geometry for a perfectly bonded interface associated with a fibre fracture or matrix crack.

A set of cylindrical polar coordinates (r, θ, z) is introduced such that the origin lies on the common axis of the two concentric cylinders (with the z-axis directed along the axis of the cylinders, i.e. in the axial loading direction, at the midpoint between two neighbouring fibre fractures or two neighbouring matrix cracks. Superscripts f and m will be used to denote parameters associated with the fibre and matrix, respectively.

For axisymmetric problems, the following equilibrium equations must be satisfied for both the fibre and matrix,

$$\frac{\partial \sigma_{rr}}{\partial r}+\frac{\partial \sigma_{rz}}{\partial z}+\frac{\sigma_{rr}-\sigma_{\theta\theta}}{r}=0, \tag{16.2}$$

$$\frac{\partial \sigma_{zz}}{\partial z}+\frac{\partial \sigma_{rz}}{\partial r}+\frac{\sigma_{rz}}{r}=0. \tag{16.3}$$

The fibre and matrix are regarded as transverse isotropic solids so that the stress-strain relations, in terms of the axial and transverse moduli E, Poisson's ratios ν, shear modulus μ and thermal expansion coefficients α are of the form

$$\varepsilon_{rr}=\frac{\partial u_r}{\partial r}=\frac{1}{E_{\mathrm{T}}}\sigma_{rr}-\frac{\nu_{\mathrm{t}}}{E_{\mathrm{T}}}\sigma_{\theta\theta}-\frac{\nu_{\mathrm{A}}}{E_{\mathrm{A}}}\sigma_{zz}+\alpha_{\mathrm{T}}\Delta T, \tag{16.4}$$

$$\varepsilon_{\theta\theta}=\frac{u_r}{r}=-\frac{\nu_{\mathrm{t}}}{E_{\mathrm{T}}}\sigma_{rr}+\frac{1}{E_{\mathrm{T}}}\sigma_{\theta\theta}-\frac{\nu_{\mathrm{A}}}{E_{\mathrm{A}}}\sigma_{zz}+\alpha_{\mathrm{T}}\Delta T, \tag{16.5}$$

$$\varepsilon_{zz}=\frac{\partial u_z}{\partial z}=-\frac{\nu_{\mathrm{A}}}{E_{\mathrm{A}}}\sigma_{rr}-\frac{\nu_{\mathrm{A}}}{E_{\mathrm{A}}}\sigma_{\theta\theta}+\frac{1}{E_{\mathrm{A}}}\sigma_{zz}+\alpha_{\mathrm{A}}\Delta T, \tag{16.6}$$

$$\varepsilon_{rz}=\frac{1}{2}\left(\frac{\partial u_r}{\partial z}+\frac{\partial u_z}{\partial r}\right)=\frac{\sigma_{rz}}{2\mu_{\mathrm{A}}},\ \varepsilon_{\theta z}=\frac{\sigma_{\theta z}}{2\mu_{\mathrm{A}}},\ \varepsilon_{r\theta}=\frac{\sigma_{r\theta}}{2\mu_{\mathrm{t}}}, \tag{16.7}$$

$$\text{where } E_{\mathrm{T}}=2\mu_{\mathrm{t}}(1+\nu_{\mathrm{t}}) \text{ but } E_{\mathrm{A}}\neq 2\mu_{\mathrm{A}}(1+\nu_{\mathrm{A}}). \tag{16.8}$$

The subscripts A and T refer the properties to the axial and transverse directions relative to the direction of the fibre axis. Following Nairn [7], when both σ_{zz} and ΔT are independent of r, the displacement u_r is compatible with the stress-strain relations (16.4) and (16.5) if the following compatibility equation for stresses is satisfied

$$(1+\nu_{\mathrm{t}})\frac{\sigma_{rr}-\sigma_{\theta\theta}}{r}=\frac{\partial}{\partial r}(\sigma_{\theta\theta}-\nu_{\mathrm{t}}\sigma_{rr}), \tag{16.9}$$

which is obtained by subtracting (16.5) from (16.4), differentiating (16.5) with respect to r, and then eliminating the displacement component u_r.

16.3 Interfacial and Radial Boundary Conditions

At the interface $r = R$ between the cylinders, the following continuity conditions must be satisfied for all values of z.

Perfect interface bonding:

$$\begin{aligned}&\sigma_{rr}^{\mathrm{f}}(R,z)=\sigma_{rr}^{\mathrm{m}}(R,z), \quad \sigma_{rz}^{\mathrm{f}}(R,z)=\sigma_{rz}^{\mathrm{m}}(R,z),\\&u_r^{\mathrm{f}}(R,z)=u_r^{\mathrm{m}}(R,z), \quad u_z^{\mathrm{f}}(R,z)=u_z^{\mathrm{m}}(R,z).\end{aligned} \tag{16.10}$$

Interface debonding:

$$\begin{aligned}&\sigma_{rr}^{\mathrm{f}}(R,z)=\sigma_{rr}^{\mathrm{m}}(R,z),\quad \sigma_{rz}^{\mathrm{f}}(R,z)=\sigma_{rz}^{\mathrm{m}}(R,z),\\ &u_r^{\mathrm{f}}(R,z)=u_r^{\mathrm{m}}(R,z),\ \sigma_{rz}^{\mathrm{f}}(R,z)=\xi\left[\theta\tau+\varphi\eta\sigma_{rr}^{\mathrm{f}}(R,z)\right].\end{aligned} \tag{16.11}$$

In (16.11), τ and η are regarded as material constants where $\theta, \varphi = 0$ or 1, and where $\xi = -1$ or $\xi = 1$. By selecting $\theta = 1$ and $\varphi = 0$ the boundary conditions (16.11) correspond to the those for an interface subject to a uniform interfacial shear stress $\pm\tau$. By selecting $\theta = 0$ and $\varphi = 1$ the boundary conditions correspond to those for an interface subject to the Coulomb law of friction where the parameter η is the friction coefficient. By selecting $\xi = 1$ the boundary conditions (16.11) correspond to those for a matrix crack, and by selecting $\xi = -1$ the conditions correspond to those for a fibre fracture. The boundary conditions (16.11) assume that on debonding at the interface the fibre and matrix remain in contact with one another and give rise to stress transfer between the cylinders where the fibre may slip relative to the matrix. If mechanical contact is lost, then clearly the stress components σ_{rr} and σ_{rz} are both zero and the displacement component u_r is discontinuous across such an interface. As no stress transfer occurs at such an interface, this case is not considered in this chapter. On the external surface $r = a$ of the outer cylinder the following boundary conditions are often imposed

$$\sigma_{rr}(a,z)=\sigma_{\mathrm{T}},\sigma_{rz}(a,z)=0, \tag{16.12}$$

where σ_{T} is a uniform transverse stress applied to the external surface of the outer cylinder. In regions away from the loading mechanism, and any matrix crack or fibre fracture, it is assumed that

$$u_z^{\mathrm{f}} \equiv u_z^{\mathrm{m}} \equiv \varepsilon_{\mathrm{A}} z, \tag{16.13}$$

where ε_{A} is the axial strain in such regions. Relation (16.13) is valid for all values of z when there is perfect fibre/matrix bonding and the system is undamaged, as described in Appendix B.

The boundary condition (16.12) for the radial stress could be replaced by a corresponding radial displacement condition, which would be appropriate when modelling fibre fractures embedded within a unidirectional composite, as considered in [6]. The radial component could be selected to be the radial displacement that would arise in an undamaged composite (see [11] for an example of this approach). The analysis will consider only boundary conditions of the type given in (16.12).

16.4 Shear-lag Theory

Stress-transfer phenomena in unidirectional composites have been examined extensively using an approximate method of stress analysis that is known as 'shear-lag theory'. The objective now is to describe the nature of this approach and to identify the various approximations that have to be made. The approach to be adopted follows the work of Nayfey [12], as in [13]. Nairn [1] has critically examined approaches to shear-lag theory, including comparison of predictions with the results of finite element analysis, and concluded that Nayfey's approach, which is now to be considered, is the most reliable shear-lag methodology, although it is by no means a satisfactory solution to stress-transfer problems in unidirectional composites.

The shear-lag approach is based on averages defined over the regions occupied by the fibre and matrix. The average of any quantity $f(r,z)$ associated with the fibre is defined by

$$\bar{f}(z) = \frac{1}{\pi R^2} \int_0^R 2\pi r\, f(r,z) \mathrm{d}r, \tag{16.14}$$

and the average of any quantity $m(r,z)$ associated with the matrix is defined by

$$\bar{m}(z) = \frac{1}{\pi\left(a^2 - R^2\right)} \int_R^a 2\pi r\, m(r,z) \mathrm{d}r. \tag{16.15}$$

Averaging (16.3) over the fibre using (16.14) leads to the result

$$\frac{\mathrm{d}\bar{\sigma}_{zz}}{\mathrm{d}z} + \frac{2\tau(z)}{R} = 0, \tag{16.16}$$

where $\tau(z) = \sigma_{rz}^{\mathrm{f}}(R,z)$. On averaging (16.3) over the matrix using (16.15) it can be shown that

$$\frac{\mathrm{d}\bar{\sigma}_{zz}}{\mathrm{d}z} - \frac{V_{\mathrm{f}}}{V_{\mathrm{m}}} \frac{2\tau(z)}{R} = 0, \tag{16.17}$$

where use has been made of the continuity condition $\sigma_{rz}^{\mathrm{f}}(R,z) = \sigma_{rz}^{\mathrm{m}}(R,z)$, and where it has been assumed that $\sigma_{rz}^{\mathrm{m}}(a,z) = 0$, i.e. the shear stress is zero on the external surface $r = a$. It follows from (16.16) and (16.17) that

$$\frac{\mathrm{d}}{\mathrm{d}z}\left(V_{\mathrm{f}}\, \bar{\sigma}_{zz}^{\mathrm{f}} + V_{\mathrm{m}}\, \bar{\sigma}_{zz}^{\mathrm{m}}\right) = 0. \tag{16.18}$$

The integrated form of (16.18) is

$$V_{\mathrm{f}}\, \bar{\sigma}_{zz}^{\mathrm{f}}(z) + V_{\mathrm{m}}\, \bar{\sigma}_{zz}^{\mathrm{m}}(z) = \sigma_{\mathrm{A}}, \tag{16.19}$$

where σ_{A} is the effective stress applied in the fibre direction. It is emphasised that relation (16.19) is valid for any value of z. The results (16.16), (16.17) and (16.19) are exact for any axisymmetric concentric cylinder model of a composite where the shear stress is everywhere zero on the external surface.

The stress-strain relation $(16.7)_1$ for shear may be written for fibre and matrix as follows:

$$\frac{\partial u_r^{\mathrm{f}}}{\partial z} + \frac{\partial u_z^{\mathrm{f}}}{\partial r} = \frac{\sigma_{rz}^{\mathrm{f}}}{\mu_{\mathrm{A}}^{\mathrm{f}}}, \qquad \frac{\partial u_r^{\mathrm{m}}}{\partial z} + \frac{\partial u_z^{\mathrm{m}}}{\partial r} = \frac{\sigma_{rz}^{\mathrm{m}}}{\mu_{\mathrm{A}}^{\mathrm{m}}}. \tag{16.20}$$

To develop a shear-lag model these relations are approximated by

$$\frac{\partial u_z^{\mathrm{f}}}{\partial r} = \frac{\sigma_{rz}^{\mathrm{f}}}{\mu_{\mathrm{A}}^{\mathrm{f}}}, \qquad \frac{\partial u_z^{\mathrm{m}}}{\partial r} = \frac{\sigma_{rz}^{\mathrm{m}}}{\mu_{\mathrm{A}}^{\mathrm{m}}}. \tag{16.21}$$

The shear-stress distribution is now assumed to have the following form

$$\sigma_{rz}^{\mathrm{f}}(r,z) = \tau(z)\frac{r}{R}, \qquad \sigma_{rz}^{\mathrm{m}}(r,z) = \tau(z)\left(\frac{R}{r} - V_{\mathrm{f}}\frac{r}{R}\right). \tag{16.22}$$

It is clear that $\sigma_{rz}^{f}(R,z) = \sigma_{rz}^{m}(R,z)$, and that $\sigma_{rz}^{m}(a,z) = 0$. On multiplying $(16.21)_1$ by r^2, on substituting the expression $(16.22)_1$ for σ_{rz} and then integrating with respect to r over the fibre, it can be shown that

$$\int_0^R r^2 \frac{\partial u_z^{\mathrm{f}}}{\partial r}\mathrm{d}r = \frac{R^3\tau(z)}{4\mu_{\mathrm{A}}^{\mathrm{f}}}. \tag{16.23}$$

On integrating the left-hand side of (16.23) by parts, it can be shown that

$$u_z^{\mathrm{f}}(R,z) - \bar{u}_z^{\mathrm{f}}(z) = \frac{R\tau(z)}{4\mu_{\mathrm{A}}^{\mathrm{f}}}. \tag{16.24}$$

On multiplying $(16.21)_2$ by $(R^2 - V_{\mathrm{f}}r^2)$, on substituting the expression $(16.22)_2$ for σ_{rz} and then integrating with respect to r over the matrix, it can be shown that

$$\int_R^a (R^2 - V_{\mathrm{f}}r^2)\frac{\partial u_z^{\mathrm{m}}}{\partial r}\mathrm{d}r = \frac{R^3\tau(z)}{2\mu_{\mathrm{A}}^{\mathrm{m}}}\left(\frac{1}{V_{\mathrm{m}}}\ln\frac{1}{V_{\mathrm{f}}} - 1 - \frac{V_{\mathrm{m}}}{2}\right). \tag{16.25}$$

On integrating the left-hand side of (16.25) by parts, it can be shown that

$$\bar{u}_z^{\mathrm{m}}(z) - u_z^{\mathrm{m}}(R,z) = \frac{R\tau(z)}{2V_{\mathrm{m}}\mu_{\mathrm{A}}^{\mathrm{m}}}\left(\frac{1}{V_{\mathrm{m}}}\ln\frac{1}{V_{\mathrm{f}}} - 1 - \frac{V_{\mathrm{m}}}{2}\right). \tag{16.26}$$

As there is perfect bonding at the fibre/matrix interface so that $u_z^{\mathrm{f}}(R,z) = u_z^{\mathrm{m}}(R,z)$, the addition of (16.24) and (16.26) leads to the important result

$$\bar{u}_z^{\mathrm{m}}(z) - \bar{u}_z^{\mathrm{f}}(z) = \frac{R}{V_{\mathrm{m}}}\eta\tau(z), \tag{16.27}$$

where

$$\eta = \frac{V_{\mathrm{m}}}{4\mu_{\mathrm{A}}^{\mathrm{f}}} + \frac{1}{2\mu_{\mathrm{A}}^{\mathrm{m}}}\left(\frac{1}{V_{\mathrm{m}}}\ln\frac{1}{V_{\mathrm{f}}} - 1 - \frac{V_{\mathrm{m}}}{2}\right). \tag{16.28}$$

Clearly the interfacial shear stress is proportional to the difference between the average axial displacements in the matrix and fibre.

The final step when developing shear-lag theory is to assume that the axial Poisson's ratio $\nu_{\mathrm{A}} = 0$ in both fibre and matrix. The stress-strain relation (16.6) is then approximated in fibre and matrix by the relations

$$\varepsilon_{zz}^{\mathrm{f}} = \frac{\partial u_z^{\mathrm{f}}}{\partial z} = \frac{1}{E_{\mathrm{A}}^{\mathrm{f}}}\sigma_{zz}^{\mathrm{f}} + \alpha_{\mathrm{A}}^{\mathrm{f}}\Delta T, \qquad \varepsilon_{zz}^{\mathrm{m}} = \frac{\partial u_z^{\mathrm{m}}}{\partial z} = \frac{1}{E_{\mathrm{A}}^{\mathrm{m}}}\sigma_{zz}^{\mathrm{m}} + \alpha_{\mathrm{A}}^{\mathrm{m}}\Delta T. \tag{16.29}$$

On using (16.14) and (16.15) the averaged forms of the stress-strain relations (16.29) may be written as

$$\frac{\partial \bar{u}_z^{\mathrm{f}}}{\partial z} = \frac{1}{E_{\mathrm{A}}^{\mathrm{f}}}\bar{\sigma}_{zz}^{\mathrm{f}} + \alpha_{\mathrm{A}}^{\mathrm{f}}\Delta T, \qquad \frac{\partial \bar{u}_z^{\mathrm{m}}}{\partial z} = \frac{1}{E_{\mathrm{A}}^{\mathrm{m}}}\bar{\sigma}_{zz}^{\mathrm{m}} + \alpha_{\mathrm{A}}^{\mathrm{m}}\Delta T. \tag{16.30}$$

Relations (16.16), (16.17), (16.29) and (16.30) then assert that

$$\frac{\partial^2 \bar{u}_z^{\mathrm{f}}}{\partial z^2} = -\frac{2}{E_{\mathrm{A}}^{\mathrm{f}}}\frac{\tau(z)}{R}, \qquad \frac{\partial^2 \bar{u}_z^{\mathrm{m}}}{\partial z^2} = \frac{2}{E_{\mathrm{A}}^{\mathrm{m}}}\frac{V_{\mathrm{f}}}{V_{\mathrm{m}}}\frac{\tau(z)}{R}. \tag{16.31}$$

On subtracting these relations and the making use of (16.27) it can be shown that the shear stress $\tau(z)$ satisfies the following second-order ordinary differential equation

$$\frac{\mathrm{d}^2\tau}{\mathrm{d}z^2} = k^2\tau, \tag{16.32}$$

where

$$k^2 = \frac{2\left(V_{\mathrm{f}}E_{\mathrm{A}}^{\mathrm{f}} + V_{\mathrm{m}}E_{\mathrm{A}}^{\mathrm{m}}\right)}{R^2\eta E_{\mathrm{A}}^{\mathrm{f}}E_{\mathrm{A}}^{\mathrm{m}}}. \tag{16.33}$$

As $\tau(0)=0$, the shear stress distribution satisfying (16.32) has the form

$$\tau(z)=A\sinh(kz), \tag{16.34}$$

where A is a constant to be determined.

For the special case $A=0$ such that the interfacial shear stress $\tau(z)\equiv 0$, i.e. there is no stress transfer between fibre and matrix, the axial strain in both the fibre and matrix have the same value ε_∞. This situation arises for isolated fibre fractures or isolated matrix cracks at large distances away from the fractures, hence the use of the subscript ∞. It follows from (16.29) that in such regions

$$\varepsilon_\infty=\frac{1}{E_{\rm A}^{\rm f}}\sigma_\infty^{\rm f}+\alpha_{\rm A}^{\rm f}\Delta T=\frac{1}{E_{\rm A}^{\rm m}}\sigma_\infty^{\rm m}+\alpha_{\rm A}^{\rm m}\Delta T, \tag{16.35}$$

and on using (16.19)

$$V_{\rm f}\,\sigma_\infty^{\rm f}+V_{\rm m}\,\sigma_\infty^{\rm m}=\sigma_{\rm A}. \tag{16.36}$$

It then follows that

$$\sigma_\infty^{\rm f}=\frac{E_{\rm A}^{\rm f}\left[\sigma_{\rm A}+V_{\rm m}E_{\rm A}^{\rm m}\left(\alpha_{\rm A}^{\rm m}-\alpha_{\rm A}^{\rm f}\right)\Delta T\right]}{V_{\rm f}E_{\rm A}^{\rm f}+V_{\rm m}E_{\rm A}^{\rm m}},\quad \sigma_\infty^{\rm m}=\frac{E_{\rm A}^{\rm m}\left[\sigma_{\rm A}+V_{\rm f}E_{\rm A}^{\rm f}\left(\alpha_{\rm A}^{\rm f}-\alpha_{\rm A}^{\rm m}\right)\Delta T\right]}{V_{\rm f}E_{\rm A}^{\rm f}+V_{\rm m}E_{\rm A}^{\rm m}}. \tag{16.37}$$

It follows from (16.16) and (16.17) that

$$\frac{{\rm d}\bar{\sigma}_{zz}}{{\rm d}z}=-\frac{2A}{R}\sinh(kz),\qquad \frac{{\rm d}\bar{\sigma}_{\rm mm}}{{\rm d}z}=\frac{V_{\rm f}}{V_{\rm m}}\frac{2A}{R}\sinh(kz). \tag{16.38}$$

On integration, and on making use of relations (16.37) that must result when $A=0$,

$$\bar{\sigma}_{zz}^{\rm f}(z)=-\frac{2A}{kR}\cosh(kz)+\sigma_{\rm f}^\infty,\qquad \bar{\sigma}_{zz}^{\rm m}(z)=\frac{V_{\rm f}}{V_{\rm m}}\frac{2A}{kR}\cosh(kz)+\sigma_{\rm m}^\infty. \tag{16.39}$$

The corresponding axial average displacement field, which satisfies the conditions $\bar{u}_z^{\rm f}(0)=\bar{u}_z^{\rm m}(0)=0$, is derived on integrating (16.30) so that

$$\begin{aligned}\bar{u}_z^{\rm f}(z)&=-\frac{2A}{k^2RE_{\rm A}^{\rm f}}\sinh(kz)+\left(\frac{\sigma_{\rm f}^\infty}{E_{\rm A}^{\rm f}}+\alpha_{\rm A}^{\rm f}\Delta T\right)z,\\ \bar{u}_z^{\rm m}(z)&=\frac{2A}{k^2RE_{\rm A}^{\rm m}}\frac{V_{\rm f}}{V_{\rm m}}\sinh(kz)+\left(\frac{\sigma_{\rm m}^\infty}{E_{\rm A}^{\rm m}}+\alpha_{\rm A}^{\rm m}\Delta T\right)z.\end{aligned} \tag{16.40}$$

This completes the development of relations for a shear-lag model that can be applied to fibre fractures or matrix cracks. It only remains to determine the value of the parameter A that characterises the distribution of the interfacial shear stress. Its value depends upon whether fibre fractures or matrix cracks are being considered.

16.4.1 Fibre Fractures

Consider now two fibre fractures separated by a distance $2L$. The axial stress must be zero on the fibre fractures so that its average is also zero. The use of result $(16.39)_1$ then leads to the following value of A

$$A=\frac{kR\sigma_{\rm f}^\infty}{2\cosh(kL)}. \tag{16.41}$$

From (16.34), (16.39) and (16.40) the average stress, shear stress and average displacement distributions are then given by

$$\bar{\sigma}_{zz}^{\mathrm{f}}(z) = \sigma_{\mathrm{f}}^{\infty}\left(1 - \frac{\cosh(kz)}{\cosh(kL)}\right), \qquad \bar{\sigma}_{zz}^{\mathrm{m}}(z) = \sigma_{\mathrm{f}}^{\infty}\frac{V_{\mathrm{f}}}{V_{\mathrm{m}}}\frac{\cosh(kz)}{\cosh(kL)} + \sigma_{\mathrm{m}}^{\infty}, \tag{16.42}$$

$$\tau(z) = \frac{1}{2}kR\sigma_{\mathrm{f}}^{\infty}\frac{\sinh(kz)}{\cosh(kL)}, \tag{16.43}$$

$$\begin{aligned} \bar{u}_z^{\mathrm{f}}(z) &= -\frac{\sigma_{\mathrm{f}}^{\infty}}{kE_{\mathrm{A}}^{\mathrm{f}}}\frac{\sinh(kz)}{\cosh(kL)} + \left(\frac{\sigma_{\mathrm{f}}^{\infty}}{E_{\mathrm{A}}^{\mathrm{f}}} + \alpha_{\mathrm{A}}^{\mathrm{f}}\Delta T\right)z, \\ \bar{u}_z^{\mathrm{m}}(z) &= \frac{\sigma_{\mathrm{f}}^{\infty}}{kE_{\mathrm{A}}^{\mathrm{m}}}\frac{V_{\mathrm{f}}}{V_{\mathrm{m}}}\frac{\sinh(kz)}{\cosh(kL)} + \left(\frac{\sigma_{\mathrm{m}}^{\infty}}{E_{\mathrm{A}}^{\mathrm{m}}} + \alpha_{\mathrm{A}}^{\mathrm{m}}\Delta T\right)z. \end{aligned} \tag{16.44}$$

It follows from (16.22) that the shear stress distributions in fibre and matrix are given by

$$\sigma_{rz}^{\mathrm{f}}(r,z) = \frac{1}{2}kr\sigma_{\mathrm{f}}^{\infty}\frac{\sinh(kz)}{\cosh(kL)}, \quad \sigma_{rz}^{\mathrm{m}}(r,z) = \frac{1}{2}k\sigma_{\mathrm{f}}^{\infty}\frac{\sinh(kz)}{\cosh(kL)}\left(\frac{R^2}{r} - V_{\mathrm{f}}r\right). \tag{16.45}$$

16.4.2 Matrix Cracks

Consider now two matrix cracks separated by a distance $2L$. The axial stress must be zero on the matrix cracks so that its average is also zero. The use of result $(16.39)_2$ then leads to the following value of A

$$A = -\frac{kR}{2}\frac{V_{\mathrm{m}}}{V_{\mathrm{f}}}\frac{\sigma_{\mathrm{m}}^{\infty}}{\cosh(kL)}. \tag{16.46}$$

From (16.34), (16.39) and (16.40) the average stress, shear stress and average displacement distributions are then given by

$$\bar{\sigma}_{zz}^{\mathrm{f}}(z) = \frac{V_{\mathrm{m}}}{V_{\mathrm{f}}}\sigma_{\mathrm{m}}^{\infty}\frac{\cosh(kz)}{\cosh(kL)} + \sigma_{\mathrm{f}}^{\infty}, \qquad \bar{\sigma}_{zz}^{\mathrm{m}}(z) = \sigma_{\mathrm{m}}^{\infty}\left(1 - \frac{\cosh(kz)}{\cosh(kL)}\right), \tag{16.47}$$

$$\tau(z) = -\frac{kR}{2}\frac{V_{\mathrm{m}}}{V_{\mathrm{f}}}\sigma_{\mathrm{m}}^{\infty}\frac{\sinh(kz)}{\cosh(kL)}, \tag{16.48}$$

$$\begin{aligned} \bar{u}_z^{\mathrm{f}}(z) &= \frac{\sigma_{\mathrm{m}}^{\infty}}{kE_{\mathrm{A}}^{\mathrm{f}}}\frac{V_{\mathrm{m}}}{V_{\mathrm{f}}}\frac{\sinh(kz)}{\cosh(kL)} + \left(\frac{\sigma_{\mathrm{f}}^{\infty}}{E_{\mathrm{A}}^{\mathrm{f}}} + \alpha_{\mathrm{A}}^{\mathrm{f}}\Delta T\right)z, \\ \bar{u}_z^{\mathrm{m}}(z) &= -\frac{\sigma_{\mathrm{m}}^{\infty}}{kE_{\mathrm{A}}^{\mathrm{m}}}\frac{\sinh(kz)}{\cosh(kL)} + \left(\frac{\sigma_{\mathrm{m}}^{\infty}}{E_{\mathrm{A}}^{\mathrm{m}}} + \alpha_{\mathrm{A}}^{\mathrm{m}}\Delta T\right)z. \end{aligned} \tag{16.49}$$

It follows from (16.22) that the shear stress distributions in fibre and matrix are given by

$$\sigma_{rz}^{\mathrm{f}}(r,z) = -\frac{kr}{2}\frac{V_{\mathrm{m}}}{V_{\mathrm{f}}}\sigma_{\mathrm{m}}^{\infty}\frac{\sinh(kz)}{\cosh(kL)}, \quad \sigma_{rz}^{\mathrm{m}}(r,z) = -\frac{k}{2}\frac{V_{\mathrm{m}}}{V_{\mathrm{f}}}\sigma_{\mathrm{m}}^{\infty}\frac{\sinh(kz)}{\cosh(kL)}\left(\frac{R^2}{r} - V_{\mathrm{f}}r\right). \tag{16.50}$$

The preferred shear-lag approach described previously is based on various approximations that are a cause for concern from a mechanics point of view. It is useful now to identify clearly these approximations. The first approximation that is made concerns the neglect in the fibre and matrix of the terms $\partial u_r / \partial z$ in the expression for the shear strain ε_{rz} (see (16.21)). This assumption is expected to lead to significant errors. The second approximation that has been made is the assumption that the axial Poisson's ratios in fibre and matrix must be zero (see (16.29)). This assumption is also expected to lead to significant errors. It is observed that the equilibrium equation (16.2) and the stress-strain relations (16.4) and (16.5) have not been considered, indicating further sources of significant error. It is concluded, that although the preferred shear-lag approach is relatively easy to develop and understand, it does suffer having to make a number of assumptions that are expected to lead to unacceptable errors.

16.5 More Accurate Stress-transfer Model

The approach to developing a solution for a damaged system is to express the solution as a sum of the undamaged solution (see Appendix B) and a perturbation solution arising from the damage. For the fibre region $0 \le r \le R$, the stress field when damage is present is assumed to be of the following form equivalent to that assumed by Nairn [7]:

$$\sigma_{zz}^{\mathrm{f}}(r,z) = \sigma_{\mathrm{f}} - C(z), \tag{16.51}$$

$$\sigma_{rz}^{\mathrm{f}}(r,z) = \frac{1}{2} C'(z) r, \tag{16.52}$$

$$\sigma_{rr}^{\mathrm{f}}(r,z) = -\frac{1}{16}\left(3+\nu_{\mathrm{t}}^{\mathrm{f}}\right) C''(z) r^2 + R_{\mathrm{f}}(z) + \sigma_{\mathrm{T}} - V_{\mathrm{m}} \frac{\phi}{R^2}, \tag{16.53}$$

$$\sigma_{\theta\theta}^{\mathrm{f}}(r,z) = -\frac{1}{16}\left(1+3\nu_{\mathrm{t}}^{\mathrm{f}}\right) C''(z) r^2 + R_{\mathrm{f}}(z) + \sigma_{\mathrm{T}} - V_{\mathrm{m}} \frac{\phi}{R^2}, \tag{16.54}$$

where σ_{f} and ϕ relate to the undamaged solution (see Appendix B), σ_{f} being the uniform axial fibre stress in an undamaged composite subject to the same loading conditions and temperature. For the matrix region $R \le r \le a$, the stress field is assumed to be of the following form, again equivalent to that used by Nairn [7]:

$$\sigma_{zz}^{\mathrm{m}}(r,z) = \sigma_{\mathrm{m}} + \frac{V_{\mathrm{f}}}{V_{\mathrm{m}}} C(z), \tag{16.55}$$

$$\sigma_{rz}^{\mathrm{m}}(r,z) = \frac{C'(z)}{2V_{\mathrm{m}}}\left(\frac{R^2}{r} - V_{\mathrm{f}} r\right), \tag{16.56}$$

$$\begin{aligned}\sigma_{rr}^{\mathrm{m}}(r,z) = {} & \left[\left(3+\nu_{\mathrm{t}}^{\mathrm{m}}\right)\frac{r^2}{a^2} - 4\left(1+\nu_{\mathrm{t}}^{\mathrm{m}}\right)\ln\frac{r}{a} - 2\left(1-\nu_{\mathrm{t}}^{\mathrm{m}}\right)\right]\frac{R^2}{16V_{\mathrm{m}}} C''(z) \\ & + R_{\mathrm{m}}(z) - \frac{S_{\mathrm{m}}(z)}{r^2} + \sigma_{\mathrm{T}} + \phi\left(\frac{1}{a^2} - \frac{1}{r^2}\right),\end{aligned} \tag{16.57}$$

$$\begin{aligned}\sigma_{\theta\theta}^{\mathrm{m}}(r,z) = {} & \left[\left(1+3\nu_{\mathrm{t}}^{\mathrm{m}}\right)\frac{r^2}{a^2} - 4\left(1+\nu_{\mathrm{t}}^{\mathrm{m}}\right)\ln\frac{r}{a} + 2\left(1-\nu_{\mathrm{T}}^{\mathrm{m}}\right)\right]\frac{R^2}{16V_{\mathrm{m}}} C''(z) \\ & + R_{\mathrm{m}}(z) + \frac{S_{\mathrm{m}}(z)}{r^2} + \sigma_T + \phi\left(\frac{1}{a^2} + \frac{1}{r^2}\right),\end{aligned} \tag{16.58}$$

where σ_{m} is the uniform axial matrix stress in an undamaged composite subject to the same loading conditions and temperature (see Appendix B). It should be noted that the interface continuity relations $(16.10)_2$ and $(16.11)_2$ are satisfied by relations (16.52) and (16.56). The functions $C(z)$ and $R_\mathrm{f}(z)$, $R_\mathrm{m}(z)$ and $S_\mathrm{m}(z)$ are regarded as being identically zero when no form of damage is present and the fibre/matrix interface is perfectly bonded. It is clear that the function $C(z)$, having the dimensions of stress, describes the stress transfer between the fibre and matrix.

On using (16.4) or (16.5), together with (16.51), (16.53) and (16.54), the corresponding representation for the displacement component u_r^f, for the fibre region $0 \le r \le R$, is given by

$$\frac{u_r^\mathrm{f}(r,z)}{r} = -\frac{1-\nu_\mathrm{t}^\mathrm{f}}{32\mu_\mathrm{t}^\mathrm{f}} C''(z) r^2 + \frac{\nu_\mathrm{A}^\mathrm{f}}{E_\mathrm{A}^\mathrm{f}} C(z) + \frac{1-\nu_\mathrm{t}^\mathrm{f}}{E_\mathrm{T}^\mathrm{f}} R_\mathrm{f}(z) + A_\mathrm{f}, \tag{16.59}$$

where A_f is defined in Appendix B. On using $(16.7)_1$, (16.52) and (16.59) it can be shown on integrating with respect to r that for $0 \le r \le R$,

$$\begin{aligned} u_z^f(r,z) = &-\frac{1}{2}\left[\left(\frac{\nu_\mathrm{A}^f}{E_\mathrm{A}^f} - \frac{1}{2\mu_\mathrm{A}^f}\right) C'(z) + \frac{1-\nu_\mathrm{t}^f}{E_\mathrm{T}^f} R_f'(z)\right]\left(r^2 - R^2\right) \\ &+ \frac{1-\nu_\mathrm{t}^f}{128\mu_\mathrm{t}^f} C'''(z)\left(r^4 - R^4\right) + H_f(z) + \varepsilon_\mathrm{A} z, \end{aligned} \tag{16.60}$$

where $H_\mathrm{f}(z) + \varepsilon_\mathrm{A} z \equiv u_z^\mathrm{f}(R,z)$ arises from the integration representing the axial displacement distribution in the fibre along the interface. The function $H_\mathrm{f}(z) \equiv 0$ when the system is in an undamaged state. Similarly, the corresponding displacement components for the matrix region $R \le r \le a$ are given by

$$\begin{aligned} \frac{u_r^\mathrm{m}(r,z)}{r} = &\left(\frac{1}{2}\frac{r^2}{a^2} - 2\ln\frac{r}{a} + 1\right)\frac{1-\nu_\mathrm{t}^\mathrm{m}}{16\mu_\mathrm{t}^\mathrm{m}}\frac{R^2}{V_\mathrm{m}} C''(z) - \frac{\nu_\mathrm{A}^\mathrm{m}}{E_\mathrm{A}^\mathrm{m}}\frac{V_f}{V_\mathrm{m}} C(z) \\ &+ \frac{1-\nu_\mathrm{t}^\mathrm{m}}{E_\mathrm{T}^\mathrm{m}} R_\mathrm{m}(z) + \frac{S_\mathrm{m}(z)}{2\mu_\mathrm{t}^\mathrm{m} r^2} + A_\mathrm{m} + \frac{\phi}{2\mu_\mathrm{t}^\mathrm{m} r^2}, \end{aligned} \tag{16.61}$$

$$\begin{aligned} u_z^\mathrm{m}(r,z) = &\frac{1-\nu_\mathrm{t}^\mathrm{m}}{16\mu_\mathrm{t}^\mathrm{m}}\frac{R^2}{V_\mathrm{m}} C'''(z) r^2 \ln\frac{r}{R} + \left(\frac{R^2}{2\mu_\mathrm{A}^\mathrm{m} V_\mathrm{m}} C'(z) - \frac{S_\mathrm{m}'(z)}{2\mu_\mathrm{t}^\mathrm{m}}\right)\ln\frac{r}{R} \\ &+\frac{1}{2}\left[\left(\frac{\nu_\mathrm{A}^\mathrm{m}}{E_\mathrm{A}^\mathrm{m}} - \frac{1}{2\mu_\mathrm{A}^\mathrm{m}}\right)\frac{V_f}{V_\mathrm{m}} C'(z) - \frac{1-\nu_\mathrm{t}^\mathrm{m}}{16\mu_\mathrm{t}^\mathrm{m}}\frac{2-\ln V_f}{V_\mathrm{m}} R^2 C'''(z) - \frac{1-\nu_\mathrm{t}^\mathrm{m}}{E_\mathrm{T}^\mathrm{m}} R_\mathrm{m}'(z)\right]\left(r^2 - R^2\right) \\ &- \frac{1-\nu_\mathrm{t}^\mathrm{m}}{128\mu_\mathrm{t}^\mathrm{m}}\frac{V_f}{V_\mathrm{m}} C'''(z)\left(r^4 - R^4\right) + H_\mathrm{m}(z) + \varepsilon_\mathrm{A} z, \end{aligned} \tag{16.62}$$

where A_m is defined in Appendix B and where $H_\mathrm{m}(z) + \varepsilon_\mathrm{A} z \equiv u_z^\mathrm{m}(R,z)$ arises from the integration representing the axial displacement distribution in the matrix along the interface. The function $H_\mathrm{m}(z) \equiv 0$ when the system is in an undamaged state. It should be noted that the displacement component u_θ is everywhere zero in both the fibre and matrix.

The stress and displacement fields specified by (16.51)–(16.62) satisfy exactly the equilibrium equations, and the compatibility equations together with the stress-strain relations (16.4), (16.5) and (16.7) for *any* function $C(z)$, and for any functions $R_\mathrm{f}(z)$, $R_\mathrm{m}(z)$, $S_\mathrm{m}(z)$, $H_\mathrm{f}(z)$ and $H_\mathrm{m}(z)$. In addition, the function $C'(z)$ is double the value of the interfacial shear stress σ_{rz} at the point z, and because from $(16.12)_2$ $\sigma_{rz}(a,z) \equiv 0$ it follows from (16.51), (16.55) and (B.44) (see Appendix B) that

$$V_{\rm f}\sigma_{zz}^{\rm f}(r,z) + V_{\rm m}\sigma_{zz}^{\rm m}(r,z) = \sigma_{\rm A}, \tag{16.63}$$

for all values of r and z. It should be noted that the axial stress-strain equation (16.6) is considered in Section 16.7.

16.6 Determination of the Integration Functions

The next objective is to determine the functions $R_{\rm f}(z)$, $R_{\rm m}(z)$ and $S_{\rm m}(z)$ that were introduced when carrying out integrations of the equilibrium equations in terms of the stress-transfer function $C(z)$. The application of the external boundary condition $(16.12)_1$ for the radial stress to relation (16.57) leads to the following expression for the function $S_{\rm m}(z)$:

$$\frac{S_{\rm m}(z)}{a^2} = \left(1+3\nu_{\rm t}^{\rm m}\right)\frac{R^2}{16V_{\rm m}}C''(z) + R_{\rm m}(z). \tag{16.64}$$

The application of the interfacial continuity condition $(16.10)_1$ or $(16.11)_1$ for the radial stress using (16.53), (16.57) and (16.64) leads to the relation

$$V_{\rm f}R_{\rm f}(z) + V_{\rm m}R_{\rm m}(z) = \alpha R^2 C''(z), \tag{16.65}$$

where α is a dimensionless parameter defined by

$$\alpha = \frac{1}{16}\left[V_{\rm f}\nu_{\rm t}^{\rm f} + V_{\rm m}\nu_{\rm t}^{\rm m} - \left(1+4\nu_{\rm t}^{\rm m}\right) - 2\left(1+\nu_{\rm t}^{\rm m}\right)\frac{V_{\rm f}}{V_{\rm m}}\ln V_{\rm f}\right]. \tag{16.66}$$

The application of the interfacial continuity condition $(16.10)_3$ or $(16.11)_3$ for the radial displacement using (16.8), (16.59), (16.61), (16.64), and relations (B.22) and (B.23) (see Appendix B) for an undamaged composite, leads to the relation

$$\begin{aligned}\left(\frac{\nu_{\rm A}^{\rm f}}{E_{\rm A}^{\rm f}} + \frac{\nu_{\rm A}^{\rm m}}{E_{\rm A}^{\rm m}}\frac{V_f}{V_{\rm m}}\right)C(z) - \left[\left(1-\nu_{\rm t}^{\rm f}\right)\frac{\mu_{\rm t}^{\rm m}}{\mu_{\rm t}^{\rm f}} + \left(1-\nu_{\rm t}^{\rm m}\right)\frac{V_f + 2(1-\ln V_f)}{V_{\rm m}} + \frac{1+3\nu_{\rm t}^{\rm m}}{V_f V_{\rm m}}\right]\frac{R^2C''(z)}{32\mu_{\rm t}^{\rm m}} \\ + \frac{1-\nu_{\rm t}^{\rm f}}{1+\nu_{\rm t}^{\rm f}}\frac{R_f(z)}{2\mu_{\rm t}^{\rm f}} - \left(\frac{1-\nu_{\rm t}^{\rm m}}{1+\nu_{\rm t}^{\rm m}} + \frac{1}{V_f}\right)\frac{R_{\rm m}(z)}{2\mu_{\rm t}^{\rm m}} = 0.\end{aligned} \tag{16.67}$$

Equations (16.65) and (16.67) are then solved simultaneously for the functions $R_{\rm f}(z)$ and $R_{\rm m}(z)$ so that

$$\begin{aligned} R_{\rm m}(z) &= \beta C(z) + \gamma R^2 C''(z), \\ R_{\rm f}(z) &= -\beta\frac{V_{\rm m}}{V_{\rm f}}C(z) + \frac{1}{V_{\rm f}}(\alpha - \gamma V_{\rm m})R^2 C''(z), \end{aligned} \tag{16.68}$$

where β, γ and ω are dimensionless parameters defined by

$$\begin{aligned} \beta &= 2\mu_{\rm t}^{\rm m}V_f\left(\frac{\nu_{\rm A}^{\rm f}}{V_f E_{\rm A}^{\rm f}} + \frac{\nu_{A}^{\rm m}}{V_{\rm m}E_{\rm A}^{\rm m}}\right)\omega, \\ \gamma &= \left[\frac{1-\nu_{\rm t}^{\rm f}}{1+\nu_{\rm t}^{\rm f}}\frac{\mu_{\rm t}^{\rm m}}{\mu_{\rm t}^{\rm f}}\left(\frac{\alpha}{V_f} - \frac{1+\nu_{\rm t}^{\rm f}}{16}\right) - \left(1-\nu_{\rm t}^{\rm m}\right)\frac{V_f + 2(1-\ln V_f)}{16\,V_{\rm m}} - \frac{1+3\nu_{\rm t}^{\rm m}}{16V_f V_{\rm m}}\right]\omega, \\ \frac{1}{\omega} &= \frac{1-\nu_{\rm t}^{\rm f}}{1+\nu_{\rm t}^{\rm f}}\frac{\mu_{\rm t}^{\rm m}}{\mu_{\rm t}^{f}}\frac{V_{\rm m}}{V_f} + \frac{1-\nu_{\rm t}^{\rm m}}{1+\nu_{\rm t}^{\rm m}} + \frac{1}{V_f}. \end{aligned} \tag{16.69}$$

16.7 Derivation of Differential Equation for a Perfectly Bonded Interface

For the stress and displacement representations derived previously it is not possible to satisfy exactly the stress-strain relations for the fibre and matrix having the form given by (16.6). However, it is possible to satisfy corresponding averaged stress-strain relations. For the fibre the averaged stress-strain relation is

$$\frac{d\bar{u}_z^{\rm f}}{dz} = -\frac{\nu_{\rm A}^{\rm f}}{E_{\rm A}^{\rm f}}\left(\bar{\sigma}_{rr}^{\rm f} + \bar{\sigma}_{\theta\theta}^{\rm f}\right) + \frac{\bar{\sigma}_{zz}^{\rm f}}{E_{\rm A}^{\rm f}} + \alpha_{\rm A}^{\rm f}\Delta T, \tag{16.70}$$

where an average value for the fibre is denoted by an overbar defined by (16.14). Similarly, on averaging (16.60), it can be shown that

$$\begin{aligned}\bar{u}_z^{\rm f}(z) = & -\frac{1-\nu_{\rm t}^{\rm f}}{192\mu_{\rm t}^{\rm f}} R^4 C''''(z) \\ & + \frac{R^2}{4}\left[\frac{1-\nu_{\rm t}^{\rm f}}{E_{\rm T}^{\rm f}} R_{\rm f}'(z) + \left(\frac{\nu_{\rm A}^{\rm f}}{E_{\rm A}^{\rm f}} - \frac{1}{2\mu_{\rm A}^{\rm f}}\right) C'(z)\right] + H_{\rm f}(z) + \varepsilon_{\rm A} z.\end{aligned} \tag{16.71}$$

For the matrix, the averaged stress-strain relation is

$$\frac{d\bar{u}_z^{\rm m}}{dz} = -\frac{\nu_{\rm A}^{\rm m}}{E_{\rm A}^{\rm m}}\left(\bar{\sigma}_{rr}^{\rm m} + \bar{\sigma}_{\theta\theta}^{\rm m}\right) + \frac{\bar{\sigma}_{zz}^{\rm m}}{E_{\rm A}^{\rm m}} + \alpha_{\rm A}^{\rm m}\Delta T, \tag{16.72}$$

where an average value for the matrix is denoted by an overbar defined by (16.15). Similarly, on averaging (16.62), it can be shown that

$$\begin{aligned}\bar{u}_z^{\rm m}(z) = & \, a_{\rm m}\frac{1-\nu_{\rm t}^{\rm m}}{16\mu_{\rm t}^{\rm m}}\frac{R^2}{V_{\rm m}} C''''(z) + b_{\rm m}\left(\frac{R^2}{2\mu_{\rm A}^{\rm m} V_{\rm m}} C'(z) - \frac{S_{\rm m}'(z)}{2\mu_{\rm t}^{\rm m}}\right) - c_{\rm m}\frac{1-\nu_{\rm t}^{\rm m}}{128\mu_{\rm t}^{\rm m}}\frac{V_{\rm f}}{V_{\rm m}} C''''(z) \\ & + \frac{d_{\rm m}}{2}\left[\left(\frac{\nu_{\rm A}^{\rm m}}{E_{\rm A}^{\rm m}} - \frac{1}{2\mu_{\rm A}^{\rm m}}\right)\frac{V_{\rm f}}{V_{\rm m}} C'(z) - \frac{1-\nu_{\rm t}^{\rm m}}{16\mu_{\rm t}^{\rm m}}\frac{2-\ln V_{\rm f}}{V_{\rm m}} R^2 C''''(z) - \frac{1-\nu_{\rm T}^{\rm m}}{E_{\rm T}^{\rm m}} R_{\rm m}'(z)\right] \\ & + H_{\rm m}(z) + \varepsilon_{\rm A} z,\end{aligned} \tag{16.73}$$

where

$$\begin{aligned} a_{\rm m} &= \frac{1}{a^2 - R^2}\int_R^a 2r^3 \ln\frac{r}{R} dr = \frac{1}{8(a^2-R^2)}\left(4a^4\ln\frac{a}{R} + R^4 - a^4\right) \\ &= \frac{R^2}{8V_{\rm f}V_{\rm m}}\left(V_{\rm f}^2 - 1 - 2\ln V_{\rm f}\right), \\ b_{\rm m} &= \frac{1}{a^2 - R^2}\int_R^a 2r\ln\frac{r}{R} dr = \frac{1}{2(a^2-R^2)}\left(2a^2\ln\frac{a}{R} + R^2 - a^2\right) = -\frac{1}{2}\left(1 + \frac{1}{V_{\rm m}}\ln V_{\rm f}\right), \\ c_{\rm m} &= \frac{1}{a^2 - R^2}\int_R^a 2r\left(r^4 - R^4\right) dr = \frac{1}{3}\left(a^2 - R^2\right)\left(a^2 + 2R^2\right) = \frac{R^4}{3}\frac{V_{\rm m}(1+2V_{\rm f})}{V_{\rm f}^2}, \\ d_{\rm m} &= \frac{1}{a^2 - R^2}\int_R^a 2r\left(r^2 - R^2\right) dr = \frac{1}{2}\left(a^2 - R^2\right) = \frac{R^2}{2}\frac{V_{\rm m}}{V_f}. \end{aligned} \tag{16.74}$$

On averaging (16.53) and (16.54), and on averaging (16.57) and (16.58), it can be shown that

$$\bar{\sigma}_{rr}^{\rm f}(z) + \bar{\sigma}_{\theta\theta}^{\rm f}(z) = 2\left(\frac{\alpha}{V_{\rm f}} - \frac{V_{\rm m}}{V_{\rm f}}\gamma - \frac{1+\nu_{\rm t}^{\rm f}}{16}\right) R^2 C''(z) - 2\beta\frac{V_{\rm m}}{V_{\rm f}} C(z) + 2\left(\sigma_{\rm T} - V_{\rm m}\frac{\phi}{R^2}\right), \tag{16.75}$$

$$\bar{\sigma}_{rr}^{\rm m}(z)+\bar{\sigma}_{\theta\theta}^{\rm m}(z)=\left[2\gamma+\frac{1+\nu_{\rm t}^{\rm m}}{8V_{\rm m}^2}\{(3+V_{\rm f})V_{\rm m}+2V_{\rm f}\ln V_{\rm f}\}\right]R^2C''(z)$$
$$+2\beta C(z)+2\left(\sigma_{\rm T}+V_{\rm f}\frac{\phi}{R^2}\right). \tag{16.76}$$

For a perfectly bonded interface, the identification $H_{\rm f}(z)\equiv H_{\rm m}(z)\equiv H(z)$ is made to satisfy the boundary condition $(16.10)_4$. The substitution of (16.51), (16.71) and (16.75) into (16.70), and of (16.55), (16.73) and (16.76) into (16.72), followed by a subtraction to eliminate the function $H(z)$, leads to the following homogeneous fourth-order ordinary differential equation that must be satisfied by the stress-transfer function $C(z)$ in the perfectly bonded region

$$FR^4C''''(z)+GR^2C''(z)+HC(z)=0, \tag{16.77}$$

where the constant coefficients F, G and H are given by

$$\begin{aligned}F&=F_0+\alpha F_\alpha+\beta F_\beta+\gamma F_\gamma,\\ G&=G_0+\alpha G_\alpha+\beta G_\beta+\gamma G_\gamma,\\ H&=H_0+\alpha H_\alpha+\beta H_\beta+\gamma H_\gamma,\end{aligned} \tag{16.78}$$

where the parameters F_0, F_α, F_β, F_γ etc. are defined by

$$F_0=\frac{\left(\nu_{\rm t}^{\rm m}\right)^2}{192V_{\rm f}V_{\rm m}^2E_{\rm T}^{\rm m}}\left[V_{\rm m}\left(V_{\rm f}^2+4V_{\rm f}-17\right)-3\left(1+V_{\rm f}\right)\left(3-V_{\rm f}\right)\ln V_{\rm f}\right]$$
$$-\frac{\nu_{\rm t}^{\rm m}}{16V_{\rm f}V_{\rm m}^2E_{\rm T}^{\rm m}}\left(V_{\rm m}+\ln V_{\rm f}\right)+\frac{1}{192V_{\rm f}E_{\rm T}^{\rm m}}\left(5+V_{\rm f}-3\ln V_{\rm f}\right)-\frac{1-\left(\nu_{\rm t}^{\rm f}\right)^2}{192E_{\rm T}^{\rm f}}, \tag{16.79}$$

$$F_\alpha=\frac{1-\nu_{\rm t}^{\rm f}}{8V_{\rm f}E_{\rm T}^{\rm f}},\ F_\beta=0,\ F_\gamma=-\frac{1+\nu_{\rm t}^{\rm m}}{8V_{\rm f}V_{\rm m}E_{\rm T}^{\rm m}}\left[(1+V_{\rm f})V_{\rm m}+2\ln V_{\rm f}\right]-\frac{\nu_{\rm t}^{\rm m}}{4E_{\rm T}^{\rm m}}\frac{V_{\rm m}}{V_{\rm f}}-\frac{1-\nu_{\rm t}^{\rm f}}{8E_{\rm T}^{\rm f}}\frac{V_{\rm m}}{V_{\rm f}},$$

$$G_0=\frac{1}{16V_{\rm m}^2\mu_{\rm A}^{\rm m}}\left[(3-V_{\rm f})V_{\rm m}+2\ln V_{\rm f}\right]-\frac{1}{16\mu_{\rm A}^{\rm f}}-\frac{\nu_{\rm A}^{\rm m}}{8E_{\rm A}^{\rm m}}$$
$$-\frac{\nu_{\rm A}^{\rm m}\left(1+\nu_{\rm t}^{\rm m}\right)}{16V_{\rm m}^2E_{\rm A}^{\rm m}}\left[(3+V_{\rm f})V_{\rm m}+2V_{\rm f}\ln V_{\rm f}\right]+\frac{\nu_{\rm A}^{\rm f}\left(1-\nu_{\rm t}^{\rm f}\right)}{16E_{\rm A}^{\rm f}}, \tag{16.80}$$

$$G_\alpha=\frac{\nu_{\rm A}^{\rm f}}{V_{\rm f}E_{\rm A}^{\rm f}},\ G_\beta=F_\gamma,\ G_\gamma=-V_{\rm m}\left(\frac{\nu_{\rm A}^{\rm f}}{V_{\rm f}E_{\rm A}^{\rm f}}+\frac{\nu_{\rm A}^{\rm m}}{V_{\rm m}E_{\rm A}^{\rm m}}\right),$$

$$H_0=\frac{V_{\rm f}}{2}\left(\frac{1}{V_{\rm f}E_{\rm A}^{\rm f}}+\frac{1}{V_{\rm m}E_{\rm A}^{\rm m}}\right),\quad H_\alpha=0,\quad H_\beta=G_\gamma,\quad H_\gamma=0. \tag{16.81}$$

It has been shown using the algebraic programming language REDUCE, that F, G and H correspond to formulae given by Nairn [7] derived using a variational technique (note that the expression for C_{35} given by Nairn should include a minus sign before the ratio $(V_2A_1)/(V_1A_2)$ that appears in his result). It is concluded that the approach being taken provides the displacement field corresponding to the stress-based variational calculation [7] that minimises the complementary energy.

16.8 The Average Axial Displacement Functions

The substitution of (16.51), (16.55), (16.75) and (16.76) into (16.70) and (16.72), followed by an integration with respect to z leads to the following expressions for the averages of the axial displacement components in the fibre and matrix, respectively:

$$\bar{u}_z^{\mathrm{f}}(z)=-\frac{2\nu_{\mathrm{A}}^{\mathrm{f}}}{E_{\mathrm{A}}^{\mathrm{f}}}\left(\frac{\alpha}{V_{\mathrm{f}}}-\frac{V_{\mathrm{m}}}{V_{\mathrm{f}}}\gamma-\frac{1+\nu_{\mathrm{t}}^{\mathrm{f}}}{16}\right)R^2C'(z)-\frac{1}{E_{\mathrm{A}}^{\mathrm{f}}}\left(1-2\beta\nu_{\mathrm{A}}^{\mathrm{f}}\frac{V_{\mathrm{m}}}{V_{\mathrm{f}}}\right)\bar{C}(z)+\varepsilon_{\mathrm{A}}z, \quad (16.82)$$

$$\begin{aligned}\bar{u}_z^{\mathrm{m}}(z)=&-\frac{2\nu_{\mathrm{A}}^{\mathrm{m}}}{E_{\mathrm{A}}^{\mathrm{m}}}\left[\gamma+\frac{1+\nu_{\mathrm{t}}^{\mathrm{m}}}{16V_{\mathrm{m}}^2}\left\{(3+V_{\mathrm{f}})V_{\mathrm{m}}+2V_{\mathrm{f}}\ln V_{\mathrm{f}}\right\}\right]R^2C'(z)\\&+\frac{V_{\mathrm{f}}}{V_{\mathrm{m}}E_{\mathrm{A}}^{\mathrm{m}}}\left(1-2\beta\nu_{\mathrm{A}}^{\mathrm{m}}\frac{V_{\mathrm{m}}}{V_{\mathrm{f}}}\right)\bar{C}(z)+\varepsilon_{\mathrm{A}}z,\end{aligned} \quad (16.83)$$

where

$$\bar{C}(z)=\int_0^z C(z')\mathrm{d}z'. \quad (16.84)$$

As, by symmetry, $C'(0)=0$, it should be noted that $\bar{u}_z^{\mathrm{f}}(0)=\bar{u}_z^{\mathrm{m}}(0)=0$.

The only remaining functions that have not been determined are the integration functions $H_{\mathrm{f}}(z)$ and $H_{\mathrm{m}}(z)$ appearing in relations (16.60) and (16.62) for the axial displacement in fibre and matrix, respectively. The function $H_{\mathrm{f}}(z)$ is obtained using the equivalent results (16.71) and (16.82) so that

$$\begin{aligned}H_{\mathrm{f}}(z)=&-\frac{1-\nu_{\mathrm{t}}^{\mathrm{f}}}{4E_{\mathrm{T}}^{\mathrm{f}}}\left(\frac{\alpha}{V_{\mathrm{f}}}-\frac{V_{\mathrm{m}}}{V_{\mathrm{f}}}\gamma-\frac{1+\nu_{\mathrm{t}}^{\mathrm{f}}}{24}\right)R^4C'''(z)+\left(\frac{1-\nu_{\mathrm{t}}^{\mathrm{f}}}{4E_{\mathrm{T}}^{\mathrm{f}}}\frac{V_{\mathrm{m}}}{V_{\mathrm{f}}}\beta+\frac{1}{8\mu_{\mathrm{A}}^{\mathrm{f}}}\right)R^2C'(z)\\&-\frac{2\nu_{\mathrm{A}}^{\mathrm{f}}}{E_{\mathrm{A}}^{\mathrm{f}}}\left(\frac{\alpha}{V_{\mathrm{f}}}-\frac{V_{\mathrm{m}}}{V_{\mathrm{f}}}\gamma+\frac{1-\nu_{\mathrm{t}}^{\mathrm{f}}}{16}\right)R^2C'(z)-\frac{1}{E_{\mathrm{A}}^{\mathrm{f}}}\left(1-2\beta\nu_{\mathrm{A}}^{\mathrm{f}}\frac{V_{\mathrm{m}}}{V_{\mathrm{f}}}\right)\bar{C}(z).\end{aligned} \quad (16.85)$$

The function $H_{\mathrm{m}}(z)$ is obtained using the equivalent results (16.73) and (16.83) so that

$$\begin{aligned}H_{\mathrm{m}}(z)=&\frac{R^4C'''(z)}{96V_{\mathrm{f}}V_{\mathrm{m}}^2E_{\mathrm{T}}^{\mathrm{m}}}\left[V_{\mathrm{m}}^2(5+V_{\mathrm{f}})-12\nu_{\mathrm{t}}^{\mathrm{m}}V_{\mathrm{m}}+\left(\nu_{\mathrm{t}}^{\mathrm{m}}\right)^2V_{\mathrm{m}}\left\{V_{\mathrm{f}}^2+4V_{\mathrm{f}}-17\right\}\right]\\&-\frac{R^4C'''(z)}{32V_{\mathrm{f}}V_{\mathrm{m}}^2E_{\mathrm{T}}^{\mathrm{m}}}\left[V_{\mathrm{m}}^2+4\nu_{\mathrm{t}}^{\mathrm{m}}+\left(\nu_{\mathrm{t}}^{\mathrm{m}}\right)^2(1+V_{\mathrm{f}})(3-V_{\mathrm{f}})\right]\ln V_{\mathrm{f}}\\&-\frac{\gamma R^4C'''(z)+\beta R^2C'(z)}{4V_{\mathrm{f}}V_{\mathrm{m}}E_{\mathrm{T}}^{\mathrm{m}}}\left[V_{\mathrm{m}}\left\{1+V_{\mathrm{f}}+\nu_{\mathrm{t}}^{\mathrm{m}}(3-V_{\mathrm{f}})\right\}+2\left(1+\nu_{\mathrm{t}}^{\mathrm{m}}\right)\ln V_{\mathrm{f}}\right]\\&-\frac{\nu_{\mathrm{A}}^{\mathrm{m}}R^2C'(z)}{8V_{\mathrm{m}}^2E_{\mathrm{A}}^{\mathrm{m}}}\left[V_{\mathrm{m}}\left\{5-V_{\mathrm{f}}+\nu_{\mathrm{t}}^{\mathrm{m}}(3+V_{\mathrm{f}})\right\}+2\left(1+\nu_{\mathrm{t}}^{\mathrm{m}}\right)V_{\mathrm{f}}\ln V_{\mathrm{f}}+16\gamma V_{\mathrm{m}}^2\right]\\&+\frac{R^2C'(z)}{8V_{\mathrm{m}}^2\mu_{\mathrm{A}}^{\mathrm{m}}}\left[V_{\mathrm{m}}(3-V_{\mathrm{f}})+2\ln V_{\mathrm{f}}\right]+\frac{V_f}{V_{\mathrm{m}}E_{\mathrm{A}}^{\mathrm{m}}}\left(1-2\beta\nu_{\mathrm{A}}^{\mathrm{m}}\frac{V_{\mathrm{m}}}{V_{\mathrm{f}}}\right)\bar{C}(z).\end{aligned} \quad (16.86)$$

For the case of perfect interface bonding, when

$$H_{\mathrm{f}}(z)\equiv H_{\mathrm{m}}(z)\equiv H(z) \quad \text{and} \quad u_z^{\mathrm{f}}(R,z)\equiv u_z^{\mathrm{m}}(R,z),$$

the solution for the stress and displacement fields in the fibre and matrix derived in Section 16.5 satisfy exactly the equilibrium equations (16.2) and (16.3), the stress-strain relations (16.4), (16.5) and (16.7), the compatibility equation (16.9), the interface conditions (16.10) and the external boundary conditions (16.12) for any stress-transfer function $C(z)$. The stress and displacement fields do not satisfy the axial stress-strain relation (16.6), but they do satisfy exactly the corresponding averaged stress-strain relations (16.70) and (16.72) provided that the stress-transfer function $C(z)$ satisfies the homogeneous fourth-order ordinary differential equation (16.77). If the interface is debonded, then $H_f(z) \neq H_m(z)$ and consequently neither the differential equation (16.77) nor the interface condition $(16.10)_4$ are satisfied, although relations (16.85) and (16.86) remain valid.

16.9 Axial Boundary Conditions

A length $2L$ of fibre and matrix are now considered where the origin of the (r, z) coordinates is at the centre of the system on the axis of the fibre. On $z = \pm L$, there are either fibre fractures or matrix cracks, and the shear stress σ_{rz} is assumed to be everywhere zero, so that the solution to the problem can be applied to the fibre fragmentation and matrix cracking problems for the special case where the crack distribution is uniform in either the fibre or the matrix. It then follows from (16.52) and (16.56) that this shear stress boundary condition is satisfied if $C'(\pm L) = 0$.

The boundary condition for the axial stress is written in the following generalised form that can be used for fibre fractures or matrix cracks

$$\begin{aligned} \sigma_{zz}^{f}(r,\pm L) &= \frac{(1+\xi)\sigma_A}{2V_f}, \quad 0 \le r \le R, \\ \sigma_{zz}^{m}(r,\pm L) &= \frac{(1-\xi)\sigma_A}{2V_m}, \quad R \le r \le a, \end{aligned} \tag{16.87}$$

where σ_A is the effective axial applied stress. It is clear that these values are consistent with relation (16.63). On setting $\xi = -1$, the boundary conditions (16.87) are valid for fibre fractures, and on setting $\xi = 1$ these boundary conditions are valid for matrix cracks. On using (16.51) and (16.55), the boundary conditions (16.87) lead to the condition

$$C(\pm L) = \sigma_f - \frac{(1+\xi)\sigma_A}{2V_f} = \frac{(1-\xi)\sigma_A}{2V_f} - \frac{V_m}{V_f}\sigma_m = F(\sigma_A, \sigma_T, \Delta T), \tag{16.88}$$

where σ_f and σ_m are the uniform axial fibre and matrix stresses for an undamaged composite. The function F depends linearly on σ_A, σ_T and ΔT, as seen from (B.37), (B.38), (B.42) and (B.49) (see Appendix B). To determine values of F, the effective properties E_A, ν_A and α_A are first calculated using (B.46)–(B.52). Given values of σ_A, σ_T and ΔT, the axial strain ε_A is first calculated using (B.49) and then the parameters ϕ, σ_f and σ_m using (B.37), (B.38) and (B.42). The value of σ_f or σ_m is then substituted into (16.88) to calculate $C(\pm L) = F(\sigma_A, \sigma_T, \Delta T)$.

16.10 Perfectly Bonded Fibre/Matrix Interfaces

Consider now a fibre fracture or matrix crack located at $z = L$, where the interface remains perfectly bonded, as illustrated in Figure 16.1. It is convenient to express the solution of the ordinary differential equation (16.77) in the following form (valid for $0 \le z \le L$)

$$C(z) = A\cosh\frac{(p+q)z}{R} + B\cosh\frac{(p-q)z}{R}, \tag{16.89}$$

where p, q, r and s are dimensionless parameters defined by

$$p = \sqrt{\frac{1}{2}(r+s)}, \quad q = \sqrt{\frac{1}{2}(r-s)} < p, \quad r = -\frac{G}{2F} > 0, \quad s = \sqrt{\frac{H}{F}}. \tag{16.90}$$

The form of solution given by (16.89) is valid only if $r > s$, a situation which is often encountered when applying the model to common composites having volume fractions which are not too large. The situation $r < s$ leads to complex values of q. The easiest way of dealing with this case is to develop computer code using complex arithmetic for variables that might be complex numbers. It should be noted that p is always real and that the situation $r = s$ needs to be considered as a special case.

In (16.89), the parameters A and B are to be determined using the boundary conditions $C'(\pm L) = 0$ and (16.88). It can be shown that

$$\begin{aligned} A &= -\frac{(p-q)\tanh\frac{(p-q)L}{R}}{\cosh\frac{(p+q)L}{R}}\Lambda F\left(\sigma_A,\sigma_T,\Delta T\right), \\ B &= \frac{(p+q)\tanh\frac{(p+q)L}{R}}{\cosh\frac{(p-q)L}{R}}\Lambda F\left(\sigma_A,\sigma_T,\Delta T\right), \end{aligned} \tag{16.91}$$

where

$$\frac{1}{\Lambda} = (p+q)\tanh\frac{(p+q)L}{R} - (p-q)\tanh\frac{(p-q)L}{R}, \tag{16.92}$$

and where the function F is defined by (16.88). The stress-transfer function for an array of equally spaced interacting cracks in the fibre or matrix having a perfectly bonded interface is, thus, determined uniquely.

The distribution of the interfacial shear stress is obtained by substituting (16.89) into (16.52) or (16.56). The interfacial shear stress will be zero at the location of the crack $z = L$, and the maximum absolute value of the interfacial shear stress occurs when $C''(z) = 0$ defining the point $z = c$ for a local maximum or minimum of the interfacial shear stress. It follows from (16.89) that the value of c must be such that

$$(p+q)^2 A\cosh\frac{(p+q)c}{R} + (p-q)^2 B\cosh\frac{(p-q)c}{R} = 0. \tag{16.93}$$

When $r > s$ so that q is a real quantity, it can be shown that the parameter c satisfies the following transcendental equation that is solved numerically

$$\tanh\frac{pc}{R}\tanh\frac{qc}{R} = \frac{p\tanh\frac{qL}{R} - q\tanh\frac{pL}{R}}{p\tanh\frac{pL}{R} - q\tanh\frac{qL}{R}}. \tag{16.94}$$

Although relation (16.94) is compact, it might cause numerical problems when q is a pure imaginary number as the tanh functions would be replaced by tan functions which periodically diverge. To avoid this potential problem, relation (16.94) is expressed in the following more expansive form that avoids the need to evaluate tanh functions

$$\sinh\frac{pc}{R}\sinh\frac{qc}{R}\left(p\sinh\frac{pL}{R}\cosh\frac{qL}{R}-q\cosh\frac{pL}{R}\sinh\frac{qL}{R}\right)$$
$$=\cosh\frac{pc}{R}\cosh\frac{qc}{R}\left(p\cosh\frac{pL}{R}\sinh\frac{qL}{R}-q\sinh\frac{pL}{R}\cosh\frac{qL}{R}\right). \tag{16.95}$$

The value $\overline{\tau}$ of the local maximum or minimum interfacial shear stress is obtained using (16.52) and (16.89) so that

$$\overline{\tau}=\frac{1}{2}RC'(c)=\frac{1}{2}A(p+q)\sinh\frac{(p+q)c}{R}+\frac{1}{2}B(p-q)\sinh\frac{(p-q)c}{R}. \tag{16.96}$$

On using (16.91)

$$\overline{\tau}=\Lambda' F(\sigma_{\mathrm{A}},\sigma_{\mathrm{T}},\Delta T), \tag{16.97}$$

where

$$\Lambda'=\frac{p+q}{2}\tanh\frac{(p-q)L}{R}\,\frac{\cosh\frac{(p+q)c}{R}}{\cosh\frac{(p+q)L}{R}}\,\Psi(c)\Lambda, \tag{16.98}$$

and

$$\Psi(x)\equiv(p+q)\tanh\frac{(p-q)x}{R}-(p-q)\tanh\frac{(p+q)x}{R}. \tag{16.99}$$

It is worth noting that when $r<s$, so that q is a pure imaginary quantity, the transcendental equation determining the location $z=c$ is

$$\tanh\frac{pc}{R}\tan\frac{|q|c}{R}=\frac{p\tan\frac{|q|L}{R}-|q|\tanh\frac{pL}{R}}{p\tanh\frac{pL}{R}+|q|\tan\frac{|q|L}{R}}. \tag{16.100}$$

For the special case of noninteracting cracks, the interfacial shear stress will be zero at the location of the crack $z=L$, and it will tend to zero far away from the crack location. It can be shown for this special case that the maximum absolute value of the interfacial shear stress occurs at the point $z=c$ where

$$\frac{c}{L}=1-\frac{R}{2Lq}\ln\frac{p+q}{p-q}. \tag{16.101}$$

The corresponding local maximum or minimum value of the interfacial shear stress is given by

$$\overline{\tau}=\frac{1}{2}\sqrt{p^2-q^2}\left(\frac{p+q}{p-q}\right)^{-\frac{p}{2q}}F(\sigma_{\mathrm{A}},\sigma_{\mathrm{T}},\Delta T). \tag{16.102}$$

On using (16.89), the function $\overline{C}(z)$ defined by (16.84) is given by

$$\overline{C}(z)=\frac{AR}{p+q}\sinh\frac{(p+q)z}{R}+\frac{BR}{p-q}\sinh\frac{(p-q)z}{R}, \tag{16.103}$$

satisfying the condition $\bar{C}(0)=0$. The average axial displacements for the fibre and matrix defined by (16.82) and (16.83) may then be calculated. As $C'(L)=0$, it follows that

$$\bar{u}_z^{\mathrm{f}}(L)=-\frac{1}{E_{\mathrm{A}}^{\mathrm{f}}}\left(1-2\beta\nu_{\mathrm{A}}^{\mathrm{f}}\frac{V_{\mathrm{m}}}{V_{\mathrm{f}}}\right)\bar{C}(L)+L\varepsilon_{\mathrm{A}}, \tag{16.104}$$

$$\bar{u}_z^{\mathrm{m}}(L)=\frac{V_{\mathrm{f}}}{V_{\mathrm{m}}E_{\mathrm{A}}^{\mathrm{m}}}\left(1-2\beta\nu_{\mathrm{A}}^{\mathrm{m}}\frac{V_{\mathrm{m}}}{V_{\mathrm{f}}}\right)\bar{C}(L)+L\varepsilon_{\mathrm{A}}. \tag{16.105}$$

When the fracture is isolated $L\to\infty$ so that from (16.91), (16.92) and (16.103),

$$\bar{C}(L)=\frac{2Rp}{p^2-q^2}F(\sigma_{\mathrm{A}},\sigma_{\mathrm{T}},\Delta T), \tag{16.106}$$

leading to the results

$$\bar{u}_z^{\mathrm{f}}(L)=-\frac{2Rp}{p^2-q^2}\frac{1}{E_{\mathrm{A}}^{\mathrm{f}}}\left(1-2\beta\nu_{\mathrm{A}}^{\mathrm{f}}\frac{V_{\mathrm{m}}}{V_{\mathrm{f}}}\right)F(\sigma_{\mathrm{A}},\sigma_{\mathrm{T}},\Delta T)+L\varepsilon_{\mathrm{A}}, \tag{16.107}$$

$$\bar{u}_z^{\mathrm{m}}(L)=\frac{2Rp}{p^2-q^2}\frac{V_{\mathrm{f}}}{V_{\mathrm{m}}E_{\mathrm{A}}^{\mathrm{m}}}\left(1-2\beta\nu_{\mathrm{A}}^{\mathrm{m}}\frac{V_{\mathrm{m}}}{V_{\mathrm{f}}}\right)F(\sigma_{\mathrm{A}},\sigma_{\mathrm{T}},\Delta T)+L\varepsilon_{\mathrm{A}}. \tag{16.108}$$

Crack opening effects now need to be taken into account. To achieve this, it is necessary to consider fibre fractures and matrix cracks separately.

16.10.1 Fibre Fractures

For fibre fractures $\xi=-1$ and it follows from (16.88) that

$$F(\sigma_{\mathrm{A}},\sigma_{\mathrm{T}},\Delta T)=\sigma_{\mathrm{f}}=\frac{\sigma_{\mathrm{A}}}{V_{\mathrm{f}}}-\frac{V_{\mathrm{m}}}{V_{\mathrm{f}}}\sigma_{\mathrm{m}}. \tag{16.109}$$

It then follows from (16.107) and (16.108) that

$$\bar{u}_z^{\mathrm{f}}(L)=-\frac{2Rp}{p^2-q^2}\frac{\sigma_{\mathrm{f}}}{E_{\mathrm{A}}^{\mathrm{f}}}\left(1-2\beta\nu_{\mathrm{A}}^{\mathrm{f}}\frac{V_{\mathrm{m}}}{V_{\mathrm{f}}}\right)+L\varepsilon_{\mathrm{A}}, \tag{16.110}$$

$$\bar{u}_z^{\mathrm{m}}(L)=\frac{2Rp}{p^2-q^2}\frac{V_{\mathrm{f}}\sigma_{\mathrm{f}}}{V_{\mathrm{m}}E_{\mathrm{A}}^{\mathrm{m}}}\left(1-2\beta\nu_{\mathrm{A}}^{\mathrm{m}}\frac{V_{\mathrm{m}}}{V_{\mathrm{f}}}\right)+L\varepsilon_{\mathrm{A}}. \tag{16.111}$$

The crack opening displacement associated with a fibre fracture Δ_{ff} is then given by

$$\Delta_{\mathrm{ff}}=\bar{u}_z^{\mathrm{m}}(L)-\bar{u}_z^{\mathrm{f}}(L)=\frac{2Rp\sigma_{\mathrm{f}}}{p^2-q^2}\left[\left(\frac{1}{V_{\mathrm{f}}E_{\mathrm{A}}^{\mathrm{f}}}+\frac{1}{V_{\mathrm{m}}E_{\mathrm{A}}^{\mathrm{m}}}\right)V_{\mathrm{f}}-2\beta\left(\frac{V_{\mathrm{m}}}{V_{\mathrm{f}}}\frac{\nu_{\mathrm{A}}^{\mathrm{f}}}{E_{\mathrm{A}}^{\mathrm{f}}}+\frac{\nu_{\mathrm{A}}^{\mathrm{m}}}{E_{\mathrm{A}}^{\mathrm{m}}}\right)\right]. \tag{16.112}$$

16.10.2 Matrix Cracks

For matrix cracks $\xi=1$ and it follows from (16.88) that

$$F(\sigma_{\mathrm{A}},\sigma_{\mathrm{T}},\Delta T)=\sigma_{\mathrm{f}}-\frac{\sigma_{\mathrm{A}}}{V_{\mathrm{f}}}=-\frac{V_{\mathrm{m}}}{V_{\mathrm{f}}}\sigma_{\mathrm{m}}. \tag{16.113}$$

It then follows from (16.107) and (16.108) that

$$\bar{u}_z^{\mathrm{f}}(L)=\frac{2Rp}{p^2-q^2}\frac{\sigma_{\mathrm{m}}}{E_{\mathrm{A}}^{\mathrm{f}}}\frac{V_{\mathrm{m}}}{V_{\mathrm{f}}}\left(1-2\beta\nu_{\mathrm{A}}^{\mathrm{f}}\frac{V_{\mathrm{m}}}{V_{\mathrm{f}}}\right)+L\varepsilon_{\mathrm{A}}, \tag{16.114}$$

$$\bar{u}_z^{\mathrm{m}}(L) = -\frac{2Rp}{p^2 - q^2}\frac{\sigma_{\mathrm{m}}}{E_{\mathrm{A}}^{\mathrm{m}}}\left(1 - 2\beta\nu_{\mathrm{A}}^{\mathrm{m}}\frac{V_{\mathrm{m}}}{V_{\mathrm{f}}}\right) + L\varepsilon_{\mathrm{A}}. \tag{16.115}$$

The crack opening displacement associated with a matrix crack is then $2\Delta_{\mathrm{mc}}$ where

$$\Delta_{\mathrm{mc}} = \bar{u}_z^{\mathrm{f}}(L) - \bar{u}_z^{\mathrm{m}}(L) = \frac{2Rp\sigma_{\mathrm{m}}}{p^2 - q^2}\frac{V_{\mathrm{m}}}{V_{\mathrm{f}}}\left[\left(\frac{1}{V_{\mathrm{f}}E_{\mathrm{A}}^{\mathrm{f}}} + \frac{1}{V_{\mathrm{m}}E_{\mathrm{A}}^{\mathrm{m}}}\right)V_{\mathrm{f}} - 2\beta\left(\frac{V_{\mathrm{m}}}{V_{\mathrm{f}}}\frac{\nu_{\mathrm{A}}^{\mathrm{f}}}{E_{\mathrm{A}}^{\mathrm{f}}} + \frac{\nu_{\mathrm{A}}^{\mathrm{m}}}{E_{\mathrm{A}}^{\mathrm{m}}}\right)\right]. \tag{16.116}$$

From (16.112) and (16.116) it is noted that

$$\Delta_{\mathrm{mc}} = \frac{V_{\mathrm{m}}\sigma_{\mathrm{m}}}{V_{\mathrm{f}}\sigma_{\mathrm{f}}}\Delta_{\mathrm{ff}}. \tag{16.117}$$

The expressions of the axial strain ε_{A} and the axial stresses σ_{f} and σ_{m} in fibre and matrix, respectively, for undamaged unidirectional composites are given in Appendix B.

16.11 Frictionally Slipping Interfaces with Uniform Interfacial Shear Stress

Consider now the situation where the fibre and matrix have debonded along the interface such that they remain in contact and are subject to frictional slip. During progressive axial loading the transverse applied stress σ_{T} is assumed to be held fixed. The first case to be considered assumes that frictional slip is characterised by a uniform interfacial shear stress τ, which is regarded as a material constant. It is useful to start this section by considering the original model of matrix cracking in the presence of debonding, which is well known in the literature as ACK theory, and was developed by Aveston et al. [2]. Their model applies to the consideration of very long matrix cracks where matrix cracking is considered in the absence of the effect of thermal residual stresses. This model was the basis of methods developed by the author to investigate crack bridging for much shorter cracks [3], and for the case where thermal stresses are present [4, 5].

16.11.1 A Simplified Model

The debonded interface is assumed to have length L and exhibit a uniform interfacial shear stress τ. It has formed because of the presence of a matrix crack on the plane $z = L$ that is normal to the fibre direction. The axial stress in the fibre is

$$\sigma_{zz}^{\mathrm{f}}(z) = \sigma_{\mathrm{f}} + \frac{2\tau}{R}z, \qquad 0 < z < L, \tag{16.118}$$

and that in the matrix is

$$\sigma_{zz}^{\mathrm{m}}(z) = \sigma_{\mathrm{m}} - \frac{2\tau}{R}\frac{V_{\mathrm{f}}}{V_{\mathrm{m}}}z, \quad 0 < z < L, \tag{16.119}$$

where σ_{f} and σ_{m} are the axial stresses in fibre and matrix, respectively, for an undamaged composite (see Appendix B). For a matrix crack, the normal traction in the matrix on the crack surface is zero so that $\sigma_{zz}^{\mathrm{m}}(L) = 0$ and it then follows from (16.119) that the debond length L is given by

$$L = \frac{R}{2\tau}\frac{V_m}{V_f}\sigma_m. \tag{16.120}$$

The stress-strain relations for fibre and matrix imply that

$$\sigma_{zz}^f(z) = E_A^f \frac{dv_f(z)}{dz} - E_A^f \alpha_A^f \Delta T, \tag{16.121}$$

$$\sigma_{zz}^m(z) = E_A^m \frac{dv_m(z)}{dz} - E_A^m \alpha_A^m \Delta T. \tag{16.122}$$

On using (16.118) and (16.119) it then follows that the axial displacement distributions are given by

$$v_f(z) = \frac{\sigma_f}{E_A^f} z + \frac{\tau}{RE_A^f} z^2 + \alpha_A^f \Delta T z, \qquad 0<z<L, \tag{16.123}$$

$$v_m(z) = \frac{\sigma_m}{E_A^m} z - \frac{\tau}{R}\frac{V_f}{V_m E_A^m} z^2 + \alpha_A^m \Delta T z, \quad 0<z<L, \tag{16.124}$$

where $v_f(0) = v_m(0) = 0$. For an undamaged composite having a uniaxial strain ε_A

$$\sigma_f = E_A^f\left(\varepsilon_A - \alpha_A^f \Delta T\right), \qquad \sigma_m = E_A^m\left(\varepsilon_A - \alpha_A^m \Delta T\right), \tag{16.125}$$

so that the displacement distributions may be written as

$$v_f(z) = \varepsilon_A z + \frac{\tau}{RE_A^f} z^2, \qquad 0<z<L, \tag{16.126}$$

$$v_m(z) = \varepsilon_A z - \frac{\tau}{R}\frac{V_f}{V_m E_A^m} z^2, \quad 0<z<L. \tag{16.127}$$

The effective axial stress applied to the composite is denoted by σ_A where

$$\sigma_A = V_f \sigma_f + V_m \sigma_m. \tag{16.128}$$

On using (16.125),

$$\sigma_A = E_A(\varepsilon_A - \alpha_A \Delta T), \tag{16.129}$$

where

$$E_A = V_f E_A^f + V_m E_A^m, \qquad E_A \alpha_A = V_f E_A^f \alpha_A^f + V_m E_A^m \alpha_A^m. \tag{16.130}$$

The parameters E_A and α_A are mixtures estimates of the effective axial Young's modulus and effective axial thermal expansion coefficient of the composite. It follows from (16.118) that the maximum value of the fibre stress occurs when $z = L$ and its value is given by

$$\sigma_f^{max} = \sigma_f + \frac{2\tau}{R} L = \sigma_f + \frac{V_m}{V_f}\sigma_m = \frac{\sigma_A}{V_f}. \tag{16.131}$$

The matrix crack opening displacement has the value $2\Delta_{mc}$ where

$$\Delta_{mc} = v_z^f(L) - v_z^m(L) = \frac{\tau}{R}\frac{E_A}{V_m E_A^m E_A^f} L^2. \tag{16.132}$$

On eliminating ε_A using $(16.125)_2$ and (16.129), it follows that

$$\sigma_m = \frac{E_A^m}{E_A}\left(\sigma_A + E_A(\alpha_A - \alpha_A^m)\Delta T\right), \tag{16.133}$$

On using (16.120) and (16.133), the result (16.132) may be written as

$$\Delta_{\rm mc} = \frac{R}{4\tau}\frac{V_{\rm m}}{V_{\rm f}^2}\frac{E_{\rm A}^{\rm m}}{E_{\rm A}E_{\rm A}^{\rm f}}\left(\sigma_{\rm A} + E_{\rm A}(\alpha_{\rm A} - \alpha_{\rm A}^{\rm m})\Delta T\right)^2. \tag{16.134}$$

On using (16.130), this result may be written as

$$\Delta_{\rm mc} = \frac{R}{4\tau}\frac{V_{\rm m}}{V_{\rm f}^2}\frac{E_{\rm A}^{\rm m}}{E_{\rm A}E_{\rm A}^{\rm f}}\left(\sigma_{\rm A} + V_{\rm f}E_{\rm A}^{\rm f}(\alpha_{\rm A}^{\rm f} - \alpha_{\rm A}^{\rm m})\Delta T\right)^2. \tag{16.135}$$

16.11.2 An Improved Stress-transfer Model for Interface Debonding

An examination of the simplified stress-transfer model described previously reveals that the interfacial shear stress is discontinuous at the boundary between the debonded zone and the perfectly bonded zone. In addition, the surface of matrix crack is subject to the shear stress τ. As these situations cannot occur in reality, it is useful to consider how these problems can be overcome. From (16.52) or (16.56) the stress-transfer function in the slip zone satisfies the first-order differential equation

$$C'(z) = \frac{2\tau}{R}. \tag{16.136}$$

The derivative $C'(z)$ governing the interfacial shear stress $\sigma_{rz}(R,z)$ is a negative quantity when stress transfer for a matrix crack ($\xi = 1$) is being analysed, but it is a positive quantity when considering stress transfer associated with a fibre fracture ($\xi = -1$). The sign of τ is set in software, selecting a negative value for matrix cracks and a positive value for fibre fractures.

Consider the geometry of the fibre, matrix and partially debonded interface shown in Figure 16.2. The fibre has radius R and there is a fibre fracture or matrix crack at the location $z = L$. As soon as debonding is initiated at the point $z = c$ on the interface at $r = R$, the solution of the stress-transfer problem is no longer governed by the differential equation (16.77) alone. The differential equation applies only in the region $0 \leq z \leq b$ where elastic stress transfer is possible as the interface has not debonded in this region. For the debonded region $b \leq z \leq c$, the interfacial shear stress is assumed to have a uniform value denoted by τ. If the parameter $c = L$, then the symmetry of the stress tensor asserts that the shear stress would have the nonzero value τ on the surface of the fibre fracture or matrix crack. This situation violates the essential condition that crack surfaces should be stress-free. To overcome this problem a transition zone is introduced in the debonded zone occupying the region $c \leq z \leq L$ in which the interfacial shear stress reduces from the specified value τ at $z = c$ to zero value on $z = L$. Fibre/matrix stress transfer will be occurring in this region, as well as in the uniform shear stress zone $b \leq z \leq c$.

For a fibre fracture on $z = L$ it is assumed that in the transition region the stress distribution in the fibre is identical to that which would arise for a perfectly bonded interface at the point of debonding. The stress distribution in the matrix would, however, be different. If the matrix were cracked on $z = L$ instead of the fibre, then the stress distribution in the matrix would then be assumed to be identical to that for a perfectly bonded interface at the point of debonding whereas that in the fibre would be different.

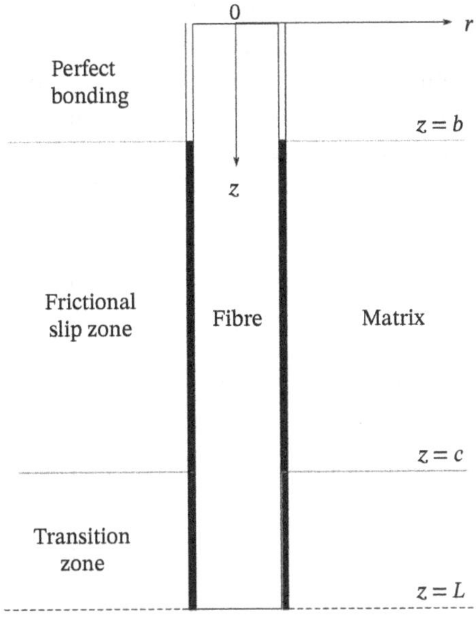

Figure 16.2 Schematic diagram showing the geometry for a partially debonded interface associated with a fibre fracture or matrix crack when the shear stress is uniform in the slipping zone.

The representation for the stress and displacement fields defined in Section 16.5 apply to all three stress-transfer regions shown in Figure 16.2. For frictionally bonded composites, the stress-transfer function $C(z)$ that is used for a debonded interface having a uniform shear stress is required to be continuous and have continuous first and second derivatives so that the axial and shear stress, and the radial displacement are continuous. For a chemically bonded interface, the debond length would be assumed to be specified in advance, and the continuity condition for the second derivative of the function $C(z)$ would need to be relaxed (a situation not considered here).

Consider now the stress-transfer function $C(z)$ for frictionally bonded interfaces having the following form.

For $0 \leq z \leq b$,

$$C(z) = A^* \cosh\frac{(p+q)z}{R} + B^* \cosh\frac{(p-q)z}{R}, \tag{16.137}$$

For $b \leq z \leq c$,

$$C(z) = \frac{2\tau}{R}(z-c) + C_0 + F(\sigma_{\mathrm{A}}, \sigma_{\mathrm{T}}, \Delta T) - F\left(\sigma_{\mathrm{A}}^{\mathrm{d}}, \sigma_{\mathrm{T}}, \Delta T\right), \tag{16.138}$$

For $c \leq z \leq L$,

$$C(z) = A\cosh\frac{(p+q)z}{R} + B\cosh\frac{(p-q)z}{R} + F(\sigma_{\mathrm{A}}, \sigma_{\mathrm{T}}, \Delta T) - F\left(\sigma_{\mathrm{A}}^{\mathrm{d}}, \sigma_{\mathrm{T}}, \Delta T\right). \tag{16.139}$$

In (16.138) and (16.139), the value of c is determined by the transcendental equation (16.94) and the function F is defined by (16.88). It is clear, from the analysis given in Section 16.7, that (16.137) satisfies the differential equation (16.77) for a perfectly bonded interface. In addition, it can be shown that (16.138) is such that the expressions (16.52) and (16.56) for the shear stress lead to the required uniform value τ on the interface $r = R$. The expression on the right-hand side of (16.139) is just one possible transition function enabling the shear stress to reduce to zero at $z = L$, thus leading to matrix cracks or fibre fractures that have a zero shear traction distribution. Given a value of the interfacial shear stress τ, it follows from (16.97) that the initial debonding stress $\sigma_{\mathrm{A}}^{\mathrm{d}}$ must satisfy the relation

$$\tau = \Lambda' F\left(\sigma_{\mathrm{A}}^{\mathrm{d}}, \sigma_{\mathrm{T}}, \Delta T\right), \tag{16.140}$$

where the value of Λ' is given by (16.98).

Consider first the boundary conditions at the end location $z = L$. The constants A and B are selected so that

$$A \cosh \frac{(p+q)L}{R} + B \cosh \frac{(p-q)L}{R} = F\left(\sigma_{\mathrm{A}}^{\mathrm{d}}, \sigma_{\mathrm{T}}, \Delta T\right), \tag{16.141}$$

$$(p+q) A \sinh \frac{(p+q)L}{R} + (p-q) B \sinh \frac{(p-q)L}{R} = 0. \tag{16.142}$$

The stress-transfer function then satisfies the conditions $C(L) = F(\sigma_{\mathrm{A}}, \sigma_{\mathrm{T}}, \Delta T)$ and $C'(L) = 0$. Consistent with (16.91), the values of A and B are given by

$$A = -\frac{(p-q)\tanh \dfrac{(p-q)L}{R}}{\cosh \dfrac{(p+q)L}{R}} \Lambda F\left(\sigma_{\mathrm{A}}^{\mathrm{d}}, \sigma_{\mathrm{T}}, \Delta T\right),$$
$$B = \frac{(p+q)\tanh \dfrac{(p+q)L}{R}}{\cosh \dfrac{(p-q)L}{R}} \Lambda F\left(\sigma_{\mathrm{A}}^{\mathrm{d}}, \sigma_{\mathrm{T}}, \Delta T\right). \tag{16.143}$$

On using (16.93), (16.98) and (16.140) it can be shown that

$$A = -\frac{p-q}{p+q} \frac{2\tau}{\Psi(c)\cosh \dfrac{(p+q)c}{R}}, \qquad B = \frac{p+q}{p-q} \frac{2\tau}{\Psi(c)\cosh \dfrac{(p-q)c}{R}}. \tag{16.144}$$

Consider now the continuity of the behaviour of the function $C(z)$ at the known location $z = c$. The continuity of $C(z)$ is assured only if the constant C_0 appearing in (16.138) is selected so that

$$C_0 = A \cosh \frac{(p+q)c}{R} + B \cosh \frac{(p-q)c}{R}. \tag{16.145}$$

On using (16.144), it can be shown that

$$C_0 = \frac{8pq}{p^2 - q^2} \frac{\tau}{\Psi(c)}. \tag{16.146}$$

The derivative $C'(z)$ must also be continuous at $z = c$, a condition that is satisfied if

$$A(p+q) \sinh \frac{(p+q)c}{R} + B(p-q) \sinh \frac{(p-q)c}{R} = 2\tau. \tag{16.147}$$

This condition is automatically satisfied because of relation (16.96). In addition, relation (16.93) implies that $C''(z)$ is continuous at $z=c$ having the value zero. It has, thus, been shown that that $C(z)$, $C'(z)$ and $C''(z)$ are all continuous at $z=c$.

Consider finally the properties of the stress-transfer function $C(z)$ in the perfectly bonded region at the point $z=b$, which is as yet an unknown quantity that determines the location of the interface between the perfectly bonded and debonded regions. There are three unknowns, namely A^*, B^* and the value of b. For consistency with the situation imposed at the point $z=c$ it is useful to apply the following conditions that ensure the continuity of $C(z)$, $C'(z)$ and $C''(z)$ at the point $z=b$. It is clear from (16.137) and (16.138) that the conditions to be satisfied are

$$\begin{aligned} A^* \cosh\frac{(p+q)b}{R} + B^* \cosh\frac{(p-q)b}{R} \\ = \frac{2\tau}{R}(b-c) + C_0 + F(\sigma_A,\sigma_T,\Delta T) - F\left(\sigma_A^d,\sigma_T,\Delta T\right), \end{aligned} \quad (16.148)$$

$$(p+q)A^* \sinh\frac{(p+q)b}{R} + (p-q)B^* \sinh\frac{(p-q)b}{R} = \frac{2\tau}{R}, \quad (16.149)$$

$$(p+q)^2 A^* \cosh\frac{(p+q)b}{R} + (p-q)^2 B^* \cosh\frac{(p-q)b}{R} = 0. \quad (16.150)$$

The easiest approach is to use the conditions (16.149) and (16.150), ensuring that $RC'(b)=2\tau$, and $C''(b)=0$, to determine the parameters A^* and B^*, with the result that

$$A^* = -\frac{p-q}{p+q}\frac{2\tau}{\Psi(b)\cosh\frac{(p+q)b}{R}}, \qquad B^* = \frac{p+q}{p-q}\frac{2\tau}{\Psi(b)\cosh\frac{(p-q)b}{R}}, \quad (16.151)$$

where the function Ψ is defined by (16.99). The values given by (16.151) ensure the continuity of $C'(z)$ and $C''(z)$ at $z=b$. It should be noted from (16.144) and (16.151) that when $b=c$ it follows that $A=A^*$ and $B=B^*$ recovering the perfect bonding distribution derived in Section 16.10.

The parameter b defining the location of the debond front has not yet been specified. Its value is determined by imposing the continuity of $C(z)$ at $z=b$. It can be shown from (16.146), (16.148) and (16.151), and that the value of b must be chosen so that

$$\left[\frac{4pq}{p^2-q^2}\left(\frac{1}{\Psi(b)} - \frac{1}{\Psi(c)}\right) - \frac{b-c}{R}\right] 2\tau = F(\sigma_A,\sigma_T,\Delta T) - F\left(\sigma_A^d,\sigma_T,\Delta T\right). \quad (16.152)$$

This result, when used in conjunction with (16.88), shows that the length of the zone having a uniform interfacial shear stress τ is determined by the value of the stress increase from the point of debond initiation. From (16.152) it is clearly seen that $b=c$ when $\sigma_A=\sigma_A^d$. The result (16.152) is a transcendental equation that must be solved numerically.

It is also worth noting that when $(p-q)b/R >> 1$ there is negligible interaction of neighbouring cracks so that

$$\tanh\frac{(p-q)b}{R} \cong \tanh\frac{(p+q)b}{R} \cong 1 \quad \text{and} \quad \Psi(x) \cong 2q. \quad (16.153)$$

Relation (16.152) determining the value of b then reduces to the simple form

$$b = c - \frac{R}{2\tau}\left[F(\sigma_A,\sigma_T,\Delta T) - F\left(\sigma_A^d,\sigma_T,\Delta T\right)\right]. \tag{16.154}$$

It should be noted that the stress-transfer function $C(z)$, defined by (16.137)–(16.139), together with its first two derivatives are continuous everywhere in the region $0 \le z \le L$, automatically ensuring the continuity of the shear stress σ_{rz}, the radial stress σ_{rr}, the axial stress σ_{zz}, the radial displacement u_r and the average axial displacement $\bar{u}_z$, in both the fibre and matrix. The axial displacement component u_z is discontinuous on $z = b$ and $z = c$ in both the fibre and matrix except at the interface. This is only continuity requirement that cannot be satisfied by the constant interfacial shear stress model. The achievement of the required continuity for all other stress and displacement components, including the average axial displacement $\bar{u}_z$, indicates that the constant interfacial shear stress model is one of high quality.

16.12 Solution for a Debonded Interface with Coulomb Friction

In the presence of a fibre fracture or matrix crack, the assumption of a constant interfacial shear stress is frequently made in the literature, although the Coulomb friction law is more acceptable from a physical point of view as the interfacial shear stress is expected to depend on the compressive normal stress at the interface. At relatively low applied loads, a frictionally bonded interface between fibre and matrix is expected to behave as a perfectly bonded interface, as described in Section 16.10. As the maximum interfacial shear stress was predicted in Section 16.10 to occur away from the crack plane, initial frictional slipping is likely to be a local event where the slip zone is wholly embedded in the interface and it gradually increases in length as load is applied. Such growth would be expected to occur at both ends of the slip zone until one of the slip zone boundaries reaches the crack plane. At first sight, the analysis of the early stages of frictional slip would appear to be very complex. However, it needs to be remembered that the stress analysis is approximate and that a more accurate solution would exhibit a shear stress singularity at the location of the interface with the crack plane, implying that frictional slip would initiate at this singularity. It is, therefore, reasonable to assume that the frictional slip zone initiates at the crack plane and progressively grows along the fibre/matrix interface. The analysis will assume that the fibre and matrix are always in mechanical contact at all points along the debonded interface, and it may not, therefore, be applicable for all states of loading. Solutions obtained will be valid only if they have been checked to ensure that the debonded interface exhibits compression or zero loading along the entire debond zone.

Consider now the geometry of the fibre, matrix and debonded interface shown in Figure 16.3. The fibre has radius R and there is a fibre fracture or matrix crack at the location $z = L$. The representation for the stress and displacement fields defined in Section 16.5 apply to both stress-transfer regions shown in Figure 16.3.

The Coulomb friction law is specified, for all z lying in the interfacial slip zone, by

$$\lambda\sigma_{rz}(R,z) = \eta\sigma_{rr}(R,z), \tag{16.155}$$

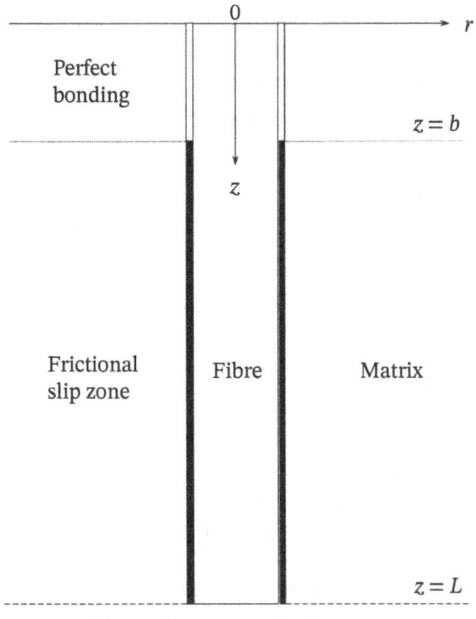

Figure 16.3 Schematic diagram showing the geometry for a partially debonded interface associated with a fibre fracture or matrix crack when Coulomb friction is assumed.

where $\eta > 0$ is the coefficient of friction and where solutions are valid only if $\sigma_{rr} \le 0$ in the debonded zone so that there is mechanical contact between the fibre and matrix in this zone. The sign of the interfacial shear stress $\sigma_{rz}(R,z)$ will depend on the type of problem being solved: negative for matrix cracks and positive for fibre fractures. In (16.155) the value of the parameter $\lambda = -1$ for fibre fractures and $\lambda = 1$ for matrix cracks. From (16.52), (16.53) and (16.68)

$$\begin{aligned}\sigma_{rz}^{\rm f}(R,z) &= \frac{1}{2} R C'(z),\\ \sigma_{rr}^{\rm f}(R,z) &= f R^2 C''(z) - h C(z) - \rho,\end{aligned} \qquad (16.156)$$

where

$$f = \frac{\alpha - \gamma V_{\rm m}}{V_{\rm f}} - \frac{3 + \nu_{\rm t}^{\rm f}}{16}, \quad h = \beta \frac{V_{\rm m}}{V_{\rm f}}, \quad \rho = V_{\rm m} \frac{\phi}{R^2} - \sigma_{\rm T}, \qquad (16.157)$$

and where α, β, γ and ϕ are defined by (16.66), (16.69) and (B.42) (see Appendix B). On substituting (16.156) into (16.155) it can be shown that in the slip zone the stress-transfer function $C(z)$ must satisfy the following second-order ordinary differential equation

$$f R^2 C''(z) - 2\lambda g R C'(z) - h C(z) = \rho, \quad \text{where} \quad g = \frac{1}{4\eta}. \qquad (16.158)$$

It should be noted that the effective axial stress $\sigma_{\rm A}$ and the axial strain $\varepsilon_{\rm A}$ for an undamaged composite are related according to the effective stress-strain relation (B.49) (see Appendix B). The differential equation (16.158) applies only in the debonded region $b \le z \le L$ where frictional slip occurs governed by the Coulomb law (16.155).

The stress-transfer function that is used for the debonded region where frictional slip is governed by the Coulomb law must satisfy the ordinary differential equation (16.158), and the boundary conditions $C'(\pm L)=0$ and (16.88) on $z = L$. The required solution of the differential equations (16.77) and (16.158) is given by:

for $0 \leq z \leq b$,

$$C(z) = A^* \cosh\frac{(p+q)z}{R} + B^* \cosh\frac{(p-q)z}{R}, \tag{16.159}$$

for $b \leq z \leq L$,

$$C(z)= \Phi\left[v \exp\left(-u\frac{L-z}{R}\right) - u \exp\left(-v\frac{L-z}{R}\right)\right] - \frac{\rho}{h}, \tag{16.160}$$

where ρ is defined by (16.157) and

$$\Phi = \frac{\frac{\rho}{h} + F(\sigma_A,\sigma_T,\Delta T)}{v-u}, \quad u = \frac{\lambda g}{f} + \sqrt{\frac{g^2}{f^2}+\frac{h}{f}}, \quad v = \frac{\lambda g}{f} - \sqrt{\frac{g^2}{f^2}+\frac{h}{f}}. \tag{16.161}$$

For applications to be considered here, it is found that $h/f>0$ so that $u>0$ and $v<0$ for all values of $g=1/(2\eta)\geq 0$. From (16.160),

$$C'(z)=\Phi uv\left[\exp\left(-u\frac{L-z}{R}\right) - \exp\left(-v\frac{L-z}{R}\right)\right], \tag{16.162}$$

and because $v-u<0$, it is clear that $C'(z)$ does not change sign in the region $b \leq z \leq L$.

The boundary conditions $C'(L)=0$ and (16.88) are sufficient to determine the function $C(z)$ uniquely for the debonded region $b \leq z \leq L$ as seen from the result (16.160). In (16.159), the parameters A^* and B^* are selected so that the function $C(z)$ and its first derivative are both continuous at the point $z = b$. On letting $\zeta = (L-b)/R$, it can be shown that

$$A^* = \Lambda^* \frac{C_1(\zeta) - (p-q)\tanh\frac{(p-q)b}{R}C_0(\zeta)}{\cosh\frac{(p+q)b}{R}},$$
$$B^* = \Lambda^* \frac{(p+q)\tanh\frac{(p+q)b}{R}C_0(\zeta) - C_1(\zeta)}{\cosh\frac{(p-q)b}{R}}, \tag{16.163}$$

where

$$\frac{1}{\Lambda^*} = (p+q)\tanh\frac{(p+q)b}{R} - (p-q)\tanh\frac{(p-q)b}{R}, \tag{16.164}$$

and where

$$C_0(\zeta) = C(b) = \Phi[v\exp(-u\zeta) - u\exp(-v\zeta)] - \frac{\rho}{h},$$
$$C_1(\zeta) = RC'(b) = \Phi uv[\exp(-u\zeta) - \exp(-v\zeta)]. \tag{16.165}$$

A criterion for the location of the debond tip at $z = b$ has not yet been given. The appropriate criterion is to expect that the Coulomb friction law (16.158) is satisfied by the stress field in the perfectly bonded zone at the point $z = b$. It is clear from the continuity of $C(z)$ and $C'(z)$ at $z = b$, and (16.52), (16.53) and (16.68), that such a condition is automatically satisfied if the function $C'(z)$ is also continuous at $z = b$, a situation that arises if the following transcendental equation determining the value of $\zeta = (L - b)/R$ is satisfied

$$\begin{aligned}(p^2 - q^2)&\left((p-q)\tanh\frac{(p+q)(L-R\zeta)}{R} - (p+q)\tanh\frac{(p-q)(L-R\zeta)}{R}\right)C_0(\zeta)\\ &-\left((p+q)\tanh\frac{(p+q)(L-R\zeta)}{R} - (p-q)\tanh\frac{(p-q)(L-R\zeta)}{R}\right)C_2(\zeta)\\ &+ 4pqC_1(\zeta) = 0,\end{aligned} \tag{16.166}$$

where

$$C_2(\zeta) = R^2C''(b) = \Phi uv[u\exp(-u\zeta) - v\exp(-v\zeta)]. \tag{16.167}$$

It should be noted that for the case of frictional slip in the debonded zone governed by the Coulomb law, the stress-transfer function $C(z)$ defined by (16.159) and (16.160), together with the first two derivatives, are continuous everywhere in the region $0 \le z \le L$, automatically ensuring the continuity of the shear stress σ_{rz}, the radial stress σ_{rr}, the axial stress σ_{zz} and the radial displacement u_r.

The axial displacement component u_z is discontinuous on $z = b$ in both the fibre and the matrix except at the interface. This is the only continuity requirement that cannot be satisfied by the Coulomb friction model. The achievement of the required continuity for all other stress and displacement components, including the average axial displacement $\bar{u}_z$, indicates that the Coulomb friction model developed is one of high quality.

Although the condition (16.166) is used to determine a value for $\zeta = (L - b)/R$ and thus the location $z = b$ of the boundary between the slip zone and the perfectly bonded region, it can be used also to determine the condition that is placed on the parameters σ_A, σ_T and ΔT such that $b = L$, i.e. such that the length of the slip zone is zero. It is easily shown using (16.165)–(16.167) that $b = L$ so that $\zeta = 0$ whenever

$$\begin{aligned}G(\sigma_A,\sigma_T,\Delta T) \equiv (p+q)&\left[\left\{(p-q)^2 + uv\right\}F(\sigma_A,\sigma_T,\Delta T) + uv\frac{\rho}{h}\right]\tanh\frac{(p+q)L}{R}\\ - (p-q)&\left[\left\{(p+q)^2 + uv\right\}F(\sigma_A,\sigma_T,\Delta T) + uv\frac{\rho}{h}\right]\tanh\frac{(p-q)L}{R} = 0.\end{aligned} \tag{16.168}$$

It follows from (16.88), (16.157), (B.37), (B.38), (B.42) and (B.49) (see Appendix B) that the function G defined by (16.168) is linear in the quantities σ_A, σ_T and ΔT so that

$$G(\sigma_A,\sigma_T,\Delta T) \equiv G_1\sigma_A + G_2\sigma_T + G_3\Delta T = 0, \tag{16.169}$$

where the coefficients are most easily found numerically using the relations

$$G_1 = G(1,0,0), \qquad G_2 = G(0,1,0), \qquad G_3 = G(0,0,1). \tag{16.170}$$

Clearly the axial stress for which $b = L$ is given by

$$\sigma_A = -\frac{G_2\sigma_T + G_3\Delta T}{G_1}. \tag{16.171}$$

It should be noted that the axial stress determined by (16.171) does not depend on the value of λ as it follows from (16.161) that $uv = -h/f$.

In Appendix B, an analysis is given that determines condition (B.77) for which contact at the interface is lost between fibre and matrix: a situation for which the radial and shear stresses at the debonded interface are zero everywhere. Relations (16.168) or (16.169) and (B.77) may be used to determine the applied axial stress range for which debonded solutions governed by the Coulomb friction law may be found using the model.

16.13 Example Predictions for Carbon-fibre Composites

Three types of model associated with fibre fractures or matrix cracks have been developed in this chapter. In Section 16.10, the interface was assumed to remain bonded for all states of loading. In Section 16.11, a debonding model was developed that assumed that the debonded interface is characterised by a uniform interfacial shear stress τ, although a transition zone had to be introduced to satisfy all traction boundary conditions on the crack surfaces. In Section 16.12, a model was developed where interfacial debonding was characterised by the Coulomb friction law. In all cases it was assumed that the undamaged composite was frictionally bonded meaning that chemical bonding between fibre and matrix is so small that it can be neglected. It is beyond the scope of this chapter to consider the application of the models to the range of materials to which they are relevant. The approach taken here is to select just one material and to investigate thoroughly the important characteristics of each type of stress-transfer model.

The material to be considered is a carbon-fibre-reinforced epoxy composite where the radius of the fibres is 3.5 μm, the volume fraction of fibres will be taken as 0.5 or 0.6 and the temperature and stress-free temperature are such that $\Delta T = -85$ °C or $\Delta T = 85$ °C. The properties assumed for the carbon fibre and the epoxy matrix are given as follows

	Fibre	*Matrix*
E_A (GPa)	208.0	3.89
E_T (GPa)	16.7	3.89
μ_A (GPa)	18.0	1.41971
ν_A	0.25	0.37
ν_t	0.35	0.37
α_A (/°C × 10^6)	1.1	55.0
α_T(/°C × 10^6)	22.1	55.0

For these properties the volume fraction of 0.5 leads to a real value of the parameter q defined by (16.90) and a volume fraction of 0.6 leads to a pure imaginary value.

Consider a RVE (concentric cylinders comprising a single fibre plus surrounding matrix) of a perfectly bonded composite, having fibre volume fraction 0.6 (so that q is an imaginary quantity), in which the fibre has fragmented into equal lengths $2L$ having a fibre crack density of 20/mm. A uniaxial axial stress of 0.03876909 GPa is applied so that the maximum interfacial stress is 30 MPa exactly. Figure 16.4 shows the axial

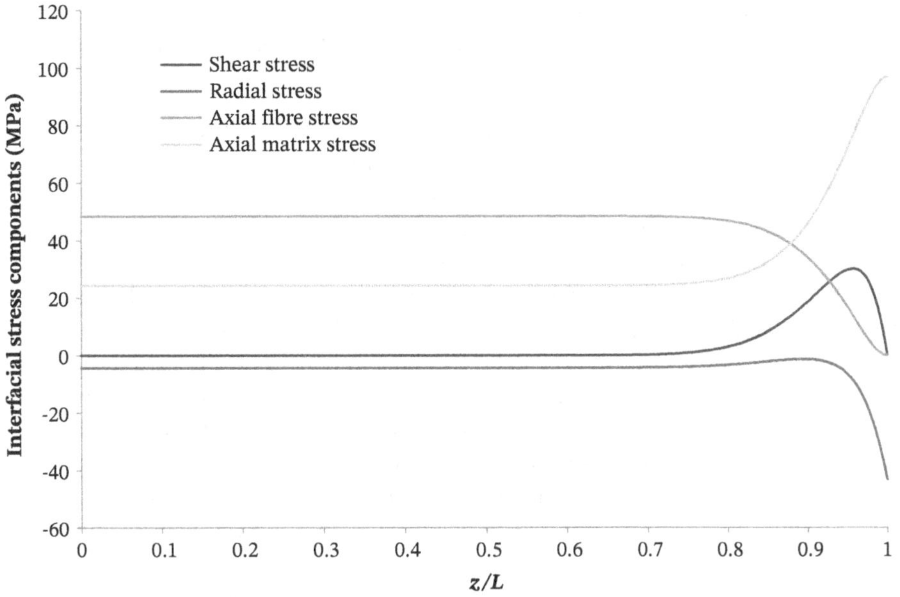

Figure 16.4 Interfacial stress distributions for a perfectly bonded interface in a CFRP composite having fibre fractures that lead to a shear stress having a maximum magnitude of 30 MPa. (Here $V_f = 0.6$, $\sigma_A = 0.0388$ GPa, $\Delta T = -85^{\circ}$C, fibre crack density = 20 mm^{-1}).

stress distributions in both the fibre and matrix as a function of z/L, together with the interfacial shear and normal stresses.

It should be noted that the stresses are all continuous and that the interfacial shear stress has an imposed maximum value of 30 MPa at the location $z = c$ where $c/L = 0.9553$, and that the normal stress is everywhere compressive. The axial fibre stress is seen to diminish to a zero value at the location $z = L$ of the fibre fracture. The axial matrix stress increases to its maximum value on the plane of fibre fracture.

It is now assumed that the stress distributions shown in Figure 16.4 is the stress state for the initiation of fibre/matrix debonding where the magnitude of the interfacial shear stress in the debonded zone is assumed to have a uniform value of 30 MPa. The same fibre crack density of 20 mm^{-1} is used. When the applied stress is increased to a value of 0.1 GPa the resulting stress distributions are shown in Figure 16.5. It should be noted that for this case the approximate formulae (16.101) and (16.154) determining the values of the parameters b and c, which define the boundaries of the constant shear stress region, are extremely accurate implying negligible interaction of neighbouring fibre fractures.

The major characteristics of the solution are as follows.

Stress distributions

- The interfacial shear stress has the uniform value 30 MPa in the region $0.7202 \leq z/L \leq 0.9553$.
- The interfacial shear stress and fibre stress tend to zero as the fibre fracture at $z = L$ is approached. In the transition zone $0.9553 \leq z/L \leq 1$, the stress distribution enables the attainment of a zero shear stress in the matrix at $z = L$.

- The interfacial radial stress is everywhere compressive.
- The interfacial shear and axial stresses are smooth, but the radial stress has sharp corners at the points $z = b$ and $z = c$.

When the fibre is intact, but the matrix has cracked instead, it is much more difficult to obtain solutions as the magnitude of the maximum interfacial shear stress for a perfectly bonded interface is much lower than for the fibre fracture case. If the fibre fracture case just described is applied instead to a matrix crack no solutions can be found as the imposed critical interfacial shear stress τ of 30 MPa is very much larger than the magnitude of the maximum shear stress (10.853 MPa) that will occur for a perfectly bonded case interface. To provide an example prediction for a matrix crack, the volume fraction is now reduced to 0.5 (so that q is a real quantity) and the value of τ is taken as 10 MPa. The resulting stress distributions are shown in Figure 16.6 for the CFRP composite having the fibre and matrix properties assumed previously. The axial fibre stress distribution is not shown as the vertical axis would need to be extended to 200 MPa. For this case, the approximate formulae (16.101) and (16.154), defining values for b and c, are again very accurate implying negligible interaction of neighbouring matrix cracks.

It is clear that the radial stress is not compressive in the whole of the transition zone, thus violating an important modelling requirement. An example has, thus, been identified indicating that the constant interfacial shear stress model for debonding is inadequate. If, however, the interfacial shear stress τ is now set to the much lower value of 2 MPa the interfacial radial stress is then always negative, thus providing a more acceptable solution. The debonded region is extensive for this case and some crack interaction occurs. To remove this interaction the matrix crack density is reduced to the value 10 mm^{-1} and the resulting stress distributions are shown in Figure 16.7.

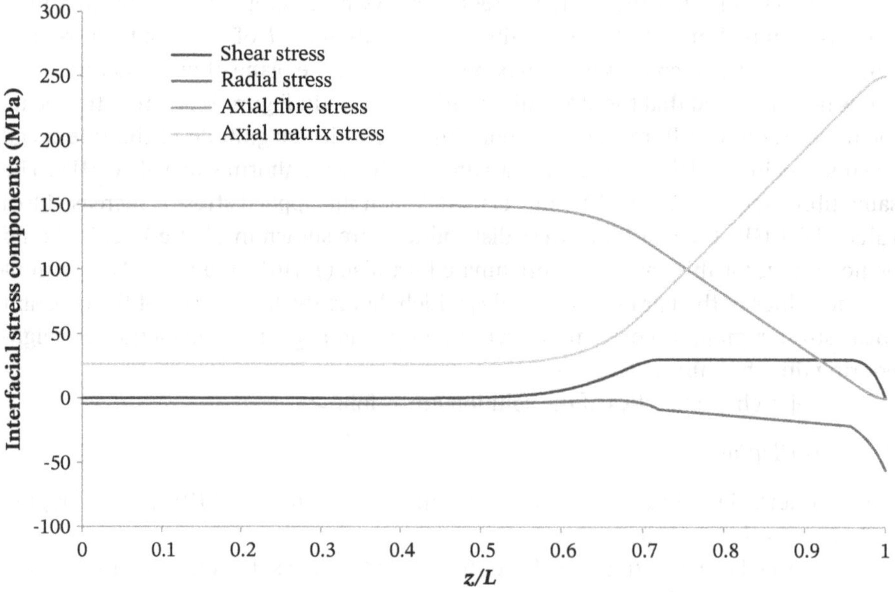

Figure 16.5 Interfacial stress distributions for a debonded interface in a CFRP composite having fibre fractures that lead to a uniform shear stress having a maximum magnitude of 30 MPa. (Here $V_f = 0.6$, $\sigma_A = 0.1$ GPa, $\Delta T = -85^\circ$C, fibre crack density = 20 mm^{-1}).

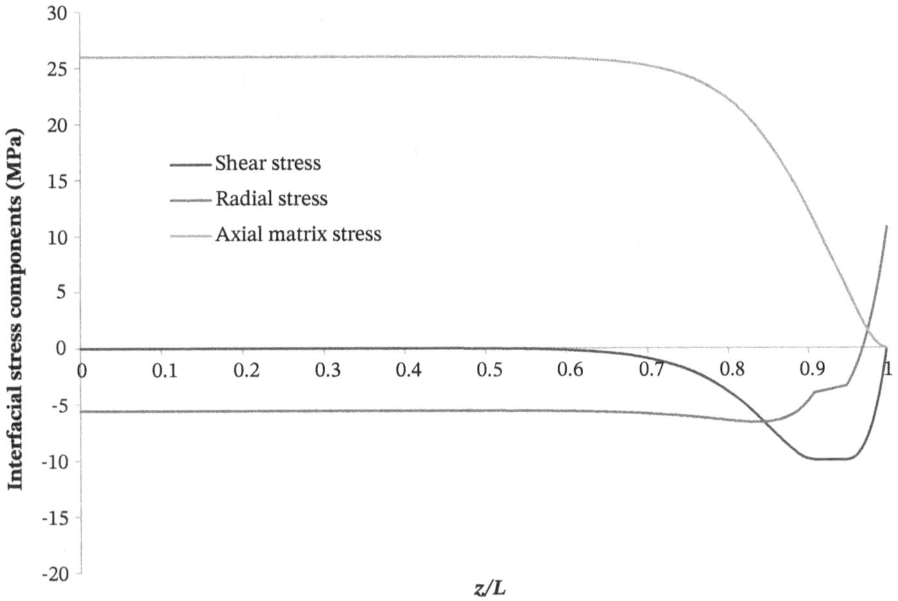

Figure 16.6 Interfacial stress distributions for a debonded interface in a CFRP composite having matrix cracks that lead to a shear stress having a maximum magnitude of 10 MPa. (Here $V_f = 0.5$, $\sigma_A = 0.1$ GPa, $\Delta T = -85^{\circ}$C, fibre crack density = 20 mm^{-1}).

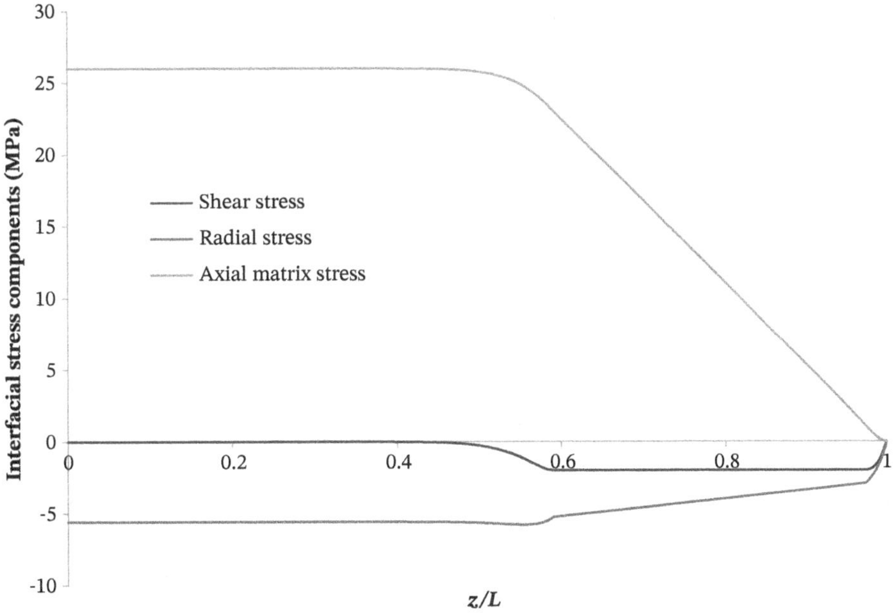

Figure 16.7 Interfacial stress distributions for a debonded interface in a CFRP composite having matrix cracks that lead to a shear stress having a maximum magnitude of 2 MPa. (Here $V_f = 0.5$, $\sigma_A = 0.1$ GPa, $\Delta T = -85^{\circ}$C, fibre crack density = 10 mm^{-1}).

It is seen that the interfacial radial stress is always negative as required.

This completes the discussion of the model of interface debonding based on the popular concept of imposing a uniform interfacial shear stress, but combined with a transition zone to prevent the violation of the zero shear stress boundary condition on the crack surfaces. The second type of debonding model described in Section 16.12 makes use of the Coulomb friction law. The previous problem that generated the results in Figure 16.7 is now solved using the Coulomb friction model. The coefficient of friction η is selected to be 0.5 and the resulting stress distributions are shown in Figure 16.8.

Again, the axial fibre stress distribution is not shown as the vertical axis would need to be extended to 200 MPa. The major characteristics of the solution are as follows.

Stress distributions

- The interfacial shear stress is nonuniform value in the debonded region $0.5909 \leq z/L \leq 1$.
- The interfacial shear stress and fibre stress tend to zero as the fibre fracture at $z = L$ is approached. No transition zone is included in the model although the stress distribution does indicate an apparent transition zone near $z = L$ that is similar to that used in the constant shear model. This apparent transition is predicted automatically when solving the differential equation (16.158).
- The interfacial radial stress is everywhere compressive.
- The interfacial shear and axial stresses are smooth, but the radial stress has a sharp corner at the point $z = b$.
- Acceptable solutions, where there is interfacial contact along the whole length of the debonded zone, are possible only if the applied axial stress lies in the range $0.424\,(\text{GPa}) \leq \sigma_A \leq 1.163\,(\text{GPa})$.

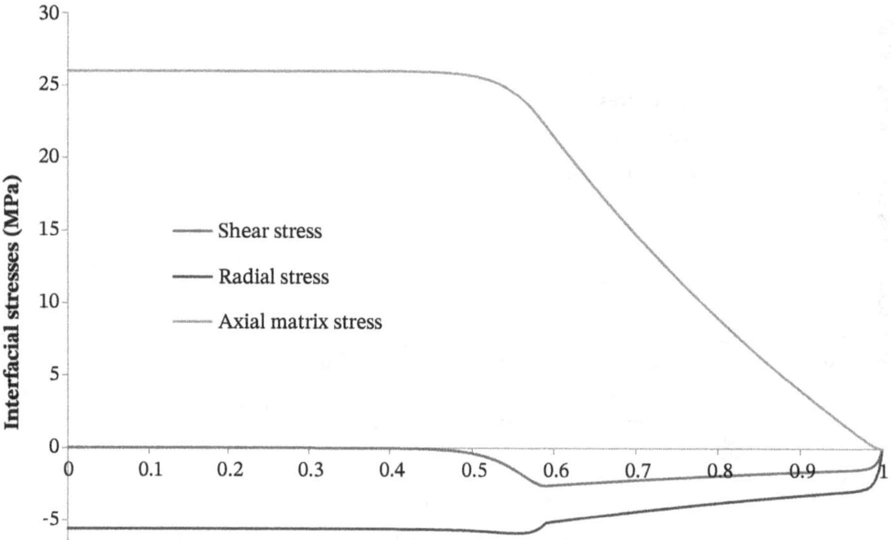

Figure 16.8 Interfacial stress distributions for a debonded interface in a CFRP composite having matrix cracks subject to Coulomb friction. (Here $V_f = 0.5$, $\sigma_A = 0.1$ GPa, $\Delta T = -85°$C, fibre crack density = 10 mm^{-1}, $\eta = 0.5$).

It should be noted that the results for the interfacial shear and normal stresses shown in Figures 16.7 and 16.8 are very similar when using a relatively large value 0.5 for the coefficient of friction. This similarity indicates that for friction problems, the values of τ used in the constant shear stress model need to be relatively small. However, it is worth mentioning that fibre fracture problems can be solved using much larger values of τ, and such values can be justified if the model is being used to predict (approximately) shear yielding phenomena associated with matrix plasticity that can occur in the matrix near fibre fractures.

When solving problems, it is useful to determine the applied stress for which the fibre and matrix are expected to separate along the debonded interface. The analysis given in Appendix B considers this aspect of debonding problems. For the example being considered where $\sigma_T = 0$ and $\Delta T = -85^\circ$C, the value of the axial stress at which separation occurs is $\sigma_A = 1.163$ GPa (not a value of practical relevance), and if this value is used in the Coulomb friction model the interfacial shear and normal stresses are zero everywhere along the interface. At this critical applied stress, the interface becomes entirely debonded because of the assumption of frictional bonding.

If the example is now applied to a fibre fracture rather than a matrix crack, keeping all the other parameters fixed, then it is not easy to obtain solutions. It is not, in fact, possible to find appropriate solutions of (16.152) unless the applied stress σ_A lies in the range $-14.7(\text{MPa}) \leq \sigma_A \leq 7.762(\text{MPa})$. At the critical value 7.762(MPa) the length of the debond zone is zero and solution for a perfectly bonded interface is obtained. Solutions obtained at other applied stresses in this range lead to tensile interfacial normal stresses in the debond zone which are not acceptable. One way of achieving negative normal stresses is to use a positive value of the temperature difference ΔT. When the fibre fracture example is solved with the value $\Delta T = 85^\circ$C at an applied stress of 12 MPa, the results shown in Figures 16.9 are obtained.

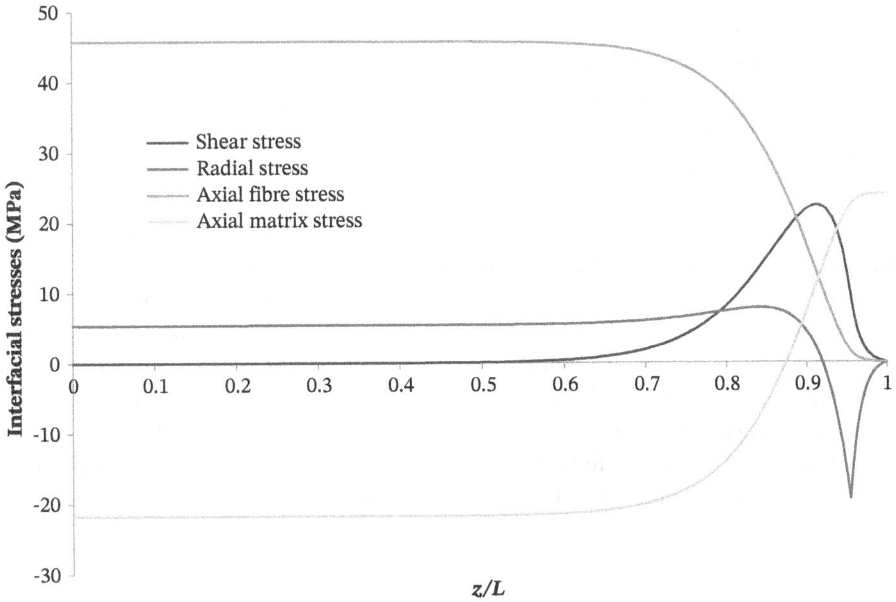

Figure 16.9 Interfacial stress distributions for a debonded interface in a CFRP composite having fibre cracks subject to Coulomb friction. (Here $V_f = 0.5$, $\sigma_A = 12$ MPa, $\Delta T = -85^\circ$C, fibre crack density = 10 mm^{-1}, $\eta = 0.5$).

The debond zone occupies the region $0.9543 \le z/L \le 1$, and it is seen that the interfacial radial stress is negative in the debonded zone as required for valid solutions. Contact is lost in the debonded zone when the applied stress σ_A exceeds the value 14.7 MPa. Clearly interfacial debonding for a frictionally bonded composite is a very limited occurrence when considering fibre fractures. In fact, acceptable solutions for this case, where there is interfacial contact along the whole length of the debonded zone, are possible only if the applied axial stress lies in the very limited range $-7.762(\text{MPa}) \le \sigma_A \le 14.7\ (\text{MPa})$. It should be noted that changing the sign of ΔT does not alter the magnitudes of the bounds found for σ_A. They are merely reversed because the example has assumed that the transverse applied stress $\sigma_T = 0$.

The results obtained using the Coulomb friction law for fibre fractures suggest that a more complex type of frictional contact must occur when the applied axial stress lies outside this range involving both a friction slip region and an open region where contact has been lost, or even loss of contact near the debond tip. An investigation of this type of stress transfer is beyond the scope of this chapter.

The analysis and results presented apply only to frictionally bonded composites, i.e. to composites where interfacial bonding is very weak so that the interfacial fracture energy for fibre/matrix debonding can be neglected. In practice, composite manufacturers attempt to achieve high levels of interface bonding, especially for polymer composites. It is remarked that for ceramic matrix composites weak interfaces are desirable so that damage will form as multiple cracking rather than as the propagation of a single dominant crack.

Beyond the scope of this chapter is the application of the debonding models to composites having a strong fibre/matrix debonding. It is, however, useful to note that much of the analysis presented applies also to this important type of composite. One approach is to regard the debond length $L-b$ as a given parameter in which case it is required to determine the stress and displacement distributions in the fibre and matrix that correspond to this debond length. This is achieved simply by removing from the analysis the requirement that either condition (16.152) or condition (16.166) are satisfied. This means that the second derivative of the stress-transfer function $C(z)$ is not continuous at the location $z=b$ of the debond tip. As a consequence, the interfacial normal stress will be discontinuous at this point where exact solutions are expected to predict a singularity. Exact solutions would also predict that the shear stress and axial stress would be singular at this location in the uncracked fibre or matrix. Such predictions are beyond the capabilities of a concentric cylinder model. However, unpublished work has been undertaken by the author that applies the stress and displacement representation derived in Sections 16.5–16.8 to an assembly of multiple concentric cylinders that enables the stress singularities at debond tips to be well represented. This is achieved by subdividing the fibre and matrix into systems of concentric cylindrical layers in which the axial stress is uniform, but which differ from layer to layer. To date, solutions have been derived only for perfectly bonded interfaces. Much further work is needed to extend the multiple cylinder model to include debonded zones.

16.14 Prediction of Matrix Cracking

It is useful to discuss crack bridging for unidirectional composites by referring to Figure 16.10. In Figure 16.10(a), a unidirectional composite having a single matrix crack with frictionally slipping fibre/matrix interfaces and subject to a uniform uniaxial applied stress σ_A is shown. Figure 16.10(b) shows the corresponding RVE for a single fibre. The simplified analysis described previously indicates that stress transfer between fibre and matrix is restricted to the regions A which are adjacent to the matrix crack. For the other regions B stress transfer is negligible so that the local stress and displacement fields in the fibre and matrix correspond to those found in an undamaged composite made of the same material and subject to the same applied stress and temperature. In regions A the stress carried by the matrix is transferred to the fibres by action of the shear stresses at the fibre/matrix interfaces. As the matrix crack is approached, the matrix stress reduces to zero whereas the stress in the fibres increases to the maximum value σ_f^{max} given by (16.131). It is assumed that all fibres remain intact during loading of the composite, so that the discontinuous tractions associated with the discrete fibre reinforcement can be approximated by a continuous distribution $p(x)$ such that

$$p(x) \equiv p = V_f \sigma_f^{max} = \sigma_A. \tag{16.172}$$

The approximation made is illustrated schematically in Figure 16.10(b) showing the discrete single-fibre model and Figure 16.10(c) showing the equivalent continuum model. It should be noted that in Figure 16.10 the stress distribution and deformation in regions B is unperturbed by the presence of the matrix crack. In the discrete fibre model shown in Figure 16.10(b) the opening of the matrix crack is denoted by $2\Delta_{mc}$

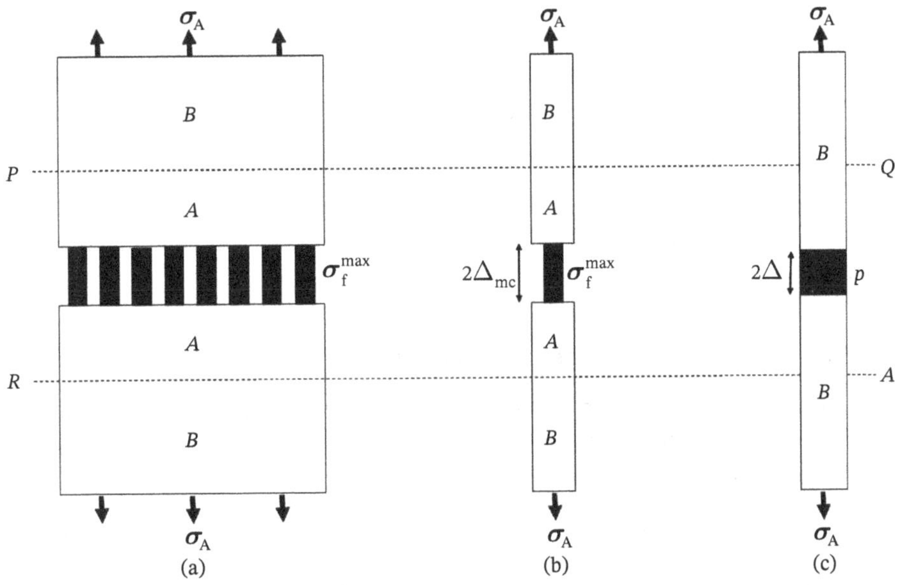

Figure 16.10 Schematic diagram of a unidirectional composite containing a matrix crack for which interfacial frictional slip has occurred: (a) composite as a whole; (b) discrete model for a single fibre; (c) corresponding continuum model.

where the value of Δ_{mc} is calculated using (16.135). In the continuum model shown in Figure 16.10(c), the stress distribution and deformation in region *B*, associated with undamaged parts of the composite, is uniform and has been extended to the matrix crack surfaces. The opening of the matrix crack, when the continuum traction applied to the crack faces is *p*, is denoted by 2Δ which differs from the value Δ_{mc} given by the discrete model. The value of Δ is determined by the axial displacements of the lines *PQ* and *RS* shown in Figure 16.10. By defining the axial displacement *V* so that it is zero on the midplane it is clear that

$$V_{PQ} = L\varepsilon_{\rm A} + \Delta, \qquad V_{RS} = -L\varepsilon_{\rm A} - \Delta, \tag{16.173}$$

where $\varepsilon_{\rm A}$ is the axial strain for an undamaged composite. It follows from (16.126) that

$$V_{PQ} = v_{\rm f}(L) = L\varepsilon_{\rm A} + \frac{\tau}{RE_{\rm A}^{\rm f}}L^2, \tag{16.174}$$

so that on using (16.120), (16.130) and (16.133)

$$\Delta = \frac{R}{4\tau}\frac{1}{E_{\rm A}^{\rm f}}\left(\frac{V_{\rm m}E_{\rm A}^{\rm m}}{V_{\rm f}E_{\rm A}}\right)^2\left(\sigma_{\rm A} + V_{\rm f}E_{\rm A}^{\rm f}\left(\alpha_{\rm A}^{\rm f} - \alpha_{\rm A}^{\rm m}\right)\Delta T\right)^2. \tag{16.175}$$

By comparing (16.135) and (16.175), it is clear that

$$\frac{\Delta}{\Delta_{\rm mc}} = \frac{V_{\rm m}E_{\rm A}^{\rm m}}{E_{\rm A}}, \tag{16.176}$$

confirming that the continuum crack opening displacement 2Δ differs from the value $2\Delta_{mc}$ obtained using the discrete model. On using (16.172) and (16.175), the crack bridging relation for the continuum model of a matrix crack may be written as

$$\Delta v = \frac{1}{\Lambda}\left(p - \sigma_{\rm th}\right)^2, \tag{16.177}$$

where

$$\Delta v \equiv 2\Delta, \qquad \Lambda = \frac{2\tau E_{\rm A}^{\rm f}}{R}\left(\frac{V_{\rm f}E_{\rm A}}{V_{\rm m}E_{\rm A}^{\rm m}}\right)^2, \qquad \sigma_{\rm th} = V_{\rm f}E_{\rm A}^{\rm f}\left(\alpha_{\rm A}^{\rm m} - \alpha_{\rm A}^{\rm f}\right)\Delta T. \tag{16.178}$$

It is noted that the thermal stress σ_{th} is zero if the temperature difference $\Delta T = 0$ or the axial thermal expansion coefficients of the fibre and matrix are equal. Result (16.177) can now be applied to the continuum bridged crack problem described in Chapter 8.

16.14.1 Solution of the Bridged Crack Problem

For matrix cracks of any length $2c$ subject only to mode I crack opening, the relationship between the effective tractions applied to the crack surfaces and the crack opening is of the form (16.177), which is written as

$$\left(p(x,c) - \sigma_{\rm th}\right)^2 = \Lambda\Delta v(x,c), \quad |x| < c \tag{16.179}$$

First, define the following parameters and the function *P*

$$t = \frac{\xi}{c}, \quad X = \frac{x}{c}, \quad \mu = \frac{4\Omega\Lambda c}{(\sigma_{\rm A} - \sigma_{\rm th})E_{\rm A}}, \quad P(X,\mu) = \frac{p(x,c) - \sigma_{\rm th}}{\sigma_{\rm A} - \sigma_{\rm th}}, \tag{16.180}$$

where the anisotropy factor is defined by (8.129) for generalised plane stress conditions or (8.130) for conditions of plane strain. Aa a unidirectional composite is being considered it is possible to assume that the homogenised undamaged material is transverse isotropic so that $E_t = E_{\mathrm{T}}, \nu_a = \nu_{\mathrm{A}}$ when using relation (8.130).

It follows from (8.145) that it is necessary to solve the following nonlinear integral equation for the function $P(X, \mu)$

$$\left(P(X,\mu)\right)^2 = \mu \int_0^1 (1 - P(t,\mu))k(t,X)\mathrm{d}t, \quad |X| < 1, \tag{16.181}$$

where

$$k(t,X) = \frac{1}{\pi}\ln\left|\frac{\sqrt{1-t^2}+\sqrt{1-X^2}}{\sqrt{1-t^2}-\sqrt{1-X^2}}\right| = k(X,t). \tag{16.182}$$

As

$$\int_0^1 k(t,X)\mathrm{d}t = \frac{1}{\pi}\int_0^1 \ln\left|\frac{\sqrt{1-t^2}+\sqrt{1-X^2}}{\sqrt{1-t^2}-\sqrt{1-X^2}}\right|\mathrm{d}t = \sqrt{1-X^2}, \quad 0 < X < 1, \tag{16.183}$$

the integral equation (16.181) may then be written as

$$\left(P(X,\mu)\right)^2 = \mu\left[\sqrt{1-X^2} - \int_0^1 P(t,\mu)k(t,X)\mathrm{d}t\right], \quad 0 < X < 1. \tag{16.184}$$

The expression (8.144) for the mode I stress intensity factor K may be written as

$$K = \sqrt{\pi c}\left(\sigma_{\mathrm{A}} - \sigma_{\mathrm{th}}\right)Y(\mu) \text{ where } Y(\mu) = \frac{2}{\pi}\int_0^1 \frac{1-P(X,\mu)}{\sqrt{1-X^2}}\mathrm{d}X. \tag{16.185}$$

As shown in [3], the integral equation (16.184) can be solved using numerical methods which are described in Appendix K to calculate the distribution $P(X,\mu)$ and the function $Y(\mu)$. Figure 16.11 shows some typical numerical results for three values of

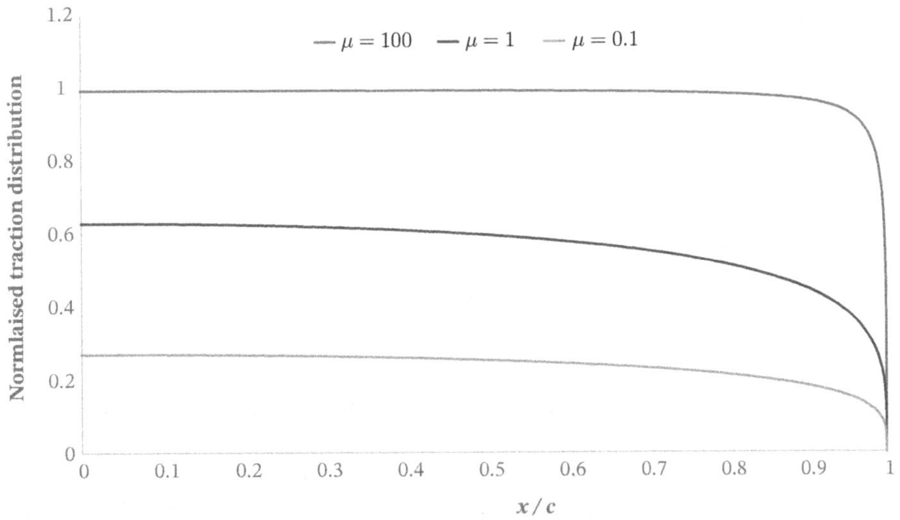

Figure 16.11 For debonded interfaces, typical numerical solutions for the distribution of tractions acting on the surface of the bridged crack. The normalised traction is $p(x)/\sigma_{\mathrm{A}}$.

$\mu = 0.1, 1$ and 100. It follows from (16.180) that the larger the crack length, the larger the value of μ. It is noted that when $\mu = 100$, the traction acting on the crack surface is approximated very well by the value σ_A for those parts of the crack that are not too close to the crack tips. This property will be exploited when considering the special case of very long cracks.

On using $(16.180)_3$ to eliminate c or $\sigma_A - \sigma_{th}$, it follows that

$$K = \sqrt{\frac{\pi E_A}{4\Omega\Lambda}}\left(\sigma_A - \sigma_{th}\right)^{3/2}\sqrt{\mu}\,Y(\mu), \tag{16.186}$$

or

$$K = 4\sqrt{\pi}\frac{\Omega\Lambda}{E_A}c^{3/2}\frac{Y(\mu)}{\mu}. \tag{16.187}$$

The value of $\sigma_A - \sigma_{th}$ for which crack growth is initiated is found by imposing the following fracture criterion (8.152) derived from energy balance considerations

$$K^2 = \frac{2E_A\gamma}{\Omega}, \tag{16.188}$$

where 2γ is the mode I effective fracture energy for matrix cracking for the homogenised material. For matrix cracks in a unidirectional composite where the fibres are normal to the crack surfaces it follows that $\gamma \equiv V_m\gamma_m$ where $2\gamma_m$ is the fracture energy of a homogeneous sample of matrix material. At the point of the initiation of crack propagation satisfying (16.188), and such that $\sigma_A = \sigma_A^{mc}$ and $\mu = \mu_c$, it follows from (16.186) and (16.188) that

$$\frac{\sigma_A^{mc} - \sigma_{th}}{\sigma_0} = \left(\frac{1}{\mu_c Y^2(\mu_c)}\right)^{1/3}, \tag{16.189}$$

where

$$\sigma_0 = \left(\frac{8\Lambda\gamma}{\pi}\right)^{1/3}. \tag{16.190}$$

From (16.187), it follows at the point of crack growth initiation, for which $c = c_{mc}$,

$$\frac{c_{mc}}{c_0} = \left(\frac{\mu_c}{Y(\mu_c)}\right)^{2/3}, \tag{16.191}$$

where

$$c_0 = \frac{E_A}{\Omega}\left(\frac{\gamma}{8\pi\Lambda^2}\right)^{1/3}. \tag{16.192}$$

It follows from (16.188), (16.190) and (16.192) that

$$\pi c_0\sigma_0^2 = \frac{2E_A\gamma}{\Omega} = K_{Ic}^2, \tag{16.193}$$

where K_{Ic} is regarded as the mode I effective fracture toughness of the homogenised material.

By solving the integral equation (16.184) for a range of values of μ and calculating the corresponding value of $Y(\mu)$ using (16.185), it is possible using (16.189) and (16.191) to generate a master curve that describes the dependence of the critical stress difference $\sigma_A^{mc} - \sigma_{th}$ on the length c of the preexisting matrix crack. This curve is

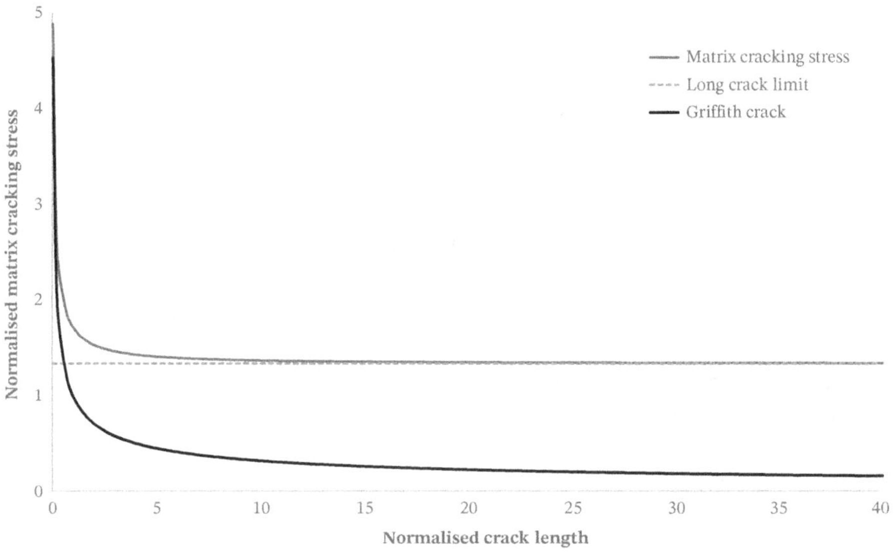

Figure 16.12 For debonded interfaces, the dependence of the critical axial applied stress for matrix cracking on the length of the preexisting matrix crack. The normalised matrix cracking stress is $(\sigma_{\rm A}^{\rm mc} - \sigma_{\rm th})/\sigma_0$ and the normalised crack length is c/c_0.

shown in Figure 16.12 which was generated by considering values of μ lying in the range $0.1 \leq \mu \leq 30$. To investigate whether or not the matrix cracking stress $\sigma_{\rm A}^{\rm mc}$ tends to an asymptotic limit as $c/c_0 \to \infty$, values of μ are considered in the range $20 \leq \mu \leq 2000$, which lead to the results listed in Table 16.1 indicating that $(\sigma_{\rm A}^{\rm mc} - \sigma_{\rm th})/\sigma \to 1.331$ as $c/c_0 \to \infty$. This result suggests that it is worth considering the special case of very long cracks in some detail.

16.14.2 Special Case of Long Matrix Cracks

On applying result (J.23) derived in Appendix J to the matrix cracking problem under consideration, it follows that at the point of matrix crack growth occurring when $\sigma_{\rm A} = \sigma_{\rm A}^{\rm mc}$

$$\int_0^c \left(\sigma_{\rm A}^{\rm mc} - p(x,c)\right)\frac{\partial \Delta v(x,c)}{\partial c}{\rm d}x + \int_0^c \frac{\partial p(x,c)}{\partial c}\Delta v(x,c){\rm d}x = 4V_{\rm m}\gamma_{\rm m}, \tag{16.194}$$

Table 16.1 Values of matrix cracking stress for various values of μ.

μ	$(\sigma_{\rm A}^{\rm mc} - \sigma_{\rm th})/\sigma_0$
0.1	2.52037997
1	1.58259402
10	1.35563328
100	1.33286170
1000	1.33091023
1500	1.33084889
2000	1.33082238

where for a matrix crack the effective continuum fracture energy 2γ is given by

$$2\gamma = 2V_{\mathrm{m}}\gamma_{\mathrm{m}}, \tag{16.195}$$

where $2\gamma_{\mathrm{m}}$ is the fracture energy for matrix material alone. It follows from (16.177) that

$$\left(p(x,c) - \sigma_{\mathrm{th}}\right)^2 = \Lambda \Delta v(x,c), \quad 0 < x < c. \tag{16.196}$$

For the case of very long cracks, the general result (16.194) may be approximated by

$$\int_{c-d}^{c} \left(\sigma_{\mathrm{A}}^{pc} - p(x-c)\right)\frac{\partial \Delta v(x-c)}{\partial c}\mathrm{d}x + \int_{c-d}^{c} \frac{\partial p(x-c)}{\partial c}\Delta v(x-c)\mathrm{d}x = 4V_{\mathrm{m}}\gamma_{\mathrm{m}}, \tag{16.197}$$

where d is the length of the region near the crack tip where the functions p and Δv are dependent on x. For the region $0 < x < c - d$ and for sufficiently long cracks $p = \sigma_{\mathrm{A}}^{pc}$ and $\Delta v = \Delta v_{\infty}$ where Δv_{∞} is independent of x and c having the following value obtained using (16.179) so that

$$\Delta v_{\infty} = \frac{1}{\Lambda}\left(\sigma_{\mathrm{A}}^{\mathrm{mc}} - \sigma_{\mathrm{th}}\right)^2. \tag{16.198}$$

The (x, c) dependence may be replaced by the $(x - c)$ dependence as the bridged crack is very long such that the local deformation in the region of the crack tip at $x = c$ is moving self-similarly as the crack tip moves. It follows that (16.197) may now be written as

$$-\sigma_{\mathrm{A}}^{\mathrm{mc}} \int_{c-d}^{c} \frac{\partial \Delta v(x-c)}{\partial x}\mathrm{d}x + \int_{c-d}^{c} p(x-c)\frac{\partial \Delta v(x-c)}{\partial x}\mathrm{d}x - \int_{c-d}^{c} \frac{\partial p(x-c)}{\partial x}\Delta v(x-c)\mathrm{d}x = 4V_{\mathrm{m}}\gamma_{\mathrm{m}}. \tag{16.199}$$

On setting $x' = x - c$ it can be shown that

$$\left(\sigma_{\mathrm{th}} - \sigma_{\mathrm{A}}^{\mathrm{mc}}\right)\int_{-d}^{0} \frac{\partial \Delta v(x')}{\partial x'}\mathrm{d}x' + \int_{-d}^{0} \left(p(x') - \sigma_{\mathrm{th}}\right)\frac{\partial \Delta v(x')}{\partial x'}\mathrm{d}x' - \int_{-d}^{0} \frac{\partial p(x')}{\partial x'}\Delta v(x')\mathrm{d}x' = 4V_{\mathrm{m}}\gamma_{\mathrm{m}}. \tag{16.200}$$

On using (16.196) for the region $c - d < x < c$,

$$2\left(p(x') - \sigma_{\mathrm{th}}\right)\frac{\partial p(x')}{\partial x'} = \Lambda \frac{\partial \Delta v(x')}{\partial x'}, \tag{16.201}$$

so that

$$\frac{\partial p(x')}{\partial x'} = \frac{1}{2}\Lambda^{1/2}\left\{\Delta v(x')\right\}^{-1/2}\frac{\partial \Delta v(x')}{\partial x'}. \tag{16.202}$$

As $\Delta v(0) = 0$ and $\Delta v(-d) = \Delta v_{\infty}$, it follows that (16.200) may then be written as

$$\left(\sigma_{\mathrm{A}}^{\mathrm{mc}} - \sigma_{\mathrm{th}}\right)\Delta v_{\infty} + \frac{1}{2}\Lambda^{1/2}\int_{-d}^{0} \left\{\Delta v(x')\right\}^{1/2}\frac{\partial \Delta v(x')}{\partial x'}\mathrm{d}x' = 4V_{\mathrm{m}}\gamma_{\mathrm{m}}, \tag{16.203}$$

so that

$$\left(\sigma_{\mathrm{A}}^{\mathrm{mc}} - \sigma_{\mathrm{th}}\right)\Delta v_{\infty} - \frac{1}{2}\Lambda^{1/2}\int_{0}^{\Delta v_{\infty}} \left\{\Delta v\right\}^{1/2}\mathrm{d}\Delta v = 4V_{\mathrm{m}}\gamma_{\mathrm{m}}, \tag{16.204}$$

i.e.

$$\left(\sigma_{\mathrm{A}}^{\mathrm{mc}} - \sigma_{\mathrm{th}}\right)\Delta v_{\infty} - \frac{1}{3}\Lambda^{1/2}\left\{\Delta v_{\infty}\right\}^{3/2} = 4V_{\mathrm{m}}\gamma_{\mathrm{m}}. \tag{16.205}$$

On using (16.198), the long crack matrix cracking stress σ_A^{mc} is given by

$$\sigma_A^{mc} - \sigma_{th} = \{6\Lambda V_m \gamma_m\}^{1/3}. \tag{16.206}$$

On using (16.190), it is clear that

$$\frac{\sigma_A^{mc} - \sigma_{th}}{\sigma_0} = \left(\frac{3\pi}{4}\right)^{1/3} = 1.33067. \tag{16.207}$$

This asymptotic limit is shown as a dashed line in Figure 16.12.

On using $(16.178)_2$, result (16.206) may be expressed as

$$\sigma_A^{mc} - \sigma_{th} = \left\{\frac{12\gamma_m \tau}{R} \frac{V_f^2 E_A^f E_A^2}{V_m (E_A^m)^2}\right\}^{1/3}. \tag{16.208}$$

When the thermal stress σ_{th} is zero this result corresponds exactly to the famous relation of ACK theory derived by Aveston et al. [2].

It is useful to consider the behaviour of matrix cracking when $\mu \to 0$ corresponding to matrix cracks of decreasing length. For this special case, it follows from (16.181) that $P(X,\mu) \to 0$ and from (16.185) that $Y(\mu) \to 1$ leading to the following expression for the stress intensity factor for a stress-free embedded crack in an infinite plate

$$K = \sqrt{\pi c}\left(\sigma_A - \sigma_{th}\right). \tag{16.209}$$

On using (16.193) the well-known Griffith fracture criterion $K = K_{Ic}$ may be expressed in the form

$$\frac{\sigma_A^{mc} - \sigma_{th}}{\sigma_0} = \sqrt{\frac{c_0}{c}}. \tag{16.210}$$

This relationship is shown in Figure 16.12 using the label 'Griffith crack'. As to be expected it approaches the bridged crack solution when $c / c_0 \to 0$, but reduces to zero as $c / c_0 \to \infty$.

It is emphasised that the bridged crack solution shown in Figure 16.12 is a master curve that applies to all materials whose deformation is governed by the simplified model. The effects of all material constants, including fracture parameters, are included in the normalising constants σ_0 and c_0 defined by (16.190) and (16.192), respectively. If an alternative stress-transfer model is used, such as that described earlier for the case of perfect fibre/matrix bonding, then a different crack bridging curve to that shown in Figure 16.12 would result. This case will be considered subsequently.

To complete the discussion of the simplified model it is also worth mentioning that, in the absence of an accurate numerical solution of the bridged crack problem, use can be made of the long crack limit and the prediction for a Griffith crack to construct an approximate approach. With reference to Figure 16.12, the long crack limiting value given by (16.207) is used in the Griffith relation (16.210) to determine the crack length c_1, denoting the point at which the long crack asymptote crosses the Griffith curve, that is given by

$$\frac{c_1}{c_0} = \left(\frac{4}{3\pi}\right)^{2/3} = 0.5648. \tag{16.211}$$

Thus, the approximate estimate of the matrix cracking stress is obtained using (16.210) when $c < c_1$ and using (16.207) when $c > c_1$. It is noted from Figure 16.12 that the approximation always underestimates the matrix cracking stress, thus providing a relatively simple conservative estimate that could be used when designing composites that must be resistant to matrix damage.

16.14.3 Consideration of Matrix Cracking for Perfectly Bonded Fibre/Matrix Interfaces

It is clear that, when considering the prediction of matrix cracking in unidirectional composites, it is required to make use of a crack bridging relation that specifies the effective normal traction distribution σ_A applied to the crack surfaces in terms of the effective crack opening displacement 2Δ that was defined previously and is illustrated in Figure 16.12. For matrix cracking problems this relationship is obtained by considering the displacement distribution in the fibre. By making use of (16.114), it is clear that

$$\Delta = \frac{2Rp}{p^2 - q^2}\frac{\sigma_m}{E_A^f}\frac{V_m}{V_f}\left(1 - 2\beta\nu_A^f\frac{V_m}{V_f}\right), \tag{16.212}$$

where p, q and β are given by (16.69) and (16.90). When the transverse stress applied to the composite is zero so that $\sigma_T = 0$, the matrix stress σ_m for an undamaged composite is obtained from relation (B.38) of Appendix B so that

$$\sigma_m = E_A^m(\varepsilon_A - \alpha_A^m \Delta T) + 2\nu_A^m V_f \frac{\phi}{R^2}. \tag{16.213}$$

The value of the parameter ϕ is obtained from (B.42) so that

$$\frac{\phi}{R^2} = \lambda\left[(\nu_A^m - \nu_A^f)\varepsilon_A + \left(\alpha_T^f + \nu_A^f\alpha_A^f\right)\Delta T - \left(\alpha_T^m + \nu_A^m\alpha_A^m\right)\Delta T\right], \tag{16.214}$$

where

$$\frac{1}{\lambda} = \frac{1}{2}\left(\frac{1}{\mu_t^m} + \frac{V_f}{k_T^m} + \frac{V_m}{k_T^f}\right). \tag{16.215}$$

When $\sigma_T = 0$ it follows from (B.49) that

$$\varepsilon_A = \frac{\sigma_A}{E_A} + \alpha_A \Delta T, \tag{16.216}$$

so that

$$\sigma_m = E_A^m\left(\frac{\sigma_A}{E_A} + (\alpha_A - \alpha_A^m)\Delta T\right) + 2\nu_A^m V_f \frac{\phi}{R^2}, \tag{16.217}$$

and

$$\frac{\phi}{R^2} = \lambda(\nu_A^m - \nu_A^f)\frac{\sigma_A}{E_A} + \lambda\left[(\nu_A^m - \nu_A^f)\alpha_A + (\alpha_T^f + \nu_A^f\alpha_A^f) - (\alpha_T^m + \nu_A^m\alpha_A^m)\right]\Delta T. \tag{16.218}$$

It then follows that

$$\begin{aligned}\sigma_m = &\left[E_A^m + 2\nu_A^m V_f\lambda(\nu_A^m - \nu_A^f)\right]\frac{\sigma_A}{E_A} \\ &+ \left[E_A^m(\alpha_A - \alpha_A^m) + 2\nu_A^m V_f\lambda\left\{(\nu_A^m - \nu_A^f)\alpha_A + (\alpha_T^f + \nu_A^f\alpha_A^f) - (\alpha_T^m + \nu_A^m\alpha_A^m)\right\}\right]\Delta T.\end{aligned} \tag{16.219}$$

The substitution of (16.219) into (16.212) then leads to the following linear crack bridging law

$$\Delta v = \frac{1}{\Lambda}(\sigma_A - \sigma_{th}), \tag{16.220}$$

where

$$\frac{1}{\Lambda} = \frac{4Rp}{p^2 - q^2} \frac{V_m}{E_A^f V_f}\left(1 - 2\beta\nu_A^f \frac{V_m}{V_f}\right)\frac{E_A^m + 2\nu_A^m V_f\lambda(\nu_A^m - \nu_A^f)}{E_A}, \tag{16.221}$$

$$\sigma_{th} = \frac{E_A^m(\alpha_A^m - \alpha_A) - 2\nu_A^m V_f\lambda\left\{(\nu_A^m - \nu_A^f)\alpha_A + (\alpha_T^f + \nu_A^f\alpha_A^f) - (\alpha_T^m + \nu_A^m\alpha_A^m)\right\}}{E_A^m + 2\nu_A^m V_f\lambda(\nu_A^m - \nu_A^f)} E_A\Delta T. \tag{16.222}$$

where $\Delta v = 2\Delta$ is the total crack opening displacement and σ_{th} is the axial thermal stress that is proportional to the temperature difference ΔT. Relation (16.220) can be used as a description of the traction/displacement law for the study of matrix cracking in unidirectional composites for the case when the fibre/matrix interfaces are perfectly bonded. This linear type of crack bridging law will arise also when considering the growth of ply cracks in laminates as described in detail in Chapter 17. The master curve that describes the dependence of the matrix cracking stress for perfectly bonded fibre/matrix interfaces on the length of the matrix crack is identical to that shown in Figure 17.2.

16.15 Conclusion

A high-quality analytical model of stress transfer has been described that can be used to predict the localised stress and displacement distributions associated with fibre fractures and matrix cracks in unidirectional composites whose fibres and matrix deform linear elastically and are subject to differing thermal expansion properties. The approach to modelling interfacial debonding has been to ensure that, wherever possible, the expressions for all relevant physical parameters are given by analytical formulae. The stress transfer associated with debonding can be modelled using two different approaches. The first assumes that the interfacial shear stress is uniform in the debonded zone. Satisfactory solutions, enabling zero shear tractions on crack surfaces, are possible only if a transition zone is included in the neighbourhood of the fibre fracture or matrix crack. The second approach assumes that stress transfer in the debonded region is governed by the Coulomb friction law: a situation that does not require the use of a transition zone. Predictions indicate that the stress-transfer models work very well for a range of composite types and loading conditions that lead to compressive interfacial normal stresses along the entire length of the debonded zone.

References

1. Nairn, J.A. (1997). On the use of shear-lag methods for analysis of stress transfer in unidirectional composites. *Mechanics of Materials* 26: 63–80.
2. Aveston, J., Cooper, G.A., and Kelly, A. (1971). Single and multiple fracture. *Conference on the Properties of Fibre Composites', National Physical Laboratory, 15–26,* IPC Science and technology Press, Guildford, Surrey, UK.
3. McCartney, L.N. (1987). Mechanics of matrix cracking in brittle-matrix fibre-reinforced composites. *Proceedings of the Royal Society of London* A409: 329.
4. McCartney, L.N. (1992). Mechanics for the growth of bridged cracks in composite materials: Parts I, basic principles. *Journal of Composites Technology and Research* 14: 133–146.
5. McCartney, L.N. (1992). Mechanics for the growth of bridged cracks in composite materials: Part II, applications. *Journal of Composites Technology and Research* 14: 147–154.
6. Hutchinson, J.W. and Jensen, H.M. (1990). 'Models of fiber debonding and pull-out in brittle composites with friction. *Mechanics of Materials* 9: 139–163.
7. Nairn, J.A. (1992). A variational mechanics analysis of the stresses around breaks in embedded fibres. *Mechanics of Materials* 13: 131–154.
8. McCartney, L.N. (2013). Analytical models for sliding interfaces associated with fibre fractures or matrix cracks. *Computers, Materials, Continua* 35 (3): 183–227.
9. McCartney, L.N. (1989). New theoretical model of stress transfer between fibre and matrix in a uniaxially fibre reinforced composite. *Proceedings of the Royal Society of London* A425: 215–244.
10. Reissner, E. (1950). On a variational theorem in elasticity. *Journal of Mathematical Physics* 29: 90–95.
11. McCartney, L.N. (1996). Stress transfer mechanics: Models that should be the basis of life prediction methodology. In: *Life Prediction Methodology for Titanium Matrix Composites*, ASTM STP 1253 (ed. W.S. Johnson, J.M. Larsen, and B.N. Cox), 85–113.
12. Nayfey, A.H. (1977). Thermomechanically induced interfacial stresses in fibrous composites. *Fibre Science and Technology* 10: 195.
13. McCartney, L.N. (1991). Analytical models of stress transfer in unidirectional composites and cross-ply laminates, and their application to the prediction of matrix/transverse cracking. *Proceedings of the IUTAM Symposium on 'Local mechanics concepts for composite materials systems'*, Blacksburg, VA, 251–282.

17

Interacting Bridged Ply Cracks in a Cross-ply Laminate

Overview: Laminates made from anisotropic materials, such as CFRP or GRP, are in common use in the aerospace, defence, leisure and general engineering sectors. When subjected to a tensile stress, these materials tend to form cracks running parallel to the fibres in the transverse plies of the laminates, affecting the effective mechanical properties of the laminates. This chapter is concerned with the development of a mathematical model to predict the instability of systems of embedded parallel transverse cracks, in the 90° ply of a 0°/90°/0° laminate, taking account of crack size and crack interaction effects. The mathematical technique used is based upon a complex variable technique for anisotropic materials that exploits the extremely useful orthogonal properties of Chebyshev polynomials when applied to crack problems arising in linear elasticity theory. Both stress-free through cracks and transverse cracks in the 90° ply bridged by the 0° plies are considered. Simple expressions are obtained for the stress intensity factors of the cracks involving coefficients that are obtained by solving a system of linear algebraic equations. The crack boundary conditions for bridged or unbridged cracks are imposed using two different techniques. The first satisfies the crack boundary conditions by exploiting the orthogonal properties of Chebyshev polynomials. The second method satisfies the crack boundary conditions exactly at a discrete number of specified points. The latter technique is more efficient for large problems where computation times are of concern. The analysis is used to confirm that well-known results for isolated cracks are recovered for both stress-free through cracks in isotropic plates and for isolated bridged cracks in a cross-ply laminate. Crack interaction effects are considered for the case of two collinear cracks in a laminate where the crack is both stress-free and bridged by 0° plies. An example is also considered of crack interaction between three equally spaced ply cracks all having the same length.

17.1 Introduction

Anisotropic materials (i.e. isotropic in a plane normal to the fibre direction), such as CFRP or GRP, are in common use in the aerospace, defence, leisure and general engineering sectors, owing to their high stiffness, strength and damage tolerance. They are often implemented as laminated structures (see Figure 9.1 for the case of a 0°/90°/0°

Properties for Design of Composite Structures: Theory and Implementation Using Software,
First Edition. Neil McCartney.

Companion Website: www.wiley.com/go/mccartney/properties

laminate), which provide tensile strength in two orthogonal directions. When subjected to an axial tensile stress, the 90° ply tends to form transverse cracks parallel to the fibres such that the crack density increases with the applied load. Such cracking may also occur during the manufacturing process itself because of thermal expansion mismatch between plies having differing orientations. Transverse cracking is the most significant mechanism contributing to the onset of damage within the laminate, and to subsequent damage growth as the load is further increased. In this chapter, one method is examined for determining the stress and displacement fields around the relatively small cracks that represent defects in the 90° ply that will grow into fully developed transverse cracks as the laminate is loaded. The method considered will be able to take account of the interaction of small defects with their neighbours, and with fully developed transverse cracks that have already formed.

The theory presented here is an extension of [1] in which *multiple angled* cracks in an *isotropic* material is considered. This chapter addresses the case of *multiple parallel* cracks embedded in a *transversely isotropic* material.

17.2 Crack-bridging for Long Isolated Cracks in Cross-ply Laminates

The following analysis relating to crack-bridging for an isolated crack in perfectly bonded cross-ply laminates is extracted from previously published work [2–4].

The thermoelastic properties of the plies of a laminate are assumed to be transverse isotropic so that they are specified by seven independent thermoelastic constants which are denoted as follows.

Elastic constants

Axial Young's moduli, $E_\mathrm{A}^0, E_\mathrm{A}^{90}, E_\mathrm{A}$,

Axial Poisson's ratios, $\nu_\mathrm{A}^0, \nu_\mathrm{A}^{90}, \nu_\mathrm{A}$,

Transverse Young's moduli, $E_\mathrm{T}^0, E_\mathrm{T}^{90}, E_\mathrm{T}$,

Transverse Poisson's ratios, $\nu_\mathrm{t}^0, \nu_\mathrm{t}^{90}, \nu_\mathrm{t}$,

Axial shear moduli, $\mu_\mathrm{A}^0, \mu_\mathrm{A}^{90}, \mu_\mathrm{A}$.

Thermal constants

Axial thermal expansion coefficients, $\alpha_\mathrm{A}^0, \alpha_\mathrm{A}^{90}, \alpha_\mathrm{A}$,

Transverse thermal expansion coefficients, $\alpha_\mathrm{T}^0, \alpha_\mathrm{T}^{90}, \alpha_\mathrm{T}$.

For cross-ply laminated composites, although it is usual to assume that each ply has the same thermomechanical properties, it is useful when developing models to allow the 0° plies to have different properties from those of the 90° plies. The performance of hybrid composites can then be readily evaluated. A superscript '0' is used to denote properties associated with the 0° plies and a superscript '90' is used to denote the properties of the 90° plies. The absence of a superscript denotes that the thermoelastic constant is an effective property associated with the laminate. For a transverse isotropic solid, there is no need to specify the transverse shear modulus as it can be calculated from other elastic constants using the following formulae,

$$\mu_\mathrm{t}^0 = \frac{E_\mathrm{T}^0}{2(1+\nu_\mathrm{t}^0)}, \quad \mu_\mathrm{t}^{90} = \frac{E_\mathrm{T}^{90}}{2(1+\nu_\mathrm{t}^{90})}, \quad \mu_\mathrm{t} = \frac{E_\mathrm{T}}{2(1+\nu_\mathrm{t})}. \tag{17.1}$$

Approximate analytical expressions have been derived [5] for the stress and displacement distributions in the 0° and 90° plies of a loaded cross-ply laminate containing transverse cracks in the 90° ply. Residual stresses arising from thermal expansion mismatch effects are taken into account and it is assumed that the interfaces between the 0° and 90° plies remain perfectly bonded even in the neighbourhood of the transverse crack. The geometry is such that $2a$ is the thickness of the 90° ply, b is the thickness of the 0° plies, and where $2h = 2(a + b)$ is the total thickness of the laminate.

It is useful to discuss crack bridging for cross-ply composites by referring to Figure 17.1. In Figure 17.1(a), a face view of a 0°/90°/0° laminate having a single-ply crack with perfectly bonded plies and subject to a uniform uniaxial applied stress σ_A is shown. Figure 17.1(b) shows the corresponding edge view of the laminate. The analysis described previously indicates that stress transfer between plies is restricted to the regions A which are adjacent to the ply crack. For the other regions B, stress transfer is negligible so that the local stress and displacement fields in the plies correspond to those found in an identical undamaged laminate subject to the same applied stress and temperature. In regions A, the stress carried by the 90° ply is transferred to the adjacent 0° plies by the action of shear stresses at the ply interfaces.

As the ply crack is approached, the axial stress in the 90° ply reduces to zero whereas the axial stress in the 0° plies increases to the maximum value σ_0^{max}. It is assumed that the discontinuous tractions on the ply crack plane associated with the ply geometry can be approximated by a continuous distribution $p(x)$ such that

$$p(x) \equiv p = b\sigma_0^{max} / h = \sigma_A. \tag{17.2}$$

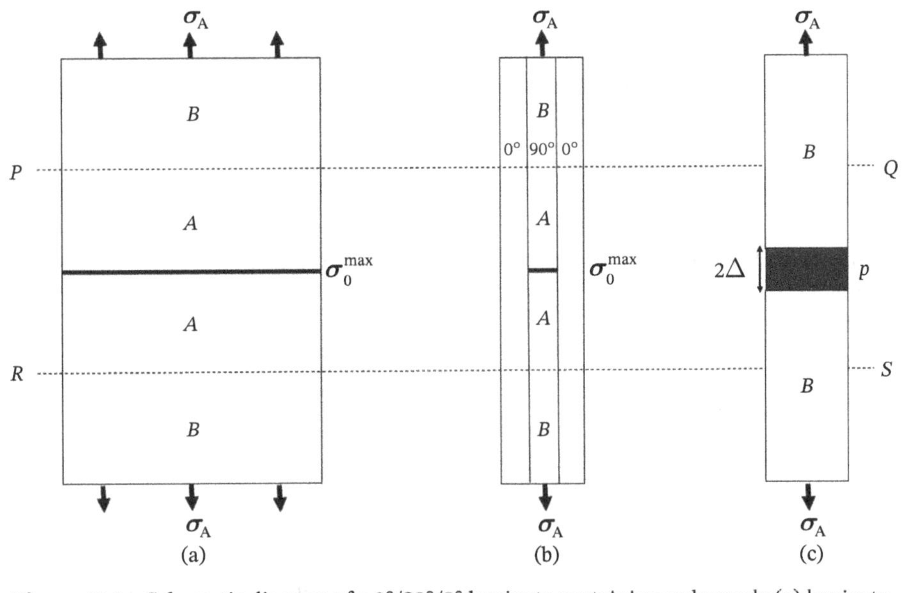

Figure 17.1 Schematic diagram of a 0°/90°/0° laminate containing a ply crack: (a) laminate as a whole; (b) discrete model; and (c) corresponding continuum model.

The approximation made is illustrated schematically in Figure 17.1(b) showing the discrete single-fibre model and Figure 17.1(c) showing the equivalent continuum model. It should be noted that in Figure 17.1, the stress distribution and deformation in regions B is unperturbed by the presence of the ply crack. In the continuum model shown in Figure 17.1(c), the stress distribution and deformation in region B, associated with undamaged parts of the laminate, is uniform and has been extended to the ply crack surfaces. The opening of the ply crack, when the continuum traction applied to the crack faces is p, is denoted by 2Δ. The value of Δ is determined by the axial displacements of the lines PQ and RS shown in Figure 17.1. By defining the axial displacement V so that it is zero on the midplane it is clear that

$$V_{PQ} = L\hat{\varepsilon}_{\mathrm{A}} + \Delta, \qquad V_{RS} = -L\hat{\varepsilon}_{\mathrm{A}} - \Delta, \tag{17.3}$$

where, as in Chapter 9, $\hat{\varepsilon}_{\mathrm{A}}$ is the axial strain for an undamaged composite. It follows from (9.109) that

$$V_{PQ} = -V_{RS} = L\hat{\varepsilon}_{\mathrm{A}} + \frac{1}{\tilde{E}_{\mathrm{A}}^{0}}\bar{C}(L) - \frac{\tilde{\nu}_{\mathrm{a}}^{0}}{6\tilde{E}_{\mathrm{A}}^{0}} b^2 C'(L), \tag{17.4}$$

where $\tilde{E}_{\mathrm{A}}^{0}$ and $\tilde{\nu}_{\mathrm{a}}^{0}$ are defined by (9.30). The stress-transfer function $C(x_1)$ for generalised plane strain conditions is defined by (9.116) such that (see (9.118) and (9.127))

$$C'(L) = 0, \quad \bar{C}(L) = \frac{a}{b}\hat{\sigma}_{11}^{90} L\Phi, \tag{17.5}$$

where $\hat{\sigma}_{11}^{90}$, calculated using (9.32) on setting $\hat{\varepsilon}_{\mathrm{T}} = \varepsilon_{\mathrm{T}}$, is the axial stress in the 90° ply for an undamaged laminate subject to the same transverse strain ε_{T} as the corresponding cracked laminate, and where Φ is the dimensionless damage-dependent parameter defined by (9.128), which may be written on using (9.123) in the form

$$\Phi = \frac{4pq}{p^2 - q^2}\frac{b}{L}\,\frac{\tanh\dfrac{(p+q)L}{b}\tanh\dfrac{(p-q)L}{b}}{(q+p)\tanh\dfrac{(p+q)L}{b} + (q-p)\tanh\dfrac{(p-q)L}{b}}, \tag{17.6}$$

where the parameters p and q are defined by (9.111) and (9.117). On using (17.3)–(17.5), it then follows that

$$\Delta = \frac{a}{b}\frac{\hat{\sigma}_{11}^{90}}{\tilde{E}_{\mathrm{A}}^{0}} L\Phi. \tag{17.7}$$

For the crack bridging problem applied to a generalised plane strain model, uniaxial loading is considered so that $\sigma_{\mathrm{A}} \neq 0$, the effective transverse stress $\sigma_{\mathrm{T}} = 0$ and through-thickness stress $\sigma_{\mathrm{t}} = 0$, so that

$$\hat{\sigma}_{11}^{90} = \frac{\tilde{E}_{\mathrm{T}}^{90}}{\xi}\left[\left(1 - \nu_{\mathrm{A}}^{90}\nu_{\mathrm{A}}^{(\mathrm{L})}\right)\frac{\sigma_{\mathrm{A}}}{E_{\mathrm{A}}^{(\mathrm{L})}} + \left(\alpha_{\mathrm{A}}^{(\mathrm{L})} + \nu_{\mathrm{A}}^{90}\alpha_{\mathrm{T}}^{(\mathrm{L})} - \tilde{\alpha}_{\mathrm{T}}^{90}\right)\Delta T\right], \tag{17.8}$$

where from (9.139)

$$\xi = 1 - \left(\nu_{\mathrm{A}}^{0}\frac{E_{\mathrm{T}}^{0}}{E_{\mathrm{A}}^{0}} - \nu_{\mathrm{A}}^{90}\right)^2 \frac{ab}{h^2}\frac{\tilde{E}_{\mathrm{A}}^{0}\tilde{E}_{\mathrm{T}}^{90}}{\tilde{E}_{\mathrm{A}}^{(L)}E_{\mathrm{T}}^{(L)}}\Phi. \tag{17.9}$$

On identifying σ_{A} with the crack face traction distribution p, the following crack bridging law is obtained from (17.7) and (17.8)

$$\Delta = \frac{a}{b}\frac{\tilde{E}_{\mathrm{T}}^{90}}{\tilde{E}_{\mathrm{A}}^{0}}\left[\frac{1-\nu_{\mathrm{A}}^{90}\nu_{\mathrm{A}}^{(\mathrm{L})}}{E_{\mathrm{A}}^{(\mathrm{L})}}p+\left(\alpha_{\mathrm{A}}^{(\mathrm{L})}+\nu_{\mathrm{A}}^{90}\alpha_{\mathrm{T}}^{(\mathrm{L})}-\tilde{\alpha}_{\mathrm{T}}^{90}\right)\Delta T\right]\frac{L\Phi}{\xi}. \tag{17.10}$$

This result can be expressed in the form

$$p = \Lambda\Delta v + \sigma_{\mathrm{th}}, \tag{17.11}$$

where $\Delta v = 2\Delta$ is the total crack opening displacement, where

$$\frac{1}{\Lambda} = \frac{2a}{b}\frac{\tilde{E}_{\mathrm{T}}^{90}}{\tilde{E}_{\mathrm{A}}^{0}}\frac{1-\nu_{\mathrm{A}}^{90}\nu_{\mathrm{A}}^{(\mathrm{L})}}{E_{\mathrm{A}}^{(\mathrm{L})}}\frac{L\Phi}{\xi}, \tag{17.12}$$

$$\sigma_{\mathrm{th}} = -E_{\mathrm{A}}^{(\mathrm{L})}\frac{\alpha_{\mathrm{A}}^{(\mathrm{L})}-\alpha_{\mathrm{T}}^{90}+\nu_{\mathrm{A}}^{90}(\alpha_{\mathrm{T}}^{(\mathrm{L})}-\alpha_{\mathrm{A}}^{90})}{1-\nu_{\mathrm{A}}^{90}\nu_{\mathrm{A}}^{(\mathrm{L})}}\Delta T = -k_1\Delta T, \tag{17.13}$$

and where use has been made of the definition (9.34) for $\tilde{\alpha}_{\mathrm{T}}^{90}$ and the definition (9.196) for the parameter k_1.

For the case of isolated ply cracks, the crack separation L tends to infinity so that it becomes necessary to consider the behaviour of the quantity $L\Phi/\xi$ in this limit. It follows from (17.6) that as $L \to \infty$

$$L\Phi \to \frac{2bp}{p^2-q^2}, \tag{17.14}$$

and from (17.9) that $\xi \to 1$ so that for an isolated ply crack

$$\frac{1}{\Lambda} \to \frac{\tilde{E}_{\mathrm{T}}^{90}}{\tilde{E}_{\mathrm{A}}^{0}}\frac{1-\nu_{\mathrm{A}}^{90}\nu_{\mathrm{A}}^{(\mathrm{L})}}{E_{\mathrm{A}}^{(\mathrm{L})}}\frac{4ap}{p^2-q^2}. \tag{17.15}$$

17.3 Method of Solution for Isolated Bridged Cracks of Any Length

For transverse cracks of any length in a cross-ply laminate subject only to mode I crack opening, it follows from (17.11) that the relationship between the effective traction distribution applied to the crack surfaces and the distribution of crack opening is of the form

$$p(x,c) = \Lambda\Delta v(x,c) + \sigma_{\mathrm{th}}, |x| < c, \tag{17.16}$$

where Λ and σ_{th} are constants defined by (17.12) and (17.13), respectively. Clearly, σ_{th} is nonzero when thermal residual stresses are present. It is useful to define, for an undamaged composite, the anisotropy factor Ω by

$$\Omega = E_{\mathrm{A}}\sqrt{a_{22}g}, \tag{17.17}$$

where g is defined by (8.125), so that the crack opening displacement distribution is obtained using (8.148) so that

$$\Delta v(x) = \frac{4\Omega}{\pi E_{\mathrm{A}}}\int_0^c \hat{p}(\xi)\ln\left|\frac{\sqrt{c^2-\xi^2}+\sqrt{c^2-x^2}}{\sqrt{c^2-\xi^2}-\sqrt{c^2-x^2}}\right| d\xi, \quad |x| < c. \tag{17.18}$$

For generalised plane stress conditions, it follows from (8.129) that

$$\Omega^2 = \frac{1}{2}\left(\sqrt{\frac{E_A}{E_T}} - \nu_A\right) + \frac{E_A}{4\mu_A}, \tag{17.19}$$

and from (8.130) that for conditions of plane strain

$$\Omega^2 = \frac{E_A}{2}\left(1-\nu_A^2\frac{E_T}{E_A}\right)\left(\sqrt{\frac{1}{E_A E_T}\left(1-\nu_A^2\frac{E_T}{E_A}\right)\left(1-\nu_t^2\frac{E_t}{E_T}\right)} - \frac{\nu_a+\nu_t\nu_A}{E_A} + \frac{1}{2\mu_A}\right). \tag{17.20}$$

If the laminate is isotropic, the parameter Ω has the value 1 for conditions of generalised plane stress and the value $1-\nu^2$ for conditions of plane strain.

Following a similar approach to that described in Chapter 16 (Section 16.14), on defining

$$X = \frac{x}{c}, \quad \mu = \frac{4\Lambda\Omega c}{E_A}, \quad P(X,\mu) = \frac{p(x,c) - \sigma_{th}}{\sigma_A - \sigma_{th}}, \tag{17.21}$$

it has been shown [2] that it is necessary to solve the following linear integral equation for the function $P(X, \mu)$

$$P(X,\mu) = \mu\int_0^1 \left[1 - P(t,\mu)\right] k(t,X)\,dt, \quad 0 \le X \le 1, \tag{17.22}$$

where

$$k(t,X) = \frac{1}{\pi}\ln\left|\frac{\sqrt{1-t^2}+\sqrt{1-X^2}}{\sqrt{1-t^2}-\sqrt{1-X^2}}\right| = k(X,t). \tag{17.23}$$

The expression for the mode I stress intensity factor K may be written as

$$K = \sqrt{\pi c}(\sigma_A - \sigma_{th})Y(\mu) \text{ where } Y(\mu) = \frac{2}{\pi}\int_0^1 \frac{1-P(X,\mu)}{\sqrt{1-X^2}}\,dX. \tag{17.24}$$

It is noted from (17.16) that, if the crack is stress-free, then $\Lambda = 0$ and $\sigma_{th} = 0$. It then follows from (17.21) and (17.24) that $P(X,\mu) \equiv 0$ so that $Y(\mu) \equiv 1$ with the result that $K = \sqrt{\pi c}\sigma_A$.

The integral equation (17.22) can be solved using numerical methods as described in that part of Appendix K concerned with a linear integral equation. The value of $\sigma_A - \sigma_{th}$ for which crack growth is initiated is found by imposing the following fracture criterion derived from energy balance considerations [2]

$$K^2 = \frac{2E_A\gamma}{\Omega}, \tag{17.25}$$

where 2γ is the mode I effective fracture energy for ply cracking for the homogenised laminate. For ply cracks in a 0°/90°/0° cross-ply laminate of total thickness $2h$ it follows that $\gamma = a\gamma_{90}/h$ where $2\gamma_{90}$ is the fracture energy for ply cracking parallel to the fibres in a 90° ply having total thickness $2a$. At the point of the initiation of crack propagation, it follows from (17.21), (17.24) and (17.25) that the ply cracking stress σ_A^{pc} is given by

$$\frac{\sigma_A^{pc} - \sigma_{th}}{\sigma_0} = \frac{1}{\sqrt{\mu}Y(\mu)}, \tag{17.26}$$

where

$$\sigma_0^2 = \frac{8\Lambda\gamma}{\pi}. \tag{17.27}$$

It is convenient to introduce the parameter c_0 defined by

$$c_0 = \frac{E_A}{4\Lambda\Omega}, \tag{17.28}$$

so that $\mu = c / c_0$. It follows from (17.25), (17.27) and (17.28) that

$$\pi c_0 \sigma_0^2 = \frac{2E_A\gamma}{\Omega} = K_{Ic}^2, \tag{17.29}$$

where K_{Ic} is regarded as the mode I effective fracture toughness for ply cracking. The fracture condition (17.26) may then be expressed in the dimensionless form

$$\frac{\sigma_A^{pc} - \sigma_{th}}{\sigma_0} = \frac{1}{\sqrt{c / c_0}Y(c / c_0)}. \tag{17.30}$$

Equation (17.30) is the relation that determines the value σ_A^{pc} of the applied stress that will initiate the growth of a preexisting transverse crack of length $2c$ when residual thermal stresses are present.

Numerical predictions of the normalised stress $(\sigma_A^{pc} - \sigma_{th}) / \sigma_0$ as a function of the normalised crack length c/c_0 are shown in Figure 17.2 using the label 'Bridged crack'. It is emphasised that these results are generating a master curve that applies to all material systems for which the linear crack bridging law (17.16) is valid.

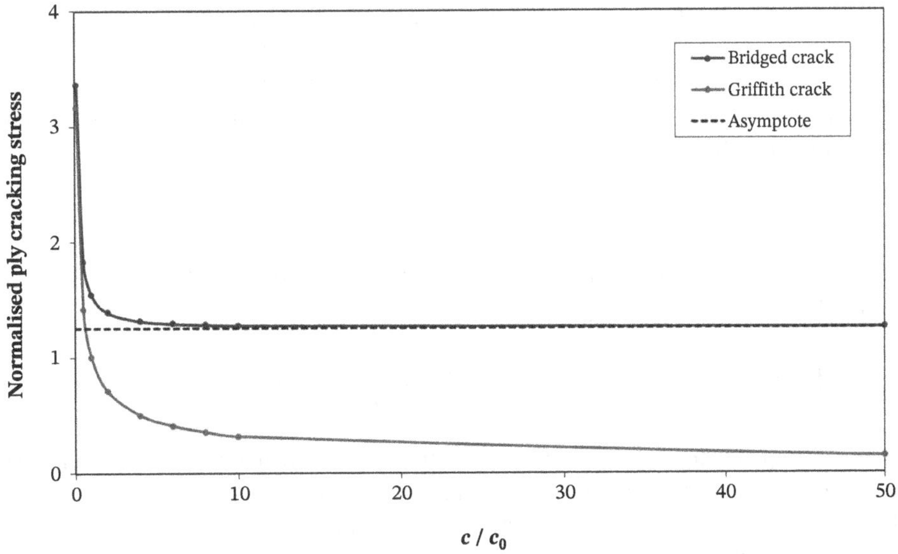

Figure 17.2 Dependence of the critical axial applied stress for ply cracking on the length of the preexisting ply crack. The normalised ply cracking stress is $(\sigma_A^{pc} - \sigma_{th}) / \sigma_0$ and the normalised crack length is c / c_0.

From (17.24) and (17.29), on setting $\mu = c/c_0$, it follows that the normalised stress intensity factor may be expressed

$$\frac{K}{K_{\mathrm{Ic}}} = \frac{\sigma_{\mathrm{A}} - \sigma_{\mathrm{th}}}{\sigma_0} \sqrt{c/c_0}\, Y(c/c_0). \tag{17.31}$$

Alternatively, on using (17.29), the stress intensity factor can be expressed in the following dimensionless form that is independent of the applied stress σ_{A}

$$\frac{K}{(\sigma_{\mathrm{A}} - \sigma_{\mathrm{th}})\sqrt{\pi c_0}} = \sqrt{\frac{c}{c_0}}\, Y\!\left(\frac{c}{c_0}\right). \tag{17.32}$$

The right-hand side of (17.32) will depend on other geometrical parameters.

17.3.1 Special Case of Long Ply Cracks

On applying the result (J.23) derived in Appendix J to the ply cracking problem under consideration and on following the approach first described at the end of Chapter 8, it follows that at the point of ply crack growth occurring when $\sigma_{\mathrm{A}} = \sigma_{\mathrm{A}}^{\mathrm{pc}}$

$$\int_0^c \left(\sigma_{\mathrm{A}}^{\mathrm{pc}} - p(x,c)\right) \frac{\partial \Delta v(x,c)}{\partial c}\,\mathrm{d}x + \int_0^c \frac{\partial p(x,c)}{\partial c} \Delta v(x,c)\,\mathrm{d}x = 4\gamma, \tag{17.33}$$

where for a ply crack the effective continuum fracture energy 2γ is given by

$$2\gamma = 2\gamma_{90} a/h, \tag{17.34}$$

where $2\gamma_{90}$ is the fracture energy for cracking 90° ply in a direction parallel to the fibres.

For the case of very long cracks ($c \to \infty$), it is useful to introduce the parameter d, which is the length of the region $c - d < x < c$ adjacent to the crack tip at $x = c$ where the functions $p(x,c)$ and $\Delta v(x,c)$ are dependent on the variables x and c. For the region $0 < x < c - d$ and for sufficiently long cracks, $p(x,c) = \sigma_{\infty}^{\mathrm{pc}}$ and $\Delta v(x,c) = \Delta v_{\infty}$ where $\sigma_{\infty}^{\mathrm{pc}}$ and Δv_{∞} are independent of x and c, the latter having the following value obtained using (17.16) so that

$$\Delta v_{\infty} = \frac{1}{\Lambda}\left(\sigma_{\infty}^{\mathrm{pc}} - \sigma_{\mathrm{th}}\right). \tag{17.35}$$

For such long-crack conditions $\sigma_{\mathrm{A}}^{\mathrm{pc}} = \sigma_{\infty}^{\mathrm{pc}}$ and the general result (17.33) may then be replaced by

$$\int_{c-d}^{c} \left(\sigma_{\infty}^{\mathrm{pc}} - p(x-c)\right) \frac{\partial \Delta v(x-c)}{\partial c}\,\mathrm{d}x + \int_{c-d}^{c} \frac{\partial p(x-c)}{\partial c} \Delta v(x-c)\,\mathrm{d}x = 4\gamma. \tag{17.36}$$

The (x, c) dependence has been replaced by the $(x - c)$ dependence shown in (17.36) because the bridged crack is very long such that the local deformation in the region of the crack tip at $x = c$ moves self-similarly as the crack propagates. It follows that (17.36) may now be written as

$$-\sigma_{\infty}^{\mathrm{pc}} \int_{c-d}^{c} \frac{\partial \Delta v(x-c)}{\partial x}\,\mathrm{d}x + \int_{c-d}^{c} p(x-c) \frac{\partial \Delta v(x-c)}{\partial x}\,\mathrm{d}x - \int_{c-d}^{c} \frac{\partial p(x-c)}{\partial x} \Delta v(x-c)\,\mathrm{d}x = 4\gamma. \tag{17.37}$$

On setting $x' = x - c$, it can then be shown that

$$\left(\sigma_{th} - \sigma_{\infty}^{pc}\right)\int_{-d}^{0} \frac{\partial \Delta v(x')}{\partial x'} dx' + \int_{-d}^{0} \left(p(x') - \sigma_{th}\right)\frac{\partial \Delta v(x')}{\partial x'} dx' - \int_{-d}^{0} \frac{\partial p(x')}{\partial x'} \Delta v(x') dx' = 4\gamma \tag{17.38}$$

where it should be noted that the thermal stress σ_{th}, which appears in relation (17.35), does, in fact, cancel in (17.38).

It follows from (17.16) that

$$p(x,c) - \sigma_{th} = \Lambda \Delta v(x,c), \ |x| < c, \tag{17.39}$$

so that for the region $c - d < x < c$, i.e. $-d < x' < 0$,

$$p(x') - \sigma_{th} = \Lambda \Delta v(x'), \quad \frac{\partial p(x')}{\partial x'} = \Lambda \frac{\partial \Delta v(x')}{\partial x'}. \tag{17.40}$$

As $\Delta v(0) = 0$ and $\Delta v(-d) = \Delta v_{\infty}$, it follows that (17.38) may now be written as

$$\left(\sigma_{\infty}^{pc} - \sigma_{th}\right)\Delta v_{\infty} = 4\gamma, \tag{17.41}$$

as the last two terms on the left-hand side of (17.38) cancel. On using (17.35) the long crack ply cracking stress σ_{∞}^{pc} is given by

$$\left(\sigma_{\infty}^{pc} - \sigma_{th}\right)^2 = 4\Lambda\gamma. \tag{17.42}$$

On using (17.27), it is clear that

$$\frac{\sigma_{\infty}^{pc} - \sigma_{th}}{\sigma_0} = \sqrt{\frac{\pi}{2}} = 1.2533. \tag{17.43}$$

It is useful to consider the behaviour of ply cracking when $\mu \to 0$ corresponding to ply cracks of decreasing length. For this special case, it follows from (17.22) that $P(X,\mu) \to 0$ and from (17.24) that $Y(\mu) \to 1$ leading to the following expression for the stress intensity factor for a stress-free embedded crack in an infinite sample of a cross-ply laminate

$$K = \sqrt{\pi c}\left(\sigma_A - \sigma_{th}\right). \tag{17.44}$$

On using (17.29), the well-known Griffith fracture criterion $K = K_{Ic}$ leads to the result

$$\frac{\sigma_A^{pc} - \sigma_{th}}{\sigma_0} = \sqrt{\frac{c_0}{c}}. \tag{17.45}$$

This relationship is shown in Figure 17.2 using the label 'Griffith crack'. As to be expected, it approaches the bridged crack solution when $c / c_0 \to 0$ but reduces to zero as $c / c_0 \to \infty$.

It is emphasised that the bridged crack solution shown in Figure 17.2 is a master curve that applies to all materials with deformation governed by the simplified model having a linear bridging relation. The effects of all material constants, including fracture parameters, are included in the normalising constants σ_0 and c_0 defined by (17.27) and (17.28). If an alternative stress-transfer model is used, then a different crack bridging curve to that shown in Figure 17.2 would result. Such a curve is easily generated using very similar numerical methods.

To complete the discussion of the simplified model it is also worth mentioning that, in the absence of an accurate numerical solution of the bridged crack problem, use can be made of the long crack limit and the prediction for a Griffith crack to construct an approximate approach. With reference to Figure 17.2, the long ply crack limiting value given by (17.43) is used in the Griffith relation (17.45) to determine the crack length c_1, denoting the point at which the long crack asymptote crosses the Griffith curve, that is given by

$$\frac{c_1}{c_0} = \frac{2}{\pi} = 0.6366. \tag{17.46}$$

Thus, the approximate estimate of the matrix cracking stress is obtained using (17.45) when $c < c_1$ and using (17.43) when $c > c_1$. It is noted from Figure 17.2 that the approximation always underestimates the ply cracking stress, thus providing a relatively simple conservative estimate that could be used when designing composites that must be resistant to ply damage.

17.4 Multiple Crack Problems

The special case of isolated cracks was considered in Chapter 8 and in Section 17.3. The objective now is to extend the approach so that interactions between multiple cracks can be considered.

17.4.1 Stress, Displacement Fields and Stress Intensity Factors

Now suppose that there is an arbitrary arrangement of m straight cracks of lengths $2c_r$, $r = 1, \ldots, m$, all parallel to the x-axis, with their tips at $z = t_1^r$ and $z = t_2^r$, $r = 1, \ldots, m$ where

$$t_1^r = a_r - c_r + ib_r,\; t_2^r = a_r + c_r + ib_r,\; r = 1,\ldots,m. \tag{17.47}$$

Corresponding to (8.87) and (8.113) complex variables are defined by

$$\left.\begin{aligned} \zeta_r &= \xi_r + i\eta_r = \frac{2z - (t_1^r + t_2^r)}{(t_2^r - t_1^r)}, \\ \zeta_j^r &= \xi_r + s_j\eta_r, \quad j = 1,2, \end{aligned}\right\} \; r = 1,\ldots,m, \tag{17.48}$$

where

$$\xi_r = \frac{x - a_r}{c_r},\; \eta_r = \frac{y - b_r}{c_r},\; r = 1,\ldots,m. \tag{17.49}$$

As linear elasticity theory is assumed, the superposition principle may be used to derive from (8.114)–(8.118) the following stress and displacement fields for m interacting cracks:

$$\sigma_{xx}(x,y) = -\sum_{r=1}^{m}\sum_{n=1}^{N_r} \mathrm{Re}\Big[\big(\alpha_n^r s_1 s_2 + \beta_n^r (s_1 + s_2)\big) H_n(\zeta_1^r) + s_2^2(\alpha_n^r s_1 + \beta_n^r)\Delta H_n(\zeta_1^r, \zeta_2^r)\Big], \tag{17.50}$$

$$\sigma_{yy}(x,y)=\sum_{r=1}^{m}\sum_{n=1}^{N_r}\text{Re}\left[\alpha_n^r H_n(\zeta_1^r)-(\alpha_n^r s_1+\beta_n^r)\Delta H_n(\zeta_1^r,\zeta_2^r)\right], \tag{17.51}$$

$$\sigma_{xy}(x,y)=\sum_{r=1}^{m}\sum_{n=1}^{N_r}\text{Re}\left[\beta_n^r H_n(\zeta_1^r)+s_2(\alpha_n^r s_1+\beta_n^r)\Delta H_n(\zeta_1^r,\zeta_2^r)\right], \tag{17.52}$$

$$\begin{aligned}u(x,y)=\sum_{r=1}^{m}c_r\sum_{n=1}^{N_r}\frac{1}{n}\text{Re}\Big[&\Big(\alpha_n^r(a_{11}s_1s_2-a_{12})+\beta_n^r a_{11}(s_1+s_2)\Big)G_n(\zeta_1^r)\\&+(a_{11}s_2^2+a_{12})(\alpha_n^r s_1+\beta_n^r)\Delta G_n(\zeta_1^r,\zeta_2^r)\Big],\end{aligned} \tag{17.53}$$

$$\begin{aligned}v(x,y)=\sum_{r=1}^{m}c_r\sum_{n=1}^{N_r}\frac{1}{n}\text{Re}\Bigg[&\left(-\alpha_n^r a_{22}\left(\frac{s_1+s_2}{s_1s_2}\right)+\beta_n^r\left(a_{12}-\frac{a_{22}}{s_1s_2}\right)\right)G_n(\zeta_1^r)\\&+\left(a_{12}s_2+\frac{a_{22}}{s_2}\right)\left(\alpha_n^r s_1+\beta_n^r\right)\Delta G_n(\zeta_1^r,\zeta_2^r)\Bigg].\end{aligned} \tag{17.54}$$

Here, N_r denotes the number of terms in the Chebyshev expansions for the rth crack of the form (8.114)–(8.118), and the values of the real coefficients α_n^r, β_n^r, $n=1,\ldots,N_r$, $r=1,\ldots,m$, that are to be determined from boundary conditions applied to the crack surfaces.

The mode I and mode II stress intensity factors for any crack in the group may also be derived, as before. For the sth crack, where $s = 1, \ldots, m$, the results corresponding to (8.133) and (8.136) are

$$K_{\text{I}}^{1,s}+iK_{\text{II}}^{1,s}=\sqrt{\pi c_s}\sum_{n=1}^{N_s}(-1)^{n+1}\left(\alpha_n^s+i\beta_n^s\right), \tag{17.55}$$

$$K_{\text{I}}^{2,s}+iK_{\text{II}}^{2,s}=\sqrt{\pi c_s}\sum_{n=1}^{N_s}\left(\alpha_n^s+i\beta_n^s\right). \tag{17.56}$$

It is worth noting that the interaction effects for multiple cracks influence only the values of the coefficients α_n^s, β_n^s, not the form of the expressions (17.55) and (17.56).

17.4.2 A Uniform Preexisting Stress Distribution with Stress-free Cracks

A uniform stress distribution having components $\sigma_{xx}^0, \sigma_{yy}^0$ and σ_{xy}^0 is considered to exist in the infinite body *before* stress-free cracks are introduced. The tractions acting at the crack locations, owing to this stress field, are given by

$$\tau_n^s+i\tau_{\text{t}}^s=\sigma_{yy}^0+i\sigma_{xy}^0,\quad s=1,\ldots,m, \tag{17.57}$$

where τ_n^s and τ_{t}^s are the normal and shear tractions at the location of the sth crack, respectively. By adding this pre-existing stress distribution to that specified by (17.51) and (17.52), the stress-free condition on the cracks may be expressed

$$S_n^s+\sigma_{yy}^0=0,\quad S_{\text{t}}^s+\sigma_{xy}^0=0,\quad s=1,\ldots,m, \tag{17.58}$$

where S_n^s and S_t^s are respectively the normal and shear tractions at the location of the sth crack arising from the representation of the stress field specified by (17.51) and (17.52). Any point $z = x_s + iy_s$ on the sth crack may be written as

$$z = \frac{1}{2}\left[(t_1^s + t_2^s) + \lambda(t_2^s - t_1^s)\right], \tag{17.59}$$

where λ is a real number lying in the range $-1 < \lambda < 1$. On substituting into (17.48), it follows that points on the sth crack may be described by

$$\zeta_{rs} = \xi_{rs} + i\eta_{rs} = \frac{(t_1^s + t_2^s) - (t_1^r + t_2^r) + \lambda(t_2^s - t_1^s)}{(t_1^r - t_2^r)}, \quad s = 1,\ldots,m, \tag{17.60}$$

which can also be expressed in the following form on using (17.47)

$$\zeta_{rs} = \xi_{rs} + i\eta_{rs} = \frac{a_s + ib_s - a_r - ib_r + \lambda c_s}{c_r}, \quad s = 1,\ldots,m, \tag{17.61}$$

so that

$$\xi_{rs} = \frac{a_s - a_r + \lambda c_s}{c_r}, \quad \eta_{rs} = \frac{b_s - b_r}{c_r}. \tag{17.62}$$

On making use of (17.51), (17.52) and (17.58) the resultant tractions acting at any point λ on the sth crack should be zero and they are given by $\tau_n^s = P_s(\lambda)$, $\tau_t^s = Q_s(\lambda)$ where

$$P_s(\lambda) = \sigma_{yy}^0 + \sum_{r=1}^{m}\sum_{n=1}^{N_r}\left(\begin{array}{l}\alpha_n^r \,\mathrm{Re}\left[H_n(\zeta_1^{rs}) - s_1\Delta H_n(\zeta_1^{rs},\zeta_2^{rs})\right] \\ \qquad + \beta_n^r \,\mathrm{Re}\left[-\Delta H_n(\zeta_1^{rs},\zeta_2^{rs})\right]\end{array}\right) = 0, \tag{17.63}$$

$$Q_s(\lambda) = \sigma_{xy}^0 + \sum_{r=1}^{m}\sum_{n=1}^{N_r}\left(\begin{array}{l}\alpha_n^r \,\mathrm{Re}\left[s_1 s_2 \Delta H_n(\zeta_1^{rs},\zeta_2^{rs})\right] \\ \qquad + \beta_n^r \,\mathrm{Re}\left[H_n(\zeta_1^{rs}) + s_2\Delta H_n(\zeta_1^{rs},\zeta_2^{rs})\right]\end{array}\right) = 0, \tag{17.64}$$

and where

$$\zeta_j^{rs} = \xi_{rs} + s_j\eta_{rs}, \quad r,s = 1,\ldots,m, \quad j = 1,2. \tag{17.65}$$

Two solution methods have been investigated and these are now described.

17.4.2.1 Method Using Orthogonality

When only one crack is present, it follows from (8.107) that the normal and tangential tractions on the crack have the form of a series of Chebyshev polynomials of the second kind. This suggests that it would be sensible to approximate the expressions (17.63) and (17.64) as follows:

$$P_s(\lambda) = \sum_{k=1}^{N_s} A_k^s U_{k-1}(\lambda), \quad s = 1,\ldots,m, \tag{17.66}$$

$$Q_s(\lambda) = \sum_{k=1}^{N_s} B_k^s U_{k-1}(\lambda), \quad s = 1,\ldots,m. \tag{17.67}$$

By making use of the following orthogonality relations

$$\int_{-1}^{1}\sqrt{1-x^2}U_n(x)U_m(x)\mathrm{d}x=\begin{cases}\pi/2 & \text{if } n=m,\\ 0 & \text{otherwise,}\end{cases} \tag{17.68}$$

it follows from (17.66) and (17.67) that for $k = 1, \ldots, N_s$ and $s = 1, \ldots, m$

$$A_k^s=\frac{2}{\pi}\int_{-1}^{1}\sqrt{1-\lambda^2}P_s(\lambda)U_{k-1}(\lambda)\mathrm{d}\lambda, \tag{17.69}$$

$$B_k^s=\frac{2}{\pi}\int_{-1}^{1}\sqrt{1-\lambda^2}Q_s(\lambda)U_{k-1}(\lambda)\mathrm{d}\lambda. \tag{17.70}$$

These coefficients may be estimated numerically using the following quadrature formula [6]

$$\int_{-1}^{1}\sqrt{1-x^2}f(x)\mathrm{d}x=\sum_{p=1}^{q}w_p f(x_p)+R_q, \tag{17.71}$$

where

$$x_p=\cos\frac{\pi p}{q+1},\qquad w_p=\frac{\pi}{q+1}(1-x_p^2),\qquad p=1,\ldots,q, \tag{17.72}$$

and where the remainder is given by

$$R_q=\frac{\pi}{(2q)!2^{2q+1}}f^{(2q)}(\xi),\quad \text{where}\quad -1<\xi<1. \tag{17.73}$$

On using (17.71), expressions (17.69) and (17.70) for the coefficients A_k^s, B_k^s are approximated by

$$A_k^s=\frac{2}{\pi}\sum_{p=1}^{q}w_pP_s(x_p)U_{k-1}(x_p),\qquad k=1,\ldots,N_s,\qquad s=1,\ldots,m, \tag{17.74}$$

$$B_k^s=\frac{2}{\pi}\sum_{p=1}^{q}w_pQ_s(x_p)U_{k-1}(x_p),\qquad k=1,\ldots,N_s,\qquad s=1,\ldots,m, \tag{17.75}$$

and their values are set to zero. On substituting (17.63) and (17.64) into (17.74) and (17.75), it then follows that

$$\sum_{r=1}^{m}\sum_{n=1}^{N_r}\alpha_n^rM_{rnsk}+\sum_{r=1}^{m}\sum_{n=1}^{N_r}\beta_n^rN_{rnsk}=K_{sk},\ k=1,\ldots,N_s,\ s=1,\ldots,m, \tag{17.76}$$

$$\sum_{r=1}^{m}\sum_{n=1}^{N_r}\alpha_n^rP_{rnsk}+\sum_{r=1}^{m}\sum_{n=1}^{N_r}\beta_n^rQ_{rnsk}=L_{sk},\ k=1,\ldots,N_s,\ s=1,\ldots,m, \tag{17.77}$$

where for $n = 1, \ldots, N_r$, $r = 1, \ldots, m$ and $k = 1, \ldots, N_s$, $s = 1, \ldots, m$,

$$K_{sk}=-\sigma_{yy}^0\sum_{p=1}^{q}w_pU_{k-1}(x_p), \tag{17.78}$$

$$L_{sk}=-\sigma_{xy}^0\sum_{p=1}^{q}w_pU_{k-1}(x_p), \tag{17.79}$$

$$M_{rnsk} = \sum_{p=1}^{q} w_p U_{k-1}(x_p) \mathrm{Re}\left[H_n(\zeta_1^{rsp}) - s_1 \Delta H_n(\zeta_1^{rsp}, \zeta_2^{rsp}) \right], \tag{17.80}$$

$$N_{rnsk} = \sum_{p=1}^{q} w_p U_{k-1}(x_p) \mathrm{Re}\left[-\Delta H_n(\zeta_1^{rsp}, \zeta_2^{rsp}) \right], \tag{17.81}$$

$$P_{rnsk} = \sum_{p=1}^{q} w_p U_{k-1}(x_p) \mathrm{Re}\left[s_1 s_2 \Delta H_n(\zeta_1^{rsp}, \zeta_2^{rsp}) \right], \tag{17.82}$$

$$Q_{rnsk} = \sum_{p=1}^{q} w_p U_{k-1}(x_p) \mathrm{Re}\left[H_n(\zeta_1^{rsp}) + s_2 \Delta H_n(\zeta_1^{rsp}, \zeta_2^{rsp}) \right], \tag{17.83}$$

where from (17.61)

$$\zeta_{rsp} = \xi_{rsp} + i\eta_{rsp} = \frac{a_s + ib_s - a_r - ib_r + x_p c_s}{c_r}, \ r, \ s = 1,\ldots,m, \ \ p = 1,\ldots,q, \tag{17.84}$$

so that

$$\xi_{rsp} = \frac{a_s - a_r + x_p c_s}{c_r}, \ \eta_{rsp} = \frac{b_s - b_r}{c_r}, \tag{17.85}$$

and where

$$\zeta_j^{rsp} = \xi_{rsp} + s_j \eta_{rsp}, \ \ r,s = 1,\ldots,m, \ \ p = 1,\ldots,q, \ \ j = 1,2. \tag{17.86}$$

If the system of linear algebraic equations (17.76) and (17.77) is solved for the unknowns α_n^r, β_n^r, use can then be made of expressions (17.55) and (17.56) to obtain values of the stress intensity factors at tips of each crack. Furthermore, by adding the uniform preexisting stress field to the stress distribution (17.50)–(17.52), the stress field everywhere in the plate is obtained. The displacement field everywhere is obtained by adding (17.53) and (17.54) to the following displacement field that generates the preexisting uniform stress field

$$u^0 = \left(a_{11}\sigma_{xx}^0 + a_{12}\sigma_{yy}^0 \right) x + \frac{1}{2} a_{66}\sigma_{xy}^0 y, \tag{17.87}$$

$$v^0 = \frac{1}{2} a_{66}\sigma_{xy}^0 x + \left(a_{12}\sigma_{xx}^0 + a_{22}\sigma_{yy}^0 \right) y. \tag{17.88}$$

17.4.2.2 Collocation Method

Relations (17.63) and (17.64) can be chosen to be satisfied at a discrete number of points on the crack surfaces. Although there are many possible choices for the locations of these points, the collocation method used here selects these points to have the following values which lead to higher densities of collocation points near the crack tips

$$x_k^s = \cos\frac{\pi(2k-1)}{2N_s}, \ \ k = 1,\ldots,N_s, \ \ s = 1,\ldots,m. \tag{17.89}$$

It follows that relations (17.63) and (17.64) will be satisfied exactly at the collocation points provided that the unknowns α_n^r, β_n^r satisfy the following linear algebraic equations

$$\sum_{r=1}^{m}\sum_{n=1}^{N_r} \alpha_n^r M_{rnsk} + \sum_{r=1}^{m}\sum_{n=1}^{N_r} \beta_n^r N_{rnsk} = K_{sk}, \quad k=1,\ldots,N_s, \quad s=1,\ldots,m, \tag{17.90}$$

$$\sum_{r=1}^{m}\sum_{n=1}^{N_r} \alpha_n^r P_{rnsk} + \sum_{r=1}^{m}\sum_{n=1}^{N_r} \beta_n^r Q_{rnsk} = L_{sk}, \quad k=1,\ldots,N_s, \quad s=1,\ldots,m, \tag{17.91}$$

where for $n = 1, \ldots, N_r$, $r = 1, \ldots, m$ and $k = 1, \ldots, N_s$, $s = 1, \ldots, m$,

$$K_{sk} = -\sigma_{yy}^0, \quad \text{independent of } s, k, \tag{17.92}$$

$$L_{sk} = -\sigma_{xy}^0, \quad \text{independent of } s, k, \tag{17.93}$$

$$M_{rnsk} = \mathrm{Re}\left[H_n(\zeta_1^{rsk}) - s_1 \Delta H_n(\zeta_1^{rsk}, \zeta_2^{rsk})\right], \tag{17.94}$$

$$N_{rnsk} = \mathrm{Re}\left[-\Delta H_n(\zeta_1^{rsk}, \zeta_2^{rsk})\right], \tag{17.95}$$

$$P_{rnsk} = \mathrm{Re}\left[s_1 s_2 \Delta H_n(\zeta_1^{rsk}, \zeta_2^{rsk})\right], \tag{17.96}$$

$$Q_{rnsk} = \mathrm{Re}\left[H_n(\zeta_1^{rsk}) + s_2 \Delta H_n(\zeta_1^{rsk}, \zeta_2^{rsk})\right], \tag{17.97}$$

where from (17.61)

$$\zeta_{rsk} = \xi_{rsk} + i\eta_{rsk} = \frac{a_s + ib_s - a_r - ib_r + x_k c_s}{c_r}, \quad k=1,\ldots,N_s, \quad r,s=1,\ldots,m, \tag{17.98}$$

so that

$$\xi_{rsk} = \frac{a_s - a_r + x_k c_s}{c_r}, \quad \eta_{rsk} = \frac{b_s - b_r}{c_r}, \tag{17.99}$$

and where

$$\zeta_j^{rsk} = \xi_{rsk} + s_j \eta_{rsk}, \quad k=1,\ldots,N_s, \quad r,s=1,\ldots,m, \quad j=1,2. \tag{17.100}$$

If the system of linear algebraic equations (17.90) and (17.91) is solved for the unknowns α_n^r, β_n^r, use can again be made of expressions (17.55) and (17.56) to obtain values of the stress intensity factors at the tips of each crack. Furthermore, by adding the uniform preexisting stress field to the stress distribution (17.50)–(17.52), the stress field everywhere in the plate is obtained. The displacement field everywhere is again obtained by adding (17.53) and (17.54) to the displacement field defined by (17.87) and (17.88) that generates the preexisting uniform stress field.

17.4.3 An Arbitrary Preexisting Stress Distribution with Bridged Cracks

A cracked transverse ply embedded in a cross-ply laminate will not behave as though it has stress-free crack surfaces because some stress will be transferred across the transverse cracks because of the presence of the undamaged neighbouring 0° plies. An attempt will now be made to model this bridging effect by the application of tensile and shear crack bridging tractions that will be linearly related to the crack opening displacement provided that there is no debonding of the interfaces. It is assumed that the linear crack bridging relation for each crack may be written in the form

$$(p+iq)(\xi)=[\,\Lambda\Delta v(\xi)+\sigma_{\text{th}}\,]+i[\,\mu\Delta u(\xi)+\tau_{\text{th}}\,], \tag{17.101}$$

where $p(\xi)$ and $q(\xi)$ are the effective (i.e. averaged across the laminate thickness) normal and shear tractions acting at the point ξ on the crack surface, respectively, where $\Delta v(\xi)$ and $\Delta u(\xi)$ are the corresponding normal and tangential displacement discontinuities, respectively, and where Λ and μ are dimensionless parameters that depend upon the properties of the laminate. The parameters σ_{th} and τ_{th} are the values of the stresses $p(\xi)$ and $q(\xi)$ needed to close the crack to overcome the effects of thermal stresses. These parameters are nonzero when account is taken of thermal expansion mismatch effects between the 0° and 90° plies. For the mode I opening mode, the parameters Λ and σ_{th} are defined in Section 17.3 in terms of the geometry and properties of the laminate plies. On using (17.57) the stress equilibrium condition (17.58) then becomes

$$S_n^s+\sigma_{yy}^0=p^s,\quad S_t^s+\sigma_{xy}^0=q^s,\quad s=1,\ldots,m, \tag{17.102}$$

and on using (8.123) and (8.124) the following new set of linear equations for the parameters α_n^r, β_n^r, is obtained:

$$\begin{aligned}P_s(\lambda)=\sigma_{yy}^0+\sum_{r=1}^{m}\sum_{n=1}^{N_r}\left\{\alpha_n^r\,\text{Re}\left[H_n(\zeta_1^r)-s_1\Delta H_n(\zeta_1^r,\zeta_2^r)\right]-\beta_n^r\,\text{Re}\left[\Delta H_n(\zeta_1^r,\zeta_2^r)\right]\right\}\\ -4\Lambda c_s\sqrt{a_{22}g}\,\sqrt{1-\lambda^2}\,\sum_{n=1}^{N_s}\frac{\alpha_n^s}{n}U_{n-1}(\lambda)-\sigma_{\text{th}}\equiv 0,\end{aligned} \tag{17.103}$$

$$\begin{aligned}Q_s(\lambda)=\sigma_{xy}^0+\sum_{r=1}^{m}\sum_{n=1}^{N_r}\left\{\alpha_n^r\,\text{Re}\left[s_1s_2\Delta H_n(\zeta_1^r,\zeta_2^r)\right]+\beta_n^r\,\text{Re}\left[H_n(\zeta_1^r)+s_2\Delta H_n(\zeta_1^r,\zeta_2^r)\right]\right\}\\ -4\mu c_s\sqrt{a_{11}g}\,\sqrt{1-\lambda^2}\,\sum_{n=1}^{N_s}\frac{\beta_n^s}{n}U_{n-1}(\lambda)-\tau_{\text{th}}\equiv 0.\end{aligned} \tag{17.104}$$

17.4.3.1 Method Using Orthogonality

By repeating the procedure used for stress-free cracks with the expressions for $P_s(\lambda)$ and $Q_s(\lambda)$ given by (17.103) and (17.104), a linear set of algebraic equations of the form (17.76) and (17.77) is again obtained, but the expressions (17.78)–(17.80) and (17.83) for the some of the coefficients must be replaced by the following:

$$K_{sk}=(\sigma_{th}-\sigma_{yy}^0)\sum_{p=1}^{q}w_pU_{k-1}(x_p),\ \text{independent of } s,k, \tag{17.105}$$

$$L_{sk}=(\tau_{th}-\sigma_{xy}^0)\sum_{p=1}^{q}w_pU_{k-1}(x_p),\ \text{independent of } s,\,k, \tag{17.106}$$

$$\begin{aligned}M_{rnsk}=\sum_{p=1}^{q}w_pU_{k-1}(x_p)\{\text{Re}\left[H_n(\zeta_1^{rsp})-s_1\Delta H_n(\zeta_1^{rsp},\zeta_2^{rsp})\right]\\ -4\delta_{rs}\Lambda c_s\sqrt{a_{22}g}\sqrt{1-x_p^2}U_{n-1}(x_p)/n\},\end{aligned} \tag{17.107}$$

$$Q_{rnsk} = \sum_{p=1}^{q} w_p U_{k-1}(x_p) \left\{ \mathrm{Re}\left[H_n(\zeta_1^{rsp}) + s_2 \Delta H_n(\zeta_1^{rsp}, \zeta_2^{rsp}) \right] - 4\delta_{rs}\mu c_s \right\}, \quad (17.108)$$

where δ_{rs} is the Kronecker delta symbol. Thus, bridged transverse cracks are readily dealt with. In subsequent numerical calculations, the parameter q introduced in the quadrature formula (17.71) is selected to be

$$q = 2\max\left(N_r,\ r = 1,\ldots,m\right). \quad (17.109)$$

This choice ensures that the remainders, defined by (17.73), related to the approximations (17.74) and (17.75) are zero when just one crack is present, i.e. the numerical analysis is exact for this special case.

17.4.3.2 Collocation Method

By repeating the procedure used for stress-free cracks with the expressions for $P_s(\lambda)$ and $Q_s(\lambda)$ given by (17.103) and (17.104), a linear set of algebraic equations of the form (17.90) and (17.91) is again obtained, but the expressions (17.92)–(17.94) and (17.97) for the some of the coefficients must be replaced by the following:

$$K_{sk} = \sigma_{\mathrm{th}} - \sigma_{yy}^0, \ \text{independent of } s, k, \quad (17.110)$$

$$L_{sk} = \tau_{th} - \sigma_{xy}^0, \ \text{independent of } s, k \quad (17.111)$$

$$M_{rnsk} = \mathrm{Re}\left[H_n(\zeta_1^{rsk}) - s_1 \Delta H_n(\zeta_1^{rsk}, \zeta_2^{rsk}) \right] - \frac{4}{n}\delta_{rs}\Lambda c_s \sqrt{a_{22}g}\sqrt{1-(x_k^s)^2}U_{n-1}(x_k^s), \quad (17.112)$$

$$Q_{rnsk} = \mathrm{Re}\left[H_n(\zeta_1^{rsk}) + s_2 \Delta H_n(\zeta_1^{rsk}, \zeta_2^{rsk}) \right] - \frac{4}{n}\delta_{rs}\mu c_s \sqrt{a_{11}g}\sqrt{1-(x_k^s)^2}U_{n-1}(x_k^s), \quad (17.113)$$

where δ_{rs} is the Kronecker delta symbol. Thus, bridged transverse cracks are again readily dealt with.

17.5 Numerical Results

In the analysis given in Section 17.4, two methods of imposing the crack boundary conditions have been described. The first method of collocation satisfies the boundary conditions at a set of discrete points on the crack surfaces. Residuals at points on the crack surfaces can be defined as the difference between the value of surface tractions predicted by the model and the values that such tractions should have if the boundary conditions are satisfied exactly. Although the collocation method leads to zero values of the residuals at a set of discrete points on the crack surfaces, the methods does not, in general, prevent significant oscillations of the residuals between the collocation points. This is not a problem for stress-free cracks where the crack boundary conditions can be satisfied almost exactly. However, for bridged cracks significant residual oscillations are encountered unless relatively large numbers of collocation points are selected. This phenomenon was the driving force for developing the orthogonal method of solving interacting bridged crack problems. Unfortunately, the orthogonality method is more cumbersome than the collocation method, leading to significantly increased computation times. In the calculations that are now described, the collocation method has been used in preference to the other method.

17.5.1 Stress-free Cracks in an Isotropic Plate

A number of checks, using unbridged stress-free cracks in a homogeneous isotropic plate, have been carried out to check the analysis and the validity of the associated software. For example, it has been checked that for an isolated stress-free crack in an isotropic material, of length $2c$ and subject to a uniaxial stress σ_A normal to the plane of the crack, the stress intensity factor has the value $K_0 = \sigma_A\sqrt{\pi c}$. The numerical results are exact for this special case.

Consider now a test example of two collinear cracks of equal length embedded in an isotropic material, where the exact solution is known and can be compared with the predictions obtained using orthogonal polynomials. Figure 17.2 shows the geometry where the two cracks have length $2c$ and the separation of the inner crack tips is $2s$. A uniaxial stress σ_A is applied in a direction normal to the crack planes. The crack problem is such that the deformation at the crack tips is mode I, i.e. the mode II stress intensity factors are zero.

The exact solution for the mode I stress intensity factors is given by [7]

$$K_I(x=s) = \sigma_A\sqrt{\pi c}\sqrt{\frac{s+2c}{\alpha c}\frac{1}{k}\left[\frac{E(k)}{K(k)} - \alpha^2\right]}, \quad k = \sqrt{1-\alpha^2}, \quad \alpha = \frac{c}{s+2c},$$

$$K_I(x=s+2c) = \sigma_A\sqrt{\pi c}\sqrt{\frac{s+2c}{c}\frac{1}{k}\left[1 - \frac{E(k)}{K(k)}\right]}. \tag{17.114}$$

For a unit applied axial stress when $c = 0.45c_0$, $s = 0.1c_0$, and on selecting $N = 100$ for any normalising crack length c_0, the methodology based on orthogonal polynomials leads to the results

$$K_I(x=s) = 1.4923379, \quad K_I(x=s+2c) = 1.2916506.$$

The corresponding values obtained from the exact solution (17.114) are

$$K_I(x=s) = 1.4923379, \quad K_I(x=s+2c) = 1.2916506,$$

which are identical to the numerical estimates, thus confirming the validity of the methodology based on orthogonal polynomials. An interaction effect between the two cracks is evident because the stress intensity factor at the inner crack tips is larger than that at the other crack tips. They would have the same value when there is no interaction, as would occur when $s \to \infty$. When the material is anisotropic, such that the crack planes are normal to the principal axis of material symmetry, the same values are obtained for the stress intensity factors.

It should be noted that relations (17.63) and (17.64) can be used to investigate the magnitude of the residual tractions on the crack surfaces at *any* points on the crack surfaces. For the example considered, the residual tractions, which should be zero, have the order of 10^{-16} indicating the very high accuracy of the methodology used.

Consider now a test example of three equally spaced vertically stacked cracks of equal length embedded in an infinite isotropic material. Figure 17.3 illustrates the geometry where the three cracks have length $2c$ and the vertical separation of the cracks is s. A unit uniaxial stress σ_A is applied in a direction normal to the crack planes. The crack problem is such that the deformation at the tips of the central crack is mode I, and the deformation at the other tips is mixed mode. The magnitudes of the model I and mode II stress intensity factors for the upper and lower cracks are expected to be the same.

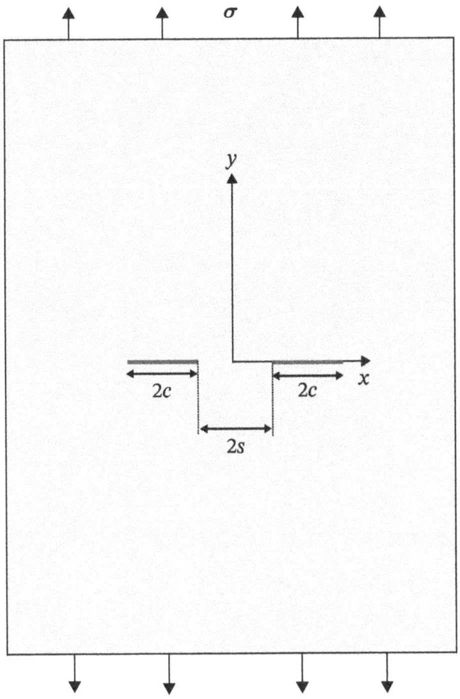

Figure 17.3 Geometry and loading of part of an infinite isotropic plate having two collinear cracks.

For a unit applied stress and when $c = 0.5c_0$, $s = 0.5c_0$ and on selecting $N = 100$ collocation points, for any normalising crack length c_0, the methodology based on orthogonal polynomials leads to the results

$$\begin{aligned}
&\text{Top crack:} \quad && K_{\text{I}} = 0.92924447, \quad K_{\text{II}} = 0.14265165, \\
&\text{Central crack:} \quad && K_{\text{I}} = 0.70031026, \quad K_{\text{II}} = 0.0, \\
&\text{Bottom crack:} \quad && K_{\text{I}} = 0.92924447, \quad K_{\text{II}} = -0.14265165.
\end{aligned}$$

It is seen that the mode I stress intensity factor for the central crack is less than that of the upper and lower cracks. This illustrates the shielding effect on the central crack because of the presence of the other two cracks. The model II stress intensity of the lowest crack is seen to be negative because the local shear stress is negative. For this example, the residual tractions, which should be zero, have the order of 10^{-16} again, indicating the very high accuracy of the methodology used. It is noted that if the material is anisotropic, such that the crack planes are normal to the principal axis of material symmetry, then the resulting stress intensity factor values differ from those given previously, which are for an isotropic plate.

17.5.2 Bridged Cracks

An isolated transverse crack in a cross-ply laminate is first considered (see Figure 17.1) where the crack bridging law (17.101) is applied for the case of mode I crack opening where $\tau_{\text{th}} = 0$. The thickness of the outer 0° plies is denoted by b and the total

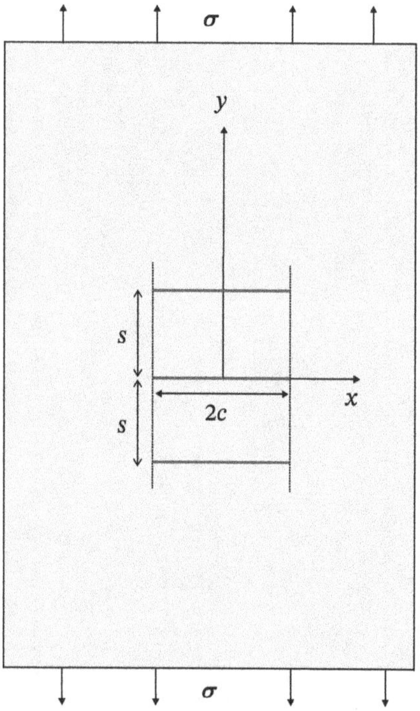

Figure 17.4 Geometry and loading of part of an infinite isotropic plate having three equally spaced vertically stacked cracks.

thickness of the inner 90° ply is $2a$ such that the total thickness of the laminate is $2h$ where $h = a + b$. The values of the thermoelastic constants for plies assumed for the calculations are given in Table 17.1 and $a = b = 1$ mm. The parameters Λ and σ_{th} appearing in (17.101) are calculated in terms of the thermoelastic constants and the temperature difference $\Delta T = 0$ °C (implying that $\sigma_{\text{th}} = 0$) using relations (17.12) and (17.13). Before any ply cracks are introduced, the stress field in the laminate is uniaxial such that a unit axial stress is applied.

As shown in Section 17.3 the problem of an isolated crack in a simple cross-ply laminate can be solved by a method involving the numerical solution of the integral equation (17.22). Such solutions have been used to generate the continuous lines shown in Figure 17.2. It is possible to solve the same problem using the analysis described in Section 17.4 applied to a single bridged crack having the same crack bridging relation or stress-free conditions. The result of applying this method is shown in Figure 17.2 as discrete points applied to both a bridged crack and an unbridged crack (i.e. a stress-free Griffith crack). There is excellent agreement between the two methods, indicating that both types of analysis are consistent with one another.

To investigate bridged crack interaction effects for a relatively simple case having relevance to the stability of ply crack fatigue damage in cross-ply laminates (see Chapter 13), the example of three stacked stress-free ply cracks is now considered for the case of bridged cracks. The crack length is selected to be $c = 0.5c_0$ and the crack separation parameter s is assumed to have values in the range $0.01 \leq s / c_0 \leq 1000$. On selecting $N = 100$ collocation points, the dependence of the normalised mode I and

Table 17.1 Elastic properties for single plies of CFRP.

Material property	Value used for calculations
E_A	140.77 GPa
E_T	8.85 GPa
μ_A	4.59 GPa
μ_t	3.0944 GPa
ν_A	0.28
ν_t	0.43
α_A	0.245×10^{-6}/°C
α_T	45.6×10^{-6}/°C

mode II stress intensity factors (defined by the left-hand side of (17.32)) on the normalised ply crack separation s / c_0 is shown in Figure 17.5. Values of s / c_0 lower than the minimum value 0.01 of the range considered, lead to numerical instabilities arising from the fact that the distinct crack tips are becoming too close. The maximum value 1000 of the range is sufficient for the model I stress intensity factors to have the same value corresponding to that of a single crack, i.e. the ply cracks do not interact. When using the collocation method, the accuracy of the analysis can be assessed by examining the magnitudes of the residuals of the boundary conditions at points lying between the collocation points where the residuals are zero. The accuracy of solutions can be increased by increasing the number of collocation points. Although all results shown have assumed $N = 100$, some solutions have been obtained using $N = 200$ collocation points to demonstrate increased accuracy and solution stability.

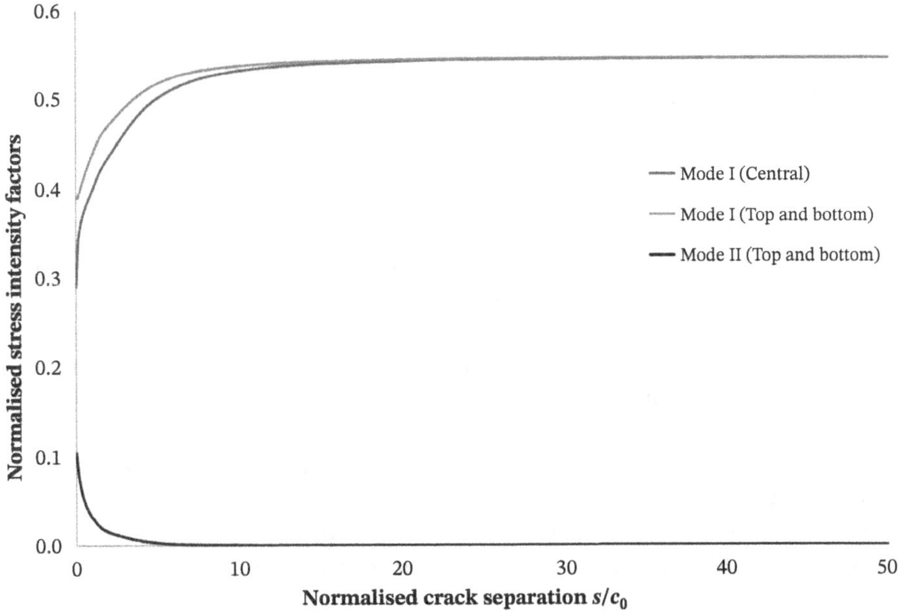

Figure 17.5 Stress intensity factors for three equally spaced stacked ply cracks (see Figure 17.4) as a function of crack separation.

It is seen from Figure 17.5 that the mode I stress intensity factors reduce as the crack separation s is reduced indicating that crack interaction is leading to a shielding effect even though the cracks are bridged by the 0° plies. Furthermore, it is seen that the mode I stress intensity factor for the central crack is lower than that of the upper and lower cracks which have equal values. This indicates an additional shielding effect arising from crack interactions. The plot in Figure 17.5 of the mode II stress intensity factor shows that crack shielding is accompanied by a small degree of mode II crack tip deformation. Mode II effects are seen to very quickly become negligible when the crack separation $s > 5c_0$. The final observation from the results is that, as the crack separation increases, the values of the mode I stress intensity factors for the central crack tends towards that of the upper and lower cracks. When $s = 1000c_0$, the mode I stress intensity factors for the three-ply cracks all have the same value. It is noted that, for the case $c = 0.5c_0$, crack interaction effects would need to be taken into account if $s < 10c_0$, when considering fatigue damage growth in cross-ply laminates.

There is only scope in this chapter to consider some simple illustrative problems concerning the interaction effects of bridged cracks in cross-ply laminates. The methodology and associated computer codes have been developed generally subject to all ply cracks being parallel (automatic in 0°/90°/0° laminate) and provided the laminate is subject to loading that generates a uniform stress field when the laminate is undamaged. The analysis and code can, therefore, be applied to quite general ply cracking damage distributions when a simple cross-ply laminate is subject to fatigue loading, as considered in Chapter 13. It is thought that the approaches made in this chapter, when analysing bridged ply cracks in simple cross-ply laminates, can be extended to multilayered general symmetric laminates provided that ply cracks occur only in plies of a single orientation.

References

1. McCartney, L.N. and Gorley, T.A.E. (1987). Complex variable method of calculating stress intensity factors for cracks in plates. *Proc. 4th Int. Conf. on Numerical Methods in Fracture*, 55–72.
2. McCartney, L.N. (1992a). Mechanics for the growth of bridged cracks in composite materials: Part I - Basic principles and Part II – Applications. *Journal of Composites Technology and Research* 14: 133–154.
3. McCartney, L.N. (1993). Mechanics of cracking in brittle matrix composites', Chapter 7 in 'Essentials of Carbon-Carbon Composites. (ed C.R. Thomas), Cambridge: Royal Society of Chemistry.
4. McCartney, L.N. (1994). The prediction of non-uniform cracking in biaxially loaded cross-ply laminates. NPL Report DMM(A)142.
5. McCartney, L.N. (1992b). *Journal of the Mechanics and Physics of Solids* 40: 27.
6. Abramowitz, M. and Stegun, I.A. (1972), Handbook of Mathematical functions. New York: Dover Publications. 889.
7. Rooke, D.P. and Cartwright, D.J. (1976). Compendium of stress intensity factors. Uxbridge, UK: Her Majesty's Stationery Office, Hillingdon Press.

18

Theoretical Basis for a Model of Ply Cracking in General Symmetric Laminates

Overview: This chapter provides the theoretical basis of an analysis to derive suitable expressions for the stress and displacement fields in general symmetric laminates having uniform distributions of ply cracks in one or more of the plies having a particular orientation. The laminate is assumed to be subject to general in-plane loading, a uniform through-thickness stress and to thermal residual stresses. The proposed stress and displacement distributions involve a set of stress-transfer functions which are determined by solving a set of fourth-order ordinary differential equations subject to a set of boundary and interface conditions. The resulting stress and displacement distributions are such that: (i) the equilibrium equations are satisfied exactly; (ii) all stress-strain relations are satisfied exactly apart form axial relations which are satisfied in an average sense; and (iii) all interface continuity conditions are satisfied exactly. It is shown how solutions can be used to estimate the effective thermoelastic constants of uniformly cracked laminates: information that is required when applying the predictive methods described in Chapter 10 to problems of practical interest. It is also shown how the effective properties of a damaged ply can be homogenised enabling the development of methods of accounting for ply crack damage in any of the plies of a general symmetric laminate.

18.1 Introduction

Laminated composites are prone to the formation of ply cracks when subjected to mechanical and thermal loading during both manufacture and service. Such ply cracking leads to a deterioration in mechanical properties, and if localised in regions of stress concentrations in composite structures, stress can be transferred from regions of stress concentration to other parts of the structure. This macroscopic stress-transfer process leads to composite structures whose strength often exceeds that which would be predicted from laboratory test data obtained for homogeneous deformation states. To understand fully such macroscopic stress-transfer processes in composite structures it is essential that a reliable methodology is developed that can be used to predict localised ply cracking for general symmetric laminates resulting from general in-plane loading, and from through-thickness loading that is encountered, for example, within bolted joints and pressure vessels. The objective of this chapter is to describe in detail

Properties for Design of Composite Structures: Theory and Implementation Using Software, First Edition. Neil McCartney.

Companion Website: www.wiley.com/go/mccartney/properties

the methodology that has been developed. This will be achieved by first showing how to predict the stress and displacement distributions associated with a uniformly spaced array of ply cracks in general symmetric laminates for conditions of multiaxial loading. These results will then be used to predict the effective thermoelastic constants of a laminate having a nonuniform distribution of ply cracks. Finally, energy methods will be used to predict progressive ply crack formation during the loading of the laminate.

Ply cracks are assumed to form when microscopic defects in the plies become unstable owing to loading so that they increase in size, being arrested when they meet an obstacle such as the nearby fibre or the interfaces between the neighbouring plies. The new types of defect formed in this way can coalesce and eventually lead to a macroscopic defect that traverses the entire thickness of the ply, but it remains of limited length in the fibre direction. Further load increases lead to the growth of this defect in the fibre direction until the entire thickness of the ply is cracked. The derivations of the required conditions for this latter stage of ply crack growth are the subject of this chapter. It should be noted that this scenario is expected whenever the thickness of the cracked ply is not too large, e.g. less than twice the total uncracked thickness. When the cracked ply is very thick, the initial instability of the microdefect may not be arrested and its growth can lead immediately to the cracking of the entire ply. The conditions for such defect instability are expected to be more demanding on the applied loading with the result that for thick plies the approach to be adopted in this chapter is expected to lead to lower bound loading conditions, and therefore pessimistic estimates for the stress required for the growth of the ply crack across the entire ply. The approach to be taken is, therefore, conservative and of utility in practical design situations. The methodology to be described in this chapter is based on the approach developed in reference [1] and requires the use of numerical techniques described in references [3, 4]. The notation has been changed so that there is consistent use of notation throughout this book.

18.2 Geometry and Basic Field Equations

When solving ply cracking problems, the first step is to rotate the laminate and loading so that the plies in which ply cracks will be allowed to form have a 90° orientation. The subsequent analysis assumes that such a transformation (if needed) has already been made. In order to achieve the challenging objectives while avoiding numerical methods such as finite element and boundary element methods, some simplifications have to be made to enable a tractable analytical development, which is nevertheless complex. The first approximation is to assume that ply cracks form in only a single orientation that must correspond to that of the fibres in a 90° ply. If the laminate does not contain a 90° ply, then the laminate and loading is rotated so that the orientation of the ply in which ply cracks are required to form becomes a 90° ply following the rotation. This approach demands that general in-plane loading must be included in the model. The second approximation is to assume that ply cracks can form in a discrete number of planes parallel to the fibres in a 90° ply. The ply crack distribution in each cracked ply must be the same, however, not all 90° plies need to be cracked. These assumptions are made so that a relatively simple representative volume can be used. It should be noted that the approach to be developed cannot consider arrays of staggered cracks, as would arise in the two 90° plies of a $[90/0]_s$ laminate where localised bending deformation could arise; a case that is not treated by the model.

The problem under consideration concerns the triaxial deformation of a symmetric, possibly unbalanced, multilayered laminate constructed of $2N + 2$ perfectly bonded layers, which can have any combination of orientations provided that there is at least one 90° ply in the laminate, and that symmetry about the midplane of the laminate is preserved. Some of the layers can have the same properties (as part of a ply refinement scheme) so that through-thickness variations in the stress and displacement fields can be modelled adequately. As laminate symmetry is assumed, it is necessary to consider only the right-hand set of $N + 1$ layers as shown in Figure 18.1. A global set of rectangular Cartesian coordinates (x_1, x_2, x_3), defined in Chapter 10, is chosen having the origin at the centre of the laminate as shown in the figure. The x_1-direction defines the longitudinal or axial direction, the x_2-direction defines the in-plane transverse direction and the x_3-direction defines the through-thickness direction.

The locations of the N interfaces in one half of the laminate ($x_3 > 0$) are specified by $x_3 = x_3^{(i)}$, $i = 1, \ldots, N$. The midplane of the laminate is specified by $x_3 = x_3^{(0)} = 0$ and the external surface by $x_3 = x_3^{(N+1)} = h$ where $2h$ is the total thickness of the laminate. The thickness of the ith layer is denoted by $h_i = x_3^{(i)} - h_3^{(i-1)}$. Stress, strain and

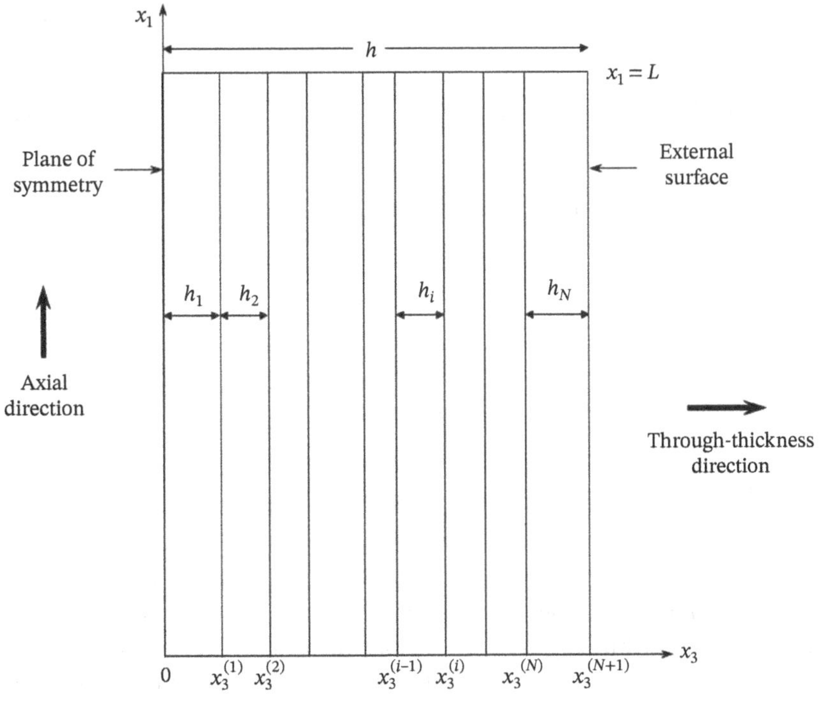

Figure 18.1 Schematic diagram illustrating the geometry of one half of a general symmetric laminate.

displacement components, and material properties associated with the ith layer are denoted by a superscript or subscript (i).

It is assumed that the ply crack distribution in damaged 90° plies is uniform having a separation $2L$, and that the cracks in each damaged 90° ply of the laminate lie in the same plane. The RVE of the laminate to be considered is specified by $0 \leq x_1 \leq L, 0 \leq x_2 \leq W, 0 \leq x_3 \leq h$. Ply cracks in some or all the 90° plies are assumed to form on the plane $x_1 = L$. The plane $x_1 = 0$ is a symmetry plane implying that neighbouring ply cracks would be on the plane $x_1 = -L$. The planes $x_2 = 0$ and $x_3 = 0$ are also planes of symmetry.

The following equilibrium equations, for the stress components $\sigma_{11}^{(i)}, \sigma_{12}^{(i)}$ etc. referred to global coordinates, must be satisfied for $i = 1, \dots, N+1$, (see (2.120)–(2.122))

$$\frac{\partial \sigma_{11}^{(i)}}{\partial x_1} + \frac{\partial \sigma_{12}^{(i)}}{\partial x_2} + \frac{\partial \sigma_{13}^{(i)}}{\partial x_3} = 0, \tag{18.1}$$

$$\frac{\partial \sigma_{12}^{(i)}}{\partial x_1} + \frac{\partial \sigma_{22}^{(i)}}{\partial x_2} + \frac{\partial \sigma_{23}^{(i)}}{\partial x_3} = 0, \tag{18.2}$$

$$\frac{\partial \sigma_{13}^{(i)}}{\partial x_1} + \frac{\partial \sigma_{23}^{(i)}}{\partial x_2} + \frac{\partial \sigma_{33}^{(i)}}{\partial x_3} = 0. \tag{18.3}$$

Based on relations (6.35) and (6.53), the required stress-strain relations referred to global coordinates, are given, for $i = 1, \dots, N+1$, by

$$\varepsilon_{11}^{(i)} \equiv \frac{\partial u_1^{(i)}}{\partial x_1} = S_{11}^{(i)}\sigma_{11}^{(i)} + S_{12}^{(i)}\sigma_{22}^{(i)} + S_{13}^{(i)}\sigma_{33}^{(i)} + S_{16}^{(i)}\sigma_{12}^{(i)} + V_1^{(i)}\Delta T, \tag{18.4}$$

$$\varepsilon_{22}^{(i)} \equiv \frac{\partial u_2^{(i)}}{\partial x_2} = S_{12}^{(i)}\sigma_{11}^{(i)} + S_{22}^{(i)}\sigma_{22}^{(i)} + S_{23}^{(i)}\sigma_{33}^{(i)} + S_{26}^{(i)}\sigma_{12}^{(i)} + V_2^{(i)}\Delta T, \tag{18.5}$$

$$\varepsilon_{33}^{(i)} \equiv \frac{\partial u_3^{(i)}}{\partial x_3} = S_{13}^{(i)}\sigma_{11}^{(i)} + S_{23}^{(i)}\sigma_{22}^{(i)} + S_{33}^{(i)}\sigma_{33}^{(i)} + S_{36}^{(i)}\sigma_{12}^{(i)} + V_3^{(i)}\Delta T, \tag{18.6}$$

$$2\varepsilon_{23}^{(i)} \equiv \frac{\partial u_2^{(i)}}{\partial x_3} + \frac{\partial u_3^{(i)}}{\partial x_2} = S_{44}^{(i)}\sigma_{23}^{(i)} + S_{45}^{(i)}\sigma_{13}^{(i)}, \tag{18.7}$$

$$2\varepsilon_{13}^{(i)} \equiv \frac{\partial u_1^{(i)}}{\partial x_3} + \frac{\partial u_3^{(i)}}{\partial x_1} = S_{45}^{(i)}\sigma_{23}^{(i)} + S_{55}^{(i)}\sigma_{13}^{(i)}, \tag{18.8}$$

$$2\varepsilon_{12}^{(i)} \equiv \frac{\partial u_1^{(i)}}{\partial x_2} + \frac{\partial u_2^{(i)}}{\partial x_1} = S_{16}^{(i)}\sigma_{11}^{(i)} + S_{26}^{(i)}\sigma_{22}^{(i)} + S_{36}^{(i)}\sigma_{33}^{(i)} + S_{66}^{(i)}\sigma_{12}^{(i)} + V_6^{(i)}\Delta T. \tag{18.9}$$

The parameters $u_j^{(i)}$, $j = 1, 2, 3$, denote the displacement components for the ith layer in the x_1, x_2 and x_3 directions, respectively. The parameter ΔT is the uniform temperature difference defined by $\Delta T = T - T_0$ for which T is the current temperature and T_0 is the temperature in the reference state for which the strain is zero and the material is everywhere stress-free with no internal or imposed external stresses. The quantities $\varepsilon_{11}^{(i)}$, $\varepsilon_{12}^{(i)}$, etc., denote the infinitesimal strain components. For individual plies, the thermoelastic constants $S_{IJ}^{(i)}$ and $V_I^{(i)}$ are defined in terms of the well-known thermoelastic properties in Chapter 6 (see (6.54)–(6.67)).

18.3 Edge Boundary Conditions for Uniformly Cracked Laminates

Boundary conditions need to be imposed on the external edges of the uniformly cracked laminate. It is assumed that biaxial and shear loads are applied to the cracked laminate by imposing displacements on the edges of the laminate that would, in an undamaged laminate, lead to a state of combined uniform biaxial strain and uniform shear strain. The laminate is assumed to have length $2L$ in the x_1-direction and width $2W$ in the x_2-direction. As assumed in Chapter 10, the edge conditions for the uncracked layers of the damaged laminate are given by

$$\begin{aligned} &\sigma_{13}=0, \quad u_1=\varepsilon_A L+\tfrac{1}{2}\gamma_A x_2, \quad u_3=\tfrac{1}{2}\gamma_A L+\varepsilon_T x_2, \quad \text{on } x_1=L, \\ &\sigma_{23}=0, \quad u_2=\varepsilon_A x_1+\tfrac{1}{2}\gamma_A W, \quad u_3=\tfrac{1}{2}\gamma_A x_1+\varepsilon_T W, \quad \text{on } x_2=W, \end{aligned} \tag{18.10}$$

where ε_A, ε_T and γ_A define the imposed in-plane axial, transverse and *engineering* shear strains for a uniformly *cracked* laminate. It is again noted that these edge conditions do not prevent the sliding of the ply crack surfaces. As a result, localised tractions will be induced on those parts of the edges near to the ply cracks.

Any cracked layer elements on $x_1 = L$ will be stress-free so that

$$\sigma_{11}^{(i)}(L,x_2,x_3)=0, \quad \sigma_{12}^{(i)}(L,x_2,x_3)=0, \quad \sigma_{13}^{(i)}(L,x_2,x_3)=0, \text{ for all } x_2,x_3. \tag{18.11}$$

By symmetry about the plane $x_1 = 0$, similar conditions apply on the other external face $x_1 = -L$ of the laminate. The external face on $x_3 = x_3^{(N+1)} = h$ of the damaged laminate is assumed to be subject to a uniform normal traction and zero shear tractions so that

$$\sigma_{13}^{(N+1)}(x_1,x_2,h)=0, \quad \sigma_{23}^{(N+1)}(x_1,x_2,h)=0, \quad \sigma_{33}^{(N+1)}(x_1,x_2,h)=\sigma_t, \text{ for all } x,z. \tag{18.12}$$

By symmetry about the plane $x_3 = 0$, similar conditions apply on the other external face $x_3 = -h$ of the laminate.

Because applied displacements of the form (18.10), involving applied strains ε_A, ε_T and γ_A, have been imposed on the edges of the cracked laminate, the corresponding stresses σ_{11}, σ_{22} and σ_{12}, resulting from an exact elasticity analysis, will be nonuniform. Corresponding effective in-plane axial, transverse and shear stresses have already been defined (see (10.4)). The effective out-of-plane strain ε_t has also been defined for a damaged laminate (see (10.5)).

The loading parameters σ_t, σ_A, τ_A and ε_T are to be applied to both cracked and undamaged laminates, whereas the values of the corresponding parameters ε_t, ε_A, γ_A and σ_T describe the response of the damaged laminate to loading. It should be noted that when the faces of the laminate are free of shear tractions, the averaging procedure defined by (10.4) is expected to nullify to a large extent the effect of the induced surface tractions that arise on $x_2 = W$ when ply crack sliding is prevented by the application of the boundary condition (18.10).

18.4 Generalised Plane Strain Conditions

The fundamental assumption for generalised plane strain conditions is that the displacement field is of the following form

$$u_1 = f_1(x_1,x_3)+\varepsilon_A^0 x_1+\frac{1}{2}\gamma_A^0 x_2,\; u_2 = f_2(x_1,x_3)+\frac{1}{2}\gamma_A^0 x_1+\varepsilon_T x_2,\;\; u_3 = f_3(x_1,x_3), \tag{18.13}$$

where ε_A^0 and γ_A^0 are the effective axial and shear strains in the undamaged laminate when the stresses σ_t, σ_A and τ_A are applied, when the in-plane transverse strain has the same value ε_T in all layers of the laminate, and when the temperature difference is ΔT. The parameters σ_t and ε_T are regarded as input values for both cracked and undamaged laminates. For an undamaged laminate, the corresponding through-thickness strain will be denoted by ε_t^0 and the corresponding effective transverse stress will be denoted by σ_T^0. It should be noted that the functions f_k, $k = 1, 2, 3$, appearing in (18.13) are all independent of x_2.

The representation (18.13) is valid only if ply cracks form in planes normal to the x_1-direction, and only if well away from the edges of the laminate where the local stress and deformation distributions depend on how the loading is applied. The regions of laminate in which the edge effects are found depends on laminate geometry, properties and loading. This issue can be examined using models of the type to be described here, and experience suggests that these effects are found in narrow strips at the edges of the laminate having thicknesses that are about the half-thickness h of the laminate. Representation (18.13) for the displacement field differs from an earlier approach [1] by a rigid rotation that does not affect the stress or strain fields.

From (18.5) and $(18.13)_2$ it follows that the in-plane transverse stress in the ith layer is given, in terms of the local values of the stress components and the transverse strain, by

$$\sigma_{22}^{(i)} = \frac{1}{S_{22}^{(i)}}\left[\varepsilon_T - S_{12}^{(i)}\sigma_{11}^{(i)} - S_{23}^{(i)}\sigma_{33}^{(i)} - S_{26}^{(i)}\sigma_{12}^{(i)} - V_2^{(i)}\Delta T\right]. \tag{18.14}$$

The motivation for deriving this expression is that it reduces by one the number of unknown stress components whose distribution needs to be found, and this approach is possible if the in-plane transverse strain ε_T is regarded as an input parameter. The stress-strain relations (18.4), (18.6) and (18.9) may then be written in the following reduced form on eliminating the stress component $\sigma_{22}^{(i)}$ using (18.14)

$$\varepsilon_{11}^{(i)} \equiv \frac{\partial u_1^{(i)}}{\partial x_1} = g_{11}^{(i)}\sigma_{11}^{(i)} + g_{12}^{(i)}\sigma_{33}^{(i)} + g_{13}^{(i)}\sigma_{12}^{(i)} + H_1^{(i)}\varepsilon_T^c + \bar{V}_1^{(i)}\Delta T, \tag{18.15}$$

$$\varepsilon_{33}^{(i)} \equiv \frac{\partial u_3^{(i)}}{\partial x_3} = g_{12}^{(i)}\sigma_{11}^{(i)} + g_{22}^{(i)}\sigma_{33}^{(i)} + g_{23}^{(i)}\sigma_{12}^{(i)} + H_2^{(i)}\varepsilon_T^c + \bar{V}_2^{(i)}\Delta T, \tag{18.16}$$

$$2\varepsilon_{12}^{(i)} \equiv \frac{\partial u_1^{(i)}}{\partial x_2} + \frac{\partial u_2^{(i)}}{\partial x_1} = g_{13}^{(i)}\sigma_{11}^{(i)} + g_{23}^{(i)}\sigma_{33}^{(i)} + g_{33}^{(i)}\sigma_{12}^{(i)} + H_3^{(i)}\varepsilon_T^c + \bar{V}_3^{(i)}\Delta T, \tag{18.17}$$

where

$$
\begin{aligned}
g_{11}^{(i)} &= S_{11}^{(i)} - \frac{S_{12}^{(i)}S_{12}^{(i)}}{S_{22}^{(i)}}, \quad g_{12}^{(i)} = S_{13}^{(i)} - \frac{S_{12}^{(i)}S_{23}^{(i)}}{S_{22}^{(i)}},\\
g_{22}^{(i)} &= S_{33}^{(i)} - \frac{S_{23}^{(i)}S_{23}^{(i)}}{S_{22}^{(i)}}, \quad g_{23}^{(i)} = S_{36}^{(i)} - \frac{S_{23}^{(i)}S_{26}^{(i)}}{S_{22}^{(i)}},\\
g_{13}^{(i)} &= S_{16}^{(i)} - \frac{S_{12}^{(i)}S_{26}^{(i)}}{S_{22}^{(i)}}, \quad g_{33}^{(i)} = S_{66}^{(i)} - \frac{S_{26}^{(i)}S_{26}^{(i)}}{S_{22}^{(i)}},\\
H_1^{(i)} &= \frac{S_{12}^{(i)}}{S_{22}^{(i)}}, \quad H_2^{(i)} = \frac{S_{23}^{(i)}}{S_{22}^{(i)}}, \quad H_3^{(i)} = \frac{S_{26}^{(i)}}{S_{22}^{(i)}},\\
\bar{V}_1^{(i)} &= V_1^{(i)} - \frac{S_{12}^{(i)}}{S_{22}^{(i)}}V_2^{(i)}, \quad \bar{V}_2^{(i)} = V_3^{(i)} - \frac{S_{23}^{(i)}}{S_{22}^{(i)}}V_2^{(i)}, \quad \bar{V}_3^{(i)} = V_6^{(i)} - \frac{S_{26}^{(i)}}{S_{22}^{(i)}}V_2^{(i)}.
\end{aligned}
\tag{18.18}
$$

18.5 Solution for Undamaged Laminates

Consider now the laminate in an undamaged state, or completely homogenised damaged state, but subject to the same loading conditions and temperature as the discretely cracked laminate such that the parameters σ_{t}, σ_{A}, τ_{A}, ε_{T} and ΔT control laminate deformation. The corresponding effective through-thickness strain, effective axial strain, effective in-plane engineering shear strain and the effective in-plane transverse stress for an undamaged or fully homogenised laminate are denoted by $\varepsilon_{\mathrm{t}}^{0}$, $\varepsilon_{\mathrm{A}}^{0}$, γ_{A}^{0} and σ_{T}^{0}, respectively. The external faces of the laminate are assumed to be subject to a uniform normal stress σ_{t} and zero shear tractions so that, for all points in the laminate away from the laminate edges

$$
\sigma_{13}^{(i)} \equiv 0, \ \sigma_{23}^{(i)} \equiv 0, \ \sigma_{33}^{(i)} \equiv \sigma_{\mathrm{t}}, \ i = 1,\ldots, n. \tag{18.19}
$$

It then follows from (18.7), (18.8) and $(18.19)_{2,3}$ that

$$
\varepsilon_{13}^{(i)} \equiv \varepsilon_{23}^{(i)} \equiv 0, \tag{18.20}
$$

and from (18.15)–(18.17) and $(18.19)_1$ that

$$
\varepsilon_{\mathrm{A}}^{0} = g_{11}^{(i)}\sigma_{\mathrm{A}}^{(i)} + g_{12}^{(i)}\sigma_{\mathrm{t}} + g_{13}^{(i)}\tau_{\mathrm{A}}^{(i)} + H_1^{(i)}\varepsilon_{\mathrm{T}} + \bar{V}_1^{(i)}\,\Delta T, \tag{18.21}
$$

$$
\varepsilon_{\mathrm{t}}^{(i)} = g_{12}^{(i)}\sigma_{\mathrm{A}}^{(i)} + g_{22}^{(i)}\sigma_{\mathrm{t}} + g_{23}^{(i)}\tau_{\mathrm{A}}^{(i)} + H_2^{(i)}\varepsilon_{\mathrm{T}} + \bar{V}_2^{(i)}\,\Delta T, \tag{18.22}
$$

$$
\gamma_{\mathrm{A}}^{0} = g_{13}^{(i)}\sigma_{\mathrm{A}}^{(i)} + g_{23}^{(i)}\sigma_{\mathrm{t}} + g_{33}^{(i)}\tau_{\mathrm{A}}^{(i)} + H_3^{(i)}\varepsilon_{\mathrm{T}} + \bar{V}_3^{(i)}\,\Delta T, \tag{18.23}
$$

where $\varepsilon_{\mathrm{t}}^{(i)}$ is the uniform out-of-plane strain, and where $\sigma_{\mathrm{A}}^{(i)}$ and $\tau_{\mathrm{A}}^{(i)}$ are the uniform axial and in-plane shear stresses in the ith layer of the undamaged laminate, respectively. From (18.14), the corresponding transverse in-plane stress component in the ith layer of the undamaged laminate is given by

$$
\sigma_{\mathrm{T}}^{(i)} = \frac{1}{S_{22}^{(i)}}\left[\varepsilon_{\mathrm{T}} - S_{12}^{(i)}\sigma_{\mathrm{A}}^{(i)} - S_{23}^{(i)}\sigma_{\mathrm{t}} - S_{26}^{(i)}\tau_{\mathrm{A}}^{(i)} - V_2^{(i)}\Delta T\right]. \tag{18.24}
$$

Relations (18.21) and (18.23) may be rearranged leading to the following expressions for the axial and shear stresses in the ith layer, for $i = 1,\ldots, n$,

$$\sigma_{\mathrm{A}}^{(i)} = \frac{1}{\lambda_2^{(i)}}\left[g_{33}^{(i)}\varepsilon_{\mathrm{A}}^{0} - g_{13}^{(i)}\gamma_{\mathrm{A}}^{0} + \lambda_1^{(i)}\sigma_{\mathrm{t}} + \frac{1}{S_{22}^{(i)}}\left(S_{26}^{(i)}g_{13}^{(i)} - S_{12}^{(i)}g_{33}^{(i)}\right)\varepsilon_{\mathrm{T}} + \left(g_{13}^{(i)}\bar{V}_3^{(i)} - g_{33}^{(i)}\bar{V}_1^{(i)}\right)\Delta T\right], \tag{18.25}$$

$$\tau_{\mathrm{A}}^{(i)} = \frac{1}{\lambda_2^{(i)}}\left[g_{11}^{(i)}\gamma_{\mathrm{A}}^{0} - g_{13}^{(i)}\varepsilon_{\mathrm{A}}^{0} + \lambda_3^{(i)}\sigma_{\mathrm{t}} + \frac{1}{S_{22}^{(i)}}\left(S_{12}^{(i)}g_{13}^{(i)} - S_{26}^{(i)}g_{11}^{(i)}\right)\varepsilon_{\mathrm{T}} + \left(g_{13}^{(i)}\bar{V}_1^{(i)} - g_{11}^{(i)}\bar{V}_3^{(i)}\right)\Delta T\right], \tag{18.26}$$

where

$$\lambda_1^{(i)} = g_{13}^{(i)}g_{23}^{(i)} - g_{12}^{(i)}g_{33}^{(i)}, \quad \lambda_2^{(i)} = g_{11}^{(i)}g_{33}^{(i)} - g_{13}^{(i)}g_{13}^{(i)}, \quad \lambda_3^{(i)} = g_{12}^{(i)}g_{13}^{(i)} - g_{11}^{(i)}g_{23}^{(i)}. \tag{18.27}$$

On substituting (18.25) and (18.26) into (18.24) and (18.22), respectively, and on using (18.18) to substitute for $H_j^{(i)}, j=1,2,3$, it can be shown that

$$\begin{aligned}\sigma_{\mathrm{T}}^{(i)} = {} & \frac{S_{26}^{(i)}g_{13}^{(i)} - S_{12}^{(i)}g_{33}^{(i)}}{\lambda_2^{(i)}S_{22}^{(i)}}\varepsilon_{\mathrm{A}}^{0} + \frac{S_{12}^{(i)}g_{13}^{(i)} - S_{26}^{(i)}g_{11}^{(i)}}{\lambda_2^{(i)}S_{22}^{(i)}}\gamma_{\mathrm{A}}^{0} - \frac{1}{\lambda_2^{(i)}S_{22}^{(i)}}\left(\lambda_1^{(i)}S_{12}^{(i)} + \lambda_2^{(i)}S_{23}^{(i)} + \lambda_3^{(i)}S_{26}^{(i)}\right)\sigma_{\mathrm{t}} \\ & + \frac{1}{\lambda_2^{(i)}S_{22}^{(i)}}\left[\lambda_2^{(i)} + \frac{S_{12}^{(i)}}{S_{22}^{(i)}}\left(S_{12}^{(i)}g_{33}^{(i)} - S_{26}^{(i)}g_{13}^{(i)}\right) + \frac{S_{26}^{(i)}}{S_{22}^{(i)}}\left(S_{26}^{(i)}g_{11}^{(i)} - S_{12}^{(i)}g_{13}^{(i)}\right)\right]\varepsilon_{\mathrm{T}} \\ & - \frac{1}{\lambda_2^{(i)}S_{22}^{(i)}}\left[\lambda_2^{(i)}V_2^{(i)} + S_{12}^{(i)}\left(g_{13}^{(i)}\bar{V}_3^{(i)} - g_{33}^{(i)}\bar{V}_1^{(i)}\right) + S_{26}^{(i)}\left(g_{13}^{(i)}\bar{V}_1^{(i)} - g_{11}^{(i)}\bar{V}_3^{(i)}\right)\right]\Delta T,\end{aligned} \tag{18.28}$$

$$\begin{aligned}\varepsilon_{\mathrm{t}}^{(i)} = {} & -\frac{\lambda_1^{(i)}}{\lambda_2^{(i)}}\varepsilon_{\mathrm{A}}^{0} - \frac{\lambda_3^{(i)}}{\lambda_2^{(i)}}\gamma_{\mathrm{A}}^{0} + \frac{1}{\lambda_2^{(i)}}\left(\lambda_1^{(i)}g_{12}^{(i)} + \lambda_2^{(i)}g_{22}^{(i)} + \lambda_3^{(i)}g_{23}^{(i)}\right)\sigma_{\mathrm{t}} \\ & + \frac{1}{\lambda_2^{(i)}S_{22}^{(i)}}\left(\lambda_1^{(i)}S_{12}^{(i)} + \lambda_2^{(i)}S_{23}^{(i)} + \lambda_3^{(i)}S_{26}^{(i)}\right)\varepsilon_{\mathrm{T}} + \frac{1}{\lambda_2^{(i)}}\left(\lambda_1^{(i)}\bar{V}_1^{(i)} + \lambda_2^{(i)}\bar{V}_2^{(i)} + \lambda_3^{(i)}\bar{V}_3^{(i)}\right)\Delta T.\end{aligned} \tag{18.29}$$

It can be shown that

$$\begin{aligned} & S_{26}^{(i)}g_{13}^{(i)} - S_{12}^{(i)}g_{33}^{(i)} = S_{16}^{(i)}S_{26}^{(i)} - S_{12}^{(i)}S_{66}^{(i)}, \qquad S_{12}^{(i)}g_{13}^{(i)} - S_{26}^{(i)}g_{11}^{(i)} = S_{12}^{(i)}S_{16}^{(i)} - S_{11}^{(i)}S_{26}^{(i)}, \\ & S_{12}^{(i)}g_{33}^{(i)} - S_{26}^{(i)}g_{13}^{(i)} = S_{12}^{(i)}S_{66}^{(i)} - S_{16}^{(i)}S_{26}^{(i)}, \qquad S_{26}^{(i)}g_{11}^{(i)} - S_{12}^{(i)}g_{13}^{(i)} = S_{11}^{(i)}S_{26}^{(i)} - S_{12}^{(i)}S_{16}^{(i)}.\end{aligned} \tag{18.30}$$

Relations (18.25), (18.26), (18.28) and (18.29) may then be written as

$$\sigma_{\mathrm{A}}^{(i)} = \frac{1}{\lambda_2^{(i)}}\left[g_{33}^{(i)}\varepsilon_{\mathrm{A}}^{0} + \frac{1}{S_{22}^{(i)}}\left(S_{16}^{(i)}S_{26}^{(i)} - S_{12}^{(i)}S_{66}^{(i)}\right)\varepsilon_{\mathrm{T}} + \lambda_1^{(i)}\sigma_{\mathrm{t}} - g_{13}^{(i)}\gamma_{\mathrm{A}}^{0} + \left(g_{13}^{(i)}\bar{V}_3^{(i)} - g_{33}^{(i)}\bar{V}_1^{(i)}\right)\Delta T\right], \tag{18.31}$$

$$\begin{aligned}\sigma_{\mathrm{T}}^{(i)} = {} & \frac{S_{16}^{(i)}S_{26}^{(i)} - S_{12}^{(i)}S_{66}^{(i)}}{\lambda_2^{(i)}S_{22}^{(i)}}\varepsilon_{\mathrm{A}}^{0} - \frac{1}{\lambda_2^{(i)}S_{22}^{(i)}}\left(\lambda_1^{(i)}S_{12}^{(i)} + \lambda_2^{(i)}S_{23}^{(i)} + \lambda_3^{(i)}S_{26}^{(i)}\right)\sigma_{\mathrm{t}} + \frac{S_{12}^{(i)}S_{16}^{(i)} - S_{11}^{(i)}S_{26}^{(i)}}{\lambda_2^{(i)}S_{22}^{(i)}}\gamma_{\mathrm{A}}^{0} \\ & + \frac{1}{\lambda_2^{(i)}S_{22}^{(i)}}\left[\lambda_2^{(i)} + \frac{S_{12}^{(i)}}{S_{22}^{(i)}}\left(S_{12}^{(i)}S_{66}^{(i)} - S_{16}^{(i)}S_{26}^{(i)}\right) + \frac{S_{26}^{(i)}}{S_{22}^{(i)}}\left(S_{11}^{(i)}S_{26}^{(i)} - S_{12}^{(i)}S_{16}^{(i)}\right)\right]\varepsilon_{\mathrm{T}} \\ & - \frac{1}{\lambda_2^{(i)}S_{22}^{(i)}}\left[\lambda_2^{(i)}V_2^{(i)} + S_{12}^{(i)}\left(g_{13}^{(i)}\bar{V}_3^{(i)} - g_{33}^{(i)}\bar{V}_1^{(i)}\right) + S_{26}^{(i)}\left(g_{13}^{(i)}\bar{V}_1^{(i)} - g_{11}^{(i)}\bar{V}_3^{(i)}\right)\right]\Delta T,\end{aligned} \tag{18.32}$$

$$\varepsilon_{\rm t}^{(i)} = -\frac{\lambda_1^{(i)}}{\lambda_2^{(i)}}\varepsilon_{\rm A}^0 + \frac{1}{\lambda_2^{(i)}S_{22}^{(i)}}\left(\lambda_1^{(i)}S_{12}^{(i)} + \lambda_2^{(i)}S_{23}^{(i)} + \lambda_3^{(i)}S_{26}^{(i)}\right)\varepsilon_{\rm T} - \frac{\lambda_3^{(i)}}{\lambda_2^{(i)}}\gamma_{\rm A}^0$$
$$+ \frac{1}{\lambda_2^{(i)}}\left(\lambda_1^{(i)}g_{12}^{(i)} + \lambda_2^{(i)}g_{22}^{(i)} + \lambda_3^{(i)}g_{23}^{(i)}\right)\sigma_{\rm t} + \frac{1}{\lambda_2^{(i)}}\left(\lambda_1^{(i)}\overline{V}_1^{(i)} + \lambda_2^{(i)}\overline{V}_2^{(i)} + \lambda_3^{(i)}\overline{V}_3^{(i)}\right)\Delta T, \tag{18.33}$$

$$\tau_{\rm A}^{(i)} = \frac{1}{\lambda_2^{(i)}}\left[-g_{13}^{(i)}\varepsilon_{\rm A}^0 + \frac{1}{S_{22}^{(i)}}\left(S_{12}^{(i)}S_{16}^{(i)} - S_{11}^{(i)}S_{26}^{(i)}\right)\varepsilon_{\rm T} + \lambda_3^{(i)}\sigma_{\rm t} + g_{11}^{(i)}\gamma_{\rm A}^0 + \left(g_{13}^{(i)}\overline{V}_1^{(i)} - g_{11}^{(i)}\overline{V}_3^{(i)}\right)\Delta T\right]. \tag{18.34}$$

The effective through-thickness strain for an undamaged or completely homogenised laminate is given by

$$\varepsilon_{\rm t}^0 = \frac{1}{h}\sum_{i=1}^{N+1} h_i\varepsilon_{\rm t}^{(i)}, \tag{18.35}$$

where $\varepsilon_{\rm t}^{(i)}$ is specified by (18.33). The effective applied axial, transverse and shear stresses defined by (10.4), and effective through-thickness strain defined by (10.5), are then given by

$$\sigma_{\rm A} = \frac{1}{h}\sum_{i=1}^{N+1} h_i\sigma_{\rm A}^{(i)} = \Omega_{11}\varepsilon_{\rm A}^0 + \Omega_{12}\varepsilon_{\rm T} + \Omega_{13}\sigma_{\rm t} + \Omega_{16}\gamma_{\rm A}^0 - \omega_1\Delta T, \tag{18.36}$$

$$\sigma_{\rm T}^0 = \frac{1}{h}\sum_{i=1}^{N+1} h_i\sigma_{\rm T}^{(i)} = \Omega_{12}\varepsilon_{\rm A}^0 + \Omega_{22}\varepsilon_{\rm T} + \Omega_{23}\sigma_{\rm t} + \Omega_{26}\gamma_{\rm A}^0 - \omega_2\Delta T, \tag{18.37}$$

$$\varepsilon_{\rm t} = -\Omega_{13}\varepsilon_{\rm A}^0 - \Omega_{23}\varepsilon_{\rm T} + \Omega_{33}\sigma_{\rm t} - \Omega_{36}\gamma_{\rm A}^0 + \omega_3\Delta T, \tag{18.38}$$

$$\tau_{\rm A} = \frac{1}{h}\sum_{i=1}^{N+1} h_i\tau_{\rm A}^{(i)} = \Omega_{16}\varepsilon_{\rm A}^0 + \Omega_{26}\varepsilon_{\rm T} + \Omega_{36}\sigma_{\rm t} + \Omega_{66}\gamma_{\rm A}^0 - \omega_6\Delta T, \tag{18.39}$$

where

$$\Omega_{11} = \frac{1}{h}\sum_{i=1}^{N+1} h_i\frac{g_{33}^{(i)}}{\lambda_2^{(i)}}, \qquad \Omega_{66} = \frac{1}{h}\sum_{i=1}^{N+1} h_i\frac{g_{11}^{(i)}}{\lambda_2^{(i)}},$$
$$\Omega_{22} = \frac{1}{h}\sum_{i=1}^{N+1}\frac{h_i}{\lambda_2^{(i)}S_{22}^{(i)}}\left[\lambda_2^{(i)} + \frac{S_{12}^{(i)}}{S_{22}^{(i)}}\left(S_{12}^{(i)}S_{66}^{(i)} - S_{16}^{(i)}S_{26}^{(i)}\right) + \frac{S_{26}^{(i)}}{S_{22}^{(i)}}\left(S_{11}^{(i)}S_{26}^{(i)} - S_{12}^{(i)}S_{16}^{(i)}\right)\right], \tag{18.40}$$
$$\Omega_{33} = \frac{1}{h}\sum_{i=1}^{N+1}\frac{h_i}{\lambda_2^{(i)}}\left(\lambda_1^{(i)}g_{12}^{(i)} + \lambda_2^{(i)}g_{22}^{(i)} + \lambda_3^{(i)}g_{23}^{(i)}\right),$$

$$\Omega_{12} = \frac{1}{h}\sum_{i=1}^{N+1} h_i\frac{S_{16}^{(i)}S_{26}^{(i)} - S_{12}^{(i)}S_{66}^{(i)}}{\lambda_2^{(i)}S_{22}^{(i)}}, \qquad \Omega_{16} = -\frac{1}{h}\sum_{i=1}^{N+1} h_i\frac{g_{13}^{(i)}}{\lambda_2^{(i)}},$$
$$\Omega_{13} = \frac{1}{h}\sum_{i=1}^{N+1} h_i\frac{\lambda_1^{(i)}}{\lambda_2^{(i)}}, \qquad \Omega_{26} = \frac{1}{h}\sum_{i=1}^{N+1} h_i\frac{S_{12}^{(i)}S_{16}^{(i)} - S_{11}^{(i)}S_{26}^{(i)}}{\lambda_2^{(i)}S_{22}^{(i)}}, \tag{18.41}$$
$$\Omega_{23} = -\frac{1}{h}\sum_{i=1}^{N+1}\frac{h_i}{\lambda_2^{(i)}S_{22}^{(i)}}\left(\lambda_1^{(i)}S_{12}^{(i)} + \lambda_2^{(i)}S_{23}^{(i)} + \lambda_3^{(i)}S_{26}^{(i)}\right), \quad \Omega_{36} = \frac{1}{h}\sum_{i=1}^{N+1} h_i\frac{\lambda_3^{(i)}}{\lambda_2^{(i)}},$$

$$\omega_1 = \frac{1}{h}\sum_{i=1}^{N+1} \frac{h_i}{\lambda_2^{(i)}}\left(g_{33}^{(i)}\overline{V}_1^{(i)} - g_{13}^{(i)}\overline{V}_3^{(i)}\right), \qquad \omega_6 = \frac{1}{h}\sum_{i=1}^{N+1} \frac{h_i}{\lambda_2^{(i)}}\left(g_{11}^{(i)}\overline{V}_3^{(i)} - g_{13}^{(i)}\overline{V}_1^{(i)}\right),$$

$$\omega_2 = \frac{1}{h}\sum_{i=1}^{N+1} \frac{h_i}{\lambda_2^{(i)} S_{22}^{(i)}}\left[\lambda_2^{(i)} V_2^{(i)} + S_{12}^{(i)}\left(g_{13}^{(i)}\overline{V}_3^{(i)} - g_{33}^{(i)}\overline{V}_1^{(i)}\right) + S_{26}^{(i)}\left(g_{13}^{(i)}\overline{V}_1^{(i)} - g_{11}^{(i)}\overline{V}_3^{(i)}\right)\right], \quad (18.42)$$

$$\omega_3 = \frac{1}{h}\sum_{i=1}^{N+1} \frac{h_i}{\lambda_2^{(i)}}\left(\lambda_1^{(i)}\overline{V}_1^{(i)} + \lambda_2^{(i)}\overline{V}_2^{(i)} + \lambda_3^{(i)}\overline{V}_3^{(i)}\right).$$

Relations (18.36)–(18.39), relating the effective applied stresses σ_A, σ_T^0, τ_A and the effective through-thickness strain ε_t^0 to the uniform strains ε_A^0, ε_T, γ_A^0 and the uniform through-thickness stress σ_t, enable the thermoelastic constants to be determined for an undamaged or completely homogenised laminate.

The undamaged laminate solution for general symmetric laminates is also relevant to laminates in which there are ply cracks in 90° plies but they are just closed, as discussed in Chapter 10. When the orientation of cracked plies is 90°, crack closure conditions occur when the condition (10.23) is satisfied. Software developments [2] have assumed that $\varepsilon_T^c = 0$. On using values of ε_A^c and ε_T^c satisfying (10.23) together with $\gamma_A^0 = 0$ and relations (18.23), (18.25) and (18.26) the ply stresses $\sigma_A^{(i)}, \sigma_T^{(i)}$ and $\tau_A^{(i)}$, $i = 1,...,N+1$ for an undamaged laminate are first calculated and then used to calculate the corresponding effective applied crack closure stresses σ_A^c, σ_T^c and τ_A^c using (18.36), (18.37) and (18.39).

18.6 Stress and Displacement Fields for Cracked Laminates

The fundamental assumption of the analysis for cracked general symmetric laminates is that the shear stress components σ_{13} and σ_{23} have the following piecewise linear forms (in x_3), for $i = 1, ..., N+1$,

$$\sigma_{13}^{(i)} = p_i'(x_1)\frac{x_3 - x_3^{(i-1)}}{h_i} - p_{i-1}'(x_1)\frac{x_3 - x_3^{(i)}}{h_i}, \ x_3^{(i-1)} \le x_3 \le x_3^{(i)}, \text{ for all } x_2, \quad (18.43)$$

$$\sigma_{23}^{(i)} = q_i'(x_1)\frac{x_3 - x_3^{(i-1)}}{h_i} - q_{i-1}'(x_1)\frac{x_3 - x_3^{(i)}}{h_i}, \ x_3^{(i-1)} \le x_3 \le x_3^{(i)}, \text{ for all } x_2, \quad (18.44)$$

where $p_i(x_1)$ and $q_i(x_1)$, $i = 1, ..., N+1$, are functions only of x_1 that are to be determined, and which have the property that they are identically zero for the case of undamaged laminates. The derivatives of these functions are denoted by primes. The form of relations (18.43) and (18.44) ensures that the stress components σ_{xy} and σ_{xz} are automatically continuous across the interfaces. From (18.43) and (18.44) it is clear that $p_i'(x_1)$ and $q_i'(x_1)$ define the distributions of the stress components $\sigma_{13}^{(i)}$ and $\sigma_{23}^{(i)}$ on the interface at $x_3 = x_3^{(i)}$. It follows from (18.3), (18.43) and (18.44) that, for $i = 1, ..., N+1$ and for all values of x_2, and all x_3 in the range $x_3^{(i-1)} \le x_3 \le x_3^{(i)}$,

$$\sigma_{33}^{(i)} = \frac{x_3 - x_3^{(i)}}{2h_i}\left[(x_3 - x_3^{(i)})p_{i-1}''(x_1) - (x_3 - x_3^{(i)} + 2h_i)p_i''(x_1)\right] + S_i(x_1) + \sigma_t, \ (18.45)$$

where $\sigma_{33}^{(i)}$ is independent of x_2 such that $\sigma_{33}^{(i)}(x_1, x_2, x_3^{(i)}) = S_i(x_1) + \sigma_t$. The functions $S_i(x_1)$, to be defined later, arise from the integration with respect to x_3 and have been chosen to be independent of x_2 and will have a zero value when the laminate is undamaged.

Assume for the moment that $\sigma_{12}^{(i)}$ and $\sigma_{22}^{(i)}$ are also independent of x_2. It follows from (18.1), (18.2), (18.43) and (18.44) that the stress components $\sigma_{11}^{(i)}$ and $\sigma_{12}^{(i)}$ are then of the following form, independent of the variables x_2 and x_3, for $x_3^{(i-1)} < x_3 < x_3^{(i)}$, $i = 1, \ldots, N+1$,

$$\sigma_{11}^{(i)} = \frac{p_{i-1}(x_1) - p_i(x_1)}{h_i} + \sigma_{\text{A}}^{(i)}, \tag{18.46}$$

$$\sigma_{12}^{(i)} = \frac{q_{i-1}(x_1) - q_i(x_1)}{h_i} + \tau_{\text{A}}^{(i)}, \tag{18.47}$$

where $\sigma_{\text{A}}^{(i)}$ and $\tau_{\text{A}}^{(i)}$, defined by (18.31) and (18.34), respectively, are the uniform axial and shear stresses that would result if the laminate were undamaged and subject to the same applied loading and thermal conditions. It is noted from (18.14) that the stress component $\sigma_{22}^{(i)}$ is specified by

$$\sigma_{22}^{(i)} = \frac{1}{S_{22}^{(i)}}\left[\varepsilon_{\text{T}} - S_{12}^{(i)}\sigma_{11}^{(i)} - S_{23}^{(i)}\sigma_{33}^{(i)} - S_{26}^{(i)}\sigma_{12}^{(i)} - V_2^{(i)}\Delta T\right], \tag{18.48}$$

and expressions (18.45)–(18.48) indicate that it is independent of x_2. The form of the results (18.47) and (18.48) confirms the validity of the assumption that the stress components $\sigma_{12}^{(i)}$ and $\sigma_{22}^{(i)}$ are independent of x_2.

It is understood that the functions $p_0(x_1)$, $p_{N+1}(x_1)$, $q_0(x_1)$ and $q_{N+1}(x_1)$ are all identically zero which means that the stress components σ_{13} and σ_{23} are automatically zero on $x_3 = 0$ and $x_3 = x_3^{(N+1)} = h$. It then follows that the value of the effective applied axial stress σ_{A} defined by $(10.4)_1$ has the same value for both cracked and uncracked laminates (see (18.36)), and similarly for the effective applied shear stress τ_{A} defined by $(10.4)_3$ and alternatively (18.39). The representation for the stress field defined by (18.14), (18.43)–(18.47) satisfies the equilibrium equations (18.1)–(18.3) for any functions $p_i(x_1)$ and $q_i(x_1)$, $i = 1, \ldots, N$, characterising stress transfer in the cracked laminate.

It follows from (18.16), (18.22) and (18.45)–(18.47) that, for $x_3^{(i-1)} < x_3 \leq x_3^{(i)}$ and $i = 1, \ldots, N+1$,

$$u_3^{(i)} = \frac{(x_3 - x_3^{(i)})^2}{6h_i} g_{22}^{(i)}\left[(x_3 - x_3^{(i)})\, p_{i-1}''(x_1) - (x_3 - x_3^{(i)} + 3h_i)\, p_i''(x_1)\right]$$
$$+ (x_3 - x_3^{(i)})\left[g_{22}^{(i)} S_i(x_1) + g_{12}^{(i)} \frac{p_{i-1}(x_1) - p_i(x_1)}{h_i} + g_{23}^{(i)} \frac{q_{i-1}(x_1) - q_i(x_1)}{h_i} + \varepsilon_{\text{t}}^{(i)}\right] + U_i(x_1), \tag{18.49}$$

where $u_3^{(i)}(x_1, x_3^{(i)}) = U_i(x_1)$, where $U_i(x_1)$, arising from the integration with respect to x_3, is the through-thickness displacement (to be defined later) at the interface $x_3 = x_3^{(i)}$, and where $\varepsilon_{\text{t}}^{(i)}$ is the through-thickness strain in the ith layer of an undamaged laminate defined by (18.33). It should be noted that the displacement component (18.49) has been defined so that it is independent of x_2 consistent with the assumption (18.13) and so that $U_i(x_1)$ will be zero for undamaged laminates.

It follows from (18.8), (18.43), (18.44) and (18.49) that the expression for $u_1^{(i)}$, for $x_3^{(i-1)} < x_3 \le x_3^{(i)}$ and $i = 1, ..., N+1$, is given by

$$\begin{aligned} u_1^{(i)} = {} & \frac{x_3 - x_3^{(i)}}{2h_i}(x_3 - x_3^{(i)} + 2h_i)\left[S_{55}^{(i)}p_i'(x_1) + S_{45}^{(i)}q_i'(x_1)\right] - \frac{(x_3 - x_3^{(i)})^2}{2h_i}\left[S_{55}^{(i)}p_{i-1}'(x_1) + S_{45}^{(i)}q_{i-1}'(x_1)\right] \\ & - \frac{(x_3 - x_3^{(i)})^3}{24h_i} g_{22}^{(i)}\left[(x_3 - x_3^{(i)})p_{i-1}'''(x_1) - (x_3 - x_3^{(i)} + 4h_i)p_i'''(x_1)\right] \\ & - \frac{(x_3 - x_3^{(i)})^2}{2}\left[g_{22}^{(i)}S_i'(x_1) + g_{12}^{(i)}\frac{p_{i-1}'(x_1) - p_i'(x_1)}{h_i} + g_{23}^{(i)}\frac{q_{i-1}'(x_1) - q_i'(x_1)}{h_i}\right] \\ & - (x_3 - x_3^{(i)})U_i'(x_1) + V_i(x_1, x_2), \end{aligned} \tag{18.50}$$

where, for $i = 1, ..., N+1$, $u_1^{(i)}(x_1, x_2, x_3^{(i)}) = V_i(x_1, x_2)$, where $V_i(x_1, x_2)$ arises from the integration with respect to x_3 subject to the assumed form $(18.13)_1$.

Similarly from (18.7), (18.43), (18.44) and (18.49), the expression for $u_2^{(i)}$, for $x_3^{(i-1)} < x_3 \le x_3^{(i)}$ and $i = 1, ..., N+1$, is given by

$$\begin{aligned} u_2^{(i)} = {} & \frac{(x_3 - x_3^{(i)})}{2h_i}(x_3 - x_3^{(i)} + 2h_i)\left[S_{45}^{(i)}p_i'(x_1) + S_{44}^{(i)}q_i'(x_1)\right] \\ & - \frac{(x_3 - x_3^{(i)})^2}{2h_i}\left[S_{45}^{(i)}p_{i-1}'(x_1) + S_{44}^{(i)}q_{i-1}'(x_1)\right] + W_i(x_1, x_2), \end{aligned} \tag{18.51}$$

where $u_2^{(i)}(x_1, x_2, x_3^{(i)}) = W_i(x_1, x_2)$, and where $W_i(x_1, x_2)$ arises from the integration with respect to x_3 subject to the assumed form $(18.13)_2$.

The functions $S_i(x_1)$, $U_i(x_1)$, $V_i(x_1, x_2)$ and $W_i(x_1, x_2)$, $i = 1, ..., N+1$, the first two of which have been defined to be identically zero for undamaged laminates, appearing in relations (18.45), (18.49)–(18.51), are determined in terms of the stress-transfer functions $p_i(x_1)$ and $q_i(x_1)$, $i = 1, ..., N$, by four sets of recurrence relations given in Appendix G. It is clear that boundary conditions prevent any x_2-dependence of the functions S_i and U_i. It is also shown in Appendix G that the stress-transfer functions must satisfy a set of fourth-order ordinary differential equations specified by (G.14) and (G.15), namely,

$$\begin{aligned} & \sum_{i=1}^{N} F_{ij}p_i^{(4)}(x_1) + \sum_{i=1}^{N} G_{ij}p_i^{(2)}(x_1) + \sum_{i=1}^{N} H_{ij}p_i(x_1) \\ & \quad + \sum_{i=1}^{N} c_{ij}q_i^{(2)}(x_1) + \sum_{i=1}^{N} d_{ij}q_i(x_1) = 0, \; j = 1, \ldots, N, \end{aligned} \tag{18.52}$$

$$\sum_{i=1}^{N} A_{ij}p_i^{(2)}(x_1) + \sum_{i=1}^{N} B_{ij}p_i(x_1) + \sum_{i=1}^{N} a_{ij}q_i^{(2)}(x_1) + \sum_{i=1}^{N} b_{ij}q_i(x_1) = 0, \; j = 1, \ldots, N, \tag{18.53}$$

where the constant coefficients are best determined numerically (see Appendix G). The superscripts '(4)' and '(2)' denote fourth- and second-order ordinary derivatives, respectively.

Provided that the recurrence relations given in Appendix G for the functions $S_i(x_1)$, $U_i(x_1)$, $\Delta V_i(x_1)$ and $\Delta W_i(x_1)$, $i = 1, ..., N+1$, are satisfied, the representation for the stress and displacement fields presented earlier in this section automatically satisfies the following boundary and interface conditions

$$\sigma_{13}^{(1)}=0,\ \sigma_{23}^{(1)}=0,\ u_3^{(1)}=0,\ \text{on } x_3=0, \tag{18.54}$$

$$\sigma_{13}^{(i)}=\sigma_{13}^{(i+1)},\ \sigma_{23}^{(i)}=\sigma_{23}^{(i+1)},\ \sigma_{33}^{(i)}=\sigma_{33}^{(i+1)},\ \text{on } x_3=x_3^{(i)},\ i=1,\ldots,N, \tag{18.55}$$

$$u_1^{(i)}=u_1^{(i+1)},\ u_2^{(i)}=u_2^{(i+1)},\ u_3^{(i)}=u_3^{(i+1)},\ \text{on } x_3=x_3^{(i)},\ i=1,\ldots,N, \tag{18.56}$$

$$\sigma_{13}^{(N+1)}=0,\ \sigma_{23}^{(N+1)}=0,\ \sigma_{33}^{(N+1)}=\sigma_t,\ \text{on } x_3=x_3^{(N+1)}=h. \tag{18.57}$$

Thus, the interface conditions between layers and the boundary conditions relating to symmetry and the boundary conditions for the external surface $x_3=h$ of the laminate are automatically satisfied.

18.7 Averaged Boundary Conditions and Stress-Strain Relations

The remaining boundary conditions, needed to ensure a unique solution of the differential equations defined by (18.52) and (18.53) are applied to the laminate edges. Any *cracked* surfaces on $x_1=L$ will be stress-free so that

$$\sigma_{11}^{(i)}(L,x_2,x_3)=0,\ \sigma_{12}^{(i)}(L,x_2,x_3)=0,\ \sigma_{13}^{(i)}(L,x_2,x_3)=0,\ \text{for all } x_2,x_3. \tag{18.58}$$

It is not possible using the representation (18.50) and (18.51) to satisfy exactly the boundary conditions (18.10) that involve the in-plane displacements. It is possible to satisfy these conditions in an average sense where the in-plane displacements are averaged through the thickness of each layer before an averaged boundary condition is applied. Any undamaged layer surfaces on $y=L$ are subject to zero shear stress, and to averaged boundary conditions derived from (18.10) as follows

$$\begin{gathered}\sigma_{13}^{(i)}(L,x_2,x_3)=0,\ \text{for all } x_2,\ x_3,\\ \bar{u}_1^{(i)}(L,x_2)=\varepsilon_A L+\frac{1}{2}\gamma_A x_2,\ \bar{u}_2^{(i)}(L,x_2)=\frac{1}{2}\gamma_A L+\varepsilon_T x_2,\ \text{for all } x_2,\end{gathered} \tag{18.59}$$

where the average in-plane displacements are given by, on integrating the through-thickness averages of (18.15) and (18.17) with respect to x_1 and x_2, respectively,

$$\begin{aligned}\bar{u}_1^{(i)}(x_1,x_2)&=\frac{1}{h_i}\int_{x_3^{(i-1)}}^{x_3^{(i)}} u_1^{(i)}(x_1,x_2,x_3)\,dx_3\\ &=\frac{h_i g_{12}^{(i)}}{6}\left[p_{i-1}'(x_1)+2p_i'(x_1)\right]+\frac{g_{11}^{(i)}}{h_i}\left[\overset{*}{p}_{i-1}(x_1)-\overset{*}{p}_i(x_1)\right]\\ &\quad+\frac{g_{13}^{(i)}}{h_i}\left[\overset{*}{q}_{i-1}(x_1)-\overset{*}{q}_i(x_1)\right]+g_{12}^{(i)}\overset{*}{S}_i(x_1)+\varepsilon_A^0 x_1+\tfrac{1}{2}\gamma_A^0 x_2,\end{aligned} \tag{18.60}$$

$$\begin{aligned}\bar{u}_2^{(i)}(x_1,x_2)&=\frac{1}{h_i}\int_{x_3^{(i-1)}}^{x_3^{(i)}} u_2^{(i)}(x_1,x_2,x_3)\,dx_3=\frac{h_i g_{23}^{(i)}}{6}\left[p_{i-1}'(x_1)+2p_i'(x_1)\right]\\ &+\frac{g_{13}^{(i)}}{h_i}\left[\overset{*}{p}_{i-1}(x_1)-\overset{*}{p}_i(x_1)\right]+\frac{g_{33}^{(i)}}{h_i}\left[\overset{*}{q}_{i-1}(x_1)-\overset{*}{q}_i(x_1)\right]+g_{23}^{(i)}\overset{*}{S}_i(x_1)+\tfrac{1}{2}\gamma_A^0 x_1+\varepsilon_T x_2,\end{aligned} \tag{18.61}$$

where ε_A^0 and γ_A^0 are defined by (18.21) and (18.23), respectively, and where

$$p_i^*(x_1) = \int_0^{x_1} p_i(s)\mathrm{d}s,\ q_i^*(x_1) = \int_0^{x_1} q_i(s)\mathrm{d}s,\ S_i^*(x_1) = \int_0^{x_1} S_i(s)\mathrm{d}s. \tag{18.62}$$

When deriving (18.60) and (18.61) the constants arising from the integrations with respect to x_1 and x_2 are selected so that the correct distributions consistent with (18.59) result when the laminate is undamaged. In Appendix G, it is shown how sets of ordinary differential equations may be derived for the unknown functions $p_i(x_1)$ and $q_i(x_1)$, $i = 1, \ldots, N$, characterising stress transfer in the cracked laminate. The method used to solve the differential equations subject to the given boundary conditions is described in Appendix H.

18.8 Calculation of Thermoelastic Constants for Cracked Laminate

The method of estimating the thermoelastic constants is based on the fact that, when deriving the stress and displacement fields from the solution of the differential equations, the parameters $\sigma_t, \sigma_A, \tau_A, \varepsilon_T$ and ΔT are specified. The analysis of Section 18.7 is then used to determine the corresponding values of $\varepsilon_t, \varepsilon_A, \gamma_A$ and σ_T for cracked laminates. If the laminate is undamaged, then these parameters have the values of $\varepsilon_t^0, \varepsilon_A^0, \gamma_A^0$ and σ_T^0. The corresponding effective through-thickness strain ε_t, the axial strain ε_A, the in-plane shear strain γ_A and the effective transverse stress σ_T, must be calculated from the solution for a uniformly cracked laminate. The axial strain ε_A and the in-plane shear strain γ_A are calculated from the displacement solution using (18.59), where use is also made of the expressions (18.60) and (18.61) applied for the special case $i = j$, where j is the label for any one of the uncracked plies.

From (10.4), the effective in-plane transverse stress is defined by

$$\sigma_T = \frac{1}{4hL}\int_{-L}^{L}\int_{-h}^{h} \sigma_{22}(x_1, W, x_3)\mathrm{d}x_1\mathrm{d}x_3 = \frac{1}{2L}\sum_{i=1}^{N+1}\frac{h_i}{h}\int_{-L}^{L}\bar{\sigma}_{22}^{(i)}(x_1)\mathrm{d}x_1, \tag{18.63}$$

where from (18.14) $\bar{\sigma}_{22}^{(i)}(x_1)$ is given by

$$\bar{\sigma}_{22}^{(i)}(x_1) = \frac{1}{S_{22}^{(i)}}\left[\varepsilon_T - S_{12}^{(i)}\bar{\sigma}_{11}^{(i)}(x_1) - S_{23}^{(i)}\bar{\sigma}_{33}^{(i)}(x_1) - S_{26}^{(i)}\bar{\sigma}_{12}^{(i)}(x_1) - V_2^{(i)}\Delta T\right]. \tag{18.64}$$

As symmetry considerations require that $\sigma_{13} = 0$ on $x_1 = \pm L$, it follows from (18.43), (G.5) and (G.6) (see Appendix G) that $p_i'(\pm L) = 0$, $i = 0, \ldots, N+1$. From (G.1) and (G.11) it can then be shown that

$$\frac{1}{2L}\int_{-L}^{L}\bar{\sigma}_{33}(x_1)\mathrm{d}x_1 = \sigma_t. \tag{18.65}$$

From (18.43), (18.44) and (18.62) it can be shown that

$$\frac{1}{2L}\int_{-L}^{L}\bar{\sigma}_{11}^{(i)}(x_1)\mathrm{d}x_1 = \frac{1}{h_i L}\left[p_{i-1}^*(x_1) - p_i^*(x_1)\right] + \sigma_A^{(i)}, \tag{18.66}$$

$$\frac{1}{2L}\int_{-L}^{L}\overline{\sigma}_{12}^{(i)}(x_1)\mathrm{d}x_1=\frac{1}{h_iL}\left[q_{i-1}^{*}(x_1)-q_i^{*}(x_1)\right]+\tau_{\mathrm{A}}^{(i)}. \tag{18.67}$$

It then follows from (18.24) (18.63)–(18.67), and the definition of the in-plane transverse stress of an undamaged laminate (18.37) that

$$\sigma_{\mathrm{T}}=\sigma_{\mathrm{T}}^{0}-\frac{1}{hL}\sum_{i=1}^{N+1}\frac{1}{S_{22}^{(i)}}\left[S_{12}^{(i)}\left(p_{i-1}^{*}(x_1)-p_i^{*}(x_1)\right)+S_{26}^{(i)}\left(q_{i-1}^{*}(x_1)-q_i^{*}(x_1)\right)\right]. \tag{18.68}$$

The effective through-thickness strain ε_{t} for a cracked laminate is defined by (10.5), namely,

$$\varepsilon_{\mathrm{t}}=\frac{1}{4hLW}\int_{-L}^{L}\int_{-W}^{W}u_3(x_1,x_2,h)\,\mathrm{d}x_1\mathrm{d}x_2. \tag{18.69}$$

As from (18.49) $u_3(x_1,x_2,h)\equiv u_3^{(N+1)}(x_1,x_2,x_3^{(N+1)})=U_{N+1}(x_1)$, and because of symmetry $U_{N+1}(x_1)=U_{N+1}(-x_1)$,

$$\varepsilon_{\mathrm{t}}=\frac{1}{hL}\int_{0}^{L}U_{N+1}(x_1)\mathrm{d}x_1. \tag{18.70}$$

It follows from the recurrence relation (G.2) (see Appendix G) that

$$U_{N+1}(x_1)=\sum_{i=1}^{N+1}\left[\begin{array}{l}h_i\varepsilon_{\mathrm{t}}^{(i)}+\frac{h_i^2}{6}g_{22}^{(i)}\left[p_{i-1}''(x_1)+2p_i''(x_1)\right]+h_ig_{22}^{(i)}S_i(x_1)\\ \quad+g_{12}^{(i)}\left[p_{i-1}(x_1)-p_i(x_1)\right]+g_{23}^{(i)}\left[q_{i-1}(x_1)-q_i(x_1)\right]\end{array}\right], \tag{18.71}$$

so that on using (G.18) the recurrence relation (G.1) and (18.62)

$$\int_{0}^{L}U_{N+1}(x_1)\mathrm{d}x_1=\sum_{i=1}^{N+1}\left[Lh_i\varepsilon_{\mathrm{t}}^{(i)}+g_{12}^{(i)}\left\{p_{i-1}^{*}(L)-p_i^{*}(L)\right\}+g_{23}^{(i)}\left\{q_{i-1}^{*}(L)-q_i^{*}(L)\right\}\right]. \tag{18.72}$$

On substituting (18.72) into (18.70) and making use of the definition for the through-thickness strain of an undamaged laminate (18.35)

$$\varepsilon_{\mathrm{t}}=\varepsilon_{\mathrm{t}}^{0}+\frac{1}{hL}\sum_{i=1}^{N+1}\left[g_{12}^{(i)}\left\{p_{i-1}^{*}(L)-p_i^{*}(L)\right\}+g_{23}^{(i)}\left\{q_{i-1}^{*}(L)-q_i^{*}(L)\right\}\right]. \tag{18.73}$$

Because linear elasticity is assumed it follows that the parameters ε_{t}, ε_{A}, γ_{A} and σ_{T} for cracked laminates are linearly related to the parameters σ_{t}, σ_{A}, τ_{A}, ε_{T} and ΔT, i.e.

$$\varepsilon_{\mathrm{t}}=A_1\sigma_{\mathrm{t}}+B_1\sigma_{\mathrm{A}}+C_1\tau_{\mathrm{A}}+D_1\varepsilon_{\mathrm{T}}+E_1\Delta T, \tag{18.74}$$

$$\varepsilon_{\mathrm{A}}=A_2\sigma_{\mathrm{t}}+B_2\sigma_{\mathrm{A}}+C_2\tau_{\mathrm{A}}+D_2\varepsilon_{\mathrm{T}}+E_2\Delta T, \tag{18.75}$$

$$\gamma_{\mathrm{A}}=A_3\sigma_{\mathrm{t}}+B_3\sigma_{\mathrm{A}}+C_3\tau_{\mathrm{A}}+D_3\varepsilon_{\mathrm{T}}+E_3\Delta T, \tag{18.76}$$

$$\sigma_{\mathrm{T}}=A_4\sigma_{\mathrm{t}}+B_4\sigma_{\mathrm{A}}+C_4\tau_{\mathrm{A}}+D_4\varepsilon_{\mathrm{T}}+E_4\Delta T, \tag{18.77}$$

The coefficients in (18.74)–(18.77) are determined numerically by calculating, from the stress and displacement distributions, the values of $\varepsilon_t, \varepsilon_A, \gamma_A$ and σ_T for each of the following five sets of values for the parameters $\sigma_t, \sigma_A, \tau_A, \varepsilon_T$ and ΔT, respectively,

{1 0 0 0 0}

{0 1 0 0 0}

{0 0 1 0 0}

{0 0 0 1 0}

{0 0 0 0 1}.

On solving (18.77) for ε_T and substituting into (18.74)–(18.76) it follows on re-ordering that

$$\begin{aligned}\varepsilon_A = &\left(A_2 - \frac{D_2}{D_4}A_4\right)\sigma_t + \left(B_2 - \frac{D_2}{D_4}B_4\right)\sigma_A + \frac{D_2}{D_4}\sigma_T \\ &+ \left(C_2 - \frac{D_2}{D_4}C_4\right)\tau_A + \left(E_2 - \frac{D_2}{D_4}E_4\right)\Delta T,\end{aligned} \tag{18.78}$$

$$\varepsilon_T = -\frac{A_4}{D_4}\sigma_t - \frac{B_4}{D_4}\sigma_A + \frac{1}{D_4}\sigma_T - \frac{C_4}{D_4}\tau_A - \frac{E_4}{D_4}\Delta T, \tag{18.79}$$

$$\begin{aligned}\varepsilon_t = &\left(A_1 - \frac{D_1}{D_4}A_4\right)\sigma_t + \left(B_1 - \frac{D_1}{D_4}B_4\right)\sigma_A + \frac{D_1}{D_4}\sigma_T \\ &\left(C_1 - \frac{D_1}{D_4}C_4\right)\tau_A + \left(E_1 - \frac{D_1}{D_4}E_4\right)\Delta T,\end{aligned} \tag{18.80}$$

$$\begin{aligned}\gamma_A = &\left(A_3 - \frac{D_3}{D_4}A_4\right)\sigma_t + \left(B_3 - \frac{D_3}{D_4}B_4\right)\sigma_A + \frac{D_3}{D_4}\sigma_T \\ &+ \left(C_3 - \frac{D_3}{D_4}C_4\right)\tau_A + \left(E_3 - \frac{D_3}{D_4}E_4\right)\Delta T.\end{aligned} \tag{18.81}$$

It is expected that the matrix formed by the coefficients of the stress terms on the right-hand sides of (18.78)–(18.81) is symmetric. Solutions confirm that such symmetry is valid (i.e. $B_1 = A_2$, $C_1 = A_3$, $D_1 = -A_4$, $C_2 = B_3$, $D_2 = -B_4$, $D_3 = -C_4$) in which case relations (18.78)–(18.81) can be written in the more convenient form used in Chapter 10 (see (10.7)–(10.10))

$$\varepsilon_A = -\frac{\nu_a}{E_A}\sigma_t + \frac{\sigma_A}{E_A} - \frac{\nu_A}{E_A}\sigma_T - \frac{\lambda_A}{\mu_A}\tau_A + \alpha_A\Delta T, \tag{18.82}$$

$$\varepsilon_T = -\frac{\nu_t}{E_T}\sigma_t - \frac{\nu_A}{E_A}\sigma_A + \frac{\sigma_T}{E_T} - \frac{\lambda_T}{\mu_A}\tau_A + \alpha_T\Delta T, \tag{18.83}$$

$$\varepsilon_t = \frac{\sigma_t}{E_t} - \frac{\nu_a}{E_A}\sigma_A - \frac{\nu_t}{E_T}\sigma_T - \frac{\lambda_t}{\mu_A}\tau_A + \alpha_t\Delta T, \tag{18.84}$$

$$\gamma_A = -\frac{\lambda_t}{\mu_A}\sigma_t - \frac{\lambda_A}{\mu_A}\sigma_A - \frac{\lambda_T}{\mu_A}\sigma_T + \frac{\tau_A}{\mu_A} + \alpha_S\Delta T, \tag{18.85}$$

where, for specified values of σ_A, σ_T, σ_t and τ_A and ΔT, the quantities ε_A, ε_T, ε_t and γ_A are the effective through-thickness, axial, in-plane transverse and shear strains of the damaged laminate, respectively, and where E_A, E_T, E_t, ν_A, ν_a, ν_t, λ_A, λ_T, λ_t, α_A, α_T, α_t and α_S denote the corresponding effective thermoelastic constants of the damaged laminate. The parameters λ_A, λ_T and λ_t are ratios indicating the degree of shear coupling. The parameter α_S is an expansion coefficient governing the amount of shear strain that can arise when the temperature is changed.

When using (10.193) to investigate crack formation, the fracture energies $2\Gamma_j$ are taken at random from a statistical distribution of fracture energies (normal assumed so that use can be made of the Box-Muller technique [5]) and then allocated at random to the various potential fracture sites in the 90° plies. The minimum fracture energy is exchanged with the fracture energy allocated to the last potential fracture site ($j = M$), it being assumed that the first site is cracked. This ensures that when the first crack forms the first and last site crack simultaneously. The values of the applied stresses are then increased gradually until the fracture condition (10.193) can again be met at one of the potential fracture sites in the laminate. This means that crack formation, for every stage of loading, must be investigated at all remaining potential cracking sites to determine the location of the next crack to form. This procedure, carried out using a computer, must be repeated for every new ply crack that forms during loading.

Figure 18.2 shows a typical distribution of ply cracking stresses when a total of six ply cracks have formed in a system having 256 potential ply cracking sites and the standard deviation of fracture energies is 10% of the mean value. The locations of the ply cracks are indicated by the spikes of the distribution (i.e. at ply locations 1, 37, 41, 91, 121 and 256 in Figure 18.2). The next crack will form at the site having the least ply cracking stress (at location 115 in Figure 18.2). It is of interest to note that at those sites near to the ply cracks the ply cracking stresses are higher than those that are more

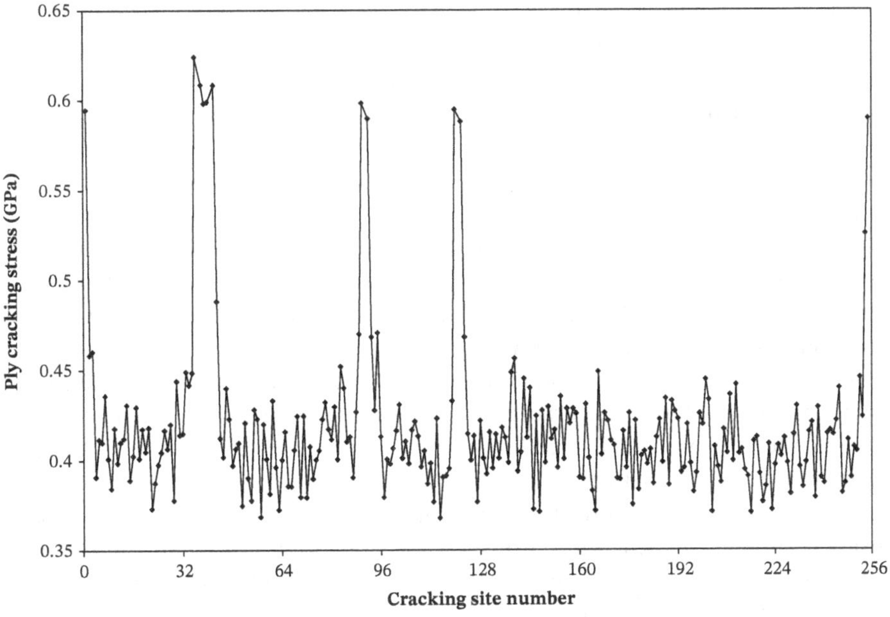

Figure 18.2 Distribution of ply cracking stresses at each potential ply cracking site when six ply cracks are present.

remote from the cracks. There is evidence of a crack shielding effect where cracks will not form very close to existing ply cracks as the stress required to do so is much larger than stresses to form cracks at other sites.

If the fracture energies allocated to potential crack formation sites are randomly distributed, then the loading of the laminate will lead to nonuniform crack spacings. Although the stress and displacement representations given in Section 8.6 are valid for such crack spacings, its application will require the values of the effective thermoelastic constants of the laminate for the case of nonuniform crack spacings. The analysis given in this chapter is valid only for uniform crack spacings (see Section 10.8). To overcome this problem an approximation can be made where, for nonuniform crack spacings, the stress and displacement fields between two neighbouring cracks are taken as those that would occur in a uniformly cracked laminate having the same crack spacing as the two neighbouring cracks that are being considered. The effective thermoelastic constants of a nonuniformly cracked laminate are then obtained by combining the effects of the effective thermoelastic constants for each region lying between two neighbouring cracks as described in Section 10.8. The important result of this approximate approach is that the key interrelationships (10.34), (10.49) and (10.62) are satisfied by the approximate thermoelastic constants for a nonuniformly cracked laminate.

The final step in the simulation procedure is to transform the predicted effective properties and ply cracking stresses so that their values correspond to the original orientation of the laminate and loading.

18.9 General Description of the Homogenisation Approach

So far in this book it has been assumed that ply cracks form in a laminate only in one or more plies that have the same orientation. Furthermore, it has been assumed that the same distribution of ply cracks is found in each of the plies that have cracked. The approach has enabled predictions of ply cracking to be made based on detailed analyses of RVEs which are defined by the locations of the ply crack planes. In practice, ply crack formation is not so orderly and it is useful to consider how more complex ply cracking distributions might be handled. The most challenging situation is to predict the effective properties of laminates having ply cracks that have formed in more than one orientation where the distribution of ply cracks in each cracked layer might be different. A tractable method of dealing with such damage states makes use of a homogenisation technique that is now described.

Before embarking on a description of the homogenisation analysis, it is useful to describe in general terms the procedure that is to be modelled. A simple example based on a $[\theta/-\theta]_s$ laminate is used for this purpose, as illustrated in Figure 18.3 (far left diagram). Assume that the plies have some fully developed ply cracks (i.e. those that traverse the entire width of the laminate and thickness of the ply) that have already been homogenised, which are shown as dotted lines in Figure 18.3(a). This means that the ply properties are homogeneous in each ply, but they have been degraded such that the properties of the laminate, estimated using modified classical laminate theory, lead to overall laminate properties that are identical to those that would be found if the ply cracks were modelled as discrete entities in one or more of

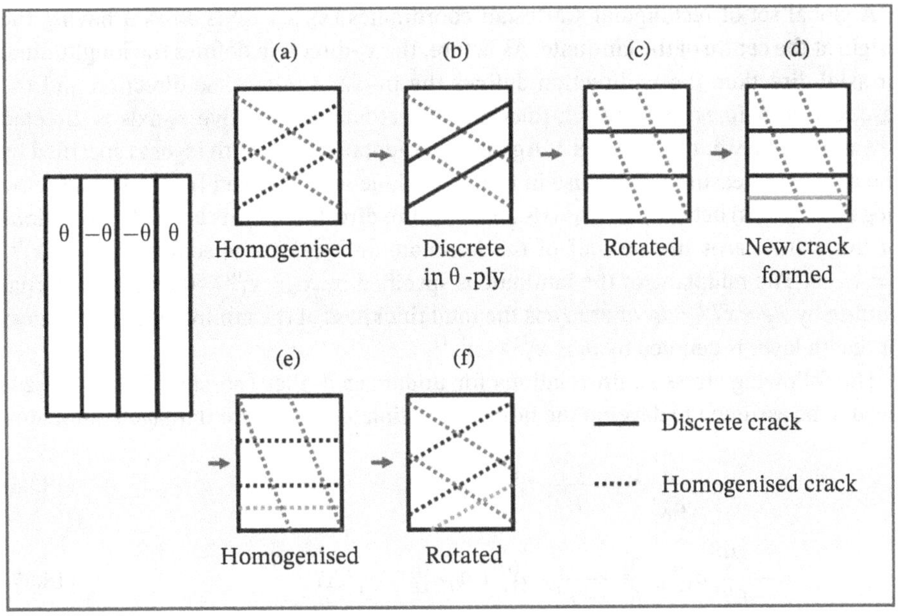

Figure 18.3 Illustration of the homogenisation procedure to account for progressive ply crack formation in an angle-ply laminate.

the plies. The next step is to describe how a new ply crack will be introduced during loading that leads to progressive ply crack formation if the procedure is applied successively. The homogenised cracks in one of the θ plies are first discretised, shown as bold continuous lines in Figure 18.3(b). The laminate and the applied in-plane stress field are then rotated so that these discrete ply cracks have a 90° orientation as shown in Figure 18.3(c). The next ply crack to form in the laminate, on increasing the loading, leads to the situation shown in Figure 18.3(d) where an additional discrete ply crack has been introduced. All ply cracks in the 90° ply of the laminate are then homogenised resulting in the damage state illustrated in Figure 18.3(e). The final stage is to rotate the fully homogenised laminate and the applied stress field back to the original orientation, as shown in Figure 18.3(f). This procedure is then repeated successively as additional cracks progressively form in the plies of the laminate.

Although the procedure has been described for the case of a simple angle-ply laminate, it is a simple matter to use the procedure to deal with cracking in any plies of a general symmetric laminate subject to general in-plane loading combined with through-thickness loading and thermal residual stresses.

18.9.1 Geometry and Basic Field Equations

As mentioned earlier, the problem under consideration concerns the triaxial deformation of a symmetric multilayered laminate of length $2L$ and width $2W$ constructed of $2n = 2(N + 1)$ perfectly bonded layers which can have any combination of orientations provided that there is symmetry about the midplane of the laminate. The analysis also applies to hybrid symmetric laminates where plies could be made of various different materials. As laminate symmetry is assumed, it is necessary to consider only the upper quadrant of the right-hand set of n layers as shown in Figure 18.1.

A global set of rectangular Cartesian coordinates (x_1, x_2, x_3) is chosen having the origin at the centre of the laminate. As before, the x_1-direction defines the longitudinal or axial direction, the x_2-direction defines the in-plane transverse direction and the x_3-direction defines the through-thickness direction. The positive x_2-axis is directed towards the viewer of Figure 18.1. Again, the orientation of the ith layer is specified by the angle ϕ_i (measured clockwise in the x_1–x_2 plane when viewed from a point on the negative x_3-axis) between the x_1-axis and the fibre direction of this layer. The locations of the n interfaces in one half of the laminate ($x_3 > 0$) are specified by $x_3 = x_3^{(i)}$, $i = 1,\dots,n$. The midplane of the laminate is specified by $x_3 = x_3^{(0)} = 0$ and the external surface by $x_3 = x_3^{(n)} = h$ where $2h$ is the total thickness of the laminate. The thickness of the ith layer is denoted by $h_i = x_3^{(i)} - x_3^{(i-1)}$.

The following stress-strain relations for undamaged plies (possibly homogenised) need to be assumed to develop the homogenisation technique for damaged laminates

$$\varepsilon_{11}^{(i)} = \frac{\sigma_{11}^{(i)}}{E_A^{(i)}} - \frac{\nu_A^{(i)}}{E_A^{(i)}}\sigma_{22}^{(i)} - \frac{\nu_a^{(i)}}{E_A^{(i)}}\sigma_{33}^{(i)} + \alpha_A^{(i)}\Delta T, \tag{18.86}$$

$$\varepsilon_{22}^{(i)} = -\frac{\nu_A^{(i)}}{E_A^{(i)}}\sigma_{11}^{(i)} + \frac{\sigma_{22}^{(i)}}{E_T^{(i)}} - \frac{\nu_t^{(i)}}{E_T^{(i)}}\sigma_{33}^{(i)} + \Phi_i\,\sigma_{12}^{(i)} + \alpha_T^{(i)}\Delta T, \tag{18.87}$$

$$\varepsilon_{33}^{(i)} = -\frac{\nu_a^{(i)}}{E_A^{(i)}}\sigma_{11}^{(i)} - \frac{\nu_t^{(i)}}{E_T^{(i)}}\sigma_{22}^{(i)} + \frac{\sigma_{33}^{(i)}}{E_t^{(i)}} + \alpha_t^{(i)}\Delta T, \tag{18.88}$$

$$2\varepsilon_{23}^{(i)} = \frac{\sigma_{23}^{(i)}}{\mu_t^{(i)}}, \tag{18.89}$$

$$2\varepsilon_{13}^{(i)} = \frac{\sigma_{13}^{(i)}}{\mu_a^{(i)}}, \tag{18.90}$$

$$2\varepsilon_{12}^{(i)} = \Phi_i\,\sigma_{22}^{(i)} + \frac{\sigma_{12}^{(i)}}{\mu_A^{(i)}}. \tag{18.91}$$

The ply properties are written here in a form where their values might vary from ply to ply because the damage, which has been homogenised, may be different in each ply. The parameter Φ_i appearing in (18.87) and (18.91) is zero when the ply is undamaged, but it will need to have nonzero values when the effect of discrete ply cracks in the ply is represented by homogenised thermoelastic constants for the damaged ply. There is no simple physical explanation of this fact, other than to emphasise that it is a mathematical construction that ensures the properties of the fully homogenised laminate are identical to those of the same laminate where one of the homogenised plies is replaced by the corresponding ply containing discrete ply cracks.

The stress-strain relations for any layer in the laminate referred to global coordinates are given, for $i = 1, \dots, N+1$, by (18.4)–(18.9) (see the Annex to this chapter for generalised relations involving the property Φ defining the elastic constants S_{ij}).

Because applied displacements of the form (18.13) need to be imposed on the edges of the cracked laminate, the corresponding stresses σ_{11}, σ_{22} and σ_{12}, resulting from an exact elasticity analysis, will be nonuniform. Effective applied axial, transverse and shear stresses are, therefore, defined by the following averages

$$\left.\begin{aligned}\sigma_{\mathrm{A}}&=\frac{1}{4hW}\int_{-W}^{W}\int_{-h}^{h}\sigma_{11}(L,x_2,x_3)\,\mathrm{d}x_2\mathrm{d}x_3\,,\quad \sigma_{\mathrm{T}}=\frac{1}{4hL}\int_{-L}^{L}\int_{-h}^{h}\sigma_{22}(x_1,W,x_3)\,\mathrm{d}x_1\mathrm{d}x_3\,,\\ \tau_{\mathrm{A}}&=\frac{1}{4hW}\int_{-W}^{W}\int_{-h}^{h}\sigma_{12}(L,x_2,x_3)\,\mathrm{d}x_2\mathrm{d}x_3=\frac{1}{4hL}\int_{-L}^{L}\int_{-h}^{h}\sigma_{12}(x_1,W,x_3)\,\mathrm{d}x_1\mathrm{d}x_3\,,\end{aligned}\right\}\tag{18.92}$$

where

$$h=\sum_{i=1}^{n}h_i.\tag{18.93}$$

The effective out-of-plane transverse strain ε_{t} is defined by

$$\varepsilon_{\mathrm{t}}=\frac{1}{4hLW}\int_{-L}^{L}\int_{-W}^{W}u_3(x_1,x_2,h)\mathrm{d}x_1\mathrm{d}x_2.\tag{18.94}$$

The equilibrium equations (18.1)–(18.3), for the stress components $\sigma_{11}^{i},\sigma_{12}^{i}$, etc., referred to the *global* coordinates, must be satisfied for $i = 1,\ldots, N + 1$.

18.9.2 Development of Homogenisation Procedure

Let Ω_{ij} and ω_i denote the thermoelastic constants appearing in stress-strain relations of the form (18.36)–(18.39) for a damaged laminate with discrete ply cracks only in the kth ply. These are regarded as known quantities resulting from a suitable ply cracking model. It is shown in the Annex to this chapter how their values may be calculated from the thermoelastic constants of a damaged laminate. Parameters $\bar{\Omega}_{ij}^{(k)}$ and $\bar{\omega}_i^{(k)}$ are defined by relations of the type

$$\bar{\Omega}_{11}^{(k)}=\frac{1}{h}\sum_{\substack{i=1\\ i\neq k}}^{n}h_i\frac{g_{33}^{(i)}}{\lambda_2^{(i)}},\quad \bar{\Omega}_{66}^{(k)}=\frac{1}{h}\sum_{\substack{i=1\\ i\neq k}}^{n}h_i\frac{g_{11}^{(i)}}{\lambda_2^{(i)}},\quad \text{etc.,}\tag{18.95}$$

that correspond exactly to the definitions given by (18.40)–(18.42) except that the kth term associated with discrete ply cracks is not included in the summations. The values of $\bar{\Omega}_{ij}^{(k)}$ and $\bar{\omega}_i^{(k)}$ are again known values as they involve only plies that are either undamaged or have already been homogenised. Now define parameters $\Omega_{ij}^{(k)}=\Omega_{ij}-\bar{\Omega}_{ij}^{(k)}$ and parameters $\omega_i^{(k)}=\omega_i-\bar{\omega}_i^{(k)}$ which are also known when applying the homogenisation procedure to the kth ply. The parameters $\Omega_{ij}^{(k)}$ and $\omega_i^{(k)}$ correspond to the kth terms that are omitted in the summations of the type (18.95). The first objective is to determine values for constants appearing in relations (18.21)–(18.23) in terms of the known values $\Omega_{ij}^{(k)}$ and $\omega_i^{(k)}$.

Using known values of $\Omega_{11}^{(k)}$, $\Omega_{16}^{(k)}$ and $\Omega_{66}^{(k)}$, it can be shown, on setting

$$\chi_k=\left(\frac{h}{h_k}\right)^2\lambda_2^{(k)}=\frac{1}{\Omega_{11}^{(k)}\Omega_{66}^{(k)}-(\Omega_{16}^{(k)})^2},\tag{18.96}$$

that the following values of the thermoelastic constants for the kth ply may be calculated

$$\begin{aligned}g_{11}^{(k)}&=S_{11}^{(k)}-\frac{S_{12}^{(k)}S_{12}^{(k)}}{S_{22}^{(k)}}=\frac{h_k}{h}\Omega_{66}^{(k)}\chi_k,\\ g_{13}^{(k)}&=S_{16}^{(k)}-\frac{S_{12}^{(k)}S_{26}^{(k)}}{S_{22}^{(k)}}=-\frac{h_k}{h}\Omega_{16}^{(k)}\chi_k,\\ g_{33}^{(k)}&=S_{66}^{(k)}-\frac{S_{26}^{(k)}S_{26}^{(k)}}{S_{22}^{(k)}}=\frac{h_k}{h}\Omega_{11}^{(k)}\chi_k.\end{aligned}\tag{18.97}$$

It is clear that

$$\begin{aligned}\Omega_{12}^{(k)} &= \frac{h_k}{\lambda_2^{(k)}h}\frac{S_{16}^{(k)}S_{26}^{(k)}-S_{12}^{(k)}S_{66}^{(k)}}{S_{22}^{(k)}} = \frac{h_k}{\lambda_2^{(k)}h}\left(g_{13}^{(k)}H_3^{(k)}-g_{33}^{(k)}H_1^{(k)}\right),\\ \Omega_{26}^{(k)} &= \frac{h_k}{\lambda_2^{(k)}h}\frac{S_{12}^{(k)}S_{16}^{(k)}-S_{11}^{(k)}S_{26}^{(k)}}{S_{22}^{(k)}} = \frac{h_k}{\lambda_2^{(k)}h}\left(g_{13}^{(k)}H_1^{(k)}-g_{11}^{(k)}H_3^{(k)}\right).\end{aligned} \tag{18.98}$$

On using (18.96) and (18.97)

$$\begin{aligned}\Omega_{11}^{(k)}H_1^{(k)}+\Omega_{16}^{(k)}H_3^{(k)} &= -\,\Omega_{12}^{(k)},\\ \Omega_{16}^{(k)}H_1^{(k)}+\Omega_{66}^{(k)}H_3^{(k)} &= -\,\Omega_{26}^{(k)},\end{aligned} \tag{18.99}$$

leading to the values

$$\begin{aligned}H_1^{(k)} &= \frac{S_{12}^{(k)}}{S_{22}^{(k)}} = \frac{\Omega_{16}^{(k)}\Omega_{26}^{(k)}-\Omega_{12}^{(k)}\Omega_{66}^{(k)}}{\Omega_{11}^{(k)}\Omega_{66}^{(k)}-\Omega_{16}^{(k)}\Omega_{16}^{(k)}} = \chi_k\left(\Omega_{16}^{(k)}\Omega_{26}^{(k)}-\Omega_{12}^{(k)}\Omega_{66}^{(k)}\right),\\ H_3^{(k)} &= \frac{S_{26}^{(k)}}{S_{22}^{(k)}} = \frac{\Omega_{12}^{(k)}\Omega_{16}^{(k)}-\Omega_{11}^{(k)}\Omega_{26}^{(k)}}{\Omega_{11}^{(k)}\Omega_{66}^{(k)}-\Omega_{16}^{(k)}\Omega_{16}^{(k)}} = \chi_k\left(\Omega_{12}^{(k)}\Omega_{16}^{(k)}-\Omega_{11}^{(k)}\Omega_{26}^{(k)}\right).\end{aligned} \tag{18.100}$$

It is also clear that

$$\Omega_{13}^{(k)} = \frac{h_k}{h}\frac{\lambda_1^{(k)}}{\lambda_2^{(k)}} = \frac{h_k}{h}\frac{g_{13}^{(k)}g_{23}^{(k)}-g_{12}^{(k)}g_{33}^{(k)}}{\lambda_2^{(k)}},\quad \Omega_{36}^{(k)} = \frac{h_k}{h}\frac{\lambda_3^{(k)}}{\lambda_2^{(k)}} = \frac{h_k}{h}\frac{g_{12}^{(k)}g_{13}^{(k)}-g_{11}^{(k)}g_{23}^{(k)}}{\lambda_2^{(k)}}, \tag{18.101}$$

which may be written in the form of linear algebraic equations for the unknowns $g_{12}^{(k)}$ and $g_{23}^{(k)}$

$$\begin{aligned}-g_{33}^{(k)}g_{12}^{(k)}+g_{13}^{(k)}g_{23}^{(k)} &= \frac{\lambda_2^{(k)}h}{h_k}\Omega_{13}^{(k)},\\ g_{13}^{(k)}g_{12}^{(k)}-g_{11}^{(k)}g_{23}^{(k)} &= \frac{\lambda_2^{(k)}h}{h_k}\Omega_{36}^{(k)}.\end{aligned} \tag{18.102}$$

On solving and making use of (18.97)

$$\begin{aligned}g_{12}^{(k)} &= -\frac{h}{h_k}\left(g_{11}^{(k)}\Omega_{13}^{(k)}+g_{13}^{(k)}\Omega_{36}^{(k)}\right) = \chi_k\left(\Omega_{16}^{(k)}\Omega_{36}^{(k)}-\Omega_{13}^{(k)}\Omega_{66}^{(k)}\right),\\ g_{23}^{(k)} &= -\frac{h}{h_k}\left(g_{33}^{(k)}\Omega_{36}^{(k)}+g_{13}^{(k)}\Omega_{13}^{(k)}\right) = \chi_k\left(\Omega_{13}^{(k)}\Omega_{16}^{(k)}-\Omega_{11}^{(k)}\Omega_{36}^{(k)}\right).\end{aligned} \tag{18.103}$$

As

$$\omega_1^{(k)} = \frac{h_k}{\lambda_2^{(k)}h}\left(g_{33}^{(k)}\bar{V}_1^{(k)}-g_{13}^{(k)}\bar{V}_3^{(k)}\right),\qquad \omega_6^{(k)} = \frac{h_k}{\lambda_2^{k}h}\left(g_{11}^{(k)}\bar{V}_3^{(k)}-g_{13}^{(k)}\bar{V}_1^{(k)}\right), \tag{18.104}$$

it follows that the values of $\bar{V}_1^{(k)}$ and $\bar{V}_3^{(k)}$ satisfy the following linear equations:

$$\begin{aligned}g_{33}^{(k)}\bar{V}_1^{(k)}-g_{13}^{(k)}\bar{V}_3^{(k)} &= \frac{\lambda_2^{(k)}h}{h_k}\omega_1^{(k)},\\ -g_{13}^{(k)}\bar{V}_1^{(k)}+g_{11}^{(k)}\bar{V}_3^{(k)} &= \frac{\lambda_2^{(k)}h}{h_k}\omega_6^{(k)}.\end{aligned} \tag{18.105}$$

On using (18.97), the solution may be expressed in the form

$$\bar{V}_1^{(k)} = \chi_k\left(\Omega_{66}^{(k)}\omega_1^{(k)} - \Omega_{16}^{(k)}\omega_6^{(k)}\right),$$
$$\bar{V}_3^{(k)} = \chi_k\left(\Omega_{11}^{(k)}\omega_6^{(k)} - \Omega_{16}^{(k)}\omega_1^{(k)}\right). \tag{18.106}$$

On writing

$$\Omega_{22}^{(k)} = \frac{h_k}{h}\frac{1}{\lambda_2^{(k)}S_{22}^{(k)}}\left[\lambda_2^{(k)} + \frac{S_{12}^{(k)}}{S_{22}^{(k)}}\left(S_{12}^{(k)}S_{66}^{(k)} - S_{16}^{(k)}S_{26}^{(k)}\right) + \frac{S_{26}^k}{S_{22}^k}\left(S_{11}^{(k)}S_{26}^{(k)} - S_{12}^{(k)}S_{16}^{(k)}\right)\right], \tag{18.107}$$

in the form, on using (18.18) and (18.30),

$$\Omega_{22}^{(k)} = \frac{h_k}{h}\left[\frac{1}{S_{22}^{(k)}} + \frac{H_1^{(k)}}{\lambda_2^{(k)}}\left(H_1^{(k)}g_{33}^{(k)} - H_3^{(k)}g_{13}^{(k)}\right) + \frac{H_3^{(k)}}{\lambda_2^{(k)}}\left(H_3^{(k)}g_{11}^{(k)} - H_1^{(k)}g_{13}^{(k)}\right)\right], \tag{18.108}$$

it can be shown using (18.96) and (18.97) that

$$\Omega_{22}^{(k)} = \frac{h_k}{h}\frac{1}{S_{22}^{(k)}} + H_1^{(k)}\left(H_1^{(k)}\Omega_{11}^{(k)} + H_3^{(k)}\Omega_{16}^{(k)}\right) + H_3^{(k)}\left(H_3^{(k)}\Omega_{66}^{(k)} + H_1^{(k)}\Omega_{16}^{(k)}\right). \tag{18.109}$$

On using (18.100),

$$\Omega_{22}^{(k)} = \frac{h_k}{h}\frac{1}{S_{22}^{(k)}} - \chi_k\left(H_1^{(k)}\Omega_{12}^{(k)} + H_3^{(k)}\Omega_{26}^{(k)}\right)\left(\Omega_{11}^{(k)}\Omega_{66}^{(k)} - \Omega_{16}^{(k)}\Omega_{16}^{(k)}\right), \tag{18.110}$$

leading to, on using (18.96),

$$\Omega_{22}^{(k)} = \frac{h_k}{h}\frac{1}{S_{22}^k} - \chi_k\left(2\Omega_{12}^{(k)}\Omega_{16}^{(k)}\Omega_{26}^{(k)} - (\Omega_{12}^{(k)})^2\Omega_{66}^{(k)} - \Omega_{11}^{(k)}(\Omega_{26}^{(k)})^2\right). \tag{18.111}$$

It then follows that

$$S_{22}^{(k)} = \frac{h_k}{h}\left[\Omega_{22}^{(k)} + \chi_k\left(2\Omega_{12}^{(k)}\Omega_{16}^{(k)}\Omega_{26}^{(k)} - \Omega_{11}^{(k)}(\Omega_{26}^{(k)})^2 - \Omega_{66}^{(k)}(\Omega_{12}^{(k)})^2\right)\right]^{-1}. \tag{18.112}$$

As

$$\Omega_{23}^{(k)} = -\frac{h_k}{h}\frac{1}{\lambda_2^{(k)}}\left(\lambda_1^{(k)}\frac{S_{12}^{(k)}}{S_{22}^{(k)}} + \lambda_2^{(k)}\frac{S_{23}^{(k)}}{S_{22}^{(k)}} + \lambda_3^{(k)}\frac{S_{26}^{(k)}}{S_{22}^{(k)}}\right)$$
$$= -\frac{h_k}{h}\left(\frac{\lambda_1^{(k)}}{\lambda_2^{(k)}}H_1^{(k)} + H_2^{(k)} + \frac{\lambda_3^{(k)}}{\lambda_2^{(k)}}H_3^{(k)}\right), \tag{18.113}$$

it follows, on using (18.27), that

$$\Omega_{23}^{(k)} = -\frac{h_k}{h}\left(\left(g_{13}^{(k)}g_{23}^{(k)} - g_{12}^{(k)}g_{33}^{(k)}\right)\frac{H_1^{(k)}}{\lambda_2^{(k)}} + H_2^{(k)} + \left(g_{12}^{(k)}g_{13}^{(k)} - g_{11}^{(k)}g_{23}^{(k)}\right)\frac{H_3^{(k)}}{\lambda_2^{(k)}}\right). \tag{18.114}$$

On using (18.96), (18.97) and (18.103),

$$\Omega_{23}^{(k)} = -\left(\frac{h_k}{h}\right)^2\chi_k\left(\Omega_{13}^{(k)}\frac{H_1^{(k)}}{\lambda_2^{(k)}} + \frac{h}{h_k\chi_k}H_2^{(k)} + \Omega_{36}^{(k)}\frac{H_3^{(k)}}{\lambda_2^{(k)}}\right). \tag{18.115}$$

The use of (18.97), (18.100) and (18.103) then leads to

$$H_2^{(k)} = -\frac{h}{h_k}\left[\Omega_{23}^{(k)} + \chi_k\left(\Omega_{13}^{(k)}\Omega_{16}^{(k)}\Omega_{26}^{(k)} - \Omega_{12}^{(k)}\Omega_{13}^{(k)}\Omega_{66}^{(k)} + \Omega_{12}^{(k)}\Omega_{16}^{(k)}\Omega_{36}^{(k)} - \Omega_{11}^{(k)}\Omega_{26}^{(k)}\Omega_{36}^{(k)}\right)\right]. \tag{18.116}$$

As

$$\Omega_{33}^{(k)} = \frac{h_k}{h}\left(\frac{\lambda_1^{(k)}}{\lambda_2^{(k)}} g_{12}^{(k)} + g_{22}^{(k)} + \frac{\lambda_3^{(k)}}{\lambda_2^{(k)}} g_{23}^{(k)}\right), \tag{18.117}$$

it follows, on using (18.101), that

$$\Omega_{33}^{(k)} = \Omega_{13}^{(k)} g_{12}^{(k)} + \frac{h_k}{h} g_{22}^{(k)} + \Omega_{36}^{(k)} g_{23}^{(k)}. \tag{18.118}$$

It then follows, on using (18.103), that

$$\Omega_{33}^{(k)} = \frac{h_k}{h} g_{22}^{(k)} + \chi_k \left(2\Omega_{13}^{(k)}\Omega_{16}^{(k)}\Omega_{36}^{(k)} - \Omega_{66}^{(k)}(\Omega_{13}^{(k)})^2 - \Omega_{11}^{(k)}(\Omega_{36}^{(k)})^2\right), \tag{18.119}$$

leading to the result

$$g_{22}^{(k)} = \frac{h}{h_k}\left[\Omega_{33}^{(k)} - \chi_k \left(2\Omega_{13}^{(k)}\Omega_{16}^{(k)}\Omega_{36}^{(k)} - \Omega_{66}^{(k)}(\Omega_{13}^{(k)})^2 - \Omega_{11}^{(k)}(\Omega_{36}^{(k)})^2\right)\right]. \tag{18.120}$$

As

$$\omega_3^{(k)} = \frac{h_k}{h}\left(\frac{\lambda_1^{(k)}}{\lambda_2^{(k)}} \bar{V}_1^{(k)} + \bar{V}_2^{(k)} + \frac{\lambda_3^{(k)}}{\lambda_2^{(k)}} \bar{V}_3^{(k)}\right), \tag{18.121}$$

it follows, on using (18.101), that

$$\omega_3^{(k)} = \Omega_{13}^{(k)} \bar{V}_1^{(k)} + \frac{h_k}{h} \bar{V}_2^{(k)} + \Omega_{36}^{(k)} \bar{V}_3^{(k)}. \tag{18.122}$$

The use of (18.106) then leads to the result

$$\bar{V}_2^{(k)} = \frac{h}{h_k}\left[\omega_3^{(k)} + \chi_k \left(\Omega_{16}^{(k)}\Omega_{36}^{(k)} - \Omega_{13}^{(k)}\Omega_{66}^{(k)}\right)\omega_1^{(k)} + \chi_k \left(\Omega_{13}^{(k)}\Omega_{16}^{(k)} - \Omega_{11}^{(k)}\Omega_{36}^{(k)}\right)\omega_6^{(k)}\right]. \tag{18.123}$$

Finally, because

$$\omega_2^{(k)} = \frac{h_k}{h}\frac{1}{\lambda_2^{(k)}}\left[\lambda_2^{(k)}\frac{V_2^{(k)}}{S_{22}^{(k)}} + \frac{S_{12}^{(k)}}{S_{22}^{(k)}}\left(g_{13}^{(k)}\bar{V}_3^{(k)} - g_{33}^{(k)}\bar{V}_1^{(k)}\right) + \frac{S_{26}^{(k)}}{S_{22}^{(k)}}\left(g_{13}^{(k)}\bar{V}_1^{(k)} - g_{11}^{(k)}\bar{V}_3^{(k)}\right)\right], \tag{18.124}$$

it follows from (18.104) that

$$\omega_2^{(k)} = \frac{h_k}{h}\frac{V_2^{(k)}}{S_{22}^{(k)}} - \frac{S_{12}^{(k)}}{S_{22}^{(k)}}\omega_1^{(k)} - \frac{S_{26}^{(k)}}{S_{22}^{(k)}}\omega_6^{(k)}. \tag{18.125}$$

On using (18.100),

$$\omega_2^{(k)} = \frac{h_k}{h}\frac{V_2^{(k)}}{S_{22}^{(k)}} - \chi_k \left(\Omega_{16}^{(k)}\Omega_{26}^{(k)} - \Omega_{12}^{(k)}\Omega_{66}^{(k)}\right)\omega_1^{(k)} - \chi_k \left(\Omega_{12}^{(k)}\Omega_{16}^{(k)} - \Omega_{11}^{(k)}\Omega_{26}^{(k)}\right)\omega_6^{(k)}, \tag{18.126}$$

leading to the result

$$V_2^{(k)} = S_{22}^{(k)}\frac{h}{h_k}\left[\omega_2^{(k)} + \chi_k \left\{\left(\Omega_{16}^{(k)}\Omega_{26}^{(k)} - \Omega_{12}^{(k)}\Omega_{66}^{(k)}\right)\omega_1^{(k)} + \left(\Omega_{12}^{(k)}\Omega_{16}^{(k)} - \Omega_{11}^{(k)}\Omega_{26}^{(k)}\right)\omega_6^{(k)}\right\}\right]. \tag{18.127}$$

To calculate the required effective thermoelastic constants of a homogenised ply from the reduced coefficients given previously, use is made of (18.18), which is rewritten in the form

$$
\begin{aligned}
&S_{12}^{(k)} = H_1^{(k)} S_{22}^{(k)}, && S_{23}^{(k)} = H_2^{(k)} S_{22}^{(k)}, && S_{26}^{(k)} = H_3^{(k)} S_{22}^{(k)},\\
&S_{11}^{(k)} = g_{11}^{(k)} + \frac{(S_{12}^{(k)})^2}{S_{22}^{(k)}}, && S_{13}^{(k)} = g_{12}^{(k)} + \frac{S_{12}^{(k)} S_{23}^{(k)}}{S_{22}^{(k)}}, && S_{16}^{(k)} = g_{13}^{(k)} + \frac{S_{12}^{(k)} S_{26}^{(k)}}{S_{22}^{(k)}},\\
&S_{33}^{(k)} = g_{22}^{(k)} + \frac{(S_{23}^{(k)})^2}{S_{22}^{(k)}}, && S_{36}^{(k)} = g_{23}^{(k)} + \frac{S_{23}^{(k)} S_{26}^{(k)}}{S_{22}^{(k)}}, && S_{66}^{(k)} = g_{33}^{(k)} + \frac{(S_{26}^{(k)})^2}{S_{22}^{(k)}},\\
&V_1^{(k)} = \bar{V}_1^{(k)} + \frac{S_{12}^{(k)}}{S_{22}^{(k)}} V_2^{(k)}, && V_3^{(k)} = \bar{V}_2^{(k)} + \frac{S_{23}^{(k)}}{S_{22}^{(k)}} V_2^{(k)}, && V_6^{(k)} = \bar{V}_3^{(k)} + \frac{S_{26}^{(k)}}{S_{22}^{(k)}} V_2^{(k)}.
\end{aligned}
\tag{18.128}
$$

As the right-hand sides of all relations in (18.128) can be calculated, using (18.112) and (18.127), it is clear that all the required thermoelastic constants for the homogenised properties of the cracked kth ply (defined for global axes) have now been derived in terms of damaged laminate properties and the properties of the other plies in an undamaged or homogenised state.

The final stage of the homogenisation procedure is to extract from the results (18.128) the corresponding values of the thermoelastic constants (appearing in (18.86)–(18.91)) of the kth ply that has now been homogenised. As this ply is in a 90° orientation following transformation it follows from relations (7)–(23) in the Annex to this chapter that on setting $\phi = \pi/2$ (i.e. $m = 0$ and $n = 1$)

$$
\begin{aligned}
&E_T = \frac{1}{S_{11}}, && E_A = \frac{1}{S_{22}}, && E_t = \frac{1}{S_{33}},\\
&\nu_a = -E_A S_{23}, && \nu_t = -E_T S_{13}, && \nu_A = -E_A S_{12},\\
&\mu_a = \frac{1}{S_{44}}, \quad \mu_t = \frac{1}{S_{55}}, && \mu_A = \frac{1}{S_{66}}, && \Phi = -S_{16},\\
&\alpha_T = V_1, && \alpha_A = V_2, && \alpha_t = V_3,
\end{aligned}
\tag{18.129}
$$

providing values of the homogenised properties (referred to local coordinates) of the ply containing the discrete set of ply cracks.

The accuracy of methods of estimating the effective thermoelastic constants of general symmetrical laminates is dependent upon accuracy of the associated stress-transfer analysis described in Sections 18.6 and 18.7. Such analyses have been validated by comparing predictions using the analysis of this chapter with those obtained through participation in international exercises (WWFE I and III, see references in Recommended reading). Model verification has been achieved by comparing predictions with those based on other methods such as variational analysis and finite element analyses, as seen in papers given in Recommended reading. This chapter, together with Chapter 10, has shown that the description of the deformation of general symmetric laminates is quite complex generating a number of thermoelastic constants associated with general in-plane loading. Rather than present graphical illustrations of the dependence of the many elastic constants on ply crack density, readers are referred to the associated software, and examples which are available at the Wiley website [6].

Annex to Chapter 18: Defining the thermoelastic constants of a homogenised laminate

For a right-handed set of reference coordinates x_1, x_2 and x_3, a right-handed second set of coordinates x_1', x_2' and x_3' is obtained by rotating the reference set of coordinates about the x_3-axis by an angle ϕ. The rotation is clockwise when viewing along the positive direction of the x_3-axis. Expressions are first derived for the stress components referred to local coordinates (defined with respect to the fibre direction) in terms of the stress components for global coordinates as follows for each of the i layers

$$\begin{aligned}
\sigma'_{11} &= m^2\sigma_{11} + n^2\sigma_{22} + 2mn\sigma_{12}, \\
\sigma'_{22} &= n^2\sigma_{11} + m^2\sigma_{22} - 2mn\sigma_{12}, \\
\sigma'_{33} &= \sigma_{33}, \\
\sigma'_{23} &= m\sigma_{23} - n\sigma_{13} = \sigma'_{32}, \\
\sigma'_{13} &= n\sigma_{23} + m\sigma_{13} = \sigma'_{31}, \\
\sigma'_{12} &= -mn\sigma_{11} + mn\sigma_{22} + \left(m^2 - n^2\right)\sigma_{12} = \sigma'_{21},
\end{aligned} \tag{1}$$

where $m = \cos\phi$ and $n = \sin\phi$. The strain components referred to local coordinates can be expressed in terms of global stress components by substituting (1) into the following homogenised stress-strain relations for the ith layer, expressed in terms of stress and strain components referred to local coordinates,

$$\varepsilon'_{11} = \frac{1}{E_A}\sigma'_{11} - \frac{\nu_A}{E_A}\sigma'_{22} - \frac{\nu_a}{E_A}\sigma'_{33} + \alpha_A\Delta T, \tag{2}$$

$$\varepsilon'_{22} = -\frac{\nu_A}{E_A}\sigma'_{11} + \frac{1}{E_T}\sigma'_{22} - \frac{\nu_t}{E_T}\sigma'_{33} + \Phi\sigma'_{12} + \alpha_T\Delta T, \tag{3}$$

$$\varepsilon'_{33} = -\frac{\nu_a}{E_A}\sigma'_{11} - \frac{\nu_t}{E_T}\sigma'_{22} + \frac{1}{E_t}\sigma'_{33} + \alpha_t\Delta T, \tag{4}$$

$$\varepsilon'_{23} = \frac{1}{2\mu_t}\sigma'_{23}, \quad \varepsilon'_{13} = \frac{1}{2\mu_a}\sigma'_{13}, \quad \varepsilon'_{12} = \frac{1}{2}\Phi\sigma'_{22} + \frac{1}{2\mu_A}\sigma'_{12}. \tag{5}$$

It should be noted that in (3) and $(5)_3$ the quantity Φ is a shear coupling parameter that has been introduced that must be included in the analysis if a cracked ply is to be homogenised exactly with respect to all thermoelastic constants. For an undamaged ply the value of Φ is zero.

The global strain components may be expressed in terms of global stress components by substituting (2)–(5) into the following relations, having made use of (1)

$$\begin{aligned}
\varepsilon_{11} &= m^2\varepsilon'_{11} + n^2\varepsilon'_{22} - 2mn\varepsilon'_{12}, \\
\varepsilon_{22} &= n^2\varepsilon'_{11} + m^2\varepsilon'_{22} + 2mn\varepsilon'_{12}, \\
\varepsilon_{33} &= \varepsilon'_{33}, \\
\varepsilon_{23} &= m\varepsilon'_{23} + n\varepsilon'_{13} = \varepsilon_{32}, \\
\varepsilon_{13} &= -n\varepsilon'_{23} + m\varepsilon'_{13} = \varepsilon_{31}, \\
\varepsilon_{12} &= mn\varepsilon'_{11} - mn\varepsilon'_{22} + \left(m^2 - n^2\right)\varepsilon'_{12} = \varepsilon_{21},
\end{aligned} \tag{6}$$

The resulting stress-strain equations relating global strain components to global stress components are given by (18.4)–(18.9) where the coefficients in the stress-strain relations are related to the thermoelastic constants of a layer as follows:

$$S_{11} = \frac{m^4}{E_A} + \frac{n^4}{E_T} + m^2n^2\left(\frac{1}{\mu_A} - \frac{2\nu_A}{E_A}\right) - 2mn^3\Phi, \tag{7}$$

$$S_{12} = m^2n^2\left(\frac{1}{E_A} + \frac{1}{E_T} - \frac{1}{\mu_A}\right) - \left(m^4 + n^4\right)\frac{\nu_A}{E_A} - mn\left(m^2 - n^2\right)\Phi, \tag{8}$$

$$S_{13} = -m^2\frac{\nu_a}{E_A} - n^2\frac{\nu_t}{E_T}, \tag{9}$$

$$S_{16} = \frac{2m^3n}{E_A} - \frac{2mn^3}{E_T} + mn\left(m^2 - n^2\right)\left(\frac{2\nu_A}{E_A} - \frac{1}{\mu_A}\right) + \left(3m^2 - n^2\right)n^2\Phi, \tag{10}$$

$$S_{22} = \frac{n^4}{E_A} + \frac{m^4}{E_T} + m^2n^2\left(\frac{1}{\mu_A} - \frac{2\nu_A}{E_A}\right) + 2m^3n\Phi, \tag{11}$$

$$S_{23} = -m^2\frac{\nu_t}{E_T} - n^2\frac{\nu_a}{E_A}, \tag{12}$$

$$S_{26} = mn\left(m^2 - n^2\right)\left(\frac{1}{\mu_A} - \frac{2\nu_A}{E_A}\right) + \frac{2mn^3}{E_A} - \frac{2m^3n}{E_T} + m^2\left(m^2 - 3n^2\right)\Phi, \tag{13}$$

$$S_{33} = \frac{1}{E_t}, \tag{14}$$

$$S_{36} = -2mn\left(\frac{\nu_a}{E_A} - \frac{\nu_t}{E_T}\right), \tag{15}$$

$$S_{44} = \frac{m^2}{\mu_t} + \frac{n^2}{\mu_a}, \tag{16}$$

$$S_{45} = mn\left(\frac{1}{\mu_a} - \frac{1}{\mu_t}\right), \tag{17}$$

$$S_{55} = \frac{m^2}{\mu_a} + \frac{n^2}{\mu_t}, \tag{18}$$

$$S_{66} = \frac{\left(m^2 - n^2\right)^2}{\mu_A} + 4m^2n^2\left(\frac{1}{E_A} + \frac{1}{E_T} + \frac{2\nu_A}{E_A}\right) - 4mn\left(m^2 - n^2\right)\Phi, \tag{19}$$

$$V_1 = m^2\alpha_A + n^2\alpha_T, \tag{20}$$

$$V_2 = m^2\alpha_T + n^2\alpha_A, \tag{21}$$

$$V_3 = \alpha_t, \tag{22}$$

$$V_6 = 2mn(\alpha_A - \alpha_T), \tag{23}$$

where $m = \cos\phi$ and $n = \sin\phi$.

Defining the thermoelastic constants of a damaged laminate

If the distribution of damage in each ply, when modelled discretely, is effectively uniform at the macroscopic level, then the effective stress-strain relations of the damaged laminate may be expressed in the same form as that for an uncracked laminate, namely

$$
\begin{aligned}
\varepsilon_A &= \frac{1}{E_A}\sigma_A - \frac{\nu_A}{E_A}\sigma_T - \frac{\nu_a}{E_A}\sigma_t - \frac{\lambda_A}{\mu_A}\tau_A + \alpha_A \Delta T, \\
\varepsilon_T &= -\frac{\nu_A}{E_A}\sigma_A + \frac{1}{E_T}\sigma_T - \frac{\nu_t}{E_T}\sigma_t - \frac{\lambda_T}{\mu_A}\tau_A + \alpha_T \Delta T, \\
\varepsilon_t &= -\frac{\nu_a}{E_A}\sigma_A - \frac{\nu_t}{E_T}\sigma_T + \frac{1}{E_t}\sigma_t - \frac{\lambda_t}{\mu_A}\tau_A + \alpha_t \Delta T, \\
\gamma_A &= -\frac{\lambda_A}{\mu_A}\sigma_A - \frac{\lambda_T}{\mu_A}\sigma_T - \frac{\lambda_t}{\mu_A}\sigma_t + \frac{1}{\mu_A}\tau_A + \alpha_S \Delta T.
\end{aligned} \tag{24}
$$

In (24), the parameters σ_A, σ_T, σ_t and τ_A denote the in-plane effective applied axial, transverse and shear stresses. For specified values of σ_A, σ_T, σ_t, τ_A and ΔT, the quantities ε_A, ε_T, ε_t and γ_A are the in-plane axial, transverse and shear strains of the damaged laminate, respectively, and E_A, E_T, E_t, ν_A, ν_a, ν_t and μ_A denote the corresponding effective thermoelastic constants of the cracked laminate. The parameters λ_A, λ_T and λ_t are ratios indicating the degree of shear coupling. These parameters are zero for the special case of cross-ply laminates. The parameter α_S is a thermal expansion coefficient governing the amount of shear strain that can arise when the temperature is changed.

For damaged laminates where there are discrete cracks in just one ply, assumed to be the kth, the values of the thermoelastic constants appearing in (24) are known, having estimated them using a suitable ply cracking model. The objective is to invert relations (24) so that they are in the form of (18.36)–(18.39), namely,

$$
\begin{aligned}
\sigma_A &= \Omega_{11}\varepsilon_A + \Omega_{12}\varepsilon_T + \Omega_{13}\sigma_t + \Omega_{16}\gamma_A - \omega_1 \Delta T, \\
\sigma_T &= \Omega_{12}\varepsilon_A + \Omega_{22}\varepsilon_T + \Omega_{23}\sigma_t + \Omega_{26}\gamma_A - \omega_2 \Delta T, \\
\varepsilon_t &= -\Omega_{13}\varepsilon_A - \Omega_{23}\varepsilon_T + \Omega_{33}\sigma_t - \Omega_{36}\gamma_A + \omega_3 \Delta T, \\
\tau_A &= \Omega_{16}\varepsilon_A + \Omega_{26}\varepsilon_T + \Omega_{36}\sigma_t + \Omega_{66}\gamma_A - \omega_6 \Delta T,
\end{aligned} \tag{25}
$$

thus enabling the calculation of the values of the thermoelastic constants Ω_{ij} that are needed to apply the homogenisation procedure.

From $(24)_4$

$$
\tau_A = \lambda_A\sigma_A + \lambda_T\sigma_T + \lambda_t\sigma_t + \mu_A\gamma_A - \mu_A\alpha_S\Delta T. \tag{26}
$$

Substitution into the other relations of (24) leads to

$$
\begin{aligned}
\varepsilon_A &= \left(\frac{1}{E_A} - \frac{\lambda_A^2}{\mu_A}\right)\sigma_A - \left(\frac{\nu_A}{E_A} + \frac{\lambda_A\lambda_T}{\mu_A}\right)\sigma_T - \left(\frac{\nu_a}{E_A} + \frac{\lambda_A\lambda_t}{\mu_A}\right)\sigma_t - \lambda_A\gamma_A + \left(\alpha_A + \lambda_A\alpha_S\right)\Delta T, \\
\varepsilon_T &= -\left(\frac{\nu_A}{E_A} + \frac{\lambda_A\lambda_T}{\mu_A}\right)\sigma_A + \left(\frac{1}{E_T} - \frac{\lambda_T^2}{\mu_A}\right)\sigma_T - \left(\frac{\nu_t}{E_T} + \frac{\lambda_T\lambda_t}{\mu_A}\right)\sigma_t - \lambda_T\gamma_A + \left(\alpha_T + \lambda_T\alpha_S\right)\Delta T, \\
\varepsilon_t &= -\left(\frac{\nu_a}{E_A} + \frac{\lambda_A\lambda_t}{\mu_A}\right)\sigma_A - \left(\frac{\nu_t}{E_T} + \frac{\lambda_T\lambda_t}{\mu_A}\right)\sigma_T + \left(\frac{1}{E_t} - \frac{\lambda_t^2}{\mu_A}\right)\sigma_t - \lambda_t\gamma_A + \left(\alpha_t + \lambda_t\alpha_S\right)\Delta T.
\end{aligned} \tag{27}
$$

The first two equations of (27) will now be solved for the applied stresses σ_A and σ_T. They are first written in the form

$$\left(\frac{1}{E_{\mathrm{A}}}-\frac{\lambda_{\mathrm{A}}^{2}}{\mu_{\mathrm{A}}}\right)\sigma_{\mathrm{A}}-\left(\frac{\nu_{\mathrm{A}}}{E_{\mathrm{A}}}+\frac{\lambda_{\mathrm{A}}\lambda_{\mathrm{T}}}{\mu_{\mathrm{A}}}\right)\sigma_{\mathrm{T}}=\varepsilon_{\mathrm{A}}+\left(\frac{\nu_{a}}{E_{\mathrm{A}}}+\frac{\lambda_{\mathrm{A}}\lambda_{\mathrm{t}}}{\mu_{\mathrm{A}}}\right)\sigma_{\mathrm{t}}+\lambda_{\mathrm{A}}\gamma_{\mathrm{A}}-\left(\alpha_{\mathrm{A}}+\lambda_{\mathrm{A}}\alpha_{S}\right)\Delta T,$$
$$-\left(\frac{\nu_{\mathrm{A}}}{E_{\mathrm{A}}}+\frac{\lambda_{\mathrm{A}}\lambda_{\mathrm{T}}}{\mu_{\mathrm{A}}}\right)\sigma_{\mathrm{A}}+\left(\frac{1}{E_{\mathrm{T}}}-\frac{\lambda_{\mathrm{T}}^{2}}{\mu_{\mathrm{A}}}\right)\sigma_{\mathrm{T}}=\varepsilon_{\mathrm{T}}+\left(\frac{\nu_{\mathrm{t}}}{E_{\mathrm{T}}}+\frac{\lambda_{\mathrm{T}}\lambda_{\mathrm{t}}}{\mu_{\mathrm{A}}}\right)\sigma_{\mathrm{t}}+\lambda_{\mathrm{T}}\gamma_{\mathrm{A}}-\left(\alpha_{\mathrm{T}}+\lambda_{\mathrm{T}}\alpha_{S}\right)\Delta T,\tag{28}$$

leading to the solution

$$\begin{aligned}\sigma_{\mathrm{A}}&=\Omega_{11}\varepsilon_{\mathrm{A}}+\Omega_{12}\varepsilon_{\mathrm{T}}+\Omega_{13}\sigma_{\mathrm{t}}+\Omega_{16}\gamma_{\mathrm{A}}-\omega_{1}\Delta T,\\ \sigma_{\mathrm{T}}&=\Omega_{12}\varepsilon_{\mathrm{A}}+\Omega_{22}\varepsilon_{\mathrm{T}}+\Omega_{23}\sigma_{\mathrm{t}}+\Omega_{26}\gamma_{\mathrm{A}}-\omega_{2}\Delta T,\end{aligned}\tag{29}$$

where

$$\begin{aligned}\Omega_{11}&=\frac{1}{\Delta}\left(\frac{1}{E_{\mathrm{T}}}-\frac{\lambda_{\mathrm{T}}^{2}}{\mu_{\mathrm{A}}}\right),\quad \Omega_{12}=\frac{1}{\Delta}\left(\frac{\nu_{\mathrm{A}}}{E_{\mathrm{A}}}+\frac{\lambda_{\mathrm{A}}\lambda_{\mathrm{T}}}{\mu_{\mathrm{A}}}\right),\quad \Omega_{16}=\frac{1}{\Delta}\left(\frac{\lambda_{\mathrm{A}}}{E_{\mathrm{T}}}+\frac{\nu_{\mathrm{A}}\lambda_{\mathrm{T}}}{E_{\mathrm{A}}}\right),\\ \Omega_{22}&=\frac{1}{\Delta}\left(\frac{1}{E_{\mathrm{A}}}-\frac{\lambda_{\mathrm{A}}^{2}}{\mu_{\mathrm{A}}}\right),\quad \Omega_{26}=\frac{1}{\Delta}\left(\frac{\lambda_{\mathrm{T}}}{E_{\mathrm{A}}}+\frac{\nu_{\mathrm{A}}\lambda_{\mathrm{A}}}{E_{\mathrm{A}}}\right),\\ \Omega_{13}&=\frac{1}{\Delta}\left[\frac{\nu_{a}}{E_{\mathrm{A}}}\left(\frac{1}{E_{\mathrm{T}}}-\frac{\lambda_{\mathrm{T}}^{2}}{\mu_{\mathrm{A}}}\right)+\frac{\nu_{\mathrm{t}}}{E_{\mathrm{T}}}\left(\frac{\nu_{\mathrm{A}}}{E_{\mathrm{A}}}+\frac{\lambda_{\mathrm{A}}\lambda_{\mathrm{T}}}{\mu_{\mathrm{A}}}\right)+\frac{\lambda_{\mathrm{t}}}{\mu_{\mathrm{A}}}\left(\frac{\lambda_{\mathrm{A}}}{E_{\mathrm{T}}}+\frac{\nu_{\mathrm{A}}\lambda_{\mathrm{T}}}{E_{\mathrm{A}}}\right)\right],\\ \Omega_{23}&=\frac{1}{\Delta}\left[\frac{\nu_{\mathrm{t}}}{E_{\mathrm{T}}}\left(\frac{1}{E_{\mathrm{A}}}-\frac{\lambda_{\mathrm{A}}^{2}}{\mu_{\mathrm{A}}}\right)+\frac{\nu_{a}}{E_{\mathrm{A}}}\left(\frac{\nu_{\mathrm{A}}}{E_{\mathrm{A}}}+\frac{\lambda_{\mathrm{A}}\lambda_{\mathrm{T}}}{\mu_{\mathrm{A}}}\right)+\frac{\lambda_{\mathrm{t}}}{\mu_{\mathrm{A}}}\frac{\lambda_{\mathrm{T}}+\nu_{\mathrm{A}}\lambda_{\mathrm{A}}}{E_{\mathrm{A}}}\right],\end{aligned}\tag{30}$$

$$\begin{aligned}\omega_{1}&=\frac{1}{\Delta}\left[\left(\frac{1}{E_{\mathrm{T}}}-\frac{\lambda_{\mathrm{T}}^{2}}{\mu_{\mathrm{A}}}\right)\alpha_{\mathrm{A}}+\left(\frac{\nu_{\mathrm{A}}}{E_{\mathrm{A}}}+\frac{\lambda_{\mathrm{A}}\lambda_{\mathrm{T}}}{\mu_{\mathrm{A}}}\right)\alpha_{\mathrm{T}}+\left(\frac{\lambda_{\mathrm{A}}}{E_{\mathrm{T}}}+\frac{\nu_{\mathrm{A}}\lambda_{\mathrm{T}}}{E_{\mathrm{A}}}\right)\alpha_{\mathrm{S}}\right],\\ \omega_{2}&=\frac{1}{\Delta}\left[\left(\frac{1}{E_{\mathrm{A}}}-\frac{\lambda_{\mathrm{A}}^{2}}{\mu_{\mathrm{A}}}\right)\alpha_{\mathrm{T}}+\left(\frac{\nu_{\mathrm{A}}}{E_{\mathrm{A}}}+\frac{\lambda_{\mathrm{A}}\lambda_{\mathrm{T}}}{\mu_{\mathrm{A}}}\right)\alpha_{\mathrm{A}}+\frac{\lambda_{\mathrm{T}}+\nu_{\mathrm{A}}\lambda_{\mathrm{A}}}{E_{\mathrm{A}}}\alpha_{\mathrm{S}}\right],\end{aligned}\tag{31}$$

and

$$\Delta=\frac{1}{E_{\mathrm{T}}}\left(\frac{1}{E_{\mathrm{A}}}-\frac{\lambda_{\mathrm{A}}^{2}}{\mu_{\mathrm{A}}}\right)-\left(\frac{\nu_{\mathrm{A}}}{E_{\mathrm{A}}}\right)^{2}-\frac{\lambda_{\mathrm{T}}}{E_{\mathrm{A}}\mu_{\mathrm{A}}}\left(\lambda_{\mathrm{T}}+2\nu_{\mathrm{A}}\lambda_{\mathrm{A}}\right).\tag{32}$$

Substitution of (29) into the third relation of (27) leads to

$$\varepsilon_{\mathrm{t}}=-\Omega_{13}\varepsilon_{\mathrm{A}}-\Omega_{23}\varepsilon_{\mathrm{T}}+\Omega_{33}\sigma_{\mathrm{t}}-\Omega_{36}\gamma_{\mathrm{A}}+\omega_{3}\Delta T,\tag{33}$$

where

$$\begin{aligned}\Omega_{13}&=\left(\frac{\nu_{a}}{E_{\mathrm{A}}}+\frac{\lambda_{\mathrm{A}}\lambda_{\mathrm{t}}}{\mu_{\mathrm{A}}}\right)\Omega_{11}+\left(\frac{\nu_{\mathrm{t}}}{E_{\mathrm{T}}}+\frac{\lambda_{\mathrm{T}}\lambda_{\mathrm{t}}}{\mu_{\mathrm{A}}}\right)\Omega_{12},\\ \Omega_{23}&=\left(\frac{\nu_{a}}{E_{\mathrm{A}}}+\frac{\lambda_{\mathrm{A}}\lambda_{\mathrm{t}}}{\mu_{\mathrm{A}}}\right)\Omega_{12}+\left(\frac{\nu_{\mathrm{t}}}{E_{\mathrm{T}}}+\frac{\lambda_{T}\lambda_{\mathrm{t}}}{\mu_{\mathrm{A}}}\right)\Omega_{22},\\ \Omega_{33}&=\left(\frac{1}{E_{\mathrm{t}}}-\frac{\lambda_{\mathrm{t}}^{2}}{\mu_{\mathrm{A}}}\right)-\left(\frac{\nu_{a}}{E_{\mathrm{A}}}+\frac{\lambda_{\mathrm{A}}\lambda_{\mathrm{t}}}{\mu_{\mathrm{A}}}\right)\Omega_{13}-\left(\frac{\nu_{\mathrm{t}}}{E_{\mathrm{T}}}+\frac{\lambda_{\mathrm{T}}\lambda_{\mathrm{t}}}{\mu_{\mathrm{A}}}\right)\Omega_{23},\\ \Omega_{36}&=\lambda_{\mathrm{t}}+\left(\frac{\nu_{a}}{E_{\mathrm{A}}}+\frac{\lambda_{\mathrm{A}}\lambda_{\mathrm{t}}}{\mu_{\mathrm{A}}}\right)\Omega_{16}+\left(\frac{\nu_{\mathrm{t}}}{E_{\mathrm{T}}}+\frac{\lambda_{\mathrm{T}}\lambda_{\mathrm{t}}}{\mu_{\mathrm{A}}}\right)\Omega_{26},\\ \omega_{3}&=\left(\alpha_{\mathrm{t}}+\lambda_{\mathrm{t}}\alpha_{S}\right)+\left(\frac{\nu_{a}}{E_{\mathrm{A}}}+\frac{\lambda_{\mathrm{A}}\lambda_{\mathrm{t}}}{\mu_{\mathrm{A}}}\right)\omega_{1}+\left(\frac{\nu_{\mathrm{t}}}{E_{\mathrm{T}}}+\frac{\lambda_{\mathrm{T}}\lambda_{\mathrm{t}}}{\mu_{\mathrm{A}}}\right)\omega_{2}.\end{aligned}\tag{34}$$

Substitution of (29) into (26) leads to

$$\tau_A = \Omega_{16}\varepsilon_A + \Omega_{26}\varepsilon_T + \Omega_{36}\sigma_t + \Omega_{66}\gamma_A - \omega_6 \Delta T, \tag{35}$$

where

$$\Omega_{16} = \lambda_A \Omega_{11} + \lambda_T \Omega_{12}, \qquad \Omega_{26} = \lambda_A \Omega_{12} + \lambda_T \Omega_{22}, \qquad \Omega_{36} = \lambda_t + \lambda_A \Omega_{13} + \lambda_T \Omega_{23},$$
$$\Omega_{66} = \mu_A + \lambda_A \Omega_{16} + \lambda_T \Omega_{26}, \quad \omega_6 = \mu_A \alpha_S + \lambda_A \omega_1 + \lambda_T \omega_2. \tag{36}$$

The most compact formulae for the symmetric quantities Ω_{ij} are

$$\begin{aligned}
\Omega_{11} &= \frac{1}{\Delta}\left(\frac{1}{E_T} - \frac{\lambda_T^2}{\mu_A}\right), \quad \Omega_{12} = \frac{1}{\Delta}\left(\frac{\nu_A}{E_A} + \frac{\lambda_A \lambda_T}{\mu_A}\right), \quad \Omega_{22} = \frac{1}{\Delta}\left(\frac{1}{E_A} - \frac{\lambda_A^2}{\mu_A}\right),\\
\Omega_{13} &= \left(\frac{\nu_a}{E_A} + \frac{\lambda_A \lambda_t}{\mu_A}\right)\Omega_{11} + \left(\frac{\nu_t}{E_T} + \frac{\lambda_T \lambda_t}{\mu_A}\right)\Omega_{12},\\
\Omega_{23} &= \left(\frac{\nu_a}{E_A} + \frac{\lambda_A \lambda_t}{\mu_A}\right)\Omega_{12} + \left(\frac{\nu_t}{E_T} + \frac{\lambda_T \lambda_t}{\mu_A}\right)\Omega_{22},\\
\Omega_{33} &= \left(\frac{1}{E_t} - \frac{\lambda_t^2}{\mu_A}\right) - \left(\frac{\nu_a}{E_A} + \frac{\lambda_A \lambda_t}{\mu_A}\right)\Omega_{13} - \left(\frac{\nu_t}{E_T} + \frac{\lambda_T \lambda_t}{\mu_A}\right)\Omega_{23},\\
\Omega_{16} &= \lambda_A \Omega_{11} + \lambda_T \Omega_{12}, \qquad \Omega_{26} = \lambda_A \Omega_{12} + \lambda_T \Omega_{22},\\
\Omega_{36} &= \lambda_t + \lambda_A \Omega_{13} + \lambda_T \Omega_{23}, \qquad \Omega_{66} = \mu_A + \lambda_A \Omega_{16} + \lambda_T \Omega_{26},
\end{aligned} \tag{37}$$

$$\begin{aligned}
\omega_1 &= \frac{1}{\Delta}\left[\left(\frac{1}{E_T} - \frac{\lambda_T^2}{\mu_A}\right)\alpha_A + \left(\frac{\nu_A}{E_A} + \frac{\lambda_A \lambda_T}{\mu_A}\right)\alpha_T + \left(\frac{\lambda_A}{E_T} + \frac{\nu_A \lambda_T}{E_A}\right)\alpha_S\right],\\
\omega_2 &= \frac{1}{\Delta}\left[\left(\frac{\nu_A}{E_A} + \frac{\lambda_A \lambda_T}{\mu_A}\right)\alpha_A + \left(\frac{1}{E_A} - \frac{\lambda_A^2}{\mu_A}\right)\alpha_T + \frac{\lambda_T + \nu_A \lambda_A}{E_A}\alpha_S\right],\\
\omega_3 &= \left(\alpha_t + \lambda_t \alpha_S\right) + \left(\frac{\nu_a}{E_A} + \frac{\lambda_A \lambda_t}{\mu_A}\right)\omega_1 + \left(\frac{\nu_t}{E_T} + \frac{\lambda_T \lambda_t}{\mu_A}\right)\omega_2,\\
\omega_6 &= \mu_A \alpha_S + \lambda_A \omega_1 + \lambda_T \omega_2.
\end{aligned} \tag{38}$$

References

1. McCartney, L.N. (2000). Model to predict effects of triaxial loading on ply cracking in general symmetric laminates. *Composites Science and Technology* 60: 2255–2279. (See errata in *Comp. Sci. Tech.* 62 (2002) 1273–1274.).
2. Software system 'PREDICT' which is a specific module of CoDA. (see http://www.npl.co.uk/npl/cmmt/cog/coda.html)
3. Hannaby, S.A. (1997). The solution of ordinary differential equations arising from stress transfer mechanics of general symmetric laminates. NPL Report CISE 13/97, April.
4. Hannaby, S.A. (1993). The solution of ordinary differential equations arising from stress transfer mechanics. NPL Report DITC 223/93, November.
5. Box, G.E.P. and Muller, M.E. (1958). A note on the generation of random normal deviates. *The Annals of Mathematical Statistics* 29: 610–611.
6. Wiley website associated with this book (www.wiley.com/go/mccartney/properties).

Recommended Reading

Hajikazemi, M. and Sadr, M.H. (2014). A variational model for stress analysis in cracked laminates with arbitrary symmetric lay-up under general in-plane loading. *International Journal of Solids and Structures* 51: 516–529. https://doi.org/10.1016/j.ijsolstr.2013.10.024

Hajikazemi, M., Sadr, M.H., and Talreja, R. (2015). Variational analysis of cracked general cross-ply laminates under bending and biaxial extension. *International Journal of Damage Mechanics* 24: 582–624. https://doi.org/10.1177/1056789514546010

Hajikazemi, M. and McCartney, L.N. (2018). Comparison of variational and generalized plane strain approaches for matrix cracking in general symmetric laminates. *International Journal of Damage Mechanics* 27: 507–540. https://doi.org/10.1177/1056789516685381

Hajikazemi, M. and van Paepegem, W. (2018). A variational model for free-edge interlaminar stress analysis in general symmetric and thin-ply composite laminates. *Composite Structures* 184: 443–451. https://doi.org/10.1016/j.compstruct.2017.10.012

Hajikazemi, M., Garoz, D., and van Paepegem, W. (2019). Model to accurately predict out-of-plane shear stiffness reduction in general cracked laminates. *Composites Science and Technology* 179: 88–96. https://doi.org/10.1016/j.compscitech.2019.05.009

Hajikazemi, M., McCartney, L.N., and van Paepegem, W. (2020). Matrix cracking initiation. propagation and laminate failure in multiple plies of general symmetric composite laminates. *Composites Part A: Applied Science and Manufacturing*, 105963. https://doi.org/10.1016/j.compositesa.2020.105963

Hajikazemi, M., McCartney, L.N., Ahmadi, H. and van Paepegem, W. (2020). Variational analysis of cracking in general composite laminates subject to triaxial and bending loads. *Composite Structures* 239: 111993. https://doi.org/10.1016/j.compstruct.2020.111993

Hajikazemi, M. (2020). Physics-based methodology for predicting ply cracking and laminate failure in symmetric composite laminates under multiaxial loading. In: *Multi-Scale Continuum Mechanics Modelling of Fibre-Reinforced Polymer Composites*, (ed. W. van Paepegem), 509–553. Woodhead. https://doi.org/10.1016/B978-0-12-818984-9.00017-2

Kaddour, A.S., Hinton, M.J., Smith. P.A., Li, S. (2013). A comparison between the predictive capability of matrix cracking, damage and failure criteria for fibre reinforced laminates: Part A of the third world-wide failure exercise (WWFE-III). *J. Comp. Mater.* 47 (20–21): 2749–2779. (Readers are warned that the assessment of the author's contribution is flawed because they did not account for all the information presented in the contribution and they wholly mis-represented the strength prediction method used. Also, readers are informed that Parts B and C of the Exercise have not been undertaken.)

Kaddour, A.S., Hinton, M.J., Smith. P.A., Li, S. (2013). The background to the third world-wide failure exercise (WWFE-III). *J. Comp. Mater.* 47 (20–21): 2417–2426.

McCartney, L.N. (1997). Prediction of ply cracking in general symmetric laminates, published in Proceedings of 4th Int. Conf. on 'Deformation & Fracture of Composites'. Manchester, UK. 24–26 March, 101–110.

McCartney, L.N., Schoeppner, G.A., and Becker, W. (2000). Validation of models for transverse ply cracks in composite laminates. *Comp. Sci. & Tech.* 60: 2347–2359.

McCartney, L.N. and Schoeppner, G.A. (2002). Predicting the effect of non-uniform ply cracking on the thermoelastic properties of cross-ply laminates. *Comp. Sci. Tech.* 62: 1841–1856.

McCartney, L.N. (2003). Physically based damage models for laminated composites. *Proc. Instn. Mech. Engrs.*, 217: Part L: J. Materials: Design & Applications, 163–199.

McCartney, L.N. (2005). Energy-based prediction of failure in general symmetric laminates. *Engng. Fract. Mech.* 72: 909–930.

McCartney, L.N. (2005). Energy-based prediction of progressive ply cracking and strength of general symmetric laminates using an homogenization method. Composites Part A 36: 119–128.

McCartney, L.N. (2008). Energy methods for fatigue damage modelling of laminates. *Comp. Sci. Tech.* 68: 2601–2615.

McCartney, L.N. (2008). Multi-scale modelling of composites using analytical methods. Chapter 7. In: *Multi-Scale Modelling and Simulation of Composite Materials and Structures*, (ed. Y.W. Kwon, D.H. Allen, and R. Talreja), Springer.

McCartney, L.N. (2013). Derivations of energy-based modelling for ply cracking in general symmetric laminates. *J. Comp. Mater.* 47 (20–21): 2641–2673.

McCartney, L.N. (2013). Energy methods for modelling damage in laminates. *J. Comp. Mater.* 47 (20–21): 2613–2640.

Soden, P.D., Hinton, M.J., Kaddour, A.S., and papers published as part of the First Worldwide Failure Exercise (WWFE-I), Part A is published in *Comp. Sci. Tech.* (1998); 58. Part B of the Exercise is published in *Comp. Sci. Tech.* (2002), 62. Part C of the Exercise is published in *Comp. Sci. Tech.* (2004).

19

Stress-transfer Mechanics for Biaxial Bending

Overview: Methodology based on an analytical model has been developed for predicting stress transfer owing to ply cracking in a cross-ply laminate subject to out-of-plane bending about two orthogonal axes. The model is valid for multiple-ply laminates that can be nonsymmetrical. Full account is taken of the combined effects of in-plane biaxial loading (leading to out-of-plane bending when the laminate is non-symmetrical) and of additional through-thickness and out-of-plane bending. Thermal residual stresses and the associated bending arising from a through-thickness variation of the temperature distribution are taken into account, although ply refinement methods are needed to deal with any through-thickness stress and displacement variations. The representations for the stress and displacement fields satisfy exactly the equilibrium equations and both traction and displacement continuity conditions at the interfaces between layers. The stress-strain equations are satisfied either exactly or in an average sense for each ply or ply layer. Stress transfer is governed by sets of fourth-order ordinary differential equations that are solved by standard methods. Solutions of the differential equations are used to determine the stress and displacement distributions at all points in a laminate subject to combined biaxial loading and out-of-plane bending about two orthogonal axes when equally spaced cracks are present in any of the 90° plies subject to tension. The stress and displacement distributions are used to predict the effective thermoelastic constants relevant to the bending of a laminate. Example predictions are cited that illustrate the capabilities of the model.

19.1 Introduction

In Chapter 11 a methodology was described, based on crack closure concepts, for understanding the effect of ply crack formation in the 90° plies of a nonsymmetric cross-ply laminate subject to in-plane biaxial loading, through-thickness loading and biaxial orthogonal bending, in the presence of thermal residual stresses. The fundamental assumption made was that the effective stress-strain relations for the damaged laminate are of exactly the same form as those for the corresponding undamaged laminate. It was not possible, using the methodology of Chapter 11, to predict how the effective

Properties for Design of Composite Structures: Theory and Implementation Using Software,
First Edition. Neil McCartney.

Companion Website: www.wiley.com/go/mccartney/properties

properties of the damaged laminate depended on the ply crack distribution (e.g. the ply crack density). To achieve this objective requires a detailed analysis of the stress and strain distributions in a laminate having ply cracks in one or more of the 90° plies. This analysis will now be described for the situation where a simplifying assumption is made where the ply crack density in the 90° plies is uniform such that ply cracks in different 90° plies are coplanar. The geometry and fundamental field equations were described in Chapter 11 (see Section 11.2).

The first objective of this chapter is to describe a model of stress transfer for unbalanced multiple-ply cross-ply laminates subject to out-of-plane bending. Account is taken of the combined effects of biaxial loading and biaxial bending, and of the effects of thermal residual stresses that also lead to out-of-plane bending when laminates are nonsymmetric, or symmetric but asymmetrically cracked. The anticlastic bending associated with thick plates is included in the analysis.

The model is based on a representation for the stress and displacement fields that satisfy exactly the equilibrium equations and both traction and displacement continuity conditions at the interfaces between plies, and ply elements if ply refinement techniques are employed. The stress-strain equations are satisfied either exactly or in an average sense for each ply or ply layer. Extensive use is made of the recurrence relations that arise by imposing the continuity of tractions and displacements between the layers representing the laminate. Stress transfer is governed by sets of fourth-order ordinary differential equations that are solved by standard methods. Solutions of the differential equations are used to determine the stress and displacement distributions at all points in a laminate subject to combined biaxial loading and out-of-plane bending when equally spaced cracks are present in any of the 90° plies.

Because the analysis is developed for laminates that may be nonsymmetrical, hybrid in nature so that plies are made of more than one type of material, subject to combined biaxial loading, out-of-plane bending and a temperature distribution, the effective stress-strain relations describing the deformation characteristics are complex in nature but predictable. The major difficulty encountered is characterising the various thermoelastic constants needed to describe behaviour. However, in Chapter 11 a variety of effective constants for damaged laminates were defined, and methods will now be developed for their prediction as a function of laminate geometry and ply crack density. The methodology is based on analysis given in references [1–3].

19.2 Representation for Stress and Displacement Fields

A multilayered model will now be described that will be useful for solving some stress-transfer problems that arise in nonsymmetric cross-ply laminates when considering the effects of ply cracking. The approach to be followed and fundamental relations have already been introduced in Section 11.3. It is seen that the in-plane transverse deformation is assumed to be characterised completely by the two strain parameters $\bar{\varepsilon}_T$, $\hat{\varepsilon}_T$, whether the laminate is damaged or not. These loading parameters apply to the transverse loading of all plies in the nonsymmetrical laminate. This assumption means that edge effects cannot be considered, and solutions will apply only in regions of the laminate which are sufficiently far away from the edges. The values of the effective transverse stress σ_T and the effective transverse bending moment per unit

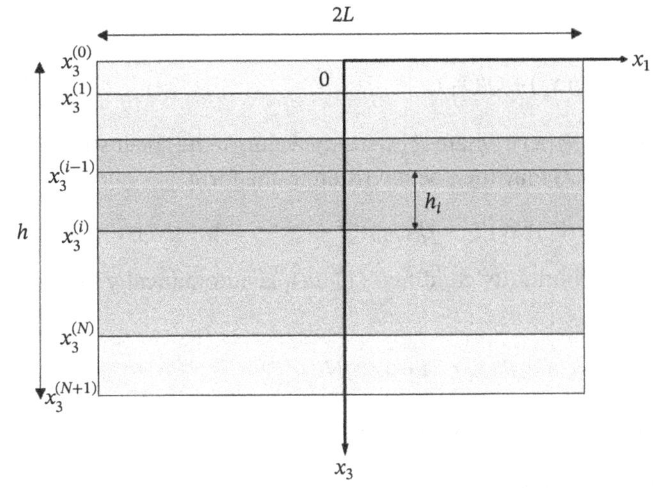

Figure 19.1 Geometry of a multilayered nonsymmetrical laminate.

cross-sectional area M_T will depend on whether the laminate is damaged. The corresponding effective axial values σ_A and M_A are assumed to have the same values for both damaged and undamaged laminates. In this chapter, a representation is first derived for the stress and corresponding displacement fields which is sufficiently general for ply cracks to be considered when applying boundary conditions. The geometry of nonsymmetric laminate to be considered is shown in Fig. 19.1.

19.2.1 The Stress Field

To allow for the anticlastic bending associated with thick laminates, bending is considered to occur orthogonally in the x_1–x_3 plane and in the x_2–x_3 plane. The bending in the x_1–x_3 plane is analysed in detail in this chapter; bending in the x_2–x_3 plane is modelled by assuming, for values $i=1,\ldots,n$, that $u_1^{(i)}$ is independent of x_2, that $u_2^{(i)}$ is linear in x_3 across the whole laminate and that there is no shear deformation in the x_2–x_3 plane throughout the laminate so that $\varepsilon_{23}^{(i)} \equiv 0$. When a multiple-ply nonsymmetric cross-ply laminate containing ply cracks is subject to combined biaxial in-plane loading and biaxial bending about the x_1- and x_2-axes, the stress component σ_{11} in each layer is assumed to be a linear function of the through-thickness coordinate x_3. It then follows from (11.26), and the assumption that $\sigma_{11}=\sigma_{11}(x_1,x_3)$, that

$$\sigma_{13}(x_1,x_3)=A(x_1)x_3^2+B(x_1)x_3+Q(x_1), \tag{19.1}$$

where $A(x_1)$, $B(x_1)$ and $Q(x_1)$ are functions of x_1. It should be noted that the stress component σ_{13} is assumed to be independent of x_2, which means that the analysis is ignoring laminate edge effects. Without loss of generality these functions can be redefined, for $x_3^{(i-1)} \le x_3 \le x_3^{(i)}$, $i=1,\ldots,N+1$, such that

$$\sigma_{13}^{(i)}(x_1,x_3)=A_i'(x_1)+B_i'(x_1)\xi_i+Q_i'(x_1)\xi_i^2, \quad \text{where} \quad \xi_i=\frac{x_3-x_3^{(i-1)}}{x_3^{(i)}-x_3^{(i-1)}}. \tag{19.2}$$

Hence, on the interface $x_3=x_3^{(i-1)}$

$$\sigma_{13}^{(i)}(x_1,x_3^{(i-1)})=A_i'(x_1), \tag{19.3}$$

and on $x_3 = x_3^{(i)}$

$$\sigma_{13}^{(i)}(x_1, x_3^{(i)}) = A_i'(x_1) + B_i'(x_1) + Q_i'(x_1). \tag{19.4}$$

A new function is introduced, $P_i(x_1)$, where $P_i'(x_1)$ is the value of the shear stress σ_{13} at the ith interface. Relation (19.2) may then be rewritten in the form

$$\sigma_{13}^{(i)}(x_1, x_3) = P_{i-1}'(x_1) + [P_i'(x_1) - P_{i-1}'(x_1)]\xi_i + Q_i'(x_1)(\xi_i^2 - \xi_i), \quad i = 1, \ldots, N+1. \tag{19.5}$$

Using this representation, the continuity condition $(11.28)_1$ is automatically satisfied such that

$$\sigma_{13}^{(i)}(x_1, x_3^{(i)}) = \sigma_{13}^{(i+1)}(x_1, x_3^{(i)}) = P_i'(x_1), \quad i = 1, \ldots, N. \tag{19.6}$$

On substituting (19.5) into (11.27), it can be shown that

$$\begin{aligned}\sigma_{33}^{(i)}(x_1, x_3) = \Bigg[& -P_{i-1}''(x_1)\xi_i + \left(P_{i-1}''(x_1) - P_i''(x_1)\right)\frac{1}{2}\xi_i^2 \\ & - Q_i''(x_1)\left(\frac{1}{3}\xi_i^3 - \frac{1}{2}\xi_i^2\right)\Bigg] h_i + S_{i-1}(x_1) + \sigma_t, \quad i = 1, \ldots, N+1,\end{aligned} \tag{19.7}$$

where $S_{i-1}(x_1) + \sigma_t$ is the through-thickness stress distribution on $x_3 = x_3^{(i-1)}$. By substitution of (19.5) into (11.26), it can be shown that

$$\sigma_{11}^{(i)}(x_1, x_3) = \frac{1}{h_i}[P_{i-1}(x_1) - P_i(x_1)] - \frac{Q_i(x_1)}{h_i}(2\xi_i - 1) + \sigma_A^{(i)}(x_3), \quad i = 1, \ldots, N+1. \tag{19.8}$$

The quantity $\sigma_A^{(i)}(x_3)$ is the longitudinal stress experienced by the ith layer of an undamaged laminate obtained using (7.202)

$$\sigma_A^{(i)}(x_3) = \tilde{E}_A^{(i)}(\bar{\varepsilon}_A^0 + \hat{\varepsilon}_A^0 x_3) + \nu_A^{(i)}\tilde{E}_T^{(i)}(\bar{\varepsilon}_T + \hat{\varepsilon}_T x_3) + \tilde{\nu}_a^{(i)}\sigma_t - \tilde{E}_A^{(i)}\tilde{\alpha}_A^{(i)}\Delta T. \tag{19.9}$$

From (11.16), the stress component $\sigma_{22}^{(i)}$ may now be calculated

$$\sigma_{22}^{(i)}(x_1, x_3) = \nu_t^{(i)}\sigma_{33}^{(i)}(x_3) + \nu_A^{(i)}\frac{E_T^{(i)}}{E_A^{(i)}}\sigma_{11}^{(i)}(x_1, x_3) - E_T^{(i)}\alpha_T^{(i)}\Delta T + E_T^{(i)}(\bar{\varepsilon}_T + \hat{\varepsilon}_T x_3). \tag{19.10}$$

On using (19.7) and (19.8), it follows that

$$\begin{aligned}\sigma_{22}^{(i)}(x_1, x_3) = & -h_i\nu_t^{(i)}P_{i-1}''(x_1)\xi_i + h_i\nu_t^{(i)}[P_{i-1}''(x_1) - P_i''(x_1)]\frac{1}{2}\xi_i^2 \\ & - h_i\nu_t^{(i)}Q_i''(x_1)\left(\frac{1}{3}\xi_i^3 - \frac{1}{2}\xi_i^2\right) + \nu_A^{(i)}\frac{E_T^{(i)}}{E_A^{(i)}}\frac{1}{h_i}[P_{i-1}(x_1) - P_i(x_1)] \\ & - \nu_A^{(i)}\frac{E_T^{(i)}}{E_A^{(i)}}\frac{Q_i(x_1)}{h_i}(2\xi_i - 1) + \nu_t^{(i)}S_{i-1}(x_1) + \sigma_T^{(i)}(x_3),\end{aligned} \tag{19.11}$$

where use has been made of relation (7.205), namely

$$\sigma_T^{(i)}(x_3) = \nu_t^{(i)}\sigma_t + \nu_A^{(i)}\frac{E_T^{(i)}}{E_A^{(i)}}\sigma_A^{(i)}(x_3) - E_T^{(i)}\alpha_T^{(i)}\Delta T + E_T^{(i)}(\bar{\varepsilon}_T + \hat{\varepsilon}_T x_3). \tag{19.12}$$

The functions $P_i(x_1)$, $Q_i(x_1)$ and $S_i(x_1)$ are stress-transfer functions to be determined, which have been configured so that they are all identically zero when the laminate is undamaged. It should be noted that $\sigma_{11}^{(i)}$, $\sigma_{22}^{(i)}$ and $\sigma_{33}^{(i)}$ are independent of x_2, as assumed previously, when demanding that the equilibrium equation (11.3) is satisfied. In addition, σ_{11} will be linear in x_3, as from (19.9) it is seen that $\sigma_{\mathrm{A}}^{(i)}(x_3)$ is linear in x_3.

From $(11.30)_{3,4}$ and (19.5) it is clear that

$$P_0'(x_1) = P_{N+1}'(x_1) \equiv 0. \tag{19.13}$$

As the functions $P_i(x_1), i = 0, \ldots, N+1$, are required to be identically zero for an undamaged laminate, on integrating (19.13) it follows that

$$P_0(x_1) = P_{N+1}(x_1) \equiv 0. \tag{19.14}$$

To complete the discussion of the stress field, it is useful to show that the bending moment in the x_1–x_3 plane per unit area M_{A} about the midplane of the laminate is uniform along the length of the laminate. First,

$$M_{\mathrm{A}} = \frac{1}{h}\sum_{i=1}^{N+1}\int_{x_3^{(i-1)}}^{x_3^{(i)}}\left(x_3 - \frac{h}{2}\right)\sigma_{11}^{(i)}\,\mathrm{d}x_3 = \frac{1}{h}\sum_{i=1}^{N+1}\int_{x_3^{(i-1)}}^{x_3^{(i)}} x_3\sigma_{11}^{(i)}\,\mathrm{d}x - \frac{1}{2}\sum_{i=1}^{N+1}\int_{x_3^{(i-1)}}^{x_3^{(i)}}\sigma_{11}^{(i)}\,\mathrm{d}x_3. \tag{19.15}$$

On integrating (19.8), it can be shown that

$$\sum_{i=1}^{N+1}\int_{x_3^{(i-1)}}^{x_3^{(i)}}\sigma_{11}^{(i)}\,\mathrm{d}x_3 = \sum_{i=1}^{N+1}\left(P_{i-1}(x_1) - P_i(x_1) + \int_{x_3^{(i-1)}}^{x_3^{(i)}}\sigma_{\mathrm{A}}^{i}(x_3)\mathrm{d}x_3\right). \tag{19.16}$$

On using (19.14) and (11.35), relation (19.16) reduces to the expected form

$$\sum_{i=1}^{N+1}\int_{x_3^{(i-1)}}^{x_3^{(i)}}\sigma_{\mathrm{A}}^{(i)}(x_3)\mathrm{d}x_3 = h\sigma_{\mathrm{A}}. \tag{19.17}$$

On substituting (19.8) and (19.17) into (19.15) and integrating the appropriate terms it can be shown that

$$M_{\mathrm{A}} = \frac{1}{h}\sum_{i=1}^{N+1}\left[\left(P_{i-1}(x_1) - P_i(x_1)\right)\left(x_3^{(i)} - \frac{h_i}{2}\right) - \frac{h_i}{6}Q_i(x_1) + \int_{x_3^{(i-1)}}^{x_3^{(i)}} x_3\sigma_{\mathrm{A}}^{i}(x_3)\mathrm{d}x_3\right] - \frac{h}{2}\sigma_{\mathrm{A}}. \tag{19.18}$$

On using the integrated form of (19.75), together with (19.18),

$$M_{\mathrm{A}} = \frac{1}{h}\left[P_0(x_1)x_3^{(0)} - P_{N+1}(x_1)x_3^{(N+1)}\right] + \frac{1}{h}\sum_{i=1}^{N+1}\int_{x_3^{(i-1)}}^{x_3^{(i)}} x_3\sigma_{\mathrm{A}}^{(i)}(x_3)\mathrm{d}x_3 - \frac{h}{2}\sigma_{\mathrm{A}}, \tag{19.19}$$

which, on using (19.14), reduces to

$$M_{\mathrm{A}} = \frac{1}{h}\sum_{i=1}^{N+1}\int_{x_3^{(i-1)}}^{x_3^{(i)}}\left(x_3 - \frac{h}{2}\right)\sigma_{\mathrm{A}}^{(i)}(x_3)\mathrm{d}x_3. \tag{19.20}$$

This expression is independent of x_1 and the degree of ply cracking, and thus the bending moment does not vary along the length of the laminate, as is expected for the case when the upper and lower surfaces of the laminate are subject only to a uniform through-thickness stress σ_t.

19.2.2 The Displacement Field

Substitution of (19.7) and (19.8) into (11.24) followed by an integration with respect to x_3 leads to

$$\begin{aligned} u_3^{(i)}(x_1,x_2,x_3) = &\frac{(x_3 - x_3^{(i-1)})^2}{12\tilde{E}_t^{(i)}}\left[2\left(P''_{i-1}(x_1) - P''_i(x_1)\right)\xi_i - 6P''_{i-1}(x_1) - Q''_i(x_1)(\xi_i^2 - 2\xi_i)\right] \\ &- \frac{\tilde{\nu}_a^{(i)}}{\tilde{E}_A^{(i)}}\left[(P_{i-1}(x_1) - P_i(x_1))\xi_i - Q_i(x_1)(\xi_i^2 - \xi_i) + \sigma_A^{(i)*}(x_3)\right] \\ &+ (x_3 - x_3^{(i-1)})\left(\tilde{\alpha}_t^{(i)}\Delta T + \frac{S_{i-1}(x_1) + \sigma_t}{\tilde{E}_t^{(i)}}\right) + U_{i-1}(x_1,x_2) - \nu_t^{(i)}\bar{\varepsilon}_T(x_3 - x_3^{(i-1)}) \\ &- \frac{1}{2}\nu_t^{(i)}\hat{\varepsilon}_T\left(x_3^2 - (x_3^{(i-1)})^2\right), \quad i = 1, \ldots, N+1, \end{aligned} \tag{19.21}$$

where $U_{i-1}(x_1,x_2) \equiv u_3^{(i)}(x_1,x_2,x_3^{(i-1)})$ represents the through-thickness displacement distribution on $x_3 = x_3^{(i-1)}$, and where

$$\sigma_A^{(i)*}(x_3) = \int_{x_3^{(i-1)}}^{x_3} \sigma_A^{(i)}(\eta)\, d\eta, \quad i = 1, \ldots, N+1. \tag{19.22}$$

On substituting (19.5) and (19.21) into (11.25), it can be shown that

$$\begin{aligned} u_1^{(i)}(x_1,x_3) = &-\frac{(x_3 - x_3^{(i-1)})^3}{120\tilde{E}_t^{(i)}}\left[5\left(P'''_{i-1}(x_1) - P'''_i(x_1)\right)\xi_i - 20P'''_{i-1}(x_1) - Q'''_i(x_1)(2\xi_i^2 - 5\xi_i)\right] \\ &+ (x_3 - x_3^{(i-1)})\left[\frac{P'_{i-1}(x_1)}{\mu_a^{(i)}} + \left(\frac{1}{\mu_a^{(i)}} - \frac{\tilde{\nu}_a^{(i)}}{\tilde{E}_A^{(i)}}\right)\left\{\left(P'_i(x_1) - P'_{i-1}(x_1)\right)\frac{1}{2}\xi_i + Q'_i(x_1)\left(\frac{1}{3}\xi_i^2 - \frac{1}{2}\xi_i\right)\right\}\right] \\ &- \frac{S'_{i-1}(x_1)}{2\tilde{E}_t^{(i)}}(x_3 - x_3^{(i-1)})^2 - \frac{\partial U_{i-1}(x_1,x_2)}{\partial x_1}(x_3 - x_3^{(i-1)}) + V_{i-1}(x_1), \quad i = 1, \ldots, N+1, \end{aligned} \tag{19.23}$$

where $V_{i-1}(x_1) \equiv u_1^{(i)}(x_1,x_2,x_3^{(i-1)})$, assumed to be independent of x_2, represents the longitudinal displacement distribution on $x_3 = x_3^{(i-1)}$. The form of (19.23) requires that $\partial U_i(x_1,x_2)/\partial x_1$, $i = 0, \ldots, N$, are independent of x_2 (see (19.61)).

19.2.3 The Recurrence Relations

By applying the relevant interfacial conditions (11.28), the following recurrence relationships can be derived on using (7.143), for $i = 1, \ldots, N$,

$$S_i(x_1) = \frac{h_i}{6}\left[Q''_i(x_1) - 3P''_{i-1}(x_1) - 3P''_i(x_1)\right] + S_{i-1}(x_1), \tag{19.24}$$

$$
\begin{aligned}
U_i(x_1,x_2) = {} & \frac{h_i^2}{12\tilde{E}_t^{(i)}}\left(Q_i''(x_1) - 2P_i''(x_1) - 4P_{i-1}''(x_1)\right) + h_i\left(\tilde{\alpha}_t^{(i)}\Delta T + \frac{S_{i-1}(x_1)+\sigma_t}{\tilde{E}_t^{(i)}}\right) \\
& - \frac{\tilde{\nu}_a^{(i)}}{\tilde{E}_A^{(i)}}\left[P_{i-1}(x_1) - P_i(x_1) + \sigma_A^{(i)*}(x_3^{(i)})\right] - h_i\nu_t^{(i)}\bar{\varepsilon}_T - \hat{h}_i\nu_t^{(i)}\hat{\varepsilon}_T + U_{i-1}(x_1,x_2),
\end{aligned}
\tag{19.25}
$$

$$
\begin{aligned}
V_i(x_1) = {} & -\frac{h_i^3}{120\tilde{E}_t^{(i)}}\left[3Q_i'''(x_1) - 5P_i'''(x_1) - 15P_{i-1}'''(x_1)\right] \\
& + \frac{h_i}{6}\left[\frac{1}{\mu_a^{(i)}}\left(3P_{i-1}'(x_1) + 3P_i'(x_1) - Q_i'(x_1)\right) + \frac{\tilde{\nu}_a^{(i)}}{\tilde{E}_A^{(i)}}\left(3P_{i-1}'(x_1) - 3P_i'(x_1) + Q_i'(x_1)\right)\right] \\
& - \frac{h_i^2}{2\tilde{E}_t^{(i)}}S_{i-1}'(x_1) - h_i\frac{\partial U_{i-1}(x_1,x_2)}{\partial x_1} + V_{i-1}(x_1),
\end{aligned}
\tag{19.26}
$$

where, on using (7.206) and (7.223),

$$
\begin{aligned}
\sigma_A^{(i)*}(x_3^{(i)}) = \int_{x_3^{(i-1)}}^{x_3^{(i)}} \sigma_A^{(i)}(\eta)\,d\eta = {} & h_i\left(\tilde{E}_A^{(i)}\bar{\varepsilon}_A^0 + \nu_A^{(i)}\tilde{E}_T^{(i)}\bar{\varepsilon}_T + \tilde{\nu}_a^{(i)}\sigma_t - \tilde{E}_A^{(i)}\tilde{\alpha}_A^{(i)}\Delta T\right) \\
& + \hat{h}_i\left(\tilde{E}_A^{(i)}\hat{\varepsilon}_A^0 + \nu_A^{(i)}\tilde{E}_T^{(i)}\hat{\varepsilon}_T\right).
\end{aligned}
\tag{19.27}
$$

Expression (19.27) may also be written in the form

$$
\sigma_A^{(i)*}(x_3^{(i)}) = \tilde{E}_A^{(i)}(h_i\bar{\varepsilon}_A^0 + \hat{h}_i\hat{\varepsilon}_A^0) + \nu_A^{(i)}\tilde{E}_T^{(i)}(h_i\bar{\varepsilon}_T + \hat{h}_i\hat{\varepsilon}_T) + h_i\,\tilde{\nu}_a^{(i)}\sigma_t - h_i\tilde{E}_A^{(i)}\,\tilde{\alpha}_A^{(i)}\Delta T.
\tag{19.28}
$$

It follows from (11.30) and (19.7) that $S_0(x_1) \equiv 0$ so that (19.24) can be used to determine the values of $S_i(x_1)$, $i = 1, \ldots, N$.

When modelling the behaviour of the *undamaged* laminate, the following identities are imposed

$$
P_i(x_1) \equiv 0, \quad \text{for} \quad i = 0, \ldots, N+1, \tag{19.29}
$$

$$
Q_i(x_1) \equiv 0, \quad \text{for} \quad i = 1, \ldots, N+1, \tag{19.30}
$$

$$
S_i(x_1) \equiv 0, \quad \text{for} \quad i = 0, \ldots, N, \tag{19.31}
$$

with the result that, from (19.7) and (19.8),

$$
\sigma_{11}^{(i)} = \sigma_A^{(i)}(x_3), \quad \sigma_{33}^{(i)} \equiv \sigma_t, \quad \text{for} \quad i = 1, \ldots, N+1. \tag{19.32}
$$

When the laminate is undamaged, the functions $U_j(x_1,x_2) = U_j^0(x_1,x_2)$, $V_j(x_1,x_2) = V_j^0(x_1)$, $j = 0, \ldots, N$, where from (19.25)–(19.31) the functions U_j^0 and V_j^0 may be determined using the recurrence relations

$$
\begin{aligned}
U_j^0(x_1,x_2) = {} & -\tilde{\nu}_a^{(j)}\left(h_j\bar{\varepsilon}_A^0 + \hat{h}_j\hat{\varepsilon}_A^0\right) - \tilde{\nu}_t^{(j)}\left(h_j\bar{\varepsilon}_T + \hat{h}_j\hat{\varepsilon}_T\right) \\
& + h_j\left(\frac{1}{\tilde{E}_t^{(j)}} - \frac{(\tilde{\nu}_a^{(j)})^2}{\tilde{E}_A^{(j)}}\right)\sigma_t + h_j\left(\tilde{\alpha}_t^{(j)} + \tilde{\nu}_a^{(j)}\tilde{\alpha}_A^{(j)}\right)\Delta T + U_{j-1}^0(x_1,x_2),
\end{aligned}
\tag{19.33}
$$

$$V_j^0(x_1) = -h_j \frac{\partial U_{j-1}^0(x_1,x_2)}{\partial x_1} + V_{j-1}^0(x_1), \tag{19.34}$$

for values $j=1,\ldots,N$, where use has been made of relations (7.207) and (7.227). It follows from (7.200) that for an undamaged laminate

$$\begin{aligned} U_0^0(x_1,x_2) &\equiv u_3^{(1)}(x_1,x_2,0) = -\frac{1}{2}\hat{\varepsilon}_A^0 x_1^2 - \frac{1}{2}\hat{\varepsilon}_T x_2^2, \\ V_0^0(x_1) &\equiv u_1^{(1)}(x_1,x_2,0) = \bar{\varepsilon}_A^0 x_1 \,. \end{aligned} \tag{19.35}$$

On operating the recurrence relations (19.33) and (19.34) using (19.35), it then follows that for the values $j=1,\ldots,N$,

$$\begin{aligned} U_j^0(x_1,x_2) &\equiv u_3^{(j)}(x_1,x_2,x_3^{(j)}) = -\sum_{i=1}^{j} \tilde{\nu}_a^{(i)}(h_i\bar{\varepsilon}_A^0 + \hat{h}_i\hat{\varepsilon}_A^0) - \sum_{i=1}^{j} \tilde{\nu}_t^{(i)}(h_i\bar{\varepsilon}_T + \hat{h}_i\hat{\varepsilon}_T) \\ &+ \sigma_t \sum_{i=1}^{j} h_i \left(\frac{1}{\tilde{E}_t^{(i)}} - \frac{(\tilde{\nu}_a^{(i)})^2}{\tilde{E}_A^{(i)}} \right) + \Delta T \sum_{i=1}^{j} h_i(\tilde{\alpha}_t^{(i)} + \tilde{\nu}_a^{(i)}\tilde{\alpha}_A^{(i)}) - \tfrac{1}{2}\hat{\varepsilon}_A^0 x_1^2 - \tfrac{1}{2}\hat{\varepsilon}_T x_2^2, \end{aligned} \tag{19.36}$$

$$V_j^0(x_1) \equiv u_1^{(j)}(x_1,x_2,x_3^{(j)}) = (\bar{\varepsilon}_A^0 + \hat{\varepsilon}_A^0 x_3^{(j)})x_1, \tag{19.37}$$

where $\bar{\varepsilon}_A^0$, $\hat{\varepsilon}_A^0$ denote the values of $\bar{\varepsilon}_A$, $\hat{\varepsilon}_A$ for an undamaged laminate, the values of $\bar{\varepsilon}_T$, $\hat{\varepsilon}_T$ denoting the in-plane transverse strain parameters for *both* damaged and undamaged laminates.

The functions U_j^0 and V_j^0 may also be calculated using results derived in Chapter 7 (see (7.201), (7.224), (7.228) and (7.229)). It is noted that for all values of $j=0,\ldots,N$, the dependence of U_j^0 on x_2 is only through the same term $\frac{1}{2}\hat{\varepsilon}_T x_2^2$.

19.3 Derivation of Differential Equations

The approach to predicting ply crack formation within a laminate centres around the assumption that the composite behaves in a manner that obeys the reduced stress-strain relations given by (11.16)–(11.18). It is now necessary to ensure that the stress-transfer functions $P_i(x_1)$ and $Q_i(x_2)$ are calculated so that they satisfy these relations as closely as possible. The axial stress-strain relation (11.17) cannot be satisfied exactly, for all x_3, by the representations that have been generated. The approach taken here is to take the average and moment of (11.17) and demand that the resulting relations, which are then independent of x_3, are satisfied exactly.

19.3.1 Averaging

The average of a variable $f_i(x_1,x_3)$ over the interval $x_3^{(i-1)} \le x_3 \le x_3^{(i)}$ of the ith element is defined to be

$$\bar{f}_i(x_1) = \frac{1}{h_i} \int_{x_3^{(i-1)}}^{x_3^{(i)}} f_i(x_1,x_3)\,\mathrm{d}x_3 \,. \tag{19.38}$$

On taking the average of (11.17),

$$\frac{d\bar{u}_1^{(i)}}{dx_1} = \frac{\bar{\sigma}_{11}^{(i)}}{\tilde{E}_A^{(i)}} - \frac{\tilde{\nu}_a^{(i)}}{\tilde{E}_A^{(i)}}\bar{\sigma}_{33}^{(i)} + \tilde{\alpha}_A^{(i)}\Delta T - \nu_A^{(i)}\frac{E_T^{(i)}}{E_A^{(i)}}\left[\bar{\varepsilon}_T + \hat{\varepsilon}_T\left(x_3^{(i)} - \frac{1}{2}h_i\right)\right], \quad i=1,\ldots,N+1. \tag{19.39}$$

It can be shown on applying (19.38) to (19.7), (19.8) and (19.23) that for $i=1,\ldots,N+1$,

$$\bar{\sigma}_{11}^{(i)} = \frac{1}{h_i}\left[P_{i-1}(x_1) - P_i(x_1) + \sigma_A^{(i)*}(x_3^{(i)})\right], \tag{19.40}$$

$$\bar{\sigma}_{33}^{(i)} = \frac{h_i}{12}\left[Q_i''(x_1) - 4P_{i-1}''(x_1) - 2P_i''(x_1)\right] + S_{i-1}(x_1) + \sigma_t, \tag{19.41}$$

$$\begin{aligned}\bar{u}_1^{(i)}(x_1) = &-\frac{h_i^3}{360\tilde{E}_t^{(i)}}\left[2Q_i'''(x_1) - 12P_{i-1}'''(x_1) - 3P_i'''(x_1)\right]\\ &+\left\{\frac{1}{12\mu_a^{(i)}}\left[4P_{i-1}'(x_1) + 2P_i'(x_1) - Q_i'(x_1)\right] + \frac{\tilde{\nu}_a^{(i)}}{12\tilde{E}_A^{(i)}}\left[2P_{i-1}'(x_1) - 2P_i'(x_1) + Q_i'(x_1)\right]\right\}h_i\\ &-\frac{h_i^2}{6\tilde{E}_t^{(i)}}S_{i-1}'(x_1) - \frac{h_i}{2}\frac{\partial U_{i-1}(x_1,x_2)}{\partial x_1} + V_{i-1}(x_1),\end{aligned} \tag{19.42}$$

where $\sigma_A^{(i)*}(x_3^{(i)})$ is given by (19.27) or (19.28). The form of (19.42) requires that $\partial U_i(x_1,x_2)/\partial x_1, i = 0, \ldots, N$, are independent of x_2 (see (19.61)). It is easily shown that

$$\frac{1}{h_i}\int_{x_3^{(i-1)}}^{x_3^{(i)}} (\bar{\varepsilon}_T + \hat{\varepsilon}_T x_3)dx_3 = \bar{\varepsilon}_T + \hat{\varepsilon}_T(x_3^{(i)} - \frac{1}{2}h_i). \tag{19.43}$$

On substituting (19.41)–(19.43) into (19.39), it can be shown that for $i=1,\ldots,N+1$,

$$\begin{aligned}&\frac{h_i^3}{360\tilde{E}_t^{(i)}}\left[12P_{i-1}''''(x_1) + 3P_i''''(x_1) - 2Q_i''''(x_1)\right] + \left[\frac{P_{i-1}''(x_1)}{6}\left(\frac{2}{\mu_a^{(i)}} - \frac{\tilde{\nu}_a^{(i)}}{\tilde{E}_A^{(i)}}\right)\right.\\ &\left.+\left(\frac{P_i''(x_1)}{6} - \frac{Q_i''(x_1)}{12}\right)\left(\frac{1}{\mu_a^{(i)}} - \frac{2\tilde{\nu}_a^{(i)}}{\tilde{E}_A^{(i)}}\right)\right]h_i + \frac{1}{h_i\tilde{E}_A^{(i)}}\left[P_i(x_1) - P_{i-1}(x_1)\right] + \frac{\tilde{\nu}_a^{(i)}}{\tilde{E}_A^{(i)}}S_{i-1}(x_1)\\ &+V_{i-1}'(x_1) - \frac{h_i}{2}\frac{\partial^2 U_{i-1}(x_1,x_2)}{\partial x_1^2} - \frac{h_i^2}{6\tilde{E}_t^{(i)}}S_{i-1}''(x_1) - \bar{\varepsilon}_A^0 - \left(x_i - \frac{h_i}{2}\right)\hat{\varepsilon}_A^0 = 0.\end{aligned} \tag{19.44}$$

It is noted that the parameters $\bar{\varepsilon}_T, \hat{\varepsilon}_T, \sigma_t$ and ΔT have cancelled during the substitutions.

19.3.2 The Integrated Moments

When the laminate is deformed, the stress and displacement fields should satisfy the integrated moment of the axial stress-strain relation (11.17), where for any variable $m_i(x_1,x_2)$ its integrated moment over the ith element is defined as

$$\breve{m}_i(x_1) = \int_{x_3^{(i-1)}}^{x_3^{(i)}} x_3 m_i(x_1,x_3)dx_3. \tag{19.45}$$

On taking moments, (11.17) may be written as

$$\frac{d\breve{u}_1^{(i)}}{dx_1} = \frac{\breve{\sigma}_{11}^{(i)}}{\tilde{E}_A^{(i)}} - \frac{\tilde{\nu}_a^{(i)}}{\tilde{E}_A^{(i)}}\breve{\sigma}_{33}^{(i)} + \tilde{\alpha}_A^{(i)}\Delta T - \nu_A^{(i)}\frac{E_T^{(i)}}{E_A^{(i)}}\int_{x_3^{(i-1)}}^{x_3^{(i)}} x_3(\bar{\varepsilon}_T + \hat{\varepsilon}_T x_3)\,dx_3, \quad i=1,\ldots,N+1. \tag{19.46}$$

A linear combination of the averaged strain equation and the integrated moment equation is defined as

$$\widehat{m}_i = \frac{1}{h_i}\breve{m}_i - x_3^{(i-1)}\bar{m}_i, \tag{19.47}$$

usefully expressed as

$$\widehat{m}_i(x_1) = \int_{x_3^{(i-1)}}^{x_3^{(i)}} \xi_i\, m_i(x_1,x_3)\,dx_3, \tag{19.48}$$

where ξ_i is defined by $(19.2)_2$. When this newly defined operator is applied to the stress-strain equation (11.17), the stress and displacement fields should satisfy the equations

$$\frac{d\widehat{u}_1^{(i)}}{dx_1} = \frac{\widehat{\sigma}_{11}^{(i)}}{\tilde{E}_A^{(i)}} - \frac{\tilde{\nu}_a^{(i)}}{\tilde{E}_A^{(i)}}\widehat{\sigma}_{33}^{(i)} + \tilde{\alpha}_A^{(i)}\Delta T - \nu_A^{(i)}\frac{E_T^{(i)}}{E_A^{(i)}}\int_{x_3^{(i-1)}}^{x_3^{(i)}} \xi_i(\bar{\varepsilon}_T + \hat{\varepsilon}_T x_3)\,dx_3, \quad i=1,\ldots,N+1, \tag{19.49}$$

which are equivalent to (19.46) provided that (19.39) is satisfied.

After the integration of the appropriate terms the following expressions are found, for $i=1,\ldots,N+1$,

$$\widehat{\sigma}_{11}^{(i)}(x_1) = \frac{1}{2}\left[P_{i-1}(x_1) - P_i(x_1)\right] - \frac{1}{6}Q_i(x_1) + \frac{h_i\tilde{E}_A^{(i)}}{6}\left(\hat{\varepsilon}_A^0 + \nu_A^{(i)}\frac{E_T^{(i)}}{E_A^{(i)}}\hat{\varepsilon}_T\right)(3x_3^{(i)} - h_i) + \frac{h_i\tilde{E}_A^{(i)}}{2}\left(\bar{\varepsilon}_A^0 + \nu_A^{(i)}\frac{E_T^{(i)}}{E_A^{(i)}}\bar{\varepsilon}_T - \tilde{\alpha}_A^{(i)}\Delta T\right), \tag{19.50}$$

$$\widehat{\sigma}_{33}^{(i)}(x_1) = \frac{h_i^2}{120}\left[7Q_i''(x_1) - 25P_{i-1}''(x_1) - 15P_i''(x_1)\right] + \frac{h_i}{2}S_{i-1}(x_1) + \frac{h_i}{2}\sigma_t, \tag{19.51}$$

$$\begin{aligned}\widehat{u}_1^{(i)}(x_1) = &-\frac{h_i^4}{5040\tilde{E}_t^{(i)}}\left[23Q_i''''(x_1) - 133P_{i-1}''''(x_1) - 35P_i''''(x_1)\right]\\ &+\frac{h_i^2}{120\mu_a^{(i)}}\left(25P_{i-1}'(x_1) + 15P_i'(x_1) - 7Q_i'(x_1)\right) + \frac{h_i^2\tilde{\nu}_a^{(i)}}{120\tilde{E}_A^{(i)}}\left(15P_{i-1}'(x_1) - 15P_i'(x_1) + 7Q_i'(x_1)\right)\\ &-\frac{h_i^3}{8\tilde{E}_t^{(i)}}S_{i-1}'(x_1) - \frac{h_i^2}{3}\frac{\partial U_{i-1}(x_1,x_2)}{\partial x_1} + \frac{h_i}{2}V_{i-1}(x_1).\end{aligned} \tag{19.52}$$

The form of (19.52) requires that $\partial U_i(x_1,x_2)/\partial x_1$, $i = 0, \ldots, N$, are independent of x_2 (see (19.61)). It can be shown that

$$\int_{x_3^{(i-1)}}^{x_3^{(i)}} \xi_i\left(\bar{\varepsilon}_T + \hat{\varepsilon}_T x_3\right)dx_3 = \frac{h_i\bar{\varepsilon}_T}{2} + \frac{h_i\hat{\varepsilon}_T}{6}(3x_3^{(i)} - h_i). \tag{19.53}$$

The substitution of (19.51)–(19.53) into (19.49) and a multiplication by $2/h_i$ lead to, for $i=1,\ldots,N+1$,

$$\begin{aligned}&\frac{h_i^3}{2520\tilde{E}_t^{(i)}}\left[133P_{i-1}''''(x_1)+35P_i''''(x_1)-23Q_i''''(x_1)\right]+\left[\frac{P_{i-1}''(x_1)}{12}\left(\frac{5}{\mu_a^{(i)}}-\frac{2\tilde{\nu}_a^{(i)}}{\tilde{E}_A^{(i)}}\right)\right.\\&+\left.\left(\frac{P_i''(x_1)}{4}-\frac{7Q_i''(x_1)}{60}\right)\left(\frac{1}{\mu_a^{(i)}}-\frac{2\tilde{\nu}_a^{(i)}}{\tilde{E}_A^{(i)}}\right)\right]h_i-\frac{1}{3h_i\tilde{E}_A^{(i)}}\left[3P_{i-1}(x_1)-3P_i(x_1)-Q_i(x_1)\right]\\&+V_{i-1}'(x_1)+\frac{\tilde{\nu}_a^{(i)}}{\tilde{E}_A^{(i)}}S_{i-1}(x_1)-\frac{2h_i}{3}\frac{\partial^2U_{i-1}(x_1,x_2)}{\partial x_1^2}-\frac{h_i^2}{4\tilde{E}_t^{(i)}}S_{i-1}''(x_1)-\bar{\varepsilon}_A^0-\left(x_3^{(i)}-\frac{h_i}{3}\right)\hat{\varepsilon}_A^0=0.\end{aligned}\tag{19.54}$$

It is noted that the parameters $\bar{\varepsilon}_T, \hat{\varepsilon}_T, \sigma_t$ and ΔT have again cancelled during the substitutions.

19.3.3 Additional Expressions Involving the Interfacial Displacements

Treating (19.44) and (19.54) as a pair of simultaneous equations in the unknowns $\partial^2U_{i-1}(x_1,x_2)/\partial x_1^2$ and $V_{i-1}'(x_1)$ it can be shown that for $i=1,\ldots,N+1$,

$$\begin{aligned}\frac{\partial^2U_{i-1}(x_1,x_2)}{\partial x_1^2}&=\frac{h_i^2}{420\tilde{E}_t^{(i)}}\left[49P_{i-1}''''(x_1)+14P_i''''(x_1)-9Q_i''''(x_1)\right]+\frac{P_{i-1}''(x_1)}{2\mu_a^{(i)}}\\&+P_i''(x_1)\left(\frac{1}{2\mu_a^{(i)}}-\frac{\tilde{\nu}_a^{(i)}}{\tilde{E}_A^{(i)}}\right)+Q_i''(x_1)\left(-\frac{1}{5\mu_a^{(i)}}+\frac{2\tilde{\nu}_a^{(i)}}{5\tilde{E}_A^{(i)}}\right)+\frac{2}{h_i^2\tilde{E}_A^{(i)}}Q_i(x_1)-\frac{h_i}{2\tilde{E}_t^{(i)}}S_{i-1}''(x_1)-\hat{\varepsilon}_A^0\end{aligned}\tag{19.55}$$

independent of x_2, and

$$\begin{aligned}V_{i-1}'(x_1)&=\frac{h_i^3}{2520\tilde{E}_t^{(i)}}\left[63P_{i-1}''''(x_1)+21P_i''''(x_1)-13Q_i''''(x_1)\right]\\&+\frac{h_i}{60}\left(\frac{1}{\mu_a^{(i)}}-\frac{2\tilde{\nu}_a^{(i)}}{\tilde{E}_A^{(i)}}\right)\left[5P_i''(x_1)-5P_{i-1}''(x_1)-Q_i''(x_1)\right]+\frac{1}{h_i\tilde{E}_A^{(i)}}\left[P_{i-1}(x_1)-P_i(x_1)+Q_i(x_1)\right]\\&-\frac{\tilde{\nu}_a^{(i)}}{\tilde{E}_A^{(i)}}S_{i-1}(x_1)-\frac{h_i^2}{12\tilde{E}_t^{(i)}}S_{i-1}''(x_1)+\bar{\varepsilon}_A^0+\hat{\varepsilon}_A^0\,x_3^{(i-1)}.\end{aligned}\tag{19.56}$$

It follows from (19.36) and (19.37) that

$$\frac{\partial^2U_{i-1}^0(x_1,x_2)}{\partial x_1^2}=-\hat{\varepsilon}_A^0,\quad V_{i-1}^{0\prime}(x_1)=\bar{\varepsilon}_A^0+\hat{\varepsilon}_A^0\,x_3^{(i-1)},\quad i=1,\ldots,N+1.\tag{19.57}$$

On defining

$$\left.\begin{aligned}&\Delta U_i(x_1)=U_i(x_1,x_2)-U_i^0(x_1,x_2),\quad\text{independent of } x_2,\\&\Delta V_i(x_1)=V_i(x_1)-V_i^0(x_1),\end{aligned}\right\}\quad i=0,\ldots,N,\tag{19.58}$$

it follows from (19.55) and (19.56) that for values $i=1,\ldots,N+1$,

$$\begin{aligned}\Delta U_{i-1}''(x_1)&=\frac{h_i^2}{420\tilde{E}_t^{(i)}}\left[49P_{i-1}''''(x_1)+14P_i''''(x_1)-9Q_i''''(x_1)\right]+\frac{P_{i-1}''(x_1)}{2\mu_a^{(i)}}\\&+P_i''(x_1)\left(\frac{1}{2\mu_a^{(i)}}-\frac{\tilde{\nu}_a^{(i)}}{\tilde{E}_A^{(i)}}\right)+Q_i''(x_1)\left(-\frac{1}{5\mu_a^{(i)}}+\frac{2\tilde{\nu}_a^{(i)}}{5\tilde{E}_A^{(i)}}\right)+\frac{2}{h_i^2\tilde{E}_A^{(i)}}Q_i(x_1)-\frac{h_i}{2\tilde{E}_t^{(i)}}S_{i-1}''(x_1),\end{aligned}\tag{19.59}$$

and

$$\Delta V'_{i-1}(x_1) = \frac{h_i^3}{2520\tilde{E}_t^{(i)}}\left[63P''''_{i-1}(x_1) + 21P''''_i(x_1) - 13Q''''_i(x_1)\right]$$
$$+\frac{h_i}{60}\left(\frac{1}{\mu_a^{(i)}} - \frac{2\tilde{\nu}_a^{(i)}}{\tilde{E}_A^{(i)}}\right)\left[5P''_i(x_1) - 5P''_{i-1}(x_1) - Q''_i(x_1)\right] + \frac{1}{h_i\tilde{E}_A^{(i)}}\left[P_{i-1}(x_1) - P_i(x_1) + Q_i(x_1)\right] \quad (19.60)$$
$$-\frac{\tilde{\nu}_a^{(i)}}{\tilde{E}_A^{(i)}}S_{i-1}(x_1) - \frac{h_i^2}{12\tilde{E}_t^{(i)}}S''_{i-1}(x_1),\ i = 1, \ldots, N+1.$$

It is noted from (19.36) and (19.58)$_1$ that

$$\frac{\partial U_i(x_1, x_2)}{\partial x_1} = \Delta U'_i(x_1) - \hat{\varepsilon}_A^0 x_1, \quad (19.61)$$

is independent of both x_2 and x_3, as assumed earlier.

19.3.4 Solving the Recurrence Relations

From the recurrence relation (19.24), the following expression can be derived for $S_i(x_1)$ on using the boundary condition (11.30)$_1$ that $\sigma_{33}^{(i)} = \sigma_t$ on $x_3 = x_3^{(0)} = 0$

$$S_i(x_1) = \frac{1}{6}\sum_{j=1}^{i} h_j\left[Q''_j(x_1) - 3P''_{j-1}(x_1) - 3P''_j(x_1)\right],\quad i = 1, \ldots, N,\quad S_0(x_1) \equiv 0. \quad (19.62)$$

During the bending of the laminate, the external surface displacements on $x_3 = 0$, $U_0(x_1, x_2)$ and $V_0(x_1, x_2)$ are nonzero. To operate the recurrence relations (19.25) and (19.26), it is useful to make use of the displacement differences in terms of the displacements on $x_3 = x_3^{(0)} = 0$, on using (19.58). First, it follows from (19.35)–(19.37) that for values $j = 1, \ldots, N$,

$$U_j^0(x_1, x_2) - U_0^0(x_1, x_2) = -\sum_{i=1}^{j}\tilde{\nu}_a^{(i)}(h_i\bar{\varepsilon}_A^0 + \hat{h}_i\hat{\varepsilon}_A^0) - \sum_{i=1}^{j}\tilde{\nu}_t^{(i)}(h_i\bar{\varepsilon}_T + \hat{h}_i\hat{\varepsilon}_T)$$
$$+\sigma_t\sum_{i=1}^{j}h_i\left(\frac{1}{\tilde{E}_t^{(i)}} - \frac{(\tilde{\nu}_a^{(i)})^2}{\tilde{E}_A^{(i)}}\right) + \Delta T\sum_{i=1}^{j}h_i(\tilde{\alpha}_t^{(i)} + \tilde{\nu}_a^{(i)}\tilde{\alpha}_A^{(i)}), \quad (19.63)$$

$$V_j^0(x_1) - V_0^0(x_1) = \hat{\varepsilon}_A^0 x_3^{(j)} x_1. \quad (19.64)$$

From (19.25) and (19.33), for values $i = 1, \ldots, N$, and on using various definitions given in Section 7.10,

$$U_i(x_1, x_2) - U_{i-1}(x_1, x_2) = \frac{h_i^2}{12\tilde{E}_t^{(i)}}(Q''_i(x_1) - 2P''_i(x_1) - 4P''_{i-1}(x_1)) + h_i\frac{S_{i-1}(x_1)}{\tilde{E}_t^{(i)}}$$
$$-\frac{\tilde{\nu}_a^{(i)}}{\tilde{E}_A^{(i)}}[P_{i-1}(x_1) - P_i(x_1)] - \tilde{\nu}_a^{(i)}(h_i\bar{\varepsilon}_A^0 + \hat{h}_i\hat{\varepsilon}_A^0) - \tilde{\nu}_t^{(i)}(h_i\bar{\varepsilon}_T + \hat{h}_i\hat{\varepsilon}_T) \quad (19.65)$$
$$+h_i\left(\frac{1}{\tilde{E}_t^{(i)}} - \frac{(\tilde{\nu}_a^{(i)})^2}{\tilde{E}_A^{(i)}}\right)\sigma_t + h_i(\tilde{\alpha}_t^{(i)} + \tilde{\nu}_a^{(i)}\tilde{\alpha}_A^{(i)})\Delta T,$$

$$U_i^0(x_1, x_2) - U_{i-1}^0(x_1, x_2) = -\tilde{\nu}_a^{(i)}(h_i\bar{\varepsilon}_A^0 + \hat{h}_i\hat{\varepsilon}_A^0) - \tilde{\nu}_t^{(i)}(h_i\bar{\varepsilon}_T + \hat{h}_i\hat{\varepsilon}_T)$$
$$+h_i\left(\frac{1}{\tilde{E}_t^{(i)}} - \frac{(\tilde{\nu}_a^{(i)})^2}{\tilde{E}_A^{(i)}}\right)\sigma_t + h_i(\tilde{\alpha}_t^{(i)} + \tilde{\nu}_a^{(i)}\tilde{\alpha}_A^{(i)})\Delta T. \quad (19.66)$$

A subtraction using (19.58) then leads to the recurrence relation

$$\Delta U_i(x_1)-\Delta U_{i-1}(x_1)=\frac{h_i^2}{12\tilde{E}_t^{(i)}}\left(Q_i''(x_1)-2P_i''(x_1)-4P_{i-1}''(x_1)\right)$$
$$+\frac{h_i}{\tilde{E}_t^{(i)}}S_{i-1}(x_1)-\frac{\tilde{\nu}_a^{(i)}}{\tilde{E}_A^{(i)}}\left[P_{i-1}(x_1)-P_i(x_1)\right],\ i=1,\dots,N,\quad \Delta U_0(x_1)\equiv 0. \tag{19.67}$$

The following expressions for $\Delta U_i(x_1)$ can then be generated:

$$\Delta U_i(x_1)=\sum_{j=1}^{i}\left[\frac{h_j^2}{12\tilde{E}_t^{(j)}}\left(Q_j''(x_1)-2P_j''(x_1)-4P_{j-1}''(x_1)\right)\right.$$
$$\left.-\frac{\tilde{\nu}_a^{(j)}}{\tilde{E}_A^{(j)}}\left[P_{j-1}(x_1)-P_j(x_1)\right]+\frac{h_j}{\tilde{E}_t^{(j)}}S_{j-1}(x_1)\right],\ i=1,\dots,N,\quad \Delta U_0(x_1)\equiv 0. \tag{19.68}$$

The expressions for $\Delta U_i''(x_1)$ are

$$\Delta U_i''(x_1)=\sum_{j=1}^{i}\left[\frac{h_j^2}{12\tilde{E}_t^{(j)}}\left(Q_j''''(x_1)-2P_j''''(x_1)-4P_{j-1}''''(x_1)\right)\right.$$
$$\left.-\frac{\tilde{\nu}_a^{(j)}}{\tilde{E}_A^{(j)}}\left(P_{j-1}''(x_1)-P_j''(x_1)\right)+\frac{h_j}{\tilde{E}_t^{(j)}}S_{j-1}''(x_1)\right],\ i=1,\dots,N. \tag{19.69}$$

On eliminating $\Delta U_i''(x_1)$ using relations (19.59) and (19.69), and on using (19.62), a set of N relations is obtained which are linear in the unknown functions $P_i(x_1)$, $i=1,\dots,N$, and $Q_i(x_1)$, $i=1,\dots,N+1$, and various even derivatives of these functions.

From (19.26) and (19.34), for values $i=1,\dots,N$,

$$V_i(x_1)-V_{i-1}(x_1)=-\frac{h_i^3}{120\tilde{E}_t^{(i)}}\left[3Q_i'''(x_1)-5P_i'''(x_1)-15P_{i-1}'''(x_1)\right]$$
$$+\frac{h_i}{6}\left[\frac{1}{\mu_a^{(i)}}\left(3P_{i-1}'(x_1)+3P_i'(x_1)-Q_i'(x_1)\right)+\frac{\tilde{\nu}_a^{(i)}}{\tilde{E}_A^{(i)}}\left(3P_{i-1}'(x_1)-3P_i'(x_1)+Q_i'(x_1)\right)\right] \tag{19.70}$$
$$-\frac{h_i^2}{2\tilde{E}_t^{(i)}}S_{i-1}'(x_1)-h_i\frac{\partial U_{i-1}(x_1,x_2)}{\partial x_1},$$

$$V_i^0(x_1)-V_{i-1}^0(x_1)=-h_i\frac{\partial U_{i-1}^0(x_1,x_2)}{\partial x_1}. \tag{19.71}$$

A subtraction using (19.58) then leads to the recurrence relation

$$\Delta V_i(x_1)-\Delta V_{i-1}(x_1)=-\frac{h_i^3}{120\tilde{E}_t^{(i)}}\left[3Q_i'''(x_1)-5P_i'''(x_1)-15P_{i-1}'''(x_1)\right]$$
$$+\frac{h_i}{6}\left[\frac{1}{\mu_a^{(i)}}\left(3P_{i-1}'(x_1)+3P_i'(x_1)-Q_i'(x_1)\right)+\frac{\tilde{\nu}_a^{(i)}}{\tilde{E}_A^{(i)}}\left(3P_{i-1}'(x_1)-3P_i'(x_1)+Q_i'(x_1)\right)\right] \tag{19.72}$$
$$-\frac{h_i^2}{2\tilde{E}_t^{(i)}}S_{i-1}'(x_1)-h_i\Delta U_{i-1}'(x_1),\ i=1,\dots,N,\quad \Delta V_0(x_1)\equiv 0,$$

which provides the following expressions for $\Delta V_i(x_1)$

$$\Delta V_i(x_1) = \sum_{j=1}^{i} \left\{ -\frac{h_j^3}{120\tilde{E}_t^{(j)}} \left(3Q_j'''(x_1) - 5P_j'''(x_1) - 15P_{j-1}'''(x_1)\right) \right.$$
$$+ \frac{h_j}{6}\left[\frac{1}{\mu_a^{(j)}}\left(3P_{j-1}'(x_1) + 3P_j'(x_1) - Q_j'(x_1)\right) + \frac{\tilde{\nu}_a^{(j)}}{\tilde{E}_A^{(j)}}\left(3P_{j-1}'(x_1) - 3P_j'(x_1) + Q_j'(x_1)\right)\right] \tag{19.73}$$
$$\left. - \frac{h_j^2}{2\tilde{E}_t^{(j)}} S_{j-1}'(x_1) - h_j \Delta U_{j-1}'(x_1) \right\}, \quad i = 1, \ldots, N, \qquad \Delta V_0(x_1) \equiv 0.$$

The expressions for $\Delta V_i'(x_1)$ follow from (19.73):

$$\Delta V_i'(x_1) = \sum_{j=1}^{i} \left\{ -\frac{h_j^3}{120\tilde{E}_t^{(j)}} \left(3Q_j''''(x_1) - 5P_j''''(x_1) - 15P_{j-1}''''(x_1)\right) \right.$$
$$+ \frac{h_j}{6}\left[\frac{1}{\mu_a^{(j)}}\left(3P_{j-1}''(x_1) + 3P_j''(x_1) - Q_j''(x_1)\right) + \frac{\tilde{\nu}_a^{(j)}}{\tilde{E}_A^{(j)}}\left(3P_{j-1}''(x_1) - 3P_j''(x_1) + Q_j''(x_1)\right)\right] \tag{19.74}$$
$$\left. - \frac{h_j^2}{2\tilde{E}_t^{(j)}} S_{j-1}''(x_1) - h_j \Delta U_{j-1}''(x_1) \right\}, \quad i = 1, \ldots, N, \qquad \Delta V_0'(x_1) \equiv 0.$$

On eliminating $\Delta V_i'(x_1)$ using relations (19.60) and (19.74), and on using (19.62), another set of N relations is obtained which are also linear in the unknown functions $P_i(x_1)$, $i = 1, \ldots, N$, and $Q_i(x_1)$, $i = 1, \ldots, N+1$, and various even derivatives of these functions. It is clear that $2N$ equations have been derived for the determination of $2N + 1$ unknown functions. A further equation is required, which is obtained by imposing the boundary conditions $(11.30)_1$ that $\sigma_{33} = \sigma_t$ on $x_3 = x_3^{(N+1)}$ for all values of x_1. From (19.13) and (19.62), it follows that

$$\sum_{j=1}^{N+1} h_j \left[Q_j''(x_1) - 3P_{j-1}''(x_1) - 3P_j''(x_1)\right] = 0. \tag{19.75}$$

Upon rearranging (19.75) and integrating twice with respect to x_1,

$$Q_{N+1}(x_1) = \frac{1}{h_{N+1}}\left[3\sum_{j=1}^{N+1} h_j\left(P_{j-1}(x_1) + P_j(x_1)\right) - \sum_{j=1}^{N} h_j Q_j(x_1)\right]. \tag{19.76}$$

This relation is used to eliminate the function $Q_{N+1}(x_1)$ in terms of the other $2N$ unknown functions, remembering from (19.14) that $P_0(x_1) = P_{N+1}(x_1) \equiv 0$. Thus, on eliminating the function $Q_{N+1}(x_1)$ in the two sets of relations obtained by eliminating the functions $\Delta U_i''(x_1)$ and $\Delta V_i'(x_1)$, the following system of fourth-order differential equations is obtained

$$\sum_{i=1}^{N} F_{ij} P_i''''(x_1) + \sum_{i=1}^{N} G_{ij} P_i''(x_1) + \sum_{i=1}^{N} H_{ij} P_i(x_1)$$
$$+ \sum_{i=1}^{N} F_{i+N,j} Q_i''''(x_1) + \sum_{i=1}^{N} G_{i+N,j} Q_i''(x_1) + \sum_{i=1}^{N} H_{i+N,j} Q_i(x_1) = 0, \quad j = 1, \ldots, N, \tag{19.77}$$

and

$$\sum_{i=1}^{N} F_{i,j+N} P_i''''(x_1) + \sum_{i=1}^{N} G_{i,j+N} P_i''(x_1) + \sum_{i=1}^{N} H_{i,j+N} P_i(x_1)$$
$$+ \sum_{i=1}^{N} F_{i+N,j+N} Q_i''''(x_1) + \sum_{i=1}^{N} G_{i+N,j+N} Q_i''(x_1) + \sum_{i=1}^{N} H_{i+N,j+N} Q_i(x_1) = 0, \ j = 1, \ldots, N. \tag{19.78}$$

These ordinary differential equations are solved for the unknown functions $P_i(x_1)$, $Q_i(x_1)$, $i=1,\ldots,N$, using standard techniques, modifying those described in [4, 5] for the cases of both axisymmetric and general symmetric laminate stress-transfer problems (see Appendix H). Having solved these differential equations subject to suitable boundary conditions (Section 19.4), the function $Q_{N+1}(x_1)$ is determined using (19.76). The various coefficients appearing in the differential equations (19.77) and (19.78) are extremely complex expressions that are too cumbersome to determine and include in this chapter. Their numerical values can, however, be easily determined using numerical methods where the residual of each the original differential equation is regarded as a function of the variables $P_i(x_1)$, $Q_i(x_1)$, $i=1,\ldots,N$, and their second- and fourth-order derivatives. The coefficients are then calculated, in turn, by setting all these variables to zero except for the variable that is associated with the coefficient that is actually being calculated.

19.4 Application of the Boundary Conditions

The functions $P_i(x_1)$ and $Q_i(x_1)$, for $i=1,\ldots,N$, that are required to specify the stress and displacement fields of a cracked laminate are now determined as the solution of the $2N$ differential equations (19.77) and (19.78). To ensure a unique solution, $8N$ boundary conditions must be imposed.

By imposing the boundary conditions $(11.31)_2$ and (11.32) in conjunction with (19.13) and (19.76), $4N$ of the boundary conditions are found. It then follows from (19.5) that $\sigma_{13}=0$ on $x_1=\pm L$ for all x_3 only if

$$P_i'(\pm L)=0, \quad i=1,\ldots,N, \tag{19.79}$$

$$Q_i'(\pm L)=0, \quad i=1,\ldots,N. \tag{19.80}$$

Suppose there are m cracked 90° layers for which $(11.31)_1$ is to be satisfied for all x_3. The application of these conditions leads to the following $4m$ boundary conditions, derived from (19.8) and (19.9),

$$Q_i(\pm L)=\frac{1}{2}h_i^2\tilde{E}_{\rm A}^{(i)}\left(\hat{\varepsilon}_{\rm A}^0+\nu_{\rm A}^{(i)}\frac{E_{\rm T}^{(i)}}{E_{\rm A}^{(i)}}\hat{\varepsilon}_{\rm T}\right), \tag{19.81}$$

$$\begin{aligned}\frac{1}{h_i\tilde{E}_{\rm A}^{(i)}}[P_i(\pm L)-P_{i-1}(\pm L)] &= \bar{\varepsilon}_{\rm A}^0+\nu_{\rm A}^{(i)}\frac{E_{\rm T}^{(i)}}{E_{\rm A}^{(i)}}\bar{\varepsilon}_{\rm T} \\ &\quad +\frac{\tilde{\nu}_{\rm a}^{(i)}}{\tilde{E}_{\rm A}^{(i)}}\sigma_{\rm t}-\tilde{\alpha}_{\rm A}^{(i)}\Delta T+\left(x_3^{(i)}-\frac{1}{2}h_i\right)\left(\hat{\varepsilon}_{\rm A}^0+\nu_{\rm A}^{(i)}\frac{E_{\rm T}^{(i)}}{E_{\rm A}^{(i)}}\hat{\varepsilon}_{\rm T}\right).\end{aligned} \tag{19.82}$$

The remaining required boundary conditions are found by demanding that the average and integrated moment of (11.33) are satisfied for the $N+1-m$ uncracked layers. For damaged laminates, it can be shown, from (11.33) and (19.38), that

$$\bar{u}_1^{(i)}(\pm L)=\pm L\left[\bar{\varepsilon}_{\rm A}+\left(x_3^{(i)}-\frac{h_i}{2}\right)\hat{\varepsilon}_{\rm A}\right], \tag{19.83}$$

and from (11.33) and (19.48) that

$$\hat{u}_1^{(i)}(\pm L) = \pm\frac{h_i L}{2}\left[\bar{\varepsilon}_{\mathrm{A}} + \left(x_3^{(i)} - \frac{h_i}{3}\right)\hat{\varepsilon}_{\mathrm{A}}\right], \tag{19.84}$$

where from (19.42), (19.44), (19.50) and (19.54), on substituting for $U_i(x_1)$ and $V_i(x_1)$ using (19.58)–(19.60), $\bar{u}_1^{(i)}(x_1)$ and $\hat{u}_1^{(i)}(x_1)$ can be expressed as

$$\begin{aligned}\bar{u}_1^{(i)}(x_1) = {} & \frac{h_i}{12}\frac{\tilde{\nu}_{\mathrm{a}}^{(i)}}{\tilde{E}_{\mathrm{A}}^{(i)}}\left[4P_{i-1}'(x_1) + 2P_i'(x_1) - Q_i'(x_1)\right] + \frac{1}{h_i\tilde{E}_{\mathrm{A}}^{(i)}}\left[P_{i-1}^*(x_1) - P_i^*(x_1)\right] \\ & - \frac{\tilde{\nu}_{\mathrm{a}}^{(i)}}{\tilde{E}_{\mathrm{A}}^{(i)}} S_{i-1}^*(x_1) + \bar{\varepsilon}_{\mathrm{A}}^0 x_1 + \left(x_3^{(i)} - \frac{h_i}{2}\right)\hat{\varepsilon}_{\mathrm{A}}^0 x_1, \quad i = 1, \ldots, N+1,\end{aligned} \tag{19.85}$$

$$\begin{aligned}\hat{u}_1^{(i)}(x_1) = {} & \frac{h_i^2}{120}\frac{\tilde{\nu}_{\mathrm{a}}^{(i)}}{\tilde{E}_{\mathrm{A}}^{(i)}}\left[25P_{i-1}'(x_1) + 15P_i'(x_1) - 7Q_i'(x_1)\right] + \frac{1}{6\tilde{E}_{\mathrm{A}}^{(i)}}\left[3P_{i-1}^*(x_1) - 3P_i^*(x_1) - Q_i^*(x_1)\right] \\ & - \frac{h_i}{2}\frac{\tilde{\nu}_{\mathrm{a}}^{(i)}}{\tilde{E}_{\mathrm{A}}^{(i)}} S_{i-1}^*(x_1) + \frac{h_i}{2}\bar{\varepsilon}_{\mathrm{A}}^0 x_1 + \frac{h_i}{6}\left(3x_3^{(i)} - h_i\right)\hat{\varepsilon}_{\mathrm{A}}^0 x_1, \quad i = 1, \ldots, N+1,\end{aligned} \tag{19.86}$$

where

$$P_i^*(x_1) = \int_0^{x_1} P_i(s)\,ds, \quad Q_i^*(x_1) = \int_0^{x_1} Q_i(s)\,ds, \quad S_i^*(x_1) = \int_0^{x_1} S_i(s)\,ds, \quad i = 0, \ldots, N+1. \tag{19.87}$$

Owing to (19.62), (19.79) and (19.80), it can be shown from (19.84)–(19.86) that the boundary conditions for uncracked layers are of the form

$$-\frac{2Q_i^*(\pm L)}{h_i^2\tilde{E}_{\mathrm{A}}^{(i)}L} = \pm\left(\hat{\varepsilon}_{\mathrm{A}} - \hat{\varepsilon}_{\mathrm{A}}^0\right), \tag{19.88}$$

$$\frac{1}{h_i\tilde{E}_{\mathrm{A}}^{(i)}L}\left[P_{i-1}^*(\pm L) - P_i^*(\pm L)\right] + \frac{2Q_i^*(\pm L)}{h_i^2\tilde{E}_{\mathrm{A}}^{(i)}L}\left(x_3^{(i)} - \frac{h_i}{2}\right) = \pm\left(\bar{\varepsilon}_{\mathrm{A}} - \bar{\varepsilon}_{\mathrm{A}}^0\right). \tag{19.89}$$

As the parameters $\bar{\varepsilon}_{\mathrm{A}}$ and $\hat{\varepsilon}_{\mathrm{A}}$, applicable to a damaged laminate, are not known in advance they need to be eliminated. This is achieved by subtracting the jth relations (19.88) and (19.89) from the others, yielding the following conditions for uncracked layers:

$$\begin{aligned}& \frac{1}{h_i\tilde{E}_{\mathrm{A}}^{(i)}}\left[P_{i-1}^*(\pm L) - P_i^*(\pm L)\right] - \frac{1}{h_j\tilde{E}_{\mathrm{A}}^{(j)}}\left[P_{j-1}^*(\pm L) - P_j^*(\pm L)\right] \\ & \qquad + \frac{Q_i^*(\pm L)}{h_i^2\tilde{E}_{\mathrm{A}}^{(i)}}\left(2x_3^{(i)} - h_i\right) - \frac{Q_j^*(\pm L)}{h_j^2\tilde{E}_{\mathrm{A}}^{(j)}}\left(2x_3^{(j)} - h_j\right) = 0,\end{aligned} \tag{19.90}$$

$$\frac{Q_i^*(\pm L)}{h_i^2\tilde{E}_{\mathrm{A}}^{(i)}} = \frac{Q_j^*(\pm L)}{h_j^2\tilde{E}_{\mathrm{A}}^{(j)}}, \tag{19.91}$$

where the jth layer is selected to be any one of the uncracked layers.

It is convenient to combine the boundary conditions for the cracked and the uncracked layers into a single set of $2N$ boundary conditions. Upon introducing a new dimensionless parameter λ_i, which takes the value one if the ith layer is cracked and zero otherwise, the conditions (19.81), (19.82), (19.90) and (19.91) are incorporated as follows for $i = 1, \ldots, N+1,\ i \neq j$,

$$\frac{1-\lambda_i}{L}\left[\frac{1}{h_i\tilde{E}_{\mathrm{A}}^{(i)}}\left[P_{i-1}^*(\pm L)-P_i^*(\pm L)\right]-\frac{1}{h_j\tilde{E}_{\mathrm{A}}^{(j)}}\left[P_{j-1}^*(\pm L)-P_j^*(\pm L)\right]\right]$$
$$+\frac{1-\lambda_i}{L}\left[\frac{Q_i^*(\pm L)}{h_i^2\tilde{E}_{\mathrm{A}}^{(i)}}\left(2x_3^{(i)}-h_i\right)-\frac{Q_j^*(\pm L)}{h_j^2\tilde{E}_{\mathrm{A}}^{(j)}}\left(2x_3^{(J)}-h_j\right)\right] \tag{19.92}$$
$$+\frac{\lambda_i}{h_i\tilde{E}_{\mathrm{A}}^{(i)}}\left[P_i(\pm L)-P_{i-1}(\pm L)\right]=\lambda_i\left[\begin{array}{l}\left(\bar{\varepsilon}_{\mathrm{A}}^0+\nu_{\mathrm{A}}^{(i)}\dfrac{E_{\mathrm{T}}^{(i)}}{E_{\mathrm{A}}^{(i)}}\bar{\varepsilon}_{\mathrm{T}}\right)-\tilde{\alpha}_{\mathrm{A}}^{(i)}\Delta T\\ +\left(x_3^{(i)}-\dfrac{h_i}{2}\right)\left(\hat{\varepsilon}_{\mathrm{A}}^0+\nu_{\mathrm{A}}^{(i)}\dfrac{E_{\mathrm{T}}^{(i)}}{E_{\mathrm{A}}^{(i)}}\hat{\varepsilon}_{\mathrm{T}}\right)\end{array}\right],$$

and for $i=1,\ldots,N$, $i\neq j$,

$$\frac{1-\lambda_i}{L}\left[\frac{Q_i^*(\pm L)}{h_i^2\tilde{E}_{\mathrm{A}}^{(i)}}-\frac{Q_j^*(\pm L)}{h_j^2\tilde{E}_{\mathrm{A}}^{(j)}}\right]+\lambda_i\frac{Q_i(\pm L)}{h_i^2\tilde{E}_{\mathrm{A}}^{(i)}}=\frac{\lambda_i}{2}\left(\hat{\varepsilon}_{\mathrm{A}}^0+\nu_{\mathrm{A}}^{(i)}\frac{E_{\mathrm{T}}^{(i)}}{E_{\mathrm{A}}^{(i)}}\hat{\varepsilon}_{\mathrm{T}}\right). \tag{19.93}$$

Relations (19.79), (19.80), (19.92) and (19.93) provide the necessary $8N$ boundary conditions for a unique solution of the differential equations (19.77) and (19.78).

The effective in-plane transverse stress applied to the laminate is defined to be

$$\sigma_{\mathrm{T}}=\frac{1}{2hL}\sum_{i=1}^{N+1}\int_{x_3^{(i-1)}}^{x_3^{(i)}}\int_{-L}^{L}\sigma_{22}^{(i)}\,\mathrm{d}x_1\mathrm{d}x_3. \tag{19.94}$$

The bending moment in the x_2–x_3 plane per unit area M_{T} about the midplane of the laminate is given by

$$M_{\mathrm{T}}=\frac{1}{2hL}\sum_{i=1}^{N+1}\int_{x_3^{(i-1)}}^{x_3^{(i)}}\left(x_3-\frac{h}{2}\right)\int_{-L}^{L}\sigma_{22}^{(i)}\,\mathrm{d}x_1\mathrm{d}x_3. \tag{19.95}$$

On using (19.94) for both a damaged and undamaged laminate, it is clear that

$$\sigma_{\mathrm{T}}=\sigma_{\mathrm{T}}^0+\frac{1}{2hL}\sum_{i=1}^{N+1}\int_{x_3^{(i-1)}}^{x_3^{(i)}}\int_{-L}^{L}\left[\sigma_{22}^{(i)}-\sigma_{\mathrm{T}}^{(i)}(x_3)\right]\mathrm{d}x_1\mathrm{d}x_3, \tag{19.96}$$

where σ_{T}^0 is the effective transverse stress applied to an undamaged laminate subject to the same loading conditions, where the stress distribution in a damaged laminate $\sigma_{22}^{(i)}$ is given by (11.16) or (19.10). Similarly, expression (19.95) for the transverse bending moment per unit area may now be written as

$$M_{\mathrm{T}}=M_{\mathrm{T}}^0+\frac{1}{2hL}\sum_{i=1}^{N+1}\int_{x_3^{(i-1)}}^{x_3^{(i)}}\left(x_3-\frac{h}{2}\right)\int_{-L}^{L}\left[\sigma_{22}^{(i)}-\sigma_{\mathrm{T}}^{(i)}(x_3)\right]\mathrm{d}x_1\mathrm{d}x_3, \tag{19.97}$$

where M_{T}^0 is the effective transverse bending moment per unit area applied to an undamaged laminate subject to the same loading conditions. On using (19.11), (19.62), (19.79), (19.80) and (19.87) it can be shown that

$$\int_{-L}^{L}\left[\sigma_{22}^{(i)}-\sigma_{\mathrm{T}}^{(i)}(x_3)\right]\mathrm{d}x_1=2\nu_{\mathrm{A}}^{(i)}\frac{E_{\mathrm{T}}^{(i)}}{E_{\mathrm{A}}^{(i)}}\left\{\frac{1}{h_i}\left[P_{i-1}^*(L)-P_i^*(L)\right]-\frac{Q_i^*(L)}{h_i}(2\xi_i-1)\right\}. \tag{19.98}$$

As, on using (19.2)$_2$,

$$\int_{x_3^{(i-1)}}^{x_3^{(i)}} (2\xi_i - 1)\,\mathrm{d}x_3 = \frac{1}{h_i}\int_{x_3^{(i-1)}}^{x_3^{(i)}} 2x_3\,\mathrm{d}x_3 - \frac{1}{h_i}\left(x_3^{(i-1)} + x_3^{(i)}\right)\int_{x_3^{(i-1)}}^{x_3^{(i)}} \mathrm{d}x_3 = 0, \tag{19.99}$$

it then follows from (19.96) and (19.98) that

$$\sigma_\mathrm{T} = \sigma_\mathrm{T}^0 + \frac{1}{hL}\sum_{i=1}^{N+1} \nu_\mathrm{A}^{(i)} \frac{E_\mathrm{T}^{(i)}}{E_\mathrm{A}^{(i)}}\left[P_{i-1}^*(L) - P_i^*(L)\right]. \tag{19.100}$$

Similarly, from (19.97) and (19.98),

$$\begin{aligned} M_\mathrm{T} = M_\mathrm{T}^0 &+ \frac{1}{hL}\sum_{i=1}^{N+1} \nu_\mathrm{A}^{(i)} \frac{E_\mathrm{T}^{(i)}}{E_\mathrm{A}^{(i)}}\frac{1}{h_i}\left[P_{i-1}^*(L) - P_i^*(L)\right]\int_{x_3^{(i-1)}}^{x_3^{(i)}} \left(x_3 - \frac{h}{2}\right)\mathrm{d}x_3 \\ &- \frac{1}{hL}\sum_{i=1}^{N+1} \nu_\mathrm{A}^{(i)} \frac{E_\mathrm{T}^{(i)}}{E_\mathrm{A}^{(i)}}\frac{Q_i^*(L)}{h_i}\int_{x_3^{(i-1)}}^{x_3^{(i)}} \left(x_3 - \frac{h}{2}\right)(2\xi_i - 1)\,\mathrm{d}x_3 . \end{aligned} \tag{19.101}$$

As

$$\begin{aligned} \int_{x_3^{(i-1)}}^{x_3^{(i)}} \left(x_3 - \frac{h}{2}\right)\mathrm{d}x_3 &= \frac{1}{2}h_i\left(x_3^{(i)} + x_3^{(i-1)}\right) - \frac{hh_i}{2} = h_i\left(x_3^{(i)} - \frac{h_i + h}{2}\right), \\ \int_{x_3^{(i-1)}}^{x_3^{(i)}} \left(x_3 - \frac{h}{2}\right)(2\xi_i - 1)\,\mathrm{d}x_3 &= \int_{x_3^{(i-1)}}^{x_3^{(i)}} x_3(2\xi_i - 1)\,\mathrm{d}x_3 \\ &= \frac{2}{h_i}\int_{x_3^{(i-1)}}^{x_3^{(i)}} x_3^2\,\mathrm{d}x_3 - \frac{x_3^{(i-1)} + x_3^{(i)}}{h_i}\int_{x_3^{(i-1)}}^{x_3^{(i)}} x_3\,\mathrm{d}x_3 = \frac{h_i^2}{6}, \end{aligned} \tag{19.102}$$

it follows from (19.101) that

$$M_\mathrm{T} = M_\mathrm{T}^0 + \frac{1}{hL}\sum_{i=1}^{N+1} \nu_\mathrm{A}^{(i)} \frac{E_\mathrm{T}^{(i)}}{E_\mathrm{A}^{(i)}}\left[\left(x_3^{(i)} - \frac{h_i + h}{2}\right)\left[P_{i-1}^*(L) - P_i^*(L)\right] - \frac{h_i}{6}Q_i^*(L)\right]. \tag{19.103}$$

The effective through-thickness strain ε_t is defined in general by (11.36), namely,

$$\varepsilon_\mathrm{t} = \frac{1}{4hLW}\int_{-W}^{W}\int_{-L}^{L}\left[u_3(x_1,x_2,h) - u_3(x_1,x_2,0)\right]\mathrm{d}x_1\mathrm{d}x_2. \tag{19.104}$$

From (19.21) and (19.58), it follows that for the ith ply the through-thickness displacement is $u_3^{(i)}(x_1,x_2,x_3^{(i-1)}) \equiv U_{i-1}(x_1,x_2)$ implying that

$$u_3(x_1,x_2,h) - u_3(x_1,x_2,0) \equiv U_N(x_1,x_2) - U_0(x_1,x_2) = \Delta U_N(x_1),$$

where the function ΔU_N is given by relation (19.68). It then follows from (19.104) that

$$\varepsilon_\mathrm{t} = \frac{1}{2hL}\int_{-L}^{L}\Delta U_N(x_1)\,dx_1 = \frac{1}{hL}\int_0^L \Delta U_N(x_1)\,\mathrm{d}x_1. \tag{19.105}$$

$$\begin{aligned} \Delta U_N(x_1) = \sum_{j=1}^{N}\Bigg[&\frac{h_j^2}{12\tilde{E}_t^{(j)}}\left(Q_j''(x_1) - 2P_j''(x_1) - 4P_{j-1}''(x_1)\right) \\ &- \frac{\tilde{\nu}_\mathrm{a}^{(j)}}{\tilde{E}_\mathrm{A}^{(j)}}\left[P_{j-1}(x_1) - P_j(x_1)\right] + \frac{h_j}{\tilde{E}_\mathrm{t}^{(j)}}S_{j-1}(x_1)\Bigg]. \end{aligned} \tag{19.106}$$

On substituting expression (19.106) into (19.105) and then making use of relations (19.62) and (19.87)$_1$, and conditions (19.79) and (19.80), it follows that

$$\varepsilon_{\mathrm{t}} = \frac{1}{2hL}\int_{-L}^{L}\frac{\tilde{\nu}_{\mathrm{a}}^{(j)}}{\tilde{E}_{\mathrm{A}}^{(j)}}\left[P_j(x_1)-P_{j-1}(x_1)\right]dx_1 = \frac{1}{hL}\frac{\tilde{\nu}_{\mathrm{a}}^{(j)}}{\tilde{E}_{\mathrm{A}}^{(j)}}\left[P_j^*(L)-P_{j-1}^*(L)\right]. \tag{19.107}$$

19.5 Determination of Effective Constants for Undamaged and Damaged Laminates

The objective now is to determine the effective thermoelastic constants of both an undamaged and damaged laminate. When first considering undamaged laminates, it will be assumed that a uniform through-thickness stress σ_{t} is acting on the external faces of the nonsymmetric cross-ply laminate. When considering the corresponding damaged laminate, it is assumed that the distribution of ply cracks in one or more of the 90° plies is uniform, and that there is an identical distribution in each of the cracked 90° plies such that the ply cracks are coplanar.

In the absence of in-plane shear deformation, the stress-strain equations (7.144), (7.145), (7.147), (7.148) and (7.156) describe the form in which undamaged laminate properties are to be expressed, and they are first rewritten in the form

$$\sigma_{\mathrm{A}}^0 = \Omega_{11}^{(L)}\bar{\varepsilon}_{\mathrm{A}}^0 + \Omega_{12}^{(L)}\bar{\varepsilon}_{\mathrm{T}}^0 + \hat{\Omega}_{11}^{(L)}\hat{\varepsilon}_{\mathrm{A}}^0 + \hat{\Omega}_{12}^{(L)}\hat{\varepsilon}_{\mathrm{T}}^0 + \Omega_{13}^{(L)}\sigma_{\mathrm{t}} - \omega_1^{(L)}\Delta T, \tag{19.108}$$

$$\sigma_{\mathrm{T}}^0 = \Omega_{12}^{(L)}\bar{\varepsilon}_{\mathrm{A}}^0 + \Omega_{22}^{(L)}\bar{\varepsilon}_{\mathrm{T}}^0 + \hat{\Omega}_{12}^{(L)}\hat{\varepsilon}_{\mathrm{A}}^0 + \hat{\Omega}_{22}^{(L)}\hat{\varepsilon}_{\mathrm{T}}^0 + \Omega_{23}^{(L)}\sigma_{\mathrm{t}} - \omega_2^{(L)}\Delta T, \tag{19.109}$$

$$M_{\mathrm{A}}^0 + \frac{1}{2}h\sigma_{\mathrm{A}}^0 = \hat{\Omega}_{11}^{(L)}\bar{\varepsilon}_{\mathrm{A}}^0 + \hat{\Omega}_{12}^{(L)}\bar{\varepsilon}_{\mathrm{T}}^0 + \tilde{\Omega}_{11}^{(L)}\hat{\varepsilon}_{\mathrm{A}}^0 + \tilde{\Omega}_{12}^{(L)}\hat{\varepsilon}_{\mathrm{T}}^0 + \hat{\Omega}_{13}^{(L)}\sigma_{\mathrm{t}} - \hat{\omega}_1^{(L)}\Delta T, \tag{19.110}$$

$$M_{\mathrm{T}}^0 + \frac{1}{2}h\sigma_{\mathrm{T}}^0 = \hat{\Omega}_{12}^{(L)}\bar{\varepsilon}_{\mathrm{A}}^0 + \hat{\Omega}_{22}^{(L)}\bar{\varepsilon}_{\mathrm{T}}^0 + \tilde{\Omega}_{12}^{(L)}\hat{\varepsilon}_{\mathrm{A}}^0 + \tilde{\Omega}_{22}^{(L)}\hat{\varepsilon}_{\mathrm{T}}^0 + \hat{\Omega}_{23}^{(L)}\sigma_{\mathrm{t}} - \hat{\omega}_2^{(L)}\Delta T, \tag{19.111}$$

$$\varepsilon_{\mathrm{t}}^0 = -\Omega_{13}^{(L)}\bar{\varepsilon}_{\mathrm{A}}^0 - \hat{\Omega}_{23}^{(L)}\bar{\varepsilon}_{\mathrm{T}}^0 - \hat{\Omega}_{13}^{(L)}\hat{\varepsilon}_{\mathrm{A}}^0 - \hat{\Omega}_{23}^{(L)}\hat{\varepsilon}_{\mathrm{T}}^0 + \Omega_{33}^{(L)}\sigma_{\mathrm{t}} + \omega_3^{(L)}\Delta T, \tag{19.112}$$

where

$$\begin{aligned}
\Omega_{11}^{(L)} &= \frac{1}{h}\sum_{i=1}^{N+1}h_i\tilde{E}_{\mathrm{A}}^{(i)}, & \Omega_{22}^{(L)} &= \frac{1}{h}\sum_{i=1}^{N+1}h_i\tilde{E}_{\mathrm{T}}^{(i)}, & \Omega_{12}^{(L)} &= \frac{1}{h}\sum_{i=1}^{N+1}h_i\nu_{\mathrm{A}}^i\tilde{E}_{T}^{(i)},\\
\hat{\Omega}_{11}^{(L)} &= \frac{1}{h}\sum_{i=1}^{N+1}\hat{h}_i\tilde{E}_{\mathrm{A}}^{(i)}, & \hat{\Omega}_{22}^{(L)} &= \frac{1}{h}\sum_{i=1}^{N+1}\hat{h}_i\tilde{E}_{\mathrm{T}}^{(i)}, & \hat{\Omega}_{12}^{(L)} &= \frac{1}{h}\sum_{i=1}^{N+1}\hat{h}_i\nu_{\mathrm{A}}^i\tilde{E}_{\mathrm{T}}^{(i)},\\
\tilde{\Omega}_{11}^{(L)} &= \frac{1}{h}\sum_{i=1}^{N+1}\tilde{h}_i\tilde{E}_{\mathrm{A}}^{(i)}, & \tilde{\Omega}_{22}^{(L)} &= \frac{1}{h}\sum_{i=1}^{N+1}\tilde{h}_i\tilde{E}_{\mathrm{T}}^{(i)}, & \tilde{\Omega}_{12}^{(L)} &= \frac{1}{h}\sum_{i=1}^{N+1}\tilde{h}_i\nu_{\mathrm{A}}^i\tilde{E}_{\mathrm{T}}^{(i)},
\end{aligned} \tag{19.113}$$

$$\begin{aligned}
\Omega_{13}^{(L)} &= \frac{1}{h}\sum_{i=1}^{N+1}h_i\tilde{\nu}_{\mathrm{a}}^{(i)}, & \Omega_{23}^{(L)} &= \frac{1}{h}\sum_{i=1}^{N+1}h_i\tilde{\nu}_{\mathrm{t}}^{(i)},\\
\hat{\Omega}_{13}^{(L)} &= \frac{1}{h}\sum_{i=1}^{N+1}\hat{h}_i\tilde{\nu}_{\mathrm{a}}^{(i)}, & \hat{\Omega}_{23}^{(L)} &= \frac{1}{h}\sum_{i=1}^{N+1}\hat{h}_i\tilde{\nu}_{\mathrm{t}}^{(i)},
\end{aligned} \tag{19.114}$$

$$\begin{aligned}
\omega_1^{(L)} &= \frac{1}{h}\sum_{i=1}^{N+1}h_i\tilde{E}_{\mathrm{A}}^{(i)}\tilde{\alpha}_{\mathrm{A}}^{(i)}, & \omega_2^{(L)} &= \frac{1}{h}\sum_{i=1}^{N+1}h_i\tilde{E}_{\mathrm{T}}^{(i)}\tilde{\alpha}_{\mathrm{T}}^{(i)},\\
\hat{\omega}_1^{(L)} &= \frac{1}{h}\sum_{i=1}^{N+1}\hat{h}_i\tilde{E}_{\mathrm{A}}^{(i)}\tilde{\alpha}_{\mathrm{A}}^{(i)}, & \hat{\omega}_2^{(L)} &= \frac{1}{h}\sum_{i=1}^{N+1}\hat{h}_i\tilde{E}_{\mathrm{T}}^{(i)}\tilde{\alpha}_{\mathrm{T}}^{(i)},
\end{aligned} \tag{19.115}$$

$$\Omega_{23}^{(\mathrm{L})} = \frac{1}{h}\sum_{i=1}^{N+1} h_i \tilde{\nu}_{\mathrm{t}}^{(i)}, \qquad \hat{\Omega}_{23}^{(\mathrm{L})} = \frac{1}{h}\sum_{i=1}^{N+1} \hat{h}_i \tilde{\nu}_{\mathrm{t}}^{(i)},$$
$$\Omega_{33}^{(\mathrm{L})} = \frac{1}{h}\sum_{i=1}^{N+1} h_i \left(\frac{1}{\tilde{E}_{\mathrm{t}}^{(i)}} - \frac{(\tilde{\nu}_{\mathrm{a}}^{(i)})^2}{\tilde{E}_{\mathrm{A}}^{(i)}}\right), \qquad \omega_3^{(\mathrm{L})} = \frac{1}{h}\sum_{i=1}^{N+1} h_i \left(\tilde{\alpha}_{\mathrm{t}}^{(i)} + \tilde{\nu}_{\mathrm{a}}^{(i)} \tilde{\alpha}_{\mathrm{A}}^{(i)}\right), \tag{19.116}$$

where

$$h_i = x_i - x_{i-1}, \qquad \hat{h}_i = \frac{1}{2}\left(x_i^2 - x_{i-1}^2\right), \qquad \tilde{h}_i = \frac{1}{3}\left(x_i^3 - x_{i-1}^3\right). \tag{19.117}$$

On using (19.108), relation (19.110) may be written as

$$M_{\mathrm{A}}^0 = \left(\hat{\Omega}_{11}^{(\mathrm{L})} - \tfrac{1}{2}h\Omega_{11}^{(\mathrm{L})}\right)\bar{\varepsilon}_{\mathrm{A}}^0 + \left(\hat{\Omega}_{12}^{(\mathrm{L})} - \tfrac{1}{2}h\Omega_{12}^{(\mathrm{L})}\right)\bar{\varepsilon}_{\mathrm{T}}^0 + \left(\tilde{\Omega}_{11}^{(\mathrm{L})} - \tfrac{1}{2}h\hat{\Omega}_{11}^{(\mathrm{L})}\right)\hat{\varepsilon}_{\mathrm{A}}^0$$
$$+ \left(\tilde{\Omega}_{12}^{(\mathrm{L})} - \tfrac{1}{2}h\hat{\Omega}_{12}^{(\mathrm{L})}\right)\hat{\varepsilon}_{\mathrm{T}}^0 + \left(\hat{\Omega}_{13}^{(\mathrm{L})} - \tfrac{1}{2}h\Omega_{13}^{(\mathrm{L})}\right)\sigma_{\mathrm{t}} - \left(\hat{\omega}_1^{(\mathrm{L})} - \tfrac{1}{2}h\omega_1^{(\mathrm{L})}\right)\Delta T. \tag{19.118}$$

Similarly, on using (19.109), relation (19.111) may be written as

$$M_{\mathrm{T}}^0 = \left(\hat{\Omega}_{12}^{(\mathrm{L})} - \tfrac{1}{2}h\Omega_{12}^{(\mathrm{L})}\right)\bar{\varepsilon}_{\mathrm{A}}^0 + \left(\hat{\Omega}_{22}^{(\mathrm{L})} - \tfrac{1}{2}h\Omega_{22}^{(\mathrm{L})}\right)\bar{\varepsilon}_{\mathrm{T}}^0 + \left(\tilde{\Omega}_{12}^{(\mathrm{L})} - \tfrac{1}{2}h\hat{\Omega}_{12}^{(\mathrm{L})}\right)\hat{\varepsilon}_{\mathrm{A}}^0$$
$$+ \left(\tilde{\Omega}_{22}^{(\mathrm{L})} - \tfrac{1}{2}h\hat{\Omega}_{22}^{(\mathrm{L})}\right)\hat{\varepsilon}_{\mathrm{T}}^0 + \left(\hat{\Omega}_{23}^{(\mathrm{L})} - \tfrac{1}{2}h\Omega_{23}^{(\mathrm{L})}\right)\sigma_{\mathrm{t}} - \left(\hat{\omega}_2^{(\mathrm{L})} - \tfrac{1}{2}h\omega_2^{(\mathrm{L})}\right)\Delta T. \tag{19.119}$$

It follows from (19.108), (19.109), (19.112), (19.118) and (19.119) that

$$\sigma_{\mathrm{A}}^0 = a_{11}\bar{\varepsilon}_{\mathrm{A}}^0 + a_{12}\bar{\varepsilon}_{\mathrm{T}}^0 + a_{13}\hat{\varepsilon}_{\mathrm{A}}^0 + a_{14}\hat{\varepsilon}_{\mathrm{T}}^0 + a_{15}\sigma_{\mathrm{t}} + a_{16}\,\Delta T, \tag{19.120}$$

$$M_{\mathrm{A}}^0 = a_{21}\bar{\varepsilon}_{\mathrm{A}}^0 + a_{22}\bar{\varepsilon}_{\mathrm{T}}^0 + a_{23}\hat{\varepsilon}_{\mathrm{A}}^0 + a_{24}\hat{\varepsilon}_{\mathrm{T}}^0 + a_{25}\sigma_{\mathrm{t}} + a_{26}\,\Delta T, \tag{19.121}$$

$$\sigma_{\mathrm{T}}^0 = a_{31}\bar{\varepsilon}_{\mathrm{A}}^0 + a_{32}\bar{\varepsilon}_{\mathrm{T}}^0 + a_{33}\hat{\varepsilon}_{\mathrm{A}}^0 + a_{34}\hat{\varepsilon}_{\mathrm{T}}^0 + a_{35}\sigma_{\mathrm{t}} + a_{36}\,\Delta T, \tag{19.122}$$

$$M_{\mathrm{T}}^0 = a_{41}\bar{\varepsilon}_{\mathrm{A}}^0 + a_{42}\bar{\varepsilon}_{\mathrm{T}}^0 + a_{43}\hat{\varepsilon}_{\mathrm{A}}^0 + a_{44}\hat{\varepsilon}_{\mathrm{T}}^0 + a_{45}\sigma_{\mathrm{t}} + a_{46}\,\Delta T, \tag{19.123}$$

$$\varepsilon_{\mathrm{t}}^0 = a_{51}\bar{\varepsilon}_{\mathrm{A}}^0 + a_{52}\bar{\varepsilon}_{\mathrm{T}}^0 + a_{53}\hat{\varepsilon}_{\mathrm{A}}^0 + a_{54}\hat{\varepsilon}_{\mathrm{T}}^0 + a_{55}\sigma_{\mathrm{t}} + a_{56}\,\Delta T, \tag{19.124}$$

where the coefficients a_{ij} are given by

$$\begin{gathered}
a_{11} = \Omega_{11}^{(L)}, \quad a_{12} = \Omega_{12}^{(L)}, \quad a_{13} = \hat{\Omega}_{11}^{(L)}, \quad a_{14} = \hat{\Omega}_{12}^{(L)}, \quad a_{15} = \Omega_{13}^{(L)}, \quad a_{16} = -\omega_1^{(L)}, \\
a_{21} = \hat{\Omega}_{11}^{(L)} - \tfrac{1}{2}h\Omega_{11}^{(L)}, \quad a_{22} = \hat{\Omega}_{12}^{(L)} - \tfrac{1}{2}h\Omega_{12}^{(L)}, \quad a_{23} = \tilde{\Omega}_{11}^{(L)} - \tfrac{1}{2}h\hat{\Omega}_{11}^{(L)}, \\
a_{24} = \tilde{\Omega}_{12}^{(L)} - \tfrac{1}{2}h\hat{\Omega}_{12}^{(L)}, \quad a_{25} = \hat{\Omega}_{13}^{(L)} - \tfrac{1}{2}h\Omega_{13}^{(L)}, \quad a_{26} = -\hat{\omega}_1^{(L)} + \tfrac{1}{2}h\omega_1^{(L)}, \\
a_{31} = \Omega_{12}^{(L)}, \quad a_{32} = \Omega_{22}^{(L)}, \quad a_{33} = \hat{\Omega}_{12}^{(L)}, \quad \hat{a}_{34} = \hat{\Omega}_{22}^{(L)}, \quad a_{35} = \Omega_{23}^{(L)}, \quad a_{36} = -\omega_2^{(L)}, \\
a_{41} = \hat{\Omega}_{12}^{(L)} - \tfrac{1}{2}h\Omega_{12}^{(L)}, \quad a_{42} = \hat{\Omega}_{22}^{(L)} - \tfrac{1}{2}h\Omega_{22}^{(L)}, \quad a_{43} = \tilde{\Omega}_{12}^{(L)} - \tfrac{1}{2}h\hat{\Omega}_{12}^{(L)}, \\
a_{44} = \tilde{\Omega}_{22}^{(L)} - \tfrac{1}{2}h\hat{\Omega}_{22}^{(L)}, \quad a_{45} = \hat{\Omega}_{23}^{(L)} - \tfrac{1}{2}h\Omega_{23}^{(L)}, \quad a_{46} = -\hat{\omega}_2^{(L)} + \tfrac{1}{2}h\omega_2^{(L)}, \\
a_{51} = -\Omega_{13}^{(L)}, \ a_{52} = -\Omega_{23}^{(L)}, \ a_{53} = -\hat{\Omega}_{13}^{(L)}, \ a_{54} = -\hat{\Omega}_{23}^{(L)}, \ a_{55} = \Omega_{33}^{(L)}, \quad a_{56} = \omega_3^{(L)}.
\end{gathered} \tag{19.125}$$

Relations (19.120) and (19.121) are now written as

$$a_{11}\bar{\varepsilon}_{\mathrm{A}}^0 + a_{13}\hat{\varepsilon}_{\mathrm{A}}^0 = \sigma_{\mathrm{A}}^0 - a_{12}\bar{\varepsilon}_{\mathrm{T}}^0 - a_{14}\hat{\varepsilon}_{\mathrm{T}}^0 - a_{15}\sigma_{\mathrm{t}} - a_{16}\,\Delta T,$$
$$a_{21}\bar{\varepsilon}_{\mathrm{A}}^0 + a_{23}\hat{\varepsilon}_{\mathrm{A}}^0 = M_{\mathrm{A}}^0 - a_{22}\bar{\varepsilon}_{\mathrm{T}}^0 - a_{24}\hat{\varepsilon}_{\mathrm{T}}^0 - a_{25}\sigma_{\mathrm{t}} - a_{26}\,\Delta T, \tag{19.126}$$

so that, on solving for $\bar{\varepsilon}_{\mathrm{A}}^0$ and $\hat{\varepsilon}_{\mathrm{A}}^0$,

$$\left(a_{11}a_{23} - a_{21}a_{13}\right)\bar{\varepsilon}_{\mathrm{A}}^0 = a_{23}\sigma_{\mathrm{A}}^0 - a_{13}M_{\mathrm{A}}^0 + \left(a_{13}a_{22} - a_{12}a_{23}\right)\bar{\varepsilon}_{\mathrm{T}}^0 + \left(a_{13}a_{24} - a_{23}a_{14}\right)\hat{\varepsilon}_{\mathrm{T}}^0$$
$$+ \left(a_{13}a_{25} - a_{23}a_{15}\right)\sigma_{\mathrm{t}} + \left(a_{13}a_{26} - a_{23}a_{16}\right)\Delta T,$$
$$\left(a_{11}a_{23} - a_{21}a_{13}\right)\hat{\varepsilon}_{\mathrm{A}}^0 = -a_{21}\sigma_{\mathrm{A}}^0 + a_{11}M_{\mathrm{A}}^0 + \left(a_{21}a_{12} - a_{11}a_{22}\right)\bar{\varepsilon}_{\mathrm{T}}^0 + \left(a_{21}a_{14} - a_{11}a_{24}\right)\hat{\varepsilon}_{\mathrm{T}}^0$$
$$+ \left(a_{21}a_{15} - a_{11}a_{25}\right)\sigma_{\mathrm{t}} + \left(a_{21}a_{16} - a_{11}a_{26}\right)\Delta T. \tag{19.127}$$

This result is now written in the form

$$\begin{aligned}\bar{\varepsilon}_A^0 &= b_{11}\sigma_A^0 + b_{12}M_A^0 + b_{13}\bar{\varepsilon}_T^0 + b_{14}\hat{\varepsilon}_T^0 + b_{15}\sigma_t + b_{16}\,\Delta T,\\ \hat{\varepsilon}_A^0 &= b_{21}\sigma_A^0 + b_{22}M_A^0 + b_{23}\bar{\varepsilon}_T^0 + b_{24}\hat{\varepsilon}_T^0 + b_{25}\sigma_t + b_{26}\,\Delta T,\end{aligned} \tag{19.128}$$

where the coefficients b_{ij} are given by

$$\begin{aligned}
b_{11} &= \frac{a_{23}}{a_{11}a_{23}-a_{21}a_{13}},\ b_{12} = -\frac{a_{13}}{a_{11}a_{23}-a_{21}a_{13}},\ b_{13} = \frac{a_{13}a_{22}-a_{12}a_{23}}{a_{11}a_{23}-a_{21}a_{13}},\ b_{14} = \frac{a_{13}a_{24}-a_{23}a_{14}}{a_{11}a_{23}-a_{21}a_{13}},\\
b_{21} &= -\frac{a_{21}}{a_{11}a_{23}-a_{21}a_{13}},\ b_{22} = \frac{a_{11}}{a_{11}a_{23}-a_{21}a_{13}},\ b_{23} = \frac{a_{21}a_{12}-a_{11}a_{22}}{a_{11}a_{23}-a_{21}a_{13}},\ b_{24} = \frac{a_{21}a_{14}-a_{11}a_{24}}{a_{11}a_{23}-a_{21}a_{13}},\\
b_{15} &= \frac{a_{13}a_{25}-a_{23}a_{15}}{a_{11}a_{23}-a_{21}a_{13}},\ b_{16} = \frac{a_{13}a_{26}-a_{23}a_{16}}{a_{11}a_{23}-a_{21}a_{13}},\ b_{25} = \frac{a_{21}a_{15}-a_{11}a_{25}}{a_{11}a_{23}-a_{21}a_{13}},\ b_{26} = \frac{a_{21}a_{16}-a_{11}a_{26}}{a_{11}a_{23}-a_{21}a_{13}}.
\end{aligned} \tag{19.129}$$

These relations apply only to an undamaged laminate.

When considering the loading of damaged laminates it is convenient to assume that the axial stress σ_A, through-thickness stress σ_t and axial bending moment per unit area M_A are specified together with the transverse strains $\bar{\varepsilon}_T$ and $\hat{\varepsilon}_T$ which are assumed to be imposed for both damaged and undamaged laminates as generalised plane strain conditions are assumed. It is required to determine the corresponding values of the axial strains $\bar{\varepsilon}_A$, ε_t, $\hat{\varepsilon}_A$ and the transverse stress σ_T and the transverse bending moment per unit area M_T. For an undamaged laminate these loading parameters lead to axial strains $\bar{\varepsilon}_A^0$, $\hat{\varepsilon}_A^0$ and to a transverse stress σ_T^0 and a transverse bending moment per unit area M_T^0. Given values of σ_A, σ_t, M_A, $\bar{\varepsilon}_T$ and $\hat{\varepsilon}_T$, relations (19.128) enable the axial strains $\bar{\varepsilon}_A^0$, $\hat{\varepsilon}_A^0$ for the corresponding undamaged laminate to be calculated.

For damaged laminates, it follows from the results (19.88), (19.89), (19.100), (19.103) and (19.107) that the following relations are valid

$$\bar{\varepsilon}_A = c_{11}\bar{\varepsilon}_A^0 + c_{12}\hat{\varepsilon}_A^0 + c_{13}\bar{\varepsilon}_T + c_{14}\hat{\varepsilon}_T + c_{15}\sigma_t + c_{16}\,\Delta T, \tag{19.130}$$

$$\hat{\varepsilon}_A = c_{21}\bar{\varepsilon}_A^0 + c_{22}\hat{\varepsilon}_A^0 + c_{23}\bar{\varepsilon}_T + c_{24}\hat{\varepsilon}_T + c_{25}\sigma_t + c_{26}\,\Delta T, \tag{19.131}$$

$$\sigma_T = c_{31}\bar{\varepsilon}_A^0 + c_{32}\hat{\varepsilon}_A^0 + c_{33}\bar{\varepsilon}_T + c_{34}\hat{\varepsilon}_T + c_{35}\sigma_t + c_{36}\,\Delta T, \tag{19.132}$$

$$M_T = c_{41}\bar{\varepsilon}_A^0 + c_{42}\hat{\varepsilon}_A^0 + c_{43}\bar{\varepsilon}_T + c_{44}\hat{\varepsilon}_T + c_{45}\sigma_t + c_{46}\,\Delta T, \tag{19.133}$$

$$\varepsilon_t = c_{51}\bar{\varepsilon}_A^0 + c_{52}\hat{\varepsilon}_A^0 + c_{53}\bar{\varepsilon}_T + c_{54}\hat{\varepsilon}_T + c_{55}\sigma_t + c_{56}\,\Delta T, \tag{19.134}$$

where values of all the coefficients c_{ij} can be determined by calculating values of $\bar{\varepsilon}_A$, $\hat{\varepsilon}_A$, σ_T, M_T, ε_t and ΔT for the following special loading cases:

$$(\bar{\varepsilon}_A^0, \hat{\varepsilon}_A^0, \bar{\varepsilon}_T, \hat{\varepsilon}_T, \sigma_t, \Delta T) = \begin{cases} (1\ \ 0\ \ 0\ \ 0\ \ 0\ \ 0),\\ (0\ \ 1\ \ 0\ \ 0\ \ 0\ \ 0),\\ (0\ \ 0\ \ 1\ \ 0\ \ 0\ \ 0),\\ (0\ \ 0\ \ 0\ \ 1\ \ 0\ \ 0),\\ (0\ \ 0\ \ 0\ \ 0\ \ 1\ \ 0),\\ (0\ \ 0\ \ 0\ \ 0\ \ 0\ \ 1). \end{cases} \tag{19.135}$$

This procedure can also be applied when the laminate is undamaged leading to a different set of coefficients c_{ij}^0. It follows from the undamaged laminate relations (19.128) that

$$\begin{aligned}\bar{\varepsilon}_A^0 &= b_{11}\sigma_A + b_{12}M_A + b_{13}\bar{\varepsilon}_T + b_{14}\hat{\varepsilon}_T + b_{15}\sigma_t + b_{16}\Delta T,\\ \hat{\varepsilon}_A^0 &= b_{21}\sigma_A + b_{22}M_A + b_{23}\bar{\varepsilon}_T + b_{24}\hat{\varepsilon}_T + b_{25}\sigma_t + b_{26}\Delta T.\end{aligned} \tag{19.136}$$

On substituting into (19.132) and (19.133),

$$\begin{aligned}d_{11}\bar{\varepsilon}_T + d_{12}\hat{\varepsilon}_T &= X_1,\\ d_{21}\bar{\varepsilon}_T + d_{22}\hat{\varepsilon}_T &= Y_1,\end{aligned} \tag{19.137}$$

where

$$\begin{aligned}d_{11} &= c_{31}b_{13} + c_{32}b_{23} + c_{33}, & d_{12} &= c_{31}b_{14} + c_{32}b_{24} + c_{34},\\ d_{21} &= c_{41}b_{13} + c_{42}b_{23} + c_{43}, & d_{22} &= c_{41}b_{14} + c_{42}b_{24} + c_{44},\end{aligned} \tag{19.138}$$

and

$$\begin{aligned}X_1 &= \sigma_T - (c_{31}b_{11} + c_{32}b_{21})\sigma_A - (c_{31}b_{12} + c_{32}b_{22})M_A\\ &\quad - (c_{31}b_{15} + c_{32}b_{25} + c_{35})\sigma_t - (c_{31}b_{16} + c_{32}b_{26} + c_{36})\Delta T,\\ Y_1 &= M_T - (c_{41}b_{11} + c_{42}b_{21})\sigma_A - (c_{41}b_{12} + c_{42}b_{22})M_A\\ &\quad - (c_{41}b_{15} + c_{42}b_{25} + c_{45})\sigma_t - (c_{41}b_{16} + c_{42}b_{26} + c_{46})\Delta T.\end{aligned} \tag{19.139}$$

On solving (19.137), the transverse strains applied to a damaged laminate are given by

$$\bar{\varepsilon}_T = \frac{d_{22}X_1 - d_{12}Y_1}{d_{11}d_{22} - d_{12}d_{21}}, \qquad \hat{\varepsilon}_T = \frac{d_{11}Y_1 - d_{21}X_1}{d_{11}d_{22} - d_{12}d_{21}}. \tag{19.140}$$

Effective strain parameters $\overset{*}{\varepsilon}_A$, $\overset{*}{\varepsilon}_T$, corresponding to the axial and transverse strains on the midplane, are defined by (11.40), namely

$$\overset{*}{\varepsilon}_A = \bar{\varepsilon}_A + \frac{1}{2}h\hat{\varepsilon}_A, \qquad \overset{*}{\varepsilon}_T = \bar{\varepsilon}_T + \frac{1}{2}h\hat{\varepsilon}_T. \tag{19.141}$$

On using (19.88), (19.89), (19.100), (19.103) and (19.107), it can then be shown that the laminate strain parameters $\overset{*}{\varepsilon}_A$, $\overset{*}{\varepsilon}_T$, $\hat{\varepsilon}_A$, $\hat{\varepsilon}_T$ and ε_t may be expressed in terms of the laminate stresses $\sigma_A, \sigma_T, \sigma_t$ and bending moments per unit area M_A, M_T as follows:

$$\overset{*}{\varepsilon}_A = \alpha_{11}\sigma_A + \alpha_{12}\sigma_T + \alpha_{13}M_A + \alpha_{14}M_T + \alpha_{15}\sigma_t + \alpha_{16}\Delta T, \tag{19.142}$$

$$\overset{*}{\varepsilon}_T = \alpha_{21}\sigma_A + \alpha_{22}\sigma_T + \alpha_{23}M_A + \alpha_{24}M_T + \alpha_{25}\sigma_t + \alpha_{26}\Delta T, \tag{19.143}$$

$$\hat{\varepsilon}_A = \alpha_{31}\sigma_A + \alpha_{32}\sigma_T + \alpha_{33}M_A + \alpha_{34}M_T + \alpha_{35}\sigma_t + \alpha_{36}\Delta T, \tag{19.144}$$

$$\hat{\varepsilon}_T = \alpha_{41}\sigma_A + \alpha_{42}\sigma_T + \alpha_{43}M_A + \alpha_{44}M_T + \alpha_{45}\sigma_t + \alpha_{46}\Delta T, \tag{19.145}$$

$$\varepsilon_t = \alpha_{51}\sigma_A + \alpha_{52}\sigma_T + \alpha_{53}M_A + \alpha_{54}M_T + \alpha_{55}\sigma_t + \alpha_{56}\Delta T. \tag{19.146}$$

The values of all the coefficients α_{ij} can be determined by calculating values of $\overset{*}{\varepsilon}_A$, $\overset{*}{\varepsilon}_T$, $\hat{\varepsilon}_A$, $\hat{\varepsilon}_T$, ε_t and ΔT for the following special mechanical or thermal loading cases

$$(\sigma_A, \sigma_T, M_A, M_T, \sigma_t, \Delta T) = \begin{cases}(1\ \ 0\ \ 0\ \ 0\ \ 0\ \ 0),\\ (0\ \ 1\ \ 0\ \ 0\ \ 0\ \ 0),\\ (0\ \ 0\ \ 1\ \ 0\ \ 0\ \ 0),\\ (0\ \ 0\ \ 0\ \ 1\ \ 0\ \ 0),\\ (0\ \ 0\ \ 0\ \ 0\ \ 1\ \ 0),\\ (0\ \ 0\ \ 0\ \ 0\ \ 0\ \ 1).\end{cases} \tag{19.147}$$

It follows from the stress-strain relations (11.41)–(11.45) for a damaged laminate that the coefficients α_{ij} can then be related to thermoelastic constants as follows:

$$\begin{aligned}
&\alpha_{11}=\frac{1}{E_A},\ \alpha_{12}=-\frac{\nu_A}{E_A},\ \alpha_{13}=-\frac{\hat{\nu}_A}{\hat{E}_A},\ \alpha_{14}=-\frac{\hat{\eta}_A}{\hat{E}_T},\ \alpha_{15}=-\frac{\nu_a}{E_A},\ \alpha_{16}=\alpha_A,\\
&\alpha_{21}=-\frac{\nu_A}{E_A},\ \alpha_{22}=\frac{1}{E_T},\ \alpha_{23}=-\frac{\hat{\nu}_T}{\hat{E}_A},\ \alpha_{24}=-\frac{\hat{\eta}_T}{\hat{E}_T},\ \alpha_{25}=-\frac{\nu_t}{E_T},\ \alpha_{26}=\alpha_T,\\
&\alpha_{31}=-\frac{\hat{\nu}_A}{\hat{E}_A},\ \alpha_{32}=-\frac{\hat{\nu}_T}{\hat{E}_A},\ \alpha_{33}=\frac{1}{\hat{E}_A},\ \alpha_{34}=-\frac{\hat{\delta}_A}{\hat{E}_A},\ \alpha_{35}=-\frac{\hat{\nu}_a}{\hat{E}_A},\ \alpha_{36}=\hat{\alpha}_A,\\
&\alpha_{41}=-\frac{\hat{\eta}_A}{\hat{E}_T},\ \alpha_{42}=-\frac{\hat{\eta}_T}{\hat{E}_T},\ \alpha_{43}=-\frac{\hat{\delta}_A}{\hat{E}_A},\ \alpha_{44}=\frac{1}{\hat{E}_T},\ \alpha_{45}=-\frac{\hat{\eta}_t}{\hat{E}_T},\ \alpha_{46}=\hat{\alpha}_T,\\
&\alpha_{51}=-\frac{\nu_a}{E_A},\ \alpha_{52}=-\frac{\nu_t}{E_T},\ \alpha_{53}=-\frac{\hat{\nu}_a}{\hat{E}_A},\ \alpha_{54}=-\frac{\hat{\eta}_t}{\hat{E}_T},\ \alpha_{55}=\frac{1}{E_t},\ \alpha_{56}=\alpha_t.
\end{aligned} \tag{19.148}$$

Knowing the values of α_{ij}, it is now straightforward to determine the values of the effective thermoelastic constants for a damaged and undamaged laminate. It is clear that

$$\begin{aligned}
&E_A=\frac{1}{\alpha_{11}},\ E_T=\frac{1}{\alpha_{22}},\ \hat{E}_A=\frac{1}{\alpha_{33}},\ \hat{E}_T=\frac{1}{\alpha_{44}},\ E_t=\frac{1}{\alpha_{55}},\\
&\nu_A=-\frac{\alpha_{12}}{\alpha_{11}}=-\frac{\alpha_{21}}{\alpha_{11}},\ \nu_a=-\frac{\alpha_{15}}{\alpha_{11}}=-\frac{\alpha_{51}}{\alpha_{11}},\ \nu_t=-\frac{\alpha_{25}}{\alpha_{22}}=-\frac{\alpha_{52}}{\alpha_{22}},\\
&\hat{\nu}_A=-\frac{\alpha_{13}}{\alpha_{33}}=-\frac{\alpha_{31}}{\alpha_{33}},\ \hat{\nu}_a=-\frac{\alpha_{35}}{\alpha_{33}}=-\frac{\alpha_{53}}{\alpha_{33}},\ \hat{\nu}_T=-\frac{\alpha_{23}}{\alpha_{33}}=-\frac{\alpha_{32}}{\alpha_{33}},\ \hat{\delta}_A=-\frac{\alpha_{34}}{\alpha_{33}}=-\frac{\alpha_{43}}{\alpha_{33}},\\
&\hat{\eta}_A=-\frac{\alpha_{14}}{\alpha_{44}}=-\frac{\alpha_{41}}{\alpha_{44}},\ \hat{\eta}_T=-\frac{\alpha_{24}}{\alpha_{44}}=-\frac{\alpha_{42}}{\alpha_{44}},\ \hat{\eta}_t=-\frac{\alpha_{45}}{\alpha_{44}}=-\frac{\alpha_{54}}{\alpha_{44}},\\
&\alpha_A=\alpha_{16},\ \alpha_T=\alpha_{26},\ \hat{\alpha}_A=\alpha_{36},\ \hat{\alpha}_T=\alpha_{46},\ \alpha_t=\alpha_{56}.
\end{aligned} \tag{19.149}$$

The accuracy of methods of estimating the effective thermoelastic constants of nonsymmetrical cross-ply laminates is dependent upon accuracy of the associated stress-transfer analysis described in Sections 19.2-4. Such analyses have been validated by comparing predictions using the analysis of this chapter with those obtained from boundary element analysis, as described in reference [6] which demonstrates excellent agreement. This chapter, together with Chapter 11, has shown that the description of the deformation of nonsymmetric cross-ply laminates is quite complex generating a number of new elastic constants associated with the orthogonal bending. Rather than present graphical illustrations of the dependence of the many elastic constants on ply crack density, readers are referred to the associated software, and examples which are available at the Wiley website [7].

While this book has focused on stress transfer methods developed by the author, mention must be made of relatively recent developments concerning variational methods applied to general laminates, as described for example in reference [8]. Variational methods have been shown to be unique in their capability of solving ply crack problems associated with nonsymmetric and unbalanced multilayer laminates having plies of any orientation. When these methods are applied to specific problems for which the methods of this book can be used, additional validation is achieved as the solutions obtained using variational methods agree exceedingly well with those obtained using the semi-analytical methods considered here.

References

1. McCartney L N (1997), Stress transfer mechanics for multiple-ply laminates subject to bending, NPL Report CMMT(A)55, February.
2. McCartney L N and Miss Pierse C (1997), Stress transfer mechanics for multiple ply laminates for axial loading and bending, Proc. 11th Int. Conf. on Composite Materials, Gold Coast, Australia, July 14–18, vol. V, 662–671.
3. McCartney L N and Byrne M J W (2001), Energy balance method for predicting cracking in cross-ply laminates during bend deformation, Proc. 10th Int. Conf. on Fracture (ICF-10). Advances in Fracture Research, Honolulu, 2–6 December.
4. Hannaby S A (1993), The solution of ordinary differential equations arising from stress transfer mechanics. NPL Report DITC 223/93, November.
5. Hannaby S A (1997), The solution of ordinary differential equations arising from stress transfer mechanics of general symmetric laminates. NPL Report CISE 13/97, April.
6. McCartney L N and Blazquez A (2009), Validating, using BEM, a model for ply cracks in cross-ply laminates subject to combined in-plane loading and out-of-plane bending, Proc. Composites 2009, Imperial College, London, April.
7. Wiley webpage (www.wiley.com/go/mccartney/properties).
8. Hajikazemi M, McCartney L N, Ahmadi H and van Paepegem W (2020), Variational analysis of cracking in general composite laminates subject to triaxial and bending loads. Composite Structures, 239, 111993. https://doi.org/10.1016/j.compstruct.2020.111993

Recommended Reading

McCartney L N (2001), Predicting ply crack formation in cross-ply laminates subject to generalised plane strain bending, Proc. 6th Int. Conf. On Deformation & Fracture of Composites, Manchester, 4–5 April, pp. 57–66.

McCartney L N (2008), Approximate method of predicting ply crack formation in general symmetric laminates subject to biaxial loading and bending, in electronic Proceedings of 13th European Conference on Composite Materials (ECCM-13), June, Stockholm.

Hajikazemi M, Sadr M H, Talreja R (2015), Variational analysis of cracked general cross-ply laminates under bending and biaxial extension. International Journal of Damage Mechanics, 24, 582–624. https://doi.org/10.1177/1056789514546010

Hajikazemi M, Sadr M H and Varna J (2017), Analysis of cracked general cross-ply laminates under general bending loads: a variational approach. Journal of Composite Materials, 51, 3089–3109. https://doi.org/10.1177/0021998316682364

Appendix A: Solution for Shear of Isolated Spherical Particle in an Infinite Matrix

1. Spherical Shell Subject to Pure Shear Loading

For a sphere of radius a, the displacement field used by Hashin [1] has the following general form (but beware of the misprint in his equation (75))

$$\mathbf{u} = A\nabla(x_1x_2) + a^5 B\nabla\left(\frac{x_1x_2}{r^5}\right) + a^{-2}C\left[r^2\nabla(x_1x_2) + \alpha x_1x_2\,\mathbf{r}\right] + a^3 D\left[r^2\nabla\left(\frac{x_1x_2}{r^5}\right) + \beta\frac{x_1x_2}{r^5}\mathbf{r}\right], \tag{A.1}$$

where A, B, C and D are dimensionless constants to be determined from boundary conditions, and where α and β are dimensionless coefficients obtained from Love [2, equation (7)] that have the values

$$\alpha = -2\frac{2\lambda+7\mu}{5\lambda+7\mu} = -2\frac{7-10\nu}{7-4\nu}, \qquad \beta = \frac{3\lambda+8\mu}{\mu} = \frac{2(4-5\nu)}{1-2\nu}, \tag{A.2}$$

where use has been made of the following well-known relations between Young's modulus E, Poisson's ratio ν, bulk modulus k and Lamé's constants λ and μ (the shear modulus)

$$\left.\begin{aligned} &E = \frac{\mu(3\lambda+2\mu)}{\lambda+\mu}, \qquad \nu = \frac{\lambda}{2(\lambda+\mu)}, \\ &\lambda = \frac{E\nu}{(1+\nu)(1-2\nu)}, \quad \mu = \frac{E}{2(1+\nu)}, \quad 3k = 3\lambda + 2\mu = \frac{E}{1-2\nu}. \end{aligned}\right\} \tag{A.3}$$

It is convenient to develop the subsequent analysis using the elastic constants ν and μ so it necessary to use the following additional relation:

$$\lambda = \frac{2\mu\nu}{1-2\nu}. \tag{A.4}$$

For spherical polar coordinates it follows that

$$x_1 = r\cos\phi\,\sin\theta, \quad x_2 = r\sin\phi\,\sin\theta, \quad x_3 = r\cos\theta. \tag{A.5}$$

Properties for Design of Composite Structures: Theory and Implementation Using Software, First Edition. Neil McCartney.

Companion Website: www.wiley.com/go/mccartney/properties

It can be shown that

$$\nabla(x_1x_2) = r\sin^2\theta\,\sin 2\phi\,\mathbf{a} + \frac{1}{2}r\sin 2\theta\,\sin 2\phi\,\mathbf{b} + r\sin\theta\,\cos 2\phi\,\mathbf{c}, \tag{A.6}$$

$$\nabla\left(\frac{x_1x_2}{r^5}\right) = -\frac{3}{2}r^{-4}\sin^2\theta\,\sin 2\phi\,\mathbf{a} + \frac{1}{2}r^{-4}\sin 2\theta\,\sin 2\phi\,\mathbf{b} + r^{-4}\sin\theta\,\cos 2\phi\,\mathbf{c}, \tag{A.7}$$

$$x_1x_2\,\mathbf{r} = \frac{1}{2}r^3\sin^2\theta\,\sin 2\phi\,\mathbf{a}, \tag{A.8}$$

$$\nabla\left(\frac{x_1x_2}{r^5}\right) = \frac{\nabla(x_1x_2)}{r^5} - 5\frac{x_1x_2}{r^7}\mathbf{r}, \tag{A.9}$$

$$r^2\nabla\left(\frac{x_1x_2}{r^5}\right) + \beta\frac{x_1x_2}{r^5}\mathbf{r} = \frac{\nabla(x_1x_2)}{r^3} + \frac{3}{1-2\nu}\frac{x_1x_2}{r^5}\mathbf{r}. \tag{A.10}$$

Substitution into (A.1) then leads to

$$\begin{aligned}\mathbf{u} = u_r\,\mathbf{a} + u_\theta\,\mathbf{b} + u_\phi\,\mathbf{c} = {} & \left[A + B\left(\frac{a}{r}\right)^5 + C\left(\frac{r}{a}\right)^2 + D\left(\frac{a}{r}\right)^3\right]\nabla(x_1x_2) \\ & - \left[5B\left(\frac{a}{r}\right)^5 - \alpha C\left(\frac{r}{a}\right)^2 - \frac{3D}{1-2\nu}\left(\frac{a}{r}\right)^3\right]\frac{x_1x_2}{r^2}\mathbf{r},\end{aligned} \tag{A.11}$$

where **a**, **b** and **c** are orthogonal unit vectors for spherical polar coordinates. It follows, on using (A.6), (A.8) and (A.11), that

$$\left.\begin{aligned} u_r &= a\left[2A\frac{r}{a} - 3B\left(\frac{a}{r}\right)^4 + 2\alpha' C\left(\frac{r}{a}\right)^3 + 2\beta' D\left(\frac{a}{r}\right)^2\right]\frac{1}{2}\sin^2\theta\,\sin 2\phi\,, \\ u_\theta &= a\left[A\frac{r}{a} + B\left(\frac{a}{r}\right)^4 + C\left(\frac{r}{a}\right)^3 + D\left(\frac{a}{r}\right)^2\right]\frac{1}{2}\sin 2\theta\,\sin 2\phi\,, \\ u_\phi &= a\left[A\frac{r}{a} + B\left(\frac{a}{r}\right)^4 + C\left(\frac{r}{a}\right)^3 + D\left(\frac{a}{r}\right)^2\right]\sin\theta\,\cos 2\phi\,, \end{aligned}\right\} \tag{A.12}$$

where, from (A.2),

$$\alpha' = 1 + \frac{1}{2}\alpha = \frac{6\nu}{7-4\nu}, \qquad \beta' = \frac{1}{2}\beta - \frac{3}{2} = \frac{5-4\nu}{2(1-2\nu)}. \tag{A.13}$$

It is useful now to redefine the coefficients C and D so that

$$\begin{aligned} \alpha' C &\to 6\nu C\,, & \beta' D &\to (5-4\nu)D\,, \\ C &\to (7-4\nu)C\,, & D &\to 2(1-2\nu)D\,. \end{aligned} \tag{A.14}$$

Relations (A.12) are then equivalent to

$$\left.\begin{aligned}
u_r &= a\left[2A\frac{r}{a} - 3B\left(\frac{a}{r}\right)^4 + 12\nu C\left(\frac{r}{a}\right)^3 + \left(10-8\nu\right)D\left(\frac{a}{r}\right)^2\right]\frac{1}{2}\sin^2\theta\,\sin 2\phi\,,\\
u_\theta &= a\left[A\frac{r}{a} + B\left(\frac{a}{r}\right)^4 + \left(7-4\nu\right)C\left(\frac{r}{a}\right)^3 + \left(2-4\nu\right)D\left(\frac{a}{r}\right)^2\right]\frac{1}{2}\sin 2\theta\,\sin 2\phi\,,\\
u_\phi &= a\left[A\frac{r}{a} + B\left(\frac{a}{r}\right)^4 + \left(7-4\nu\right)C\left(\frac{r}{a}\right)^3 + \left(2-4\nu\right)D\left(\frac{a}{r}\right)^2\right]\sin\theta\,\cos 2\phi\,.
\end{aligned}\right\} \quad (A.15)$$

The corresponding strain field is given by (see Eringen [3, p. 209] and (2.142))

$$\left.\begin{aligned}
&\varepsilon_{rr} \equiv \frac{\partial u_r}{\partial r}, \quad \varepsilon_{\theta\theta} \equiv \frac{1}{r}\frac{\partial u_\theta}{\partial \theta} + \frac{u_r}{r}, \quad \varepsilon_{\phi\phi} \equiv \frac{1}{r\sin\theta}\frac{\partial u_\phi}{\partial \phi} + \frac{u_r}{r} + \frac{u_\theta}{r}\cot\theta,\\
&\varepsilon_{r\theta} \equiv \frac{1}{2}\left(\frac{1}{r}\frac{\partial u_r}{\partial \theta} + \frac{\partial u_\theta}{\partial r} - \frac{u_\theta}{r}\right), \quad \varepsilon_{r\phi} \equiv \frac{1}{2}\left(\frac{1}{r\sin\theta}\frac{\partial u_r}{\partial \phi} + \frac{\partial u_\phi}{\partial r} - \frac{u_\phi}{r}\right),\\
&\varepsilon_{\theta\phi} \equiv \frac{1}{2}\left(\frac{1}{r}\frac{\partial u_\phi}{\partial \theta} + \frac{1}{r\sin\theta}\frac{\partial u_\theta}{\partial \phi} - \frac{u_\phi}{r}\cot\theta\right).
\end{aligned}\right\} \quad (A.16)$$

The stress-strain relations for an isotropic material, in the absence of thermal expansion effects, are written in the form

$$\left.\begin{aligned}
\sigma_{rr} &= \lambda\left(\varepsilon_{rr} + \varepsilon_{\theta\theta} + \varepsilon_{\phi\phi}\right) + 2\mu\varepsilon_{rr}\,,\\
\sigma_{\theta\theta} &= \lambda\left(\varepsilon_{rr} + \varepsilon_{\theta\theta} + \varepsilon_{\phi\phi}\right) + 2\mu\varepsilon_{\theta\theta}\,,\\
\sigma_{\phi\phi} &= \lambda\left(\varepsilon_{rr} + \varepsilon_{\theta\theta} + \varepsilon_{\phi\phi}\right) + 2\mu\varepsilon_{\phi\phi}\,,
\end{aligned}\right\} \quad (A.17)$$

so that $\sigma_{rr} + \sigma_{\theta\theta} + \sigma_{\phi\phi} = \dfrac{2\mu(1+\nu)}{1-2\nu}\left(\varepsilon_{rr} + \varepsilon_{\theta\theta} + \varepsilon_{\phi\phi}\right).$ (A.18)

2. Deformation and Stress Fields in a Spherical Particle

The results are now applied to the interior region of a spherical particle having radius denoted by a. Consider the displacement and stress fields in a single spherical particle referred to spherical polar coordinates where the origin is located at the particle centre. The displacement and stress fields must be bounded at the origin $r = 0$ so that from (A.15) the displacement field to be considered has the form

$$\left.\begin{aligned}
u_r^{\mathrm{p}} &= a\left[2A_{\mathrm{p}}\frac{r}{a} + 12\nu_{\mathrm{p}}C_{\mathrm{p}}\left(\frac{r}{a}\right)^3\right]\frac{1}{2}\sin^2\theta\,\sin 2\phi\,,\\
u_\theta^{\mathrm{p}} &= a\left[A_{\mathrm{p}}\frac{r}{a} + \left(7-4\nu_{\mathrm{p}}\right)C_{\mathrm{p}}\left(\frac{r}{a}\right)^3\right]\frac{1}{2}\sin 2\theta\,\sin 2\phi\,,\\
u_\phi^{\mathrm{p}} &= a\left[A_{\mathrm{p}}\frac{r}{a} + \left(7-4\nu_{\mathrm{p}}\right)C_{\mathrm{p}}\left(\frac{r}{a}\right)^3\right]\sin\theta\,\cos 2\phi\,.
\end{aligned}\right\} \quad (A.19)$$

It follows from (A.16) and (A.19) that the corresponding strain field in the particle is given by

$$\varepsilon_{rr}^{\mathrm{p}} \equiv \frac{\partial u_r^{\mathrm{p}}}{\partial r} = \left[\mathrm{A_p} + 18\nu_{\mathrm{p}}C_{\mathrm{p}}\left(\frac{r}{a}\right)^2\right]\sin^2\theta\,\sin 2\phi, \tag{A.20}$$

$$\begin{aligned}\varepsilon_{\theta\theta}^{\mathrm{p}} \equiv \frac{1}{r}\frac{\partial u_\theta^{\mathrm{p}}}{\partial \theta} + \frac{u_r^{\mathrm{p}}}{r} &= \left[\mathrm{A_p} + (7-4\nu_{\mathrm{p}})C_{\mathrm{p}}\left(\frac{r}{a}\right)^2\right]\sin 2\phi \\ &- \left[\mathrm{A_p} + 14(1-\nu_{\mathrm{p}})C_{\mathrm{p}}\left(\frac{r}{a}\right)^2\right]\sin^2\theta\,\sin 2\phi,\end{aligned} \tag{A.21}$$

$$\begin{aligned}\varepsilon_{\phi\phi}^{\mathrm{p}} \equiv \frac{1}{r\sin\theta}\frac{\partial u_\phi^{\mathrm{p}}}{\partial \phi} + \frac{u_r^{\mathrm{p}}}{r} + \frac{u_\theta^{\mathrm{p}}}{r}\cot\theta &= -\left[A_{\mathrm{p}} + (7-4\nu_{\mathrm{p}})C_{\mathrm{p}}\left(\frac{r}{a}\right)^2\right]\sin 2\phi \\ &- (7-10\nu_{\mathrm{p}})C_{\mathrm{p}}\left(\frac{r}{a}\right)^2\sin^2\theta\,\sin 2\phi,\end{aligned} \tag{A.22}$$

$$\varepsilon_{r\theta}^{\mathrm{p}} \equiv \frac{1}{2}\left(\frac{1}{r}\frac{\partial u_r^{\mathrm{p}}}{\partial \theta} + \frac{\partial u_\theta^{\mathrm{p}}}{\partial r} - \frac{u_\theta^{\mathrm{p}}}{r}\right) = \left[\mathrm{A_p} + (7+2\nu_{\mathrm{p}})C_{\mathrm{p}}\left(\frac{r}{a}\right)^2\right]\frac{1}{2}\sin 2\theta\,\sin 2\phi, \tag{A.23}$$

$$\varepsilon_{r\phi}^{\mathrm{p}} \equiv \frac{1}{2}\left(\frac{1}{r\sin\theta}\frac{\partial u_r^{\mathrm{p}}}{\partial \phi} + \frac{\partial u_\phi^{\mathrm{p}}}{\partial r} - \frac{u_\phi^{\mathrm{p}}}{r}\right) = \left[\mathrm{A_p} + (7+2\nu_{\mathrm{p}})C_{\mathrm{p}}\left(\frac{r}{a}\right)^2\right]\sin\theta\,\cos 2\phi, \tag{A.24}$$

$$\varepsilon_{\theta\phi}^{\mathrm{p}} \equiv \frac{1}{2}\left(\frac{1}{r}\frac{\partial u_\phi^{\mathrm{p}}}{\partial \theta} + \frac{1}{r\sin\theta}\frac{\partial u_\theta^{\mathrm{p}}}{\partial \phi} - \frac{u_\phi^{\mathrm{p}}}{r}\cot\theta\right) = \left[\mathrm{A_p} + (7-4\nu_{\mathrm{p}})C_{\mathrm{p}}\left(\frac{r}{a}\right)^2\right]\cos\theta\,\cos 2\phi. \tag{A.25}$$

It can be shown from (A.20)–(A.22) and (A.17) that

$$\varepsilon_{rr}^{\mathrm{p}} + \varepsilon_{\theta\theta}^{\mathrm{p}} + \varepsilon_{\phi\phi}^{\mathrm{p}} = -21(1-2\nu_{\mathrm{p}})\,C_{\mathrm{p}}\left(\frac{r}{a}\right)^2\sin^2\theta\,\sin 2\phi, \tag{A.26}$$

$$\left.\begin{aligned}\sigma_{rr}^{\mathrm{p}} &= \lambda_{\mathrm{p}}\left(\varepsilon_{rr}^{\mathrm{p}} + \varepsilon_{\theta\theta}^{\mathrm{p}} + \varepsilon_{\phi\phi}^{\mathrm{p}}\right) + 2\mu_{\mathrm{p}}\,\varepsilon_{rr}^{\mathrm{p}} = 2\mu_{\mathrm{p}}\left[A_{\mathrm{p}} - 3\nu_{\mathrm{p}}C_{\mathrm{p}}\left(\frac{r}{a}\right)^2\right]\sin^2\theta\,\sin 2\phi, \\ \sigma_{r\theta}^{\mathrm{p}} &= \mu_{\mathrm{p}}\left[A_{\mathrm{p}} + (7+2\nu_{\mathrm{p}})C_{\mathrm{p}}\left(\frac{r}{a}\right)^2\right]\sin 2\theta\,\sin 2\phi, \\ \sigma_{r\phi}^{\mathrm{p}} &= 2\mu_{\mathrm{p}}\left[A_{\mathrm{p}} + (7+2\nu_{\mathrm{p}})C_{\mathrm{p}}\left(\frac{r}{a}\right)^2\right]\sin\theta\,\cos 2\phi,\end{aligned}\right\} \tag{A.27}$$

where use has been made of the following relationship $\lambda_{\mathrm{p}}(1-2\nu_{\mathrm{p}}) = 2\mu_{\mathrm{p}}\nu_{\mathrm{p}}$.

3. Deformation and Stress Fields in an Infinite Matrix

Consider the stress field in the infinite matrix surrounding a single spherical particle of radius a where the origin of spherical polar coordinates is at the particle centre. The stress field must be bounded as $r \to \infty$ so that from (A.15) the displacement field to be considered has the form

$$\left.\begin{aligned} u_r^m &= a\left[2A_m\frac{r}{a} - 3B_m\left(\frac{a}{r}\right)^4 + (10-8\nu_m)D_m\left(\frac{a}{r}\right)^2\right]\frac{1}{2}\sin^2\theta\,\sin 2\phi,\\ u_\theta^m &= a\left[A_m\frac{r}{a} + B_m\left(\frac{a}{r}\right)^4 + (2-4\nu_m)D_m\left(\frac{a}{r}\right)^2\right]\frac{1}{2}\sin 2\theta\,\sin 2\phi,\\ u_\phi^m &= a\left[A_m\frac{r}{a} + B_m\left(\frac{a}{r}\right)^4 + (2-4\nu_m)D_m\left(\frac{a}{r}\right)^2\right]\sin\theta\,\cos 2\phi. \end{aligned}\right\} \tag{A.28}$$

It follows from (A.16) and (A.28) that the corresponding strain field in the particle is given by

$$\varepsilon_{rr}^{\mathrm{m}} \equiv \frac{\partial u_r^{\mathrm{m}}}{\partial r} = \left[\mathrm{A_m} + 6B_{\mathrm{m}}\left(\frac{a}{r}\right)^5 - (10-8\nu_{\mathrm{m}})D_{\mathrm{m}}\left(\frac{a}{r}\right)^3\right]\sin^2\theta\,\sin 2\phi, \tag{A.29}$$

$$\begin{aligned} \varepsilon_{\theta\theta}^{\mathrm{m}} \equiv \frac{1}{r}\frac{\partial u_\theta^{\mathrm{m}}}{\partial\theta} + \frac{u_r^{\mathrm{m}}}{r} &= \left[A_{\mathrm{m}} + B_{\mathrm{m}}\left(\frac{a}{r}\right)^5 + (2-4\nu_{\mathrm{m}})D_{\mathrm{m}}\left(\frac{a}{r}\right)^3\right]\sin 2\phi\\ &\quad - \left[A_{\mathrm{m}} + \frac{7}{2}B_{\mathrm{m}}\left(\frac{a}{r}\right)^5 - (1+4\nu_{\mathrm{m}})D_{\mathrm{m}}\left(\frac{a}{r}\right)^3\right]\sin^2\theta\,\sin 2\phi, \end{aligned} \tag{A.30}$$

$$\begin{aligned} \varepsilon_{\phi\phi}^{\mathrm{m}} \equiv \frac{1}{r\sin\theta}\frac{\partial u_\phi^{\mathrm{m}}}{\partial\phi} + \frac{u_r^{\mathrm{m}}}{r} + \frac{u_\theta^{\mathrm{m}}}{r}\cot\theta &= -\left[A_{\mathrm{m}} + B_{\mathrm{m}}\left(\frac{a}{r}\right)^5 + (2-4\nu_{\mathrm{m}})D_{\mathrm{m}}\left(\frac{a}{r}\right)^3\right]\sin 2\phi\\ &\quad - \left[\frac{5}{2}B_{\mathrm{m}}\left(\frac{a}{r}\right)^5 - 3D_{\mathrm{m}}\left(\frac{a}{r}\right)^3\right]\sin^2\theta\,\sin 2\phi, \end{aligned} \tag{A.31}$$

$$\varepsilon_{r\theta}^{\mathrm{m}} \equiv \frac{1}{2}\left(\frac{1}{r}\frac{\partial u_r^{\mathrm{m}}}{\partial\theta} + \frac{\partial u_\theta^{\mathrm{m}}}{\partial r} - \frac{u_\theta^{\mathrm{m}}}{r}\right) = \left[\mathrm{A_m} - 4B_{\mathrm{m}}\left(\frac{a}{r}\right)^5 + (2+2\nu_{\mathrm{m}})D_{\mathrm{m}}\left(\frac{a}{r}\right)^3\right]\frac{1}{2}\sin 2\theta\sin 2\phi, \tag{A.32}$$

$$\varepsilon_{r\phi}^{\mathrm{m}} \equiv \frac{1}{2}\left(\frac{1}{r\sin\theta}\frac{\partial u_r^{\mathrm{m}}}{\partial\phi} + \frac{\partial u_\phi^{\mathrm{m}}}{\partial r} - \frac{u_\phi^{\mathrm{m}}}{r}\right) = \left[\mathrm{A_m} - 4B_{\mathrm{m}}\left(\frac{a}{r}\right)^5 + (2+2\nu_{\mathrm{m}})D_{\mathrm{m}}\left(\frac{a}{r}\right)^3\right]\sin\theta\cos 2\phi, \tag{A.33}$$

$$\begin{aligned} \varepsilon_{\theta\phi}^{\mathrm{m}} &\equiv \frac{1}{2}\left(\frac{1}{r}\frac{\partial u_\phi^{\mathrm{m}}}{\partial\theta} + \frac{1}{r\sin\theta}\frac{\partial u_\theta^{\mathrm{m}}}{\partial\phi} - \frac{u_\phi^{\mathrm{m}}}{r}\cot\theta\right) =\\ &\left[\mathrm{A_m} + B_{\mathrm{m}}\left(\frac{a}{r}\right)^5 + (2-4\nu_{\mathrm{m}})D_{\mathrm{m}}\left(\frac{a}{r}\right)^3\right]\cos\theta\,\cos 2\phi. \end{aligned} \tag{A.34}$$

It follows from (A.29)–(A.31) that

$$\varepsilon_{rr}^{\mathrm{m}} + \varepsilon_{\theta\theta}^{\mathrm{m}} + \varepsilon_{\phi\phi}^{\mathrm{m}} = -6(1-2\nu_{\mathrm{m}})D_{\mathrm{m}}(a/r)^3\sin^2\theta\,\sin 2\phi. \tag{A.35}$$

From (A.17)

$$\left.\begin{aligned} \sigma_{rr}^{\mathrm{m}} &= 2\mu_{\mathrm{m}}\left[A_{\mathrm{m}} + 6B_{\mathrm{m}}(a/r)^5 - 2(5-\nu_{\mathrm{m}})\,D_{\mathrm{m}}(a/r)^3\right]\sin^2\theta\,\sin 2\phi,\\ \sigma_{r\theta}^{\mathrm{m}} &= \mu_{\mathrm{m}}\left[A_{\mathrm{m}} - 4B_{\mathrm{m}}(a/r)^5 + (2+2\nu_{\mathrm{m}})D_{\mathrm{m}}(a/r)^3\right]\sin 2\theta\,\sin 2\phi,\\ \sigma_{r\phi}^{\mathrm{m}} &= 2\mu_{\mathrm{m}}\left[A_{\mathrm{m}} - 4B_{\mathrm{m}}(a/r)^5 + (2+2\nu_{\mathrm{m}})D_{\mathrm{m}}(a/r)^3\right]\sin\theta\,\cos 2\phi, \end{aligned}\right\} \tag{A.36}$$

where use has been made of the relationship

$$6(1-2\nu_{\mathrm{m}})\lambda_{\mathrm{m}}+4(5-4\nu_{\mathrm{m}})\mu_{\mathrm{m}}=4(5-\nu_{\mathrm{m}})\mu_{\mathrm{m}}.$$

4. Isolated Spherical Particle Embedded in an Infinite Matrix

A single spherical particle of radius a is placed in an infinite matrix and the origin of spherical polar coordinates (r, θ, ϕ) is taken at the centre of the particle. The system is then subject only to a shear stress applied at infinity. At the particle/matrix interface, the following boundary conditions must be satisfied:

$$\left.\begin{aligned} u_r^{\mathrm{p}}(a,\theta,\phi)&=u_r^{\mathrm{m}}(a,\theta,\phi), & \sigma_{rr}^{\mathrm{p}}(a,\theta,\phi)&=\sigma_{rr}^{\mathrm{m}}(a,\theta,\phi),\\ u_\theta^{\mathrm{p}}(a,\theta,\phi)&=u_\theta^{\mathrm{m}}(a,\theta,\phi), & \sigma_{\theta\theta}^{\mathrm{p}}(a,\theta,\phi)&=\sigma_{\theta\theta}^{\mathrm{m}}(a,\theta,\phi),\\ u_\theta^{\mathrm{p}}(a,\theta,\phi)&=u_\theta^{\mathrm{m}}(a,\theta,\phi), & \sigma_{\phi\phi}^{\mathrm{p}}(a,\theta,\phi)&=\sigma_{\phi\phi}^{\mathrm{m}}(a,\theta,\phi).\end{aligned}\right\} \tag{A.37}$$

It follows from (A.19), (A.27), (A.28) and (A.36) that the continuity conditions (A.37) are satisfied if the following four independent relations are satisfied

$$\left.\begin{aligned} &2A_{\mathrm{p}}+12\nu_{\mathrm{p}}C_{\mathrm{p}}=2A_{\mathrm{m}}-3B_{\mathrm{m}}+(10-8\nu_{\mathrm{m}})D_{\mathrm{m}},\\ &A_{\mathrm{p}}+(7-4\nu_{\mathrm{p}})C_{\mathrm{p}}=A_{\mathrm{m}}+B_{\mathrm{m}}+(2-4\nu_{\mathrm{m}})D_{\mathrm{m}},\\ &\mu_{\mathrm{p}}\left[A_{\mathrm{p}}-3\nu_{\mathrm{p}}C_{\mathrm{p}}\right]=\mu_{\mathrm{m}}\left[A_{\mathrm{m}}+6B_{\mathrm{m}}-2(5-\nu_{\mathrm{m}})D_{\mathrm{m}}\right],\\ &\mu_{\mathrm{p}}\left[A_{\mathrm{p}}+(7+2\nu_{\mathrm{p}})C_{\mathrm{p}}\right]=\mu_{\mathrm{m}}\left[A_{\mathrm{m}}-4B_{\mathrm{m}}+(2+2\nu_{\mathrm{m}})D_{\mathrm{m}}\right].\end{aligned}\right\} \tag{A.38}$$

In the matrix at infinity, the displacement and stress field should tend to the state of homogenous shear. As the displacement field must tend, as $r\to\infty$, to that of uniform shear, it follows that the coefficient $\mathrm{A_m}$ has the value γ which is the applied shear strain. Substitution into (A.38) then leads to the following equations determining the unknown coefficients $\mathrm{A_p}$, B_{m}, C_{p} and D_{m}

$$\left.\begin{aligned} &A_{\mathrm{p}}+\frac{3}{2}B_{\mathrm{m}}+6\nu_{\mathrm{p}}C_{\mathrm{p}}-(5-4\nu_{\mathrm{m}})D_{\mathrm{m}}=\gamma,\\ &A_{\mathrm{p}}-B_{\mathrm{m}}+(7-4\nu_{\mathrm{p}})C_{\mathrm{p}}-(2-4\nu_{\mathrm{m}})D_{\mathrm{m}}=\gamma,\\ &\frac{\mu_{\mathrm{p}}}{\mu_{\mathrm{m}}}A_{\mathrm{p}}-6B_{\mathrm{m}}-3\nu_{\mathrm{p}}\frac{\mu_{\mathrm{p}}}{\mu_{\mathrm{m}}}C_{\mathrm{p}}+(10-2\nu_{\mathrm{m}})D_{\mathrm{m}}=\gamma,\\ &\frac{\mu_{\mathrm{p}}}{\mu_{\mathrm{m}}}A_{\mathrm{p}}+4B_{\mathrm{m}}+(7+2\nu_{\mathrm{p}})\frac{\mu_{\mathrm{p}}}{\mu_{\mathrm{m}}}C_{\mathrm{p}}-(2+2\nu_{\mathrm{m}})D_{\mathrm{m}}=\gamma.\end{aligned}\right\} \tag{A.39}$$

It can then be shown that

$$\left.\begin{aligned} A_{\mathrm{p}}&=\frac{15(1-\nu_{\mathrm{m}})\gamma}{2(4-5\nu_{\mathrm{m}})\mu_{\mathrm{p}}/\mu_{\mathrm{m}}+(7-5\nu_{\mathrm{m}})}, & B_{\mathrm{m}}&=\frac{3(1-\mu_{\mathrm{p}}/\mu_{\mathrm{m}})\gamma}{2(4-5\nu_{\mathrm{m}})\mu_{\mathrm{p}}/\mu_{\mathrm{m}}+(7-5\nu_{\mathrm{m}})},\\ C_{\mathrm{p}}&=0, & D_{\mathrm{m}}&=\frac{5(1-\mu_{\mathrm{p}}/\mu_{\mathrm{m}})\gamma}{4(4-5\nu_{\mathrm{m}})\mu_{\mathrm{p}}/\mu_{\mathrm{m}}+2(7-5\nu_{\mathrm{m}})}.\end{aligned}\right\} \tag{A.40}$$

References

1. Hashin, Z. (1962). The elastic moduli of heterogeneous materials. *Journal of Applied Mechanics* 29: 143–150.
2. Love, A.E.H. (1944). *A Treatise on the Mathematical Theory of Elasticity*. Chapter XI, 4th ed. New York: Dover Publications.
3. Eringen, A.C. (1967). *Mechanics of Continua*. New York - London - Sydney: John Wiley & Sons Inc.

Appendix B: Elasticity Analysis of Two Concentric Cylinders

The following analysis applies to two concentric perfectly bonded cylinders having external radii R and $a > R$, subject to a uniform temperature change ΔT, that is subject to a uniform axial strain $\varepsilon_{\rm A}$ and uniform transverse stress $\sigma_{\rm T}$ on the external surface of the outer cylinder. A set of cylindrical polar coordinates (r,θ,z) will be used where the origin lies on the axis of the inner cylinder.

The fibre is regarded as a homogeneous transverse isotropic solid so that the stress-strain relations are of the form

$$\varepsilon_{rr}^{\rm f} = \frac{1}{E_{\rm T}^{\rm f}}\sigma_{rr}^{\rm f} - \frac{\nu_{\rm t}^{\rm f}}{E_{\rm T}^{\rm f}}\sigma_{\theta\theta}^{\rm f} - \frac{\nu_{\rm A}^{\rm f}}{E_{\rm A}^{\rm f}}\sigma_{zz}^{\rm f} + \alpha_{\rm T}^{\rm f}\Delta T, \tag{B.1}$$

$$\varepsilon_{\theta\theta}^{\rm f} = -\frac{\nu_{\rm t}^{\rm f}}{E_{\rm T}^{\rm f}}\sigma_{rr}^{\rm f} + \frac{1}{E_{\rm T}^{\rm f}}\sigma_{\theta\theta}^{\rm f} - \frac{\nu_{\rm A}^{\rm f}}{E_{\rm A}^{\rm f}}\sigma_{zz}^{\rm f} + \alpha_{\rm T}^{\rm f}\Delta T, \tag{B.2}$$

$$\varepsilon_{zz}^{\rm f} = -\frac{\nu_{\rm A}^{\rm f}}{E_{\rm A}^{\rm f}}\sigma_{rr}^{\rm f} - \frac{\nu_{\rm A}^{\rm f}}{E_{\rm A}^{\rm f}}\sigma_{\theta\theta}^{\rm f} + \frac{1}{E_{\rm A}^{\rm f}}\sigma_{zz}^{\rm f} + \alpha_{\rm A}^{\rm f}\Delta T\,, \tag{B.3}$$

$$\varepsilon_{rz}^{\rm f} = \frac{\sigma_{rz}^{\rm f}}{2\mu_{\rm A}^{\rm f}},\ \varepsilon_{\theta z}^{\rm f} = \frac{\sigma_{\theta z}^{\rm f}}{2\mu_{\rm A}^{\rm f}},\ \varepsilon_{r\theta}^{\rm f} = \frac{\sigma_{r\theta}^{\rm f}}{2\mu_{\rm t}^{\rm f}}, \tag{B.4}$$

$$\text{where } E_{\rm T}^{\rm f} = 2\mu_{\rm t}^{\rm f}(1+\nu_{\rm t}^{\rm f}) \text{ but } E_{\rm A}^{\rm f} \neq 2\mu_{\rm A}^{\rm f}(1+\nu_{\rm A}^{\rm f}). \tag{B.5}$$

The matrix is also regarded as a homogeneous transverse isotropic solid so that

$$\varepsilon_{rr}^{\rm m} = \frac{1}{E_{\rm T}^{\rm m}}\sigma_{rr}^{\rm m} - \frac{\nu_{\rm t}^{\rm m}}{E_{\rm T}^{\rm m}}\sigma_{\theta\theta}^{\rm f} - \frac{\nu_{\rm A}^{\rm m}}{E_{\rm A}^{\rm m}}\sigma_{zz}^{\rm m} + \alpha_{T}^{\rm m}\Delta T, \tag{B.6}$$

$$\varepsilon_{\theta\theta}^{\rm m} = -\frac{\nu_{\rm t}^{\rm m}}{E_{\rm T}^{\rm m}}\sigma_{rr}^{\rm m} + \frac{1}{E_{\rm T}^{\rm m}}\sigma_{\theta\theta}^{\rm m} - \frac{\nu_{\rm A}^{\rm m}}{E_{\rm A}^{\rm m}}\sigma_{zz}^{\rm m} + \alpha_{T}^{\rm m}\Delta T, \tag{B.7}$$

$$\varepsilon_{zz}^{\rm m} = -\frac{\nu_{\rm A}^{\rm m}}{E_{\rm A}^{\rm m}}\sigma_{rr}^{\rm m} - \frac{\nu_{\rm A}^{\rm m}}{E_{\rm A}^{\rm m}}\sigma_{\theta\theta}^{\rm m} + \frac{1}{E_{\rm A}^{\rm m}}\sigma_{zz}^{\rm m} + \alpha_{\rm A}^{\rm m}\Delta T, \tag{B.8}$$

$$\varepsilon_{rz}^{\rm m} = \frac{\sigma_{rz}^{\rm m}}{2\mu_{\rm A}^{\rm m}},\ \varepsilon_{\theta z}^{\rm f} = \frac{\sigma_{\theta z}^{\rm m}}{2\mu_{\rm A}^{\rm m}},\ \varepsilon_{r\theta}^{\rm m} = \frac{\sigma_{r\theta}^{\rm m}}{2\mu_{\rm t}^{\rm m}}, \tag{B.9}$$

$$\text{where } E_{\mathrm{T}}^{\mathrm{m}} = 2\mu_{\mathrm{t}}^{\mathrm{m}}(1+\nu_{\mathrm{t}}^{\mathrm{m}}) \text{ but } E_{\mathrm{A}}^{\mathrm{m}} \neq 2\mu_{\mathrm{A}}^{\mathrm{m}}(1+\nu_{\mathrm{A}}^{\mathrm{m}}). \tag{B.10}$$

For isotropic fibres and matrix

$$E_{\mathrm{A}}^{\mathrm{f}} = E_{\mathrm{T}}^{\mathrm{f}} = E_{\mathrm{f}},\ \mu_A^f = \mu_{\mathrm{t}}^f = \mu_{\mathrm{f}},\ \nu_{\mathrm{A}}^{\mathrm{f}} = \nu_{\mathrm{t}}^{\mathrm{f}} = \nu_{\mathrm{f}},\ \alpha_{\mathrm{A}}^{\mathrm{f}} = \alpha_{\mathrm{T}}^{\mathrm{f}} = \alpha_{\mathrm{f}},$$
$$E_{\mathrm{A}}^{\mathrm{m}} = E_{\mathrm{T}}^{\mathrm{m}} = E_{\mathrm{m}},\ \mu_{\mathrm{A}}^{\mathrm{m}} = \mu_{\mathrm{t}}^{\mathrm{m}} = \mu_{\mathrm{m}},\ \nu_{\mathrm{A}}^{\mathrm{m}} = \nu_{\mathrm{t}}^{\mathrm{m}} = \nu_{\mathrm{m}},\ \alpha_{\mathrm{A}}^{\mathrm{m}} = \alpha_{\mathrm{T}}^{\mathrm{m}} = \alpha_{\mathrm{m}}.$$

The equilibrium equations for the fibre, assuming symmetry about the fibre axis, are

$$\frac{\partial \sigma_{rr}^{\mathrm{f}}}{\partial r} + \frac{\partial \sigma_{rz}^{\mathrm{f}}}{\partial z} + \frac{\sigma_{rr}^{\mathrm{f}} - \sigma_{\theta\theta}^{\mathrm{f}}}{r} = 0, \tag{B.11}$$

$$\frac{\partial \sigma_{zz}^{\mathrm{f}}}{\partial z} + \frac{\partial \sigma_{rz}^{\mathrm{f}}}{\partial r} + \frac{\sigma_{rz}^{\mathrm{f}}}{r} = 0. \tag{B.12}$$

Similarly, for the matrix

$$\frac{\partial \sigma_{rr}^{\mathrm{m}}}{\partial r} + \frac{\partial \sigma_{rz}^{\mathrm{m}}}{\partial z} + \frac{\sigma_{rr}^{\mathrm{m}} - \sigma_{\theta\theta}^{\mathrm{m}}}{r} = 0, \tag{B.13}$$

$$\frac{\partial \sigma_{zz}^{\mathrm{m}}}{\partial z} + \frac{\partial \sigma_{rz}^{\mathrm{m}}}{\partial r} + \frac{\sigma_{rz}^{\mathrm{m}}}{r} = 0. \tag{B.14}$$

The boundary and interface conditions that must be satisfied are

$$\sigma_{rr}^{\mathrm{m}}(a,z) = \sigma_{\mathrm{T}}, \tag{B.15}$$

$$\sigma_{rz}^{\mathrm{m}}(a,z) = 0, \tag{B.16}$$

$$\sigma_{rr}^{\mathrm{m}}(R,z) = \sigma_{rr}^{\mathrm{f}}(R,z), \tag{B.17}$$

$$\sigma_{rz}^{\mathrm{m}}(R,z) = \sigma_{rz}^{\mathrm{f}}(R,z), \tag{B.18}$$

$$u_r^{\mathrm{m}}(R,z) = u_r^{\mathrm{f}}(R,z), \tag{B.19}$$

$$u_z^{\mathrm{m}}(R,z) = u_z^{\mathrm{f}}(R,z). \tag{B.20}$$

In regions away from the loading mechanism it is reasonable to assume that

$$u_z^{\mathrm{f}} \equiv u_z^{\mathrm{m}} \equiv \varepsilon_{\mathrm{A}}\, z, \tag{B.21}$$

where $\varepsilon_{\mathrm{A}} = \alpha_{\mathrm{A}}\Delta T$ is the axial strain applied to the composite and α_{A} is the effective axial thermal expansion coefficient that is defined later in terms of fibre and matrix properties. A solution is now sought of the following classical Lamé form

$$u_r^{\mathrm{f}} = \mathrm{A}_{\mathrm{f}}\, r,\ u_\theta^{\mathrm{f}} \equiv 0, \tag{B.22}$$

$$u_r^{\mathrm{m}} = \mathrm{A}_{\mathrm{m}} r + \frac{\phi}{2\mu_{\mathrm{t}}^{\mathrm{m}} r},\ u_\theta^{\mathrm{m}} \equiv 0, \tag{B.23}$$

where A_f, A_m and ϕ are constants to be determined. It is assumed that the outer radius a of the cylindrical composite is given by

$$a = \frac{R}{\sqrt{V_{\mathrm{f}}}}, \tag{B.24}$$

where V_f is the volume fraction of the composite represented by the concentric cylinder model. This selection ensures that the fibre volume fraction in the concentric cylinder model is the same as that in the unidirectional composite being modelled. On differentiating the displacement field, it follows that

$$\varepsilon_{rr}^{\mathrm{f}} = \frac{\partial u_r^{\mathrm{f}}}{\partial r} = \mathrm{A_f}, \; \varepsilon_{\theta\theta}^{\mathrm{f}} = \frac{u_r^{\mathrm{f}}}{r} = \mathrm{A_f}, \; \varepsilon_{zz}^{\mathrm{f}} = \varepsilon_{\mathrm{A}}, \tag{B.25}$$

$$\varepsilon_{rz}^{\mathrm{f}} = \varepsilon_{\theta z}^{\mathrm{f}} = \varepsilon_{r\theta}^{\mathrm{f}} \equiv 0, \tag{B.26}$$

$$\varepsilon_{rr}^{\mathrm{m}} = \frac{\partial u_r^{\mathrm{m}}}{\partial r} = \mathrm{A_m} - \frac{\phi}{2\mu_{\mathrm{t}}^{\mathrm{m}} r^2}, \; \varepsilon_{\theta\theta}^{\mathrm{m}} = \frac{u_r^{\mathrm{m}}}{r} = \mathrm{A_m} + \frac{\phi}{2\mu_{\mathrm{t}}^{\mathrm{m}} r^2}, \; \varepsilon_{zz}^{\mathrm{m}} = \varepsilon_{\mathrm{A}}, \tag{B.27}$$

$$\varepsilon_{rz}^{\mathrm{m}} = \varepsilon_{\theta z}^{\mathrm{m}} = \varepsilon_{r\theta}^{\mathrm{m}} \equiv 0. \tag{B.28}$$

It follows directly from (B.4), (B.9), (B.26) and (B.28) that

$$\begin{aligned} \sigma_{rz}^{\mathrm{f}} &= \sigma_{\theta z}^{\mathrm{f}} = \sigma_{r\theta}^{\mathrm{f}} \equiv 0, \\ \sigma_{rz}^{\mathrm{m}} &= \sigma_{\theta z}^{\mathrm{m}} = \sigma_{r\theta}^{\mathrm{m}} \equiv 0. \end{aligned} \tag{B.29}$$

On subtracting (B.1) and (B.2), and (B.6) and (B.7), it follows that

$$\begin{aligned} \varepsilon_{\theta\theta}^{\mathrm{f}} - \varepsilon_{rr}^{\mathrm{f}} &= \frac{1}{2\mu_{\mathrm{t}}^{\mathrm{f}}}(\sigma_{\theta\theta}^{\mathrm{f}} - \sigma_{rr}^{\mathrm{f}}), \\ \varepsilon_{\theta\theta}^{\mathrm{m}} - \varepsilon_{rr}^{\mathrm{m}} &= \frac{1}{2\mu_{\mathrm{t}}^{\mathrm{m}}}(\sigma_{\theta\theta}^{\mathrm{m}} - \sigma_{rr}^{\mathrm{m}}). \end{aligned}$$

Relations (B.25) and (B.27) then assert that

$$\begin{aligned} \sigma_{\theta\theta}^{\mathrm{f}} &= \sigma_{rr}^{\mathrm{f}}, \\ \sigma_{\theta\theta}^{\mathrm{m}} &= \sigma_{rr}^{\mathrm{m}} + \frac{2\phi}{r^2}. \end{aligned} \tag{B.30}$$

On substituting (B.29) and (B.30) into the equilibrium equations (B.11)–(B.14), it follows that

$$\frac{\partial \sigma_{rr}^{\mathrm{f}}}{\partial r} = 0, \; \frac{\partial \sigma_{zz}^{\mathrm{f}}}{\partial z} = 0, \tag{B.31}$$

$$\frac{\partial \sigma_{rr}^{\mathrm{m}}}{\partial r} = \frac{2\phi}{r^3}, \; \frac{\partial \sigma_{zz}^{\mathrm{m}}}{\partial z} = 0. \tag{B.32}$$

On integrating $(\mathrm{B.32})_1$ subject to condition (B.15),

$$\sigma_{rr}^{\mathrm{m}} = \sigma_{\mathrm{T}} + \phi\left(\frac{1}{a^2} - \frac{1}{r^2}\right), \tag{B.33}$$

and from (B.30) it follows that

$$\sigma_{\theta\theta}^{\mathrm{m}} = \sigma_{\mathrm{T}} + \phi\left(\frac{1}{a^2} + \frac{1}{r^2}\right). \tag{B.34}$$

On integrating $(\mathrm{B.31})_1$ using (B.33) together with the continuity condition (B.17)

$$\sigma_{rr}^{\mathrm{f}} = \sigma_{\mathrm{T}} + \phi\left(\frac{1}{a^2} - \frac{1}{R^2}\right) = \sigma_{\mathrm{T}} - V_{\mathrm{m}}\frac{\phi}{R^2}, \tag{B.35}$$

and from (B.30) it follows that

$$\sigma_{\theta\theta}^{\mathrm{f}} = \sigma_{\mathrm{T}} + \phi\left(\frac{1}{a^2} - \frac{1}{R^2}\right) = \sigma_{\mathrm{T}} - V_{\mathrm{m}}\frac{\phi}{R^2}, \tag{B.36}$$

where use has been made of (B.24) and the relation $V_{\mathrm{m}} = 1 - V_{\mathrm{f}}$. The substitution of (B.25)$_3$, (B.35) and (B.36) into (B.3) leads to

$$\sigma_{zz}^{\mathrm{f}} = \sigma_{\mathrm{f}} = E_{\mathrm{A}}^{\mathrm{f}}(\varepsilon_{\mathrm{A}} - \alpha_{\mathrm{A}}^{\mathrm{f}}\Delta T) + 2\nu_{\mathrm{A}}^{\mathrm{f}}\left(\sigma_{\mathrm{T}} - V_{\mathrm{m}}\frac{\phi}{R^2}\right). \tag{B.37}$$

It is clear from (B.37) that σ_{zz}^{f} is a constant, automatically satisfying (B.31)$_2$. The substitution of (B.27)$_3$, (B.33) and (B.34) into (B.8) leads to

$$\sigma_{zz}^{\mathrm{m}} = \sigma_{\mathrm{m}} = E_{\mathrm{A}}^{\mathrm{m}}(\varepsilon_{\mathrm{A}} - \alpha_{\mathrm{A}}^{\mathrm{m}}\Delta T) + 2\nu_{\mathrm{A}}^{\mathrm{m}}\left(\sigma_{\mathrm{T}} + V_{\mathrm{f}}\frac{\phi}{R^2}\right), \tag{B.38}$$

which is also a constant, thus automatically satisfying (B.32)$_2$. The substitution of (B.25), (B.27), (B.33)–(B.38) into relations (B.1) and (B.6) leads to

$$\mathrm{A}_{\mathrm{f}} = -\nu_{\mathrm{A}}^{\mathrm{f}}\varepsilon_{\mathrm{A}} + \left(\alpha_{\mathrm{T}}^{\mathrm{f}} + \nu_{\mathrm{A}}^{\mathrm{f}}\alpha_{\mathrm{A}}^{\mathrm{f}}\right)\Delta T + \frac{1}{2k_{\mathrm{T}}^{\mathrm{f}}}\left(\sigma_{\mathrm{T}} - V_{\mathrm{m}}\frac{\phi}{R^2}\right), \tag{B.39}$$

$$\mathrm{A}_{\mathrm{m}} = -\nu_{\mathrm{A}}^{\mathrm{m}}\varepsilon_{\mathrm{A}} + \left(\alpha_{\mathrm{T}}^{\mathrm{m}} + \nu_{\mathrm{A}}^{\mathrm{m}}\alpha_{\mathrm{A}}^{\mathrm{m}}\right)\Delta T + \frac{1}{2k_{\mathrm{T}}^{\mathrm{m}}}\left(\sigma_{\mathrm{T}} + V_{\mathrm{f}}\frac{\phi}{R^2}\right), \tag{B.40}$$

where

$$\frac{1}{k_{\mathrm{T}}^{\mathrm{f}}} = \frac{2\left(1-\nu_{\mathrm{t}}^{\mathrm{f}}\right)}{E_{\mathrm{T}}^{\mathrm{f}}} - \frac{4\left(\nu_{\mathrm{A}}^{\mathrm{f}}\right)^2}{E_{\mathrm{A}}^{\mathrm{f}}}, \quad \frac{1}{k_{\mathrm{T}}^{\mathrm{m}}} = \frac{2\left(1-\nu_{\mathrm{t}}^{\mathrm{m}}\right)}{E_{\mathrm{T}}^{\mathrm{m}}} - \frac{4\left(\nu_{\mathrm{A}}^{\mathrm{m}}\right)^2}{E_{\mathrm{A}}^{\mathrm{m}}}. \tag{B.41}$$

It now only remains to determine the constant ϕ which can be specified on applying the remaining condition (B.19), because conditions (B.16), (B.18) and (B.20) are automatically satisfied by (B.21) and (B.29). It follows from (B.19), (B.22), (B.23), (B.39) and (B.40) that

$$\frac{\phi}{R^2} = \lambda\left[(\nu_{\mathrm{A}}^{\mathrm{m}} - \nu_{\mathrm{A}}^{\mathrm{f}})\varepsilon_{\mathrm{A}} + \left(\alpha_{\mathrm{T}}^{\mathrm{f}} + \nu_{\mathrm{A}}^{\mathrm{f}}\alpha_{\mathrm{A}}^{\mathrm{f}}\right)\Delta T - \left(\alpha_{\mathrm{T}}^{\mathrm{m}} + \nu_{\mathrm{A}}^{\mathrm{m}}\alpha_{\mathrm{A}}^{\mathrm{m}}\right)\Delta T + \frac{1}{2}\left(\frac{1}{k_{\mathrm{T}}^{\mathrm{f}}} - \frac{1}{k_{\mathrm{T}}^{\mathrm{m}}}\right)\sigma_{\mathrm{T}}\right], \tag{B.42}$$

where

$$\frac{1}{\lambda} = \frac{1}{2}\left(\frac{1}{\mu_{\mathrm{t}}^{\mathrm{m}}} + \frac{V_{\mathrm{f}}}{k_{\mathrm{T}}^{\mathrm{m}}} + \frac{V_{\mathrm{m}}}{k_{\mathrm{T}}^{\mathrm{f}}}\right). \tag{B.43}$$

The displacement distribution is specified by (B.21)–(B.23), and the corresponding stress distribution is specified by (B.29) and (B.33)–(B.38). The stress-strain relations (B.1)–(B.4), (B.6)–(B.9) and the equilibrium equations (B.11)–(B.14) are satisfied exactly. The boundary and interface conditions (B.15)–(B.20) are also satisfied exactly. In addition the displacement distribution satisfies the compatibility equations although they have not been considered in this analysis.

As

$$\sigma_A = V_f \sigma_{zz}^f + V_m \sigma_{zz}^m, \tag{B.44}$$

it follows from (B.37) and (B.38) that

$$\sigma_A = E_A^* \varepsilon_A - E_A^* \alpha_A^* \Delta T + 2\nu_A^* \sigma_T + 2(\nu_A^m - \nu_A^f) V_f V_m \frac{\phi}{R^2}, \tag{B.45}$$

where

$$E_A^* = V_f E_A^f + V_m E_A^m, \tag{B.46}$$

$$E_A^* \alpha_A^* = V_f E_A^f \alpha_A^f + V_m E_A^m \alpha_A^m, \tag{B.47}$$

$$\nu_A^* = V_f \nu_A^f + V_m \nu_A^m. \tag{B.48}$$

Substitution of (B.42) into (B.45) leads to

$$\sigma_A = \left[E_A^* + 2\lambda \left(\nu_A^m - \nu_A^f\right)^2 V_f V_m \right] \varepsilon_A + 2\left[\nu_A^* - \frac{\lambda}{2}\left(\nu_A^m - \nu_A^f\right)\left(\frac{1}{k_T^m} - \frac{1}{k_T^f}\right) V_f V_m \right] \sigma_T$$
$$- \left[E_A^* \alpha_A^* + 2\lambda \left(\nu_A^m - \nu_A^f\right)\left(\alpha_T^m + \nu_A^m \alpha_A^m - \alpha_T^f - \nu_A^f \alpha_A^f\right) V_f V_m \right] \Delta T,$$

$$\text{i.e. } \sigma_A = E_A \varepsilon_A + 2\nu_A \sigma_T - E_A \alpha_A \Delta T, \tag{B.49}$$

where E_A, ν_A and α_A are the effective axial modulus, Poisson's ratio and axial thermal expansion coefficient, respectively, for the concentric cylinder system representing the unidirectional composite defined by

$$E_A = E_A^* + 2\lambda \left(\nu_A^m - \nu_A^f\right)^2 V_f V_m, \tag{B.50}$$

$$\nu_A = \nu_A^* - \frac{\lambda}{2}\left(\frac{1}{k_T^m} - \frac{1}{k_T^f}\right)\left(\nu_A^m - \nu_A^f\right) V_f V_m, \tag{B.51}$$

$$E_A \alpha_A = E_A^* \alpha_A^* + 2\lambda \left(\nu_A^m - \nu_A^f\right)\left(\alpha_T^m + \nu_A^m \alpha_A^m - \alpha_T^f - \nu_A^f \alpha_A^f\right) V_f V_m. \tag{B.52}$$

From (B.24) and (B.27)

$$\varepsilon_T = \frac{u_r(r=a)}{a} = A_m + \frac{V_f \phi}{2\mu_t^m R^2}. \tag{B.53}$$

On using (B.40) and (B.42) it can be shown that

$$\varepsilon_T = -\left[\nu_A^m + \frac{\lambda V_f}{2}\left(\frac{1}{\mu_t^m} + \frac{1}{k_T^m}\right)\left(\nu_A^f - \nu_A^m\right) \right] \varepsilon_A$$
$$+ \left[\frac{1}{2k_T^m} + \frac{\lambda V_f}{4}\left(\frac{1}{\mu_t^m} + \frac{1}{k_T^m}\right)\left(\frac{1}{k_T^f} - \frac{1}{k_T^m}\right) \right] \sigma_T \tag{B.54}$$
$$+ \left[\left(\alpha_T^m + \nu_A^m \alpha_A^m\right) + \frac{\lambda V_f}{2}\left(\frac{1}{\mu_t^m} + \frac{1}{k_T^m}\right)\left\{\left(\alpha_T^f + \nu_A^f \alpha_A^f\right) - \left(\alpha_T^m + \nu_A^m \alpha_A^m\right)\right\} \right] \Delta T.$$

For the composite regarded as a homogeneous medium, the stress-strain relations may be written as

$$\varepsilon_{\mathrm{A}} = \frac{\sigma_{\mathrm{A}}}{E_{\mathrm{A}}} - \frac{2\nu_{\mathrm{A}}}{E_{\mathrm{A}}}\sigma_{\mathrm{T}} + \alpha_{\mathrm{A}}\Delta T\,, \qquad \varepsilon_{\mathrm{T}} = -\frac{\nu_{\mathrm{A}}}{E_{\mathrm{A}}}\sigma_{\mathrm{T}} + \frac{1-\nu_{\mathrm{t}}}{E_{\mathrm{T}}}\sigma_{\mathrm{A}} + \alpha_{\mathrm{T}}\Delta T. \tag{B.55}$$

On eliminating the axial stress, it can be shown that

$$\varepsilon_{\mathrm{T}} = -\,\nu_{\mathrm{A}}\,\varepsilon_{\mathrm{A}} + \frac{\sigma_{\mathrm{T}}}{2k_{\mathrm{T}}} + (\alpha_{\mathrm{T}} + \nu_{\mathrm{A}}\alpha_{\mathrm{A}})\Delta T. \tag{B.56}$$

By comparing (B.54) with (B.56) it is deduced that for the quantity ψ representing ν_{A}, $1/k_{\mathrm{T}}$ or $\alpha_{\mathrm{T}} + \nu_{\mathrm{A}}\alpha_{\mathrm{A}}$

$$\psi = \psi_{\mathrm{m}} + \frac{\lambda V_{\mathrm{f}}}{2}\left(\frac{1}{\mu_{\mathrm{t}}^{\mathrm{m}}} + \frac{1}{k_{\mathrm{T}}^{\mathrm{m}}}\right)\left(\psi_{\mathrm{f}} - \psi_{\mathrm{m}}\right). \tag{B.57}$$

On using (B.43), relation (B.57) may be written in the following form, analogous to that of relations (B.50)–(B.52) expressing the properties as the sum of a mixtures rule term together with a correction term,

$$\psi = V_{\mathrm{f}}\psi_{\mathrm{f}} + V_{\mathrm{m}}\psi_{\mathrm{m}} - \frac{\lambda}{2}\left(\frac{1}{k_{\mathrm{T}}^{\mathrm{f}}} - \frac{1}{k_{\mathrm{T}}^{\mathrm{m}}}\right)\left(\psi_{\mathrm{f}} - \psi_{\mathrm{m}}\right)V_{\mathrm{f}}V_{\mathrm{m}}. \tag{B.58}$$

It then follows directly that

$$\frac{1}{k_{\mathrm{T}}} = \frac{V_{\mathrm{f}}}{k_{\mathrm{T}}^{\mathrm{f}}} + \frac{V_{\mathrm{m}}}{k_{\mathrm{T}}^{\mathrm{m}}} - \frac{\lambda}{2}\left(\frac{1}{k_{\mathrm{T}}^{\mathrm{f}}} - \frac{1}{k_{\mathrm{T}}^{\mathrm{m}}}\right)^2 V_{\mathrm{f}}V_{\mathrm{m}}, \tag{B.59}$$

$$\begin{aligned}\alpha_{\mathrm{T}} = &-\nu_{\mathrm{A}}\alpha_{\mathrm{A}} + (\alpha_{\mathrm{T}}^{\mathrm{f}} + \nu_{\mathrm{A}}^{\mathrm{f}}\alpha_{\mathrm{A}}^{\mathrm{f}})V_{\mathrm{f}} + (\alpha_{\mathrm{T}}^{\mathrm{m}} + \nu_{\mathrm{A}}^{\mathrm{m}}\alpha_{\mathrm{A}}^{\mathrm{m}})V_{\mathrm{m}} \\ &+ \frac{\lambda}{2}\left(\frac{1}{k_{\mathrm{T}}^{\mathrm{f}}} - \frac{1}{k_{\mathrm{T}}^{\mathrm{m}}}\right)\left[(\alpha_{\mathrm{T}}^{\mathrm{m}} + \nu_{\mathrm{A}}^{\mathrm{m}}\alpha_{\mathrm{A}}^{\mathrm{m}}) - (\alpha_{\mathrm{T}}^{\mathrm{f}} + \nu_{\mathrm{A}}^{\mathrm{f}}\alpha_{\mathrm{A}}^{\mathrm{f}})\right]V_{\mathrm{f}}V_{\mathrm{m}}\,.\end{aligned} \tag{B.60}$$

In addition, result (B.58), when applied with $\psi = \nu_{\mathrm{A}}$, leads to result (B.51) that has already been derived.

Separation Condition for a Sliding Interface

Consider the concentric cylinder model of a composite where the fibre is broken or matrix is cracked. Following fracture, interface sliding is assumed to have occurred such that extensive sliding has resulted either when fracture first occurred or following subsequent loading of the composite. For such large-scale sliding conditions, it is of interest to determine the critical loading conditions that correspond to the separation of the fibre and matrix at the interface in the regions close to the fibre fracture or matrix crack when stress transfer is governed by the Coulomb friction law. Interface separation takes place when $\sigma_{rr}^{\mathrm{f}} = \sigma_{rr}^{\mathrm{m}} = 0$ on $r = R$.

The displacement field is assumed to have the following form

$$u_r^{\mathrm{f}} = \mathrm{A}_{\mathrm{f}}' r, u_\theta^{\mathrm{f}} \equiv 0, \quad u_z^{\mathrm{f}} = \varepsilon_{\mathrm{A}}^{\mathrm{f}} z, \tag{B.61}$$

$$u_r^{\mathrm{m}} = \mathrm{A}_{\mathrm{m}}' r + \frac{\phi'}{2\mu_{\mathrm{t}}^{\mathrm{m}} r}, \quad u_\theta^{\mathrm{m}} \equiv 0, \quad u_z^{\mathrm{m}} = \varepsilon_{\mathrm{A}}^{\mathrm{m}} z, \tag{B.62}$$

where A'_f, A'_m and ϕ' are constants to be determined and ε_A^f and ε_A^m are the uniform axial strain in the fibre and matrix, respectively. On differentiating the displacement field, it follows that the nonzero strain components are

$$\varepsilon_{rr}^f = \frac{\partial u_r^f}{\partial r} = A'_f, \quad \varepsilon_{\theta\theta}^f = \frac{u_r^f}{r} = A'_f, \quad \varepsilon_{zz}^f = \varepsilon_A^f, \tag{B.63}$$

$$\varepsilon_{rr}^m = \frac{\partial u_r^m}{\partial r} = A'_m - \frac{\phi'}{2\mu_t^m r^2}, \quad \varepsilon_{\theta\theta}^m = \frac{u_r^m}{r} = A'_m + \frac{\phi'}{2\mu_t^m r^2}, \quad \varepsilon_{zz}^m = \varepsilon_A^m. \tag{B.64}$$

As from the stress-strain relations

$$\varepsilon_{\theta\theta}^f - \varepsilon_{rr}^f = \frac{1}{2\mu_t^f}(\sigma_{\theta\theta}^f - \sigma_{rr}^f), \qquad \varepsilon_{\theta\theta}^m - \varepsilon_{rr}^m = \frac{1}{2\mu_t^m}(\sigma_{\theta\theta}^m - \sigma_{rr}^m), \tag{B.65}$$

it can be shown that

$$\sigma_{rr}^f = \sigma_{\theta\theta}^f = \sigma_T - V_m \frac{\phi'}{R^2}, \tag{B.66}$$

$$\sigma_{rr}^m = \sigma_T + \phi'\left(\frac{1}{a^2} - \frac{1}{r^2}\right), \qquad \sigma_{\theta\theta}^m = \sigma_T + \phi'\left(\frac{1}{a^2} + \frac{1}{r^2}\right). \tag{B.67}$$

In the axial direction it is assumed that

$$\sigma_A^f = \frac{(1+\xi)\sigma_A}{2V_f}, \qquad \sigma_A^m = \frac{(1-\xi)\sigma_A}{2V_m}. \tag{B.68}$$

When $\xi = 1$ a matrix crack is being considered as $\sigma_A^m = 0$, and when $\xi = -1$ a fibre fracture is being considered as $\sigma_A^f = 0$. The axial stress-strain relations then lead to

$$\sigma_A^f = \frac{(1+\xi)\sigma_A}{2V_f} = E_A^f(\varepsilon_A^f - \alpha_A^f \Delta T) + 2\nu_A^f\left(\sigma_T - V_m \frac{\phi'}{R^2}\right), \tag{B.69}$$

$$\sigma_A^m = \frac{(1-\xi)\sigma_A}{2V_m} = E_A^m(\varepsilon_A^m - \alpha_A^m \Delta T) + 2\nu_A^m\left(\sigma_T + V_f \frac{\phi'}{R^2}\right). \tag{B.70}$$

Substitution of (B.63), (B.64) and (B.66)–(B.70) into the radial stress-strain equations leads to

$$A'_f = \frac{1-\nu_t^f}{E_T^f}\left(\sigma_T - V_m \frac{\phi'}{R^2}\right) - \frac{\nu_A^f}{E_A^f}\frac{(1+\xi)\sigma_A}{2V_f} + \alpha_T^f \Delta T, \tag{B.71}$$

$$A'_m = \frac{1-\nu_t^m}{E_T^m}\left(\sigma_T + V_f \frac{\phi'}{R^2}\right) - \frac{\nu_A^m}{E_A^m}\frac{(1-\xi)\sigma_A}{2V_m} + \alpha_T^m \Delta T. \tag{B.72}$$

It now only remains to determine the constant ϕ' which can be specified on applying the continuity condition for the radial displacement. It follows from (B.61) and (B.62) that

$$A'_f = A'_m + \frac{\phi'}{2\mu_t^m R^2}. \tag{B.73}$$

On using (B.71) and (B.72),

$$\frac{\phi'}{R^2}=\lambda'\left[\left(\frac{1-\nu_t^f}{E_T^f}-\frac{1-\nu_t^m}{E_T^m}\right)\sigma_T+\frac{\nu_A^m}{E_A^m}\frac{(1-\xi)\sigma_A}{2V_m}-\frac{\nu_A^f}{E_A^f}\frac{(1+\xi)\sigma_A}{2V_f}+\left(\alpha_T^f-\alpha_T^m\right)\Delta T\right], \quad (B.74)$$

where

$$\frac{1}{\lambda'}=\frac{1-\nu_t^m}{E_T^m}V_f+\frac{1-\nu_t^f}{E_T^f}V_m+\frac{1}{2\mu_t^m}. \quad (B.75)$$

It should be noted from (B.66) that the interfacial radial stress is zero when

$$\frac{\phi'}{R^2}=\frac{\sigma_T}{V_m}. \quad (B.76)$$

As $E_T^m=2\mu_t^m(1+\nu_t^m)$, it then follows from (B.74) that this occurs when

$$\left(\frac{\nu_A^f}{E_A^f}\frac{1+\xi}{2V_f}-\frac{\nu_A^m}{E_A^m}\frac{1-\xi}{2V_m}\right)\sigma_A=-\frac{2}{V_mE_T^m}\sigma_T+\left(\alpha_T^f-\alpha_T^m\right)\Delta T. \quad (B.77)$$

Appendix C: Gibbs Energy per Unit Volume for a Cracked Laminate

Consider a macroscopic region V of a multiple-ply general symmetric laminate that contains ply cracks in the 90° plies. The cracks may be nonuniformly distributed, and they need not extend across the entire width or length of the laminate. The Gibbs energy per unit volume of a homogeneous linear elastic solid has the form

$$g = -\frac{1}{2} S_{ijkl}\sigma_{ij}\sigma_{kl} - \sigma_{ij}\alpha_{ij}\Delta T + g_0(\Delta T), \tag{C.1}$$

where $g_0(\Delta T)$ is the Gibbs energy per unit volume when the stress field is zero, and where $S_{ijkl} = S_{klij}$. The stress-strain relations are obtained by differentiation as follows:

$$\varepsilon_{ij} = -\frac{\partial g}{\partial \sigma_{ij}} = S_{ijkl}\sigma_{kl} + \alpha_{ij}\,\Delta T. \tag{C.2}$$

The parameters S_{ijkl} denote the elastic compliances and the parameters α_{ij} denote the thermal expansion coefficients. On using (C.2), it follows that (C.1) may be expressed in the form

$$g = -\frac{1}{2}\sigma_{ij}\left(\varepsilon_{ij} + \alpha_{ij}\Delta T\right) + g_0(\Delta T). \tag{C.3}$$

Thus, the average Gibbs energy density for the region V, in which the stress and strain fields will be nonuniform owing to the presence of ply cracks, is written as

$$\bar{g} = \frac{1}{V}\int_V g\,dV = -\frac{1}{2V}\int_V \sigma_{ij}\varepsilon_{ij}\,dV - \frac{1}{2V}\int_V \sigma_{ij}\alpha_{ij}\,\Delta T\,dV + g_0(\Delta T). \tag{C.4}$$

By making use of the strain–displacement relations, the symmetry relations $\sigma_{ij} = \sigma_{ji}$, the equilibrium equations and the Gauss theorem, it can be shown that

$$\int_V \sigma_{ij}\varepsilon_{ij}\,dV = \int_S n_i\sigma_{ij}u_j\,dS, \tag{C.5}$$

where S is the surface describing the external surface of the laminate having outward unit normal n_i and where u_j are the surface displacements.

1. In-plane Biaxial Plus Shear Deformation

An assumption is made that the crack distribution does not lead to any asymmetrical property that could induce in-plane shear and out-of-plane bending deformation in a multiaxially loaded laminate. The external faces of the laminate are subject to a uniform applied normal stress σ_t and the edges of undamaged plies are subject to the boundary conditions (10.1) defining the effective applied in-plane strains for the cracked laminate.

As a uniform through-thickness stress σ_t is applied to the laminate faces, and because displacements are applied on the edges of the laminate such that the effective in-plane strains are given by (10.1), it follows from (C.5) that

$$\frac{1}{V}\int_V \sigma_{ij}\varepsilon_{ij}\, dV = \sigma_A\varepsilon_A + \sigma_T\varepsilon_T + \sigma_t\varepsilon_t + \tau_A\gamma_A, \tag{C.6}$$

where σ_A, σ_T and τ_A are the effective in-plane applied stresses defined by (10.4). The effective out-of-plane strain ε_t is defined by (10.5). Result (C.6) demonstrates that the value of the integral for a cracked laminate, having nonuniform stress and strain distributions, has the same form as that which arises for undamaged laminates.

Let σ_{ij}^c and ε_{ij}^c denote the state of stress and strain when the cracked laminate is subject to multiaxial stresses that just close all the cracks in the 90° plies for the case when the shear strain $\gamma_A = \gamma_A^c = 0$. It follows from (C.2) that

$$\varepsilon_{ij}^c = S_{ijkl}\sigma_{kl}^c + \alpha_{ij}\Delta T. \tag{C.7}$$

Consider now the quantity I defined by

$$I = \int_V \sigma_{ij}\,\alpha_{ij}\,\Delta T dV. \tag{C.8}$$

It follows on using (C.7) that (C.8) can be written in the form

$$I = \int_V \sigma_{ij}\left[\varepsilon_{ij}^c - S_{ijkl}\sigma_{kl}^c\right]dV. \tag{C.9}$$

The symmetry relations $S_{ijkl} = S_{klij}$ enable (C.9) to be written as

$$I = \int_V \sigma_{ij}\varepsilon_{ij}^c dV - \int_V \sigma_{ij}^c S_{ijkl}\sigma_{kl} dV. \tag{C.10}$$

On using the stress-strain relations (C.2) the result (C.10) may be expressed

$$I = \int_V \sigma_{ij}\varepsilon_{ij}^c dV - \int_V \sigma_{ij}^c\left[\varepsilon_{ij} - \alpha_{ij}\Delta T\right]dV. \tag{C.11}$$

The strain–displacement relations, the equilibrium equations and the symmetry relations $\sigma_{ij} = \sigma_{ji}$ and $\sigma_{ij}^c = \sigma_{ji}^c$ lead to the expression

$$I = \int_V \frac{\partial}{\partial x_i}\left(\sigma_{ij}u_j^c\right)dV - \int_V \frac{\partial}{\partial x_i}\left(\sigma_{ij}^c u_j\right)dV + \int_V \sigma_{ij}^c\alpha_{ij}\Delta T dV. \tag{C.12}$$

The use of Gauss' theorem leads directly to

$$I = \int_S n_i\sigma_{ij}u_j^c\, dS - \int_S n_i\sigma_{ij}^c u_j dS + \int_V \sigma_{ij}^c\alpha_{ij}\Delta T dV, \tag{C.13}$$

where S denotes the surface describing the external surface of the laminate. On applying the boundary conditions, and using the definitions (10.1) and (10.5) for the applied strains, it follows from (C.8), (C.13) and the fact that $\gamma_A^c = 0$ that

$$\begin{aligned}\frac{1}{V}\int_V \left(\sigma_{ij} - \sigma_{ij}^c\right)\alpha_{ij}\,\Delta T dV = {} & \sigma_A\varepsilon_A^c + \sigma_T\varepsilon_T^c + \sigma_t\varepsilon_t^c \\ & -\sigma_A^c\varepsilon_A - \sigma_T^c\varepsilon_T - \sigma_t^c\varepsilon_t - \tau_A^c\gamma_A,\end{aligned} \tag{C.14}$$

where $\sigma_A^c, \sigma_T^c, \sigma_t^c, \tau_A^c, \varepsilon_A^c, \varepsilon_T^c, \varepsilon_t^c$ and γ_A^c are the stresses and strains which arise at the point of crack closure. On eliminating the strains in (C.14) using the stress-strain relations it can be shown that

$$\frac{1}{V}\int_V \left(\sigma_{ij}-\sigma_{ij}^c\right)\alpha_{ij}\,\Delta T dV = \left(\sigma_A-\sigma_A^c\right)\alpha_A + \left(\sigma_T-\sigma_T^c\right)\alpha_T + \left(\sigma_t-\sigma_t^c\right)\alpha_t + \left(\tau_A-\tau_A^c\right)\alpha_S. \tag{C.15}$$

On substituting (C.6) and (C.15) into (C.4) the following expression for the average Gibbs energy in region V is obtained

$$\begin{aligned}\bar{g} = \frac{1}{V}\int_V g\,dV &= -\frac{1}{2}\left[\sigma_A\varepsilon_A + \sigma_T\varepsilon_T + \sigma_t\varepsilon_t + \tau_A\gamma_A\right]\\ &-\frac{1}{2}\left[\left(\sigma_A-\sigma_A^c\right)\alpha_A + \left(\sigma_T-\sigma_T^c\right)\alpha_T + \left(\sigma_t-\sigma_t^c\right)\alpha_t + \left(\tau_A-\tau_A^c\right)\alpha_S\right]\Delta T\\ &-\frac{1}{2V}\int_V \sigma_{ij}^c\alpha_{ij}\Delta T\,dV + g_0(\Delta T).\end{aligned} \tag{C.16}$$

On eliminating the strains using (10.7)–(10.10) it can be shown that

$$\begin{aligned}\bar{g} = &-\frac{\sigma_A^2}{2E_A} - \frac{\sigma_T^2}{2E_T} - \frac{\sigma_t^2}{2E_t} - \frac{\tau_A^2}{2\mu_A} + \frac{\nu_A}{E_A}\sigma_A\sigma_T + \frac{\nu_a}{E_A}\sigma_A\sigma_t + \frac{\nu_t}{E_T}\sigma_T\sigma_t\\ &+ \frac{\lambda_A}{E_A}\sigma_A\tau_A + \frac{\lambda_T}{E_A}\sigma_T\tau_A + \frac{\lambda_t}{E_A}\sigma_t\tau_A - \left[\sigma_A\alpha_A + \sigma_T\alpha_T + \sigma_t\alpha_t + \tau_A\alpha_S\right]\Delta T\\ &+\frac{1}{2}\left[\sigma_A^c\alpha_A + \sigma_T^c\alpha_T + \sigma_t^c\alpha_t + \tau_A^c\alpha_S\right]\Delta T - \frac{1}{2V}\int_V \sigma_{ij}^c\alpha_{ij}\,\Delta T\,dV + g_0(\Delta T).\end{aligned} \tag{C.17}$$

It should be noted that the stress-strain relations (10.7)–(10.10) can be recovered from (C.17) using the relations

$$\varepsilon_A = -\frac{\partial\bar{g}}{\partial\sigma_A},\quad \varepsilon_T = -\frac{\partial\bar{g}}{\partial\sigma_T},\quad \varepsilon_t = -\frac{\partial\bar{g}}{\partial\sigma_t},\quad \gamma_A = -\frac{\partial\bar{g}}{\partial\tau_A}. \tag{C.18}$$

This proves that the form assumed for the stress-strain relations (10.7)–(10.10) is valid. In addition, the average Gibbs energy for a cracked laminate with nonuniform stress and displacement distributions obeys the same type of relation as that which applies for homogeneous stress and strain states (see (C.2)).

2. Bend Deformation

On using (11.14), (11.36) and (11.38) it can be shown that

$$\begin{aligned}\int_S n_i\sigma_{ij}u_j\,dS = &\,2L\bar{\varepsilon}_A\int_0^h\int_{-W}^W \sigma_{11}(L,x_2,x_3)\,dx_2dx_3 + 2L\hat{\varepsilon}_A\int_0^h\int_{-W}^W x_3\sigma_{11}(L,x_2,x_3)\,dx_2dx_3\\ &+2W\bar{\varepsilon}_T\int_0^h\int_{-L}^L \sigma_{22}(x_1,W,x_3)\,dx_1dx_3 + 2W\hat{\varepsilon}_T\int_0^h\int_{-L}^L x_3\sigma_{22}(x_1,W,x_3)\,dx_1dx_3 + 4hLW\sigma_t\varepsilon_t,\end{aligned} \tag{C.19}$$

where use has been made of the symmetry relations

$$\sigma_{11}(L,x_2,x_3) \equiv \sigma_{11}(-L,x_2,x_3), \ \sigma_{22}(x_1,W,x_3) \equiv \sigma_{22}(x_1,-W,x_3). \tag{C.20}$$

On using (11.34) and (11.35) it follows from (C.19) that

$$\begin{aligned}\int_S n_i\sigma_{ij}u_j\, dS &= 4hLW\sigma_A\left(\bar{\varepsilon}_A + \frac{1}{2}h\hat{\varepsilon}_A\right) + 4hLW\sigma_T\left(\bar{\varepsilon}_T + \frac{1}{2}h\hat{\varepsilon}_T\right)\\ &+ 4hLW\, M_A\hat{\varepsilon}_A + 4hLW\, M_T\hat{\varepsilon}_T + 4hLW\sigma_t\varepsilon_t.\end{aligned} \tag{C.21}$$

As the volume of the laminate is $V = 4hLW$, the use of (C.5) leads to

$$\frac{1}{V}\int_V \sigma_{ij}\,\varepsilon_{ij}\, dV = \sigma_A\,\bar{\varepsilon}_A + \sigma_T\bar{\varepsilon}_T + \sigma_t\,\varepsilon_t + M_A\hat{\varepsilon}_A + M_T\hat{\varepsilon}_T. \tag{C.22}$$

The effective out-of-plane strain ε_t is defined by (11.36). Result (C.22) demonstrates that the value of the integral for a cracked laminate, having nonuniform stress and strain distributions, has the same form as that which arises for undamaged laminates.

Let σ_{ij}^c and ε_{ij}^c denote the state of stress and strain when the cracked laminate is subject to multiaxial stresses that just close all the ply cracks in the 90° plies. It follows from (C.2) that

$$\varepsilon_{ij}^c = S_{ijkl}\,\sigma_{kl}^c + \alpha_{ij}\,\Delta T. \tag{C.23}$$

Consider now the quantity I defined by

$$I = \int_V \sigma_{ij}\,\alpha_{ij}\,\Delta T\, dV. \tag{C.24}$$

It follows on using (C.23) that (C.24) can be written in the form

$$I = \int_V \sigma_{ij}\left[\varepsilon_{ij}^c - S_{ijkl}\,\sigma_{kl}^c\right] dV. \tag{C.25}$$

The symmetry relations $S_{ijkl} = S_{klij}$ enable (C.25) to be written as

$$I = \int_V \sigma_{ij}\,\varepsilon_{ij}^c\, dV - \int_V \sigma_{ij}^c\, S_{ijkl}\,\sigma_{kl}\, dV. \tag{C.26}$$

On using the stress-strain relations (C.2), the result (C.26) may be expressed

$$I = \int_V \sigma_{ij}\,\varepsilon_{ij}^c\, dV - \int_V \sigma_{ij}^c\left(\varepsilon_{ij} - \alpha_{ij}\,\Delta T\right) dV. \tag{C.27}$$

The strain–displacement relations, the equilibrium equations and the symmetry relations $\sigma_{ij} = \sigma_{ji}$ and $\sigma_{ij}^c = \sigma_{ji}^c$ lead to the expression

$$I = \int_V \frac{\partial}{\partial x_i}\left(\sigma_{ij}\, u_{ij}^c\right) dV - \int_V \frac{\partial}{\partial x_i}\left(\sigma_{ij}^c\, u_j\right) dV + \int_V \sigma_{ij}^c\,\alpha_{ij}\,\Delta T\, dV. \tag{C.28}$$

The use of Gauss' theorem leads directly to

$$I = \int_S n_i\sigma_{ij}\, u_j^c\, dS - \int_S n_i\sigma_{ij}^c\, u_j\, dS + \int_V \sigma_{ij}^c\,\alpha_{ij}\,\Delta T\, dV, \tag{C.29}$$

where S denotes the surface describing the external surface of the laminate. On applying the boundary conditions, and using the definitions (11.36) and (11.39) for the applied strains, it follows from (C.24) and (C.29) that

$$\frac{1}{V}\int_V \left(\sigma_{ij}-\sigma_{ij}^{c}\right)\alpha_{ij}\Delta T\,dV = \sigma_A\,\bar{\varepsilon}_A^{c} + \sigma_T\bar{\varepsilon}_T^{c} + \sigma_t\varepsilon_t^{c} + M_A\,\hat{\varepsilon}_A^{c} + M_T\hat{\varepsilon}_T^{c} - \sigma_A^{c}\bar{\varepsilon}_A - \sigma_T^{c}\bar{\varepsilon}_T - \sigma_t^{c}\varepsilon_t - M_A^{c}\hat{\varepsilon}_A - M_T^{c}\hat{\varepsilon}_T, \tag{C.30}$$

where σ_A^c, σ_T^c, σ_t^c, M_A^c, M_T^c, ε_A^c, ε_T^c, ε_t^c, $\hat{\varepsilon}_A^c$ and $\hat{\varepsilon}_T^c$ are the stresses, moments and strains which arise at the point of crack closure (see Appendix D). On eliminating the strains in (C.30) using the stress-strain relations it can be shown that

$$\frac{1}{V}\int_V \left(\sigma_{ij}-\sigma_{ij}^{c}\right)\alpha_{ij}\Delta T\,dV = \begin{bmatrix} \left(\sigma_A-\sigma_A^{c}\right)\alpha_A + \left(\sigma_T-\sigma_T^{c}\right)\alpha_T + \left(\sigma_T-\sigma_t^{c}\right)\alpha_T \\ +\left(M_A-M_A^{c}\right)\hat{\alpha}_A + \left(M_T-M_T^{c}\right)\hat{\alpha}_T \end{bmatrix}\Delta T. \tag{C.31}$$

On substituting (C.22) and (C.31) into (C.4), the following expression for the average Gibbs energy in region V is obtained:

$$\begin{aligned}\bar{g} = &-\frac{1}{2}\left(\sigma_A\bar{\varepsilon}_A + \sigma_T\bar{\varepsilon}_T + \sigma_t\varepsilon_t + M_A\hat{\varepsilon}_A + M_T\hat{\varepsilon}_T\right) \\ &-\frac{1}{2}\Big[\left(\sigma_A-\sigma_A^{c}\right)\alpha_A + \left(\sigma_T-\sigma_T^{c}\right)\alpha_T + \left(\sigma_t-\sigma_t^{c}\right)\alpha_t \\ &+\left(M-M_A^{c}\right)\hat{\alpha}_A + \left(M_T-M_T^{c}\right)\hat{\alpha}_T\Big]\Delta T + \frac{1}{2V}\int_V \sigma_{ij}^{c}\alpha_{ij}\Delta T\,dV + g_0(\Delta T).\end{aligned} \tag{C.32}$$

On eliminating the strains using the stress-strain relations it can be shown that

$$\begin{aligned}\bar{g} = &-\frac{\sigma_A^2}{2E_A} - \frac{\sigma_T^2}{2E_T} - \frac{\sigma_t^2}{2E_t} - \frac{M_A^2}{2\hat{E}_A} - \frac{M_T^2}{2\hat{E}_T} \\ &+\frac{\nu_A}{E_A}\sigma_A\sigma_T + \frac{\nu_a}{E_A}\sigma_A\sigma_t + \frac{\nu_t}{E_T}\sigma_T\sigma_t + \frac{\hat{\nu}_A}{\hat{E}_A}\sigma_A M_A + \frac{\hat{\nu}_t}{\hat{E}_A}\sigma_T M_A + \frac{\hat{\nu}_a}{\hat{E}_A}\sigma_t M_A \\ &+\frac{\hat{\eta}_A}{\hat{E}_T}\sigma_A M_T + \frac{\hat{\eta}_T}{\hat{E}_T}\sigma_T M_T + \frac{\hat{\eta}_t}{\hat{E}_T}\sigma_t M_T + \frac{\hat{\delta}_A}{\hat{E}_A}M_A M_T \\ &-\left[\sigma_A\alpha_A + \sigma_T\alpha_T + \sigma_t\alpha_t + M_A\hat{\alpha}_A + M_T\hat{\alpha}_T\right]\Delta T \\ &+\frac{1}{2}\left[\sigma_A^{c}\alpha_A + \sigma_T^{c}\alpha_T + \sigma_t^{c}\alpha_t + M_A^{c}\hat{\alpha}_A + M_T^{c}\hat{\alpha}_T\right]\Delta T - \frac{1}{2V}\int_V \sigma_{ij}^{c}\alpha_{ij}\Delta T\,dV + g_0(\Delta T).\end{aligned} \tag{C.33}$$

It should be noted that the stress-strain relations (11.41)–(11.45) can be recovered from (C.33) using the relations

$$\varepsilon_A^* = -\frac{\partial \bar{g}}{\partial \sigma_A},\ \varepsilon_T^* = -\frac{\partial \bar{g}}{\partial \sigma_T},\ \varepsilon_t = -\frac{\partial \bar{g}}{\partial \sigma_t},\ \hat{\varepsilon}_A = -\frac{\partial \bar{g}}{\partial M_A},\ \hat{\varepsilon}_T = -\frac{\partial \bar{g}}{\partial M_T}. \tag{C.34}$$

The average Gibbs energy for a cracked laminate with nonuniform stress and displacement distributions obeys the same type of relation as that which applies for homogeneous stress and strain states (see (C.2)).

Appendix D: Crack Closure Conditions for Laminates

The conditions of ply crack closure in a laminate can easily be found from an analysis of an undamaged laminate provided that a stress state can be found for which the tractions acting at the locations of the planes of the ply cracks in the corresponding damaged laminate are all zero. In other words, at the point of crack closure, the stress field in both a damaged and corresponding undamaged laminate are identical such that the tractions at the locations of the ply cracks are identically zero. The global coordinate system uses Cartesian coordinates (x_1, x_2, x_3) such that the x_1-axis is directed in the axial direction, the x_2-axis is directed in the in-plane transverse direction whereas the x_3-axis is directed in the through-thickness direction of a ply.

1. Crack Closure Analysis for Cross-ply Laminates (in the Presence of Bending)

Consider a symmetric cross-ply laminate having multiple layers of 0° and 90° plies. The stress-strain relations for a 0° ply for a given material may be written as

$$\begin{aligned}
\varepsilon_{11} \equiv \varepsilon_{\mathrm{A}} &= \frac{1}{E_{\mathrm{A}}}\sigma_{\mathrm{A}} - \frac{\nu_{\mathrm{A}}}{E_{\mathrm{A}}}\sigma_{\mathrm{T}} - \frac{\nu_{\mathrm{a}}}{E_{\mathrm{A}}}\sigma_{\mathrm{t}} + \alpha_{\mathrm{A}}\,\Delta T, \\
\varepsilon_{22} \equiv \varepsilon_{\mathrm{T}} &= -\frac{\nu_{\mathrm{A}}}{E_{\mathrm{A}}}\sigma_{\mathrm{A}} + \frac{1}{E_{\mathrm{T}}}\sigma_{\mathrm{T}} - \frac{\nu_{\mathrm{t}}}{E_{\mathrm{T}}}\sigma_{\mathrm{t}} + \alpha_{\mathrm{T}}\Delta T, \\
\varepsilon_{33} \equiv \varepsilon_{\mathrm{t}} &= -\frac{\nu_{\mathrm{a}}}{E_{\mathrm{A}}}\sigma_{\mathrm{A}} - \frac{\nu_{\mathrm{t}}}{E_{\mathrm{T}}}\sigma_{\mathrm{T}} + \frac{1}{E_{\mathrm{t}}}\sigma_{\mathrm{t}} + \alpha_{\mathrm{t}}\Delta T, \\
2\varepsilon_{12} \equiv \gamma_{\mathrm{A}} = \frac{1}{\mu_{\mathrm{A}}}\tau_{\mathrm{A}}, &\quad 2\varepsilon_{13} \equiv \gamma_{\mathrm{a}} = \frac{1}{\mu_{\mathrm{a}}}\tau_{\mathrm{a}}, \quad 2\varepsilon_{23} \equiv \gamma_{\mathrm{t}} = \frac{1}{\mu_{\mathrm{t}}}\tau_{\mathrm{t}},
\end{aligned} \tag{D.1}$$

whereas for a 90° ply made of the same material the stress-strain relations are

$$\begin{aligned}
\varepsilon_{11} \equiv \varepsilon_{\mathrm{A}} &= \frac{1}{E_{\mathrm{T}}}\sigma_{\mathrm{A}} - \frac{\nu_{\mathrm{A}}}{E_{\mathrm{A}}}\sigma_{\mathrm{T}} - \frac{\nu_{\mathrm{t}}}{E_{\mathrm{T}}}\sigma_{\mathrm{t}} + \alpha_{\mathrm{T}}\,\Delta T, \\
\varepsilon_{22} \equiv \varepsilon_{\mathrm{T}} &= -\frac{\nu_{\mathrm{A}}}{E_{\mathrm{A}}}\sigma_{\mathrm{A}} + \frac{1}{E_{\mathrm{A}}}\sigma_{\mathrm{T}} - \frac{\nu_{\mathrm{a}}}{E_{\mathrm{A}}}\sigma_{\mathrm{t}} + \alpha_{\mathrm{A}}\,\Delta T, \\
\varepsilon_{33} \equiv \varepsilon_{\mathrm{t}} &= -\frac{\nu_{\mathrm{t}}}{E_{\mathrm{T}}}\sigma_{\mathrm{A}} - \frac{\nu_{\mathrm{a}}}{E_{\mathrm{A}}}\sigma_{\mathrm{T}} + \frac{1}{E_{\mathrm{t}}}\sigma_{\mathrm{t}} + \alpha_{\mathrm{t}}\,\Delta T, \\
2\varepsilon_{12} \equiv \gamma_{\mathrm{A}} = \frac{1}{\mu_{\mathrm{A}}}\tau_{\mathrm{A}}, &\quad 2\varepsilon_{13} \equiv \gamma_{\mathrm{a}} = \frac{1}{\mu_{\mathrm{t}}}\tau_{a}, \quad 2\varepsilon_{23} \equiv \gamma_{\mathrm{t}} = \frac{1}{\mu_{\mathrm{a}}}\tau_{t}.
\end{aligned} \tag{D.2}$$

Relations (D.1) and (D.2) correspond to the stress-strain relations (6.99) and (6.100).

For a cross-ply laminate that need not be symmetric, the longitudinal displacement u_1 and the in-plane transverse displacement u_2 in any ply of the undamaged laminate are given by

$$u_1(x_1,x_3)=\left(\bar{\varepsilon}_A+\hat{\varepsilon}_A x_3\right)x_1,\quad u_2(x_2,x_3)=\left(\bar{\varepsilon}_T+\hat{\varepsilon}_T x_3\right)x_2, \tag{D.3}$$

where $\bar{\varepsilon}_A$, $\hat{\varepsilon}_A$ and $\bar{\varepsilon}_T$, $\hat{\varepsilon}_T$ are uniform strain parameters which have the same value in each layer of the laminate (well away from laminate edges). It follows from (D.3) that

$$\varepsilon_{11}=\bar{\varepsilon}_A+\hat{\varepsilon}_A x_3,\quad \varepsilon_{22}=\bar{\varepsilon}_T+\hat{\varepsilon}_T x_3. \tag{D.4}$$

Assuming that the 0° plies and 90° plies might be made of different materials, superscripts '(0)' and '(90)' will now be attached to the ply properties, the in-plane stresses and the out-of-plane strain. When bend deformation occurs in the absence of shear deformation, it then follows from (D.1) and (D.4) that the stress-strain relations for the 0° plies in the laminate subject to a uniform through-thickness stress σ_t are given by

$$\begin{aligned}
\varepsilon_t^{(0)} &= -\frac{\nu_a^{(0)}}{E_A^{(0)}}\sigma_A^{(0)}(x_3)-\frac{\nu_t^{(0)}}{E_T^{(0)}}\sigma_T^{(0)}(x_3)+\frac{1}{E_t^{(0)}}\sigma_t+\alpha_t^{(0)}\Delta T,\\
\bar{\varepsilon}_A+\hat{\varepsilon}_A x_3 &= \frac{1}{E_A^{(0)}}\sigma_A^{(0)}(x_3)-\frac{\nu_A^{(0)}}{E_A^{(0)}}\sigma_T^{(0)}(x_3)-\frac{\nu_a^{(0)}}{E_A^{(0)}}\sigma_t+\alpha_A^{(0)}\Delta T,\\
\bar{\varepsilon}_T+\hat{\varepsilon}_T x_3 &= -\frac{\nu_A^{(0)}}{E_A^{(0)}}\sigma_A^{(0)}(x_3)+\frac{1}{E_T^{(0)}}\sigma_T^{(0)}(x_3)-\frac{\nu_t^{(0)}}{E_T^{(0)}}\sigma_t+\alpha_T^{(0)}\Delta T,
\end{aligned} \tag{D.5}$$

where $\sigma_A^{(0)}(x_3)$ and $\sigma_T^{(0)}(x_3)$ are the axial and in-plane transverse stress distributions in the 0° ply. From (D.2) and (D.4) it follows that the corresponding stress-strain relations for the 90° plies are

$$\begin{aligned}
\varepsilon_t^{(90)} &= -\frac{\nu_t^{(90)}}{E_T^{(90)}}\sigma_A^{(90)}(x_3)-\frac{\nu_a^{(90)}}{E_A^{(90)}}\sigma_T^{(90)}(x_3)+\frac{1}{E_t^{(90)}}\sigma_t+\alpha_t^{(90)}\Delta T,\\
\bar{\varepsilon}_A+\hat{\varepsilon}_A x_3 &= \frac{1}{E_T^{(90)}}\sigma_A^{(90)}(x_3)-\frac{\nu_A^{(90)}}{E_A^{(90)}}\sigma_T^{(90)}(x_3)-\frac{\nu_t^{(90)}}{E_T^{(90)}}\sigma_t+\alpha_T^{(90)}\Delta T,\\
\bar{\varepsilon}_T+\hat{\varepsilon}_T x_3 &= -\frac{\nu_A^{(90)}}{E_A^{(90)}}\sigma_A^{(90)}(x_3)+\frac{1}{E_A^{(90)}}\sigma_T^{(90)}(x_3)-\frac{\nu_a^{(90)}}{E_A^{(90)}}\sigma_t+\alpha_A^{(90)}\Delta T,
\end{aligned} \tag{D.6}$$

where $\sigma_A^{(90)}(x_3)$ and $\sigma_T^{(90)}(x_3)$ are the axial and in-plane transverse stress distributions in the 90° ply. From $(D.6)_3$

$$\sigma_T^{(90)}(x_3)=E_A^{(90)}\left(\bar{\varepsilon}_T+\hat{\varepsilon}_T x_3\right)+\nu_A^{(90)}\sigma_T+\nu_A^{(90)}\sigma_A^{(90)}(x_3)-E_A^{(90)}\alpha_A^{(90)}\Delta T. \tag{D.7}$$

Substitution into $(D.6)_2$ leads to

$$\begin{aligned}
\bar{\varepsilon}_A+\hat{\varepsilon}_A x_3 = &\left(\frac{1}{E_T^{(90)}}-\frac{(\nu_A^{(90)})^2}{E_A^{(90)}}\right)\sigma_A^{(90)}(x_3)-\nu_A^{(90)}\left(\bar{\varepsilon}_T+\hat{\varepsilon}_T x_3\right)\\
&-\left(\frac{\nu_t^{(90)}}{E_T^{(90)}}+\frac{\nu_a^{(90)}\nu_A^{(90)}}{E_A^{(90)}}\right)\sigma_t+\left(\alpha_T^{(90)}+\nu_A^{(90)}\alpha_A^{(90)}\right)\Delta T.
\end{aligned} \tag{D.8}$$

At the point of crack closure for the ply cracks in the 90° plies $\sigma_A^{(90)}(x_3)=0$, $\sigma_t=\sigma_t^c$, whereas $\bar{\varepsilon}_A=\bar{\varepsilon}_A^c$, $\hat{\varepsilon}_A=\hat{\varepsilon}_A^c$ and $\bar{\varepsilon}_T=\bar{\varepsilon}_T^c$, $\hat{\varepsilon}_T=\hat{\varepsilon}_T^c$. It then follows from (D.8) that

$$\begin{aligned}\bar{\varepsilon}_A^c &= -\nu_A^{(90)}\bar{\varepsilon}_T^c-\left(\frac{\nu_t^{(90)}}{E_T^{(90)}}+\frac{\nu_a^{(90)}\nu_A^{(90)}}{E_A^{(90)}}\right)\sigma_t^c+\left(\alpha_T^{(90)}+\nu_A^{(90)}\alpha_A^{(90)}\right)\Delta T,\\ \hat{\varepsilon}_A^c &= -\nu_A^{(90)}\hat{\varepsilon}_T^c.\end{aligned} \tag{D.9}$$

It is noted that the derivation of the result (D.9) does not make use of the stress-strain relations (D.5), and that relation $(D.9)_1$ is identical to that derived when considering previously the closure of ply cracks in general symmetric laminates in the absence of bending. Relation $(D.9)_2$ is relevant only when some bending is required to achieve the ply crack closure condition, as would be the case if the cross-ply laminate was nonsymmetrical. Result (D.9) is used in Chapter 11 to derive sets of very useful interrelationships between the effective thermoelastic constants of damaged possibly nonsymmetric cross-ply laminates subject to bend deformation.

2. Crack Closure Analysis for General Symmetric Laminates (in the Absence of Bending)

Consider now the case of a general symmetric laminate in which there is at least one 90° ply. As it is now assumed that the plies in the laminate might be made of different materials, the superscript '(90)' will be attached to the properties, the in-plane and shear ply stresses and the in-plane and out-of-plane strains of the 90° plies. With reference to the global axes of the laminate and in the absence of bend deformation, it follows from (D.2) that the stress-strain relations for a 90° ply are given by

$$\begin{aligned}\varepsilon_t^{(90)} &= \frac{1}{E_t^{(90)}}\sigma_t-\frac{\nu_t^{(90)}}{E_T^{(90)}}\sigma_A^{(90)}-\frac{\nu_a^{(90)}}{E_A^{(90)}}\sigma_T^{(90)}+\alpha_t^{(90)}\Delta T,\\ \varepsilon_A &= -\frac{\nu_t^{(90)}}{E_T^{(90)}}\sigma_t+\frac{1}{E_T^{(90)}}\sigma_A^{(90)}-\frac{\nu_A^{(90)}}{E_A^{(90)}}\sigma_T^{(90)}+\alpha_T^{(90)}\Delta T,\\ \varepsilon_T &= -\frac{\nu_a^{(90)}}{E_A^{(90)}}\sigma_t-\frac{\nu_A^{(90)}}{E_A^{(90)}}\sigma_A^{(90)}+\frac{1}{E_A^{(90)}}\sigma_T^{(90)}+\alpha_A^{(90)}\Delta T,\\ \gamma_a^{(90)} &= \frac{1}{\mu_t^{(90)}}\tau_a,\quad \gamma_t^{(90)}=\frac{1}{\mu_a^{(90)}}\tau_t,\quad \gamma_A=\frac{1}{\mu_A^{(90)}}\tau_A^{(90)},\end{aligned} \tag{D.10}$$

where ε_A, ε_T and γ_A are in-plane laminate strains which have the same value in all plies of the laminate. The parameter σ_t is the though-thickness stress which also has the same value in all plies of the laminate.

For any off-axis ply in the laminate labelled i, it follows from (18.4)–(18.9) that the stress-strain relations have the form

$$\begin{aligned}\varepsilon_A &= S_{11}^{(i)}\sigma_A^{(i)}+S_{12}^{(i)}\sigma_T^{(i)}+S_{13}^{(i)}\sigma_t+S_{16}^{(i)}\tau_A^{(i)}+\alpha_1^{(i)}\Delta T,\\ \varepsilon_T &= S_{12}^{(i)}\sigma_A^{(i)}+S_{22}^{(i)}\sigma_T^{(i)}+S_{23}^{(i)}\sigma_t+S_{26}^{(i)}\tau_A^{(i)}+\alpha_2^{(i)}\Delta T,\\ \varepsilon_t^{(i)} &= S_{13}^{(i)}\sigma_A^{(i)}+S_{23}^{(i)}\sigma_T^{(i)}+S_{33}^{(i)}\sigma_t+s_{36}^{(i)}\tau_A^{(i)}+\alpha_3^{(i)}\Delta T,\\ \gamma_t^{(i)} &= S_{44}^{(i)}\tau_t+S_{45}^{(i)}\tau_a,\\ \gamma_a^{(i)} &= S_{45}^{(i)}\tau_t+S_{55}^{(i)}\tau_a,\\ \gamma_A &= S_{16}^{(i)}\sigma_A^{(i)}+S_{26}^{(i)}\sigma_T^{(i)}+S_{36}^{(i)}\sigma_t+S_{66}^{(i)}\tau_A^{(i)}+\alpha_6^{(i)}\Delta T.\end{aligned} \tag{D.11}$$

It should be noted from (D.10) and (D.11) that the strains $\varepsilon_{\rm A}, \varepsilon_{\rm T}, \gamma_{\rm A}$ and the stresses $\sigma_{\rm t}, \tau_{\rm t}, \tau_{\rm a}$ have the same value in every ply of the laminate.

At the point where the ply cracks in a 90° ply are just closed, for cases where there is no bend deformation, it is clear that $\sigma_{\rm A}^{(90)} = 0$ and $\tau_{\rm A}^{(90)} = 0$. The variables describing laminate behaviour are such that $\varepsilon_{\rm A} = \varepsilon_{\rm A}^{\rm c}$, $\varepsilon_{\rm T} = \varepsilon_{\rm T}^{\rm c}$, $\sigma_{\rm T} = \sigma_{\rm t}^{\rm c}$, $\gamma_{\rm A} = \gamma_{\rm A}^{\rm c}$ where the superscript denotes a critical value at the point of ply crack closure. It then follows from (D.10) that

$$\begin{aligned} \varepsilon_{\rm A}^{\rm c} &= -\frac{\nu_{\rm A}^{(90)}}{E_{\rm A}^{(90)}}\sigma_{\rm T}^{(90)} - \frac{\nu_{\rm t}^{(90)}}{E_{\rm T}^{(90)}}\sigma_{\rm t}^{\rm c} + \alpha_{\rm T}^{(90)}\Delta T, \\ \varepsilon_{\rm T}^{\rm c} &= \frac{1}{E_{\rm A}^{(90)}}\sigma_{\rm T}^{(90)} - \frac{\nu_{\rm a}^{(90)}}{E_{\rm A}^{(90)}}\sigma_{\rm t}^{\rm c} + \alpha_{\rm A}^{(90)}\Delta T, \qquad \gamma_{\rm A}^{\rm c} = 0\,. \end{aligned} \tag{D.12}$$

On eliminating $\sigma_{\rm T}^{(90)}$ it follows that

$$\varepsilon_{\rm A}^{\rm c} + \nu_{\rm A}^{(90)}\varepsilon_{\rm T}^{\rm c} = -\left(\frac{\nu_{\rm t}^{(90)}}{E_{\rm T}^{(90)}} + \frac{\nu_{\rm a}^{(90)}\nu_{\rm A}^{(90)}}{E_{\rm A}^{(90)}}\right)\sigma_{\rm t}^{\rm c} + \left(\alpha_{\rm T}^{(90)} + \nu_{\rm A}^{(90)}\alpha_{\rm A}^{(90)}\right)\Delta T, \tag{D.13}$$

indicating the most general condition of laminate loading that ensures ply cracks just close. The crack closure value of $\sigma_{\rm T}^{(90)}$ is given by

$$\sigma_{\rm T,c}^{(90)} = E_{\rm A}^{(90)}\varepsilon_{\rm T}^{\rm c} + \nu_{\rm a}^{(90)}\sigma_{\rm t}^{\rm c} - E_{\rm A}^{(90)}\alpha_{\rm A}^{(90)}\Delta T, \tag{D.14}$$

and the through-thickness strain in the 90° ply at the point of crack closure is given by

$$\varepsilon_{\rm t,c}^{(90)} = -\nu_{\rm a}^{(90)}\varepsilon_{\rm T}^{\rm c} + \left(1 - (\nu_{\rm A}^{(90)})^2\frac{E_{\rm t}^{(90)}}{E_{\rm A}^{(90)}}\right)\frac{\sigma_{\rm t}^{\rm c}}{E_{\rm t}^{(90)}} + \left(\alpha_{\rm t}^{(90)} + \nu_{\rm a}^{(90)}\alpha_{\rm A}^{(90)}\right)\Delta T. \tag{D.15}$$

It is noted that the derivation of the crack closure condition (D.13) does not, in fact, make use of the stress-strain relations (D.11). In addition, this condition does not depend upon the value of the in-plane shear strain $\gamma_{\rm A}$. From (D.10), because $\tau_{\rm A}^{(90)} = \mu_{\rm A}^{(90)}\gamma_{\rm A}$, the shear stress $\tau_{\rm A}^{(90)}$ can be zero, as required for ply crack closure, only if $\gamma_{\rm A}^{\rm c} = 0$.

On using (7.81) and (7.82), the stress-strain relations for a general symmetric laminate having ply cracks in the 90° plies are of the form

$$\begin{aligned} \varepsilon_{\rm A} &= \frac{1}{E_{\rm A}}\sigma_{\rm A} - \frac{\nu_{\rm A}}{E_{\rm A}}\sigma_{\rm T} - \frac{\nu_{\rm a}}{E_{\rm A}}\sigma_{\rm t} - \frac{\lambda_{\rm A}}{\mu_{\rm A}}\tau_{\rm A} + \alpha_{\rm A}\Delta T, \\ \varepsilon_{\rm T} &= -\frac{\nu_{\rm A}}{E_{\rm A}}\sigma_{\rm A} + \frac{1}{E_{\rm T}}\sigma_{\rm T} - \frac{\nu_{\rm t}}{E_{\rm T}}\sigma_{\rm t} - \frac{\lambda_{\rm T}}{\mu_{\rm A}}\tau_{\rm A} + \alpha_{\rm T}\Delta T, \\ \varepsilon_{\rm t} &= -\frac{\nu_{\rm a}}{E_{\rm A}}\sigma_{\rm A} - \frac{\nu_{\rm t}}{E_{\rm T}}\sigma_{\rm T} + \frac{1}{E_{\rm t}}\sigma_{\rm t} - \frac{\lambda_{\rm t}}{\mu_{\rm A}}\tau_{\rm A} + \alpha_{\rm t}\Delta T, \\ \gamma_{\rm t} &= \frac{1}{\mu_{\rm t}}\tau_{\rm t} + \Phi\tau_{\rm a}, \\ \gamma_{\rm a} &= \Phi\tau_{\rm t} + \frac{1}{\mu_{\rm a}}\tau_{\rm a}, \\ \gamma_{\rm A} &= -\frac{\lambda_{\rm A}}{\mu_{\rm A}}\sigma_{\rm A} - \frac{\lambda_{\rm T}}{\mu_{\rm A}}\sigma_{\rm T} - \frac{\lambda_{\rm t}}{\mu_{\rm A}}\sigma_{\rm t} + \frac{1}{\mu_{\rm A}}\tau_{\rm A} + \alpha_{\rm S}\Delta T. \end{aligned} \tag{D.16}$$

In order that the ply crack closure condition $\tau_{\mathrm{A}}^{(90)}=0$ is satisfied, it is required that the in-plane shear stress $\gamma_{\mathrm{A}}=0$. It then follows from the last of relations (D.16) that this condition is satisfied for any values of σ_{A}, σ_{T}, σ_{t} and ΔT if the applied in-plane shear stress has the value

$$\tau_{\mathrm{A}}=\tau_{\mathrm{A}}^{\mathrm{c}}=\lambda_{\mathrm{A}}\sigma_{\mathrm{A}}+\lambda_{\mathrm{T}}\sigma_{\mathrm{T}}+\lambda_{\mathrm{t}}\sigma_{\mathrm{t}}-\mu_{\mathrm{A}}\alpha_{\mathrm{S}}\Delta T. \tag{D.17}$$

On substituting this value into the first three of relations (D.16), it follows that

$$\begin{aligned}
\varepsilon_{\mathrm{A}} &= \frac{1}{\tilde{E}_{\mathrm{A}}}\sigma_{\mathrm{A}}-\frac{\tilde{\nu}_{\mathrm{A}}}{\tilde{E}_{\mathrm{A}}}\sigma_{\mathrm{T}}-\frac{\tilde{\nu}_{\mathrm{a}}}{\tilde{E}_{\mathrm{A}}}\sigma_{\mathrm{t}}+\tilde{\alpha}_{\mathrm{A}}\Delta T,\\
\varepsilon_{\mathrm{T}} &= -\frac{\tilde{\nu}_{\mathrm{A}}}{\tilde{E}_{\mathrm{A}}}\sigma_{\mathrm{A}}+\frac{1}{\tilde{E}_{\mathrm{T}}}\sigma_{\mathrm{T}}-\frac{\tilde{\nu}_{\mathrm{t}}}{\tilde{E}_{\mathrm{T}}}\sigma_{\mathrm{t}}+\tilde{\alpha}_{\mathrm{T}}\Delta T,\\
\varepsilon_{\mathrm{t}} &= -\frac{\tilde{\nu}_{\mathrm{a}}}{\tilde{E}_{\mathrm{A}}}\sigma_{\mathrm{A}}-\frac{\tilde{\nu}_{\mathrm{t}}}{\tilde{E}_{\mathrm{T}}}\sigma_{\mathrm{T}}+\frac{1}{\tilde{E}_{\mathrm{t}}}\sigma_{\mathrm{t}}+\tilde{\alpha}_{\mathrm{t}}\Delta T,
\end{aligned} \tag{D.18}$$

where

$$\begin{aligned}
&\frac{1}{\tilde{E}_{\mathrm{A}}}=\frac{1}{E_{\mathrm{A}}}-\frac{\lambda_{\mathrm{A}}^2}{\mu_{\mathrm{A}}},\quad \frac{1}{\tilde{E}_{\mathrm{T}}}=\frac{1}{E_{\mathrm{T}}}-\frac{\lambda_{\mathrm{T}}^2}{\mu_{\mathrm{A}}},\quad \frac{1}{\tilde{E}_{\mathrm{t}}}=\frac{1}{E_{\mathrm{t}}}-\frac{\lambda_{\mathrm{t}}^2}{\mu_{\mathrm{A}}},\\
&\frac{\tilde{\nu}_{\mathrm{A}}}{\tilde{E}_{\mathrm{A}}}=\frac{\nu_{\mathrm{A}}}{E_{\mathrm{A}}}+\frac{\lambda_{\mathrm{A}}\lambda_{\mathrm{T}}}{\mu_{\mathrm{A}}},\quad \frac{\tilde{\nu}_{\mathrm{a}}}{\tilde{E}_{\mathrm{A}}}=\frac{\nu_{\mathrm{a}}}{E_{\mathrm{A}}}+\frac{\lambda_{\mathrm{A}}\lambda_{\mathrm{t}}}{\mu_{\mathrm{A}}},\quad \frac{\tilde{\nu}_{\mathrm{t}}}{\tilde{E}_{\mathrm{T}}}=\frac{\nu_{\mathrm{t}}}{E_{\mathrm{T}}}+\frac{\lambda_{\mathrm{T}}\lambda_{\mathrm{t}}}{\mu_{\mathrm{A}}},\\
&\tilde{\alpha}_{\mathrm{A}}=\alpha_{\mathrm{A}}+\lambda_{\mathrm{A}}\alpha_{\mathrm{S}},\quad \tilde{\alpha}_{\mathrm{T}}=\alpha_{\mathrm{T}}+\lambda_{\mathrm{T}}\alpha_{\mathrm{S}},\quad \tilde{\alpha}_{\mathrm{t}}=\alpha_{\mathrm{t}}+\lambda_{\mathrm{t}}\alpha_{\mathrm{S}}.
\end{aligned} \tag{D.19}$$

The stress-strain relations (D.18) describe laminate deformation when it is constrained so that the in-plane shear strain γ_{A} is zero, as required for ply crack closure. At the point of crack closure, the stress-strain relations (D.18) may be written as

$$\begin{aligned}
\varepsilon_{\mathrm{A}}^{\mathrm{c}} &= \frac{1}{\tilde{E}_{\mathrm{A}}}\sigma_{\mathrm{A}}^{\mathrm{c}}-\frac{\tilde{\nu}_{\mathrm{A}}}{\tilde{E}_{\mathrm{A}}}\sigma_{\mathrm{T}}^{\mathrm{c}}-\frac{\tilde{\nu}_{\mathrm{a}}}{\tilde{E}_{\mathrm{A}}}\sigma_{\mathrm{t}}^{\mathrm{c}}+\tilde{\alpha}_{\mathrm{A}}\Delta T,\\
\varepsilon_{\mathrm{T}}^{\mathrm{c}} &= -\frac{\tilde{\nu}_{\mathrm{A}}}{\tilde{E}_{\mathrm{A}}}\sigma_{\mathrm{A}}^{\mathrm{c}}+\frac{1}{\tilde{E}_{\mathrm{T}}}\sigma_{\mathrm{T}}^{\mathrm{c}}-\frac{\tilde{\nu}_{\mathrm{t}}}{\tilde{E}_{\mathrm{T}}}\sigma_{\mathrm{t}}^{\mathrm{c}}+\tilde{\alpha}_{\mathrm{T}}\Delta T,\\
\varepsilon_{\mathrm{t}}^{\mathrm{c}} &= -\frac{\tilde{\nu}_{\mathrm{a}}}{\tilde{E}_{\mathrm{A}}}\sigma_{\mathrm{A}}^{\mathrm{c}}-\frac{\tilde{\nu}_{\mathrm{t}}}{\tilde{E}_{\mathrm{T}}}\sigma_{\mathrm{T}}^{\mathrm{c}}+\frac{1}{\tilde{E}_{\mathrm{t}}}\sigma_{\mathrm{t}}^{\mathrm{c}}+\tilde{\alpha}_{\mathrm{t}}\Delta T,
\end{aligned} \tag{D.20}$$

where the closure strains $\varepsilon_{\mathrm{A}}^{\mathrm{c}}$, $\varepsilon_{\mathrm{T}}^{\mathrm{c}}$ and the through-thickness stress at closure $\sigma_{\mathrm{t}}^{\mathrm{c}}$ must satisfy the condition (D.13).

The corresponding stress-strain relations for an undamaged laminate are of the form

$$\begin{aligned}
\varepsilon_{\mathrm{A}}^{\mathrm{c}} &= \frac{1}{\tilde{E}_{\mathrm{A}}^{0}}\sigma_{\mathrm{A}}^{\mathrm{c}}-\frac{\tilde{\nu}_{\mathrm{A}}^{0}}{\tilde{E}_{\mathrm{A}}^{0}}\sigma_{\mathrm{T}}^{\mathrm{c}}-\frac{\tilde{\nu}_{\mathrm{a}}^{0}}{\tilde{E}_{\mathrm{A}}^{0}}\sigma_{\mathrm{t}}^{\mathrm{c}}+\tilde{\alpha}_{\mathrm{A}}^{0}\Delta T,\\
\varepsilon_{\mathrm{T}}^{\mathrm{c}} &= -\frac{\tilde{\nu}_{\mathrm{A}}^{0}}{\tilde{E}_{\mathrm{A}}^{0}}\sigma_{\mathrm{A}}^{\mathrm{c}}+\frac{1}{\tilde{E}_{\mathrm{T}}^{0}}\sigma_{\mathrm{T}}^{\mathrm{c}}-\frac{\tilde{\nu}_{\mathrm{t}}^{0}}{\tilde{E}_{\mathrm{T}}^{0}}\sigma_{\mathrm{t}}^{\mathrm{c}}+\tilde{\alpha}_{\mathrm{T}}^{0}\Delta T,\\
\varepsilon_{\mathrm{t}}^{\mathrm{c}} &= -\frac{\tilde{\nu}_{\mathrm{a}}^{0}}{\tilde{E}_{\mathrm{A}}^{0}}\sigma_{\mathrm{A}}^{\mathrm{c}}-\frac{\tilde{\nu}_{\mathrm{t}}^{0}}{\tilde{E}_{\mathrm{T}}^{0}}\sigma_{\mathrm{T}}^{\mathrm{c}}+\frac{1}{\tilde{E}_{\mathrm{t}}^{0}}\sigma_{\mathrm{t}}^{\mathrm{c}}+\tilde{\alpha}_{\mathrm{t}}^{0}\Delta T.
\end{aligned} \tag{D.21}$$

Results (D.20) and (D.21) are used to derive in Chapter 10 sets of very useful interrelationships between the effective thermoelastic constants of damaged general symmetric laminates.

3. Crack Closure Analysis for Cross-ply Laminates with Delaminations but Without Bending

Consider a simple undamaged cross-ply laminate where both the 0° and 90° plies are made of the same material. The through-thickness stress is zero throughout the laminate. The stress-strain relations for the 0° ply are given by

$$\varepsilon_{\mathrm{A}} = \frac{1}{\widehat{E}_{\mathrm{A}}}\sigma_{\mathrm{A}}^{(0)} - \frac{\widehat{\nu}_{\mathrm{A}}}{\widehat{E}_{\mathrm{A}}}\sigma_{T}^{(0)} + \widehat{\alpha}_{\mathrm{A}}\Delta T, \tag{D.22}$$

$$\varepsilon_{\mathrm{T}} = -\frac{\widehat{\nu}_{\mathrm{A}}}{\widehat{E}_{\mathrm{A}}}\sigma_{\mathrm{A}}^{(0)} + \frac{1}{\widehat{E}_{\mathrm{T}}}\sigma_{T}^{(0)} + \widehat{\alpha}_{\mathrm{T}}\Delta T, \tag{D.23}$$

and those for the 90° ply are given by

$$\varepsilon_{\mathrm{A}} = \frac{1}{\widehat{E}_{\mathrm{T}}}\sigma_{\mathrm{A}}^{(90)} - \frac{\widehat{\nu}_{\mathrm{A}}}{\widehat{E}_{\mathrm{A}}}\sigma_{T}^{(90)} + \widehat{\alpha}_{\mathrm{T}}\Delta T, \tag{D.24}$$

$$\varepsilon_{\mathrm{T}} = -\frac{\widehat{\nu}_{\mathrm{A}}}{\widehat{E}_{\mathrm{A}}}\sigma_{\mathrm{A}}^{(90)} + \frac{1}{\widehat{E}_{\mathrm{A}}}\sigma_{T}^{(90)} + \widehat{\alpha}_{\mathrm{A}}\Delta T, \tag{D.25}$$

where for any ply property p the symbol $\widehat{p}$ denotes that the same transverse isotropic material is being used to make both the 0° and the 90° plies. The axial and in-plane transverse stresses in the 0° plies have been denoted by $\sigma_{\mathrm{A}}^{(0)}$ and $\sigma_{\mathrm{T}}^{(0)}$, respectively, and the axial and in-plane transverse stresses in the 90° plies have been denoted by $\sigma_{\mathrm{A}}^{(90)}$ and $\sigma_{\mathrm{T}}^{(90)}$, respectively.

3.1 Uniaxial Axial Loading

The laminate is first loaded by a uniaxial axial applied stress such that each ply in the laminate experiences the same values, $\varepsilon_{\mathrm{A}}^{\mathrm{c}}$ and $\varepsilon_{\mathrm{T}}^{\mathrm{c}}$, for the axial and transverse in-plane strains, respectively, and that $\sigma_{\mathrm{A}}^{(90)} = 0$ for a given value of the temperature difference ΔT. The resulting stress state in the damaged laminate corresponds to the case when ply cracks in the 90° plies and delaminations just close. For this special case mechanical equilibrium asserts that

$$\sigma_{\mathrm{A}}^{(0)} = \frac{h\sigma_{\mathrm{A}}^{\mathrm{c}}}{h^{(0)}},\ \sigma_{\mathrm{A}}^{(90)} = 0,\ h^{(0)}\sigma_{\mathrm{T}}^{(0)} + h^{(90)}\sigma_{\mathrm{T}}^{(90)} = 0, \tag{D.26}$$

where $h^{(0)}$ and $h^{(90)}$ denote the total thicknesses in the laminate of all 0° plies and all 90° plies, respectively, and where $\sigma_{\mathrm{A}}^{\mathrm{c}}$ is the effective applied axial stress at the point of crack closure.

Substitution of the first two of relations (D.26) into the stress-strain equations (D.22)–(D.25) for the 0° and 90° plies leads to

$$\varepsilon_{\mathrm{A}}^{\mathrm{c}} = \frac{1}{\widehat{E}_{\mathrm{A}}}\frac{h\sigma_{\mathrm{A}}^{\mathrm{c}}}{h^{(0)}} - \frac{\widehat{\nu}_{\mathrm{A}}}{\widehat{E}_{\mathrm{A}}}\sigma_{\mathrm{T}}^{(0)} + \widehat{\alpha}_{\mathrm{A}}\Delta T, \tag{D.27}$$

$$\varepsilon_{\mathrm{T}}^{\mathrm{c}} = -\frac{\widehat{\nu}_{\mathrm{A}}}{\widehat{E}_{\mathrm{A}}}\frac{h\sigma_{\mathrm{A}}^{\mathrm{c}}}{h^{(0)}} + \frac{1}{\widehat{E}_{\mathrm{T}}}\sigma_{\mathrm{T}}^{(0)} + \widehat{\alpha}_{\mathrm{T}}\Delta T, \tag{D.28}$$

$$\varepsilon_{\mathrm{A}}^{\mathrm{c}} = -\frac{\widehat{\nu}_{\mathrm{A}}}{\widehat{E}_{\mathrm{A}}}\sigma_{\mathrm{T}}^{(90)} + \widehat{\alpha}_{\mathrm{T}}\Delta T, \tag{D.29}$$

$$\varepsilon_{\mathrm{T}}^{\mathrm{c}} = \frac{1}{\widehat{E}_{\mathrm{A}}} \sigma_{\mathrm{T}}^{(90)} + \widehat{\alpha}_{\mathrm{A}} \Delta T. \tag{D.30}$$

It follows from (D.27) and (D.28), on eliminating $\sigma_{\mathrm{T}}^{(0)}$, that

$$\varepsilon_{\mathrm{A}}^{\mathrm{c}} + \widehat{\nu}_{\mathrm{A}} \frac{\widehat{E}_{\mathrm{T}}}{\widehat{E}_{\mathrm{A}}} \varepsilon_{\mathrm{T}}^{\mathrm{c}} = \frac{1}{\widehat{E}_{\mathrm{A}}} \left(1 - \widehat{\nu}_{\mathrm{A}}^2 \frac{\widehat{E}_{\mathrm{T}}}{\widehat{E}_{\mathrm{A}}} \right) \frac{h \sigma_{\mathrm{A}}^{\mathrm{c}}}{h^{(0)}} + \left(\widehat{\alpha}_{\mathrm{A}} + \widehat{\nu}_{\mathrm{A}} \frac{\widehat{E}_{\mathrm{T}}}{\widehat{E}_{\mathrm{A}}} \widehat{\alpha}_{\mathrm{T}} \right) \Delta T, \tag{D.31}$$

and from (D.29) and (D.30), on eliminating $\sigma_{\mathrm{T}}^{(90)}$, that

$$\varepsilon_{\mathrm{A}}^{\mathrm{c}} + \widehat{\nu}_{\mathrm{A}} \varepsilon_{\mathrm{T}}^{\mathrm{c}} = \left(\widehat{\alpha}_{\mathrm{T}} + \widehat{\nu}_{\mathrm{A}} \widehat{\alpha}_{\mathrm{A}} \right) \Delta T. \tag{D.32}$$

From (D.26)₃, (D.28) and (D.30), on eliminating $\sigma_{\mathrm{T}}^{(0)}$ and $\sigma_{\mathrm{T}}^{(90)}$,

$$\varepsilon_{\mathrm{T}}^{\mathrm{c}} = -\widehat{\nu}_{\mathrm{A}} \frac{\widehat{E}_{\mathrm{T}}}{\widehat{E}_{\mathrm{A}}} \frac{h \sigma_{\mathrm{A}}^{\mathrm{c}}}{h^{(0)} \widehat{E}_{\mathrm{T}} + h^{(90)} \widehat{E}_{\mathrm{A}}} + \frac{h^{(0)} \widehat{E}_{\mathrm{T}} \widehat{\alpha}_{\mathrm{T}} + h^{(90)} \widehat{E}_{\mathrm{A}} \widehat{\alpha}_{\mathrm{A}}}{h^{(0)} \widehat{E}_{\mathrm{T}} + h^{(90)} \widehat{E}_{\mathrm{A}}} \Delta T. \tag{D.33}$$

From (D.32) and (D.33)

$$\varepsilon_{\mathrm{A}}^{\mathrm{c}} = \widehat{\nu}_{\mathrm{A}}^2 \frac{\widehat{E}_{\mathrm{T}}}{\widehat{E}_{\mathrm{A}}} \frac{h \sigma_{\mathrm{A}}^{\mathrm{c}}}{h^{(0)} \widehat{E}_{\mathrm{T}} + h^{(90)} \widehat{E}_{\mathrm{A}}} - \widehat{\nu}_{\mathrm{A}} \frac{h^{(0)} \widehat{E}_{\mathrm{T}} \widehat{\alpha}_{\mathrm{T}} + h^{(90)} \widehat{E}_{\mathrm{A}} \widehat{\alpha}_{\mathrm{A}}}{h^{(0)} \widehat{E}_{\mathrm{T}} + h^{(90)} \widehat{E}_{\mathrm{A}}} \Delta T + \left(\widehat{\alpha}_{\mathrm{T}} + \widehat{\nu}_{\mathrm{A}} \widehat{\alpha}_{\mathrm{A}} \right) \Delta T. \tag{D.34}$$

It can then be shown from (D.31), (D.33) and (D.34) that the crack closure stress for uniaxial axial loading is given by

$$\sigma_{\mathrm{A}}^{\mathrm{c}} = \frac{h^{(0)} \widehat{E}_{\mathrm{A}}}{h} \frac{h^{(0)} \widehat{E}_{\mathrm{T}} + h^{(90)} \widehat{E}_{\mathrm{A}} - h \widehat{\nu}_{\mathrm{A}} \widehat{E}_{\mathrm{T}}}{h^{(0)} \widehat{E}_{\mathrm{T}} + h^{(90)} \widehat{E}_{\mathrm{A}} - h \widehat{\nu}_{\mathrm{A}}^2 \widehat{E}_{\mathrm{T}}} \left(\widehat{\alpha}_{\mathrm{T}} - \widehat{\alpha}_{\mathrm{A}} \right) \Delta T = -k_1 \Delta T, \tag{D.35}$$

where

$$k_1 = \frac{h^{(0)} \widehat{E}_{\mathrm{A}}}{h} \frac{h^{(0)} \widehat{E}_{\mathrm{T}} + h^{(90)} \widehat{E}_{\mathrm{A}} - h \widehat{\nu}_{\mathrm{A}} \widehat{E}_{\mathrm{T}}}{h^{(0)} \widehat{E}_{\mathrm{T}} + h^{(90)} \widehat{E}_{\mathrm{A}} - h \widehat{\nu}_{\mathrm{A}}^2 \widehat{E}_{\mathrm{T}}} \left(\widehat{\alpha}_{\mathrm{A}} - \widehat{\alpha}_{\mathrm{T}} \right). \tag{D.36}$$

3.2 Uniaxial in-plane Transverse Loading

The laminate is now loaded by a uniaxial applied stress for the in-plane transverse direction such that each ply in the laminate experiences the same values, $\bar{\varepsilon}_{\mathrm{A}}^{\mathrm{c}}$ and $\bar{\varepsilon}_{\mathrm{T}}^{\mathrm{c}}$, for the axial and in-plane transverse strains, respectively, and that $\sigma_{\mathrm{A}}^{(90)} = 0$ for a given value of the temperature difference ΔT. The resulting stress state in the damaged laminate corresponds to the case when ply cracks in the 90° plies and delaminations just close. For this special case mechanical equilibrium asserts that

$$\sigma_{\mathrm{A}}^{(0)} = \sigma_{\mathrm{A}}^{(90)} = 0, \; h^{(0)} \sigma_{\mathrm{T}}^{(0)} + h^{(90)} \sigma_{\mathrm{T}}^{(90)} = h \sigma_{\mathrm{T}}^{\mathrm{c}}, \tag{D.37}$$

where $\sigma_{\mathrm{T}}^{\mathrm{c}}$ is the effective applied in-plane transverse stress at the point of crack closure. Substitution of (D.37) into the stress-strain equations (D.22)–(D.25) for both 0° and 90° plies leads to the relations

$$\bar{\varepsilon}_{\mathrm{A}}^{\mathrm{c}} = -\frac{\widehat{\nu}_{\mathrm{A}}}{\widehat{E}_{\mathrm{A}}} \sigma_{\mathrm{T}}^{(0)} + \widehat{\alpha}_{\mathrm{A}} \Delta T, \tag{D.38}$$

$$\bar{\varepsilon}_{\mathrm{T}}^{\mathrm{c}} = \frac{1}{\widehat{E}_{\mathrm{T}}} \sigma_{\mathrm{T}}^{(0)} + \widehat{\alpha}_{\mathrm{T}} \Delta T, \tag{D.39}$$

$$\bar{\varepsilon}_{\mathrm{A}}^{\mathrm{c}} = -\frac{\widehat{\nu}_{\mathrm{A}}}{\widehat{E}_{\mathrm{A}}}\sigma_{\mathrm{T}}^{(90)} + \widehat{\alpha}_{\mathrm{T}}\Delta T, \tag{D.40}$$

$$\bar{\varepsilon}_{\mathrm{T}}^{\mathrm{c}} = \frac{1}{\widehat{E}_{\mathrm{A}}}\sigma_{\mathrm{T}}^{(90)} + \widehat{\alpha}_{\mathrm{A}}\Delta T. \tag{D.41}$$

It follows from (D.38) and (D.39), on eliminating $\sigma_{\mathrm{T}}^{(0)}$, that

$$\bar{\varepsilon}_{\mathrm{A}}^{\mathrm{c}} + \widehat{\nu}_{\mathrm{A}}\frac{\widehat{E}_{\mathrm{T}}}{\widehat{E}_{\mathrm{A}}}\bar{\varepsilon}_{\mathrm{T}}^{\mathrm{c}} = \left(\widehat{\alpha}_{\mathrm{A}} + \widehat{\nu}_{\mathrm{A}}\frac{\widehat{E}_{\mathrm{T}}}{\widehat{E}_{\mathrm{A}}}\widehat{\alpha}_{\mathrm{T}}\right)\Delta T, \tag{D.42}$$

and from (D.40) and (D.41), on eliminating $\sigma_{\mathrm{T}}^{(90)}$, that

$$\bar{\varepsilon}_{\mathrm{A}}^{\mathrm{c}} + \widehat{\nu}_{\mathrm{A}}\bar{\varepsilon}_{\mathrm{T}}^{\mathrm{c}} = \left(\widehat{\alpha}_{\mathrm{T}} + \widehat{\nu}_{\mathrm{A}}\widehat{\alpha}_{\mathrm{A}}\right)\Delta T. \tag{D.43}$$

A subtraction then yields

$$\widehat{\nu}_{\mathrm{A}}\left(1 - \frac{\widehat{E}_{\mathrm{T}}}{\widehat{E}_{\mathrm{A}}}\right)\bar{\varepsilon}_{\mathrm{T}}^{\mathrm{c}} = \left[\left(1 - \widehat{\nu}_{\mathrm{A}}\frac{\widehat{E}_{\mathrm{T}}}{\widehat{E}_{\mathrm{A}}}\right)\widehat{\alpha}_{\mathrm{T}} - \left(1 - \widehat{\nu}_{\mathrm{A}}\right)\widehat{\alpha}_{\mathrm{A}}\right]\Delta T, \tag{D.44}$$

and from (D.43)

$$\left(1 - \frac{\widehat{E}_{\mathrm{T}}}{\widehat{E}_{\mathrm{A}}}\right)\bar{\varepsilon}_{\mathrm{A}}^{\mathrm{c}} = \left[\left(1 - \widehat{\nu}_{\mathrm{A}}\frac{\widehat{E}_{\mathrm{T}}}{\widehat{E}_{\mathrm{A}}}\right)\widehat{\alpha}_{\mathrm{A}} - \frac{\widehat{E}_{\mathrm{T}}}{\widehat{E}_{\mathrm{A}}}\left(1 - \widehat{\nu}_{\mathrm{A}}\right)\widehat{\alpha}_{\mathrm{T}}\right]\Delta T. \tag{D.45}$$

From (D.37), (D.39) and (D.41), on eliminating $\sigma_{\mathrm{T}}^{(0)}$ and $\sigma_{\mathrm{T}}^{(90)}$, the closure stress applied in the in-plane transverse direction is given by

$$h\sigma_{\mathrm{T}}^{\mathrm{c}} = \left(h^{(0)}\widehat{E}_{\mathrm{T}} + h^{(90)}\widehat{E}_{\mathrm{A}}\right)\bar{\varepsilon}_{\mathrm{T}}^{\mathrm{c}} - \left(h^{(0)}\widehat{E}_{\mathrm{T}}\widehat{\alpha}_{\mathrm{T}} + h^{(90)}\widehat{E}_{\mathrm{A}}\widehat{\alpha}_{\mathrm{A}}\right)\Delta T. \tag{D.46}$$

On using (D.44), the crack closure stress for uniaxial in-plane transverse loading is then given by

$$\sigma_{\mathrm{T}}^{\mathrm{c}} = -\left[h^{(0)}\widehat{E}_{\mathrm{T}} + h^{(90)}\widehat{E}_{\mathrm{A}} - h\widehat{\nu}_{\mathrm{A}}\widehat{E}_{\mathrm{T}}\right]\frac{\widehat{E}_{\mathrm{A}}\left(\widehat{\alpha}_{\mathrm{A}} - \widehat{\alpha}_{\mathrm{T}}\right)}{h\widehat{\nu}_{\mathrm{A}}\left(\widehat{E}_{\mathrm{A}} - \widehat{E}_{\mathrm{T}}\right)}\Delta T = -k_2\,\Delta T, \tag{D.47}$$

where

$$k_2 = \left[h^{(0)}\widehat{E}_{\mathrm{T}} + h^{(90)}\widehat{E}_{\mathrm{A}} - h\widehat{\nu}_{\mathrm{A}}\widehat{E}_{\mathrm{T}}\right]\frac{\widehat{E}_{\mathrm{A}}\left(\widehat{\alpha}_{\mathrm{A}} - \widehat{\alpha}_{\mathrm{T}}\right)}{h\widehat{\nu}_{\mathrm{A}}\left(\widehat{E}_{\mathrm{A}} - \widehat{E}_{\mathrm{T}}\right)}. \tag{D.48}$$

Appendix E: Derivation of the Solution of Nonlinear Equations

In Chapter 15 it is required to solve (15.66)–(15.68), which may be written in the form

$$L_1\bar{P} + L_2\bar{Q} + L_3\bar{R} + \Psi\Omega + 9k_m\mu_m + \frac{3k_m - 2\mu_m}{2B}\bar{P} - \frac{\mu_m}{B}\bar{Q} = \frac{S_{2211}}{B}\Omega, \tag{E.1}$$

$$L_1\bar{P} + L_2\bar{Q} + L_3\bar{R} + \Psi\Omega + 9k_m\mu_m - \frac{\mu_m}{C}\bar{Q} - \frac{3k_m - 2\mu_m}{C}\bar{R} = \frac{S_{1122}}{C}\Omega, \tag{E.2}$$

$$L_1\bar{P} + L_2\bar{Q} + L_3\bar{R} + \Psi\Omega + 9k_m\mu_m + \frac{2\mu_m}{D}\bar{Q} - \frac{3k_m + 4\mu_m}{D}\bar{R} = \frac{S_{1111}}{D}\Omega, \tag{E.3}$$

where Ω is defined by (15.61). On subtracting (E.1) and (E.2),

$$\frac{3k_m - 2\mu_m}{2B}\bar{P} + \left(\frac{1}{C} - \frac{1}{B}\right)\mu_m\bar{Q} + \frac{3k_m - 2\mu_m}{C}\bar{R} = \left(\frac{S_{2211}}{B} - \frac{S_{1122}}{C}\right)\Omega. \tag{E.4}$$

On subtracting (E.1) and (E.3),

$$\frac{3k_m - 2\mu_m}{2B}\bar{P} - \left(\frac{1}{B} + \frac{2}{D}\right)\mu_m\bar{Q} + \frac{3k_m + 4\mu_m}{D}\bar{R} = -\left(\frac{S_{1111}}{D} - \frac{S_{2211}}{B}\right)\Omega. \tag{E.5}$$

On subtracting (E.2) and (E.3),

$$\left(\frac{1}{C} + \frac{2}{D}\right)\mu_m\bar{Q} + \left(\frac{3k_m - 2\mu_m}{C} - \frac{3k_m + 4\mu_m}{D}\right)\bar{R} = \left(\frac{S_{1111}}{D} - \frac{S_{1122}}{C}\right)\Omega. \tag{E.6}$$

Relations (E.4)–(E.6) are now written as

$$(3k_m - 2\mu_m)C\bar{P} + 2\mu_m(B - C)\bar{Q} + 2(3k_m - 2\mu_m)B\bar{R} = 2(CS_{2211} - BS_{1122})\Omega, \tag{E.7}$$

$$(3k_m - 2\mu_m)D\bar{P} - 2\mu_m(2B + D)\bar{Q} + 2(3k_m + 4\mu_m)B\bar{R} = -2(BS_{1111} - DS_{2211})\Omega, \tag{E.8}$$

$$(2C + D)\mu_m\bar{Q} + \left[(3k_m - 2\mu_m)D - (3k_m + 4\mu_m)C\right]\bar{R} = (CS_{1111} - DS_{1122})\Omega. \tag{E.9}$$

The elimination of $\bar{R}$ in (E.7) and (E.8) leads to the relation

$$\begin{aligned}&(3k_{\mathrm{m}}-2\mu_{\mathrm{m}})\left[(3k_{\mathrm{m}}+4\mu_{\mathrm{m}})C-(3k_{\mathrm{m}}-2\mu_{\mathrm{m}})D\right]\bar{P}\\&\quad+\left[9k_{\mathrm{m}}B-(3k_{\mathrm{m}}+4\mu_{\mathrm{m}})C+(3k_{\mathrm{m}}-2\mu_{\mathrm{m}})D\right]2\mu_{\mathrm{m}}\bar{Q}\\&\qquad=\begin{bmatrix}\left\{(3k_{\mathrm{m}}+4\mu_{\mathrm{m}})C-(3k_{\mathrm{m}}-2\mu_{\mathrm{m}})D\right\}S_{2211}\\-\left\{(3k_{\mathrm{m}}+4\mu_{\mathrm{m}})S_{1122}-(3k_{\mathrm{m}}-2\mu_{\mathrm{m}})S_{1111}\right\}B\end{bmatrix}2\Omega.\end{aligned} \tag{E.10}$$

Using (15.63) and (15.64), define parameters Γ and λ such that

$$\Gamma=(3k_{\mathrm{m}}+4\mu_{\mathrm{m}})C-(3k_{\mathrm{m}}-2\mu_{\mathrm{m}})D=2(3k_{\mathrm{m}}+\mu_{\mathrm{m}})B-(3k_{\mathrm{m}}-2\mu_{\mathrm{m}})A, \tag{E.11}$$

and

$$\begin{aligned}\lambda&=(3k_{\mathrm{m}}+4\mu_{\mathrm{m}})S_{1122}-(3k_{\mathrm{m}}-2\mu_{\mathrm{m}})S_{1111}\\&=2(3k_{\mathrm{m}}+\mu_{\mathrm{m}})S_{2211}-(3k_{\mathrm{m}}-2\mu_{\mathrm{m}})(S_{2222}+S_{2233}).\end{aligned} \tag{E.12}$$

Substitution into (E.10) then leads to

$$\Gamma\bar{P}+(A+B)\,2\mu_{\mathrm{m}}\bar{Q}=2\left[(S_{2222}+S_{2233})B-S_{2211}A\right]\Omega. \tag{E.13}$$

The result (E.9) can then be written as

$$\Gamma\bar{R}-(2C+D)\mu_{\mathrm{m}}\bar{Q}=(DS_{1122}-CS_{1111})\,\Omega. \tag{E.14}$$

From relations (15.53) and (15.56), the parameters A, B, C and D are given by

$$\begin{aligned}A&=\sum_{i=1}^{N}\left[(Q_iR_i-P_iS_i)(S_{2222}+S_{2233})-(3k_{\mathrm{m}}+\mu_{\mathrm{m}})P_i-\mu_{\mathrm{m}}Q_i\right]\frac{V_{\mathrm{p}}^i}{\Delta_i},\\B&=\sum_{i=1}^{N}\left[(Q_iR_i-P_iS_i)S_{2211}-\tfrac{1}{2}(3k_{\mathrm{m}}-2\mu_{\mathrm{m}})P_i+\mu_{\mathrm{m}}Q_i\right]\frac{V_{\mathrm{p}}^i}{\Delta_i},\\C&=\sum_{i=1}^{N}\left[(Q_iR_i-P_iS_i)S_{1122}+\mu_{\mathrm{m}}Q_i+(3k_{\mathrm{m}}-2\mu_{\mathrm{m}})R_i\right]\frac{V_{\mathrm{p}}^i}{\Delta_i},\\D&=\sum_{i=1}^{N}\left[(Q_iR_i-P_iS_i)S_{1111}-2\mu_{\mathrm{m}}Q_i+(3k_{\mathrm{m}}+4\mu_{\mathrm{m}})R_i\right]\frac{V_{\mathrm{p}}^i}{\Delta_i}.\end{aligned} \tag{E.15}$$

Clearly,

$$\begin{aligned}A+B&=\sum_{i=1}^{N}\left[(Q_iR_i-P_iS_i)(S_{2211}+S_{2222}+S_{2233})-\frac{9}{2}k_{\mathrm{m}}P_i\right]\frac{V_p^i}{\Delta_i},\\2C+D&=\sum_{i=1}^{N}\left[(Q_iR_i-P_iS_i)(S_{1111}+2S_{1122})+9k_{\mathrm{m}}R_i\right]\frac{V_p^i}{\Delta_i},\end{aligned} \tag{E.16}$$

and

$$\begin{aligned}(S_{2222}+S_{2233})B-S_{2211}A&=\sum_{i=1}^{N}\left[\frac{1}{2}\lambda P_i+\mu_{\mathrm{m}}(S_{2211}+S_{2222}+S_{2233})Q_i\right]\frac{V_p^i}{\Delta_i},\\DS_{1122}-CS_{1111}&=\sum_{i=1}^{N}\left[\lambda R_i-\mu_{\mathrm{m}}(S_{1111}+2S_{1122})Q_i\right]\frac{V_p^i}{\Delta_i}.\end{aligned} \tag{E.17}$$

Relations (E.1)–(E.3) are now written as

$$L_1B\overline{P} + L_2B\overline{Q} + L_3B\overline{R} + \Psi B\Omega + 9Bk_{\rm m}\mu_{\rm m} + \frac{1}{2}(3k_{\rm m} - 2\mu_{\rm m})\overline{P} - \mu_{\rm m}\overline{Q} = S_{2211}\Omega\,, \quad \text{(E.18)}$$

$$L_1C\overline{P} + L_2C\overline{Q} + L_3C\overline{R} + \Psi C\Omega + 9Ck_{\rm m}\mu_{\rm m} - \mu_{\rm m}\overline{Q} - (3k_{\rm m} - 2\mu_{\rm m})\overline{R} = S_{1122}\Omega\,, \quad \text{(E.19)}$$

$$L_1D\overline{P} + L_2D\overline{Q} + L_3D\overline{R} + \Psi D\Omega + 9Dk_{\rm m}\mu_{\rm m} + 2\mu_{\rm m}\overline{Q} - (3k_{\rm m} + 4\mu_{\rm m})\overline{R} = S_{1111}\Omega. \quad \text{(E.20)}$$

On multiplying (E.19) by 2, adding to (E.20) and then dividing by $(2C + D)$

$$L_1\overline{P} + L_2\overline{Q} + \left(L_3 - \frac{9k_{\rm m}}{2C + D}\right)\overline{R} = \left(\frac{S_{1111} + 2S_{1122}}{2C + D} - \Psi\right)\Omega - 9k_{\rm m}\mu_{\rm m}. \quad \text{(E.21)}$$

On using (E.13) and (E.14) to eliminate $\overline{P}$ and $\overline{R}$ in (E.21), it follows that

$$\begin{aligned}
&\left[(2C + D)\mu_{\rm m}L_3 - 2(A + B)\mu_{\rm m}L_1 + \Lambda L_2 - 9k_{\rm m}\mu_{\rm m}\right]\overline{Q} \\
&= \left[\left(\frac{S_{1111} + 2S_{1122}}{2C + D} - \Psi\right)\Gamma - 2\left\{(S_{2222} + S_{2233})B - S_{2211}A\right\}L_1\right. \\
&\qquad \left. - \frac{S_{1122}D - S_{1111}C}{2C + D}\left\{(2C + D)L_3 - 9k_{\rm m}\right\}\right]\Omega - 9k_{\rm m}\mu_{\rm m}\Gamma\,.
\end{aligned} \quad \text{(E.22)}$$

On using (15.46), it follows that (E.22) may be written as

$$\begin{aligned}
&(2C + D)\mu_{\rm m}L_3 - 2(A + B)\mu_{\rm m}L_1 + \Gamma L_2 - 9k_{\rm m}\mu_{\rm m} \\
&= 9k_{\rm m}\mu_{\rm m}\left[2S_{2211}C - S_{1111}A + 2S_{1122}B - (S_{2222} + S_{2233})D - 1\right],
\end{aligned} \quad \text{(E.23)}$$

and that

$$\begin{aligned}
&\left(\frac{S_{1111} + 2S_{1122}}{2C + D} - \Psi\right)\Gamma - 2\left\{(S_{2222} + S_{2233})B - S_{2211}A\right\}L_1 \\
&\qquad - \frac{S_{1122}D - S_{1111}C}{2C + D}\left\{(2C + D)L_3 - 9k_{\rm m}\right\} \\
&= \lambda\left[1 - 2S_{2211}C + (S_{2222} + S_{2233})D + S_{1111}A - 2S_{1122}B\right] \\
&\quad + \left[(3k_{\rm m} - 2\mu_{\rm m})A - 2(3k_{\rm m} + \mu_{\rm m})B\right]\left[S_{1111}(S_{2222} + S_{2233}) - 2S_{1122}S_{2211}\right].
\end{aligned} \quad \text{(E.24)}$$

It then follows from (E.22)–(E.24) that

$$\Phi\overline{Q} = (\Gamma\Psi - \lambda\Phi)\frac{\Omega}{9k_{\rm m}\mu_{\rm m}} - \Gamma, \quad \text{(E.25)}$$

where

$$\Phi = 2S_{2211}C - (S_{2222} + S_{2233})D - S_{1111}A + 2S_{1122}B - 1. \quad \text{(E.26)}$$

On using (E.15) and (E.26),

$$\Phi = \sum_{i=1}^{N}\left[\begin{aligned}
&2(Q_iR_i - P_iS_i)\Psi + \left\{S_{1111}(3k_{\rm m} + \mu_{\rm m}) - S_{1122}(3k_{\rm m} - 2\mu_{\rm m})\right\}P_i \\
&\quad + \left\{(S_{1111} + 2S_{1122}) + 2(S_{2211} + S_{2222} + S_{2233})\right\}\mu_{\rm m}Q_i \\
&+ \left\{2S_{2211}(3k_{\rm m} - 2\mu_{\rm m}) - (S_{2222} + S_{2233})(3k_{\rm m} + 4\mu_{\rm m})\right\}R_i
\end{aligned}\right]\frac{V_{\rm p}^i}{\Delta_i} - 1. \quad \text{(E.27)}$$

From (E.13) and (E.14),

$$\Gamma\overline{P} = 2\left[(S_{2222} + S_{2233})B - S_{2211}A\right]\Omega - 2(A + B)\mu_{\rm m}\overline{Q}\,, \quad \text{(E.28)}$$

$$\Gamma\bar{R} = (S_{1122}D - S_{1111}C)\Omega + (2C+D)\mu_{\mathrm{m}}\bar{Q}. \tag{E.29}$$

On substituting for $\bar{Q}$ using (E.25),

$$\Phi\bar{P} = \left[2\{(S_{2222}+S_{2233})B - S_{2211}A\}\Phi - \frac{2}{9k_{\mathrm{m}}}(\Gamma\Psi - \lambda\Phi)(A+B)\right]\frac{\Omega}{\Gamma} + 2\mu_{\mathrm{m}}(A+B), \tag{E.30}$$

$$\Phi\bar{R} = \left[\{S_{1122}D - S_{1111}C\}\Phi + \frac{1}{9k_{\mathrm{m}}}(\Gamma\Psi - \lambda\Phi)(2C+D)\right]\frac{\Omega}{\Gamma} - \mu_{\mathrm{m}}(2C+D). \tag{E.31}$$

It follows from (E.15) that

$$\begin{aligned}
A &= (S_{2222}+S_{2233})W - (3k_{\mathrm{m}}+\mu_{\mathrm{m}})X - \mu_{\mathrm{m}}Y,\\
B &= S_{2211}W - \frac{1}{2}(3k_{\mathrm{m}}-2\mu_{\mathrm{m}})X + \mu_{\mathrm{m}}Y,\\
C &= S_{1122}W + \mu_{\mathrm{m}}Y + (3k_{\mathrm{m}}-2\mu_{\mathrm{m}})Z,\\
D &= S_{1111}W - 2\mu_{\mathrm{m}}Y + (3k_{\mathrm{m}}+4\mu_{\mathrm{m}})Z,
\end{aligned} \tag{E.32}$$

where

$$W = \sum_{i=1}^{N} V_p^i \frac{Q_iR_i - P_iS_i}{\Delta_i}, \quad X = \sum_{i=1}^{N} V_p^i \frac{P_i}{\Delta_i}, \quad Y = \sum_{i=1}^{N} V_p^i \frac{Q_i}{\Delta_i}, \quad Z = \sum_{i=1}^{N} V_p^i \frac{R_i}{\Delta_i}. \tag{E.33}$$

It follows from (E.16) and (E.17) that

$$\begin{aligned}
A + B &= (S_{2211}+S_{2222}+S_{2233})W - \frac{9}{2}k_{\mathrm{m}}X,\\
2C + D &= (S_{1111}+2S_{1122})W + 9k_{\mathrm{m}}Z,
\end{aligned} \tag{E.34}$$

$$\begin{aligned}
(S_{2222}+S_{2233})B - S_{2211}A &= \frac{1}{2}\lambda X + \mu_{\mathrm{m}}(S_{2211}+S_{2222}+S_{2233})Y,\\
DS_{1122} - CS_{1111} &= \lambda Z - \mu_{\mathrm{m}}(S_{1111}+2S_{1122})Y,
\end{aligned} \tag{E.35}$$

and

$$\Gamma = \lambda W + 9k_{\mathrm{m}}\mu_{\mathrm{m}}Y, \tag{E.36}$$

and from (E.26)

$$\begin{aligned}
\Phi = 2\Psi W &+ \left[(3k_{\mathrm{m}}+\mu_{\mathrm{m}})S_{1111} - (3k_{\mathrm{m}}-2\mu_{\mathrm{m}})S_{1122}\right]X\\
&+ \left[S_{1111} + 2(S_{1122}+S_{2211}+S_{2222}+S_{2233})\right]\mu_{\mathrm{m}}Y\\
&+ \left[2(3k_{\mathrm{m}}-2\mu_{\mathrm{m}})S_{2211} - (3k_{\mathrm{m}}+4\mu_{\mathrm{m}})(S_{2222}+S_{2233})\right]Z - 1.
\end{aligned} \tag{E.37}$$

Consider now

$$\begin{aligned}
\lambda\Phi - \Gamma\Psi = \lambda\Psi W &+ \lambda\left[(3k_{\mathrm{m}}+\mu_{\mathrm{m}})S_{1111} - (3k_{\mathrm{m}}-2\mu_{\mathrm{m}})S_{1122}\right]X\\
&+ \left[\lambda S_{1111} + 2\lambda(S_{1122}+S_{2211}+S_{2222}+S_{2233}) - 9k_{\mathrm{m}}\Psi\right]\mu_{\mathrm{m}}Y\\
&+ \lambda\left[2(3k_{\mathrm{m}}-2\mu_{\mathrm{m}})S_{2211} - (3k_{\mathrm{m}}+4\mu_{\mathrm{m}})(S_{2222}+S_{2233})\right]Z - \lambda.
\end{aligned} \tag{E.38}$$

Substitution into (E.25), (E.30) and (E.31) leads to

$$\begin{aligned}
\Phi\bar{P} = &\left[\begin{array}{l}\{9k_{\mathrm{m}}\mu_{\mathrm{m}}X - 2\mu_{\mathrm{m}}(S_{2211}+S_{2222}+S_{2233})W\}\Psi\\ \quad + 2\mu_{\mathrm{m}}(S_{2211}+S_{2222}+S_{2233})\Phi\end{array}\right]\frac{\Omega}{9k_{\mathrm{m}}\mu_{\mathrm{m}}}\\
&- \{9k_{\mathrm{m}}\mu_{\mathrm{m}}X - 2\mu_{\mathrm{m}}(S_{2211}+S_{2222}+S_{2233})W\},
\end{aligned} \tag{E.39}$$

$$\Phi\bar{Q} = \left[\{9k_{\mathrm{m}}\mu_{\mathrm{m}}Y + \lambda W\}\Psi - \lambda\Phi\right]\frac{\Omega}{9k_{\mathrm{m}}\mu_{\mathrm{m}}} - \{9k_{\mathrm{m}}\mu_{\mathrm{m}}Y + \lambda W\}, \tag{E.40}$$

$$\begin{aligned}\Phi\bar{R} = &\left[\{9k_{\mathrm{m}}\mu_{\mathrm{m}}Z + \mu_{\mathrm{m}}(S_{1111} + 2S_{1122})W\}\Psi - \mu_{\mathrm{m}}(S_{1111} + 2S_{1122})\Phi\right]\frac{\Omega}{9k_{\mathrm{m}}\mu_{\mathrm{m}}} \\ &- \{9k_{\mathrm{m}}\mu_{\mathrm{m}}Z + \mu_{\mathrm{m}}(S_{1111} + 2S_{1122})W\}.\end{aligned} \tag{E.41}$$

These results are of the form (15.71)–(15.77) given in Chapter 15.

Appendix F: Analysis for Transversely Isotropic Cylindrical Inclusions

As $S_{1111} = S_{1122} = 0$ for the fibre case, it follows from (15.46), (15.49) and (15.52) that $\Psi = 0$ and

$$\Delta_i = \mu_{\mathrm{m}} \frac{1+\nu_{\mathrm{m}}}{1-\nu_{\mathrm{m}}} (Q_i - 2R_i) - 3k_{\mathrm{m}} \frac{1-2\nu_{\mathrm{m}}}{2(1-\nu_{\mathrm{m}})} R_i + 9k_{\mathrm{m}}\mu_{\mathrm{m}} ,$$
$$\bar{\Delta} = \mu_{\mathrm{m}} \frac{1+\nu_{\mathrm{m}}}{1-\nu_{\mathrm{m}}} (\bar{Q} - 2\bar{R}) - 3k_{\mathrm{m}} \frac{1-2\nu_{\mathrm{m}}}{2(1-\nu_{\mathrm{m}})} \bar{R} + 9k_{\mathrm{m}}\mu_{\mathrm{m}} . \qquad (F.1)$$

Equations (15.57)–(15.60) reduce to

$$\frac{\Omega}{2(1-\nu_{\mathrm{m}})} - (3k_{\mathrm{m}} + \mu_{\mathrm{m}})\bar{P} - \mu_{\mathrm{m}}\bar{Q} = A\bar{\Delta}, \qquad (F.2)$$

$$\frac{\nu_{\mathrm{m}}\Omega}{2(1-\nu_{\mathrm{m}})} - \frac{1}{2}(3k_{\mathrm{m}} - 2\mu_{\mathrm{m}})\bar{P} + \mu_{\mathrm{m}}\bar{Q} = B\bar{\Delta}, \qquad (F.3)$$

$$\mu_{\mathrm{m}}\bar{Q} + (3k_{\mathrm{m}} - 2\mu_{\mathrm{m}})\bar{R} = C\bar{\Delta}, \qquad (F.4)$$

$$-2\mu_{\mathrm{m}}\bar{Q} + (3k_{\mathrm{m}} + 4\mu_{\mathrm{m}})\bar{R} = D\bar{\Delta}. \qquad (F.5)$$

From (F.4) and (F.5),

$$9k_{\mathrm{m}}\bar{R} = (2C + D)\bar{\Delta}, \qquad (F.6)$$

$$3\mu_{\mathrm{m}}\left(\bar{Q} - 2\bar{R}\right) = (C - D)\bar{\Delta}. \qquad (F.7)$$

Substitution into (F.1) leads to the following expression for $\bar{\Delta}$

$$\frac{1}{\bar{\Delta}} = \frac{1-\nu_{\mathrm{m}} - \nu_{\mathrm{m}}C + \frac{1}{2}D}{9k_{\mathrm{m}}\mu_{\mathrm{m}}(1-\nu_{\mathrm{m}})} = \frac{1}{9k_{\mathrm{m}}\mu_{\mathrm{m}}}\left(1 + \frac{D - 2\nu_{\mathrm{m}}C}{2(1-\nu_{\mathrm{m}})}\right). \qquad (F.8)$$

Consider now the constants C and D defined by (15.56). As $S_{1111} = S_{1122} = 0$, it follows using (F.1) that

$$\Delta_i = \frac{9k_{\mathrm{m}}\mu_{\mathrm{m}}}{3k_{\mathrm{m}} + 4\mu_{\mathrm{m}}}(Q_i - 3R_i + 3k_{\mathrm{m}} + 4\mu_{\mathrm{m}}), \qquad (F.9)$$

$$C=\sum_{i=1}^{N}\left[\mu_{\mathrm{m}}Q_i+(3k_{\mathrm{m}}-2\mu_{\mathrm{m}})R_i\right]\frac{V_p^i}{\Delta_i}, \tag{F.10}$$

$$D=\sum_{i=1}^{N}\left[-2\mu_{\mathrm{m}}Q_i+(3k_{\mathrm{m}}+4\mu_{\mathrm{m}})R_i\right]\frac{V_p^i}{\Delta_i}. \tag{F.11}$$

On using (15.48), it then follows that

$$\Delta_i=27k_{\mathrm{m}}\mu_{\mathrm{m}}\frac{k_{\mathrm{T}}^i+\mu_{\mathrm{m}}}{3k_{\mathrm{m}}+4\mu_{\mathrm{m}}}, \tag{F.12}$$

$$C=\frac{3k_{\mathrm{m}}+4\mu_{\mathrm{m}}}{27k_{\mathrm{m}}\mu_{\mathrm{m}}}\sum_{i=1}^{N}V_{\mathrm{p}}^i\left[2\left(3k_{\mathrm{m}}+\mu_{\mathrm{m}}\right)\nu_{\mathrm{A}}^i-\left(3k_{\mathrm{m}}-2\mu_{\mathrm{m}}\right)\right]\frac{k_{\mathrm{T}}^i}{k_{\mathrm{T}}^i+\mu_{\mathrm{m}}}, \tag{F.13}$$

$$D=\frac{3k_{\mathrm{m}}+4\mu_{\mathrm{m}}}{27k_{\mathrm{m}}\mu_{\mathrm{m}}}\sum_{i=1}^{N}V_{\mathrm{p}}^i\left[2\left(3k_{\mathrm{m}}-2\mu_{\mathrm{m}}\right)\nu_{\mathrm{A}}^i-\left(3k_{\mathrm{m}}+4\mu_{\mathrm{m}}\right)+\frac{9k_{\mathrm{m}}\mu_{\mathrm{m}}}{k_{\mathrm{T}}^i}\right]\frac{k_{\mathrm{T}}^i}{k_{\mathrm{T}}^i+\mu_{\mathrm{m}}}. \tag{F.14}$$

It can be shown that

$$2C+D=\frac{3k_{\mathrm{m}}+4\mu_{\mathrm{m}}}{3\mu_{\mathrm{m}}}\sum_{i=1}^{N}V_{\mathrm{p}}^i\left(2\nu_{\mathrm{A}}^i-1+\frac{\mu_{\mathrm{m}}}{k_{\mathrm{T}}^i}\right)\frac{k_{\mathrm{T}}^i}{k_{\mathrm{T}}^i+\mu_{\mathrm{m}}}, \tag{F.15}$$

$$C-D=\frac{3k_{\mathrm{m}}+4\mu_{\mathrm{m}}}{9k_{\mathrm{m}}}\sum_{i=1}^{N}V_{\mathrm{p}}^i\left[2\left(1+\nu_{\mathrm{A}}^i\right)-\frac{3k_{\mathrm{m}}}{k_{\mathrm{T}}^i}\right]\frac{k_{\mathrm{T}}^i}{k_{\mathrm{T}}^i+\mu_{\mathrm{m}}}. \tag{F.16}$$

It then follows from (F.6) and (F.7) that

$$\bar{Q}=\bar{\Delta}\frac{3k_{\mathrm{m}}+4\mu_{\mathrm{m}}}{27k_{\mathrm{m}}\mu_{\mathrm{m}}}\sum_{i=1}^{N}V_{\mathrm{p}}^i\left[6\nu_{\mathrm{A}}^i-\frac{\left(3k_{\mathrm{m}}-2\mu_{\mathrm{m}}\right)}{k_{\mathrm{T}}^i}\right]\frac{k_{\mathrm{T}}^i}{k_{\mathrm{T}}^i+\mu_{\mathrm{m}}}, \tag{F.17}$$

$$\bar{R}=\bar{\Delta}\frac{3k_{\mathrm{m}}+4\mu_{\mathrm{m}}}{27k_{\mathrm{m}}\mu_{\mathrm{m}}}\sum_{i=1}^{N}V_{\mathrm{p}}^i\left(2\nu_{\mathrm{A}}^i-1+\frac{\mu_{\mathrm{m}}}{k_{\mathrm{T}}^i}\right)\frac{k_{\mathrm{T}}^i}{k_{\mathrm{T}}^i+\mu_{\mathrm{m}}}. \tag{F.18}$$

From (F.8)

$$\frac{1}{\bar{\Delta}}=\left(\frac{3k_{\mathrm{m}}+4\mu_{\mathrm{m}}}{9k_{\mathrm{m}}\mu_{\mathrm{m}}}+\frac{2C+D}{9k_{\mathrm{m}}}-\frac{C-D}{3\mu_{\mathrm{m}}}\right)\frac{1}{3k_{\mathrm{m}}+4\mu_{\mathrm{m}}}, \tag{F.19}$$

so that on using (F.15) and (F.16)

$$\frac{1}{\bar{\Delta}}=\frac{1}{9k_{\mathrm{m}}\mu_{\mathrm{m}}}\left(V_{\mathrm{m}}+\left(k_{\mathrm{m}}+\frac{4}{3}\mu_{\mathrm{m}}\right)\sum_{i=1}^{N}\frac{V_{\mathrm{p}}^i}{k_{\mathrm{T}}^i+\mu_{\mathrm{m}}}\right). \tag{F.20}$$

It then follows from (15.51), (F.17) and (F.18) that

$$\bar{Q}\equiv 6\nu_{\mathrm{A}}^{\mathrm{eff}}k_{\mathrm{T}}^{\mathrm{eff}}-3k_{\mathrm{m}}+2\mu_{\mathrm{m}}=\frac{\sum_{i=1}^{N}\left[6\nu_{\mathrm{A}}^ik_{\mathrm{T}}^i-3k_{\mathrm{m}}+2\mu_{\mathrm{m}}\right]\frac{V_{\mathrm{p}}^i}{k_{\mathrm{T}}^i+\mu_{\mathrm{m}}}}{\sum_{i=1}^{N}\frac{V_{\mathrm{p}}^i}{k_{\mathrm{T}}^i+\mu_{\mathrm{m}}}+\frac{3V_{\mathrm{m}}}{3k_{\mathrm{m}}+4\mu_{\mathrm{m}}}}, \tag{F.21}$$

$$\bar{R} \equiv \left(2\nu_{\mathrm{A}}^{\mathrm{eff}} - 1\right)k_{\mathrm{T}}^{\mathrm{eff}} + \mu_{\mathrm{m}} = \frac{\sum_{i=1}^{N}\left[\left(2\nu_{\mathrm{A}}^{i} - 1\right)k_{\mathrm{T}}^{i} + \mu_{\mathrm{m}}\right]\dfrac{V_{\mathrm{p}}^{i}}{k_{\mathrm{T}}^{i} + \mu_{\mathrm{m}}}}{\sum_{i=1}^{N}\dfrac{V_{\mathrm{p}}^{i}}{k_{\mathrm{T}}^{i} + \mu_{\mathrm{m}}} + \dfrac{3V_{\mathrm{m}}}{3k_{\mathrm{m}} + 4\mu_{\mathrm{m}}}}. \tag{F.22}$$

On eliminating $\nu_{\mathrm{A}}^{\mathrm{eff}}$ in (F.21) and (F.22),

$$3\left(k_{\mathrm{T}}^{\mathrm{eff}} + \mu_{\mathrm{m}}\right) - \left(3k_{\mathrm{m}} + 4\mu_{\mathrm{m}}\right) = \frac{\sum_{i=1}^{N}\left[3\left(k_{\mathrm{T}}^{i} + \mu_{\mathrm{m}}\right) - \left(3k_{\mathrm{m}} + 4\mu_{\mathrm{m}}\right)\right]\dfrac{V_{\mathrm{p}}^{i}}{k_{\mathrm{T}}^{i} + \mu_{\mathrm{m}}}}{\sum_{i=1}^{N}\dfrac{V_{\mathrm{p}}^{i}}{k_{\mathrm{T}}^{i} + \mu_{\mathrm{m}}} + \dfrac{3V_{\mathrm{m}}}{3k_{\mathrm{m}} + 4\mu_{\mathrm{m}}}}. \tag{F.23}$$

It then follows, on using the relation $k_{\mathrm{T}}^{\mathrm{m}} = k_{\mathrm{m}} + \frac{1}{3}\mu_{\mathrm{m}}$, that the result (F.23) may be written as the following mixtures relationship:

$$\frac{1}{k_{\mathrm{T}}^{\mathrm{eff}} + \mu_{\mathrm{m}}} = \sum_{i=1}^{N}\frac{V_{\mathrm{p}}^{i}}{k_{\mathrm{T}}^{i} + \mu_{\mathrm{m}}} + \frac{V_{\mathrm{m}}}{k_{\mathrm{T}}^{\mathrm{m}} + \mu_{\mathrm{m}}}. \tag{F.24}$$

From (F.21),

$$\nu_{\mathrm{A}}^{\mathrm{eff}}k_{\mathrm{T}}^{\mathrm{eff}} = \frac{\dfrac{\left(3k_{\mathrm{m}} - 2\mu_{\mathrm{m}}\right)}{2\left(3k_{\mathrm{m}} + 4\mu_{\mathrm{m}}\right)}V_{\mathrm{m}} + \sum_{i=1}^{N}V_{\mathrm{p}}^{i}\dfrac{\nu_{\mathrm{A}}^{i}k_{\mathrm{T}}^{i}}{k_{\mathrm{T}}^{i} + \mu_{\mathrm{m}}}}{\sum_{i=1}^{N}\dfrac{V_{\mathrm{p}}^{i}}{k_{\mathrm{T}}^{i} + \mu_{\mathrm{m}}} + \dfrac{V_{\mathrm{m}}}{k_{\mathrm{T}}^{\mathrm{m}} + \mu_{\mathrm{m}}}}. \tag{F.25}$$

As

$$\frac{\nu_{\mathrm{m}}k_{\mathrm{T}}^{\mathrm{m}}}{k_{\mathrm{T}}^{\mathrm{m}} + \mu_{\mathrm{m}}} = \frac{3k_{\mathrm{m}} - 2\mu_{\mathrm{m}}}{2\left(3k_{\mathrm{m}} + 4\mu_{\mathrm{m}}\right)}, \tag{F.26}$$

it follows that

$$\nu_{\mathrm{A}}^{\mathrm{eff}}k_{\mathrm{T}}^{\mathrm{eff}} = \frac{\sum_{i=1}^{N}V_{\mathrm{p}}^{i}\dfrac{\nu_{\mathrm{A}}^{i}k_{\mathrm{T}}^{i}}{k_{\mathrm{T}}^{i} + \mu_{\mathrm{m}}} + \dfrac{V_{\mathrm{m}}\nu_{\mathrm{m}}k_{\mathrm{T}}^{\mathrm{m}}}{k_{\mathrm{T}}^{\mathrm{m}} + \mu_{\mathrm{m}}}}{\sum_{i=1}^{N}\dfrac{V_{\mathrm{p}}^{i}}{k_{\mathrm{T}}^{i} + \mu_{\mathrm{m}}} + \dfrac{V_{\mathrm{m}}}{k_{\mathrm{T}}^{\mathrm{m}} + \mu_{\mathrm{m}}}}. \tag{F.27}$$

On using (F.24), the following mixtures relation is obtained

$$\frac{\nu_{\mathrm{A}}^{\mathrm{eff}}k_{\mathrm{T}}^{\mathrm{eff}}}{k_{\mathrm{T}}^{\mathrm{eff}} + \mu_{\mathrm{m}}} = \sum_{i=1}^{N}V_{\mathrm{p}}^{i}\frac{\nu_{\mathrm{A}}^{i}k_{\mathrm{T}}^{i}}{k_{\mathrm{T}}^{i} + \mu_{\mathrm{m}}} + \frac{V_{\mathrm{m}}\nu_{\mathrm{m}}k_{\mathrm{T}}^{\mathrm{m}}}{k_{\mathrm{T}}^{\mathrm{m}} + \mu_{\mathrm{m}}}. \tag{F.28}$$

On subtracting (F.2) and (F.3) to obtain a value for Ω and then substituting into (15.61),

$$\left[\left(3k_{\mathrm{m}} + 4\mu_{\mathrm{m}}\right) + \bar{Q} - 3\bar{R}\right]\bar{P} = \bar{Q}\bar{R} - 4\mu_{\mathrm{m}}\bar{Q} - 2(A - B)\bar{\Delta}. \tag{F.29}$$

Thus, on using (15.51),

$$3\left(k_{\mathrm{T}}^{\mathrm{eff}} + \mu_{\mathrm{m}}\right)\bar{P} = \left[\left(2\nu_{\mathrm{A}}^{\mathrm{eff}} - 1\right)k_{\mathrm{T}}^{\mathrm{eff}} - 3\mu_{\mathrm{m}}\right]\left(6\nu_{\mathrm{A}}^{\mathrm{eff}}k_{\mathrm{T}}^{\mathrm{eff}} - 3k_{\mathrm{m}} + 2\mu_{\mathrm{m}}\right) - 2\left(A - B\right)\bar{\Delta}. \tag{F.30}$$

From (15.53) and on using the appropriate values of S_{ijkl} relevant to cylinders

$$A=\sum_{i=1}^{N}\left[\frac{Q_iR_i-P_iS_i}{2(1-\nu_{\rm m})}-(3k_{\rm m}+\mu_{\rm m})P_i-\mu_{\rm m}Q_i\right]\frac{V_p^i}{\Delta_i}, \tag{F.31}$$

$$B=\sum_{i=1}^{N}\left[\frac{(Q_iR_i-P_iS_i)\nu_{\rm m}}{2(1-\nu_{\rm m})}-\frac{1}{2}(3k_{\rm m}-2\mu_{\rm m})P_i+\mu_{\rm m}Q_i\right]\frac{V_p^i}{\Delta_i}. \tag{F.32}$$

It then follows that

$$2(A-B)=\sum_{i=1}^{N}\left[(Q_iR_i-P_iS_i)-(3k_{\rm m}+4\mu_{\rm m})P_i-4\mu_{\rm m}Q_i\right]\frac{V_p^i}{\Delta_i}. \tag{F.33}$$

This relation is now written, on using (15.48) and (F.12), as

$$\begin{aligned}2(A-B)\frac{9k_{\rm m}\mu_{\rm m}}{3k_{\rm m}+4\mu_{\rm m}}&=2\mu_{\rm m}\left(k_{\rm m}-\tfrac{2}{3}\mu_{\rm m}\right)\sum_{i=1}^{N}\frac{V_{\rm p}^i}{k_{\rm T}^i+\mu_{\rm m}}-2\left(k_{\rm m}+\tfrac{4}{3}\mu_{\rm m}\right)\sum_{i=1}^{N}\frac{V_{\rm p}^ik_{\rm T}^i\nu_{\rm A}^i}{k_{\rm T}^i+\mu_{\rm m}}\\&-\sum_{i=1}^{N}V_{\rm p}^iE_{\rm A}^i-4\mu_{\rm m}\sum_{i=1}^{N}V_{\rm p}^i\left(\nu_{\rm A}^i\right)^2+\left(k_{\rm m}+\tfrac{4}{3}\mu_{\rm m}\right)\sum_{i=1}^{N}V_{\rm p}^i+4\mu_{\rm m}^2\sum_{i=1}^{N}V_{\rm p}^i\frac{\left(\nu_{\rm A}^i\right)^2}{k_{\rm T}^i+\mu_{\rm m}}.\end{aligned} \tag{F.34}$$

It should be noted from (F.20) and (F.24) that

$$\bar{\Delta}=27k_{\rm m}\mu_{\rm m}\frac{k_{\rm T}^{\rm eff}+\mu_{\rm m}}{3k_{\rm m}+4\mu_{\rm m}}. \tag{F.35}$$

Substitution into (F.30) leads to

$$\bar{P}=\left[\left(2\nu_{\rm A}^{\rm eff}-1\right)k_{\rm T}^{\rm eff}-3\mu_{\rm m}\right]\frac{6\nu_{\rm A}^{\rm eff}k_{\rm T}^{\rm eff}-(3k_{\rm m}-2\mu_{\rm m})}{3\left(k_{\rm T}^{\rm eff}+\mu_{\rm m}\right)}-2(A-B)\frac{9k_{\rm m}\mu_{\rm m}}{3k_{\rm m}+4\mu_{\rm m}}. \tag{F.36}$$

On using (15.51) it follows that

$$\begin{aligned}E_{\rm A}^{\rm eff}=2\mu_{\rm m}\left(1-3\nu_{\rm A}^{\rm eff}\right)-\left[\left(2\nu_{\rm A}^{\rm eff}-1\right)k_{\rm T}^{\rm eff}-3\mu_{\rm m}\right]\frac{6\nu_{\rm A}^{\rm eff}\mu_{\rm m}+(3k_{\rm m}-2\mu_{\rm m})}{3\left(k_{\rm T}^{\rm eff}+\mu_{\rm m}\right)}\\-2(A-B)\frac{9k_{\rm m}\mu_{\rm m}}{3k_{\rm m}+4\mu_{\rm m}}.\end{aligned} \tag{F.37}$$

From (F.34) and (F.37)

$$\begin{aligned}E_{\rm A}^{\rm eff}&=\left(1-2\nu_{\rm A}^{\rm eff}\right)k_{\rm m}+\tfrac{4}{3}\mu_{\rm m}\left(1+\nu_{\rm A}^{\rm eff}\right)\left(1-3\nu_{\rm A}^{\rm eff}\right)+\frac{2\mu_{\rm m}\left(1+\nu_{\rm A}^{\rm eff}\right)}{k_{\rm T}^{\rm eff}+\mu_{\rm m}}\left[2\nu_{\rm A}^{\rm eff}\mu_{\rm m}+\left(k_{\rm m}-\tfrac{2}{3}\mu_{\rm m}\right)\right]\\&-2\mu_{\rm m}\left(k_{\rm m}-\tfrac{2}{3}\mu_{\rm m}\right)\sum_{i=1}^{N}\frac{V_{\rm p}^i}{k_{\rm T}^i+\mu_{\rm m}}+2\left(k_{\rm m}+\tfrac{4}{3}\mu_{\rm m}\right)\sum_{i=1}^{N}\frac{V_{\rm p}^ik_{\rm T}^i\nu_{\rm A}^i}{k_{\rm T}^i+\mu_{\rm m}}\\&+\sum_{i=1}^{N}V_{\rm p}^iE_{\rm A}^i+4\mu_{\rm m}\sum_{i=1}^{N}V_{\rm p}^i\left(\nu_{\rm A}^i\right)^2-\left(k_{\rm m}+\tfrac{4}{3}\mu_{\rm m}\right)\sum_{i=1}^{N}V_{\rm p}^i-4\mu_{\rm m}^2\sum_{i=1}^{N}V_{\rm p}^i\frac{\left(\nu_{\rm A}^i\right)^2}{k_{\rm T}^i+\mu_{\rm m}}.\end{aligned} \tag{F.38}$$

On using (F.24) and (F.28), it can be shown after some calculation that the following mixtures relationship is valid that determines the effective axial modulus E_A^{eff} of the unidirectional multiphase composite

$$E_A^{eff} + \frac{4k_T^{eff}\left(\nu_A^{eff}\right)^2 \mu_m}{k_T^{eff} + \mu_m} = \sum_{i=1}^{N} V_p^i \left(E_A^i + \frac{4k_T^i\left(\nu_A^i\right)^2 \mu_m}{k_T^i + \mu_m} \right) + V_m \left(E_m + \frac{4k_T^m \nu_m^2 \mu_m}{k_T^m + \mu_m} \right), \quad \text{(F.39)}$$

where

$$k_T^m = k_m + \frac{1}{3}\mu_m. \quad \text{(F.40)}$$

Appendix G: Recurrence Relations, Differential Equations and Boundary Conditions

1. Recurrence Relations

The functions S_i, U_i, V_i and W_i appearing in the expressions (18.45) and (18.49)–(18.51), respectively, are the values of $\sigma_{33}^{(i)}$, $u_1^{(i)}$, $u_2^{(i)}$ and $u_3^{(i)}$, respectively, at the interfaces $x_3 = x_3^{(i)}$, for $i = 1, \ldots, N$ and at the external surface $x_3 = x_3^{(N+1)}$. These definitions can be extended without loss of generality so that S_0, U_0, V_0 and W_0 are the values of these variables at $x_3 = x_3^{(0)} = 0$. It follows on substituting $x_3 = x_3^{(i-1)}$ into (18.45) that

$$S_{i-1}(x_1) = \frac{h_i}{2}\left[p''_{i-1}(x_1) + p''_i(x_1)\right] + S_i(x_1), \text{ for } i = N+1,\ldots,1, \text{where } S_{N+1}(x_1) \equiv 0, \quad \text{(G.1)}$$

because $\sigma_{33} = \sigma_t$ on $x_3 = h$ implying that the functions S_i are all independent of x_2. On substituting $x_3 = x_3^{(i-1)}$ into (18.49) the following recurrence relation is derived for the functions $U_i(x_1)$

$$\begin{aligned} U_i(x_1) &= \frac{h_i^2}{6} g_{22}^{(i)}\left[p''_{i-1}(x_1) + 2p''_i(x_1)\right] \\ &+ h_i\left[g_{22}^{(i)} S_i(x_1) + g_{12}^{(i)} \frac{p_{i-1}(x_1) - p_i(x_1)}{h_i} + g_{23}^{(i)} \frac{q_{i-1}(x_1) - q_i(x_1)}{h_i} + \varepsilon_t^{(i)}\right] + U_{i-1}(x_1), \quad \text{(G.2)} \\ &\text{for } i = 1, \ldots, N+1, \quad \text{where } U_0(x_1) \equiv 0, \end{aligned}$$

because from $(18.54)_3$ $u_3 = 0$ on $x_3 = 0$ implying that the functions U_i are all independent of x_2. On substituting $x_3 = x_3^{(i-1)}$ into (18.50) and then eliminating x_2 by subtraction, the following recurrence relation can be derived for the functions $\Delta V_i(x_1) \equiv V_i(x_1,x_2) - V_{N+1}(x_1,x_2)$, which must be independent of x_2,

$$\begin{aligned} \Delta V_{i-1}(x_1) &= -\frac{h_i}{2}\left[S_{55}^{(i)} p'_i(x_1) + S_{45}^{(i)} q'_i(x_1)\right] - \frac{h_i}{2}\left[S_{55}^{(i)} p'_{i-1}(x_1) + S_{45}^{(i)} q'_{i-1}(x_1)\right] \\ &- \frac{h_i^2}{2}\left[g_{22}^{(i)} S'_i(x_1) + g_{12}^{(i)} \frac{p'_{i-1}(x_1) - p'_i(x_1)}{h_i} + g_{23}^{(i)} \frac{q'_{i-1}(x_1) - q'_i(x_1)}{h_i}\right] \\ &- \frac{h_i^3}{24} g_{22}^{(i)}\left[p'''_{i-1}(x_1) + 3p'''_i(x_1)\right] + h_i U'_i(x_1) + \Delta V_i(x_1), \text{ for } i = N+1\ldots2, \end{aligned} \quad \text{(G.3)}$$

where $\Delta V_{N+1}(x_1) \equiv 0$.

On substituting $x_3 = x_3^{(i-1)}$ into relation (18.51) and then eliminating x_2 by subtraction, the following recurrence relation for the functions

$$\Delta W_i(x_1) \equiv W_i(x_1, x_2) - W_{N+1}(x_1, x_2),$$

which must be independent of x_2, can be derived

$$\Delta W_{i-1}(x_1) = -\frac{h_i}{2}\left[S_{45}^{(i)} p_i'(x_1) + S_{44}^{(i)} q_i'(x_1)\right] - \frac{h_i}{2}\left[S_{45}^{(i)} p_{i-1}'(x_1) + S_{44}^{(i)} q_{i-1}'(x_1)\right] + \Delta W_i(x_1), \quad \text{(G.4)}$$
$$\text{for } i = N+1, \ldots, 2, \text{ where } \Delta W_{N+1}(x_1) \equiv 0.$$

Because symmetry is assumed about $x_3 = 0$, and because $p_0(x_1)$ and $q_0(x_1)$ must have zero values for uncracked laminates, relations (18.43) and (18.44) and the boundary conditions (18.54) assert that the functions p_0 and q_0 must be such that

$$p_0(x_1) \equiv 0, \; q_0(x_1) \equiv 0. \quad \text{(G.5)}$$

Similarly, relations (18.43) and (18.44) and the boundary conditions (18.57) assert that

$$p_{N+1}(x_1) \equiv 0, \; q_{N+1}(x_1) \equiv 0. \quad \text{(G.6)}$$

The expressions derived for the stress and displacement components satisfy exactly the equilibrium equations (18.1)–(18.3) together with the stress-strain relations (18.4), (18.6), (18.8) and (18.9) for *any* functions $p_i(x_1)$ and $q_i(x_1)$, $i = 1, \ldots, N$. In addition, provided that the functions $S_i(x_1)$, $U_i(x_1)$, $V_i(x_1, x_2)$ and $W_i(x_1, x_2)$ are generated by the recurrence relations (G.1)–(G.4), respectively, the stress and displacement fields automatically satisfy the interface conditions (18.55) and (18.56) and some of the other boundary conditions. The remaining stress-strain relations (18.5) and (18.7) (and (18.15) and (18.17)) can only be satisfied in an average sense that is now described.

2. Derivation of Differential Equations

When deriving the displacement distributions given by (18.49)–(18.51), use has been made of the stress-strain relations (18.7), (18.8) and (18.16) in such a way that these relations are automatically satisfied by the displacement field. It is necessary, therefore, to consider the remaining stress-strain relations given by (18.15) and (18.17) for the in-plane strains ε_{11} and ε_{12}. As it is not possible to satisfy these relations exactly using the stress and displacement representation, an attempt is made to satisfy them approximately by averaging these stress-strain relations through the thickness of each layer. By making use of ply refinement techniques, where each ply is subdivided into layers to resolve through-thickness variations of stresses and displacements, the thickness of each layer can be made small enough for there to be minimal error when replacing (18.15) and (18.17) by their through-thickness averages. The average of any quantity $\phi_i(x_1, x_2, x_1)$ through the thickness of the ith layer is denoted by an overbar and defined by

$$\bar{\phi}_i(x_1, x_2) = \frac{1}{h_i} \int_{x_3^{(i-1)}}^{x_3^{(i)}} \phi_i(x_1, x_2, x_3)\, dx_3, \; h_i = x_3^{(i)} - x_3^{(i-1)}. \quad \text{(G.7)}$$

The through-thickness averages of the stress-strain relations (18.15) and (18.17) are written, for $i = 1, \ldots, N + 1$, as

$$\bar{\varepsilon}_{11}^{(i)} \equiv \frac{\partial \bar{u}_1^{(i)}}{\partial x_1} = g_{11}^{(i)} \bar{\sigma}_{11}^{(i)} + g_{12}^{(i)} \bar{\sigma}_{33}^{(i)} + g_{13}^{(i)} \bar{\sigma}_{12}^{(i)} + H_1^{(i)} \varepsilon_{\mathrm{T}}^{\mathrm{c}} + \bar{V}_1^{(i)} \Delta T, \tag{G.8}$$

$$2\bar{\varepsilon}_{12}^{(i)} \equiv \frac{\partial \bar{u}_1^{(i)}}{\partial x_2} + \frac{\partial \bar{u}_2^{(i)}}{\partial x_1} = g_{13}^{(i)} \bar{\sigma}_{11}^{(i)} + g_{23}^{(i)} \bar{\sigma}_{33}^{(i)} + g_{33}^{(i)} \bar{\sigma}_{12}^{(i)} + H_3^{(i)} \varepsilon_{\mathrm{T}}^{\mathrm{c}} + \bar{V}_3^{(i)} \Delta T. \tag{G.9}$$

From (18.45)–(18.47) it is clear that, on averaging,

$$\bar{\sigma}_{11}^{(i)} \equiv \sigma_{11}^{(i)} = \frac{p_{i-1}(x_1) - p_i(x_1)}{h_i} + \sigma_{\mathrm{A}}^i, \; \bar{\sigma}_{12}^{(i)} \equiv \sigma_{12}^{(i)} = \frac{q_{i-1}(x_1) - q_i(x_1)}{h_i} + \tau_{\mathrm{A}}^i. \tag{G.10}$$

$$\bar{\sigma}_{33}^{(i)} = \frac{h_i}{6}\left[p_{i-1}''(x_1) + 2p_i''(x_1)\right] + S_i(x_1) + \sigma_{\mathrm{t}}. \tag{G.11}$$

From (18.50), the average of the displacement component $u_1^{(i)}$ is given by

$$\begin{aligned}\bar{u}_1^{(i)}(x_1, x_2) = &-\frac{h_i}{3}\left[S_{55}^{(i)} p_i'(x_1) + S_{45}^{(i)} q_i'(x_1)\right] - \frac{h_i}{6}\left[S_{55}^{(i)} p_{i-1}'(x_1) + S_{45}^{(i)} q_{i-1}'(x_1)\right] \\ &-\frac{h_i^2}{6}\left[g_{22}^{(i)} S_i'(x_1) + g_{12}^{(i)} \frac{p_{i-1}'(x_1) - p_i'(x_1)}{h_i} + g_{23}^{(i)} \frac{q_{i-1}'(x_1) - q_i'(x_1)}{h_i}\right] \\ &-\frac{h_i^3}{120} g_{22}^{(i)}\left[p_{i-1}'''(x_1) + 4p_i'''(x_1)\right] + \frac{h_i}{2} U_i'(x_1) + \Delta V_i(x_1) + V_{N+1}(x_1, x_2).\end{aligned} \tag{G.12}$$

From (18.51) the average of the displacement component $u_2^{(i)}$ is given by

$$\begin{aligned}\bar{u}_2^{(i)}(x_1, x_2) = &-\frac{h_i}{3}\left[S_{45}^{(i)} p_i'(x_1) + S_{44}^{(i)} q_i'(x_1)\right] - \frac{h_i}{6}\left[S_{45}^{(i)} p_{i-1}'(x_1) + S_{44}^{(i)} q_{i-1}'(x_1)\right] \\ &+ \Delta W_i(x_1) + W_{N+1}(x_1, x_2).\end{aligned} \tag{G.13}$$

The required governing differential equations are obtained by substituting (G.12) and (G.13) into (G.8) and (G.9) using (G.11) and (G.10) having determined the functions $S_i(x_1)$, $U_i(x_1)$, $\Delta V_i(x_1)$ and $\Delta W_i(x_1)$ using (G.1)–(G.4), and then eliminating by subtraction the functions $V_{N+1}(x_1, x_2)$ and $W_{N+1}(x_1, x_2)$. It can be shown that the resulting differential equations are of the following form:

$$\begin{aligned}&\sum_{i=1}^{N} F_{ij} p_i^{(4)}(x_1) + \sum_{i=1}^{N} G_{ij} p_i^{(2)}(x_1) + \sum_{i=1}^{N} H_{ij} p_i(x_1) \\ &\quad + \sum_{i=1}^{N} c_{ij} q_i^{(2)}(x_1) + \sum_{i=1}^{N} d_{ij} q_i(x_1) = 0, \; j = 1, \ldots, N,\end{aligned} \tag{G.14}$$

$$\sum_{i=1}^{N} A_{ij} p_i^{(2)}(x_1) + \sum_{i=1}^{N} B_{ij} p_i(x_1) + \sum_{i=1}^{N} a_{ij} q_i^{(2)}(x_1) + \sum_{i=1}^{N} b_{ij} q_i(x_1) = 0, \; j = 1, \ldots, N, \tag{G.15}$$

where the coefficients in (G.14) and (G.15) are best determined numerically. In (G.14) and (G.15), $p^{(n)}(x_1)$ and $q^{(n)}(x_1)$, $n \geq 1$, replace the prime notation to denote the nth derivatives of the functions $p(x_1)$ and $q(x_1)$, respectively. It should be noted that the results (18.60) and (18.61) are equivalent to (G.12) and (G.13) because the differential equations (G.14) and (G.15) are satisfied.

3. Calculation of the Coefficients

The left-hand side of the jth equation of the set (G.14) is denoted by $\Phi_j(p, p^{(2)}, p^{(4)}, q, q^{(2)})$ where $p \equiv \{p_i,\ i = 1, ..., N\}$, having second and fourth derivatives denoted by $p^{(2)} \equiv \left\{p_i^{(2)},\ i=1,\ldots,N\right\}$ and $p^{(4)} \equiv \left\{p_i^{(4)},\ i=1,\ldots,N\right\}$, respectively, and similarly for functions q.

The numerical values of the coefficients F_{ij}, G_{ij} and H_{ij}, $i = 1, ..., N$, $j = 1, ..., N$, are then calculated simply by obtaining values of Φ_j, $j = 1, ..., N$, using suitable sets of values for the functions $p, p^{(2)}, p^{(4)}, q, q^{(2)}$. Thus,

$$\begin{aligned} F_{ij} &\equiv \Phi_j(0,0,I_i,0,0), \\ G_{ij} &\equiv \Phi_j(0,I_i,0,0,0), \quad i=1,\ldots,N,\ j=1,\ldots,N, \\ H_{ij} &\equiv \Phi_j(I_i,0,0,0,0), \end{aligned} \tag{G.16}$$

with $I_i \equiv \{\delta_{ik},\ k = 1, ..., N\}$ where δ_{ik} is the Kronecker delta, and with 0 denoting a corresponding set of zero values. Similarly, the values of the coefficients c_{ij}, d_{ij}, $j = 1, ..., N$, are obtained using

$$c_{ij} \equiv \Phi_j(0,0,0,0,I_i),\quad d_{ij} \equiv \Phi_j(0,0,0,I_i,0),\quad i=1,\ldots,N,\ j=1,\ldots,N. \tag{G.17}$$

A similar method is used to determine coefficients appearing in the differential equations (G.15).

4. Boundary Conditions for the Differential Equations

Figure 18.1 shows the region under consideration in the $x_1 - x_3$ plane for stress transfer between the $N + 1$ perfectly bonded layers. The objective now is to specify boundary conditions that will lead to a unique solution of the system of differential equations specified by (G.14) and (G.15). Such a solution can then be used to calculate the stress and displacement fields everywhere in the laminate using relations (18.43)–(18.51). Solutions symmetric about $x_1 = 0$ and $x_3 = 0$ are considered in the region $0 \le x_1 \le L$, $0 \le x_3 \le h$, so that only $3N$ boundary conditions are required. The functions $p_i(x_1)$ and $q_i(x_1)$ are then symmetric about $x_1 = 0$ and it is necessary to impose the loading boundary conditions only at $x_1 = L$.

As the shear stress σ_{13} is zero everywhere on $x_1 = 0$ and $x_1 = L$, it follows from (18.43) that

$$p_i'(0) = 0,\ p_i'(L) = 0,\ i = 1,\ldots,N. \tag{G.18}$$

On $x_1 = L$ it follows from (18.11) that for *cracked layers* the normal stress σ_{11} and shear stress σ_{12} are zero so that from (18.46) and (18.47) the following boundary conditions are obtained

$$p_i(L) - p_{i-1}(L) = h_i \sigma_{\mathrm{A}}^{(i)}, \tag{G.19}$$

$$q_i(L) - q_{i-1}(L) = h_i \tau_{\mathrm{A}}^{(i)}, \tag{G.20}$$

where $\sigma_{\mathrm{A}}^{(i)}$ and $\tau_{\mathrm{A}}^{(i)}$ are given by (18.25) and (18.26) and where i (taking values appropriate for cracked layers) lies in the range $1 \le i \le N+1$.

Consider now the boundary conditions that must be applied to *uncracked layers* in the laminate. Use is made of the boundary conditions (18.10) relating to displacement components for uncracked layers. As these conditions cannot be satisfied exactly by relations (18.50) and (18.51) because of the x_3 dependence, the following less-restrictive equivalent average conditions for uncracked layers are imposed on $x_1 = L$, namely

$$\bar{u}_1^{(i)}(L,x_2)=\varepsilon_A L+\frac{1}{2}\gamma_A x_2,\ \bar{u}_2^{(i)}(L,x_2)=\frac{1}{2}\gamma_A L+\varepsilon_T x_2, \tag{G.21}$$

where $\bar{u}_1^{(i)}$ and $\bar{u}_2^{(i)}$ are specified by (18.60) and (18.61), and where i (taking values appropriate for uncracked layers) lies in range $1 \le i \le N+1$.

When use is made of (G.1), (G.18), (18.60) and (18.61) together with the symmetry of the functions $p_i(x_1)$ and $q_i(x_1)$ about $x_1 = 0$, conditions (G.21) for *uncracked layers* may be expressed as

$$\varepsilon_A = \frac{g_{11}^{(i)}}{h_i L}\left[\overset{*}{p}_{i-1}(L)-\overset{*}{p}_i(L)\right]+\frac{g_{13}^{(i)}}{h_i L}\left[\overset{*}{q}_{i-1}(L)-\overset{*}{q}_i(L)\right]+\varepsilon_A^0, \tag{G.22}$$

$$\gamma_A = \frac{g_{13}^{(i)}}{h_i L}\left[\overset{*}{p}_{i-1}(L)-\overset{*}{p}_i(L)\right]+\frac{g_{33}^{(i)}}{h_i L}\left[\overset{*}{q}_{i-1}(L)-\overset{*}{q}_i(L)\right]+\gamma_A^0, \tag{G.23}$$

where i (taking values appropriate for uncracked layers) lies in the range $1 \le i \le N+1$. The parameters ε_A and γ_A are not yet known and therefore must be eliminated from these boundary conditions. It is convenient to let j be the label of the nearest uncracked layer to the centre plane of the laminate. It then follows by subtraction that the required boundary conditions for *uncracked layers* derived from (G.22) and (G.23) are of the form

$$\begin{aligned}&\frac{g_{11}^{(i)}}{h_i L}\left[\overset{*}{p}_{i-1}(L)-\overset{*}{p}_i(L)\right]-\frac{g_{11}^{(j)}}{h_j L}\left[\overset{*}{p}_{j-1}(L)-\overset{*}{p}_j(L)\right]\\&\quad+\frac{g_{13}^{(i)}}{h_i L}\left[\overset{*}{q}_{i-1}(L)-\overset{*}{q}_i(L)\right]-\frac{g_{13}^{(j)}}{h_j L}\left[\overset{*}{q}_{j-1}(L)-\overset{*}{q}_j(L)\right]=0,\end{aligned} \tag{G.24}$$

$$\begin{aligned}&\frac{g_{13}^{(i)}}{h_i L}\left[\overset{*}{p}_{i-1}(L)-\overset{*}{p}_i(L)\right]-\frac{g_{13}^{(j)}}{h_j L}\left[\overset{*}{p}_{j-1}(L)-\overset{*}{p}_j(L)\right]\\&\quad+\frac{g_{33}^{(i)}}{h_i L}\left[\overset{*}{q}_{i-1}(L)-\overset{*}{q}_i(L)\right]-\frac{g_{33}^{(j)}}{h_j L}\left[\overset{*}{q}_{j-1}(L)-\overset{*}{q}_j(L)\right]=0,\end{aligned} \tag{G.25}$$

where i and j (taking values appropriate for uncracked layers such that $i \neq j$) lie in the range $1 \le i, j \le N+1$.

Conditions (G.18)–(G.20), (G.24) and (G.25) provide the $3N$ boundary conditions required to ensure that the system of differential equations given by (G.14) and (G.15) has a unique solution that is symmetric about $x_1 = 0$. Once the solution of these differential equations is known satisfying the appropriate boundary conditions, the functions $V_{N+1}(x_1, x_2)$ and $W_{N+1}(x_1, x_2)$ are calculated using (G.8), (G.9), (G.12) and (G.13). The functions $V_i(x_1, x_2)$, $W_i(x_1, x_2)$, $i = 1, \ldots, N$, can then be found having used the recurrence relations (G.3) and (G.4) to determine values of the functions $\Delta V_i(x_1)$, $\Delta W_i(x_1)$, $i = 1, \ldots, N$. The method of solving numerically this system of differential equations is described in Appendix H.

Appendix H: Solution of Differential Equations

Semi-analytical numerical methods have been developed in Chapter 18 (see [3, 4]) and are used to solve the differential equations (G.14) and (G.15) involving the unknown functions $p_i(x_1)$ and $q_i(x_1)$, $i = 1, ..., N$. Solutions symmetric about $y = 0$ will now be obtained for values $-L \le x_1 \le L$ subject to the boundary conditions (G.18)–(G.25). The differential equations (G.14) require $4N$ boundary conditions and (G.15) require $2N$ boundary conditions, making $6N$ in total. As solutions are to be restricted so that they are symmetrical about $x_1 = 0$, only $3N$ conditions need to be imposed.

The first N boundary conditions are given by (G.18). Suppose there are c cracked and u uncracked ply elements. Clearly, $c+u=N+1$ as N is the number of interfaces between elements. Let $I_{\mathrm{p}}=\{1,2,\ldots,N+1\}$ denote all the ply elements in one half of the laminate, let $I_{\mathrm{c}}(\subset I_{\mathrm{p}})$ denote the indices for all the cracked elements and let $I_{\mathrm{u}}(\subset I_{\mathrm{p}})$ denote the set of indices for of the uncracked elements. It follows that $I_{\mathrm{p}}=I_{\mathrm{c}}\cup I_{\mathrm{u}}$ and $I_{\mathrm{c}}\cap I_{\mathrm{u}}=0$. For each of the cracked ply elements, conditions (G.19) and (G.20) are applied. There are clearly $2c$ of these conditions. The nearest uncracked ply element to the mid-plane of the laminate is identified and its ply element number is assigned to the integer variable j. For each of the uncracked ply elements apart from the jth, the boundary conditions (G.24) and (G.25) are imposed for all values of $i \ne j$. There are $2u-2$ of these boundary conditions leading to $N+2c+(2u-2)=N+2(c+u-1)=3N$, conditions in total, as required.

It should be noted that when the laminate is a cross-ply laminate the coefficients $c_{ij}, d_{ij}, \mathrm{A}_{ij}$ and B_{ij} all vanish with the result that the differential equations given by (G.14) and (G.15) separate into two independent systems of differential equations. The numerical solution to be described cannot be used when the laminate is a cross-ply, although very good approximations for cross-ply laminates can be obtained by offsetting the 0° plies by fractions of one degree. Alternatively, and more accurately, a solution technique especially designed for cross-ply laminates can be used.

1. The Auxiliary/Characteristic Equation

Using a matrix and differential operator notation, the differential equations (G.14) and (G.15) may be expressed as

$$\begin{aligned} &\boldsymbol{F}^{\mathrm{T}}\mathrm{D}^4\boldsymbol{p}+\boldsymbol{G}^{\mathrm{T}}\mathrm{D}^2\boldsymbol{p}+\boldsymbol{H}^{\mathrm{T}}\boldsymbol{p}+\boldsymbol{c}^{\mathrm{T}}\mathrm{D}^2\boldsymbol{q}+\boldsymbol{d}^{\mathrm{T}}\boldsymbol{q}=0,\\ &\boldsymbol{A}^{\mathrm{T}}\mathrm{D}^2\boldsymbol{p}+\boldsymbol{B}^{\mathrm{T}}\boldsymbol{p}+\boldsymbol{a}^{\mathrm{T}}\mathrm{D}^2\boldsymbol{q}+\boldsymbol{b}^{\mathrm{T}}\boldsymbol{q}=0, \end{aligned} \tag{H.1}$$

where $\boldsymbol{p} = (p_1(x_1), \ldots, p_N(x_1))^{\mathrm{T}}$ and $\boldsymbol{q} = (q_1(x_1), \ldots, q_N(x_1))^{\mathrm{T}}$ and D^i is the operator denoting the ith derivative. Combining these relations leads to following coupled system

$$\begin{pmatrix} \boldsymbol{F}^{\mathrm{T}}\mathrm{D}^4 + \boldsymbol{G}^{\mathrm{T}}\mathrm{D}^2 + \boldsymbol{H}^{\mathrm{T}} & \boldsymbol{c}^{\mathrm{T}}\mathrm{D}^2 + \boldsymbol{d}^{\mathrm{T}} \\ \boldsymbol{A}^{\mathrm{T}}\mathrm{D}^2 + \boldsymbol{B}^{\mathrm{T}} & \boldsymbol{a}^{\mathrm{T}}\mathrm{D}^2 + \boldsymbol{b}^{\mathrm{T}} \end{pmatrix} \begin{pmatrix} \boldsymbol{p} \\ \boldsymbol{q} \end{pmatrix} = 0. \tag{H.2}$$

Let the differential operator on the left-hand side of (H.2) be denoted by $\boldsymbol{M}(\mathrm{D})$, i.e.

$$\boldsymbol{M}(\mathrm{D}) \equiv \begin{pmatrix} \boldsymbol{F}^{\mathrm{T}}\mathrm{D}^4 + \boldsymbol{G}^{\mathrm{T}}\mathrm{D}^2 + \boldsymbol{H}^{\mathrm{T}} & \boldsymbol{c}^{\mathrm{T}}\mathrm{D}^2 + \boldsymbol{d}^{\mathrm{T}} \\ \boldsymbol{A}^{\mathrm{T}}\mathrm{D}^2 + \boldsymbol{B}^{\mathrm{T}} & \boldsymbol{a}^{\mathrm{T}}\mathrm{D}^2 + \boldsymbol{b}^{\mathrm{T}} \end{pmatrix}. \tag{H.3}$$

A solution of the system (H.2) is sought of the form

$$p_k(x_1) = P_k e^{\lambda x_1}, \quad q_k(x_1) = Q_k e^{\lambda x_1}, \quad k = 1, \ldots, N. \tag{H.4}$$

Substitution into (H.2) then leads to the linear algebraic equations

$$\begin{pmatrix} \boldsymbol{F}^{\mathrm{T}}\lambda^4 + \boldsymbol{G}^{\mathrm{T}}\lambda^2 + \boldsymbol{H}^{\mathrm{T}} & \boldsymbol{c}^{\mathrm{T}}\lambda^2 + \boldsymbol{d}^{\mathrm{T}} \\ \boldsymbol{A}^{\mathrm{T}}\lambda^2 + \boldsymbol{B}^{\mathrm{T}} & \boldsymbol{a}^{\mathrm{T}}\lambda^2 + \boldsymbol{b}^{\mathrm{T}} \end{pmatrix} \begin{pmatrix} \boldsymbol{P} \\ \boldsymbol{Q} \end{pmatrix} = \boldsymbol{0}, \tag{H.5}$$

where $\boldsymbol{P} = (P_1, \ldots, P_N)^{\mathrm{T}}$ and $\boldsymbol{Q} = (Q_1, \ldots, Q_N)^{\mathrm{T}}$. Non-trivial solutions of (H.5) exist only if the following characteristic equation for the coupled equations is satisfied

$$\left|\boldsymbol{M}(\lambda)\right| \equiv \det \begin{pmatrix} \boldsymbol{F}^{\mathrm{T}}\lambda^4 + \boldsymbol{G}^{\mathrm{T}}\lambda^2 + \boldsymbol{H}^{\mathrm{T}} & \boldsymbol{c}^{\mathrm{T}}\lambda^2 + \boldsymbol{d}^{\mathrm{T}} \\ \boldsymbol{A}^{\mathrm{T}}\lambda^2 + \boldsymbol{B}^{\mathrm{T}} & \boldsymbol{a}^{\mathrm{T}}\lambda^2 + \boldsymbol{b}^{\mathrm{T}} \end{pmatrix} = 0. \tag{H.6}$$

It now follows that more general solutions of (H.2) may be taken of the form

$$p_i(x_1) = \sum_{k=1}^{6N} P_k^i e^{\lambda_k x_1}, \quad q_i(x_1) = \sum_{k=1}^{6N} Q_k^i e^{\lambda_k x_1}, \quad i = 1, \ldots, N. \tag{H.7}$$

The characteristic equation (H.6) is a polynomial of degree $6N$ in λ. As the numerical solution of a polynomial having degree $6N$ is often unstable, a better approach is now described.

As the derivatives appearing in the coupled system are all even, the roots λ_k, $k = 1, \ldots, 6N$, appear as positive- and negative-valued pairs that may be arranged so that $\lambda_k = -\lambda_{k+3N}$, $k = 1, \ldots, 3N$. Having arranged the roots, which could be complex valued, in this way, parameters $\xi_k \equiv \lambda_k^2$ for $i=1, \ldots, 3N$, are introduced. The linear equations (H.5) may then be written as

$$\begin{pmatrix} \boldsymbol{F}^{\mathrm{T}}\xi^2 + \boldsymbol{G}^{\mathrm{T}}\xi + \boldsymbol{H}^{\mathrm{T}} & \boldsymbol{c}^{\mathrm{T}}\xi + \boldsymbol{d}^{\mathrm{T}} \\ \boldsymbol{A}^{\mathrm{T}}\xi + \boldsymbol{B}^{\mathrm{T}} & \boldsymbol{a}^{\mathrm{T}}\xi + \boldsymbol{b}^{\mathrm{T}} \end{pmatrix} \begin{pmatrix} \boldsymbol{P} \\ \boldsymbol{Q} \end{pmatrix} = \boldsymbol{0}. \tag{H.8}$$

Now define a vector R by

$$R = \xi P, \tag{H.9}$$

so that from (H.8)

$$\begin{aligned} \boldsymbol{H}^{\mathrm{T}}\boldsymbol{P}+\boldsymbol{d}^{\mathrm{T}}\boldsymbol{Q} &= -\xi\left(\boldsymbol{F}^{\mathrm{T}}\boldsymbol{R}+\boldsymbol{G}^{\mathrm{T}}\boldsymbol{P}+\boldsymbol{c}^{\mathrm{T}}\boldsymbol{Q}\right), \\ \boldsymbol{B}^{\mathrm{T}}\boldsymbol{P}+\boldsymbol{b}^{\mathrm{T}}\boldsymbol{Q} &= -\xi\left(\boldsymbol{A}^{\mathrm{T}}\boldsymbol{P}+\boldsymbol{a}^{\mathrm{T}}\boldsymbol{Q}\right). \end{aligned} \tag{H.10}$$

These equations are now written in the equivalent form

$$\begin{pmatrix} \boldsymbol{H}^{\mathrm{T}} & \boldsymbol{d}^{\mathrm{T}} & \boldsymbol{0} \\ \boldsymbol{B}^{\mathrm{T}} & \boldsymbol{b}^{\mathrm{T}} & \boldsymbol{0} \\ \boldsymbol{0} & \boldsymbol{0} & \mathbf{I} \end{pmatrix}\begin{pmatrix} \boldsymbol{P} \\ \boldsymbol{Q} \\ \boldsymbol{R} \end{pmatrix} = \xi\begin{pmatrix} -\boldsymbol{G}^{\mathrm{T}} & -\boldsymbol{c}^{\mathrm{T}} & -\boldsymbol{F}^{\mathrm{T}} \\ -\boldsymbol{A}^{\mathrm{T}} & -\boldsymbol{a}^{\mathrm{T}} & \boldsymbol{0} \\ \mathbf{I} & \boldsymbol{0} & \boldsymbol{0} \end{pmatrix}\begin{pmatrix} \boldsymbol{P} \\ \boldsymbol{Q} \\ \boldsymbol{R} \end{pmatrix}, \tag{H.11}$$

where **I** is the identity matrix. Non-trivial solutions exist only if

$$\det\left[\begin{pmatrix} \boldsymbol{H}^{\mathrm{T}} & \boldsymbol{d}^{\mathrm{T}} & \boldsymbol{0} \\ \boldsymbol{B}^{\mathrm{T}} & \boldsymbol{b}^{\mathrm{T}} & \boldsymbol{0} \\ \boldsymbol{0} & \boldsymbol{0} & \mathbf{I} \end{pmatrix} - \xi\begin{pmatrix} -\boldsymbol{G}^{\mathrm{T}} & -\boldsymbol{c}^{\mathrm{T}} & -\boldsymbol{F}^{\mathrm{T}} \\ -\boldsymbol{A}^{\mathrm{T}} & -\boldsymbol{a}^{\mathrm{T}} & \boldsymbol{0} \\ \mathbf{I} & \boldsymbol{0} & \boldsymbol{0} \end{pmatrix}\right] = 0, \tag{H.12}$$

which leads to a polynomial in ξ of degree $3N$. This is the generalised eigenvalue problem form of the characteristic equation that is solved numerically.

On solving the polynomial (H.12) to provide the roots ξ_k, $k = 1,\ldots,3N$, and hence values λ_k, $k = 1,\ldots,6N$, where $\lambda_k = \pm\left(\xi_k\right)^{1/2}$, solutions of the differential equations (G.14) and (G.15) may be taken having the form given by (H.7) provided that there are no coincident eigenvalues. This condition on eigenvalues is checked in software after solving (H.12) and has been found to be satisfied by input data for all simulations so far undertaken.

As only $3N$ of the eigenvalues λ_k are independent (the complete set of $6N$ appearing in positive and negative pairs), and from the symmetry of the solution about $x_1 = 0$, it follows that only $3N$ of the constants P_k^i and Q_k^i are independent. In (H.7), the coefficient of $\mathrm{e}^{\lambda_k x_1}$ must match that of $\mathrm{e}^{\lambda_{k+3N} x_1}$ for each of $k = 1, \ldots, 3N$. Therefore, the reduced series expansions of the solution may be written, on setting $Q_k^i = P_k^{N+i}$, as

$$p_i(x_1) = 2\sum_{k=1}^{3N} P_k^i \cosh(\lambda_k x_1), \quad q_i(x_1) = 2\sum_{k=1}^{3N} P_k^{N+i} \cosh(\lambda_k x_1), \quad i = 1,\ldots,N. \tag{H.13}$$

It should be noted that complex eigenvalues must appear as conjugate pairs.

Vectors X_k, $k = 1, \ldots, 3N$, are now introduced defined by

$$\boldsymbol{X}_k = \left(P_k^1, \ldots, P_k^{2N}\right)^{\mathrm{T}}, \quad k = 1, \ldots, 3N, \tag{H.14}$$

and the following $2N \times 2N$ matrices are evaluated

$$\boldsymbol{M}_k = \boldsymbol{M}\left(\lambda_k\right) = \begin{pmatrix} \boldsymbol{F}^{\mathrm{T}}\lambda_k^4 + \boldsymbol{G}^{\mathrm{T}}\lambda_k^2 + \boldsymbol{H}^{\mathrm{T}} & \boldsymbol{c}^{\mathrm{T}}\lambda_k^2 + \boldsymbol{d}^{\mathrm{T}} \\ \boldsymbol{A}^{\mathrm{T}}\lambda_k^2 + \boldsymbol{B}^{\mathrm{T}} & \boldsymbol{a}^{\mathrm{T}}\lambda_k^2 + \boldsymbol{b}^{\mathrm{T}} \end{pmatrix}, \quad k = 1, \ldots, 3N. \tag{H.15}$$

It then follows from the linear equations (H.5) that

$$\boldsymbol{M}_k \boldsymbol{X}_k = \boldsymbol{0}, \quad k = 1, \ldots, 3N. \tag{H.16}$$

The LU decomposition of the matrix M_k into lower and upper diagonal forms is now sought such that (H.16) may be written as

$$\boldsymbol{L}_k\boldsymbol{U}_k\boldsymbol{X}_k = \boldsymbol{0}, \quad k = 1, ..., 3N. \tag{H.17}$$

As L_k is always invertible and the system is homogeneous, (H.17) can be pre-multiplied by the inverse of L_k so that

$$\boldsymbol{U}_k\boldsymbol{X}_k = \boldsymbol{0}, \quad k = 1, ..., 3N. \tag{H.18}$$

For convenience, the components of the matrix U_k are denoted by $U_{i,j}^k$, $i, j = 1, ..., 2N$. The upper triangular form of U_k provides the following linear relationships between the constants

$$\sum_{j=i}^{2N} U_{i,j}^k P_k^j = 0, \; i = 1, ..., 2N-1, \; k = 1, ..., 3N. \tag{H.19}$$

The constants P_k^{2N} are for the moment arbitrary constants to be determined from boundary conditions. All other constants can be determined in terms of P_k^{2N} by back-substitution, as follows:

$$P_k^i = -\frac{1}{U_{i,i}^k}\sum_{j=i+1}^{2N} U_{i,j}^k P_k^j, \quad i = 2N-1, ...,1, \; k = 1, ..., 3N. \tag{H.20}$$

Thus, when $i = 2N - 1$,

$$P_k^{2N-1} = -\frac{U_{2N-1,2N}^k}{U_{2N-1,2N-1}^k} P_k^{2N} = \mu_{k,1} P_k^{2N}, \text{ where } \mu_{k,1} = -\frac{U_{2N-1,2N}^k}{U_{2N-1,2N-1}^k}, \quad k = 1, ..., 3N, \tag{H.21}$$

and, when $i = 2N - 2$,

$$P_k^{2N-2} = -\frac{1}{U_{2N-2,2N-2}^k}\left[U_{2N-2,2N-1}^k \mu_{k,1} + U_{2N-2,2N}^k\right] P_k^{2N} = \mu_{k,2} P_k^{2N}, \; k = 1,\ldots,N, \tag{H.22}$$

where

$$\mu_{k,2} = -\frac{1}{U_{2N-2,2N-2}^k}\left[U_{2N-2,2N-1}^k \mu_{k,1} + U_{2N-2,2N}^k\right], \quad k = 1,\ldots,3N. \tag{H.23}$$

The continuation of this procedure leads to the following recurrence relation

$$P_k^i = -\frac{1}{U_{i,i}^k}\left(\sum_{j=i+1}^{2N} U_{i,j}^k \mu_{k,2N-j}\right) P_k^{2N} = \mu_{k,2N-i} P_k^{2N}, \; i = 2N-1,\ldots,1, k = 1,\ldots, 3N, \tag{H.24}$$

and to the recurrence relation

$$\mu_{k,2N-i} = -\frac{1}{U_{i,i}^k}\left(\sum_{j=i+1}^{2N} U_{i,j}^k \mu_{k,2N-j}\right), i = 2N-1,\ldots,1, \text{ where } \mu_{k,0} = 1, k = 1,\ldots,3N. \tag{H.25}$$

The solution of the linear equations (H.16) has now been found where, for $k = 1, ..., 3N$, the $P_k^i, i = 1, ..., 2N-1$, have been expressed in terms of the arbitrary constants P_k^{2N} which are to be determined from boundary conditions.

2. Series Solution of the Differential Equations

On using (H.24), the unknown functions of the coupled differential equations (G.14) and (G.15) may now be represented by the following series that are derived from (H.13)

$$p_i(x_1) = 2\sum_{k=1}^{3N} R_k \mu_{k,2N-i} \cosh(\lambda_k x_1),\ q_i(x_1) = 2\sum_{k=1}^{3N} R_k \mu_{k,N-i} \cosh(\lambda_k x_1),\ i = 1,\ldots,N, \quad (H.26)$$

where R_k are arbitrary constants, which will be determined from boundary conditions, such that

$$R_k = P_k^{2N},\ k = 1,\ldots,3N. \quad (H.27)$$

Situations can arise where large eigenvalues, when combined with the use of the cosh function, can lead to exponential overflows during numerical analysis. To avoid such problems the representations (H.26) are first written as

$$p_i(x_1) = \sum_{k=1}^{3N} R_k \mu_{k,2N-i} \left(e^{\lambda_k x_1} + e^{-\lambda_k x_1} \right),\ q_i(x_1) = \sum_{k=1}^{3N} R_k \mu_{k,N-i} \left(e^{\lambda_k x_1} + e^{-\lambda_k x_1} \right),\ i = 1,\ldots,N. \quad (H.28)$$

As solutions are sought in the region $-L \le x_1 \le L$, it is useful to replace the arbitrary constants R_k by R_k^* where

$$R_k^* = R_k e^{\lambda_k L},\quad k = 1,\ldots,3N. \quad (H.29)$$

It then follows that

$$\left.\begin{aligned} p_i(x_1) &= \sum_{k=1}^{3N} R_k^* \mu_{k,2N-i} \left(e^{\lambda_k (x_1 - L)} + e^{-\lambda_k (x_1 + L)} \right), \\ q_i(x_1) &= \sum_{k=1}^{3N} R_k^* \mu_{k,N-i} \left(e^{\lambda_k (x_1 - L)} + e^{-\lambda_k (x_1 + L)} \right), \end{aligned}\right\} \quad i = 1,\ldots,N. \quad (H.30)$$

The nth derivatives of these functions are given by

$$\left.\begin{aligned} p_i^{(n)}(x_1) &= \sum_{k=1}^{3N} R_k^* \mu_{k,2N-i} \lambda_k^n \left(e^{\lambda_k (x_1 - L)} + (-1)^n e^{-\lambda_k (x_1 + L)} \right), \\ q_i^{(n)}(x_1) &= \sum_{k=1}^{3N} R_k^* \mu_{k,N-i} \lambda_k^n \left(e^{\lambda_k (x_1 - L)} + (-1)^n e^{-\lambda_k (x_1 + L)} \right), \end{aligned}\right\} \quad i = 1,\ldots,N, \quad (H.31)$$

and the following functions are required when imposing boundary conditions (see (18.62), (G.24) and (G.25))

$$\left.\begin{aligned} p_i^*(x_1) &= \sum_{k=1}^{3N} R_k^* \frac{\mu_{k,2N-i}}{\lambda_k} \left(e^{\lambda_k (x_1 - L)} - e^{-\lambda_k (x_1 + L)} \right), \\ q_i^*(x_1) &= \sum_{k=1}^{3N} R_k^* \frac{\mu_{k,N-i}}{\lambda_k} \left(e^{\lambda_k (x_1 - L)} - e^{-\lambda_k (x_1 + L)} \right), \end{aligned}\right\} \quad i = 1,\ldots,N. \quad (H.32)$$

Relations (H.30)–(H.32) are then substituted into the boundary conditions (G.18)–(G.20), (G.24) and (G.25), leading to $3N$ linear algebraic equations for the unknowns R_k^*, $k = 1,\ldots,3N$, which are solved numerically. The stress transfer functions may then be calculated using (H.30) enabling the distributions of displacement and stress components to be determined.

Appendix I: Energy Balance Equation for Delamination Growth

It is assumed that the ply crack density $\rho = 1/(2L)$ is fixed, and that the uniform temperature difference ΔT is prescribed and remains fixed. It is also assumed that the cracks are always open, or have frictionless contacts if there is some closure. For a region V of the damaged laminate, bounded by the surface S having dimensions $2L \times 2W \times 2h$ associated with just one ply crack and its delaminations, the energy balance equation at time t, in the absence of inertia effects, may be written as

$$16\gamma W \dot{c}(t) + \frac{\mathrm{d}}{\mathrm{d}t}\int\limits_V f(x_1,x_2,x_3,t)\mathrm{d}V = \int\limits_S \tau_j(x_1,x_2,x_3,t)\frac{\partial u_j(x_1,x_2,x_3,t)}{\partial t}\mathrm{d}S, \quad \text{(I.1)}$$

where $\dot{c}(t)$ is the speed of delamination growth, $f(x_1,x_2,x_3,t)$ is the non-uniform distribution at time t of the local Helmholtz energy per unit volume, $\tau_j(x_1,x_2,x_3,t)$ is the distribution of the jth applied traction component on the external surface S of the region V and $u_j(x_1,x_2,x_3,t)$ is the corresponding distribution of the displacement component. The first term on the left-hand side arises because account has to be taken of energy changes that occur when a new crack surface is formed as a result of delamination. The factor 16 appears in (I.1) because the region of the laminate considered, having width $2W$, has four delamination cracks each having fracture energy 2γ, where the factor 2 is included because two free surfaces are generated as each crack grows. The energy balance equation (I.1) is valid only for quasi-static conditions of delamination growth.

The boundary conditions applied on the external surface S are such that on the edges of the laminate the shear tractions are zero and the applied axial displacement $\delta_A(t)$ is uniform leading to an effective applied axial strain $\varepsilon_A(t)$ and a corresponding effective applied axial stress $\sigma_A(t)$, which is the average of the non-uniform traction distribution on laminate edges, as specified by (12.1). The applied transverse displacement $\delta_T(t)$ on the edges of the laminate is also uniform leading to an effective applied transverse strain $\varepsilon_T(t)$, and a corresponding effective applied transverse traction $\sigma_T(t)$ that is the average of the non-uniform local transverse traction distribution on laminate edges, as specified by (12.2). The faces of the laminate are assumed to be traction-free. Because uniform normal displacements $\delta_A(t)$, $\delta_T(t)$ and zero shear tractions have been applied on the laminate edges, and on using (12.1) and (12.2) together with the fact that the laminate faces are stress-free, the global energy balance equation (I.1) may be expressed in the form

$$\frac{16\gamma W}{V}\dot{c}(t) + \frac{dF}{dt} = \sigma_{\mathrm{A}}\frac{d\varepsilon_{\mathrm{A}}}{dt} + \sigma_{\mathrm{T}}\frac{d\varepsilon_{\mathrm{T}}}{dt}, \quad (\Delta T, \rho \text{ held fixed}), \tag{I.2}$$

where the global Helmholtz energy averaged per unit volume at time t is denoted by $F(t)$ and is defined by

$$F(t) = \frac{1}{V}\int_V f(x_1,x_2,x_3,t)\mathrm{d}V. \tag{I.3}$$

The local Helmholtz energy per unit volume $f(x_1,x_2,x_3,t)$ is a function of the local strain field $\varepsilon_{ij}(x_1,x_2,x_3,t)$ and the uniform fixed temperature T such that

$$f\left(x_1,x_2,x_3,t\right) = \hat{f}\left(\varepsilon_{ij}(x_1,x_2,x_3,t), T\right), \tag{I.4}$$

where

$$\mathrm{d}\hat{f} = -s\,\mathrm{d}T + \sigma_{ij}\,\mathrm{d}\varepsilon_{ij}, \quad s = -\frac{\partial\hat{f}(\varepsilon_{ij},T)}{\partial T}, \quad \sigma_{ij} = \frac{\partial\hat{f}(\varepsilon_{ij},T)}{\partial\varepsilon_{ij}}, \tag{I.5}$$

where σ_{ij} is the stress field corresponding to the strain field ε_{ij}, and where s is the local entropy per unit volume. Both σ_{ij} and s depend on the coordinates, time and the temperature.

On introducing the local Gibbs energy per unit volume $\hat{g}$ defined by

$$\hat{g} = \hat{f} - \sigma_{ij}\varepsilon_{ij}, \tag{I.6}$$

it follows from (I.5) that

$$\mathrm{d}\hat{g} = -s\,\mathrm{d}T - \varepsilon_{ij}\,\mathrm{d}\sigma_{ij}, \quad \text{where} \quad s = -\frac{\partial\hat{g}}{\partial T}, \quad \varepsilon_{ij} = -\frac{\partial\hat{g}}{\partial\sigma_{ij}}. \tag{I.7}$$

From (I.3) and (I.6), the global Gibbs energy averaged per unit volume at time t is denoted by $G(t)$ and is defined by

$$G(t) = \frac{1}{V}\int_V g(x_1,x_2,x_3,t)\mathrm{d}V = F(t) - \frac{1}{V}\int_V \sigma_{ij}(x_1,x_2,x_3,t)\varepsilon_{ij}(x_1,x_2,x_3,t)\mathrm{d}V. \tag{I.8}$$

On using the symmetry of the stress tensor, the strain–displacement relations, the equilibrium equations and the divergence theorem, it can be shown that

$$\int_V \sigma_{ij}\varepsilon_{ij}\,\mathrm{d}V = \int_S \tau_j u_j\,\mathrm{d}S = V\left[\sigma_{\mathrm{A}}(t)\varepsilon_{\mathrm{A}}(t) + \sigma_{\mathrm{T}}(t)\varepsilon_{\mathrm{T}}(t)\right]. \tag{I.9}$$

Relation (I.8) may then be expressed in the form

$$G(t) = F(t) - \sigma_{\mathrm{A}}(t)\varepsilon_{\mathrm{A}}(t) - \sigma_{\mathrm{T}}(t)\varepsilon_{\mathrm{T}}(t), \tag{I.10}$$

and substitution into (I.2) leads to the following alternative statement of the energy balance equation (I.1)

$$\frac{16\gamma W}{V}\dot{c}(t) + \frac{dG}{dt} = -\varepsilon_{\mathrm{A}}(t)\frac{d\sigma_{\mathrm{A}}}{dt} - \varepsilon_{\mathrm{T}}(t)\frac{d\sigma_{\mathrm{T}}}{dt}, \quad (\Delta T,\ \rho \text{ held fixed}). \tag{I.11}$$

For the assumed conditions of quasi-static delamination growth, the delamination length $c(t)$ is the only parameter that is directly dependent on time. The external loading parameters might depend on delamination length and, thus, be time-dependent indirectly. Solutions of the delamination problem, subject to boundary conditions where the prescribed loading parameters are independent of the delamination length, will provide a value for the Gibbs energy per unit volume that is of the form

$$G(t) = \widehat{G}\big(\sigma_A(c(t)), \sigma_T(c(t)), \Delta T, \rho, c(t)\big). \tag{I.12}$$

This form of the function $\widehat{G}$ may be derived using methods described in Appendix C on making use of the stress-strain relations (12.3) and (12.4) that result from the following relations

$$\varepsilon_A = -\frac{\partial \widehat{G}}{\partial \sigma_A}, \qquad \varepsilon_T = -\frac{\partial \widehat{G}}{\partial \sigma_T}. \tag{I.13}$$

For delamination growth, when both ρ and ΔT are fixed, it then follows from (I.12) that the energy balance equation (I.11) reduces to the form

$$\left(\frac{16\gamma W}{V} + \frac{\partial \widehat{G}}{\partial c}\right)\dot{c}(t) = 0\,, \quad (\Delta T,\ \rho \text{ held fixed}). \tag{I.14}$$

Delamination growth (i.e. $\dot{c}(t) > 0$) is possible only if

$$\frac{16\gamma W}{V} + \frac{\partial \widehat{G}}{\partial c} = 0\,, \quad (\Delta T,\ \rho \text{ held fixed}). \tag{I.15}$$

This relation, which is valid for all delamination lengths and ply crack densities, will be the basis of a study of delamination governed by energy balance considerations for conditions of self-similar delamination growth.

Appendix J: Derivation of Energy-based Fracture Criterion for Bridged Cracks

Consider a large plate of uniform thickness $2h$ at time t, which is subject to fixed complex loading, having an embedded plane crack of variable length $2c(t)$ that is small enough for interactions with the external boundary to be negligible. The geometry is shown in Figure J.1 where part of the external boundary S, denoted by S_T, is subject to a prescribed nonuniform distribution $\hat{\tau}_j$ of tractions whereas the remainder of the surface S, denoted by S_u, is subject to a nonuniform prescribed displacement distribution $\hat{u}_j$. The external surface S bounds the plate of arbitrary shape forming a region that is denoted by V. The origin of a rectangular Cartesian system of coordinates (x_1, x_2, x_3) is located at some point of the crack. At any time t, the location of the right-hand crack tip is $x_1 = c^+(t)$ whereas the left-hand tip is at $x_1 = -c^-(t)$. Thus, the crack occupies the planar region $-c^-(t) \leq x_1 \leq c^+(t)$, $x_2 = 0$, $|x_3| \leq h$ and its total length is $2c(t) = c^+(t) + c^-(t)$. The crack surfaces are assumed to be subject to applied normal and shear tractions described by the functions $p(x_1)$ and $q(x_1)$, respectively, such that they are both continuous across the crack at all points. These tractions can represent the bridging effects of intact fibres that cross the crack or of cohesive forces, or of friction.

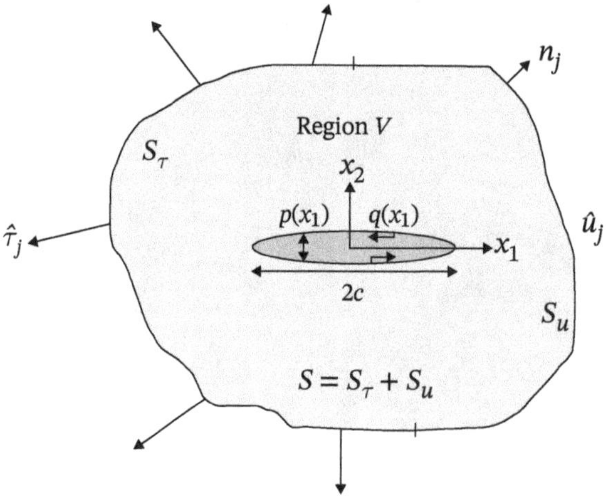

Figure J1 Schematic diagram of a bridged crack embedded in a plate of arbitrary shape subject to normal and shear crack tractions and tractions/displacements applied externally.

In the absence of inertia effects, the global energy balance equation has the form

$$4\gamma\dot{c}(t) + \frac{\mathrm{d}}{\mathrm{d}t}\int_V f\,\mathrm{d}V = \int_{S_\tau}\hat{\tau}_j\frac{\partial u_j}{\partial t}\mathrm{d}S - \int_{-c^-(t)}^{c^+(t)} p\frac{\partial\Delta u_2}{\partial t}\mathrm{d}x_1 - \int_{-c^-(t)}^{c^+(t)} q\frac{\partial\Delta u_1}{\partial t}\mathrm{d}x_1, \quad (\mathrm{J.1})$$

where

2γ = effective fracture energy,
$\dot{c}(t)$ = rate of increase of half crack length,
f = Helmholtz energy per unit volume,
$\Delta u_1 = u_1(x_1,0+,t) - u_1(x_1,0-,t)$,
$\Delta u_2 = u_2(x_1,0+,t) - u_2(x_1,0-,t)$.

The first term on the left-hand side of (J.1) accounts for the energy that is required to form new fracture surface, the factor 4 appearing as there are two crack tips that extend creating new crack surface at four different locations. The second term accounts for the change in stored energy which is written in terms of Helmholtz energy rather than strain energy as thermal residual stresses are included in the analysis. The first term on the right-hand side is the rate of working of the applied tractions $\hat{\tau}_j$ which are applied on the part S_τ of the external boundary S. The displacements $\hat{u}_j$ applied to the remainder of the surface S are fixed and do no work. The second and third terms on the right-hand side of (J.1) take into account the rate of working of the tractions applied to the crack surfaces which result from the mechanical effect of bridging ligaments or cohesive forces. The form of the energy balance equation assumes that the temperature is held fixed at all points in the region V.

For a linear anisotropic thermoelastic solid, the Helmholtz energy per unit volume (assuming infinitesimal deformations) has the form

$$f = \frac{1}{2}C_{ijkl}\,\varepsilon_{ij}\,\varepsilon_{kl} - \alpha_{ij}\,\varepsilon_{ij}\,\Delta T, \quad (\mathrm{J.2})$$

where ε_{ij} is the infinitesimal strain tensor derived from the displacement vector u_j using the relation

$$\varepsilon_{ij} = \frac{1}{2}\left(\frac{\partial u_i}{\partial x_j} + \frac{\partial u_j}{\partial x_i}\right) = \varepsilon_{ji}, \quad (\mathrm{J.3})$$

and where C_{ijkl} and α_{ij} are the elastic constants and thermal expansion coefficients of an anisotropic material, respectively. The parameter $\Delta T = T - T_0$ is the difference between the temperature T of the system and a reference temperature T_0 for which the strains are everywhere zero when the material is unloaded. Strictly speaking, relation (J.2) should include a term that depends only on the temperature difference ΔT, but, as the temperature is held fixed during crack growth, there is no loss of generality in assuming that the additional term is zero. The stress tensor σ_{ij}, corresponding to the strain field ε_{ij} at temperature ΔT, is obtained by differentiating (J.2) so that

$$\sigma_{ij} \equiv \frac{\partial f}{\partial\varepsilon_{ij}} = C_{ijkl}\,\varepsilon_{kl} - \alpha_{ij}\,\Delta T, \quad (\mathrm{J.4})$$

which is a general form for the linear anisotropic stress-strain relations.

When deriving (J.4), use has been made of the well-known symmetry properties of the tensors σ_{ij} and ε_{ij}, and of the thermoelastic constants, namely $C_{ijkl} = C_{klij}$ and $\alpha_{ij} = \alpha_{ji}$. The stress tensor must satisfy the equilibrium equations (for conditions of quasistatic crack growth where inertia effects are neglected, and in the absence of body forces such as gravity) so that

$$\frac{\partial \sigma_{ij}}{\partial x_i} = 0. \tag{J.5}$$

The objective now is to derive a local form of the energy balance equation (J.1) assuming that the prescribed applied tractions $\hat{\tau}_j$ and displacements $\hat{u}_j$ are held fixed during crack growth. On using (J.4), relation (J.2) can be written in the form

$$f = \frac{1}{2}\sigma_{ij}\,\varepsilon_{ij} - \frac{1}{2}\alpha_{ij}\,\varepsilon_{ij}\,\Delta T. \tag{J.6}$$

On using (J.3), the stress-strain product in the first term of (J.6) is now expressed as

$$\sigma_{ij}\,\varepsilon_{ij} = \frac{1}{2}\sigma_{ij}\left(\frac{\partial u_i}{\partial x_j} + \frac{\partial u_j}{\partial x_i}\right) = \sigma_{ij}\frac{\partial u_j}{\partial x_i} = \frac{\partial}{\partial x_i}(\sigma_{ij}u_j) - u_j\frac{\partial \sigma_{ij}}{\partial x_i} = \frac{\partial}{\partial x_i}(\sigma_{ij}u_j), \tag{J.7}$$

where use has been made of the symmetry of the stress tensor and the equilibrium equations (J.5).

On integration over the region V, it can be shown, on using the divergence theorem, that

$$\begin{aligned}\int_V \sigma_{ij}\,\varepsilon_{ij}\,dV = \int_V \frac{\partial}{\partial x_i}(\sigma_{ij}u_j)\,dV = \int_{S_\tau} \hat{\tau}_j u_j\,dS + \int_{S_u} \tau_j \hat{u}_j\,dS \\ - \int_{-c^-(t)}^{c^+(t)} p\Delta u_2\,dx_1 - \int_{-c^-(t)}^{c^+(t)} q\Delta u_1\,dx_1,\end{aligned} \tag{J.8}$$

and that

$$\begin{aligned}\int_V \alpha_{ij}\,\varepsilon_{ij}\,dV = \int_V \frac{\partial}{\partial x_i}(\alpha_{ij}u_j)\,dV = \int_{S_\tau} n_i\alpha_{ij}u_j\,dS + \int_{S_u} n_i\alpha_{ij}\hat{u}_j\,dS \\ - \int_{-c^-(t)}^{c^+(t)} \alpha_{22}\Delta u_2\,dx_1 - \int_{-c^-(t)}^{c^+(t)} \alpha_{12}\Delta u_1\,dx_1,\end{aligned} \tag{J.9}$$

where n_i is the outward unit normal to the bounding surface S and crack surfaces, and where $\tau_j = n_i\,\sigma_{ij}$. On integrating (J.6) over the region V using (J.8) and (J.9), it then follows that

$$\begin{aligned}\int_V f\,dV = \frac{1}{2}\int_{S_\tau} \hat{\tau}_j u_j\,dS + \frac{1}{2}\int_{S_u} \tau_j \hat{u}_j\,dS - \frac{1}{2}\int_{-c^-(t)}^{c^+(t)} p\Delta u_2 dx_1 - \frac{1}{2}\int_{-c^-(t)}^{c^+(t)} q\Delta u_1 dx_1 \\ -\frac{1}{2}\int_{S_\tau} n_i\alpha_{ij}\Delta T u_j\,dS - \frac{1}{2}\int_{S_u} n_i\alpha_{ij}\Delta T\,\hat{u}_j\,dS + \frac{1}{2}\int_{-c^-(t)}^{c^+(t)} \alpha_{22}\Delta T\,\Delta u_2\,dx_1 + \frac{1}{2}\int_{-c^-(t)}^{c^+(t)} \alpha_{12}\Delta T\,\Delta u_1 dx_1.\end{aligned} \tag{J.10}$$

The terms in (J.10), which are integrals over the crack, arise because the displacement field is discontinuous across the crack whereas the corresponding stress field is continuous. On making use of the strain–displacement relations, the symmetry of the stress tensor and the divergence theorem, it can also be shown that

$$\int_V \left(\sigma_{ij} + \alpha_{ij}\Delta T\right)\varepsilon_{ij}^0 \,\mathrm{d}V = \int_{S_\tau} \left(\hat{\tau}_j + n_i\alpha_{ij}\Delta T\right)u_j^0 \,\mathrm{d}S + \int_{S_u} \left(\tau_j + n_i\alpha_{ij}\Delta T\right)\hat{u}_j \,\mathrm{d}S, \quad \text{(J.11)}$$

$$\begin{aligned}\int_V \left(\sigma_{ij}^0 + \alpha_{ij}\Delta T\right)\varepsilon_{ij}\,\mathrm{d}V &= \int_{S_\tau} \left(\hat{\tau}_j + n_i\alpha_{ij}\Delta T\right)u_j\,\mathrm{d}S + \int_{S_u} \left(\tau_j^0 + n_i\alpha_{ij}\Delta T\right)\hat{u}_j\,\mathrm{d}S \\ &- \int_{-c^-(t)}^{c^+(t)} \left(p + \alpha_{22}\Delta T\right)\Delta u_2\,\mathrm{d}x_1 - \int_{-c^-(t)}^{c^+(t)} \left(q + \alpha_{12}\Delta T\right)\Delta u_1\,\mathrm{d}x_1,\end{aligned} \quad \text{(J.12)}$$

where $\tau_j^0 \equiv n_i\sigma_{ij}^0$, ε_{ij}^0 and u_j^0 are the traction, strain and displacement distributions for the special case when the crack is absent and the plate is subject to the same applied tractions $\hat{\tau}_j$ and displacements $\hat{u}_j$ as for the cracked plate. Owing to the symmetry relation $C_{ijkl} = C_{klij}$, it follows that

$$\left(\sigma_{ij}^0 + \alpha_{ij}\Delta T\right)\varepsilon_{ij} = \left(\sigma_{ij} + \alpha_{ij}\Delta T\right)\varepsilon_{ij}^0, \quad \text{(J.13)}$$

and a subtraction of (J.11) and (J.12) leads to

$$\begin{aligned}\int_{S_\tau} \left(\hat{\tau}_j + n_i\alpha_{ij}\Delta T\right)\left(u_j - u_j^0\right)\mathrm{d}S - \int_{S_u} \left(\tau_j - \tau_j^0\right)\hat{u}_j\,\mathrm{d}S \\ = \int_{-c^-(t)}^{c^+(t)} \left(\sigma_{22}^0 + \alpha_{22}\Delta T\right)\Delta u_2\,\mathrm{d}x_1 + \int_{-c^-(t)}^{c^+(t)} \left(\sigma_{12}^0 + \alpha_{12}\Delta T\right)\Delta u_1\,\mathrm{d}x_1.\end{aligned} \quad \text{(J.14)}$$

Relations (J.10) and (J.14) then assert that

$$\begin{aligned}\int_V f\mathrm{d}V - \int_{S_\tau} \hat{\tau}_j u_j\,\mathrm{d}S &= \frac{1}{2}\int_{S_u} \left(\tau_j^0 - n_i\alpha_{ij}\Delta T\right)\hat{u}_j\,\mathrm{d}S - \frac{1}{2}\int_{S_\tau} \left(\hat{\tau}_j + n_i\alpha_{ij}\Delta T\right)u_j^0\mathrm{d}S \\ &- \frac{1}{2}\int_{-c^-(t)}^{c^+(t)} \left(\sigma_{22}^0 + p\right)\Delta u_2\mathrm{d}x_1 - \frac{1}{2}\int_{-c^-(t)}^{c^+(t)} \left(\sigma_{12}^0 + q\right)\Delta u_1\mathrm{d}x_1.\end{aligned} \quad \text{(J.15)}$$

As the parameters $\hat{\tau}_j, \hat{u}_j, \tau_j^0, u_j^0, \sigma_{12}^0, \sigma_{22}^0$ and ΔT are all independent of time when considering time-dependent crack growth, and because the crack is closed at its ends $x_1 = \pm c^{\pm}(t)$, it follows on differentiating (J.15) that

$$\begin{aligned}\frac{\mathrm{d}}{\mathrm{d}t}\int_V f\mathrm{d}V - \int_{S_\tau} \hat{\tau}_j \frac{\partial u_j}{\partial t}\mathrm{d}S &= -\frac{1}{2}\int_{-c^-(t)}^{c^+(t)} \left(\sigma_{22}^0 + p\right)\frac{\partial \Delta u_2}{\partial t}\mathrm{d}x_1 - \frac{1}{2}\int_{-c^-(t)}^{c^+(t)} \left(\sigma_{12}^0 + q\right)\frac{\partial \Delta u_1}{\partial t}\mathrm{d}x_1 \\ &- \frac{1}{2}\int_{-c^-(t)}^{c^+(t)} \frac{\partial p}{\partial t}\Delta u_2\,\mathrm{d}x_1 - \frac{1}{2}\int_{-c^-(t)}^{c^+(t)} \frac{\partial q}{\partial t}\Delta u_1\,\mathrm{d}x_1.\end{aligned} \quad \text{(J.16)}$$

Substitution into the energy balance equation (J.1) then leads to the result

$$\begin{aligned}8\gamma\dot{c}(t) &= \int_{-c^-(t)}^{c^+(t)} \left(\sigma_{22}^0 - p\right)\frac{\partial \Delta u_2}{\partial t}\mathrm{d}x_1 + \int_{-c^-(t)}^{c^+(t)} \left(\sigma_{12}^0 - q\right)\frac{\partial \Delta u_1}{\partial t}\mathrm{d}x_1 \\ &+ \int_{-c^-(t)}^{c^+(t)} \frac{\partial p}{\partial t}\Delta u_2\,\mathrm{d}x_1 + \int_{-c^-(t)}^{c^+(t)} \frac{\partial q}{\partial t}\Delta u_1\,\mathrm{d}x_1.\end{aligned} \quad \text{(J.17)}$$

The global energy balance equation (J.1) has, thus, been reduced to the form (J.17) involving only integrals over the crack surfaces, the volume and surface integrals having been eliminated by introducing the stress and displacement distributions for the plate in an uncracked state, but subject to the same loading conditions as the cracked plate.

To proceed further it is necessary to investigate the dependence of the functions p, q, Δu_1 and Δu_2 on the location x_1 in the crack and on the time t. Because the applied tractions $\hat{\tau}_j$, displacements $\hat{u}_j$ and temperature difference ΔT are held fixed during crack growth, and on assuming that inertia effects are negligible, it follows that

$$\begin{aligned} p &= p\left(x_1, c^+(t), c^-(t)\right), & q &= q\left(x_1, c^+(t), c^-(t)\right), \\ \Delta u_1 &= \Delta u_1\left(x_1, c^+(t), c^-(t)\right), & \Delta u_2 &= \Delta u_2\left(x_1, c^+(t), c^-(t)\right). \end{aligned}$$

As $2\dot{c}(t) = \dot{c}^+(t) + \dot{c}^-(t)$, result (J.17) may then be expressed as

$$\begin{aligned} &\left[4\gamma - \int_{-c^-(t)}^{c^+(t)} \left(\sigma_{22}^0 - p\right)\frac{\partial \Delta u_2}{\partial c^+}\,\mathrm{d}x_1 - \int_{-c^-(t)}^{c^+(t)} \left(\sigma_{22}^0 - q\right)\frac{\partial \Delta u_1}{\partial c^+}\,\mathrm{d}x_1 \right.\\ &\qquad \left. - \int_{-c^-(t)}^{c^+(t)} \frac{\partial p}{\partial c^+}\Delta u_2\,\mathrm{d}x_1 - \int_{-c^-(t)}^{c^+(t)} \frac{\partial q}{\partial c^+}\Delta u_1\,\mathrm{d}x_1 \right]\dot{c}^+(t) \\ &+ \left[4\gamma - \int_{-c^-(t)}^{c^+(t)} \left(\sigma_{22}^0 - p\right)\frac{\partial \Delta u_2}{\partial c^-}\,\mathrm{d}x_1 - \int_{-c^-(t)}^{c^+(t)} \left(\sigma_{22}^0 - q\right)\frac{\partial \Delta u_1}{\partial c^-}\,\mathrm{d}x_1 \right.\\ &\qquad \left. - \int_{-c^-(t)}^{c^+(t)} \frac{\partial p}{\partial c^-}\Delta u_2\,\mathrm{d}x_1 - \int_{-c^-(t)}^{c^+(t)} \frac{\partial q}{\partial c^-}\Delta u_1\,\mathrm{d}x_1 \right]\dot{c}^-(t) = 0\,. \end{aligned} \tag{J.18}$$

It then follows that energy balance principles allow crack growth at the right-hand tip, i.e. $\dot{c}^+(t) > 0$, $\dot{c}^-(t) = 0$, only if

$$\begin{aligned} &\int_{-c^-(t)}^{c^+(t)} \left(\sigma_{22}^0 - p\right)\frac{\partial \Delta u_2}{\partial c^+}\,\mathrm{d}x_1 + \int_{-c^-(t)}^{c^+(t)} \left(\sigma_{22}^0 - q\right)\frac{\partial \Delta u_1}{\partial c^+}\,\mathrm{d}x_1 \\ &\qquad + \int_{-c^-(t)}^{c^+(t)} \frac{\partial p}{\partial c^+}\Delta u_2\,\mathrm{d}x_1 + \int_{-c^-(t)}^{c^+(t)} \frac{\partial q}{\partial c^+}\Delta u_1\,\mathrm{d}x_1 = 4\gamma\,. \end{aligned} \tag{J.19}$$

Similarly, crack growth at the left-hand tip, i.e. $\dot{c}^+(t) = 0$, $\dot{c}^-(t) > 0$, can occur only if

$$\begin{aligned} &\int_{-c^-(t)}^{c^+(t)} \left(\sigma_{22}^0 - p\right)\frac{\partial \Delta u_2}{\partial c^-}\,\mathrm{d}x_1 + \int_{-c^-(t)}^{c^+(t)} \left(\sigma_{22}^0 - q\right)\frac{\partial \Delta u_1}{\partial c^-}\,\mathrm{d}x_1 \\ &\qquad + \int_{-c^-(t)}^{c^+(t)} \frac{\partial p}{\partial c^-}\Delta u_2\,\mathrm{d}x_1 + \int_{-c^-(t)}^{c^+(t)} \frac{\partial q}{\partial c^-}\Delta u_1\,\mathrm{d}x_1 = 4\gamma\,. \end{aligned} \tag{J.20}$$

For bridged crack problems which are geometrically symmetrical about the plane $x_1 = 0$, $c^+(t) = c^-(t) = c(t)$ and, provided the loading is also symmetrical about this plane,

$$p = p(x_1, c(t)), \qquad q = q(x_1, c(t)),$$
$$\Delta u_1 = \Delta u_1(x_1, c(t)), \quad \Delta u_2 = \Delta u_2(x_1, c(t)).$$

It follows from (J.17) that

$$\left[4\gamma - \int_0^{c(t)} \left(\sigma_{22}^0 - p\right) \frac{\partial \Delta u_2}{\partial c} \mathrm{d}x_1 - \int_0^{c(t)} \left(\sigma_{22}^0 - q\right) \frac{\partial \Delta u_1}{\partial c} \mathrm{d}x_1 \right.$$
$$\left. - \int_0^{c(t)} \frac{\partial p}{\partial c} \Delta u_2 \,\mathrm{d}x_1 - \int_0^{c(t)} \frac{\partial q}{\partial c} \Delta u_1 \,\mathrm{d}x_1 \right] \dot{c}(t) = 0. \tag{J.21}$$

Crack growth, i.e. $\dot{c}(t) > 0$, can occur only if

$$\int_0^{c(t)} \left(\sigma_{22}^0 - p\right) \frac{\partial \Delta u_2}{\partial c} \mathrm{d}x_1 + \int_0^{c(t)} \left(\sigma_{12}^0 - q\right) \frac{\partial \Delta u_1}{\partial c} \mathrm{d}x_1 + \int_0^{c(t)} \frac{\partial p}{\partial c} \Delta u_2 \,\mathrm{d}x_1 + \int_0^{c(t)} \frac{\partial q}{\partial c} \Delta u_1 \,\mathrm{d}x_1 = 4\gamma. \tag{J.22}$$

For many applications considered in this book, shear loading is absent and crack opening occurs only in mode I so that $\Delta u_1 \equiv 0$, $q \equiv 0$ and $\sigma_{12}^0 = 0$. The fracture criterion (J.22) then reduces to the form

$$\int_0^{c(t)} \left(\sigma_{22}^0 - p\right) \frac{\partial \Delta u_2}{\partial c} \mathrm{d}x_1 + \int_0^{c(t)} \frac{\partial p}{\partial c} \Delta u_2 \,\mathrm{d}x_1 = 4\gamma. \tag{J.23}$$

Appendix K: Numerical Solution of Integral Equations for Bridged Cracks

Two types of integral equation are encountered in this book. The first is nonlinear and arises from a simplified model of fibre/matrix debonding (a weak interface) in the presence of a matrix crack embedded in a uniaxial fibre reinforced composite. The second type is linear and arises when there is perfect fibre/matrix bonding (i.e. a very strong interface). These two types will be considered separately as the solution approach differs.

1. Nonlinear Case

For a given value of μ, and on writing $P(X,\mu)=P(X)$, the nonlinear integral equation (16.184) that governs crack bridging for a matrix crack in a unidirectional composite may be written as

$$\left(P(X)\right)^2=\mu\left[\sqrt{1-X^2}-\int_0^1 P(t)k(t,X)\mathrm{d}t\right],\ 0<X<1. \tag{K.1}$$

By making use of relations (16.182) and (16.183), this integral equation is first written in the following form for values $0<X<1$

$$(P(X))^2=\mu\sqrt{1-X^2}\,(1-P(X))-\frac{\mu}{\pi}\int_0^1\{P(t)-P(X)\}\ln\left|\frac{\sqrt{1-t^2}+\sqrt{1-X^2}}{\sqrt{1-t^2}-\sqrt{1-X^2}}\right|\mathrm{d}t, \tag{K.2}$$

which removes the logarithmic singularity in the integrand at the point $t=X$. It is clear on dividing (K.2) by μ and then taking the limit $\mu\to\infty$, the solution of the integral equation is asymptotic to

$$P(X)=1,\qquad \text{for}\qquad 0<X<1. \tag{K.3}$$

It is convenient to introduce the function G defined by

$$G(X)\equiv 1-P(X), \tag{K.4}$$

and to consider the following integral equation that follows from (K.2) for values $0<X<1$

$$\left(G(X)\right)^2-\left\{2+\mu\sqrt{1-X^2}\right\}G(X)+1=\frac{\mu}{\pi}\int_0^1\left\{G(t)-G(X)\right\}\ln\left|\frac{\sqrt{1-t^2}+\sqrt{1-X^2}}{\sqrt{1-t^2}-\sqrt{1-X^2}}\right|\mathrm{d}t. \quad \text{(K.5)}$$

The following substitutions into (K.5) are now made

$$t=\frac{1-u^4}{1+u^4}, \qquad X=\frac{1-v^4}{1+v^4}, \qquad G(t)\equiv g(u), \quad \text{(K.6)}$$

so that for values $0<v<1$

$$\left(g(v)\right)^2-\left\{2+\mu S(v)\right\}g(v)+1=\frac{\mu}{\pi}\int_0^1\left\{g(u)-g(v)\right\}\ln\left|\frac{S(u)+S(v)}{S(u)-S(v)}\right|\frac{u^3\,\mathrm{d}u}{(1+u^4)^2}, \quad \text{(K.7)}$$

where

$$S(u)=\frac{2u^2}{1+u^4}\equiv\sqrt{1-t^2}. \quad \text{(K.8)}$$

The function $Y(\mu)$ defined by (16.185) associated with the stress intensity factor for a bridged crack may be written as

$$Y=\frac{8}{\pi}\int_0^1\frac{u\,g(u)\mathrm{d}u}{1+u^4}. \quad \text{(K.9)}$$

The integral equation (K.7) is solved numerically by iteration using the following discrete formulation:

$$\left(g_i^{(N+1)}\right)^2-\left\{2+\mu S_i\right\}g_i^{(N+1)}+1=\mu\sum_{\substack{j=0\\ j\neq i}}^{n}\left\{g_j^{(N)}-g_i^{(N)}\right\}K_{ij},\ i=0,\ldots,n,\ N\geq 0, \quad \text{(K.10)}$$

where

$$g_i=g(i/n), \quad S_i=\frac{2(i/n)^2}{1+(i/n)^4}, \quad g_i^{(0)}=0, \quad i=0,\ldots,n, \quad \text{(K.11)}$$

and

$$K_{ij}=\frac{8\delta_j}{\pi n^4}\ln\left|\frac{S_j+S_i}{S_j-S_i}\right|\frac{j^3}{\left(1+(j/n)^4\right)^2}, \quad \text{(K.12)}$$

with

$$\delta_0=\delta_n=\frac{1}{3}, \quad \delta_j=\left\{\begin{array}{l}\frac{4}{3}\ \text{if } j \text{ is odd}\\[4pt] \frac{2}{3}\ \text{if } j \text{ is even}\end{array}\right\}, \quad j=1,\ldots,n-1. \quad \text{(K.13)}$$

This discrete form of the integral equation is based on Simpson's integration formula and it assumes that n is even.

When calculating a new solution $g_i^{(N+1)}$, $i=0,\ldots,n$, using (K.10) it is necessary to solve a quadratic equation of the form

$$g^2 - (2 + \mu s)g + 1 - \kappa = 0. \tag{K.14}$$

The required solution such that $0 < g < 1$ is given by

$$g = 1 + \frac{1}{2}\mu s - \sqrt{\mu s + (\tfrac{1}{2}\mu s)^2 + \kappa}. \tag{K.15}$$

The convergence criterion for the iteration procedure if given by

$$\sqrt{\frac{1}{n+1}\sum_{i=0}^{n}\left\{g_i^{(N+1)} - g_i^{(N)}\right\}^2} < \varepsilon, \tag{K.16}$$

where the parameter $\varepsilon = 10^{-8}$.

2. Linear Case

For a given value of μ, and on writing $P(X,\mu) = P(X)$, the linear integral equation (17.22) that governs crack bridging for a matrix crack in a unidirectional composite may be written as

$$P(X) = \mu\left[\sqrt{1-X^2} - \int_0^1 P(t)k(t,X)\mathrm{d}t\right], 0 < X < 1. \tag{K.17}$$

By making use of relations (16.182) and (16.183), this integral equation is first written in the following form for values $0 < X < 1$

$$P(X) = \mu\sqrt{1-X^2}\,(1-P(X)) - \frac{\mu}{\pi}\int_0^1\{P(t)-P(X)\}\ln\left|\frac{\sqrt{1-t^2}+\sqrt{1-X^2}}{\sqrt{1-t^2}-\sqrt{1-X^2}}\right|\mathrm{d}t, \tag{K.18}$$

which removes the logarithmic singularity in the integrand at the point. As before, it is clear on dividing (K.2) by μ and then taking the limit $\mu \to \infty$, the solution of the integral equation is asymptotic to

$$P(X) = 1, \quad \text{for} \quad 0 < X < 1. \tag{K.19}$$

It is convenient to introduce the function G defined by

$$G(X) \equiv 1 - P(X), \tag{K.20}$$

and to consider the following integral equation that follows from (K.2) for values $0 < X < 1$

$$1 - \left\{1 + \mu\sqrt{1-X^2}\right\}G(X) = \frac{\mu}{\pi}\int_0^1\{G(t) - G(X)\}\ln\left|\frac{\sqrt{1-t^2}+\sqrt{1-X^2}}{\sqrt{1-t^2}-\sqrt{1-X^2}}\right|\mathrm{d}t. \tag{K.21}$$

The substitutions (K.6) are again made so that for values $0 < v < 1$

$$1 - \{1 + \mu S(v)\}g(v) = \frac{\mu}{\pi}\int_0^1\{g(u) - g(v)\}\ln\left|\frac{S(u)+S(v)}{S(u)-S(v)}\right|\frac{u^3\,\mathrm{d}u}{(1+u^4)^2}. \tag{K.22}$$

where the function $S(u)$ is defined by (K.8). The function $Y(\mu)$ defined by (16.185) associated with the stress intensity factor for a bridged crack is again given by (K.9).

The integral equation (K.22) is solved numerically by iteration using the following discrete formulation:

$$1-\left\{1+\mu S_i\right\}g_i^{(N+1)}=\mu\sum_{\substack{j=0\\ j\neq i}}^{n}\left\{g_j^{(N)}-g_i^{(N)}\right\}K_{ij},\; i=0,\ldots,n,\; N\geq 0, \tag{K.23}$$

and relations (K.11)–(K.13). The convergence criterion for the iteration procedure is again given by (K.16) where the parameter $\varepsilon=10^{-8}$.

Index

- Index entries appearing in Figures and their Captions have italic page numbers. Index entries appearing in Tables have bold page numbers. This convention is overridden if the entry appears also in the text.
- Context to some entries is provided after the '/' symbol.
- The symbol '–' indicates either the range of pages for which an index entry is relevant (e.g. a Section or Chapter), or a range of pages where the index entry appears.

Properties for Design of Composite Structures: Theory and Implementation Using Software, First Edition. Neil McCartney.

Companion Website: www.wiley.com/go/mccartney/properties

S

y